FUNDAMENTALS OF DAIRY CHEMISTRY

Second Edition

Some other AVI books

Dairy Science and Technology

BYPRODUCTS FROM MILK, 2ND EDITION *Webb and Whittier*
DAIRY LIPIDS AND LIPID METABOLISM *Brink and Kritchevsky*
DRYING OF MILK AND MILK PRODUCTS, 2ND EDITION *Hall and Hedrick*
FLUID MILK INDUSTRY, 3RD EDITION *Henderson*
ICE CREAM, 2ND EDITION *Arbuckle*
JUDGING DAIRY PRODUCTS *Nelson and Trout*
MILK PASTEURIZATION *Hall and Trout*

Food Science and Technology

BAKERY TECHNOLOGY AND ENGINEERING, 2ND EDITION *Matz*
BREAD SCIENCE AND TECHNOLOGY *Pomeranz and Shellenberger*
CHOCOLATE, COCOA AND CONFECTIONERY *Minifie*
CONFECTIONERY AND CHOCOLATE PROGRESS *Pratt*
COOKIE AND CRACKER TECHNOLOGY *Matz*
ECONOMICS OF FOOD PROCESSING *Grieg*
ECONOMICS OF NEW FOOD PRODUCT DEVELOPMENT *Desrosier and Desrosier*
EGG SCIENCE AND TECHNOLOGY *Stadelman and Cotterill*
ENCYCLOPEDIA OF FOOD ENGINEERING *Hall, Farrall, and Rippen*
FOOD ANALYSIS: THEORY AND PRACTICE *Pomeranz and Meloan*
FOOD DEHYDRATION, 2ND EDITION, VOLS. 1 AND 2 *Van Arsdel, Copley, and Morgan*
FOOD FLAVORINGS, 2ND EDITION *Merory*
FOOD OILS AND THEIR USES *Weiss*
FOOD PACKAGING *Sacharow and Griffin*
FOOD SANITATION *Guthrie*
FOOD SCIENCE, 2ND EDITION *Potter*
FOOD TEXTURE *Matz*
FUNDAMENTALS OF FOOD ENGINEERING, 2ND EDITION *Charm*
FUNDAMENTALS OF FOOD PROCESSING OPERATIONS *Heid and Joslyn*
HANDBOOK OF SUGARS *Junk and Pancoast*
LABORATORY MANUAL FOR FOOD CANNERS AND PROCESSORS, 3RD EDITION, VOLS. 1 AND 2 *National Canners Association*
MICROBIOLOGY OF FOOD FERMENTATIONS *Pederson*
PHOSPHATES IN FOOD PROCESSING *DeMan and Melnychyn*
PRACTICAL BAKING, 2ND EDITION *Sultan*
PRACTICAL FOOD MICROBIOLOGY, 2ND EDITION *Weiser, Mountney, and Gould*
PRINCIPALS OF PACKAGE DEVELOPMENT *Griffin and Sacharow*
QUALITY CONTROL FOR THE FOOD INDUSTRY, 3RD EDITION, VOLS. 1 AND 2 *Kramer and Twigg*
THE FREEZING PRESERVATION OF FOODS, 4TH EDITION, VOLS. 1, 2, 3, AND 4 *Tressler, Van Arsdel, and Copley*

Nutrition and Biochemistry

CARBOHYDRATES AND THEIR ROLES *Schultz, Cain, and Wrolstad*
CHEMISTRY AND PHYSIOLOGY OF FLAVORS *Schultz, Day, and Libbey*
LIPIDS AND THEIR OXIDATION *Schultz, Day, and Sinnhuber*
NUTRITIONAL EVALUATION OF FOOD PROCESSING *Harris and Von Loesecke*
PROTEINS AND THEIR REACTIONS *Schultz and Anglemier*

FUNDAMENTALS OF DAIRY CHEMISTRY

Second Edition

Edited by

BYRON H. WEBB
Collaborator and Retired Chief,
Dairy Products Laboratory,
Consultant, Dairy Research, Inc.,
Agricultural Research Service
U.S. Department of Agriculture
Washington, D.C.

ARNOLD H. JOHNSON
Consultant and Retired Director of Research,
Research and Development Division,
Kraftco, Inc.,
Glenview, Illinois

JOHN A. ALFORD
Chief,
Dairy Foods Nutrition Laboratory,
Nutrition Institute,
Agricultural Research Service,
U.S. Department of Agriculture,
Beltsville, Maryland

WESTPORT, CONNECTICUT

THE AVI PUBLISHING COMPANY, INC.

1974

Library of Congress Catalog Card Number: 73–85050
ISBN-0-87055-143-4

Printed in the United States of America

Dedication

To Lore A. Rogers, George E. Holm, Earle O. Whittier and associates in the Research Laboratories of the Bureau of Dairy Industry, whose pioneering research in the first half of this century provided a basis for their two editions of Fundamentals of Dairy Science and for its successor, Fundamentals of Dairy Chemistry.

Preface

A decade has passed since material was gathered for the 1st Edition of Fundamentals of Dairy Chemistry (1965), successor to Fundamentals of Dairy Science, 1928 and 1935. All chapters in this 2nd Edition have been revised and updated. Sections of many chapters have been completely re-written to keep up with rapid research progress. Of 23 authors writing for this volume 10 are new. Thirteen authors are from the Agricultural Research Service, USDA, 9 from 7 State Universities and 1 from industry.

Even with writing deadlines, delay in manuscript completion seems inevitable. The completion date was set for October 1971, but it was actually the end of 1972 before all chapters were in the hands of the editors. Hence the literature is covered into 1971 but with some chapters having 1972 references.

The 2nd Edition remains primarily a text and reference book. The basic chemistry of milk and dairy products is emphasized but related aspects of microbiology and technology are included.

While the 1928 edition of Fundamentals of Dairy Science contained a complete bibliography for each chapter, the great increase in research reports during the last 40 years has made a careful selection of references necessary. Present emphasis has been on reviews and on the most important papers in each field. Some 4,465 references are cited in this Edition. Dairy Science Abstracts (DSA) and Chemical Abstracts (CA) have been used in literature searches. When the abstract number is cited it is placed in a bracket; when the reference is to a journal page number, no bracket is used. Figures and tables are identified by chapter number, then numerically as Figure 9.2 for Figure 2 in Chapter 9.

The editors acknowledge with appreciation the generous assistance of the authors of the various chapters; it is they who have made possible publication of a reference book of this kind. This revision was produced with the encouragement of Ivan A. Wolff, Area Director, Eastern Regional Research Laboratory, and Michael J. Pallansch, Chief, Dairy Products Laboratory, who generously made available the time and facilities of their groups including the cheerful assistance of our two

Agricultural Research Service secretaries, Carolyn Taylor, and Frances Lowe. The senior editor completed his editing duties as a USDA Collaborator and a member of the Washington Staff of Dairy Research, Inc., an arrangement for which he was most appreciative.

BYRON H. WEBB
ARNOLD H. JOHNSON
JOHN A. ALFORD

November 1972

Contents

CHAPTER 1

Arnold H. Johnson †

The Composition of Milk*

HISTORICAL

Since the milk of animals was used as food in prehistoric times, it is not possible to state precisely when man first became aware of the fact that milk contains more than one substance. It is reasonable to suppose that the use of sour milks and cheeses as food followed very closely the use of fresh milk. The mention of milk, butter, and cheese in the earlier books of the Old Testament is in every instance incidental, and implies their previous use for an extended time. The use of butter for food and for sacrificial purposes is mentioned in the Hindoo Vedas, written between 1400 and 2000 B.C. These facts indicate that even in prehistoric times man knew that milk contains fat and protein. He cannot be credited with devising any precise analytical process, since the separation of these components was imperfect and occurred without application of effort on his part.

Curd, butterfat, and whey were mentioned as the only components of milk by Bartolettus of Mantua[38] in 1619, but in 1633[39] he wrote of a crude milk sugar that he obtained by evaporation of whey. Scheele[336] about 1780 proved that lactose is a true sugar and listed the components of milk as butterfat, casein, lactose, a small proportion of extractive matter, a little salt, and water.

Lactalbumin as a component of milk was first definitely established by Bouchardat[52] in 1857, though earlier investigators had separated the protein of milk into several fractions. Lactoglobulin was identified in whey by Sebelien[343] in 1885, and its presence in whole milk and colostrum was established soon after. The first of the trace elements, iron, was noted in milk by Bunge[60] in 1889. The enzymes of milk received attention after Arnold[25] demonstrated the presence of peroxidase in milk in 1881, and Babcock and Russell[29] reported on the protease activity in milk in 1897.

Since 1900, the elucidation of milk composition has been most rapid. The milk proteins, casein, lactalbumin, and lactoglobulin have been

* Revised from Chapter 1 by E. A. Corbin and E. O. Whittier in the first edition.

† Arnold H. Johnson is the retired Director of Research & Development Division, Kraftco Corporation, Glenview, Illinois.

shown to be mixtures of proteins rather than homogeneous entities. The components of milkfat have been identified and studied, and the large number of mineral elements, vitamins, and other minor components of milk is growing still larger. Although these recent advances have been most enlightening, knowledge of the compostiton of milk has not yet reached the stage of complete clarity.

DEFINITION OF MILK

Milk is defined in the Milk Ordinance and Code recommended by the United States Public Health Service[391] as "the lacteal secretion, practically free from colostrum, obtained by the complete milking of one or more healthy cows, which contains not less that 8¼% of milk-solids-not-fat and not less than 3¼% of milkfat." Since this definition was promulgated as a basis for the enforcement of regulatory laws, it limits milk to that of the cow and prescribes only minimal percentages of two chief components. Minimal standards in the states vary from 8.0 to 8.5% for milk-solids-not-fat and from 3.0 to 3.8% for milkfat.[107] The udder, immediately after parturition, secretes a fluid known as colostrum, which differs considerably in composition from the later secretion; hence the exclusion of colostrum in the above definition.

GROSS COMPOSITION OF MILK

Milk is the liquid food secreted by the mammary gland for the nourishment of the newly born, containing water, fat, proteins, lactose, and minerals (ash). An average gross composition of cow's milk would be as follows:[407] water, 87%; fat, 3.5–3.7%; lactose, 4.9%; proteins, 3.5% and ash (minerals), 0.7%. In 60 specified marketing areas under U.S. Federal and State regulations in 1970, the average fat content of the market milk (about 90,000 samples) was 3.47%, varying between 3.22 and 3.84%.[363]

Water

Milk is a natural liquid food containing a high percentage of water (87%). Milk is actually a concentrated food designed to produce rapid growth in young mammals and contains more solid material than many of our other common foods. Water is the medium in which all the other components of milk (total solids) are dissolved or suspended. A small percentage of the water in milk is hydrated to the lactose and salts, and some is bound in the proteins.

Fat

Milkfat is the most variable of all milk components; and any average value of 3.5 to 3.7% means little when applied to one breed, one animal, or one source of supply. The mixture of mixed triglycerides which makes up 98 to 99% of milkfat is peculiar to milk; though quite bland in taste, it imparts smoothness and palatability to fat-containing dairy products. The remaining 1 to 2% of milkfat is composed of phospholipids, sterols, carotenoids, the fat-soluble vitamins A, D, E, and K, and some traces of free fatty acids. Milk lipids are treated in detail in Chapters 4 and 5.

Lactose

This disaccharide is the characteristic sugar of milk and for most purposes can be considered as the only carbohydrate present. Trace amounts of other carbohydrates found in milk will be mentioned later. Lactose is the predominant solid in milk, except in the case of some high-yielding cows that may produce more fat than lactose; and even then the amount does not vary much from the average value of 4.9%. Chapter 6 is devoted to a detailed discussion of lactose.

Proteins

The classic nomenclature for milk proteins identifies them as casein, lactalbumin, and lactoglobulin. The principal protein of milk, casein, makes up 80% of the total, while lactalbumin and lactoglobulin, also known as the serum proteins, make up the remaining 20%. The classic nomenclature is in the process of continual revision, since these three fractions have been shown to be heterogeneous and to consist of many proteins.[328] Skimmilk protein is designated to consist of 45 to 56% α_s-casein, 8 to 15% κ -casein, 25 to 35% β-casein, 3 to 7 % γ casein, 2 to 5% lactalbumin, 7 to 12% β-lactoglobulin, 0.7 to 1.3% blood serum albumin, 1 to 2% Ig G1 and ≃ 0.2 to 0.5% Ig G2 immunoglobulins, ≃ 0.1 to 0.2% IgM immunoglobulin, ≃ 0.05 to 0.10% IgA immunoglobulin, and 2 to 6% proteose-peptone fraction.

The protein fraction of milk also includes the enzymes of milk, and perhaps some unidentified minor proteins. Chapter 3 is devoted to a comprehensive presentation of the proteins of milk.

Kiermeier[179A] has published an extensive review of the nonprotein components of milk, including the influence of many factors of milk production upon them. More than 600 reports were evaluated of which 352 were cited. In view of the mass of data being reported, Kiermeier suggests that some system of cataloging the data be set up.

Ash

The ash content of milk is an analytical value indicating the amount of noncombustible matter in milk. In normal milk the value remains rather constant at about 0.7%; a value much higher than this is indicative of abnormal conditions in the secreting gland. The ash content gives an idea of the total mineral content, but it can in no way indicate how the minerals were originally distributed in the milk. The ash content of Holstein Friesen milk in Japan[260] over a yearly period averaged 0.712% ± 0.015. Ash contents both higher[243] and lower[128] than this have been reported.

Acidity

The acidity of milk is discussed in Chapter 8. Fresh milk is slightly acid in reaction, having a pH of 6.5 to 6.7.[198,224] Because of many variables it is difficult to give a definite value for titratable acidity, but normal milk usually varies between 0.15 and 0.18% expressed as lactic acid. However, the titratable acidity represents only a hypothetical "endpoint" which is the amount of alkali necessary to shift the protein and other buffer systems to that pH value at which the indicator used (phenolphthalein) changes color. Acidity in milk may be expressed by other means; i.e., Soxhlet-Henkel, Thörner or Dornic.[271] Brusilovskii[57] has developed a formula linking pH with titratable acidity by which acidity can be accurately calculated from the pH value.

Extensive analyses of milk in Germany[380] showed an average acidity of 6.85°SH ± 1.35°SH and in Yugoslavia[90] acidities between 7.0° and 10.4°SH (average 8.44°SH) were determined. Milks with high acidities are usually high in total solids. Soxhlet-Henkel degrees are the number of ml of 0.25*N* NaOH required to neutralize 100 ml milk, and are convertible to percent lactic acid by dividing by 44.4.

MAJOR DIFFERENCES IN GROSS COMPOSITION

Milks differ widely in composition, the greatest differences being among milks of different species of mammals. The variations in milks within a species depend on so many factors that it is difficult to determine accurately the relationship between a specific factor and composition. Differences in the composition of milks may be compared in two ways: by comparing the percentages of individual components as they occur in whole milk, or by comparing the percentages of individual components as they occur in the total solids of milk. It should be borne in mind that neither the percentage of each component in milk nor its percentage in the total milk solids is an indication of the amount of a component produced by an animal in a stated time. Duclaux has stated[304] that there is no such entity as *milk,* but only *milks.* Since there is such a wide variation in the composition of individual samples of milk, the data in the accompanying tables are, for the most part, averages of analyses of a large number of samples produced under varying conditions. These values are, therefore, statistical in nature; they indicate tendencies only, are true in general, but are untrue in a particular instance.

One general statement may be made regarding the composition of milk from individuals. The osmotic pressure of milk is equal to that of the blood of the mammal secreting it; hence, the percentage of lactose and ash found in milks varies between rather narrow limits and is nearly constant in milks for a single species. Any increase or decrease in lactose content is compensated by a decrease or increase in other soluble components. This osmotic relationship helps explain why human milk, with a high percentage of lactose, is very low in ash content, while reindeer milk, with a low percentage of lactose, has a very high ash content.

Gross Composition of Milks from Different Species of Mammals

All milks contain the same gross components, but in widely varying proportions. The gross composition of many different species has been determined; and it is of interest to know the composition of whales' milk, hippopotamus' milk, rats' milk, and others. Hence, Jenness and Sloan[161] have reviewed in a quantitative manner the composition of milks from 144 species representing 13 orders of mammals. However, most of the detailed examination of milks has been centered on those used for human consumption. Table 1.1 gives the average gross composition of milks which fall into this category.

The values for human milk are averages of 1150 determinations covering the entire lactation period. Whittle[412] found human milk to vary within the following percentage ranges: fat, 1.14 to 4.44; nonfat solids, 8.16 to 10.59; protein, 2.18 to 4.43; lactose, 5.0 to 7.16; and ash, 0.20 to 0.40. Selected analyses of human milk from the following respective countries are generally within these ranges: India,[26] Morocco,[83] Japan,[112] Russia,[227] Hungary,[276] New Guinea,[44] Bulgaria,[45] and the United States.[161]

The first set of figures for cow's milk in Table 1.1 are from many sources and are considered to represent the average composition of milk produced in the United States over a period of a year. The second

Table 1.1

AVERAGE COMPOSITION (PERCENT) OF MILKS OF VARIOUS MAMMALS[1]

Species	in Milk							in Total Solids				
	Water	Fat	Protein	Lactose	Ash	Nonfat Solids	Total Solids	Fat	Protein	Lactose	Ash	Nonfat Solids
Woman[115]	87.43	3.75	1.63	6.98	0.21	8.82	12.57	29.83	12.97	55.53	1.67	70.17
Cow[407]	87.2	3.7	3.5	4.9	0.7	9.1	12.8	28.9	27.34	38.28	5.47	71.1
Cow[277]	86.61	4.14	3.58	4.96	0.71	9.25	13.39	30.91	26.76	37.04	5.30	69.09
Goat[32,231]	87.00	4.25	3.52	4.27	0.86	8.75	13.00	32.69	27.08	32.85	6.62	67.31
Ewe[272]	80.71	7.90	5.23	4.81	0.90	11.39	19.29	40.96	27.11	24.94	4.67	59.05
Egyptian buffalo[282]	82.09	7.96	4.16	4.86	0.78	9.95	17.91	44.44	23.23	27.14	4.36	55.56
Chinese buffalo[63]	76.80	12.60	6.04	3.70	0.86	10.60	23.20	54.31	26.03	15.94	3.71	45.69
Philippine cara-bao[93]	78.46	10.35	5.88	4.32	0.84	11.19	21.54	48.05	27.30	20.06	3.90	51.95
Indian buffalo[346]	82.76	7.38	3.60	5.48	0.78	9.86	17.24	42.81	20.88	31.78	4.52	57.19
Camel[36]	87.61	5.38	2.98	3.26	0.70	7.01	12.39	43.42	24.05	26.31	5.65	56.58
Mare[221]	89.04	1.59	2.69	6.14	0.51	9.37	10.96	14.51	24.54	56.02	4.65	85.49
Ass[123]	89.03	2.53	2.01	6.07	0.41	8.44	10.97	23.06	18.32	55.33	3.74	76.94
Reindeer[37]	63.30	22.46	10.30	2.50	1.44	14.24	36.70	61.20	28.06	6.81	3.92	38.80
Llama[145]	86.55	3.15	3.90	5.60	0.80	10.30	13.45	23.42	29.00	41.63	5.95	76.58

[1] Ed. note. An unpublished survey (1973) of over one million commercial cow milk samples indicates an average composition of: fat 3.68%, nonfat solids 8.48%, protein 3.14%, lactose 4.64%, ash .7%.

set of values are from 2426 samples and include milk from Ayrshire, Brown Swiss, Guernsey, Holstein, Jersey, and Guernsey-Holstein cows.[227] Similar tabulation of the composition of cow's milk have also been made.[76,237]

The other mammals listed in Table 1.1 are domesticated and supply milk for human consumption in some part of the world. Differences in composition are rather large; the least variable value appears to be the percentage of protein in the total solids. The differences between cow's milk and human milk are quite apparent; and, because of these differences, cow's milk is usually "modified" to bring its composition closer to human milk before it is given to infants. Parkash and Jenness[284] have reviewed the composition of goats' milk, and Laxminarayana and Dastur[213] have done likewise for buffalo milk.

Growth rate is an inherited trait of newborn mammals. The composition of the milk of each species is designed to provide for this natural rate of growth. Abderhalden[1] studied the milk composition of various species from this standpoint, and Table 1.2 shows his result. The values are self-explanatory; the faster the rate of growth, the more concentrated are the milk components needed for this growth.

Cow's Milk

Differences in Composition Due to Breed.—All the milk produced in the United States and marketed in the fluid condition is from the cow with the exception of comparatively small quantities of goats' milk. Cow's milk is not a uniform article of commerce; its composition varies in response to a long list of physiological, inherited, and environ-

Table 1.2

COMPOSITION OF MAMMAL'S MILK AS RELATED TO THE RATE OF GROWTH OF THE YOUNG MAMMAL[1]

Species	Days Required to Double Birth Weight	Percent in Milk			
		Protein	Ash	Lime	Phosphoric Acid
Man	180	1.60	0.20	0.033	0.047
Horse	60	2.00	0.40	0.124	0.131
Cow	47	3.50	0.70	0.160	0.197
Goat	22	3.67	0.77	0.197	0.284
Sheep	15	4.88	0.84	0.245	2.293
Pig	14	5.21	0.81	0.249	0.308
Cat	9.5	7.00	1.02	—	—
Dog	9	7.44	1.33	0.454	0.508
Rabbit	6	10.38	2.50	0.891	0.997

mental factors. The one indisputable fact regarding the composition of milk can be stated in one word: variation. Those factors responsible for the never-ending variations in milk composition will be presented by limiting the data to cow's milk.

The first factor is the breed of cow. Typical analyses of the milks of the six principal breeds of cows in the United States are given in Table 1.3. These values are averages of herd milks; they would not be valid for individual strains, individual herds, or individual animals. For example, the range of fat percentages within a breed may be as much as 2%. The percentage of lactose remains relatively constant for all breeds in spite of changes in the percentage of other solids. The percentages of protein in the total solids show no definite relationship to the percentages of other components, but remain within somewhat narrow limits. Armstrong,[24] in reviewing the composition of milk from the various breeds, concluded that the average composition of milk seems to show increases in both fat and solids-not-fat over the period from 1900 to 1957. There were some indications of an increase in protein during this same period. He also concluded that the percentages of fat, solids-not-fat, protein, lactose, and ash show a tendency to be slightly higher in individual cow samples than in herd samples for the same breed of cows. Adachi[4] has also reported on changes in the composition of milk in Japan over the last 80 years.

Wilcox, Gaunt, and Farthing[414] recently conducted nationwide studies on the composition of milk of the principal dairy breeds in the United States with a goal of 6,000 cows of each breed, the goal being attained for Holsteins, Jerseys, and Guernseys. The data are given in Table 1.4. It is of interest to compare the data of Overman[277] (1945) with the more recent data.

A vast amount of information has been accumulated on the composition of milk produced in many countries of the world. Selected data are given in Table 1.5. Average values for composition were taken from the published data. In some cases the averages represent results from large numbers of milks. In addition to the gross composition data set down in Table 1.5, certain of the references include other factors such as acidity, pH, Ca, P_2O_5, vitamins and trace minerals.

The data in Table 1.5 show substantial variation that exists in milk composition, resulting in difficulty in setting standards.

Minima for milk composition as set by regulatory agencies differ, and it is pointed out in some of the references that, no matter what these standards are, milk as produced by the cow does not always meet them.

Comprehensive reviews of dairy research in Australia[225,319] and

Table 1.3

TYPICAL COMPOSITION (PERCENT) OF THE MILKS OF COWS OF SIX BREEDS

	in Milk							in Total Solids				
Breed	Water	Fat	Protein	Lactose	Ash	Nonfat Solids	Total Solids	Fat	Protein	Lactose	Ash	Nonfat Solids
Guernsey[277]	85.35	5.05	3.90	4.96	0.74	9.60	14.65	34.47	26.62	33.86	5.05	65.53
Jersey[277]	85.47	5.05	3.78	5.00	0.70	9.48	14.53	34.75	26.02	34.41	4.82	65.25
Ayrshire[277]	86.97	4.03	3.51	4.81	0.68	9.00	13.03	30.93	26.94	36.91	5.22	69.07
Brown Swiss[277]	86.87	3.85	3.48	5.08	0.72	9.28	13.13	29.32	26.50	38.69	5.48	70.68
Shorthorn[98]	87.43	3.63	3.32	4.89	0.73	8.94	12.57	28.88	26.41	38.82	5.81	71.12
Holstein[277]	87.72	3.41	3.32	4.87	0.68	8.87	12.28	27.77	27.03	39.66	5.54	72.23

Table 1.4

MILK EQUIVALENT LACTATION RECORDS[414]
(THIRTY THOUSAND TWICE A DAY, 305 DAY MILKINGS)

	Weight in Pounds										Percent							
	Yield		Fat		Nonfat Solids		Total Solids		Protein		Fat		Nonfat Solids		Total Solids		Protein	
	$\overline{X}$	SD*	$\overline{X}$	SD	$\overline{X}$	SD	$\overline{X}$	SD	$\overline{X}$	SD	$\overline{X}$	SD	$\overline{X}$	SD	$\overline{X}$	SD	$\overline{X}$	SD
Ayrshire	11567	2378	466	100	990	208	1456	301	390	81	3.99	0.33	8.52	0.47	12.55	0.62	3.34	0.29
Guernsey	10601	2414	521	123	961	222	1482	340	390	93	4.87	0.45	9.01	0.29	13.94	0.64	3.62	0.29
Holstein	15594	3142	583	128	1325	270	1908	389	499	104	3.70	0.39	8.45	0.32	12.19	0.59	3.11	0.25
Jersey	9798	2492	507	137	907	234	1415	366	386	98	5.13	0.54	9.21	0.37	14.39	0.77	3.80	0.30
Brown Swiss	12814	3133	539	139	1159	288	1698	423	463	114	4.16	0.35	8.99	0.32	13.20	0.57	3.53	0.26

* Standard deviation.

Table 1.5

WORLD-WIDE DATA ON MILK COMPOSITION

Country	Breed	Total Solids %	Fat %	Protein %	Lactose %	Ash %
Australia[388]	Shorthorn	12.4	3.8			
	Friesian	12.2	3.7			
	Jersey	13.7	4.7			
Bulgaria[128]	Red Danish	11.97	3.58	3.52		0.71
Bulgaria[129]	Bulgarian Brown	11.63	3.15	3.34		0.68
England[168]	All Breeds	12.52	3.82			
	Channel Island	13.75	4.75			
Finland[291]	Bulk Milk	13.05	4.32	3.34	4.76	
France[125]	Many Breeds—5 Regions		3.71	3.24		
Germany[74]	Angeln	14.56	5.52	3.63	4.98	
Holland[155]	Friesian	12.34	3.65	0.522*	4.63	
Ireland[116]	Dublin Milk Supply	12.23	3.63	3.19	4.46	
Italy[12]	Cremona Province Area	12.26	3.62	3.08		0.71
Italy[84]	Friesian (Milan Area)	12.30	3.65	3.10		
Italy[341]	Pesaro Province	12.28	3.30			
Japan[420]	Holstein-Friesian	11.53	3.42	2.87	4.54	0.70
	Jersey	13.63	4.98	3.40	4.50	0.72
Japan[260]	Tohoka (Holstein-Friesian)	11.52	3.56	2.96	4.34	0.71
India[246]	Native Cows (Orissa)	14.04	4.65	3.35		0.66
Poland[160]	Polish Reds	12.68	3.90	3.29		
	Polish Reds x Jerseys	13.52	4.62	3.45		
Poland[180]	College Friesians	12.35	3.65	3.33		
Poland[59]	Polish Black Pied	11.68	3.34	3.09		
Russia[51]	Ayrshire	13.67	4.44	3.76	4.77	
	Brown Latvian	12.99	3.90	3.70	4.66	
	Kholmogor	12.38	3.48	3.48	4.68	
Russia[2]	Rostov Milk Supply	12.42	3.70	3.29	4.85	0.66
Russia[395]	Kula	13.00	4.06	3.46	4.87	0.72
Russia[378]	City Market Milk	12.33	3.75	3.30		
Yugoslavia[30]	Native Spotted Cows	12.20	3.80	3.27		
Yugoslavia[31]	Simmental	12.22	3.81			
Sweden[229]	Central Dairy Milk Supply	12.65	3.92	3.15	4.92	

* Total nitrogen

Germany[176] give data on the composition of milk in those countries.

Differences in Composition Between One Milking and Another.—The composition of milk is not markedly affected by different intervals between milkings except for the percentage of fat. When the intervals are not equal, the highest percentage of fat is found in the milk drawn after the shortest interval. The milk yield during such intervals may vary inversely with fat percentage.[324] Using oxytocin injections to obtain the residual milk, Schmidt[338] studied the milk yields for various milking intervals and found that the rates of milk, fat, and solids-not-fat secretion were linear for intervals up to and including 12 hr, but that significant reductions in rates of secretion of these components were found at 16- and 20-hr intervals. Extensive data exist on milking intervals of 4, 8, 12, 16, 20, and 24 hr[220] and on 6, 12, 18, 24, 30, and 36 hr.[409] In the latter instance it was observed that the individual components decreased curvilinearly with duration of the interval, but the degree of curvilinearity differed among the components.

Varying results are reported when the milking intervals are the same or nearly so. Thus for 12−12, 14−10, and 16−8 hr[219,344,400] no significant differences in milk yield or in fat content were observed. However, other data[344,400] show evening milk to be higher in percentage of fat, total solids, and yield than morning milk. It is commented that tests on morning milk or evening milk may not be representative of the milk for 24 hr. The weight of evidence as to whether there is a difference in the composition of milk as obtained after equal intervals between milking is about equal.[99,273,324,338,389]

Higher fat and milk yields[9] were obtained after cows were milked at 4 equally spaced intervals rather than after 2 equal intervals. Eliminating one milking per week[69,169] reduced fat and milk yield by 5 to 10%. Another milking schedule[207,347,416] involving two 18-hr milking intervals and one 12-hr interval over the 48-hr period usually resulted in reducing milk yield and milk composition as compared with four 12-hr intervals. Radcliff[308] and coworkers noted that omission of one and two consecutive milkings per week for a year reduced milk yield by 3 and 14%, protein by 2 and 12%, fat by 2 and 12%, and solids-not-fat by 3 and 14%, respectively. Nielsen and Sorensen[268], on the other hand noted very little difference in average daily milk yield or percentages of fat and protein on omitting one milking per week.

A group of cows,[285] 6 months into the lactation period, were milked once daily. This resulted in a 35% reduction in milk yield but did not change the composition of the milk. If milk is permitted to accumulate in the udder long enough to exert maximum pressure, some resorption takes place; the percentages of lactose, fat, calcium, and phosphorus decrease; the percentages of protein and chloride increase; and the milk becomes alkaline.[298]

Changes in Composition of Milk During Milking.—The percentage of fat increases markedly during milking; hence, the udder must be effectively stripped if the highest percentage of fat possible is desired in the milk. If high-fat milk is left in the udder at one milking, it

Table 1.6

CHANGES IN THE PERCENTAGES OF FAT IN MILK DURING MILKING[316]

	Cow No. 1	Cow No. 2	Cow No. 3
First portion	0.57	2.09	1.73
Second portion	1.82	2.66	2.65
Third portion	4.15	3.66	3.82
Fourth portion, strippings	5.56	6.42	4.80

may be obtained by stripping at a subsequent milking. No appreciable differences other than in fat have been found between foremilk and strippings. Table 1.6 contains data which show typical variations in fat during milking.[316] These findings are confirmed in more recent work.[218,253] Sundaresan *et al.*[367] found no difference in milk yield, quality, or composition whether the complete milking was conducted by hand or by machine.

Variability of Milks from Different Quarters of the Udder.—The four quarters of the udder are separate units physiologically as well as anatomically. Proks[305] has shown that there are distinct differences among samples from different quarters of the udder, not only in gross composition but also in the percentage of calcium in the ash and in the composition of the fat. Benton[46] found that if quarters are milked separately, the samples may be no more alike than if they had been obtained from different cows, and that milk from any one quarter differs from day to day just as does the total pooled sample. Turner[389] has explained that the order of milking the quarters influences the fat test of the milk from each quarter. The quarters milked first yield milk having the highest fat content; milk from the last quarters milked has the lowest fat content. If a milking machine is used and all four quarters are milked simultaneously, the milk from the different quarters will be very similar in fat percentage. Confirmatory observations have been reported by other investigators.[183,288,294,401,422]

King[184] has examined the influence of the side on which the cow lies on the yield and composition of milk. Milk samples were taken separately from the lower and upper halves of the udder from tied cows which lay on one side for 2 hr before milking. The lower sides of the udder yielded less milk than the upper sides when the cows were giving > 12 lb per milking, but no corresponding difference when the cows were giving < 12 lb. Similar results were obtained for percentage of solids-not-fat, but there were no significant differences in the fat percentage.

Table 1.7

TRANSITION FROM COLOSTRUM TO NORMAL MILK[103]

Time after Calving, Hr	Total Protein [a]%	Casein %	Albumin [b]%	Fat %	Lactose %	Ash %	Total Solids, %
0	17.57	5.08	11.34	5.10	2.19	1.01	26.99
6	10.00	3.51	6.30	6.85	2.71	0.91	20.46
12	6.05	3.00	2.96	3.80	3.71	0.89	14.53
24	4.52	2.76	1.48	3.40	3.98	0.86	12.77
30	4.01	2.56	1.20	4.90	4.27	0.83	13.63
36	3.98	2.77	1.03	3.55	3.97	0.84	12.22
48	3.74	2.63	0.99	2.80	3.97	0.83	11.46
72	3.86	2.70	0.97	3.10	4.37	0.84	11.86
96	3.76	2.68	0.82	2.80	4.72	0.83	11.85
120	3.86	2.68	0.87	3.75	4.76	0.85	12.67
168	3.31	2.42	0.69	3.45	4.96	0.84	12.13

a 6.37 × nitrogen.
b This is undoubtedly almost entirely globulins.

Changes in Composition During the Lactation Period.—The secretion of the mammary glands for the first few days of lactation is known as colostrum; it differs considerably in composition from the later secretion known as normal milk. Colostrum has a strong odor, a bitter taste, a slight reddish-yellow color, and contains a remarkably large percentage of immune globulins. It is richer in practically all milk components except lactose, potassium, pantothenic acid, and of course, water.[237] Table 1.7[103] shows the gradual change from colostrum to normal milk. Other analyses[101,117] have shown an upsurge of fat to between 8 and 15% a few hours after calving. This high level does not persist for more than a few hours. The composition of colostrum and the rate of its change probably depends, to a considerable extent, on the individuality of the cow. About 5 days after parturition the secretion of the udder is considered to be normal milk, suitable for human consumption, although slight changes in composition will still occur for some time.

The percentage of solids-not-fat in the milk of barren cows shows no consistent succession of changes during the lactation period, except that is may decrease slightly as lactation progresses. With pregnant cows, the behavior of the solids-not-fat content of the milk during lactation is rather consistent regardless of breed. The amount of solids-not-fat is high shortly after parturition, drops to a lactation low at 2 to 3 months, increases slowly to six months, and then increases rapidly to the end of lactation. Pregnancy effects are noticeable 2 to

5 months after a successful service and account for most of the increase in solids-not-fat in the late stages of lactation.[33,413] Total protein varies during lactation in the same manner as the solids-not-fat.[97]

The fat content of milk varies widely during the lactation period, but the general trend of the fat percentage is similar to the trend of the solids-not-fat. Fat content is high after parturition, decreases to a low during the first or second month of lactation, and then gradually increases during the remainder of the lactation period.[165] The only change in lactose percentage attributable to stage of lactation is a slight decrease toward the end.

Other investigators[58,172,181,209,226,369,408,410] report results somewhat similar to those in Table 1.7. Thus Buecher[58] reported that the fat content dropped 25% during the first 6 months post partum, but gradually increased to a level just below that at the beginning of lactation. The nitrogen content decreased first, but increased toward the end of lactation. The calcium and sodium contents increased and the potassium content decreased toward the end of lactation. Sebela and Klicnik[342] reported that the protein increased steadily over the lactation period.

Wheelock *et al.* and Rathore studied the effects of no dry periods[410] or varying dry periods[311] on milk composition and the milk production following lactation. Economically the optimum dry period was found to be 60 days, but for milk yield in the next lactation it was found to be 121 to 180 days, for butterfat percentage 31 to 60 days, and for solids-not-fat percentage 0 to 30 days.

Influence of Season on the Composition of Milk.—In Table 1.8 are shown seasonal differences in the composition of milk. The values given are averages derived from analyses of milks of 6 herds of cows: Ayrshire, Jersey, Guernsey, Brown Swiss, Holstein, and Guernsey-Holstein cross. Because the number of samples was large (2426), and from individual cows of the various breeds during all stages of lactation, the variations shown should truly represent seasonal changes.

The most pronounced change is the sharp decrease in the percentage of total solids in the milk from May to June.[226] This decrease is divided among the fat, protein, and ash, and is often attributed to a change in feed or temperature. Neither of these is a primary factor; it has been shown that cows fed the same ration month after month and with only slight temperature variation still exhibited this seasonal low in total solids content.[165] Lactose is the least variable of the gross components on a percentage basis; but because of its osmotic relationship in milk, it increases slightly as the other solids decrease, and decreases as the other solids increase. Although the percentage of total

TABLE 1.8

INFLUENCE OF SEASON ON COMPOSITION OF MILK[277] (PERCENT)

Month	No. Cows	No. Samples	in Milk						in Total Solids				
			Fat	Protein	Lactose	Ash	Nonfat Solids	Total Solids	Fat	Protein	Lactose	Ash	Nonfat Solids
Jan.	130	227	4.31	3.67	4.87	0.72	9.26	13.57	31.80	27.04	35.87	5.33	68.24
Feb.	127	199	4.22	3.62	4.89	0.72	9.23	13.45	31.39	26.90	36.37	5.35	68.62
Mar.	134	228	4.16	3.56	4.98	0.71	9.25	13.41	31.02	26.56	37.12	5.30	68.91
Apr.	128	210	4.10	3.54	5.01	0.71	9.27	13.37	30.69	26.49	37.43	5.33	69.28
May	132	208	4.10	3.53	5.04	0.71	9.27	13.37	30.69	26.41	37.71	5.32	69.28
June	124	201	3.96	3.45	5.02	0.70	9.17	13.13	30.16	26.26	38.22	5.32	69.80
July	123	195	3.95	3.46	5.02	0.70	9.16	13.12	30.12	26.39	38.23	5.30	69.62
Aug.	116	173	3.95	3.54	5.00	0.69	9.24	13.18	29.95	26.86	37.92	5.25	70.03
Sept.	109	176	4.10	3.62	4.96	0.70	9.28	13.38	30.64	27.02	37.06	5.22	69.30
Oct.	112	182	4.24	3.66	4.92	0.71	9.29	13.53	31.33	27.07	36.35	5.25	68.67
Nov.	119	207	4.27	3.69	4.88	0.72	9.28	13.55	31.48	27.21	36.02	5.28	68.51
Dec.	128	220	4.30	3.65	4.92	0.72	9.29	13.59	31.64	26.87	36.22	5.20	68.39
Average			4.14	3.58	4.96	0.71	9.25	13.39	30.91	26.76	37.04	5.30	69.09
Range of values from average			+0.17 −0.19	+0.11 −0.13	+0.08 −0.09	+0.01 −0.02	+0.04 −0.09	+ 0.20 − 0.27	+ 0.89 − 0.96	+ 0.45 − 0.50	+ 1.19 − 1.17	+0.05 −0.10	+ 0.94 − 0.85
Range of values as per cent of average			+4.11 −4.59	+3.07 −3.63	+1.61 −1.82	+1.41 −2.82	+0.43 −0.97	+ 1.49 − 2.02	+ 2.88 − 3.11	+ 1.68 − 1.87	+ 3.21 − 3.16	+0.94 −1.89	+ 1.36 − 1.23

solids in milk does show this definite seasonal low in late spring or early summer, it is remarkable that the solids-not-fat are so constant throughout the year.

Table 1.8 also shows the percentages of the components in the total solids. When calculated on this basis the protein and ash give the most constant values for the season.

Since Overman's report[277] was published, many studies have been made on seasonal variations in the composition of milk. Reference is made to the work in India,[120] Australia,[75] the United States,[81,359,408] France,[58 82] the Netherlands,[209] Russia,[131] Italy,[248] England,[156 418] Czechoslovakia,[89] Japan,[275] and New Zealand.[88] In certain instances[131] differences in the melting point of the fat were observed to occur seasonally

Composition of Milk as Affected by Feeds and Nutritional Level.—Changing the feed of cows from full to half production ration has been shown to cause a reduction in yield of milk and a decrease of 0.3 to 0.5 in the percentage of solids-not-fat, but to have no consistent effect on the percentage of fat.[323] The percentage of solids-not-fat returned to normal when the full ration was restored. Mature cows well fed before calving gave milk which averaged 0.28% higher in fat and 0.11% higher in solids-not-fat during the first 3 months of lactation than cows poorly fed before calving. Underfeeding cows after calving caused an increase in the percentage of fat and a decrease in the percentage of solids-not-fat, the latter decrease being accounted for by a decrease in protein. Underfeeding both before and after calving had a severe effect on the percentage of solids-not-fat in milk; it was lowered by as much as 0.5 or 0.6% in some instances.[110] In general, the results from underfeeding have been in agreement in showing that the protein content is suppressed somewhat more than the lactose content.[215]

Feeding more than the normal amount of feed, or feeding additional concentrates, has been reported to increase the solids-not-fat slightly. Feeding excess protein does not appreciably alter the protein content of milk, but the nonprotein nitrogen content may be enhanced. When cows are allowed fresh pasturage the solids-not-fat increases, mainly because of an increase in protein. The lactose content is not changed.[215]

In further investigations of feed requirements Voigtländer[399] fed cows at three levels of protein and energy intakes. In the well-fed groups, fat, protein, and minerals were higher and chloride lower than in the average or poorly fed cows. Seidler and Petkow[345] observed that when three comparable groups of cows were fed recommended standard rations and that ration was increased or reduced by 15%

of digestible protein, the milk yield increased or decreased accordingly, with only slight effect on the fat content. On similarly increasing or decreasing the oat-units in the standard ration by 30%, the increase resulted in greater yield of milk of higher fat content, while the reduction brought about only a lower milk yield.

Other investigators[64,121,122,157,189,233,289,289B,310,333,383] report the effect of ration composition on milk yield and milk composition. Usually, but not in all cases, a reduction of the protein regardless of its nature affected milk composition.

Lizal and Opletalova[223] found that urea could be substituted for 50% of the protein without deleterious effect. Virtanen[398] fed protein-free feeds using urea and ammonia as sources of nitrogen. Kaplan and associates[171] noted that fasting for 24 hr decreased both fat content and milk yield.

Burt and Dunton[61] observed that 10 feedings a day of equal composition caused a reduction in milk yield but no change in fat content as compared with 4 feedings (2 concentrate + 2 roughage).

Effect of Environmental Temperature on the Composition of Milk.—At the Missouri Experiment Station,[71] 43 cows of Jersey, Holstein, Brown Swiss, and Brahman breeds were exposed to temperatures ranging from − 15° to 40°C, and their milk analyzed. The data show that environmental temperatures from − 1° to 23°C do not materially influence milk yield or milk composition in dairy cattle of European ancestry. Temperatures below or above this range (−1° to 23°C) were found to cause substantial changes in the composition and yield of their milk. The composition of the milk of Brahman cows tested at 10° to 40°C was not affected by the high environmental temperatures, nor was the yield. This breed was not subjected to low environmental temperatures (below 10°C).

Other investigators[136,234,313] also noted the effect of temperatures higher than 25°C in lowering milk yield and the composition of milk, with Hafez[136] reviewing the effect of both high temperature and humidity on production of animal products. Roussel and Beatty[329] noted that cooling the head and neck of the cows significantly increased milk production as compared with control cows. Composition of the milk was not affected. Himmel[147] supplied water to cows at 23° and at 3°C. Milk production was higher with water at the higher temperature but milk composition was not affected. The cows preferred the warmer water.

Several investigators[217,306,372] have studied the effect of protective housing versus loose-housing or outside maintenance on milk yield and composition. Protective housing usually resulted in increased milk production with variable effects on milk composition. Climatic factors were considered operative in the differences that occurred.

Effect of Disease on the Composition of Milk.—Pronounced changes in the composition of milk are associated with an inflamed condition in the udder. Studies of milks from cows affected by mastitis made by several investigators have led to the following general conclusions. The composition of milk from the affected quarters of the udder tends to approach that of blood. The percentages of fat, solids-not-fat, lactose, and casein are reduced, while the percentages of chloride and milk-serum proteins are increased.[27,185,192,200,264,296,330,364,404] Stefanovic and co-workers[364] have related change in milk composition to five increasing degrees of mastitis, as determined by the Zagreb test. Because milk from inflamed udders does contain a relatively high percentage of chlorides, this fact has been employed as a means of detecting mastitis. The chloride-lactose number (discussed in the Salts and Ash section) is also used to detect abnormal or diseased conditions in the udder.

At the onset of foot and mouth disease there is a marked reduction in the volume of milk secreted, whether the udder is inflamed or not. If the udder is not inflamed, moderate increases in the percentages of fat, protein, and ash occur, with a decrease in the percentage of lactose. If the quarters are inflamed, the changes are much greater and resemble those of severe mastitis.[195]

Other Factors Affecting the Composition of Milk.—*Age of the Cow.*—The average fat percentage of milk decreases slightly with the age of the cow.[389] A gradual, though irregular, decrease in solids-not-fat content is also noted with age.[413] The decline in solids-not-fat percentage with age is approximately twice the magnitude of the decline in fat percentage. Lactose and casein are the solids-not-fat components that are most affected by the aging of the cow. Declines in solids-not-fat ranging from 0.21 to 0.45% during the first 7 years of lactation have been reported.[215] Other investigators show similar results.[113,174,286,393,400] Thus maximum milk yields may occur from the fifth to the ninth lactation. Gacula[113] found that the lactation year of maximum yield varied for the five breeds which he studied. He noted that fat content of the milk decreased from the second to the tenth lactation.

Weather Conditions.—In dry years the total yield of milk tends to decrease, together with a decrease in the solids-not-fat and an increase in fat percentage. In wet years there is a tendency for both the solids-not-fat and the fat percentage to decrease, with or without an increase in the pounds of milk, depending on the level of nutrition.[211] Controlled humidity, as opposed to natural weather conditions, does not appear to influence milk composition when the temperature is kept below 24°C.[215]

Johnstone, Stone, and Frye[167] reported that during a 120-day hot period in Louisiana, cows in a shaded pasture or in a covered shed showed losses in total milk production and in fat and solids-not-fat content of the milk. Akhmetov[5] noted that cows in the sunshine for 4 hr per day (temperature 39.2°C) suffered losses in milk yield and in the fat, protein, and lactose contents of the milk, as compared with cows in the shade on the same days where the temperature was 36.2°C.

Heat or Oestrum.—The effect of oestrum on milk composition is real but inconsistent. There is usually much variation in the fat test, together with a slight decrease in yield during this period. These changes have been attributed to increased excitability and nervousness causing the cow to either hold up some of the milk or to secrete less milk.[92,182,199,289,366A]

Gestation.—This factor has been touched on briefly in the section dealing with composition during the lactation period. Gestation can affect the composition of milk in an indirect manner by hastening the end of lactation which, as noted previously, causes marked changes in the composition of milk.[279] Aside from this indirect effect, it is accepted that gestation does cause a direct change in the composition of milk. An increase in milk solids, especially solids-not-fat, starts at about the fourth month of pregnancy and continues, increasing to the end of lactation. Cows which are open show no rise or a negligible rise in solids content during a comparable period in their lactation.[215,286,410,413]

Exercise.—Dairy cows are benefited by mild exercise. During a period of exercise feed consumption is increased, milk production is maintained, and the percentage of fat is increased over a corresponding period of rest.[389]

Breeding and the Composition of Milk.—Because butterfat has always been the most expensive component of milk, dairy cattle are usually selected and bred on the basis of their fat-producing ability. The selection and breeding for increased fat has been successful, and the trend of herd averages over the last 30 or 40 years in most countries where dairying is a major enterprise has been an increase of about 2 lb fat per year.[230] This trend is only partly caused by genetic improvement; environmental and nutritional conditions must also be considered as factors.

At present there is a trend to place more economic value on the solids-not-fat, and even to price milk with the solids-not-fat receiving more consideration. If such pricing is adopted, dairy cattle breeders will undoubtedly strive to increase the solids-not-fat portion of milk. There is enough evidence available to indicate that such breeding would be successful.[164,215,230,390] Von Krosigk, as reported by Lush,[230] has estimated that in 19 years of selecting solely to increase the percentage of protein in milk, the protein would increase by 0.28 percentage point, the solids-not-fat by 0.37, and fat by 0.48. If, on the other hand, selection was directly and wholly for total solids, in the same time protein would increase by 0.25, solids-not-fat by 0.34, and fat by 0.64.

It is evident that selective breeding for one component of milk is impossible; the breeder in selecting for one characteristic is automatically selecting for many others at the same time.[164] However, since part of the variations found in milk composition are genetic, selection for solids-not-fat or basic components should be effective in changing the composition of milk of the future dairy cow population.[390] Comparable results in changing the composition of the milk, especially its protein content, have been reported by other investigators.[62,119,252,339]

MINOR COMPONENTS OF MILK

The analyses of milk previously given have included only what are considered the major components. Succeeding chapters are devoted to the composition of milk products and to the characteristics of each of the major components. Changes in the chemical components of milk due to processing, storage, or bacterial fermentation will be discussed when these main topics are presented. But there are in milk a number of inorganic and organic substances in small percentages or traces, some apparently unimportant and others highly important because of effects they produce, which are out of all proportion to their concentration. These are considered in the remainder of this chapter.

Salts and Ash

The salts of milk and the ash of milk are not identical. The ash of milk denotes the white residue left after milk has been incinerated at a low red heat. Since the metallic elements are in excess of the nonmetallic, the ash is always alkaline in character. The composition of the ash does not represent the state of the salts as they occur in milk, since there is considerable alteration due to the chemical reactions taking place during incineration. The ash contains substances derived from both the organic and inorganic compounds of the milk. The CO_2 of the carbonates is formed mostly from the organic components; the SO_3 of the sulfates is considered to be a decomposition product of the proteins; and part of the P_2O_5 must be from the casein, since this protein contains phosphorus equivalent to about 1.62% P_2O_5. Chlorides are frequently volatilized and lost by the use of excessively high temperatures of ashing.[302] The loss may be as much as 45%,[272]

Table 1.9

PERCENTAGE COMPOSITION OF THE ASH OF MILK

Investigator	K_2O	CaO	Na_2O	MgO	Fe_2O_3	P_2O_5	C1	SO_3
Babcock[214]	25.02	20.01	10.01	2.42	0.13	24.29	14.28	3.84
Schepang[337]	30.33	20.90	8.41	2.25	0.05	24.80	14.55	2.55
Orla-Jensen[272]	23.63	27.32	5.82	2.42	—	26.89	13.57	2.96
Richmond[322]	28.71	20.27	6.67	2.80	0.40	29.33	14.00	Trace
Schrodt and Hansen[a]	25.42	21.45	10.94	2.54	0.11	24.11	14.60	4.11
Fleischmann[a]	25.71	24.68	11.92	3.12	0.31	21.57	16.38	—
Storch[a]	25.31	21.93	9.44	2.87	—	28.69	13.73	—

[a] As reported by Allen.[7]

but is minimized by keeping the temperature below 600°C. The citrate content is completely lost during incineration. Average normal milk is considered to contain 0.70% ash, and this amount represents a salt content of about 0.90%.

The salt components as they are found in the ash of milk are given in Table 1.9. Table 1.10 gives an average value for each salt component on a whole-milk basis.

The salts of milk are considered to be the chlorides, phosphates, and citrates of potassium, sodium, calcium, and magnesium. Although the salts comprise less than 1% of milk, they influence the heat stability of milk, the stability of milk toward alcohol coagulation, the age thickening of sweetened condensed milk, the feathering of cream in

Table 1.10

AVERAGE VALUES FOR MILK SALT CONSTITUENTS

Constituent	Content in Whole Milk, Mg/100 Ml	Number of Samples	References
Calcium	123	824	237,266
Magnesium	12	759	237,266
Phosphorus	95	829	237,266
Sodium	58	491	237
Potassium	141	472	237
Chlorine	119	1579	237
Sulfur	30	80	237
Citric acid	160	307	237,266

Table 1.11

DISTRIBUTION OF CALCIUM, MAGNESIUM, PHOSPHORUS, AND CITRIC ACID IN MILK[396]
(15 Samples)

	Total Mg/100 Ml	Soluble Mg/100 Ml	% Soluble
Calcium	132.1	51.8	39.2
Magnesium	10.8	7.9	73.2
Phosphorus	95.8	36.3	37.9
Citric acid	156.6	141.6	90.4

coffee, the coagulation of milk by rennin, and the clumping of the fat globules on homogenization.[396]

With regard to the chemical combinations and physical states in which the salt components exist in milk, knowledge is incomplete and uncertain. Freshly drawn milk is not in a state of stable equilibrium; it changes slowly merely upon standing and changes faster during handling and processing. Potassium, sodium, and chlorine are entirely in solution and presumably entirely ionized. Phosphates, calcium, magnesium, and citric acid are partly in solution and partly in suspended combinations. Table 1.11 gives the soluble and insoluble distribution of these four components. The salts of milk exist in a complex, easily altered equilibrium that has not been completely formulated either from a gross heterogeneous standpoint or from the ionic standpoint. Milk equilibria are discussed in detail in later chapters.

The percentage of salt and ash components in milk varies with the usual factors: individuality of the animal, the breed, the feed consumed, the season, stage of lactation, and disease. The first four factors produce definite but rather small variations in composition. Stage of lactation and disease produce more noticeable effects.

Lactation.—The total ash content is higher in colostrum and at the end of lactation than in the intervening period. Chloride increases quite markedly toward the end of lactation. Potassium, the only salt constituent that is lower in colostrum than in normal milk, increases quickly to a normal value, reaches a maximum at about 2 months, and then declines again. The other salt constituents—phosphorus, calcium, magnesium, and sodium—are higher in colostrum, quickly decrease to a normal level, and then increase again toward the end of lactation.[386] Tokovoi and Lapshina[384] made similar observations on the ash content as well as on trace-element variation during lactation and season, as did also Comberg and Meyer.[73]

Disease.—A diseased udder produces milk which is abnormal in many respects, but a general conclusion has evolved concerning the salt and other constituents. In disease, the proportion of albumin, globulin, chloride, sodium, and sulfate increase, while a decrease is noted in lactose, potassium, magnesium, calcium, and phosphate. The increase in sodium and chloride is quite large, is easily detected, and caused some earlier investigators to label abnormal milk of this type as "salt milk".[7] The inverse relationship between chloride and lactose in abnormal milk has led to the chloride-lactose number which is equal to (% chloride / % lactose) × 100. The value for normal milk will fall within the range 1.5 to 3.5. A value larger than 3.5 indicates abnormal milk from a diseased udder or milk from a cow in late lactation. Other investigators[158,190] show similar relationships in the minerals of milk from mastitis-affected cows.

Citric acid, present as the citrate radical, is an important component of milk because of the part it plays in the salt equilibria. Earlier literature values for citric acid varied widely, but more recent work has shown that the content in milk does not vary much from the average value of 160 mg per 100 ml. Sherwood and Hammer[352] studied citric acid in 335 samples and concluded that the content in milk is not affected by the breed, by the stage of lactation, or a change from summer to winter rations. Other investigators[42,381,421] reporting on the citric acid content of milk have found variation with season and with stage of lactation. Thus Tikhomirova[381] gives citric acid values of 264 mg/100 ml for winter milk and 180 mg/100 ml for summer milk.

Trace Elements

In addition to the elements that occur in milk in relatively large proportions (K, Ca, Na, Mg, P, Cl, and S) there are a large number of elements usually measured in ppm or μg per liter, and referred to as trace elements. The fact that they are present in trace amounts does not detract from their importance, since many of these elements are known to possess remarkable physiological and nutritional qualities. The trace elements in milk come from the feeds eaten by the animal, and ultimately from the soil in which the feeds were grown. Most soils and feeds contain a sufficiency of these elements; but there are agricultural areas where cobalt, copper, or iodine is lacking, and a supplement of the particular element must be fed to prevent deficiency diseases from occurring. At the other extreme are the seleniferous areas of the West North Central States where selenium is present in such quantities that feeds from these soils are toxic to livestock.

Although feedstuffs are the primary source of trace elements in milk, a certain quantity could be obtained from the water supply (bromine

Table 1.12

TRACE ELEMENTS IN COW'S MILK (μG PER L)

Element	Cow receiving Normal Ration	Cow receiving Supplement of Element in Ration	Reference
Aluminum	460	810	20,126,188
Arsenic	50	450	15
Barium	Qualitative	—	—
Boron	270	660	109,188
Bromine	600	Increases	105,210
Bromine (coastal area)	2800	—	105
Cadmium	26	No Increase	245,258
Chromium	15	—	187,188
Cobalt	0.6	2.4	17
Copper	130	No Increase	15,259
Fluorine	150	Increases	8,188
Iodine	43	Up to 2700	15,188
Iron	450	No Increase	15,259
Lead	40	Increases	15,256,259,332
Lithium	Qualitative	—	
Manganese	22	64	23,188,259
Molybdenum	73	371	14,19
Nickel	27	No Increase	18,188
Rubidium	2000	Increases	256
Selenium (nonseleniferous area)	40	Increases	124,134,135
Selenium (seleniferous area)	Up to 1270	—	355
Silicon	1430	No Increase	22,126
Silver	47	—	258
Strontium	171	—	259,316A
Tin	Qualitative	—	—
Titanium	Qualitative	—	126
Vanadium	0.092	—	356
Zinc	3900	5100	16,259

and/or fluorine), from insecticide and spray residues (arsenic and/or lead), from glass containers (silicon), or from metal cans and processing equipment of the dairy (copper, iron, nickel and/or zinc). Table 1.12 lists the trace elements found in cow's milk. An accepted quantitative value for the usual concentration of each of the trace elements in milk is not available. For many of them there is no quantitative data at all; they have been shown to be in milk ash qualitatively only by spectrographic analysis. The values in Table 1.12 have been selected to give a general idea of the proportions encountered in milk. Since the feeding of an element in large amounts as a supplement may increase its content in milk, values are given for a normal ration and a ration with the element added.

TABLE 1.13

TRACE ELEMENTS IN COW'S COLOSTRUM (μG PER LITER)

Element	Normal Milk	Colostrum
Chromium	15 ± 5	57 ± 28
Cobalt	1.3 ± 0.4	3.6 ± 1.9
Copper	100	670
Iodine	70	260
Manganese	30 [15]	130
Zinc	3700 [15]	13,500

Colostrum from all species invariably contains a higher proportion of trace elements than normal milk. Information is meager in this particular area, but Table 1.13 gives the values of Kirchgessner[187] as reported in a review by Archibald.[15]

The data of Murthy *et al.*[259] in Table 1.14 give recent findings for the content of some of the trace elements in milk and the review of Ammerman and others[10] reports recent developments in Co, Cu, Fe, Mn, Zn, Se, and I in relation to ruminant nutrition and possible milk relationships.

Cobalt.—This element has only recently been accepted as a trace element in milk because it is present in very minute percentages, and it is not detected spectrographically in milk ash. Cobalt forms the center of vitamin B_{12} molecule. As such it is important in animal nutrition. Vitamin B_{12} functions in blood regeneration, a lack of it resulting in anemia. With ruminants it is only necessary to supply the cobalt; the ruminant bacteria are capable of synthesizing the vitamin. The cobalt needed by cows (less than 1 mg per day) if not available in feed, is supplied from trace-mineralized salt blocks.

Cobalt was first reported by Bertrand and Mâcheboeuf[48] in 1925. Values for normal milk are usually less than 1 μg per liter.[17,100] The amount present can easily be increased by supplemental feeding, and the cobalt content is always higher in colostrum. Other investigators[68,201,216,281,188,373] have confirmed the levels of cobalt given in Table

Table 1.14

RANGE[259] IN SOME TRACE MINERALS FOR MARKET MILKS COLLECTED FOR A 1-YEAR PERIOD IN 62 CITIES OF THE UNITED STATES

Copper	0.044	to 0.190,	avg.	0.086	ppm
Iron	0.29	to 0.151,	avg.	0.64	ppm
Manganese	0.033	to 0.211,	avg.	0.091	ppm
Strontium	0.040	to 0.480,	avg.	0.171	ppm
Zinc	2.30	to 5.10,	avg.	3.28	ppm

1.12, but with somewhat higher levels in a few cases. It was noted[216] that when cobalt and iodine were added to a silage-corn diet, the cobalt content in the milk increased by 100%, the nickel and iron by 50%, and the manganese by 10%.

Chromium.—Kirchgessner[188] and co-workers report a chromium content of 15 μg/l milk. Lamport[208] suggests that chromium may be needed to make insulin, in promoting glucose metabolism, especially in some diabetics.

Copper.—Much attention has been given to the copper content of milk because of its catalytic effect on the development of oxidized flavor in milk and milk products. It is a normal component of milk, present in amounts of about 20 to 200μg per liter,[15,47,102,206,253] but this may increase during processing and storage in metal containers. Some of the high values reported for copper[70,368] probably include an amount dissolved from metal equipment. Recent investigations[259] have shown, however, that milk can vary considerably in its copper content without involving contamination. Early investigators[188] reported that the feeding of supplemental copper did not increase the percentage of copper in the milk, but a deficiency of copper in the pasture or feed resulted in a lower copper content in the milk.[253] Recently[95 104] it has been noted that feeds with added $CuSO_4$ or copper-EDTA resulted in increasing the copper content of the milk. Thus by drenching cows with a 1% solution of $CuSO^4$, King and Dunkley[181A] were able to raise the copper content of the milk by a few μ g per liter. Copper is essential in the formation of hemoglobin; a deficiency of copper produces anemia and other physiological disorders in animals.

Iodine.—Iodine has only one known physiological function namely as a constituent of thyroxine, the hormone secreted by the thyroid gland. A deficiency of iodine causes an enlargement of the thyroid, the enlarged gland being known as a goiter.

Iodine is always present in milk, but in variable quantities. Since sea water is so rich in iodine, milk from coastal agricultural areas contains more iodine (and bromine) than milk from inland areas. The inland areas also vary in the percentage of iodine present in soils and feeds. Areas of the world where endemic goiter is prevalent are deficient in iodine and are known as goiter "belts." Iodine is unique in that it is easily transmitted to milk from the animal's feed.[6,15,49,53,72,108,232,188,299,366] Thus by adding iodine to the feed, the iodine content of the milk can be increased 200 times as compared to the iodine content of the milk from cows on the original feed. A salt such as potassium iodide may be added to the feed, or the supplement can be organically combined iodine.[55] Organic iodine in proportions as great as 20 ppm imparts no unpleasant taste or odor to milk.

The iodine content of colostrum is about three times that of normal milk. In milk, iodine appears to be in organic combination and associated with the milk serum. Scharrer and Schwaibold[335] found 4.5% of the total iodine from normal cow's milk with the fat, 31.0% with the protein, 60.5% in organic combination in the serum, and 4.0% in inorganic combination in the serum.

Iron.—The presence of iron in milk was noted by Bunge in 1889[60] long before any of the other trace elements were detected. Iron was also the first of the trace elements to be recognized as essential to life functions. It is always present in milk, the usual range being from 100 to 900μ g per liter.[14,15,77,166,203,206,254,331] Values much higher than these [78,80,149,323] are probably from milks which have been contaminated with iron from metal containers after milking.

The feeding of supplemental iron to cows does not increase the iron content of the milk. Iron is important as a constituent of hemoglobin and as a component of some respiratory enzymes. As such it is an essential element and must be present in the animal diet. The amount needed is easily obtained since many foods contain considerable iron; but for humans, cow's milk is a poor source of dietary iron. Iron in milk is mostly in organic combination and associated with the fat globule membrane. Iron is present in the enzymes, xanthine oxidase and lactoperoxidase. It is also bound to other proteins.[15] Only a small amount of iron can be dialyzed from milk. The data in Table 1.14 and those of Kirchgessner[188] indicate the range of the iron content of milk as found over wide geographical areas.

Manganese.—The analytical values[54] for this element deserve special mention because of the agreement between so many workers. Almost all values reported for cow's milk are between 20 and 30 μ g per liter.[15,21,23,149,173,354] Broek *et al.*[54] reported a lower value, 7μ g per liter; and Krauss[202] reported slightly higher, 34 μ g per liter. The percentage of manganese in milk can be increased by supplemental feeding,[23] and the percentage in colostrum is substantially higher than in normal milk.[187] Manganese is considered an essential element in animal nutrition since it is an integral part of the liver enzyme arginase.

Since this earlier work was done, somewhat variable results have been obtained for the manganese content of milk[11,14,191,259,276,377,402,415] which show values both higher and lower than those previously reported. Thus Murthy and Rhea[259] report the values given for Mn in Table 1.14 for market milk samples collected for a period of a year in 62 cities in the United States. Other investigators[11,415] found no increase in Mn in milk when the feed contained added radioactive manganese.

Molybdenum.—Milk is a good source of the enzyme xanthine oxidase. Molybdenum is a component part of this enzyme. Richert and Westerfeld[321] found that xanthine oxidase from milk contained 0.03% molybdenum. In normal milk molybdenum is present in higher proportion than most of the other elements, and the percentage can be increased about 5 times by feeding ammonium molybdate as a supplement.[17] Recent data[14,142,362,387] show molybdenum values in milk of 24 to 60 μg/l, the content varying with intake in the feed.

Nickel.—Nickel was earlier not found as a normal component of cow's milk. Even when the cow was fed a supplement of nickel it did not appear in the milk.[18] Commercial milk which has been processed and stored in metal containers may contain a variable percentage of nickel. Kirchgessner *et al.*,[188] however, report nickel in milk as 26 (18–36) μ g/l and up to 100 μ g/l in colostrum.

Zinc.—of all the trace elements found in milk, zinc is consistently present in the largest proportion. An average value for normal milk would be 3500 μ g per l.[14,16,40,54,188,206,244,284,334] By feeding the cow zinc oxide the content in milk can be almost doubled.[16] Colostral milk contains a much larger percentage, yet both Sato and Murata[334] and Kirchgessner[187] reported as much as 13,500 μg per liter. Zinc is present in the enzyme carbonic anhydrase, and is therefore an essential nutrient for animals. However, a zinc deficiency among farm animals receiving a normal ration is unknown. Additional data are given in Table 1.14.

Radioactive Trace Elements.—The presence of naturally occurring radiation in the environment has been known for some time; but the relatively recent detonations of man-made nuclear devices, together with the knowledge that ionizing radiation is damaging to living tissue, has stimulated an interest in the significance and quantity of radionuclides in the air, water, and food supply. Milk has been shown to contain radionuclides, and as a major dietary ingredient has received much study. Naturally-occurring radionuclides in milk include potassium^{40}, carbon^{14}, radium^{226}, and in some isolated localities, certain members of the thorium decay series, such as radium^{228}, radium^{224}, thorium^{228}, lead^{212}, bismuth^{212}, and actinium^{228}. Doubtless other radionuclides, not yet identified, also occur naturally in milk. Thus Murthy[257] has shown that the average Rh^{87} content of milk varied from 14.2 to 84.8 pCi/kg for samples collected from various sections of the United States. The abundance of the radioactive form in the naturally occurring element was 27.85%.

Radionuclides reported in milk following nuclear testing are: strontium (Sr^{89},Sr^{90}) yttrium^{90} (daughter of Sr^{90}), barium^{140}, lanthanum^{140} (daughter of Ba^{140}), cerium^{144}, praseodymium^{144} (daughter of Ce^{144}), cesium^{137}, zirconium^{95}, niobium^{95} (daughter of Zr^{95}), iodine^{131}, and ruthenium^{105} with its daughter rhodium^{106}. In the future, as nuclear industry expands and applications of radioisotopes become widespread, it is probable that other radionuclides will be found in milk. For example, Perkins *et al.*[292] found zinc^{65} and chromium^{51} in milk produced on a farm irrigated with water taken downstream from the same river used as coolant for nuclear reactors.

Of the above radionuclides, those generally of most concern to public health authorities are I^{131}, Sr^{89}, Sr^{90}, Y^{90}, Cs^{137}, and Ra^{226}. Iodine^{131} concentrates in the thyroid gland. Radiostrontium and radium, being related to calcium, accumulate in bone which, of course, is adjacent to blood-forming tissue. Cesium^{137} is related to potassium and is found generally distributed throughout soft tissues.

Radionuclides are normally present in milk in extremely small traces. Their concentrations are usually measured in picocuries (micromicrocuries) per liter. One picocurie represents 2.22 radioactive decay events per minute and represents, typically, 10^{-14} gm or less of the nuclide in question. The levels of the various radionuclides are subject to wide variation. "Typical" values obtained from various surveillance programs, including those of the U. S. Public Health Service, individual states, and other countries, are available in the periodical Radiological Health Data and Reports.[309] These reports give data from the continuous collection of milk samples by a network of cooperating stations sponsored by the U. S. Public Health service. For understanding the significance and implications of radioactivity in milk, see the Staff Report No. 2 of the Federal Radiation Council.[106]

Although the concentrations of I^{131} and Sr^{90} in milk are low, it may be advisable under certain conditions to remove them. Thus certain farm practices such as land fertilization[138] and large-scale ion-exchange systems[242] permit control of these radionuclides in the milk. Paakkola[278] reports the Sr^{90} and Cs^{137} as they occur in some foods in Scandinavia and the uptake of these elements by children. Nakanishi and Sugawara[261] reported the average levels of Sr^{90} and Cs^{137} in milk of the Tohoku district in Japan to be 18.6 and 46.6 pc/l in summer and 10.9 and 20.3 pc/l in winter, respectively. These values show a substantial decline in comparison with results two years earlier.

Gases

Milk contains dissolved oxygen, nitrogen, and carbon dioxide. This is true whether the milk is obtained anaerobically or with access to the air as shown in Table 1.15. The proportions of these gases in milk change easily depending on the treatment given the milk. The carbon dioxide content, which is high immediately after milking, decreases when exposed to air and agitation, while the oxygen and nitrogen content increase. Heating, cooling, aeration, or vacuum treatment

Table 1.15

GAS CONTENT (VOLUME %) OF MILK IMMEDIATELY AFTER MILKING[a]

	Oxygen	Nitrogen	Carbon Dioxide	Total Gas
Anaerobically drawn	0.12–0.14	0.82–1.02	3.44–4.96	4.40–6.10
Regular milking	0.86	1.71–2.19	2.93–3.78	5.50–6.84

[a] Data of Marshall[241] as recalculated to volume % by Noll and Supplee.[269]

affect the gas content of milk in accordance with the gas laws governing the solubility of gases in liquids. Commercial raw milk as it arrives at the dairy plant will have established a gaseous equilibrium and will contain the three gases in proportions similar to those in Table 1.16.

Dissolved nitrogen is an inert and passive component of milk.

Carbon dioxide can exist in milk as carbonic acid or as bicarbonate.[394] As carbonic acid it contributes to the acidity of milk. Milk exposed to the air after milking decreases in titratable acidity for a time due to loss of carbonic acid.

Oxygen is known for its undesirable effects in milk, such as oxidized flavor development and the loss of ascorbic acid.

Sheets and Lopez[351] report the average total gas, CO_2, O_2, and N_2 for the milk layer and the cream layer for commercial pasteurized milk as well as for pasteurized homogenized milk. Kreula and Moisio[205] determined N_2, O_2, and CO_2 by a chromatographic procedure and report results in terms of mg/l.

ENZYMES

Milk contains a number of enzymes. An enzyme is a biologic catalyst elaborated by a living cell, milk enzymes being elaborated by the cells of the mammary tissue. It is not yet clear whether the enzymes in milk serve some purpose or whether they should be considered as extraneous material introduced into milk during the secretory process. The enzymes produced by bacteria in the milk cannot be considered as normal components of milk. Enzymes are proteins. They are denatured and inactivated by high temperatures, they possess a pH of optimum activity, and they exhibit specificity for certain substrates. Comprehensive reviews have been published including quantitative data on the role and significance of the enzymes of milk (Shahani,[349] McKenzie,[235] and Syväoja and Virtanen[371]).

Table 1.16

GAS CONTENT (VOLUME %) OF COMMERCIAL MIXED RAW MILK

Source of Data		Oxygen	Nitrogen	Carbon Dioxide	Total Gas
Noll and Supplee[269]	Minimum	0.30	1.18	3.44	4.92
(63 samples)	Maximum	0.59	1.63	6.28	8.50
	Average	0.47	1.29	4.45	6.21
Sheets and Lopez[351]	Pasteurized Milk (Cream Layer)	0.19	1.70	1.19	3.08
	Pasteurized Milk (Milk Layer)	0.29	2.80	1.29	4.30
	Pasteurized, Homogenized Milk	0.07	2.21	0.43	2.70
Kreula and Moisio[205]	Fresh Raw Milk	2.9	12.0	72	
(mg/l)	Processed Milk	8.7	15.0	27	

Lipase

A lipase is an enzyme which catalyzes the hydrolysis of fats to glycerol and fatty acids. Rogers[326] in 1904 found lipase to be the cause of increased acidity in canned butter, and Maasz[236] in 1909 offered convincing evidence for the presence of lipase in milk. Though bitter, rancid flavors were observed earlier in products made from homogenized milk, it was not until 1931 that it was shown that raw milk became rancid as a result of homogenization and that the lipase causing this development of objectionable flavors could be inactivated by holding the milk at 55°C for 30 min.[91]

Milk lipase is a mixture of several enzymes and is capable of hydrolyzing many types of fats. Tarassuk and Frankel[375] agree that these lipases are present in the plasma of fresh milk, but differentiate between "plasma lipase" which remains in the plasma when the milk is cooled, and "membrane lipase" which is irreversibly adsorbed on the fat globule membrane as fresh milk is cooled. The "plasma lipase" requires an activation treatment (homogenization or agitation) before it produces rancidity, whereas the "membrane lipase" requires cooling to place it in contact with the substrate. Cows late in lactation and on dry feed contain more of this "membrane lipase," and the milk is subject to spontaneous lipolysis[375]; but many investigators do not agree that lipolytic activity increases toward the end of lactation.[146,325] Lipase has not yet been crystallized or obtained in pure form, and the percentage in milk is still in doubt. Downey and Andrews[94] have evaluated the evidence for the presence of several lipases in milk. Wallander and Swanson[403] have investigated the effect of heat treatment on lipase activity in milk and noted a decrease as the temperature was increased. However, complete inactivation did not occur at 73.9 or 85°C for 30 min.

Chandan, Shahani, and co-workers[66,67] have developed quantitative methods for determining the lipase activity of milk, expressing the activity in units (μ equivalents of acid liberated/minute) per ml of milk. They report values of 1.0 to 1.7 units/ml for fresh raw cow's milk. Finnish workers[228] report substantially equivalent values. Several excellent reviews[162,212,360,387] have been published on lipase phenomena.

The milk lipase system is discussed in Chapter 5.

Esterase

An esterase is an enzyme which catalyzes the hydrolysis of esters. Lipase is an esterase, but the two words are not considered to be synonymous. Milk exhibits considerable esterase activity in addition to lipase activity. Forster *et al.*[111] identified three esterases in milk and classified them as A, B, or C, similar to a classification used for blood esterases.

A-esterases (Arylesterases).—Typical aromatic esterases which hydrolyze phenyl acetate at a higher rate than phenylbutyrate. Aliphatic esters are normally not attacked. It is resistant to both organophosphorus and physostigmine.

B-esterases (Aliphatic Esterases, Lipases).—Both aliphatic and aromatic esters, but not choline esters, are hydrolyzed. It is sensitive to organophosphorus compounds but resistant to 10^{-5} physostigmine. Some B-esterases are sensitive to the latter compound.

C-esterases (Cholinesterases).—Choline esters are split at a higher rate than both aliphatic and aromatic esters, the latter usually being hydrolyzed at a lower rate than aliphatic esters or not at all. They are sensitive to organophosphorus compounds and to physostigmine, the latter giving complete inhibition at 10^{-5} or lower concentrations.

Activities as measured by the Warburg manometric techniques are expressed as A_{10}, that is, the μ liters of CO_2 evolved by 1 ml of milk during 30 min of incubation under prescribed conditions of temperature, pH, substrate, and inhibitor. Typical values for the A-, B-, and C-esterases are 60, 234, and 20, respectively. Esterase-A was high in milk from diseased udders. Esterase-B showed considerable variation in the milk from different cows.

Other investigators[239,240,249,307] have sought to expedite the method for determining the esterases, particularly A-esterase, also observing intercow variations, effects of lactation stage, and the higher values in colostrum and in mastitic milk.

Alkaline Phosphatase

This enzyme, also known as alkaline phosphomonoesterase, catalyzes the hydrolysis of organic phosphates, yielding an alcohol or a phenol and phosphoric acid. It is a native enzyme of milk, always present, but is destroyed during pasteurization; hence, its absence is an index to the effectiveness of pasteurization. The well-known "phosphatase test" is based on this characteristic. Alkaline phosphatase has not been prepared in crystalline form, but highly purified preparations have been studied. From 30 to 40% of the enzyme is concentrated in the cream adsorbed on the fat globule, and the balance is dispersed throughout the skimmilk, probably in the lipoprotein particles.[411] The pH of optimum activity is about 9.6. Wüthrich and co-workers[419] report the average alkaline phosphatase activity of milk to be 160 IU for normal milk and 30 units for colostrum. The data of Kitchen, Taylor, and White[193] in Table 1.17 show comparable results but with a considerable range in values. Quantitative data on several other milk enzymes are given in this table.

Table 1.17

RANGE OF ENZYME ACTIVITIES IN NORMAL AND IN MASTITIC WHOLE MILK[193]

Enzyme	Activity, units/ml[a]
Catalase (normal milk)	7.5–36.0
Catalase (mastitic milk)	175.0[b]
Xanthine oxidase (normal milk)	15.6–21.4
Xanthine oxidase (mastitic milk)	9.4[b]
Aldolase (normal milk)	5.04–8.7
Aldolase (mastitic milk)	56.5[b]
Ribonuclease	19.2–35.4
Acid phosphatase	0.0026–0.0037
Alkaline phosphatase	0.18–0.27

[a] Catalase, xanthine oxidase, acid phosphatase, and alkaline phosphatase activities are expressed in μ moles product formed/min. Aldolase and ribonuclease activities are expressed in units as described in the text. The activities expressed for ribonuclease, acid phosphatase, and alkaline phosphates are from normal milk.
[b] Bulked milk from infected animals.

Alkaline phosphatase exhibits the characteristic of reappearing or becoming reactivated in heated milk that has been held. Kresheck and Harper,[204] and Battiati and Corso[41] have studied this phenomenon.

Since the "phosphatase test" is used in determining the effectiveness of milk pasteurization, extensive investigations have been conducted in improving, modifying, accelerating, or automating the procedure.[194,197,262,267,270,318,361]

Acid Phosphatase

An acid phosphomonoesterase has been reported in milk in low concentrations.[255] It is found in the skimmilk, and its optimum pH is about 4.0. It is unstable when exposed to sunlight or ultraviolet radiation but is very heat-resistant, requiring 5 min at 96°C for complete destruction. Zittle[423] has reviewed the properties of acid phosphatase. Wuthrich[419] reports the activity of acid phosphatase to be 72 IU for normal cow's milk and 14 units for colostrum. Kitchen *et al.* give quantitative data for the phosphatases.

Xanthine Oxidase

Milk is the best-known source of the enzyme xanthine oxidase. An oxidase is an enzyme which catalyzes the addition of oxygen to a substance or the removal of hydrogen from it. When the emphasis is on removal of hydrogen, the term dehydrogenase is sometimes used; when the point of view is that of the donor of oxygen, the term reductase is applied.

The presence of an oxidation catalyst in milk was announced in 1902 by Schardinger,[334A] by whose name it is frequently designated.

In the presence of a suitable oxidant, Schardinger's enzyme catalyzes the oxidation of aldehydes; hence the names aldehyde oxidase and aldehydrase are sometimes encountered. This enzyme was found in milk by Koning,[196] Harden and Lane-Claypon,[139] and Stetter,[365] all of whom called it reductase because of its ability to cause reduction of methylene blue to its leuco base in the presence of an aldehyde. They stated that it is more abundant in fat and cream than in skimmilk, and according to Koning is not present in separator slime. It is more abundant in the last part of the milking and in milk from diseased udders. It is not destroyed by heating for 30 min at 65°C. Piccard and Rising[297] showed that it is active up to the coagulating temperature of albumin, 72 to 80°C. They also reported that some of the enzyme remains after the fat and casein have been removed from milk.

Morgan, Stewart, and Hopkins[250] in 1922 found in milk an enzyme capable of oxidizing the purine bases, hypoxanthine and xanthine, to uric acid. Shortly afterward, Haas and Hill[133] reported in milk an adenine oxidase and a nitrate-reducing enzyme active in the presence of an aldehyde. Subsequently, Dixon and Thurlow[86] confirmed both these findings, but presented evidence which suggested that xanthine oxidase, adenine oxidase, and nitrate-reducing enzymes are identical with aldehyde oxidase. At present, all these enzymes are considered to be one and the same; and the name xanthine oxidase is used for all of them.

Xanthine oxidase was crystallized in a high degree of purity by Avis, Bergel, and Bray[28] in 1955. The structure and properties of the enzyme have been reviewed by Whitney.[411] The structure consists of protein, flavin adenine dinucleotide, molybdenum, and iron in the molar ratio of 1:2:1.4:8. The enzyme has a molecular weight of about 300,000, and the isoelectric point is at pH 5.3–5.4. Greenbank[127] has estimated that the xanthine oxidase content of milk averages 160 mg per liter.

Other investigators[142,179,193,423] have also characterized xanthine oxidase. Hart and co-workers[142] noted that the xanthine oxidase content correlated well with the molybdenum content of the milk, but that when the molybdenum content of the feed was increased by adding molybdenum salts, there was no corresponding increase in the milk xanthine oxidase activity. Table 1.17 from data of Kitchen *et al.* gives quantitative data on the range of xanthine oxidase activities in normal and mastitic milks.

Lactoperoxidase

A peroxidase is an enzyme which catalyzes the transfer of oxygen from peroxides, especially hydrogen peroxide, to other substances. The presence of a peroxidase in milk was first demonstrated by Arnold[25] in 1881 and later by others.[139,196,301,397] All milk contains peroxidase. The content varies somewhat, but milk is one of the best known sources of peroxidase. Milk-peroxidase is known as lactoperoxidase; it is a heme protein with an iron content of about 0.07%.[301] It has been crystallized in pure form by Polis and Shmukler[301] and found to have a molecular weight of about 82,000. Its optimum pH is 6.8, and pasteurization temperatures do not inactivate it.

Woerner[417] has updated the characterization of lactoperoxidase, and other investigators[85,327,370] have added confirmatory data on the molecular weight, conditions of activity, iron content, and resistance to heat. Wüthrich and co-workers[419] have reported the average peroxidase activity of milk as 22,000 IU with a standard deviation of 1612. For colostrum the values were 29,200 (SD 2,000). As indicated previously, an IU is equal to the μM substrate/minute/1,000 ml milk at 37°C.

Protease

A protease is an enzyme which catalyzes the hydrolysis of the peptide linkages of proteins to produce smaller protein fragments, i.e., peptones, proteoses, peptides, amino acids, and ammonia. The presence of a proteolytic enzyme in milk was first reported by Babcock and Russell[29] in 1897 and given the misleading name of "galactase." Thatcher and Dahlberg[379] in 1917 also demonstrated the presence of protease in normal milk, and since then it has been fully established that protease is a normal component of milk.

Warner and Polis[406] found that almost all the protease of milk is precipitated with the casein, and they developed a procedure for concentrating it to obtain a 150-fold increase in activity. Milk protease has not yet been isolated in pure form. It acts on casein in a neutral or slightly alkaline medium; its action is retarded in an acid medium and is inhibited by 1% chloroform or by 15% sodium chloride.[279] Its optimal activity is at pH 8.5.[406] It is inactivated by heat at 75 to 80°C, and in acid solution is destroyed at 72°C in 10 min.

Harper and associates[141] incubated aseptically drawn milk in 0.2*M* borax buffer solutions at pH 8.5 for 6 to 8 hr at 38°C. They found tryosine to be the only amino acid produced and designated the protease activity of milk in terms of increases in tyrosine. For whole raw milk, proteolytic activity showed a range of 4 to 126 μg/5 ml, with an average of 57.0. It was concluded that the protease activity of milk was not likely to have any serious effect in milk processing.

Other investigators[170,222] have been concerned with the heat stability, variability, distribution, and determination of protease in milk and the subject has been reviewed by Zittle.[423]

Amylase

Amylases are enzymes which catalyze the hydrolysis of starch to dextrin or maltose. An amylase that catalyzes the hydrolysis of central glucosidic linkages in the starch molecule, thus producing dextrinization, is designated as alpha-amylase; one that catalyzes saccharification is called beta-amylase. In practice the distinction appears not to be sharp. Richardson and Hankinson[320] have shown that cow's milk possesses both dextrinizing and saccharifying activity, and hence believe that it contains both alpha- and beta-amylase. Alpha-amylase is considerably inactivated by heating milk at 55°C for 30 min; beta-amylase withstands 65°C for 30 min without loss of activity. Guy and Jenness[132] prepared a highly concentrated alpha-amylase system from the protein fraction precipitated from whey. It appears to be a single enzyme with a single optimum pH at 7.4 at 34°C. Both calcium and chloride ions must be present for alpha-amylase activity. Alpha-amylase is definitely a normal, native constituent of cow's milk, and it is accepted that some milks contain a low percentage of beta-amylase.

Posch-Czubik[303] has presented a comprehensive study on milk amylases. Pavel[290] has pointed out the changes in amylase activity of milk as affected by individuality of the cow, lactation period, and diet. Wüthrich and Richterich[419] noted the effects of heat treatment on the enzyme and its distribution between skimmilk, cream, and sediment on centrifugation of milk. Syväoja and Virtanen[371] have measured the alpha-amylase activity in milk of cows on normal and modified rations. With the normal or control ration the alpha-amylase activity of the milk was 1029 IU (μM/minute/1000 ml/37°C). Some data are also given by Syväoja and Virtanen in their respective units for aldolase, alkaline phosphatase, catalase, peroxidase, and xanthine oxidase.

Catalase

A catalase is an enzyme that catalyzes the decomposition of hydrogen peroxide to water and inactive oxygen. The presence of catalase in normal cow's milk was first demonstrated by Raudnitz.[312] That milk catalase is a secreted enzyme has been shown by Harden and Lane-Claypon.[139]

The percentage of catalase in milk differs with the breed of the cow, with individual animals, and with different intervals between milkings. The last portion of milk drawn contains the largest porportion of catalase, but milk containing a high percentage of fat does not necessarily contain a high proportion of catalase. Colostral milk and milk from cows in the last stage of lactation contain high percentages of catalase. The feeding of green feeds increases the catalase content of milk. An increased catalase content accompanies appreciable numbers of bacteria or leucocytes in milk. For this reason the catalase content of milk has been proposed as a means of determining the quality of milk and of detecting milk from diseased udders. The maximal activity of the enzyme at 0°C is at pH 6.8 to 7.0.[34] Its activity is retarded in acid solution. Heating milk for a half hour at 65 to 75°C completely destroys the catalase, and its activity is greatly impaired if it is heated for the same length of time at somewhat lower temperatures. This fact is the basis of a test for determining whether milk has been heated. Cream contains more catalase than the milk from which it has been separated, and a still greater proportion is to be found in the separator slime. On coagulation of milk the catalase is precipitated with the casein. Recent data on the activity of catalase in normal and mastitic milk is found in Table 1.17.

Aldolase

The enzyme aldolase reversibly splits fructose 1,6-diphosphate into dihydroxyacetone phosphate and phosphoglyceric aldehyde. Polis and Shmukler[300] have found this enzyme to be present in fresh milk in the same concentration range as in blood serum. It is present in greater concentration in cream than in milk. It is unstable when in milk, but its stability is enhanced by purification. It is completely inactivated in milk by heating at 45°C for 20 min. Aldolase activity in normal and mastitic milk is given in Table 1.17, and was determined by use of 1.0 ml of enzyme solution, 0.6 ml of 0.5*M* tris buffer of pH 8.6, and 0.2 ml of 0.56*M* hydrazine of pH 8.6, the reaction being started by addition of 0.2 ml of 0.05*M* fructose 1,6-diphosphate. One unit of activity was equivalent to 0.0445 μ moles of fructose 1,6-diphosphate split/hour at 37°C.

Ribonuclease

This enzyme hydrolyzes nucleic acid to its component nucleotides. It is present in milk in relatively large amounts. The enzyme is quite heat-stable, there being little loss when heated to 90°C for 20 min at pH 3.5; however, under the same conditions at pH 7, all the activity is lost.[179,425] Kiermeier and Hundt[179] found optimum activity at pH 6.9 and 60°C. Ribonuclease activity is not reduced by pasteurization. Chandan and co-workers[67] give the ribonuclease content of milk as 1100 μ g/100 ml. Additional values are expressed as 5 × crystallized ribonuclease, given in Table 1.17., in units of which one unit of activity is equivalent to that amount of acid-soluble oligonucleotide which causes an increase in absorbance of 1.0 at 260 mμ.

Lysozyme

The possibility that milk had natural antibacterial property was observed many years ago, but its quantitative assay has been possible only recently. The assay method involves preparation of milk whey as the lysozyme source, mixing it with a substrate consisting of dead cells of *Micrococcus lysodeikticus,* and observing the extent of lysis spectrophotometrically. The cell suspension is prepared in a Sörensen buffer at pH 6.24 to yield a light transmission of 10 or 30% at 540 mμ with distilled water at 100% transmission. The change in optical density before and after 20 min incubation at 37°C is a measure of lysozyme concentration. Shahani and co-workers[350] found the lysozyme content of milk from 67 cows to vary between 0 and 260 μg/100 ml. The average lysozyme contents of Brown Swiss, Guernsey, Holstein, and Jersey milks were 21, 15, 11, and 5μg/100ml, respectively. Stage of lactation or milk yield did not affect lysozyme secretion. In a later publication Chandan and co-workers[67] report the lysozyme content of cow's milk as 13 μ g/100 ml. For human milk it is 40,000 μ g/100 ml.

Carbonic Anhydrase

This enzyme catalyzes the hydration of carbon dioxide and the reverse reaction, the dehydration of carbonic acid. It is a zinc-containing protein associated with the red blood cells and is an important enzyme in the animal body. Its purpose or value in milk is unknown; and, as is the case with other milk enzymes, its presence in milk probably occurs unintentionally during the secretory process. It has been reported in cow, goat, and sheep milk.[114] Kitchen *et al.*[193] found no carbonic anhydrase in milk they tested.

Salolase

A salolase is an enzyme which catalyzes the hydrolysis of a salicylate, specifically, phenyl salicylate. The presence of salolase in milk has been reported by Vandevelde.[392] Grimmer[130] showed its presence in the mammary glands of cows.

Rhodonase

This enzyme catalyzes the conversion of cyanide into thiocyanate. Rhodonase activity has been reported in cow's milk and somewhat higher activity in the milk of goats and sheep.[114]

Beta-Galactosidase (Lactase)

Beta-galactosidase is an enzyme which catalyzes the hydrolysis of lactose to glucose and galactose. Some investigators have reported low percentages of lactase in milk; others have been unable to demonstrate its presence. Whether normal to milk or of bacterial origin, the presence or absence of a lactase in milk has not been definitely established.

Other Enzymes

By the very nature of milk, other enzymes will be identified and their action quantitatively determined. Mention can be made at this time of transaminase and sorbitol anhydrase,[265] phosphohexose isomerase,[148] sulfhydryl oxidase,[177] lactate dehydrogenase,[144] beta-glucuronidase,[178] thromboplastin,[424] and *p*-diamine oxidase.[419] Quantitative data are already available on some of these. Thus Wüthrich and co-workers[419] report 3180 IU of *p*-diamine oxidase for normal cow's milk and SD 1070.

NONPROTEIN NITROGENOUS SUBSTANCES

The total nonprotein nitrogen of milk has been reported as 25 to 30 mg per 100 ml of milk,[237,266,348] which constitutes 5 to 6% of the total nitrogen in milk. The nonprotein nitrogen components listed in Table 1.18 are known to be the end products of nitrogen metabolism in the animal body and are presumably introduced into milk directly from the blood. The amount in milk will vary with the individual animal, the breed, and the feed being consumed; but Nickerson[266] found no significant change due to season of the year. The unaccounted nitrogen in Table 1.18 includes such compounds as hippuric acid, 5.1 mg per 100 ml skimmilk[289]; orotic acid, 5 to 10 mg per 100 ml milk[137]; indican, 0.124 mg per 100 ml milk[357]; and the compounds identified by Schwartz and Pallansch,[340] phosphoglyceroethanolamine, *o*-phosphoethanolamine, and phenylacetylglutamine.

TABLE 1.18

NONPROTEIN NITROGEN COMPONENTS IN COW'S MILK[348]
(MG PER 100 ML MILK)

Component	Average of Milk Samples	
	14 Individual	14 Mixed
Total nonprotein N	28.1	23.8
Ammonia N	0.59	0.67
Urea N	13.1	8.38
Creatinine[a]	0.87	0.49
Creatine[a]	3.72	3.93
Uric Acid[a]	2.32	2.28
Alpha-Amino N	4.82	3.74
Unaccounted N	7.41	8.81

[a] Reported as such.

Milk serum contains a number of free amino acids and peptides in very small proportions. Block[50] has reported glutamic acid, glycine, alanine, valine, leucines, aspartic acid, and serine. The peptides studied contained mostly glutamic acid and glycine. Schwartz and Pallansch[340] found the same amino acids as Block, plus threonine, sarcosine, proline, histidine, arginine, and they tentatively identified alpha-amino-*n*-butyric acid, phenylalanine, ornithine, and lysine.

OTHER COMPONENTS OF MILK

Flavoring Substances

Milk cannot be considered a highly flavored food; yet it does possess a delicate and, to most people, a pleasing flavor. Feed or weed flavors very often obscure the normal flavor of fresh milk. The substances, other than the major components, that give flavor to fresh milk have not been identified in detail as yet. Harper and Huber[140] report carbonyl compounds such as acetaldehyde, acetone, and formaldehyde in fresh raw milk, which are accepted as being flavor components of many foods. Patton *et al.*[289A] have identified methyl sulfide in fresh raw milk and attribute the faint "cowy" flavor of fresh milk to this substance.

Kinsella[186] has discussed the flavor chemistry of milk lipids, and Tamsma *et al.*[374] have reviewed the contribution of milk fat to the flavor of milk. Pangborn and Dunkley[280] have evaluated salt and sweetness characteristics of milk. Honkanen *et al.*[153] have noted that the presence of ketones, alcohols, and some esters in the feed of the cows can have a deleterious effect on milk flavor.

Phospholipids

Lecithin, cephalin, and sphingomyelin, which are present in milk, are known collectively as phospholipids or phosphatides. They are fat-like substances, contain phosphorus and nitrogen, are associated with the protein of milk, and are easily oxidized. Values of 0.028, 0.031, 0.034, and 0.037% for total phospholipid content are believed to be true estimates for whole milk; values in a higher range probably include other phosphorus compounds.[150,154,247,317] The trace amounts of cerebrosides found in milk are usually included with the phosphatides.

Walstra[405] found the phospholipid content of milk to be 0.024%. On producing higher fat content products up to a 40% cream, the phospholipid content increased to 0.204%. Patton *et al.*[288A] and co-workers observed that the same classes of phospholipids (lecithin, cephalin, sphingomyelin, etc.) with a similar fatty acid composition were present in similar proportions in milk, skimmilk, cream, and buttermilk. This is taken to indicate that phospholipids occur in units of relatively homogeneous composition.

Other investigators[87,251,358,376] have confirmed the results of earlier investigators for the percentage of phopholipids in milk but have added extensive information on constituents of the total phosphatide, and on the effects of feed, pasture, and lactation stage on milk phosphatide content. The phospholipids are discussed in detail in Chapter 4.

Sterols

The dominant sterol in milk is cholesterol. Low percentages of lanosterol and vitamin D are also present. Cholesterol in milk has been reported to be present to the extent of 120,[13] 110,[263] and 140[85] ppm. It has been stated that the percentage of cholesterol in milk varies in direct proportion to that of fat present.[85] Undoubtedly, this is not precisely true since, although the greater the percentage of fat in milk, the greater the percentage of cholesterol,[263] not all the cholesterol is associated with the fat, approximately 18% being associated with the proteins.[13] Recent results[151,159,238,382] for cholesterol in milk are in the range reported earlier. DeMan[238] noted that cholesterol was present in milk as free cholesterol. Intrieri and Badolato,[159] however, found a certain portion of it to be in ester form. These workers also showed decreasing contents in whole milk, partially skimmed milk, and skimmilk. Timmen[382] has reported on the major hydroxy compounds in milk lipids. Brewington, Caress, and Schwartz[56] have found dihydrolanosterol in milk.

Carbohydrates Other Than Lactose

Lactose is no longer considered to be the only carbohydrate in milk. Free glucose and galactose have been shown to be present in fresh cow's milk. Limited data indicate a glucose content of about 7 mg per 100 ml milk[10A,152] and about 2 mg per 100 ml milk for galactose.[152] Phosphate esters of glucose, galactose, and lactose are present in milk as well as carbohydrate-protein combinations, which on hydrolysis yield one or more of the following: glucosamine, galactosamine, neuraminic acid, mannose, galactose, or fucose.[3] Oligosaccharides are also known to be present in milk. Trucco *et al.*[385] studied 7 different ones from cow's milk and on hydrolysis found them to be made up of 2, 3, or 4 of the following: lactose, glucose, galactose, neuraminic acid, mannose, and acetyl-glucosamine. The *Lactobacillus bifidus* growth factors in milk have been shown to be substances of oligosaccharide nature.[3]

Barker and Stacey[35] have reviewed the carbohydrates of milk and expanded on the information previously reported. Reineccius and co-workers[314] have noted the following saccharides in milk as indicated by gas-liquid chromatographic patterns: alpha- and beta-glucose, alpha-, beta-, and gamma-galactose, and alpha- and beta-lactose. Free glucose and galactose contents were 13.8 and 11.7 mg/100 ml. These were the only free sugars found. Clamp and Dawson[70] report further on the saccharide residues existing in milk glycoproteins. Gatschew[118] also studied the monosaccharide content of milk, particularly glucose.

Vitamins

The vitamins listed in Table 1.19 are normally present in fresh milk. The values given are representative and were selected particularly by Hartman and Dryden[143] from those in the literature. A detailed discussion of the vitamins in milk is given in Chapter 7. Macy *et al.*[237] have reported somewhat comparable results. Recent data on pantothenic acid and vitamins B_6 and B_{12} are reported by Orr.[274]

Pigments

Milk contains fat-soluble and water-soluble pigments which are largely the carotenes and riboflavin, called earlier lactoflavin. Since

Table 1.19

VITAMINS IN FRESH MILK[143]

	Mg/100 ml	Range
Vitamin A[a]	159	136–176
Carotenoids[213]	0.030	0.025–0.060
Vitamin D[b,188]	2.21	0–10.9
Vitamin E	0.100	0.02–0.18
Vitamin K	0.00467	0.0–0.0160
Vitamin C	2.09	1.57–2.75
Biotin	0.003	0.0012–0.0060
Choline	13.7	4.3–28.5
Folacin	0.0059	0.0038–0.0090
Myo-Inositol (total)[c]	11.0	6.0–18.0
Niacin	0.09	0.03–0.20
Pantothenic acid	0.34	0.26–0.49
Riboflavin	0.17	0.08–0.26
Thiamine	0.04	0.02–0.08
Vitamin B-6	0.06	0.02–0.08
Vitamin B-12	0.00042	0.00024–0.00074
p-Aminobenzoic acid	0.01	0.004–0.015

[a] Expressed as IU/100 ml (0.048 mg/100 ml[213]).
[b] Expressed as IU/100 ml
[c] About ¾ of it exists as free inositol.

these pigments are also important vitamins they are discussed in Chapter 7. Hartman and Dryden[143] have reviewed the factors affecting the color of milk and the components of milk that affect the color of dairy products.

The yellow appearance of milkfat is caused by fat-soluble carotene. Water-soluble riboflavin produces a slight yellow or greenish tint in skimmilk which becomes a distinct green fluorescence in whey after removal of the light-reflecting components in cheese-making.

The white or "milky" appearance of cow's milk is caused by the scattering of reflected light by the fat globules, the colloidal calcium caseinate, and the colloidal calcium phosphate in the milk. Dispersions of each of these ingredients are milky when prepared separately in concentrations similar to that occurring in milk, and, in the case the fat, with the particles the size of the globules in milk.

REFERENCES

1. ABDERHALDEN, E., Z. physiol Chem., *27*, 594 (1889).
2. ABRAMOVA, O., Molochn. Prom., *26*, 43 (1965). Cited in DSA *28*, 39 (1966).
3. ADACHI, S., and PATTON, S., J. Dairy Sci., *44*, 1375 (1961).
4. ADACHI, S., Rept. Tohoku Brch Japan. Soc. zootech. Sci., *17*, 36 (1967). Cited in DSA *31*, 398 (1969).
5. AKHMETOV, I. Z., Uzbeksk. Biol. Zh., *8*, 28 (1964). Cited in DSA *27*, 555 (1965).
6. ALDERMAN, G., and STRANKS, M. H., J. Sci. Food Agr., *18*, 151 (1967). Cited in DSA *29*, 521 (1967).
7. ALLEN, L. A., J. Dairy Res., *3*, 1 (1931).
8. ALLCROFT, R., BURNS, K. N., and HERBERT, N. C., Animal Diseases Survey Rept., Ministry Agr. Fish Food, No. 2, Part 2 (1965), H. M. Stationery Office, London.
9. AL-SAFAR, T. A., and RAMZY, M., Iraq J. Agr. Sci., *2*, 27 (1967). Cited in DSA *31*, 121 (1969).
10. AMMERMAN, C. B., THOMAS, J. W., MILLER, W. J., HOGUE, D. E., and HEMKEN, R. W., J. Dairy Sci., *53*, 1097 (1970).
10A. ANANTAKRISHNAN, C. P., and HERRINGTON, B. L., Arch. Biochem., *18*, 327 (1948).
11. ANKE, M., DIETTRICH, M., HOFFMANN, G., and JEROCH, H., Arch. Tierernähr., *17*, 81 (1967). Cited in DSA *29*, 501 (1967).
12. ANON., Latte, *44*, 19 (1970). Cited in DSA *32*, 508 (1970).
13. ANSBACHER, S., and SUPPLEE, G. C., J. Biol. Chem., *105*, 391 (1934).
14. ANTILA, P., and ANTILA, V., Intern. Dairy Congr., 18th, Sydney, 1E, 94 (1970). Cited in DSA *32*, 723 (1970).
15. ARCHIBALD, J. G., Dairy Sci. (Abst.), *20*, 711 (1958).
16. ARCHIBALD, J. G., J. Diary Sci., *27*, 257 (1944).
17. ARCHIBALD, J. G., J. Dairy Sci., *30*, 293 (1947).
18. ARCHIBALD, J. G., J. Dairy Sci., *32*, 877 (1949).
19. ARCHIBALD, J. G., J. Dairy Sci., *34*, 1026 (1951).
20. ARCHIBALD, J. G., J. Dairy Sci., *38*, 159 (1955).
21. ARCHIBALD, J. G., Milk Plant Monthly, *30*, No. 9, 36 (1941).
22. ARCHIBALD, J. G., and FENNER, H., J. Dairy Sci., *40*, 703 (1957).
23. ARCHIBALD, J. G., and LINDQUIST, H. G., J. Dairy Sci., *26*, 325 (1943).
24. ARMSTRONG, T. V., J. Dairy Sci., *42*, 1 (1959).

25. ARNOLD, C., Arch. Pharm., XVI, No. 1, 41 (1881).
26. ASHDIR, S., and PURI, B., Indian J. Pediat., *29* (171), 99 (1962). Cited in DSA *27,* 136 (1965).
27. ASHWORTH, U. S., and BLOSSER, T. H., J. Dairy Sci., *47,* 696 (1964).
28. AVIS, P. G., BERGEL, F., and BRAY, R. C., J. Chem. Soc., *1955,* 1100.
29. BABCOCK, S. M., and RUSSELL, H. L., Wis. Agr. Expt. Sta., 14th Ann. Rept., p. 161 (1897).
30. BAČIĆ, B., and VUJIČIĆ, I., Savremena Poljopriv., *3,* 213 (1963). Cited in DSA 25, 291 (1963).
31. BAČIĆ, B., and VUJIČIĆ, I., Let. naučn. Rad. Poljopriv. Fak. Novom Sadu, *8,* reprint (1964). Cited in DSA *27,* 187 (1965).
32. BAGNALL, H. H., Analyst, *68,* 148 (1943).
33. BAILEY, G. L., J Dairy Res., *19,* 102 (1952).
34. BALLS, A. K., and HALE, W. S., J. Biol. Chem., *107,* 767 (1934).
35. BARKER, S. A., and STACEY, M., Dairy Sci. Abst., *25,* 445 (1963).
36. BARTHE, L., J. pharm. chim. [6], *21,* 386 (1905).
37. BARTHEL, C., and BERGMAN, A. M., Z. Nahr. Genussm., *26,* 238 (1913).
38. BARTOLETTUS, FABRITIUS, "Enzyclopaedia hermetico-dogmatics," p. 168, Bononiae (1619).
39. BARTOLETTUS, FABRITIUS, "Methodus in Dyspnoeam seu de Respirationibus," Liber V, p. 400, Bononiae (1633).
40. BAS, J. M. VAN DER, and MULDER, H., Neth. Milk Dairy J., *18,* 103 (1964). Cited in DSA *26,* 601 (1964).
41. BATTIATI, G., and CORSO, G., Genoa Ig. Med. Prev., *5,* 326 (1964). Cited in DSA *28,* 212 (1966).
42. BATTISTOTTI, B., Lattee, *37,* 754 (1964). Cited in DSA *27,* 32 (1965).
43. BECK, A.B., Australian J. Exptl. Biol. Med. Sci., *19,* 145 (1941).
44. BECROFT, T. C., Med. J. Australia, 1967-II, 598. Cited in DSA *30,* 432 (1968).
45. BELCHEVA, M., MATROVA, TS., and SEPETLIEV, D., Akusherstvo Ginekol., Sofia, *5,* 157 (1966). Cited in DSA *31,* 530 (1969).
46. BENTON, A. G., J. Dairy Sci., *12,* 481 (1929).
47. BERTOLANI, G., and GAVIOLI, E., Scienza Tecnica latt-casear. Parma., *20,* 315 (1969). Cited in DSA *32,* 258 (1970).
48. BERTRAND, G., and MÂCHEBOEUF, M., C. R. Acad. Sci., Paris, *180,* 1380 (1925).
49. BINNERTS, W. T., Neth. Milk Dairy J., *21,* 3 (1967). Cited in DSA *29,* 521 (1967).
50. BLOCK, R. J., J. Dairy Sci., *34,* 1 (1951).
51. BONDAREVSKII, N. A., Zhivotnovodstvo, Mosk., *27,* 63 (1965). Cited in DSA 27, 604 (1965).
52. BOUCHARDAT, A., and QUEVENNE, T. A., "Du Lait," Paris (1957).
53. BRAND, N., and GEDALIA, I., J. S. African Vet. Med. Assoc., *34,* 33 (1963). Cited in DSA *26,* 30 (1964).
54. BROEK, A., and WOLFF, L. K., Acta Brevia Neerland, Physiol, Pharmacol., Microbiol., *5,* 80 (1935).
55. BROWN, H. E., Am. Creamery & Poultry Rev., *73,* 1048 (1932).
56. BREWINGTON, C. R., CARESS, E. A., and SCHWARTZ, D. P., J. Lipid Res., *11,* 355 (1970). Cited in DSA *32,* 663 (1970).
57. BRUSILOVSKII, L. P., Tr. Vses. Nauchn.-Issled. Inst. Molochn. Prom., *27,* 180 (1970). Cited in DSA *32,* 761 (1970).
58. BUECHER, M., Ecol. Natl. Vet. Lyon: Thèse No. 9 (1965). Cited in DSA *28,* 262 (1966).
59. BUDSLAWSKI, J., DAMICZ, W., TOMASIK, M., and RUSIECKI, M., Zeszyty Nauk. Wyzsz. Szk. roln. Olsztyn., *21,* 365 (1966). Cited in DSA *29,* 112 (1967).
60. BUNGE, G., Z. physiol. Chem., *13,* 399 (1889).
61. BURT, A. W. A., and DUNTON, C. R., Proc. Nutr. Soc., *26,* 181 (1967). Cited in DSA *30,* 137 (1968).
62. BUTCHER, K. R., SARGENT, F. D., and LEGATES, J. E., J. Dairy Sci., *50,* 185 (1967).

63. CADBURY, W. W., Am. J. Diseases Children, *19,* 38 (1920).
64. CAMPBELL, C., Dairy Farmers Ipswich, *13,* 55 (1966). Cited in DSA *28,* 124 (1966).
65. CARLSTRÖM, A., Acta Chem. Scand., *23,* 171, 185, 203 (1969). Cited in DSA *31,* 397 (1969).
66. CHANDAN, R. C., and SHAHANI, K. M., J. Dairy Sci., *47,* 471 (1964).
67. CHANDAN, R. C., PARRY, R. M., JR., and SHAHANI, K. M., J. Dairy Sci., *51,* 606 (1968).
68. CHESHEV, K. S., and GERASIMOVA, L. K., Vopr. Pitaniya, *28,* 70 (1969). Cited in DSA *31,* 536 (1969).
69. CLAESSON, O., Dairy Farmers Ipswich, *9* (9), 36 (1962). Cited in DSA 25, 9 (1963).
70. CLAMP, J. R., DAWSON, G., and HOUGH, L., Biochim. Biophys. Acta, *148,* 342 (1967). Cited in DSA *30,* 129 (1968).
71. COBBLE, J. W., and HERMAN, H. A., Mo. Agr. Expt. Sta. Research Bull. 485 (1951).
72. COLAGHIS, S., Atti III Simp. Intern. Zootech., p. 781 (1968). Cited in DSA *31,* 495 (1969).
73. COMBERG, G., and MEYER, H., Züchtungskunde, *35,* 363 (1963). Cited in DSA *26,* 190 (1964).
74. COMBERG, G., and GRÖNING, M., Züchtungskunde, *38,* 337 (1966). Cited in DSA *29,* 112 (1967).
75. CULLITY, M., and NEEDHAM, K., Mimeograph, Perth, Western Australia Dept. Agric. (1964). Cited in DSA *27,* 182 (1965).
76. Dairy Council Dig., *42,* No. 1 (1971).
77. DAHLBERG, A. C., and CARPENTER, D. C., J. Dairy Sci., *19,* 541 (1936).
78. DAVIDSON, L. S. P., and LEITCH, I., Nutrition Abst. Rev., *3,* 901 (1934).
79. DAVIES, W. L., J. Dairy Res., *3,* 86 (1931).
80. DAVIES, W. L., J. Dairy Res., *6,* 363 (1935).
81. DAVIS, A. V., and WOODWARD, R. S., J. Dairy Sci., *49,* 744 (1966).
82. DECAEN, C., and JOURNET, M., Ann. Zootech., *15,* 259 (1966). Cited in DSA *29,* 224 (1967).
83. DELON, J., and BELLE, G., Pr. Méd., *69* (49), 2189 (1961). Cited in DSA *24,* 406 (1962).
84. DELFORNO, G., FROSIO, A., and PAGANI, F., Sci. Aliment., *15,* 283 (1969). Cited in DSA *32,* 314 (1970).
85. DENIS, W., and MINOT, A. S., J. Biol. Chem., *36,* 59 (1918).
86. DIXON, M., and THURLOW, S., Biochem. J., *18,* 989 (1924).
87. DOI, T., MORI, S., NIKI, T., and MINO, K., Japan J. Zootech. Sci., *38,* 181 (1966). Cited in DSA *29,* 528 (1967).
88. DOLBY, R. M., CREAMER, L. K., and ELLEY, E. R., N. Z. J. Dairy Technol., *4,* 18 (1969). Cited in DSA *31,* 454 (1969).
89. DOLEŽÁLEK, J., FORMAN, L., and HODAŇOVÁ, I., Sb. Vysoke Školy Chem.-Technol. Praze, *16,* 19 (1967). Cited in DSA *30,* 346 (1968).
90. DORDEVIĆ, J., Zborn. Rad. Poljopriv. Fak. Univ. Beogr., *17,* No. 490 (1969). Cited in DSA *32,* 511 (1970).
91. DORNER, W., and WIDMER, A., Lait, *11,* 545 (1931).
92. DOROTYUK, É. N., Dokl. Mosk. sel'.-khoz. Akad. K. A. Timiryazeva, *1963* (85), 138. Cited in DSA *27,* 183 (1965).
93. DOVEY, E. R., Philippine J. Sci., *8,* Section A, 151 (1913).
94. DOWNEY, W. K., and ANDREWS, P., Biochem. J., *112,* 559 (1969). Cited in DSA *31,* 453 (1969).
95. DUNKLEY, W. L., FRANKE, A. A., ROBB, J., and RONNING, M., J. Dairy Sci., *51,* 863 (1968).
96. ECHAVE, D., Rev. farm. (Buenos Aires), *78,* 337 (1936).
97. ECKLES, C. H., and SHAW, R. H., U.S. Dept. Agr., Bur. An. Ind., Bull. 155 (1913).

98. ECKLES, C. H., and SHAW, R. H., U.S. Dept. Agr., Bur. An. Ind., Bull. 156 (1913).
99. ECKLES, C. H., and SHAW, R. H., U.S. Dept. Agr., Bur. An. Ind., Bull. 157 (1913).
100. ELLIS, G. H., and THOMPSON, J. F., Ind. Eng. Chem., Anal. Ed., *17,* 254 (1945).
101. ELSDON, G. D., Analyst, *59,* 665 (1934).
102. ELVEHJEM, C. A., STEENBOCK, H., and HART, E. B., J. Biol. Chem., *83,* 27 (1929).
103. ENGEL, H., and SCHLAG, H., Milchw. Forsch., *2,* 1 (1924).
104. ENGEL, R. W., HARDISON, W. A., MILLER, R. F., PRICE, N. O., and HUBER, J. T., J. Animal Sci., *23,* 1160 (1964). Cited in DSA *27,* 124 (1965).
105. ERÄMETSÄ, O., and HEIKONEN, M., Suomen Kemist., *39B,* 101 (1966). Cited in DSA *29,* 51 (1967).
106. Federal Radiation Council, Background material for the development of radiation protection standards, Rept. No. 2, reprinted by the U. S. Department of Health, Education and Welfare, Public Health Service, Washington 25, D. C. (1961).
107. Federal and State Standards for the Composition of Milk Products, Handbook No. 51 (1971).
108. FELLENBERG, T. von, Biochem. Z., *152,* 141 (1924).
109. FENNER, H., and ARCHIBALD, J. G., J. Dairy Sci., *41,* 803 (1958).
110. FLUX, D. S., PATCHELL, M. R., CAMPBELL, I. L., and McDOWALL, F. H., The Dairy Research Inst. (N.Z.), Pub. No. 277 (1955).
111. FORSTER, T. L., MONTGOMERY, M. W., and MONTOURE, J. E., J. Dairy Sci., *44,* 1420 (1961).
112. FURUICHI, E., DOI, T., and IMAMURA, M., *et al.,* J. Japan. Soc. Food Nutr., *15,* 135 (1962); *ibid., 16,* 327 (1963). Cited in DSA *25,* 484 (1963); *ibid.,* 26, 392 (1964).
113. GACULA, M. C., JR., GAUNT, S. N., and DAMON, R. A., JR., J. Dairy Sci., *48,* 803 (1965).
114. GARCIA ALFONSO, C., and CASTELLA BERTRÁN, E., Lait, *33,* 386 (1953).
115. GARDNER, J. A., and FOX, F. W., The Practitioner, *114,* 153 (1925).
116. GARDINER, K. D., and McGANN, T. C. A., Irish J. Agr. Res., *7,* 37 (1968). Cited in DSA 30, 613 (1968).
117. GARRETT, O. F., and OVERMAN, O. R., J. Dairy Sci., *23,* 13 (1940).
118. GATSCHEW, E. P., Compt. Rend. Acad. Bulgare Sci., *19,* 751 (1966). Cited in DSA *29,* 370 (1967).
119. GAUNT, S. N., WILCOX, C. J., FARTHING, B. R., and THOMPSON, N. R., J. Dairy Sci., *51,* 1396 (1968).
120. GHOSH, S. N., and ANANTAKRISHNAN, C. P., Indian J. Dairy Sci., *16,* 190 (1963). Cited in DSA *27,* 311 (1965).
121. GLEESON, P., Farm Res. News, *9,* 107 (1968). Cited in DSA *31,* 3 (1969).
122. GOLOVNINA, A. I., Dokl. Mosk. sel'.-khoz. Akad. K. A. Timiryazeva, No. *78,* 111 (1962). Cited in DSA *28,* 124 (1966).
123. GONZALEZ-DIAZ, C., and CRAVIOTO, R. O., Anales escuela nacl. cience. biol. (Mex.), *4,* 371 (1947).
124. GRANT, A. B., and WILSON, G. F., New Zealand J. Agr. Res., *11,* 733 (1968). Cited in DSA *31,* 153 (1969).
125. GRAPPIN, R., RICORDEAU, G., MOCQUOT, G., JEUNET, R., and TASSENCOURT, L., Rev. Lait. fr. "Industrie Lait," No. *240,* 71 (1967). Cited in DSA *29,* 365 (1967).
126. GREBENNIKOV, E. P., and SOROKA, V. R., Pediatriya, *42,* 16–18 (1963). Cited in DSA *26,* 393 (1964).
127. GREENBANK, G. R., J. Dairy Sci., *37,* 644 (1954).
128. GRIGOROV, KH., Vet. Med. Nauk. Sofia, *5,* 59 (1968). Cited in DSA *31,* 96 (1969).
129. GRIGOROV, KH., Vet. Med. Nauk. Sofia, *6,* 77 (1969). Cited in DSA *32,* 129 (1970).

130. GRIMMER, W., Biochem. Z., *53*, 429 (1913).
131. GULYAEV-ZAITSEV, S., and TVERDOKHLEB, G., Molochn. Prom., *26*, 7 (1965). Cited in DSA *28*, 154 (1966).
132. GUY, E. J., and JENNESS, R., J. Dairy Sci., *41*, 13 (1958).
133. HAAS, P., and HILL, T. G., Biochem. J., *17*, 671 (1923).
134. HADJIMARKOS, D. M., and BONHORST, C. W., J. Pediat., *59*, 256 (1961).
135. HADJIMARKOS, D. M., J. Pediat., *68*, 470 (1966). Cited in DSA *28*, 325 (1966).
136. HAFEZ, E. S. E., World Rev. Animal Prod., *3*, 22 (1967). Cited in DSA *30*, 363 (1968).
137. HALLANGER, L. E., LAAKSO, J. W., and SCHULTZE, M. O., J. Biol. Chem., *202*, 83 (1953).
138. HANSEN, W. G., CAMPBELL, J. E., FOOKS, J. H., MITCHELL, H. C., and ELLER, C. H., U.S. Public Health Service Publ. 999-R-6 (1964).
139. HARDEN, A., and LANE-CLAYPON, J. E., J. Hyg., *12*, 144 (1912).
140. HARPER, W. J., and HUBER, R. M., J. Dairy Sci., *39*, 1609 (1956).
141. HARPER, W. J., ROBERTSON, J. A., JR., and GOULD, I. A., J. Dairy Sci., *43*, 1850 (1960).
142. HART, L. I., OWEN, E. C., and PROUDFOOT, R., Brit. J. Nutr., *21*, 617 (1967). Cited in DSA *30*, 56 (1968).
143. HARTMAN, A. M., and DRYDEN, L. P., Am. Dairy Sci. Assoc., Champaign, Ill., 117 pp. (1965). Also Personal Communication (1972) and Chapter 7.
144. HELLUNG-LARSEN, P., Comp. Biochem. Physiol., *27*, 703 (1968). Cited in DSA *31*, 340 (1969).
145. HERRINGTON, B. L., "Milk and Milk Processing," 1st Ed., p. 25, McGraw-Hill Book Company, Inc., New York (1948).
146. HERRINGTON, B. L., and KRUKOVSKY, V. N., J. Dairy Sci., *22*, 127 (1939).
147. HIMMEL, U., Tierzucht, *18*, 133 (1964). Cited in DSA *28*, 496 (1966).
148. HEYNDRICKX, G. V., Enzymologia, *27*, 209 (1964). Cited in DSA *27*, 31 (1965).
149. HODGES, M. A., and PETERSON, W. H., J. Am. Dietetic Assoc., *7*, 6 (1931).
150. HOLM, G. E., WRIGHT, P. A., and DEYSHER, E. F., J. Dairy Sci., *19*, 631 (1936).
151. HOMER, D. R., and VIRTANEN, A. I., Milchwiss., *22*, 1 (1967). Cited in DSA *29*, 366 (1967).
152. HONER, C. J., and TUCKEY, S. L., J. Dairy Sci., *36*, 559 (1953).
153. HONKANEN, E., KARVONEN, P., and VIRTANEN, A. I., Acta Chem. Scand., *18*, 612 (1964). Cited in DSA *26*, 540 (1964).
154. HORRALL, B. E., Purdue Agr. Expt. Sta., Bull. 401 (1935).
155. HORST, MARIA G. TER, Neth. Milk Dairy J., *17*, 162 (1963).
156. HOUSE, M. A., PIM, F. B., and SMELLIE, T. J., Intern. Dairy Congr. 18th, Sydney, 1E, 664 (1970).
157. HUBER, J. T., and BOMAN, R. L., J. Dairy Sci., *49*, 395 (1966).
158. IMAMURA, T., KATAOKA, K., and ISHII, R., Nippon Chikusan Gakkai-ho, *35*, 117 (1964). Cited in CA *63*, 2313f (1965).
159. INTRIERI, F., and BADOLATO, F., Acta Med. Vet., Napoli, *11*, 473 (1965). Cited in DSA *28*, 436 (1966).
160. JASIOROWSKI, H., and POCZYNAJLO, S., Intern. Dairy Congr. 17th, Munich, A 21 (1966).
161. JENNESS, R., and SLOAN, R. E., Dairy Sci. Abst., *32*, 599 (1970).
162. JENSEN, R. G., J. Dairy Sci., *47*, 210 (1964).
163. JOHNSON, A. H., Proceedings of the Perkin Centennial 161–9 (1956).
164. JOHNSON, K. R., J. Dairy Sci., *40*, 723 (1957).
165. JOHNSON, K. R., FOURT, D. L., HIBBS, R. A., and ROSS, R. H., J. Dairy Sci., *44*, 658 (1961).
166. JOHNSTON, F. A., Food Research, *9*, 212 (1944).
167. JOHNSTON, J. E., STONE, E. J., and FRYE, J. B., JR., Bull. Louisianna Agr. Expt. Sta. 608 (1966). Cited in DSA *29*, 143 (1967).
168. Joint Milk Quality Committee, England and Wales, Progress Report (1968).

169. JORGENSEN, K. E., and CARUOLO, E. V., J. Dairy Sci., *46,* 624 (1963).
170. KAMINOGAWA, S., YAMAUCHI, K., and TSUGO, T., Japan. J. Zootech. Sci., *40,* 559 (1969); also XVIII Intern. Dairy Congr., Sydney, 1E, *60,* (1970). Cited in DSA *32,* 379, 728 (1970).
171. KAPLAN, V. A., and BOĬKO, O. A., Molochn.-myas. Skotovod. Respub. mezhved. temat. nauch. Sb., No. *4,* 53 (1966). Cited in DSA *30,* 482 (1968).
172. KÄSTLI, P., and GERBER, H., Mitt. Lebensm. Hyg. Bern, *54,* 258 (1963). Cited in DSA *27,* 84 (1965).
173. KEMMERER, A. R., and TODD, W. R., J. Biol. Chem., *94,* 317 (1931).
174. KECSKÉS, S., Allattenyész., *12,* 101 (1963). Cited in DSA *26,* 63 (1964).
175. KIERMEIER, F., and HAISCH, K. H., Z. Tierphysiol. Tierernähr., *18,* 186 (1963). Cited in DSA *26,* 144 (1964).
176. KIERMEIER, F., and RENNER, E., Dairy Sci. Abst., *28,* 331 (1966).
177. KIERMEIER, F., and PETZ, E., Z. Lebensm.-Untersuch. Forsch., *134,* 97 (1967). Cited in DSA *29,* 636 (1967).
178. KIERMEIER, F., and GÜLL, J., Z. Lebensm.-Untersuch. Forsch., *138,* 205 (1968). Cited in DSA *31,* 96 (1969).
179. KIERMEIER, F., and HUNDT, D., Z. Lebensm.-Untersuch. Forsch., *141,* 76 (1969). Cited in DSA *32,* 314 (1970).
179A. KIERMEIER, F., 18th Intern. Dairy Congress (Sydney) Proc., II, 206 (1970).
180. KIJAK, Z., Przegl. Hodowl., *32,* 18 (1963). Cited in DSA *27,* 252 (1965).
181. KIN, Y., ARIMA, S., and HASHIMOTO, Y., J. Fac. Agr. Hokkaido Univ., *53,* 228 (1963). Cited in DSA *26,* 600 (1964).
181A. KING, R. L., and DUNKLEY, W. L., J. Dairy Sci., *42,* 420 (1959).
182. KING, J. O. L., Res. Vet. Sci., *4,* 526 (1963). Cited in DSA *26,* 43 (1964).
183. KING, J. O. L., Vet. Record, *79,* 480 (1966). Cited in DSA *29,* 148 (1967).
184. KING, J. O. L., Brit. Vet. J., *123,* 358 (1967). Cited in DSA *29,* 615 (1967).
185. KING, J. O. L., Brit. Vet. J., *125,* 63 (1969). Cited in DSA *31,* 214 (1969).
186. KINSELLA, J. E., Chem. Ind., *1969,* 36. Cited in DSA *31,* 274 (1969).
187. KIRCHGESSNER, M., Schriftenreihe Mangelkrankheiten, Heft, 6, 61 and 105 (1955).
188. KIRCHGESSNER, M., FRIESECKE, H., and KOCH, G., "Nutrition and the Composition of Milk," Lippincott and Co., Philadelphia (1967).
189. KIRCHGESSNER, M., FRIESECKE, H., and KOCH, G., "Fütterung und Milchzusammensetzung," BLV Bayerischer Landwirtschaftsverlag, Munich (1965).
190. KISZA, J., KARWOWICZ, E., and SOBINA, A., Milchwiss., *19,* 437 (1964).
191. KISZA, J., and BATURA, K., Milchwiss., *24,* 281 (1969). Cited in DSA *31,* 529 (1969).
192. KISZA, J., and BATURA, K., Milchwiss. *24,* 465 (1969). Cited in DSA *31,* 663 (1969).
193. KITCHEN, B. J., TAYLOR, G. C., and WHITE, I. C., J. Dairy Res., *37,* 279 (1970).
194. KLEYN, D. H., and LIN, S. H. C., J. Assoc. Offic. Anal. Chem., *51,* 802 (1968). Cited in DSA *30,* 562 (1968).
195. KOESTLER, G., and ELSER, E., Landw. Jahrb. Schweiz., *36,* 133 (1922).
196. KONING, C. J., Milchw. Zentr., *4,* 156 (1908).
197. KOSIKOWSKI, F. V., J. Milk Food Technol., *27,* 268 (1964). Cited in DSA *27,* 318 (1965).
198. KOSSILA, V., SEPPÄLÄ, A., and PAULAMÄKI, M., Karjantuote, *50,* 450 (1967). Cited in DSA *30,* 438 (1968).
199. KOSTOV, L., Vet. Med. Nauki, Sofia, *4,* 29 (1967). Cited in DSA *29,* 520 (1967).
200. KOSTOV, L., and DZHUROV, TS., Hig. Mlek. Zwalcz. Schorzén Grucz. mlecz., *46,* (1968). Cited in DSA *31,* 214 (1969).
201. KOTELYANSKAYA, L. I., Vopr. Pitaniya, *22,* 71 (1963). Cited in DSA *26,* 87 (1964).
202. KRAUSS, W. E., Ohio Agr. Expt. Sta. Bull. 477 (1931).
203. KRAUSS, W. E., and WASHBURN, R. G., J. Biol. Chem., *114,* 247 (1936).

204. KRESHECK, G. C., and HARPER, W. J., Milchwiss., *22,* 72 (1967). Cited in DSA *29,* 302 (1967).
205. KREULA, M., and MOISIO, T., Suomen Kemistilehti, *43B,* 51 (1970). Cited in DSA *32,* 592 (1970).
206. KREULA, M., and HEIKONEN, M., Karjantuote, *52,* 336 (1969). Cited in DSA *32,* 583 (1970).
207. LABUSSIÈRE, J., and COINDET, J., Ann. Zootech., *17,* 231 (1968). Cited in DSA *31,* 237 (1969).
208. LAMPERT, L. M., "Modern Dairy Products," Chemical Publishing Company, N.Y. (1970).
209. LAMPO, PH., WILLEMS, A., and VANSCHOUBROEK, F., Neth. Milk Dairy J., *20,* 17 (1966).
210. LAUE, W., KRETZSCHMANN, F., SCHÜTZE, P., and HORNAWSKY, G., Mh. Vet. Med., *24,* 526 (1969). Cited in DSA *32,* 658 (1970).
211. LARSON, B. L., J. Dairy Sci., *41,* 440 (1958).
212. LAWRENCE, R. C., Dairy Sci. (Abst.), *29,* 1, 59 (1967).
213. LAXMINARAYANA, H., and DASTUR, N. N., Dairy Sci. (Abst.), *30,* 177 (1968).
214. LEACH, A. E., "Food Inspection and Analysis," p. 111, John Wiley and Sons, Inc., New York (1920).
215. LEGATES, J. E., J. Dairy Sci., *43,* 1527 (1960).
216. LENOV, V. A., and TERENT'EV, M. V., Mikroelementy v Sel'sk. Khoz. i Med., Ukr. nauchno-issled. Inst. Fiziol. Rast., Akad. Nauk Ukr. SSR, Materialy 4-go (Chetvertogo) vses. Soveshch., Kiev, *1962,* 499. Cited in DSA *28,* 180 (1966).
217. LEVIN, A. B., Dokl. Mosk. Sel'skokhoz. Akad., No. *110,* 53 (1965). Cited in DSA *29,* 77 (1967).
218. LEWIS, J., and DAVIES, J., J. Soc. Dairy Technol., *18,* 218 (1965).
219. LINNERUD, A. C., WILLIAMS, J. B., and DONKER, J. D., J. Dairy Sci., *47,* 766 (1964).
220. LINNERUD, A. C., Diss. Abst. Univ. Minn., *25,* 2129 (1964). Cited in DSA *27,* 166 (1965).
221. LINTON, R. G., J. Agr. Sci., *21,* 669 (1931).
222. LIVREA, G., CAMPANELLA, S., and FAMÀ-CAMBRIA, M., Quaderni Nutr., Bologna, *24,* 1 (1964). Cited in DSA *27,* 311 (1965).
223. LÍZAL, F., and OPLETALOVA L., Živočišná Výroba, *7,* 781 (1962). Cited in DSA *25,* 141 (1963).
224. LJUNGGREN, B., Intern. Dairy Congr. 17th, Munich, *A 317* (1966).
225. LOFTUS-HILLS, G., and BROWN, B. M., Dairy Sci. Abst., *28,* 163, 219 (1966).
226. LOGANATHAN, S., and THOMPSON, N. R., J. Dairy Sci., *50,* 1009 (1967).
227. LUGININA-KOVALEVSKAYA, N. M., Gigiena i Sanit., *29* (10), 54 (1964). Cited in DSA *27,* 32 (1965).
228. LUHTALA, A., and ANTILA, M., Maataloust. Aikakausk., *40,* 171 (1968). Cited in DSA *31,* 157 (1969).
229. LJUNGGREN, B., Svenska Mejeritidn., *59,* 103 (1967). Cited in DSA *29,* 365 (1967).
230. LUSH, J. L., J. Dairy Sci., *43,* 702 (1960).
231. LYTHGOE, H. C., J. Dairy Sci., *23,* 1097 (1940).
232. McCLENDON, J. F., REMINGTON, R. E., KOLNITZ, H. von, and RUFE, R., J. Am. Chem. Soc., *52,* 541 (1930).
233. McCOY, G. C., THURMON, H. S., OLSON, H. H., and REED, A., J. Dairy Sci., *49,* 1058 (1966).
234. McDOWELL, R. E., MOODY, E. G., VAN SOEST, P. J., LEHMANN, R. P., and FORD, G. L., J. Dairy Sci., *52,* 188 (1969).
235. McKENZIE, H. A., Adv. in Protein Chem., *22,* 55 (1967).
236. MAASZ, C., Milchw. Zentr., *5,* 329 (1909).
237. MACY, I. G., KELLY, H. J., and SLOAN, R. E., Natl. Res. Council, Publication 254 (1953).
238. MAN, J. M. De, Ernährungswiss., *5,* 1 (1964). Cited in DSA *26,* 445 (1964).
239. MARQUARDT, R. R., and FORSTER, T. L., J. Dairy Sci., *48,* 1526 (1965).

240. MARQUARDT, R. R., and FORSTER, T. L., J. Dairy Sci., *48,* 1602 (1965).
241. MARSHALL, C. E., Mich. Agr. Coll. Expt. Sta. Special Bull., 16 (1902).
242. MARSHALL, R. O., SPARLING, E. M., HEINEMANN, B., and BALES, R. E., J. Dairy Sci., *51,* 673 (1968).
243. MARTIN, T. G., and FRAZEUR, D. R., J. Dairy Sci., *50,* 1009 (1967).
244. MILLER, W. J., CLIFTON, C. M., FOWLER, P. R., and PERKINS, H. F., J. Dairy Sci., *48,* 450 (1965).
245. MILLER, W. J., LAMPP, B., POWELL, G. W., SALOTTI, C. A., and BLACKMON, D. M., J. Dairy Sci., *50,* 1404 (1967).
246. MISHRA, M., and NAYAK, N. C., Indian Vet. J., *39,* 203 (1962). Cited in DSA *25,* 248 (1963).
247. MOHR, W., BROCKMANN, C., and MÜLLER, W., Molkerei Ztg., *46,* 633 (1932).
248. MONTEMURRO, O., SALERNO, A., and CIANCI, D., Annali Fac. Agr. Bari, *17,* 25 (1963). Cited in DSA *28,* 5 (1966).
249. MONTGOMERY, M. W., and FORSTER, T. L., J. Dairy Sci., *44,* 1165 (1961).
250. MORGAN, E. J., STEWART, C. P., and HOPKINS, F. G., Proc. Roy. Soc. (London), *94B,* 109 (1922).
251. MORRISON, W. R., Lipids, *3,* 101 (1968). Cited in DSA *30,* 348 (1968).
252. MUDRA, K., and SCHÖNMUTH, G., Arch. Tierzücht, *8,* 355 (1965). Cited in DSA *28,* 61 (1966).
253. MULDER, H., MENGER, J. W., and MEIJERS, P., Neth. Milk Dairy J., *18,* 52 (1964). Cited in DSA *26,* 334 (1964).
254. MULDER, H., MEIJERS, P., and MENGER, J. W., Neth. Milk Dairy J., *18,* 93 (1964). Cited in DSA *26,* 601 (1964).
255. MULLEN, J. E. C., J. Dairy Res., *17,* 288 (1950).
256. MURTHY, G. K., RHEA, U., and PEELER, J. T., J. Dairy Sci., *50,* 651 (1967).
257. MURTHY, G. K., J. Dairy Sci., *50,* 818 (1967).
258. MURTHY, G. K., and RHEA, U., J. Dairy Sci., *51,* 610 (1968).
259. MURTHY, G. K., and RHEA, U., J. Dairy Sci., *54,* 1001 (1971). Also reference 18 in reference 259, unpublished data.
260. NAKANISHI, T., YAMAJI, A., and SUGAWARA, H., Japan. J. Dairy Sci., *17,* A83 (1968). Cited in DSA *31,* 398 (1969).
261. NAKANISHI, T., and SUGAWARA, H., Japan. J. Dairy Sci., *16,* A85 (1967). Cited in DSA *29,* 633 (1967).
262. NASR, S., EL-SAWAH, H., and YOUSSEF, M. H., J. Arabian Vet. Med. Assoc., *26,* 131 (1966). Cited in DSA *29,* 476 (1967).
263. NATAF, B., MICKELSEN, O., KEYS, A., and PETERSEN, W. E., J. Nutr., *36,* 495 (1948).
264. NATZKE, R. P., SCHULZ, H. L., BARR, G. R., and HOLTMANN, W. B., J. Dairy Sci., *48,* 1295 (1965).
265. NEUMAN, V., KUDĚLKA, E., and ŠINDELÁŘOVÁ, K., Vet. Med. Praha, *10,* 349. (1965). Cited in DSA *27,* 536 (1965).
266. NICKERSON, T. A., J. Dairy Sci., *43,* 598 (1960).
267. NIELSEN, E. W., Maelkeritidende, *80,* 821 (1967). Cited in DSA *30,* 62 (1968).
268. NIELSEN, S. M., and SØRENSEN, A. N., Intern. Dairy Congr. 18th, Sydney, 1E, 606 (1970).
269. NOLL, C. I., and SUPPLEE, G. C., J. Dairy Sci., *24,* 993 (1941).
270. OBRIEN, J. E., J. Dairy Sci., *49,* 1482 (1966).
271. O'CONNOR, C. B., Irish Agr. Cream Rev., *23,* 23 (1970). Cited in DSA *32,* 516 (1970).
272. ORLA-JENSEN, S., Ann. Agr. Suisse, *5–6,* 125, 291 (1904–1905).
273. ORMISTON, E. E., SPAHR, S. L., TOUCHBERRY, R. W., and ALBRIGHT, J. L., J. Dairy Sci., *50,* 1597 (1967).
274. ORR, M. L., Home Economics Res. Rept. No. 36, USDA (1969).
275. OTAKE, Y., NAGAI, K., KANAMORI, T., KAMOSHIDA, I., KOBAYASHI, T., and TANAKA, A., Japan. J. Dairy Sci., *17,* A132 (1968). Cited in DSA *31,* 340 (1969).

276. OVERBY, A. J., and VIGH-LARSEN, M., Årsskr. K. Vet. Landbohøjsk., p. 101 (1966). Cited in DSA *28,* 375 (1966).
277. OVERMAN, O. R., J. Dairy Sci., *28,* 305 (1945).
278. PAAKKOLA, O., Beih. Zentr. Vet. Med., *11,* 88 (1970). Cited in DSA *32,* 376 (1970).
279. PALMER, L. S., and ECKLES, C. H., J. Dairy Sci., *1,* 185 (1917).
280. PANGBORN, R. M., and DUNKLEY, W. L., J. Dairy Sci., *48,* 762 (1965).
281. PANIĆ, B., STOŠIĆ, D., and SOTIROVA-PEŠEVSKA, V., Zbornik Radova Poljopriv. Fak. Univ. Beogr., *15* (451), 1–6 (1967). Cited in DSA *31,* 98 (1969).
282. PAPPEL, A., and HOGAN, G., Egypt Dept. Pub. Health, Hygienic Inst., Pub. No. 4, Cairo, Govt. Press (1914).
283. PARKASH, S., and JENNESS, R., Dairy Sci. Abst., *30,* 67 (1968).
284. PARKASH, S., and JENNESS, R., J. Dairy Sci., *50,* 127 (1967).
285. PARKER, O. F., Proc. Ruakura Farmers' Conference Week, p. 236 (1965). Cited in DSA *28,* 125 (1966).
286. PARKHIE, M. R., GILMORE, L. O., and FECHHEIMER, N. S., J. Dairy Sci., *49,* 1410 (1966).
287. PARRY, R. M., JR., CHANDAN, R. C., and SHAHANI, K. M., J. Dairy Sci., *49,* 356 (1966).
288. PATEL, R. D., and PATEL, B. M., Indian J. Dairy Sci., *16,* 76 (1963). Cited in DSA *26,* 66 (1964).
288A. PATTON, S., DURDAN, N., and McCARTHY, R. D., J. Dairy Sci., *47,* 489 (1964).
289. PATTON, S., J. Dairy Sci., *36,* 943 (1953).
289A. PATTON, S., FORSS, D. A., and DAY, E. A., J. Dairy Sci., *39,* 1469 (1956).
289B. PAULA ASSIS, F. De, ROCHA, G. L. Da, MEDINA, P., BECKER, M., and CINTRA, B., Bol. Indústr. Animal, *19,* 13 (1961). Cited in DSA *26,* 208 (1964).
290. PAVEL, J., Sborn. čsl. Akad. zeměd. Věd (Živočišná Výroba), *33,* 945 (1960). Cited in DSA *23,* 294 (1961).
291. PELTOLA, E., ANTILA, P., and MÄLKKI, Y., Meijerit. Aikakausk., 24, 3 (1963). Cited in DSA *25,* 336 (1963).
292. PERKINS, R. W., NIELSEN, J. M., ROESCH, W. C., and McCALL, R. C., Sci., *132,* 1895 (1960).
293. PETERSEN, W. E., and RIGOR, T. V., Proc. Soc. Exptl. Biol. Med., *30,* 257 (1932).
294. PETRAĬTIS, I. P., and NORKUS, A. Yu., Dokl. Vses. Akad. Sel'skokhoz. Nauk, No. *5,* 25 (1965). Cited in DSA *29,* 133 (1967).
295. PFEFFER, J. C., JACKSON, H. C., and WECKEL, K. G., J. Dairy Sci. (Abst.), *21,* 143 (1938).
296. PHILPOT, W. N., J. Dairy Sci., *50,* 978 (1967).
297. PICCARD, J., and RISING, M., J. Am. Chem. Soc., *40,* 1275 (1918).
298. PIERCE, A. W., Australian J. Exptl. Biol. Med. Sci., *12,* 7 (1934).
299. PODGÓRSKI, W., Roczniki Nauk Rolniczych (Ser. Zootech.), *87,* 87 (1965). Cited in DSA *28,* 210 (1966).
300. POLIS, B. D., and SHMUKLER, H. W., J. Dairy Sci., *33,* 619 (1950).
301. POLIS, B. D., and SHMUKLER, H. W., J. Biol. Chem., *201,* 475 (1953).
302. PORCHER, C., and CHEVALLIER, A., Lait, *3,* 97 (1923).
303. POSCH-CZUBIK, H., Oesterr. Milchwirtsch., *15,* 200 (1960). Cited in CA *59,* 2102h (1963).
304. POURIAU, A. F., and AMMANN, L., "La Laiterie," Maison Rustique, Paris (1936).
305. PROKS, J., Lait, *8,* 553 (1928).
306. PUHAČ, I., PEČAR, S., and BEŠLIN, R., Vet. Glasn., *23,* 11 (1969). Cited in DSA *32,* 271 (1970).
307. PURR, A., MATHIES, P., and KOTTER, L., Deut. Lebensm.-Rundschau, *65,* 105 (1969). Cited in DSA *31,* 404 (1969).
308. RADCLIFF, J. C., BAILEY, L. F., and HORNE, M. L., Intern. Dairy Congr. 18th, Sydney, 1E, 607 (1970).

309. Radiological Health Data and Reports. Published monthly by U.S. Public Health Service, Supt. of Documents, U.S. Government Printing Office, Washington, D.C.
310. RAHMAN, M. S., and LEIGHTON, R. E., J. Dairy Sci., *51,* 1667 (1968).
311. RATHORE, A. K., Intern. Dairy Congr. 18th, Sydney, 1E, 581 (1970).
312. RAUDNITZ. R. W., Monatssch, Kinderheilk., zehntes sammelreferate, Band 6, 281 (1907–08); elftes sammelreferate, Band 6, 579 (1907–08); zwölftes sammelreferate, Band 7, 369 (1908–09); dreizehntes and vierzehntes sammelreferate, Band 8, 233 (1909); fünfzehntes sammelreferate, Band 9, 82 (1910–11).
313. REES, H. V., Res. Bull. Dept. Agr. Tasmania, *4,* (1964). Cited in DSA *26,* 524 (1964).
314. REINECCIUS, G. A., KAVANAGH, T. E., and KEENEY, P. G., J. Dairy Sci., *53,* 1018 (1970).
315. REIS, F., and CHAKMAKJIAN, H. H., J. Biol. Chem., *98,* 237 (1932).
316. REISSIG, G., Milchw Forsch., *19,* 273 (1938).
316A. REHNBERG, G. L., STRONG, A. B., PORTER, C. R., and CARTER, M. W., Envir. Sci. Technol., *3,* 171 (1969). Cited in DSA *32,* 192 (1970).
317. REWALD, B., Lait, *17,* 225 (1937).
318. REYNOLDS, R. G., and TELFORD, W. J. P., J. Milk Food Technol., *30,* 21 (1967).
319. RICE, E. B., Dairy Sci. Abst., *32,* 525 (1970).
320. RICHARDSON, G. A., and HANKINSON, C. L., J. Dairy Sci., *19,* 761 (1936).
321. RICHERT, D. A., and WESTERFELD, W. W., J. Biol. Chem., *203,* 915 (1953).
322. RICHMOND, H. D., "Dairy Chemistry," Charles Griffin & Co., London (1920).
323. RIDDET, W., CAMPBELL, I. L., McDOWALL, F. H., and COX, G. A., New Zealand J. Sci. Tech., *23A,* 80, 99 (1942).
324. ROADHOUSE, C. L., and HENDERSON, J. L., J. Dairy Sci., *15,* 1 (1932).
325. ROAHEN, D. C., and SOMMER, H. H., J. Dairy Sci., *23,* 831 (1940).
326. ROGERS, L. A., U.S. Dept. Agr., Bur. An. Ind., Bull. 57 (1904).
327. ROMBAUTS, W. A., SCHROEDER, W. A., and MORRISON, M., Biochem., *6,* 2965 (1967). Cited in CA *67,* 105407z (1967); see also CA *67,* 113921c (1967).
328. ROSE, D., BRUNNER, J. R., KALAN, E. B., LARSON, B. L., MELNYCHYN, P., SWAISGOOD, H. E., and WAUGH, D. F., J. Dairy Sci., *53,* 1 (1970).
329. ROUSSEL, J. D., and BEATTY, J. F., J. Dairy Sci., *53,* 1085 (1970).
330. ROWLAND, S. J., J. Dairy Res., *9,* 47 (1938).
331. RUEGAMER, W. R., MICHAUD, L., and ELVEHJEM, C. A., J. Biol. Chem., *158,* 573 (1945).
332. RUSSEL, H., Arch. Lebensmittelhyg., *16,* 82 (1965). Cited in DSA *27,* 417 (1965).
333. SALERNO, A., MALOSSINI, F., and PILLA, A. M., Annali 1st Sper. Zootech. Roma, 11, 225 (1967). Cited in DSA *31,* 414 (1969).
334. SATO, M., and MURATA, K., J. Dairy Sci., *15,* 451 (1932).
334A. SCHARDINGER, F., Z. Nabr. Genussm., *5,* 1113 (1902).
335. SCHARRER, K., and SCHWAIBOLD, J., Biochem. Z., *207,* 332 (1929).
336. SCHEELE, C. W., "Kongl. Vetenskaps Akademiens nya handlingar," tom I, 116, 269 (1780).
337. SCHEPANG, W., Dissertation, Leipzig (1917).
338. SCHMIDT, G. H., J. Dairy Sci., *43,* 213 (1960).
339. SCHNEIDER, P., Arch. Tierzücht, *8,* 419 (1965). Cited in DSA *28,* 119 (1966).
340. SCHWARTZ, D. P., and PALLANSCH, M. J., J. Agr. Food Chem., *10,* 86 (1962).
341. SCRIMA, M., Boll. Lab. Chim. Prov., 19, 13 (1968). Cited in DSA *31,* 589 (1969).
342. ŠEBELA, F., and KLÍČNÍK, V., Intern. Dairy Congr. 18th, Sydney, 1E, 31 (1970). Cited in DSA *32,* 722 (1970).
343. SEBELIEN, J., Z. Physiol. Chem., *9,* 445 (1885).
344. SEDLÁKOVÁ, L., Živočišná Výroba, *14,* 573 (1969). Cited in DSA *32,* 151 (1970).
345. SEIDLER, S., and PETKOW, K., Zeszyty Nauk. Wyzszej Szkoly Rolniczej, Szezec., No. *15,* 36 (1964). Cited in DSA *28,* 450 (1966).
346. SEN, K. C., and DASTUR, N. N., 11th Intern. Congr. Pure and Applied Chem., London, Proc., 1947, *3,* 271 (1951).

347. SENFT, B., Mitt. Deut. Landwirtsch. Ges., *81,* 1146 (1966). Cited in DSA *29,* 542 (1967).
348. SHAHANI, K. M., and SOMMER, H. H., J. Dairy Sci., *34,* 1010 (1951).
349. SHAHANI, K. M., J. Dairy Sci., *49,* 907 (1966).
350. SHAHANI, K. M., CHANDAN, R. C., KELLY, P. L., and MACQUIDDY, E. L., Intern. Dairy Congr. 16th, Copenhagen, C285 (1962).
351. SHEETS, E. H., and LOPEZ, A., Food Technol., *16* (10), 143 (1962).
352. SHERWOOD, F. F., and HAMMER, B. W., Iowa State Coll., Agr. Expt. Sta., Res. Bull. 90 (1926).
353. SÎRBULESCU, V., Lucrarile Stiint. Inst. Agron. Nicolae Bălescu, Ser. C (Zootech. Med. Vet.) *10,* 129 (1967). Cited in DSA *30,* 651 (1968).
354. SKINNER, J. T., PETERSON, W. H., and STEENBOCK, H., J. Biol. Chem., *90,* 65 (1931).
355. SMITH, M. I., and WESTFALL, B. B., U.S. Pub. Health Rept. 52, 1375 (1937).
356. SÖREMARK, R., J. Nutr., *92,* 183 (1967). Cited in DSA *29,* 579 (1967).
357. SPINELLI, F., Boll. Soc. Italian Biol. Sper., *22,* 211 (1946).
358. SPRECHER, H. W., Diss. Abst. Univ. Wisconsin, *24* (9), 3530 (1964). Cited in DSA *26,* 389 (1964).
359. SLACK, N. H., Diss. Abst. New Jersey State Univ., *25* (3), 1466 (1964). Cited in DSA *27,* 55 (1965).
360. STADHOUDERS, J., Offic. Organ. Konink. Ned Zuivelb, *56,* 1141 (1964). Cited in DSA *27,* 256 (1965).
361. Standards Institution of Israel, Israel Stand., S.I. 563 (1965). Cited in DSA *28,* 363 (1966).
362. STANTON, R. E., and HARDWICK, A. J., Analyst, London, *93* (1004), 193 (1968). Cited in DSA *30,* 349 (1968).
363. Statistical Reporting Service, Fluid Milk and Cream, USDA, March 19 (1971).
364. STEFANOVIĆ, R., BEŠLIN, R., and DORDEVIĆ, J., Sborn. Rad. Poljopriv. Fak. Univ. Beogr., *16* (1968). Cited in DSA *32,* 374 (1970).
365. STETTER, A., Milchw. Zentr., *43,* 369 (1914).
366. STÖCKI, W., and LESKOVA, R., Milchwiss., *22,* 692 (1967). Cited in DSA *30,* 168 (1968).
366A. STRȲBAK, G. YA., Sborn Dokl. vses. Knof. po molochn. Delu posvyashch. 100-letiyu so Dnya Rozhd. Prof. A. A. Kalantara, Erevan, *1960,* 98–100 (1961). Cited in DSA *25,* 291 (1963).
367. SUNDARESAN, D., MALIK, S. S., and TIWARI, M. P., Indian J. Dairy Sci., *17,* 75 (1964). Cited in DSA *28,* 9 (1966).
368. SUPPLEE, G. C., and BELLIS, B., J. Dairy Sci., *5,* 455 (1922).
369. SCHWARK, H.-J., and JÄHNE, M., Arch. Tierzücht, *8,* 441 (1965). Cited in DSA *28,* 61 (1966).
370. SWOPE, F. C., KOLAR, C. W., JR., and BRUNNER, J. R., J. Dairy Sci., *49,* 1279 (1966).
371. SYVÄOJA, E.-L., and VIRTANEN, A. I., Milchwiss., *23,* 200 (1968). Cited in DSA *30,* 400 (1968).
372. SZÉP, I., Agrártud. Egyet. Mezögazdtud. Kar. Közlemen. Gödöllö, 315 (1965). Cited in DSA *29,* 132 (1967).
373. TAKTAKISHVILI, S. D., Vopr. Pitaniya, *22,* 73 (1963). Cited in DSA *26,* 87 (1964).
374. TAMSMA, A., KURTZ, F. E., BRIGHT, R. S., and PALLANSCH, M. J., J. Dairy Sci., *52,* 1910 (1969).
375. TARASSUK, N. P., and FRANKEL, E. N., J. Dairy Sci., *40,* 418 (1957).
376. TARJÁN, R., KRÁMER, M., SZÖKE, K., LINDNER, K., SZARVAS, T., and DWORSCHÁK, E., Nutritio Dieta, 7, 136 (1965). Cited in DSA *27,* 555 (1965).
377. TAUCINŠ, E., and SVILANE, A., Latvijas PSR Zinatnu Akad. Vestis, No. *10,* 71 (1966). Cited in DSA *29,* 521 (1967).
378. TEREKHINA, E., Izv. Kuibyshev. Sel'skokhoz. Inst., *20,* 212 (1967). Cited in DSA *30,* 345 (1968).
379. THATCHER, R. W., and DALHLBERG, A. C., J. Agr. Res., *11,* 437 (1917).

380. THIEME, D., Mh. Vet. Med., *23,* 109 (1968). Cited in DSA *31,* 591 (1969).
381. TIKHOMIROVA, T. V., Izv. Timiryazev. Sel'skokhoz. Akad., No. *3,* 215 (1961). Cited in DSA *25,* 120 (1963).
382. TIMMEN, H., Intern. Dairy Congr. 18th, Sydney, 1E, 77 (1970). Cited in DSA *32,* 725 (1970).
383. TKHAKAKHOV, KH. KH., Dokl. Mosk. Sel'skokohoz. Akad. Timiryazeva (Zootekhniya), No. *65,* 133 (1961). Cited in DSA *26,* 208 (1964).
384. TOKOVOI, N., and LAPSHINA, L., Tr. Krasnoyarsk. Sel'sokohoz. Inst. *13,* 69 (1962). Cited in CA *60,* 14923h (1964).
385. TRUCCO, R. E., VERDIER, P., and REGA, A., Bichim. Biophys. Acta, *15,* 582 (1954).
386. TRUNZ, A., Z. Physiol. Chem., *40,* 263 (1903).
387. TSVETKOVA, I. N., Gigiena Sanit., *34,* 92 (1969). Cited in DSA *31,* 594 (1969).
388. TUCKER, V. C., Australian J. Dairy Technol., *23,* 42 (1968). Cited in DSA *30,* 445 (1968).
389. TURNER, C. W., Univ. Mo. Agr. Expt. Sta. Bull. 365 (1936).
390. TYLER, W. J., J. Dairy Sci., *41,* 447 (1958).
391. United States Dept. of Pub. Health, Publication No. 229 (1965).
392. VANDEVELDE, A. J. J., Biochem. Z., *11,* 61 (1908).
393. VANSCHOUBROEK, F., WILLEMS, A., and LAMPO, PH., Neth. Milk Dairy J., *18,* 79 (1964). Cited in DSA *27,* 2 (1965).
394. VAN SLYKE, L. L., and BAKER, J. C., J. Biol. Chem., *40,* 335 (1919).
395. VELEV, S., MIKHAĬLOVA, T., and RACHEV, R., Izv. nauchno-izsled. Inst. mlechn. Prom., Vidin, *1,* 73 (1966). Cited in DSA *29,* 244 (1967).
396. VERMA, I. S., and SOMMER, H. H., J. Dairy Sci., *40,* 331 (1957).
397. VIOLLE, H., Compt. Rend., *169,* 248 (1919).
398. VIRTANEN, A. I., Neth. Milk Dairy J., *21,* 223 (1967). Cited in DSA *30,* 312 (1968).
399. VOIGTLÄNDER, K. H., Arch. Tierzücht, *6,* 186 (1963). Cited in DSA *25,* 434 (1963).
400. VOIGTLÄNDER, K. H., Arch. Tierzücht, *7,* 237 (1964). Cited in DSA *26,* 391 (1964).
401. VOIGTLÄNDER, K. H., Arch. Tierzücht, *9,* 213 (1966). Cited in DSA *29,* 50 (1967).
402. VOTH, J. L., Ind. Eng. Chem., Anal. Ed., *35,* 1957 (1963).
403. WALLANDER, J. F., and SWANSON, A. M., J. Dairy Sci., *50,* 949 (1967).
404. WALSH, J. P., and NEAVE, F. K., Irish J. Agr. Res., *7,* 81 (1968). Cited in DSA *30,* 438 (1968).
405. WALSTRA, P., and DE GRAAF, J. J., Neth. Milk Dairy J., *16,* 283 (1962). Cited in DSA *25,* 170 (1963).
406. WARNER, R. C., and POLIS, E., J. Am. Chem. Soc., *67,* 529 (1945).
407. WATT, B. K., and MERRILL, A. L., U.S. Dept. Agr., Agriculture Handbook No. 8, 189 (1963).
408. WEESE, S. J., BUTCHER, D. F., and THOMAS, R. O., W. Va. Univ. Agr. Expt. Sta., Bull. 544 (1967). Cited in DSA *29,* 520 (1967).
409. WHEELOCK, J. V., ROOK, J. A. F., DODD, F. H., and GRIFFIN, T. K., J. Dairy Res., *33,* 161 (1966).
410. WHEELOCK, J. V., ROOK, J. A. F., and DODD, F. H., J. Dairy Res. *32,* 249 (1965).
411. WHITNEY, R. MCL., J. Dairy Sci., *41,* 1303 (1958).
412. WHITTLE, E. G., Analyst, *68,* 247 (1943).
413. WILCOX, C. J., PFAU, K. O., MATHER, R. E., and BARTLETT, J. W., J. Dairy Sci., *42,* 1132 (1959).
414. WILCOX, J. C., GAUNT, S. N., and FARTHING, B. R., Southern Cooperative Series Bull. No. 155 (1971).
415. WILSON, D. W., and WARD, G. M., J. Dairy Sci., *50,* 592 (1967).
416. WITT, M., and SENFT, B., Mitt, Deut. Landwirtsch. Ges., *78,* 551 (1963). Cited in DSA *28,* 66 (1966).

417. WOERNER, F., Kiel. Milchw. Forsch., *13,* 361 (1961). Cited in CA *57,* 8973b (1962).

418. WRIGHT, J. A., ROOK, J. A. F., and PANES, J. J., J. Dairy Res., *36,* 399 (1969).

419. WÜTHRICH, S., RICHTERICH, R., and HOSTETTLER, H., Z. Lebensm.-Untersuch. Forsch., *124,* 336 (1964). Cited in DSA *26,* 601 (1964).

420. YAMAMOTO, T., HAMADA, H., TAKAHASHI, K., CHUKUMA, G., and KOISHIKAWA, T., Bull. Nat. Inst. Animal Ind., Chiba, *14,* 11 (1967). Cited in DSA *30,* 218 (1968).

421. YUSA, K., ANDO, K., and ONODERA, Y., Japan J. Zootech. Sci., *40,* 32 (1969). Cited in DSA *31,* 341 (1969).

422. YUSA, K., ANDO, K., and ONODERA, Y., Animal Husb. Tokyo, *22,* 1611 (1969). Cited in DSA in *31,* 96 (1969).

423. ZITTLE, C. A., J. Dairy Sci., *47,* 202 (1964).

424. ZUBOVA, V. A., Pediatriya, Mosk., *47* (10), 43 (1968). Cited in DSA *31,* 75 (1969).

CHAPTER 2

Robert E. Hargrove
John A. Alford†

Composition of Milk Products*

INTRODUCTION

Milk is a dynamically balanced mixture of proteins, fats, carbohydrates, salts and water co-existing as emulsions, colloidal suspensions, and true solutions. Milk products are prepared by alteration of these relationships either by removal of, or by changing the ratio of, one or more of the components. The process may range from simple removal of water to yield evaporated or dry whole milk to selective changes in relative concentrations of several components to yield one of several varieties of cheese. For example, in most cheeses the casein : fat ratio remains essentially unaltered while the lactose disappears and the ratios of other components are changed considerably. In many cheeses and in fermented milks the actual chemical composition is changed by microbial action, and in cheeses of the *pasta filata* type, heat alters the physical characteristics of the product.

Usually, the fraction containing most of the fat is the primary milk product sought (e.g., cheese) while the remaining fraction is referred to as a by-product (whey). However, the widespread use of nonfat dry milk as compared to the surplus of butterfat raises some question as to which is the primary product. This chapter will make no attempt to distinguish between milk product and milk by-product. It will be primarily concerned with brief descriptions of the more traditional products, although some new products will be described to illustrate the kinds of development necessary for the dairy industry to remain competitive in the food industry.

Milk products have basically the same composition from one country to another, but some variation does occur. International standards proposed by the International Dairy Federation, the Food and Agriculture Organization of the United Nations, and the World Health Organization have done much to standardize dairy products throughout the world. In the United States the composition of most dairy products

---

* Revised from Chapter 2 in the first edition by R. W. Bell and E. O. Whittier.

† Robert E. Hargrove and John A. Alford, Dairy Foods Nutrition Laboratory, Nutrition Institute, Agricultural Research Service, U.S. Department of Agriculture, Beltsville, Maryland 20705.

is regulated by Federal and State Standards.[29] In instances where there are no Federal standards regulating interstate commerce for a particular product, the state standard prevails. Legal standards, where applicable, are given within the discussion of each product. Typical chemical analyses are given in Tables 2.2 through 2.7.

Table 2.1 shows the approximate percentages of the total milk supply utilized for various products in the United States and three other countries selected to indicate the diversity of use around the world. The figures do not include amounts used for stock feed or other nonfood uses, and reflect export markets as well as the eating habits of the countries indicated.

MARKET MILK

The term "market milk" usually refers to fluid whole milk that is sold directly to the consumer. Market milk products include homogenized whole milk, Vitamin D whole milk, fortified milk, flavored milks, low-fat milks, skimmilk and fat-free milk. In the United States the composition of market milk is regulated primarily by state standards. The sanitary quality is usually regulated by sanitary codes established by state and local health departments, although the U.S. Public Health Service has established standards for milk served on interstate carriers. Approximately 50% of the milk produced in the United States is consumed as market milk (Table 2.1). Typical analyses of fluid milk products are given in Table 2.2.

Whole Milk

Fluid Grade A raw milk received by the dairy is usually clarified, standardized to a uniform milk fat content, pasteurized, and homogenized. It may or may not contain added vitamin D. In the United States, all states require a minimum fat content for whole milk, which may vary from 3.0 to 3.8%, 3.25% being the most prevalent.[29] Most require a minimum solids nonfat content of at least 8.25%. Standards in other countries are near the 3.25% fat and 8.25% solids nonfat content.

Flavored Milk

Flavored milks are whole milks that contain added flavoring materials, such as chocolate, vanilla, and fruit flavors. If the product is to contain less than the milk fat content of whole milk, it must be

Table 2.1

UTILIZATION OF MILK PRODUCED IN SELECTED COUNTRIES[5,21,28]

	Whole Milk Products						
	Market Milk (and Cream)	Butter[a] Butteroil Nonfat Dry Milk	Cheese	Concentrated Milks	Dried Whole Milk	Frozen Desserts	Fermented and Miscellaneous Products
	%	%	%	%	%	%	%
United States	48	20	17	3	*[b]	10	*
The Netherlands	25	15	35	16	6	*	*
New Zealand	9	72	15	*	*	*	*
United Kingdom	70	11	10	5	2	*	*

Butter and nonfat dry milk are grouped together since both products usually come from the same milk.
Individual percentages too small to be accurately indicated.
Data provided by Agricultural Counselor, Royal Netherlands Embassy, Washington, D.C.

Table 2.2

TYPICAL COMPOSITION OF FLUID AND FERMENTED MILKS[3,23,24,33]

	Moisture	Protein	Fat	Lactose	Ash	Calcium	Phosphorus	Lactic acid	Ethyl alcohol
	%	%	%	%	%	%	%	%	%
Whole milk	87.4	3.5	3.5	4.8	0.7	0.1	0.09		
Chocolate milk	81.5	3.4	3.4	11.0[a]	0.7	0.11	0.09		
Chocolate drink	82.8	3.3	2.3	10.9[a]	0.7	0.11	0.09		
10—2 Milk	87.0	4.2	2.0	6.0	0.8	0.14	0.11		
Low fat, 1%, milk	89.5	3.5	1.0	4.9	0.7	0.12	0.09		
Skim milk	90.5	3.6	0.1	5.1	0.7	0.12	0.09		
Cultured buttermilk	90.5	3.6	0.1[b]	4.3	0.7	0.12	0.10	0.8	
Sour cream	74.5	2.8	18.0	3.4	0.5	0.10	0.08	0.6	
Acidophilus skim milk	90.1	3.5	0.5	4.4	0.7	0.12	0.09	0.7	
Kefir, part skim	89.4	3.5	2.0	4.0	0.7	0.10	0.09	0.6	1.0
Yoghurt, plain	87.2	3.4	3.4	4.1	0.6	0.12	0.09	0.9	
Yoghurt, solids added	83.1	5.0	4.8	6.0	0.8	0.18	—	0.9	
Yoghurt, part skim	89.0	3.4	1.7	5.2	0.7	0.11	0.09	0.9	
Yoghurt, full skim	91.0	3.4	—	4.0	0.7	0.12	0.09	0.9	
Yoghurt, fruit	c	3.4	1.7	12.5[b]	—	0.14	0.10	0.8	

a Carbohydrate other than lactose added.
b Additional fat may be added.
c Varies with solids content of added fruit.

labeled as a "Milk Drink," "Flavored Milk Drink," or "Dairy Drink."

Chocolate milk is the most popular flavored milk in the United States. In making chocolate milk about 1% cocoa and 6% sucrose are added to whole milk. Where permitted, a small percentage of stabilizer, such as a vegetable gum, is added to keep the cocoa in suspension. As with whole milk, the minimum fat content required in flavored milks varies with the state standard.

Low-Fat Milk

Low-fat milks are processed much like whole milk except that the fat content is reduced to 1 to 2% milk fat. Fanciful names and terms, such as 99% fat-free, have been given to low-fat milks containing 1% fat. In some areas a fortified low-fat milk is available which contains 2% milk fat and 1 to 2% additional skim-milk solids. Such a product may be referred to as "10–2 milk."

Skimmilk

Skimmilk is processed by removing most or all of the milk fat from whole milk. The percentages of all components, with the exception of milk fat and vitamins associated with the fat, are proportionally greater than in the whole milk from which it was made. Vitamins A and D may be used to fortify skimmilks in states where permitted. State standards vary in their requirements for skimmilk ranging from a maximum fat content of 0.1 to 0.5% and a minimum solids nonfat content from 8.0 to 8.25%.[29]

Multivitamin-Mineral Milk

Specially fortified milks containing multiple vitamins and minerals are available in some states. The vitamin and mineral content per quart is regulated by state law, but is usually the minimum daily requirement for an adult as established by the Food and Drug Administration. A typical fortified milk may contain the following:

Vitamin A	4000	USP units
Vitamin D	400	USP units
Thiamine (B_1)	1	mg
Riboflavin (B_2)	2	mg
Niacin	10	mg
Iron	10	mg
Iodine	10	mg

Low-sodium Milk

Low-sodium milk is available in some areas as a specialty product for consumers who require a diet of low-sodium foods. In the processing procedure normal milk is usually passed over an ion-exchange resin in which the sodium of the milk is replaced by potassium ions. The normal sodium content of milk is reduced from about 50 mg of sodium/100 ml to approximately 3 mg/100 ml. The other components of milk remain essentially the same after processing.

FERMENTED MILKS[34]

Fermented milks are cultured dairy products made from skim, whole or slightly concentrated milk that require specific lactic acid bacteria to develop their characteristic flavor and texture. Fermented milks are usually fluid or semifluid in consistency, and all contain lactic acid in varying proportions. There are no Federal standards for fermented milks in the United States, but these products may be regulated by state standards.[29] Compositional standards for fermented milks have been proposed by the International Dairy Federation.[13] Typical analyses of fermented milks are given in Table 2.2.

Cultured Buttermilk

Cultured buttermilk is a skimmed or partly skimmed milk that has been fermented with lactic acid bacteria. Most commercial buttermilk is made with mixed-strain bacterial starters containing lactic streptococci and leuconostocs. A product called "Bulgarian buttermilk" is made by culturing the milk with *Lactobacillus bulgaricus*. About 0.1% salt is usually added to the cultured milk, and butter granules or cream may or may not be added. Buttermilks are similar to skimmilk in composition, except that total acid calculated as lactic acid is present up to 0.9%. The percentage of lactose is reduced in proportion to the percentage of actual lactic acid. The fat content may be as great as 1%, sometimes in the form of small flakes which cause it to simulate churned buttermilk, a product prepared from churning cream to produce butter. Very little of this type of product is available for retail consumption; it is usually condensed and dried for commercial use.

A product equivalent to cultured buttermilk may be prepared by direct acidification. Acid or acid anhydrides are added to unfermented milk to provide controlled acidity and desirable body characteristics. Flavoring material may also be added to simulate the cultured product. Acidity induced similarly has been used in the production of other products such as sour cream and Cottage cheese.

Sour Cream

Sour cream or cultured cream is a cream containing about 18% milk fat that has been cultured with lactic acid bacteria similar to those used in preparing cultured buttermilk. There are no Federal standards, but most states require 18% fat, and nonfat dry milk solids may be added to the cream to obtain a thick body. Where permitted by law, a small amount of stabilizer or rennet may be added to give a smooth appearance and thick body.

Yoghurt

Yoghurt is a fermented milk product that has characteristics intermediate between cultured buttermilk and unripened cheeses such as Cottage, and the basic medium may contain whole milk, partly skimmed, skimmed evaporated or dried milk. The texture may vary from a rennet-like custard to a creamy, highly viscous liquid depending upon the milk solids and fat content. Fruits or fruit flavors may be added to yoghurt before or after fermentation. Two microorganisms *Streptococcus thermophilus* and *Lactobacillus bulgaricus*, growing together symbiotically, are responsible for the lactic acid fermentation and yoghurt flavor. There are no Federal standards, but several states specify fat (1.0 to 3.5%) and milk solids not fat (8.25 to 8.5%). FAO proposed minimum standards are 0.5% fat and 8.5% MSNF.[7]

Acidophilus Milk

Acidophilus milk is a fermented whole or skimmilk product that contains an active culture of *Lactobacillus acidophilus*. Nutrients such as honey, glucose and tomato juice may be added to stimulate bacterial growth. Plain acidophilus milk would have the same composition as whole or skimmilk with the exception of replacement of part of the lactose with lactic acid. From 0.6 to 1% lactic acid may be found in acidophilus milk.

Kefir

Kefir is a self-carbonated beverage which may be made with whole, part skim or skimmilk to contain about 1% lactic acid and 1% alcohol. Kefir is made with a fermenting agent called "Kefir Grains," the grains consisting of casein and gelatinous colonies of microorganisms growing together symbiotically. Kefir milk has essentially the same composition as the original milk, some of the lactose being converted into lactic acid, alcohol, and CO_2. There is limited proteolysis in the milk, with development of a definite yeasty aroma. The quantities of lactic acid fermented to alcohol and CO_2 in kefir milk may be regulated by the temperature of incubation.

Concentrated Fermented Milks

The Danish fermented milk, "Ymer," and the Swedish product, "Lactofil," are concentrated fermented milks that may be made from skimmilk or whole milk. Skimmilk is usually cultured with an active lactic butter culture containing *Streptococcus diacetilactis* and *Leuconostoc* species. After coagulation, about 50% of the whey is removed and then cream is added to the curd to give a fat content of about 5% and a total solids content of about 15%. The cream and curd are homogenized to a smooth creamy consistency.

MARKET CREAM

Market cream refers to all fluid cream that is processed and sold directly to the consumer. It is prepared by separating the cream from whole milk and standardizing to a prescribed fat content. Creams may be prepared to have from 12 to 80% fat, the percentage depending upon the speed of the separator, the temperature of the milk, the rate of milk flow and the number of times the product is passed through the separator. It is usually processed in such a manner as to increase its viscosity and thickness. The whippability of cream depends upon the fat content, the age of the cream, the type of beater and cream temperature. Adding nonfat dry milk solids usually improves whipping ability.

Market creams throughout the world fall into discrete classes, depending upon the fat content; these are: 10 to 12%, 18 to 20%, 25 to 30%, 34 to 36%, and 48%, although not all countries market all types. The official designations for these products are not always the same, however. For example, a product in the 25 to 30% range is called "Medium Cream" in the United States and "Reduced Cream" in Australia, while the term "Cream" officially varies from 18% (Food and Agriculture Organization) to 40% (New Zealand).

In the United States fluid creams are offered to the consumer as Half and Half, Light, Medium, Whipping and Heavy. A product made for manufacturing purposes containing about 80% fat is known as "Plastic Cream." Federal standards and range of state standards for minimum fat content of specific products as well as similar standards proposed for the United Kingdom are given in Table 2.3. Typical compositions of these products are listed in Table 2.4.

Table 2.3

MINIMUM FAT STANDARDS FOR MARKET CREAMS IN THE UNITED STATES AND THE UNITED KINGDOM[29,32]

Type of Cream	Minimum Fat Required Federal	States
	%	%
Half and half	—	10–12
Light, table, coffee	18	16–20
Medium	—	24–30
Whipping	30	30–36
Heavy whipping	36	34–36
United Kingdom		
Half and half	12	
Sterilized half and half	12	
Cream	20	
Sterilized cream	23	
Whipped cream	35	
Whipping cream	35	
Double cream	48	
Clotted cream	48	

BUTTER, BUTTER OIL, SPREADS

Butter

Most creamery butter is produced by churning sweet cream so that the fat globules coalesce into a soft mass. The Federal standard for butter,[29] set by an act of Congress in 1923, requires that butter contain not less than 80% milk fat. This fat level is universal, and most states and other countries also specify a maximum moisture level of 16%.[8,28] A typical analysis of butter is given in Table 2.4. Whey butter has a similar composition, but is derived from the milk fat recovered from cheese whey.

Butter oil

Butter oil or anhydrous milk fat is a refined product prepared by centrifuging melted butter or by separating the milk fat from high-fat cream. There are no Federal standards in the United States, but FAO has suggested standards of 99.3% fat and 0.5% moisture for butter oil, and 99.6% fat and 0.2% moisture for anhydrous butter oil.[7]

Ghee

Ghee is a nearly anhydrous butterfat used in many parts of India and Egypt. It is usually made from buffalo milk and much of the

Table 2.4

TYPICAL COMPOSITION OF MARKET CREAMS, BUTTER, AND FROZEN DESSERTS[a,3,12,33]

	Moisture	Protein	Fat	Lactose	Ash	Calcium	Phosphorus
	%	%	%	%	%	%	%
Market creams							
Half and half	80.0	3.1	11.6	4.5	0.7	0.10	0.08
Light cream	73.0	2.9	19.3	4.2	0.6	0.10	0.08
Whipping, light	62.9	2.5	30.5	3.6	0.5	0.08	0.06
Whipping, heavy	57.3	2.2	36.8	3.2	0.5	0.07	0.05
Plastic	18.2	0.7	80.0	1.0	0.1	0.03	0.02
Butter, butter oil, ghee							
Butter	16.5	0.6	80.5	0.4	2.5	0.02	0.02
Butter oil	0.2	0.3	99.5	0.0	0.0	—	—
Ghee	0.1	0.1	99.8	0.0	0.0	—	—
Frozen desserts							
Ice cream	62.1	4.0	12.5	20.3[b]	0.8	0.12	0.10
Ice cream, low fat	63.2	4.5	10.6	20.8[b]	0.9	0.15	0.12
Ice cream, high fat	62.8	2.6	16.1	18.0[b]	0.5	0.08	0.06
Ice milk	66.7	4.8	5.1	22.4[b]	1.0	0.16	0.12
Sherbet	67.0	0.9	1.2	30.8[b]	0.1	0.02	0.01

[a] Salt concentration in butter ranges from 0.8–2.3%. The lower value is typical of most European countries, the higher value of the United States, New Zealand, an Australia

[b] Carbohydrate other than lactose added.

typical flavor comes from the burned nonfat solids remaining in the product. Ghee is made in the United States from butter, and recently a procedure has been developed for its production from cream.[33]

Miscellaneous Spreads

A dry product referred to as Butter Powder has been produced in Australia.[8,17] It has the same fat content as butter and only 0.6% water, but contains other components (added nonfat dry milk, emulsifier, etc.) which affect its legal status.

Dairy spreads and products simulating butter have emerged in the past few years. "Butterine," a product available in Wisconsin, is composed of 40% milk fat, 40% vegetable fat, 1% milk solids, salt and added vitamins A and D. Other lower-calorie spreads containing about 50% moisture and 40% milk fat have been developed in the United States, Canada, Eire, and Sweden.[16] A product developed at the University of South Dakota contains 44% moisture, 40% milk fat, 14 to 16% nonfat dry milk, synthetic butter flavor, high-acid starter distillate, salt, butter coloring, and a combination of gelatin and sodium carboxymethylcellulose as a stabilizer.

A spread-type product called "Bregott" in which 15% of the total fat is soybean oil is marketed in Sweden. The oil is added to cream, and then churned with minor adjustments in temperature and time. Although it is not competitive with the best margarines in price, after one year it had captured 4% of the total spread market.[8]

CONCENTRATED MILK PRODUCTS[34]

Whole milk, skimmilk, and buttermilk are concentrated by removal of water and may be preserved by heat, addition of sugar, or refrigeration. Typical analyses of these products are given in Table 2.5.

Evaporated Milk

This product is made by the evaporation of water from whole milk under a vacuum. Low percentages of sodium phosphate, sodium citrate, calcium chloride, and/or carrageenan may be added to improve stability.[30] Vitamin D is usually added and the concentrate is homogenized and canned, then sterilized under pressure at 117°C for 15 min, or 127°C for 2 min. Ultra-high temperatures (130 to 150°C for a few seconds), followed by aseptic packaging, have been applied with some success, but have found limited commercial application.[16]

United States standards of identity require that evaporated milk contain not less than 7.9% milk fat and 25.9% total milk solids,[29] and most other countries, including the proposed standards by FAO, are similar.[2,7] Exceptions are Australia, where requirements of 8% milk fat and 28% total solids are found,[35] and Great Britain where 9% fat and 31% TMS are required.[23a]

Plain Condensed Milk

Plain condensed milk or concentrated milk has the same standard of identity in the United States as evaporated milk, except that it is not given additional heat processing after concentration, and the optional ingredients are not used.[30] This product is shipped in bulk containers and is perishable. Technology is available to produce it in a sterile or almost sterile manner, and its extended shelf life gives it a potential, but as yet undeveloped, market as a source of beverage milk. Whole milk can be successfully concentrated up to 45% total solids, and these higher concentrations have found some use in the bulk product market.

Sweetened Condensed Whole Milk

Sweetened condensed milk is made by the addition of approximately 18% sugar to whole milk, followed by concentration under vacuum to approximately one-half the volume. The product is canned without sterilizing, the sugar acting as a preservative.

Federal standards of identity require 8.5% fat, 28.0% total milk solids and sufficient sugar to prevent spoilage.[30] State standards range from 7.7 to 8.5% fat and 25.9 to 28% total milk solids. The proposed standards by FAO of 8% fat and 28% solids are fairly common worldwide, although Australia and New Zealand require 9% fat and 31% total solids.[20,35]

Condensed Skimmilk

Plain Condensed Skimmilk.—This product is usually sold in bulk in the United States for increasing milk solids in ice cream, bakery goods, and many other foods. It is usually less expensive, though more perishable, than nonfat dry milk. There are no Federal standards, but several states require 20% total solids-not-fat and one requires 27%. FAO proposes a MSNF of 20%, and Australian states and New Zealand require 26.5%.

Sweetened Condensed Skimmilk.—This product is prepared from skimmilk in a process similar to that used for whole milk. The final product contains at least 60% sugar and 72 to 74% solids. There are no Federal standards in the United States, and state requirements range from 18 to 28% total milk solids. FAO proposed standards require 24% milk solids; 26.5% is required in Australian states and New Zealand.

Condensed Skimmilk, Acid.—This a product manufactured primarily for animal feed. It is made from skimmilk by developing about 2% acidity by means of a *Lactobacillus* culture and a mycoderm, then concentrating the milk to about ⅓ its weight.[34]

Condensed Buttermilk

Condensed or semisolid buttermilk is a creamery buttermilk (usually from sweet cream) which is allowed to ripen to an acidity of 1.6% or more, then condensed. It has found limited use in the baking industry. There are no Federal standards, but a typical product contains about 28% total solids.

DRIED MILK PRODUCTS[1,9,34]

Typical analyses of dried milk products are given in Table 2.5.

Nonfat Dry Milk

Nonfat dry milk (NDM) has become an increasingly important product in the dairy industry during the past twenty years, and in 1970 it utilized about 20% of the total milk supply. Since most of the fat from milk used for NDM goes into butter, percentages on distribution of the milk supply necessarily overlap for these two products.

Nonfat dry milk is produced from skimmilk by condensing with conventional equipment followed by drum or spray drying. The drum-dried product is relatively insoluble and is used principally in animal feeds. Over 85% of the NDM in the United States is used for human foods and is produced by spray drying. Most instant NDM is made by rewetting the conventionally spray-dried product, allowing the particles to agglomerate, then reducing the moisture content with added heat. Foam spray drying by spray drying a pressurized concentrated milk also gives a very acceptable product.

NDM has only fat and water removed, and standards in the United States and proposed by FAO allow a maximum of 5% moisture and less than 1.5% fat.[6,29]

Dried Whole Milk

Dried whole milk is prepared by conventional spray or roller drying with some modifications of the preheat treatment of the milk. The

Table 2.5

TYPICAL ANALYSES OF CONCENTRATED MILKS AND DRIED PRODUCTS[1,3,12,33]

Milk Products	Protein	Fat	Moisture	Carbohydrate Lactose	Sucrose	Ash	Calcium	Phosphorus	Lactic acid
Concentrated	%	%	%	%	%	%	%	%	%
Evaporated milk	7.0	7.9	73.8	9.7	0	1.6	0.252	0.205	0
Sweetened condensed, whole	8.1	8.7	27.1	11.4	44.3	1.8	0.262	0.206	0
Plain condensed skim	10.0	0.3	73.0	14.7	0	2.3	0.250	0.200	0
Sweetened condensed skim	10.0	0.3	28.4	16.3	42.0	2.3	0.300	0.230	0
Condensed buttermilk (acid)	9.9	1.5	72.0	12.0	0	2.2	—	—	5.7
Condensed skim (acid)	10.19	0.17	72.0	9.43	0	2.13	—	—	6.08
Condensed whey	7.0	2.4	48.1	38.5	0	4.0	—	—	2.4
Sweetened condensed whey	5.0	1.7	24.0	28.5	38.0	2.8	—	—	0
Dried									
Whole milk	26.4	27.5	2.0	38.2	0	5.9	0.909	0.708	0
Skim (conventional)	35.9	0.8	3.0	52.3	0	8.0	1.308	1.016	0
Skim (instant)	35.8	0.7	4.0	51.6	0	7.9	1.293	1.005	0
Buttermilk (sweet)	34.3	5.3	2.8	50.0	0	7.6	1.248	0.970	0
Buttermilk (acid)	37.6	5.7	4.8	38.8	0	7.4	—	—	5.7
Malted milk	14.7	8.3	2.6	20.0	50.5[a]	3.6	0.288	0.380	0
Cream	13.4	65.0	0.8	18.0	0	2.91			0
Whey (sweet) Cheddar	12.9	0.9	4.5	73.5	0	8.0	0.646	0.589	2.3
Whey (acid) cottage	13.0	0	3.2	66.5	0	10.2	1.44	1.17	8.6
Casein (commercial)	88.5	0.2	7.0	0	0	3.8	—	—	—
Casein (co-precipitate)	83.0	1.5	4.0	1.0	0	10.5	2.5	—	—

50.5% = maltose and dextrin.

product is usually stored under nitrogen to delay oxidative off-flavor development. In spite of these processing changes, flavor defects and short storage life have limited the market for dried whole milk; most of that produced is utilized in the confectionery and baking industries.

Federal standards require a minimum fat content of 26% and a moisture content of 1.5% to 3.0%, depending upon the grade. Most states, as well as proposed FAO standards and those of several other countries, permit up to 5% moisture with 26% fat.[6,29]

Dry Buttermilk

Most dry buttermilk is prepared from sweet cream buttermilk, which is processed in a manner similar to that for nonfat dry milk. Dry buttermilk has a higher phospholipid content than other dry-milk products, and therefore is a natural emulsifier for use in baking, dry mixes, and other foods. A dry, high acid buttermilk can be produced from milk fermented by *Lactobacillus bulgaricus*. It is difficult to dry, however, and has found only limited use in the baking industry. There are no United States or FAO standards for this product, although typically the moisture content is less than 5%.

Dry Cream

Dry cream may be produced by spray drying or foam drying a good quality standardized cream.[34] Higher heat treatments make the product more resistant to oxidation, and gas packaging is used to reduce the oxygen in the head space to 0.75% or less. There are no standards as to composition, although the usual range is from 40 to 70% fat and a moisture content less than 2%. The solids-not-fat content is usually higher than is normal for market creams. A foam spray-dried sour cream has also been manufactured. Cream tablets have been produced containing added lactose to aid in tableting, but the commercial acceptance of this product has been negligible.

Malted Milk Powder

Malted milk powder is made by concentrating a mixture of milk and an extract from a mash of ground barley malt and wheat flour, to obtain a solid which is ground to powder. It usually contains not less than 7.5% milkfat and not more than 3.5% moisture. One pound is considered equivalent to 2.65 lb of fluid milk on the basis of fat content and 4.4 lb on the basis of solids-not-fat content. This difference in equivalents results from using milk containing approximately 2.0% fat in making malted milk.

CHEESE

Cheese is a concentrated dairy product that requires select microorganisms and their enzyme systems to develop characteristic flavor and texture. Cheeses consist of varying ratios of milk proteins, fat and moisture; they may be made from whole milk, partly skimmed, skimmilk or whey, and may or may not contain added cream or nonfat dry milk solids. Casein is usually coagulated with rennet and lactic acid or with other suitable enzymes and acids. Ripening agents may vary from none to select strains of bacteria, yeasts and molds. Factors having the greatest effect on the composition of the finished cheese are source of milk (cow, ewe, goat, buffalo), fat content, ratio of casein to milk solids, and method of manufacture.

Cheeses contain in concentrated form much of the nutrients of milk. The protein content of cheese consists mainly of casein.

The water-soluble lactalbumin and lactoglobulin which represent about 1/5 of the total protein in hard cheese do not coagulate at the acidity and the temperatures used to make most cheeses. The extent to which whey is retained in the cheese will determine to a great extent the amount of water-soluble nutrients left in the finished cheese.

About 4/5 of the milk calcium, ⅔ of the phosphorus, and 5/6 of vitamin A remain in a Cheddar type cheese. Varying quantities of lactose are found in freshly prepared cheeses; however, the lactose content decreases rapidly with ripening, and may completely disappear after 4 to 6 weeks. The rate and extent of protein and fat decomposition in a cheese during ripening will vary markedly with the ripening agent and the enzymes used in its manufacture. This is discussed more fully in Chapter 12. The vitamins in cheese are known to vary with the fat, solids and moisture content;[11] they are discussed in Chapter 11.

Classification

Over 400 varieties of cheese have been described throughout the world;[26] however, the manufacturing processes for many are quite similar if not identical. Cheeses may be classified according to texture, moisture, ripening agent, and method of manufacture; most cheeses fall within the groupings described below, which are based on these factors. Cheeses obviously have characteristics related to more than one group. The most significant and distinguishing characteristic has been used in classifying them. Representative cheeses in each class or group are given.

1. Soft unripened; low fat
   (a) Cottage
   (b) Quarg
   (c) Bakers
2. Soft unripened; high fat
   (a) Cream
   (b) Neufchatel (US)
   (c) Petit Suisse
3. Soft; ripened by surface bacteria and yeast
   (a) Limburger
   (b) Liederkranz
   (c) Romadur
   (d) Bel Paese
4. Soft; ripened by external molds
   (a) Camembert
   (b) Brie
   (c) Coulommiers
5. Soft; ripened by bacteria, preserved by salt
   (a) Feta
   (b) Domiati
   (c) Teleme
6. Semi-soft; ripened by internal bacteria, with surface growth
   (a) Brick
   (b) Muenster
   (c) Tilsiter
7. Semi-soft; ripened by internal mold
   (a) Blue
   (b) Roquefort
   (c) Stilton
   (d) Gorgonzola
8. Hard; ripened by bacteria
   (a) Cheddar
   (b) Colby
   (c) Cheshire
9. Hard; ripened by eye-forming bacteria
   (a) Swiss
   (b) Edam
   (c) Gouda
10. Very hard; ripened by bacteria
    (a) Parmesan

   (b) Romano
   (c) Asiago
11. Pasta Filata or plastic cheese
   (a) Provolone
   (b) Mozzarella
   (c) Cacciocavallo
12. Skim milk and low fat cheese (ripened)
   (a) Sapsago
   (b) Euda
13. Whey cheese
   (a) Ricotta
   (b) Mysost
   (c) Primost
14. Process cheese
   (a) Pasteurized process cheese
   (b) Process cheese food
   (c) Process cheese spread

Brief descriptions are given below of selected cheeses from each group and typical compositions are given in Table 2.6. Federal standards of identity where established are given in Table 2.7.

Cottage Cheese

Cottage cheese is a soft, unripened, acid cheese made primarily in the United States, Canada, and England from various combinations of pasteurized skimmilk, partly condensed skimmilk and reconstituted low-heat nonfat dry milk. In conventional Cottage cheese-making, curd coagulation is caused by lactic acid formed by lactic starter bacteria and the optional use of a coagulator. The curds are firmed by cooking, then washed and cooled with cold water. Creamed Cottage cheese must contain 4% milk fat and its addition is usually in the form of a cream dressing added to the Cottage curd. Cottage cheese with its high moisture content has a shelf life of 2 to 3 weeks.

Cream Cheese

Cream cheese is a soft, unripened, high fat, lactic acid type of cheese made from a homogenized milk and cream mixture containing about 16% milk fat. The curd is precipitated with lactic acid formed by lactic starter bacteria and separated from the whey by centrifugal separators or drained into muslin bags. United States Federal Standards allow the use of up to 0.5% stabilizer to prevent whey leakage.

Table 2.6

TYPICAL ANALYSES OF CHEESE[3,12,26,33]

Type	Cheese	Moisture	Fat	Protein	Fat in dry-matter	Salt	Ash	Lactose	Calcium	Phosphorus
		%	%	%	%	%	%	%	%	%
Soft-unripened Low-fat	Cottage	79.0	0.4	16.9	1.9	1.0	0.8	2.7	0.09	0.05
	Creamed cottage	78.3	4.2	13.6	19.3	1.0	0.8	3.3	0.09	0.05
	Quarg	72.0	8.0	18.0	28.5			3.0	0.30	0.35
	Quarg (high fat)	59.0	18.0	19.0				3.0	0.30	0.35
Soft-unripened high	Cream	51	37	8.8	75.5	1.0	1.2	1.5–2.1	0.08	0.06
	Neufchatel	55	23	18.0	51.1	1.0	2.0			
Soft-ripened by surface bacteria	Limburger	46	27	21.5	50.0	2.0	3.6	0–2.2	0.5	0.4
	Liederkranz	52	28	16.5	58.3	1.5	3.5	0	0.3	0.25
		53	25.5	16.8	54.2	1.7	3.9			
Soft-ripened by external molds	Camembert	51	26	20.0	53.0	2.5	3.8	0–1.8	0.6	0.5
	Brie	45	30	21.6	54.5	2.0	4.0	0–2.0	0.6	0.4
Soft-ripened by bacteria, preserved by salt	Feta	57	24	20	55.8	5.0				
	Domiati	55	25	20.5	55.5	4.8				

Semi-soft, ripened	Brick	42	31	21	53.4	2.0	4.2	0–1.9	0.6	0.4
by bacteria with	Muenster	44	28	25	50.0	1.8			0.5	0.35
surface growth										
Semi-soft, ripened	Blue	41.5	30.5	21.5	52.1	4.0	6.0	0–2.0	0.7	0.5
by internal	Roquefort	40.0	31.0	22.0	50.1	4.2	6.0		0.65	0.45
molds	Gorganzola	36.0	32.0	26.0	50.0	2.4	5.0			
Hard, ripened by	Cheddar	37.0	32.0	22	50.8	1.6	3.7	0–2.1	0.7	0.5
bacteria	Colby	39.0	31.0	21	50.8	1.7	3.6	0	.7	.5
Hard, ripened by	Swiss	37	28	27.5	44.4	1.3	3.8	0–1.7	1.0	0.6
eye-forming	Edam	39	25	28.0	40.9	2.0	4.4	0–1.0	0.75	0.45
bacteria	Gouda	36.5	29	25.0	45.6	1.7		0–1.0	0.60	0.38
Very hard, ripened	Parmesan	30.0	26.0	36.0	37.1	1.8	5.1	0–2.9	1.1	0.8
by bacteria	Romano	32.0	30.0		44.1	4.6	5.4	0		
Pasta filata										
(stretch cheese)	Provolone	38.	28	28	45.1	3.0	4.0	0	0.7	0.6
	Mozarella	53	18	22	38.3	1.0		0.3		
Low-fat or skim	Euda	56.5	6.5	30.0		2.6		1.0		
milk cheese	Sapsago	37.0	7.4	41.0		4.5				
(ripened)										
Whey cheese	Ricotta	72.0	10.0	12.5	35.7	1.2	3.6	3.0		
	Primost	13.8	30.2	10.9	35.0	—		36.6		
Process cheese	Process Cheddar	39.5	31.5	22.2	52.0	1.7	4.9	0	0.7	0.7
	Process Cheese food	43.0	24.0	20.5	42.1	1.0		7.0	0.6	0.6
	Process cheese spread	48.5	21.5	16.0	41.1	1.0		7.0	0.8	0.8

Table 2.7
FEDERAL STANDARDS OF IDENTITY FOR CHEESE[29,30]

Cheese Type	Moisture (maximum)	Milk fat[a] (minimum in solids)	Milk fat[b] (minimum in cheese
	%	%	%
Cottage curd	80	—	—
Creamed cottage	80	—(20)	4
Cream	55	—(73.3)	33
Limburger	50	50	—(25)
Camembert	—	50[c]	—
Feta	—	50[c]	—
Brick	44	50	—(28)
Blue	46	50	—(27)
Cheddar	39	50	—(30.5)
Swiss	41	43	—(25.4)
Parmesan	32	32	—(21.8)
Provalone	45	45	—(24.8)
Ricotta	80[d]	11[e]	11[f]
Process Cheddar	39	50	—(30.5)
Process Swiss	41	43	—(25.4)
Cheese food	44	—(42.6)	23
Cheese spread	50 (44 minimum)	—(50)	20

[a] Federal Standards set on milk fat in solids. Figures in ( ) calculated from standard.
[b] Federal Standards set on milk fat in finished cheese. Figures in ( ) calculated from standard.
[c] Federal Standards for cheese class only.
[d] No Federal Standards; standard for California, Minnesota, and New York.
[e] No Federal Standards; standard for California.
[f] No Federal Standards; standard for Minnesota and New York.

Limburger

Limburger is typical of a soft, surface-ripened variety of cheese. It is usually made from cow's milk and requires about 2 months to develop the characteristic flavor and aroma. The bacterial growth of *Bacterium linens* in association with yeast is primarily responsible for ripening, which occurs from the surface to the interior. Soft cheeses, such as Limburger, must be small in size to provide the optimum amount of surface in relation to mass. During ripening there is extensive protein decomposition accompanied by a strong odor and flavor.

Feta

Feta cheese is a white, soft-ripened variety of cheese that originated in Greece; it belongs to a type of cheese called "pickled cheese" because of its high salt content. High levels of salt were formerly used as a means of preservation. Feta is usually made from ewes' or goat's milk, but in the United States it is made from cow's milk. Lactic acid bacteria and rennet are used to coagulate the curd. The curd is well drained, matted, cut into 1-in. slices, and heavily salted. The salted curd is ripened about 1 month before eating. The cheese is soft in body and resembles Cottage cheese in flavor, although it is very salty and acid.

Camembert

Camembert cheese is typical of the group of soft cheeses that is ripened from the surface by molds. The distinctive characteristic of Camembert is its leathery exterior and creamy semi-liquid interior. Ripening is due primarily to the growth of the mold, *Penicillium camemberti*. Other microorganisms found in association with this mold are yeast and *Bacterium linens*. Softening of the cheese is accompanied by casein hydrolysis and an increase in water-soluble proteins. Characteristic body texture and flavor are evident after 4 to 5 weeks curing.

Brick

Brick cheese, which is similar to Muenster, was developed in the United States. Its flavor and texture are intermediate between Limburger and Cheddar types. Surface microorganisms of yeast and *Bacterium linens* develop on the surface of the cheese and contribute to its flavor; however, most of the ripening takes place within the cheese. Relatively little proteolysis occurs during the 2 to 3-month ripening process.

Blue-veined

Blue-veined or Blue cheese is a semi-soft, blue-veined variety of cheese made in the United States from cow's milk. It is similar to many other varieties that are internally ripened by the mold *Penicillium roqueforti*, such as Stilton, Gorgonzola, and Danablu. The term "Roquefort" is officially limited to the original blue-veined cheese manufactured only from sheep's milk in a small area in southeastern France near the town of Roquefort. During the curing, the cheeses are punctured with slender needles to permit air to enter and favor the growth of mold. The mold, *P. roqueforti*, is mainly responsible for ripening and characteristic flavor development. The flavor in part has been attributed to the hydrolysis of milk fat and the formation of methyl ketones during ripening.[22]

Cheddar

Cheddar is a hard, close-textured, bacteria-ripened variety of cheese that requires several months curing at low temperatures to develop the characteristic flavor. It is made from warmed and pressed curd obtained by the action of rennet and lactic acid bacteria on whole milk. The milk is usually standardized to a definite ratio of fat to casein. Lactic acid bacteria added as bacterial starters are primarily responsible for ripening and mild flavor. Freshly made Cheddar has a firm elastic body and a mild acid flavor. Reduction in firmness and an increase in flavor are accompanied by ripening and conversion of part of the protein to water-soluble compounds such as proteoses, peptones and amino acids.

Swiss

Swiss cheese made in the United States is the counterpart of Emmentaler cheese made in Switzerland. It is a hard bacteria-ripened variety that develops large characteristic eyes or gas holes during the ripening process. Although lactic acid bacteria, such as *Lactobacillus bulgaricus* and *Streptococcus thermophilus,* are essential in the making process, eye formation and the typical sweet flavor are due to the growth of *Propionibacterium shermanii.* The cheese milk is usually standardized to a definite ratio of fat to casein. Casein coagulation is due to rennet and lactic acid. Cheese curing usually requires about 3 months.

Provolone

Provolone is a hard, bacteria-ripened cheese that is typical of the "Pasta Filata" or heat-stretch curd variety. Pasta Filata (plastic curd) is a class of Italian cheeses characterized by the fact that the curd is immersed in hot water or whey and is worked, stretched and molded while hot in a plastic condition. Provolone is representative of this group in that it is acid-ripened by bacteria, cooked to a relatively high temperature, and the curd pulled and molded while hot into long cylindrical shapes, chilled, salted in brine, smoked, waxed, and ripened similar to Cheddar. Much of the typical flavor results from the lipolysis of milk fat achieved by adding special mammalian lipases.

Parmesan

Parmesan or Grana cheese, as it is known in Italy, is a very hard, granular textured cheese made from partly defatted cow's milk. In addition to rennet, special lipolytic enzyme preparations of animal origin are usually used to develop the characteristic rancid flavor. After making and salting, which takes about 40 days, the cheese is stored in cooled, ventilated rooms for 1 to 2 years to ripen. The texture and hardness of the cured cheese are such that it must be grated and used for seasoning. As a group, these cheeses are characterized by low moisture and fat content, both of which contribute to their hardness.

Skimmilk or Low-fat Cheese

Sapsago, made chiefly in Switzerland, is a very hard, cone-shaped, greenish, grating cheese which derives its color and flavor from added dried leaves of aromatic clover. It has a sharp pungent aroma and pleasing flavor but must be used as a grating cheese because of its hardness. In contrast, Euda cheese, developed by the U.S. Department of Agriculture[10,31] is a ripened, low-fat, semisoft, skimmilk cheese that has a very mild flavor and soft body resembling Colby cheese in appearance. Lactic acid bacteria are primarily responsible for ripening; however, much of the flavor is attributed to the predevelopment of lipolysis in the small amount of milk fat used.

Ricotta

About 50% of the solids in milk remain in the whey after making most cheeses. Cheese-like products may be made from the lactalbumin and lactose remaining in the whey. Wheys in which the noncasein protein has not been coagulated may be made into whey cheese by either of two methods. In one, whey is concentrated by boiling until it has a firm sugary consistency on cooling. Primost and Gjetost are manufactured by this procedure.

Ricotta cheese, made by the other method, is made from whey to which 10% skimmilk has been added or whey to which 10% whole milk has been added. The milk and whey proteins are coagulated by high heat (80 to 100°C) and acid. Fresh Ricotta is bland in flavor and resembles Cottage cheese in appearance. There are no Federal standards of identity for Ricotta cheese; however, some states require that Ricotta cheese made from whole milk must contain a minimum of 11% fat and not more than 80% moisture.

Process Cheese

Process cheeses are modified from their natural state by comminuting, heating, and blending one or more lots of cheese with a suitable emulsifying agent into a homogeneous mass. Heating temperatures above pasteurization are sufficient to stop ripening and destroy most bacteria. Process cheese which contains only Cheddar cheese is called "Pasteurized Process Cheddar Cheese."

Pasteurized Process Cheese Foods may contain optional dairy ingredients other than cheese such as skimmilk, cream, cheese whey and lactalbumin.

Pasteurized Process Cheese Spreads may contain the same optional ingredients as cheese foods, additional moisture, and stabilizing agents such as gums, gelatin and algin.

FROZEN DESSERTS

Frozen desserts containing milk products include ice cream, frozen custard, ice milk, and sherbet. Federal standards for ice cream require a minimum of 10% fat in ice cream and frozen custard, 2% in ice milk, and 1% in fruit sherbets.[29] Although consumption of ice cream and other frozen desserts is increasing around the world, only in the United States does a significant part of the milk supply go into ice cream (Table 2.1). Typical analyses are given in Table 2.4. Detailed compositions and manufacturing procedures for frozen milk products are given in Chapter 14.

CASEIN

Commercial casein is usually manufactured from skim milk, precipitation being accomplished by acidification or rennet.[19] Casein exists in milk as a calcium caseinate-calcium phosphate complex. When acid is added, the complex is dissociated and at pH 4.7, the isoelectric point of casein, maximum precipitation occurs. Relatively little commercial casein is produced in the United States, but imports amounted to well over 100 million pounds in 1970. Casein is widely used in food products as a protein supplement. Industrial uses include paper coatings, glues, plastics and man-made fibers. Casein is typed according to the process used to precipitate it from milk, such as hydrochloric acid casein, sulfuric acid casein, lactic acid casein, coprecipitated casein, rennet casein and low-viscosity casein. Differences in composition of casein are mostly by differences in the manufacturing process and the care taken in precipitation and washing.

In the United States a classification of edible casein by grade has been established.[4]

Specifications for standard and extra grades are as follows:

	Extra grade	*Standard grade*
Moisture %	10	12
Fat, max. %	1.5	2

Free acid, max.	0.20 ml	0.27 ml
Ash, max.	2.2	2.2
Protein content %	95	95
Plate count/gm, max.	30,000	100,000
Coliform count, max. 0.1gm	0	2

Australian standards have been established for both acid and rennet caseins. The standards for acid casein are much the same as those for United States casein. Rennet casein usually has between 7.0 and 8.3 ash. The fact that rennet casein is essentially a calcium caseinate accounts for this comparatively large value.

Sodium Caseinate

Sodium caseinate, edible grade, is made from isoelectric casein which has been prepared to meet the sanitary standards for edible casein. Casein is solubilized with food-grade caustic soda and the resulting soluble product (20 to 25% solids) is spray-dried. Spray-drying procedures are adjusted to obtain a product with 5% or less moisture content. Dry sodium caseinate usually contains about 90 to 94% protein, 3 to 5% moisture, 6 to 7% ash, and 0.7 to 1% fat. The best flavor in dried sodium caseinate is obtained when the product is made directly from fresh wet curd. The calcium and lactose content and moisture in fresh curd should be as low as possible, since all three adversely affect the resulting dried product. Isoelectric casein usually has better keeping qualities than sodium caseinate. The uses for sodium caseinate are much the same as those of commercial casein. Increasing quantities are being used as a protein supplement in dietetic and bakery products, in stews, soup, and imitation milk.

LACTOSE

Lactose is the characteristic carbohydrate of milk, averaging about 4.9% for fluid whole cow's milk and 4.8% for sheep and goat's milk. The commercial source of lactose today is almost exclusively from sweet whey, the by-product from cheese making. Details of its production are given in Chapter 6.

Tentative standards for anhydrous lactose have been proposed by a Joint Codex Alimentarius Commission of the Food and Agriculture Organization of the United Nations.[14] Lactose for export and import would have these tentative specifications:

Lactose, anhydrous	99% min. (on dry basis)
Sulfated ash	0.3% max (on dry basis)
Loss on drying (16 hr at 120°C)	6% max.
pH, 10% solution	4.5–7.0
Arsenic	1 mg/kg max.
Lead	2 mg/kg max.
Copper	2 mg/kg max.

WHEY

Whey is the liquid remaining after removal of the casein and fat from milk in the process of cheesemaking; it contains most of the salts, lactose and water-soluble proteins of the milk. Whey from Cottage cheese manufacture, called "acid whey," has a lower pH and consequently longer storage life than "sweet whey" from Cheddar and other kinds of cheese. Only about half the total quantity of whey produced is utilized today. However, as an increased concern about environmental pollution combines with newly developed uses for whey, the amount of whey utilized should be substantially increased in the future.

Whey is the major source of lactose (Chapter 6), although the total quantity utilized for this purpose is small. The major use of whey is in the dry form as a supplement for animal feeds. Of increasing importance are wheys partially desalted by electrodialysis and whey fractions produced by ultrafiltration.[18] Protein fractions produced by the latter method show considerable promise as a protein supplement for human foods.

Yeast fermentation of whey, particularly by *Saccharomyces fragilis,* is used commercially to produce a feed supplement useful as milk replacers in calves. It also is finding increased use in human foods.[27]

SPECIALTY PRODUCTS

Recent progress in scientific and technological research, coupled with the recognized nutritive value of milk, has led to a new concept of milk as a readily processable raw material for the development of new dairy foods. These foods differ widely in chemical and physical characteristics from traditional dairy products,[17] and new dairy foods containing substantial quantities of milk and its components are constantly being developed. For example, a milk-orange drink containing 37% orange juice, 57% milk, plus sugar and emulsifying agent has recently been marketed.[25] Whey spreads, whey ices, and unique protein concentrates from whey are or soon should be on the market. Whey-based beverages such as Rivella are popular in Europe, and similar products have received some attention in the United States.[15,17]

REFERENCES

1. AMERICAN DRY MILK INSTITUTE Bull. 916, Chicago (1961).
2. CANADA DEPT. AGRICULTURE, Dairy Products Act of 1951, as amended. Ottawa (1971).
3. FACIUS, W., "Handbuch der Nährwert-Kontrolle Nahrstoff-Calorien- und Vitamin-Tabellen". 3rd ed. Koln, Heyman (1968).
4. Federal Register, July 20, 1968.
5. Federation of UK Milk Marketing Boards. "UK Dairy Facts and Figures. 1970". Milk Mkting Bd., T. Ditton, Surrey (1971).
6. Food and Agriculture Organization, United Nations. "Code of Principles Concerning Milk and Milk Products and Associated Standards". 6th ed. FAO, Rome (1968).
7. Food and Agriculture Organization, United Nations. CX 5/70, 13th session, FAO, Rome (1970).
8. GUNNIS, L. F., Milk Ind., *68,* 17 (1971).
9. HALL, C. W., and HEDRICK, T. I., "Drying Milk and Milk Products", Avi Publishing Co., Westport, Conn., 1966.
10. HARGROVE, R. E., and McDONOUGH, R. E., J. Dairy Sci., *49,* 796 (1966).
11. HARTMAN, A. M., and DRYDEN, L. P., "Vitamins in Milk and Milk Products", Amer. Dairy Sci. Assoc., Champaign, Ill., 1965.
12. HARTOG, C. den, Insert in Voeding, *30,* (1) (1969).
13. International Dairy Federation, Intern. Stand. FIL-IDF 47 (1969). Cited in Dairy Sci. Abstr., *32,* 1086 (1970).
14. Joint FAO/WHO Codex Alimentarious Commission. Comm. on Sugars, Rpt of 4th Mtg., London, April 18–21, (1967).
15. KOSIKOWSKI, F., "Cheese and Fermented Milk Foods", publ. by author, Cornell Univ., Ithaca, N.Y., 1966.
16. LAMPERT, L. M., "Modern Dairy Products", Chem. Publ. Co., New York, 1970.
17. MANN, E. J., Dairy Sci. Abstr., *33,* 1 (1971).
18. McDONOUGH, F. E., MATTINGLY, W. A., and VESTAL, J. H., J. Dairy Sci., *54,* 1406 (1971).
19. MULLER, L. L., Dairy Sci. Abstr., *33,* 659 (1971).
20. New Zealand Govt, "Food and Drug Regulations", 1946, as amended, Wellington, 1966.
21. New Zealand Dairy Board. Annual Report 1970, Wellington, 1971.
22. PATTON, S., J. Dairy Sci., *33,* 680 (1950).
23. PEARSON, D., Rev. Nutr. Food Sci., *14,* 10 (1969).
23a. PEARSON, D., The Chemical Analysis of Foods, 6th ed. J. & A. Churchill, London, 1970.
24. PORTER, J. W. G., J. Dairy Res., *27,* 321 (1960).
25. SHENKENBERG, D. R., CHANG, J. C., and EDMONDSON, L. F., Food Eng., *43* (4), 97 (1971).
26. U. S. Department of Agriculture, USDA Agr. Handbook 54, 1953.
27. U. S. Department of Agriculture, USDA, ARS 73–69, 1970.
28. U. S. Department of Agriculture, USDA, SRS Bull. Da 2–1(71), 1971.
29. U. S. Department of Agriculture, USDA Agr. Handbook 51, 1971.
30. U. S. Office of the Federal Register, "Code of Federal Regulations", Title 21, Parts 1–119, 1971.
31. WALTER, H. E., SADLER, A. M., and MATTINGLY, W. A., J. Dairy Sci., *52,* 1133 (1969).

32. WARD, A. G., "Food Standards Committee Report on Cream", H. M. Stationery Office, London, 1967.
33. WATT, B. K., and MERRILL, A. L., USDA Agr. Handbook 8, 1963.
34. WEBB, B. H., and WHITTIER, E. O., "Byproducts from Milk", 2nd ed., Avi Publishing Co., Westport, Conn., 1970.
35. Western Australia Dept. of Health, "Standards for Milk and Milk Products", as amended, Perth, 1971.

CHAPTER 3

William G. Gordon
Edwin B. Kalan†

Proteins of Milk*

INTRODUCTION

Proteins are of great importance in biochemistry, since they form a large proportion of the food and the body tissues of man and other animals. Research on proteins is extremely difficult because of the large size and complex structure of their molecules and aggregates, their insolubility in most common solvents, their usually noncrystalline form, and the readiness with which they undergo decomposition or other chemical change when attempts are made to isolate and purify them or to subject them to the usual methods of characterization. Proteins may be defined as substances composed principally of amino acids chemically combined. Carbon, hydrogen, nitrogen, oxygen, sulfur, and, in a few instances, phosphorus are the elements present in proteins. Cow's milk contains approximately 3.5% protein, or, as distributed, 2.9% casein and 0.6% whey proteins.

The chemistry of casein was reviewed by Jollès[113] in 1966, of the whey proteins by Garnier[66] in 1965, and of minor components of casein by Groves[86] in 1969. A comprehensive review of the chemistry of milk proteins by McKenzie[137] appeared in 1967, and a two-volume treatise on the chemistry and molecular biology of the milk proteins, edited by the same author,[138] was published in 1970 and 1971.

NOMENCLATURE

In discussing the proteins of milk, it is necessary to distinguish between the proteins *in* milk and those obtained *from* milk by various chemical and physical fractionation procedures. Because of the ease with which casein can be isolated from milk, the earliest subdivision of milk proteins was to casein and whey proteins. However, this implies

* Revised from Chapter 3 in the 1st edition by W.G. Gordon and E.O. Whittier.

† William G. Gordon and Edwin B. Kalan, Eastern Regional Research Laboratory, Agricultural Research Service, U.S. Department of Agriculture, Philadelphia, Pennsylvania 19118.

that casein exists in milk in the same form as it does in the isolated state, which is not true when casein is precipitated by acid, as is customary. The casein in milk is in the form of a complex or micelle, consisting of calcium caseinate plus phosphate, additional calcium, magnesium, and citrate. *Casein* may be defined most simply as the protein precipitated by acidifying skimmilk to a pH value near 4.6 at 20°C. The proteins remaining after casein has been removed from skimmilk are known as *whey proteins* or *milk-serum proteins*. They have been fractionated by salting-out methods to produce a lactalbumin fraction and a lactoglobulin fraction. Each of these—casein, lactalbumin and lactoglobulin—was generally considered to be a single chemical entity until Linderstrøm-Lang[130,131] in 1925 reported the fractionation of casein, and Palmer[162] in 1934 the further fractionation of the whey proteins. Since then research has resulted in the isolation of at least nine main components of milk proteins, several of which are still heterogeneous, as indicated by electrophoretic measurements.

Subdivisions of milk-serum proteins sometimes mentioned in milk protein literature are the *heat-labile milk-serum proteins* and the *proteose-peptone fraction,* the first being the portion of the milk-serum proteins rendered acid-precipitable at pH 4.6 to 4.7 by previous heat treatment of the milk or whey, the second being the portion not made precipitable by this means. *Commercial lactalbumin* is a mixture of the heat-labile, acid-precipitable milk-serum proteins. The so-called classic *lactalbumin fraction* is the portion of the milk-serum proteins that is soluble in neutral half-saturated ammonium sulfate solution or neutral saturated magnesium sulfate solution. In a classification advocated by Rowland,[189] *lactalbumin* is the portion of the heat-labile milk-serum protein that is soluble in saturated magnesium sulfate solution. The classic *lactoglobulin fraction* is the portion of the milk-serum proteins that is insoluble in neutral half-saturated ammonium sulfate solution or saturated neutral magnesium sulfate solution. In the Rowland classification, *lactoglobulin* is the portion of the heat-labile milk-serum protein that is insoluble in saturated magnesium sulfate solution. A complex of proteins and enzymes adsorbed on the surface of the fat globules of milk is usually designated as the *membrane proteins of the fat globules*.

The contemporary nomenclature of the principal proteins found in cow's milk, as reported[188] by a Committee of the American Dairy Science Association, appears in Table 3.1, in which there are listed distinguishing characteristics of the proteins. Some of the problems that have arisen in the naming of newly discovered milk proteins have been dealt with in previous reports of the Committee.[110,34,214] The present status is discussed cogently by Jenness.[109]

DETERMINATION OF PROTEIN FRACTIONS

The separation of the proteins of skimmilk into the casein, lactalbumin, and lactoglobulin fractions that was employed for many years was preparatory to the determination, characterization, and study of each of these fractions. The methods employed were precipitation of casein by acidification to its isoelectric point, and salting-out methods for the lactalbumin and lactoglobulin fractions. Rowland[189] devised an analytical scheme to determine these fractions, one fraction consisting of proteoses and peptones, and the other consisting of nonprotein nitrogen compounds. The results are usually expressed in terms of the percentage distribution of nitrogen among these five fractions. The preparation of samples for study is incidental. The following is an outline of the Rowland scheme.

Total N(I) of the milk is determined by the Kjeldahl procedure. Noncasein N(II) is determined by precipitating casein by a mixture of acetic acid and sodium acetate, filtering, and analyzing the filtrate for N. Nonprotein N(III) is determined by precipitating the total protein by means of 15% trichloroacetic acid, filtering, and analyzing the filtrate for N. Nonprotein N plus proteose-peptone N(IV) is determined by boiling an aliquot of the filtrate II, adjusting the reaction to pH 4.75, filtering, and analyzing the filtrate for N. Lactoglobulin N(V) is determined by adjusting an aliquot of the filtrate II to pH 6.8–7.2, saturating it with magnesium sulfate, filtering off the lactoglobulin fraction after several hours, washing it with saturated magnesium sulfate solution, and analyzing it for N. Then, casein N = I − II, lactalbumin N = II − (IV + V), lactoglobulin N = V, proteose-peptone N = IV − III, and nonprotein N = III. Rowland, by use of his method, found the average nitrogen distribution in normal milk to be: 78.3% casein N; 9.1% lactalbumin N; 3.5% lactoglobulin N; 4.1% proteose-peptone N; and 5.0% nonprotein N.

The classic fractions obtained by the Rowland procedure were investigated electrophoretically by Larson and Rolleri.[125,186] The electrophoretic method not only reveals the complexity of the fractions produced by chemical partition, but gives somewhat different results, particularly in the distribution of the albumin and globulin fractions. The distribution percentages shown in Table 3.1 were compiled from both types of study which, incidentally, agree in assigning 2 to 6% of total skimmilk protein to the proteose-peptone fraction.

A newer chemical partition method employing sodium sulfate at controlled pH values for the preparation of fractions was developed by Aschaffenburg and Drewry.[11] This procedure gives results in substantial agreement with the electrophoretic data and also permits the

Table 3.1

SOME PROPERTIES OF PROTEINS ISOLATED FROM COW'S MILK[a]

Protein or Protein Fraction	Approx. % of Skimmilk Protein[b]	Occurrence in Electrophoretic Pattern (Peak Number)[e]	Reference to Preparation[g]	Electrophoretic Mobility[j]	pI[l]	Sedimentation Constant (S_{20})[n]	Molecular Weight[o]	Components
I Casein (Precipitated from skimmilk by acid at pH 4.6)	76–86		101,224			1.3, 6.0–7.5[m] [226]	15,000[222] 33,600[35]	
α-Casein	53–70	1	101,224	−6.7[101]	4.1[224]	3.99[204]	27,000[154] 23,000[193]	
α_s-Casein	45–55	1[f]	213,226	−6.7[f] [101]	4.1[110]	3.99[204]	23,000[p]	α_{s1}-Variants A, B, C, D α_{s0}-, α_{s2}-, α_{s3}-, α_{s4}-, α_{s5}-[s]
β-Casein	25–35	2	101,224	−3.1[101]	4.5[224]	1.57[204]	24,000[89,173,176,184,204]	Variants A¹, A², A³, B, C D, B_z
κ-Casein	8–15	1[f]	209,226,231	−6.7[f] [101]	4.1[209]	1.4[209]	19,000[q] [230]	Variants A, B, sub-variants containing 0 to 5 carbohydrate chains
γ-Casein	3–7	3	101	−2.0[101]	5.8–6.0[101]	1.55[151]	21,000[r]	Variants A¹, A², A³, B components R, S, TS (TS has several variants)
II Noncasein Proteins	14–24							
A—Lactalbumin (soluble in ½ saturated $(NH_4)_2SO_4$ solution)								
β-Lactoglobulin	7–12	6	10,162	−5.3[188]	5.3[188]	2.7[216,217]	36,000[215,217]	Variants A, A_{DR}, B, B_{DR}, C, D
α-Lactalbumin	2–5	4	10,72,75	−4.2[72]	4.2–4.5[123]	1.75[72]	14,440[30]	Variants A, B in Zebu
Blood Serum Albumin	0.7–1.3	7	177	−6.7[177]	4.7[177]	4.0[55]	69,000[177]	
B—Lactoglobulin (insoluble in ½ saturated $(NH_4)_2SO_4$ solution)								
Euglobulin[c]	0.8–1.7	1	197	−1.8[151,197]	6.0[151,197]	8.77[151]	180,000[197] 252,000[151]	
Pseudoglobulin[d]	0.6–1.4	2	197	−2.0 to 2.2[k]	5.6[151,197]	8.06[151]	180,000[197] 289,000[151]	

IgG Immunoglobulins[h]								
IgG1	1–2	1 and 2		−2.0 to 2.2[k]		6.3[188]	150,000 to 170,000[188]	A1 and A2 allotypes recognized in serum IgG1
IgG2	~ 0.2–0.5	1		−1.1[197]		6.6[188]		
IgM Immunoglobulin	~ 0.1–0.2	2				18 to 19[188]	900,000 to 1,000,000[188]	Insufficient data
IgA Immunoglobulin	~ 0.05–0.1	2 (?)				10 to 12[188]	300,000 to 500,000[188]	Insufficient data
C—Proteose-Peptone Fraction[i] (not precipitated at pH 4.6 from skimmilk previously heated to 95–100°C, 30 min)	2–6	3,5,8	12,110,125	−3.8 to 9.3[188]	3.3–3.7[188]	0.8 to 4.0[188]	4,100 to 20,000[188]	Multiple, including glycoproteins

[a] Adapted from the reports of Thompson *et al.*[214] and Rose *et al.*[188]

[b] Values compiled or calculated from Rowland nitrogen distribution data, relative areas of electrophoretic patterns, and protein yield studies.[101,110,124,125,186,189,195]

[c] Euglobulin is not well characterized and is not included in the most recent revision of the nomenclature of the proteins of cow's milk.[188]

[d] Pseudoglobulin is considered to be primarily IgG1.[188]

[e] Free-boundary electrophoresis in veronal buffer at pH 8.6, $\Gamma/2 = 0.1$. Casein components designated in descending order of mobility in casein pattern; whey proteins, designated in ascending order of mobility in acid whey pattern.[125]

[f] Value for whole α-casein (i.e., α_s- and κ-casein complex).

[g] See Reference 138 for other references concerning preparation of milk proteins.

[h] Nomenclature and properties of immunoglobulins of the cow have recently been reviewed.[37]

[i] Composition and physiocochemical properties of proteose-peptone fraction are summarized in Reference 188.

[j] Electrophoretic mobility = 10^{-5} cm^2 volts^{-1} sec^{-1} in Tiselius moving-boundary method, 2°C. in veronal buffer, pH 8.6, $\Gamma/2$ 0.1; descending pattern.

[k] Average of values from Smith[197] and Murthy and Whitney.[151]

[l] Isoelectric point, or pH of zero electrophoretic mobility.

[m] From bimodal sedimentation pattern.

[n] Sedimentation coefficient, $S_{20} = (dx/dt)(\omega^2 x)$ in Svedberg units ($S = 1 \times 10^{-13}$) corrected to 20°C.

[o] Refer to original literature for methods and conditions of determination.

[p] From amino-acid sequence.[147]

[q] Value for carbohydrate-free monomer; approximately 600 should be added for each carbohydrate chain.

[r] From amino-acid analyses. Groves, M. L., and Gordon, W. G., unpublished data. For discussions of the minor proteins (γ-, R-, S-, TS-casein) and their relationship to β-casein see References 71, 86, 188.

[s] See References 3 and 105 for a discussion of these proteins.

further differentiation of the classic albumin fraction into α-lactalbumin and β-lactoglobulin fractions.

The free-boundary electrophoretic method, when applied in 1939 by Mellander[145] to casein itself, showed that the protein was made up of at least three components, which were designated α-, β-, and γ-casein in the order of decreasing mobility in a phosphate buffer. The more discriminating method of zonal electrophoresis, for example, starch gel electrophoresis as used by Wake and Baldwin,[223] reveals that casein is in fact a much more complex mixture of proteins.

Another method which has yielded important information about the proteins in milk is ultracentrifugation. It is mentioned here primarily in connection with the names of the individual whey proteins, α-lactalbumin, β-lactoglobulin, and γ- or immuno-lactoglobulin. In the sedimentation diagrams of decalcified skimmilk described by Pedersen,[168] the peaks were identified by Greek letters in the order of increasing sedimentation constants, the smallest whey protein being designated α-lactalbumin, the next larger β-lactoglobulin (see Cannan *et al.*[39]), and the largest γ-lactoglobulin. Peaks with even larger constants were ascribed to components of casein.

The components of the milk protein fractions will be dealt with as individual proteins in later sections of this chapter.

THE CASEINATE COMPLEX

The principal protein in milk, casein, is present in the form of micelles, or particles of macromolecular sizes. These are made up of the various components of casein bonded together as "calcium caseinate," and complexed further with calcium phosphate and magnesium and citrate ions. The casein complex is considered in detail elsewhere in this volume.

CASEIN

Elementary Composition

In the early years of protein chemistry the determination of the elementary constitution of isolated proteins, as of other organic chemicals, was an important method of characterization. With the recognition that proteins are very large molecules, composed essentially of the same amino acids, it became evident that differences in carbon and hydrogen content could not be great, and determinations of these elements are seldom made on proteins. Nevertheless, because casein was one of the first proteins to be prepared in purified form, the following results of an analysis of acid-precipitated casein reported in 1883 by Hammarsten[93] are at least of historical interest: C, 53.0; H, 7.05; N, 15.65; S, 0.76; and P, 0.85%. The determinations of phosphorus, sulfur and nitrogen compare favorably with modern analyses. Estimations of these elements are still useful in the characterization of casein and other proteins, since phosphorus is an uncommon constituent of proteins, sulfur provides a measure of the total sulfur-containing amino acids present, and nitrogen percentage is a convenient rough index of the content of protein in foods, feeds, and other biological materials. It has been customary for many years to estimate the protein content of biological materials by determining their nitrogen content, usually by the Kjeldahl method, and multiplying by the factor 6.25, because purified proteins contain about 16% nitrogen. In the case of dairy products, the factor used is 6.38, this figure reflecting more accurately the nitrogen content of milk proteins.

Amino Acid Composition

Many important discoveries regarding proteins and their amino acid composition were made in the nineteenth and early twentieth centuries by investigators who used the readily prepared casein as a typical protein. For example, a few amino acids were discovered in hydrolyzates of casein and were recognized later to be common constituents of all proteins.[221] Casein was also widely used in the development of methods for the quantitative amino acid analysis of proteins. Actually, however, it was another milk protein, β-lactoglobulin, that was the first large protein to have its complete amino acid composition established through the classic research of Brand *et al.*[28] in 1945. A few years later casein and its main components were also analyzed by similar methods.[74] These results, as well as later analyses of purified casein components and other milk proteins, are summarized in Table 3.2. Compilations from the older literature of the many quantitative determinations of amino acids in milk proteins have been made by Block and Weiss[26] and Orr and Watt.[159] More up-to-date tabulations may be found in the treatise by McKenzie.[138]

Milk and milk products provide food proteins of excellent quality for the nutrition of man and animals. Casein, the dominant protein, is a good source of amino acids indispensable to man. Commercial lactalbumin, or the heat-coagulable whey proteins, has an even higher biological value. Table 3.3 lists the content of essential amino acids (also cystine and tyrosine) in these proteins and in dried skimmilk and dried whey. The data are taken from Table 2 of the Orr and Watt report.[159] The nutritive value of milk proteins is also reviewed by Henry.[96]

Table 3.2

AMINO-ACID COMPOSITION OF COW'S MILK PROTEINS[f]
(GRAMS PER 100 GM PROTEIN)

Constituent	α-Lactalbumin[77]	β-Lactoglobulin[69,175]	Blood Serum Albumin[203]	Immune Globulin[94,95,197]	Casein[73,74]	α-Casein[73,74]	α_{s1}-Casein B[d147]	κ-Casein[210]	β-Casein[73,74]	γ-Casein[73,74]
Total N	15.9	15.6[28]	16.1	15.3–16.1	15.6	15.5	15.4	15.3	15.3	15.4
Total P	0.0	0.0	0.0	0.0	0.9	0.99	1.05	0.16[192]	0.61	0.11
Total S	1.9	1.6[28]	1.9	1.0	0.8	0.72[101]	0.68	0.70	0.86[101]	1.03[101]
Glycine	3.2	1.4	1.8	5.2	2.0	2.3	2.9	1.2	1.6	1.5
Alanine	2.1	7.0	6.3	4.8	3.2	3.8	3.4	5.4	2.0	2.3
Valine	4.7	6.1	5.9	9.6	7.2	6.3	5.5	6.3	10.2	10.5
Leucine	11.5	15.5	12.3	9.6	9.2	7.9	9.4	6.1	11.6	12.0
Isoleucine	6.8	6.9	2.6	3.0	6.1	6.4	6.1	7.1	5.5	4.4
Proline	1.5	5.1	4.8	10.0	10.6	7.5	8.3	11.0	15.1	17.0
Phenylalanine	4.5	3.5	6.6	3.9	5.0	4.6	5.6	3.9	5.8	5.8
Tyrosine	5.4	3.7	5.1	6.7	6.3	8.1	7.7	7.6	3.2	3.7
Tryptophan	(7.0)[a] 5.3[b]	2.7	0.58	2.7	1.7	2.2	1.7	1.0[201]	0.83	1.2
Serine	4.8	4.0	4.2	11.5	6.3	6.3	7.1	5.0	6.8	5.5
Threonine	5.5	5.0	5.8	10.5	4.9	4.9	2.5	6.7	5.1	4.4
Cystine + Cysteine	6.4[c]	3.4	6.5	3.2	0.34	0.43	0.0	1.2[e230]	0.0	0.0
Methionine	0.95	3.2	0.81	0.9	2.8	2.5	3.2	1.7	3.4	4.1
Arginine	1.2	2.8	5.9	4.1	4.1	4.3	4.4	4.0	3.4	1.9
Histidine	2.9	1.6	4.0	2.1	3.1	2.9	3.2	2.4	3.1	3.7
Lysine	11.5	11.8	12.8	6.8	8.2	8.9	8.7	6.5	6.5	6.2
Aspartic acid	18.7	11.4	10.9	9.4	7.1	8.4	8.5	7.7	4.9	4.0
Glutamic acid	12.9	19.3	16.5	12.3	22.4	22.5	24.3	19.8	23.2	22.9
Amide N	1.4	1.1[28]	0.78	—	1.1	1.6	1.3	1.9	1.6	1.6

[a] By method of Spies and Chambers.[202]
[b] By method of Spies;[201] more reliable value.
[c] No cysteine present.
[d] Based on amino acid sequence of protein.[147]
[e] No cysteine present.
[f] Table 3.2 does not include analyses of many other milk proteins whose amino acid compositions have been reported. The reader is referred to the treatise edited by McKenzie[138] for more inclusive compositional data.

Table 3.3

AMINO ACIDS ESSENTIAL TO MAN IN COW'S MILK PRODUCTS
(GRAMS PER 100 GM)

	Casein	Lactalbumin	Dried Nonfat Milk	Dried Whey
Tryptophan	1.3	2.2	0.50	0.15
Threonine	4.3	5.2	1.6	0.68
Isoleucine	6.6	6.2	2.3	0.73
Leucine	10.0	12.3	3.5	1.04
Lysine	8.0	9.1	2.8	0.77
Methionine	3.1	2.3	0.87	0.19
Cystine	0.38	3.4	0.32	0.25
Phenylalanine	5.4	4.4	1.7	0.32
Tyrosine	5.8	3.8	1.8	0.13
Valine	7.4	5.7	2.4	0.64

Sulfur

Casein contains 0.78% sulfur; the amino acids cystine and methionine account for 0.09 and 0.69%, respectively, cysteine being absent.[119] When casein is fractionated, it is found that α-casein contains all the cystine in casein.[74] On fractionation of α-casein, most of the cystine is found in κ-casein,[225,115] some in the minor components α_{s3}-, α_{s4}- and α_{s5}-caseins.[105]

Phosphorus

When casein is treated with proteolytic enzymes, rather large polypeptides, some of which contain phosphorus and resist further enzymatic degradation, are formed. It has been thought that these products, often called phosphopeptones, may play an important role in the nutrition of the young mammal. Thus, Mellander suggested that phosphopeptones can combine with other nutrients, such as calcium and iron, in this way favoring absorption of these elements, and, furthermore, that during subsequent digestion a unique mixture of amino acids is provided at the proper time for optimal utilization of protein.[146]

In 1927, Posternak[179] isolated a phosphopeptone from tryptic digests of casein and found that it contained 5.9% phosphorus, 11.9% nitrogen, and glutamic and aspartic acids, serine and isoleucine. He believed that all the phosphoric acid found had been linked to serine, presumably because serine was the only hydroxy-amino acid present in the hydrolyzate. That some phosphoric acid is indeed bound to serine was proved by the isolation of phosphoserine from a weak acid hydrolyzate of casein by Lipmann[132] in 1933. At about the same time Levene and Hill hydrolyzed further a phosphopeptone from casein and isolated a dipeptide made up of phosphoserine and glutamic acid.[128] Twenty years later, in 1953, de Verdier[220] was able to prepare phosphothreonine from casein hydrolyzed by weak acid. From the preceding evidence and many other studies of a similar nature, it may be concluded that the phosphorus in casein is bound chiefly, if not entirely, in ester linkages with the hydroxyl groups of serine and threonine.

As to the nature of the ester linkages, Perlmann has postulated that not only orthomonophosphate ester linkages, but also diester

$$\begin{array}{rcl} & OH & \\ & | & \\ -CH_2-O- & P & -O-CH_2- \\ & \| & \\ & O & \end{array} \quad \text{and pyrophosphate} \quad \begin{array}{rcccl} & OH & & OH & \\ & | & & | & \\ -CH_2-O- & P & -O- & P & -O-CH_2- \\ & \| & & \| & \\ & O & & O & \end{array} \quad \text{linkages}$$

occur in casein, and that the principal electrophoretic components of casein, α- and β-caseins, differ in the types of linkage present.[169] However, a considerable number of more recent investigations indicate that phosphorus is bonded in the same way in whole casein, α-casein, or β-casein and that the bond is most likely the orthomonophosphate ester linkage.[2,107,118,161,191]

Work on the isolation of phosphopeptides and phosphopeptones has continued to provide important information concerning the structure of casein. This research was facilitated by the fractionation of casein into its several components and by the development of newer, more powerful methods of separating and characterizing products of partial hydrolysis of the components. For example, Peterson *et al.*[174] isolated a relatively large, electrophoretically homogeneous phosphopeptone from tryptic digests of β-casein. The phosphopeptone, with a molecular weight of about 3,000, consisted of 24 amino acid residues of 10 different amino acids, and 5 phosphoric acid groups presumably attached to the 4 serine and 1 threonine residues present. Essentially all the phosphorus of β-casein appeared to be concentrated in this portion of the molecule. In later work on the primary structure of β-casein A^2, one of the genetic variants of this protein, the same phosphopeptide was isolated by Ribadeau-Dumas *et al.*[184] as tryptic peptide, T1. Subsequently, the same authors reported much of the primary structure of the entire β-casein molecule with peptide T1 positioned at the N-terminal end.[184] Similarly, in the case of α-casein, a tryptic polypeptide of 35 amino acid residues with 7 phosphate groups was isolated and partly sequenced by Österberg in 1964.[161] A 21-residue phosphopeptide, probably a portion of the Österberg peptide, was prepared from α-casein by Schormüller *et al.*[194] in 1948. Presumably the same larger peptide was isolated later by Grosclaude *et al.*[84] from $_{\alpha s1}$-casein B. Now designated Tm1, the peptide was found to contain 8 of the 9 phosphate groups in this α_S-casein polymorph. In 1970 Mercier *et al.*[147] worked out the partial sequence of the phosphopeptide and its position in the primary structure of $_{\alpha s1}$-casein B.

The localization of phosphorylated amino acids in limited regions of both the α_{s1}- and β-casein molecules is an important outgrowth of this research. Such information will lead to a fuller understanding of the interactions among components of casein and the structure and behavior of casein micelles.

Carbohydrate in Casein and Other Milk Proteins

The primary carbohydrate of whole unfractionated cow's milk is, of course, lactose, which occurs to the extent of 50 gm per l. Very little carbohydrate is associated with milk proteins through covalent linkage; however, the carbohydrate which is, seems to play an important role in stabilizing the casein micelle. The presence of small proportions of carbohydrate in casein has been detected by various investigators. For example, using the orcinol reaction, Sørensen and Haugaard[199] found highly purified, acid-precipitated casein to contain 0.31% hexose, believed to be galactose but later shown to be a mixture of hexose, hexosamine and sialic acid.[48] Nitschmann *et al.*[158] demonstrated that a large peptide, which was split off from casein by the action of rennet, contained considerable percentages of galactose, galactosamine and neuraminic acid. The peptide was called a glycomacropeptide. The other major product of the rennet reaction is para-casein which is essentially devoid of carbohydrate. It was subsequently shown that the carbohydrate is concentrated largely in the α-fraction of casein and more particularly in the κ-portion of α-casein, from which the glycomacropeptide may be prepared.[1,115] The carbohydrate is found in the carboxyl-terminal portion of the κ-casein molecule and is linked to the protein by an O-glycosidic linkage between the –OH groups of serine or threonine or both, and N-acetyl-galactosamine.[60,218] The other carbohydrate moieties are linked to the galactosamine, N-acetylneuraminic acid being terminal.[115,116,218] Variability in carbohydrate content of κ-casein is responsible for one type of heterogeneity of the protein, since from 0 to 5 carbohydrate chains may occur in the molecule.[230]

Other milk proteins are now known to have covalently bound carbohydrate. Aschaffenburg isolated a minor component of crystalline α-lactalbumin B preparations which was analyzed by Gordon and found to have the exact amino acid composition of α-lactalbumin B. Brew and Hill showed that this component contains 1 mole of hexosamine per mole of protein.[68] Barman[13] observed that as much as 7% of an α-lactalbumin preparation can be recovered as a glyco-α-lactalbumin after column chromatography. This heterogeneous glycoprotein has up to 15% carbohydrate, consisting of mannose, galactose, glucosamine, galactosamine, and N-acetylneuraminic acid. Hindle and Wheelock have reported similar findings.[99] A variant of β-lactoglobulin isolated[17,18] from Australian Droughtmaster beef cattle was shown to contain N-acetylneuraminic acid, hexosamine and hexose in a ratio of 1 : 4.3 : 2.7; otherwise, the protein is identical with β-lactoglobulin A.[140] The red protein, lactoferrin, which binds iron, has been shown to contain hexose, hexosamine and N-acetylneuraminic acid.[70] Other proteins containing carbohydrate are fat globule membrane proteins,[108 211] glycoprotein-a,[88] M-1 glycoproteins,[20,21,22] and the proteosepeptone fractions.[188] In addition, several enzymes appearing in milk are glycoproteins; these include lactoperoxidase[187] and ribonuclease.[23,24] The carbohydrate, though found in relatively low percentages, seems to be distributed over a wide number and variety of proteins, and it may be inferred that, where it does occur in covalent linkage to protein, it plays an important biological role.

Solubility

Casein is dissolved by aqueous solutions of acids, alkalies, and alkaline salts. The amount dissolved in a definite weight of solvent depends on the pH value of the solvent. In acid solutions of a pH on the acid side of the isoelectric point of casein, compounds are formed with the nonmetallic element or radical of the acid, and, in alkaline solutions, compounds are formed with the metal of the alkali or alkaline salt. Hence, such solutions are not solutions of casein in a strict sense. Solubility of casein, strictly speaking, is the solubility of chemically unaltered hydrogen caseinate, and hence is the solubility of casein in solutions at its isoelectric point. That point is usually considered to be pH 4.6, but it is shifted by the presence of neutral salts in solution and may be a point in a zone extending from pH 4.0 to pH 4.8, approximately.[81] Much of the literature on solubility of casein does not state whether the determinations were carried out under isoelectric conditions and consequently must be accepted with reservations. Because the amount of casein dissolved by many solvents varies with the quantity of casein added to a definite quantity of solvent, many available data have limited usefulness. This is a necessary consequence of the fact that the casein preparations studied do not consist of a single molecular species, and therefore no true solubility constant for casein can be expected. The solubility of casein in water at the isoelectric point has been reported as 0.05 gm/l at 5°C,[170] and 0.11 gm/l at 25°C.[43] Its solubility in various organic acids, mixtures of water and certain organic solvents, and aqueous solutions of a variety of salts has been investigated. Appropriate references to these studies in the older literature may be found in the preceding edition (1965) of this book.

Optical Rotation

Solutions of casein rotate polarized light in the levo direction. The specific rotation of casein varies with samples prepared in different ways and with the solvent. Gould[78] found $[\alpha]_{D}^{30°C} = -81.7°$ for 2% Hammarsten casein in 10% sodium acetate solution; values for commercial caseins under the same conditions differed over the range of $-70°$ to $-90°$. Hipp *et al.*[101] determined $[\alpha]_{D}^{25°C}$ for casein and purified fractions in 1% solutions in veronal buffer, pH 8.4, ionic strength 0.1; the values were $-105°$ for whole casein, -87.4 to $-90.5°$ for α-casein prepared by different methods, $-125°$ for β-casein, and $-132°$ for γ-casein.

Combining Capacity with Acids and Bases

Much effort has been spent in determining the combining capacity of casein as an acid and as a base. Electrometric titration,[44] determination of the minimal amount of acid or base required to dissolve a given weight of casein,[170] and conductivity represent some of the methods employed. The results obtained for the combining capacity of casein have varied greatly depending on the method of preparing the casein and the method used to determine its combining capacity. Variations in the combining capacity of casein can in part be explained by the now known heterogeneity of the casein and its lability.

From the acid- and base-combining capacity of casein Cohn and Berggren[44] calculated the number of dissociating groups. They found that this number could be correlated with the amino acid constitution of the casein. More recently, Hipp *et al.*[102] determined the acid- and base-binding capacities of α-, β- and γ-caseins from their titration curves (Fig. 3.1).

By selecting the pH values where the ionic groups of the amino acid residues would be expected to dissociate, the number of ionizing groups was estimated from the titration curve. Thus, at pH 6.35, carboxyl groups plus one equivalent of phosphoric acid are considered to be dissociated and were estimated by the equivalents of alkali combined at this pH. Similar estimates for the remaining dissociating groups in casein were made at other pH values. The estimates of the

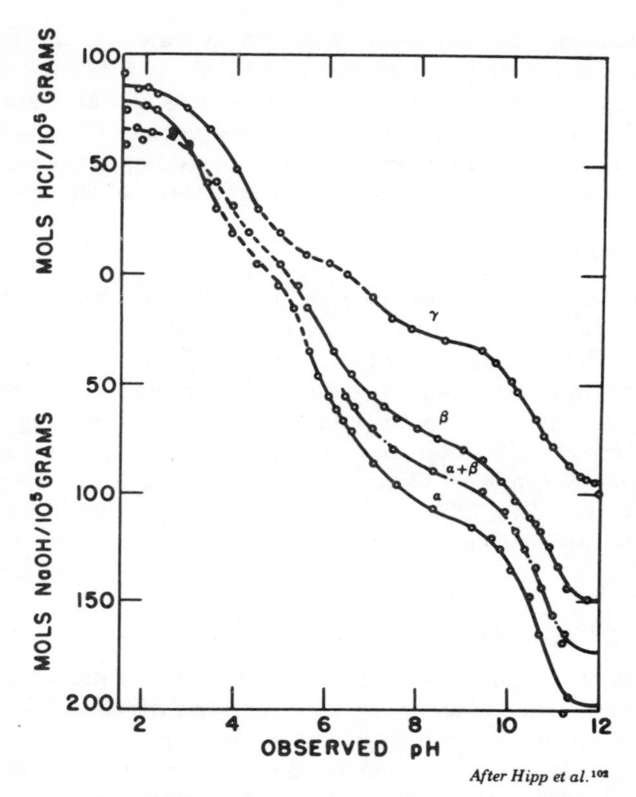

After Hipp et al.[102]

FIG. 3.1. ACID BASE TITRATION CURVES FOR α-, β-, and γ-CASEIN AND AN EQUAL MIXTURE OF α- and β-CASEIN AT PROTEIN CONCENTRATION OF 1% 0.05 IONIC STRENGTH AND 25°C

Ordinates are moles bound/10^5 gm. The dotted line is the pH region where, under the conditions of the experiment, all the protein was not soluble. For clarity, only a portion of the curve for the mixture of α- and β-casein is given.

number of ionic groups in α-, β-, or γ-casein arising from its amino acid residues were consistent with the number of ionic groups calculated to be present from the amino acid compositions of these proteins, as reported by Gordon *et al.*[73,74] and given in Table 3.2. The maximum acid- and base-combining capacities, as determined from the titration curves (Fig. 3.1) of α-, β-, and γ-caseins, were calculated to be 78, 66, and 85 moles of acid and 198, 150, and 96 moles of base per 10^5 gm, respectively. These values for the base-binding capacity are somewhat higher than the values of 176 and 128 moles per 10^5 gm for α- and β-casein calculated from their amino acid constitutions. However, the values determined for the base-combining capacity of

α-, β-, and γ-casein by titration (Fig. 3.1) lead to a calculated value of 183 moles per 10^5 gm of casein based on the constitution of unfractionated casein being 16 parts α-, 4 parts β-, and 1 part γ-casein, which is in agreement with the value reported by Cohn and Berggren[44] for the base-combining capacity of several casein preparations.

Derivatives of Casein

Earlier editions of this volume dealt at some length with many different compounds made from casein. In numerous studies, casein, because of its ready availability in purified from, was used as a model protein for the preparation of various derivatives. Furthermore, since large quantities of the protein were manufactured, some of the unique properties of the native protein or its derivatives were exploited for diverse industrial applications. Derivatives that were prepared include halogenated caseins, desaminocasein, nitrocasein, methylated casein, a number of acylated caseins, formaldehyde casein, carbamidocasein, casein treated with sulfur-containing reagents and metal compounds of casein. References to original papers are listed in the preceding edition (1965) and in the book by Sutermeister and Browne[205] published in 1939. Most of this work has been outdated but the reaction of casein with formaldehyde is still of some importance, as are also certain industrial uses of casein. These topics will be discussed presently.

Formaldehyde Casein

The reaction of formaldehyde with casein is of practical as well as of theoretical interest, since formaldehyde is used as a hardening agent for plastic casein in industry. It is agreed that the free amino groups of casein are the points initially attacked, but several ideas have been advanced as to the reactions involved. Blum[27] believed that methylene caseins, $R\text{-}N = CH_2$ and $R\text{-}NH\text{-}CH_2\text{-}NH\text{-}R_1$, were formed with loss of water. Benedicenti[19] suggested that formaldehyde might be added, without elimination of water, to form a compound of the formula $R\text{-}NH\text{-}CH_2OH$. The probability that this reaction takes place is supported by the fact that much of the formaldehyde can be removed from formaldehyde casein by steam distillation, but the fact that heat treatment fixes part of the formaldehyde on the casein indicates that other reactions take place.[155,157,208] According to Wormell and Kaye,[229] formaldehyde, under neutral conditions, first becomes attached to each lysine side chain, and, in a second stage, cyclic compounds may be formed when two NH_2 groups have been linked by a methylene group in the first step:

$$R\text{-}NH\text{-}CH_2\text{-}NH\text{-}R_1 + 2CH_2O \rightarrow \underbrace{R\text{-}N\text{-}CH_2\text{-}O\text{-}CH_2\text{-}N\text{-}R}_{CH_2} + H_2O$$

Under acid conditions, amino and amide groups are linked through methylene groups. Fraenkel-Conrat and Olcott[64] have offered evidence that, in neutral or slightly acid solutions, CH_2O introduces $-CH_2-$ bridges between amino groups and reactive $=CH$ groups of phenolic and imidazole rings of proteins, and that the hardening action of formaldehyde is not due to a primary reaction, such as that suggested by Benedicenti, but is due to a secondary reaction which transforms methylol, $-CH_2OH$, groups into cross-linking methylene bridges that link amino and primary amide or guanidyl groups:

$$R\text{-}NH\text{-}CH_2OH + H_2N\text{-}COR_1 \rightarrow R\text{-}NH\text{-}CH_2\text{-}NH\text{-}COR_1 + H_2O$$

$$\text{and } R\text{-}NH\text{-}CH_2OH + H_2N\text{-}C\begin{matrix} /\!\!/ NH \\ \diagdown NHR_1 \end{matrix} \rightarrow R\text{-}NH\text{-}CH_2NH\text{-}C\begin{matrix} /\!\!/ NH \\ \diagdown NHR_1 \end{matrix} + H_2O$$

Browning Reaction

Concentrated and dried forms of milk, when held for considerable time in storage, gradually develop a brown color. Very often, even under conditions when browning does not occur, the forerunner of discoloration, especially in high moisture dry milks, is off-flavor. In either case, whether browning takes place or not, the products are usually unfit for human consumption. The reaction involved has been studied by many investigators.[45,126,127,149,163] Considerable evidence[47,165] has been presented in support of the view that browning in milk is caused by a Maillard-type interaction[135] between the free amino groups of milk proteins and the aldehyde group of lactose. Early formation of a lactose-casein complex has been postulated[79,166] and the involvement of the ϵ-amino group of lysine was demonstrated by the experiments of Henry *et al.*[97] and Lea and Hannan[127]; more recently, a lysine derivative was isolated from heated milk powders.[61,32] Participation of other amino acids (arginine, histidine, methionine, and tyrosine) has also been documented.[127,163,56,164] The browning reaction is accelerated by increases in temperature, pH value, and water content of the protein-containing material, or in the relative humidity to which a solid protein is exposed. The reactions appear to be similar to those between casein and formaldehyde, not only in that the primary reaction is between aldehyde and amino groups, but also in that there is a secondary reaction, which, in this instance, produces color and decreases the dispersibility of the protein.

The isolation and identification of products of browning, such as maltol, 5-hydroxy-methyl-2-furfuraldehyde and furfuryl alcohol, have been reported.[165,167] These have been shown to arise from the casein-catalyzed degradation of lactose. The isolation and identification of 40 compounds among the products arising from the browning reaction of casein-lactose in the dry state were reported.[58,59] These compounds have not as yet been evaluated for flavor, but such studies will be important in correlating the development of off-flavor with manifestations of the browning reaction. The chemistry of the browning reaction was reviewed by Danehy and Pigman,[49] by Hodge[106] and, with particular reference to milk and dairy products, by Patton.[165]

One interest of the dairy industry in the browning reaction is its relation to nutrition. The economic interest in browning and production of color and concomitant off-flavor is obvious. Browning results in lowered consumption of the foods involved because of poor palatability, appearance and physical properties, loss of nutritional value from destruction of essential amino acids and vitamins, and loss of biological value and digestibility of protein; finally, browning may produce toxic substances and metabolic inhibitors. The nutritive value of milk and milk products has been reviewed.[122,96]

Precipitation of Casein

It might be expected that all the casein in a sample of milk would be precipitated simply by adding sufficient acid to bring the pH value to approximately 4.6. However, the reaction of acid with caseinate complex is not instantaneous and the pH will tend to rise slowly with time. If a considerable excess of strong acid, such as hydrochloric, is added, the complex will be broken down more rapidly, but some of the separated casein will dissolve as chloride and may not reprecipitate completely on readjustment of the pH to the isoelectric point. To stabilize the pH, a buffer mixture of acetic acid and sodium acetate may be used. In order to increase the rate of precipitation, the acetic acid is added alone to give a pH value slightly on the acid side of the isoelectric point. After a few minutes, the sodium acetate is added and thus the pH is brought to the isoelectric point and stabilized there. Because of the buffer action of the acetate mixture, the same proportions of acid and salt are equally effective with milks of different percentages of caseinate complex. According to Rowland[189] maximal precipitation of casein from milk is effected by adding for each 10 ml milk, 80 ml water at 40°C, then 1.0 ml 10% acetic acid, and, after 10 min, 1.0 ml normal sodium acetate.

Preparation of Casein for Research Purposes

The earliest accepted method for the preparation of so-called "pure casein" was that of Hammarsten.[93] Many modifications of this method and a few somewhat different methods have been proposed. Brief reference is made below to several of these, but the original publications should be consulted for details.

Casein "nach Hammarsten" is prepared by the following method. Skimmilk is diluted with 4 times its volume of water and the casein precipitated by the addition of dilute acetic acid. The casein is then repeatedly dissolved in water containing the least amount of alkali that will dissolve it, the solution filtered to remove insoluble substances, the casein reprecipitated with dilute acetic acid and washed with water. The casein must not be exposed to high concentrations of hydroxyl ion nor of hydrogen ion because of the danger of racemization or hydrolysis under these conditions. Without previous drying, the casein is triturated with absolute ethanol, the ethanol removed by vacuum, the casein treated with anhydrous ether, the ether removed, and the product dried either over sulfuric acid at reduced pressure or over calcium chloride at atmospheric pressure. The first method of drying yields a product that is difficult to wet with water; the second gives a product that contains traces of water and is therefore readily wetted.

Dunn[51] precipitated casein at pH 4.8 by adding 0.5*N* hydrochloric acid to dilute skimmilk, and then washed the precipitate with water, ethanol, and ether. Warner[224] obtained raw milk at time of milking, added toluene as preservative, and chilled the milk immediately to 2°C. All subsequent operations were carried out at that temperature. The milk was skimmed and the casein precipitated at pH 4.6 by adding 0.1*N* hydrochloric acid. It was then washed with ice-water, dissolved with sodium hydroxide to pH 6.5, and the solution extracted with ether. The solution was diluted and the casein reprecipitated with 0.01*N* hydrochloric acid. Finally, it was washed thoroughly with water. A dry product was also made using ethanol and ether. This casein contained 0.86% phosphorus.

Precipitation of casein at pH 4.1, as in the commercial grain-curd process described later, followed by thorough washing with slightly acidulated water, removes all but about 0.20% calcium oxide and all the phosphorus pentoxide in excess of 1.80% (0.79% phosphorus). For most purposes, the small quantity of extraneous ash remaining—less than 0.50%—is not objectionable, and, what is of considerable importance, the casein is not exposed to alkalinities greater than that of milk itself.

Casein that is practically free of both ash and vitamins may be prepared by a method of Block and Howard,[25] which avoids exposure of casein to alkaline pH values. Purified sulfur dioxide is bubbled through skimmilk held at approximately 36°C until the pH value is 4.5 to 4.6. The precipitated casein is removed by filtration and washed with water at pH 4.5 to 4.6. The casein is redissolved by stirring it in suspension in water and passing in sulfur dioxide until the pH value is 1.8 to 1.9. It is reprecipitated by adding dilute alkali until the pH value is again 4.5 to 4.6; it is then separated by filtration, washed with water, and dried by use of anhydrous solvents as described above.

Preparation of Casein for Industrial Use

Commercial casein is made by either of two general methods—precipitation by acid or coagulation by rennin. The latter method is used almost exclusively for producing casein for the manufacture of plastics, since rennin casein has the peculiar properties considered essential for this product. Acid precipitation is employed for producing casein for its other commercial uses, for most of which adhesive properties are the most important requirement.

In the manufacture of rennin casein,[152] a fresh, low-fat, skimmilk is warmed to 35.5°C and curdled with about 4 oz rennin extract to each 100 gal milk. The coagulation should be complete in 15 to 20 min., after which the curd is broken up and gradually heated to 55 to 65°C. After it has settled for 10 min, the curd is drained of whey and washed several times with water at 26 to 32°C. It is then pressed for an hour, shredded, and dried in thin layers at 43 to 46°C. Rapid drying at a low temperature is essential if the product is to be of a light color.

Acidification methods[227] may be of the self-sour type, in which the acid is formed in the skimmilk by bacterial fermentation of lactose to lactic acid, or of the type in which acids are added to the milk in sufficient quantity to precipitate the casein and attain the desired degree of acidity. There are numerous modifications, depending on the acid and the temperatures used and on the mechanical equipment. Several procedures employ continuous precipitation, washing and drying, which, as is to be expected, give a product of remarkable uniformity in composition, color, and adhesiveness.

In the self-sour processes, the milk is inoculated with lactic acid bacteria and allowed to stand at a favorable temperature until curdling takes place. The curd at this acidity—about pH 4.7—is soft and fine; in order to agglomerate it, heat is employed. Excessively high temperatures produce a rubbery curd that is impossible to wash successfully unless it is chopped. After draining, the curd is washed several times, drained, pressed overnight, shredded, and dried at 54°C. Casein of this type, unless exposed to acidity greater than pH 4.7, contains a somewhat greater percentage of ash than the grain-curd casein described below.

The above-described process may be carried out more rapidly if the curd is precipitated by adding dilute acid with moderate stirring; but if a mineral acid, such as hydrochloric or sulfuric, is added to the skimmilk at above 35°C in sufficient quantity to bring the reaction to pH 4.1 (apparently pH 4.6, if methyl red is used as indicator), a curd of coarse, granular texture is formed. This is known as grain-curd; because of its open texture and the fact that calcium phosphate is completely in solution at this acidity, it is easily washed to produce a low-ash, low-acid casein. In the continuous adaptation of this process, temperatures of coagulation as high as 43°C are employed and the chewing gum-textured curd produced is chopped before washing. The washing, pressing, shredding, and drying operations are the same for all types of curd. Casein for paper coating should be ground to 20 to 30 mesh, the proportion of finer particles being kept as small as possible.

Short descriptions of more up-to-date methods for preparing caseins of these types, as well as coprecipitated casein (casein + whey protein), low-viscosity casein and sodium caseinate, are given by Fox.[62] Coprecipitate manufacture is also discussed by Beeby *et al.*[14] A comprehensive review of the manufacture and uses of casein and coprecipitate, with special emphasis on developments since 1953, has been published by Muller.[150]

Uses of Casein

The most extensive nonfood use of casein in the United States, some 34 million pounds[62] in 1967, is in coating paper for books and magazines. For this use it is dissolved in alkali, mineral pigments are added, and the suspension is applied to the paper by means of rollers or brushes. The casein binds the pigment to the surface of the paper and renders the surface smooth and nonabsorptive, thus making the paper suitable for fine printing. Formaldehyde may be used to make the casein-bound coatings waterproof, as on playing cards and wall papers. Casein is used in the paper industry also as a dispersing agent for the rosin used for sizing paper.

Casein glues are used chiefly in the woodworking industries. They consist of casein, a solution of alkali as solvent and a calcium compound. The calcium caseinate formed becomes insoluble on drying and causes the glue to become waterproof. About 10 million lb of casein were used in 1967 in this manner[62] and smaller quantities in the following ways.

Plastic casein is made by stirring dry rennin casein, pigments, and a small proportion of water into a heavy dough and extruding this dough through dies under pressure. Formaldehyde is used as an insolubilizer and hardening agent, being applied usually to the finished article.

Casein-containing, water-thinned paints may consist of pigments, an insolubilizer and a solution of casein in alkali, or may contain oil in the form of an oil-in-water emulsion. The first type, of which whitewash made from skimmilk and lime is the simplest example, is used mostly on exterior surfaces; the second type is popular for use on interior walls and ceilings.

Casein is used in the textile industry for many purposes, such as fixing colors, loading, sizing, softening, and waterproofing. In the leather industry it is used in solution in a minimum amount of alkali to give a gloss to light leathers. A mixture of lime and casein is used as an adhesive and spreader in applying insecticides. Many more uses of minor importance are listed in Sutermeister and Browne.[205]

Ground casein can be converted into fibrous forms by extruding an alkaline solution of the protein into an acid coagulating bath, or by extruding a heated mixture of casein and water into air. The term casein fiber is reserved for the fine wool-like filaments obtained by the first method. The coarser product of the second method is called casein bristle. Casein fiber was produced commercially in the United States for a number of years as a wool substitute during war-time scarcity. As an example of the research which was carried out with the goal of improving the properties of the fiber, the description of continuous-filament casein yarn by Peterson *et al.* in 1948 may be cited.[172] The preparation of casein fibers and bristles was reported by Whittier *et al.*[227A] and by McMeekin *et al.*[103,144,171] and these attained limited commercial production.

Nonfood uses of casein have been described at somewhat greater length by Salzberg.[190]

Of the approximately 100 million pounds of casein and sodium caseinate consumed annually in the United States, about one-third is used by the food industry in the manufacture of a variety of products. This topic is dealt with in the chapters by Fox[62] and Beeby *et al.*,[14] and in the review by Muller.[150]

Fractions and Components of Casein

Casein occurs in milk as a colloidal calcium phosphate complex which contains about 7% inorganic material.[182] It has been realized for some time that although casein can be prepared and purified in a readily reproducible fashion, the product cannot be regarded, chemically, as a single protein; rather it is made up of a number of different proteins which can be separated and shown to have distinctive properties. The starting point for the fractionation of casein is, of course, the so-called whole casein preparation. This material is generally obtained in any one of three ways. The earliest and most widely-used method is that of acid precipitation under controlled conditions of temperature and pH. These methods eventually find their antecedent in the method of Hammarsten.[93] A second method employs centrifugation of skimmilk in the presence of added $CaCl_2$, with appropriate consideration of conditions of temperature, pH and calcium-ion concentration.[222,226] Finally, a method of salt precipitation using 264 gm $(NH_4)_2SO_4$ per l of skimmilk held at 2°C has been described.[137,139] Each of the three methods of preparation is carefully considered by McKenzie.[137,139]

An early observation of the heterogeneity of casein was made by Osborne and Wakeman[160], who isolated a small amount of casein with unique properties from alcohol extracts obtained in drying casein. Casein has since been fractionated on the basis of differential solubility in solutions of hydrochloric acid,[131] acidified ethanol,[130] and 50% ethanol,[160] by precipitation from solutions containing ammonium chloride by means of acetone,[41,42] and by successive precipitations by other methods.[82,112] Fractions of varying phosphorus, tyrosine, and tryptophan content were obtained, but no evidence of homogeneity of the fractions was presented, nor was complete separation into distinct components claimed. In 1939, Mellander, using the method of electrophoresis developed by Tiselius, showed the presence of three components in casein which he designated α-, β-, and γ-casein in decreasing order of mobility.[145] The method enabled him to isolate small amounts of α- and γ-caseins, which were found to be quite different in their phosphorus-to-nitrogen ratios.

The separation of casein into α- and β-caseins, its principal components, was achieved by Warner[224] by preparing a very dilute solution of acid-precipitated casein, 0.12 to 0.3%, at pH 3.5 and 2°C and then adding 0.01N sodium hydroxide to pH 4.4. Under these conditions, β-casein is more soluble and remains in solution, whereas α-casein is precipitated. By reprecipitating α-casein in like manner at least 6 times, it is obtained free of β-casein as shown by electrophoresis.

From the supernatant solution β-casein is precipitated by adjusting the pH to 4.9 at room temperatures. It too can be readily purified so as to be free of other electrophoretic components. Warner's preparations were not electrophoretically homogeneous under all conditions, but neither component contained any of the other.

Two important methods for the separation of α- and β-casein, and for the preparation of γ-casein, have been described by Hipp *et al.* In the first,[101] the fractionation is carried out with solutions of casein in 50% ethanol by varying temperature, pH, and ionic strength. In the second,[101] casein is dissolved in 6.6*M* urea, and the fractionation of components is achieved by the addition of water. The order of precipitation of the casein fractions by both methods is α-, β-, and γ-casein, indicating that the charge on the casein component is the solubility-determining factor under these conditions.

The development of chemical methods for the preparation of the principal electrophoretic components of casein has enabled numerous investigators to study their properties and composition. A few of the distinctive properties are listed in Table 3.1, and amino acid compositions in Table 3.2.

That α-casein is not always homogeneous in its electrophoretic behavior has already been mentioned. Evidence had accumulated that the protective colloid for the stabilization of the micelles in milk is α-casein, and that it is the component on which the enzyme rennin acted.[40] Nitschmann and Keller[156] found that nonprotein nitrogen is produced from only the α-casein component by the action of rennin. In 1955 and in subsequent publications, von Hippel and Waugh[222,225,226] described the separation of casein from milk by high-speed centrifugation and its fractionation by means of calcium ion. They showed that α-casein is really made up of subfractions: α-casein, precipitable by calcium ion under certain conditions and also called "calcium-sensitive casein" and κ-casein (calcium-insensitive casein) not precipitable by calcium ion. They concluded[226] that the κ-casein fraction is the one acted upon by rennin and is responsible for micelle stability. The two points of view concerning the action of rennin were reconciled by McKenzie and Wake[141] on the premise that the α-casein samples examined prior to 1956 contained both α_s-casein and κ-casein. They showed that κ-casein is concentrated with α_s-casein in fraction A during the alcohol fractionation procedure of Hipp *et al.*[101] On the other hand, fraction B contains α_s-casein essentially free of κ-casein. The α-casein obtained in the urea fractionation procedure[101] was found to consist of a mixture of α_s- and κ-caseins. Thus, only alcohol fraction B was a suitable source for preparing α_s-casein. Those α-casein fractions containing α_s- and κ-casein contain the protective colloid, whereas those containing only α_s-casein have no protective properties.

The heterogeneity of the α-casein fraction was also investigated by McMeekin *et al.*[100,104,142] using calcium chloride and ammonium sulfate to separate the components. These investigators isolated three fractions designated α_1-, α_2, and α_3-casein which, in their properties, appear to be similar to Waugh's α_s-casein, λ-casein,[133] and κ-casein, respectively.

Mellander,[145] as has already been mentioned, demonstrated the existence of three electrophoretic fractions of casein, designated α-, β-, and γ-casein. Hipp *et al.*[101] obtained a γ-casein preparation which was electrophoretically homogeneous on the alkaline side of the isoelectric point, pH 5.8 to 6.0. The preparation, low in phosphorus and high in sulfur relative to α- and β-caseins, was thought to be similar to the casein isolated by Osborne and Wakeman.[160] Groves *et al.*[91] obtained essentially homogeneous γ-casein by the method of column chromatography, using DEAE-cellulose columns and phosphate buffers. γ-Casein was eluted with 0.02M phosphate, pH 8.3. A further examination of this γ-casein fraction[90] has revealed the existence of at least three other minor proteins designated S-, R-, and TS-caseins. The isolation, characterization and some properties of the components of the γ-casein fraction have been reviewed.[86,139,188] Further, the interesting relationship between β-casein and the members of the γ-casein fraction has been reported.[71]

There are other α_S-caseins (i.e., caseins which are calcium-sensitive and stabilized by κ-casein) which have been designated α_{S0}-, α_{S2}-, α_{S3}-, α_{S4}-, α_{S5}-caseins,[3,105] to distinguish them from the genetically variable α_{S1}-casein.[213] Annan and Manson[3] isolated some of these α_S-caseins by use of SE-Sephadex chromatography in 8*M* urea at pH 4.0. They described some of the physical chemical properties of these proteins. The relationship of α_{S3}-, α_{S4}-, and α_{S5}-caseins has been partially clarified by the report of Hoagland *et al.*[105] It appears that α_{S5}-casein gives rise to α_{S3}- and α_{S4}-caseins upon reduction, and that α_{S3}- and α_{S4}-caseins are very similar, based on amino acid molar ratios. More information is to be anticipated regarding these calcium-sensitive proteins.

A thorough discussion of casein components can be found in two recent publications concerned with milk proteins.[137,139] It is evident that whole casein is an extraordinarily complicated chemical mixture. This is readily apparent, as mentioned previously, from the analyses of zone electrophoresis in concentrated urea solutions described by Wake and Baldwin.[223] These authors report that about 20 components can be resolved in a single starch-gel analysis of casein. It is equally apparent from what has been written here that many investigators are making important contributions to clarify the complexity of whole casein. Still another aspect of the complexity of casein was brought to light by Aschaffenburg.[6] Caseins prepared from the milks of individual cows and inspected by electrophoretic techniques were found to show genetically controlled polymorphism in the β- and γ-caseins. Polymorphism in α_{S1}-caseins was demonstrated by Thompson and Kiddy[213] and genetic variation has now been confirmed in all the major and some of the minor casein components.[86,139,134,212] The biological significance of milk protein polymorphism has also been considered.[57]

Individual casein components have been isolated and purified to such a degree that the complete linear sequence of the polypeptide chain can be determined. Such studies have been undertaken on α-, β-, γ-, and κ-caseins. In fact, the complete sequence for α_{S1}-casein has been reported,[147] and the exact location of amino-acid substitutions and deletions is now known, as well as the monoester linkage of each phosphate group.[84,147] Similarly, the French workers have also determined the position of each amino acid in the primary structure of β-casein, and the amino-acid substitutions of several of the polymorphic forms of the protein.[31,183,184] The partial sequence of some of the γ-casein variants, as well as of the minor components of the γ-casein fraction, has also been reported,[71] and the possible relationship of these proteins to β-casein is discussed in the same report. Finally, the partial sequence of κ-casein has also been published.[114] The rennin-sensitive bond was shown to be Phe-Met.[50,114] It would seem that the sequences of all the major caseins will be known, and such information will be of great value in a rational approach to the study of the biological and physicochemical properties and industrial uses of milk and its protein components.

WHEY PROTEINS

Prior to 1934, it was generally considered that whey protein consisted of two main components, named lactalbumin and lactoglobulin according to contemporary terminology, and comparatively small proportions of other protein-like substances. But in 1934, Palmer[162] reported the isolation from the lactalbumin fraction of a protein having the characteristics of a globulin. This protein, which comprises more than half the lactalbumin fraction, was at first called Palmer's globulin, but is now named β-lactoglobulin. This and several other proteins have been isolated from the classic fractions by methods to be described.

Preparation of Whey Protein

The proteins of cheese whey, which include some casein in addition to globulins and albumins, may be precipitated together and recovered efficiently by procedures that require careful control of temperature and acidity.[36,92,136] If a product free from casein is desired, the whey should be from the acid precipitation of casein from skimmilk, as previously described. The whey is put through a cream separator to remove as much fat as possible, its reaction adjusted to pH 6.3 to 6.5 by addition of 10*N* sodium hydroxide, and the whey heated to above 90°C with constant stirring to render the protein precipitable by acid. Then, 100 ml 33% acetic acid is added rapidly for each 100 lb whey and the stirring stopped as soon as the acid is thoroughly distributed. When the coagulum has collected, it is removed by filtration and, if desired, washed, drained, and dried. Solutions of other acids may be employed as precipitants in place of the acetic acid, the reaction being brought to pH 4.8 to 5.3. The whey protein prepared in this way is suitable for use in food products.

A soluble whey protein concentrate with excellent nutritional properties can be prepared following precipitation of the proteins by polymeric phosphate.[228] The product is available commercially.

Other methods for the preparation of whey proteins in both denatured and undenatured form, and in the form of coprecipitate with casein, have been proposed. Some of these have been described by Fox,[62] by Beeby *et al.*,[14] and by Muller.[150]

Preparation of the Lactoglobulin and Lactalbumin Fractions

A method typical of older procedures used for the preparation of the lactoglobulin fraction is that of Rowland. In his scheme for the partition of milk proteins, Rowland[189] obtained precipitation of the lactoglobulin fraction as follows. Casein was removed from milk by means of acetic acid and sodium acetate, as described earlier in this chapter, the reaction of the filtrate was adjusted to pH 6.8 to 7.2, and the liquid was then saturated with magnesium sulfate to precipitate the lactoglobulin.

The method of Sjögren and Svedberg[196] for preparing the classic lactalbumin fraction is, similarly, illustrative of older procedures. It consisted of half-saturating skimmilk with ammonium sulfate, adding acetic acid to give a pH value of 5.2, removing the precipitated casein and lactoglobulin fractions by filtration, and increasing the ammonium sulfate concentration to 80% of saturation. The resulting precipitate was removed by centrifuging. It was dissolved in water and reprecipitated by adding ammonium sulfate and sulfuric acid. The precipitate was again dissolved in water, salts were removed by dialysis, and the protein was dried.

β-Lactoglobulin

β-Lactoglobulin, a protein of molecular weight 36,000, occurs in milk to the extent of 3.0 g/l. In 1934, Palmer[162] discovered that a crystalline protein, insoluble in water, could be prepared from the classic lactalbumin fraction. The protein was named β-lactoglobulin and was shown to be the most abundant of the whey proteins. It was widely used in protein chemistry in research requiring a pure, crystalline protein. It was used for the first complete amino acid analysis of a fairly large protein; for the calculation of correlation between composition and titration curves of proteins with acids and bases; for the determination of the hydration of a protein crystal; for the investigation of the diffusion of electrolytes and nonelectrolytes into protein crystals; and for other important physicochemical studies. The extensive literature concerning the chemical and physical properties of β-lactoglobulin has been summarized.[137,140,215] A few pertinent data are given in Tables 3.1 and 3.2.

Palmer's original method for preparing β-lactoglobulin had as its starting point whey, which was obtained by adjusting skimmilk to pH 4.6 by adding hydrochloric acid and subsequently removing the precipitated casein. Crystalline β-lactoglobulin was obtained after fractionation of the whey with ammonium sulfate at pH 6.0 and dialysis at pH 5.2 of the precipitate obtained when the whey was saturated with ammonium sulfate. In 1957, Aschaffenburg and Drewry[10] prepared β-lactoglobulin in a somewhat different manner. Whey is obtained after precipitation of casein, globulins, other protein and fat by means of sodium sulfate (200 g/l). α-Lactalbumin, serum albumin and other proteins are removed by acidification of the filtrate to pH 2.0. The β-lactoglobulin is salted out from the filtrate by adjusting the pH to 6.0 and adding ammonium sulfate (200 g/l). The protein is then crystallized by dialysis. Other methods reported[5,7,63,185] are variations of the general scheme of Aschaffenburg and Drewry. These methods have been compared and evaluated.[137,140] Modifications generally are related to whether β-lactoglobulin or α-lactalbumin is the protein desired. Chromatographic methods[4,175,180] for the separation of whey proteins have also been utilized.

That even well-purified β-lactoglobulin, recrystallized many times, a single substance in both electrophoresis and ultracentrifugation,[168] was not truly homogeneous became apparent from the solubility studies of Grönwall[83] and electrophoretic experiments of Li[129] and McMeekin *et al.*[143] Polis *et al.*[178] isolated a homogeneous, crystalline component, β_1-lactoglobulin, upon fractionation of crystalline β-lactoglobulin. An understanding of the heterogeneity came with the discovery by Aschaffenburg and Drewry[8,9] that individual cows produce either a mixture of two electrophoretically distinct β-lactoglobulins or only one or the other of these. The capacity to produce the different types was genetically controlled, and the two forms of the protein became known as β-lactoglobulins A and B.

At the present time 6 genetic variants of β-lactoglobulin are known. The C variant, found in Jersey cows, was discovered by Bell;[15,16] the D polymorph was shown to be present in Montbeliarde cattle in France.[85] The final two variants are identical with β-lactoglobulins A and B, but contain covalently bound carbohydrate.[17,18,137] These have so far been found only in Australian Droughtmaster beef cattle; hence the designation β-lactoglobulin A_{DR} and B_{DR}.[188] The distribution of the various β-lactoglobulins in different breeds is considered by McKenzie,[140] who has also listed the amino-acid compositions of the proteins and the differences in number of amino-acid residues per monomer. As an example, Gordon *et al.*[69] and Piez *et al.*[175] found that β-lactoglobulin A had one more residue of aspartic acid and valine and one less residue of glycine and alanine than β-lactoglobulin B in a total of 160 amino acids per monomer (18,500). The amino acids are substituted as aspartic acid (β-A) → glycine (β-B) and valine (β-A) → alanine (β-B).[117] A partial sequence for β-lactoglobulins A and B has been published.[65] Many investigations have been carried out on β-lactoglobulin concerning the location and reactivity of cystine, cysteine, tyrosine, and tryptophan residues.[140] In addition, much research has been reviewed[137,140] on its electrochemical properties, such as pH titration curves, isoionic points and ion binding, and electrophoresis; its solution properties, such as molecular size and conformation; the effect of heat, detergents and other denaturants; the interaction of the protein with κ-casein; and its X-ray crystallographic structure.

β-Lactoglobulins have proved of great interest in protein chemistry because they are a family of simply prepared, crystalline proteins. Their interactions with κ-casein in milk are of technological importance, and these reactions may also be relevant to problems of allergenicity. In milk, β-lactoglobulin is the most abundant protein having a free sulfhydryl group in the form of a cysteine residue.

α-Lactalbumin

α-Lactalbumin, second in concentration to β-lactoglobulin among the whey proteins, can also be crystallized from the lactalbumin fraction. The name was first used in connection with a "lactalbumin" isolated by Kekwick and with its behavior in the ultracentrifugal investigations of Svedberg and Pederson.[206,207] It is most likely that the same protein was crystallized by Sørensen and Sørensen[200] in 1939 and designated "crystalline insoluble substance" because of its insolubility in water at pH 4.6. A method for the isolation of the protein, based on the observations of Sørensen and Sørensen, was reported by Gordon and Semmett,[72] who also characterized the protein and proposed that it be known as α-lactalbumin. The method was later modified[75] and described in detail.[76] In principle, the original Palmer method is used to remove casein and the lactoglobulin fraction from skimmilk and to crystallize β-lactoglobulin from the lactalbumin fraction. The mother liquor from the crystallization is adjusted to pH 4 and ammonium sulfate is added to precipitate crude α-lactalbumin. It is dissolved at pH 8, reprecipitated at pH 4, dissolved again, and crystallized by the addition of ammonium sulfate to ⅔ saturation at pH 6.6; it is recrystallized by repeating these steps. Other more convenient procedures which have been developed for the preparation of α-lactalbumin were listed and described by Gordon.[67]

Table 3.1 shows some of the chemical properties of α-lactalbumin and Table 3.2 its amino-acid composition. It is of interest that α-lactalbumin contains no free sulfhydryl groups, though its content of cystine is high. It is also rich in tryptophan. α-Lactalbumin occurs in two forms, genetic polymorphs A and B. In milks from Western breeds of dairy cattle the variant found is B, while in milks from African and Indian Zebu and Australian Droughtmaster cattle, both forms occur. Chemically, A differs from B by the simple substitution of glutamine for arginine.[67]

Until 1966 α-lactalbumin was considered to be a protein of good nutritive value but of little importance because of its low concentration in milk. However, in that year Ebner *et al.*[54] reported that an enzymatic role could be assigned to α-lactalbumin, since it was identified as one of the two subunits of lactose synthetase. Beginning with this discovery, much research was devoted not only toward elucidation of the reactions involved in the biosynthesis of lactose, but also to the chemical structure of α-lactalbumin and to the correlation of its structure with its biological function. Thus the complete amino-acid sequence of the protein was worked out by Brew *et al.*,[29] and the positions of the disulfide bonds were established by Vanaman *et al.*[219] Surprising similarities in primary structure and possibly in conformation of α-lactalbumin and chicken egg white lysozyme were brought to light. Furthermore, it was found that the "A protein" subunit of lactose synthetase acts primarily as a galactosyl transferase to produce N-acetyllactosamine; the function of α-lactalbumin (the "B protein" subunit) is to modify the enzyme so as to increase greatly its activity as a lactose synthetase. Many of the developments in this very active field of research have been reviewed by Ebner,[52,53] Hill *et al.*[98] and Gordon.[67]

Bovine Immunoglobulins

These proteins are present in ordinary milk in low concentration but they occur in colostrum—the milk secreted for a few days after parturition—in much larger percentages. They are of unique importance to the new-born calf because they are absorbed into its circulation, where they fulfill, temporarily, the immunological functions of blood gamma-globulin. The previous designation of immune globulins of milk as eu- and pseudo-globulins has now been superseded by a more general nomenclature.[37 188] The early work of Smith *et al.*[197,198] has also been reassessed in light of more recent developments in the field of immunology. These developments have been reported in a symposium on the bovine immune system.[38]

Three distinct classes of bovine immunoglobulins occur in milk; they are designated IgM (γM), IgA (γA), and IgG (γG). IgG is subdivided into IgG1 and IgG2. IgG1 is selectively accumulated in the colostrum and milk of cows. IgM is a 19S protein with 12.3% carbohydrate which has been partially characterized after isolation from a fraction of whey insoluble in 33% ammonium sulfate.[80] The isolation of bovine IgA from milk has been reported by many investigators. It contains 8 to 9% carbohydrate, has a sedimentation coefficient of 10 to 12S, and can be eluted from Sephadex G-200 in the fractionation of whey. Glycoprotein-a[88] occurs free and bound to lacteal IgA. Hence, glycoprotein-a and lacteal IgA are probably homologous to the "secretory piece" and "secretory IgA," respectively, described for other species. IgG1 and IgG2 have about 2 to 4% carbohydrate and a sedimentation constant of about 7S. IgG2 molecules are somewhat more basic than IgG1 and are not retained on DEAE-cellulose columns at low ionic strength (pH 8.3), in contrast to IgG1 which is the more abundant protein in lacteal secretions. The two subclasses also differ in amino-acid composition.[88,120,148]

The two subclasses have been correlated[37] with the early preparations of Smith. Smith's pseudoglobulin contains mostly IgG1 and also "secretory IgA." The euglobulin consists of IgG2-like globulins, slower IgG1 globulins, IgA and IgM. Amino-acid analyses of older γ-globulin preparations are shown in Table 3.2. Butler has extensively reviewed the immunoglobulins of the cow.[37,188]

Blood Serum Albumin

By repeated fractionations of the mother liquor remaining after crystallization of β-lactoglobulin from the crude lactalbumin fraction, Polis *et al.*[177] succeeded in isolating and crystallizing a true, water-soluble milk albumin. When this was compared with crystalline bovine serum albumin, it was found that the two are identical in physical properties and composition; some of these data are shown in Tables 3.1 and 3.2. It was demonstrated by Coulson and Stevens that the proteins are immunologically indistinguishable.[46]

Other Milk Proteins

Table 3.1 makes reference to a proteose-peptone fraction, amounting to about 2 to 6% of the total protein present in milk. The fraction is operationally defined as that portion of the protein system that is not precipitated by heating at 95 to 100°C for 20 min and subsequent acidification to pH 4.7, but is precipitated by 12% (w/v) trichloroacetic acid.[189] Various workers have prepared substances of this nature from milk under such names as minor-protein fraction, sigma-proteose, and milk component 5, as well as proteose-peptone. The principal components of the proteose-peptone fraction have been designated milk-serum components "3," "5," and "8," in ascending order of mobility in moving-boundary electrophoresis.[125] More recently, the proteose-peptone fraction has been separated into four principal components designated "3," "5," and "8-slow," and "8-fast."[121,153] Each component is a glycoprotein and has been isolated from both heated and unheated skimmilk. The properties of the proteose-peptone components have been summarized.[188]

The interphase between fat globules and milk plasma, the fat globule membrane, is of great interest in milk chemistry and has been the subject of several review articles.[33,111,181] The reader is also referred to Chapter 10 and to the review and discussion by Groves[87] of the proteins of the fat globule membrane. In addition, the same author[87] has considered the great variety of minor proteins and enzymes which occur in milk. Among these are lactoferrin (red protein), blood serum transferrin, lactollin, kininogen, M-1 glycoproteins; also nucleases, lactoperoxidase, xanthine oxidase, lipases, esterases, amylases, phosphatases, lysozyme and other enzymes. More refined methods of isolation, fractionation, analysis, and characterization have stimulated interest in proteins occurring in almost trace amounts. This is true because many of these proteins are of great importance in catalyzing reactions of biological interest, in binding minerals and vitamins, and in affecting the stability and flavor of milk and milk products.

REFERENCES

1. ALAIS, C., and JOLLÈS, P., Biochim. Biophys. Acta, *51,* 315 (1961).
2. ANDERSON, L., and KELLEY, J. J., J. Am. Chem. Soc., *81,* 2275 (1959).
3. ANNAN, W. D., and MANSON, W., J. Dairy Res., *36,* 259 (1969).
4. ARMSTRONG, J. McD., HOPPER, K. E., McKENZIE, H. A., and MURPHY, W. H., Biochim. Biophys. Acta, *214,* 419 (1970).
5. ARMSTRONG, J. McD., McKENZIE, H. A., and SAWYER, W. H., Biochim. Biophys. Acta, *147,* 60 (1967).
6. ASCHAFFENBURG, R., Nature, *192,* 431 (1961).
7. ASCHAFFENBURG, R., J. Dairy Sci., *51,* 1295 (1968).
8. ASCHAFFENBURG, R., and DREWRY, J., Nature, *176,* 218 (1955)
9. ASCHAFFENBURG, R., and DREWRY, J., Nature, *180,* 376 (1957).
10. ASCHAFFENBURG, R., and DREWRY, J., Biochem. J., *65,* 273 (1957).
11. ASCHAFFENBURG, R., and DREWRY, J., 15th Intern. Dairy Congr. Proc., *3* 1631 (1959).
12. ASCHAFFENBURG, R., and OGSTON, A. G., J. Dairy Res., *14,* 316 (1946).
13. BARMAN, T. E., Biochim. Biophys. Acta, *214,* 242 (1970).
14. BEEBY, R., HILL, R. D., and SNOW, N. S., in "Milk Proteins. Chemistry and Molecular Biology," Vol. 2, Chap. 17, Academic Press, New York, 1971.
15. BELL, K., Nature, *195,* 705 (1962).
16. BELL, K., Biochim. Biophys. Acta, *147,* 100 (1967).
17. BELL, K., McKENZIE, H. A., and MURPHY, W. H., Austral. J. Sci., *29,* 87 (1966).
18. BELL, K., McKENZIE, H. A., MURPHY, W. H., and SHAW, D. C., Biochim. Biophys. Acta, *214,* 427 (1970).
19. BENEDICENTI, A., Arch. Anat. Physiol., Physiol. Abt., 210 (1897).
20. BEZKOROVAINY, A., Arch. Biochem. Biophys., *110,* 558 (1965).
21. BEZKOROVAINY, A., J. Dairy Sci., *50,* 1368 (1967).
22. BEZKOROVAINY, A., and GROHLICH, D., Biochem. J., *115,* 817 (1969).
23. BINGHAM, E. W., and ZITTLE, C. A., Arch. Biochem. Biophys., *106,* 235 (1964).
24. BINGHAM, E. W., and KALAN, E. B., Arch. Biochem. Biophys., *121,* 317 (1967).
25. BLOCK, R. J., and HOWARD, H. W. (to the Borden Co.), U. S. Patent 2,468,730 (May 3, 1949).
26. BLOCK, R. J., and WEISS, K. W., "Amino Acid Handbook," p. 266, C. C. Thomas, Springfield, Illinois, 1956.
27. BLUM, F., Z. Physiol. Chem., *22,* 127 (1896).
28. BRAND, E., SAIDEL, L. J., GOLDWATER, W. H., KASSEL, B., and RYAN, F. J., J. Am. Chem. Soc., *67,* 1524 (1945).
29. BREW, K., CASTELLINO, F. J., VANAMAN, T. C., and HILL, R. L., J. Biol. Chem., *245,* 4570 (1970).
30. BREW, K., VANAMAN, T. C., and HILL, R. L., J. Biol. Chem., *242,* 3747 (1967).
31. BRIGNON, G., RIBADEAU-DUMAS, B., GROSCLAUDE, F., and MERCIER, J.-C., Eur. J. Biochem., *22,* 179 (1971).
32. BRÜGGEMANN, J., and ERBERSDOBLER, H., Z. Lebensm.-Unters. Forsch., *137,* 137 (1968).
33. BRUNNER, J. R., in "Fundamentals of Dairy Chemistry," Chap. 10, Avi Publishing Co., Westport, Connecticut, 1965.

34. BRUNNER, J. R., ERNSTROM, C. A., HOLLIS, R. A., LARSON, B. L., WHITNEY, R. McL., and ZITTLE, C. A., J. Dairy Sci., *43,* 901 (1960).
35. BURK, N. F., and GREENBERG, D. M., J. Biol. Chem., *87,* 197 (1930).
36. BURKEY, L. A., and WALTER, H. E., U.S. Dept. Agr. Bur. Dairy Ind. BDIM-Inf.-46, (1947).
37. BUTLER, J. E., J. Dairy Sci., *52,* 1895 (1969).
38. BUTLER, J. E., WINTER, A. J., and WAGNER, G. G., J. Dairy Sci., *54,* 1309 (1971).
39. CANNAN, R. K., PALMER, A. H., and KIBRICK, A. C., J. Biol. Chem., *142,* 803 (1943).
40. CHERBULIEZ, E., and BAUDET, P., Helv. Chim. Acta, *33,* 398, 1673 (1950).
41. CHERBULIEZ, E., and MEYER, F., Helv. Chim. Acta, *16,* 600 (1933).
42. CHERBULIEZ, E., and SCHNEIDER, M. L., Helv. Chim. Acta, *15,* 597 (1932); Lait, *13,* 264 (1933).
43. COHN, E. J., J. Gen. Physiol., *4,* 697 (1922).
44. COHN, E. J., and BERGGREN, R. E. L., J. Gen. Physiol., *7,* 45 (1924).
45. COOK, B. B., FRAENKEL-CONRAT, J., SINGER, B., MORGAN, A. F., BUELL, R., and MOISES, J. G., J. Nutr., *44,* 217 (1951).
46. COULSON, E. J., and STEVENS, H., J. Biol. Chem., *187,* 355 (1950).
47. COULTER, S. T., JENNESS, R., and GEDDES, W. F., Advances in Food Res., *3,* 45 (1951).
48. CRAVIOTO-MUNOZ, J., JOHANSSON, B., and SVENNERHOLM, L., Acta Chem. Scand., *9,* 1033 (1955).
49. DANEHY, J. P., and PIGMAN, W. W., Advances in Food Res. *3,* 241 (1951).
50. DELFOUR, A., JOLLÈS, J., ALAIS, C., and JOLLÈS, P., Biochem. Biophys. Res. Commun., *19,* 452 (1965).
51. DUNN, M. S., Biochem. Preparations, *1,* 22 (1949).
52. EBNER, K. E., Accts. Chem. Res. *3,* 41 (1970).
53. EBNER, K. E., and BRODBECK, U., J. Dairy Sci., *51,* 317 (1968).
54. EBNER, K. E., DENTON, W. L., and BRODBECK, U., Biochem. Biophys. Res. Commun., *24,* 232 (1966).
55. EHRENPREIS, S., MAURER, P. H., and RAM, J. S., Arch. Biochem. Biophys., *67,* 178 (1957).
56. EVANS, R. J., and BUTTS, H. A., Food Research, *16,* 415 (1951).
57. FARRELL, H. M., JR., and THOMPSON, M. P., J. Dairy Sci., *54,* 1219 (1971).
58. FERRETTI, A., and FLANAGAN, V. P., J. Agri. Food Chem., *19,* 245 (1971).
59. FERRETTI, A., FLANAGAN, V. P., and RUTH, J. M., J. Agri. Food Chem., *18,* 13 (1970).
60. FIAT, A. M., ALAIS, C., and JOLLÈS, P., Chimia, *22,* 137 (1968).
61. FINOT, P. A., BRICOUT, J., VIANI, R., and MAURON, J., Experientia, *24,* 1097 (1968).
62. FOX, K. K., in "Byproducts from Milk," 2nd Edition, Chap. 11, Avi Publishing Co., Westport, Connecticut, 1970.
63. FOX, K. K., HOLSINGER, V. H., POSATI, L. P., and PALLANSCH, M. J., J. Dairy Sci., *50,* 1363 (1967).
64. FRAENKEL-CONRAT, H., and OLCOTT, H. S., J. Biol. Chem., *174,* 827 (1948); J. Am. Chem. Soc., *70,* 2673 (1948).
65. FRANK, G., and BRAUNITZER, G., Z. Physiol. Chem., *348,* 1691 (1967).
66. GARNIER, J., Lait, *45,* 519 (1965).
67. GORDON, W. G., in "Milk Proteins. Chemistry and Molecular Biology," Vol. 2, Chap. 15, Academic Press, New York, 1971.
68. GORDON, W. G., ASCHAFFENBURG, R., SEN, A., and GHOSH, S. K., J. Dairy Sci., *51,* 947 (1968).
69. GORDON, W. G., BASCH, J. J., and KALAN, E. B., J. Biol. Chem., *236,* 2908 (1961).
70. GORDON, W. G., GROVES, M. L., and BASCH, J. J., Biochemistry, *2,* 817 (1963).

71. GORDON, W. G., GROVES, M. L., GREENBERG, R., JONES, S. B., KALAN, E. B., PETERSON, R. F., and TOWNEND, R. E., J. Dairy Sci., *55*, 261 (1972).
72. GORDON, W. G., and SEMMETT, W. F., J. Am. Chem. Soc., *75*, 328 (1953).
73. GORDON, W. G., SEMMETT, W. F., and BENDER, M., J. Am. Chem. Soc., *72*, 4282 (1950); *75*, 1678 (1953).
74. GORDON, W. G., SEMMETT, W. F., CABLE, R. S., and MORRIS, M., J. Am. Chem. Soc., *71*, 3293 (1949).
75. GORDON, W. G., SEMMETT, W. F., and ZIEGLER, J., J. Am. Chem. Soc., *76*, 287 (1954).
76. GORDON, W. G., and ZIEGLER, J., Biochem. Preparations, *4*, 16 (1955).
77. GORDON, W. G., and ZIEGLER, J., Arch. Biochem. Biophys., *57*, 80 (1955).
78. GOULD, S. P., Ind. Eng. Chem., *24*, 1077 (1932).
79. GOULDEN, J. D. S., Nature, *177*, 85 (1956).
80. GOUGH, P. M., JENNESS, R., and ANDERSON, R. K., J. Dairy Sci., *49*, 718 (1966).
81. GREEN, A. A., J. Biol. Chem., *93*, 517 (1931).
82. GROH, J., KARDOS, E., DÉNES, K., and SERÉNYI, V., Z. physiol. Chem., *226* 32 (1934).
83. GRÖNWALL, A., Compt. rend. trav. lab. Carlsberg, Ser. Chim., *24*, 185 (1942).
84. GROSCLAUDE, F., MERCIER, J.-C., RIBADEAU-DUMAS, B., Eur. J. Biochem., *14*, 98 (1970); *16*, 447 (1970).
85. GROSCLAUDE, F., PUJOLLE, J., GARNIER, J., and RIBADEAU-DUMAS, B., Ann. Biol. Anim., Biochim., Biophys., *6*, 215 (1966).
86. GROVES, M. L., J. Dairy Sci., *52*, 1155 (1969).
87. GROVES, M. L., in "Milk Proteins. Chemistry and Molecular Biology," Vol. 2, Chap. 16, Academic Press, New York, 1971.
88. GROVES, M. L., and GORDON, W. G., Biochemistry, *6*, 2388 (1967).
89. GROVES, M. L., and GORDON, W. G., Biochim. Biophys. Acta, *194*, 421 (1969).
90. GROVES, M. L., GORDON, W. G., and KIDDY, C. A., J. Dairy Sci., *51*, 946, (1968).
91. GROVES, M. L., McMEEKIN, T. L., HIPP, N. J., and GORDON, W. G., Biochim. Biophys. Acta, *57*, 197 (1962).
92. GUY, E. J., VETTEL, H. E., and PALLANSCH, M. J., J. Dairy Sci., *50*, 828 (1967).
93. HAMMARSTEN, O., Z. Physiol. Chem., *7*, 227 (1883); *9*, 273 (1885).
94. HANSEN, R. G., and CARLSON, D. M., J. Dairy Sci., *39*, 663 (1956).
95. HANSEN, R. G., POTTER, R. L., and PHILLIPS, P. H., J. Biol. Chem., *171*, 229 (1947).
96. HENRY, K. H., Dairy Sci. Abs., *19*, 603, 691 (1957).
97. HENRY, K. H., KON, S. K., LEA, C. H., and WHITE, J. C. D., J. Dairy Res., *15*, 292 (1948).
98. HILL, R. L., BREW, K., VANAMAN, T. C., TRAYER, I. P., and MATTOCK, P., Brookhaven Symp. Biol., Vol. 1, *21*, 139 (1968).
99. HINDLE, E. J., and WHEELOCK, J. V., Biochem. J., *119* (3) 14P (1970).
100. HIPP, N. J., BASCH, J. J., and GORDON, W. G., Arch. Biochem. Biophys., *94*, 35 (1961).
101. HIPP, N. J., GROVES, M. L., CUSTER, J. H., and McMEEKIN, T. L., J. Am. Chem. Soc., *72*, 4928 (1950); J. Dairy Sci., *35*, 272 (1952).
102. HIPP, N. J., GROVES, M. L., and McMEEKIN, T. L., J. Am. Chem. Soc., *74*, 4822 (1952).
103. HIPP, N. J., GROVES, M. L., and McMEEKIN, T. L., Textile Res. J., *24*, 618 (1954).
104. HIPP, N. J., GROVES, M. L., McMEEKIN, T. L., Arch. Biochem. Biophys., *93*, 245 (1961).
105. HOAGLAND, P. D., THOMPSON, M. P., and KALAN, E. B., J. Dairy Sci., *54*, 1103 (1971).

106. HODGE, J. E., J. Agri. Food Chem., *1,* 928 (1953).
107. HOFMAN, T., Biochem. J., *69,* 139 (1958).
108. JACKSON, R. H., COULSON, E. J., and CLARK, W. R., Arch. Biochem. Biophys., *97,* 373 (1962).
109. JENNESS, R., in "Milk Proteins. Chemistry and Molecular Biology," Vol. 1, Chap. 2, Academic Press, New York, 1970.
110. JENNESS, R., LARSON, B. L., McMEEKIN, T. L., SWANSON, A. M., WHITNAH, C. H., and WHITNEY, R. McL., J. Dairy Sci., *39,* 536 (1956).
111. JENNESS, R., and PATTON, S., "Principles of Dairy Chemistry," John Wiley and Sons, New York, 1959.
112. JIRGENSONS, B., Biochem. Z., *268,* 414 (1934).
113. JOLLÈS, P., Angew. Chem., *5,* 558 (1966).
114. JOLLÈS, J., ALAIS, C., and JOLLÈS, P., Biochim Biophys. Acta, *168,* 591 (1968); Helv. Chim. Acta. *53.* 1918 (1970).
115. JOLLÈS, P., ALAIS, C., and JOLLÈS, J., Biochim. Biophys. Acta, *51,* 309 (1961); Arch. Biochem. Biophys., *98,* 56 (1962).
116. JOLLES, P., ALAIS, C., ADAM, A., DELFOUR, A., JOLLÈS, J., Chimia, *18,* 357 (1964).
117. KALAN, E. B., GORDON, W. G., BASCH, J. J., and TOWNEND, R. E., Arch, Biochem. Biophys., *96,* 376 (1962).
118. KALAN, E. B., and TELKA, M., Arch. Biochem. Biophys., *79,* 275 (1959); *85,* 273 (1959).
119. KASSEL. B., and BRAND. E., J. Biol. Chem., *125,* 435 (1938).
120. KICKHÖFEN, B., HAMMER, D. K., and SCHEEL, D., Z. Physiol. Chem., *349,* 1755 (1968).
121. KOLAR, C. W., and BRUNNER, J. R., J. Dairy Sci., *53,* 997 (1970).
122. KON, S. K., and HENRY, S. M., J. Dairy Res. *16,* 68 (1949).
123. KRONMAN, M. J., and ANDREOTTI, R. E., Biochemistry *3,* 1145 (1964); KRONMAN, J. J., ANDREOTTI, R. E., and VITOLS, R., Biochemistry *3,* 1152 (1964).
124. LARSON, B. L., and KENDALL, K. A., J. Dairy Sci., *40,* 377 (1957).
125. LARSON, B. L., and ROLLERI, G. D., J. Dairy Sci., *38,* 351 (1955).
126. LEA, C. H., J. Dairy Res. *15,* 369 (1948).
127. LEA, C. H., and HANNAN, R. S., Biochim. Biophys. Acta, *3,* 313 (1949); *4,* 518 (1950); *5,* 433 (1950); *7,* 366 (1951).
128. LEVENE, P. A., and HILL, D. W., J. Biol. Chem., *101,* 711 (1933).
129. LI, C. H., J. Am. Chem. Soc., *68,* 2746 (1946).
130. LINDERSTRØM-LANG, K., Compt. rend. trav. lab. Carlsberg, *17,* 9 (1929).
131. LINDERSTRØM-LANG, K., and KODAMA, S., Compt. rend. trav. lab. Carlsberg, *16,* No. 1 (1925).
132. LIPMANN, F., Biochem. Z., *262,* 3 (1933).
133. LONG, J., VAN WINKLE, Q., and GOULD, I, A., J. Dairy Sci., *41,* 317 (1958).
134. MACKINLEY, A. G., and WAKE, R. G., in "Milk Proteins. Chemistry and Molecular Biology," Vol. 2, Chap. 12, Academic Press, New York, 1971.
135. MAILLARD, L. C., Compt. rend., *154,* 66 (1912); Ann. Chim. (9), *5,* 258 (1916).
136. MALKAMES, J. P., JR., WALTER, H. E., and SAGER, O. S., Bur. Dairy Ind., U.S. Dept. Agri., BDI-Inf-118 (1951).
137. McKENZIE, H. A., Advances in Protein Chem., *22,* 55 (1967).
138. McKENZIE, H. A., "Milk Proteins. Chemistry and Molecular Biology," Vol. 1 and 2, Academic Press, New York, 1970, 1971.
139. McKENZIE, H. A., in "Milk Proteins. Chemistry and Molecular Biology," Vol. 2, Chap. 10, Academic Press, New York, 1971.
140. McKENZIE, H. A., in "Milk Proteins. Chemistry and Molecular Biology," Vol. 2, Chap. 14, Academic Press, New York, 1971.
141. McKENZIE, H. A., and WAKE, R. G., Austral. J. Chem., *12,* 712, 723, 734 (1959).
142. McMEEKIN, T. L., HIPP, N. J., and GROVES, M. L., Arch. Biochem. Biophys., *83,* 35 (1959).

143. McMEEKIN, T. L., POLIS, B. D., DELLAMONICA, E. S., and CUSTER, J. H., J. Am. Chem. Soc., *70,* 881 (1948).
144. McMEEKIN, T. L., REID, T. S., WARNER, R. C., and JACKSON, R. W., Ind. Eng. Chem., *37,* 685 (1945); U.S. Patent No. 2,521,738 (September 12, 1950).
145. MELLANDER, O., Biochem. Z., *300,* 240 (1939).
146. MELLANDER, O., Nutr. Revs., *13,* 161 (1955).
147. MERCIER, J. C., GROSCLAUDE, F., and RIBADEAU-DUMAS, B., Eur. J. Biochem., *14,* 108 (1970); *16,* 453 (1970); *23,* 41 (1971).
148. MILSTEIN, C. P., and FEINSTEIN, A., Biochem. J., *107,* 559 (1968).
149. MOHAMMAD, A., FRAENKEL-CONRAT, H., and OLCOTT, H. S., Arch. Biochem. *24,* 157 (1949).
150. MULLER, L. L., Dairy Sci. Abs., *33,* 659 (1971).
151. MURTHY, G. K., and WHITNEY, R. McL., J. Dairy Sci., *41,* 1 (1958).
152. NABENHAUER, F. P., Ind. Eng. Chem., *22,* 54 (1930).
153. NG, W. C., BRUNNER, J. R., and RHEE, K. C., J. Dairy Sci., *53,* 987 (1970).
154. NIELSEN, H. C., and LILLEVIK, H. A., J. Dairy Sci., *40,* 598 (1957).
155. NITSCHMANN, H., and HADORN, H., Helv. Chim, Acta, *27,* 299 (1944).
156. NITSCHMANN, H., and KELLER, W., Helv. Chim. Acta, *38,* 942 (1955).
157. NITSCHMANN, H., and LAUENER, H., Helv. Chim. Acta, *29,* 174 (1946).
158. NITSCHMANN, H., WISSMANN, H., and HENZI, R., Chimia, *11,* 76 (1957).
159. ORR, M. L., and WATT, B. K., Home Economics Research Report, No. 4, U.S. Dept. Agri. (1957).
160. OSBORNE, T. B., and WAKEMAN, A. J., J. Biol. Chem., *33,* 243 (1918).
161. ÖSTERBERG, R., Biochim. Biophys. Acta, *54,* 424 (1961); Acta Chem. Scand., *18,* 795 (1964).
162. PALMER, A. H., J. Biol. Chem., *104,* 359 (1934).
163. PATTON, A. R., HILL, E. G., and FOREMAN, E. M., Science, *197,* 623 (1948).
164. PATTON, A. R., SALANDER, R. C., and PIANO, M., Food Res., *19,* 444 (1954).
165. PATTON, S., J. Dairy Sci., *33,* 102, 324, 904 (1950); *38,* 457 (1955).
166. PATTON, S., and FLIPSE, R. J., J. Dairy Sci., *36,* 766 (1953).
167. PATTON, S., amd JOSEPHSON, D. V., J. Dairy Sci., *32,* 222 (1949).
168. PEDERSEN, K. O., Biochem. J., *30,* 948, 961 (1936).
169. PERLMANN, G. E., Advances in Protein Chem., *10,* 1 (1955).
170. PERTZOFF, V., J. Gen. Physiol., *10,* 961 (1927).
171. PETERSON, R. F., MCDOWELL, R. L., and HARRINGTON, B. J., Textile Res. J., *24,* 747 (1954).
172. PETERSON, R. F., McDOWELL, R. L., and HOOVER, S. R., Textile Res. J., *18,* 744 (1948).
173. PETERSON, R. F., NAUMAN, L. W., and HAMILTON, D. F., J. Dairy Sci., *49,* 601 (1966).
174. PETERSON, R. F., NAUMAN, L. W., and McMEEKIN, T. L., J. Am. Chem. Soc., *80,* 95 (1958).
175. PIEZ, K. A., DAVIE, E. W., FOLK, J. E., and GLADNER, J. A., J. Biol. Chem., *236,* 2912 (1961).
176. PION, R., GARNIER, J., RIBADEAU-DUMAS, B., DEKONING, P. F., and VAN ROOYEN, P. J., Biochem. Biophys. Res. Commun. *20,* 246 (1965).
177. POLIS, B. D., SHMUKLER, H. W., and CUSTER, J. H., J. Biol. Chem., *187,* 349 (1950).
178. POLIS, B. D., SHMULKER, H. W., CUSTER, J. H., and McMEEKIN, T. L., J. Am. Chem. Soc., *72,* 4965 (1950).
179. POSTERNAK, S., Biochem. J., *21,* 289 (1927).
180. PRÉAUX, G., and LONTIE, R. in "Protides of the Biological Fluids: Proceedings of the 9th Colloquium," (H. Peeters, Ed.), Elsevier, Amsterdam, 1962.
181. PRENTICE, J. H., Dairy Sci. Abs., *31,* 353 (1969).
182. RAMSDELL, G. A., and WHITTIER, E. O., J. Biol. Chem., *154,* 413 (1944).
183. RIBADEAU-DUMAS, B., BRIGNON, B., GROSCLAUDE, F., and MERCIER, J.-C., Eur. J. Biochem., *20,* 258, 264 (1971); *25,* 505 (1972).

184. RIBADEAU-DUMAS, B., GROSCLAUDE, F., and MERCIER, J.-C., Eur. J. Biochem., *14,* 451 (1970); *18,* 252 (1971).
185. ROBBINS, F. M., and KRONMAN, M. J., Biochim. Biophys. Acta, *82,* 186 (1964).
186. ROLLERI, G. D., LARSON, B. L., and TOUCHBERRY, R. W., J. Dairy Sci., *39,* 1683 (1956).
187. ROMBAUTS, W. A., SCHROEDER, W. A., and MORRISON, M., Biochemistry, *6,* 2965 (1967).
188. ROSE, D., BRUNNER, J. R., KALAN, E. B., LARSON, B. L., MELNYCHYN, P., SWAISGOOD, H. E., and WAUGH, D. F., J. Dairy Res., *53,* 1 (1970).
189. ROWLAND, S. J., J. Dairy Sci., *9,* 30, 35, 38, 40, 42, 47 (1938).
190. SALZBERG, H. K., in "Encyclopedia of Polymer Science and Technology," Vol. 2, p. 859, Interscience Publications, John Wiley and Sons, New York, 1965.
191. SAMPATH KUMAR, K. S. V., SUNDARARAJAN, T. A., and SARMA, P. S., Enzymologia, *18,* 228, 234 (1957).
192. SCHMIDT, D. G., BOTH, P., and DEKONING, P. J., J. Dairy Sci., *49,* 776 (1966).
193. SCHMIDT, D. G., PAYENS, T. A. J., VANMARKWIJK, B. W., and BRINKHUIS, J. A., Biochem. Biophys. Res. Commun., *27,* 448 (1967).
194. SCHORMÜLLER, J., HANS, R., and BELITZ, H.-D., Z. Lebensm. Untersuch. u. Forsch., *131,* 65 (1966).
195. SHAHANI, K. M., and SOMMER, H. H., J. Dairy Sci., *34,* 1010 (1951).
196. SJÖGREN, B., and SVEDBERG, T., J. Am. Chem. Soc., *52,* 3650 (1930).
197. SMITH, E. L., J. Biol. Chem., *164,* 354 (1946); *165,* 665 (1946); J. Dairy Sci., *31,* 127 (1948).
198. SMITH, E. L., and HOLM, A., J. Biol. Chem., *175,* 349 (1948).
199. SØRENSEN, M., and HAUGAARD, G., Biochem. Z., *260,* 247 (1933).
200. SØRENSEN, M., and SØRENSEN, S. P. L., Compt. rend, trav. lab. Carlsberg, Ser. Chim., *23,* 55 (1939).
201. SPIES, J. R., Anal. Chem., *39,* 1412 (1967).
202. SPIES, J. R., and CHAMBERS, D. C., Anal. Chem., *21,* 1249 (1949).
203. STEIN, W. H., and MOORE, S., J. Biol. Chem., *178,* 79 (1949).
204. SULLIVAN, R. A., FITZPATRICK, M. M., STANTON, E. K., ANNINO, R., KISSEL, G., and PALERMITE, F., Arch. Biochem. Biophys., *55,* 455 (1955).
205. SUTERMEISTER, E., and BROWNE, F. L., "Casein and Its Industrial Applications," 2nd Edition, Reinhold Publishing Corp., New York, 1959.
206. SVEDBERG, T., Nature, *139,* 1051 (1937).
207. SVEDBERG, T., and PEDERSEN, K. O., "The Ultracentrifuge," p. 379, Oxford, University Press, London, 1940.
208. SWAIN, A. P., KOKES, E. L., HIPP, N. J., WOOD, J. L., and JACKSON, R. W. Ind. Eng. Chem., *40,* 465 (1948).
209. SWAISGOOD, H. E., and BRUNNER, J. R., J. Dairy Sci., *45,* 1 (1962).
210. SWAISGOOD, H. E., BRUNNER, J. R., and LILLEVIK, H. A., Biochemistry, *3,* 1616 (1964).
211. SWOPE, F. C., and BRUNNER, J. R., Milchwiss., *23,* 470 (1968).
212. THOMPSON, M. P., in "Milk Proteins. Chemistry and Molecular Biology," Vol. 2, Chap. 11, Academic Press, New York, 1971.
213. THOMPSON, M. P., and KIDDY, C. A., J. Dairy Sci., *47,* 626 (1964)
214. THOMPSON, M. P., TARASSUK, N. P., JENNESS, R., LILLEVIK, H. A., ASHWORTH, U. S., and ROSE, D., J. Dairy Sci., *48,* 159 (1965).
215. TILLEY, J. M. A., Dairy Sci. Abs., *22,* 111 (1960).
216. TIMASHEFF, S. N., and TOWNEND, R. E., J. Am. Chem. Soc., *80,* 4433 (1958).
217. TOWNEND, R. E., HERSKOVITZ, T. T., SWAISGOOD, H. E., and TIMASHEFF, S. N., J. Biol. Chem., *239,* 4196 (1964).
218. TRAN, V. D., and BAKER, B. E., J. Dairy Sci., *53,* 1009 (1970).
219. VANAMAN, T. C., BREW, J., and HILL, R. L., J. Biol. Chem., *245,* 4583 (1970).
220. VERDIER, C. H. DE, Acta Chem. Scand., *7,* 196 (1953).
221. VICKERY, H. B., and SCHMIDT, C. L. A., Chem. Revs, *9,* 169 (1931).
222. VONHIPPEL, P. H., and WAUGH, D. F., J. Am. Chem. Soc., *77,* 4311 (1955).

223. WAKE, R. G., and BALDWIN, R. L., Biochim. Biophys. Acta, *47,* 225 (1961).
224. WARNER, R. C., J. Am. Chem. Soc., *66,* 1725 (1944).
225. WAUGH, D. F., Discussions Faraday Soc., No. 25, 186 (1958); J. Phys. Chem., *65,* 1793 (1961).
226. WAUGH, D. F., and VONHIPPEL, P. H., J. Am. Chem. Soc., *78,* 4576 (1956).
227. WHITTIER, E. O., U.S. Dept. Agri. Circ. 279 (rev. 1942).
227A. WHITTIER, E. O., and GOULD, S. P., Ind Eng. Chem., *32,* 906 (1940).
228. WINGERD, W. H., (to the Borden Co.), Canadian Patent 790,580 (July 23, 1968); WINGERD, W. H., SAPERSTEIN, S., and LUTWAK, L., Food Techn. *24,* 758 (1970).
229. WORMELL, R. L., and KAYE, M. A. G., J. Soc. Chem. Ind., *64,* 75 (1945).
230. WOYCHIK, J. H., KALAN, E. B., and NOELKEN, M. E., Biochemistry *5,* 2276 (1966).
231. ZITTLE, C. A., and CUSTER, J. H., J. Dairy Sci., *46,* 1183 (1963).

CHAPTER 4

Floyd E. Kurtz

The Lipids of Milk: Composition and Properties

INTRODUCTION

Milkfat consists chiefly of triglycerides of fatty acids. It also contains varying quantities of other compounds as indicated in Table 4.1. Some additional materials are included in the text. Others are undoubtedly present but still unidentified. The relative proportions of these compounds in milkfat are affected by such factors in its production as the type and amount of feed, stage of lactation, and species of animal or breed of cow. Constituents not normally present may be incorporated

Table 4.1

COMPOSITION OF BOVINE MILK LIPIDS

Class of Lipid	% Total Milk Lipids
Triglycerides of fatty acids	95–96
Diglycerides	1.26–1.59
Monoglycerides	0.016–0.038
Keto acid glycerides (total)	0.85–1.28
Ketonogenic glycerides	0.03–0.13
Hydroxy acid glycerides (total)	0.60–0.78
Lactonogenic glycerides	0.06
Neutral glyceryl ethers	0.016–0.020
Neutral plasmalogens	0.04
Free fatty acids	0.10–0.44
Phospholipids (total)	0.80–1.00
Sphingolipids (less sphingomyelin)	0.06
Sterols	0.22–0.41
Squalene	0.007
Carotenoids	0.0007–0.0009
Vitamin A [a]	0.0006–0.0009
Vitamin D	0.00000085–0.0000021
Vitamin E	0.0024
Vitamin K	0.0001

[a] Based on the free alcohol.

Floyd E. Kurtz, Dairy Products Laboratory, Agricultural Research Service, U.S. Department of Agriculture, Washington, D.C. 20250. Retired.

into milkfat under unusual feeding conditions. The proportions of some of these components may vary widely with dairy processing. Butterfat prepared from churned cream, for example, may be devoid of detectable phospholipids. It is, therefore, incorrect to use the terms "milkfat" and "butterfat" interchangeably.

GLYCERIDES—GENERAL

Structure

The glycerides dealt with in this chapter are compounds in which 1, 2, or 3 hydroxyl groups of glycerol have reacted to form esters or ethers. Considering the substituents in a general sense as a, b, and c, the various possible isomers are shown below.

$$(1)\quad \begin{array}{c} H_2CO\text{-}a \\ | \\ HCOH \\ | \\ H_2COH \end{array} \qquad (2)\quad \begin{array}{c} H_2COH \\ | \\ HCO\text{-}a \\ | \\ H_2COH \end{array}$$

monoglycerides

$$(3)\quad \begin{array}{c} H_2CO\text{-}a \\ | \\ HCOH \\ | \\ H_2CO\text{-}a \end{array} \qquad (4)\quad \begin{array}{c} H_2CO\text{-}a \\ | \\ HCO\text{-}a \\ | \\ H_2COH \end{array}$$

diglycerides with 2 like substituents

$$(5)\quad \begin{array}{c} H2CO\text{-}a \\ | \\ HCOH \\ | \\ H_2CO\text{-}b \end{array} \qquad (6)\quad \begin{array}{c} H_2CO\text{-}a \\ | \\ HCO\text{-}b \\ | \\ H_2COH \end{array} \qquad (7)\quad \begin{array}{c} H_2CO\text{-}b \\ | \\ HCO\text{-}a \\ | \\ H_2COH \end{array}$$

diglycerides with 2 unlike substituents

```
        H2CO-a              H2CO-a
          |                   |
(8)     HCO-b       (9)     HCO-a
          |                   |
        H2CO-a              H2CO-b
```

triglycerides with 2 like and 1 unlike substituents

```
      H2CO-a             H2CO-b              H2CO-a
        |                  |                   |
(10)  HCO-b      (11)    HCO-a       (12)    HCO-c
        |                  |                   |
      H2CO-c             H2CO-c              H2CO-b
```

triglycerides with 3 unlike substituents

The positions of attachment of the substituents, from top to bottom, have been variously known as external, internal, external; α, ß, α'; α, ß, α; α, ß, γ; or 1, 2, 3. It is apparent from these examples that the following number of positional isomers are possible: monoglycerides, 2; diglycerides with 2 like substituents, 2; diglycerides with 2 unlike substituents, 3; triglycerides with 3 like substituents (not shown), 1; triglycerides with 2 like and 1 unlike substituents, 2; triglycerides with 3 unlike substituents, 3. Such positional isomers may show differences in physical properties, e.g., melting point, and in chemical properties, e.g., rate of hydrolysis.

Certain of the above forms are capable of an additional type of isomerism, namely, optical isomerism. Optical isomers are identical in most chemical and physical properties. An exception is their ability to rotate the plane of polarized light; enantiomorphic forms rotate the plane of polarized light an equal amount but in opposite directions. Such isomers also frequently show differences in chemical reactions with other optically active compounds and in those physical, chemical, and biological reactions in which adsorption on an optically active surface is an essential feature. The structural requirement for optical isomerism is an asymmetric molecule. Of the structures given above, 1, 4, 5, 6, 7, 9, 10, 11, and 12 satisfy this requirement, so each of these should be capable of existing in enantiomorphic forms. Optically active mono- and diglycerides were synthesized in early work. However, except for those containing an optically active acid, fatty acid triglycerides with demonstrable optical activity could neither be isolated from natural sources nor synthesized.[80] Proceeding on the hypothesis that this resulted from a combination of high molecular weight and insufficient disparity between the terminal acyl groups, Schlenk[285] was able to synthesize a number of triglycerides with easily demonstrable optical activity.

The triglycerides are represented above as having all the substituents lying on the same side of the glyceryl group. Actually, in the liquid state there is considerable freedom for groups to rotate about their bonds. In crystals, however, the orientation would be organized and uniform for a particular crystal form. With reference to fatty acid triglycerides, Clarkson and Malkin[48] have concluded from molecular models that the presence of all three fatty acid radicals on the same side of the glyceryl group would introduce considerable strain through steric interference. They proposed instead a "tuning fork" arrangement of the fatty acid groups:

```
          H
          |
     H —  C — OR
         /
RO — C — H
         \
     H —  C — OR
          |
          H
```

This arrangement is also consistent with the x-ray data showing the length of the triglyceride unit cell to correspond to a double-[48] and sometimes to a triple-chain[46,79,209] structure:

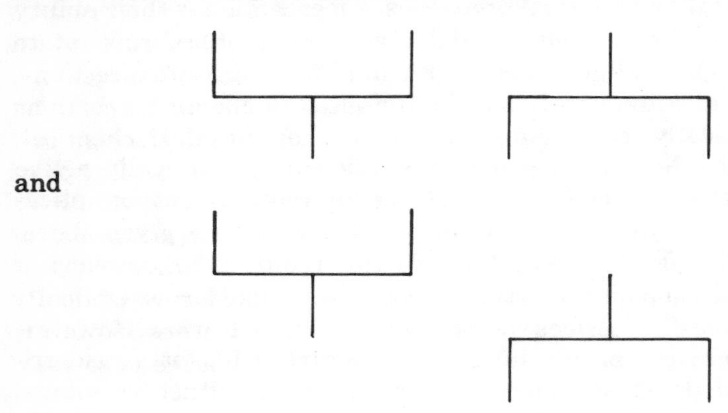

and

Lutton[209] believed that, in certain instances, the triple-chain glyceride was probably represented by a "chair" structure,

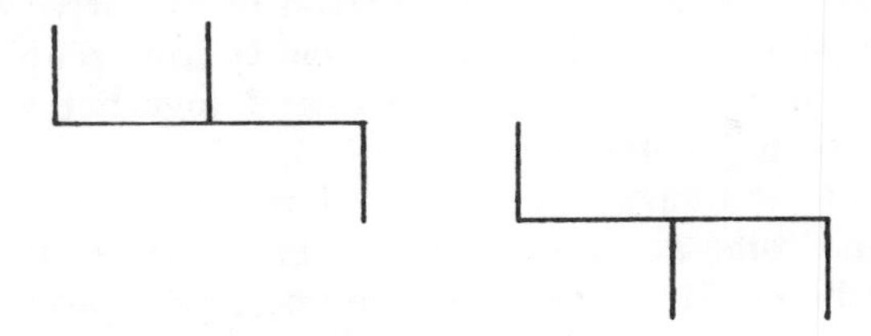

rather than by the "tuning fork" structure.

He found both double- and triple-chain structures for the glyceride, C_{18}-C_{12}-C_{18}. Normally a glyceride would appear in only one of these structures; the particular one it appears in would depend upon its fatty acid components. The individual triglycerides can show other differences of molecular packing which produce different crystalline forms. This phenomenon is known as polymorphism. As a result, the unit cells of different polymorphic forms may have different cell dimensions and facial angles and the crystals different physical properties, such as density and melting point. Bailey[21] has stated that at least 2 polymorphic forms have been observed for all fatty acids and glycerides and, for some, as many as 3 or 4. This phenomenon is of interest insofar as it affects the behavior of a fat during melting or solidification. It must also be taken into account in determining the melting point of a pure compound.

The crystalline form obtained from a melt depends upon the temperature at which crystallization takes place. On the basis of existing experimental data, largely on simple saturated triglycerides, it appears that the melt can always be supercooled by rapid cooling to a temperature slightly below the melting point of the least stable form. From these crystals those of intermediate forms, if any, can be obtained by rapidly heating to a temperature only slightly above the melting point of the next higher-melting form, and then cooling. Crystallization from solvents usually results in formation of the most stable form. Transformation of an unstable to a stable form on heating usually occurs at a measurable rate at a temperature appreciably below the melting point of the unstable form. Whether or not this form will be detected during the melting of the material depends upon the rate of transformation and upon the rate of heating. The transformation is slow enough in simple saturated triglycerides to permit easy observation of double melting. In contrast, while two melting points have been obtained for oleic acid, the transformation of saturated fatty acids is so rapid that this phenomenon has never been observed.[21]

Chemical Properties.—The majority of the many chemical reactions which glycerides are capable of undergoing are inappropriate for discussion in this book. Those involved in oxidative deterioration and in the development of lipolytic rancidity are treated in another chapter. Other reactions will be considered together with the lipid classes to which they pertain.

With regard to the various glyceride structures listed above, it was pointed out that some are positional isomers which differ in chemical properties. Although the differences may be small, positional isomers exhibit different reaction rates with either symmetric or asymmetric reagents. A second class of structures is made up of the optically active glycerides which differ from their optical isomers, but only in reactions involving asymmetric reagents. An example of compounds exhibiting this type of chemical differentiation is the monoglyceride, structure 1 above, and its optical isomer in which the substituent "a" is transposed to the other primary hydroxyl group.

There is still a third class of glycerides, namely, compounds which exhibit an internal specificity between like groups in reactions with asymmetric reagents. This class consists of glycerol itself, any triglyceride having 3 identical substituents, and the following glycerides illustrated above: the monoglyceride, structure 2; the diglyceride, structure 3; and the triglyceride, structure 8. All these are examples of the structural type C_{aabc}, which represents a molecule in which the central carbon atom is substituted with 4 symmetrical groups, of which two (a) are identical, but different from the two dissimilar groups (b and c). Such a molecule has a plane of symmetry and therefore is optically inactive. The two halves of the molecule are not identical, however, but have a left hand-right hand relationship. This can be seen in Figure 4.1. Because of the similarity with other compounds, e.g., *meso*-tartaric acid, in which the optical inactivity results from the left hand-right hand relationship of its two halves, the central carbon atom of glycerol and other molecules of the C_{aabc} type has been referred to as a *meso*-carbon atom.[295] In this type of molecule, asymmetric reagents or reactions, such as enzymatically catalyzed reactions which involve intermediate complexes with asymmetric molecules, discriminate between the identical groups.

The possibility of such discrimination was foreseen by Ogston[248] who showed that it could be effected if the reaction involved a 3-point attachment of the substrate to an enzyme. This possibility was later confirmed by Schambye *et al.*[283] and by Swick and Nakao,[296] who demonstrated discrimination between the two primary hydroxyl groups of glycerol in biological systems. Schwartz and Carter[295] showed that this type of discrimination does not require orientation of the substrate on an enzyme. They reacted 1-α-phenethylamine with ß-phenylglutaric anhydride and found a preferential reaction with one of its two like groups.

The fundamental reason for discrimination between the like groups in C_{aabc}-type molecules by asymmetric reagents might be stated in the generalization that unlike molecules, except optical isomers, differ in energy content and in probability of formation. Reaction of a symmetric reagent with one "a" group (other than a reagent which would alter an "a" group to a "b" or "c" group) forms a mixture of two optically active antipodes. Since the energy content of optical isomers is equal, the mixture contains equal quantities of each antipode. Reaction with an asymmetric reagent, however, produces an unequal mixture of diastereoisomers, since diastereoisomers differ in energy content.

The extent of discrimination in these reactions differs widely. In the Schwartz and Carter reactions with simple compounds, the ratio of diastereoisomers varied from an undetectable difference to 3 to 2 (in the 1-α-phenethylamine-ß-phenylglutaric anhydride reaction). Van Deenen and de Haas[58] hydrolyzed synthetic (α,γ-disteroyl)-ß-lecithin with phospholipase A. In this reaction discrimination appeared to be complete, as one equivalent only of fatty acid was removed to form an optically active lyso lecithin, and no further hydrolysis could be detected after prolonged incubation. In contrast, no discrimination could be detected in the hydrolysis of symmetric triglycerides with pancreatic lipase.[161, 338]

To distinguish between its two primary carbinol groups, several systems for numbering the carbon atoms of glycerol have been proposed.

Figure 4.1 illustrates the system in most general current use.[143] Glycerol is oriented with the secondary carbinol group to the observer's

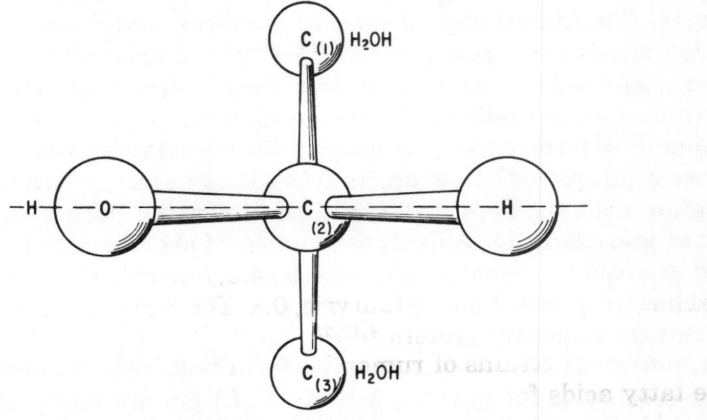

FIG. 4.1. sn-GLYCEROL

left in a Fischer projection in which the substituents above and below the central carbon atom are farther from the observer than those shown to the right and left.[133] The carbon atoms are then numbered as shown in the figure. To distinguish the numbering in this system from that in other systems, the prefix *sn*- designating "stereospecific numbering," is used. This term is printed in the lower case italics and, in names containing glycerol or its equivalent, immediately precedes the term signifying glycerol and is separated from it by a hyphen. For further details the reader is referred to the published recommendations on lipid nomenclature.[143]

Biosynthesis

Before considering the composition of milkfat and its variations with species, diet, and other conditions, it seems appropriate to review information regarding the bio-origins of the substances that are incorporated into the glycerides.

Rumen Activity.—Fatty acids may originate in the lipids of the animal's food or may be formed, in ruminants, by microbial action in the rumen on the nonlipid portion of the diet. To a lesser degree microbial action can alter dietary materials in nonruminants. It is at a minimum in carnivorous species, such as the cat and the dog. The horse and the rabbit represent species with enlarged intestines in which an intermediate degree of activity takes place.

Fatty acids from grass lipids are typically long-chain and contain a high proportion of polyunsaturated fatty acids.[311] Before leaving the rumen the dietary glycerides are largely hydrolyzed to free acids.[92] Microbial activity causes extensive hydrogenation of the unsaturated acids.[93,118,271,311,312] Both saturated and partially saturated acids are formed. The hydrogenation process produces considerable rearrangement of double bonds[4,93,118,307,310-312] both geometrically and positionally, so that one can expect the remaining unsaturated acids to contain an abundance of both geometrical and positional isomers.

Microbial fermentation of dietary carbohydrates in the rumen produces large proportions of acetic acid with smaller proportions of other volatile acids.[8,74] Annison[8] found the following percentage composition of volatile fatty acids in the rumen of sheep which had been fed hay: acetic, 81; propionic, 13; butyric, 4.3; valeric, 0.5; isobutyric, 0.6; isovaleric, 0.3; and 2-methylbutyric, 0.3. The branched-chain acids are derived from dietary protein.[8,10,75]

A number of strains of rumen bacteria require branched-chain volatile fatty acids for growth. Allison *et al.*[5] found that these acids are utilized in forming both branched-chain fatty acids and branched-chain aldehydes. Isovaleric acid was incorporated into 15-carbon acid, 17-carbon acid, and 15-carbon aldehyde. Isobutyric acid was incorporated into lipids mainly as 14-carbon and 16-carbon fatty acids. Of that incorporated into the lipids, 11% was found in 14- and 16-carbon carbonyl compounds, which presumably were branched-chain aldehydes. The aldehydes were found in the phospholipid fraction and were presumed to be constituents of plasmalogens.

Keeney *et al.*[169] studied the fatty-acid composition of the lipids of the rumen bacteria and protozoa of a cow. The acids were mostly long-chain and included substantial percentages of branched- and odd-carbon acids. The odd- and branched-chain acids were also found in the cow's blood serum lipids, thus showing that they became available to the gland for incorporation into milkfat. The authors concluded, from the nature and percentages of the acids isolated from the microbial lipids, that significant proportions of milkfat acids originate from rumen microbial synthesis of long-chain acids from volatile acids, and that this is the major source of odd-carbon and branched-chain acids.

Circulatory Transport.—Fatty acids that have entered the general circulation, regardless of their origin, are subject to the metabolic requirement of the animal. Utilization in milkfat is but one of these requirements. For example, long-chain acids may be oxidized in the liver to β-hydroxybutyric acid, or may be stored in the animal's depot fat and subsequently be released into the circulatory system over a long period of time. In this connection, conjugated acids have been found in milkfat at elevated levels for a prolonged period after discontinuing the addition of linseed oil to the animal's diet.[134] Likewise, radioactivity from tritium-labeled stearic acid was detected in milkfat for more than a month after feeding.[99]

Volatile fatty acids of the rumen pass directly into the blood and are transported as free acids to the mammary gland by way of the liver. Large quantities of acetic acid reach the gland and are removed by it from the blood. Only a small portion of the propionic and butyric acids, as such, reach the mammary gland. Butyric acid is partly oxidized to β-hydroxy butyric acid by the ruminal epithelia[9,262,340] and both butyric and propionic acids undergo extensive metabolic changes in the liver.[221] Esterified fatty acids, as components of triglycerides, phospholipids, and cholesteryl esters, are carried by the blood to the mammary gland in the form of lipoprotein complexes. Nonesterified medium- and long-chain fatty acids are transported in combination with albumin.

Mammary Gland Activity.—As raw materials for milkfat synthesis, the mammary gland removes both lipids and nonlipids from the blood.

Utilization of Blood Lipids.—It is known that triglycerides are removed from the blood in large quantities and are believed to be a principal source of long-chain acids in milkfat. Albumin-bound nonesterified fatty acids can also be utilized by the gland;[200] it has been postulated that this source of milkfat might become important during fasting. The utilization of fatty acids from other lipid classes is more controversial. This subject is treated in reviews by, for example, Luick[206] and McCarthy.[221] It appears likely that, prior to incorporation into milkfat, the fatty acids of blood glycerides are either hydrolyzed or rearranged.[222,256,259]

The fact that dietary lipids can contribute fatty acids to milkfat has been established by adding unusual fatty acids to the diet and noting their subsequent appearance in the milkfat. Examples are erucic acid from the addition of rapeseed oil to the diet,[131] iodinated fatty acids,[17] tritium-labeled stearic acid,[99] and synthetic odd-carbon acids.[12] Coconut oil cake added to the diet produced elevated levels of myristic and lauric acids in the milkfat.[130] Measurable radioactivity which appeared in the milkfat 4 hr. after feeding tritium-labeled stearic acid[99] could be detected in the fat for 37 days.

The average milk cow secretes about twice as much fat in her milk as she takes in from her diet.[206] This sets an absolute maximum of about 50% for that part of milkfat originating in dietary lipids. Glascock *et al.*,[99] from studies based on the incorporation of radioactive stearic acid, concluded that the actual contribution of dietary fat to milkfat is in the range of 18 to 25%.

Fat Synthesis.—It has been proposed[123] that milkfat is derived entirely from blood fat, the short-chain fatty acids being formed by an *omega* oxidation of fatty acids in the intact triglyceride. It is now known that this view is incorrect. Synthesis of fat in the mammary gland was first recognized from respiration studies and later confirmed with isotopically labeled metabolites. Graham *et al.*[104] found that the respiratory quotient of the lactating goat udder *in vivo* was greater than unity. This indicated utilization of an oxygen-rich substrate for the synthesis of fat. Later, in a series of publications,[85] Folley and French reported their studies of the respiration of tissue slices, taken from the lactating mammary glands of various species, in the presence of glucose or of acetate. The animals studied were mouse, rat, guinea pig, rabbit, sheep, goat, and cow. The ruminant tissue utilized acetate but not glucose, with a respiratory quotient above unity. This situation was reversed with nonruminant tissue. It was concluded that when the respiratory quotient was high, net fatty acid synthesis from an oxygen-rich substrate (acetate for ruminant tissue and glucose for nonruminant tissue) was occurring.

Synthesis of milkfat by the mammary gland has been confirmed and elucidated by experiments utilizing C^{14}-labeled metabolites. Reviews are available covering much of this research.[82,83,84,261,262] It has been found that acetic acid makes the most important contribution to the synthesis of fatty acids in the mammary gland[275] and is utilized by various species.[53,86,89,146,180,265,266,267] Acetic and butyric acids act as precursors of both fatty acids and glycerol.[275] Although propionic acid is utilized for the synthesis of minor amounts of longer-chain fatty acids with an odd number of carbon atoms,[146] it is relatively unimportant as a precursor of fatty acids.[275] It is, however, indirectly the main glycerol precursor in the cow by conversion into glucose outside the gland.[207,275] Within the mammary gland, glucose is converted into glycerol, which is incorporated into the triglyceride.[207] It appears that valeric acid is utilized for fatty-acid synthesis after being broken down to acetate and propionate.[94] Butyric acid is absorbed into the mammary gland chiefly as β-hydroxybutyrate, which is either reduced directly to butyric acid or first broken down into two-carbon units before participating in fatty-acid synthesis.[9,187,261,305]

The synthetic route by which fatty acids are formed from acetate in the mammary gland has been established by Popják and his associates.[264,265,266] A lactating goat was injected intravenously with $CH_3C^{14}OONa$ and the goat milked at frequent intervals up to 48 hr after the injection.[265] It was found that, at the time of their maximum isotope content, the milk fatty acids contained several hundred times as high a concentration of C^{14} as did the plasma fatty acids at any time during the experiment. This indicated that the milk fatty acids must have been synthesized to a large extent within the udder from small molecules and not transferred as such from the blood. The particularly high C^{14} content of the steam-volatile acids indicated that they were synthesized independently rather than being derived from long-chain fatty acids of mammary gland fat. It was calculated that, of the total acetate present in the body of the goat at any time, about 80% would be oxidized within 6 hr, and about 50% of the remainder would participate in the fat metabolism of the udder.

Popják *et al.*[266] resolved the fatty acids obtained from the above-described experiment into two groups: one group contained the acids from all fat samples taken during a period 0 to 12 hr after injection of the isotopically labeled acetate; the other contained acids from samples taken during a 12- to 48-hr period. In the first group the radioactivity per unit of carbon increased to a maximum with decanoic acid, fell off to myristic and palmitic acids and then dropped precipitously to oleic and stearic acids. In the second group the activity increased uniformly to myristic and palmitic acids and then again dropped precipitously to oleic and stearic acids.

These results were interpreted as indicating that in the goat's udder the intracellular pools of the short-chain acids up to and including decanoic acid are very small, but that they are increasingly larger from lauric acid upward. The very low concentration of C^{14} in stearic and oleic acids indicated that their origin is different from that of the other acids in milkfat. They may be absorbed directly from the blood instead of being synthesized from small molecules. An alternative explanation is that the pools of these acids in the mammary gland tissue are so very large that the concentration of any newly synthesized acids would remain small in comparison to that of the pre-existing acids.

Acetic, butyric, caproic,[266] and caprylic[264] acids were degraded by methods which permitted measurement of the radioactivity of each carbon atom of each acid, together with positive identification of the position within the acid from which the sample was derived. The investigation of acetic acid showed that no redistribution of the C^{14} isotope occurred in the body, since only the carboxyl carbon contained this isotope.

In the other acids, each of the odd-numbered carbon atoms (designating the carboxyl carbon as 1) was radioactive and each of the even-numbered carbon atoms was nonradioactive. Carbon atoms 1 and 3 of butyric acid had equal radioactivity. Carbon atoms 3 and 5 of caproic acid had equal radioactivity, and this was the same as that of carbon atoms 1 and 3 of butyric acid. Carbon atom 1 of caproic acid had a radioactivity approximately 2.5 times that of carbon atoms 3 and 5. These results demonstrate conclusively that butyric acid is synthesized by the combination of two acetate units, and that the synthesis of caproic acid involves the addition of an acetate unit to the carboxyl end of butyric acid. It was likewise shown[264] that synthesis of caprylic acid involves the addition of an acetate unit to the carboxyl end of caproic acid.

The lower activity of the carbon atoms derived from butyric acid was interpreted as indicating that the butyric acid synthesized from acetate was diluted 2.5 times with butyric acid or a butyric acid derivative present in the blood.[266] This compound was not identified, but was believed to be β-hydroxybutyric acid. Such a dilution of the acetate-derived butyric acid was not invariably found with other animals. The fact that in the lactating goat the butyric acid was labeled by having a lower C^{14} content than the acetate provided the basis for the unequivocal demonstration that synthesis always involves the addition of an acetate unit to the carboxyl end of the acid that is being lengthened. Without this circumstance the evidence would not have precluded condensation of larger units nor combination of the carboxyl end of the acetate with the hydrocarbon end of the acid undergoing enlargement.

This synthetic pathway, as described, is incomplete in the sense that a chain elongation by one acetate unit actually involves a series of steps. These intermediate reactions and their requirements of enzymes and cofactors are treated in several general reviews.[52,105,210,353] The review of Folley and McNaught[86] deals more specifically with synthesis by the mammary gland.

At least two distinct pathways may be followed in the biosynthesis of saturated fatty acids. Additional pathways, or at least different enzyme systems, are indicated in some instances. The first step always seems to be the activation of the fatty acids[183] (both components involved in the elongation) by conversion to the thiol ester of coenzyme A:

$$RCOOH + CoASH + ATP \rightleftharpoons RCOSCoA + AMP + PP_i^*$$

The energy necessary for this reaction is furnished by the cleavage of adenosine triphosphate to adenosine monophosphate and inorganic pyrophosphate.

One of the pathways is a reversal of that followed in the *beta* oxidation of fatty acids and probably involves one or more enzymes in addition to those of the *beta*-oxidation sequence.[301] Wakil[353] has suggested that this pathway may be followed in the further elongation of such long-chain acids as palmitic. The steps in a reversal of *beta* oxidation are:

(1) $RCOSCoA + CH_3COSCoA \rightleftharpoons CoASH + RCOCH_2COSCoA$
(2) $RCOCH_2COSCoA \rightleftharpoons RCHOHCH_2COSCoA$
(3) $RCHOHCH_2COSCoA \rightleftharpoons RCH{=}CHCOSCoA$
(4) $RCH{=}CHCOSCoA \rightleftharpoons RCH_2CH_2COSCoA$

While these are equilibrium reactions the equilibrium for step 1 is extremely unfavorable to synthesis. However, an enzyme not involved in the *beta*-oxidation sequence, which can irreversibly reduce the unsaturated ester in step 4, has been discovered.[198,301] This reduction leads, through each of the four steps, to a continuous removal of the product in step 1, making the overall reaction possible.

Another pathway utilizes malonyl coenzyme A which is formed in the first step[352] of the synthesis:

$$CH_3COSCoA + CO_2 + ATP \xrightarrow{Mn^{++}} HOOCCH_2COSCoA + ADP + P_i$$

This reaction is catalyzed by an enzyme containing tightly bound biotin. The major products of the overall synthesis using the enzyme system

*Abbreviations used: CoASH or CoA, coenzyme A; TPNH and TPN+, reduced and oxidized forms of triphosphopyridine nucleotide; ATP, adenosine triphosphate; ADP, adenosine diphosphate; AMP, adenosine monophosphate; P_i, inorganic phosphate; PP_i, pyrophosphate.

isolated from pigeon liver can be represented as being formed in accordance with the following equation:

$$RCOSCoA + 7HOOCCH_2COSCoA + 14\ TPNH + 14\ H^+ \rightarrow$$

$$R(CH_2)_{14}COOH + 14TPN^+ + 7CO_2 + 8CoASH + 6H_2O$$

When malonyl CoA was reacted with acetyl CoA, palmitic acid formed 80% of the synthesized acids. When propionyl CoA was substituted for acetyl CoA, the synthesis was much slower and heptadecanoic acid formed 70% of the synthesized acids. When butyryl CoA was used, the reaction rate was even slower; stearic acid formed 85% and palmitic acid 10% of the synthesized acids.

The exact mechanism of this chain elongation and the nature of the intermediate reductive steps have not yet been established. On the basis of available evidence the following scheme has been proposed[35] for the synthesis of long-chain fatty acids by the soluble pigeon liver system:

(1) Both malonyl CoA and acetyl CoA (when the synthesis is started with acetic acid) become bound to the enzymes of fatty acid synthesis by an exchange reaction in which coenzyme A is freed and thiol esters are formed with enzyme sulfhydryl groups.

(2) The enzyme-bound acyl groups undergo a decarboxylation-condensation reaction, which may be followed either by reduction, dehydration, and reduction again to form an enzyme-bound saturated C_4 compound, or by repeated condensations of multiple malonyl groups on the acetyl moiety to form an enzyme-bound keto compound. The carbonyl groups of the ketoacyl compound are then reduced by TPNH, dehydrated, and reduced again to the saturated acyl derivative, which is deacylated to the free fatty acid. In the absence of TPNH, the ketoacyl compound may be deacylated and set free without reduction.

While the mammary gland enzymes are incompletely characterized, there is some evidence that the malonyl-coenzyme A pathway is an important route for the synthesis of fatty acids in the gland. Ganguly[91a] found, using a soluble enzyme system isolated from cow mammary gland, that the rate of synthesis from malonyl CoA is much faster than from acetyl CoA. The C_4, C_6, C_8, C_{10}, C_{12} acids, and a group of longer-chain acids which Ganguly did not resolve, were synthesized by this enzyme system. Dils and Popják[62] studied the synthesis of fatty acids by a soluble enzyme system prepared with rat mammary gland. They concluded that the major route for the synthesis of fatty acids is by way of malonyl coenzyme A. The main products of synthesis with this preparation are the even-numbered fatty acids from C_8 to

C_{18}. These results, in contrast with those from nonmammary tissue enzymes, showed the synthesis of significant amounts of shorter-chain fatty acids.

Avena and Kumar[15] studied the formation of fatty acids by mammary gland slices from the rabbit and the goat. The C^{14}-labeled metabolites studied were acetate, butyrate, and β-hydroxybutyrate. The synthesized acids were isolated as their coenzyme A derivatives and separated into two groups: C_8 and over, and the shorter-chain acids. With both species acetate was incorporated to a greater extent in the longer-chain acids than in the shorter-chain acids; no butyrate was formed from acetate by rabbit slices and none initially by goat slices. About 90% of the β-hydroxybutyric acid incorporated into the fatty acids appeared in the short-chain group. Butyric acid, likewise, but to a smaller extent, was incorporated preferentially in the short-chain acids. These studies showed that in the ruminants, butyric acid and longer-chain acids may be synthesized by two independent processes. Acetate is used to a great extent for chain elongation, while β-hydroxybutyrate is converted predominantly to butyric acid.

It is known that triglycerides are synthesized in the mammary gland from fatty acids and glycerol.[207] The following pathway has been established for this synthesis in nonmammary gland tissues:

(1) $2RCOSCoA + CH_2OHCHOHCH_2OPO_3^{=} \rightarrow$
$CH_2O(OCR)CHO(OCR)CH_2OPO_3^{=}$

(2) $CH_2O(OCR)CHO(OCR)CH_2OPO_3^{=} \rightarrow$
$CH_2O(OCR)CHO(OCR)CH_2OH$

(3) $CH_2O(OCR)CHO(OCR)CH_2OH + RCOSCoA \rightarrow$
$CH_2O(OCR)CHO(OCR)CH_2O(OCR)$

In the first step two molecules of fatty acids, in the active form of coenzyme A esters, react with L-α-glycerophosphate to form a phosphatidic acid.[184] In the second step this is converted to an α,β-diglyceride by the action of a specific phosphatase.[320] The diglyceride is then esterified by a third fatty acyl CoA to form the triglyceride.[356]

The synthesis of triglycerides by mammary gland enzymes has been studied by Kumar and Pynadath,[188] using a mitochondrial-microsomal particulate fraction from lactating goat mammary glands. The glyceride synthetic activity resides in this fraction. C^{14}-labeled palmitate and palmityl CoA are incorporated into glycerides in proportions comparable to those found in intestinal mucosa, adipose tissue, and liver. Besides palmityl CoA, octanyl CoA and hexanoyl CoA are incorporated in relatively large percentages. Butyryl CoA is incorporated to a much lesser extent than in goat milkfat. The authors were unable to explain this. The fatty acid acceptors are α-glycerophosphate and 1,2-diglycerides. The results are in agreement with the scheme proposed by Weiss *et al.*[356]

Unsaturated Acids.—Most of the experimental material cited in this section does not establish the body sites where unsaturated acids are formed. Even though synthesized outside the mammary gland, they may be transported in the general circulation to the gland and thus become available for incorporation into milkfat.

There is no proof that unsaturated acids can be formed in animal tissues by biohydrogenation of more highly unsaturated acids. C^{14}-labeled palmitoleic acid fed to rats was converted in small amounts to radioactive palmitic acid.[227] The extent of this biohydrogenation was of relatively minor importance; the possibility that even this occurred in the intestines from microbial action could not be excluded. In contrast, it has been demonstrated that, by dehydrogenation, double bonds can be introduced into saturated acids and into certain unsaturated acids.

Schoenheimer and Rittenberg fed mice the methyl esters of linseed oil which had been reduced catalytically with deuterium.[287] The unsaturated acids from the body carcass fat contained more deuterium than the body water. Stetten and Schoenheimer,[329] by isotopic exchange, prepared a palmitic acid containing 22 atom per cent deuterium. This was fed to rats as the methyl ester. The unsaturated acids, after isolation from the acids of the carcass fat, were hydrogenated. Palmitic acid containing deuterium was found. This demonstrated a desaturation of palmitic acid. In a similar experiment[7], in which C^{14}-labeled myristic acid was included in the diet of rats, desaturation of myristic acid was demonstrated.

These experiments are sometimes cited as demonstrating the formation of oleic, palmitoleic, and myristoleic acids from their saturated counterparts. They did not, in fact, show either the number or the position of double bonds introduced into the saturated acids. Glascock and Reinius[100] provided more definite evidence on these points. Tritium-labeled stearic acid was fed to a goat and oleic acid was isolated from the milkfat after conversion by oxidation to a characteristic derivative, dihydroxystearic acid. The activity of this derivative was such as to indicate that it was derived from an oleic acid which had been formed by desaturation of the labeled stearic acid.

Lauryssens *et al.*[200] have subsequently shown that desaturation can take place in the mammary gland. More than one-half of the activity incorporated into udder tissue glycerides from 1-C^{14}-stearate in a perfusion experiment was found in the oleic acid fraction of the chromatographically separated fatty acids.

Enzyme systems and pathways in the formation of monounsaturated acids in yeasts[26] and in an anaerobic bacterium[284] have been discussed by Bloch and his associates.

Mead, in a review article[225], has made a remarkably clear presentation of the principles underlying the formation of polyunsaturated fatty acids in animals. The steps in the conversion of linoleic to arachidonic acid are shown below:

$$CH_3{-}(CH_2)_4{-}CH{=}CH{-}CH_2{-}CH{=}CH{-}(CH_2)_7{-}COOH$$

(linoleic acid)

$$\downarrow$$

$$CH_3{-}(CH_2)_4{-}CH{=}CH{-}CH_2{-}CH{=}CH{-}CH_2{-}CH{=}CH{-}(CH_2)_4{-}COOH$$

(gamma-linolenic acid)

$$\downarrow$$

$$CH_3{-}(CH_2)_4{-}CH{=}CH{-}CH_2{-}CH{=}CH{-}CH_2{-}CH{=}CH{-}(CH_2)_4{-}CH_2{-}CH_2{-}COOH$$

(8,11,14-eicosatrienoic acid)

$$\downarrow$$

$$CH_3{-}(CH_2)_4{-}CH{=}CH{-}CH_2{-}CH{=}CH{-}CH_2{-}CH{=}CH{-}CH_2{-}CH{=}CH{-}CH_2{-}CH_2{-}CH_2{-}COOH$$

(arachidonic acid)

This pathway was elucidated by a series of experiments[140,226,228,327] involving the administration to rats of C^{14}-labeled acids, followed by determination of the location of the C^{14} in arachidonic acid. Following the administration of acetate-1-C^{14}, activity appeared only in the carboxyl group of arachidonic acid. Linoleate-1-C^{14} activity appeared in carbon 1, as a result of its breakdown to acetate, and in carbon 3. Gamma-linolenate-1-C^{14} activity appeared almost exclusively in carbon 3, indicating a very rapid conversion to arachidonic acid. The activity from 8,11,14-eicosatrienoic-2,3-C^{14} acid appeared in carbons 2 and 3 of arachidonic acid.

It was likewise shown[181,224,328] that linolenic acid is converted by a series of similar steps to the end products, 5,8,11,14,17-eicosapentaenoic acid, 7,10,13,16,19-docosapentaenoic acid, and 4,7,10,13,16,19-docosahexaenoic acid. Oleic acid was shown[90] to be the precursor of 5,8,11-eicosatrienoic acid in fat-deficient rats. A pathway analogous to those from linoleic and linolenic acids was indicated both by the constitution of the trienoic acid itself, and by the isolation of 8,11-eicosadienoic acid, a postulated intermediate. In this same study another triene, 7,10,13-eicosatrienoic acid, was isolated. This acid presumably would be formed in a similar manner from palmitoleic acid.

These results can be generalized[225] as follows: Starting from the existing double bonds of the precursor acid, additional double bonds are introduced in a 1,4-relationship toward the carboxyl group. This dehydrogenation is limited by the restriction that a double bond cannot be introduced closer to the carboxyl than the 4-position. Chain lengthening by the addition of acetate units may provide one or more intermediate precursors which can be dehydrogenated further.

It is notable that each of the initial precursor acids in these studies has a *cis* double bond in the 9-position. Intermediates, with *cis* double bonds nearer the carboxyl, but with definite relationships to the 9-position as shown above, also served as precursors of more highly unsaturated acids. Holman[136] has shown that *trans* isomers can be dehydrogenated. If the positional isomer of oleic acid, *cis*-12-octadecenoic acid, could serve as a precursor, it would be converted initially to linoleic acid; but it has been reported to be ineffective as an essential fatty acid.[344]

Mead[225] has suggested that the formation of polyunsaturated acids requires three enzyme systems: a polydehydrogenase which dehydrogenates mono- or polyenoic acids in a 1,4-relationship from the existing double bonds toward the carboxyl group; an acyl transferase, which adds two carbons when chain-lengthening is necessary; and a system which disposes of unsaturations in the 2- or 3-positions.

Mead has also discussed the special physiological functions of the polyunsaturated acids in relation to their shapes and polar centers. Kishimoto and Radin[178] determined the structure of a number of monoenoic and several dienoic acids isolated from brain sphingolipids. The chain lengths and positions of unsaturations of some of these acids indicated the likelihood that they were formed, via an α-oxidation system, by the removal of one carbon atom from the next higher homolog.

NEUTRAL GLYCEROLIPIDS

Fatty Acid Triglycerides

Chemical Properties.—Pertinent reactions of the ester linkage can be generalized as follows:

$$\mathrm{G{-}O{-}\overset{\overset{\displaystyle O}{\|}}{C}{-}R + R'\,OH \rightleftharpoons R{-}\overset{\overset{\displaystyle O}{\|}}{C}{-}O{-}R' + G{-}OH}$$

When R′ is hydrogen, the hydrolysis forms fatty acids and, successively, diglycerides, monoglycerides, and glycerol. The reaction with water alone takes place slowly at room temperature and, in the presence of both reaction products, is incomplete. It can be speeded at elevated temperatures and in the presence of a catalyst. The reaction velocity is also affected by certain substituents in the acyl moiety of the glyceride. It can be driven to completion by removing either of the reaction products as formed.

Fatty acids with an odd number of carbon atoms were reported to be liberated from stored butter more rapidly than acids with an even number of carbon atoms.[166] The increase in free fatty acids in sterile concentrated milk during storage appeared to result from random hydrolysis of the milk lipids.[203]

When R′ is an alkyl group, the acyl groups of the glyceride equilibrate between the hydroxyls of the glycerol and those of the added alcohol. If a suitable catalyst is used, e.g., an alkali-metal alcoholate with a sufficient excess of anhydrous alcohol, the reaction is complete enough to serve as a practical means of obtaining esters without the more laborious intermediate step of isolating the free acids. If the glyceride is heated and another ester and a catalyst, such as sodium ethylate or sodium-potassium alloy, or with the catalyst alone, the resultant reaction is known as interesterification; it consists of random rearrangement of the acyl groups among all the hydroxyl groups.

General Properties of Fatty Acid Components.—About 85% of the total weight of the milk lipids occurs in the fatty acid moiety of the triglycerides. The properties of milkfat, accordingly, are influenced by those of their component acids and by the relative proportions and the inter- and intramolecular distribution of these acids.

The lower molecular weight saturated acids and the 9-*cis* unsaturated acids up to the C_{20} acid, which melts at 25°C, are liquids at normal room temperatures. Factors affecting the melting point include the molecular weight, the number and position of alkyl substituents, and the number, position, and geometrical configuration of unsaturations.

The series of fatty acids with an even number of carbon atoms and that with an odd number both show a nearly linear increase of melting point with increased molecular weight, but the curve of the latter is displaced downward on the melting-point scale. The melting point of an acid is also a function of its degree of symmetry. For a given molecular weight the melting point is highest for the straight-chain isomer. The melting point is progressively lowered with increased branching, and the effect is the greatest for branching located near the center of the molecule.

The boiling points of fatty acids also increase with increased molecular weight, but without distinction as to whether the acid has an odd or an even number of carbon atoms. The boiling point of an acid decreases with increased symmetry and with increased unsaturation, but the effect is much smaller than in the case of the melting point. Treatises are available[60,216] covering these and other physical properties of the fatty acids.

The fatty acids are among those compounds which can associate by means of hydrogen bonding. In this instance a dimer is formed:

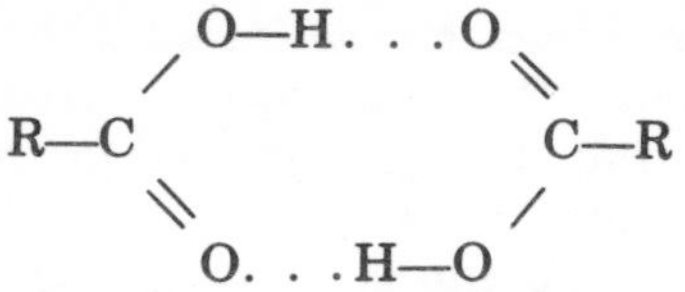

$$\begin{array}{ccccc} & O\text{—}H & \cdots & O & \\ / & & & & \backslash\backslash \\ R\text{—}C & & & & C\text{—}R \\ \backslash\backslash & & & & / \\ & O & \cdots & H\text{—}O & \end{array}$$

X-ray analysis showing crystals having layers two molecules thick has substantiated this structure. This double-molecule structure has the effect of raising the melting point of fatty acids considerably above values found for nonassociated compounds of equal molecular weight. The ester derivatives of fatty acids are, for the most part, unassociated, and consequently have considerably lower melting points than acids of the same molecular weight. For example, the melting point of stearic acid has been reported as 69.6°C, of ethyl stearate as 34.0°C, and of ethyl palmitate (same molecular weight as stearic acid) as 25.0°C. The methyl esters are apparently partially associated. This would account for the fact that, in general, they have somewhat higher melting points than the corresponding ethyl esters. The effect of polymorphism on the melting point of fatty acids has been mentioned in the section on glycerides.

Hydrogen bonds are weakened with an increase in temperature. At the boiling points of formic and acetic acids they are still strong enough to maintain the integrity of the dimeric form. This phenomenon causes the fatty acids to boil at higher temperatures than unassociated molecules of the same molecular weight. It is, therefore, common practice to use the methyl or ethyl esters in effecting separations of the fatty acids by distillation. It is also common practice to use ester derivatives in effecting separations of the fatty acids by gas-liquid chromatography. Separations of the free acids are best accomplished with a polar stationary phase. Under these conditions association of the acids, with its resultant complications, is minimized.

Contrary to a widespread belief, the fatty acids do not ionize appreciably less as their molecular weight increases above that of butyric acid.[172] This is of interest in connection with the titration of acids. If one considers the following equilibrium,

$$RC(=O)\text{—}OH + OH^- \rightleftharpoons RC(=O)\text{—}O^- + H_2O,$$

he will find that, regardless of molecular weight, the fatty acids can be determined quantitatively by titrating to a suitable end point if the reactants are maintained in solution. It is well known that the salt of a higher fatty acid hydrolyzes to a considerable extent in water to give an alkaline solution, e.g.,

$$C_{17}H_{35}\overset{\overset{\displaystyle O}{\|}}{C}—O^- + H_2O \rightleftharpoons C_{17}H_{35}\overset{\overset{\displaystyle O}{\|}}{C}—OH + OH^-.$$

This reaction is not primarily due to a low ionization constant of stearic acid, however, but rather to its insolubility in water. The alcoholic alkali solution generally used for determining fatty acids provides sufficient solvent action to prevent excessive hydrolysis.

Determination of Fatty-acid Composition.—Hilditch[124] has described the methods by which he and his associates, as well as others, accumulated much of the earlier data on the fatty-acid composition of triglycerides. The methods employed were aimed at separating the complex mixture of fatty acids into groups containing no more than two acids each. Steam distillation of the acids, fractional vacuum distillation of the esters, and differential solubilities of the salts, chiefly the lead salts, were utilized toward this end. Determinations of the saponification and iodine values, and the weights of material in each fraction, gave the information needed for a quantitative estimation of the acid composition of the fat. If the preliminary separations failed to limit a fraction to two acids, as often occurred in the group of C_{18} acids, additional separations or analyses were necessary.

Chromatographic methods have proved to have outstanding utility in recent investigations of fatty-acid composition. Adsorption chromatography has been applied successfully[190,274] to the difficulty separable group of C_{18} acids. This form of chromatography has been largely superseded, however, by partition chromatography, both liquid-liquid and gas-liquid. Other useful techniques are low-temperature (down to -65 to -70°C) crystallization from solvents and countercurrent distribution. Of these various techniques low-temperature crystallization is probably the least effective in separating individual fatty acids. It has the advantage of being applicable, without damage to the fatty acids, to large quantities of material. These characteristics have made it useful, as a preliminary step, in obtaining concentrated samples of a particular type of acid, e.g., the polyethenoic acids. Fractional vacuum distillation is also applicable to large samples of fatty acids (usually as the methyl esters). It is effective in separating acids on the basis of chain length—much less so on the basis of unsaturation. It has the disadvantages of being relatively slow and, as a result of prolonged exposure to elevated temperatures, of subjecting the acids to possible thermal damage. Countercurrent distribution is applicable to samples of intermediate size. The apparatus is rather bulky and the procedure is time-consuming. However, the material is not damaged, and it is sometimes possible to obtain effective separations by proper selection of solvents and application of a sufficiently large number of equilibria.

Chromatographic methods, particularly those employing gas-liquid chromatography, permit the use of columns with very high theoretical plate ratings. An interesting example of the high efficiency of gas-liquid chromatography is the resolution of the four geometrical isomers of methyl linoleate.[202] In general, the amounts of material separable by gas-liquid chromatography are small, although a distinction is made between preparative and nonpreparative columns. By using the former and by combining material collected from identical portions of a series of separations, samples of sufficient size for further investigation can be obtained. The elevated temperatures employed in gas-liquid chromatography may cause structural alternations in certain esters.[233] Conjugated trienoates can be isomerized geometrically, and esters with a hydroxyl or hydroperoxide group *alpha* to a double bond can undergo dehydration. In general, other types of acid are unaffected.[233,332] This technique requires expensive equipment and considerable experience in order to avoid misinterpretations of the data. Columnar liquid-liquid chromatography, in contrast, is an elegantly simple method, in regard to both the required apparatus and its operation, and is suitable for larger samples. Paper and thin-layer chromatography are modifications adapted to the resolution and quantitative estimation of small samples. The required apparatus and its operation are both very simple. These methods are eminently suited to the routine analysis of a large number of samples.

The methods employed in identifying and estimating the acids in the various fractions separated from a fat depend upon the complexity of its fatty acid composition and upon whether or not acids of unknown structure are present. In the simpler situations it may be sufficient to determine the weight, saponification value, and iodine value of a distillation fraction, the position and area under a peak in a gas chromatogram, or the position and titration value of a fraction obtained by columnar partition chromatography. In more complex situations additional tests or procedures are necessary. Reviews are available on the investigation of fatty acid structure by chemical means,[326] by nuclear magnetic resonance,[138] mass spectrometry,[71] infrared spectroscopy,[246] and ultraviolet spectroscopy.[245]

In absorption spectra the following relation exists between the wavelength of absorbed radiation and the energy change involved in its associated transition: $\Delta E = hc\lambda$. In this equation E is the absorbed energy, h is Planck's constant, c is the speed of light and λ is the wavelength of absorbed radiation. It is apparent that the energy change becomes smaller as the wavelength increases.

Ultraviolet absorption spectra arise from transitions of valence electrons between different energy levels. Fatty acids with isolated double bonds show absorption maxima in the region just below 200 mμ. Conjugation introduces transition levels with smaller energy differences, thus shifting the absorption maxima to longer wavelengths. Two conjugated double bonds shift the absorption band to about 234 mμ, 3 conjugated double bonds to about 268 mμ, 4 to 315 mμ, 5 to 346 mμ, 6 to 374 mμ, etc. Compounds with a sufficiently large number of conjugations, e.g., carotene, absorb in the region of longer wavelengths associated with visible light, and are therefore colored. On the basis of data obtained from known compounds, the absorption spectra of solutions can be used to detect the presence of conjugated acids and to estimate their concentrations from, respectively, the positions of absorption maxima and their intensities.

Alkali isomerization rearranges the isolated double bonds into a conjugated system. Under standardized conditions the absorption spectra before and after isomerization afford data suitable for determining the identity of nonconjugated acids (with respect to the number of double bonds present) and their concentrations. Fig. 4.2 shows the ultraviolet absorption spectra of linoleic and linolenic acids both before and after alkali isomerization. It is to be noted that isomerized linolenic acid shows absorption maxima for systems of both 3 conjugated double bonds and of 2 double bonds. The latter, of course, is part of the total system. The determination of unsaturated acids by ultraviolet absorption spectroscopy is described in detail by Mehlenbacher.[229]

Absorption in the infrared region concerns radiation of considerably longer wavelengths than that of the ultraviolet, and accordingly involves much smaller energy changes. These spectra involve vibrations of atoms or groups between different energy levels within a molecule. O'Connor[245] has assembled a collection of infrared absorption bands applicable to fatty acids and other lipids. Absorption at appropriate wavelengths is evidence of the presence of definite groups. Fig. 4.3 illustrates the distinction between absorption due to *cis* double bonds and that characteristic of *trans* double bonds. The latter has been developed into a quantitative determination.[247] Many additional structural features can be recognized from the presence of absorption at specific wavelengths. Comparison of infrared spectra with those of known compounds is an effective means of establishing the identity of unknown compounds or of those whose identity is doubtful. Infrared absorption spectroscopy has become an important tool in lipid research.

Spectroscopic data on lipids are usually reported as the wavelength (λ) in angstroms (A), millimicrons (mμ), or microns (μ), or as wavenumbers, i.e., the number of waves per unit length. For absorption

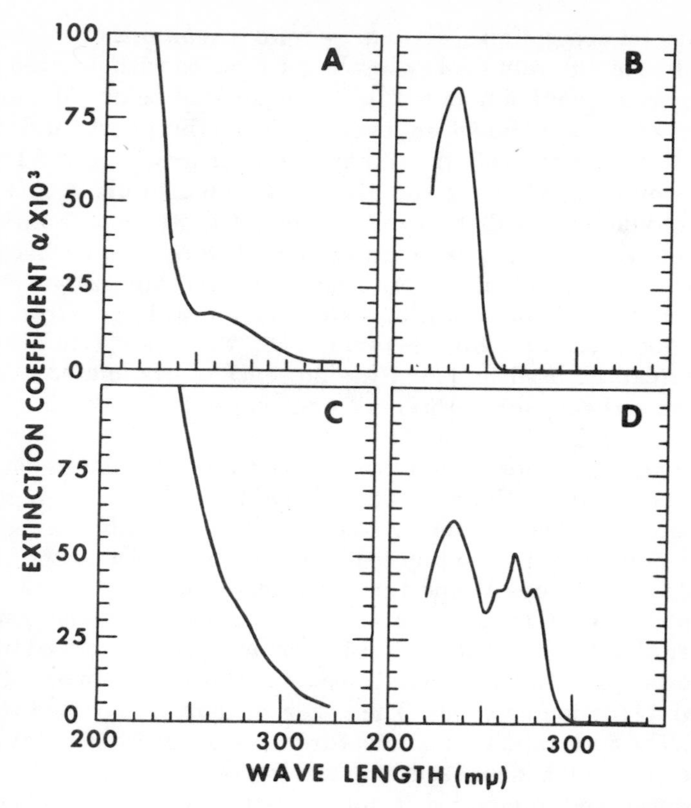

Adapted from O'Connor[244]

FIG. 4.2. ULTRAVIOLET ABSORPTION SPECTRA OF POLY-UNSATURATED FATTY ACIDS BEFORE AND AFTER ALKALI ISOMERIZATION

A—Linoleic acid before. B—After. C—Lineolic acid before. D—After.

in the infrared region it is usually expressed in reciprocal centimeters (cm^{-1}). The relationships between these units are given here for the convenience of the reader who may need to interconvert data for the sake of making comparisons in the same units:

1 meter = 10^2 centimeters = 10^6 microns = 10^9 millimicrons = 10^{10} angstroms

10^4/(wavenumber in cm^{-1}) = wavelength in microns.

It was stated above that the methods employed in an analysis depend upon the complexity of the material. It should be emphasized here that they also depend upon the purpose of the investigation, which may, for example, be satisfied with a determination of only the known major component acids. In this instance a simple analytical procedure

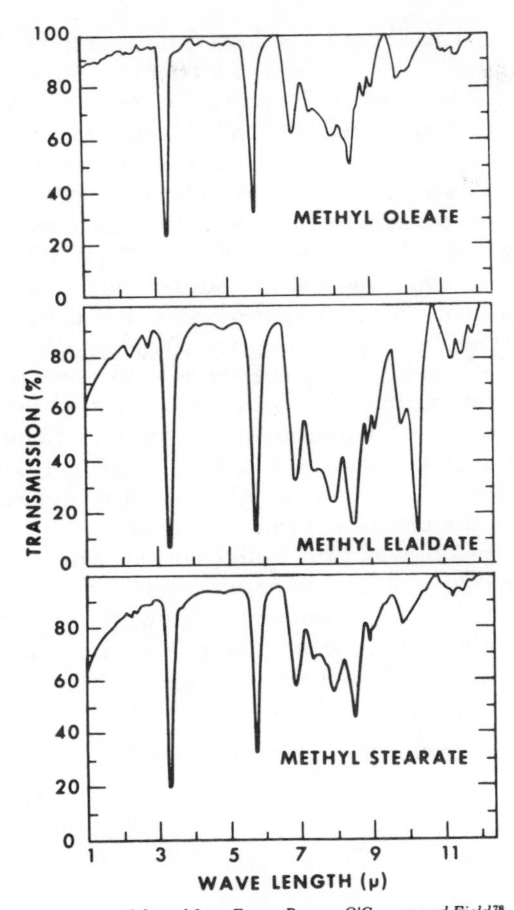

Adapted from Feuge, Pepper, O'Connor and Field[78]

FIG. 4.3. COMPARISON OF INFRARED ABSORPTION SPECTRA OF *cis*-UNSATURATED, *trans*-UNSATURATED, AND SATURATED FATTY ACID ESTERS

might well be adequate. This section is being concluded, however, with an illustration of the use of various techniques in the investigation of a complex oil in which it was felt necessary to obtain positive identification of the component acids.

In this investigation[331] it was desired to characterize completely and with certainty the unsaturated fatty acids of menhaden body oil. The acids from 500 gm of oil were fractionally crystallized from acetone at temperatures ranging down to −70°C. Those soluble at this temperature (137 gm) were esterified with methanol, and 100 gm of esters was fractionally vacuum-distilled. The fractions rich in C_{16} acids were combined and redistilled into 9 fractions. Small samples of each fraction were hydrogenated and checked for homogeneity by gas-liquid chromatography. Fractions 4 to 7 were exclusively C_{16}. Their ultraviolet spectra showed no conjugated acids; after isomerization of small samples the presence of dienoic, trienoic, and tetraenoic acids was shown.

Six gm of C_{16} esters, free from other chain lengths, was saponified and the free acids separated by countercurrent distribution into groups of acids differing in number of double bonds. Using a 200-tube apparatus and 375 equilibria, the acids were separated into 4 groups, as shown in Fig. 4.4. Aliquots of each peak were isomerized with alkali. Ultraviolet spectra showed the identity of each peak in terms of the number of double bonds. These spectra also showed that each peak was free from contamination by more highly unsaturated acids. Superimposition of these experimental curves on theoretical curves furnished additional evidence of purity. Ultraviolet analysis of the unisomerized material showed the absence of conjugated acids, and infrared analysis the absence of *trans* isomers.

Each of these peaks was thus homogeneous with respect to chain length, to the number of double bonds, and to geometrical isomers, but might contain positional isomers needing characterization of the double bond patterns. This was accomplished as follows: Oxidative ozonolysis degraded the unsaturated acids by attack at the double

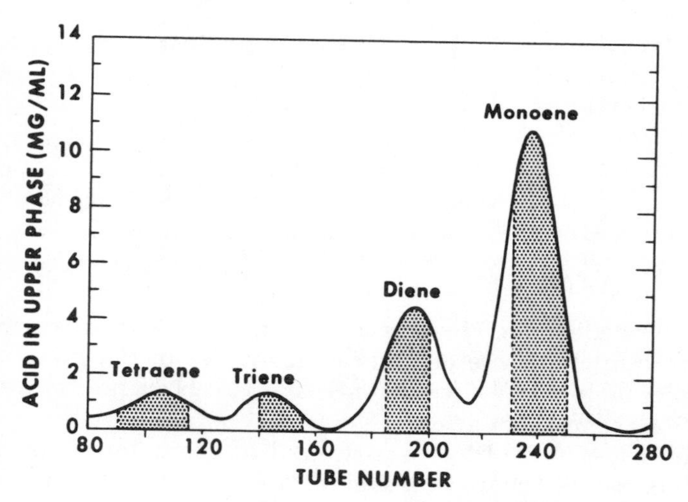

Adapted from Stoffel and Ahrens[331]

FIG. 4.4. SEPARATION BY COUNTERCURRENT DISTRIBUTION OF UNSATURATED FATTY ACIDS ON BASIS OF NUMBER OF DOUBLE BONDS

System: heptane/methyl alcohol, formamide, acetic acid.

bonds into mixtures of mono- and dicarboxylic acids. Utilizing columnar, paper, and gas-liquid chromatography, these fragments were identified and quantitated. From these data it was possible to identify the original structures. The presence of terminal unsaturations was verified by infrared spectra showing absorption maxima at 910 and 990 cm^{-1}.

Techniques additional to those mentioned above have been and are being made available for lipid research. One of particular utility is the use of silica gel or silicic acid modified by impregnation with silver nitrate. This adsorbent has been widely used in both columnar and thin-layer partition chromatography, e.g., for separating *cis-* from *trans*-fatty acid isomers,[61] methyl esters of different degrees of unsaturation,[178] and conjugatable from *cis, cis*-nonconjugatable dienoates.[156] Mass spectrometry is now used in many structural investigations. In determining the structure of the hydroxy acid components of milkfat sphingolipids, Morrison and Hay,[236] prior to mass spectrometry, converted the acids to the trimethylsilyl ethers. The resultant spectra, unique for ethers in the 2-position, identified the structure of the hydroxy acids unambiguously.

Component Fatty Acids of Cow's Milkfat.—Prior to 1951, the fatty acids found in natural glycerides had been characterized, almost without exception, by a straight chain and an even number of carbon atoms. The isolation by Hansen and Shorland[110] of 14-methylhexadecanoic acid and 15-methylhexadecanoic acid from milkfat provided acids having neither of these characteristics. These were but the first of what has become a long series of branched-chain and odd number-carbon atom acids isolated from natural fats. The following branched-chain saturated acids have also been reported to be present in milkfat: 10- and 11-methyldodecanoic acids,[309] 12-methyltridecanoic acid,[116] 12- and 13-methyltetradecanoic acids,[115,308] 14-methylpentadecanoic acid,[114] a methylheptadecanoic acid of uncertain structure,[113] and a multibranched C_{20} acid of uncertain structure.[111] Such an acid has recently been identified[23,324] as 3,7,11,15-tetramethylhexadecanoic acid. It is noted that most of these acids are branched on either the penultimate or the antepenultimate carbon atom. Those with branching on the antepenultimate carbon atom have structures meeting the essential requirement for optical activity. In each instance the form isolated has been the dextrorotatory isomer.

Unsaturated Acids.—Oleic, the most abundant of this group, is *cis*-9-octadecenoic acid. C_{10}, C_{12}, C_{14}, and C_{16} acids with a double bond at the 9-position have also been identified.[129,299] The C_{12} and C_{14} acids each had an isomer with a terminal unsaturation; but the positional isomer with internal unsaturation predominated and had both the *cis* and *trans* configuration,[211] the *cis* form predominating.[299] The C_{16} and C_{18} acids, likewise, were shown to have both *cis* and *trans* isomers.[299] *Cis*-9-heptadecenoic acid has been isolated from butterfat in a concentration of approximately 0.06%.[117] Vaccenic acid (*trans*-11-octadecenoic acid) was reported by Bertram[24] to be present in milkfat to the extent of 0.01%. More recent investigations have confirmed its presence in substantially higher concentrations.[18,96] Positional isomers (4-octadecenoic acid,[147] 13-octadecenoic acid,[147] *trans*-16-octadecenoic acid,[18] and the series of 7-to 14-octadecenoic acids[164]) have also been reported.

The presence of linoleic acid (*cis, cis*-9-12-octadecadienoic acid) in milkfat has been a subject of controversy. Hilditch and Jasperson[128] failed to obtain tetrabromo- or tetrahydroxystearic acid from the octadecadienoic acid which they identified spectrophotometrically and by other means. From chemical studies they concluded that the acid is *cis*-9, *trans*-12- and/or *trans*-9, *cis*-12-octadecadienoic acid. Shorland,[306] using what he considered a sensitive technique for detecting the tetrabromo-derivative of linoelic acid, also failed to find any evidence of linoleic acid in New Zealand milkfats. These findings differ from those of White and Brown[358] who, after a preliminary concentration of the acid by low-temperature crystallization, obtained the tetrabromo-derivative of linoleic acid in such amounts as would have been obtained if ⅔ to ¾ of the octadecadienoic acid were linoleic acid.

Shorland[306] found that the octadecatrienoic acid content of New Zealand milkfat consists almost entirely of linolenic acid, i.e., *cis, cis, cis*-9,12,15-octadecatrienoic acid. He has suggested that this finding—in contrast with that of Hilditch and co-workers, who could not isolate the hexabromo- derivative of linolenic acid in any appreciable quantity—might be explained on the basis of direct incorporation of linolenic acid present in the diet into the milkfat. Bosworth and Sisson[29] isolated octabromoarachidic acid and concluded that arachidonic acid (*cis, cis, cis, cis*-5,8,11,14-eicosatetraenoic acid) is present in milkfat.

Sambasivarao and Brown[280] have presented impressive evidence, based on the use of a variety of techniques, that both linoleic and linolenic acids are indeed present in milkfat. Fractions containing concentrates of the polyunsaturated C_{18}-acids were analyzed. The linoleic acid content of the dienoic acids averaged about 62%, that of conjugated acids about 10%, and the remainder consisted of *cis, trans* or *trans, cis* isomers of which the greater part consisted of positional isomers with more widely separated double bonds. The linolenic acid content of the C_{18}-trienoic acids averaged 75%.

Since the above material was written for the 1965 edition of this book, the list of known acids has been tremendously expanded. With only a few exceptions, Tables 4.2 to 4.6 contain all the unsubstituted fatty acids (other than the keto- and hydroxy acids) known by this author to have been reported as constituents of the neutral glycerolipids of cow's milk. The following acids also have been reported: the dibasic acids, glutaric, adipic, pimelic, and azeleic,[38] which were identified on the basis of chromatographic retention data, and 4-octadecenoic acid,[147] which was not found in a recent detailed study of the octadecenoic acids of milkfat.[121]

Identification of the acids involved a number of techniques, among which gas-liquid chromatography was invariably prominent. Because of the complexity of the total fatty-acid mixture, identification always was preceded by treatments designed to simplify the mixtures. These preliminary operations included fractional crystallization, fractional distillation, urea-complex fractionation, columnar chromatography, preparative gas-liquid chromatography, thin-layer chromatography, and argentation chromatography.

In addition to chromatographic retention data, infrared, ultraviolet, and mass spectrometry were used in final identification. Isomeric mixtures of unsaturated acids were separated into *cis* and *trans* groups by means of argentation chromatography, and the double bond positions located by oxidative cleavage followed by chromatographic analysis of the fragments.

Distribution of Fatty Acids in Triglycerides.—The manner in which these acids are associated in the triglycerides has been the subject of a vast amount of research effort. The monoacid or simple triglyceride theory postulated that each triglyceride contains only one kind of acid,

Table 4.2
NORMAL SATURATED FATTY ACIDS OF BOVINE MILKFAT [a]

Number of Carbon Atoms	Wt. %	Number of Carbon Atoms	Wt. %
2	0.09	16	23.8
4	2.79	17	0.70
5	0.01	18	13.2
6	2.34	19	0.27
7	0.02	20	0.28
8	1.06	21	0.04
9	0.03	22	0.11
10	3.04	23	0.03
11	0.03	24	0.07
12	2.87	25	0.01
13	0.06	26	0.07
14	8.94	27	< 0.01
15	0.79	28	< 0.01

[a] Data for C_2 acid from reference 119 and for C_{27} and C_{28} acids from reference 144. All other data from reference 122.

Table 4.3

MONOBRANCHED SATURATED FATTY ACIDS OF BOVINE MILKFAT

Number of Carbon Atoms	Structure	Ref.	Wt. %[a]	Ref.
9	Branched	166	< 0.01	166
11	Branched	166	< 0.01	166
12	9-Methylundecanoic	278	0.1 mole %	278
12	10-Methylundecanoic	144	0.01	144
13	10-Methyldodecanoic	309	0.05 mole %	278
13	11-Methyldodecanoic	309	0.04	144
14	12-Methyltridecanoic	116	0.03	144
15	7-Methyltetradecanoic	278	—	—
15	8-Methyltetradecanoic	278	—	—
15	9-Methyltetradecanoic	278	—	—
15	10-Methyltetradecanoic	278	—	—
15	11-Methyltetradecanoic	278	—	—
15	12-Methyltetradecanoic	115, 308	0.23	94
15	13-Methyltetradecanoic	115, 308	0.14	122
16	14-Methylpentadecanoic	114	0.20	144
17	8-Methylhexadecanoic	278	—	—
17	9-Methylhexadecanoic	278	—	—
17	10-Methylhexadecanoic	278	—	—
17	11-Methylhexadecanoic	278	—	—
17	12-Methylhexadecanoic	278	—	—
17	14-Methylhexadecanoic	110	0.23	144
17	15-Methylhexadecanoic	110	0.36	144
17	11-Cyclohexylundecanoic	288	≥0.01	288
18	16-Methylheptadecanoic	144	0.02	144
19	16-Methyloctadecanoic	278	0.3 mole %	278
20	Branched	122, 144	0.01	144
21	Branched	144	0.01	144
22	Branched	144	0.02	144
23	Branched	144	0.01	144
24	Branched	144	0.02	144
25	Branched	144	0.004	144
26	Branched	144	0.004	144

[a] The seven C_{15} acids comprised 1.20 mole % and the eight C_{17} acids 0.70 mole % of the total acids of the neutral glycerolipids. Data from reference 278.

the number of different triglycerides being equal to the number of acids. Based more on assumption than experiment, this theory is of historical interest only.

Details of earlier studies of glyceride structure have been described by Hilditch.[124] This work, to a large extent, was based on determining the distribution of fatty acids among the main glyceride types based

Table 4.4

MULTIBRANCHED SATURATED FATTY ACIDS OF BOVINE MILKFAT [a]

Number of Carbon Atoms	Structure	Wt. %
16	D,D-4,8,12-Trimethyltridecanoic	0.005
17	3-Side chains	0.01
18	3-Side chains	0.16
19	2,6,10,14-Tetramethylpentadecanoic	0.01
20	3,7,11,15-Tetramethylhexadecanoic	0.12 [b], 0.18 [c]
21	4 Side chains	0.2
22	4 Side chains	0.02
23	4 Side chains	0.01
24	4 Side chains	0.10
25	4 Side chains	0.10
26	3 Side chains	0.01
26	4 Side chains	0.04
27	4 Side chains	0.04
28	3 Side chains	0.02
28	4 Side chains	0.12
28	5 Side chains	0.01

[a] Data for C_{16} acid from reference 108, for C_{19} acid from reference 109, and for C_{20} acid from references 16, 23, and 324. All other data from reference 144.
[b] Spring fat.
[c] Summer fat.

Table 4.5

MONOENOIC FATTY ACIDS OF BOVINE MILKFAT

Acid [a,b]	Structure	Wt. %
10:1	9-Decenoic	0.27
12:1	11-Dodecenoic	} 0.14
12:1	9-*cis*-Dodecenoic	
12:1	9-*trans*-Dodecenoic	
14:1	5-*cis*-Tetradecenoic	0.0075
14:1	6-*cis*-Tetradecenoic	0.0060
14:1	7-*cis*-Tetradecenoic	0.0068
14:1	8-*cis*-Tetradecenoic	0.0045
14:1	9-*cis*-Tetradecenoic	0.72
14:1	*trans*-Tetradecenoic	0.0076
14:1	13-*trans*-Tetradecenoic	—
15:1	Pentadecenoic	0.07
16:1	5-*cis*-Hexadecenoic	trace
16:1	6-*cis*-Hexadecenoic	0.021
16:1	7-*cis*-Hexadecenoic	0.092
16:1	8-*cis*-Hexadecenoic	trace
16:1	9-*cis*-Hexadecenoic	1.46
16:1	10-*cis*-Hexadecenoic	trace
16:1	11-*cis*-Hexadecenoic	0.042

Table 4.5 (*Continued*)

Acid[a,b]	Structure	Wt. %
16:1	12-*cis*-Hexadecenoic	trace
16:1	13-*cis*-Hexadecenoic	0.0099
16:1	5-*trans*-Hexadecenoic	0.0038
16:1	6-*trans*-Hexadecenoic	0.014
16:1	7-*trans*-Hexadecenoic	0.012
16:1	8-*trans*-Hexadecenoic	0.0087
16:1	9-*trans*-Hexadecenoic	0.057
16:1	10-*trans*-Hexadecenoic	0.0013
16:1	11-*trans*-Hexadecenoic	0.0078
16:1	12-*trans*-Hexadecenoic	0.022
16:1	13-*trans*-Hexadecenoic	0.017
16:1	14-*trans*-Hexadecenoic	0.0078
17:1	6-*cis*-Heptadecenoic	0.0092
17:1	7-*cis*-Heptadecenoic	0.0057
17:1	8-*cis*-Heptadecenoic	0.054
17:1	9-*cis*-Heptadecenoic	0.19
17:1	10-*cis*-Heptadecenoic	trace
17:1	11-*cis*-Heptadecenoic	0.0078
17:1	12-*cis*-Heptadecenoic	trace
17:1	*trans*-Heptadecenoic	0.018
18:1	7-*cis*-Octadecenoic	trace
18:1	8-*cis*-Octadecenoic	0.45
18:1	9-*cis*-Octadecenoic	25.5
18:1	10-*cis*-Octadecenoic	trace
18:1	11-*cis*-Octadecenoic	0.67
18:1	6-*trans*-Octadecenoic	0.030
18:1	7-*trans*-Octadecenoic	0.024
18:1	8-*trans*-Octadecenoic	0.097
18:1	9-*trans*-Octadecenoic	0.31
18:1	10-*trans*-Octadecenoic	0.32
18:1	11-*trans*-Octadecenoic	1.08
18:1	12-*trans*-Octadecenoic	0.12
18:1	13-*trans*-Octadecenoic	0.32
18:1	14-*trans*-Octadecenoic	0.27
18:1	15-*trans*-Octadecenoic	0.21
18:1	16-*trans*-Octadecenoic	0.23
19:1	—	0.06
20:1	—	0.22
21:1	—	0.02
22:1	—	0.03
23:1	—	0.03
24:1	—	0.01
25:1	—	0.0008
26:1	—	0.0008

[a] Acid designated by number of carbon atoms, colon, number of double bonds.
[b] Data for C_{10}, three C_{12}, 13-*trans*-tetradecenoic, C_{15}, and C_{19-24} acids taken from reference 122. Data for C_{25} and C_{26} acids taken from reference 144. All other data taken from reference 121.

Table 4.6

POLYENOIC FATTY ACIDS OF BOVINE MILKFAT

Acid [a]	Structure	Ref.	Wt. %	Ref.
14:2	—	144	0.04	144
16:2	—	144	0.02	144
18:2	*cis,cis*-9,12-acid	280	2.11 [b]	122,280
18:2	*cis,cis*-9,15-acid	156		156
18:2	*cis,cis*-10,15-acid	156		156
18:2	*cis,cis*-11,15-acid	156	0.02	156
18:2	*cis,cis*-8,15 and/or 8,12-acid	156		156
18:2	*cis,cis*-7,15 and/or 7,12-acid	156		156
18:2	*cis,cis*-6,15 and/or 6,12-acid	156		156
18:2	*cis,trans* (or *trans,cis*)-11,16 and/or 11,15-acid	357		
18:2	*cis,trans* (or *trans,cis*)-10,16 and/or 10,15-acid	357		
18:2	*cis,trans* (or *trans,cis*)-9,15 and/or 9,16-acid	357		
18:2	*cis,trans* (or *trans,cis*)-8,16 and/or 8,15 and/or 8,12 acid	357[c]	0.44[b]	122,280
18:2	*trans,trans*-12,16-acid	357		
18:2	*trans,trans*-11,16 and/or 11,15-acid	357		
18:2	*trans,trans*-10,16 and/or 10,15-acid	357		
18:2	*trans,trans*-9,16 and/or 9,15 and/or 9,13-acid	357		
18:2	conjugated	122,280	0.28 [b]	122,280
18:3	*cis,cis,cis*-9,12,15-acid	280	0.38[b]	122,280
18:4	—	144	0.10	144
20:2	—	122	0.05	122
20:3	—	122	0.11	122
20:4	*cis,cis,cis,cis*-acid	122,299	0.14	122
20:5	*cis,cis,cis,cis,cis*-acid	122,299	0.04	122
22:2	—	122	0.01	122
22:3	—	122	0.02	122
22:4	*cis,cis,cis,cis*-acid	122,299	0.05	122
22:5	*cis,cis,cis,cis,cis*-acid	122,299	0.06	122
24:2	—	144	0.02	144
26:2	—	144	0.0004	144

[a] Acid designated by number of carbon atoms, colon, number of double bonds.
[b] Calculated roughly from data in references 122 and 280.
[c] Calculated roughly from data in this table and references 122 and 280.

on degree of unsaturation. These are GS_3 (trisaturated), GS_2U (disaturated, monounsaturated, GSU_2 (monosaturated, diunsaturated), and GU_3 (triunsaturated). Hilditch generalized the experimental data in a theory of "even" distribution which in essence stated that an acid tends to have as wide a distribution as possible among the triglycerides. A strict application of this theory, which Hilditch later said should not be attempted,[125] would require any acid to represent more than 33 mole% of the total acids to occur twice in a triglyceride, and more than 67 mole % to form a simple triglyceride. Sommer,[322] vigorously opposing this theory, pointed out that while the monoacid theory implies a chemical incompatibility of unlike fatty acid molecules, the theory of "even" distribution implies the converse, or an incompatibility of like fatty acid molecules. Hilditch later emphasized[125] that he never intended this rule to be more than an expression of the major tendency. More precisely, in most natural fats the content of glycerides containing one group of a particular acid reaches a maximum when that acid is exactly one-third of the total acids and, likewise, the content of glycerides containing two groups of the acid reaches a maximum when the acid is exactly two-thirds of the total acids. At these maxima about 85 to 90% of the whole fat consists of glycerides with, respectively, one or two groups of the particular acid. Glycerides containing one group of a particular acid persist, in decreasing amounts, until that acid forms about 75% of the total acids, and those containing two groups of the acid start appearing when the acid forms 20 to 25% of the total acids. Simple triglycerides start to appear when the acid forms 60% or more of the total acids.

A third theory, that of random distribution, has been defended by Sommer[322] and by many others. According to this theory, distribution of fatty acids among the glycerides is determined solely by chance and, therefore, follows the law of probability. The total number of triglycerides, including both positional and optical isomers, present in a fat is equal to the third power of the number of fatty acids present. This figure can be subdivided in such a manner as to show the different types of triglycerides included in it, thus:

(1) $G(AAA) = X$
(2) $G(AAB) = 2X(X - 1)$
(3) $G(ABA) = X(X - 1)$
(4) $G(ABC) = X(X - 1)(X - 2)$

in which, as before, X is the number of fatty acids present. The formula in (4) includes the positional isomers G(ACB) and G(CAB). The molecules represented in (2) and (4) are asymmetric and, of the number of different glycerides calculated for each of these types, one-half consists of optical isomers of the other half. X^3 represents the total number of different glycerides if both optical forms of asymmetric molecules are present; $(X^3 + X^2)/2$ represents the number of glycerides if only one optical form of each asymmetric molecule is present. Earlier calculations (This Book, 1965 edition) were made using a figure of 60 fatty acids. On this earlier basis, the 60 fatty acids would be distributed among 216,000 different glycerides including both forms of optical isomers, or 124,800 different glycerides including but one form. The random theory also permits calculation of the quantitative occurrence of the various glycerides. In the following equations *a, b,* and *c* represent, respectively, the mole fractions of acids A, B, and C. By the use of these formulas the mole fraction of any glyceride can thus be calculated. This figure multiplied by 100 gives the mole percent of the glyceride in the fat.

$$G(AAA) = a^3$$

$$\left.\begin{array}{l} G(AAB) = 2a^2b \\ G(ABA) = a^2b \end{array}\right\} = 3a^2b$$

$$\left.\begin{array}{l} G(ABC) = 2abc \\ G(ACB) = 2abc \\ G(CAB) = 2abc \end{array}\right\} = 6abc$$

Doerschuk and Daubert[68] have shown that the glycerides of corn oil do not fit the pattern of either an even or a random distribution of the fatty acids. The partial random theory of these investigators permits an acid present in less than 33 mole % to appear twice in some molecules. It also permits an acid present in 33 to 67 mole % to be absent from some molecules.

Kartha[162] analyzed for the GS_3, GS_2U, GSU_2, and the GU_3 content of 27 natural fats and studied the analyses of 46 natural fats published by others. He concluded that each natural fat has a limit, which cannot be exceeded, for trisaturated glycerides and that its glyceride composition corresponds to that calculated from a random deposition of fatty acids until this limit is reached. Any excess saturated acids are utilized in the random formation of GS_2U and GSU_2 at the expense of GSU_2 and GU_3, respectively. Hilditch[125] has criticized these conclusions. Without obtaining evidence for an even distribution, Quimby *et al.*[268] have nevertheless concluded that the fatty acids are nonrandomly distributed in animal fats. They concluded that lard is composed largely of 2-palmityl glycerides and tallows probably of 1-palmityl glycerides. Vander Wal[346] and Coleman and Fulton[51] developed the 1, 3-random-2-random theory to account for the known deviations from random theory. Jensen and Sampugna[153] have discussed the applicability of the random and 1, 3-random-2-random theories to milkfat. Thus it appears that in many instances—perhaps in all—a completely random distribution of the fatty acids has been modified biologically.

It has been shown that the fatty acids of synthetic fats are randomly distributed. Similarly, if a nonrandomly distributed natural fat is heated with a catalyst, such as sodium methoxide, the fat is resynthesized by an interesterification process into a new material in which the fatty acids are randomly distributed among its glycerides. Norris and Mattil[242] have used this process as the basis of a test for the presence or absence of random distribution in natural fats and oils. If the characteristics of the fat before and after treatment are identical, it is presumed that the natural fat follows a pattern of random distribution. If the characteristics differ, the original distribution must have been nonrandom. Interesterification of butterfat has been found to change its refractive index,[25] its melting point,[355] and other physical characteristics. It thus appears that the distribution of fatty acids in the glycerides of milkfat is nonrandom.

While the earlier studies of glyceride structure concerned the intermolecular distribution of fatty acids, that of Quimby *et al.* pertained to their intramolecular distribution. Their conclusions were based on data obtained by thermal and x-ray diffraction studies. An additional tool for the study of glyceride structure has been provided by the discovery that pancreatic lipase initially catalyzes the hydrolysis of only primary ester linkages.[217,282,337,338] This lipase has been used in studies of the positional distribution of fatty acids in cow's milk glycerides. The results show considerable conflict of detail, but support the view that at least some of the fatty acids are nonrandomly distributed.

Kumar and his associates[187,189] found that butyric acid is preponderantly in the α-positions. Entressangles *et al.*,[77] on the basis of their findings that pancreatic lipase hydrolyzes butyric acid linkages more rapidly than those of long-chain acids, questioned the validity of results obtained by applying this method to milkfat. Jack *et al.*[145] studied the method and felt that it could be validly applied to milkfat by using conditions which minimize the preferential hydrolysis of butyric acid linkages. They, like Kumar's group, found a large part of the butyric acid in the α-positions.

Several groups[14,222,256] have found a greater than random percentage of saturated acids in the ß-position of cow's milkfat. In this respect these results resemble those found for hog fat and contrast with those found for beef fat,[268] other animal and vegetable fats,[218] and seed fats.[219] Jack *et al.*,[145] however, found a less than random percentage of the saturated acids in the ß-position; palmitic acid, in contrast to the group as a whole, was preferentially located in this position.

Jensen *et al.*[155] found that pancreatic lipase shows no preference for the butyric acid linkage as such. For example, using short digestion periods, it released butyric and palmitic acids in equimolar quantities from glyceryl 1-palmitate-2, 3-dibutyrate. However, this enzyme does discriminate, with respect to the speed with which acids are released from the α-position, between molecular species. These investigators felt, accordingly, that results obtained on milk triglycerides should be viewed with reservation; those obtained with short digestion periods yield information essentially about the more rapidly hydrolyzed triglycerides; those obtained with prolonged digestion are vitiated by acyl migration, which obscures positional specificity.

A number of others[31,49,132,154,155,197,317] have confirmed the preferential location of butyric acid in the terminal positions of glycerol. In each instance, however, it was either not possible to demonstrate to the contrary, or else positive evidence was obtained for placement of part of the butyric acid in the 2-position.

Caproic acid, also, has been found in the terminal positions in greater than random proportions in milkfat[31,132] and in the milkfat fraction soluble in pentane at −60°C.[317]

Both stearic and oleic acids were found in the terminal positions in greater than random proportions in milkfat[31,132,197] and in various fractions obtained by fractionally crystallizing milkfat.[197]

Myristic acid was found in the 2-position in greater than random proportions in milkfat,[31,64,132,155,197] in various milkfat fractions obtained by crystallization,[197] by silicic acid fractionation,[64] and in the fraction soluble in pentane at - 60°C.[317]

Palmitic acid was found in the 2-position in greater than random concentrations in milkfat,[31,132,155,197] in the fraction soluble in pentane at −60°C,[317] and in various fractionally crystallized samples.[197] In one study,[64] palmitic acid was distributed almost randomly between the 2-position and the terminal positions. This fat was fractionated on silicic acid into a series of samples in which the triglycerides varied from high to low molecular weights. In the high molecular-weight samples the palmitic acid was more than random in the 2-position; in the low molecular weight samples it was less than random in the 2-position, its distribution in the 2-position decreasing continuously from high to low in the samples of intermediate molecular weight.

The distribution of fatty acids in the terminal positions has been studied further by a stereospecific analysis[37] which discriminates between positions 1 and 3 of *sn*-glycerol. In this analysis the triglycerides are partially hydrolyzed with pancreatic lipase and the resultant diglycerides isolated chromatographically. The diglycerides, consisting of a mixture of 1, 2-diacyl-*sn*-glycerol and 2, 3-diacyl-*sn*-glycerol, are treated with phenyldichlorophosphate, which reacts with the free hydroxyl groups to form phosphorylphenols. The mixture is reacted with phospholipase A, which can attack only the 1,2-diacyl-*sn*-glycero-3-phosphorylphenol to liberate the fatty acids from the 2-position. The resultant mixture thus contains unchanged 2, 3-diacylphospholipids, lysophospholipids with acids in the 1-position, and free fatty acids. After separation, analysis of these fractions and the original triglycerides gives sufficient information for designating the distribution of the acids among the three positions of *sn*-glycerol.

Pitas *et al.*[263] have stereospecifically analyzed milkfat. Breckenridge and Kuksis[32,33] have stereospecifically analyzed two fractions obtained by molecular distillation of butterfat. The so-called short-chain triglyceride fraction contained triglycerides with 34 to 44 acyl carbon atoms per molecule and most of the butyric acid in milkfat. The other fraction—the long-chain triglyceride fraction—contained triglycerides with 36 to 54 acyl carbon atoms per molecule. The fatty-acid composition of each fraction is shown in Table 4.7. The results of these stereospecific analyses have been assembled in Table 4.8.

It is obvious that, in general, the fatty acids in the terminal positions are not divided equally between the 1- and 3-positions. Of all the fatty acids, the greatest discrimination is shown in the preferential placement of butyric acid in the 3-position of *sn*-glycerol.

Stull and Brown[334] made a comparative study of the fatty acids in the milkfats of some common dairy breeds. Composite samples from each of three herds (Guernsey, Holstein-Friesian, and Jersey) were taken monthly for a period of two years while the cows were on a uniform diet at a constant energy level according to production. The Holstein-Friesian group differed significantly from the other two with

Table 4.7

FATTY-ACID COMPOSITION OF SHORT- AND LONG-CHAIN TRIGLYCERIDE FRACTIONS [a]

	Mole %	
Fatty Acid [b]	Short-Chain	Long-Chain
4:0	18.3	1.5
6:0	7.5	2.2
8:0	2.0	1.6
10:0	3.5	2.9
12:0	3.1	3.5
14:0	11.0	11.4
14:1	1.0	trace
15:0	1.8	3.8
16:0	27.8	28.2
16:1	1.6	3.0
17:0	1.3	0.9
18:0	6.7	13.1
18:1	13.0	26.1
18:2	1.0	1.2

a Data from reference 33.
b Designated by number of carbon atoms:number of double bonds.

Table 4.8

POSITIONAL DISTRIBUTION OF FATTY ACIDS IN TRIGLYCERIDES [a]

Fatty Acid [b]	1-Position			2-Position			3-Position		
	Total Fat [c]	Short-chain [d]	Long-chain [e]	Total Fat [c]	Short-chain [d]	Long-chain [e]	Total Fat [c]	Short-chain [d]	Long-chain [e]
4:0	9.8	0.0	0.0	5.6	0.0	0.0	84.6	100	100
6:0	16.3	0.0	0.0	25.8	0.0	0.0	58.0	100	100
8:0	17.2	0.0	trace	42.1	15.0	trace	40.7	85.0	<100
10:0	20.7	8.6	13.8	49.7	41.0	32.2	29.6	50.5	54.0
12:0	24.6	33.3	16.2	47.5	69.9	37.1	28.0	– 3.2	46.7
14:0	27.6	32.8	18.5	53.8	69.1	46.8	18.6	– 1.8	34.7
14:1	—	17.1	trace	—	85.7	trace	—	– 3.6	—
15:0	—	42.6	16.7	—	81.5	32.5	—	–24.1	50.9
16:0	45.5	59.3	44.8	41.7	44.8	45.7	12.8	5.9	9.5
16:1	48.5	39.6	31.1	35.6	64.6	42.2	15.9	– 4.2	26.7
17:0	—	84.6	76.1	—	33.3	38.1	—	–17.9	–14.3
18:0	58.2	73.6	48.1	25.9	17.4	21.6	16.0	9.0	30.3
18:1	41.8	50.8	32.8	27.9	30.3	24.8	30.3	19.0	42.3
18:2	36.7	33.3	27.0	77.9	40.0	44.4	–14.6	26.7	27.8

a Percentage of each acid in designated position of *sn*-glycerol.
b Designated by number of carbon atoms:number of double bonds per molecule.
c From reference 263.
d Calculated from mole % data in reference 32.
e Calculated from mole % data in reference 33.

respect to five acids. The milkfat of this group had smaller proportions of capric and lauric acids and larger proportions of palmitic, palmitoleic, and oleic acids. The magnitude of these differences, while significant, was not considered large enough to affect the gross characteristics of the milkfats.

Effect of Species Variation on Fatty Acids.—The various species in Table 4.9 are grouped so as to reflect the extent to which ingested food is subjected to microbial activity in the digestive tract. This is greatest in the ruminants and decreases in subsequent groups; it is probably at a minimum in carnivores. Odd-carbon acids, branched-chain acids, *trans* unsaturations, and conjugated double bond systems are all formed by microbial activity. Such acids in an animal's milkfat may have been formed by activity in the animal itself, or may have been ingested in the animal's food.

Fatty acids of grass lipids are chiefly C_{16} and C_{18}.[311] The fatty acids of the blood glycerides in a cow were found[205] to be chiefly C_{16} and C_{18}. The dietary fatty acids in a standard ration fed rats (Table 4.9) were chiefly C_{16} and C_{18}. As a basis for discussion, we can consider the group of acids from C_4 to C_{14} inclusive to have resulted primarily from mammary gland synthesis. Taking the average composition of the milkfat of each group in Table 4.9, these mammary gland-synthesized acids form the following percentages of the total acids: ruminants, 24.5; nonruminant herbivore, 22.0; omnivores, 18.8. Many individuals species show a greater similarity to species in other groups than do the group averages. It is apparent that the greatest intragroup differences occur among the omnivores. It is also apparent that individual fatty acids vary more widely between species and groups than do the synthesized acids as a whole. The C_4 and C_6 acids are most abundant in the ruminants and decrease in subsequent groups.

The unusually high polyunsaturated acid content of horse milkfat may be due to the availability of an abundance of dietary polyunsaturated acids which are subjected to only limited microbial hydrogenation. Some interesting studies of variations in milkfat-acid composition might be based on the circumstance that the horse has less microbial activity than ruminants, while most other variables such as diet, climate, freedom of m ovement, etc., could be controlled to be either the same, or to vary independently.

The data of Glass *et al.*[101,102] cover the component acids of the milkfat of 91 species representing 14 orders. Additional species not included in these studies are the Eland antelope (*Taurotragus oryx*),[50] the grey (Atlantic) seal (*Halichoerus grypus*),[1] the collared peccary (*Pecari tajacu*),[333] the kangaroo rat (*Dipodomys merriami*),[333] the bats (*Laptomycteris nivalis* and *Tadarida brasiliensis*),[333] the fin whale

Table 4.9

COMPONENT ACIDS (WEIGHT PERCENT) OF THE MILKFAT OF VARIOUS SPECIES

Group Species	Saturated								Odd Carbon	Branched Chain	Mono-ethenoid	Poly-ethenoid	*Trans*	Conju-gated	Ref.
	C_4	C_6	C_8	C_{10}	C_{12}	C_{14}	C_{16}	C_{18}							
Herbivore															
Ruminant															
Cow [a]	2.8	2.3	1.1	3.0	2.9	9.0	24.0	13.2	3.7	1.5	33.3	3.8	5.	1.1	122
Sheep [b]	4.0	2.1	2.0	6.0	2.8	5.3	17.5	15.6	1.5[i]	—	38.6	5.6	12.1	2.7	95
Goat [c]	3.0	2.5	2.8	10.0	6.0	12.3	27.9	6.0	0.9[l]	—	25.1	3.8	—	—	127
Buffalo [d]	3.0	0.2	1.4	1.7	2.6	9.2	37.7	11.5	1.1[l]	—	30.5	0.9	—	—	6
Nonruminant															
Horse [e]	0.4	0.9	2.6	5.5	5.6	7.0	16.1	2.9	trace[i]	—	29.9	28.8	—	—	127
Omnivore															
Rat [f]	—	—	0.4	8.3	11.5	16.1	27.1	3.0	0.4[i]	—	13.9	9.4	—	0.9[j]	22
Hog [g]	[		1.3		]	1.5	26.9	6.5	<0.8[i]	—	45.0	18.8	—	—	59
Human [h]	[	0.3	]	1.63	6.89	8.5	20.9	7.3	2.7	1.2	39.5	10.1	1.0–4.0[k]	—	141
Carnivore															
Cat				1.0	1.9	7.2	24.6	10.4	2.0	—	42.5	10.5	—	—	102
Dog				0–1.5	trace to 1.0	4.0	27.3	4.4	trace to 0.8	—	48.2	14.5	—	—	102

[a] Ayrshire/Philadelphia, Pa./May/grain-bran-molasses supplement (1 lb/3 lb milk produced) + timothy hay ad libitum, 4-hr pasture daily.
[b] Romney/Palmerston North, New Zealand/Sept. (early spring) 4 weeks postpartum/pasture of ryegrass and white clover.
[c] Norfolk, England/hay, kale, beet, 1–1.5 lb peanut cake, 2-hr pasture.
[d] Murrah/India/wheat bran, grain, peanut cake, grain husk (4.0:2.0:1.5:2.5) 1 lb/2 lb milk + 70 lb green grass + 3 lb straw.
[e] Shire/Cheshire, England/summer pasture.
[f] Wistar rats/Ottawa, Canada/12–15 days post partum/fox cubes with 3% fat (fatty acids:some 14:0, 18:0, 18:3; chiefly 16:0, 18:1, 18:2).
[g] No background information given.
[h] Regular hospital and home food taken ad libitum/11 mothers from 9 to 180 days post partum.
[i] Reference 102. The rat was Norway rat.
[j] Reference 149.
[k] Reference 165.
[l] Reference 88. Expressed in mole %.

(*Balaenoptera physalus*),[2] the African elephant (*Loxodonta africana*),[223] the baboons (*Papio anubis, P. cynocephalus,* and *P. papio*),[42] the rabbit (Comune Migliorata breed),[281] and the hooded seal (*Cystophora crista*).[148]

The milkfat of the Eland antelope contrasted to that of other ruminants by having only 0.9% butyric acid and 0.5% caproic acid. The milkfat of the African elephant had the unusually high caprice acid level of 64.5%. The milkfat of the baboon was unusual in containing 37% linoleic acid; this was attributed to the high level of corn oil in its diet.

Freeman *et al.*[88] examined the milkfats of the cow, sheep, Indian buffalo, goat, and human. They found that butyric and caproic acids were predominantly in the terminal positions of glycerol in the milkfat of all these species. Myristic and pentadecanoic acids were preferentially in the 2-position and stearic acid in the terminal positions of all fats. Human milkfat was notable in having a higher proportion of palmitic acid in the 2-position and of stearic acid and oleic acid in the terminal positions than did the milkfat of any of the ruminants.

Dimick *et al.*[64] found that in both cow's and goats' milkfat palmitic acid was placed preferentially in the 2-position in the high molecular-weight triglycerides and preferentially in the terminal positions in the low molecular-weight triglycerides.

Duncan and Garton[70] examined the milkfat of the cow and found palmitic acid preferentially in the 2-position and stearic, oleic, and linoleic acids preferentially in the 1,3-positions.

Breckenridge *et al.*[34] found that the saturated triglycerides of human milkfat consisted chiefly of *sn*-glycerol-1-stearate-2-palmitate esterified at the 3-position with C_{14} to C_{18} fatty acids.

Marai *et al.*[212] stereospecifically analyzed goat and sheep milkfats after separation into short-chain (34 to 44 acyl carbon atoms) and long-chain (40 to 54 acyl carbon atoms) fractions by means of silicic acid chromatography. Each fraction formed about 50% of the total fat. Of the 20 to 23 mole % of C_4 to C_8 fatty acids present in the short-chain fraction, at least 95% were in the 3-position of sn-glycerol in each species. The distribution of the other acids showed a less marked specificity for either the 1- or the 2-position.

The specific placement of the fatty acids could best be accounted for by assuming a common pool of long-chain 1,2-diglycerides which served as precursors of the bulk of both short- and long-chain triglycerides during milkfat synthesis. This concept is in agreement with the results of Kinsella and McCarthy,[174] which suggested that the milkfat triglycerides were produced by the action of 1,2-diglycerides with endogenously synthesized fatty acids. The mechanism of biosynthesis of the 1,2-diglycerides could not be inferred from these results. It is noted in this connection that myristic acid, an endogenously synthesized acid, was found preferentially in the 2-position in the milk fat of both these species as well as in that of the cow.[32,33]

Effect of Composition of Diet on Fatty Acids.—Kaufmann *et al.*[165] found that the *trans* unsaturated fatty acids in human milkfat vary considerably with the nature of the diet. *Trans* acids could no longer be detected after 9 to 11 days of a *trans* acid-free diet in milkfat containing 2% *trans* acids on the fifth day of the diet. In experiments with five cows the milkfat produced from grass in September contained 6.3 to 9.0% *trans* acids, but in winter when beet leaves, beet, hay, and concentrates were fed the content decreased to 0 to 1.7%.

Hilditch[124] has summarized the results of a number of investigations in which specific oils have been added to the diet of cows. Other studies have concerned the effect of added oils on the composition of the milkfat of buffaloes[6] and of rats.[22] The daily inclusion in the diet of 7 lb coconut oil cake, rich in lauric and myristic acids, almost doubled these acids in cow's milkfat. Buffaloes' milkfat was affected in a similar but less pronounced fashion. A variety of oils rich in linoleic acid or linolenic acid has failed to appreciably increase the content of either of these acids in the milkfat of cows or buffaloes.

Rapeseed oil contains about 15% linoleic acid and as much as 50% erucic (13-docosenoic) acid. Addition to the diet of 4 oz daily resulted in the incorporation of 3.7% erucic acid and no detectable linoleic acid into cow's milkfat. In contrast, rats, fed a diet of fox cubes supplemented with 20% rapeseed oil, produced milk with a substantial transfer of acids from the oil into the milkfat.[22] The following figures designate, in order, the acid, its percentage content before, and that after the addition of rapeseed oil to the rats' diet: 18:1, 14—35; 18:2, 9—21; 18:3, 0.5—5; 20:1, 0—5; and 22:1, 0—7. When their diet was supplemented with 20% corn oil rich in linoleic acid, rats produced milkfat in which the linoleic acid content was raised from 9 to 40.5%.

The following percentage composition of cod-liver oil unsaturated acids has been reported: 16:1, 10—20; 18:1 + 18:2, 25—31; 20:3, 25—32; 22:3 + 22:4, 10—20. When 4 oz daily was added to cow's diet the fat content of the milk was reduced, the lower fatty acids were reduced to about half their normal proportions, oleic acid was proportionately increased, and the C_{20} and C_{22} unsaturated acids were greatly increased from a normal value of about 1% to about 5%.

A linseed oil supplement, increased from 260 gm per day to 610 gm per day over a 30-day period, decreased the percentage of C_4 to C_{16} acids and increased that of stearic, oleic, linoleic, and linolenic acids in cow's milkfat.[279]

Kuzdzal[194] established that the levels of conjugated dienes in cow's milkfat are directly related to the levels of dietary linolenic acid.

The addition of 15% safflower oil (containing 81% linoleic acid and no palmitoleic acid) to the diet of cows depressed the production of milkfat and changed its composition.[254] The percentage of all saturated acids was decreased and that of unsaturated acids, as a group, nearly doubled, with individual acids (designated as chain length:number of unsaturations) as follows: 18:1 from 27.4 to 46.6%; 18.2 from 1.9 to 2.9%; 18:3 from 1.1 to 2.9%; 16:1 (absent in oil) from 1.7 to 2.8%.

Many of these findings can be explained on the basis of the extensive hydrogenation of unsaturated acids known to take place in the rumens of ruminants. The unusual effects of cod-liver oil supplements are in accord with the hypothesis that the presence of this oil can alter normal ruminal microbial activity. The substantial incorporation of C_{20} and C_{22} polyunsaturated acids suggests less active hydrogenation by ruminal microbes, while the lower content of volatile acids in the milkfat suggests an interference with the normal production of volatile acids in the rumen.

Shaw and Ensor[304] found that when cows were fed a normal diet supplemented with 300 ml daily of cod-liver oil, oleic acid, or linoleic acid, the ratio of propionic to acetic acid in the rumen was greatly increased. The absolute amount of acetic acid was only slightly decreased by the cod-liver oil and linoleic acid supplements and was actually increased over 75% by the oleic acid supplement. In each instance, however, the amount of propionic acid was greatly increased. The volatile fatty acid components of a cow's milkfat were substantially reduced by the inclusion of this same oleic acid supplement. The reduction was greatest in butyric and caproic acids, which were reduced to about one-half the amounts present when the diet did not include an oil supplement. In each of these instances there was also a reduction in the percentage of milkfat in the milk.

Similar changes in fat content, fat composition, and ruminal volatile acids have been found when cows are fed special diets composed chiefly of cooked concentrates supplemented with small amounts of ground and pelleted hay.[47,76,304] The fineness to which hay is ground can in itself affect the proportions of fatty acids in milkfat.[319] Smith *et al.*[315] have reported on the effect of a variety of hays on some of the major fatty acids in milkfat.

Virtanen[351] has described "zero" milk from cows fed on synthetic, protein-free diets in which all the nitrogen is supplied by urea and ammonium salts. Although the total blood lipids are less than one-third the level in cows on normal diets, "zero" milk contains at least as much fat as does normal milk. The fatty-acid composition of this fat, however, is quite distinctive. From Table 4.10 it is clear that "zero" milkfat has unusually high levels of lauric, palmitic, and palmitoleic acids and unusually low levels of stearic and oleic acids. The levels of branched acids are also distinctly higher in "zero" milkfat.

These milk samples were taken from cows which had undergone a period of adaptation during which the microbial population of the rumen became adapted to the synthesis of the essential amino acids. After adaptation, protozoa had disappeared from the rumen while the number of bacteria had increased enormously to about 50 times that of cows on normal diets.

Table 4.10

FATTY ACID COMPOSITION OF MILKFAT FROM COWS ON NORMAL AND PROTEIN-FREE FEEDS[a]

	Protein-free Feed			Normal Feed
	Cow 1[c]	Cow 2[d]	Cow 2[e]	Herd
Fatty acid[b]	% of Total Fatty Acids			
4:0	2.58	2.63	3.05	3.05
5:0	0.12	0.13	0.05	< 0.02
6:0	1.88	1.90	1.75	1.76
7:0	0.14	0.12	0.05	< 0.02
8:0	1.30	1.08	1.12	1.02
9:0	0.18	0.12	0.06	< 0.02
10:0	3.63	2.38	2.73	2.13
10:1	0.54	0.44	0.40	0.25
11:0	0.30	0.22	0.14	< 0.02
12:0	5.41	3.64	3.74	2.65
12:1	0.20	0.19	0.18	0.08
13:0 br	0.20	0.15	0.17	0.03
13:0	0.55	0.51	0.26	0.04
14:0 br	0.18	0.32	0.21	0.09
14:0	11.99	10.17	10.95	987
14:1	1.63	2.63	1.95	1.01
15:0 br	0.80	0.52	0.82	0.50
15:0	4.09	4.88	2.70	0.90
16:0 br	0.11	0.48	0.33	0.19
16:0	40.29	48.16	38.90	28.41
16:1	3.54	4.29	4.22	2.03
17:0 br	1.08	0.94	0.96	0.58
17:0	1.21	1.57	0.90	0.46
17:1	0.77	1.13	0.77	0.39
18:0	1.93	0.98	2.58	12.29
18:1	13.18	8.35	19.14	29.37
18:2	1.48	1.40	1.15	1.79
18:3	0.68	0.67	0.74	1.11

[a] From reference 351.
[b] Designated by number of carbon atoms:number of double bonds; br is branched.
[c] Daily diet included 36.6 gm olive oil, 18.4 gm corn oil, and 18.6 gm linseed oil.
[d] Daily diet included 0.0 gm olive oil, 18.4 gm corn oil, and 18.6 gm linseed oil.
[e] Daily diet included 36.6 gm olive oil, 46.0 gm corn oil, and 46.5 gm linseed oil. Different lactation cycle from preceding.

Changes in ruminal volatile acids are partly due to changes in the proportions of various species of ruminal microbes, but may also be partly due to an altered ruminal environment affecting the proportions of the end products formed by individual species.[40] The reasons for these effects have not been completely elucidated. Certain factors, such as the concentration of iron, the presence of oxidizing salts, and the pH are known to affect the proportions of end products.

Leffel[201] incubated whole rumen fluids with various concentrations of C^{14}-labeled glucose. When the glucose was added in trace amounts its label was distributed uniformly among the volatile fatty acids. When the glucose was added in large amounts (20 to 40μ-moles/ml of incubate) the resultant propionic acid contained from 40 to 60% of the activity present in the volatile fatty acids. This was true whether the rumen fluids were obtained from animals on a high roughage, acetate-producing diet, or on a high concentrate, propionate-producing diet. From these results it was concluded that the concentration of glucose at any particular time influences the volatile fatty acids formed, and that the high production of propionic acid observed in animals on high concentrate diets can be accounted for, at least in part, by the rapid rate at which concentrates are fermented.

Van Soest has reviewed[347] theories attempting to explain how these special diets depress the milkfat concentration. Several of these theories lead to situations resulting in an eventual shortage of β-hydroxybutyric acid, the acid which Avena and Kumar[15] have shown to be especially important in the production of butyric and caproic acids within the mammary gland. It was noted just above that these are the acids that are especially depressed by a dietary oleic acid supplement. Higher levels of propionic acid lead to the formation of more glucose by the animal's metabolism. High levels of glucose are known to depress the oxidation of fat by the liver. Thus, less β-hydroxybutyric acid is formed and less of this important material is available to the mammary gland for synthesis of the lower fatty acids.

Plane of Nutrition.—The plane of nutrition can have a pronounced effect on the composition of milkfat. In Table 4.11 a comparison is made of milkfats produced just before and during a period in which food was withheld completely from the animals. Cow number 1 seemed to present a case of simple inanition. The milkfat yield, on the 11th and 12th days of fasting, was 40% of the original value. With cow number 2 the data are complicated by the development, on the second day of fasting, of severe symptoms of "milk fever" which, however, responded rapidly to treatment. During this complication the secretion of milk and of milkfat almost ceased. By the end of the experiment the fat production had risen to 20% of its original value. These data show that fasting produces effects which are similar to, but more pronounced than those of the special diets described above. All the lower acids, butyric through myristic, i.e., the group previously referred to as arising chiefly from mammary gland synthesis, are greatly reduced in percentage.

Effect of Miscellaneous Factors on Fatty Acids.—Analyses of English milkfats have shown typical changes when the animals go from winter stall feeding to pasture grazing. These changes normally include a sudden increase in the iodine value and, as shown in Table 4.12, an increase in the oleic acid content of the fat. The stall-fed animals referred to in this table had received the usual winter diet of that period consisting largely of hay, roots, and farm-grown concentrates. The silage-fed group had been receiving only hay and silage. The other two groups had been feeding on fresh grass of early and late summer. From the fact that the acid composition of the hay and silage-fed group was similar to that of typically winter-fed animals, Hilditch and Jasperson[126] concluded that the observed differences between winter and summer milkfats cannot be directly related to

Table 4.11

EFFECT OF FASTING ON COMPOSITION (WEIGHT PERCENT) OF MILKFAT ACIDS[a]

Acid	Cow No. 1		Cow No. 2
	Before Fasting	During Fasting[b]	During Fasting[c]
Butyric	3.5	1.2	2.7
Caproic	0.6	—	0.1
Caprylic	1.0	0.1	0.1
Capric	1.8	0.2	1.0
Lauric	2.5	0.1	0.6
Myristic	11.9	2.8	3.8
Palmitic	23.5	20.0	22.1
Stearic	11.6	14.3	9.9
Arachidic	1.1	0.9	0.9
9-Decenoic	0.2	—	0.2[d]
9-Dodecenoic	0.2	—	0.2[d]
9-Tetradecenoic	0.9	0.4[d]	0.4[d]
9-Hexadecenoic	3.2	1.4	2.0
Oleic	35.9	52.8	51.7
Octadecadienoic	1.2	2.5	0.8
C_{20} unsaturated	0.8	3.3	3.5

[a] Data from reference 314.
[b] Sample from mixed fat secreted on 11th and 12th days of fasting.
[c] Sample from mixed fat secreted on 7th to 12th days of fasting.
[d] Considered by Smith and Dastur to be somewhat higher than true values.

Table 4.12

SEASONAL VARIATION (WEIGHT PERCENT) OF COMPONENT ACIDS OF MILKFAT

Acid	Stall-Fed Winter[a]	Silage-Fed Winter[a]	June Pasture[b]	August Pasture[b]
Butyric	3.0	3.6	3.7	3.5
Caproic	1.4	2.0	1.7	1.9
Caprylic	1.5	0.5	1.0	0.7
Capric	2.7	2.3	1.9	2.1
Lauric	3.7	2.5	2.8	1.9
Myristic	12.1	11.1	8.1	7.9
Palmitic	25.3	29.0	25.9	25.8
Stearic	9.2	9.2	11.2	12.7
As Arachidic	1.3	2.4	1.2	1.5
9-Decenoic	0.3	0.1	0.1	0.1
9-Dodecenoic	0.4	0.1	0.2	0.2
9-Tetradecenoic	1.6	0.9	0.6	0.6
9-Hexadecenoic	4.0	4.6	3.4	2.4
Oleic	29.6	26.7	32.8	34.0
As Octadecadienoic	3.6	3.6	3.7	3.7
As C_{20-22} unsaturated	0.3	1.4	1.7	1.0

[a] Data from reference 129.
[b] Data from reference 126.

dietary changes. They postulated that these seasonal differences may result from differences in freedom of movement and exercise between stall-fed and pasture-grazing animals.

Similar changes in fatty-acid composition have been observed in other parts of the northern hemisphere. Jensen *et al.*[152] found that Connecticut milkfat was richer in oleic and stearic acids and poorer in myristic acid in summer than in winter. Patton *et al.*[260] compared the acid composition of late winter (March) and early summer (June) milkfats. They found the following percentage increases in the summer fat: 18:0, 4.2; *cis*-18:1, 4.4; *trans*-18:1, 3.9; 18:2 + 3, 0.4. These increases were mainly at the expense of palmitic acid (9.0% loss) and myristic acid (2.9% loss).

Milkfat samples from both southern and central Alberta were examined by Wood and Haab.[360] The iodine value reached a maximum in May in the southern fat and in September in the northern fat. It reached minimum values in November to December in both regions. Oleic acid reached maximum values in May and minimum values in November in both regions. The nonconjugated polyunsaturated acids did not vary geographically. The dienoic fraction did not vary seasonally. The trienoic fraction reached a maximum in July. The conjugated polyunsatuated acids had maximum values during the grazing season, May to September, and minimum values during the stall-feeding period.

The iodine value of butterfat in France[196] was found to be higher throughout the year in evening milk (36.9 average) than in morning milk (34.2 average). The nonconjugated diene and tetraene acids were relatively constant throughout the year. They averaged 1.1 and 0.34 gm/100 gm fat, respectively. The conjugated diene acids increased during the grazing period, late April to November. They averaged, in the winter, 0.5 and in the summer, 1.6 gm/100 gm fat. Nonconjugated triene, i.e., linolenic acid, averaged in the summer 0.64, and in the winter 0.40 gm/100 gm fat. This acid was also slightly higher in evening than in morning milkfats.

The relationships between season of the year and milkfat composition in New Zealand are quite different from those recorded in the northern hemisphere. Hansen and Shorland[112] examined bimonthly samples of milkfat from cows grazing throughout the year on pasture which reaches its maximum lushness within the months of September to January and slowly declines thereafter. The lactation period starts about the middle of July, reaches a peak in November, and ends in May. During the winter the cow's diet is normally supplemented with hay or silage.

The analyses of Hansen and Shorland showed that the iodine values were highest in July and September and dropped sharply to a minimum in November. The greatest changes in fatty-acid composition also occurred in the November samples. The C_{18} unsaturated acids dropped from 29.5 mole % at the beginning of lactation to 23.4 mole % in November. The C_6 to C_{14} acids increased during this same period, from 18.6 to 26.8 mole %. The seasonal changes were interpreted as resulting from changes in the plane of nutrition and the stage of lactation. Later analyses of sheep milkfat[95] from sheep grazing under similar conditions showed that similar fatty-acid compositional changes took place at the same time of year as those in cow's milkfat. Since the stages of lactation were completely different in the two species, it was concluded that the plane of nutrition is the most important factor in producing seasonal changes in milkfat composition.

In French butterfat,[195] 2.3% of the total acids were *trans* in March (in stalls) and 7.0% were *trans* in May (on pasture). Vaccenic was the major *trans* acid. Pentadecenoic acid was mostly *trans*, hexadecenoic acid was 15%, and myristoleic and heptadecenoic acids each less than 15% *trans*. Antilla[11] has reported on the seasonal variation of conjugated, *trans*, and branched acids in Finnish butterfat.

No significant difference was found in the fatty-acid composition of morning and evening milks.[272,335] A number of studies on cow's milk[3,57,272,335,336] and one on human milk[270] showed significant increases in the short-chain acids from colostrum through at least 8 weeks of lactation. The results of Hosogai[139] were in conflict with these findings.

The effect of high temperatures and humidity on milkfat composition was studied by Richardson *et al.*[273] The experiments were designed to maintain all other conditions constant. It was found that typical changes in fatty-acid composition took place when any combination of high temperature and high relative humidity resulted in a rectal temperature of 103 to 105°F. This situation was marked by a relative decrease in the low-melting acids, short-chain and oleic, and a relative increase in the higher-melting acids, chiefly palmitic and stearic.

Similar effects were found by Moody *et al.*[231] Comparisons were made between one group of cows held two weeks at 15 to 24°C and another group at a constant 32°C and 60% relative humidity. The milkfat of the latter group had depressed levels of 12:0, 14:0, 14:1 + 15:0, 18:1, and 18:2 acids. The lower acids were not determined.

In concluding this section on factors affecting the acid composition of triglycerides, it is noted that apparently unrelated data may sometimes be unified or what seem to be contradictory results resolved if coordinating data are available on the animal's metabolism and that of the rumen flora. Two examples come to mind. In the first, both starvation and high-concentrate diets, representing the two extremes of the plane of nutrition, lower the relative proportion of the C_4 to C_{14} acids in milkfat. Because of the important relationship of β-hydroxybutyric acid to the C_4 and C_6 acids,[15] a shortage of this precursor might adequately explain the preferential depression of these acids in the milkfat of cows on a high concentrate diet. During starvation there is no relative shortage of β-hydroxybutyric acid. However, in addition to a shortage of acetic acid, there appear to be metabolic changes which interfere with the mammary gland's ability to use either of these substrates normally for fatty-acid synthesis.[208]

The second example concerns the varying extent to which polyunsaturated acids have been found to be incorporated into milkfat. It is well known that in normal diets extensive ruminal hydrogenation acts as a barrier to the passage of these acids into milkfat. Since neither these acids nor their parent oils are water-miscible, the extent to which they undergo ruminal hydrogenation must be influenced by the manner in which they are presented to ruminal action. For example, a single massive dose of an oil probably would be less completely hydrogenated than the same amount administered in a number of smaller portions.

Mochrie and Tove[230] have presented data demonstrating the

extremes of this principle. On the one hand, soybean oil was fed as the whole bean so that the oil, by a slow release from the bean, would be subjected to a maximum of microbial action. On the other hand, cottonseed oil was infused into an animal intravenously so that ruminal action would be completely bypassed. Linoleic acid from the soybean feeding increased in the milkfat from 2.1 to 3.2%; that from the cottonseed oil infusion increased to nearly 15%.

Scott *et al*[297,298] have attempted to bypass ruminal hydrogenation in a way that would be practical for producing milk with a softer fat and higher levels of polyunsaturated acids. Particles containing a core of polyunsaturated oil encapsulated with protein were prepared by spray drying a homogenate of sodium caseinate and the desired oil. The particles were then treated with formalin to protect the protein coat from microbial proteolysis. When these protected oils were fed to ruminants they largely escaped hydrogenation in the rumen; the formaldehyde-protein complex was hydrolyzed in the acidic conditions of the abomasum, and the fatty acids were absorbed from the small intestine. The levels of octadecatrienoic acid were raised to from 12 to 22% in the milkfat of cows and to from 21 to 25% in that of goats with encapsulated linseed oil. The levels of octadecadienoic acid in the milkfat of cows were raised to 25% with sunflower oil, 20% with corn oil, 21% with peanut oil, and 35% with safflower oil. Safflower oil also raised the level of this acid to 35% in the milkfat of a goat.

McCarthy[221] has stated that milkfat from a cow on a high-concentrate diet contained the unusually high levels of 8.6% linoleic acid and 2.3% linolenic acid. Both high-concentrate diets and added oils have been shown to affect the formation of volatile fatty acids in the rumen. The extent to which this is associated with metabolic changes that also affect the capacity for hydrogenating fatty acids is not known. It should be considered in relation to the effects of both high-concentrate diets and added oils.

Keto Acid-containing Triglycerides

β-Keto Acid Glycerides.—A homologous series of methyl ketones with an odd number of carbon atoms, C_3 to C_{15} inclusive, has been identified in the unsaponifiable matter of milkfat.[260a] The concentration of these ketones is increased by heating. This circumstance has led to the postulation of glyceride-bound β-keto acids as their precursor.[28] Support for this hypothesis was obtained by the isolation of pyrazolones from butterfat treated with Girard T reagent.[349] Also, synthetically prepared glycerides containing β-keto acids reacted similarly, forming methyl ketones on heating and pyrazolones on treatment with Girard T reagent. Parks *et al.*[249] demonstrated the validity of this hypothesis by isolating the precursor and identifying it as a triglyceride containing one β-keto acid.

The formation of methyl ketones from precursors is a two-step process: (a) hydrolysis of the ester linkage binding the β-keto acid in the triglyceride, and (b) decarboxylation of the unstable free acid. The hydrolytic step requires water,[199,293,349] and is completely inhibited if the milkfat is adequately dried, as with calcium carbide. Both steps are accelerated at elevated temperatures. Langler and Day[199] obtained maximum ketone formation by heating the fat for 3 hr at 140°C. Obviously, the β-keto ester linkage is much less stable than that of the fatty acids, since complete hydrolysis is obtained under conditions which leave the bulk of the milkfat intact.

Parks *et al.*[249] found that the β-keto acid-containing glycerides represent 0.045% of the milkfat and have fatty-acid components typical of those in milkfat, except for a slightly higher palmitic acid content. The β-keto acid moieties had the following composition, expressed in mole %: C_6, 11.5; C_8, 25.7; C_{10}, 13.8; C_{12}, 11.4; C_{14}, 13.2; C_{16}, 24.4.

The methyl ketone potential of milkfat can vary widely. Parks *et al.*[249] found 0.58 μmoles, Schwartz *et al.*[292a] 0.4 to 1.2 μmoles, and Langler and Day[199] 1.7 μmoles per gm fat.

Dimick and Walker[66] studied the effect of a number of factors on the monocarbonyl potential. As the methyl ketones generated by heating the moist fat were found to compose from 70 to 90% of the monocarbonyls, the conclusions regarding monocarbonyls can, in a general way, be applied to the methyl ketones, and hence to the β-keto acid-containing glycerides.

Samples taken weekly from herd milk over a period of one year showed that the monocarbonyl potential followed a seasonal trend, being higher in the winter (while on barn feeding) and lower in the summer (while on pasture feeding). The monocarbonyl potential correlated, in this respect, with the lactone potential[63] and the levels of short-chain fatty acids.

Studies on three groups of cows, Holstein, Brown Swiss, and Jersey-Guernsey, showed no significant correlation between monocarbonyl potential and breed of cow. This was also true of the lactone potential.[63]

The effect of the stage of lactation on monocarbonyl potential was studied with individual cows throughout a complete lactation period. Both monocarbonyls and methyl ketones (which were assayed separately in this study) increased with the progress of lactation. The lactone potential likewise increased.[63]

Analyses of the milkfat from high fat- and low fat-producing cows showed no significant difference in the monocarbonyl potential.

Ketosis produced low levels of the C_5 to C_{15} methyl ketones.

Other Keto Acid Glycerides.—Vander Ven[348] removed the β-keto acids from milkfat by steam deodorization and then extracted a mixture of other keto acids with Girard T reagent. After reduction of the keto groups to hydroxyl groups, the 4- and 5-hydroxy acids were converted to lactones, which were identified by gas liquid chromatography. The lactones, corresponding to 2.5 ppm in the fat, consisted mainly of δ-octa, δ-deca, and δ-dodecalactones with lesser percentages of γ-deca, γ-undeca, and γ-dodecalactones. From these results the presence of the corresponding 5- and 4-keto acid-containing triglycerides can be inferred.

Keeney *et al.*[170] studied the total keto acids of milkfat and found them to consist chiefly of isomeric ketostearic acids with low percentages of C_{10} to C_{16} saturated keto acids. Unsaturated C_{18} keto acids were also found. The percentage composition of the ketostearic acid fraction was as follows: 8-ketostearic, 1.2; 9-ketostearic, 39.4; 10-ketostearic, 27.5; 11-ketostearic, 10.1; 12-ketostearic, 4.6; and 13-ketostearic acid, 17.3. The levels of keto acid-containing triglycerides in Table 4.1 were calculated from the data obtained in this study.[168] These triglycerides, with the exception of the diglycerides and possibly of the phospholipids, are the most abundant, on a weight basis, of the minor components of milkfat. The keto acid components of these glycerides form about 0.3 to 0.5% of the component acids of milkfat.

Hydroxy Acid-containing Triglycerides

Lactonogenic 4- and 5-Hydroxy Acid Glycerides.—In 1956 Keeney and Patton[171] identified δ-decalactone and in 1960 Tharp and Patton[339] identified δ-dodecalactone in butterfat. The list has been expanded to where these became recognized as members of a homologous series, all the members of which have been identified from C_6 through C_{16}. These, along with a similar series of γ-lactones, have been assembled in Table 4.13. Additional lactones, not members of either of these series, have unsaturations and branchings. They are: 5-hydroxy-9-dodecenoic acid lactone,[364] 4-hydroxy-6-dodecenoic acid lactone,[364] 5-hydroxy-9-tetradecenoic acid lactone (indicated but not proved),[364] 2,3-dimethyl-4-hydroxy-2,4-nonadienoic acid lactone at a level of 0.00005% in butterfat,[28] and *trans*-4-methyl-5-hydroxy-hexanoic acid lactone.[137] The last-named lactone was found "free" at a level of 0.00000005% in butterfat. This will be discussed below.

The fact that heating butterfat increases the yield of lactones led to the postulate that the lactones are derived from nonvolatile precursors, namely, triglycerides containing lactonogenic hydroxy acid components. Boldingh and Taylor[28] lent weight to this hypothesis by synthesizing δ-hydroxy acid-containing triglycerides and finding that they

Table 4.13

SATURATED n-ALIPHATIC LACTONES IN BUTTERFAT

Number of Carbon Atoms	δ-Lactones Levels in Butterfat ppm	Ref.	γ-Lactones Levels in Butterfat ppm	Ref.
6	2.0	252	trace	67
7	trace	67	trace	67
8	3.8[a]	65	0.5	175
9	trace	67	0.2	175
10	17.7[a]	65	1.2	158
11	0.7	158	0.5	158
12	28.6[a]	65	1.6	158
13	1.5	158	0.5	158
14	40.7[b]	65	1.4	158
15	6.4	158	1.3	158
16	32.3[b]	65	1.3	158
18	35	362	—	—
19	5[c]	361	—	—

[a] Average values of lactones obtained by steam deodorization.
[b] Average values of lactones obtained by silicic acid chromatography.
[c] δ(?) – C_{19}

did generate lactones on heating. In later work[175,253,362], the lactone precursors in butterfat were concentrated by silicic acid chromatography prior to further study.

Parliment *et al.*[253] concentrated all the lactone precursors in a polar fraction that contained 0.2% of its weight in bound δ-decalactone and 0.8% in bound dodecalactone. Based on chromatographic behavior and molecular weight considerations, they concluded that the precursors are triglycerides containing one esterified hydroxy acid moiety.

Kinsella *et al.*[175] concentrated all the lactone precursors in a polar fraction which represented 0.75% of the butterfat. From this fraction, after heating, they identified most of the saturated lactones known to be derivable from butterfat. After pancreatic lipase hydrolysis of the lactone-precursor fraction, the resultant 2-monoglycerides did not form lactones on heating. From this they concluded that the hydroxy acids are esterified in the α-positions.

Wyatt *et al.*[362] resolved a polar fraction which, from their description, appeared to have a high concentration of lactone precursors. They found that its infrared spectrum strongly suggested that the precursors are hydroxy acid-containing triglycerides.

The level of free lactones in fresh butterfat must be low, since their odor could not be detected in a concentrated percursor fraction.[175] However, during storage of butterfat or dried whole milk, even under refrigeration, lactones are formed until their presence can be detected by smell and by taste. On the contrary, in sterile concentrated milk[203] the δ-C_{10} and δ-C_{12} lactones decreased during storage at all temperatures studied, while the δ-C_{14} lactone decreased at 4°C and 37°C. Lactone formation is a two-step process: hydrolysis of the glycerol-hydroxy acid ester linkage and lactonization of the free hydroxy acid. The first step requires a low concentration of water for its initiation.[362] It is relatively slow at room temperature or below, usually requiring a period of weeks before the lactones become noticeable. It can be speeded up enormously at higher temperatures. Boldingh and Taylor[28] reported nearly 100% formation of lactone from synthetic triglycerides containing δ-hydroxy decanoic acid when heated at 145°C for about 85 min.

The lactonization step is generally reported as rapid, the expression "spontaneous lactonization" often being encountered. This step does, however, depend on the pH. Under sufficiently alkaline conditions, salts of the hydroxy acids are formed. Kontson *et al.*[182] found that at room temperature a pH well below neutrality was needed for complete and rapid lactonization. Studies of the rate of lactonization in butterfat or in milk at room temperature or lower, if reported, are unknown to this author. Possibly, the anomalous behavior of lactones in sterile concentrated milk involved an equilibration with hydroxy acid salts.

Lactones originating in milkfat are optically active,[27,28] indicating a biological origin. However, they are not optically pure,[350] which suggests that both optical antipodes are formed in milkfat in unequal proportions, by different pathways.

Flavor thresholds of the lactones have been included in a review by Kinsella *et al.*[176] and have been determined and compared with those of other flavor compounds by Siek *et al.*[313] The lactone of 2,3-dimethyl-4-hydroxy-2,4-nonadienoic acid has been described as celery-like,[28] δ-decalactone as coconut-like, and δ-dodecalactone as peach-like. The last two are the most important contributors to the flavor complex arising from the decomposition of lactone precursors in milkfat. These flavors, in themselves, are pleasant. At very low levels they have been considered by some to be important contributors to the desirable flavor of fresh milk and at rather high levels to the desirable flavor imparted by butter to candy and baked products. In this connection it has been claimed in patent applications[345] that lactones impart a buttery flavor to margarine.

High levels of lactones in fresh milk, however, are foreign to its natural flavor and are generally considered a flavor defect. The development of lactones in stored whole milk powder, therefore, has been one factor which has mitigated against its acceptance as a replacement for fresh milk. Patton was issued a patent[255] for stabilizing the flavor of milkfat against an excessive formation of lactones. The process utilized vacuum steam deodorization, which decomposes the lactone precursors and volatilizes the resultant lactones. Powders with an initial flavor equal to that of fresh whole milk, and with excellent storage qualities, have been prepared from a homogenate of skim milk and steam-deodorized butterfat.[193]

Dimick and Horner[63] examined the effect of several factors on the lactone potential of milkfat. On barn feed, in the winter, butterfat samples averaged 96 ppm lactones (range 59 to 139), while on pasture feed, in the summer, the average was 67 ppm lactones (range 48 to 100). During a 310-day lactation period of one Holstein, the concentration was about 25 to 30 ppm following parturition. It increased to from 170 to 180 ppm at 150 days, and then decreased to final values of about 120 ppm. The effect of breed was examined with groups of Holsteins, Brown Swiss, and Jersey-Guernsey cows. The Holsteins showed a somewhat higher lactone potential than did the other groups. Ketosis lowered the potential markedly. There was a high positive correlation between the lactone potential and the levels of short-chain fatty acids, suggesting a connection in the biosynthesis of the two classes of acids.

"Zero" milk[351] had concentrations of C_6 to C_{10} δ-lactones similar to those in normal milk while the C_{12} and C_{14} δ-lactones were higher than in normal milk. It is not clear whether the reference is to free lactones or to their precursors. However, *trans*-4-methyl-5-hydroxyhexanoic acid lactone[137] was removed by sweeping butterfat warmed to 50°C with carbon dioxide under vacuum. Under these conditions, designed to recover only free lactones, "zero" milk had 20 times the level of this lactone found in normal milk (10 μg vs. 0.5 μg/kg butterfat).

The level of lactones shown in Table 4.13 corresponds to about 0.0167% of the butterfat. With an average molecular weight of 212 for the lactones and 251 for the fatty acids, the lactonogenic hydroxy acid-containing triglycerides have an average molecular weight of 770 and are equivalent to about 0.061% of the butterfat.

Other Hydroxy Acid Glycerides.—Although a number of investigators have reported the presence of nonlactonogenic hydroxy acid triglycerides, there seem to have been no studies of their composition. Schwartz[291] found levels of hydroxy acid triglycerides corresponding to 0.60 to 0.66% of the butterfat. This is in close agreement with the values of 0.61 to 0.78% found in total milk lipids by Timmen.[341] These values, in conjunction with those for lactonogenic triglycerides, indicate that the nonlactonogenic glycerides constitute about 90% of the total hydroxy acid-containing triglycerides.

Neutral Glyceryl Ethers

$$\begin{array}{ccc} CH_2 \text{———} & CH \text{———} & CH_2 \\ | & | & | \\ OCH_2CH_2R & OCOR & OCOR \end{array}$$

It can be conjectured that the naturally occurring glyceryl ethers are triglycerides containing two esterified fatty acids. However, investigations of this class of compounds have been made on material isolated from the unsaponifiable fraction of milkfat; hence any fatty acids originally present would have been removed by the preliminary saponification. The presence of the ether linkage on the α-carbon of glycerol has been demonstrated by Davies *et al.*[54] and by Karnovsky and Brumm.[160]

The glyceryl ether content of cow's milkfat has been reported to be 0.0045% by Schogt *et al.*[289] and 0.01% by Hallgren and Larsson.[106] Both groups isolated their material from the unsaponifiable fraction and based their calculations on fatty acid-free compounds. Schogt *et al.* assumed, for their calculations, a palmityl glyceryl ether; the alcohol components found by Hallgren and Larsson are listed just below.

Hallgren and Larsson[106] have investigated the glyceryl ethers of both human and cow's milk. The lipids of human milk contained 0.1% glyceryl ethers and, as stated above, those of cow's milk 0.01%. The monohydric alcohol components of cow's milk glyceryl ethers were found to be palmityl, stearyl, and oleyl in the ratios 1.15:1.00:0.73, respectively. It was felt by these investigators that the composition is more complex than this, the findings being limited by the small amount of available material.

Following are the alcoholic components and their percentage occurrence in the glyceryl ethers of human milk: 16:0—23.9; 16:1—trace; 17 (including both normal and branched)—3.6; 18:0—22.8; 18:1—33.8; 18:2—1.4; 19 (including both normal and branched)—2.4; 20:0—1.6; 20:1—2.3; 22.0—0.7; 22:1—3.4; 24—2.1; unidentified components, 2.0.

As the data of Hallgren and Larsson were obtained from an examination of the total lipids of cream, the possibility that phosphatidic glyceryl ethers were involved is not excluded. However, Schwartz and Weihrauch[291] found the same glyceryl ether level, 0.016 to 0.020% in butterfat, as in the total milk lipids. These levels, based on triglycerides containing two fatty-acid components of average molecular weight, are recorded in Table 4.1. They identified $C_{10,11,12,13,14,15,16,17 \text{ and } 18}$ saturated alcohols and indications of both higher- and lower-molecular weight saturated alcohols; 60 unsaturated alcohols were detected, of which only $C_{18:1}$ *and* $C_{18:2}$ were identified with certainty.[291,294]

Neutral Plasmalogens

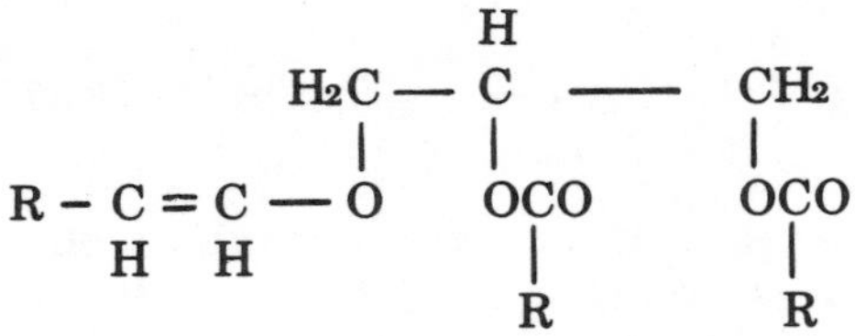

```
                          H
                 H2C ---  C  -----  CH2
                  |       |          |
R - C = C --- O          OCO        OCO
    H   H                 |          |
                          R          R
```

In this structure, the presence of the ether linkage on the α-carbon of glycerol is indicated by the structural investigations of Eichberg *et al.*[72] on material isolated from the common starfish. The work of Schogt *et al.*[289] on the neutral plasmalogens of milkfat indicated that the major portion of this material likewise had the vinyl ether linkage attached to the α-carbon atom of glycerol.

The aldehydes formed by hydrolysis of the neutral plasmalogens of milkfat have been examined by Schogt *et al.*[290] and by Parks *et al.*[250] Each group found a mixture of normal and branched-chain aldehydes, the branched C_{14} and branched C_{15} being by far the most abundant. Schogt *et al.* found all chain lengths from C_{13} through C_{18}, while Parks *et al.* found all chain lengths from C_9 through C_{18}. The values in Table 4.1 are calculated from these studies.

Diglycerides

The reported levels of diglycerides in fresh milk fat vary considerably: 0.25 to 0.48%,[154a] 1.26 to 1.59%,[341] 4.4 to 6.6%,[30] and 4 to 6 mole %.[220]

The free hydroxyl group is in one of the terminal positions of the diglyceride.[342,343]

Boudreau and deMan[30] reported a fatty-acid composition similar to that of diglycerides formed by partial hydrolysis of butterfat with pancreatic lipase. Their finding of relatively high concentrations of short-chain acids in the diglycerides is unfavorable to a proposal[258] that short-chain acids are esterified in the final step of triglyceride synthesis.

In contrast, Timmen *et al.*[342] have reported a fatty-acid composition which favors this sequence of synthesis. Following is a typical analysis: C_4, $< 0.1\%$; C_6 $\leq 0.1\%$; C_8, 0.4%; C_{10}, 0.4%; C_{12}, 1.2%; C_{14}, 9.2%; $C_{14:1}$, 1.5%; C_{16}, 36.5%; $C_{16:1}$, 1.8%; C_{18}, 13.9%; $C_{18:1}$, 27.7%; $C_{18:2}$, 2.3%; $C_{18:3}$, 1.2%.

It would appear that these conflicting results are a consequence of the methodologies used in the various studies. The levels of diglycerides recorded in Table 4.1 are those of Timmen and his associates.[341,342] These values, from mixed herd milks, had no distinct seasonal trends. From individual cows, values within the same range were obtained, with occasional lower values down to 0.91%. In the milk of a single cow, the diglyceride content increased steadily from 0.84% of the fat immediately after parturition to 1.40% after 2 months.

Monoglycerides

The content of monoglycerides, presented in Table 4.1, represents measurements on completely fresh milkfat.[151] Milk taken directly from the animals, without cooling, was treated at once to inactivate lipolytic enzymes.

MISCELLANEOUS NEUTRAL LIPIDS

Steroids

Cholesterol, General Properties.—Cholesterol is insoluble in water, sparingly soluble in cold alcohol or petroleum ether, and soluble in hot alcohol and most other organic solvents. Its solubility in cold ethyl acetate differentiates it from many other lipids. The presence of asymmetric carbon atoms at positions 3, 8, 9, 10, 13, 14, 17, and 20 imparts optical activity to the molecule. Values of −34.3° in chloroform and of both −31.1° and −29.0° in ether have been reported. Cholesterol melts at 148.5°C; it can be sublimed, and distilled under high vacuum.

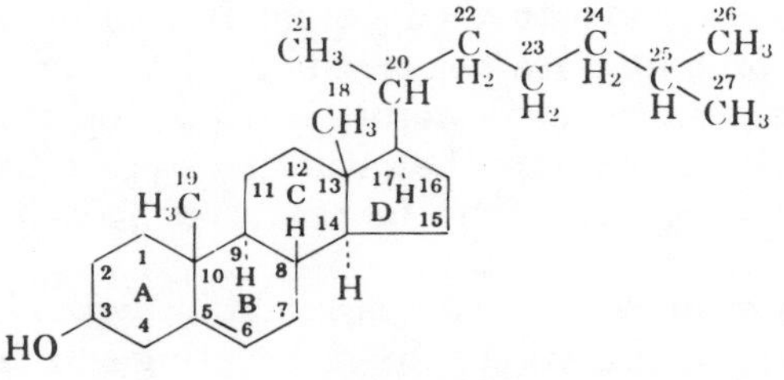

Cholesterol

The formula of cholesterol, above, shows the system used for designating the various rings and carbon atoms. The use of a solid line for showing the attachment of hydrogen or an alkyl group to a nuclear carbon indicates that the atom or group extends above the general plane of the steroid nuclear ring system, while the use of a dotted line indicates an extension below this plane. Thus, it is seen that in cholesterol the B and C rings and the C and D rings have a *trans* relationship. This is common to all the steroids. Steroids with the configuration at C_3, as shown for cholesterol, are referred to as β-compounds.

Cholesterol appears to be a component of all higher animal cells and is particularly abundant in nervous tissue. A number of suggestions have been made regarding its function in the animal organism. It has been suggested as a protective agent for nervous matter, as controlling cell permeability, as a precursor of the bile acids and some of the hormones, including the sex hormones, and of provitamin D_3. Attempts have been made to correlate the incidence of such disorders as atherosclerosis with a high level of cholesterol in the blood.

For a discussion of its biological functions as well as its biosynthesis and its chemistry the reader is referred to the monograph by Kritchevsky.[185]

In addition to many reactions unique to itself and related steroids, cholesterol shows a number of reactions to be expected from the presence of a carbon-carbon unsaturation and of a secondary alcohol group. A reaction of particular interest, because of its use in the isolation and determination of cholesterol, is that with digitonin to form the molecular addition compound, $C_{56}H_{92}O_{29}C_{27}H_{46}O$. This substance has a solubility of 14 mg in 100 ml 95% ethanol at 18°C. An essential structural feature for the occurrence of this reaction is the presence of the C_3 hydroxyl group in a β-configuration. Alteration of either the group or its configuration prevents the formation of a digitonide. Since most, if not all, naturally occurring free sterols meet these two requirements, it is obvious that the cholesterol digitonide would be contaminated with any other sterols present in the original solution. Schoenheimer[286] developed a method for separating cholesterol or other unsaturated sterols from saturated sterols. It is based on the observation that sterol bromides are not precipitated by digitonin. Thus, by brominating the mixture, cholesterol adds bromine at C_5 and C_6 and digitonin is used to precipitate the saturated sterols. Cholesterol, free from saturated sterols, can then be recovered from the solution by debromination.

The detection of plant sterols in association with cholesterol is of practical interest from the standpoint of detecting adulteration of milkfat with vegetable fat. In one method[13] the sterols are isolated from the glycerides and acetylated. The acetylated sterols are then subjected to repeated crystallization from alcohol or aqueous alcohol, and the melting point determined. That of pure cholesteryl acetate is 114°C while that of phytosteryl acetate (a mixture) varies from 125 to 137°C. The presence of plant sterols will usually be apparent from the melting point of the first crop of crystals, but confirmation can be obtained by further recrystallizations. If phytosteryl acetate is present, the melting point will rise. Chromatographic methods have also been applied successfully to the detection of foreign sterols in butterfat.[73]

When treated with strong acid under dehydrating conditions cholesterol gives a series of color reactions, the details of which depend upon the conditions employed. An example of such a test is the Liebermann-Burchard reaction, which is obtained by dissolving a crystal of cholesterol in 2 ml chloroform and adding 10 drops acetic anhydride and 1 drop concentrated sulfuric acid. A violet-pink coloration which changes to blue is produced. These reactions require nuclear unsaturation in the steroid; they can be obtained with other unsaturated steroids besides cholesterol, and with a number of other polynuclear hydroaromatic compounds which are either unsaturated or become unsaturated under the conditions of the test.

Cholesterol in Milk.—The quantitative determination of cholesterol in milk has usually involved precipitation with digitonin to separate the cholesterol from glycerolipids and other associated materials not precipitated by this reagent. The steroid can be calculated from the weight of the precipitated complex and this complex can be treated further. A number of micro methods are based on the use of reagents which react with the cholesterol to produce color which is measured with a colorimeter or with a spectrophotometer at selected wavelengths.

Both free cholesterol and cholesteryl esters can be determined by analyses made both before and after saponification of the milk lipids. That made before saponification represents the free cholesterol and that made after, the total cholesterol. Chromatographic methods are now available and can be expected to assume increasing importance.

Cholesterol in milkfat has been reported to be both completely esterified and, by Nataf *et al.*[240], completely free. Others have found it to be partially esterified. Patton and McCarthy[257] have reported concentrations of esters corresponding to from 10 to 15% of the total cholesterol; Nieman and Groot[241] found a variation with season—the highest concentration, in May, was about one-third of the total cholesterol. Intrieri and Badolato[142] found the ratio of free to esterified cholesterol to vary with the fat content of the milk as follows: 3.3% fat milk, 2.0; 1.8% fat milk, 2.9; skimmilk, 9.0.

Patton and McCarthy[257] separated the cholesteryl esters chromatographically from other milk lipids and found the following fatty acid composition (the chain length, colon, number of double bonds, dash, and percentage of the total acids are recorded in sequence): 10:0—1.3, 12:0—1.6, 14:0—4.4, 15:0—2.8, 16:0—17.9, 16:1—3.1, 18:0—8.8, 18:1—15.9, 18:2—23.9, 18:3—4.5, 20 to 26—4.5, and 11 unidentified acids—11.3.

Some data regarding the quantitative occurrence of cholesterol in milkfat are given in Table 4.14.

Although cows fed on a special protein-free diet had one-third the

Table 4.14

CHOLESTEROL CONTENT OF MILKFAT[a]

Breed of Cow	Fat Content of Milk, %	Total Cholesterol, %
Holstein	2.4	0.413
Holstein	3.3	0.288
Jersey	4.9	0.288
Jersey	5.0	0.300
Jersey	3.9	0.297
Guernsey	3.0	0.346
Guernsey	3.7	0.278

[a] Data calculated from reference 240.

blood cholesterol of cows on normal diet, the total cholesterol in "zero" milkfat was higher—413 mg/100gm compared with 317 mg/100 gm .[351]

Other Steroids.—Lanosterol, previously obtained only from wool fat and yeast, has been isolated by Morice.[232] This compound is synthesized biologically from squalene and is, itself, an intermediate in the biosynthesis of cholesterol. Its skeleton formula is shown below:

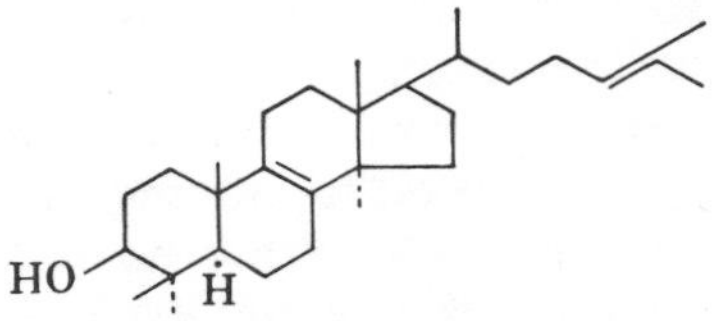

HO H

Lanosterol

Schwartz *et al.*[292] also found dihydrolanosterol in butterfat. Together, these sterols constituted 0.0046 to 0.0054% of the fat. Brewington *et al.*[36] identified β-sitosterol, a plant sterol, which was present in butterfat at a much lower level than the lanosterols—so low that it would present no complication in the detection of vegetable fat adulterants in butterfat. Parks *et al.*[251] isolated Δ^7 -cholesten-3-one from butterfat and identified it by a variety of means including infrared spectra, mass spectra, and melting point. This work was described as the first report of its occurrence in nature.

Vitamins and Carotenoids.—Milkfat contains vitamins A,D,E, and K and the following carotenoids: α-carotene, β-carotene, γ-carotene, ζ-carotene, lycopene, cryptoxanthin, lutein, zeaxanthine, and neo-β-carotene. The presence of γ-carotene has been inferred but not definitely proved. An extended discussion of the vitamins is given in Chapter 7.

The structures of the various carotenoids can be described on the basis of the structures of β-carotene and lycopene.

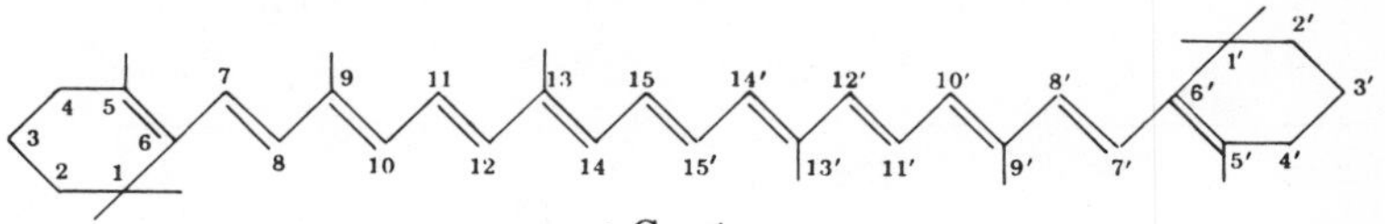

β-Carotene

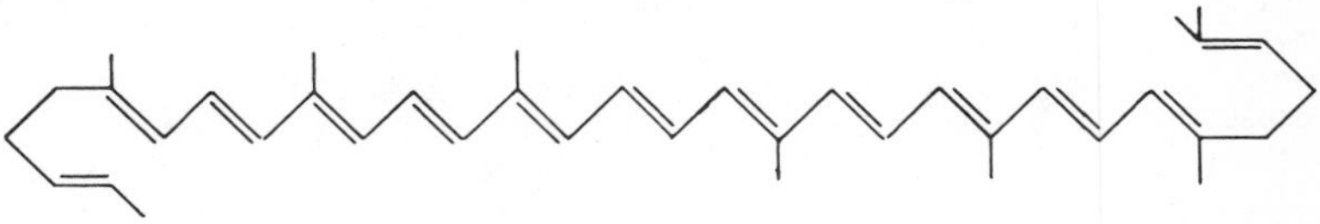

Lycopene

In these structures all the double bonds have the *trans* configuration. The carbon skeleton is considered as being filled out with hydrogen atoms as needed for satisfying the valency requirements of carbon. The lines extended from the skeleton represent attached methyl groups.

β-Carotene is a provitamin A; it is the most active of the carotenoids in this respect. Other provitamins A are α-carotene, γ-carotene, neo-β-carotene, and cryptoxanthin. α-Carotene has the same structure as β-carotene, except that the double bond at the 5′-position is shifted to the 4′-position. γ-Carotene is the same as β-carotene except that one of the rings is open to form a terminal structure as in lycopene. Neo-β-carotene is a mixture of geometrical isomers of β-carotene. The most abundant of these in milkfat appears to be the one with *cis* double bonds at positions 9 and 15. Cryptoxanthin is 3′-hydroxy-β-carotene. All these provitamins A are discussed in Chapter 7.

ζ-Carotene, found in milkfat by Nash and Zscheile[239] is 7,8,7′,8′-tetrahydrolycopene. Lycopene was first reported in milkfat by Gillam and Heilbron.[97] Lutein, found in milkfat by Gillam *et al.*,[98] is 3,3′-dihydroxy-α-carotene. Zeaxanthine, found in milkfat by Frankel *et al.*,[87] is 3,3′-dihydroxy-β-carotene.

Miscellaneous

Squalene, a liquid acyclic triterpene, was isolated from milkfat by Fitelson[81] in an amount corresponding to 7 mg squalene per 100 gm milkfat. This compound is an intermediate in the biosynthesis of cholesterol. Its formula is as follows:

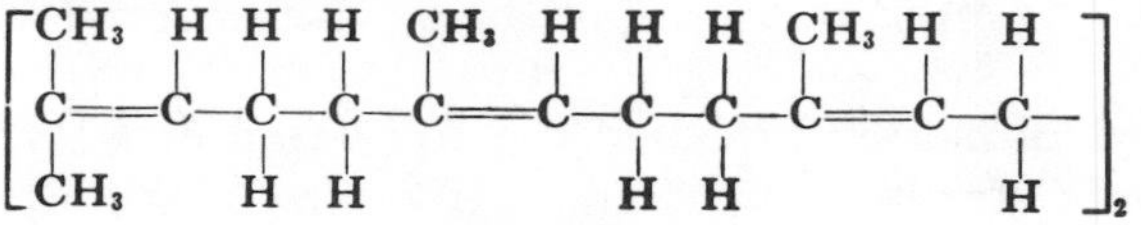

$$\left[(CH_3)_2C{=}CH{-}CH_2{-}CH_2{-}C(CH_3){=}CH{-}CH_2{-}CH_2{-}C(CH_3){=}CH{-}CH_2{-}\right]_2$$

Squalene

Lopez Lorenzo *et al.*[204] found ubiquinone at a level of 0.817 to 1.17 μg per gm butterfat, and provisionally identified solanesol and dolichol in all samples. Ubiquinone has the following formula:

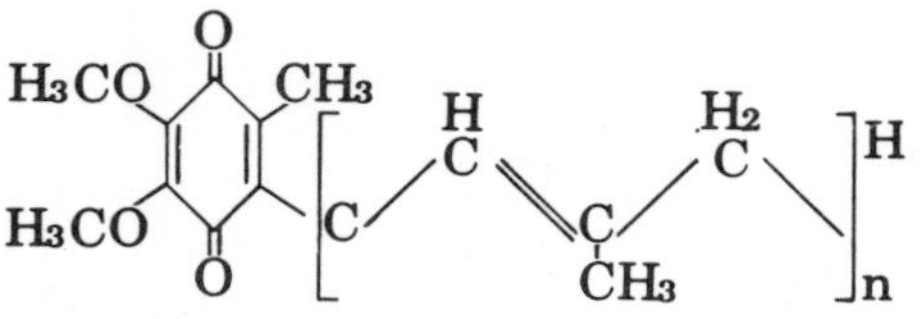

Ubiquinone (n = 6 − 10)

Solanesol, $H(CH_2\overset{CH_3}{C}=CHCH_2)_9OH$, a constituent of tobacco leaf, has been discussed by Stedman.[325] Dolichol is a long-chain isoprenoid alcohol for which the formula, $H \cdot [CH_2 \cdot C(CH_3){:}CH \cdot CH_2]_{19} \cdot CH_2$ $CH(CH_3) \cdot CH_2 \cdot CH_2OH$, has been proposed.[41]

Sulfur Compounds.—Although organic esters of sulfuric acid are widely distributed in nature, it is not known whether such compounds are present in milkfat. Trace quantities of organically bound sulfur have been reported.[330] These compounds have not been identified.

The biosynthesis of a number of the compounds discussed in the preceding sections has been reviewed by Goodwin.[103]

FREE FATTY ACIDS

Small proportions of free fatty acids are always present in fresh milkfat; larger percentages are found in the fat isolated from milk or cream that has been subjected to bacterial action or in fat that has been stored for some time. Lipolysis of milkfat, resulting in the production of free fatty acids, is discussed in another chapter. The presence of free acids in unlipolyzed fat may be the result of an incomplete esterification in the mammary gland.

Kintner and Day[177] found the composition of the free fatty acids in fresh milk to be similar to that of the acids bound in the milk glycerides.

Kawashiro *et al.*[166] reported that during storage of butter relatively more of the odd-numbered than of the even-numbered fatty acids are liberated.

Loney and Bassette[203] studied the liberation of fatty acids during the storage of sterile concentrated milk. Enzymatic lipolysis should not be a factor in this product, so the liberated acids can be considered a consequence of uncatalyzed hydrolysis of the milk lipids. The rate of fatty acid liberation varied directly with temperature, no detectable lipid hydrolysis taking place during storage at 4°C. The composition of free acids, after storage, was similar to that in the parent lipids.

PHOSPHOLIPIDS AND SPHINGOLIPIDS

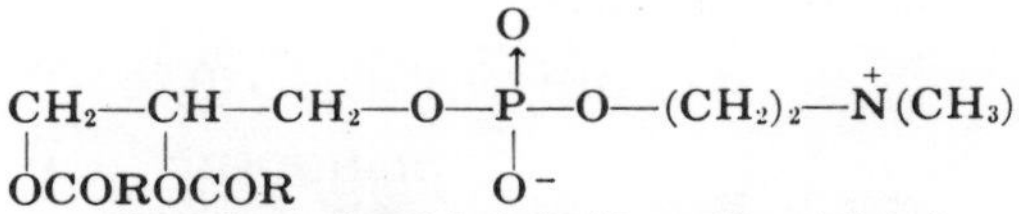

$$\underset{\text{OCOR}}{CH_2}—\underset{\text{OCOR}}{CH}—CH_2—O—\overset{O}{\overset{\uparrow}{\underset{O^-}{P}}}—O—(CH_2)_2—\overset{+}{N}(CH_3)_3$$

(I) Phosphatidylcholine (Lecithin)

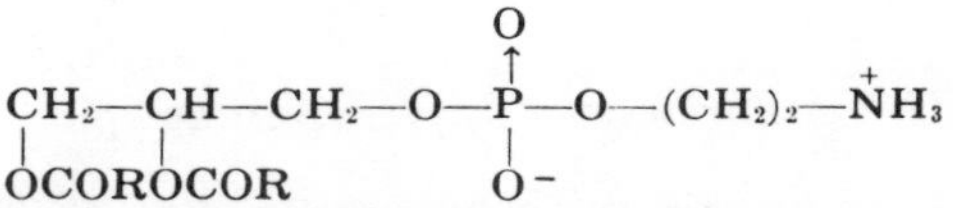

$$\underset{\text{OCOR}}{CH_2}—\underset{\text{OCOR}}{CH}—CH_2—O—\overset{O}{\overset{\uparrow}{\underset{O^-}{P}}}—O—(CH_2)_2—\overset{+}{N}H_3$$

(II) Phosphatidylethanolamine (Cephalin)

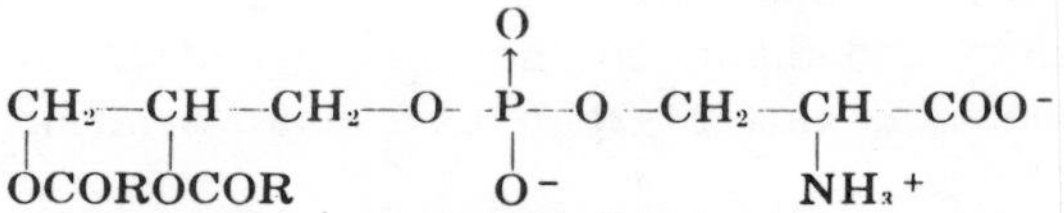

$$\underset{\text{OCOR}}{CH_2}—\underset{\text{OCOR}}{CH}—CH_2—O—\overset{O}{\overset{\uparrow}{\underset{O^-}{P}}}—O—CH_2—\underset{NH_3^+}{CH}—COO^-$$

(III) Phosphatidylserine

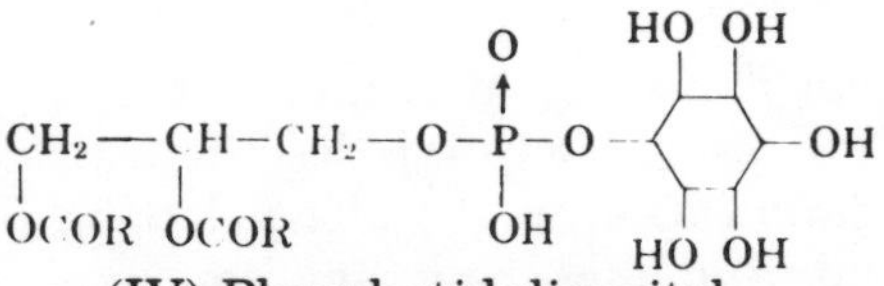

$$\underset{\text{OCOR}}{CH_2}—\underset{\text{OCOR}}{CH}—CH_2—O—\overset{O}{\overset{\uparrow}{\underset{OH}{P}}}—O—C_6H_6(OH)_5 \quad (\text{inositol ring: HO, OH, OH, HO, OH})$$

(IV) Phosphatidylinositol

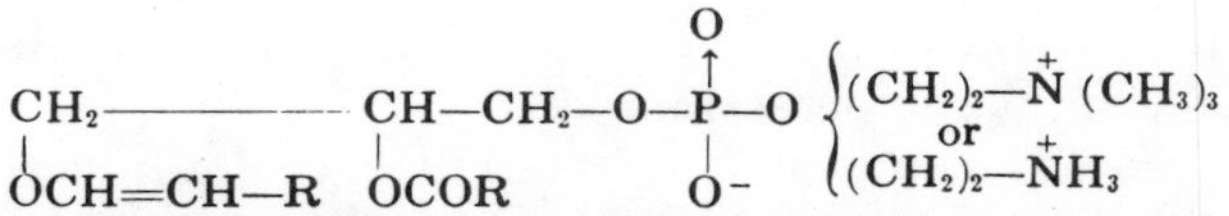

$$\underset{\text{OCH=CH—R}}{CH_2}—\underset{\text{OCOR}}{CH}—CH_2—O—\overset{O}{\overset{\uparrow}{\underset{O^-}{P}}}—O\begin{cases}(CH_2)_2—\overset{+}{N}(CH_3)_3\\ \text{or}\\ (CH_2)_2—\overset{+}{N}H_3\end{cases}$$

(V) Phosphatidalcholine or ethanolamine (Plasmalogen)

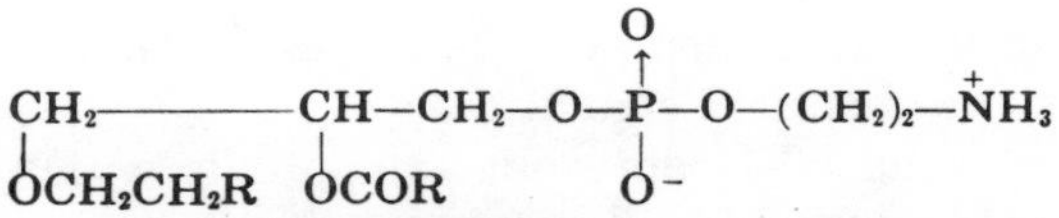

$$\underset{\text{OCH}_2\text{CH}_2\text{R}}{CH_2}—\underset{\text{OCOR}}{CH}—CH_2—O—\overset{O}{\overset{\uparrow}{\underset{O^-}{P}}}—O—(CH_2)_2—\overset{+}{N}H_3$$

(VI) Phosphatidic glyceryl ether

```
                                              O
                                              ↑                      +
CH3(CH2)12CH=CH—CH—CH—CH2—O—P—O—(CH2)2—N(CH3)3
                |   |                 |
               OH HNCOR               O-
```

(VII) Sphingomyelin

```
                                        H
                                        |
CH3(CH2)12CH=CH—CH—CH—O—C———————┐
                |   |      |        |
               OH  NH    HCOH       |
                    |      |        |
                   COR  HOCH        O
                           |        |
                        HOCH        |
                           |        |
                         HC—————————┘
                           |
                          CH2OH
```

(VIII) Cerebroside—a glycosphingolipid

Formulas I-VII are phospholipids and VII-VIII are sphingolipids, sphingomyelin being a member of each group. In addition to variation in their acyl components, 31 long-chain bases in bovine milk sphingomyelin[235] and 33 in bovine milk glycosphingolipids[236] have been identified. The latter group also has variation in its sugar moiety.

General

The phospholipids are widely distributed; they appear to be components of all cells. Proposals regarding their biological functions include the following:[107] implication in the blood-clotting process; as storage forms for fatty acids and phosphate; as a source of choline in nervous tissue; as essential structural elements in living cells; as integral components in biological oxidations; as intermediates in the transport, absorption, and metabolism of fatty acids; and in the transport and utilization of sodium and potassium ions.

For many years the substance referred to as cephalin was thought to be represented by Formula II, above. Folch,[83] however, showed that brain cephalin is a mixture. In addition to phosphatidylethanolamine, he isolated phosphatidylserine and an inositol-containing phospholipid. The term "cephalin" is, accordingly, less specific than nomenclature which indicates the chemical structure.

The above formulas I-VI represent compounds which, if partially hydrolyzed, would produce α-glycerophosphoric acid, in which the phosphoric acid is attached to one of the terminal carbon atoms of glycerol. α-Glycerophosphoric acid contains the elements essential for optical activity; β-glycerophosphoric acid, in which the attachment is through the middle carbon atom of glycerol, is optically inactive. Although partial hydrolysis of phospholipids has yielded both α- and β-glycerophosphoric acid, this cannot be considered proof that both structures are present in the naturally occurring phospholipids, since hydrolysis of phospholipids is accompanied by migration of the phosphoric acid radical to form an equilibrium mixture of the α- and β-glycerophosphoric acids.[19] Evidence from other sources has been obtained for the occurrence of the α-attachment in the natural phospholipids, but not for the β-attachment.

Almost without exception the natural phospholipids are the isomers in which the phosphoric-acid moiety is attached in the 3-position of *sn*-glycerol. The known exception[157,163] occurs in phospholipid components of the extremely halophylic bacterium, *Halobacterium cutirubrum,* in which the phosphoric-acid moiety is attached in the 1-position of *sn*-glycerol. Another unusual feature of these phospholipids is that they contain no acyl groups but, instead, are diethers.

It is noted that the base, sphingosine, in the sphingomyelin molecule has two hydroxyl groups through which the phosphoric acid radical could conceivably be attached. Evidence for the attachment as shown through the terminal hydroxyl (α-attachment) has been furnished by Rouser *et al.*[277] and by Marinetti *et al.*[214] Carter and Fugino[44] have reviewed the evidence that sphingosine is Ds-erythro-1,3-dihydroxy-2-amino-4-*trans*-octadecene.

The plasmalogens are lipids characteristic of the animal kingdom which generate aldehydes on acid hydrolysis. The structure of phosphatidic plasmalogens is shown by formula V, above. The linkage to glycerol as an α,β-unsaturated ether was established by hydrogenation[269] and ozonolysis.[55] Hydrogenation formed a saturated ether which no longer had aldehydogenic properties. Ozonolysis produced long-chain aldehydes, a result consistent with a vinyl ether structure.

The position of attachment of the aldehydogenic group has been the subject of considerable controversy. Some evidence is available indicating that both positions of attachment may exist in natural plasmalogens; the predominance of current evidence, however, favors α-attachment. Part of the earlier confusion has resulted from hydrolysis studies employing lecithinase A. This enzyme had been thought to be specific in hydrolyzing fatty acids from the α-position of phospholipids. Tattrie,[337] however, has presented what appears to be unequivocal evidence that the opposite is true, and that lecithinase A, now known as phospholipase A, is specific in hydrolyzing fatty acids from the β-position. Plasmalogens have been hydrogenated and the resultant saturated ether hydrolyzed to form an acid-free glyceryl ether. Ethers formed in this manner have been shown to be attached through the α-position of glycerol. This has been done by the formation of ethers of known structure,[215] as well as by infrared spectra and periodate-oxidation analysis.[56]

Norton *et al.*[243] have prepared the α-1-alkenyl glyceryl ether from pure phosphatidalcholine. The infrared spectrum indicated that the double bond has a *cis* configuration.

Formula VI, above, represents a saturated counterpart of phosphatidalethanolamine. Such a compound has been isolated[45] from egg yolk. The α-attachment was deomonstrated by periodate-oxidation analysis and proton magnetic resonance studies.

Another type of structural variation, differing from those just discussed, is that involved in acid-base equilibria. In the formulas above, the dipolar ionic forms have been used for those compounds containing both acidic and basic elements. In compounds such as phosphatidylethanolamine, the contribution of the dipolar ionic form is a maximum at the isoelectric point. The presence of excess hydroxyl or hydrogen ions leads to an increased representation of those forms containing a nonionized basic or acidic group, respectively.

However, in choline-containing compounds, such as lecithin, the nitrogen atom is not attached to an ionizable hydrogen atom which could be removed by hydroxyl ions to form a nonionized base. Also, according to the Pauli exclusion principle, this nitrogen atom is not capable of attaching an additional atom or group through a covalent bond. It cannot, therefore, combine with a hydroxyl ion to form a nonionized base. It would thus seem that the choline group must remain ionized, regardless of variations in acidity. Ionization of the phosphate radical is in all instances influenced by the acidity.

Baer and Maurukas[20] obtained analyses of synthetic lecithin which they felt were consistent with the nonionized forms of both the phosphate and choline radicals. The same analyses would be expected, however, if the ionized form contained one molecule of water of crystallization.

Properties.—The phospholipids are not truly soluble in water but can be emulsified to form colloidal solutions or suspensions. Lecithin and cephalin, in addition, are hygroscopic. In general, phospholipids are soluble in fat solvents. There are several exceptions to this statement; for example, they have low solubility in acetone. This property has been widely used in separating phospholipids from fats. Also, sphingomyelin has low solubility in ether, and this fact is utilized in its isolation from other phospholipids. Classically, cephalin has been isolated by its limited solubility in alcohol. It is now known that pure phosphatidylethanolamine is quite soluble in alcohol, phosphatidylserine is less so, and the inositol phospholipids are the least soluble in this solvent.

It should be pointed out here that these solubility relationships are influenced considerably by both the nature of the fatty-acid constituents of the molecule and by the presence of impurities. Thus, while lecithin is ordinarily soluble in ether, if its fatty acids are completely saturated it is insoluble in this solvent.[359] The presence of triglycerides increases the solubility of phospholipids in acetone. The presence of inositol phospholipids decreases the solubility of phosphatidylethanolamine in alcohol.

Chemically, the reactions of phospholipids are those to be expected from a consideration of their constituents. It is notable that unsaturated acids combined in phospholipids are more susceptible to oxidation than if combined in a triglyceride.

Preparation and Determination.—The preparation and purification of phospholipids in earlier work have been based largely on the solubility relationships mentioned above, together with the use of complexes with inorganic salts such as cadmium chloride. These relationships are still occasionally employed in preliminary steps, but today chromatographic procedures are in general use. Marinetti[213] has reviewed in some detail the problems involved in isolating phospholipids from natural sources, and in their chromatographic separation, identification, and analysis.

The identification and estimation of phospholipids may be based on a determination of characteristic constituents such as lipid phosphorus, lipid nitrogen, choline, or ethanolamine. Spectral methods are being used increasingly for identifying phospholipids and for determining their purity. These methods are also being applied to their quantitative estimation. Details are given elsewhere.[213,359] Figure 4.5, taken from the work of Smith and Freeman,[316] shows infrared spectra of a cerebroside and of three classes of phospholipids. A correlation between chemical groups and absorption peaks has been illustrated in another publication.[318] Cerebrosides and the phospholipid, sphingomyelin, contain an amide linkage. They can be distinguished from other phospholipids by the strong absorption bands at 6.1 and 6.5 μ that are characteristic of monosubstituted amides. The spectra of both lecithin and cephalin show absorption peaks at 5.8 μ due to an ester carbonyl group, which is lacking in sphingomyelin. Lecithin has a strong absorption peak at 10.3 μ which serves to distinguish it from cephalin.

A suggestion of the complexity of this fraction of the milk lipids

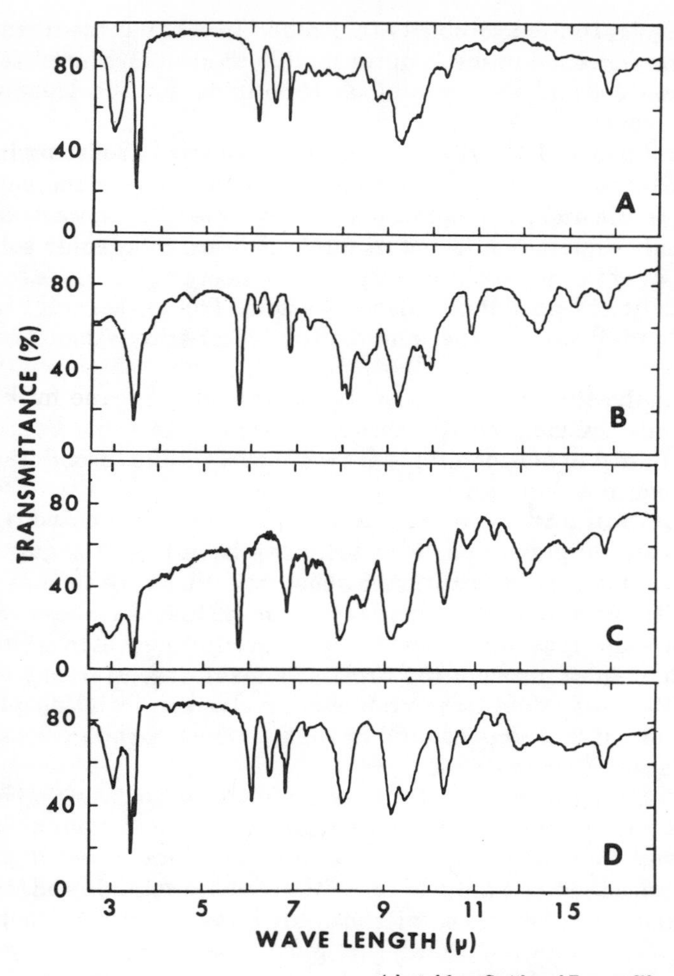

Adapted from Smith and Freeman[316]

FIG. 4.5. INFRARED ABSORPTION SPECTRA OF A CEREBROSIDE AND VARIOUS PHOSPHOLIPIDS

A—Phrenosine. B—Cephalin. C—Lecithin. D—Sphingomyelin.

appeared in early work,[191,192] which indicated the presence of lecithin, cephalin, sphingomyelin, and cerebrosides. Morrison *et al.*,[237] in a summary of later work, reported the following distribution of the major classes of compounds present in this fraction of bovine milk lipids: 30% phosphatidylethanolamine, 28% phosphatidylcholine, 8% phosphatidylserine, 5% phosphatidylinositol, 19% sphingomyelin, 1% phosphatidalethanolamine, 3% phosphatidalcholine, and 3% each of the sphingolipids, ceramidemonohexoside and ceramidedihexoside. Comparative distribution data are also available[234] for the phospholipids of the cow, ewe, Indian buffalo, camel, ass, sow, and human.

The phospholipids have been reported as generally existing in milk combined by semilabile and labile linkages to protein and carbohydrates.[91] The carbohydrates, which can be retained by proper methods of isolation, are reported to be very easily broken by polar solvents.

Phospholipids in Milk

The data in Table 4.15 are indicative of the levels of phospholipids in milk and some milk products. Approximate levels of individual phospholipid and sphingolipid classes in the total bovine milk lipids can be calculated from the figure 0.87% of this table and the distribution data given above.

Morrison and associates[234,237,238] have studied the effect of species on the fatty-acid composition of milk phospholipids. Data on the acids of the bovine milk glycerophospholipids are given in Table 4.16 and of bovine milk sphingomyelin in Table 4.17. Morrison[234] summarized their findings in the following manner. The ruminant herbivores (camel, cow, sheep, Indian buffalo) have branched-chain acids in all phospholipid classes, they have only a few percent of fatty acids with more than two double bonds, and their sphingomyelin contains about 30% of 23:0 and only a small percentage of 24:1. The nonruminant herbivore (ass) has larger percentages of 18:3, and its sphingomyelin has only a low percentage of 23:0 but about 16% of 24:1. The nonherbivores (pig and human) have less 18:3 than the ass and larger proportions of long-chain polyunsaturated acids. Camel-milk phospholipids are somewhat atypical of their group and contain the exceptionally high level of 15% plasmalogens.

Table 4.15

PHOSPHOLIPID CONTENT (PERCENT) OF MILK AND MILK PRODUCTS[a]

Product	Phospholipids in Product	Fat in Product	Phospholipids in Fat
Whole milk	0.0337	3.88	0.87[b]
Skim milk	0.0169	0.090	17.29
Cream	0.1816	41.13	0.442
Buttermilk	0.1819	1.94	9.378
Butter	0.1872	84.8	0.2207

[a] Data calculated or taken from reference 135.
[b] The reported value of 0.0869 appears to be a misprint.

Table 4.16

COMPONENT ACIDS OF BOVINE MILK GLYCEROPHOSPHOLIPIDS: THEIR LEVELS AND SPECIFIC DISTRIBUTION[a,b]

Acid[c]	Phosphatidylethanolamine		Phosphatidylserine		Phosphatidylcholine	
	1-position	2-position	1-position	2-position	1-position	2-position
10:0					0.2	0.7
11:0	0.1	0.2	0.5	0.1	trace	0.1
12:0	0.3	0.6	1.1	0.5	0.3	0.8
13:0	trace	0.1	0.4	0.1	0.1	trace
14:0i	0.2	0.1	0.1	0.1	0.2	0.1
14:0	1.9	1.3	4.0	2.2	5.6	10.8
15:0i	0.1	0.1	0.3	0.2	0.3	0.4
15:0ai	0.3	0.3	0.5	0.5	0.3	0.6
15:0	0.6	0.4	0.6	0.5	1.6	2.4
15:1	0.1	0.1	0.4	0.3	0.2	0.3
16:0i	trace	0.1	—	0.2	0.4	1.0
16:0	19.7	4.7	24.5	7.6	41.9	30.6
16:1	1.2	2.2	1.7	2.3	0.6	1.2
17:0i	0.4	0.2	0.3	0.2	0.4	0.5
17:0ai	0.4	0.3	1.3	0.5	1.0	1.0
17:0	1.3	0.9	1.3	0.7	1.3	0.6

17:1	0.3	0.6	0.7	0.6	0.2	0.3
18:0i	0.1	trace	0.1	0.1	0.1	0.1
18:0	19.0	1.3	28.0	3.9	17.5	2.4
18:1	45.8	47.8	25.8	46.2	20.3	27.8
18:2	2.9	21.4	1.6	18.3	2.7	9.2
18:2c,tt	0.5	0.2	—	0.4	0.3	0.5
18:3	1.1	4.5	1.2	3.0	0.8	1.8
18:3c,ttt	0.4	1.8	—	1.0	1.1	1.0
19:0	0.4	0.4	0.7	0.4	0.6	0.2
20:0i	0.1	0.1	—	0.2	0.1	0.1
20:0	0.7	0.1	1.0	0.2	0.3	0.2
20:1 ?	0.1	0.2	—	0.1	0.1	0.2
20:3	0.2	2.2	1.2	2.0	—	1.6
20:4	0.2	3.0	1.1	2.7	0.2	1.2
21:0	0.1	0.1	—	0.4	—	—
21:1	—	0.3	1.5	0.6	0.2	0.3
22:0	0.1	0.1	—	0.4	—	—
Unknowns	1.1	3.7	—	4.0	1.1	2.0
Total Saturated	46	11	65	19	72	53
Total Unsaturated	54	89	35	81	28	47

[a] Data from reference 237.
[b] Levels expressed as mole % of total acids in 1- and 2-positions of *sn*-glycerol.
[c] Acid designated by number of carbon atoms, colon, number of unsaturations. Abbreviations: i-iso, ai-anteiso, c-conjugated, t-*trans*.

Table 4.17

COMPONENT ACIDS OF BOVINE MILK SPHINGOLIPIDS[a]

Acid[b]	Percent Composition[c]					
	Sphingomyelin		Glucosylceramide		Lactosylceramide	
	Fatty Acid	2-Hydroxy Acid	Fatty Acid	2-Hydroxy Acid	Fatty Acid	2-Hydroxy Acid
12:0	0.1				0.1	
14:0	0.4	1.6	1.0	2.8	0.3	0.6
15:0	0.1		0.3		0.1	
16:0	7.8	9.2	9.3	12.6	7.7	10.5
16:1		0.8	1.4		0.3	
17:0	0.3	1.1	1.3	1.3	0.2	1.1
18:0	1.6	6.2	13.7	8.4	3.3	3.7
18:1	0.2	0.7	12.2		1.3	
18:2	0.2		2.0		0.2	
19:0	0.2	0.6	1.3		0.2	
20:0	0.6	0.6	0.9	4.5	1.1	1.2
21:0	0.9	2.0	1.2	1.3	1.4	0.9
21:1			0.1			
22:0	20.7	17.2	17.0	16.7	24.9	15.4
22:1	0.7		0.9		0.6	
23:0	30.4	31.5	22.0	31.0	29.5	26.9
23:1	5.0	2.5	3.4	1.7	6.6	6.3
24:0	22.8	21.8	9.9	19.7	16.5	29.5
24:1	4.0	1.5	2.1		3.7	4.1
25:0	1.6	1.9			0.7	
25:1	1.6	1.0			1.4	
26:0	0.8					
Total Unsaturates	11.7	6.5	22.1	1.7	14.1	10.4

[a] From reference 236
[b] Acid designated by number of carbon atoms, colon, number of double bonds.
[c] With reference to each type of acid. The 2-hydroxy were less than 1% of total acids in each lipid class.

The positional distribution of fatty acids in phosphatidylcholine and phosphatidylethanolamine was studied. The acids generally fit a pattern of increasing preference for the α-position with increasing chain length of the saturated acids, and an increasing preference for the β-position with increasing unsaturation. The fatty acids in the β-position of milk phospholipids differed significantly in composition from those in the β-position of the corresponding neutral triglycerides. This was considered support for the idea that the biosynthesis of neutral triglycerides and phospholipids does not proceed by simple direct routes from a common precursor.

Phosphatidic Glyceryl Ethers.—On questionable evidence,[91] these compounds were reported present and were thought to represent the entire glyceryl ether fraction of milkfat. However, glyceryl ethers have been found in butterfat (which is free from phospholipids) at the same level as in the total milk lipids.[291,294]

Phosphatidic Plasmalogens.—Various plasmalogen levels, up to 4% of bovine milk phospholipids, have been reported. Duin[69] found 0.4 to 0.8% bound aldehydes in the phospholipid fraction. This corresponds approximately to 1.3 to 2.5% of the phospholipids. Duin found that the aldehydes were a complicated mixture with chain lengths of 14 to 18 carbon atoms. Besides saturated aldehydes he found unsaturated and probably branched-chain aldehydes. Stearaldehyde and palmitaldehyde were identified with certainty. In addition, there were indications of the following: an α,β-unsaturated C_{18} aldehyde, a C_{18} aldehyde with unsaturation in some other position, a normal or branched C_{17} aldehyde, an α-β-unsaturated or branched C_{15} aldehyde, and a C_{14} aldehyde.

The biosynthesis of phospholipids will not be discussed here except to mention that phosphatidic acid, α′,β-diacyl-α-glycerophosphoric acid, is a common intermediate from which either triglycerides or phospholipids may be formed. It is of interest in this connection that there are two series of naturally occurring compounds which differ only in the nature of the substituent on the α-carbon atom of glycerol. These are the neutral series and the phospholipid series. Examples of the neutral series and their phosphatidic analogues are: (1) triglycerides and the phospholipids represented by formulas I, II, III, and IV; (2) neutral plasmalogens and phosphatidic plasmalogens, formula V; and (3) neutral glyceryl ethers and phosphatidic glyceryl ethers, formula VI. As discussed above, however, there is evidence[234] that if the two classes of compounds are derived from a common precursor, their biosynthesis does not proceed by simple direct routes.

For further information the reader is referred to works discussing phospholipids with reference to their structure,[107] their isolation and analysis,[107,213] their biosynthesis,[173,276] and their function.[276]

Sphingolipids in Milk

In addition to a variation in component fatty acids, sphingolipids are characterized by a great variation in their long-chain base moiety. Karlsson[159] has reviewed the sphingolipid long-chain bases with regard to their nomenclature, distribution, metabolism, biologic properties, chemistry, and methods of characterization. His review includes 60 long-chain bases which have been identified or proposed.

Morrison[235] identified 31 long-chain bases in bovine milk sphingomyelin. They consisted of both saturated and unsaturated dihydroxy bases. The saturated bases had normal 12 to 20 carbon atom chains, *iso* 17 to 20 carbon atom chains, and *anteiso* 17 to 19 carbon atom chains. The unsaturated bases had normal 12 to 20 carbon atom chains, *iso* 14 to 19 carbon atom chains, and *anteiso* 15, 17, and 19 carbon atom chains. The unsaturated bases formed 82% of the total.

The long-chain bases in glucosylceramide and lactosylceramide[236] included all those in the sphingomyelin fraction plus two additional: an *iso* saturated and an *iso* monounsaturated 16 carbon atom base.

The fatty-acid and substituted fatty-acid components of bovine milk sphingolipids are shown in Table 4.17. Table 4.18 gives information on the structure of the monoenoic acids of sphingomyelin. Long-chain acids are formed by chain elongation from the carboxyl end of the molecule.[179] The data in this table show a close relationship between the *cis* monoenoic acids of the neutral triglycerides and the longer-chain *cis* monoenoic acids of sphingomyelin. No such similarity is evident between the *trans* monoenoic acids of the two lipids. Table 4.19 shows the high *trans* content of the sphingolipid monoenoic acids.

Obvious differences between the fatty-acid composition of the neutral triglycerides and that of the phospholipid-sphingolipid fraction of milkfat are the absence of short-chain acids in the latter group and the abundance of chain lengths greater than 18 in the sphingolipids. There are also a number of acids present in the sphingolipids which either are absent from the neutral triglycerides or have not been as completely characterized. Notable in this respect are the 2-hydroxy acids listed in Table 4.17 and the numerous positional and *cis, trans* isomers in the 23:1, 24:1, and 25:1 acids of sphingomyelin. Additionally, a conjugated *trans, trans, trans*- 18:3 acid[237] was tentatively identified, and a 22:6 acid[120] was reported present in the phospholipids.

PHYSICAL PROPERTIES OF MILKFAT

Structure

The size of the milkfat globules has been reported to vary from 0.1 to 22 μ in diameter.[59] The majority, however, are within the size range of 1 to 5μ.

Table 4.18

COMPARISON OF DISTRIBUTION OF DOUBLE BONDS IN CERTAIN *cis* AND *trans* MONOENOIC FATTY ACIDS OF BOVINE MILK SPHINGOMYELIN AND NEUTRAL TRIGLYCERIDES[a]

Position of Double Bond from Terminal Methyl Group	*cis*-Isomers				*trans*-Isomers		
	Sphingomyelin			Neutral Triglycerides	Sphingomyelin		Neutral Triglycerides
	23:1	24:1	25:1	18:1	23:1	24:1	18:1
12		0.3				0.6	
11	1.2	0.1	0.7		1.2	0.8	1.2
10	2.3	1.0	3.0	1.0	2.1	7.5	2.3
9	79.0	96.5	92.5	96.0	59.6	21.1	3.6
8	5.2	1.0	2.8	0.5	9.3	6.5	3.6
7	8.6	—	1.0	2.5	1.7	24.4	45.3
6	2.3	0.3			8.9	5.7	4.7
5	0.7				2.5	11.3	8.8
4	0.7				12.4	19.6	8.8
3					3.2	3.2	8.4
2							13.0

[a] From reference 236. Acids designated by number of carbon atoms, colon, number of double bonds.

Table 4.19

PERCENT *trans* CONTENT OF MONOENOIC FATTY ACIDS IN BOVINE MILK SPHINGOLIPIDS[a]

Acid[b]	Sphingomyelin	Glucosylceramide	Lactosylceramide
18:1	94	44	57
22:1	98	84	84
23:1	64, 70[c]	49	64
24:1	23, 17[c]	20	19
25:1	7, 5[c]	1	2
Total monoenes	44, 46[c]	47, 43[c]	46, 51[c]

[a] From reference 236
[b] Acid designated by number of carbon atoms, colon, number of double bonds.
[c] Values found by different methods of analysis.

The fat globules are stabilized by a surface film containing, among other components, protein and phospholipids. The composition of this film will not be further discussed here as it has been treated in detail in a later chapter. Sommer[323] has pictured the phospholipids as the primary stabilizing force in preventing further growth of the fat globules. According to his conception, triglyceride molecules secreted by the mammary gland, being insoluble in the aqueous phase, combine with similar molecules previously or concurrently formed and thus build up a fat globule. At the same time phospholipids, present in the fluid or simultaneously secreted, deposit in a strictly random fashion on the growing globules. The phospholipid molecules, having both non-polar and strongly polar groups, orient on the surface of the fat globules with their polar groups in the aqueous phase. When, as a result of this random deposition, the surface of a fat globule is completely covered with phospholipids, it is sealed against the further deposition of triglyceride molecules and so ceases to grow. The protein or proteins of the surface film are then added by complexing with the polar groups of the phospholipids. Sommer has concluded that the phospholipid content of milk is, therefore, just sufficient to seal off the globules against further growth by forming a monolayer on each globule. He has pointed out that this concept affords an explanation of the relationship between globule size and breed of cow and, at the same time, explains the small globules of late lactation. With a rapid formation of triglycerides in high concentration, the globules would grow more than under less favorable conditions before becoming sealed off with phospholipids.

This would explain the comparatively large globules in Jersey milk. In the later stages of lactation, while the concentration of fat in the milk is high, its formation is slow and the phospholipids are relatively more abundant. Under these conditions the surface would become covered while the globule was still small. Regardless of the correctness of Sommer's views, it is a fact that the milkfat is capable of existence in a state of much greater subdivision than occurs naturally. By homogenization the number of globules has been increased more than 1000 times.

Plastic Properties

Whereas a viscous material will flow under stress—the rate of flow being proportional to the force applied—a plastic material will react to sufficiently small forces by an elastic deformation. Only upon the application of forces greater than a critical value, known as the yield point or yield value, will the material flow. The relationship between the rate of flow and the pressure, while nonlinear, approaches linearity with increased pressure. If a straight line is fitted to this curve in the proper manner it can be used to define the plastic properties of a material such as milkfat.[321] The straight line must fulfill the dual requirements, as closely as possible, of intercepting the pressure axis at 4/3 the yield value and of being asymptotic to the curve in the region of 5/3 the yield value. The slope of such a line, expressed in centipoises, has been referred to by Søltoft[321] as the viscosity, and the intercept, expressed in dynes/cm^2, as the firmness. He used the term "consistency" to express the total plastic properties of a fat.

Various types of instruments have been used to measure the plastic properties of fats. With one type, in which a fat is forced through a capillary tube, measurements such as those pertaining to firmness and viscosity can be expressed in fundamental scientific units. With another type the data can be expressed only in arbitrary units; examples of this type are penetrometers, such as the falling ball or falling needle, and plunger or cutting-wire plastometers.

The response of a plastic material to stresses less than the yield value is at least partially elastic in nature. Thereafter the flow becomes viscous. Since elasticity and viscosity have different dimensions, $ML^{-1}T^{-2}$ and $ML^{-1}T^{-1}$, respectively, they cannot be compared quantitatively nor expressed in a single scientific unit. Søltoft, however, has shown that since a certain relationship usually does exist between firmness and viscosity, it is possible to calculate one of them approximately from the other, and to express the consistency in a single arbitrary unit. This is of practical interest in connection with the use of the instruments mentioned above.

It is also of considerable interest that Scott Blair and co-workers[300] have found that there is no psychological barrier to making comparisons between viscosity and firmness. In experiments in which rubber balls (showing only elastic behavior) and balls of bitumen (showing only viscous behavior) were manually squeezed for a definite time, the subjects were able to reproduce their descriptions of the relative firmness of the two materials. The materials were rated as equally firm neither when the total deformations at the end of the squeezing time nor when the rates of change of deformation were equal, but rather when some intermediate entity was the same.

Bailey[21] has reprinted photomicrographs of plastic fats showing them to consist of crystals of fat enmeshing a considerable proportion of liquid oil. He has stated that in the range in which they are form-retaining, yet easily workable, fats appear to contain usually about 10 to 30% of solid material. It can be seen in Table 4.20 that milkfat meets these requirements for solids content within the temperature range of about 15 to 25°C.

The plasticity of milkfat and its variation under different conditions are a matter of interest in several ways. Of prime importance is their effect upon the spreadability and palatability of butter and upon its use as a shortening agent in baking.

Of the various factors which affect the plasticity of a fat, perhaps the most important is the proportion of solids present. Søltoft[321] has reported that, on varying the temperature, both the firmness and the viscosity of fats increase logarithmically with an increase in solids content. The solids contents of the fats were measured dilatometrically. This method depends upon the fact that, upon changing the temperature of a material, the volume change associated with a change of phase is greater than that of either the liquid or the solid. So, by heating the material in an apparatus designed to give either a measure of the volume changes at constant pressure or the pressure changes at constant volume, data can be obtained from which it is possible to calculate the proportion of liquid and solid materials. Bailey[21] has discussed the techniques of dilatometry, and has pointed out that this method involves uncertainties which limit its accuracy when applied to unknown mixtures. From dilatometric data, obtained with a Danish butterfat and an American butterfat, he has calculated the percentages of solids in these butterfats at different temperatures. The data are reproduced in Table 4.20.

The plastic properties of a fat depend not only on the proportion of solids but also on the physical properties of its crystals. These, in turn, are to a certain extent dependent upon the history of the fat as regards its thermal treatment and mechanical handling. Bailey[21] has stated his belief that crystals of high-melting glycerides would impart greater plasticity to a fat than would those of lower-melting

Table 4.20

VARIATION OF BUTTERFAT SOLIDS WITH TEMPERATURE[a]

Temperature, °C	Percent Solids	
	American Butterfat[b]	Danish Butterfat[c]
40	0	0
35	2.0	1.5
30	7.5	5.5
25	13.0	9.5
20	22.0	21.0
15	36.0	34.0
10	43.0	40.0
5	47.0	43.0

[a] Data from Table 63, reference 21.
[b] M.p. 37.7°C.
[c] M.p. 38.0°C.

glycerides. He has also pointed out that fats containing large crystals, produced by slow cooling or by other means, are in general softer than the same fats at the same solids content, but with the solids in the form of smaller crystals.

In addition to the factors mentioned, Søltoft[321] has concluded that the effect of temperature on the plastic properties of a fat is due not only to its effect on the solids content of the fat, but also to its effect on the viscosity of the liquid phase. He has attributed 30% of the temperature change of firmness and 50% of the temperature change of viscosity to a change in the viscosity of the liquid phase. While the effect of the viscosity of the liquid phase upon that of the fat might be expected, his conclusions regarding its effect upon firmness appear to be questionable.

Density

The density of milkfat varies with the fatty-acid composition of the triglycerides, with the nature and amounts of nontriglyceride material present in the fat, with the temperature, and with the physical state of the fat. In the introductory paragraph of this chapter reference was made to the fact that the composition of milkfat is altered by dairy processing. In this discussion a distinction will be made between milkfat as it occurs naturally in milk; as it is obtained in bulk by churning, with or without purification beyond that necessary to remove nonlipid materials; and, because of its commercial importance, as it occurs in the fat layer of a Babcock test.

Sommer[323] has disputed reports in the literature that the density of the fat in milk is related to the size of the globule. He noted that smaller globules have a higher proportion of surface-adsorbed phospholipids than do larger ones, and considered that this probably accounts for the reported density differences. Determination of the total solids or of the milk-solids-not-fat content of milk by means of a fat determination and a lactometer reading made on the whole milk involves the use of an equation containing, in one of its terms, a value for the density of milkfat. Many equations proposed for this purpose were found to be inaccurate if applied to samples other than those used for obtaining the data upon which their derivation was based. Sharp and Hart[303] concluded that a major source of error lay in the determination of the specific gravity of the milk before the fat had reached its phase equilibrium at the temperature of measurement. At the temperatures 15°C and 60°F, one of which is usually specified for specific gravity measurements, the establishment of phase equilibrium in milkfat is especially slow. It is even slower in the fat globules in whole milk than in bulk fat because, due to the isolation of individual fat globules from one another, fat crystals in one particle cannot act as seeds to induce crystallization in the remainder of the fat.

Bulk fat extracted from the milk by appropriate solvents contains the phospholipids associated with it in the whole milk. That obtained by churning has a lower content of phospholipids. Jenness *et al.*[150] conducted an extensive and careful study of the density of milkfat. They obtained the fat by churning, washed it with water, clarified it by centrifuging, dissolved it in acetone, and filtered it. They obtained an average density value of 0.8892 gm/ml at 60°C and an average coefficient of expansion of 78.34×10^{-5} ml/ml/°C for the temperature range 30 to 60°C. This corresponds to a density change of 0.00070 gm/ml/°C for this same temperature range. They were able to conclude that the density of the purified milkfat is relatively constant and is not affected to any marked extent by breed, season, or feed. The same investigators found that the material in the fat column of Babcock tests had an average density of 0.8918 at 60°C and an average coefficient of expansion of 75.58×10^{-5} ml/ml/°C for the temperature range 30 to 60°C.

The data of Sharp[302] (Fig. 4.6) show the density of milkfat for the temperature range 5 to 80°C. Sharp did not describe his method of preparing the milkfat. It is presumed to be similar to that of Jenness *et al.* A literature survey[354] of milkfat densities determined throughout Europe and the United States showed the values to be generally within 0.002 gm/ml of those of Sharp for that part of the curve represented by liquid fat. Values reported at 15°C or 60°F, however, varied over a range of 0.029 gm/ml. It seems probable that this was because measurements were made before phase equilibrium was established.[303]

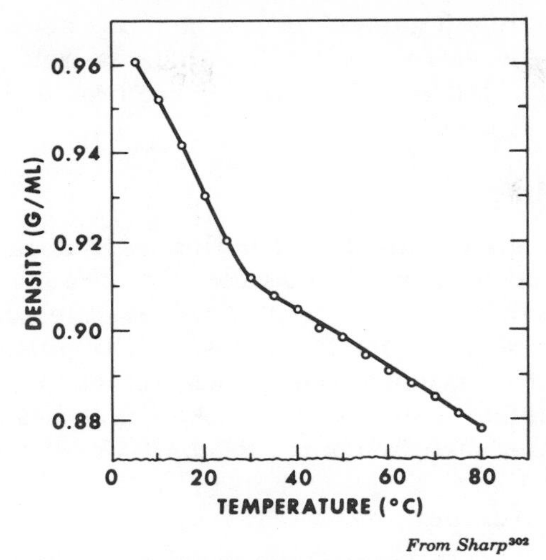

From Sharp[302]

FIG. 4.6. DENSITY OF MILKFAT

METHODS OF EXAMINATION

Milkfat, like other fats, can be characterized by the values obtained from chemical and physical tests. Specific procedures for many of these have been published elsewhere.[229] Some of these tests may be useful in detecting deterioration of the fat; this subject is treated in another chapter. Other tests may indicate the presence of adulterants.

Refractive Index

The refractive index of a fat is influenced by both the molecular weight and the degree of saturation of the component fatty acids. Its determination may be useful as an indication of adulteration but would seldom, in itself, be conclusive.

Iodine Number

The absorption of IBr (Hanus method) and ICl (Wijs method) gives a measure of the unsaturated acids present in a fat. One molecule of halogen compound is absorbed by each unsaturated linkage, and the absorption is expressed as the equivalent number of grams of iodine absorbed by 100 gm fat.

Reichert-Meissl Number

This value is equivalent to the number of milliliters of decinormal sodium hydroxide solution required to neutralize the soluble, volatile acids obtainable from 5 gm fat. Milkfat contains more of these acids than does any of the fats from which it might be desirable to distinguish it. The Canadian standard for milkfat[43] requires a Reichert-Meissl number of not less than 24.

Polenske Number

This value equals the number of milliliters of decinormal sodium hydroxide solution required to neutralize the insoluble volatile acids obtainable from 5 gm fat. As most of the volatile acids obtainable from milkfat are soluble, this test helps to distinguish between this fat and one of high caprylic acid content (e.g., coconut oil). The Canadian standard for milkfat[43] requires a Polenske number of not more than 3.5 and not greater than 10% of the Reichert-Meissl number.

Saponification Number

The saponification number is equivalent to the number of milligrams of potassium hydroxide required to saponify 1 gm fat. A butterfat that yields a high Reichert-Meissl value also gives a high saponification value, though the latter may drop considerably in the last stages of lactation. The two tests should be collated. A high saponification value and a low Reichert-Meissl value are grounds for suspicion of adulteration.

Some characteristic values of milkfat are given in Table 4.21. It is realized that data not covered by the author, if included, might alter the variation limits shown in this table.

Detection of Adulterants in Milkfat

There are two general ways in which adulteration of milkfat can be established. One involves the identification of a foreign material in the fat. The other involves the demonstration that a normal component of milkfat is present in proportions not encountered in authentic samples. The first method seems to be restricted to the detection of adulteration by vegetable fats.

The difference between the sterols of animal and vegetable fats affords a positive means for identifying adulteration by a vegetable fat. Phytosterol is not a single compound, however, but a mixture. Some of its forms give acetates with chemical and physical properties much closer to those of cholesteryl acetate than do the others, and may not be detectable by the crystallization method with certainty unless the adulterant is present in relatively high proportions. A much more sensitive method for detecting plant sterols has been described

Table 4.21

RANGE OF CHARACTERISTIC MILKFAT VALUES

Refractive index	1.4527–1.4566
Iodine no.	26–28
Saponification no.	220–241
Reichert-Meissl no.	23–33
Polenske no.	1–3

by Eisner *et al.*[73] The sterols of butter and of margarine oils were isolated and fractionated by chromatography of the unsaponifiable fractions on Florisil columns followed by gas-liquid chromatography of the sterol fractions. Cholesterol was the only sterol component found in any butter sample. Stigmasterol, β-sitosterol, and γ-sitosterol were found in all margarine samples. This method was successful in detecting as little as 1% margarine in butter.

Vitamin E exists in several modifications. Brown[39] found only α-tocopherol in cow's milk, while most of the vegetable oils he examined also had β-, γ- or δ-tocopherol. The Canadian standard[43] requires that the tocopherol content shall not exceed 50 μgm/gm of fat.

The detection of animal fat adulterants is less positive in that it depends upon establishing a quantitative change in the composition or the physical constants of the milkfat. The high proportion of the soluble, volatile fatty acids, particularly butyric acid, is the most prominent difference between that of milkfat and other animal fats. Tests such as the Reichert-Meissl number, the Kirschner number, and the direct determination of butyric acid can be used to establish gross adulteration of milkfat. With smaller proportions of adulterants the quantitative differences in these values of course become less pronounced. Legal proof, in such instances, is made still more difficult by the occasional reports in the literature of extreme values for butyric acid content and for such tests as the Reichert-Meissl number.

Of great interest in connection with the detection of adulterants in milkfat are the surveys of United States butterfat constants by Zehren and Jackson[363] and by Keeney.[167] Zehren and Jackson determined the Reichert-Meissl, Polenske, and refractive index values monthly for one year on butterfat obtained from 42 sources from 6 geographical areas having a coverage of 25 states. They analyzed a total of 500 samples. The Reichert-Meissl values ranged from 24.24 to 33.55, with an average value of 28.99. The Polenske values varied from 1.12 to 2.95, with an average value of 1.93. The refractive index varied from 1.4531 to 1.4557, with an average value of 1.4540. Keeney analyzed the samples for butyric acid content. He found that this value varied from 9.6 to 11.3 mole %, with a mean value of 10.41 mole %.

High-concentrate, low-roughage diets, as pointed out earlier, may decrease the proportion of butyric acid quite markedly. This introduces a further complication into the problem of detecting adulterants, as authentic milkfat from cows on such diets may have uncharacteristically low butyric acid values. Similarly, the high levels of linoleic acid that can now be introduced into milkfat may further distort the characteristic pattern of fatty acid composition. Bhalerao and Kummerow[25] have approached the adulteration problem from the viewpoint that it is more difficult to simulate the glyceride structure of milkfat than its fatty-acid composition. They proposed a method based upon a preliminary fractionation of the glycerides. Kuksis and McCarthy[186] have investigated analysis of the triglycerides by gas liquid chromatography as a means of detecting adulteration.

REFERENCES

1. ACKMAN, R. G., and BURGHER, R. D., Can. J. Biochem. Physiol., *41,* 2501 (1963).
2. ACKMAN, R. G., EATON, C. A., and HOOPER, S. N., Can. J. Biochem. Physiol., *46,* 197 (1968).
3. ADDA, J., Kieler milchw. ForschBer., *16,* 379 (1964).
4. ALLEN, R. R., and KIESS, A. A., J. Am. Oil. Chem. Soc., *32,* 400 (1955).
5. ALLISON, M. J., BRYANT, M. P., KATZ, I., and KEENEY, M., J. Bacteriol., *83,* 1084 (1962).
6. ANANTAKRISHNAN, C. P., BHALERAO, V. R., and PAUL, T. M., Arch. Biochem., *13,* 389 (1947).
7. ANKER, H. S., J. Biol. Chem., *194,* 177 (1952).
8. ANNISON, E. F., Biochem. J., *57,* 400 (1954).
9. ANNISON, E. F., HILL, K. J., and LEWIS, D., Biochem. J., *66,* 592 (1957).
10. ANNISON, E. F., and LEWIS, D., "Metabolism in the Rumen"., p. 68, John Wiley and Sons, Inc., New York, 1959.
11. ANTILLA, V., Meijeritiet. Aikakausk., *27,* (1) (1966); Dairy Sci. Abstr., *28,* [3250] (1966).
12. APPEL, H., BÖHM. H., KEIL, W., and SCHILLER, G., Z. Physiol. Chem., *282,* 220 (1947).
13. Association of Official Agricultural Chemists, "Official Methods of Analysis," 8th ed., p. 471, Washington, D.C., Association of Official Agricultural Chemists, 1955.
14. AST, H. J., and VANDER WAL, R. J., J. Am. Oil. Chem. Soc. *38,* 67 (1961).
15. AVENA, R. M., and KUMAR, S., Personal communication, 1962.
16. AVIGNAN, J., Biochim. Biophys. Acta, *125,* 607 (1966).
17. AYLWARD, F. X., BLACKWOOD, J. H., and SMITH, J. A. B., Biochem. J., *31,* 130 (1937).
18. BACKDERF, R. H., and BROWN, J. B., Arch. Biochem. Biophys., *76,* 15 (1958).
19. BAER, E., and KATES, M., J. Biol. Chem., *175,* 79 (1948); *185,* 615 (1950).
20. BAER, E., and MAURUKAS, J., J. Am. Chem. Soc., *74,* 158 (1952).
21. BAILEY, A. E., "Melting and Solidification of Fats," Interscience Publishers, Inc. New York, 1950.
22. BEARE, J. L., GREGORY, E. R. W., SMITH, D. M., and CAMPBELL, J. A., Can. J. Biochem. Physiol., *39,* 195 (1961).

23. BEERS, G. J. VAN, KEUNING, R., and SCHOGT, J. C. M., Abstr. Papers, Congr. Intern. Soc. Fat Res., 5th Gdansk, Poland, 1960.
24. BERTRAM, S. H., Biochem. Z., *197,* 433 (1928).
25. BHALERAO, V. R., and KUMMEROW, F. A., J. Dairy Sci., *39,* 956 (1956).
26. BLOOMFIELD, D. K., and BLOCH, K., Biochim. Biophys. Acta, *30,* 220 (1958); J. Biol. Chem., *235,* 337 (1960).
27. BOLDINGH, J., BEGEMANN, P. H., JONG, A. P. DE, and TAYLOR, R. J., Revue fr. Cps. gras, *13,* 327 (1966).
28. BOLDINGH, J., and TAYLOR, R. J., Nature, *194,* 909 (1962).
29. BOSWORTH, A. W., and SISSON, E. W., J. Biol. Chem., *107,* 489 (1934).
30. BOUDREAU, A., and MAN, J. M. DE, Biochim. Biophys. Acta, *98,* 47 (1965).
31. BOUDREAU, A., and MAN, J. M. DE, Can. J. Biochem. Physiol. *43,* 1799 (1965).
32. BRECKENRIDGE, W. C., and KUKSIS, A., J. Lipid Res., *9,* 388 (1968).
33. BRECKENRIDGE, W. C., and KUKSIS, A., Lipids, *4,* 197 (1969).
34. BRECKENRIDGE, W. C., MARAI, L., and KUKSIS, A., Can. J. Biochem. Physiol., *47,* 761 (1969).
35. BRESSLER, R., and WAKIL, S. J., J. Biol. Chem., *237,* 1441 (1962).
36. BREWINGTON, C. R., CARESS, E. A., and SCHWARTZ, D. P., J. Lipid Res., *11,* 355 (1970).
37. BROCKERHOFF, H., J. Lipid Res., *6,* 10 (1965).
38. BROGIONI, M., and BASI, G., Olearia, *20,* 80 (1966).
39. BROWN, F., Biochem. J., *51,* 237 (1952).
40. BRYANT, M. P., Personal communication, 1964.
41. BURGOS, J., HEMMING, F. W., PENNOCK, J. F., and MORTON, R. A., Biochem. J., *88,* 470 (1963).
42. BUSS, D. H., Lipids, *4,* 152 (1969).
43. Canada Dairy Products Regulations 2 (1)(j) (p. 7 of reprint "Canada Dairy Products Act-The Canada Dairy Products Regulations," extracted from The Canada Gazette, *88,* Part II, No. 23, Dec. 8, 1954).
44. CARTER, H. E., and FUJINO, Y., J. Biol. Chem., *221,* 879 (1956).
45. CARTER, H. E., SMITH, D. B., and JONES, D. N., J. Biol. Chem., *232,* 681 (1958).
46. CARTER, M. G. R., and MALKIN, T., J. Chem. Soc., *1947,* 554.
47. CHILSON, W. H., and SOMMER, H. H., J. Dairy Sci., *36,* 561 (1953).
48. CLARKSON, C. E., and MALKIN, T., J. Chem. Soc. *1934,* 666.
49. CLÉMENT, G., CLÉMENT, J., BEZARD, J., DI CASTANZO, G., and PARIS, R., Archs Sci. physiol., *16,* 237 (1962); J. Dairy Sci., *46,* 1423 (1963).
50. CMELIK, S. H. W., J. Sci. Fd. Agric., *13,* 662 (1962).
51. COLEMAN, M. H., and FULTON, W. C., *In* Enzymes of Lipid Metabolism, p. 127 (1961). P. DESNUELLE, ed., Pergamon Press, New York.
52. CORNFORTH, J. W., J. Lipid Res., *1,* 3 (1959).
53. COWIE, A. T., DUNCOMBE, W. G., FOLLEY, S. J., FRENCH, T. H., GLASCOCK, R. F., MASSART, L., PEETERS, G. J., and POPJÁK, G., Biochem. J., *48,* XXXIX; *49,* 610 (1951).
54. DAVIES, W. H., HEILBRON, I. M., and JONES, W. E., J. Chem. Soc., *1933,* 165.
55. DEBUCH, H., Biochem. J., *67,* 27P (1957).
56. DEBUCH, H., Z. Physiol. Chem., *314,* 49 (1959); *ibid.; 317,* 182 (1959).
57. DECAEN, C., and ADDA, J., 17th Int. Dairy Congr. *A,* 161 (1966).
58. DEENEN, L. L. M. VAN, and HAAS, G. H. DE, Biochim. Biophys. Acta, *70,* 538 (1963).
59. DE LA MARE, P. B. D., and SHORLAND, F. B., Nature, *153,* 380 (1944).
60. DEUEL, H. J., JR., "The Lipids," Vol. 1, Interscience Publishers, Inc., New York, 1951.
61. DEVRIES, B., J. Am. Oil Chem. Soc. *40,* 184 (1963).
62. DILS, R., and POPJÁK, G., Biochem. J., *83,* 41 (1962).
63. DIMICK, P. S., and HARNER, J. L., J. Dairy Sci., *51,* 22 (1962).
64. DIMICK, P. S., McCARTHY, R. D., and PATTON, S., J. Dairy Sci., *48,* 735 (1965).

65. DIMICK, P. S., and WALKER, N. J., J. Dairy Sci., *50,* 97 (1967).
66. DIMICK, P. S., and WALKER, H. M., J. Dairy Sci., *51,* 478 (1968).
67. DIMICK, P. S., WALKER, N. J., and PATTON, S., Agr. Fd. Chem., *47,* 649 (1969).
68. DOERSCHUK, A. P., and DAUBERT, B. F., J. Am. Oil Chem. Soc., *25,* 425 (1948).
69. DUIN, H. VAN, Neth. Milk Dairy J., *12,* 90 (1958).
70. DUNCAN, W. R. H., and GARTON, G. A., J. Dairy Res., *33,* 255 (1966).
71. DUTTON, H. J., J. Am. Oil Chem. Soc., *38,* 660 (1961).
72. EICHBERG, J., GILBERTSON, J. R., and KARNOVSKY, M. L., J. Biol. Chem., *236,* PC 15 (1961).
73. EISNER, J., WONG, N. P., FIRESTONE, D., and BOND, J., J. Assoc. Offic. Agr. Chemists, *45,* 337 (1962).
74. EL-SHAZLY, K., Biochem. J., *51,* 640 (1952).
75. EL-SHAZLY, K., Biochem. J., *51,* 647 (1952).
76. ENSOR, W. L., SHAW, J. C., and TELLECHEA, H. F., J. Dairy Sci., *42,* 189 (1959).
77. ENTRESSANGLES, B., PASERO, L., SAVARY, P., SARDA, L., and DESNUELLE, P., Bull. Soc. Chim. Biol., *43,* 581 (1961).
78. FEUGE, R. O., PEPPER, M. B., JR., O'CONNOR, R. T., and FIELD, E. T., J. Am. Oil. Chem. Soc., *28,* 420 (1951).
79. FILER, L. J., JR., SIDHU, S. S., DAUBERT, B. F., and LONGENECKER, H. E., J. Am. Chem. Soc., *68,* 167 (1946).
80. FISCHER, H. O. L., and BAER, E., Chem. Rev., *29,* 287 (1941).
81. FITELSON, J., J. Assoc. Offic. Agr. Chemists, *26,* 506 (1943).
82. FLEISCHMANN, W., "Lehrbuch der Milchwirtschaft" (rev. by H. WEIGMANN), 7th ed., p. 89, Paul Parey, Berlin (1932).
83. FOLCH, J., J. Biol. Chem., *146,* 35 (1942).
84. FOLLEY, S. J., Biochem. Soc. Symp. (Cambridge, Engl.), No. *9,* 52 (1952).
85. FOLLEY, S. J., and FRENCH, T. H., Nature, *161,* 933 (1948); Biochem. J., *42,* XLVII (1948); Biochem. J., *43,* LV (1948); Nature, *163,* 174 (1949); Biochem. J., *45,* 117 (1949); Biochem. J., *45,* 270 (1949); Biochem. J., *46,* 465 (1950).
86. FOLLEY, S. J., and McNAUGHT, M. L., "Milk: the Mammary Gland and Its Secretion," Vol. 1, Chapt. 12, S. K. KON and A. T. COWIE, editors, Academic Press, New York, 1961.
87. FRANKEL, E. N., SMITH, L. M., and JACK, E. L., J. Dairy Sci., *41,* 483 (1958).
88. FREEMAN, C. P., JACK, E. L., and SMITH, L. M., J. Dairy Sci., *48,* 853 (1965).
89. FRENCH, T. H., and POPJÁK, G., Biochem. J., *49,* III (1951).
90. FULCO, A. J., and MEAD, J. F., J. Biol. Chem., *234,* 1411 (1959).
91. GALANOS, D. S., and KAPOULAS, V. M., Biochim. Biophys. Acta, *98,* 293, 313 (1965).
91a. GANGULY J., Biochim. Biophys. Acta, *40,* 110 (1960).
92. GARTON, G. A., HOBSON, P. N., and LOUGH, A. K., Nature, *182,* 1511 (1958).
93. GARTON, G. A., LOUGH, A. K., and VIOQUE, E., Biochem. J., *73,* 46P (1959).
94. GERSON, T., HAWKE, J. C., SHORLAND, F. B., and MELHUISH, W. H., Biochem. J., *74,* 366 (1960).
95. GERSON, T., SHORLAND, F. B., and BARNICOAT, C. R., Biochem. J., *68,* 644 (1958).
96. GEYER, R. P., NATH, H., BARKI, V. H., ELVEHJEM, C. A., and HART, E. B., J. Biol. Chem., *169,* 227 (1947).
97. GILLAM, A. E., and HEILBRON, I. M., Biochem. J., *29,* 834 (1935).
98. GILLAM, A. E., HEILBRON, I. M., MORTON, R. A., BISHOP, G., and DRUMMOND, J. C., Biochem. J., *27,* 878 (1933).
99. GLASCOCK, R. F., DUNCOMBE, W. G., and REINIUS, L. R., Biochem. J., *62,* 535 (1956).
100. GLASCOCK, R. F., and REINIUS, L. R., Biochem J., *62,* 529 (1956).
101. GLASS, R. L., and JENNESS, R., Comp. Biochem. Physiol., *38B,* 353 (1971).
102. GLASS, R. L., TROOLIN, H. A., and JENNESS, R., Comp. Biochem. Physiol., *22,* 415 (1967).

103. GOODWIN, T. W., "Recent Advances in Biochemistry," Little, Brown and Company, Boston, 1960.
104. GRAHAM, W. R., JR., HOUCHIN, O. B., PETERSON, V. E., and TURNER, C. W., Am. J. Physiol., *122,* 150 (1938).
105. GREEN, D. E., and WAKIL, S. J., "Lipide Metabolism," K. BLOCH, editor, John Wiley and Sons, Inc., New York, 1960.
106. HALLGREN, B., and LARSSON, S., J. Lipid Res., *3,* 39 (1962).
107. HANAHAN, D. J., "Lipide Chemistry," John Wiley and Sons, Inc., New York, 1960.
108. HANSEN, R. P., J. Dairy Res., *36,* 177 (1969).
109. HANSEN, R. P., and MORRISON, J. D., Biochem. J., *93,* 225 (1964).
110. HANSEN, R. P., and SHORLAND, F. B., Biochem. J., *50,* 207 (1951).
111. HANSEN, R. P., and SHORLAND, F. B., Biochem. J., *50,* 358 (1952).
112. HANSEN, R. P., and SHORLAND, F. B., Biochem. J., *52,* 207 (1952).
113. HANSEN, R. P., SHORLAND, F. B., and COOKE, N. J., Chem. Ind. (London), *1951,* 839.
114. HANSEN, R. P., SHORLAND, F. B., and COOKE, N. J., Chem. Ind. (London), *1959,* 124.
115. HANSEN, R. P., SHORLAND, F. B., and COOKE, N. J., Biochem. J., *57,* 297 (1954).
116. HANSEN, R. P., SHORLAND, F. B., and COOKE, N. J., Biochem. J., *58,* 358 (1954).
117. HANSEN, R. P., SHORLAND, F. B., and COOKE, N. J., Biochem. J., *77,* 64 (1960).
118. HARTMAN, L., SHORLAND, F. B., and McDONALD, I. R. C., Nature, *174,* 185 (1954); Biochem. J., *61,* 603 (1955).
119. HAWKE, J. C., J. Dairy Res., *24,* 366 (1957).
120. HAWKE, J. C., J. Lipid Res., *4,* 255 (1963).
121. HAY, J. D., and MORRISON, W. R., Biochim. Biophys. Acta, *202,* 237 (1970).
122. HERB, S. F., MAGIDMAN, P., LUDDY, F. E., and RIEMENSCHNEIDER, R. W., J. Am. Oil Chem. Soc., *39,* 142 (1962).
123. HILDITCH, T. P., "The Chemical Constitution of Natural Fats," 2nd ed., p. 306, Chapman and Hall, Ltd., London, 1947.
124. HILDITCH, T. P., "The Chemical Constitution of Natural Fats," 3rd ed., John Wiley and Sons, Inc., New York, 1956.
125. HILDITCH, T. P., J. Am. Oil Chem. Soc., *31,* 433 (1954).
126. HILDITCH, T. P., and JASPERSON, H., J. Soc. Chem. Ind. (London), *60,* 305 (1941).
127. HILDITCH, T. P., and JASPERSON, H., Biochem. J., *38,* 443 (1944).
128. HILDITCH, T. P., and JASPERSON, H., J. Soc. Chem. Ind. (London), *64,* 109 (1945).
129. HILDITCH, T. P., and LONGENECKER, H. E., J. Biol. Chem., *122,* 497 (1938).
130. HILDITCH, T. P., and SLEIGHTHOLME, J. J., Biochem. J., *24,* 1098 (1930).
131. HILDITCH, T. P., and THOMPSON, H. M., Biochem. J., *30,* 677 (1936).
132. HIRAYAMA, O., and NAKAE, T., Agr. Biol. Chem., *28,* 201 (1964).
133. HIRSCHMANN, H., J. Biol. Chem., *235,* 2762 (1960).
134. HOFLUND, S., HOLMBERG, J., and SELLMANN, G., Cornell Vet., *45,* 254 (1955).
135. HOLM, G. E., WRIGHT, P. A., and DEYSHER, E. F., J. Dairy Sci., *19,* 631 (1936).
136. HOLMAN, R. T., Proc. Soc. Exp. Biol. Med., *76,* 100 (1951).
137. HONKANEN, E., MOISIO, T., KARVONEN, P., VIRTANEN, A. I., and PAASIVIRTA, J., Acta Chem. Scand., *22,* 2041 (1968).
138. HOPKINS, C. Y., J. Am. Oil Chem. Soc., *38,* 664 (1961).
139. HOSOGAI, Y., Jap. J. Zootech. Sci., *35,* 137 (1964); Dairy Sci. Abstr., *26* [3666] (1964).
140. HOWTON, D. R., and MEAD, J. F., J. Biol. Chem., *235,* 3385 (1960).
141. INSULL, W., JR., and AHRENS, E. H., JR., Biochem. J., *72,* 27 (1959).
142. INTRIERI, F., and BADOLATO, F., Acta Med. Vet. Napoli, *11,* 473 (1965).

143. IUPAC-IUB Commission on Biochemical Nomenclature, European J. Biochem., *2*, 127 (1967); J. Lipid Res., *8*, 523 (1967).
144. IVERSON, J. L., EISNER, J., and FIRESTONE, D., J. Am. Oil Chem. Soc., *42*, 1063 (1965).
145. JACK, E. L., FREEMAN, C. P., SMITH, L. M., and MICKLE, J. B., J. Dairy Sci., *46*, 284 (1963).
146. JAMES, A. T., PEETERS, G., and LAURYSSENS, M., Biochem. J., *64*, 726 (1956).
147. JAMES, A. T., and WEBB, J., Biochem. J., *66*, 515 (1957).
148. JANGAARD, P. M., and KE, P. J., J. Fisheries Res. Board Can., *25*, 2419 (1968).
149. JART, A., FUNCH, J. P., and DAM, H., Acta Chem. Scand., *13*, 1910 (1959).
150. JENNESS, R., HERREID, E. O., CAULFIELD, W. J., BURGWALD, L. H., JACK, E. L., and TUCKEY, S. L., J. Dairy Sci., *25*, 949 (1942).
151. JENSEN, R. G., GANDER, G. W., and DUTHIE, A. H., J. Dairy Sci., *42*, 1913 (1959).
152. JENSEN, R. G., GANDER, G. W., and SAMPUGNA, J., J. Dairy Sci., *45*, 329 (1962).
153. JENSEN, R. G., and SAMPUGNA, J., J. Dairy Sci., *49*, 460 (1966).
154. JENSEN, R. G., SAMPUGNA, J., CARPENTER, D. L., and PITAS, R. E., J. Dairy Sci., *50*, 231 (1967).
154a. JENSEN, R. G., SAMPUGNA, J., and GANDER, G. W., J. Dairy Sci., *44*, 1983 (1961).
155. JENSEN, R. G., SAMPUGNA, J., and PEREIRA, R. L., J. Dairy Sci., *47*, 727 (1964).
156. JONG, K. DE, and WEL, H. VAN DER, Nature, *202*, 553 (1964).
157. JOO, C. N., and KATES, J., Biochim. Biophys. Acta, *176*, 278 (1969).
158. JURRIENS, G., and OELE, J. M., J. Am. Oil Chem. Soc., *42*, 857 (1965).
159. KARLSSON, K-A., Lipids, *5*, 878 (1970).
160. KARNOVSKY, M. L., and BRUMM, A. F., J. Biol. Chem., *216*, 689 (1955).
161. KARNOVSKY, M. L., and WOLFF, D., 4th Intern. Congr. Biochem., Vienna 1–6 Sept. 1958. Proc. *15*, 208 (1960).
162. KARTHA, A. R. S., J. Am. Oil Chem. Soc., *30*, 326 (1953).
163. KATES, M., PALAMETA, B., and YENGOYAN, L. S., Biochem., *4*, 1595 (1965).
164. KATZ, I., and KEENEY, M., J. Dairy Sci., *46*, 605 (1963).
165. KAUFMANN, H. P., VOLBERT, F., and MANKEL, G., Fette, Seifen, Anstrichmittel, *63*, 261 (1961).
166. KAWASHIRO, I., TANABE, H., and ISHII, A., Shokuhin Eiseigaku Zasshi, *1*, 78 (1960); Chem. Abstr., 2629 (1961).
167. KEENEY, M., J. Assoc. Offic. Agr. Chemists, *39*, 212 (1956).
168. KEENEY, M., Personal communication, 1964.
169. KEENEY, M., KATZ, I., and ALLISON, M. J., J. Am. Oil Chem. Soc., *39*, 198 (1962).
170. KEENEY, M., KATZ, I., and SCHWARTZ, D. P., Biochim. Biophys. Acta, *62*, 615 (1962).
171. KEENEY, P. G., and PATTON, S., J. Dairy Sci., *39*, 1104, 1114 (1956).
172. KENDALL, J., *In* "International Critical Tables," Vol. VI, p. 259, McGraw-Hill, New York (1929).
173. KENNEDY, E. P., Federation Proc., *20*, 934 (1961).
174. KINSELLA, J. E., and McCARTHY, R. D., Biochim. Biophys. Acta, *164*, 518, 530, 540 (1968).
175. KINSELLA, J. E., PATTON, S., and DIMICK, P. S., J. Am. Oil Chem. Soc., *44*, 202 (1967).
176. KINSELLA, J. E., PATTON, S., and DIMICK, P. S., J. Am. Oil Chem. Soc., *44*, 449 (1967).
177. KINTNER, J. A., and DAY, E. A., J. Dairy Sci., *48*, 1575 (1965).
178. KISHIMOTO, Y., and RADIN, N. S., J. Lipid Res., *4*, 437 (1963).
179. KISHIMOTO, Y., and RADIN, N. S., Lipids, *1*, 47 (1966).

180. KLEIBER, M., SMITH, A. H., BLACK, A. L., BROWN, M. A., and TOLBERT, B. M., J. Biol. Chem., *197,* 371 (1952).
181. KLENK, E., and MOHRHAUER, H., Z. Physiol. Chem. *320,* 218 (1960).
182. KONTSON, A., TAMSMA, A., KURTZ, F. E., and PALLANSCH, M. J., J. Dairy Sci., *53,* 410 (1970).
183. KORNBERG, A., and PRICER, W. E., JR., J. Biol. Chem., *204,* 329 (1953).
184. KORNBERG, A., and PRICER, W. E., JR., J. Biol. Chem., *204,* 345 (1953).
185. KRITCHEVSKY, D., "Cholesterol," John Wiley and Sons, Inc., New York, 1958.
186. KUKSIS, A., and McCARTHY, M. J., J. Am. Oil Chem. Soc., *41,* 17 (1964).
187. KUMAR, S., and LALKA, K., Federation Proc., *19,* 228 (1960).
188. KUMAR, S., and PYNADATH, T. I., Personal communication, 1964.
189. KUMAR, S., PYNADATH, T. I., and LALKA, K., Biochim. Biophys. Acta, *42,* 373 (1960).
190. KURTZ, F. E., J. Am. Chem. Soc. *74,* 1902 (1952).
191. KURTZ, F. E., and HOLM, G. E., J. Dairy Sci. *22,* 1011 (1939).
192. KURTZ, F. E., JAMIESON, G. S., and HOLM, G. E., J. Biol. Chem., *106,* 717 (1934).
193. KURTZ, F. E., TAMSMA, A., and PALLANSCH, M. J., J. Dairy Sci., *54,* 173 (1971).
194. KUZDZAL, S., Fac. Sci. Univ. Paris: Thèses (1964).
195. KUZDZAL-SAVOIE, S., and KUZDZAL-SAVOIE, R., Ann. Biol. Anim. Biochim. Biophys., *5,* 497 (1965).
196. KUZDZAL-SAVOIE, S., and KUZDZAL, W., Ann. Technol. Agr., *10,* 73 (1961).
197. LANE, C. B., and KEENEY, M., J. Dairy Sci., *47,* 665 (1964).
198. LANGDON, R. G., J. Am. Chem. Soc., *77,* 5190 (1955); J. Biol. Chem., *226,* 615 (1957).
199. LANGLER, J. E., and DAY, E. A., J. Dairy Sci., *47,* 1291 (1964).
200. LAURYSSENS, M., VERBEKE, R., and PEETERS, G., J. Lipid Res., *2,* 383 (1961).
201. LEFFEL, E. C., Proc. Nutr. Conf. Feed Mfrs., Univ. Maryland, 1960, 93.
202. LITCHFIELD, C., ISBELL, A. F., and REISER, R., J. Am. Oil Chem. Soc., *39,* 330 (1962).
203. LONEY, B. E., and BASSETTE, R., J. Dairy Sci., *54,* 343 (1971).
204. LOPÉZ LARENZO, P., SANZ PÉREZ, B., and BURGOS, J., Anales Bromatol., *14,* 205 (1962); Dairy Sci. Abstr., *25* [2963] (1963).
205. LOUGH, A. K., and GARTON, G. A., Biochem. J., *67,* 345 (1957).
206. LUICK, J. R., J. Dairy Sci., *43,* 1344 (1960).
207. LUICK, J. R., J. Dairy Sci., *44,* 652 (1961).
208. LUICK, J. R., and SMITH, L. M., J. Dairy Sci., *46,* 1251 (1963).
209. LUTTON, E. S., J. Am. Chem. Soc., *70,* 248 (1948).
210. LYNEN, F., Fèderation Proc., *20,* 941 (1961).
211. MAGIDMAN, P., HERB, S. F., BARFORD, R. A., and RIEMENSCHNEIDER, R. W., J. Am. Oil Chemists' Soc., *39,* 137 (1962).
212. MARAI, L., BRECKENRIDGE, W. C., and KUKSIS, A., Lipids, *4,* 562 (1969).
213. MARINETTI, G. V., J. Lipid Res., *3,* 1 (1962).
214. MARINETTI, G., BERRY, J. F., ROUSER, G., and STOTZ, E., J. Am. Chem. Soc., *75,* 313 (1953).
215. MARINETTI, G. V., ERBLAND, J., and STOTZ, E., J. Am. Chem. Soc., *80,* 1624 (1958); *ibid., 81,* 861 (1959).
216. MARKLEY, K. S., "Fatty Acids," 2nd ed., Part 1, Interscience Publishers, Inc., New York, 1960.
217. MATTSON, F. H., and BECK, L. W., J. Biol. Chem., *214,* 115 (1955).
218. MATTSON, F. H., and LUTTON, E. S., J. Biol. Chem., *233,* 868 (1958).
219. MATTSON, F. H., and VOLPENHEIN, R. A., J. Biol. Chem., *236,* 1891 (1961).
220. MATTSSON, S., Rep. Milk Dairy Res. Alnarp, *74* (1966).
221. McCARTHY, R. D., J. Agr. Food Chem., *10,* 126 (1962).
222. McCARTHY, R. D., PATTON, S., and EVANS, L. J. Dairy Sci., *43,* 1196 (1960).

223. McCULLAGH, K. A., LINCOLN, H. G., and SOUTHGATE, D. A. T., Nature, Lond., *222,* 493 (1969).
224. MEAD, J. F., J. Biol. Chem., *227,* 1025 (1957).
225. MEAD, J. F., Federation Proc., *20,* 952 (1961).
226. MEAD, J. F., and HOWTON, D. R., J. Biol. Chem., *229,* 575 (1957).
227. MEAD, J. F., and NEVENZEL, J. C., J. Lipid Res. *1,* 305 (1960).
228. MEAD, J. F., STEINBERG, G., and HOWTON, D. R., J. Biol. Chem., *205,* 683 (1953).
229. MEHLENBACHER, V. C., "The Analysis of Fats and Oils," The Garrard Press, Champaign, Illinois, 1960.
230. MOCHRIE, R. D., and TOVE, S. B., Cited by S. B. TOVE, J. Dairy Sci., *43,* 1354 (1960).
231. MOODY, E. G., VAN SOEST, P. J., McDOWELL, R. E., and FORD, G. L., J. Dairy Sci., *54,* 1457 (1971).
232. MORICE, I. M., J. Chem. Soc., *1951,* 1200.
233. MORRIS, L. J., HOLMAN, R. T., and FONTELL, K., J. Lipid Res., *1,* 412 (1960).
234. MORRISON, W. R., Lipids, *3,* 107 (1968).
235. MORRISON, W. R., Biochim. Biophys. Acta, *176,* 537 (1969).
236. MORRISON, W. R., and HAY, J. D., Biochim. Biophys. Acta, *202,* 460 (1970).
237. MORRISON, W. R., JACK, E. L., and SMITH, L. M., J. Am. Oil Chem. Soc., *42,* 1142 (1965).
238. MORRISON, W. R., and SMITH, L. M., J. Dairy Sci., *49,* 701 (1966); Lipids, *2,* 178 (1967).
239. NASH, H. A., and ZSCHEILE, F. P. Arch. Biochem., *7,* 305 (1945).
240. NATAF, B., MICKELSEN, O., KEYS, A., and PETERSEN, W. E., J. Nutr., *36,* 495 (1948).
241. NIEMAN, C., GROOT, E. H., and ROOSELAAR, W. J., Acta Physiol. Pharmacol. Neerl., *1,* 488 (1950).
242. NORRIS, F. A., and MATTIL, K. F., J. Am. Oil Chem. Soc., *24,* 274 (1947).
243. NORTON, W. T., GOTTFRIED, E. L., and RAPPORT, N. M., J. Lipid Res., *3,* 456 (1962).
244. O'CONNOR, R. T., J. Am. Oil Chem. Soc., *32,* 616 (1955).
245. O'CONNOR, R. T., "Fatty Acids," 2nd ed., Part I, I. S. MARKLEY, Editor, Interscience Publishers, Inc., New York, 1960.
246. O'CONNOR, R. T., J. Am. Oil Chem. Soc., *38,* 641,648 (1961).
247. O'CONNOR, R. T., (chairman), *et al.,* J. Am. Oil Chem. Soc., *36,* 627 (1959).
248. OGSTON, A. G., Nature, Lond., *162,* 963 (1948).
249. PARKS, O. W., KEENEY, M., KATZ, I., and SCHWARTZ, D. P., J. Lipid Res., *5,* 232 (1964).
250. PARKS, O. W., KEENEY, M., and SCHWARTZ, D. P., J. Dairy Sci., *44,* 1940 (1961).
251. PARKS, O. W., SCHWARTZ, D. P., KEENEY, M., and DAMICO, J. N., Nature, Lond., *210,* 417 (1966).
252. PARLIMENT, T. H., NAWAR, W. W., and FAGERSON, I. S., J. Dairy Sci., *48,* 615 (1965).
253. PARLIMENT, T. H., NAWAR, W. W., and FAGERSON, I. S., J. Dairy Sci., *49,* 1109 (1966).
254. PARRY, R. M., JR., SAMPUGNA, J., and JENSEN, R. G., J. Dairy Sci., *47,* 37 (1964).
255. PATTON, S., U. S. Patent 3,127,275 (1964).
256. PATTON, S., EVANS, L., and McCARTHY, R. D., J. Dairy Sci., *43,* 95 (1960).
257. PATTON, S., and McCARTHY, R. D., J. Dairy Sci., *46,* 396 (1963).
258. PATTON, S., and McCARTHY, R. D., J. Dairy Sci., *46,* 916 (1963).
259. PATTON, S., McCARTHY, R. D., EVANS, L., JENSEN, R. G., and GANDER, G. W., J. Dairy Sci., *45,* 248 (1962).
260. PATTON, S., McCARTHY, R. D., EVANS, L., and LYNN, T. R., J. Dairy Sci., *43,* 1187 (1960).

260a. PATTON, S., and THARP, B. W., J. Dairy Sci., *42,* 49 (1959).
261. PEETERS, G., COUSSENS, R., and SIERENS, G., Arch. Intern. Pharmacodyn., *95,* 153 (1953).
262. PENNINGTON, R. J., Biochem. J., *51,* 251 (1952).
263. PITAS, R. E., SAMPUGNA, J., and JENSEN, R. G., J. Dairy Sci., *50,* 1332 (1967).
264. POPJAK, G., Nutr. Abstr. & Rev., *21,* 535 (1952).
265. POPJÁK, G., FRENCH, T. H., and FOLLEY, S. J., Biochem. J., *46,* XXVIII (1950); Biochem. J., *48,* 411 (1951).
266. POPJÁK, G., FRENCH, T. H., HUNTER, G. D., and MARTIN, A. J. P., Biochem. J., *48,* 612 (1951).
267. POPJÁK, G., HUNTER, G. D., and FRENCH, T. H., Biochem. J., *54,* 238 (1953).
268. QUIMBY, O. T., WILLE, R. L., and LUTTON, E. S., J. Am. Oil Chem. Soc., 30, 186 (1953).
269. RAPPORT, M. M., LERNER, B., ALONZO, N., and FRANZL, R. E., Abstr. Papers, Am. Chem. Soc., Meeting, 130th, Atlantic City, 1956, 66c; J. Biol. Chem., *225,* 859 (1957).
270. READ, W. C., and SARRIF, A., Am. J. Clin. Nutr., *17,* 177 (1965).
271. REISER, R., and REDDY, H. G. R., J. Am. Oil Chem. Soc., *33,* 155 (1956).
272. RENNER, E., and SENFT, B., Zuchtungskunde *43,* 26 (1971).
273. RICHARDSON, C. W., JOHNSON, H. D., GEHRKE, C. W., and GOERLITZ, D. F., J. Dairy Sci., *44,* 1937 (1961).
274. RIEMENSCHNEIDER, R. W., HERB, S. F., and NICHOLS, P. L., JR., J. Am. Oil Chem. Soc., *26,* 371 (1949).
275. ROGERS, T. A., and KLEIBER, M., Biochim. Biophys. Acta, *22,* 284 (1956).
276. ROSSITER, R. J., and STRICKLAND, K. P., "Lipide Metabolism," K. BLOCH, editor, John Wiley and Sons, Inc., New York (1960).
277. ROUSER, G., BERRY, J. F., MARINETTI, G., and STOTZ, E., J. Am. Chem. Soc., *75,* 310 (1953).
278. RYHAGE, R., J. Dairy Res., *34,* 115 (1967).
279. SAITO, T., and ONUMA, T., Nakanishi Jap. J. Dairy Sci., *19,* A121 (1970); Dairy Sci. Abstr., *33* [2609] (1971).
280. SAMBASIVARAO, K., and BROWN, J. B., J. Am. Oil Chem. Soc., *39,* 340 (1962).
281. SAMPIERI, M., VOLFRE, F., and MALETTO, S., Riv. Zooteck. Agr. Vet., *7,* 164 (1969).
282. SAVARY, P., and DESNUELLE, P., Biochim. Biophys. Acta, *21,* 349 (1956).
283. SCHAMBYE, P., WOOD, H. G., and POPJÁK, G., J. Biol. Chem. *206,* 875 (1954).
284. SCHEUERBRANDT, G., GOLDFINE, H., BARONOWSKY, P. E., and BLOCH, K., J. Biol. Chem., *236,* PC70 (1961).
285. SCHLENK, W., JR., J. Am. Oil Chem. Soc., *42,* 945 (1965).
286. SCHOENHEIMER, R., Z. Physiol. Chem., *192,* 77 (1930).
287. SCHOENHEIMER, R., and RITTENBERG, D., J. Biol. Chem., *113,* 505 (1936).
288. SCHOGT, J. C. M., and BEGEMANN, P. H., J. Lipid Res., *6,* 466 (1965).
289. SCHOGT, J. C. M., HAVERKAMP BEGEMANN, P., and KOSTER, J., J. Lipid Res. *1,* 446 (1960).
290. SCHOGT, J. C. M., HAVERKAMP BEGEMANN, P., and RECOURT, J. H., J. Lipid Res., *2,* 142 (1961).
291. SCHWARTZ, D. P., Personal communication, 1971.
292. SCHWARTZ, D. P., BURGWALD, L. H., SHAMEY, J., and BREWINGTON, C. R., J. Dairy Sci., *51,* 929 (1968).
292a. SCHWARTZ, D. P., PARKS, O. W., and YONCOSKIE, R. A., J. Am. Oil Chem. Soc., *43,* 128 (1966).
293. SCHWARTZ, D. P., SPIEGLER, P. S., and PARKS, O. W., J. Dairy Sci., *48,* 1387 (1965).
294. SCHWARTZ, D. P., and WEIRAUCH, J. L., J. Dairy Sci., *53,* 637 (1970).
295. SCHWARTZ, P., and CARTER, H. E., Proc. Nat. Acad. Sci. U.S.A., *40,* 499 (1954).
296. SCHWIK, R. W., and NAKAO, A., J. Biol. Chem., *206,* 883 (1954).

297. SCOTT, T. W., COOK, L. J., FERGUSON, K. A., McDONALD, I. W., BUCHANAN, R. A., and LOFTUS HILLS, G., Australian J. Sci., *32,* 291 (1970).
298. SCOTT, T. W., COOK, L. J., and MILLS, S. C., J. Am. Oil Chem. Soc., *48,* 358 (1971).
299. SCOTT, W. E., HERB, S. F., MAGIDMAN, P., and RIEMENSCHNEIDER, R. W., J. Agr. Food Chem., *7,* 125 (1959).
300. SCOTT BLAIR, G. W., and BARON, M., Intern. Dairy Congr., Proc., 12th, Stockholm 1949, *2,* 49.
301. SEUBERT, W., GREULL, G., and LYNEN, F., Angew. Chem., *69,* 359 (1957).
302. SHARP, P. F., J. Dairy Sci., *11,* 259 (1928).
303. SHARP, P. F., and HART, R. G., J. Dairy Sci., *19,* 683 (1936).
304. SHAW, J. C., and ENSOR, W. L., J. Dairy Sci., *42,* 1238 (1959).
305. SHAW, J. C., and KNODT, C. B., J. Biol. Chem., *138,* 287 (1941).
306. SHORLAND, F. B., Nature, *166,* 745 (1950).
307. SHORLAND, F. B., Ann. Rev. Biochem., *25,* 101 (1956).
308. SHORLAND, F. B., GERSON, T., and HANSEN, R. P., Biochem. J., *59,* 350 (1955).
309. SHORLAND, F. B., GERSON, T., and HANSEN, R. P., Biochem. J., *61,* 702 (1955).
310. SHORLAND, F. B., and HANSEN, R. P., Dairy Sci. Abstr., *19,* 167 (1957).
311. SHORLAND, F. B., WEENINK, R. O., and JOHNS, A. T., Nature, *175,* 1129 (1955).
312. SHORLAND, F. B., WEENINK, R. O., JOHNS, A. T., and McDONALD, I. R. C., Biochem. J., *67,* 328 (1957).
313. SIEK, T. J., ALBIN, I. A., SATHER, L. A., and LINDSAY, R. C., J. Dairy Sci., *54,* 1 (1971).
314. SMITH, J. A. B., and DASTUR, N. N., Biochem. J., *32,* 1868 (1938).
315. SMITH, L. M., DAIRIKI, T., DUNKLEY, W. L., and RONNING, M., Intern. Dairy Congr., 17th, A199 (1966).
316. SMITH, L. M., and FREEMAN, N. K., J. Dairy Sci., *42,* 1450 (1959).
317. SMITH, L. M., FREEMAN, C. P., and JACK, E. L., J. Dairy Sci., *47,* 665 (1964); *ibid.*, *48,* 531 (1965).
318. SMITH, L. M., and LOWRY, R. R., J. Dairy Sci. *45,* 581 (1962).
319. SMITH, L. M., and RONNING, M., J. Dairy Sci., *44,* 1170 (1961).
320. SMITH, S. W., WEISS, S. B., and KENNEDY, E. P., J. Biol. Chem., *228,* 915 (1957).
321. SØLTOFT, P., "On the Consistency of Mixtures of Hardened Fats," (trans. from the Danish by E. CHRISTENSEN) Bjarne Kristensen-Bogtrykkeri, Copenhagen (1947).
322. SOMMER, H. H., Cherry-Burrell Circle, *37,* (5–6), 3 (1952).
323. SOMMER, H. H., Cherry-Burrell Circle, *37* (7–8), 7 (1952).
324. SONNEVELD, W., HAVERKAMP BEGEMANN, P., BEERS, G. J. VAN, KEUNING, R., and SCHOGT, J. C. M., J. Lipid Res., *3,* 351 (1962).
325. STEDMAN, R. L., Chem. Rev., *68,* 153 (1968).
326. STEIN, R. A., J. Am. Oil Chem. Soc., *38,* 636 (1961).
327. STEINBERG, G., SLATON, W. H., JR., HOWTON, D. R., and MEAD, J. F., J. Biol. Chem., *220,* 257 (1956).
328. STEINBERG, G., SLATON, W. H., JR., HOWTON, D. R., and MEAD, J. F., J. Biol. Chem., *224,* 841 (1957).
329. STETTEN, D., JR., and SCHOENHEIMER, R., J. Biol. Chem., *133,* 329 (1940).
330. STINE, C. M., and PATTON, S., J. Dairy Sci., *36,* 516 (1953).
331. STOFFEL, W., and AHRENS, E. H., JR., J. Am. Chem. Soc., *80,* 6604 (1958).
332. STOFFEL, W., INSULL, W., JR., and AHRENS, E. H., JR., Proc. Soc. Exp. Biol. Med., *99,* 238 (1958).
333. STULL, J. W., and BROWN, W. H., J. Dairy Sci., *47,* 676 (1964).
334. STULL, J. W., and BROWN, W. H., J. Dairy Sci., *47,* 1412 (1964).
335. STULL, J. W., and BROWN, W. H., J. Dairy Sci., *48,* 802 (1965).
336. STULL, J. W., BROWN, W. H., VALDEZ, C., and TUCKER, H., J. Dairy Sci., *49,* 1401 (1966).

337. TATTRIE, N. H., J. Lipid Res. *1,* 60 (1959).
338. TATTRIE, N. H., BAILEY, R. A., and KATES, M., Arch. Biochem. Biophys., *78,* 319 (1958).
339. THARP, B. W., and PATTON, S., J. Dairy Sci., *43,* 475 (1960).
340. THIN, C., and ROBERTSON, A., J. Comp. Pathol. Therap., *63,* 184 (1953); Biochem. J., *51,* 218 (1952).
341. TIMMEN, H., Intern. Dairy Congr., 18th, *1E,* 77 (1970).
342. TIMMEN, H., DIMICK, P. S., and PATTON, S., Intern. Dairy Congr., 18th, *1E,* 78 (1970).
343. TIMMEN, H., DIMICK, P. S., PATTON, S., and POHANKA, D. S. Milchwiss., *25,* 217 (1970).
344. TURPEINEN, O., J. Nutr., *15,* 351 (1938).
345. UNILEVER LTD., U.S. Patent 2,819,169 (1958); Margarinbologet Aktiebolag, U.S. Patent 2,903,364 (1959).
346. VAN DER WAL, R. J., *In* Advances in Lipid Research, *2,* 1 (1964). PAOLETTI, R., and KRITCHEVSKY, D., editors. Academic Press, New York.
347. VAN SOEST, P. J. J. Dairy Sci., *46,* 204 (1963).
348. VEN, B. VAN DER, Rec. Trav. Chim. Pay-Bas, *83,* 976 (1964).
349. VEN, B. VAN DER, BEGEMANN, P. H., and SCHOGT, J. C., J. Lipid Res., *4,* 91 (1963).
350. VEN, B. VAN DER, and JONG, K. DE, J. Am. Oil Chem. Soc., *47,* 299 (1970).
351. VIRTANEN, A. I., Neth. Milk Dairy J., *21,* 223 (1967).
352. WAKIL, S. J., J. Am. Chem. Soc., *80,* 6465 (1958).
353. WAKIL, S. J., J. Lipid Res. *2,* 1 (1961).
354. WATSON, P. D., Personal communication, 1960.
355. WEIHE, H. D., J. Dairy Sci., *44,* 944 (1961).
356. WEISS, S. B., KENNEDY, E. P., and KIYASU, J. Y., J. Biol. Chem., *235,* 40 (1960).
357. WEL, H. VAN DER, and JONG, K. DE, Fette, Seifen, Anstrichmittel, *69,* 279 (1967).
358. WHITE, M. F., and BROWN, J. B., J. Am. Oil Chem. Soc., *26,* 385 (1949).
359. WITTCOFF, H., "The Phosphatides,", p. 58, Reinhold Publishing Corp., New York, 1951.
360. WOOD, F. W., and HAAB, W., Can. J. Animal Sci., *37,* 1 (1957).
361. WYATT, C. J., PEREIRA, R. L., and DAY, E. A., Lipids, *2,* 208 (1967).
362. WYATT, C. J., PEREIRA, R. L., and DAY, E. A., J. Dairy Sci., *50,* 1760 (1967).
363. ZEHREN, V. L., and JACKSON, H. C., J. Assoc. Offic. Agr. Chemists, *39,* 194 (1956).
364. ZIJDEN, A. S. M. VAN DER, JONG, K. DE, SLOOT, D., CLIFFORD, J., and TAYLOR, R. J., Rev. Franc. Corps Gras, *13,* 731 (1966).

CHAPTER 5

Daniel P. Schwartz*
Owen W. Parks†

The Lipids of Milk: Deterioration

Daniel P. Schwartz

Part I. Lipolysis and Rancidity

Market milk and some products manufactured from milk at times possess a flavor described as rancid. This term as used in the dairy industry denoted implicitly the flavor due to the accumulation of the proper concentration and types of free fatty acids hydrolytically cleaved from milkfat under the catalytic influence of the lipases normally present in milk.

The development of rancid flavor in milk and some other fluid products is usually undesirable and detracts from their market value. In contrast, the popularity of certain dairy products, notably some cheese varieties, and also some confectionery items containing milk as an ingredient, is thought to be partially due to the proper intensity of the rancid flavor. Hence, knowledge of the factors involved in the development of rancidity is of great practical importance to several industries.

The literature on the subject is quite large. The present review has been limited to milk lipases, but good reviews on this, other dairy products, milk esterases, and on microorganisms are available.[11,92,145,147,162,163,195,223,313,318]

GENERAL

A lipase has been defined as an enzyme hydrolyzing the esters from emulsified glycerides at an oil/water interface.[61] Adherence to this definition has been maintained in this review; as a consequence, investigations which involved water-soluble substrates or substrates containing an alcoholic moiety other than glycerol have not been included.

* Daniel P. Schwartz, Dairy Products Laboratory, Agricultural Research Service, U.S. Department of Agriculture, Washington, D.C. 20250
† Owen W. Parks, Dairy Products Laboratory, Agricultural Research Service, U.S. Department of Agriculture, Washington, D.C. 20250

The term enzyme system has been proposed to include multiple enzymes.[154] It has been suggested that the term "lipase system" be used in dairy literature to express the lipase multiplicity in milk.[108] This expression will be adopted here and will be used synonymously with lipases and lipolytic enzymes.

It should be mentioned at the outset that a number of variables affect lipase activity. Unlike most enzymic reactions, lipolysis takes place at an oil/water interface. This rather unique situation gives rise to variables not ordinarily encountered in enzyme reactions. Factors such as the amount of surface area available, the permeability of the emulsion, the type of glyceride employed, the physical state of the substrate (complete solid, complete liquid or liquid-solid), and the degree of agitation of the reaction medium must be taken into account for the results to be meaningful. Other variables common to all enzymic reactions—such as pH, temperature, presence of inhibitors and activators, concentration of enzyme and substrate, light, and the duration of the incubation period—will affect activity and the subsequent interpretation of the results. Many of these variables have not been standardized in milk lipase studies, and it is difficult to assess results and draw concrete conclusions in a number of instances.

Enzymes are produced and elaborated by living cells—a fact that has prompted some investigations into the origin of milk lipases. It is only recently that synthesis of glycerides by milk lipases has been demonstrated.[205,229,231,232] Using tripalmitin isotopically labeled in both the glycerol and fatty-acid moieties, Koskinen *et al.*[205] demonstrated that glyceride synthesis occurs in freshly drawn milk, and that synthesis and hydrolysis occur simultaneously.[229] Luhtala *et al.*[231] showed that intracellular enzymes isolated from homogenized somatic cells of milk were capable of synthesis and lipolysis of milk triglycerides. However, there has been no evidence that any of the lipases of milk are identical with the lipase(s) involved in milkfat synthesis *in vivo*. It is of interest in this line of investigation to note that Morton[245-247] has shown that milk phosphatase is derived from mammary gland microsomes released into the milk during the normal secretory process.

Bovine blood serum is lipolytically active, but cows producing milk which goes rancid quickly do not have sera more lipolytically active than those producing normal milk.[284] Leucocytes, which are present in large numbers in milk, are especially high in mastitic milk; they are the source of milk catalase, but are apparently not the source of milk lipases.[252]

The lipases of milk are apparently inactive in the udder and at the time of milking.[145] Milk always contains relatively large proportions of unesterified fatty acids,[360] but these may be left over from the metabolic pool. The milk lipases are unusually slow-acting unless some physical or thermal treatment is applied to the milk. This may account for the inactivity in the udder, but no experiments have been conducted to substantiate this.

FARM FACTORS AND LIPOLYSIS

Spontaneous Rancidity

Studies have been undertaken to determine how widespread rancidity really is. Hemingway *et al.*[141] examined 12 herds and reported that about 50% of the herd samples showed some initial rancidity and 21% of samples from 15 cows were rancid. Differences in degree of rancidity were marked. Another report contends that from 2 to 22% of cows in a herd produce milk which goes rancid quickly.[150] Milk which inherently possesses the quality of high susceptibility toward rancidity has been variously termed "naturally rancid milk"[63], "bitter milk of advanced lactation"[258], "naturally active" or "naturally lipolytically active",[214,343,345,348] "normally active",[360] and "spontaneous".[345,348] The latter term has been more or less adopted in recent years. These various designations were introduced in an effort to distinguish such milk from "non-spontaneous" (normal) milk.

Lipolysis in freshly drawn milk normally proceeds at a very slow rate even upon prolonged incubation, unless proper thermal or mechanical treatment is applied to the milk. This, of course, always occurs in practice as raw, warm milk is never consumed on the market. It is through these necessary practices that lipolysis in normal milk is accelerated. As a consequence, milk may be made rancid either deliberately or accidentally. The so-called spontaneous type of milk needs no treatment other than cooling to 15 to 20°C when drawn or shortly afterward to hasten lipolysis.[354] Once the milk has been cooled, lipolysis is not materially affected whether the milk is aged in the cold or rewarmed to 20, 30, or 37°C and aged at these temperatures. Lipolysis in normal milk is not accelerated to the same degree by cooling and aging.

The reason that rancid milk is not more prevalent in market milk is due to the fortuitous fact that spontaneous rancidity can be prevented or reduced by mixing such milk within 1 hr after milking with 4 to 5 times its volume of normal milk.[348] Since usually only about 1 out of 5 cows in a herd produces spontaneous milk, this defect is thus almost automatically eliminated or reduced. It is clear, however, that farmers with only a few cows are likely to encounter spontaneously rancid milk during the lactation period.

The dilution of normal milk which has been activated by thermal or mechanical treatment does not diminish activity of the lipases.[121,220,244]

Feed.—The cow's feed has been shown to be an important practical factor in influencing the susceptibility of the milk to rancidity. Feeding experiments and practical observations have demonstrated that green pasture decreases and dry feed increases the incidence of rancidity.[44] Poor quality ration fed at reduced energy levels can significantly increase the incidence of rancidity[115], as can the feeding of a high-carbohydrate diet.[197]

Lactation.—Individual cows maintained under identical conditions seem to vary markedly in the susceptibility of their milk to rancidity.[176,206] An increased incidence of rancidity has also been associated with advanced lactation, particularly during long lactation periods.[8,16,48,70,83,102,150,199,258] There are reports, however, which fail to show a correlation between rancidity and advanced lactation.[149,268]

Mastitis.—Mastitis has been implicated in rancidity[16,127,338,358]; according to Guthrie and Herrington[127] and Tarassuk *et al.*[358], it may be more important than late lactation. Luhtala and Antila[230], however, found lower lipolytic activity in mastitic milks. They also reported that lipase activity was higher in foremilk than in strippings.

Estrous

The effect of the estrous period on rancidity has also been investigated. According to Wells *et al.*[370], who studied lipase activity in the milk and blood of cows throughout their lactation periods, the peak blood plasma lipase values occur about 24 hr before the onset of observed estrous. Changes in the blood lipase activity were reflected and magnified in the milk, although it was noted that the increase in the lipase level in milk occurred 9 to 15 hr after it was observed in the blood. Bachmann[16] also has indicated that hormonal disturbances are linked to rancidity. He differentiates between rancidity produced by cows in late lactation and rancidity due to hormonal disturbances on the basis of an increase in lipase concentration in the latter.

Pipeline Milkers

The increased use of pipeline milkers and farm tanks on dairy farms has coincided with a noticeable increase in rancidity.[114,146,167,178,257,327,359] About 6 times as much rancid milk has been reported from pipeline milkers as compared to nonpipeline systems.[168] The trouble has been traced to risers in the pipelines, that is, vertical sections connecting one pipeline to another at a higher level.

Air leaking excessively into the milk lines primarily at the claw, teat cups, milk hose and loose line joints causes considerable foaming of the warm, raw milk lifted in the risers under reduced pressure.[44] The formation of foam due to air agitation was found to be an important feature of the mechanism involved in the acceleration of lipolysis and the resultant appearance of rancid flavor in milk from pipeline milkers. Optimal conditions for activation by air agitation appear to be foaming with the continuous mixing of foam and milk at temperatures that keep the milkfat liquid.[346]

Remedial measures that suppress foaming and agitation in pipeline milkers have been recommended. The use of a pipeline located below the cow was reported to virtually eliminate rancidity or to significantly reduce the acid degree value which is defined as ml N KOH required to neutralize the free fatty acids in 100 gm fat.[114] Shortening of the main pipeline and minimizing the number of risers, joints, and sharp bends will also reduce foam formation and subsequent rancidity.[160,243]

DISTRIBUTION AND PURIFICATION OF MILK LIPASES

Sufficient evidence now exists to support the view that a number of lipases exist in milk. This was first proposed by Herrington and Krukovsky[149] on the basis of the effect of formaldehyde on lipase activity. Schwartz *et al.*[309] presented kinetic data and conducted more detailed studies on formaldehyde inhibition,[310] the results of which were interpreted in favor of the existence of two or three lipases in skimmilk. Albrecht and Jaynes[4] and Nelson[251] obtained multiple pH optima on tributyrin hydrolysis with their enzyme source. The relative specificity of the milk lipase system toward a variety of substrates was shown to be different for individual milks; and this observation, together with data obtained using a number of selective inhibitors, also led to the conclusion that more than one lipase is normal to milk.[106-108,347]

There seems to be little doubt that the majority of the cell-free lipases of milk are bound in some manner to the casein complex,[65,66,67,68,111,139,295,318] although it has not been firmly established that the lipase activity is associated with any particular casein species. Lipase activity has been reported to be in the α-casein fraction,[320] the β-casein fraction,[296] and the κ-casein fraction.[104,380] The association of lipases with the casein fraction of milk can be rationalized from the fact that lipases in general show a strong affinity for interfaces. Baskys *et al.*[21] took advantage of this fact and used adsorption of the enzyme at an ether/water interface as their initial purification step. A recent report[275] has shown that lipoprotein lipase as well as pancreatic lipase will attack triolein impregnated onto "Celite", demonstrating that creation of this type of interface will attract the enzyme. This experiment also may indicate that emulsification of the substrate is not essential. It may offer a means for simple standardization of lipolytic substrates.

Downey and Andrews[66] showed that, upon addition of pancreatic lipase to milk, a decrease in activity is manifested due to binding of the enzyme to both micellar and soluble casein complexes. They demonstrated that the binding of the lipase to casein was not dependent upon the presence of colloidal phosphate; hence, complete micellar structure is not essential for association of lipases with casein. Downey and Murphy[68] intimated that lipases in milk appear to be involved in the equilibrium between micellar and soluble casein in that activity is apparently influenced by this equilibrium.

The binding of milk lipases to casein micelles apparently imparts some stability to the enzymes, for as purification progresses the milk lipases become less stable, and more so as the concentration of casein decreases.[66]

Lipases associated with the casein micelles in skimmilk are not fully active, but both dilution and the addition of sodium chloride stimulate or restore activity, presumably by dissociating the micelle-lipase complex. Sodium chloride is an inhibitor of lipolysis (see p. 232), but the proper dilution and addition of this salt can elicit maximal activity.[66]

Other lipases are present in milk but are associated with cells or cell debris. Somatic cells isolated from separator slime were found to contain lipase activity[110,231] and lipases have been isolated and purified to a high degree from separator slime.[10,285] The cell-bound lipases are ostensibly distinct entities and differ from the lipases normally associated with casein.

Milk also contains a lipoprotein lipase similar to that found in mammalian adipose and heart tissue.[204a] The enzyme is highly specific for lipoproteins and probably is not involved in rancidity development to any great extent.

In order to explain the phenomenon of spontaneous rancidity, Tarassuk and Frankel[347] presented evidence to show that when freshly drawn milk is cooled, irreversible adsorption of a lipase on the membrane surrounding the fat globules occurs. This enzyme, termed "membrane lipase", was thought to occur in very low concentration in normal milk and in relatively high concentration in spontaneous milk. Thus, the accelerated lipolysis in spontaneous milk would be due, according to Tarassuk and Frankel[347], to the higher concentration of "membrane lipase". These workers acknowledged that at least one other lipase is present in milk; it is associated with casein, and is not active to any extent in spontaneous lipolysis, but can be activated by proper physical means (see below).

The appearance of more elegant techniques for purification of proteins has resulted in an increase in the number of attempts to purify the lipases of milk. Purification of cell-free lipases usually begins with the casein complex. At least 5 distinct lipases all capable of catalyzing the hydrolysis of tributyrin are associated with casein.[67] These enzymes can be dissociated from casein with 0.75*M* sodium chloride and subsequently separated to some degree by gel filtration.[65,67] All the lipases separated in this manner show the characteristic property of lipases, namely, that they hydrolyze emulsified triglycerides at a greater rate than they hydrolyze the dissolved triglyceride, triacetin. Molecular weights of the lipases were estimated as between 35,000 and 180,000.

Curd from rennet casein was used by Gaffney *et al.*[111] as a starting point for fractionating the lipases on DEAE cellulose and on Sephadex G-35. Their results indicated that 11 of the fractions showed lipase activity.

Fox and Tarassuk,[103] also starting from rennet curd, isolated a lipase with a specific activity 500 times that of the original skimmilk. The molecular weight of their preparation was 210,000. The lipase had a pH optimum at 9.2 on milkfat and a temperature optimum at 37°C.

Other lipases have been isolated from the separator slime deposited from milk. Chandan and Shahani[41] purified a lipase which had a molecular weight of about 7,000, a pH optimum of 9.0 to 9.2, and a temperature optimum at 37°C. It hydrolyzed both milkfat and tributyrin and was inhibited by sulfhydryl reagents. The lipase showed little activity toward simple esters.

Foissy,[92] working with separator slime, showed that three proteins isolated by gel electrophoresis exhibited lipolytic activity.

A recent report by Richter and Randolph[285] described the isolation of a lipase from separator slime with a molecular weight of 8,500 which contained carbohydrate, had a pH optimum at 9.2, and also had a temperature optimum at 37°C. It hydrolyzed simple short-chain fatty-acid triglycerides more rapidly than long-chain glycerides, but had little specificity for natural oil emulsions. Whether this is the same lipase reported by Chandan and Shahani was not unequivocally established.

ACTIVATION OF LIPASES

Homogenization and Agitation

All methods of agitation of milk appear to increase the rate of lipolysis. The increased incidence of rancidity in pipeline milkers as opposed to conventional milking procedures due to foaming and agitation has already been discussed. Homogenization (a more violent form of agitation) of raw milk when conducted at temperatures between 37.7°C and 54.4°C will render milk rancid within a very short time in some cases in only a few minutes.[145,366] The length of time of homogenization as well as the homogenization pressure[253] will influence subsequent lipase activity, lipolysis increasing within limits, as the magnitude of these variables increases.[230,253]

Other forms of agitation, including shaking raw milk containing liquid fat,[51,60,213,319] the churning of raw milk or cream, and pumping[372] will accelerate lipolysis. The severity of agitation and the temperature at which it is conducted are of prime importance.

Foaming due to agitation also promotes lipolysis, but the increased activity in foam is probably independent of the accelerated lipolysis due to agitation. According to Tarassuk and Frankel,[346] foaming promotes lipolysis by providing (a) greatly increased surface area, (b) selective concentration of enzyme at the air-liquid interface, (c) "activation" of the substrate by surface denaturation of the membrane materials around the fat globules, and (d) intimate contact of the lipases and the "activated" substrate.

All forms of agitation, with the exception of churning, increase the surface area of the substrate, and this is the foremost reason for the increase in lipase activity. However, agitation produces other effects which are conducive to lipase action. The process of diffusion, which has been shown to be very important,[239] is speeded up. Diffusion permits the lipases to migrate more readily to the oil/water interface while simultaneously allowing the fatty acids produced in lipolysis to leave the interface.

Thermal Manipulation

Unlike spontaneous milk, normal (nonspontaneous) milk requires additional thermal "shocking" beyond the first cooling to activate the milk lipase system. Krukovsky and Herrington[212] were the first to demonstrate that lipolysis in normal milk could be hastened by warming cold milk to 29.4°C, and then recooling beyond the solidifying point of the fat. Most samples of milk subjected to this treatment will become rancid within 24 hr.[145] The temperature of approximately 30°C is critical, and heating below or appreciably beyond that point diminishes the degree of activation that can be obtained. This type of activation is of great practical importance because it can happen accidentally. For example, if warm morning milk is added to a can of milk refrigerated from the night before and the whole cooled again, the milk may be rancid by the time it is ready for processing.

Milk containing fat globules with a natural fat globule membrane can be activated, deactivated, and reactivated by proper changes in temperature. The phenomenon of temperature activation is found only when the fat globules have their natural layer of adsorbed materials. Neither homogenized milk, nor emulsions of tributyrin, nor of butter oil emulsified in skim milk can be activated in this manner.

Several hypotheses have been advanced to explain the peculiar phenomenon of temperature activation. These include the attainment of a favorable liquid-to-solid glyceride ratio[144,215], an increase in the permeability of the fat globule membrane to the lipases,[353] and reorientation of glycerides more susceptible to lipolysis toward the fat/water interface.[283] However, the first and latter hypotheses would seem to be inconsistent with the fact that homogenized milk cannot be temperature-activated.

The freezing of raw milk followed by thawing to 4°C causes an increase in lipolysis compared to unfrozen controls stored at 4°C, but the increase in activity varies considerably. Repeated freezing and thawing also causes a notable increase in lipolytic activity. The temperature of freezing has a marked effect, the increase in lipolysis being most pronounced when the temperature is lowered from −10 to −20°C; little further increase in activity occurs between −20 and −33°C. Slow freezing causes greater lipolysis than does rapid freezing.[371]

Chemical Activation

Downey and Andrews' experiments[66] have indicated that there is a bivalent cation requirement for full milk lipase activity. Dunkley and Smith[81] had previously stated that small amounts of $CaCl_2$ accelerate lipolysis. These observations are in keeping with those made on lipases from other sources where Ca^{++} was found to stimulate activity.[376]

Pitocin, a hormone, was reported to increase lipolysis,[179,180] and another hormone, diethylstilbestrol, is said to increase lipase activity toward tributyrin but not toward milkfat.[84] Shahani and Chandan[314] found that purified lipase from separator slime can be activated by the proper concentration of pseudoglobulin, euglobulin, lactalbumin, blood serum albumin, and bovine plasma albumin.

The milk lipase system is reported to be activated by mercuric chloride. Raw milk preserved with corrosive sublimate contains, in some instances, a much larger concentration of free fatty acids than unpreserved samples. Pasteurized milk preserved in a similar fashion does not show an increase in free fatty acids.[235]

INHIBITION OF LIPASES

Thermal Inhibition

Heat-treatment of milk is the most important practical means for inactivation of its lipases. The temperature-time relationship necessary for partial or complete inactivation has been extensively studied, but a number of discrepancies have been apparent. These are probably due to several factors, among which might be mentioned the sensitivity of the assay procedure, length of the incubation period following heating, the presence and concentration of fat and solids-not-fat in the milk at the time of heating, and the type and condition of the substrate. In view of these variables, references to a number of early studies on heat inactivation have been omitted.

The data of Nilsson and Willart[254] indicate that heating at 80°C for 20 sec is sufficient to destroy all lipases in normal milk. Their studies included assays after 48 hr incubation following the heat-treatment. At lower temperatures for 20 sec, some lipolysis was detected after the 48 hr incubation period after heating. Thus, 10% residual activity remained at 73°C. Below a temperature of 68°C the amount of residual activity was enough to render the milk rancid in 3 hr, and temperatures below 60°C had no appreciable effect on lipolysis. With holding times of 30 min, 40°C produced only slight inactivation, and at 55°C, 80% inactivation was reported. The effect of other temperature-time relationships was studied; the results are given in Table 5.1.

Table 5.1

INFLUENCE OF HEATING TIME ON THE HEAT-INACTIVATION OF MILK LIPASE(S)[a]

Temperature °C	Heating Time Sec	% Inactivation Determined after Incubation for	
		3 Hr	24 Hr
65	5	23	20
	22	52	50
	36	70	57
	74	100	81
72	5	85	30
	22	96	91
	36	100	96
	72	100	100

[a] Data of Nilsson and Willart.[254]

The data of Harper and Gould[138] are essentially in agreement with those of Nilsson and Willart. They also detected no inactivation until a temperature of 60°C for 17.6 sec was reached. At 87.7°C (17.6 sec) some lipase still survived.

Schwartz[308] investigated the effect of various heat treatments on the pH-activity curve of the lipase system in raw skimmilk powder. For this purpose the powder was reconstituted and the milk exposed to the temperature and times listed in Table 5.2.

Fat apparently protects the lipases to some extent from heat inactivation, 1° to 2°C higher temperatures being necessary for whole milk than for skimmilk.[109,137,138,254,297] The influence of the fat content of milk on heat inactivation of the milk lipase system is given in Table 5.3.

Harper and Gould[138] indicate that besides the protective effect of fat on lipase inactivation, the solids-not-fat content is also a factor. Higher solids-not-fat concentration, within limits, afforded some protection.

Inhibition by Light and Ionizing Irradiation

The milk lipase system shows a remarkable sensitivity to light. Kay[175] exposed fresh milk in glass vessels to bright summer sunshine for 10 min and found that 40% of the lipolytic activity was destroyed. Exposure for 30 min resulted in a loss of 80%, and exposure to an 800-watt quartz mercury-vapor lamp at a distance of 15 cm destroyed 75% of the activity. He noted, however, that if oxygen was first removed from the system before exposure to sunlight, the effect of the light was greatly diminished. Kannan and Basu[172] observed that in some cases exposure to ultraviolet light destroyed the lipase system and diffused daylight brought about a partial inactivation.

Table 5.2

EFFECT OF VARIOUS HEATING PROCEDURES ON LIPOLYSIS[a]

Temp. °C	Time Sec[b]	pH Levels 6.2	6.6	7.0	7.5	7.9	8.5	9.5
		Percent Inactivation						
60.0	17.4	73.5	67.4	55.0	45.8	49.7	41.3	53.5
66.8	14.2	65.1	61.4	65.7	65.4	47.8	48.4	90.7
72.0	14.4	91.0	90.1	84.4	83.5	82.1	83.8	91.3

[a] Data of Schwartz.[308]
[b] Milk attained temperature at or within these times.

Table 5.3

INFLUENCE OF FAT CONTENT ON THE HEAT-INACTIVATION OF MILK LIPASES[a]

Fat	% Inactivation after Heating at 55°C	
	5 min	15 min
0.1	50	70
5	40	63
10	35	60
20	29	57

[a] Data of Nilsson and Willart.[254]

Frankel and Tarassuk[109] exposed a layer of raw skimmilk 1 cm thick to direct sunlight at room temperature and noted a loss in lipase activity of 84% in 5 min and 96% after 10 min. In diffuse daylight inactivation was less, but 71% was lost in 1 hr. The loss of activity by light was independent of the temperature of the milk, equal losses being observed at 0°C and at 37°C. The enzymes are markedly protected against light inactivation by the presence of fat.

Stadhouders and Mulder[329] confirmed Kay's observation that the shorter wavelengths (about 4300 Å) of the spectrum are most destructive to milk lipases. The destructive effect of light could be repressed by addition of reducing agents such as metol, hydroquinone, and especially by hydrogen sulfide. Ascorbic acid and methionine had no effect, but cysteine afforded a significant protection. Lipases which had been inactivated by light were not reactivated by treating milk with hydrogen sulfide.

Irradiation by ionizing radiation and its effect on milk lipase activity has also been studied.[367] Irradiation doses of 6.6×10^4 rads destroyed 70% of the activity. The udders of lactating cows when exposed to cobalt-60 gamma rays gave milk with decreased lipase and esterase activity.[233]

Chemical Inhibition

A large variety of chemical compounds have been added to milk or purified lipase preparations in order to determine their effect on lipase activity. The conditions under which the inhibitor is studied are very important. Factors such as pH, temperature, time of addition of the chemical, sequence of addition of reactants and the presence or absence of substrate are undoubtedly involved. The presence of substrate appears to offer some degree of protection to the enzymes. Consequently, in lipase studies the surface area of the emulsified substrate is probably also important.

Heavy metals usually affect enzymes adversely, and milk lipases are no exception. Copper, cobalt, nickel, iron, chromium, manganese, and silver are inhibitors.[57] Raw skimmilk treated with 5 to 20 ppm Cu^{++} for 15 min at room temperature caused 7 to 17% loss of lipolytic activity, whereas 5 ppm at 37°C for 1 hr resulted in a 69% loss. There was less inhibition in the presence of substrate.[109] Earlier, however, Krukovsky and Sharp[216] showed that Cu^{++} was ineffective as a lipase inhibitor in nonhomogenized milk if oxygen was absent. At the same time they also found that oxygen alone is an active inhibitor, its effect being magnified by the presence of low percentages of copper.

A number of salts inhibit lipolysis, the most effective being sodium chloride.[39,117,269,373] Lipolysis in cream was found to be insignificant in the presence of 4% sodium chloride and in homogenized milk containing 5 to 8% of this salt.[117] It should be noted that Downey and Andrews[63] used 0.75*M* sodium chloride (about 4.3%) to dissociate the lipases from the casein micelles as the first step in their fractionation procedure, but still went on to isolate 5 active lipases. It is not known whether loss of part of the activity of their lipases was due directly to the salt treatment, or whether activity was restored by subsequent purification.

Phosphate buffer (0.6*M*) slightly inhibited lipolysis, but the same concentration of borate and barbiturate buffers was without effect.[266]

Zinc chloride, potassium cyanide, manganese sulfate, cysteine, and magnesium chloride retarded milk lipase activity to various degrees. All these compounds were tested at pH 8.5 with tributyrin as substrate during a 30-min incubation period.[266]

N-Ethyl maleimide inhibits lipase activity in milk activated by shaking, temperature fluctuations and homogenization, 0.02*M* being completely inhibitory.[357] An equimolar concentration of glutathione markedly reduces inhibition by *N*-ethyl maleimide. This reagent can also completely inhibit lipolysis in spontaneous milk.[357] It was concluded on the basis of these experiments that sulfhydryl groups are essential sites of activity on milk lipases. This is supported by the ability of reducing agents such as glutathione, hydroquinone, and potassium thiocyanate to offer stability to the milk lipase system during storage.[109] It was also concluded that purified lipase from separator slime contains 1 free and 1 masked sulfhydryl group, which are essential for activity. Sulfhydryl reagents such as *p*-chloromercuribenzoate, iodoacetic acid, formamidine disulfide, and *N*-ethyl maleimide are potent inhibitors of slime lipase.[43]

Other chemicals which inhibit the milk lipase system include hydrogen peroxide, animal cephalin,[37] sodium arsenite, diisopropyl fluorophosphate, 2,4-dinitro-1-fluorobenzene, *p*-hydroxymercuribenzoate,[290] potassium dichromate,[375] lauryl dimethyl benzyl ammonium chloride,[296] aureomycin, penicillin, streptomycin, and terramycin.[40]

The most studied chemical inhibitor of lipolysis has been formaldehyde, for the reason that formaldehyde had been widely used as a preservative in milk lipase studies without knowledge of its effect. After the existence of a lipase in milk had been established, Herrington and Krukovsky[148] postulated that there was more than one lipase in milk, since experiments had suggested that there was a formaldehyde-sensitive and a formaldehyde-tolerant lipase in milk. Other investigators[226,289] published data which essentially substantiated this, but there was also information obtained under different experimental conditions in which formaldehyde was shown to be noninhibitory.[117,344]

An extensive study of the effects of formaldehyde in milk lipase inhibition showed that formaldehyde acts as a competitive inhibitor and also, under the proper conditions, selectively inhibits the lipases of raw skimmilk.[310] It was shown in this study that the inhibitory effect of formaldehyde was dependent on such factors as pH, time of addition of inhibitor, length of incubation period, concentration and availability of substrate, and concentration of inhibitor. Many of the conflicting results encountered with formaldehyde can be explained on the basis of dependence on one or more of these factors.

PROPERTIES OF MILK LIPASES

Specificity

A review of the specificity of a purified slime lipase from milk and also of pancreatic lipase has been published by Jensen,[163] who has been one of the foremost investigators in this field. Pancreatic lipase and the purified slime lipase exhibit similar characteristics in their specificities, and although slime lipase represents only one of a number of lipases in milk, it is probable that other lipases in milk will show similar specificities as they are isolated and purified.

A study of lipase specificity requires that the enzyme and substrates be virtually pure. Contamination of the lipase preparation with esterases will give rise to misleading results. Pure, synthetic substrates of known configuration are essential, and the same available surface area should be present after emulsification for meaningful data to be obtained. Since most of the earlier workers have disregarded one or more of these variables, their data are useless for all practical purposes, and will not be included here.

There are four types of specificity which may be considered: (1) a specificity between glycerides, i.e., intermolecular specificity; (2) a specificity between different fatty acids in the same position of a glyceride, that is, the 1- and 3-positions, also called intramolecular specificity; (3) positional specificity, i.e., a specificity for the 2-position over the 1- and 3-positions, or vice versa; and (4) stereospecific specificity, i.e., whether lipases preferentially differentiate between the 1- and 3-positions when they are occupied by the same fatty acid.

Purified slime lipase shows an apparent specificity for short-chain fatty acid-containing triglycerides relative to longer chain-containing glycerides including milkfat. This was shown by Jensen *et al.*[165] It also shows a specificity for triolein over saturated triglycerides such as trilaurin or trimyristin, but this may have been because the latter two are solids and consequently are more difficult to emulsify to the same degree as liquid glycerides.

Purified slime and pancreatic lipases, among others, have been shown to have no intramolecular specificity, hydrolyzing both short- and long-chain acids at the same rate when the acids occupy the 1- and 3-positions in the same molecule.[163] This observation refuted earlier work, which had indicated that pancreatic lipase shows a specificity for shorter-chain acids in a similar situation.

Positional specificity is exhibited by slime and pancreatic lipases. Fatty acids in the 1- and 3-positions are hydrolyzed at a greatly accelerated rate compared to fatty acids in the 2-position. In fact, there is some question whether the 2-position is attacked at all, since acyl migration from the 2-position to the 1-position occurs to some extent.[163]

Stereospecificity has not been studied on milk lipases, although it has been demonstrated to be absent for pancreatic lipase.[174]

pH Optimum

Enzymes usually exert their catalytic influence over a somewhat restricted pH range. Within this range the activity passes through a maximum, commonly called the pH optimum, and then falls off again. Although the pH optimum and the pH range are generally characteristic of a given enzyme, they may in some instances be altered by such factors as type and strength of buffer, ionic strength, temperature, type of substrate employed, and, in the case of lipases, the condition of the interface where lipolysis must proceed.

Lipases are sensitive to extremes in pH, and even in the vicinity of the pH optimum, where enzymes are supposedly more stable, marked inhibition may occur.[108] Thus, it must be borne in mind also that the length of the incubation period and the prior history of the preparation can influence the range and perhaps the shape of the pH-activity curve.

The incubation of raw skimmilk at pH 6.0 and at pH 8.9 for 1 hr at 37°C in the absence of substrate was subsequently shown to cause a 47% and 40% decrease, respectively, in lipase activity when the milk was later incubated with milkfat. When tributyrin was the substrate the inhibition was even more marked. Although some of the inactivation was due to temperature, the majority of it is attributable to pH exposure. Stadhouders and Mulder[330] have also demonstrated that milk lipase subjected to incubation at pH 5.0 is almost completely destroyed.

The point on the acid side of the pH curve where milk lipase activity ceases is of considerable practical importance, but there is still constoversy regarding it. Bosco[33] found that milk lipase is active down to pH 4.7, whereas Schwartz *et al.*[309] could detect no activity at pH 5.2 on butterfat. Although Peterson *et al.*[267] found no milk lipase activity on tributyrin at pH 7.0, activity has been reported on this substrate at pH 5.0 and even at pH 4.7 when 24-hr incubation periods were used.[330] Willart and Sjostrom[374] have also observed activity in the range pH 4.1 to 5.7.

It has been reported that there are lipases and esterases operating in the range of pH 5.0 to 6.6 on tributyrin and simple esters. These lipases were stated to have pH optima at 5.4 and 6.3 and to be inhibited by formaldehyde.[4] For two enzymes to operate in such a restricted range is unusual, and may indicate that nonenzymatic entities are present in milk which can promote hydrolysis of esters.

The pH activity curves for the milk lipase system were shown by Schwartz *et al.*[309] to vary with the concentration of milkfat used as substrate. They observed activity between pH 5.2 and 9.8. At high substrate concentrations (16.5% milkfat), optima were observed at pH 8.5, 6.5 to 7.0, and 7.9; at a very low concentration of substrate (0.92% milkfat), a broad optimum between pH 7.0 and 8.0 was noted.

In general, when the substrate concentration is not limiting, the pH optimum most frequently observed for the milk lipase system lies between pH 8.5 and 9.0. The lipase isolated and purified from separator slime of milk shows an optimum at 9.0 to 9.2.[42]

Apparent Temperature Optimum

A rise in temperature has a dual effect upon an enzyme-catalyzed reaction: it increases the rate of the reaction, but it also increases the rate of thermal inactivation of the enzyme itself. Like the pH optimum, the temperature optimum may in certain instances be altered by environmental conditions, e.g., pH, type and strength of buffer, etc. The term "temperature optimum," therefore, is useless unless the incubation time and other conditions are specified. A more enlightening nomenclature is "apparent temperature optimum," which indicates that the optimum has been obtained under a certain set of conditions, and may or may not hold when these conditions are changed.

The apparent temperature optimum for the milk lipase system is reported to be around 37°C both on milkfat and on tributyrin.[106,289] This temperature has been recorded both at pH 8.9 and pH 6.6 for milkfat[106,289] and at pH 8.0 and pH 6.6 on tributyrin.[106]

Stadhouders and Mulder,[328] however, using tributyrin and milkfat together as substrates, found that at pH 9.1 under their conditions poorer activity was apparent at 37°C compared to 25°C or 15°C after 1, 4, and 24 hr of incubation. In fact, 90% less activity occurred at 37°C at 24 hr of incubation than at 15°C. The activity at 15°C was quite constant, indicating complete stability of the lipase system under the conditions used. Downey and Andrews[66] have stated in this regard that the lipases which they isolated are unstable even at 37°C, but are more active at this temperature than at 25°C. Their results suggest that it is unwise to assay at 37°C.

Stability

Some discussion regarding stability of milk lipases was presented in the preceding section. Other workers have also studied the stability of the milk lipase system under different circumstances. Schwartz[308] found that raw skimmilk showed no loss of activity at pH 6.2, 6.6, 7.0, 7.5, and 8.5 after the milk had been stored at pH 6.6 for 6 hr in the dark at 4°C prior to assay. Frankel and Tarassuk[109] permitted raw skimmilk to stand 1 hr at 37°C at pH levels ranging from 5.2 to 9.8, and found greatest activity in the range pH 6.6 to pH 7.6 when the milk was subsequently incubated with the substrate at pH 8.9. Hence it appears that normal milk is in an optimum pH range for stability of the lipases. The same pH range of optimum stability was found in whole milk.

The stability of the lipase system in raw, lyophilized skimmilk has also been investigated.[308] This powder stored at 4°C showed no loss in activity over a 7-month period when assayed at pH 6.6, 7.9, and 8.5.

SOME EFFECTS OF LIPOLYSIS

The most serious effect of lipolysis is the appearance of the so-called rancid flavor. The fatty acids and their soaps which are thought to be implicated in the flavor have been studied in an effort to assess the role of the individual acids in the overall rancid flavor picture.

Scanlan *et al.*[305] reported that only the even-numbered fatty acids from C_4 to C_{12} account for the contribution of fatty acids to the flavor, but that no single acid exerts a predominating influence. Another study has implicated the sodium and/or calcium salts of capric and lauric acids as major contributors to the rancid flavor.[5] Butyric acid, assumed to be the compound most intimately associated with the flavor, was not singled out in either study as being especially involved.

Besides changing the natural flavor of milk, lipolysis may produce a variety of other effects. One of the most noticeable of these is the lowering of surface tension as lipolysis proceeds.[39,72,81,204,353-356] Fatty acids, and more especially their salts, and mono- and diglycerides, being good surface-active agents, depress the surface tension of milk (see under Methods). Milkfat obtained from milk subject to lipase action also gives lower interfacial tensions with water than does milkfat obtained from nonlipolyzed milk.[31,82]

Rancid milk decreases the quality of cream, butter, and buttermilk made from it, and a limit on the acid degree value of the fat of milk from which butter is eventually to be made has been proposed.[160]

The higher saturated fatty acids have been noted to inhibit rennet action, whereas the lower fatty acids enhance it. The inhibitory effect of the higher acids could be nullified by $CaCl_2$.[354]

As little as 0.1% rancid milkfat proved to be a very effective foam depressant during the condensing of skimmilk and whey.[35] This effect was attributed to the mono- and diglycerides.

Lipolytic action has been observed to occur in composite samples preserved with mercuric chloride and decreased the reading of the Babcock test as much as 0.15%.[235]

An inhibitory effect of rancid milk on the growth of *S. lactis* has been reported. Koestler *et al.*[198] claim that rancid milk significantly inhibits the growth of bacteria in general and of *S. lactis* in particular. It has been stated that rancidity in milk may reach such a degree as to actually render the product sterile.[8] Tarassuk and Smith[356] attributed the inhibitory effect of rancid milk to changes in surface tension, but Costilow and Speck[49,50] believe that the inhibition is due to the toxic effect of the individual fatty acids.

Although rancidity is a serious defect in market milk, it has also been utilized profitably. Whole milk powder made from lipase-modified milk has generally met acceptance among chocolate manufacturers. It is used as a partial replacement for whole milk because it imparts a rich, distinctive flavor to milk chocolate, other chocolate products like fudge and compound coatings, caramels, toffees, and butter creams.[381]

METHODS FOR DETERMINING LIPASE ACTIVITY

A number of methods are available for following lipase activity. Although numerous modifications and variations have been introduced, the basic methods may be listed as: (1) titration of the liberated fatty acids, (2) changes in surface tension, (3) colorimetric determination of the fatty acids, (4) use of gas-liquid chromatography, and (5) use of radioactive substrates.

Titration

Titration of the fatty acids formed by the action of the milk lipase system has been the most widely used procedure. Titration has been conducted directly on the reaction medium either manually[120] or automatically,[263] in the presence of added organic solvents,[81,266,267] and after separation of the lipid phase by extraction,[106] distillation,[101,289] churning,[101,354] or adsorption of the medium followed by elution of the fatty acids.[140] All these techniques have their shortcomings. The most widely used laboratory methods appear to be the silica gel extraction[140] and the pH-stat methods,[263] and in the field, the method of Thomas *et al.*[359a]

Surface Tension

Efforts have been expended to apply surface-tension measurements for following lipolysis in milk.[72,81,348,352,354-356] As mentioned earlier, the hydrolysis products resulting from lipase action are strongly surface-active. Tarassuk and Regan[352] have stated that the lowering of surface tension resulting from lipolysis is the most distinct change differentiating rancid from nonrancid milk. However, a great many variables influence the surface tension of milk, such as the elaboration of structurally different mono- and diglycerides and their concentration.

Colorimetry

Copper[71] or cobalt[255] soaps of long-chain fatty acids ($\geq C_{12}$) are soluble in chloroform and can be determined quantitatively by colorimetric determination of the extracted metal. This method has been used in some nondairy lipase investigations, but could conceivably be used as such or modified for use in milk lipase work. It is very sensitive.

Another sensitive colorimetric procedure is that of Mackenzie *et al.*,[234] which utilizes the dye Rhodamine B to form benzene-soluble complexes with fatty acids. Nakai *et al.*[250] developed a rapid, simple method for screening rancid milk based on the above procedure. The test is said to detect rancid milk with an acid degree value above 1.2. Like the copper or cobalt soap method, the Rhodamine B reagent is also limited to the longer-chain fatty acids.

Gas-Liquid Chromatography

Gas-liquid chromatography (GLC) affords both a qualitative and, if adequate internal standards are used, a quantitative analysis of the products of lipolysis. It is necessary, however, first to isolate the acids by a suitable method and then inject them as free acids or as esters. The partial glycerides can be isolated by thin-layer chromatography and can also be determined by GLC of suitable derivatives. The acid(s) remaining in the partial glycerides can be identified readily by GLC following transesterification. Jensen and his co-workers have utilized these techniques in their studies on lipase specificity.[164,166]

Radioactive Substrates

Koskinen *et al.*,[205] Luhtala *et al.*,[231-232] and Scott[311] have used labeled triglycerides as substrates for milk lipases. This method, which is extremely sensitive, requires that the acids released by lipase action be isolated uncontaminated with any tagged glycerides. It also requires the preparation of labeled substrate, and, of course, counting equipment.

Miscellaneous

A manometric technique utilizing a Warburg apparatus has been used to follow esterase activity. The carbon dioxide liberated from sodium bicarbonate by the fatty acids is measured.[373]

An agar diffusion procedure has been utilized for screening microorganisms for lipolytic enzymes. The presence of lipase is indicated by clear zones in the turbid media.[224]

Owen W. Parks

Part II. Autoxidation

Lipid autoxidation in fluid milk and a number of its products has been a concern of the dairy industry for a number of years. The need for low-temperature refrigeration of butter and butter oil, and inert-gas or vacuum packing of dry whole milks to prevent or retard lipid deterioration, in addition to the loss of fluid and condensed milks as a result of oxidative deterioration have been major problems of the industry.

The autoxidation of milk lipids is not unlike that of lipids in other edible products. However, the complex composition of dairy products, physical state of the product (liquid, solid, emulsion, etc.) presence of natural anti- or pro-oxidants, as well as processing, manufacturing, and storage conditions tend to influence both the rate of autoxidation and the composition and percentage of autoxidation products formed.

The literature dealing with the autoxidative mechanism involved in lipid deterioration has been concerned with investigations on pure unsaturated fatty acids and their esters. The reactions involved, however, are representative of those occurring in lipids and lipid-containing food products.

AUTOXIDATION MECHANISM

The initial step in the autoxidation of unsaturated fatty acids and their esters is the formation of free radicals. Although the initiation of such radicals is not completely understood, the resulting free-radical chain reaction has been elucidated in the investigations of Farmer and Sutton,[89] and others.[23,32] In the case of monounsaturated and nonconjugated polyene fatty acids—the acids of significance in milkfat—the reaction is initiated by the removal of a hydrogen atom from the methylene (α-methylene) group adjacent to the double bond (I). The resulting free radical, stabilized by resonance, adds oxygen to form peroxide-containing free radicals (II); these in turn react with another mole of unsaturated compound to produce 2 isomeric hydroperoxides in addition to free radicals (III) capable of continuing the chain reaction.

$$-CH_2-CH=CH-CH_2- \xrightarrow{H} -CH\cdot-CH=CH-CH_2- \leftrightarrow -CH=CH-CH\cdot-CH_2-$$

(I)

$$-CH\cdot-CH=CH-CH_2- \xrightarrow{+O_2} -CH(OO\cdot)-CH=CH-CH_2-$$
$$\updownarrow$$
$$-CH=CH-CH\cdot-CH_2- \xrightarrow{+O_2} -CH=CH-CH(OO\cdot)-CH_2-$$

(II)

$$-CH(OO\cdot)-CH=CH-CH_2- + -CH_2-CH=CH-CH_2- \rightarrow -CH(OOH)-CH=CH-CH_2- + -CH\cdot-CH=CH-CH_2-$$

$$-CH=CH-CH(OO\cdot)-CH_2- + -CH_2-CH=CH-CH_2- \rightarrow -CH=CH-CH(OOH)-CH_2- + -CH\cdot-CH=CH-CH_2-$$

(III)

Oleic acid, having two α-methylene groups, gives rise to 4 isomeric hydroperoxides which have been isolated in equal amounts by various workers.[89,279] The preferential points of attack in polyene nonconjugated systems are the α-methylene groups located between the double bonds. Hence the autoxidation of linoleic acid and linolenic acid can lead to the formation of 3 and 6 isomeric hydroperoxides, respectively, as a result of the attack on the C_{11} methylene group of linoleic acid and on the C_{11} and C_{14} methylene groups of linolenic acid. However, a characteristic of hydroperoxide formation is the shifting of double bonds to form the conjugated system,[38,278] and the existence of an 11-linoleate hydroperoxide or an 11- or 14-linolenate hydroperoxide has not been established.[20] The α-methylene groups of polyunsaturated acids other than those located between double bonds are also subject to attack, but to a lesser degree. In all, 7 hydroperoxides from linoleic acid and 10 hydroperoxides from linolenic acid are theoretically possible during the autoxidation of these acids.

In addition to the formation of hydroperoxides, other reactions are known to occur simultaneously. The formation of polyperoxides, carbon-to-carbon polymerization, and the formation of epoxides and cyclic peroxides have been proposed or demonstrated in lipid oxidation—subjects which are not within the scope of this book.

PRODUCTS OF OXIDATION

The hydroperoxides formed in the autoxidation of unsaturated fatty acids are unstable and readily decompose. The main products of hydroperoxide decomposition are saturated and unsaturated aldehydes. The mechanism suggested for the formation of aldehydes involves cleavage of the isomeric hydroperoxide (I) to the alkoxyl radical (II), which undergoes carbon-to-carbon fission to form the aldehyde (III).[105]

$$\underset{\text{(I)}}{\underset{\text{O:OH}}{\text{R—CH—R}^1}} \rightarrow \underset{\text{(II)}}{\underset{\text{O}\cdot}{\text{R—CH}}\text{—R}^1} \rightarrow \underset{\text{(III)}}{\text{R—CHO} + \text{R}^1\cdot}$$

Other products, such as unsaturated ketones,[331] saturated and unsaturated alcohols,[153,332,333] saturated and unsaturated hydrocarbons,[94,157,184] and semi-aldehydes,[105] have been observed in the decomposition of hydroperoxides of oxidized lipid systems.

A comprehensive review and study by Badings[20] includes a listing of the carbonyls which can result from the dismutation of the theoretical hydroperoxides formed in the autoxidation of the major unsaturated acids of butterfat and those which have been observed. In addition to those carbonyls theoretically possible, various others have been isolated and identified in the autoxidation of pure fatty acids or their esters. Their presence suggests that migration of double bonds,[18] further oxidation of the unsaturated aldehydes initially formed,[17] and/or isomerization of the theoretical geometric form[20] may occur during autoxidation.

In addition to the major fatty acids, milk also contains many minor polyunsaturated acids;[219] hence the autoxidation of dairy products can lead to a multitude of saturated and unsaturated aldehydes.

OXIDATION AND OFF-FLAVORS

The overwhelming consideration in regard to lipid deterioration is the resulting off-flavors. Aldehydes, both saturated and unsaturated, impart characteristic off-flavors at minute concentrations. Terms such as painty, nutty, melon-like, grassy, tallowy, oily, cardboard, fishy. cucumber, etc. have been used to characterize the flavors imparted by individual saturated and unsaturated aldehydes, as well as by mixtures of these compounds. Moreover, the concentration necessary to impart off-flavors is so low that oxidative deterioration need not progress substantially before the off-flavors are detectable. For example, Patton *et al.*[265] reported that 2, 4-decadienal, which imparts a deep-fried fat or oily flavor, is detectable in aqueous solution at levels approaching 0.5 part per billion.

In addition to aldehydes, other secondary products of lipid oxidation, such as unsaturated ketones and alcohols, impart characteristic flavors, and their presence in oxidized milk systems has been established.[20,331,332]

Generally speaking, the flavor threshold values for aldehydes are governed to varying degrees by the number of carbon atoms; degree of unsaturation; location of unsaturation in the chain; form of geometric isomer; additive and/or antagonistic effects of mixtures of compounds; and the medium in which the flavor compounds are present.[59,240] With respect to the latter point, the flavor potency of many aldehydes identified in oxidized lipids is up to 100 times greater in an aqueous medium than in a fat or oil. Hence, the extent of oxidative deterioration of fluid milk need not progress to the same point as that in butter oil before the onset of off-flavors in the fluid product.

The off-flavors which develop in dairy products as a result of oxidative deterioration are collectively referred to as the "oxidized flavor". However, the organoleptic properties of the off-flavor differ between products as well as in the same product, depending on the degree of deterioration. Descriptive terms, such as cappy and cardboard have been used to characterize the off-flavor in fluid milk, and the off-flavor in dry whole milk and butter oil has been referred to as oily or tallowy. Butter undergoes a continuous change in flavor defects during storage which usually develop in an order described as metallic, fatty, oily, trainy, and tallowy.[20]

Although the conditions under which the above-mentioned products are normally stored undoubtedly influence the extent of deterioration and hence the character of off-flavor, the lipid constituents involved in the reaction also influence the resulting flavor. The site of oxidative deterioration in fluid milk and cream is the highly unsaturated phospholipid fraction associated with the fat globule membrane material.[20,326,337] On the other hand, in products such as butter and dry whole milk, both the phospholipids and the triglycerides are subject to oxidative deterioration.[20] The off-flavor appearing in butter oil is understandably the result of triglyceride deterioration.

MEASUREMENT OF FAT OXIDATION

Various methods have been employed to measure the extent of autoxidation in lipids and lipid-containing food products. For obvious reasons, such methods should be capable of detecting the autoxidative process before the onset of off-flavor. Milk and its products, which develop characteristic off-flavors at low levels of oxidation, require procedures that are extremely sensitive to oxidation. Thus methods of measuring the decrease in unsaturation (Iodine No.) or increase in diene conjugation as a result of the reaction do not lend themselves to quality control procedures, although they have been used successfully in determining the extent of autoxidation in model systems.[128,274]

Several methods have been introduced which express the degree of oxidative deterioration in terms of hydroperoxides per unit weight of fat. The Modified Stamm Method,[134] the most sensitive of the peroxide determinations, is based on the reaction of oxidized fat and 1,5-diphenylcarbohydrazide to yield a red color. The Lea Method[6,225] depends on the liberation of iodine from potassium iodide, wherein the amount of iodine liberated by the hydroperoxides is used as the criterion of the extent of oxidative deterioration. The colorimetric ferric thiocyanate procedure adapted to dairy products by Loftus-Hills and Thiel,[15] with modifications by various workers,[271,335] involves conversion of the ferrous ion to the ferric state in the presence of ammonium thiocyanate, presumably by the hydroperoxides present, to yield the red pigment ferric thiocyanate. Hamm and Hammond[132] have shown that the results of these three methods can be interrelated by use of the proper correction factors. However, those methods based on the direct or indirect determination of hydroperoxides which do not consider previous dismutations of these primary reaction products are not necessarily indicative of the extent of the reaction, nor do they tend to correlate well with the degree of off-flavors in the product.[196]

Two variations of the Thiobarbituric Acid Method have been widely used to determine the degree of lipid oxidation in dairy products.[78,187] The methods of approximately equal sensitivity are based on the condensation of two molecules of thiobarbituric acid with one of malonaldehyde,[307] resulting in the formation of a red color complex with an absorption maximum at 532 to 540 m_{μ}. King[187] has shown (Table 5.4) that a correlation exists between the determined TBA values and the intensity of the oxidized flavor in fluid milks. Similar observations have been reported by others in fluid milks[85] and ultra-high temperature creams.[64] The TBA method of Dunley and Jennings[78] has been reported to be more applicable than the King method in determining the extent of autoxidation in butter, although no correlation with the

Table 5.4

RELATION BETWEEN ORGANOLEPTIC AND TBA VALUES OF FLUID MILK[187]

Flavor Score	Description	Range of Optical Density (432 mμ)
0	No oxidized flavor	0.010–0.023
1	Questionable to very slight	0.024–0.029
2	Slight but consistently detectable	0.030–0.040
3	Distinct or strong	0.041–0.055
4	Very strong	> 0.056

extent of the off-flavor is apparent.[64] Both methods have been used extensively in studies of the autoxidation of extracted milk components and model lipid systems.[112,128,187] Lillard and Day reported[228] a significant correlation between a modified TBA test and the reciprocal of the Average Flavor Threshold of oxidized butterfat. A similar correlation also existed between the Peroxide Value and the reciprocal of the Average Flavor Threshold of butterfat.

In addition to the previously mentioned chemical tests, methods based on the carbonyl content of oxidized fats have also been suggested[143,228] as a measure of oxidative deterioration. The procedures determine the secondary products of autoxidation and have been reported to correlate significantly with the degree of off-flavor in butter oil[228] The methods, however, are cumbersome and are not suited for routine analysis.

ANTIOXIDANTS

The use of synthetic antioxidants in the prevention or retardation of autoxidation in lipids and lipid-containing food products has been the subject of numerous investigations. Although the present U.S. standards do not permit antioxidants in dairy products, and hence the question of their effectiveness is one of only theoretical interest, they are of practical interest in countries where their use is permitted. Many compounds containing two or more phenolic hydroxy groups, such as esters of gallic acid, butylated hydroxyanisole, norhydroguaiaretic acid, hydroxyquinone and dihydroquercitin, have been employed as antioxidants in studies of dairy products. These compounds apparently exert their influence by interrupting the chain reaction in autoxidation by capture of the free radicals necessary for continuation of hydroperoxide formation.[18]

Considerations, other than legal, that must be taken into account regarding use of antioxidants in dairy products include off-flavors imparted by the antioxidant itself,[113,294] ease of incorporation into the product,[135] and effectiveness of the antioxidant in different mediums. With regard to the latter point, studies of the use of antioxidants in dairy products reveal variations in their antioxidative properties in different products. Norhydroguaiaretic acid is effective in preventing the development of an oxidized flavor in fluid milk, but tends to increase the rate of autoxidation in milkfat.[135] The tocopherols, while of little value in dry whole milks[3] and butter oil,[272] are highly effective in preventing spontaneous or copper-induced oxidation in fluid milk.[79,189] Compounds reported to be among the most antioxidative in specific dairy products include: dodecyl gallate in spray-dried whole milks,[2,341] ascorbyl palmitate in cold storage-cultured butter,[201] sodium gentisate in frozen whole milk,[113] and quercitin and propyl gallate in butter oil.[379]

Synergists, such as the polybasic acids citric and phosphoric, have been used in conjunction with antioxidants. These compounds have no antioxidative value in themselves, but increase the effectiveness of antioxidants. Their synergistic influence on antioxidants may be due to the sequestering of metallic ions,[18,161] inhibiting the antioxidant catalysis of peroxide decomposition,[280] or regenerating the antioxidant in the system.[323] It has been reported that these synergists, like the phenolic antioxidants, are capable of performing the dual role of retarding autoxidation at low levels and accelerating it at higher levels.[280]

In addition to antioxidants alone or in the presence of synergists, metal chelating compounds, such as the various salts of ethylenediaminetetraacetic acid,[10,192] neocuproine[325] among others,[301] have also proven their effectiveness as inhibitors of autoxidation.

OXIDATIVE DETERIORATION IN FLUID MILK

Fluid milks have been classified by Thurston[362] into three categories with regard to their ability to undergo oxidative deterioration: (a) spontaneous, for those milks that spontaneously develop off-flavor within 48 hr after milking; (b) susceptible, for those milks that develop off-flavor within 48 hr after contamination with cupric ion; and (c) resistant, for those milks that exhibit no flavor, even after contamination with copper and storage for 48 hr. A similar classification has been employed by Dunkley and Franke.[73]

With the advent of noncorrodible dairy equipment, oxidative deterioration in fluid milks as a result of copper contamination has decreased significantly, although it has not been completely eliminated.[293] However, the incidence of spontaneous oxidation remains a major problem of the dairy industry. For example, Bruhn and Franke[34] have shown that 38% of samples produced in the Los Angeles milkshed are susceptible to spontaneous oxidation; and Potter and Hankinson reported[276] that 23.1% of almost 3000 individual samples tasted were criticized for oxidized flavor after 24 to 48 hr storage. Significantly, certain animals consistently produce milk which develops oxidized flavor spontaneously, others occasionally, and still others not at all.[260] Differences have even been observed in the milk from the different quarters of the same animal.[126]

Greenbank[122] attributed the resistance of certain milks to oxidation, even in the presence of added copper, to its poising action, i.e., the resistance of milk to a change in oxidation-reduction potential. That a correlation exists between the appearance of an oxidized flavor and conditions favoring a mild oxidation, as measured by the oxidation-reduction potential, was shown by Tracey *et al.*[365] and by Greenbank.[122] This apparent correlation, as well as other factors, tends to discredit theories as to the role of enzymes as catalytic agents in the development of oxidized flavor. Such a theory had been proposed initially by Kende,[181] who claimed that milk contains "oleinase" which catalyzes the oxidation of oleic acid to produce the characteristic off-flavor. More recently, xanthine oxidase has been proposed[12,13,15] as the catalytic agent in the development of spontaneously oxidized milk. The studies of Smith and Dunkley,[322] among others,[282] do not corroborate these studies, and they conclude that xanthine oxidase is itself not a limiting factor in the off-flavor.

Despite literature reports of anomalous behavior in several aspects, sufficient evidence has been accumulated in recent years to establish that the susceptibility or resistance of milk to oxidative deterioration is dependent on the percentage and/or distribution of naturally occurring pro- and antioxidants.

METALS

Metal-catalyzed lipid oxidative reactions were recognized in dairy products as early as 1905.[116] Investigations through the years have shown that copper and iron are the important metal catalysts in the development of oxidized flavors. Of these two metals, copper exerts the greater catalytic effect, while ferrous ion is more influential than ferric ion.[122]

Both copper and iron are normal components of milk. Disregarding variations due to individuality, stage of lactation, and contamination, the former is present at average levels of 20 to 40 μg/liter[157,203,227] and the latter at 100 to 250 μg/liter. Despite the greater abundance of iron in milk, copper has been shown by specific chelating agents to be the catalytic agent in the development of oxidized fluid milk.[325]

The natural copper content of milk originates in the cow's food, and is transmitted to the milk by way of the blood stream.[131] The studies of Dunkley and co-workers[75,287] suggest that an animal's feed can influence the natural copper content of its milk—a view which is not shared by others.[249] Nevertheless, the total natural copper content of a milk is not the overall deciding factor in the spontaneous development of an oxidized flavor in fluid milk.

Poulsen and Jensen[277] reported that "neither the absolute amount nor the range in content of naturally occurring copper during the lactation period has any significant influence on the tendency of milk to acquire oxidized flavor." Samuelsson[300] investigated milks from cows of low and high yield production ranging in copper content from 0.023 to 0.204 ppm. He concluded that oxidation may occur irrespective of the copper content, but no oxidation faults have been observed in milks with a copper content less than 0.060 ppm. Similar results have been reported by others.[191]

Natural copper and iron exist in milk in the form of complexes with proteins and as such are not dialyzable at normal pH of milk.[193,302] Copper and iron added to milk are, however, slightly dialyzable, the ease of dialysis of added copper increasing with a decrease in pH.[302] The latter observation suggests that the copper-protein bond of added copper is different from that of natural copper. King *et al.*[193] reported that 10 to 35% of the natural copper and 20 to 47% of the natural iron are associated with the fat globule membrane material. Only 2 to 3% of added copper and negligible percentages of added iron, however, become associated with the fat globule membrane. Similar trends in the distribution of natural and added copper in milk have been reported by others;[248] the subject has recently been reviewed by Haase and Dunkley.[131]

Samuelsson observed[300] that most of the natural copper associated with the cream phase can be removed by washing with water, and that the actual fat globule membrane proteins contain approximately 4% of the total natural copper content. Nevertheless, the value represents the highest concentration of copper per gram of protein in the milk system. Koops stated[203] that "although the amount of natural copper in early lactation may be very high, the concentration of copper (average 11.0 μg/100 g fat globules) in the membrane does not deviate substantially from that of normal uncontaminated milk." King[186] observed that milks which developed oxidized flavor spontaneously had a higher total copper concentration in the fat globule membrane than did milks classified as susceptible or resistant.

Samuelsson concluded[300] on the basis of his studies that the close proximity of a copper-protein complex to the phospholipids which are also associated with the fat globule membrane is an important consideration in the development of an oxidized flavor in fluid milks. Haas and Dunkley[131] stated that although "some aspects of catalysis of oxidative reactions in milk by copper still appear anomalous . . . the mechanism of oxidized flavor development with copper as catalyst involves a specific grouping of lipoprotein-metal complexes in which the spatial orientation is a critical factor."

ROLE OF ASCORBIC ACID

That copper, naturally occurring or present as a contaminant, accelerates the development of oxidative deteriorations in fluid milk is evident. However, its presence is not the only consideration as to whether or not oxidative deterioration occurs. Olson and Brown[256] showed that washed cream (free of ascorbic acid) from susceptible milk did not develop an oxidized flavor when contaminated with copper and stored for 3 days. Subsequently, the addition of ascorbic acid to washed cream, even in the absence of added copper, was observed[270] to promote the development of an oxidized flavor. Krukovsky and Guthrie[210] and Krukovsky[209] reported that 0.1 ppm of added copper did not promote oxidative flavors in milk or butter depleted of their vitamin C content by quick and complete oxidation of ascorbic acid to dehydroascorbic acid. Krukovsky and Guthrie[208,210] further showed that the oxidative reaction in ascorbic acid-free milk could be initiated by the addition of ascorbic acid to such milk. Accordingly, these workers and others have concluded that ascorbic acid is an essential link in a chain of reactions resulting in the development of an oxidized flavor in fluid milk.

Various workers[56,136,369] have observed a correlation between the oxidation of ascorbic acid to dehydroascorbic acid and the development of an oxidized flavor. Smith and Dunkley[323] concluded, however, that ascorbic acid oxidation cannot be used as a criterion for lipid oxidation. Their studies showed that although ascorbic acid oxidation curves for homogenized and pasteurized milk were similar, the homogenized samples were significantly more resistant to oxidized flavor. Furthermore, whereas pasteurization caused an appreciable decrease in the rate of ascorbic acid oxidation compared to raw milk, the pasteurized samples were more susceptible to oxidation.

Haase and Dunkley[129,130] reported, as a result of studies on model systems of potassium linoleate, that ascorbic acid functioned as a true catalyst, i.e., it accelerated the oxidation of linoleate but it itself was not oxidized. When copper was added to the system, however, the oxidation of ascorbic acid occurred simultaneously with the linoleate. In this respect, Smith and Dunkley[324] reported that a significant correlation exists between the rate of ascorbic acid oxidation and the natural copper content of milk. Furthermore, King reported[188] a positive relation between lipid oxidation and ascorbic acid oxidation in model systems containing fat globule membrane material, the component of uncontaminated milk having the highest concentration of copper per gram of lipid. Although ascorbic acid alone in model systems of linoleate has been observed to be pro-oxidant, low concentrations of ascorbic acid in combination with copper exhibited greater catalytic activity than the additive activity of the two catalysts individually.[130] Possible explanations for the enhanced catalysis include reduction of copper by ascorbic acid to the more pro-oxidative cuprous form,[24,130,325] increased concentration of a semidehydroascorbic acid radical,[24,130] and the formation of a metal-ascorbic acid-oxygen complex.[130]

The behavior of ascorbic acid in the oxidative reaction, however, is anomalous, as evidenced by the studies of several workers.[27,28,45,211] Their results indicate that concentrations normal to milk (10 to 20 mg per l) promote oxidative deterioration, while higher concentrations (50 to 200 mg per l) inhibit the development of off-flavors.

Various researchers have proposed explanations for the inhibitory behavior of high concentrations of ascorbic acid in fluid milk. Chilson[45] reported that added ascorbic acid acts as a reducing agent which oxidizes more readily than milkfat. This either prevents or prolongs the time required for fat oxidation and the development of an oxidized flavor. Bell *et al.*[27] concluded that the addition of L-ascorbic acid to concentrated sweet cream lowers its oxidation-reduction potential and thus produces a medium less conducive to oxidation. In this respect, Campbell *et al.*[36] reported that the O-R potential of milk is entirely dependent on its vitamin C content, and Greenbank[123] has shown that the oxidation of ascorbic acid to dehydroascorbic acid is reflected in gradual increases in Eh. Krukovsky[209] reported that the oxidative reaction is initiated more rapidly in milk when the ratio of ascorbic to dehydroascorbic acid is approximately 1 to 1 or lower. He states "that an unfavorable proportion of dehydroascorbic acid could not be accumulated if the rate of its oxidation to non-reducible substances surpassed that of ascorbic acid to dehydroascorbic acid. Consequently, the protective influence of ascorbic acid added in large but variable quantities to milk could be attributed to the exhaustion of occluded oxygen prior to the establishment of a favorable equilibrium between these two forms of vitamin C". Smith and Dunkley[325] disputed this theory and suggested that the results were influenced by higher than normal ascorbic acid contents when the ratio of ascorbic acid to dehydroascorbic acid was greater than 1 to 1 in the experimental milks. In this regard, King[186] was not able to duplicate Krukovsky's results in milks with normal ascorbic acid levels.

King theorized[188] that when the initial concentration of ascorbic acid increases beyond that necessary to saturate the copper in the system, the oxidation of ascorbic acid becomes so rapid and the products of the reaction accumulate so rapidly that they either block the reaction involving the lipids in the system or prevent the copper from acting as a catalyst.

Haase and Dunkley[129] reported that, although high concentrations of ascorbic acid in model systems of potassium linoleate were prooxidant, a decrease in the rate of oxidation was observed. They further noted[130] that certain concentrations of ascorbic acid and copper inhibited the formation of conjugated dienes, but not the oxidation of ascorbic acid, and caused a rapid loss of part of the conjugated dienes already present in the system. They theorized that certain combination concentrations of ascorbic acid and copper inhibit oxidation by the formation of free-radical inhibitors which terminate free-radical chain reactions, and that the inhibitors are complexes that include the free radicals.

ROLE OF α-TOCOPHEROL

The literature[123,286] appears to be in general agreement that the use of green feeds tends to inhibit and that of dry feeds to promote the development of oxidized flavors in dairy products. Furthermore, the observation[52,237] that milks produced during the winter months are more susceptible to oxidative deteriorations is the result, no doubt, of differences in feeding practices.

Investigations concerned with variations in the oxidative stability of milk as a result of feeding practices have centered on the transfer to milk of natural antioxidants. Although Kanno *et al.*[173] have reported the presence of γ-tocopherol, the only known natural antioxidant of consequence in milk is α-tocopherol.

Milk contains on the average approximately 25 μg of α-tocopherol per gram milkfat.[34,86,173] Dicks[62] has assembled a comprehensive bibliography of the literature on the α-tocopherol content of milk and its products, including data on the numerous variables which influence

Vitamin E content. Foremost among these variables is the feed of the animal as influenced by season of the year. Kanno *et al.*[173] reported that milk produced from May to October on pasture feeding averaged 33.8 μgα-tocopherol per gram fat, while that produced by dry-lot feeding from November to April contained an average of 21.6 μ g α-tocopherol per gram fat. Similar results have been reported by others.[190,217,312]

Krukovsky *et al.*[218] found a significant correlation between the tocopherol content of milkfat and the ability of milk to resist autoxidation. A high proportion of samples which contained less than 25 μgα tocopherol per gram fat were unstable and developed oxidized flavors during storage. Erickson *et al.*[87] reported that the tocopherol concentration in the fat globule membrane lipids correlated more closely with oxidative stability of the milk than did the tocopherol content of the butter oil. Dunkley *et al.*[74] stated, however, that the concentration of α-tocopherol in milk is not satisfactory as a sole criterion for predicting oxidative stability, and that the concentration of copper must also be considered. In this regard, King *et al.*[194] found a direct relationship between the tocopherol level and the percentage of copper tolerated by milk. Spontaneous milk oxidation was reported by Bruhn and Franke[34] to be directly proportional to the copper content and inversely proportional to the α-tocopherol content of milk.

Erickson *et al.*[88] observed that, although containing only 8% of the total tocopherols in milk, the fat globule membrane contains the highest concentration of α-tocopherol per gram fat in milk (44.0 μ g/g). Erickson and co-workers had previously concluded[87] that since "the lipids in the fat globule membrane are most susceptible to oxidation because of their unsaturation and their close association with the pro-oxidants copper and ascorbic acid, the α-tocopherol in the membrane is more important in inhibiting oxidation than that inside the fat globule". A similar conclusion has also been reached by King.[189]

Recently, several studies have been concerned with increasing the α-tocopherol levels of milk to prevent the development of oxidized flavors when tocopherol-rich forages are not available for feed. Dunkley *et al.*,[79,80] King *et al.*[194] and Merk and Crasemann[241] have reported increases in the α-tocopherol content of milk and increased resistance to spontaneous and copper-induced oxidation by supplementing the cow's ration with varying proportions of α-tocopherol acetate. Dunkley *et al.*[80] reported that supplementing the ration of an animal with 500 mg *d*-α-tocopherol acetate increased the total milk tocopherol content by 28.6 μ g/g lipid; and King *et al.*[190] reported that supplementing the feed to achieve a total intake of 1.0 g α-tocopherol/cow/day provides an effective control against oxidation in milk containing 0.1 ppm copper contamination. Several reports[74,194] have shown that approximately

2% of the total α-tocopherol intake is transferred to milk and as such, supplementing the ration with α-tocopherol acetate is a relatively inefficient procedure. King[189] has reported, however, that the direct addition of d-α-tocopherol in an emulsified form at a concentration of 25 μg/g milkfat would prevent the development of oxidized flavor in milk containing 0.1 ppm added copper—the same α-tocopherol concentration found to be effective when the ration was supplemented with α-tocopherol acetate. Control of oxidized flavor by direct addition of emulsfied α-tocopherol to milk can be achieved with only 1% of the amount required by ration supplementation.

FACTORS AFFECTING OXIDATIVE DETERIORATION IN MILK AND ITS PRODUCTS

Storage Temperature

The role of storage temperature in the oxidative deterioration of dairy products is anomalous. Tracey[364] recognized that fluid milk was more susceptible to oxidized flavor when stored at 4°C than at 20°C. Dunkley and Franke[73] also observed more intense oxidized flavors and higher TBA values in fluid milks stored at 0°C than at 4°C and 8°C. The flavor intensity and the TBA values decreased with increasing storage temperature. Bell[26] reported that, other conditions being equal, condensed milk stored at −17°C is more susceptible to the development of oxidized flavor than is condensed milk maintained at −7°C.

In contrast to the above, low storage temperatures tend to decrease the rate of light-induced oxidative deterioration[76] and to decrease or inhibit oxidative deterioration in other dairy products. Pyenson and Tracey[281] reported that storage temperatures of 2°C retarded the development of oxidative deterioration in dry whole milk, as determined by O_2 absorption and flavor scores, in comparison with samples stored at 38°C in an atmosphere of air. Downey[64] reported that oxidative deterioration in UHT cream occurred 2 to 3 times more rapidly at 18°C than at 10°C, while little or no oxidation occurred at 4°C. Holm *et al.*[156] showed that in the case of butter, approximately 4 times more storage time was necessary at −10°C to obtain the same 2-point decrease in flavor score given products held at 10°C. Sattler-Dornbacher[304] reported an increase in the O-R potential of butter as the storage temperature increased with a corresponding increase in the rate of flavor deterioration. Hamm *et al.*[133] demonstrated the rates of oxidative deterioration in butter oils during storage at temperatures ranging from −10 to 50°C. Despite dramatic differences in the rate of oxidation, increasing rates with increasing temperatures; they concluded that the same flavors were formed on storage, and that the reaction sequence for flavor formation was the same at all temperatures.

Oxygen Levels

The inhibition of oxidative deterioration in fluid milk held at higher storage temperatures has been attributed by various workers[56,122,361] to a lowering of the oxygen content as a result of bacterial activity. In this respect, it has been noted that the increase in incidence of oxidized flavor in milk has paralleled the bacteriologically improved milk supply.[161] Collins and Dunkley[47] have reported, however, that although large numbers of bacteria slightly retard development of oxidized flavor, the relatively small numbers of bacteria normally found in market milk are of no practical consequence in determining whether or not milk will develop off-flavor. Furthermore, Sharp *et al.*[316] stated that the number of bacteria necessary to reduce the oxygen content materially would be sufficient to cause other types of deterioration.

Removal of the dissolved oxygen in fluid milk or its replacement with nitrogen was shown by Dahle and Palmer[56] to inhibit the development of oxidized flavors. Sharp *et al.*[315] further showed that deaeration would inhibit the appearance of off-flavor even in the presence of 0.1 mg copper per l milk. Singleton *et al.*[317] confirmed previous observations that oxygen was required for the development of light-induced off-flavors. Schaffer *et al.*,[306] applying deaeration to products other than fluid milk, concluded that, to prevent the production of tallowy flavor in butter oil, the available oxygen should be less than 0.8% of the volume of the fat. Similar storage conditions were also proposed by Lea *et al.*[226] Although the deaeration of these products is of significance only from a scientific standpoint, the deaeration of dry milk products has practical applications.

Vacuum treatment or replacement of available oxygen with an inert gas has proved its reliability in preventing or retarding the onset of oxidation in dry whole milk for extended periods of storage. Greenbank *et al.*[125] showed that inert gas-packing to an oxygen level of 3 to 4% increased the storage life of whole milk powder 2 to 3 times that of air-packed samples, the length of storage being dependent on the initial quality of the product. Lea *et al.*[226] showed that, whereas oxidative deterioration in milk powders packed at the 3 to 6% oxygen level was retarded significantly, inert gas containing 0.5 to 1.0% oxygen would prevent the development of recognizable tallowy flavors for an indefinite period. Tamsma *et al.*[342] showed statistically a highly significant improvement in storage stability of whole milk powders packed in inert gases containing 0.1% oxygen over those packed at a 1% oxygen level. Schaffer *et al.*[306] concluded that the time required for the production of a tallowy flavor is inversely proportional to the oxygen concentration.

Several dearation techniques other than mechanical methods have been utilized to inhibit or retard the development of tallowy flavors in dry milks. Meyer and Jokay[242] reported that milk powders packed in the presence of an oxygen scavenger (glucose oxidase-catalase) and desiccant (calcium oxide) were comparable flavorwise to samples stored in the presence of an inert gas, the enzymes demonstrating the ability to reduce oxygen levels to 0.5% in one week. Jackson and Loo,[159] employing an oxygen-absorbing mixture (0.5 g Na_2SO_3 and 0.75 g $CuSO_4 \cdot 5H_2O$) enclosed in porous paper pouches, demonstrated keeping qualities equal to those of dry milks stored in the presence of an inert gas. Abbot and Waite[1] reported favorable results in the keeping quality of dry whole milk by using a mixture of 90% nitrogen and 10% hydrogen in the presence of a palladium catalyst. The metal catalyzes the formation of water from the hydrogen and residual oxygen to produce an almost oxygen-free atmosphere in the pack. Tamsma *et al.*[339] reported obtaining within 24 hr a pack containing less than 0.001% oxygen by use of an oxygen-scavenging system consisting of 95% nitrogen, 5% hydrogen and a platinum catalyst. Marked improvements in the keeping quality of milk powders packed in the scavenging system were reported.

Heat Treatment

Pasteurization of fluid milk leads to increased susceptibility to spontaneous,[31] copper-induced,[263,323] and light-induced oxidized flavor.[90] Heating to higher temperatures, however, reduces the susceptibility.[30,323] A possible explanation for the increased incidence of oxidized flavor as a result of pasteurization temperatures is suggested by several studies. Sargent and Stine[303] reported a substantial migration of added copper to the cream phase of milk at temperatures higher than 60°C. Van Duin and Bruns[69] also observed an increase in the copper content of creams prepared from pasteurized milk. Samuelsson[301] reported that washed cream made from milk heated to 80°C for 10 min contained twice as much copper as that prepared from unheated milk. The migration of the additional copper to the cream phase, which also contains the readily oxidized phospholipids, increased the potential of the system toward oxidative deterioration. Tarassuk *et al.*[350] also observed that washed cream is very sensitive to the development of trainy (fishy) flavor when heated to temperatures between 60 and 90°C. The effect of previous heat treatment on the copper content of butter was reported by Van Duin and Bruns.[69] They observed that pasteurization of cream at 78°C for 15 to 30 sec gave high copper concentrations in butter and low values in the buttermilk, the reverse being true when the cream was heated to above 82°C. They recommended that creams prepared from pasteurized milks should be heated to the higher temperatures to decrease the susceptibility of butter to oxidative deterioration during storage.

The inhibitory effect of high heat treatment on oxidative deterioration in fluid milk and its products has been reported by various workers.[56,122,340] Gould and Sommer,[119] in conjunction with studies on the development of a cooked flavor in heated milks, noted a decrease in the oxidation-reduction potential of the product. They attributed the cooked flavor to the formation of sulfhydryl compounds and correlated the liberation of these compounds to the heat retardation and prevention of oxidized flavor. The work of Josephson and Doan[171] conducted simultaneously with these workers confirmed the relationship between sulfhydryl compounds, cooked flavor, decreased Eh, and inhibition of oxidized flavor. They further reported that most heated products do not become tallowy or oxidized until the sulfhydryls are first oxidized and the cooked flavor has disappeared. Wilson and Herreid[342] prolonged substantially the onset of oxidative deterioration of 30% sterilized cream by increasing to 13% the solids-not-fat content of the cream prior to sterilization, presumably by increasing the potential sulfhydryl content of the finished product. Gould and Keeney[118] showed that oxidized flavor occurred in heated cream to which copper had been added when the active sulfhydryl compounds had decreased to a level approximating 3 mg per l of cystine HCl.

β-Lactoglobulin has been shown by Larson and Jenness[222]—and this finding was confirmed by Hutton and Patton[158]—to be the major source of sulfhydryl groups in milk, while the fat globule membrane material contributes a minor portion of these reducing compounds.

Time-temperature relationships have been established by various workers as being optimum for preventing or retarding the development of oxidized flavors in dairy products: cream, 88°C for 5 min;[118] condensed milk, 76.5°C for 8 min;[26] dry whole milk, preheat at 76.5°C for 20 min;[46] frozen whole milk, 76.5°C for 1 min.[29] Few, if any, instances of a tallowy flavor have been reported in evaporated milk; undoubtedly a major reason for its stability toward oxidation can be attributed to the sterilization temperatures employed in its manufacture.

Josephson reported[170] that butterfat prepared from butter heated to 149, 177, and 204.5°C was extremely stable to oxidation, while that heated to 121°C oxidized readily when stored at 60°C. When butter oil itself was heated from 121 to 204.5°C it also oxidized rapidly. The addition of 1% skimmilk powder to butter oil, however, prior to heating at 204.5°C for 10 min also resulted in a significant antioxidative effect, which Josephson concluded was the result of a protein-lactose reaction (carmelization). Wyatt and Day[379] reported that the addition of 0.5% nonfat milk solids to butter oil followed by heating at 200°C and 15 mm Hg for 15 min caused the formation of antioxidants which protected the butter oil against oxidative deterioration for one year, thus surpassing the effectiveness of many synthetic antioxidants tested.

Exposure To Light

The catalytic effect of natural light in promoting off-flavor development in fluid milk has been recognized for some years. The extent of deterioration appears to be dependent on the wavelengths involved, intensity of the source, and the length of exposure.[14,76,122,236] Off-flavors have also been reported to develop in butterfat which has been exposed to the action of natural light.[334] In addition to natural light, incandescent or fluorescent lights employed in storage coolers may promote deteriorative reactions,[321] while the development of off-flavors is the limiting factor in the preservation of dairy products by high-energy radiation.[58,152] Efforts to inhibit or retard the onset of off-flavors as a result of exposure to sunlight has led to the introduction of doorstep coolers and, in certain cases, of amber-colored milk bottles.

Two distinct flavors may develop in milk exposed to light:[299,368] a burnt, activated or sunlight flavor which develops rapidly, and a typically oxidized flavor which develops on prolonged exposure.[336] It is possible that the presence of contradictory statements in the literature regarding deterioration on exposure to light may be attributed to the failure of various investigators to recognize the existence of more than one off-flavor.

Studies[264,368] have shown that riboflavin plays a significant role in the development of the activated flavor. Although removal of riboflavin from milk by passing through Florisil prevented the development of activated flavor, such treatments did not prevent the development of the oxidized flavor. The later observation does not agree with the reports of other workers,[14,136,369] which indicate that riboflavin plays a significant role in the oxidized flavor. Ascorbic acid has also been implicated in the development of off-flavors in fluid milks exposed to light.[14,77,299] The exact nature of its involvement, however, is not clear.

Limited studies have been conducted on the lipid components oxidized in milk exposed to sunlight. Finley *et al.*[90] observed a decrease in the oleic and linoleic acid contents of an isolated low-density lipoprotein from milk, and implicated the lipoprotein as a major substrate for the photoxidation reaction. Although previous studies[378] suggested that the monoene fatty acids are important oxidizing substrates in milk exposed to sunlight, Wishner noted[377] that photoxidation of methyl linoleate in the presence of photosensitizers produces significant percentages of the less stable 11-hydroperoxide,[183] which on decomposition forms alk-2-enals, the significant carbonyls found in milk exposed to sunlight.

The sunlight flavor has been shown[91,264] to originate in the proteins of milk. Hendrick[142] concluded that the serum proteins are the main source of activated flavor in milk, with riboflavin as the photosensitizer. Similar results have been reported by Storgards and Ljungren.[336] Singleton *et al.*[317] demonstrated a relationship between riboflavin destruction, tryptophan destruction, and the intensity of the sunlight flavor in milk, and implicated a tryptophan-containing protein rather than a single low molecular weight compound as one of the reactants. Finley *et al.* reported[90] that a low-density lipoprotein fraction associated with the fat globule membrane served as a carrier and a precursor for the light-induced off-flavor. Studies of the degradation of the lipoprotein on exposure to light showed that both the lipid and protein portions of the lipoprotein were degraded. In addition to tryptophan, they observed the destruction of methionine, tyrosine, cysteine, and lysine in the lipoprotein on exposure to light in the presence of riboflavin. The photoxidation of amino acids other than tryptophan has been observed in enzymes exposed to sunlight.[377]

Methional, formed by the degradation of the amino acid methionine, has been reported[264,368] to be the principal contributor to the activated flavor. Samuelsson[298] reported, in studies of di- and tripeptides containing methionine, that irradiation did not result in any hydrolysis of the peptides, and the presence of methional in the reaction products could not be demonstrated. He concluded that methional can only occur in irradiated milks from the free methionine in the milk serum. Thiols, sulfides and disulfides observed as products of the irradiated peptides may be of greater significance in the activated flavor.

Acidity

The development of a fishy flavor in butter is well known, and its association with salted butter made from acid cream was first demonstrated by Rogers in 1909.[291] Cream acidities ranging from 0.20 to 0.30% appear to represent those levels where flavor development is marginal.[156,292] Although the development of fishy flavors in unsalted butters is rarely encountered,[292] it is not restricted to those products containing salt. Pont *et al.*[273] induced the development of a fishy flavor in commercial butterfat by the addition of nordihydroguaiaretic acid and citric or lactic acid. In addition, Tarassuk *et al.*[350] reported the development of fishy flavors in washed cream adjusted to pH 4.6.

Koops[202] conducted a comprehensive study of the development of trainy (fishy) flavor which occurs in butter prepared from cultured cream (pH 4.6) during cold storage. He observed[203] that, although the acidification of milk or cream to pH 4.6 did not result in a transfer of natural copper from the plasma proteins to the fat globule membrane, 30 to 40% of added copper migrated to the membrane proteins at pH 4.6. He concluded[202] that the development of a trainy flavor in cultured butter is the result of the migration of the plasma-bound added (contaminated) copper to the fat globule membrane and the enhanced interaction between the cephalin fraction of the membrane phospholipids, which is highly susceptible to oxygen,[200] and the copper-containing membrane protein.

Although not studied extensively, reports on other dairy products suggest that titratable acidity as well as hydrogen-ion concentration tend to influence the development of oxidative deteriorations. Anderson[7] found a relationship between the titratable acidity and the development of an oxidized flavor in milk. Furthermore, his results showed that, while milks developed an oxidized flavor at a titratable acidity of 0.19%, the deteriorative mechanism was inhibited when the milks were neutralized to acidities of 0.145% or lower. Greenbank[122] found that an increase in pH of 0.1 was sufficient to inhibit the development of oxidized flavors in fluid milks for 24 hr. Anderson[7] reported similar results. In addition to fluid milk, Dahle and Folkes[54] attributed the development of oxidized flavors in strawberry ice cream to the presence of copper and the acid content of the fruit.

Homogenization

Homogenization was found in 1933 by Tracey *et al.*[365] to inhibit the development of an oxidized flavor in fluid milk. Subsequently, similar observations were reported on cream,[350] ice cream,[55] dry whole milk,[155] and frozen condensed milk.[26] The inhibitory effect, however, is not absolute. Roadhouse and Henderson[288] found that the absolute pressure required varies with different milks contaminated with the same concentration of cupric ion. The results of Larsen *et al.*[220] and Smith and Dunkley[323] indicate that the inhibitory effect of homogenization is dependent on the degree of metallic contamination.

Various workers have proposed explanations for the inhibitory effect of homogenization on oxidative deterioration. Tracey *et al.*[365] considered it to be apparent rather than actual, resulting from changes in the physical consistency of the milk, which may alter the taste. These workers based their proposal on the observation that homogenization has no apparent effect on the Eh of milk. Similar observations have been noted by others.[221] Still others have proposed that the inhibition is real, and is due to migration of the phospholipids into either the serum phase[363] or interior of the fat globule,[207] to general redistribution of the phospholipids in the milk proper,[124] or to denaturation of proteins resulting in an increase in the number of available -SH groups.[100] King[186] proposed that homogenization effects an irreversible change in the structural configuration of the copper-protein complex in such a way that ascorbic acid is no longer able to initiate the formation of lipid free radicals. Smith and Dunkley[323] theorized that homogenization causes a change in the copper-protein binding by the formation of a chelate that is less active in ascorbic acid oxidation and inactive in lipid peroxidation. Tarassuk and Koops[349] stated that "the decrease in concentration of phospholipids and the copper-protein complex per unit of newly formed fat globule surface appears to be the most important factor, if not the only one, that retards the development of oxidized flavor in homogenized milk."

Dunkley *et al.*[77] demonstrated, by the use of TBA values and a highly trained taste panel, that although homogenization inhibits light-induced lipid oxidation, the process increases the susceptibility of milk to development of the activated flavor. An increase in the intensity of off-flavors in homogenized milks exposed to sunlight has been reported by several workers.[53,177] Finley concluded[90] as a result of his studies that any treatment (e.g., homogenization) which affects the fat globule membrane increases the susceptibility of milk to light-induced off-flavors. It is evident from the literature that homogenization affords a degree of protection against oxidative deterioration in fluid milks provided excessive metallic contamination and undue exposure to light are avoided.

CARBONYL CONTENT OF OXIDIZED DAIRY PRODUCTS

Considerable effort has been expended in recent years on the odorous compounds formed in autoxidized dairy products. Although some of the early identification studies lack present-day sophisticated methodology, may be incomplete, and do not differentiate between isomeric forms of the various compounds, their contribution to the knowledge of the products of autoxidation in dairy products is invaluable.

Table 5.5 summarizes the carbonyls that have been identified in several selected dairy products. Despite the general similarity in the qualitative carbonyl content of oxidized dairy products, flavor differences are apparent. Attempts to correlate the off-flavors with specific compounds or groups of compounds, however, are made difficult for several reasons. These include: (a) the multitude of compounds produced; (b) difficulties arising in the quantitative analyses of oxidized dairy products; (c) differences in threshold values of individual compounds; (d) similarity of flavors imparted by individual compounds near threshold; (e) a possible additive and/or antagonistic effect, flavorwise and with regard to threshold values of mixtures of compounds; (f) the possible existence of a compound or group of compounds heretofore not identified; and (g) the difficulties involved in adding pure compounds to dairy products as a means of evaluating their flavor characteristics.

Several individual compounds formed by the autoxidation of milk lipids, however, have been implicated in specific off-flavors. Stark and Forss[331] have identified 1-octen-3-one as the compound responsible for the metallic flavor which develops in dairy products. This compound has also been shown to be an integral part of other oxidized flavor defects.[20,95,96]

4-*cis*-Heptenal, responsible for the creamy flavor of butter,[25] results

Table 5.5

CARBONYLS IDENTIFIED IN AUTOXIDIZED DAIRY PRODUCTS

Product	Alkanal	Alk-2-enal	Alk-2,4-dienal
Skimmilk, copper-induced[98,99a]	C_2, C_6	C_4 to C_{11}	C_6 to C_{11}
Whole milk, spontaneous oxidation[260]	C_5 to C_{10}	C_6 to C_{11}	C_6 to C_{12}
Whole milk, light-induced[378]	—	C_4, C_6 to C_{11}	—
Dry whole milk air-packed[261]	C_1 to C_3 C_5 to C_{10}	C_5 to C_{11}	traces
Butter oil, exposed to air[58a,85a]	C_1 to C_{10}	C_4 to C_{11}	C_7, C_{10}
Butter, cold storage[b] defects[20]	C_5 to C_{12}	C_5 to C_{11}	C_7[c], C_9, C_{10}[c], C_{11}

[a]References

[b] Miscellaneous carbonyls: 4-heptenal [c]; 2,6-nonadienal [c]; 2,5-octadienal [c]; 2,4,6-nonatrienal [c]; 2,4,7-decatrienal [c]; 1-penten-3-one; 1-octen-3-one; 3,5-octadien-3-one; 3,5-undecadien-3-one.

[c] Includes *cis/trans* geometric isomers.

from autoxidation of minor isolinoleic acids in butterfat.[169] At higher concentrations this compound has also been implicated in the trainy flavor which develops in cold storage butter.[19] 6-*trans*-Nonenal has been identified as the compound responsible for the "drier" flavor[262] which frequently appears in freshly prepared foam spray-dried milks—an off-flavor which is peculiar to this particular product. Although the evidence suggests that it is formed in foam spray-dried milk by trace ozonolysis of minor milk lipids, it has also been identified in stored sterile milks.[259] The latter observation suggests it may also appear in dairy products as a result of autoxidation reactions.[182]

The findings of other studies suggest that the preponderance of certain carbonyls or groups of carbonyls is involved in the off-flavors of various dairy products. Forss *et al.*[98,99] reported that the C_6 to C_{11} 2-enals and the C_6 to C_{11} 2,4-dienals—and more specifically 2-octenal, 2-nonenal, 2,4-heptadienal, and 2,4-nonadienal—constitute a basic and characteristic factor in copper-induced cardboard flavor in skimmilk. The same workers concluded that "while these compounds in milk closely simulate the cardboard flavor, the resemblance is not complete" and that "the defect contains further subsidiary flavor elements".

Bassette and Keeney[22] ascribed the cereal-type flavor in dry skimmilk to a homologous series of saturated aldehydes resulting from lipid oxidation in conjunction with products of the browning reaction. The results of Parks and Patton[261] suggest that saturated and unsaturated aldehydes at levels near threshold may impart an off-flavor suggestive of staleness in dry whole milk. Wishner and Keeney[378] concluded from studies on milk exposed to sunlight that C_6 to C_{11} alk-2-enals are important contributors to the oxidized flavor in this product. Parks *et al.*[260] concluded, as a result of quantitative carbonyl analysis and flavor studies, that alk-2,4-dienals, especially 2,4-decadienal, constitute a major portion of the off-flavor associated with spontaneously oxidized fluid milk. Forss *et al.*[95,96] reported that the fishy flavor in butterfat and washed cream is in reality a mixture of an oily fraction in addition to 1-octene-3-one, the compound responsible for the metallic flavor. *n*-Heptanal, *n*-hexanal, and 2-hexenal were found to be constituents of the oily fraction in washed cream, and these three carbonyls plus heptanone-2 were constituents of the oily fraction isolated from fishy butterfat. Badings[20] identified 40 volatile compounds in cold storage cultured butter which had a trainy (fishy) off-flavor. Included among the 14 compounds which were present in above-threshold levels were: 4-*cis*-heptenal; 2-*trans,* 4-*cis*-decadienal; 2-*trans,* 6-*cis*-nonadienal; 2,4,7-decatrienal; 3-*trans,* 5-*cis*-octadien-2-one; 1-octene-3-one; and 1-octen-3-ol.

Comparative studies by Forss and co-workers[95,97] on the fishy, tallowy, and painty flavors of butterfat tend to emphasize the importance of the relative and total carbonyl contents in dairy products with different off-flavors. These researchers showed that three factors distinguished painty and tallowy butterfat from fishy flavored butterfat. First, there was a relative increase in the *n*-heptanal, *n*-octanal, *n*-nonanal, heptanone-2, 2-heptenal, and 2-nonenal in the tallowy butterfat, and a relative increase in the *n*-pentanal, and the C_5 to C_{10} alk-2-enals in the painty butterfat. Secondly, 1-octen-3-one was present in such low concentrations in both the tallowy and painty butterfats as to have no effect on the flavor. Thirdly, the total weight of the volatile carbonyl compounds was about ten times greater in the tallowy and 100 times greater in the painty butterfat than in the fishy-flavored butterfat.

REFERENCES

1. ABBOT, J., and WAITE, R., J. Dairy Res., *28,* 285 (1961).
2. ABBOT, J., and WAITE, R., J. Dairy Res., *29,* 55 (1962).
3. ABBOT, J., and WAITE, R., J. Dairy Res., *32,* 143 (1965).
4. ALBRECHT, T. W., and JAYNES, H. O., J. Dairy Sci., *38,* 137 (1955).
5. AL-SHABIBI, E. H., LANGNER, E. H., TOBIAS, J., and TUCKEY, E. H., J. Dairy Sci., *47,* 295 (1964).
6. American Oil Chemists' Society, "Official and Tentative Methods", Official Method Cd 8-53 (1960).
7. ANDERSON, E. O., Intern. Assoc. Milk Dealers, 30th Ann. Conv. Lab. Sect. Proc., 153 (1937).
8. ANDERSON, J. A., Milk Dealer, *37,* No. 2, 90 (1937).
9. ANDERSON, K. P., and JENSEN, S. G. K., Beretn. Forsgsm. Kbh., *136,* 58 (1962).
10. ARRINGTON, L. R., and KRIENKE, W. A., J. Dairy Sci., *37,* 819 (1954).
11. ASCHAFFENBURG, R., J. Dairy Res., *23,* 134 (1956).
12. ASTRUP, H., J. Dairy Sci., *46,* 1425 (1963).
13. AURAND, L. W., CHU, T. M., SINGLETON, J. A., and SHEN, R., J. Dairy Sci., *50,* 465 (1967).
14. AURAND, L. W., SINGLETON, J. A., and NOBLE, B. W., J. Dairy Sci., *49,* 138 (1966).
15. AURAND, L. W., WOODS, A. E., and ROBERTS, W. M., J. Dairy Sci., *42,* 1111 (1959).
16. BACHMAN, M., Schweiz, Milchztg., *87,* 629 (1961).
17. BADINGS, H. T., J. Am. Oil Chemists Soc., *36,* 648 (1959).
18. BADINGS, H. T., Neth. Milk Dairy J., *14,* 215 (1960).
19. BADINGS, H. T., Neth. Milk Dairy J., *19,* 69 (1965).
20. BADINGS, H. T., Ph.D. Thesis, Vageningen, The Netherlands (1970).
21. BASKYS, B., KLEIN, E., and LEVER, W., Arch. Biochem. Biophys., *102,* 201 (1963).
22. BASSETTE, R., and KEENEY, M., J. Dairy Sci., *43,* 1744 (1960).
23. BATEMAN, L., Quart. Rev. (London), *8,* 147 (1954).
24. BAUERNFEIND, J. C., and PINKERT, D. M., Adv. Food Res., *18,* 219 (1970).
25. BEGEMAN, P. HAVERKAMP, and KOSTER, J. C., Nature, *202,* 552 (1964).
26. BELL, R. W., J. Dairy Sci., *22,* 89 (1939).
27. BELL, R. W., ANDERSON, H. A., and TITTSLER, R. P., J. Dairy Sci., *45,* 1019 (1962).

28. BELL, R. W., and MUCHA, T. J., J. Dairy Sci., *32,* 833 (1949).
29. BELL, R. W., and MUCHA, T. J., J. Dairy Sci., *34,* 432 (1951).
30. BERGMAN, T., BERGOLF, A., and KJELL, S., 16th Intern. Dairy Cong. Proc., Vol. *A,* Sect. II:1, 675, Copenhagen (1962).
31. BERGMAN, T., BEETELSEN, E., BERGOLD, A., and LARSSON, S., 16th Intern. Dairy Cong. Proc.; Vol *A,* Sect. II:1, 579, Copenhagen (1962).
32. BOLLAND, J. L., Quart. Rev. (London), *3,* 1 (1949).
33. BOSCO, J., 11th World's Dairy Cong. Proc. *2,* 3, Berlin (1937).
34. BRUHN, J. C., and FRANKE, A. A., J. Dairy Sci., *54,* 761 (1971).
35. BRUNNER, J. R., J. Dairy Sci., *33,* 741 (1950).
36. CAMPBELL, J. J. R., PHELPS, R. H., and KEUR, L. B., J. Milk Food Tech., *22,* 346 (1959).
37. CAMPBELL, L. B., WATROUS, G. H., JR., and KEENEY, P. G., J. Dairy Sci., *51,* 910 (1968).
38. CANNON, J. A., ZILCH, K. T., BURKET, S. C., and DUTTON, H. J., J. Am. Oil Chemists Soc., *29,* 447 (1952).
39. CASTELL, C. H., J. Milk Technol., *5,* 195 (1942).
40. CHANDRAN, R. C., and SHAHANI, K. M., J. Dairy Sci., *43,* 841 (1960).
41. CHANDRAN, R. C., and SHAHANI, K. M., J. Dairy Sci., *46,* 275 (1963).
42. CHANDRAN, R. C., and SHAHANI, K. M., J. Dairy Sci., *46,* 503 (1963).
43. CHANDRAN, R. C., and SHAHANI, K. M., J. Dairy Sci., *48,* 1417 (1965).
44. CHEN, J. H. S., and BATES, C. R., J. Milk Food Technol., *25,* 176 (1962).
45. CHILSON, W. H., Milk Plant Monthly, *24,* No. 11, 24; *24,* No. 12, 30 (1935).
46. CHRISTENSEN, L. J., DECKER, C. W., and ASHWORTH, U. S., J. Dairy Sci., *34,* 404 (1951).
47. COLLINS, E. B., and DUNKLEY, W. L., J. Dairy Sci., *40,* 603 (1957).
48. COLMEY, J. C., DEMOTT, B. J., and WARD, G. M., J. Dairy Sci., *40,* 608 (1957).
49. COSTILOW, R. N., and SPECK, M. L., J. Dairy Sci., *34,* 1104 (1951).
50. COSTILOW, R. N., and SPECK, M. L., J. Dairy Sci., *34,* 1119 (1951).
51. CROWE, L. K., J. Dairy Sci., *38,* 969 (1955).
52. DAHLE, C. D., Penna. Agr. Expt. Sta. Bull. *320,* 2 (1935).
53. DAHLE, C. D., Milk Dealer, *27,* No. 5, 68 (1938).
54. DAHLE, C. D., and FOLKERS, E. C., J. Dairy Sci., *16,* 529 (1933).
55. DAHLE, C. D., and JOSEPHSON, D. V., Ice Cream Review, *20,* 31 (1937).
56. DAHLE, C. D., and PALMER, L. S., Penna. Agr. Expt. Sta. Bull. *347,* 3 (1937).
57. DAVIES, W. D., J. Dairy Res., *3,* 264 (1932).
58. DAY, E. A., FORSS, D. A., and PATTON, S., J. Dairy Sci. *40,* 922 (1957).
58a. DAY, E. A., and LILLARD, D. A., J. Dairy Sci., *43,* 585 (1960).
59. DAY, E. A., LILLARD, D. A., and MONTGOMERY, M. W., J. Dairy Sci., *46,* 291 (1963).
60. DEMOTT, B. J., J. Dairy Sci., *43,* 436 (1960).
61. DESNUELLE, P., Adv. Enzymology, *23,* 129 (1961).
62. DICKS, M. W., Wyoming Agr. Expt. Sta. Bull. 435 (1965).
63. DORNER, W., and WIDMER, A., Lait., *11,* 545 (1931).
64. DOWNEY, W. K., J. Soc. Dairy Technol., *22,* 154 (1969).
65. DOWNEY, W. K., and ANDREWS, P., Biochem. J., *94,* 642 (1965).
66. DOWNEY, W. K., and ANDREWS, P., Biochem. J., *101,* 651 (1966).
67. DOWNEY, W. K., and ANDREWS, P., Biochem. J., *112,* 559 (1969).
68. DOWNEY, W. K., and MURPHY, R. F., J. Dairy Res., *37,* 47 (1970).
69. DUIN, H. VAN and BRONS, C., Alg. Zuivelbl., *60,* 37 (1967).
70. DUKMAN, A. J., and SCHIPPER, C. J., Veet-en Zuivelbericht *7,* No. 11, 525 (1964); Dairy Sci. Abst., *27* [626] (1965).
71. DUNCOMBE, W. G., Biochem. J., *88,* 7 (1963).
72. DUNKLEY, W. L., J. Dairy Sci., *34,* 515 (1951).
73. DUNKLEY, W. L., and FRANKE, A. A., J. Dairy Sci., *50,* 1 (1967).
74. DUNKLEY, W. L., FRANKE, A. A., and ROBB, J., J. Dairy Sci., *51,* 531 (1968).
75. DUNKLEY, W. L., FRANKE, A. A., ROBB, J., and RONNING, M., J. Dairy Sci., *5,* 863 (1968).

76. DUNKLEY, W. L., FRANKLIN, J. D., and PANGBORN, R. M., Food Technol. *16,* 112 (1962).
77. DUNKLEY, W. L., FRANKLIN, J. D., and PANGBORN, R. M., J. Dairy Sci., *45,* 1040 (1962).
78. DUNKLEY, W. L., and JENNINGS, W. G., J. Dairy Sci., *34,* 1064 (1951).
79. DUNKLEY, W. L., RONNING, M., FRANKE, A. A., and ROBB, J., J. Dairy Sci., *50,* 492 (1967).
80. DUNKLEY, W. L., RONNING, M., and SMITH, L. M., 17th Intern. Dairy Cong. Proc., Sect. A:2., 223, Munich (1964).
81. DUNKLEY, W. L., and SMITH, L. M., J. Dairy Sci., *34,* 935 (1951).
82. DUTHIE, A. H., JENSEN, R. G., and GANDER, G. W., J. Dairy Sci., *44,* 401 (1961).
83. ECKLES, C. H., and SHAW, R., U.S. Dept. Agr. Bur. Animal Ind., Bull. 155 (1913).
84. EL-NAHTRA, A., Milchwiss. Berichte, Wolf Passing, *13,* 139 (1963).
85. EL-NEGOUMY, A. M., J. Dairy Sci., *48,* 1406 (1965)
85a. EL-NEGOUMY, A. M., MILES, D. M., and HAMMOND, E. G., J. Dairy Sci., *44,* 1047 (1961).
86. ERICKSON, D. R., and DUNKLEY, W. L., Anal. Chem., *36,* 1055 (1964).
87. ERICKSON, D. R., DUNKLEY, W. L., and RONNING, M., J. Dairy Sci., *46,* 911 (1963).
88. ERICKSON, D. R., DUNKLEY, W. L., and SMITH, L. M., J. Food Sci., *29,* 269 (1964).
89. FARMER, E. H., and SUTTON, D. A., J. Chem. Soc., *1943,* 119 (1943).
90. FINLEY, J. W., Ph.D. Thesis. Cornell Univ. (1968).
91. FLAKE, J. C., JACKSON, H. C., and WECKEL, K. G., J. Dairy Sci., *23,* 1087 (1940).
92. FOISSY, H., Öst. Milch., *25,* 217 (1970); Dairy Sci. Abst., *32,* [3878] (1970).
93. FORD, J. E., J. Dairy Res., *34,* 239 (1967).
94. FORSS, D. A., ANGELINI, P., BAZINET, M. L., and MERRITT, C., J. Am. Oil Chemists Soc., *44,* 141 (1967).
95. FORSS, D. A., DUNSTONE, E. A., and STARK, W., J. Dairy Res., *27,* 211 (1960).
96. FORSS, D. A., DUNSTONE, E. A., and STARK, W., J. Dairy Res., *27,* 373 (1960).
97. FORSS, D. A., DUNSTONE, E. A., and STARK, W., J. Dairy Res., *27,* 381 (1960).
98. FORSS, D. A., PONT, E. G., and STARK, W., J. Dairy Res., *22,* 91 (1955).
99. FORSS, D. A., PONT, E. G., and STARK, W., J. Dairy Res., *22,* 345 (1955).
100. FORSTER, T. L., and SOMMER, H. H., J. Dairy Sci., *34,* 992 (1951).
101. FOUTS, E. L., J. Dairy Sci., *23,* 173 (1940).
102. FOUTS, E. L., and WEAVER, E., J. Dairy Sci., *19,* 482 (1936).
103. FOX, P. F., and TARASSUK, N. P., J. Dairy Sci., *51,* 826 (1968).
104. FOX, P. F., YAGUCHI, M., and TARASSUK, N. P., J. Dairy Sci., *50,* 307 (1967).
105. FRANKEL, E. N., NOWAKOWSKA, J., and EVANS, C. D., Am. Oil Chemists Soc., *38,* 161 (1961).
106. FRANKEL, E. N., and TARASSUK, N. P., J. Dairy Sci., *39,* 1506 (1956).
107. FRANKEL, E. N., and TARASSUK, N. P., J. Dairy Sci., *39,* 1517 (1956).
108. FRANKEL, E. N., and TARASSUK, N. P., J. Dairy Sci., *39,* 1532 (1956).
109. FRANKEL, E. N., and TARASSUK, N. P., J. Dairy Sci., *42,* 409 (1959).
110. GAFFNEY, P. J., JR., and HARPER, W. J., J. Dairy Sci., *48,* 613 (1965).
111. GAFFNEY, P. J., JR., HARPER, W. J., and GOULD, I. A., J. Dairy Sci., *49,* 921 (1966).
112. GAWEL, J., and PIJANOWSKI, E., Nahrung, *14,* 469 (1970).
113. GELPI, P. J., RUSOFF, L. L., and PINEIRO, E., J. Agr. Food Chem., *10,* 89 (1962).
114. GHOLSON, J. H., GELPI, A. J., JR., and FAYE, J. B., JR., J. Milk Food Technol., *29,* 248 (1966).
115. GHOLSON, J. H., SCHEXNAILDER, R. H., and RUSOFF, L. L., J. Dairy Sci., *49,* 1136 (1966).
116. GOLDING, J., and FEILMAN, E., J. Soc. Chem. Ind., *24,* 1285 (1905).
117. GOULD, I. A., J. Dairy Sci., *24,* 779 (1941).

118. GOULD, I. A., and KEENEY, P. G., J. Dairy Sci., *40,* 297 (1957).
119. GOULD, I. A., and SOMMER, H. H., Mich. Agr. Expt. Sta. Tech. Bull. *164,* (1939).
120. GOULD, I. A., and TROUT, G. M., J. Agr. Res., *52,* 49 (1936).
121. GOULD, I. A., and TROUT, G. M., Mich. Agr. Expt. Sta. Quart. Bull. *22,* 101 (1939).
122. GREENBANK, G. R., J. Dairy Sci., *23,* 725 (1940).
123. GREENBANK, G. R., J. Dairy Sci., *31,* 913 (1948).
124. GREENBANK, G. R., and PALLANSCH, M. J., J. Dairy Sci., *44,* 1547 (1961).
125. GREENBANK, G. R., WRIGHT, P. A., DEYSHER, E. F., and HOLM, G. E., J. Dairy Sci., *29,* 55 (1946).
126. GUTHRIE, E. S., and BRUECKNER, H. J., N.Y. Agr. Expt. Sta. Bull. *606* (1934).
127. GUTHRIE, E. S., and HERRINGTON, B. L., J. Dairy Sci., *43,* 843 (1960).
128. HAASE, G., and DUNKLEY, W. L., J. Lipid Res., *10,* 555 (1969).
129. HAASE, G., and DUNKLEY, W. L., J. Lipid Res., *10,* 561 (1969).
130. HAASE, G., and DUNKLEY, W. L., J. Lipid Res., *10,* 568 (1969).
131. HAASE, G., and DUNKLEY, W. J., Milchwiss. *25,* 656 (1970).
132. HAMM, D. L., and HAMMOND, E. G., J. Dairy Sci., *50,* 1166 (1967).
133. HAMM, D. L., HAMMOND, E. G., and HOTCHKISS, D. K., J. Dairy Sci., *51,* 483 (1968).
134. HAMM, D. L., HAMMOND, E. G., PARVANAH, V., and SNYDER, H. E., J. Am. Oil Chemists Soc., *42,* 920 (1965).
135. HAMMOND, E. G., Am. Dairy Rev., *32,* No. 6, 40 (1970).
136. HAND, D. B., GUTHRIE, E. S., and SHARP, P. F., Science *87,* 439 (1938).
137. HAND, D. B., and SHARP, P. F., Intern. Assoc. Milk Dealers Bull., *33,* No. 17, 460 (1941).
138. HARPER, W. J., and GOULD, I. A., 15th Intern. Dairy Cong. Proc., *6,* 455, London (1959).
139. HARPER, W. J., GOULD, I. A., and BADAMI, M., J. Dairy Sci., *39,* 910 (1956).
140. HARPER, W. J., SCHWARTZ, D. P., and EL-HAGARAWY, I. S., J. Dairy Sci., *39,* 46 (1956).
141. HEMINGWAY, E. B., SMITH, G. H., and ROOK, J. A. F., J. Soc. Dairy Technol., *23,* 44 (1970).
142. HENDRICK, H., MOOR, H. de, and DEVOGELAERE, R., Mededel-Landbov-whogeschool Opzockingsstn Staat. Gent. *29,* 119 (1964); Chem. Abs. *64* [10318d] (1966).
143. HENICK, A. S., BENCA, M. F., and MITCHELL, J. H., JR., J. Am. Oil Chemists Soc., *31,* 88 (1954).
144. HENNINGSON, R. W., and ADAMS, J. B., J. Dairy Sci., *50,* 961 (1967).
145. HERRINGTON, B. L., 43rd Ann. Meeting Milk Ind. Foundation Proc., Lab. Sect. *1950,* 30 (1950).
146. HERRINGTON, B. L., J. Dairy Sci., *37,* 775 (1954).
147. HERRINGTON, B. L., and GUTHRIE, E. S., J. Dairy Sci., *41,* 707 (1958).
148. HERRINGTON, B. L., and KRUKOVSKY, V. N., J. Dairy Sci., *22,* 127 (1939).
149. HERRINGTON, B. L., and KRUKOVSKY, V. N., J. Dairy Sci., *22,* 149 (1939).
150. HILEMAN, J. L., and COURTNEY, E., J. Dairy Sci., *18,* 247 (1935).
151. HILLS, G. LOFTUS, and THIEL, C. C., J. Dairy Res., *14,* 340 (1946).
152. HOFF, J. E., WERTHEIM, J. H., and PROCTOR, B. E., J. Dairy Sci., *42,* 468 (1959).
153. HOFFMANN, G., J. Am. Oil Chemists Soc., *39,* 439 (1962).
154. HOFFMAN-OSTENHOF, O., Adv. Enzymology, *14,* 219 (1953).
155. HOLM, G. E., GREENBANK, G. R., and DEYSHER, E. F., J. Dairy Sci., *8,* 515 (1925).
156. HOLM, G. E., WRIGHT, P. A., WHITE, W., and DEYSHER, E. F., J. Dairy Sci., *21,* 385 (1938).
157. HOEVAT, R. J., McFADDAN, W. H., NG, H., BLACK, D. R., LANE, W. G., and TEETER, R. M., J. Am. Oil Chemists Soc., *42,* 1112 (1965).
158. HUTTON, J. T., and PATTON, S., J. Dairy Sci., *35,* 699 (1952).

159. JACKSON, W. P., and LOO, C. C., J. Dairy Sci., *42,* 912 (1959).
160. JAMOTTE, P., Publs. Stn. Lait Etat Gem Bloug, *1968;* Dairy Sci. Abst. *31* [3997] (1969).
161. JENNESS, R., and PATTON, S., "Principles of Dairy Chemistry", John Wiley and Sons, Inc., New York, 1959.
162. JENSEN, R. G., J. Dairy Sci., *47,* 210 (1964).
163. JENSEN, R. G., Progress in the Chemistry of Fats and Other Lipids, *9,* Part 3, 349. Pergamon Press, New York and London, 1971.
164. JENSEN, R. G., and SAMPUGNA, J., J. Dairy Sci., *45,* 646 (1962).
165. JENSEN, R. G., SAMPUGNA, J., PARRY, R. M., JR., SHAHANI, K. M., and CHANDRAN, R. G., J. Dairy Sci., *45,* 1527 (1962).
166. JENSEN, R. G., SAMPUGNA, J., and PEREIRA, R. L., J. Dairy Sci., *47,* 727 (1964).
167. JENSEN, R. G., SMITH, A. C., MACLEOD, PATRICIA, and DOWD, L. R., J. Milk Food Technol., *20,* 352 (1957).
168. JOHNSON, P. E., and GUNTEN, R. L., von, Okla. Ag. Expt. Sta. Bull. *B-593* (1962).
169. JONG, K. de, and VAN der WEL, H., Nature, *202,* 553 (1964).
170. JOSEPHSON, D. V., Abst. of Doctoral Diss. *6,* Penna. State College (1943).
171. JOSEPHSON, D. V., and DOAN, F. J., Milk Dealer, *29,* No. 2, 35 (1939).
172. KANNAN, A., and BASU, K. P., Indian J. Dairy Sci., *4,* 8 (1951).
173. KANNO, C., YAMAUCHI, K., and TSUGO, T., J. Dairy Sci., *51,* 1713 (1968).
174. KARNOVSKY, M. L., and WOLFF, D., Biochemistry of Lipids, *5,* 53 Pergamon Press, New York and London, 1960.
175. KAY, H. D., Nature, *157,* 511 (1946).
176. KEITH, J. I., and FOUTS, E. L., Intern. Assoc. Milk Dealers, 30th Ann. Conv., Lab. Sect. Proc., *172* (1937).
177. KELLEY, E., "Report of Chief, Div. of Market Milk Investigations", Bur. Dairy Indus., U.S. Dept. Agr. (1942).
178. KELLEY, L. A., and DUNKLEY, W. L., J. Milk Food Technol., *17,* 306 (1954).
179. KELLEY, P. L., J. Dairy Sci., *26,* 385 (1943).
180. KELLEY, P. L., J. Dairy Sci., *28,* 793 (1945).
181. KENDE, S., Milchw. Forsch., *13,* 111 (1932).
182. KEPPLER, J. G., HORIKX, M. M., MEIJBOOM, P. W., and FEENSTRA, W. H., J. Am. Oil Chemists Soc., *44,* 543 (1967).
183. KHAN, N. A., LUNDBERG, W. O., and HOLMAN, R. T., J. Am. Chem. Soc., *76,* 1779 (1954).
184. KHATEI, L. L., Diss. Abst. *26,* No. 11, 6638 (1966).
185. KIERMEIER, F., and WAIBLINGER, W., Z. Lebensmittel-Untersuch. u. Forsch., *142,* 36 (1970).
186. KING, R. L., Ph.D. Thesis, University of California, Davis, Cal., 1958.
187. KING, R. L., J. Dairy Sci., *45,* 1165 (1962).
188. KING, R. L., J. Dairy Sci., *46,* 267 (1963).
189. KING, R. L., J. Dairy Sci., *51,* 1705 (1968).
190. KING, R. L., BURROWS, F. A., HEMKEN, R. W., and BASHORE, D. L., J. Dairy Sci., *50,* 943 (1967).
191. KING, R. L., and DUNKLEY, W. L., J. Dairy Sci., *42,* 420 (1959).
192. KING, R. L., and DUNKLEY, W. L., J. Dairy Sci., *42,* 897 (1959).
193. KING, R. L., LUICK, J. R., LITMAN, I. I., JENNINGS, W. G., and DUNKLEY, W. L., J. Dairy Sci., *42,* 780 (1959).
194. KING, R. L., TIKEITI, H. H., and OSKARSSON, M., J. Dairy Sci., *49,* 1574 (1966).
195. KITCHEN, B. J., J. Dairy Res., *38,* 171 (1971).
196. KLIMAN, P. G., TAMSMA, A., and PALLANSCH, M. J., J. Agr. Food Chem., *10,* 496 (1962).
197. KODGEV, A., and RACHEV, R., 18th Intern. Dairy Congress, Proc., *IE,* 200, Sydney (1970).
198. KOESTLER, G., Schweiz. Mitchztg., *103* (1928).

199. KOESTLER, G., ROADHOUSE, C. L., and LORTSCHER, W., Landw. Jaheb. Schweiz., *42,* 937 (1928).
200. KOOPS, J., Verslag. Ned. Inst. Zuivelonderzoek, No. 80 (1963); Chem. Abs., *60* [2255c] (1964).
201. KOOPS, J., Neth. Milk and Dairy J., *18,* 38 (1964).
202. KOOPS, J., Neth. Milk and Dairy J., *18,* 220 (1964).
203. KOOPS, J., Neth. Milk and Dairy J., *23,* 200 (1969).
204. KOPACZEWSKI, W., J. Dairy Sci., *20,* A25 (1937).
204a. KORN, E. D., J. Lipid Res., *3,* 246 (1962).
205. KOSKINEN, E. H., LUHTALA, A., and ANTILA, M., Milchwiss., *24,* 20 (1969).
206. KRIENKE, W. A., J. Dairy Sci., *27,* 683 (1944).
207. KRUKOVSKY, V. N., J. Dairy Sci., *35,* 21 (1952).
208. KRUKOVSKY, V. N., J. Dairy Sci., *38,* 595 (1955).
209. KRUKOVSKY, V. N., J. Agr. Food Chem., *9,* 439 (1961).
210. KRUKOVSKY, V. N., and GUTHRIE, E. S., J. Dairy Sci., *28,* 565 (1945).
211. KRUKOVSKY, V. N., and GUTHRIE, E. S., J. Dairy Sci., *29,* 293 (1946).
212. KRUKOVSKY, V. N., and HERRINGTON, B. L., J. Dairy Sci., *22,* 137 (1939).
213. KRUKOVSKY, V. N., and SHARP, P. F., J. Dairy Sci., *19,* 279 (1936).
214. KRUKOVSKY, V. N., and SHARP, P. F., J. Dairy Sci., *21,* 671 (1938).
215. KRUKOVSKY, V. N., and SHARP, P. F., J. Dairy Sci., *23,* 1109 (1940).
216. KRUKOVSKY, V. N., and SHARP, P. F., J. Dairy Sci., 23, 1119 (1940).
217. KRUKOVSKY, V. N., and WHITING, F., J. Dairy Sci., *31,* 666 (1948).
218. KRUKOVSKY, V. N., WHITING, F., and LOOSLI, J. K., J. Dairy Sci., *33,* 791 (1950).
219. KURTZ, F. E., "Fundamentals of Dairy Chemistry *2nd Edition*", Chapter IV, Avi Publ. Co., Westport, Conn. 1974.
220. LARSEN, P. B., GOULD, I. A., and TROUT, G. M., J. Dairy Sci., *24,* 789 (1941).
221. LARSEN, P. B., TROUT, G. M., and GOULD, I. A., J. Dairy Sci., *24,* 771 (1941).
222. LARSSON, B. L., and JENNESS, R., J. Dairy Sci., *33,* 896 (1950).
223. LAWRENCE, R. C., Dairy Sci. Abst., *29,* 59 (1967).
224. LAWRENCE, R. C., FRYER, T. F., and REITER, B., Nature, *213,* 1264 (1967).
225. LEA, C. H., "Rancidity in Edible Fats", Chemical Publ. Co., New York, 1939.
226. LEA, C. H., MORAN, T., and SMITH, J. A. B., J. Dairy Res., *13,* 162 (1943).
227. LEMBKE, A., and FRAHM, H., Kiel. Milchw. Forsch., *16,* 427 (1964); Chem. Abs. *66* [166ON] (1967).
228. LILLARD, D. A., and DAY, E. A., J. Dairy Sci., *44,* 623 (1961).
229. LUHTALA, A., MEIJERTIET. AIKAKAVSK., *29,* 7 (1969); Dairy Sci. Abst. *32,* 1327 (1970).
230. LUHTALA, A., and ANTILA, M., Fette, Seifen. Anstrichmittel, *70,* 280 (1968).
231. LUHTALA, A., KORHONEN, H., KOSKINEN, E. H., and ANTILA, M., 18th Intern. Dairy Congr., Proc., *1E,* 80, Sidney (1970).
232. LUHTALA, A., KOSKINEN, E. H., and ANTILA, M., 18th Intern. Dairy Congr., Proc., *1E,* 79, Sydney (1970).
233. LUICK, J. R., and MAZRIMAS, J. A., J. Dairy Sci., *49,* 1500 (1966).
234. MACKENZIE, R. D., BLOHM, T. R., AUXIER, E. M., and LUTHER, A. C., J. Lipid Res., *8*. 589 (1967).
235. MANUS, L. J., and BENDIXEN, H. A., J. Dairy Sci., *39,* 508 (1956).
236. MARQUARDT, J. C., Milk Dealer, *22,* No. 12, 39 (1932).
237. MATTICK, A. T. R., J. Agr. Sci., *17,* 388 (1927).
238. MATTICK, E. C. V., and KAY, H. D., J. Dairy Res., *9,* 58 (1938).
239. MATTSON, F. H., and VOLPENHEIM, R. A., J. Am. Oil Chemists Soc., *43,* 286 (1966).
240. MEIJBOOM, P. W., J. Am. Oil Chemists Soc., *41,* 326 (1964).
241. MERK, W., and CRASEMANN, E., Z. Tierphysiol. Tierernahr. Futtermittelk, *16,* 197 (1961).
242. MEYER, R. I., and JOKAY, L., J. Dairy Sci., *43,* 844 (1960).
243. MOLLER-MADSEN. A. A., and HORVATH, Z., Beretn. St. Forsøgsmejeri, *145,* 20 (1964); Dairy Sci. Abst., *27,* [1938] (1965).

244. MOORE, A. V., and TROUT, G. M., Can. Dairy Ice Cream J., *25*, No. 7, 33 (1946).
245. MORTON, R. K., Nature, *171*, 734 (1953).
246. MORTON, R. K., Biochem. J., *57*, 231 (1954).
247. MORTON, R. K., Biochem. J., *60*, 573 (1955).
248. MULDER, H., and KOPPEJAN, C. A., 13th Intern. Dairy Congr., Proc. *3*, 1402, The Hague (1953).
249. MULDER, M., MENGER, J. W., and MEIJERS, P., Neth. Milk and Dairy J., *18*, 52 (1964).
250. NAKAI, S., PERRIN, J. J., and WRIGHT, V., J. Dairy Sci., *53*, 537 (1970).
251. NELSON, H. G., Ph.D. Thesis, Univ. of Minnesota (1952).
252. NELSON, H. G., and JEZESKI, J. J., J. Dairy Sci., *38*, 479 (1955).
253. NILSSON, R., and WILLART, S., "Milk and Dairy Research (Alnarp) Report", *60* (1960).
254. NILSSON, R., and WILLART, S., "Milk and Dairy Research (Alnarp) Report", *64* (1961).
255. NOVAK, M., J. Lipid Res., *6*, 431 (1965).
256. OLSON, F. C., and BROWN, W. C., J. Dairy Sci., *25*, 1027 (1942).
257. OLSON, J. C., THOMAS, E. L., and NIELSEN, A. J., Am. Milk Rev., *18*, No. 10, 98 (1956).
258. PALMER, L. S., J. Dairy Sci., *5*, 201 (1922).
259. PARKS, O. W., and ALLEN, C. A., Unpublished data (1972).
260. PARKS, O. W., KEENEY, M., and SCHWARTZ, D. P., J. Dairy Sci., *46*, 295 (1963).
261. PARKS, O. W., and PATTON, S., J. Dairy Sci., *44*, 1 (1961).
262. PARKS, O. W., WONG, N. P., ALLEN, C. A., and SCHWARTZ, D. P., J. Dairy Sci., *52*, 953 (1969).
263. PARRY, R. M., JR., CHANDRAN, R. C., and SHAHANI, K. M., J. Dairy Sci., *49*, 356 (1966).
264. PATTON, S., J. Dairy Sci., *37*, 446 (1954).
265. PATTON, S., BARNES, I. J., and EVANS, L. E., J. Am. Oil Chemists Soc., *36*, 280 (1959).
266. PETERSON, M. H., JOHNSON, M. J., and PRICE W. V., J. Dairy Sci., *26*, 233 (1943).
267. PETERSON, M. H., JOHNSON, M. J., and PRICE W. V., J. Dairy Sci., *31*, 31 (1948).
268. PFEFFER, J. C., JACKSON, H. C., and WECKEL, K. G., J. Dairy Sci., *21*, 143 (1938).
269. PIJANOWSKI, E., WOJTOWICZ, M., and LOCHOWSKA, H., 16th Intern. Dairy Congr., Proc., Vol. A, Sect. II:1, 633, Copenhagen (1962).
270. PONT, E. G., J. Dairy Res., *19*, 316 (1952).
271. PONT, E. G., Aust. J. Dairy Technol., *10*, 72 (1955).
272. PONT, E. G., Aust. J. Dairy Technol., *19*, 108 (1964).
273. PONT, E. G., FORSS, D. A., DUNSTONE, E. A., and GUNNIS, L. F., J. Dairy Res., *27*, 205 (1960).
274. PONT, E. G., and HOLLOWAY, G. L., J. Dairy Res., *34*, 231 (1967).
275. POSNER, I., and BASCH, V., J. Lipid Res., *12*, 768 (1971).
276. POTTER, F. E., and HANKINSON, D. J., J. Dairy Sci., *43*, 1887 (1960).
277. POULSEN, P. R., and JENSEN, G. K., 17th Intern. Dairy Congr., Proc., Vol. A, Sect. A:2, 229, Munich (1966).
278. PRIVETT, O. S., LUNDBERG, W. O., KHAN, N. A., TOLBERG, W. E., and WHEELER, D. H., J. Am. Oil Chemists Soc., *30*, 61 (1953).
279. PRIVETT, O. S., and NICKELL, E. C., Fette, Seifen, Anstrichmittel, *61*, 842 (1959).
280. PRIVETT, O. S., and QUACKENBUSH, F. W., J. Am. Oil Chemists Soc., *31*, 321 (1954).
281. PYENSON, H., and TRACEY, P. H., J. Dairy Sci., *29*, 1 (1946).
282. RAJAN, T. S., RICHARDSON, G. A., and STEIN, R. W., J. Dairy Sci., *45*, 933 (1962).
283. RAO, S. R., Ph.D. Thesis, Univ. of Wisconsin (1951).

284. REDER, R., J. Dairy Sci., *21,* 475 (1938).
285. RICHTER, R. L., and RANDOLPH, H. E., J. Dairy Sci., *54,* 1275 (1971).
286. RIEL, R. R., Ph.D. Thesis, Univ. of Wisconsin (1952).
287. RIEST, U., RONNING, M., DUNKLEY, W. L., and FRANKE, A. A., Milchwiss. *22,* 551 (1967).
288. ROADHOUSE, C. L., and HENDERSON, J. L., Revised Edition, "The Market-Milk Industry", McGraw-Hill Book Co., Inc., New York, 1950.
289. ROAHEN, D. C., and SOMMER, H. H., J. Dairy Sci., *23,* 831 (1940).
290. ROBERTSON, J. A., HARPER, W. J., and GOULD, I. A., J. Dairy Sci., *49,* 1386 (1966).
291. ROGERS, L. A., U.S. Bur. Animal Industry, Cir. *146* (1909).
292. ROGERS, L. A., Milk Dealer, *10,* 10 (1914).
293. ROGERS, W. P., and PONT, E. G., Aust. J. Dairy Technol., *20,* 200 (1965).
294. ROMANSKAYA, N. N., and VALEEVA, A. N., Tr. Frunzensk. Politekhn. Inst., *1962,* 17 (1962); Chem. Abs. *62* [8311E] (1965).
295. SAITO, Z., Jap. J. Zootech. Sci., *34,* 94 (1963).
296. SAITO, Z., and HASHIMOTO, Y., Jap. J. Zootech. Sci., *34,* 393 (1963).
297. SAITO, Z., NAKAMURA, S., and IGARASHI, Y., Jap. J. Dairy Sci., *18,* A176 (1969); Dairy Sci. Abs., *32* [3081] 1970.
298. SAMUELSSON, E. G., 16th Intern. Dairy Congr., Proc., Vol. A, Sect. II:1, 552, Copenhagen (1962).
299. SAMUELSSON, E. G., Svenska Mejeritidn., *54,* 511 (1962); Dairy Sci. Abs., *25,* [2413] (1963).
300. SAMUELSSON, E. G., Milchwiss., *21,* 335 (1966).
301. SAMUELSSON, E. G., Milk Dairy Res. (Alnarp) Report No. 77. Chem. Abs., *67* [42654E] (1967).
302. SAMUELSSON, E. G., Milk Dairy Res. (Alnarp) Report No. 77. Cited in Milchwiss. *25,* 656 (1970).
303. SARGENT, J. S. E., and STINE, C. M., J. Dairy Sci., *47,* 662 (1964).
304. SATTLER-DORNBACHER, S., Milchwiss. Ber., *13,* 53 (1963).
305. SCANLAN, R. A., SATHER, LOIS A., and DAY, E. A., J. Dairy Sci., *48,* 1582 (1965).
306. SCHAFFER, P. S., GREENBANK, G. R., and HOLM, G. E., J. Dairy Sci., *29,* 145 (1946).
307. SCHMIDT, H., Fette, Seifen Anstrimittel, *61,* 881 (1959).
308. SCHWARTZ, D. P., Ph.D. Thesis, Ohio State Univ. (1954).
309. SCHWARTZ, D. P., GOULD, I. A., and HARPER, W. J., J. Dairy Sci., *39,* 1364 (1956).
310. SCHWARTZ, D. P., GOULD, I. A., and HARPER, W. J., J. Dairy Sci., *39,* 1375 (1956).
311. SCOTT, K., Aust. J. Dairy Technol., *20,* 36 (1965).
312. SEARLES, S. K., and ARMSTRONG, J. G., J. Dairy Sci., *53,* 150 (1970).
313. SHAHANI, K. M., J. Dairy Sci., *49,* 907 (1966).
314. SHAHANI, K. M., and CHANDRAN, R. C., Arc. Biochem. Biophys., *111,* 257 (1965).
315. SHARP, P. F., Intern. Assoc. Milk Dealers Bull., *20,* 523 (1941).
316. SHARP, P. F., HAND, D. B., and GUTHRIE, E. S., Intern. Assoc. Milk Dealers Bull., *34,* No. 17, 365 (1942).
317. SINGLETON, J. A., AURAND, L. W., and LANCASTER, F. W., J. Dairy Sci., *46,* 1050 (1963).
318. SJÖSTRÖM, G., Milk Dairy Research (Alnarp) Report No. *58* (1959).
319. SJÖSTRÖM, G., and WILLART, S., Svenska Mejeritidn., *48,* 421, 435, 441 (1956).
320. SKEAN, J. D., and OVERCAST, W. W., J. Dairy Sci., *44,* 823 (1961).
321. SMITH, A. C., and MacLEOD, PATRICIA, J. Dairy Sci., *38,* 870 (1955).
322. SMITH, G. J., and DUNKLEY, W. L., J. Dairy Sci., *43,* 278 (1960).
323. SMITH, G. J., and DUNKLEY, W. L., 16th Intern. Dairy Congr., Proc., Vol. A, Sect. II:1, 625, Copenhagen (1962).

324. SMITH, G. J., and DUNKLEY, W. L., J. Food Sci., *27,* 127 (1962).
325. SMITH, G. J., and DUNKLEY, W. L., J. Dairy Sci., *45,* 170 (1962).
326. SMITH, L. M., and DUNKLEY, W. L., J. Dairy Sci., *42,* 896 (1959).
327. SPEER, J. F., WATROUS, G. H., and KESLER, E. M., J. Milk Food Technol., *21,* 33 (1958).
328. STADHOUDERS, J., and MULDER, H., Neth. Milk Dairy J., *12,* 117 (1958).
329. STADHOUDERS, J., and MULDER, H., Neth. Milk Dairy J., *13,* 122 (1959).
330. STADHOUDERS, J., and MULDER, H., Neth. Milk Dairy J., *18,* 30 (1964).
331. STARK, W., and FORSS, D. A., J. Dairy Res., *29,* 173 (1962).
332. STARK, W., and FORSS, D. A., J. Dairy Res., *31,* 253 (1964).
333. STARK, W., and FORSS, D. A., J. Dairy Res., *33,* 31 (1966).
334. STEBNITZ, V. C., and SOMMER, A. H., J. Dairy Sci., *20,* 181 (1937).
335. STINE, C. M., HARLAND, H. A., COULTER, S. T., and JENNESS, R., J. Dairy Sci., *37,* 202 (1954).
336. STORGARDS, T., and LJUNGREN, B., Milchwiss., *17,* 406 (1962).
337. SWANSON, A. M., and SOMMER, H. H., J. Dairy Sci., *23,* 201 (1940).
338. TALLAMY, P. T., and RANDOLPH, H. E., J. Dairy Sci., *52,* 1569 (1969).
339. TAMSMA, A., KURTZ, F. E., and PALLANSCH, M. J., J. Dairy Sci., *50,* 1562 (1967).
340. TAMSMA, A., MUCHA, T. J., and PALLANSCH, M. J., J. Dairy Sci., *45,* 1435 (1962).
341. TAMSMA, A., MUCHA, T. J., and PALLANSCH, M. J., J. Dairy Sci., *46,* 114 (1963).
342. TAMSMA, A., PALLANSCH, M. J., and MUCHA, T. J., J. Dairy Sci., *44,* 1644 (1961).
343. TARASSUK, N. P., Assoc. Bull. (Intern. Assoc. Milk Dealers), *32,* 153 (1939).
344. TARASSUK, N. P., Can. Dairy Ice Cream J., *19,* 32 (1940).
345. TARASSUK, N. P., Milk Plant Monthly, *31,* No. 4, 24 (1942).
346. TARASSUK, N. P., and FRANKEL, E. N., J. Dairy Sci., *38,* 438 (1955).
347. TARASSUK, N. P., and FRANKEL, E. N., J. Dairy Sci., *40,* 418 (1957).
348. TARASSUK, N. P., and HENDERSON, J. L., J. Dairy Sci., *25,* 801 (1942).
349. TARASSUK, N. P., and KOOPS, J., J. Dairy Sci., *43,* 93 (1960).
350. TARASSUK, N. P., KOOPS, J., and PETTE, J. W., Neth. Milk Dairy J., *13,* 258 (1959).
351. TARASSUK, N. P., and PALMER, L. S., J. Dairy Sci., *22,* 543 (1939).
352. TARASSUK, N. P., and REGAN, W. M., J. Dairy Sci., *26,* 987 (1943).
353. TARASSUK, N. P., and RICHARDSON, G. A., Science, *93,* 310 (1941).
354. TARASSUK, N. P., and RICHARDSON, G. A., J. Dairy Sci., *24,* 667 (1941).
355. TARASSUK, N. P., and SMITH, F. R., J. Dairy Sci., *22,* 415 (1939).
356. TARASSUK, N. P., and SMITH, F. R., J. Dairy Sci., *23,* 1163 (1940).
357. TARASSUK, N. P., and YAGUCHI, M., J. Dairy Sci., *41,* 708 (1958).
358. TARASSUK, N. P., YAGUCHI, M., and NOORLANDER, D., Western Div., Am. Dairy Sci., Assoc. Proc., *39,* 566 (1958).
359. THOMAS, E. L., NIELSEN, A. J., and OLSON, J. C., JR., J. Dairy Sci., *38,* 596 (1955).
359a. THOMAS, E. L., NIELSEN, A. J., and OLSON, J. C., JR., Am. Milk Rev., *17,* 50 (1955).
360. THOMAS, W. R., HARPER, W. J., and GOULD, I. A., J. Dairy Sci., *38,* 315 (1955).
361. THORNTON, H. R., and HASTINGS, E. G., J. Dairy Sci., *13,* 221 (1930).
362. THURSTON, L. M., Intern. Assoc. Milk Dealers, Proc., *30,* Lab. Sect. 143 (1937).
363. THURSTON, L. M., BROWN, W. C., and DUSTMAN, R. B., J. Dairy Sci., *19,* 671 (1936).
364. TRACEY, P. H., Milk Dealer, *21,* 68 (1931).
365. TRACEY, P. H., RAMSEY, R. J., and RUEHE, H. A., Ill. Agr. Expt. Sta. Bull. *389* (1933).
366. TROUT, G. M., HALLORAN, C. P., and GOULD, I. A., Mich. Agr. Expt. Sta. Tech. Bull. *145* (1935).

367. TSUGO, T., and HAYASHI, T., Jap. J. Zootech. Sci., *33,* 125 (1962).
368. VELANDER, H. J., and PATTON, S., J. Dairy Sci., *38,* 593 (1955).
369. VLEESCHAUWER, A. de, TIJTGAT, B., HENDRICKX, H., and DESCHACHT, W., Meded LangHogesch. Gent., *25,* 939 (1960); Dairy Sci. Abst., *23,* 146 (1961).
370. WELLS, M. E., PRYOR, O. P., HAGGERTY, D. M., PICKETT, H. C., and MICKLE, J. B., J. Dairy Sci., *52,* 1110 (1969).
371. WILLART, S., Svenska Mejeritidn., *49,* 177 (1957).
372. WILLART, S., and ARPH, S. O., Svenska, Mejeritidn., *49,* 177 (1957).
373. WILLART, S., and SJÖSTRÖM, G., 15th Intern. Dairy Congr., Proc., 3, 1482, London (1959).
374. WILLART, S., and SJÖSTRÖM, G., 16th Intern. Dairy Congr., Proc., Sect. *A,* 669, Copenhagen (1962).
375. WILLART, S., and SJÖSTRÖM, G., Svenska, Mejeritidn., *56,* 561 (1964).
376. WILLS, E. D., Adv. Lipid Research, *3,* 197 (1965). Academic Press, New York and London.
377. WISHNER, L. A., J. Dairy Sci., *47,* 216 (1964).
378. WISHNER, L. A., and KEENEY, M., J. Dairy Sci., *46,* 785 (1963).
379. WYATT, J. C., and DAY, E. A., J. Dairy Sci., *48,* 682 (1965).
380. YAGUCHI, M., TARASSUK, N. P., and ABE, N., J. Dairy Sci., *47,* 1167 (1964).
381. ZAMBIA, J. V., Food Eng., *41,* 105 (1969).

CHAPTER 6

T. A. Nickerson

Lactose

OCCURRENCE

The characteristic carbohydrate of milk is lactose (milk sugar). It was formerly thought to occur only in milk, to be the only sugar in milk, and to be found in the milk of all species. Those beliefs are substantially correct, but with recognized exceptions.

True, lactose is found in high concentrations only in milk and the mammary gland, but low percentages appear also in the blood and urine, especially during pregnancy and lactation, the result of escape of lactose formed in the mammary gland. Lactose in the urine, a condition known as lactosuria, is found in about 9% of healthy humans of either sex on a normal diet; its origin is probably alimentary.[66] Other sources are extremely rare, e.g., as a constituent of various oligosaccharides,[210] in *Forsythia* flowers,[128] and in *Sapotacea.*[177]

Lactose has been thought to be present in the milk of all mammals, although the milk of all species has not been analyzed for it specifically. In many cases the analyses have been based solely on reducing power or by difference between total solids and the sum of the fat, protein, and ash, which, of course, would not be indicative of lactose alone. Now chromatographic procedures are showing the presence of lactose specifically, as well as the monosaccharides glucose and galactose, and a variety of oligosaccharides; thus knowledge of milk composition is becoming more precise. Milks from a few species, e.g., sea lion,[171] some seals,[170] and opossum,[18] have been found totally lacking in lactose, while other milks contain only low levels.[114,172,196] In some milks, other carbohydrates occur in higher concentrations than lactose.[113] This is a particularly interesting area for mammologists or those concerned with evolutionary biochemistry, and a recent review serves as an excellent reference.[114]

The first record of isolation of lactose was in 1633, by Bartolettus, by evaporation of whey. During the 18th century, lactose became a commercial commodity, used principally in medicine. Whey had been

T. A. Nickerson, Department of Food Science and Technology, University of California, Davis, CA 95616.

used by physicians since the time of Hippocrates to utilize the unique biochemical functions and properties of lactose.

The literature on lactose is voluminous and no attempt is made here to review it comprehensively. Those interested in further details and additional references may consult the general review articles on lactose and the specialized reviews referred to in the various sections of this chapter.[11,39,44,222,223,225,226]

The lactose in normal cow's milk generally ranges from 4.4 to 5.2%, averaging 4.8% anhydrous lactose. This usually amounts to 50 to 52% of the total solids in skimmilk. Human milk contains a much greater percentage of lactose (av. 7.0%); hence the practice of adding lactose or some other carbohydrate to cow's milk for use in feeding human infants. The lactose content of milk from different mammals varies greatly, as shown in Table 6.1.

Molecular Structure

Lactose is a disaccharide that yields D-glucose and D-galactose on hydrolysis. It may be designated as 4-0-β-D-galactopyranosyl-D-

Table 6.1

MILK COMPOSITION OF DOMESTICATED AND EXPERIMENTAL MAMMALS

Mammal	Fat %	Protein %	Lactose %	Total solids %
Cow	3.7	3.4	4.8	12.7
Man	3.8	1.0	7.0	12.4
Sheep	7.4	5.5	4.8	19.3
Goat	4.5	2.9	4.1	13.2
Water buffalo	7.4	3.8	4.8	17.2
Dromedary	4.5	3.6	5.0	13.6
Horse	1.9	2.5	6.2	11.2
Llama	2.4	7.3	6.0	16.2
Reindeer	16.9	11.5	2.8	33.1
Yak	6.5	5.8	4.6	17.3
Indian elephant	11.6	4.9	4.7	21.9
Dog	12.9	7.9	3.1	23.5
Cat	4.8	7.0	4.8	17.6
Pig	6.8	4.8	5.5	18.8
Norway rat	10.3	8.4	2.6	21.0
Golden hamster	4.9	9.4	4.9	22.6
Guinea pig	3.9	8.1	3.0	16.4
Rhesus monkey	4.0	1.6	7.0	15.4
Baboon	5.0	1.6	7.3	14.4
Rabbit	18.3	13.9	2.1	32.8
Mink	3.4	7.5	2.0	21.2

[a] Large differences within species are reported by various workers.
from Jenness and Sloan[114]

glucopyranose, and occurs in both alpha and beta forms. The conclusive evidence establishing this structure has been reviewed[44,226] in detail, so only a brief description is given here.

By hydrolyzing lactosone and lactobionic acid and obtaining free D-galactose plus glucosone and gluconic acid, respectively, it has been shown that the two monosaccharides are linked through the aldehyde group of D-galactose. Thus the aldehydic portion of lactose is on the glucose residue. That the configuration of the D-galactose residue is of the beta form was shown by use of an enzyme, β-D-galactosidase, that would hydrolyze lactose and would also hydrolyze a methyl β-D-galactopyranoside, but not the α anomer. Conversely, it was shown that an enzyme that hydrolyzed an α-D-galactoside but not a β-D-galactoside would not hydrolyze lactose. That the D-galactose in the lactose molecule is the beta form was also shown by its synthesis from D-glucose and D-galactose. The point of union of the two monosaccharides was established through products of hydrolysis of methylated lactose. The alpha or beta configurations of lactose are easily distinguished, since the designation alpha is arbitrarily assigned to the form having the greater rotation in the dextro direction. The structural formula for lactose may be written in either the straight-chain or Haworth forms, as illustrated in Fig. 6.1. Beta-lactose is depicted by interchanging the OH and H on the reducing group.

A series of rare carbohydrates is known whose structures differ only slightly from that of lactose. Lactose has been prepared from epilactose, 4-0-β-D-galactopyranosyl-D-mannose, from which it differs in configuration at C-2 of the D-glucose residue.[86] Lactulose, 4-0-β-D-galactopyranosyl-D-fructose, is produced from lactose during the heat processing and storage of certain dairy products. It is also one of the products resulting from the action of lime-water on lactose.[146] Its properties, presence, and significance in milk products have been reviewed.[2] A number of oligosaccharides that may be derivatives of lactose have been found in milks of various species.[126,127,181,210] They may differ considerably in composition in the various milks and, on hydrolysis, yield such compounds as fucose, glucosamine, galactosamine, neuraminic acid, and D-mannose, in addition to D-glucose and D-galactose. Many exhibit bifidus growth activity for *Lactobacillus bifidus* var. *pennsylvanicus (L. bifidus* var. *penn.)*[78] and are of interest because of nutritional and physiological significance.[232] Milk and especially colostrum are rich sources of complex oligosaccharides.[74] However, metabolic significance is obscure. A review of the literature covers the isolation, identification, and structure of these oligosaccharides.[44] Further discussion of oligosaccharides is presented later, in conjunction with hydrolysis of lactose.

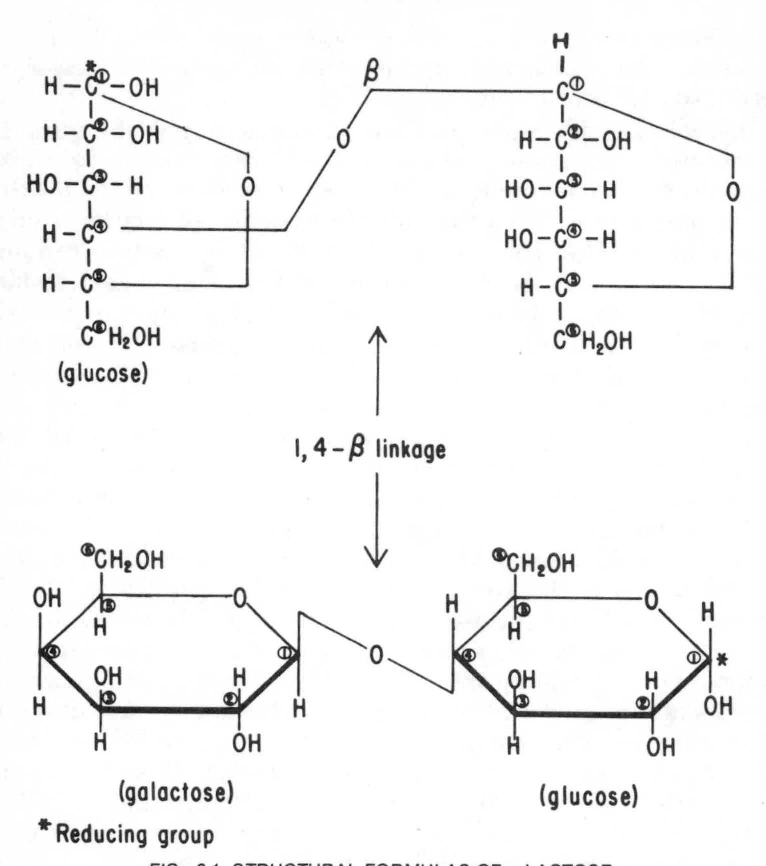

FIG. 6.1. STRUCTURAL FORMULAS OF α-LACTOSE

Conventional numbering of carbons indicated by circled numerals.

BIOSYNTHESIS

Biochemistry of the mammary gland leading to lactose synthesis has been the subject of a considerable volume of published work and has been generally reviewed.[56,117]

It is well established that blood D-glucose is the principal precursor of lactose.[121,176] Arterio-venous differences show considerable uptake of D-glucose by the mammary gland. It was logical to assume that blood D-glucose is the source of the glucose moiety of lactose. Tissue cultures of lactating mammary glands readily converted S-glucose to lactose but did not convert the other hexoses tested.[75,140] Later, the use of carbon-14-labeled compounds confirmed the view that

D-glucose is the most important precursor of lactose. It was calculated that about 80% of the lactose carbon is derived directly from the plasma glucose pool.[13,121] Other studies showed that the D-glucose residue is incorporated into lactose in a quite straightforward manner. The main pathway (about 70%) for D-galactose synthesis was also shown to be from blood D-glucose without rupture of the carbon chain. However, additional D-galactose appears to arise by way of intermediates that are in equilibrium with the many possible metabolic pathways and cycles. As a result, there is a dilution and randomization of the label in D-galactose, as compared with the D-glucose residue, when D-glucose-1-C^{14} is fed. It appears that the glucosyl moiety of lactose originates from D-glucose in equilibrium with blood D-glucose, and that the galactosyl donor is uridine diphosphate galactose formed in the mammary gland.[85,186,220]

The pathway of lactose synthesis is now accepted as involving three enzymes, which catalyze critical reactions as follows:

I. UDPG pyrophosphorylase (EC 2.7.7.9) catalyzes uridine triphosphate (UTP) + glucose-1-P $\rightleftharpoons$ uridine diphosphate glucose (UDPG) + pyrophosphate (PP).

II. UDPGal-4-epimerase (EC 5.1.3.2) catalyzes UDP-glucose $\rightleftharpoons$ UDP-galactose.

III. Lactose synthetase (EC 2.4.1.22) catalyzes UDP-galactose + glucose $\rightarrow$ lactose + UDP.

The systematic name for lactose synthetase under the International Union of Biochemistry, Commission on Enzymes, is UDP-galactose-d-glucose 1-galactosyltransferase.

Studies of lactose synthesis eventually led to discovery of the biological role of the protein α-lactalbumin, a major component of the whey proteins (see Chapter 3). It is now recognized as a subunit of the lactose synthetase system catalyzing the final reaction linking the glucose and galactose moieties.

Lactose synthetase activity is dependent on the presence of two proteins, named A and B protein subunits.[29,30] The B protein subunit was found in high concentrations in milk. Subsequently it was isolated in the crystalline state and identified as α-lactalbumin.[28] The A protein subunit was found to be bound chiefly to the cellular particulate matter of the mammary gland, but has not been isolated in a pure state. It is recognized, however, as a more general galactosyltransferase[27] found normally in a variety of tissues. It will transfer galactose from UDPGalactose to a number of compounds, but not to glucose in the absence of B protein subunits. In the presence of B protein subunits (now identified as α-lactalbumin), the A protein subunit transfers galactose from UDPGalactose to glucose to form the lactose molecule. It has been postulated that secretion of α-lactalbumin in the mammary gland controls the level of lactose synthesis,[26] indeed, a high correlation has been found between the lactose and α-lactalbumin contents of milks from 6 species.[138] This role of α-lactalbumin is a very interesting control mechanism whereby the activity already present in the mammary gland is redirected to make a product specific to lactation. The excellent review of α-lactalbumin, including structural and immunological species differences, should be consulted for more details and historical development of the subject.[56]

Lactose content of milk is affected by inheritance, age, stage of lactation, and interquarter differences.[216] Such variations in lactose content constitute the major factor in variations observed in the SNF content of milk.[217]

PHYSICAL PROPERTIES

Lactose is normally found in nature or in dairy products in either of two crystalline forms—alpha-hydrate and anhydrous beta—or as an amorphous "glass" mixture of alpha- and beta-lactose. Several other forms may be produced under special conditions.

Alpha-Hydrate

Ordinary commercial lactose is alpha-lactose monohydrate ($C_{12}H_{22}O_{11} \cdot H_2O$), or simply alpha-hydrate. It is prepared by concentrating an aqueous lactose solution to supersaturation and allowing crystallization to take place at a moderate rate below 93.5°C. That alpha-hydrate is the stable solid form at ordinary temperatures is indicated by the fact that the other solid forms change to the hydrate in the presence of a small amount of water below 93.5°C. It has a specific optical rotation in water of $[\alpha]^{20}_{D} = 89.4°$ (anhydrous-weight basis), and a melting point of 201.6°C. A study of the crystalline structure by X-ray diffraction[33] has given the following constants: a + 7.98 Å; b = 21.68 Å; c = 4.836 Å; β = 109°47′, which are in close agreement with previously reported constants.[122,189] These values refer to the dimensions of the unit cell and one of the axial angles. The value of ρ exp. = 1.497, indicating a z value of 2.03 molecules per unit cell.

Alpha-hydrate may form a number of crystal shapes, depending on the conditions of crystallization, but the most familiar forms are the prism and tomahawk shapes.[92 110] As the crystals are hard and not very soluble, they feel gritty when placed in the mouth, similar to sand particles. This is the origin of the term "sandy" to describe the defect in texture of ice cream, condensed milk, or processed cheese spread that contains perceptible α-hydrate crystals. This defect depends on both the size and the numbers of the crystals. Crystals that are 10 μ or smaller are undetectable in the mouth. Above 16 μ, however, the larger the crystals the fewer can be tolerated without affecting the texture. When they are as large as 30 μ, only a few crystals are sufficient to cause sandiness in the several products.[101,109,151]

Crystalline Habit.—Alpha-lactose hydrate crystals are observed in a wide variety of shapes, depending on conditions of crystallization. The principal factor governing the crystalline habit of lactose is the precipitation of pressure, the ratio of actual concentration to solubility.[92] When the pressure is high and crystallization is forced rapidly, only prisms form. As precipitation pressure lessens, the dominant crystal form changes to diamond-shaped plates, then to pyramids and tomahawks, and, finally, in slow crystallization, to the fully developed crystal. These types of crystals are illustrated in Fig. 6.2.

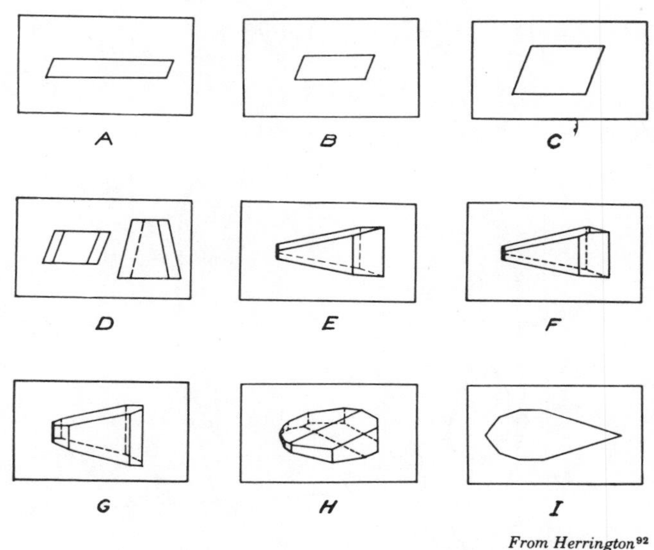

From Herrington[92]

FIG 6.2. THE CRYSTALLINE HABIT OF LACTOSE α-HYDRATE

A—Prism formed when velocity of growth is very high. B—Prism, formed more slowly than prism A. C—Diamond-shaped plates, transition between prism and pyramid. D—Pyramids resulting from increase in thickness of diamond. E—Tomahawk, a tall pyramid with bevel faces at the base. F—Tomahawk, showing another face which sometimes appears. G—The form most commonly described as "fully developed." H—A crystal having 13 faces. The face shown in F is not present. I—A profile view of H with the tomahawk blade sharpened.

Recently detailed studies on the growth rates of the individual faces of α-lactose crystals have appreciably increased understanding of the crystallization process. All the habits of lactose crystals found in dairy products are crystallographically equivalent to the tomahawk form; different relative growth rates on the crystal faces account for the various shapes observed.[92,124] The axes and faces of the tomahawk crystal are depicted in Fig. 6.3. Some typical lactose crystals are shown in Fig. 6.4.

The rate of crystal growth increases rapidly as supersaturation (precipitation pressure) is increased. Data from several studies have shown that the growth rate increases with a power of the supersaturation greater than 1.[124,212] Again, the rate is different for the different faces, thus altering the shape of the crystals. It is observed that the more the faces are oriented toward the b direction the less they grow. The (0$\bar{1}$0) face does not grow at all, the (011) face does not grow at

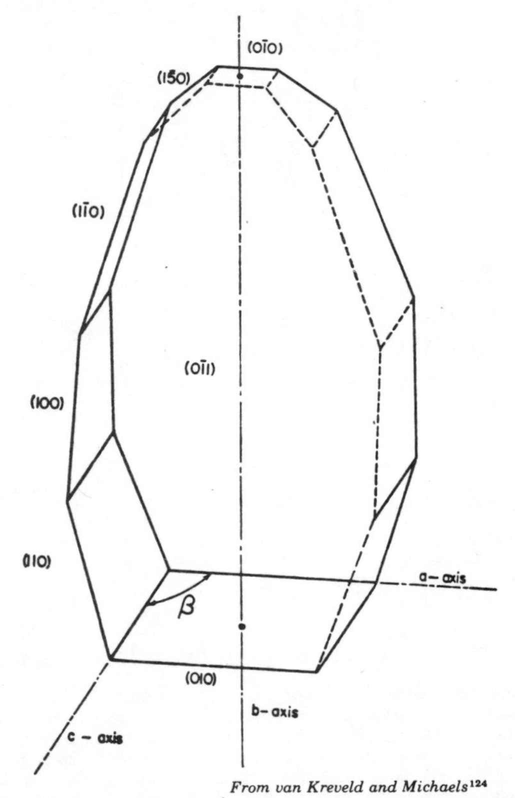

From van Kreveld and Michaels[124]

FIG. 6.3. TOMAHAWK CRYSTAL OF α-LACTOSE MONOHYDRATE

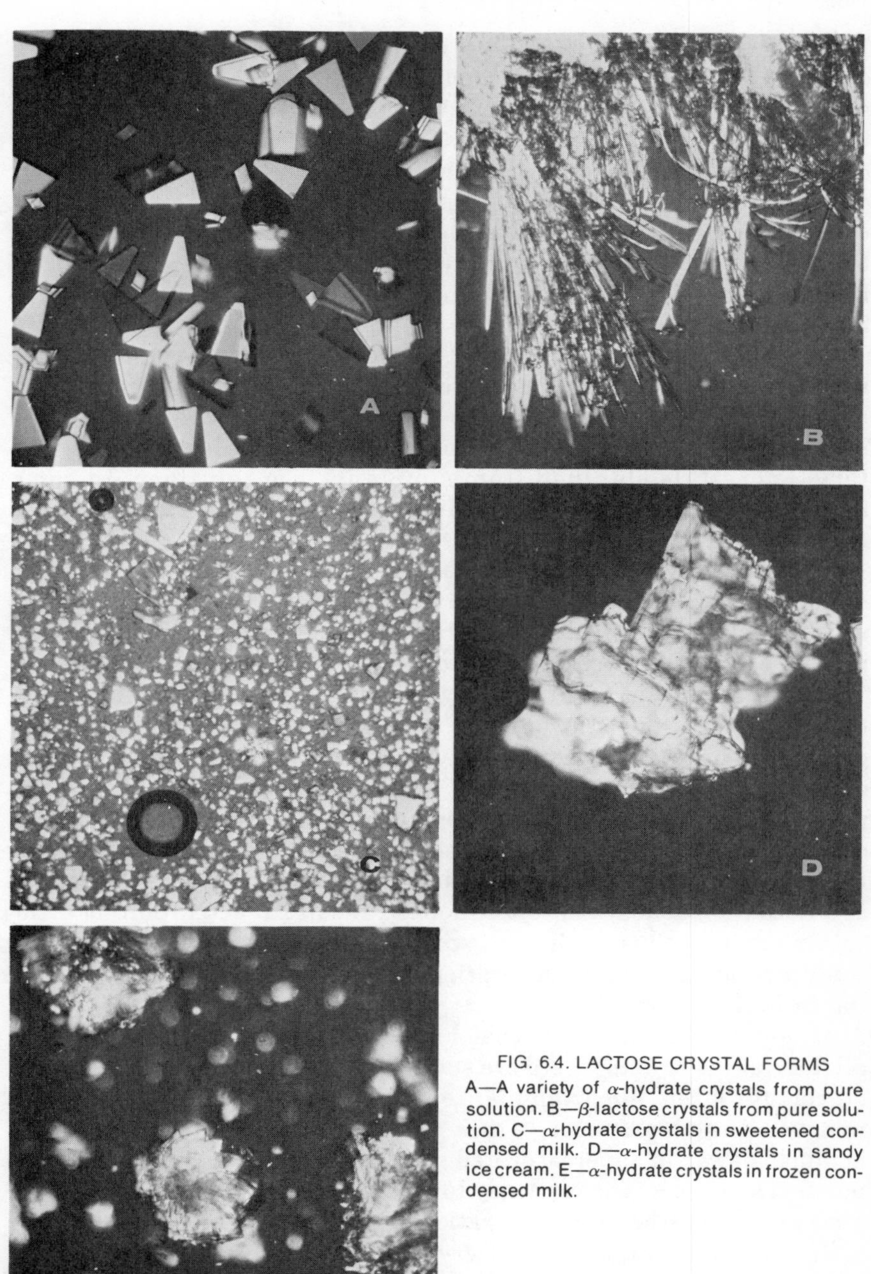

FIG. 6.4. LACTOSE CRYSTAL FORMS

A—A variety of α-hydrate crystals from pure solution. B—β-lactose crystals from pure solution. C—α-hydrate crystals in sweetened condensed milk. D—α-hydrate crystals in sandy ice cream. E—α-hydrate crystals in frozen condensed milk.

low supersaturation but grows slightly at high supersaturation; the $(1\bar{1}0)$ face grows slightly at low and moderately at high supersaturations; the (100) face always takes an intermediate position, and the (110) and (010) always grow fastest. Growth on the (010) face can vary enormously, whereas the $(0\bar{1}1)$ and $(1\bar{1}0)$ faces are fairly constant in growth rate.

Growth studies of broken crystals, as well as studies of the individual faces, have shown that lactose crystals grow only in one direction of its principal axis and therefore have their nucleus in the apex of the tomahawk.[124]

In dairy products, crystallization is often more complex. The impurities (e.g., other milk components) as far as lactose is concerned, may interfere with the crystalline habit. As a result, the crystals tend to be irregularly shaped and clumped instead of yielding the characteristic crystals obtained from simple lactose solutions. In some instances, the impurities may inhibit the formation of nuclei and thus retard or prevent lactose crystallization.[153]

The influence of a number of additives on growth rates has been studied; some additives resulted in marked retardation, whereas others accelerated growth on specific crystal faces.[143] Alterations in the growth process by additives are assumed to involve two opposing mechanisms: (1) acceleration of crystallization by reducing the edge energy at dislocation centers on the crystal face, thereby favoring more rapid step-generation rate by permitting a higher curvature of steps near a dislocation; and (2) inhibition of crystallization by retarding step-propagation by adsorption of the additive on the crystal face. The concentration of the additive can influence the relative importance of these two reactions. For example, low concentrations of a surface-active agent, sodium dodecylbenzenesulfonate, result in "activation" of dislocation centers, thus leading to accelerated crystal growth; at higher concentrations, however, adsorption on the crystal face is rapid, resulting in inhibited growth.[143]

Although most additives that have been studied retard growth on all faces of the crystal, there are some which definitely promote growth on certain faces. For example, repeated recrystallization of lactose removes growth-promoting trace substances, so that crystal growth is much slower in supersaturated solutions of this lactose than in less purified solutions. The tendency to spontaneous nucleation is also lowered upon repeated recrystallization.[143]

Gelatin is an example of a crystallization inhibitor that reduces the rate to 1/3 to 3/4 of normal, even at low gelatin concentrations.[143] In highly supersaturated lactose solutions, however, gelatin cannot suppress nucleation, which explains its ineffectiveness in preventing sandiness in ice cream.[153]

Various marine and vegetable gums are currently in wide use in ice-cream formulations. Shown to inhibit the formation of lactose crystal nuclei, they have been the principal factor responsible for the reduced incidence of sandiness in ice cream in recent years.[153] These gums should prove useful for inhibiting lactose crystallization in other milk-containing foods where crystals give an undesirable texture.

Both methanol and ethanol accelerate crystallization even at low (1%) concentration.[143] The mechanism is unexplained, but several factors seem to be involved. Although the solubility of lactose is depressed by alcohol, it does not seem to be depressed enough to account for the observed acceleration. It is more likely that the effects are due to promotion of step-generation by adsorption of alcohol on the steps. Since alcohol promotes spontaneous nucleation, this may be another factor involved.[143]

The rate of lactose crystallization is also markedly increased at low pH's (<1). Since the acceleration could not be explained by mutarotation, it was suggested that the effect of low pH may be influencing the crystal-surface reaction.[212]

Some carbohydrates actively inhibit the crystallization of lactose, whereas others do not. Carbohydrates that are active contain either the β-galactosyl or the 4-O-glucose group of lactose, so that adsorption can occur specifically at certain crystal faces.[123] Beta-lactose, which is always formed in any solution containing lactose (see mutarotation), has been shown to retard the crystallization process greatly, and incidentally is responsible for a much slower crystallization of lactose than of sucrose, which does not have an isomeric form to interfere with the crystallization process. The retarding action of β-lactose during crystallization of alpha-hydrate lactose is ascribed to the fact that the β-galactosyl part of its molecule is the same as in α-lactose. The β-lactose molecules, along with α molecules, become attached to certain crystal faces which are acceptors of β-galactosyl groups. Once the β-lactose molecules are incorporated on the crystal, they impede further growth, because of their β-glucose group, which is foreign to the crystal structure. Likewise, carbohydrates with the same 4-O-glucose group of α-lactose inhibit the growth of faces that are acceptors for this group.

The retarding action of certain additives is more apparent in solutions where crystallization is slow (low supersaturation), since, under conditions of rapid growth, there is no time for the additive to be adsorbed on the surface of the crystal.

Riboflavin also may be adsorbed on growing lactose crystals and alter the crystalline habit. Since it is naturally present in the whey from which lactose hydrate is made and is present in all dairy foods, its influence on lactose crystallization may be of special interest. Adsorption is dependent upon concentration of riboflavin in solution, on degree of lactose supersaturation and on temperature.[134,135,143] No adsorption occurs below a certain minimum (critical) concentration of riboflavin (2.5 μg per ml), but adsorption increases linearly with riboflavin concentration above this critical level. Increasing the temperature of crystallization increases the amount of riboflavin left in solution; therefore, less riboflavin is adsorbed at higher than at lower temperatures. Adsorption is favored at lower supersaturation levels of lactose where crystallization is slow, in keeping with the action of additives in general. By proper control of these variables, concentrations of 200 to 300 μg of riboflavin per gram of lactose are practical.

Forms of anhydrous α-lactose.—The water of crystallization may be removed from alpha-hydrate crystals under various conditions to produce two types of anhydrous lactose. Technically it is more correct to call them anhydrous lactose rather than lactose anhydrides, as is often done in the dairy literature, since the constituents of water have not actually been extracted from the molecular structure, but only water of crystallization removed.

Regular (unstable) anhydrous α-lactose.—An anhydrous form of alpha-lactose is produced by heating alpha-hydrate above 100°C *in vacuo*. The loss of moisture is negligible at 85°C, becomes significant at 90°C, and rises steadily with increasing temperature, being rapid at 120–125°C.[89,93] Its melting point is 222.8°C. Workers have had difficulty preparing this type of lactose in high purity; for example, drying at 110°C for 48 hr yielded a product of only 90 to 95% alpha, with the remainder beta.[33] This regular anhydrous lactose is stable in dry air, but is highly hygroscopic and therefore unstable when exposed to normal atmospheric conditions. In the presence of water it apparently forms the hydrate without first dissolving. In a solution saturated as to α-lactose hydrate, however, it will not dissolve. This behavior suggests little change in the crystalline structure other than removal of the water of crystallization.[97] Even so, the alpha-hydrate crystals are extensively fractured by the heat and vacuum treatment required to remove the water of crystallization, as shown by electron-microscope micrographs (Fig. 6.8). These fractures remain even when the anhydrous lactose is rehydrated by absorption of 5% moisture from a moist environment.[139]

Stable anhydrous α-lactose.—A stable form of anhydrous α-lactose (not hygroscopic) can be prepared by heating alpha-hydrate crystals at temperatures high enough to drive off the water of crystallization (100 to 190°C) while maintaining the atmospheric environment of the crystals at a water-vapor pressure between 6 and 80 cm mercury.[192] This environment is intermediate between rapid removal of vapor, by which the regular (unstable) anhydrous lactose is formed, and heavier vapor-pressure conditions under which beta-lactose is formed (see section on Manufacture of β-Lactose). Laboratory-scale experiments[123] have been successful with various heat treatments: α-hydrate crystals in a covered Petri dish were heated for various periods from 2 to 3 hr at 130°C to ½ to 1 hr at 160°C.

This same stable anhydrous α-lactose may also be prepared by crystallizing from suitable solvents, such as dry methyl alcohol or 95% or absolute ethanol. With methanol, the reaction occurs more rapidly and at lower temperatures; a product containing 99% anhydrous α-lactose was produced by boiling hydrate crystals for 1 hr with absolute methanol in a ratio 1 : 10.[33] At refluxing temperature the anhydrous form is produced in high yields (98 to 99%) in 1 hr at all ratios of alpha-lactose hydrate to dry methanol. At room temperature the reaction is slower, the rate decreasing with increasing proportions of alpha-hydrate in dry methanol (Fig. 6.5). The decreasing rate has been attributed[139] to the increase in water in the system with increased alpha-lactose hydrate.

Using dilute rather than dry methanol causes a lag period before production of stable anhydrous lactose starts (Fig. 6.6); the more dilute the methanol the longer the induction period. This lag time, needed for the development of nuclei for a new crystal form, is eliminated when nuclei are furnished to the lactose-methanol system by adding stable anhydrous crystals (Fig. 6.7). In the presence of moisture the formation of nuclei is inhibited, and conversion to the anhydrous form is prevented until the new nuclei are formed.[139] The difference in crystal shape can be seen in the electron-microscope micrographs (Fig. 6.8). The new crystalline structure reduces the tendency of the product to take up water of crystallization, so this anhydrous form remains stable even in an environment of 50% relative humidity.

The water of crystallization can be removed from alpha-hydrate by refluxing it in a high-boiling organic solvent that is immiscible with water. For example, the moisture in lactose hydrate has been determined by the toluene-distillation method that is often used to determine moisture in milk powder; but with lactose prolonged distillation (5 hr) is necessary to remove the hydrate moisture.[118,231] The powder remaining after distillation is the stable anhydrous form.[157]

The stable anhydrous α-lactose differs from the regular anhydrous form produced by heating the hydrate under vacuum in that it has greater density, is not appreciably hygroscopic, must dissolve in water before forming the hydrate, and dissolves readily in a solution that is already saturated as far as α-lactose hydrate is concerned.[97] This solution is unstable, to be sure, and soon deposits crystals of α-lactose hydrate. Stable anhydrous α-lactose is more soluble in water than

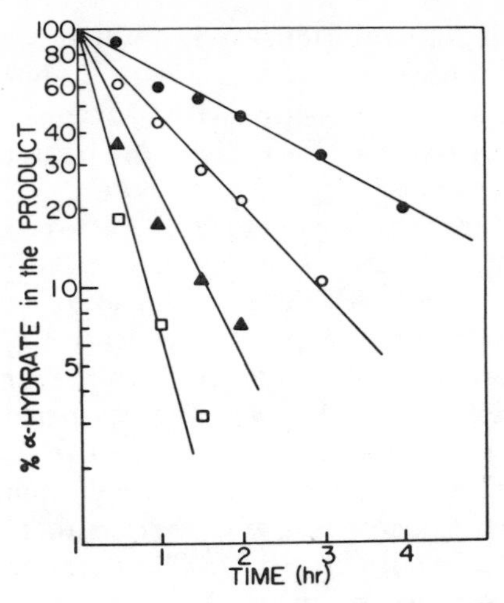

From Lim and Nickerson[139]

FIG. 6.5. EFFECT OF RATIO OF α-LACTOSE HYDRATE TO METHANOL ON THE RATE OF REACTION AT ROOM TEMPERATURE

●—Ratio 1:2. ○—Ratio 1:5. ▲—Ratio 1:10. □—Ratio 1:15 and 1:20.

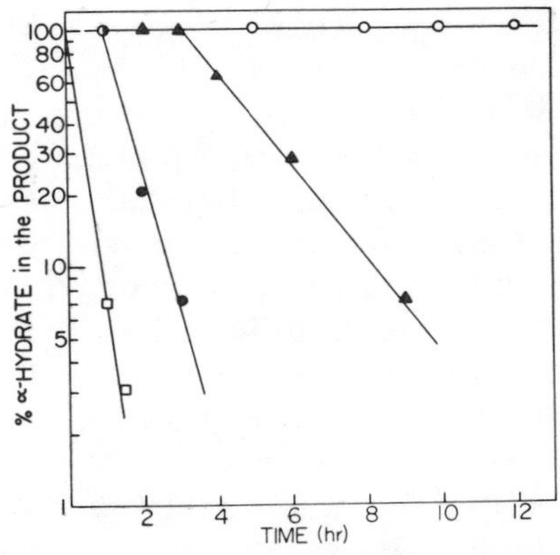

From Lim and Nickerson[139]

FIG. 6.6. EFFECT OF MOISTURE IN SOLVENT ON THE RATE OF REACTION AT ROOM TEMPERATURE

○—2% moisture in solvent. ▲—1.5% moisture in solvent. ●—1.0% moisture in solvent. □—0.5% moisture in solvent.

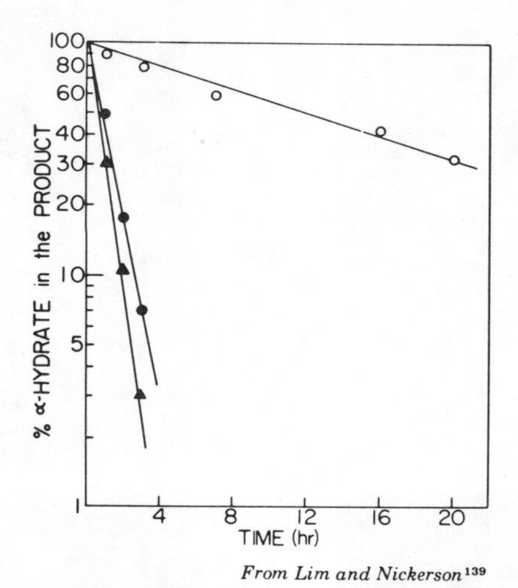

From Lim and Nickerson[139]

FIG. 6.7. EFFECT OF MOISTURE ON THE RATE OF REACTION AFTER SEEDING AT ROOM TEMPERATURE

○—2.4% moisture in solvent. ●—1.5% moisture in solvent. ▲—1.0% moisture in solvent.

either α-hydrate or β-lactose. This form of lactose has been used to prepare solutions that are highly supersaturated in α but low in β.[123]

Neither the regular (hygroscopic) nor the stable form of anhydrous lactose is available commercially, but their production may be warranted in the future since their properties are appreciably different from those of α-hydrate, beta, or the glass form. The anhydrous forms have occasioned increased interest recently; they have been used in lactose investigations and are the subject of recent patents. The lattice constants of the anhydrous α-forms have not been reported.

A third type of anhydrous lactose crystal can be prepared by shaking finely powdered α-hydrate crystals at room temperature in 10 times their weight of methanol containing 1 to 5% anhydrous HCl.[98] The characteristic crystals of lactose gradually disappear, and tiny needles form. They contain a mixture of anhydrous α- and β-lactose in a ratio of 5: 3.

Anhydrous Lactose Glass (Amorphous Noncrystalline Glass)

When a lactose solution is dried rapidly, the viscosity increases so quickly that crystallization cannot take place.[87] The dry lactose is essentially in the same condition as it was in solution except for removal of the water. This is spoken of as a concentrated syrup or an amorphous (noncrystalline) glass. Various workers have shown conclusively that

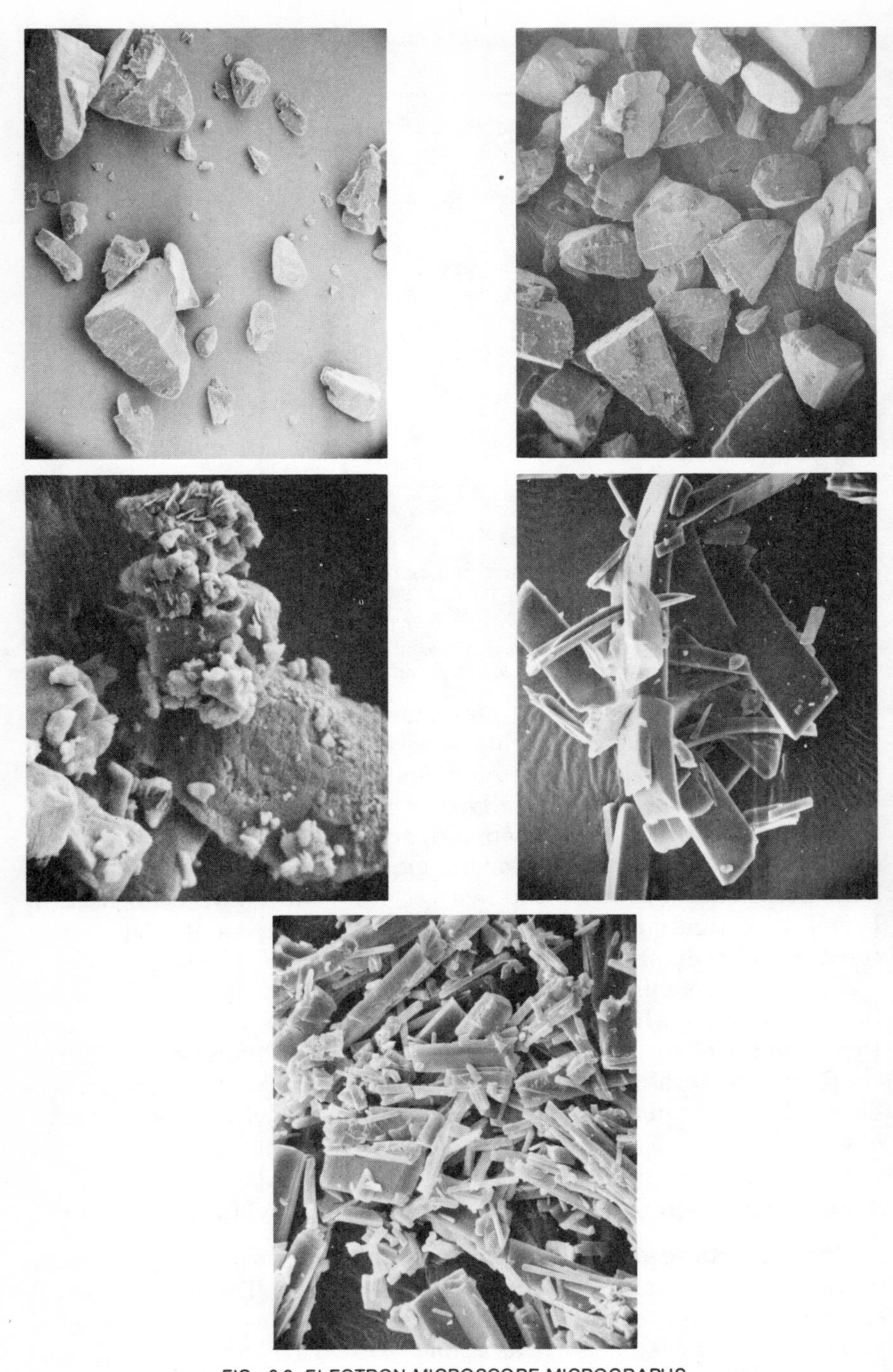

FIG. 6.8. ELECTRON MICROSCOPE MICROGRAPHS

Top Left—α-hydrate crystals at 120X. Top Right—Regular anhydrous α-lactose (hygroscopic) at 100X. Middle Left—Regular anhydrous α-lactose (hygroscopic) at 2300X. Middle Right—Stable anhydrous α-lactose from methanol at 1150X. Bottom Left—β-lactose crystals at 2000X.

lactose in milk powder (spray, roller, or freeze-dried) is noncrystalline and exists in the same equilibrium mixture of α- and β-lactose as existed in the milk prior to drying.[40,154,193,194,209]

In vacuum-oven methods for moisture determination, such as the official method of AOAC, lactose solutions are dried quickly. The result in the dried product is amorphous lactose glass. Since lactose glass is very hygroscopic, the dried sample must be protected from moisture until final weighing. If alpha-hydrate crystals are present in the product to be analyzed, the sample is diluted with water to dissolve the crystals, since slow removal of the water of crystallization under the temperature and vacuum conditions of the moisture test unduly prolongs the moisture determination.

Lactose glass is stable if protected from moisture, but since it is very hygroscopic, it rapidly takes up moisture from the air and becomes sticky. When the moisture content reaches about 8% or relative vapor pressure near 0.5, the lactose molecules are diluted enough to become mobile and can orient into a crystal lattice.[19,31,187] Alpha-hydrate crystals develop at all temperatures below 93.5°C, and as they grow the crystals bind adjacent powder particles together. Dry milk products containing lactose glass therefore tend to become lumpy or cake together during storage unless protected from moisture absorption. When moisture is absorbed, part will be incorporated as water of crystallization in the α-hydrate and the remainder will be released, since crystalline α-hydrate is not hygroscopic.[205]

Beta-lactose

When lactose crystallization occurs above 93.5°C, the crystals formed are anhydrous and have a specific rotation of $[\alpha]^{20}_{D} = 35.0°$ and a melting point of 252.2°C. They are composed of anhydrous beta-lactose, which is sweeter and considerably more soluble than α-hydrate (see Tables 6.2 and 6.6). The common form of crystal is an uneven-sided diamond when crystallized from water, and curved needle-like prisms when from alcohol. The lattice constants for the crystalline structure as determined by X-ray diffraction[33] are:
a = 10.81 Å; b = 13.34 Å; c = 4.84 Å; β = 91°15′.

Equilibrium in Solution (Mutarotation)

As mentioned previously, lactose exists in two forms, α and β. By definition, α is the form having the greater optical rotation in the dextro direction. The specific rotation of a substance is characteristic of that substance and is defined as the rotation in angular degrees produced by a length of 1 dm of a solution containing 1 gm of substance per ml. Therefore the specific rotation may be represented by the formula $[\alpha] = 100a/lc$, in which α = specific rotation, a = degrees of angular rotation, l = length of tube in dm, and c = concentration of substance in gm per 100 ml of solution.

Also important, besides the variables of the equation, are temperature of the solution, wavelength of the light source, and concentration of the solution. The standard light source used to measure optical rotation has been the bright yellow D lines of the sodium spectrum, but the single mercury line, λ5461 Å, is now used frequently for precision measurements. Generally, the specific rotation is reported at 20°C and expressed as:

$$[\alpha]_{D}^{20} \text{ or } [\alpha]_{Hg}^{20} .$$

The following formulas[79,103] express variations in specific rotation in terms of these variables:

$[\alpha]_{D}^{t} = 55.23 - 0.01688C - 0.07283\,(t-25)$
where C is gm anhydrous lactose/100 ml solution
and t is degrees centigrade

$[\alpha]_{Hg}^{t} = 61.77 - 0.007C - 0.076\,(t-20)$
where C is gm lactose monohydrate/100 ml solution

The values given earlier for specific rotations of α- and β-lactose are the initial values. When either form is dissolved in water, however, there is a gradual change-over of one form to the other until equilibrium is established. Regardless of the form used in preparing a solution, the rotation will change (mutarotation) until $[\alpha]_{D}^{20} = 55.3°$ at equilibrium (anhydrous weight basis). This is equivalent to 37.3% in the α form and 62.7% in the β form, since the equilibrium rotation is the sum of the individual rotations of the α and β forms. The equilibrium ratio of β to α at 20°C, therefore, is 62.7/37.3 = 1.68. This value is affected slightly by differences in temperature, but not by differences in pH.[163] The proportion of lactose in the α form increases gradually and at a constant rate as the temperature rises. The equilibrium constant (β/α) consequently decreases with rising temperatures (Fig. 6.9).

Mutarotation has been shown to be a first-order reaction, the velocity constant being independent of reaction time and concentration of reactant. The rate of mutarotation increases 2.8 times with a 10°C rise in temperature. By applying the law of mass action, equations have been developed to measure the rate of the reversible reaction between the α and β forms of lactose. If a dilute lactose solution at constant temperature contains a moles of α and b moles of β, then the amount of β formed (x) per unit of time is

$$\frac{dx}{dt} = k_1\,(a - x) - k_2\,(b + x)\cdot$$

The mutarotation coefficient ($k_1 + \kappa_2$) can be determined by the change in optical rotation with time:

$$k_1 + k_2 = \frac{1}{t} \log \frac{r_0 - r_\infty}{r_t - r_\infty}$$

where r_0 is the optical rotation at zero time, r_t is the rotation at time t, and r_∞ is the equilibrium (final) rotation. The equation expresses a first-order reaction. Plotting the difference in rotation at time t and equilibrium ($r_t - r_\infty$) against time gives a straight line with a slope equivalent to the mutarotation coefficient.

Procedures based on rotation have been used for quantitative measurement of the amounts of α- and β-lactose in dry milk products and in ice cream.[152,194]

The rate of mutarotation is influenced greatly by both temperature and pH. The rate is slow at low temperature but increases as the temperature rises, becoming almost instantaneous at about 75°C. The rate of change from α to β is given by Hudson[105] as: 51.1% complete in 1 hr at 25°C, 17.5% complete in 1 hr at 15°C, and 3.4% complete in 1 hr at 0°C. The rate of mutarotation is minimum at about pH 5.0, increasing with changes in pH on either side of this value (Fig. 6.10). The rate is rapid at very low pH values but increases most rapidly in alkaline solutions, establishing equilibrium within a few minutes at pH 9.

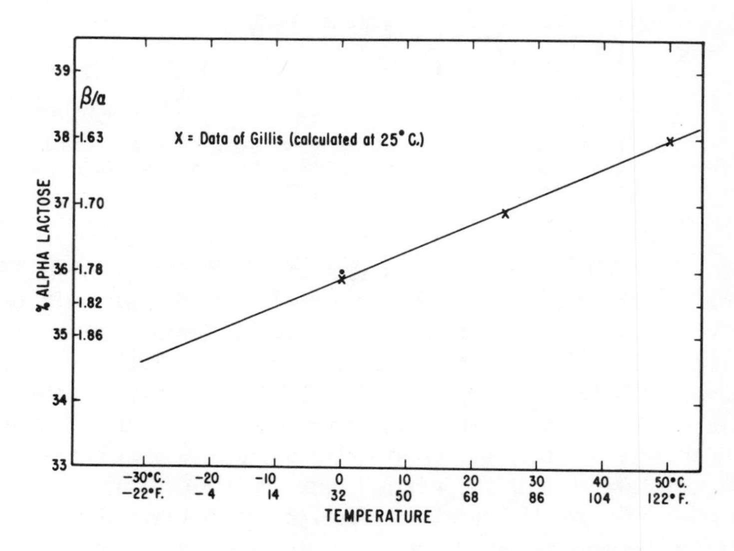

From Nickerson[152]

FIG. 6.9. EFFECT OF TEMPERATURE UPON THE EQUILIBRIUM RATIO OF β TO α-LACTOSE

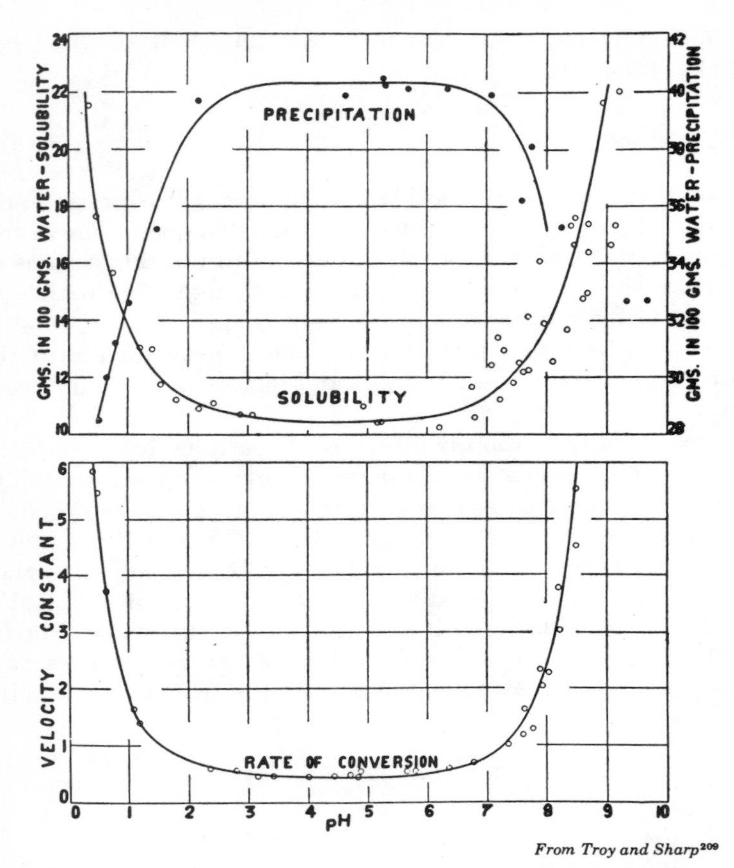

From Troy and Sharp[209]

FIG. 6.10. THE EFFECT OF pH ON THE RATE OF CHANGE OF THE FORMS OF LACTOSE INTO EACH OTHER AS INFLUENCING THE RATE OF SOLUTION AND PRECIPITATION OF LACTOSE

The presence of sugars and salts also can affect the rate of mutarotation. Although the effect is small in dilute solutions,[94,95] a combination of salts equal to that found in solution in milk nearly doubles the rate of mutarotation.[79] This catalytic effect was attributed primarily to the citrates and phosphates of milk.[164] The presence of high levels of sucrose, on the other hand, has the opposite effect. The effect of sucrose is only slight at concentrations up to 40%, but as the concentration is increased above this level mutarotation is rapidly decreased to about half the normal rate. A level of 30% sucrose or more also eliminates the catalytic effect of citrates and phosphates.[164] Some of the data suggest an interaction between the salts and the sugars. Some physicochemical properties have been reported for a complex of lactose and calcium chloride that has an empirical formula of α-lactose · $CaCl_2$ · $7H_2O$. However, a study of the calcium distribution in milk indicated that no soluble compounds between calcium and lactose exist at concentrations normally occurring in milk.[199]

The specific rotation of lactose varies with the solvent. It is higher in glycerol than in aqueous solutions,[95] but lower in alcoholic or acetone solutions.[208] Not only the specific rotation but also the equilibrium ratio of α to β may be changed by the nature of the solvent. For example, upon dilution with water, concentrated solutions of lactose in methanolic calcium chloride show a high (1.3) initial to final rotation, regardless of whether the α- or β-isomer was used originally in preparing the solution.[53]

Solubility

Mutarotation also manifests itself in the solubility behavior of lactose. When α-lactose hydrate is added in excess to water, with agitation, a definite amount dissolves rapidly, after which an additional amount dissolves slowly until final solubility is attained.

Equations similar to those for mutarotation have been derived, expressing the relationship between the solubility behavior of the two forms of lactose and the equilibrium or rate constants.[104] The constants derived by both mutarotation and solubility methods are in agreement. The solubility equations have been used to develop procedures for measuring α- and β-lactose in dry milk.[40,41]

This initial solubility is the true solubility of the alpha form. The increasing solubility with time is due to mutarotation. As some of the α is converted to β, the solution becomes unsaturated with respect to α, and more α-hydrate dissolves. This process continues until equilibrium is established between α and β in solution and no more α-hydrate can dissolve, thus establishing the final solubility. This solution is saturated with respect to α, but a great deal of β-lactose powder can be dissolved in it because of the greater initial solubility of the β form. The solution becomes saturated with α long before the saturation point of β is reached. However, additional β dissolving in such a solution upsets the equilibrium and mutarotation takes place. Since the solution was already saturated with α, α formed by mutarotation will crystallize to re-establish equilibrium. Since β-lactose is much more soluble and mutarotation is slow, it is possible to form more highly concentrated solutions by dissolving β- rather than α-lactose hydrate. In either case the final solubility of the lactose in solution will be the same. Solubility values for lactose are shown in Table 6.2.

The solvent and the presence of salts or sucrose influence the solubility of lactose, as was the case with mutarotation. The solubility of lactose increases with increasing concentrations of several calcium

Table 6.2

SOLUBILITIES OF LACTOSE (GM PER 100 GM WATER)

°C	Initial α	Initial β	Final	Super-solubility
0	5.0	45.1	11.9	25
10.0	5.8	—	15.1	—
15.0	7.1	—	16.9	38
25.0	8.6	—	21.6	50
30.0	9.7	—	24.8	—
39.0	(12.6)[a]	—	31.5	74
49.0	(17.8)	—	42.4	—
50.0	17.4	—	43.7	—
59.1	—	—	59.1	—
63.9	—	—	64.2	—
64.0	(26.2)	—	65.8	—
73.5	—	—	84.5	—
74.0	(34.4)	—	86.2	—
79.1	—	—	98.4	—
87.2	—	—	122.5	—
88.2	—	—	127.3	—
89.0	(55.7)	—	139.2	—
90.0	60.0	—	143.9	—
100.0	—	(94.7)	157.6	—
107.0	—	—	177.0	—
121.5	—	—	227.0	—
133.6	—	—	273.0	—
138.8	—	—	306.0	—

[a] Calculated values assuming K = 1.50 and solubility of one form is independent of the other.
Data from Whittier[226] and Herrington[97]

salts—chloride, bromide, or nitrate—and exceedingly stable concentrated solutions are formed.[96] One explanation for the increased solubility is the complex formation previously mentioned between the lactose and the salt.

It has been shown that calcium chloride also markedly increases the solubility of lactose in methanol.[53] From the highly concentrated viscous solutions formed there slowly crystallizes a complex of β-lactose, calcium chloride, and methanol in a molecular ratio of 1 : 1 : 4. On addition of water to the concentrated solution, the complex previously described (α-lactose · $CaCl_2$ · $7H_2O$) soon crystallizes.

Only limited studies have been made on the effects of other sugars on the solubility of lactose. At 10 to 18°C a sucrose concentration in water of 14%, comparable to that in ice-cream mix, reduces lactose solubility only slightly. However, the data in Table 6.3 show that concentrations of 40 to 70% sucrose reduce the solubility of lactose appreciably—to 40 to 80% of normal. It has also been shown that

Table 6.3

RELATIVE SOLUBILITY OF LACTOSE IN SUCROSE SOLUTIONS[a],[156]

Solution	Temperature, °C					
	25	40	50	60	80	85
	Percent[a]					
40% Sucrose	74.5	76.7	75.5	81.9	89.4	80.5
50% Sucrose	63.0	64.8	64.9	71.9	76.7	73.0
60% Sucrose	50.9	53.5	53.3	57.8	70.2	66.4
70% Sucrose	42.1	44.3	43.2	54.3	63.9	62.7

[a] % of lactose solubility in distilled water at the same temperature.

at temperatures near 0°C, the solubility of lactose is reduced about one-half by saturating the solution with sucrose.[169]

As mentioned previously, alcohol greatly reduces the solubility of lactose, but the glass or amorphous form dissolves in alcoholic solutions to form supersaturated solutions. This has been used to extract lactose from whey or skimmilk powder with methanol or ethanol. A high-grade lactose subsequently crystallizes from the alcoholic solution. Methanol is the better solvent and allows recovery of soluble proteins in addition to the lactose.[136,137]

Ethanol and methanol (preferred with less than 3% moisture) have recently been used to extract lactose from skimmilk or whey powders.[129] The dried lactose powder that crystallized from the alcoholic extract was believed to be anhydrous α-lactose, but work in the author's laboratory indicates that the product is a mixture of anhydrous α-and β-lactose.[139]

Since alcohol greatly reduces the solubility of lactose, addition of alcohol accelerates crystallization and therefore influences crystal habit, as discussed earlier. When alcohol is added to a lactose solution, the mixture becomes milky white for a few seconds, and then clears up.[92] After a few minutes a permanent precipitate of lactose crystals appears. The composition of the precipitate may vary a great deal with the percentage of alcohol added.[91] Only α-hydrate is precipitated at low concentration, but β is also included at higher concentrations. Stable anhydrous α-lactose is produced when α-hydrate is treated with alcohol (see section beginning on page 285). Unlike α-hydrate, β-lactose is not altered by methanol, either at room or at refluxing temperatures.

Other organic solvents have been employed to facilitate the isolation of lactose from whey. The principle is to add a suitable solvent to concentrated whey in an amount sufficient to precipitate some of the impurities and then to add more solvent to lower the solubility of lactose. Ethyl, propyl, and butyl alcohols have been used, and ketones have been suggested.[60,133]

Acetone also reduces the solubility of lactose, and a procedure to recover lactose from whey is based on that fact.[119] Acetone is added to concentrated whey (18 to 20% lactose) in amounts sufficient to precipitate some of the impurities. After these are filtered off, the gradual addition of acetone to over 65% allows recovery of 85% of the lactose during a 3½ hr period. The yield of lactose and rapidity of crystallization are influenced by the rate of addition of the acetone.

Crystallization

Solutions of lactose are capable of being highly supersaturated before spontaneous crystallization occurs.[91] Even then, crystallization may occur only after a considerable period. In general, the supersolubility at any temperature is equal to the saturation value at a temperature 30°C higher. This is shown by the lactose solubility curves of Fig. 6.11.

Ostwald[159] is credited with introducing the concept of supersaturation and extending it into "metastable" and "labile" areas. The metastable area occurs in the first stages of supersaturation produced by cooling a saturated solution or by continued evaporation beyond the saturation point. Crystallization does not occur readily in this supersaturation range. The labile area is found at higher levels of supersaturation, where crystallization occurs readily.

The true picture[188] is far more complex than indicated in Fig. 6.11. In reality, a series of supersolubility curves should be pictured whose locations will depend on specific seed surface, rate of supersaturation production (e.g., cooling rate, evaporation rate), and mechanical disturbances (e.g., agitation). The concept of "regions" of supersolubility is correct qualitatively, but not quantitatively. The significant points of the concept are that (a) neither growth nor nucleation can take place in the unsaturated region; (b) growth of a crystal can take place in both the metastable and labile areas; (c) nucleation can take place in the metastable area only if seeds (centers for crystal growth) are added; and (d) spontaneous nucleation (crystallization) can take place in the labile area without addition of seeding materials.

These principles have been used to detect crystals of α-hydrate, β-lactose or both in various products.[106,193,209] A supersaturated solution in the metastable zone prepared with respect to the form being tested is undersaturated with respect to the other. When the product in question contains crystals, the solution will become cloudy with newly-formed crystals as a result of seeding. If crystals are not present in the material the solution will remain stable and clear.

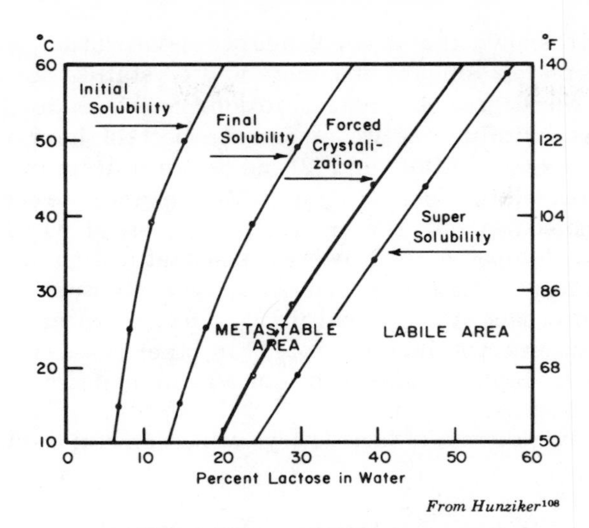

From Hunziker[108]

FIG. 6.11. LACTOSE SOLUBILITY CURVES

Crystallization in general is a two-step process involving (1) nucleation and (2) growth of the nucleus to a macro-size. Nucleation involves the activation of small unstable particles with sufficient excess surface energy to form a new stable phase. This may occur in supersaturated solutions as a result of mechanical shock, the introduction of small crystals of the desired type, or the presence of certain impurities that can act as centers for growth.

Early work indicated that a certain minimum-size fragment is required to induce crystallization. More recent work[213] indicates the size of such a critical nucleus to be of the order of 100 molecules (a diameter of 100 Å).

With increasing concentration, the probability of nucleus formation increases to a maximum and then quickly decreases to zero.[91] The stability of lactose glass is apparently due to the small probability that nuclei will form at such high concentrations. When lactose glass absorbs moisture, as milk powder does, great numbers of nuclei form, since the concentration has been reduced to the region of maximum nucleus formation. After a nucleus forms, subsequent growth of any crystal depends on the rate of transfer of solute to the crystal surface and the rate of orientation of these molecules at the surface. Thus, the rate of crystal growth is controlled by the degree of supersaturation, the surface area available to deposition, and the diffusion rate to the crystal surface, which depends upon viscosity, agitation, and temperature of the solution. With lactose there is the additional factor of the rate of mutarotation of β to α. As mentioned previously, this rate is very rapid above 75°C but is slow at low temperatures.

It has been shown that the optimum crystallization temperature (where the greatest amount of lactose will crystallize per unit time) varies with the degree of supersaturation.[212] This is due to the fact that temperature influences two important aspects of the crystallization process: (1) supersaturation, and (2) the crystallization rate constant, including rate of diffusion, rate of mutarotation, and rate of orientation of lactose molecules into the crystal lattice, all of which probably increase with temperature. However, supersaturation will decrease with temperature. These two factors oppose each other and in some instances cancel each other, so that changes in temperature have practically no effect on crystallization rate.[80] In other cases crystallization is accelerated at higher temperatures as a result of decreased viscosity and increased kinetic activity.[227]

Thus, the overall rate of lactose crystallization can be summarized by the reaction:

$$\beta\text{-lactose} \underset{}{\overset{\text{Step 1}}{\rightleftharpoons}} \alpha\text{-lactose} \overset{\text{Step 2}}{\rightleftharpoons} \alpha\text{-lactose hydrate crystals}$$

If mutarotation (step 1) is slower than crystallization (step 2), it will determine the overall reaction, and the α-lactose level will be lower than the mutarotatory equilibrium value (37.3% α at 20°C). Conversely, if crystallization is slower, then the α- and β-lactose isomers in solution will be close to their equilibrium value. It has been shown that mutarotation occurs more rapidly under conditions normally found in milk products; thus crystallization becomes the rate-determining step.[80] However, under conditions of very rapid crystallization where the supersaturated α-lactose is being deposited on a large surface area of nuclei, the percentage of α in solution will drop below its equilibrium value. Under these conditions neither mutarotation nor surface orientation appears to be completely rate-limiting.[212]

Figs. 6.12 and 6.13 show the effects of a few variables on lactose crystallization rate.

Crystallization of the lactose in concentrated skimmilk (40 to 54% TS) can result in rapid increases in viscosity even during the short interval between concentration and spray-drying.[12]

Other Physical Properties

Density.—The densities of the various lactose crystals differ slightly from each other: α-hydrate is 1.540, β-anhydride is 1.589, α-anhydride formed by dehydration under vacuum is 1.544, and α-anhydride crystallized from alcohol is 1.575. Densities of lactose solutions are not straight-line functions of concentration. Equations have been developed[142] relating the percentage (p) by weight of the lactose to density. The equations differ, depending on whether they are based

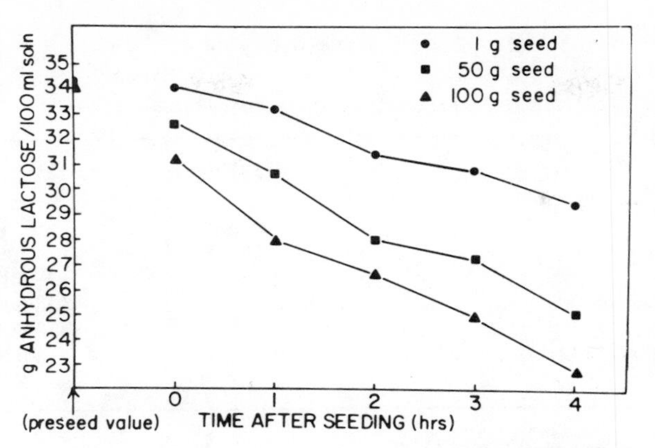

From Twieg and Nickerson[212]

FIG. 6.12. EFFECT OF SURFACE AREA ON CRYSTALLIZATION RATE (24°C, pH 4.0)

on the hydrated or anhydrous form, temperature of solution, and range of concentration, e.g., $D_4^{20} = 0.99823 + 0.003739p + 0.00001281p^2$, where the hydrate is present between 0 and 16% and equilibrium established. Similarly, equations have been developed relating such variables as concentration, form of lactose, and temperature to refractive index: $n_D^{20} = 1.33299 + 0.001409p + 0.00000498p^2$, where hydrate is present at less than 20% concentration and equilibrium established. Tables are available containing precise data on the density and refractive indexes of lactose solutions.[142 233] Other physical properties of lactose are presented in Table 6.4.

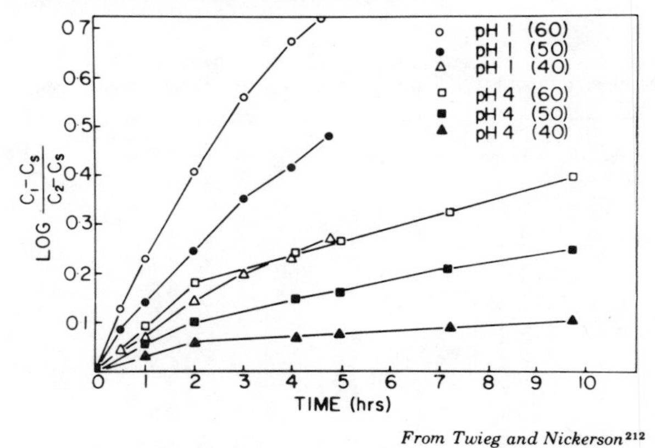

From Twieg and Nickerson[212]

FIG. 6.13. SEMILOG PLOT SHOWING EFFECT OF SUPERSATURATION AND pH ON CRYSTALLIZATION RATE

Figures in parentheses indicate initial lactose concentration in grams of α-hydrate per 100 ml water.

Table 6.4

PHYSICAL PROPERTIES OF LACTOSE[6,32,226]

Property	α-Hydrate	β
Specific heat	0.299	0.2895
$S_{298,16}$°K (E.u./mole)	99.1	92.3
$\Delta S_{298\ 16}$°K (E.u.)	−586.2	−537.2
$\Delta H^{\circ}_{298,16}$°K (cal./mole)	−592,900	−533,800
$\Delta F^{\circ}_{298,16}$°K (cal./mole)	−418,200	−373,700
Heat of combustion (cal./gm.)	3,761.6	3,932.7
Melting point	201.6° (d.)	252.2° (d.)

Relative Sweetness.—It has been amply demonstrated that the relative sweetness of sugars changes with concentration. Therefore it is misleading to say that one sugar is so many times as sweet as another, because this will be true only at certain concentrations. Table 6.5 summarizes results of several investigators on the relative sweetness of some common sugars. It should be noted that lactose is relatively sweeter at higher than at lower concentrations and is sweeter than is usually reported in the literature.

β-Lactose is sweeter than α (Table 6.6) but β is not appreciably sweeter than the equilibrium mixture except when the concentration of lactose solution equals or is greater than 7%.[162] Since there is approximately 63% β in the equilibrium mixture, a β-lactose solution differs

Table 6.5

RELATIVE SWEETNESS OF SUGARS
(PERCENT CONCENTRATION TO GIVE EQUIVALENT SWEETNESS)

Sucrose	Glucose	Fructose	Lactose	Reference
0.5	0.9	0.4	1.9	160
1.0	1.8	0.8	3.5	160
2.0	3.6	1.7	6.5	160
2.0	3.8	—	6.5	36
2.0	3.2	—	6.0	47
5.0	8.3	4.2	15.7	160
5.0	8.3	4.6	14.9	36
5.0	7.2	4.5	13.1	47
10.0	13.9	8.6	25.9	160
10.0	14.6	—	0	36
10.0	12.7	8.7	20.7	47
15.0	17.2	12.8	27.8	47
15.0	20.0	13.0	34.6	160
20.0	21.8	16.7	33.3	47

Table 6.6

RELATIVE SWEETNESS OF α- AND β-LACTOSE[162]

	Concentration %	No. Evaluations	Sec. after Hydration	Percent Response Considering		
				α Sweeter	Equilibrium Sweeter	β Sweeter
α-lactose	5.0	40	270	40.0	60.0	—
	7.0	44	222	13.6	86.4[c]	—
β-lactose	5.0	40	124	—	45.0	55.0
	7.0	40	140	—	35.0	65.0[a]
α-versus	5.0	40	184	25.0	—	75.0[b]
β-lactose	7.0	40	266	12.5	—	87.5[c]

[a] Significant at P = 0.05.
[b] Significant at P = 0.01.
[c] Significant at P = 0.001.

less in sweetness from a solution in equilibrium than does α-lactose solution. However, for practical purposes there is little advantage in using β for sweetness in preference to an equilibrium solution at these concentrations, since the small difference is quickly eliminated by mutarotation.

APPLICATION IN DAIRY TECHNOLOGY

These crystallization principles are applied in processing several dairy products, e.g., sweetened condensed milk, instant milk powder, stabilized whey powders, and lactose. Control of lactose crystallization in ice cream is discussed in Chapter 14.

Sweetened Condensed Milk

Processing of sweetened condensed milk has recently been reviewed.[222] Small crystals are desired to assure a smooth-bodied product. Since, in general, rapid crystallization produces small crystals, it is desirable to crystallize at the optimum temperature for rapid crystallization. This varies with the concentration of the original solution, as mentioned previously. The "forced crystallization" curve in Fig. 6.11 represents the optimum crystallization temperatures for the various concentrations of lactose.[108] This curve is in the metastable area, where crystallization will not occur spontaneously. Therefore, one does not depend upon spontaneous production of an adequate number of uniformly small crystals; instead, one controls crystallization by inoculating with many crystal centers, i.e., seed lactose.

Recent advances in this field have been concerned with supplying adequate numbers of crystal centers. If the seeding material and conditions of crystallization do not produce crystals of uniform size, large crystals grow at the expense of small ones during storage. Sweetened condensed milk may be subjected to rather wide temperature fluctuations. As a result, some lactose dissolves when the temperature increases, and is redeposited when it goes down again. Since small crystals have a greater surface per unit volume, they dissolve more rapidly, and redeposition takes place on the remaining surfaces of the larger crystals. Over a long period the small crystals become smaller, but (more important) the large crystals grow larger and cause a defective product, e.g., mealy or sandy texture. On the other hand, if crystals are of uniform size they will be affected uniformly by temperature fluctuations. Uniformity of crystal size can be improved by passing the seeded concentrate through a colloid mill or homogenizer to break up the crystals.[148]

Instead of relying on the production of crystal centers by conventional cooling and agitating procedures, new seeding techniques[34,59] add all the nuclei that are needed (about 400,000/ml). To allow for growth, their diameter must be about 1μ. Normal seed lactose is produced by grinding α-hydrate powder in a hammer mill equipped with an air separator that automatically regrinds the large particles until all are reduced to the desired fineness. Such a mechanical procedure, however, cannot in practice produce uniform crystals as small as 1μ. For that purpose the unstable state of lactose glass is utilized. As explained previously, the lactose glass formed by rapid drying of milk, whey, or lactose solutions is unstable when it is slightly moistened, and crystallization occurs spontaneously as a result of the nuclei formed upon dilution of the supersaturated solution (glass). The crystals produced are exceedingly small and ideally suited for seeding material. When this is used to supply the required 400,000 nuclei per ml, the lactose in the supersaturated solution can be deposited without producing large crystals. It is not necessary to produce additional crystals by cooling slowly with vigorous agitation. Continuous vacuum cooling can be used to reduce cooling time from several hours to less than 1 min. In fact this new seeding method gives the best results when it is used in conjunction with rapid cooling. No explanations are given as to why this should be true, but it would not be surprising if these conditions initiate further spontaneous crystallization. Similarly, when ice-cream mix is seeded with these minute crystals while being cooled and agitated rapidly in the freezer, tremendous numbers of very small crystals are produced, and sandiness is prevented.[151]

Instant Dry Milk

Lumping and caking during storage is a problem if milk powder is not protected from moisture. The absorption of moisture dilutes the glass to the stage where molecular orientation is possible and crystallization occurs. The hard α-hydrate crystals that gradually form cement the milk particles together, producing lumpiness and then caking. The caseinate system of the powder also becomes insoluble. Similarly, caseinate insolubility, which develops during storage of frozen milk, is associated with lactose crystallization (see Chapter 14). It appears that in both products lactose crystallization is observed prior to loss of solubility.[52,147]

The instantizing process for producing readily-soluble powder is somewhat similar. The surfaces of milk powder particles are humidified, or, alternatively, the particles are only partially dried during manufacture, so that the surface is tacky and partial crystallization occurs before the particles are redried.[168] This produces a clustering of the particles in loose, spongy aggregates of low density. The aggregates are free-flowing and readily dispersible in water. Water quickly penetrates the spongy structure of the aggregates, allowing them to sink and disperse, whereas normal milk particles tend to clump and ball up, making it difficult for water to disintegrate them.

The lactose equilibrium has been shifted in such instant milk to about 3:2 α and β, instead of the usual 2:3.[21] Conditions of rehydrating, holding to allow agglomeration, and redrying must be carefully controlled to preserve the solubility of the powder. Unfortunately, little technical information is available on the process, though it is well known that powder can be rendered insoluble almost instantly by exposing it to elevated temperatures at high humidities.[230] More detailed descriptions of milk-drying methods are available in recent publications.[83,222]

Stabilized Dry Whey

Lactose, being about 70% of the solids in whey, understandably plays a dominant role in determining the properties of whey and whey products. It is often difficult to dry by normal methods but can be dried by modifying the processes used to dry milk, i.e., roller or spray methods.[83,222] Modifications usually involve procedures to cope with the sticky hygroscopic glass that may be formed with either method. Most special processes cause a considerable portion of the lactose to crystallize by holding the product at some stage in the presence of sufficient moisture. This may be induced in the condensed product or at some stage where the product is only partially dried, or by rehumidifying the dry product.

Crystallizing conditions generally favor the production of α-hydrate crystals, but if crystallization occurs above 93.5°C, especially under pressure, anhydrous β crystals will be formed. After crystallization has occurred and the product finally dried, the resulting whey powder is granular and free-flowing, and does not tend to become sticky and caked.

Foam-spray drying was introduced successfully for drying high-solids condensed whey.[83,84,222] Normally, the high acidity of many such wheys causes difficulty in drying; lumps form and clog the dryer. By introducing compressed air into the whey just prior to the spray nozzle, a foam structure is produced that dries rapidly, forming free-flowing particles. These are hygroscopic, of course, and apparently more research is needed to overcome this difficulty. Claimed to be another method for prevention of caking is the addition of calcium silicate to whey, either before or after drying.[43]

The high level of milk salts (over 10% of the solids) limits the whey solids that can be used in many applications. Electrodialysis is being used to remove most of the electrolytes without appreciable loss of diffusible nonionic substances or damage to the whey proteins.[4,10,145,229] Daniel describes a dialysis procedure to remove 80% of the milk salts and half of the lactose from whey or skimmilk, yielding special products for use in ice cream and other foods.[49] Reviews on whey utilization should be consulted for additional details of processing and uses of whey.[3,221]

Manufacture of α-Lactose

Until about 1880, lactose was produced only in Switzerland; the source material was cheese whey. Shortly after that, factories were established in Germany and the United States, and lactose is now produced in other European countries, South America, and New Zealand. Until World War II, a starting material of importance in the United States was hydrochloric acid whey from the manufacture of casein. At present, however, the chief source of lactose is cheese whey.

For many years, only the highly refined U.S.P. grade of lactose was marketed, but now that lactose is used in diverse products, a variety of grades are available. Specifications for some of these grades are shown in Table 6.7. In addition, small quantities of β-lactose are produced for certain uses where its higher solubility or sweetness may be an advantage. Only the basic procedures of manufacture are described herein; further details are in several reviews.[39,222,223] Schematic representation of a process for the manufacture of crude and refined lactose is shown in Fig. 6.14.

Table 6.7

TYPICAL PHYSICAL AND CHEMICAL DATA FOR VARIOUS GRADES OF LACTOSE

Analysis	Fermen-tation	Crude	Edible	U.S.P.	U.S.P. Spray Process
Lactose, %	98.0	98.4	99.0	99.85	99.4
Moisture, nonhydrate %	0.35	0.3	0.5	0.1	0.5
Protein (N × 6.38), %	1.0	0.8	0.1	0.01	0.05
Ash, %	0.45	0.4	0.2	0.03	0.09
Lipids, %	0.2	0.1	0.1	0.001	0.01
Acidity, as lactic, %	0.4	0.4	0.06	0.04	0.03
Heavy metals, as Pb, ppm	*a*	*a*	<2	<1	<2
Specific rotation $[\alpha]^{25}_{D}$	*a*	*a*	+52.4°	+52.4°	+52.4°
Turbidity, ppm	*a*	*a*	<5	<5	<5
Other sugars, mg.	*a*	*a*	15	5	10
Color, ppm	*a*	*a*	10	5	5
Bacterial estimate					
Standard plate count, per gm.	*a*	*a*	<100	<30	<30
Coliforms in 10 mg.	*a*	*a*	Neg.	Neg.	Neg.
Sporeformers in 10 mg.	*a*	*a*	Neg.	Neg.	Neg.
Molds in 10 mg.	*a*	*a*	Neg.	Neg.	Neg.
Yeasts in 10 mg.	*a*	*a*	Neg.	Neg.	Neg.

[a] Not determined routinely for this grade.

Sweet fluid whey contains about 5% lactose, but differs according to the amount fermented during the setting of cheese. There is usually 0.85% protein and 0.75% ash in acid whey, and 1.10% protein and 0.50% ash in rennet whey. The difference in ash is attributable mainly to calcium phosphate, which is present in acid whey in appreciable proportions, but is lacking in rennet whey. Rennet whey contains a small percentage of acid-precipitable protein, which accounts for the greater protein percentage than in acid whey, and adds to the difficulty in isolating the lactose. The degree to which protein and salts are removed from whey prior to concentration and crystallization largely determines the purity of the lactose. Protein and salts present during crystallization may result in contamination of the crystals and may cause additional difficulties. They may greatly increase the viscosity of concentrated whey and thus make separation of the crystallized lactose exceedingly difficult, in extreme instances even preventing its crystallization. The protein and salts are controlled by adjusting pH, by heat treatments, by use of proteolytic enzymes, or by addition of tetraphosphates.

In a commercial process, whey is limed, heated and filtered to remove the proteins and calcium phosphate. The clear whey is then concen-

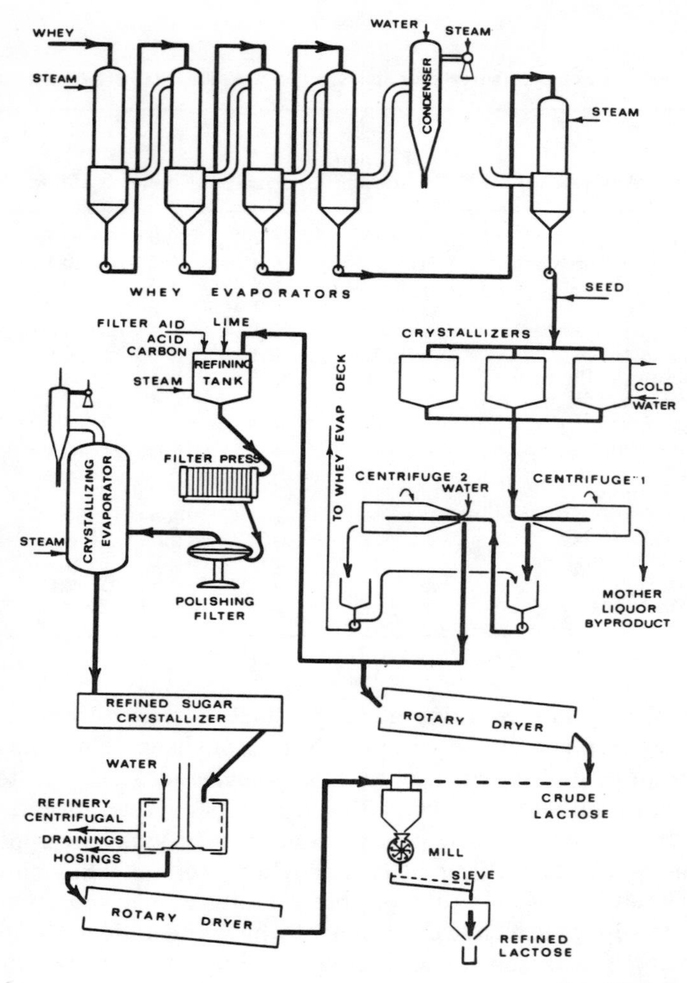

From Wilson[228]

FIG. 6.14. PROCESS FOR THE MANUFACTURE OF CRUDE AND REFINED LACTOSE FROM SWEET WHEY

trated to approximately 30% solids and refiltered to remove any proteins and salts that separate. After further concentration, crystallization is induced. The greater the degree of concentration, the greater the yield, but if the concentration is too great the crystals stick together, making the occluded material difficult to remove. The slurry of syrup and crystals is drawn to perforated basket centrifuges, which revolve at high speed to separate the crystals from the mother liquor. While the centrifuge continues to spin, the crystals are sprayed with clear water to remove the adhering liquor. The crystals may be dissolved for further purification, or they may be dried to form "crude lactose." Refining to produce better grades of lactose consists of redissolving in water, treating with activated carbon to decolorize the solution, filtering, concentrating, and either recrystallizing or spray-drying the solution. Commercial practice makes use of numerous modifications of this basic procedure.

A general recent trend has been to apply the principle of ion exchange to the purification of whey or lactose solutions. Anionic and cationic exchange resins are used to remove impurities from the solution, which can then be condensed and crystallized or spray-dried directly. Weisberg[223] described a number of applications of ion exchange to lactose production.

In another commercial process[5], whey is treated with sodium tetraphosphate to prevent heat coagulation and hydration of the protein, and is then condensed to a total solids content suitable for lactose graining and crystallization. The lactose crystals are then separated from the concentrated whey, the protein and salts remaining in solution.

A process has been described for producing a high-quality crude lactose by methanol extraction of whey or nonfat milk powders.[136] The powder is dispersed in the solvent under controlled conditions of concentration and temperature. The resulting supersaturated lactose solution is stable enough to allow separation of the coagulated proteins before crystallization starts. The purity obtained by a single crystallization is comparable to that of refined commercial lactose, and, in addition, practically all the milk proteins are recovered as a good-quality product. This process has not yet been used commercially.

Manufacture of β-Lactose

Because β-lactose has a much higher initial solubility and is sweeter than α-lactose, there is a demand for a limited amount of it. The processes for the manufacture of β-lactose are based on the fact that β is the stable form precipitated from lactose solutions above 93.5°C.[17]

One process consists of drying a lactose solution as a film on a heated surface above 100°C and removing this film while it is still a paste containing at least 2% moisture.[206] The lactose crystallizes in the β-form, and the heat remaining in the paste completes the drying.

In another process, α-lactose is added to a saturated solution maintained above 93.5°C, and an equivalent quantity of β-lactose is removed.[191] Since equilibrium between the two forms is established rapidly at this temperature, the α dissolves and β crystals form readily. Using a heated centrifuge to recover the crystals yields a sugar that is high in β-lactose.

β-Lactose can also be produced by heating lactose powder in an atmosphere of water vapor at a temperature above the transition point (93.5°C) so that a temporary solution is formed on the surface of the lactose particles.[195] The β-lactose then crystallizes from this minute solution because above 93.5°C it is the more insoluble form. The moisture can be derived from the water of crystallization in α-hydrate, or if another form of lactose powder is used the moisture enabling the reaction to proceed can be supplied from an external source. After the conversion to β-lactose is complete, the surplus moisture is permitted to escape, leaving the dry crystalline β-lactose. During the conversion process, the supplemental moisture is so limited that the lactose always remains in a pulverulent condition, which eliminates the problems normally encountered in recovery and drying.

The old procedure of Verschuur[214] is of academic interest because of the solvent system used for crystallization as a means of preventing contamination with α-lactose. Pyridine, b.p. 115°C, is added to a boiling lactose solution. Since the aqueous phase distills off more rapidly, pyridine becomes the liquid medium in which the β crystals are deposited as the lactose solution is concentrated. The crystals are recovered, washed with boiling pyridine and hot ethyl alcohol, and dried. The method is not suitable for production of a food product, because of the disagreeable odor and moderate toxicity of pyridine.

CHEMICAL REACTIONS

In its chemical behavior, lactose is like a number of similar carbohydrates and reacts according to the general rules of carbohydrate chemistry. Thus the reactions involve such groups as: (a) glycosidic linkage between the two monosaccharides; (b) the reducing group of the glucose; (c) the free hydroxyl groups; and (d) the carbon-carbon bonds. A more complete discussion is available on the chemical properties of lactose and its early chemical history.[44,71]

Hydrolysis

Lactose may be hydrolyzed by the enzyme β-D-galactosidase, also called lactase, and by dilute solutions of strong acids. The immediate products of hydrolysis are glucose and galactose, in equal proportions, but side reactions may alter this picture through the production of oligosaccharides.

Lactose is quite resistant to acid hydrolysis. In fact, organic acids, such as citric, that easily hydrolyze sucrose are unable to hydrolyze lactose under the same conditions. This is useful in analyzing a mixture of these two sugars, because the quantity of sucrose can be measured by the extent of changes in optical rotation of reducing power as a result of mild acid hydrolysis. The speed of hydrolysis of lactose[174] varies with time, temperature, and concentration of reactant, as shown in Table 6.8. During hydrolysis there is a progressive reduction in lactose and the newly formed hexoses. These in turn may recombine through condensation reactions to form oligosaccharides by a process known as "reversion".[141] Acid hydrolysis of lactose in whey is reported to occur much more rapidly than in pure lactose solutions, considerable hydrolysis occurring at 90°C and pH 4.7, for example; such conditions leave pure aqueous solutions of lactose unchanged. Whey protein is suggested as responsible for this catalytic effect.[150]

A recent innovation has been the use of ion-exchange resins in the hydrogen form to hydrolyze lactose solutions at elevated temperatures. This is the same principle as acid hydrolysis, but the product formed is claimed to be superior.[20]

β-D-Galactosidase (lactase) activity is widely distributed in nature, having been found in many tissues of higher animals, in yeasts, in bacteria, in birds, and in plants. General reviews are available on the occurrence and properties of β-galactosidases.[173,215] Current interest in the enzyme is discussed further in the section beginning on page 314. The presence of galactose inhibits the enzymatic hydrolysis of lactose, but glucose does not. Increasing the enzyme concentration or the temperature increases the rate of hydrolysis. The reaction, however, is not limited to a simple hydrolysis with the production of glucose

Table 6.8

HYDROLYSIS OF LACTOSE BY ACID

Lactose in Solution	HCl/ 1000 gm Lactose Solution	Heating Conditions		Reaction after Heating,	Lactose Hydro- lyzed,	Velocity Constant (K^a ×	Calcu- lated Time to Invert 99.5%,
		Temp.,	Time,				
%	Mole	°C.	Min.	pH	%	10^4)	Min.
33.6	0.034	130	36.0	1.23	82.0	476	111.3
29.0	0.023	130	58.8	1.46	79.7	271	195.4
28.4	0.023	140	30.0	1.47	84.5	622	85.3
23.2	0.019	165	8.2	1.60	79.0	1904	27.8

[a] $K = 1/t \times 2.303 \times \log a/(a-x)$ where x represents the amount of hydrolysis attained in time t, and a the initial concentration of lactose. Data from Ramsdell and Webb.[174]

and galactose, but is accompanied by the immediate formation of various oligosaccharides.[9] These remain unchanged for at least 12 hr after all the lactose has been hydrolyzed. Only traces of oligosaccharides are formed in dilute lactose solutions, but large percentages are formed in higher (up to 35%) concentrations.[180]

The formation of oligosaccharides during the enzymatic hydrolysis of other disaccharides also is known. The products of hydrolysis may be transferred to a suitable acceptor, which may be another disaccharide molecule or to another hydrolyzed monosaccharide. Oligosaccharides also form during acid hydrolysis of lactose, but some appear to differ from those produced enzymatically, possibly in the α or β form of the sugar or in the glycosidic linkages.[9]

At least ten oligosaccharides have been detected during β-D-galactosidase hydrolysis of lactose.[179] Three of these have been identified[167] as 3-0-β-D-galactopyranosyl-D-glucose, 6-0-β-D-galactopyranosyl-D-glucose, and 6-0-β-D-galactopyranosyl-D-galactose. Galactose is primarily involved in the formation of the oligosaccharides, which accounts for the lower concentration of free galactose than of free glucose during hydrolysis. Similar oligosaccharides are found in the cecal contents of rats fed a high-lactose diet.[179] Their formation *in vivo* suggests they are of physiological importance, or that they may be a means of removing excess free galactose from the system.

The use of hydrolyzing enzymes has the important advantage of reducing the lactose content of milk or milk products without affecting the proteins, which is impossible, of course, with acid A number of patents describe processes using enzymes to hydrolyze lactose to improve product stability, prevent lactose crystallization, or decrease lactose content for feed purposes.[200-202]

Hydrolysis of a portion of the lactose has been shown to be an effective method of retarding lactose crystallization and consequently of improving the stability of frozen concentrated milk[115,211] (also see Chapter 14). One part lactase from a specially cultured *Saccharomyces fragilis* yeast strain added to 40 parts lactose in condensed skimmilk yields 75% hydrolysis, usually in 4 hr at 38-43°C. When this is used to standardize untreated condensed whole milk, the useful storage life of the frozen product is greatly increased: by about double if 10% of the lactose is hydrolyzed, or a fivefold increase in storage life if 30% of the lactose is hydrolyzed.

Pyrolysis

There is considerable disagreement in the literature as to the temperature to which α-hydrate can safely be heated without losing its molecule of water (see forms of anhydrous α-lactose, p. 284). Vapor-pressure measurements indicate that water may be lost at 85°C (14.0 mm Hg) and even at 80°C (10.4 mm Hg) under very low humidity.[40] Specifications of U.S.P. XIV require 2 hr drying at 80°C in sampling α-hydrate. After several days at 100°C the water of crystallization is almost entirely lost from the powder. At 130°C the water is lost quickly, giving rise to an anhydrous powder.[142] At 150 to 165°C lactose becomes yellow, and at 175°C it becomes brown, emits a characteristic odor, and loses about 13% of its original weight. The anhydro sugars or glycosans are produced by elimination of water from lactose, glucose, and galactose as a result of heating under vacuum or destructive distillation.

The most important heat-induced changes in dairy products that involve lactose are the changes associated with browning. An extensive review is available of the browning of milk and associated changes.[165] Other pertinent reviews discuss the Maillard reaction[58] and the Amadori rearrangement.[99] The Maillard-type browning, sugar-amino type, is the most prevalent, since it requires a relatively low energy of activation and is autocatalytic. Direct caramelization, on the other hand, has a rather high energy of activation and therefore is of lesser importance. Nevertheless, some of the carbon dioxide formed by sterilization of milk can be traced to caramelization of lactose.[55]

Lactose and casein are the two principal reactants in the browning of dairy products, but whey proteins are involved in some circumstances.[166] A number of components have been shown to influence the reaction in model systems, if not in milk, including phosphates, oxygen, metals, and such compounds as acetaldehyde, 5-hydroxymethylfurfural, and methyl glyoxal.

In addition to accounting for the major part of browning, the protein-carbohydrate complex or its decomposition products also result in the production of reducing substances, fluorescent substances, and disagreeable flavor materials. Indeed, many products are possible (Fig. 6.15). For example, 40 compounds were isolated and identified from a model system of casein and lactose that had been stored at 80°C and 75% R.H. for 8 days to accelerate browning. On the basis of gas chromatographic, infrared, and mass spectroscopic data in comparison with authentic samples, 13 furanics, 9 lactones, 5 pyrozines, 2 pyridines, 2-acetyl pyrrole, 2 amides, pyrrolidinone, succinamide, glutarimide, two carboxylic acids, acetone, 2-heptanone, and maltol were identified in the brown mixture, as well as D-galactose, D-tagatose, and lactulose.[62] Nearly 40 additional compounds were found in a later study with more sensitive techniques.[61]

A number of compounds have been shown to inhibit the browning

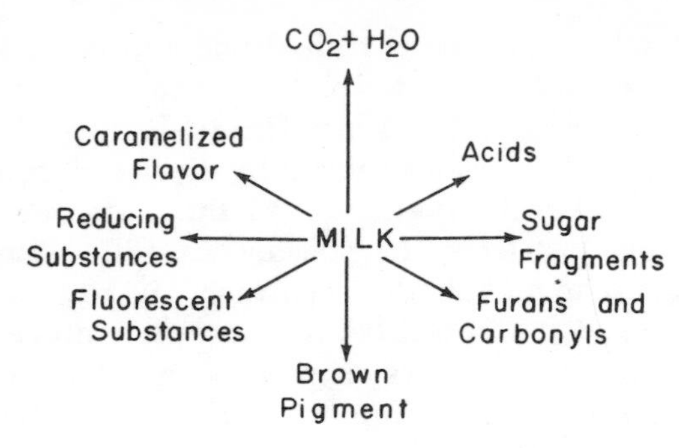

From Patton[165]

FIG. 6.15. A SCHEME OF THE MORE SIGNIFICANT MANIFESTATIONS OF BROWNING IN MILK

reaction. In milk products, active sulfhydryl groups serve as natural inhibitors in retarding heat-induced browning, but the mechanism is not understood. Sodium bisulfite, sulfur dioxide, and formaldehyde also inhibit browning in milk systems as well as in simpler amino acid-sugar solutions. In actual practice, browning is controlled in dairy products by limiting heat treatments, moisture content, and time and temperature of storage.

Browning has a detrimental effect on the nutritive value of milk products through interaction of the protein and carbohydrate and the resulting rearrangement products. Destruction of essential amino acids, particularly lysine and probably histidine, has been shown to occur during the storage and browning of nonfat dry milk of high (7.6%) moisture content.[90] Similar powders of low (3.0%) moisture did not deteriorate in nutritive value during storage. Reduction of β-lactoglobulin with lactose in the "dry" (10% moisture) state resulted in various degrees of lysine destruction, depending upon temperature and heating times. Neither arginine, histidine, nor the acidic and neutral amino acids were damaged by the thermal treatments (0 to 90°C) in the presence of lactose.[70]

The reaction of sugar with protein becomes irreversible. For example, in a model system, after glucose reacted with casein none of the glucose could be oxidized enzymatically with glucose oxidase, nor could any glucose be regenerated by dilute acid or alkali hydrolysis.[130]

Oxidation

The products of oxidation of lactose, as with other sugars, vary with the oxidizing agent used and the conditions under which the reaction is carried out. Lactose does not react in a unique manner, but seems to follow the general oxidative reactions of sugars.

The carbonyl group may be oxidized selectively to a carboxyl group to form its aldobionic acid, lactobionic acid. This important oxidation reaction occurs, for example, when the reducing power of lactose is measured, i.e., reduction of alkaline copper solutions such as with Fehling's or Benedict's solution. Oxidation with bromine in the presence of a suitable buffer yields salts of lactobionic acid. The electrolytic oxidation of lactose with bromine in the presence of sodium bicarbonate (buffering agent) is reported to improve the yield and purity of lactobionate.[190] Without buffer, the hydrobromic acid formed causes hydrolysis with subsequent oxidation, yielding D-gluconic and D-galactonic acids. Likewise, hypoiodite may be used to produce lactobionic acid.

Certain aerobic organisms, notably of the genus *Pseudomonas* but also algae and yeasts, are capable of oxidizing lactose to lactobionic acid without hydrolysis to monosaccharides.[203] Lactose dehydrogenase stoichiometrically oxidizes lactose to lactobionic-δ-lactone in the presence of an appropriate hydrogen acceptor. The lactobionic-δ-lactone is subsequently hydrolyzed to lactobionic acid by another enzyme, lactonase.[158]

Treating with nitric acid oxidizes the D-glucose and D-galactose portions to their respective dicarboxylic acids, D-glucaric (saccharic) and D-galactaric (mucic) acids. If the acid is sufficiently concentrated or hot, it may cause further oxidation to tartaric, racemic, oxalic, and carbonic acids. Complete oxidation to carbon dioxide and water can be accomplished in alkaline solution with potassium permanganate, or with such catalysts as cerous hydroxide, ferrous sulfate, or sodium sulfite. Biological oxidation is also capable of degrading lactose to the end products carbon dioxide and water.

Reduction

The primary effect of hydrogenation (reduction) of an aldose sugar is conversion of the terminal aldehyde group to an alcohol group. Unless preventive measures are taken, hydrolytic and other degradative changes follow, resulting in the formation of simpler alcohols and related substances. Thus, using sodium amalgam as the reducing agent gives a mixture containing dulcitol, sodium lactate, isopropanol, ethanol, and hexanol.[25] Calcium amalgam, used in an atmosphere of carbon dioxide to prevent secondary reactions, gives the crystalline lactose alcohol, lactitol (4-0-β-D-galactopyranosyl-D-glucitol).[149] It is possible to produce lactitol by electrolytic reduction of an acidified aqueous solution of lactose, a procedure similar to that used to produce sorbitol from glucose.[82,88]

NUTRITIONAL AND PHYSIOLOGICAL EFFECTS OF LACTOSE

Milk and its derivatives contribute lactose to the diet, the only major carbohydrate of animal origin in man's food supply. Lactose differs in some respects from other sugars in physiological behavior. Some differences are attributable to the lactose molecule itself but others are due to the galactose moiety liberated on hydrolysis.[54]

In the digestive tract, lactose may be fermented by bacteria; in the intestine it may be absorbed directly or hydrolyzed by β-D-galactosidase (lactase) and its components adsorbed. β-D-galactosidase is an introcellular enzyme[35] which in man is found within the cells of the intestinal mucous membrane.[48,218] Hydrolysis, therefore, occurs during transport through the intestinal wall.

A great deal of interest has recently centered on the enzyme lactase not only because it brings about certain diarrheal syndromes in infants, but also because of its widespread deficiency in non-Caucasian adults, who normally consume little or no milk after weaning.[8] Lactose intolerance, causing abdominal cramps, gaseous distention, or diarrhea in severe cases, is attributed to a deficiency of lactase in the intestinal mucosa. Lactase deficiency is detected by a biopsy of the mucosa or by feeding lactose to the subject after a period of fasting and measuring the rate of increase in blood-sugar level.[15] Current studies of adults indicate lactase deficiency in many groups; for example, it is probably present in at least 8% of U.S. whites, 70% of U.S. Negroes, 95% of African Bantus, and over 90% of Formosans, Filipinos, Japanese, Thais, and South American (Colombian) Indians.[14,23,50,64,76]

The question has arisen whether this condition is inherited or acquired. In most mammals, lactase declines to low levels, or is entirely absent after weaning. It would seem entirely feasible that it could be an acquired characteristic in peoples who customarily do not drink milk. Infants of the populations discussed above can be nursed or fed milk, so they are not congenitally lactase-deficient, but studies with Thais and Ugandan Bantu children indicate that intolerance becomes increasingly prevalent with age regardless of milk ingestion, at least in the small samples tested.

A review of medical research concerning lactose intolerance has led to the conclusion that ethnic differences are largely genetic in origin.[197] A culture historical hypothesis has recently been offered to explain the present-day occurrence of various Old World groups with high and low incidences of lactose intolerance based on milk use.[198]

Conversely, the adaptive nature of enzyme systems to substrate is well known. The adaptation of intestinal lactase to lactose in the diet has been demonstrated in the rat, although it required considerable time.[24] In human beings experimental evidence suggests that withdrawal of lactose from the diet for 2 to 5 yr is required before lactase deficiency occurs.[22] All attempts to reestablish the activity in lactase-deficient subjects by feeding either milk or lactose have been unsuccessful.[22] It therefore appears that continual consumption of milk in the diet is essential if elevation of lactase activity is to be maintained, but this has not been convincingly demonstrated. Lack of correlation between lactose intake and lactose tolerance in several population studies also casts doubt on the adaptive nature of lactase in man.[73]

The significance of this subject is obvious because of its implications on the suitability of milk as a food after weaning in countries where there is a high incidence of lactase deficiency in the population. Even so, aid programs which use milk products to improve the diet have been welcome in such areas because of the improved health that results. Apparently these programs have not encountered ill effects that can be attributed to lactase deficiency. However, the controversy stimulated the Protein Advisory Group of the United Nations to emphasize in a recent report that low lactase activity or reaction to high lactose under experimental conditions is not synonymous with milk-intolerance. The report continues that it would be highly inappropriate, on present evidence, to discourage programs aimed at improving milk supplies and increasing milk consumption among children for fear of milk-intolerance.

Several long-term studies are presently in progress which should resolve the etiology of lactase deficiency in segments of the world's population by clarifying the genetic and environmental factors.[22,73] Those interested in further details should refer to the various reviews available.[65,182,197]

Lactose may also be transferred to the blood or urine (lactosuria) without being hydrolyzed, particularly after consumption of a large quantity of lactose or during lactation.[67] Under normal milking conditions only a small portion of the lactose produced in the mammary gland escapes to the blood plasma. When milking is suspended or the milking interval is prolonged, however, appreciable lactose is lost to the blood[125] and, in turn, is excreted mainly in the urine.[100,224] Most, though not all, of the lactose given intravenously is eliminated through the urine, only a small percentage being metabolized.[37] The liver also shows considerable lactase activity.[51]

Since lactose is absorbed slowly, a portion usually reaches the ileum, where it is utilized by bacterial flora, with the production of acid. A detailed discussion of lactic acid production is given in Chapter 13. Lactose inhibits putrefaction by promoting the growth of aciduric bacteria in the intestine.[107] This reduces the intestinal pH, which was formerly given as the explanation for the increased calcium absorption.[120] It has long been recognized that inclusion of lactose in a diet improves the utilization of calcium and other minerals, though the mechanism by which this is accomplished remains obscure.[38] It has been suggested that lactose makes for more efficient utilization of the absorbed calcium or that the lactose effect is a nonspecific membrane phenomenon, resulting in increased permeability to alkaline-earth cation.[132,144,219] Magnesium, barium, strontium, radium, zinc, and phosphorus, in addition to calcium, are absorbed more efficiently in the presence of lactose.[69,131] Cholesterol absorption and catabolism also seem to be influenced by lactose.[111]

Lactose is considered by many to be the preferred carbohydrate for modifying cow's milk and formulating infant foods, which is probably its greatest single use. The nutritional significance of variations in human and cow's milk compositions has been reviewed.[68] Lactose appears to give rise to a characteristic body composition, with less fat and firmer tissues.[72,112] Rat-feeding studies have confirmed that lactose in the diet increases the activity of the disaccharidases in the intestinal mucosa.[116] The effect has been attributed to the presence of galactose. By promoting a more desirable flora in the lower digestive tract, it is effective in combating gastrointestinal disturbances caused by putrefactive bacteria as well as in promoting synthesis of the B-vitamins for absorption by the host.[38,81] It may also be used in therapy against constipation and diarrhea. It may be used to advantage in both animal and human feeding, though its nutritive value varies with the experimental animal used.[11] Excessive lactose in the diet is undesirable, however, since it results in diarrhea and poor growth.[63] Diarrhea in lactase-deficient subjects results, in large part, from combination of net fluid secretion by the small intestine in response to the osmotic load caused by the presence of unmetabolized lactose, plus interference with subsequent fluid adsorption. Products of bacterial fermentation of lactose in the lower intestinal tract may play a role in the decreased colonic adsorption.[42] In rats, excess lactose results in poor growth and in cataracts, due primarily to their inability to metabolize large percentages of galactose.[51,102,178] Other components in high-lactose-containing diets affect the time of appearance and the severity of the cataracts. Excellent reviews of the nutritional and physiological effects of lactose should be consulted for more details and references.[11,54]

USES

As mentioned previously, lactose has unique physicochemical properties that make it useful in a number of products.[7,175] In addition to its extensive use in baby foods, it is listed as an optional ingredient in many food standards, but for economic reasons it has not been used widely by the food industry. However, with the present concern about waste disposal of whey, which is primarily a problem of disposal of lactose, the major component of whey solids (67 to 73%), there is likely to be increased research effort on the technological and nutritional properties of this interesting carbohydrate.[3]

Being less sweet than other commercial sugars makes lactose useful in processing many foods. It may be added to increase osmotic pressure or viscosity or improve texture without making the product too sweet. It is added in the manufacture of beer in some instances because it is not fermented by the yeast and remains in the product to improve flavor and contribute to viscosity and mouth-feel. It has similar uses in other beverages and low-calorie foods.[1] Toppings, icings, and various types of pie fillings are examples of uses where its inclusion in the formulation can improve quality.

Lactose is a major contributor to the acceptability of milk as a beverage, and variations of 0.33% lactose are readily detected by taste test.[161] Lactose may not be present at optimum levels, however, so that supplementation of milk products such as buttermilk or chocolate drink effectively improves acceptance, apparent richness, and smoothness. Consequently, lactose is included as an optional ingredient in standards of identity of such foods.

The candy industry uses lactose to achieve desirable characteristics in certain types of candies. It changes the crystallization habits of other sugars present and improves body, texture, chewiness, or shelf-life.

Lactose excels in absorbing flavors, aromas, and coloring materials. As a result it has found application as a carrier for flavorings or volatile aromas. It is used to trap such materials during their preparation, or in filters to remove undesirable volatiles. The anhydrous forms of lactose have recently been shown to have greater adsorption capacity for certain odors than do other sugars or other forms of lactose.[155] Lactose anhydride may be mixed with ethanol to produce a powder which dissolves in water, yielding ethanol and lactose.[184] To retard flavor losses, lactose may be added to various foods during processing. Likewise, it is used to carry fragrances where a gradual release of odor is desired over a period of time, such as in sachet wafers, or as a carrier for seasonings. For example, wine can be incorporated into cake mixes by absorbing the wine on an anhydrous lactose.[185] It is used in conjunction with saccharine or cyclamate to produce a sweetening agent.[57] It gives a better color to some foods. In other cases it is used as a carrier for colors because it dissolves slowly, releasing the color for uniform dispersion. It is used in flavoring mixtures.[46,204]

Lactose in the glass state may be used as a protective coating on certain materials, either to seal in components or to protect the material from the environment.[207] Materials may be coated with lactose solution and dried, or a solution containing the material and lactose can be spray-dried.[183] This latter application has been used to preserve enzymatic activity during spray drying and storage.[16]

The pharmaceutical industry has used lactose for many years because of its properties that aid flow characteristics and tablet or pill formation. The drug is distributed uniformly in lactose powder, which is easily molded or compressed into tablets. These have good dispersing characteristics, similar in some respects to the properties of "instantized" products. Other tablets are given a shell by first moistening the surface of the tablets with a small amount of coating syrup, and then tumbling them in lactose powder. This process is repeated for as many coats as desired. Such a coating procedure should be very useful in producing certain food products where the coating could seal in the contents, but the product would be easily handled and readily dispersible. Although the potential of this type of application seems very promising, it has not been given the consideration or study it deserves.

Another expanding area where lactose is useful is in "instantizing" or increasing the dispersibility of certain foods. Products are prepared containing 15 to 50% lactose, spray-dried, and then "instantized" by moistening and redrying. This allows some of the lactose to crystallize; the particles then agglomerate, becoming free-flowing and capable of dispersing rapidly, as described under Instant Dry Milk (p. 303). Such materials are well suited for items to be dispensed from vending machines, or in convenience foods for easy preparation. It would appear that this method could be used to prepare in more usable form ingredients that normally disperse with difficulty. Lactose powder is also used as a dispensing aid in powdered products, because it is nonhydroscopic and thus helps to maintain the product in a free-flowing condition.

The desirable properties of lactose, that are important to the baking industry have been reviewed recently.[45,77] Being a reducing sugar, it readily reacts with proteins by the Maillard reaction to form the highly flavored golden-brown materials commonly found in the crusts of baked goods. Caramelization by heat during baking also contributes flavor and color. Lactose is not fermented by bakers' yeast, so its functional properties are effective throughout baking and during storage. Its emulsifying properties aid in creaming and promote greater efficiency from shortening. Thus, lactose can contribute to the flavor, texture, appearance, shelf-life, and toasting qualities of baked goods.

A variety of other uses for lactose have been reported[39,175,221,223] but the examples cited illustrate the functional properties of lactose and how they can be applied.

REFERENCES

1. Abbott Laboratories. British Patent 1,148,395 (1969). Cited in Food Sci. and Tech. Abs. *1,* 9T275 (1969).
2. ADACHI, S., and PATTON, S., J. Dairy Sci. *44,* 1375 (1961).
3. Agric. Research. Ser. USDA Publ. 3340 (1970).
4. AL, J., and WIECHERS, S. G., Research (London) *5,* 256 (1952).
5. ALMY, E. F., and HULL, M. E., U.S. Patent 2,467,453 (1949).
6. ANDERSON, A. G., and STEGEMAN, G., J. Am. Chem. Soc. *63,* 2119 (1941).
7. Anon. Oil, Paint, and Drug Reporter *181,* 36 (1962).
8. Anon. Nutrition Rev. *28,* 138 (1970).
9. ARONSON, M., Arch. Biochem. Biophys. *39,* 370 (1952).
10. ATEN, A. H. W., WEGELIN, E., and WIECHERS, S. G., 12th Internat. Dairy Congr., Proc. *3,* 381 (1949).
11. ATKINSON, R. L., KRATZER, F. H., and STEWART, G. F., J. Dairy Sci. *40,* 1114 (1957).
12. BAUCKE, A. G., and SANDERSON, W. B., 18th Internat. Dairy Congr., Proc. *1E,* 256 (1970).
13. BAXTER, C. F., KLEIBER, M., and BLACK, A. L., Biochim Biophys. Acta *21,* 277 (1956).
14. BAYLESS, T. M., HUANG, S. S., and FERRY, G. D., Am. J. Clin. Nutr. *23,* 659 (1970).
15. BAYLESS, T. M., ROSENWEIG, N. S., CHRISTOPHER, N., and HUANG, S. S., Gastroenterology *54,* 475 (1968).
16. BEEK, M. J. VAN DE, and GERLSMA, S. Y., Neth. Milk and Dairy J. *23,* 46 (1969).
17. BELL, R. W., Ind. Eng. Chem. *22,* 51 (1930).
18. BERGMAN, H. C., and HOUSLEY, C., Comp. Biochem. Physiol. *25,* 213 (1968).
19. BERLIN, E., ANDERSON, B. A., and PALLANSCH, M. J., J. Dairy Sci. *51,* 1912 (1968).
20. BLOCK, R. J., U.S. Patent 2,592,509 (1952).
21. BOCKIAN, A. H., STEWART, G. F., and TAPPEL, A. L., Food Research *22,* 69 (1957).
22. BOLIN, T. D., Gastroenterology *60,* 347 (1971).
23. BOLIN, T. D., DAVIS, A. E., SEAH, C. S., CHUA, K. L., YONG, V., KHO, K. M., SIAK, C. L., and JACOB, E., Gastroenterology *59,* 76 (1970).
24. BOLIN, T. D., PIROLA, R. C., and DAVIS, A. E., Gastroenterology *57,* 406 (1969).
25. BOUCHARDAT, G., Ann. chim. phys. [4] *27,* 68 (1872).
26. BREW, K., Nature *222,* 671 (1969).
27. BREW, K., VANAMAN, T. C., and HILL, R. L., Proc. Natl. Acad. Sci. *59,* 491 (1968).
28. BRODBECK, U., DENTON, W. L., TANAHASHI, N., and EBNER, K. E., J. Biol. Chem. *242,* 1391 (1967).
29. BRODBECK, U., and EBNER, K. E., J. Biol. Chem. *241,* 762 (1966).
30. BRODBECK, U., and EBNER, K. E., J. Biol. Chem. *241,* 5526 (1966).
31. BUMA, T. J., Neth. Milk Dairy J. *20,* 91 (1966).
32. BUMA, T. J., and MEERSTRA, J., Neth. Milk Dairy J. *23,* 124 (1969).
33. BUMA, T. J., and WIEGERS, G. A., Neth. Milk Dairy J. *21,* 208 (1967).
34. BUYZE, H. G., Neth. Milk Dairy J. *6,* 218 (1952).
35. CAJORI, F. A., J. Biol. Chem. *109,* 159 (1935).
36. CAMERON, A. T., Sugar Research Foundation, Inc. Scientific Report Series No. 9 (1947).

37. CARLETON, F. J., MISLER, S., and ROBERTS, H. R., J. Biol. Chem. *214*, 427 (1955).
38. CHANEY, M. S., and ROSS, M. L., "Nutrition," 7th Ed., Houghton Mifflin Co., Boston, 1966.
39. CHOI, R. P., DOAN, F. J., HULL, M. E., and CALL, A. O., J. Dairy Sci. *41*, 319 (1958).
40. CHOI, R. P., TATTER, C. W., O'MALLEY, C. M., and FAIRBANKS, B. W., J. Dairy Sci. *32*, 391 (1949).
41. CHOI, R. P., TATTER, C. W., and O'MALLEY, C. M., Anal. Chem. *23*, 888 (1951).
42. CHRISTOPHER, N. L., and BAYLESS, T. M., Gastroenterology *60*, 845 (1971).
43. CIPOLLA, R. H., DAVIS, D. W., and VANDER LINDEN, C. R., U.S. Patent 2,995,447 (1961).
44. CLAMP, J. R., HOUGH, L., HICKSON, J. L., and WHISTLER, R. L., Advances in Carbohydrate Chem. *16*, 159 (1961).
45. CRAIG, T. W., and COLMEY, J. C., Bakers Digest *45*(1), 36 (1971).
46. Cumberland Packing Co. British Patent 1,119,490 (1968). Cited in Food Sci. and Tech. Abs. *1*, 1T24 (1969).
47. DAHLBERG, A. C., and PENCZEK, E. S., N.Y. Agric. Expt. Sta. Tech. Bull. 258 (1941).
48. DAHLQVIST, A., and BORGSTRÖM, B., Biochem. J. *81*, 411 (1961).
49. DANIEL, F. K., U.S. Patent 2,437,080 (1948).
50. DAVIS, A. E., and BOLIN, T. D., Nature *216*, 1244 (1967).
51. DE GROOT, A. P., and HOOGENDOORN, P., Neth. Milk Dairy J. *11*, 290 (1957).
52. DESAI, I. D., NICKERSON, T. A., and JENNINGS, W. G., J. Dairy Sci. *44*, 215 (1961).
53. DOMOVS, K. B., and FREUND, E. H., J. Dairy Sci. *43*, 1216 (1960).
54. DUNCAN, D. L., Nutrition Abs. and Rev. *25*, 309 (1955).
55. DUTRA, R. C., TARASSUK, N. P., and KLEIBER, M., J. Dairy Sci. *41*, 1017 (1958).
56. EBNER, K. E., and BRODBECK, U., J. Dairy Sci. *51*, 317 (1968).
57. EISENSTADT, B., Canadian Patent 824,100 (1969). Cited in Food Sci. and Tech. Abs. *2*, 3T75 (1970).
58. ELLIS, G. P., Advances in Carbohydrate Chem. *14*, 63 (1959).
59. EVENHUIS, N., and DE VRIES, Th. R., Neth. Milk Dairy J. *11*, 184 (1957).
60. EVERETT, S. M., and PEERS, A. E., British Patent 546,447 (1942). Cited in DSA, *5*, 12 (1943-44).
61. FERRETTI, A., and FLANAGAN, V. P., J. Agr. Food Chem. *19*, 245 (1971).
62. FERRETTI, A., FLANAGAN, V. P., and RUTH, J. M., J. Agr. Food Chem. *18*, 13 (1970).
63. FISCHER, J. E., and SUTTON, T. S., J. Dairy Sci. *32*, 139 (1949).
64. FLATZ, G., SAENGUDOM, C., and SANGUANBHOKHAI, T., Nature *221*, 758 (1969).
65. FLOCH, M. H., Am. J. Clin. Nutr. *22*, 327 (1969).
66. FLYNN, F. V., and HARPER, C., Lancet *265*, 698 (1953).
67. FOLIN, O., and BERGLUND, H., J. Biol. Chem. *51*, 213 (1922).
68. FOMON, S. J., CLEMENT, D. H., FORBES, G. B., FRASER, D., HANSEN, A. E., LOWE, C. U., MAY, C. D., SMITH, C. A., and SMITH, N. J., Pediatrics *26*, 1039 (1960).
69. FOURNIER, P., and DIGUAD, A., C. r. hebd. Séanc. Acad. Sci., Sér. D, Paris *269* (20), 2001 (1969). Cited in DSA. *32*, 1623 (1970).
70. FREIMUTH, U., and TRÜBSBACK, A., Nahrung *13*, 199 (1969).
71. FUNCK, E., Milchwiss. *3*, 152 (1948).
72. GERSTLEY, J. R., COHN, D. J., and LAWRENCE, G., J. Pediat. *27*, 521 (1945).
73. GILAT, T., Gastroenterology *60*, 346 (1971).
74. GOTTSCHALK, A., "Glycoproteins, Their Composition, Structure and Function," Elsevier Publ. Co., Amsterdam, 1966.

75. GRANT, G. A., Biochem. J. *29,* 1905 (1935).
76. GUDMAND-HOYER, E., and JARNUM, S., Acta med. scand. *186*(3), 235 (1969). Cited in DSA. *32*, 1658 (1970).
77. GUY, E. J., Bakers Digest *45*(2), 34 (1971).
78. GYÖRGY, P., Pediatrics *11*, 98 (1953).
79. HAASE, G., and NICKERSON, T. A., J. Dairy Sci. *49,* 127 (1966).
80. HAASE, G., and NICKERSON, T. A., J. Dairy Sci. *49,* 757 (1966).
81. HAENEL, H., RUTTLOFF, H., and ACKERMANN, H., Biochem. Z. *331,* 209 (1959).
82. HALES, R. A., U.S. Patent 2,300,218 (1942).
83. HALL, C. W., and HEDRICK, T. I., "Drying Milk and Milk Products," 2nd Edition. Avi Publishing Co., Westport, Conn., 1971.
84. HANRAHAN, F. P., and WEBB, B. H., Food Eng. *33*(8), 37 (1961).
85. HANSEN, R. G., WOOD, H. G., PEETERS, G. J., JACOBSON, B., and WILKEN, J., J. Biol. Chem. *237,* 1034 (1962).
86. HASKINS, W. T., HANN, R. M., and HUDSON, C. S., J. Am. Chem. Soc. *64,* 1852 (1942).
87. HAUSER, E. A., and HERING, H., Le Lait *4,* 388 (1924).
88. HEFTI, H. R., and KOLB, W., U.S. Patent 2,507,973 (1950).
89. HEINRICH, C., Milchwiss. *25,* 387 (1970).
90. HENRY, K. M., KON, S. K., LEA, C. H., and WHITE, J. C. D., J. Dairy Res. *15,* 292 (1948).
91. HERRINGTON, B. L., J. Dairy Sci. *17,* 501 (1934).
92. HERRINGTON, B. L., J. Dairy Sci. *17,* 533 (1934).
93. HERRINGTON, B. L., J. Dairy Sci. *17,* 595 (1934).
94. HERRINGTON, B. L., J. Dairy Sci. *17,* 659 (1934).
95. HERRINGTON, B. L., J. Dairy Sci. *17,* 701 (1934).
96. HERRINGTON, B. L., J. Dairy Sci. *17,* 805 (1934).
97. HERRINGTON, B. L., "Milk and Milk Processing," McGraw-Hill Book Co., Inc., New York, 1948.
98. HOCKETT, R. C., and HUDSON, C. S., J. Am. Chem. Soc. *53,* 4455 (1931).
99. HODGE, J. E., Advances in Carbohydrate Chem. *10,* 169 (1955).
100. HOGAN, A. G., J. Biol. Chem. *18,* 485 (1914).
101. HOMBERGER, R. E., and COLE, W. C., Ice Cream Rev. *17*(4), 29 (1933).
102. HÖRMANN, E., Albrecht von Graefe's Arch. Ophthalmol. *154,* 561 (1954).
103. HORST, M. G. ter, Rec. trav. chim. *72,* 878 (1953).
104. HUDSON, C. S., J. Am. Chem. Soc. *26,* 1065 (1904).
105. HUDSON, C. S., J. Am. Chem. Soc. *30,* 1767 (1908).
106. HUDSON, C. S., and BROWN, F. C., J. Am. Chem. Soc. *30,* 960 (1908).
107. HULL, T. G., and RETTGER, L., J. Bacteriol. *2,* 47 (1917).
108. HUNZIKER, O. F., "Condensed Milk and Milk Powder," 4th Ed., Pub. by Author, La Grange, Ill., 1926.
109. HUNZIKER, O. F., "Condensed Milk and Milk Products," 6th Ed., Pub. by Author, La Grange, Ill., 1946.
110. HUNZIKER, O. F., and NISSEN, B. H., J. Dairy Sci. *10,* 139 (1927).
111. IRITANI, N., and TAKEUCHI, N., J. Atheroscler. Res. *10,* 207 (1969).
112. JARVIS, B. W., Am. J. Diseases Children *40,* 993 (1930).
113. JENNESS, R., REGEHR, E. A., and SLOAN, R. E., Comp. Biochem. Physiol. *13,* 339 (1964).
114. JENNESS, R., and SLOAN, R. E., DSA. *32,* p. 599 (1970).
115. JOHNSON, A. H., and TUMERMAN, L., 17th Internat. Dairy Congr., Proc. *B,* 1057 (1962).
116. JONES, D. P., and SKROMAK, E., Am. J. Clin. Nutr. *23,* 665 (1970).
117. JONES, E. A., J. Dairy Res. *36,* 145 (1969).
118. JONES, J. M., and McLACHLAN, T., Analyst *52,* 383 (1927).
119. KERKKONEN, H. K., KÄRKKÄINEN, V. J., and ANTILA, M., Finnish J. Dairy Sci. *24,* 61 (1963).

120. KESSEL, J. F., Proc. Soc. Exptl. Biol. Med. *27,* 113 (1929).
121. KLEIBER, M., BLACK, A. L., BROWN, M. A., BAXTER, C. F., LUICK, J. R., and STADTMAN, F. H., Biochim. Biophys. Acta *17,* 252 (1955).
122. KNOOP, E., and SAMHAMMER, E., Milchwiss. *17,* 129 (1962).
123. KREVELD, A. VAN, Neth. Milk Dairy J. *23,* 258 (1969).
124. KREVELD, A. VAN, and MICHAELS, A. S., J. Dairy Sci. *48,* 259 (1965).
125. KUHN, N. J., and LINZELL, J. L., J. Dairy Res. *37,* 203 (1970).
126. KUHN, R., Angew. Chem. *69,* 23 (1957).
127. KUHN, R., GAUHE, A., and BAER, H. H., Chem. Ber. *86,* 827 (1953).
128. KUHN, R., and LÖW, I., Chem. Ber. *82,* 479 (1949).
129. KYLE, R. C., and HENDERSON, R. J., U.S. Patent 3,511,226 (1970).
130. LEA, C. H., and HANNAN, R. S., Biochim. Biophys. Acta *4,* 518 (1950).
131. LENGEMANN, F. W., J. Nutr. *69,* 23 (1959).
132. LENGEMANN, F. W., WASSERMAN, R. H., and COMAR, C. L., J. Nutr. *68,* 443 (1959).
133. LEVITON, A., U.S. Patent 2,116,931 (1938).
134. LEVITON, A., Ind. Eng. Chem. *35,* 589 (1943).
135. LEVITON, A., Ind. Eng. Chem. *36,* 744 (1944).
136. LEVITON, A., Ind. Eng. Chem. *41,* 1351 (1949).
137. LEVITON, A., and LEIGHTON, A., Ind. Eng. Chem. *30,* 1305 (1938).
138. LEY, J. M., and JENNESS, R., Arch. Biochem. Biophys. *138,* 464 (1970).
139. LIM, S. G., and NICKERSON, T. A., J. Dairy Sci. *56,* 843 (1973).
140. MALPRESS, F. H., and MORRISON, A. B., Biochem. J. *46,* 307 (1950).
141. MALYOTH, G., and STEIN, H. W., Angew. Chem. *64,* 399 (1952).
142. McDONALD, E. J., and TURCOTTE, A. L., J. Research Nat. Bur. Standards *41,* 63 (1948).
143. MICHAELS, A. S., and KREVELD, A. VAN, Neth. Milk Dairy J. *20,* 163 (1966).
144. MILLS, R., BREITER, H., KEMPSTER, E., McKEY, B., PICKENS, M., and OUTHOUSE, J., J. Nutr. *20,* 467 (1940).
145. MOHR, W., and CORTE, H., Kieler Milchwirtschaft. Forsch. Berichte *2,* 13 (1950).
146. MONTGOMERY, E. M., and HUDSON, C. S., J. Am. Chem. Soc. *52,* 2101 (1930).
147. NAKANISHI, T., and ITOH, T., 18th Internat. Dairy Congr., Proc. *1E,* 268 (1970).
148. Nestle's Products Limited. British Patent 972,220 (1964). Cited in DSA. *28,* 774 (1966).
149. NEUBERG, C., and MARX, F., Biochem. Z. *3,* 539 (1907).
150. NEY, K. H., and WIROTAMA, I. P. G., Zts. Lebensmitteluntersuchung und Forschung *143*(2), 93 (1970). Cited in DSA. *32,* 5323 (1970).
151. NICKERSON, T. A., J. Dairy Sci. *37,* 1099 (1954).
152. NICKERSON, T. A., J. Dairy Sci. *39,* 1342 (1956).
153. NICKERSON, T. A., J. Dairy Sci. *45,* 354 (1962).
154. NICKERSON, T. A., COULTER, S. T., and JENNESS, R., J. Dairy Sci. *35,* 77 (1952).
155. NICKERSON, T. A., and DOLBY, R. M., J. Dairy Sci. *54,* 1212 (1971).
156. NICKERSON, T. A., and MOORE, E. E., J. Food Sci. *37,* 60 (1972).
157. NICKERSON, T. A., and MOORE, E. E., Unpublished data.
158. NISHIZUKA, Y., and HAYAISHI, O., J. Biol. Chem. *237,* 2721 (1962).
159. OSTWALD, W., Z. Physik. Chem. *22,* 289 (1897).
160. PANGBORN, R. M., J. Food Sci. *28,* 726 (1963).
161. PANGBORN, R. M., and DUNKLEY, W. L., J. Dairy Sci. *49,* 1 (1966).
162. PANGBORN, R. M., and GEE, S. C., Nature *191,* 810 (1961).
163. PARISI, P., Giorn. Chim. ind. applicata *12,* 225 (1930).
164. PATEL, K. N., and NICKERSON, T. A., J. Dairy Sci. *53,* 1654 (1970).
165. PATTON, S., J. Dairy Sci. *38,* 457 (1955).
166. PATTON, S., and JOSEPHSON, D. V., J. Dairy Sci. *32,* 398 (1949).
167. PAZUR, J. H., TIPTON, C. L., BUDOVICH, T., and MARSH, J. M., J. Amer. Chem. Soc. *80,* 119 (1958).
168. PEEBLES, D. D., Food Technol. *10,* 64 (1956).
169. PETER, P. N., J. Phys. Chem. *32,* 1856 (1928).

170. PILSON, M. E. Q., Am. Zool. *5,* 220 (1965).
171. PILSON, M. E. Q., and KELLY, A. L., Science *135,* 104 (1962).
172. PILSON, M. E. Q., and WALLER, D. W., J. Mammal. *51,* 74 (1970).
173. POMERANZ, Y., Food Technol. *18,* 682 (1964).
174. RAMSDELL, G. A., and WEBB, B. H., J. Dairy Sci. *28,* 677 (1945).
175. REGER, J. V., Cereal Science Today *3,* 270 (1958).
176. REISS, O. K., and BARRY, J. M., Biochem. J. *55,* 783 (1953).
177. REITHEL, F. J., and VENKATARAMAN, R., Science *123,* 1083 (1956).
178. RIGGS, L. K., Arch. Ophthalom. *51,* 415 (1954).
179. ROBERTS, H. R., and McFARREN, E. F., J. Dairy Sci. *36,* 620 (1953).
180. ROBERTS, H. R., and PETTINATI, J. D., J. Agri. Fd. Chem. *5,* 130 (1957).
181. ROBERTS, H. R., PETTINATI, J. D., and BUCEK, W., J. Dairy Sci. *37,* 538 (1954).
182. ROSENSWEIG, N. S., J. Dairy Sci. *52,* 585 (1969).
183. Sanei Kagaku Kogyo Co. Ltd. Japanese Patent 22 505/70 (1970). Cited in Food Sci. Tech. Abs. *3,* 5J634 (1971).
184. SATO, J., British Patent 1,138,124 (1969). Cited in Food Sci. Tech. Abs. *2,* 3H285 (1970).
185. SATO, J., Japanese Patent 21 615/70 (1970). Cited in Food Sci. Tech. Abs. *3,* 5M611 (1971).
186. SCHAMBYE, P., WOOD, H. G., and KLEIBER, M., J. Biol. Chem. *226,* 1011 (1957).
187. SCHMOEGER, M., Ber. *14,* 2121 (1881).
188. SCHOEN, H. M., Ind. Eng. Chem. *53,* 607 (1961).
189. SEIFERT, H., and LABROT, G., Naturwiss. *48,* 691 (1961).
190. SEN GUPTA, M. L., BHATTACHARYYA, N., and BASU, U. P., Indian J. Technol. *6,* 146 (1968).
191. SHARP, P. F., U.S. Patent 1,956,811 (1934).
192. SHARP, P. F., U.S. Patent 2,319,562 (1943).
193. SHARP, P. F., J. Dairy Sci. *21,* 445 (1938).
194. SHARP, P. F., and DOOB, H., JR., J. Dairy Sci. *24,* 589 (1941).
195. SHARP, P. F., and HAND, D. B., U.S. Patents 2,182,618 and 2,182,619 (1939).
196. SHERGIN, A. L., SHERGINA, V. P., and BELOBORODOV, A. G., Trudy vologod. molock. Inst. *55,* 13 (1967). Cited in DSA. *31,* 1086 (1969).
197. SIMOONS, F. J., Am. J. Dig. Diseases *14,* 819 (1969).
198. SIMOONS, F. J., Am. J. Dig. Diseases *15,* 695 (1970).
199. SMEETS, W. T. G. M., Neth. Milk Dairy J. *9,* 249 (1955).
200. STIMPSON, E. G., U.S. Patent 2,668,765 (1954).
201. STIMPSON, E. G., U.S. Patent 2,681,858 (1954).
202. STIMPSON, E. G., U.S. Patent 2,781,266 (1957).
203. STODOLA, F. H., and LOCKWOOD, L. B., J. Biol. Chem. *171,* 213 (1947).
204. SUKEGAWA, T., Japanese Patent 16 767/70 (1970). Cited in Food Sci. Tech. Abs. *3,* 1T51 (1971).
205. SUPPLEE, G. C., J. Dairy Sci. *9,* 50 (1926).
206. SUPPLEE, G. C., and FLANIGAN, G. E., U.S. Patent 1,954,602 (1934).
207. TEMPLETON, R. A. S., West German Patent Application 1 492 698 (1970). Cited in Food Sci. Tech. Abs. *3,* 5J668 (1971).
208. TREY, H., Z. Physik Chem. *46,* 620 (1903).
209. TROY, H. C., and SHARP, P. F., J. Dairy Sci. *13,* 140 (1930).
210. TRUCCO, R. E., VERDIER, P., and REGA, A., Biochim. Biophys. Acta *15,* 582 (1954).
211. TUMERMAN, L., FRAM, H., and CORNELY, K. W., J. Dairy Sci. *37,* 830 (1954).
212. TWIEG, W. C., and NICKERSON, T. A., J. Dairy Sci. *51,* 1720 (1968).
213. VAN HOOK, A., "*Crystallization: Theory and Practice.*" Reinhold Publishing Corp., New York, 1961.
214. VERSCHUUR, R., Rec. trav. chim. *47,* 123 (1928).
215. WALLENFELS, K., and MALHOTRA, O. P., "The Enzymes," Vol. 4, Ch. 24, Academic Press, New York, 1960.

216. WALSH, J. P., ROOK, J. A. F., and DODD, F. H., J. Dairy Res. *35,* 91 (1968).
217. WALSH, J. P., ROOK, J. A. F., and DODD, F. H., J. Dairy Res. *35,* 107 (1968).
218. WASSERMAN, R. H., J. Histochem. Cytochem. *9,* 452 (1961).
219. WASSERMAN, R. H., and LENGEMANN, F. W., J. Nutr. *70,* 377 (1960).
220. WATKINS, W. M., and HASSID, W. Z., J. Biol. Chem. *237,* 1432 (1962).
221. WEBB, B. H., and WHITTIER, E. O., J. Dairy Sci. *31,* 139 (1948).
222. WEBB, B. H., and WHITTIER, E. O., "Byproducts from Milk," 2nd Ed., Avi Publishing Co., Westport, Conn., 1970.
223. WEISBERG, S. M., J. Dairy Sci. *37,* 1106 (1954).
224. WHEELOCK, J. V., and ROOK, J. A. F., J. Dairy Res. *33,* 37 (1966).
225. WHITTIER, E. O., Chem. Rev. *2,* 85 (1925).
226. WHITTIER, E. O., J. Dairy Sci. *27,* 505 (1944).
227. WHITTIER, E. O., and GOULD, S. P., Ind. Eng. Chem. *23,* 670 (1931).
228. WILSON, A. D., personal communication. Edendale, New Zealand (1969).
229. WINGERD, W. H., and BLOCK, R. J., J. Dairy Sci. *37,* 932 (1954).
230. WRIGHT, N. C., J. Dairy Res. *4,* 123 (1932-33).
231. WRIGHT, P. A., J. Dairy Sci. *11,* 240 (1928).
232. YOSHIOKA, Y., IWASAKI, T., and KANAUCHI, T., J. Agr. Chem. Soc. Japan *42,* 651 (1968).
233. ZERBAN, F. W., and MARTIN, J., J. Assoc. Offic. Agric. Chemists *32,* 709 (1949).

CHAPTER 7

Arthur M. Hartman
Leslie P. Dryden*

The Vitamins in Milk and Milk Products

INTRODUCTION

Experiments carried out during the latter part of the 19th and early part of the 20th centuries led to the realization that a diet consisting only of protein, carbohydrate, fat, minerals, and water would not suffice to maintain life and that certain other organic components present in foods in very small amounts are essential for normal well-being. Casimir Funk in 1912, speculating on the possible existence of separate factors capable of curing various human dietary deficiency diseases, coined the term "vitamine," signifying an amine essential to life, to describe these accessory food factors. This term, although erroneous in its implication as to the general chemical nature of these substances, has continued to be used in its shortened form "vitamin."

The exact biochemical functions of all the vitamins are still by no means completely understood. It is clear, however, that the B vitamins, at least, by participating in specific enzyme systems, act as catalysts for particular metabolic processes that occur in the body. The activity of a vitamin is not always confined to a single chemical substance but may in some cases be exhibited by a number of different, though structurally related, compounds. These compounds are not necessarily, however, equal in biological potency.

Some species of animals need not have all the vitamins supplied directly in their food, for they may obtain an ample supply of some of them by synthesis, either by the microflora in their alimentary tracts or by their own tissues. Thus, if the original concept of an essential food substance is rigidly adhered to, it is evident that a particular substance may be a vitamin for one species but not for another.

Milk played an important part in the recognition and early development of knowledge of the vitamins. Lunin as early as 1881 recognized that experimental animals would not grow normally on purified diets composed of the principal components of milk, and concluded that this food must contain "small quantities of other and unknown substances

* Arthur M. Hartman, Retired, Animal Science Research Division and Leslie P. Dryden, Animal Physiology and Genetics Institute, Agricultural Research Service, U.S. Department of Agriculture, Beltsville, Maryland, 20705

essential to life." Purified diets were shown to be satisfactory when supplemented by insignificantly small additions of milk in the work of Pekelharing and in more extensive studies by Hopkins and by Osborne and Mendel. In 1913, these latter workers expressed the idea that milk contains all the factors essential for growth and maintenance. This concept has proved to be largely correct insofar as the vitamins are concerned, for all the recognized vitamins are to be found in milk. Some are not present in important quantities, but for others, milk is a very rich source—indeed one of the best available.

As might be expected, the so-called fat-soluble vitamins A, D, E, and K are associated with the fat component of milk, while the water-soluble vitamins—the B vitamins and vitamin C—are found, with minor exceptions, exclusively in the nonfat portion. The quantities of most of the fat-soluble vitamins in milk are principally dependent upon those present in the diet of the cow; the water-soluble vitamins and vitamin K are largely, if not altogether, independent of the diet, since they are synthesized by bacteria in the cow's rumen or by her tissues.

VITAMIN A

Vitamin A is not found in plants but is solely of animal origin. Although the vitamin occurs in more than one form, it is generally found as retinol (vitamin A_1) in nature, being present as such in all animals and fish. Dehydroretinol (vitamin A_2) also occurs in the bodies of freshwater fish but not in land animals unless they consume these fish.

```
   16     17            19                 20
  CH3   CH3  H  H     CH3  H   H   H     CH3  H
    \   /    |  |       |  |   |   |       |  |
      C1     |  |       |  |   |   |       |  |
H2C2     6C—C=C—C=C—C=C—C=C—CH2OH
  |       ||  7  8     9  10  11  12  13  14  15
H2C3  4  5C—CH3
    \   /       18
      C
      |
      H2
```

Retinol

Dehydroretinol contains an additional double bond in the ring structure, between carbons 3 and 4. Its biological activity is only about 40% that of retinol. For the most part, interconversion of the two compounds seems to be quite limited.

Stereoisomers of vitamin A also exist, since either *cis* or *trans* configurations can occur about the double bonds in the straight-chain part of the molecule. In addition to all-*trans* form, 5 other stereoisomers have been characterized chemically and physically (13-*cis*, 11-*cis*, 9-*cis*,

11, 13-di-*cis* and 9, 13-di-*cis*). Three of these (13-*cis*, 9-*cis*, and 9,13-di-*cis*), together with all-*trans* retinol, represent the maximum number of sterically unhindered forms. These are more stable than the sterically hindered isomers. The 13-*cis* isomer always accompanies all-*trans* retinol. The other stereoisomers may also be present in small amounts. The 13-*cis* isomer has only 75% of the biological potency of the all-*trans* form, while the other isomers have much smaller potencies.

Deficiency of vitamin A results in cessation of growth, a keratinizing of the epithelium in various parts of the body, such as the respiratory, alimentary, reproductive and genito-urinary tracts, defects in the teeth, disturbances in bone growth, and changes in the eye, which becomes hemorrhagic, keratinized, and eventually completely degenerated.

Primary effects of the deficiency are complicated by secondary effects due to infections, to which vitamin A-deficient animals are prone, and inanition.

Vitamin A plays an essential role in regard to the process of normal vision;[162] however, its more general role in regard to cellular metabolism has still to be clarified.

Forms of Vitamin A in Milk

Because of its alcohol group, retinol readily forms esters. In normal milk and colostrum, almost all the vitamin occurs in the ester form. Only about 2 to 6% is present as the alcohol. Breed of cow or stage of lactation has little or no effect on the relative proportions of ester and alcohol present in milk. Retinyl esters in milk apparently arise both from the absorption of the ester present in the blood and from synthesis by the mammary gland from the retinol in the blood.

Carotenoids in Milk

The vitamin A activity of milk and milk products is due not only to the vitamin as such but also to provitamin A carotenoids, which can be converted to the vitamin in the animal body. The carotenoids are the pigments in milkfat that are responsible for the natural yellow color of butter. They occur quite abundantly in plant material and appear in milk as a result of their ingestion by the cow. Not all of them, however, act as precursors of the vitamin.

Carotenoids make up from 11 to 50% of the total vitamin A activity of milk, the exact percentage depending upon the breed of the cow and the level of carotenoid intake. Since the latter is generally higher during the summer than during the winter, the fraction of total vitamin A activity due to the provitamin is greater in summer milk. Moreover, the fluctuations in the carotenoid content of milk as between winter and summer are greater than the variations in vitamin A.

The most active of the provitamins and the one most widely distributed in nature is β-carotene. The molecule of β-carotene is symmetrical, containing two β-ionone rings connected by a chain of 4 isoprene units. The isoprene units are reversed in the center of the molecule.

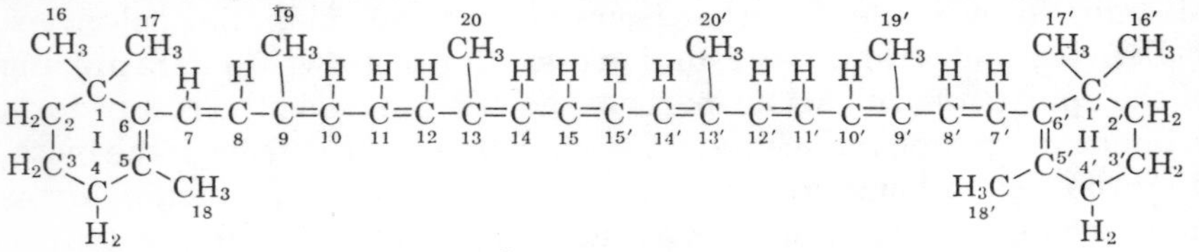

```
 16    17          19           20                20'          19'         17'   16'
CH3   CH3         CH3          CH3               CH3          CH3         CH3   CH3
   \C/       H  H  |  H  H  H   |   H  H  H  H    |  H  H  H   |  H  H       \C/
H2C2 1 6C-C=C-C=C-C=C-C=C-C=C-C=C-C=C-C=C-C=C-C6' 1' 2'CH2
 |   I  ||  7  8  9 10 11 12 13 14 15 15' 14' 13' 12' 11' 10' 9' 8' 7' ||  II  |
H2C3 4 5C                                                       C5' 4' 3'CH2
   \C/   CH3                                                H3C/   \C/
    |    18                                                 18'     |
    H2                                                              H2
```

β-carotene

Changes in the structure of the β-carotene molecule involve changes in biological activity. β-Carotene is isomeric with α-carotene, γ-carotene, and lycopene, all of them being hydrocarbons with a composition of $C_{40}H_{56}$ but differing in the structure of one or both of the rings. In α-carotene, the double bond in ring II connects carbons 4′ and 5′ rather than carbons 5′ and 6′, thus forming an α-ionone ring. In γ-carotene, ring II opens between carbons 1′ and 6′. Each of these two isomers possesses only half or less of the biological activity of β-carotene. In lycopene, both rings are open; it has no provitamin A activity.

Among other important naturally occurring carotenoids are some having oxygen in the molecule. These include carotenols which contain oxygen in the form of one or more hydroxyl groups. Among these substances are cryptoxanthin (β-carotene with hydroxyl group attached to carbon 3′), which has a biological potency somewhat in excess of 50% of that of β-carotene, and lutein (xanthophyll) (α-carotene with hydroxyl groups attached to carbons 3 and 3′), which has no biological potency. Carotenols are widely distributed in plants.

β-Carotene makes up by far the greatest fraction of the carotenoids occurring in milk. If the feed of the cow contains other carotenoids, however, some of them will also appear in milk to a greater or lesser extent. For example, α-carotene is generally absent from butterfat; but if cows are fed carrots, which contain about 25% of their total carotenoids in the form of α-carotene, this form of carotene will also appear in the milk. Cryptoxanthin has also been found in butter. Neo-β-carotenes, stereoisomers of β-carotene containing *cis* bonds, have been reported[175] in butterfat. The *cis* isomer present was believed to be largely neo-β-carotene B, an isomer in which *cis* bonds occur between carbons 9 and 10 and carbons 15 and 15′. However, there is some question as to whether this *cis* isomer occurs naturally in butter or was formed during the processes of clarification and saponification used in analysis.

A part of the total carotenoids in milk may consist of compounds that are completely inactive as provitamin A. Estimates of the inactive substances generally range from 5 to 25% of the total carotenoids, but in winter milkfat, the proportion has at times been found to be higher—as much as 40% with ordinary feeds and as high as 75% when cows are fed large amounts of yellow turnips. These inactive carotenoids consist of lycopene, lutein (xanthophyll), ζ-carotene (lycopene with carbons 7, 8, 7′, and 8′ hydrogenated) and zeaxanthin (β-carotene with hydroxyl groups attached to carbons 3 and 3′). When cows are fed certain types of silage, pigments formed as decomposition products of xanthophyll or carotene, especially by the action of mineral acids on the silage, may also appear in the milk in low amounts.

Transfer of Vitamin A and Carotene to Milk

The conversion of β-carotene to vitamin A takes place mainly in the intestinal mucosa. Cleavage, catalyzed by an enzyme, β-carotene, 15, 15′-dioxygenase, takes place in the center of the molecule to yield two molecules of retinal (vitamin A aldehyde). Cattle are less efficient at this process than many other species, but they absorb and circulate carotene and other carotenoids as well as vitamin A in their blood.

The retinal formed from carotene is reduced to retinol. The retinol formed in this manner, as well as by hydrolysis of retinyl esters taken in the food, are absorbed as such through the mucosal cell membrane and then reesterified with long-chain fatty acids, primarily palmitate. These retinyl esters are incorporated into mucosal chylomicrons and thus transported through the lymphatic system. The major portion of the vitamin in the blood consists of retinol, though the ester form increases rapidly in the postabsorptive state after vitamin A intake.

Under normal conditions of adequate carotene intake, the level of the carotenoids and vitamin A in milk is derived through the uptake by the mammary gland of vitamin A ester and carotenoids of immediate dietary origin.

There is evidence, however, of uptake from the blood of some of the retinol and carotenoids associated with plasma protein. The mammary gland undoubtedly dissociates most of these substances from protein and transfers them to the fat. It has been suggested, however, that in all milk, a small proportion of carotenoids and retinol remains associated with milk serum proteins.

When the level of vitamin A in the blood was increased by intravenous injection of an aqueous vitamin A dispersion, the increase in the vitamin A level in the milk was 15 times greater than when the same dose was administered orally in an oily solution, and about 3 times as high as when it was administered orally in an aqueous dispersion.

The maximum transfer of the vitamin to the milk took place much earlier with intravenous than with oral administration.

Absorption of Vitamin A from Milk

As indicated above, vitamin A is better absorbed by the body from an aqueous medium than from an oily solution, particle size apparently being the most important factor. Tests with infants indicated that the vitamin A in milk was absorbed almost as well as when the vitamin was given as an aqueous dispersion. These results were apparently due to the dispersion of the butterfat globules in milk, since the vitamin was much less well absorbed when it was fed in the form of liquefied butterfat.[103] It appears that vitamin A is satisfactorily absorbed by infants from vitamin A-fortified nonfat dry milk.[39] Although a minimum level of dietary fat has been suggested as necessary for adequate vitamin A absorption to occur, tests with rats indicated that added vitamin A was absorbed better from fortified nonfat dry milk than from a stock diet containing 5% fat.[135]

Effect of Season and Diet on Vitamin A Content of Milk

In the northern hemisphere, milk produced during winter months contains less vitamin A activity than that produced during summer months. The highest concentrations generally occur during late spring and early fall, while the lowest prevail during late winter. Such changes in the vitamin A content of milk were quite early shown to be due not to season as such, but to changes in the diet of the cow associated with season, particularly to differences between winter feeds and summer pasture. Summer milk contains from 1½ to as much as 15 times the amount of total vitamin A contained in winter milk. The smaller differences are more common, the larger differences being found only when winter rations are experimentally designed to yield low vitamin A values in milk. From the results of a nationwide survey carried out by the U.S. Department of Agriculture,[169] summer milk was estimated to contain, on the average, about 1.6 times as much vitamin A as winter milk.

Within certain limits, the amount of vitamin A secreted into the milk and butterfat, either as vitamin A itself or as carotene, depends directly upon the level of carotene in the ration of the cow. Wiseman *et al.*[169] expressed the relation between carotene intake and the potency of the butter produced by the following equation: $Y = -12912 + 8937X$ where Y is the potency of vitamin A in the butter in terms of I.U. (International Units) per lb and X is the logarithm of carotene intake in mg per day. From this equation, they calculated that to produce butter with a potency of average winter butter, a cow requires an intake of approximately 480 mg carotene daily in the feed, while to produce average summer butter, an intake of about 2.5 gm a day is needed. Other estimates have indicated that somewhat smaller amounts are required.[155]

When carotene is sharply increased in the diet, responses as measured by the amount of vitamin A secreted in the milk are very rapid, large differences being observed within one to two weeks.

Among natural foods, pasture provides the most plentiful source of carotene for the cow. With the feeding of abundant pasture, some workers have obtained a maximum beyond which further increases in carotene intake do not increase the vitamin A content of the milk. If the intake of carotene by a cow already receiving enough to provide for summer levels is suddenly increased, a marked rise in vitamin A potency may occur, but it is transient in nature.

When cows are transferred from pasture to winter rations in the fall, a decline in the vitamin A content of the milk begins; it continues slowly through the winter until the cows are once more put on pasture in the spring. Winter feeds vary considerably in carotene content, however, and with proper selection, it is possible to increase the vitamin A content of winter milk and butter to a level that approaches, but in most cases does not equal, that obtained on summer pasture. Hays contain more carotene than grains, but vary among themselves; alfalfa generally contains more carotene than timothy, and artifically dried hays are superior to corresponding field-cured crops. Silages generally provide more of the provitamin than hay, alfalfa silage usually but not always containing more than corn silage. Feeding corn silage containing a high level of nitrate had no harmful effect upon the vitamin A content of milk.[78]

Large increases in the vitamin A content of milk can be obtained by supplementing the ration of the cow with concentrated sources of vitamin A, such as fish-liver oils. Since these oils contain only true vitamin A and no carotene, the increase in the total vitamin A activity of the milk is all in the form of true vitamin A itself.

The efficiency of utilization of vitamin A or carotene in the diet of the cow in regard to secretion in milk is quite low when large amounts are fed, but larger when smaller amounts are given. Most figures indicate an apparent transfer from the diet to the milk of between 0.09 and 6% for high or relatively high intakes, whereas with low or very low intakes values of from 10 to 47% have been observed. Some of the higher percentages probably represent apparent rather than actual utilization, however, since under conditions of very low intake, the cow may draw from its vitamin A body stores to increase the vitamin A content of its milk. A considerable portion of the carotene or administered vitamin A is destroyed in the rumen and retuculum or carried into the feces.[115,168]

Relation Between Vitamin A and Vitamin E (Tocopherol)

In the north island of New Zealand, pasture feeding is possible all year. Even under these conditions, however, seasonal fluctuations were noticed in the vitamin A content of milk and butter, the low point occurring during the summer season. These seasonal variations were not related to the carotene content of the pasture but to its stage of maturity and the consequent differences in the percentage and nature of the lipids present in the pasture grasses.[106] Supplementation with carotene did not increase the level of vitamin A in the milk, but such an increase was obtained by feeding the cows supplemental tocopherols[172] even though there was no evidence of a seasonal trend in the tocopherol content of the pasture.[106] Increases secured by the use of tocopherol appear to be due to its antioxidant properties, resulting in an improvement in the absorption and utilization of carotene and vitamin A. A similar trend for values lower in mid-summer than in late spring or early autumn was observed in Great Britain.[157]

Effect of Breed

Guernsey and Jersey milk contain more total vitamin A than does Holstein or Ayrshire milk, but this difference is due primarily, if not altogether, to the higher fat content. Guernsey cows convert a smaller proportion of the carotene from their feed into vitamin A than do Holsteins and Ayrshires. Thus cows of the Guernsey breed secrete a much higher proportion of their total vitamin A activity as carotene, and their butterfat is accordingly more highly pigmented. On the other hand, more of the vitamin A activity in Holstein and Ayrshire milk consists of the true vitamin. In these comparisons, the Jersey and Brown Swiss breeds occupy an intermediate position. In experiments at Beltsville,[16] the proportion of the total vitamin A potency of milk contributed by carotene was on the average 18% for Holsteins and 36% for Jerseys, or, recalculated according to the international standard since adopted, 21 and 40%, respectively.

Breed differences in this respect are a reflection of relative efficiency in converting carotene to vitamin A. Breeds that secrete a greater percentage of the total vitamin A in their milk as carotene are also less selective in the absorption and secretion of carotenoids.

Change During Lactation

The total vitamin A content of colostrum is highest immediately after calving, being from 4 to 25 times that of milk. This concentration declines quickly thereafter, however, so that by the 3rd to 10th day after parturition, it has reached the level in normal milk. Except for the colostral period, the total vitamin A activity of milk seems to be largely independent of the phase of lactation.

The same factors that affect the vitamin A content of milk also affect that of colostrum. Such factors include breed of cow and the carotene content of the ration, especially during the prepartum period.

The high concentration of vitamin A in the colostrum may be attributed, at least partially, to an accumulation of the vitamin in the tissues during the latter stages of pregnancy when milk flow has ceased. Presumably, near parturition, body stores of vitamin A and carotene are released into the blood. The amounts of the vitamin in the blood plasma rise rapidly, beginning 3 or 4 days prepartum, and then decrease sharply at about the time of parturition or shortly before, reaching a minimum level at about 3 days to 1 week following parturition. These amounts subsequently rise, but do not reach normal levels until about 2 to 4 weeks after calving.

Association of Vitamin A and Carotenoids with Fat Particles

The concentration of carotenoids in butter obtained by churning is less than might be expected from its fat content. In the making of butter, one study showed that, although 93 to 100% of the vitamin A went into the butter, 2 to 4% into the skimmilk, and 0.4 to 1% into the buttermilk, the carotenoids were distributed 89 to 94% in the butter, 10 to 14% in the skimmilk, and 0.8 to 2% in the buttermilk. In another study, the fat of skimmilk obtained in making butter and the fat of separated cheese whey contained, respectively, 7 and 11 times as high a concentration of carotenoids as the original milkfat. The relative amounts of these substances in the fat of buttermilk and in the fat of whey itself were also somewhat higher than in the original milkfat, although the differences were not as great. The content of carotenoids per gm of fat in a particular fraction was negatively correlated with the average size of the fat globules in that fraction.

The carotene and vitamin A contents of successive portions of a single milking were studied by Eaton *et al.*[35] using colostrum, and by White *et al.*[165] using later milk. In both studies, the carotenoids and vitamin A per gm of milk as well as the fat percentage increased with successive increments of milk. No significant trends were observed between vitamin A per unit of fat and stage of milking. In one of these studies,[35, see 95] carotenoids per gm fat showed a decreasing trend with successive increments of milk throughout the entire milking, together with an increase in the size of the fat globules. In the other instances,[165] there was no correlation between stage of milking and fat globule size; the carotenoids per gm fat tended to decline to a minimum at a point approximately midway through the milking, and to increase thereafter. In this work, vitamin A and carotenoid content per gm fat were both found to be negatively correlated with the size of the fat globules, although the differences were not as great with vitamin A as with the carotenoids. Since smaller globules have greater surface areas relative to their volumes than do larger globules, these results were interpreted to indicate that vitamin A and carotene are not concentrated in the interiors of fat globules but rather on their surfaces, possibly in a dilute solution or loose chemical complex.

The centrifugation of milk at increasing speeds produced fractions containing smaller and smaller fat globules, and these fractions became progressively richer in carotenoids although unchanged in concentration of total vitamin A.[156] The traces of fat in skimmilk contained 15% of the vitamin A as retinol in comparison to 1 to 2% in the fat of whole milk.

In other work,[109] however, when fat was obtained from milk by centrifugation, its carotenoid and vitamin A contents appeared to be almost independent of globule size. Moreover, some vitamin A and carotene remained in the virtually fat-free serum obtained from milk and were precipitated with the various serum proteins, particularly the globulin fraction. In extracted fat, retinol varied in the same way as the carotenoids, but retinyl ester did not. It was suggested that the difference in concentration as the size of the fat globule changed was not due to a surface layer of carotenoids and retinol on the fat globule but to the presence in milk of protein-bound forms of these substances. The protein-bound materials, extracted with the fat, would in effect increase the concentration of carotenoids and retinol in fat obtained from milk and milk products by solvent extraction rather than by centrifugation. This apparent increase in concentration in the extracted fat would be progressively greater as the size of the globule, and hence the quantity of the fat relative to the volume of milk from which it was obtained, decreased.

Ling *et al.*[104] suggested that both explanations may be partly correct —that the carotenoids and retinol are present both in the fat globule membrane and as a protein complex in the milk serum.

Effect of Heat Treatment

According to most workers, procedures such as pasteurization, sterilization,[13] drying by the spray or roller processes or evaporation cause little or no loss of the vitamin A or carotene. However, prolonged heating of milk, butter or butterfat at high temperatures in the presence of air results in a considerable decrease in vitamin A activity.

Influence of Light and Radiation

Losses of vitamin A and carotene when milkfat is exposed to daylight depend to some extent on the previous history of the fat. Thus, with fresh milkfat, exposure for 8 hr to northern daylight passing through window glass destroyed only about 2% of both vitamin A and carotene. On the other hand, when milkfat which had been stored in the dark at 5°C for 4 to 7 months was similarly exposed for 8 to 10 hr, decreases of 60 to 70% in vitamin A and of 15 to 20% in carotene were observed. Milk in amber-glass bottles exposed to light equivalent to 3 hr sunlight lost about 10% of the vitamin.[56] Ultraviolet radiation for about 2 sec to enrich its vitamin D content caused no diminution in vitamin A.[170] Fluorescent light for 12 hr brought about no net loss of retinol, but about 30 to 40% decrease in carotene, according to one study;[147] in another,[155] progressive loss of the vitamin occurred with increasing exposure.

The use of X-rays to sterilize milk caused no decrease in vitamin A. Irradiation with gamma rays from Co^{60} (80,000 roentgens/hr), however, brought about the disappearance of 85% of the vitamin A and 45% of the carotenoids in 12 hr.[100] Large losses were also observed with evaporated milk, cream, butter, and cheese. According to Pederson,[130] 20,000 rads of gamma radiation are needed to produce losses of carotene and vitamin A in liquid milk. Exposure of milk to one megarad destroyed 64 and 68%, respectively, of its vitamin A and carotene; replacement of air in the milk by nitrogen reduced these losses to 51 and 59%. Processing of milk to remove cationic fission products brought about no loss of vitamin A or carotene.[49;76]

Content in Milk and Butter

Obviously, the vitamin A content of milk and milk products varies considerably, depending upon the factors previously discussed. During World War II, a nationwide survey to determine the vitamin A potency of milk and butter produced and marketed in different regions of the United States was conducted by the U.S. Department of Agriculture and 21 cooperating state agricultural experiment stations. According to this survey,[16] the average potency of the total output of creamery butter in the United States was approximately 15,700 I.U. per lb, while the average annual potency of milk was calculated to be 1540 I.U. per qt. These values were calculated on the basis that 1 I.U. is equal to

0.25 μg retinol or 0.6 μg β-carotene. Since the publication of the results of this survey, the Committee on Biological Standardization of the World Health Organization has set, as an international standard, 1 I.U. equal to 0.30 μg retinol. Judged by this standard, the above value for butter is about 15% too high. Results of the survey indicate that butter contains on the average 2.7 μg carotene and 5.0 μg retinol per gm during the winter, and 6.1 μg carotene and 7.5 μg retinol per gm during the summer. More detailed data are given in Tables 7.1 and 7.2 along with roughly comparable figures from other countries.

In the United States, the trend in recent years has been to fortify cattle feeds with synthetic vitamin A and to increase the intake of silage as well. As a consequence, the annual average for vitamin A in butter has very likely increased over that found in the above survey.

Little or no change has been found in the carotene or vitamin A content of butter maintained for periods and under conditions that commercial butters are ordinarily stored. Insofar as the effect of storage is concerned, the average vitamin A potency of butter sold on the retail market is not significantly different from that of butter produced in creameries.

The dietary allowance of vitamin A recommended by the National Research Council is 5000 I.U. daily for most adults. A quart of milk would supply on the average about 25 to 30% of this requirement. Milk and milk products, excluding butter, contribute 12% of the total vitamin A in the American diet. Fats and oils, including butter, supply an additional 10%, but in recent years there has been a shift away from butter toward vitamin A-fortified margarine. Estimates for Great Britain during the 1957-1959 period indicated that milk and milk

Table 7.1

CONTENT (I.U./QT) OF VITAMIN A IN FLUID WHOLE MILK

Country	Winter Milk	Summer Milk	Annual	Ref.
United States of America:[a]				
Figures as published	1,120	1,820	1,540	16
Calculated according to new standard[b]	1,025	1,690	1,425	—
Great Britain[a]	1,060	1,520	—	—
Netherlands[c]	1,220	1,870	1,620	98
New Zealand[a]	1,990	1,600	1,670	107
India (Bangalore)	—	—	1,535	139

[a] Calculated from butter or butterfat data on basis of milk containing 4% fat.
[b] Set by Committee on Biological Standardization, World Health Organization, 1949, equating 1 I.U. to 0.30 instead of 0.25 μg retinol.
[c] Average fat content, 3.6 to 3.7%.

Table 7.2

AVERAGE (I.U./LB.) VITAMIN A POTENCY OF BUTTER[a]

	Winter Butter	Summer Butter	Annual	Ref.
United States of America				
North Central States	10,880	18,120	15,390	16
South Atlantic and South Central States	11,270	17,360	15,807	16
Rocky Mountain and Pacific States	13,900	19,080	17,710	16
Average				
As published	11,170	18,160	15,680	16
Calculated according to new standard[b]	9,650	15,900	13,700	—
Canada				
Ontario	9,915	14,702	13,269	149[c]
Manitoba[d]	8,320	13,470	10,490	137
Denmark	7,500	15,000	—	157
Great Britain	10,500	15,000	—	157
Netherlands	12,900	20,200	17,400	98
New Zealand	19,010	15,330	16,740	107
Sweden	6,800	11,000	—	157

[a] Corrected for losses in analysis: 7% vitamin A in references 107 (for annual figure) and 16 and 8% in ref. 98; others not corrected for such losses.
[b] Set by Committee on Biological Standardization, World Health Organization, in 1949, equating 1 I.U. to 0.30 instead of 0.25 μg vitamin A alcohol.
[c] Quoted in ref. 137.
[d] Calculated from butterfat data, assuming butter containing 80% fat.

products contributed about 20 to 30% of the vitamin A in the British diet.[93]

Content in Milk Products

The vitamin A content of milk products depends to a large extent upon their fat content. It also varies with the season of year in which these products are manufactured, since their potency depends upon the vitamin A content of the original milk.

In the manufacture of cheese, between 80 and 95% of the vitamin A and carotene in the original milk is recovered in the curd. In many instances little or no change in the content of the vitamin has been observed during ripening or storage of cheese for a year.

Losses in making whole milk powder are less than 10%. No further decreases have been found during 6 months storage of this product at room temperature under inert gas, and only a small diminution when the powder was air-packed. Indeed, dried milk prepared under modern conditions can be maintained in air-pack for 6 to 12 months without a decrease in vitamin A or carotene. However, storage of dried milk or sterilized milk for prolonged periods at elevated temperatures or in sunlight causes extensive losses.

Estimates of the vitamin A content of some milk products, as found in the literature, are given in Table 7.3.

Addition of Vitamin A to Dairy Products

Tests have shown that nonfat dry milk can be satisfactorily enriched by either (1) homogenizing a coconut fat carrier of synthetic retinyl palmitate and vitamin D in condensed milk before drying, or (2) blending a dry beadlet carrier of synthetic retinyl palmitate and vitamin D into the dry milk just before packaging. The latter process tends to yield slightly higher vitamin A retention values under prolonged storage. Since the vitamin is somewhat less stable in nonfat dry milk products than in whole milk products, some oxidized flavor may develop.

Fortification of nonfat dairy products with vitamin A was recommended by the World Health Organization for reconstituted nonfat milk products used as fresh milk substitutes in undeveloped countries. Starting in 1965, dry milk enriched with vitamin A (5000 I.U./100 gm) and vitamin D (500 I.U./100 gm) was purchased for export in connection with the Food for Peace Program.[23]

THE B VITAMINS

Historically, the various members of the vitamin B complex have frequently been considered as a group. This tendency can no doubt be attributed more to their common occurrence in animal tissues and in foods such as liver and yeast than to any likeness in structure or function. A further point of similarity can be found in ruminants where synthesis of the B vitamins by the rumen microflora occurs at a position in the alimentary tract whence they become available for absorption into the system. For this reason, ruminants, unlike monogastric animals, do not require a dietary supply of these food essentials.

Evidence pointing toward synthesis of the B vitamins in the cow was obtained at an early date. Much of this evidence was based on the comparative concentrations of the vitamins in rumen contents and in the feed ingested. The concentrations of thiamin, riboflavin, vitamin B-6, biotin, pantothenic acid, niacin, and folacin have been observed to be greater in the rumen contents than in the food eaten by the cow, the differences being quite large in some cases. Niacin is undoubt-

Table 7.3

VITAMIN A CONTENT OF MILK PRODUCTS[a] IN (I.U./100 GM)[b]

Milk or Milk Product	Average	Range	Cheese Variety[c]	Average	Range
Whole milk			Very hard		
Fluid	159	136– 176[d]	Parmesan	1,410	— (1)[e]
Condensed	276	141– 352(4)[e]	Hard		
Evaporated	369	342– 464(4)	Cheddar	1,169[f]	750–1985(10)
Dried	1,110	600–1600(6)	Edam	1,203	733–1788(4)
Skimmilk			Gruyère	822	267–1333(3)
Fluid	9	4– 18(4)	Swiss	1,592	954–2680(3)
Dried	68	25– 140(3)	Semisoft		
Buttermilk			Blue[g]	1,935	1,000–3502(6)
Fluid	12	4– 20(2)	Brick	1,626	853–2400(2)
Yoghurt	69	— (1)	Limburger	1,280	— (1)
Cream			Roquefort[h]	1,971	900–4012(3)
Half and half	480	— (1)	Stilton	1,235	— (1)
Light table	880	— (1)	Soft		
Medium whipping	1,336	— (1)	Ripened		
Heavy whipping	1,598	— (1)	Brie	667	— (1)
Butter	2,917	1,962–3836[i]	Unripened		
Ice cream	523	425– 600(4)	Cottage		
Whey			Creamed	291	185– 397(2)
Fluid	11	10– 12(2)	Uncreamed	42	9– 60(3)
Dried	50	— (1)	Cream	2,194	1,552–2819(3)
			Neufchâtel	1,495	— (1)
			Processed		
			Cheddar	1,705	1,250–2160(2)
			Swiss	1,680	1,390–1970(2)

[a] Mean and range of average values obtained from the literature.
[b] I.U. per 100 ml for products designated "fluid".
[c] Classified primarily according to Sanders.[140]
[d] Calculated from data in Table 7.1, as average and range of annual figures for the various countries listed (winter-summer average where annual figure not given).
[e] Figures in parentheses indicate the number of references consulted.
[f] One high value (5,500 I.U./100 gm) omitted.
[g] May be made from milk of species other than the cow.
[h] Made from ewe's milk.
[i] Calculated from data in Table 7.2, as average and range of annual figures for the various countries listed (winter-summer average where annual figure not given).

edly synthesized from tryptophan by the tissues of the animal as well as by bacteria in the rumen. Myo-inositol is also probably synthesized in the rumen. The nature of the feed ingested by the cow has some influence on the amounts of the vitamins synthesized by the rumen microflora. Larger amounts of most vitamins were found in rumen fluid from steers on all-concentrate diets than in that from animals fed lucerne hay.[63] Synthesis in the rumen of lactating cows fed urea and ammonium salts as the sole sources of nitrogen is normal for thiamin, vitamin B-6, folacin, biotin and vitamin B-12, and somewhat higher than normal for riboflavin, niacin and pantothenic acid.[161]

Comparisons were made[96] of the thiamine, riboflavin, niacin, pantothenic acid, vitamin B-6, biotin, folacin, and vitamin B-12 concentrations in the diets of fistulated yearling steers and in their rumen contents when the animals were fed either natural diets or semisynthetic low-vitamin diets. The tendency was for the vitamin concentrations to remain the same in the rumen contents regardless of type of diet fed, so that if the natural diets contained the vitamin, synthesis was apparently much less, where it occurred at all, on the natural diets than on the low-vitamin diets. Synthesis begins at a relatively early age—by 1 to 3 months for thiamine, riboflavin, and vitamin B-12.

There is relatively little information available on the specific microorganisms responsible for the synthesis of the B vitamins in the rumen. Porter[132] studied the synthesis by five strains of *Escherichia coli* and five strains of *Streptococcus bovis* isolated from the rumen. *E. coli* synthesized appreciable quantities of a number of the vitamins, while *S. bovis* apparently synthesized only niacin, vitamin B-6, and pantothenic acid. Biosynthesis of vitamin B-12 will be further considered in the section covering that vitamin.

As would be expected from the relatively constant levels of the various B vitamins in rumen contents, their concentrations in milk do not vary with changes in the vitamin content of the ration of the cow. Even when the ration is devoid, or nearly so, of the vitamins, the milk from cows fed such a ration contains as much of these vitamins as that from normal cows.

Thiamine (Vitamin B-1)[1]

Thiamine is essential for the growth and metabolism of all animals as well as of many plants and microorganisms. In coenzyme form, it occurs as thiamine pyrophosphate (cocarboxylase). Unphosphorylated thiamine and cocarboxylase have the same biological activity for higher animals. Thiamine pyrophosphate serves as a coenzyme for a number of biochemical reactions concerned with the

[1] "Thiamin" is the alternate spelling recommended by the IUNS Committee on Nomenclature.

intermediary metabolism of carbohydrates. Slight modifications in the structure of the thiamine molecule lead to extensive or complete loss of biological activity.

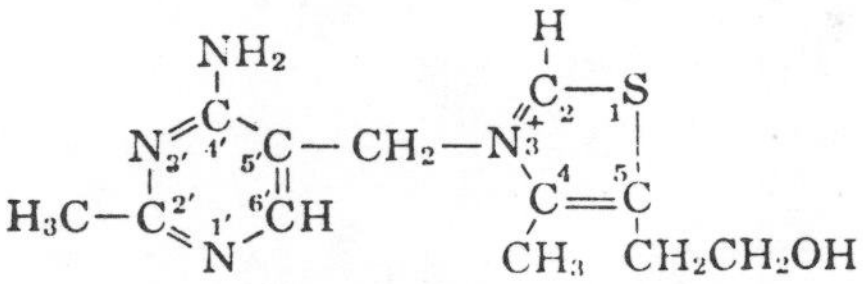

```
            NH2                      H
             |                       |
             C                       C—S
          N3' 4' 5'C — CH2 — N3+  2  1  |
          |       ||          4    5'   |
H3C — C2'  1'  6'CH             C == C
          N                    CH3   CH2CH2OH
```

Thiamin

Thiamine deficiency may produce such symptoms as loss of appetite and weight, general weakness, gastrointestinal disturbances, edema, polyneuritis, neuralgia, degenerative changes in the nervous system, muscular incoordination, and heart failure resulting from an abnormal enlargement of the heart. The body seems to have little capacity to store the vitamin, the tissues maintaining at best a few weeks reserve.

Forms in Milk.—Thiamine occurs in milk not only as free thiamine but also in phosphorylated form and complexed with protein. In normal milk, the total thiamine consists of 50 to 70% free thiamine, 18 to 45% phosphorylated thiamine, and 5 to 17% protein-bound thiamine. DeJong[28] found no cocarboxylase present and concluded that the phosphorylated form was thiamine monophosphate. Gregory and Kon[53] found only thiamine monophosphate present in cow's milk during the second week of lactation, and only free thiamine in mid- or late-lactation milk.

Content in Milk and Milk products.—The average thiamine content of fluid whole milk is about 0.43 mg per l (Table 7.4). The daily dietary allowance of thiamine for adults recommended by the National Research Council varies between 1.0 and 1.4 mg, depending upon age, sex, and other factors. One quart of milk per day would supply about 30 to 40% of these recommended amounts. In the United States and western Europe, milk and milk products contribute about 10 to 15% of the total thiamine in the diet. Figures from the literature for the thiamine content of various milk products are given in Table 7.4.

Effect of Season, Diet of Cow, and Breed.—The content of thiamine in milk remains fairly constant. In most cases, variations in the nature of the feed of the cow have no effect at all on the content of the vitamin in milk, while in the few instances where differences have been observed, they have generally been quite small. Factors such as season and breed likewise have a minor effect, if any at all.

Change During Lactation.—Colostrum and early milk contain 1½ to 2 times as much total thiamine as later milk. This decreases very rapidly during the first 10 days postpartum, and then more gradually

Table 7.4

CONTENT (MG/KG[a]) OF THIAMINE, RIBOFLAVIN, NIACIN, AND PANTOTHENIC ACID IN MILK AND MILK PRODUCTS[b]

Milk or Milk Product	Thiamine		Riboflavin		Niacin		Pantothenic Acid	
	Average	Range	Average	Range	Average	Range	Average	Range
Whole milk								
Fluid	0.43	0.20–0.80(54)[c]	1.74	0.81– 2.58(77)	0.93[d]	0.30– 2.00(38)	3.39	2.58– 4.90(27)
Condensed	1.1	0.8 –1.5 (5)	3.6	2.6 – 4.0 (4)	2.1	1.6 – 2.4 (3)	8.7	7.5 –10.4 (3)
Evaporated	0.56	0.40–0.82(6)	3.8	2.8 – 4.8 (6)	2.0	1.8 – 2.3 (5)	7.0	5.8 – 8.0 (4)
Dried	3.3	2.5 –5.1 (10)	15.5	9.8 –25.6 (9)	7.3	6.1 – 9.0 (7)	27.3	22.7 –39.0 (7)
Skim milk								
Fluid	0.40	0.20–0.53(6)	1.7	1.5 – 1.8 (4)	0.86	0.74– 1.1 (4)	3.6	2.8 – 4.0 (6)
Dried	3.6	2.2 –4.6 (14)	18.9	13.0 –25.4 (25)	10.6	8.2 –18.3 (8)	38.8	22.9 –77.0 (12)
Buttermilk								
Fluid	0.42	— (1)	1.7	1.6 – 1.8 (2)	0.55	0.27– 0.82(2)	3.8	2.9 – 4.7 (2)
Dried	3.5	— (1)	32.0	29 –35 (10)	8.6	— (1)	28.0	27.0 –30.1 (3)
Yoghurt	0.37	— (1)	1.4	0.8 – 1.8 (3)	1.3	0.8 – 1.9 (2)	—	—
Cream								
Half and half	0.3	— (1)	1.5	— (1)	0.4	— (1)	—	—
Light table	0.3	— (1)	1.4	1.4 – 1.5 (2)	0.4	— (1)	—	—
Medium whipping	0.25	— (1)	1.3	— (1)	0.4	— (1)	—	—
Heavy whipping	0.2	— (1)	1.2	— (1)	0.4	— (1)	2.6	— (1)
Butter	0.03	— (1)	0.16	0.08– 0.37(4)	0.5	0 – 1.0 (2)	2.3	0 – 4.6 (2)
Ice cream	0.48	0.38–0.65(3)	2.3	2.0 – 2.6 (6)	1.1	1.0 – 1.2 (2)	—	—
Whey								
Fluid	0.4	— (1)	1.2	0.5 – 1.6 (3)	0.85	0.72– 1.03(3)	3.4	2.1 – 4.1 (5)
Condensed	3.3	— (1)	16.3	— (1)	3.5	— (1)	15.1	— (1)
Dried	3.7	1.7 –4.9 (5)	23.4	20.0 –29.7 (8)	9.6	8.0 –11.2 (2)	47.3	42.4 –56.0 (4)

[a] Mg per l for products designated "fluid."
[b] Mean and range of average values obtained from the literature.
[c] Figures in parentheses indicate number of references consulted.
[d] This average is based on determinations made by both microbiological and chemical assays, but some of the early values obtained by chemical means were extremely high (3.0, 4.5, and 8.2 mg per l) and have been omitted.

until about 30 to 60 days after parturition, when it reaches normal levels. During mid-lactation, the amount remains practically constant. Toward the end of lactation, a slight increase, a slight decrease, or no change at all have been variously reported.

More marked than the changes in total thiamine are the changes in the proportions of the various forms in which thiamine occurs in milk. At about the second week postpartum, the content of free thiamine shows a minimum value, while the contents of phosphorylated and protein-bound thiamine attain a maximum value. Thereafter, throughout lactation, the proportion of free thiamine tends to increase, while that of phosphorylated thiamine and, to a lesser extent, of protein-bound thiamine tend to decrease.[19,75] The changes in the various forms of thiamine are correlated with simultaneous changes in the phosphatase content of milk. Increases in phosphatase are associated with increases in free thiamine and decreases in phosphorylated and protein-bound thiamine.

Effect of Heat and Other Treatments.—Thiamine in milk is subject to more or less destruction by heat, the amount of destruction increasing with the frequency and severity of the heat treatment, but being essentially independent of the presence of oxygen.

Losses through pasteurization by the holding method amount to between 5.5 and 25%, the lower values being more commonly observed. When milk is pasteurized by the high temperature-short time process, decreases in thiamine content of only about 3 to 4% are encountered. Destruction by sterilization of milk is much greater, amounting to from 20 to 45% or more. Here again, use of a brief heating period, by the ultrahigh-temperature process, can prevent the thiamine from being lowered by more than a few percent if at all. Milk can be boiled quickly with but little disappearance of thiamine, but at times decreases of 4.5 to 8% have been observed.

Simulated milk systems containing some major components of milk and labeled thiamine were sterilized in cans and the ensuing losses of thiamine compared with control solutions containing thiamine alone. Destruction was decreased in the presence of lactose, sodium caseinate and milk fat and increased by lactose and sodium caseinate together. A major part of sulfur-containing degradation products were bound to protein; thiamine was bound reversibly to protein components.[118]

In the preparation of evaporated milk, destruction of between 20 and 60% of the thiamine originally present has been found. Losses of as little as 3.5% and as much as 75% have been reported in the preparation of condensed milk; recent figures place them at about 14 to 20%.

In the manufacture of dried milk powder, loss of 10% and at times much more has been observed; but under optimum and modern conditions of manufacture appreciable lowering of the thiamine content is avoided. None is destroyed during the preparation of instant nonfat dried milk.

The comparative losses of thiamine resulting when a number of these different processing treatments were applied by British workers to one bulk of raw milk are shown in Table 7.5.

Exposure of powdered milk to 2.79 megarads or 5.58 megarads of gamma radiation led to no decrease in thiamine, in marked contrast to 70 to 95% losses of the vitamin in certain other foods tested. This relative stability of the vitamin in powdered milk to gamma radiation was attributed to the lack of moisture in this product.[173] According to Pedersen,[130] 50,000 rads will produce losses of thiamine in liquid milk. In another investigation, exposure of liquid milk to one megarad of gamma radiation led to the disappearance of all the thiamine; replacement of air in the milk by nitrogen afforded no protection. In work by Vincent,[160] exposure to 0.147 megarad destroyed 35% of the vitamin in pasturized milk and 85% of that in sweetened condensed milk. Fortification of milk with ascorbic acid appeared to protect thiamine from the destructive effect of gamma radiation. Processing milk with cationic resins to remove radionuclides decreased the thiamine content by 50% or more.[49,76] On the other hand, treatment with anionic resins did not affect the thiamine content of milk.[5]

Effect of Storage.—When milk was kept in dark bottles for 72 hr at different temperatures, it lost 24% of its thiamine at 4°C, 14% at 12°C and 16% at 20°C. The apparent decreases were smaller at the higher temperatures because of synthesis of thiamine in the milk by lactic acid bacteria, which counteracted in part the losses due to storage.

Storage of evaporated milk for 12 months at 4°C led to no further loss of thiamine, but at room temperature and at 37°C losses increased considerably.[52] Evaporated and sterilized milk declined in potency between 15 and 50% after long periods. The thiamine content of condensed milk kept at 8 to 15°C for 2 or up to 4 years decreased by 33 to 47%.

Spray-dried whole milk did not undergo any change in thiamine potency during 6 or 12 months storage. Some decline in potency was noted after 15 months, but little change was found in roller-dried or evaporated milk after 18 months.

Content in Cheese.—Since certain differences in processing at various stages in cheesemaking give rise to different varieties of cheese, such variations might be expected to exert some influence on the relative levels of the vitamins in the finished products. This is particularly

Table 7.5

EFFECT ON SOME OF THE B VITAMINS OF DIFFERENT PROCESSING TREATMENTS APPLIED TO A SINGLE LOT OF RAW MILK[a]

Treatment of Milk	Percent Loss on Processing						
	Thiamine	Riboflavin	Niacin	B_{-6}	Pantothenic Acid	Biotin	B_{-12}
Sterilized milk							
In-bottle process							
After homogenization	<10	0	0	0	<10	<10	10
After autoclave sterilization in-bottle	30–50	<10	0	0	<10	<10	90–100
Ultrahigh temperature process							
After U.H.T.sterilization in-bulk	<10	<10	0	0	<10	<10	15–20
After U.H.T. + resterilization in-bottle	40–50	<10	0	0	10–15	<10	90–100
Evaporated milk							
After evaporation	<10	0	0	<10	<10	<10	70
After sterilization in-can	30–40	0	0	0	<10	10–15	90
Sweetened condensed milk							
After preheating with sugar	<10	0	0	<10	<10	<10	30
After condensation	10–20	<10	0	10–15	<10	10–15	40
Dried whole milk							
Roller dried	20–30	10–15	0	0	<10	10–15	20
Spray dried							
Low temperature	<10	<10	0	0	<10	10–15	35
High temperature	10–15	<10	0	0	<10	10–15	35

[a] Data from Chapman *et al.*[18]

true with the B vitamins, because of their water-soluble character and because of their synthesis by some and their utilization by other microorganisms involved in the manufacture and curing of most cheeses. As with other components, changes in the B vitamin content of natural cheese from that of milk may usually be considered as occurring in two principal stages, separation of the whey and ripening of the curd.

The retention of thiamine during the manufacture of various kinds of cheese and subsequent alterations in the content of this vitamin during ripening have been studied by several groups of workers.

In the studies of Dearden *et al.*,[27] the main decreases in thiamine took place at the drawing of the first whey. Losses in the second and third wheys were quite small, the final recovery in the young cheese being 13% for Cheddar, 12% for Cheshire, and about 14 to 17% for Stilton. No significant changes were noted during ripening of these cheeses.

Evans *et al.*,[38] however, noted a progressive decrease in thiamine content during the maturation of Cheddar cheese. For cheese ripened at 4.4°C, the average loss in 12 months amounted to from 43 to 52%, while for that cured at 14.4°C, it was 60 to 73%.

Burkholder *et al.*[11] also studied the changes in thiamine content of certain types of cheese (Camembert, Liederkranz, Brie, and Limburger) during curing. With these cheeses, relatively little change occurred in the central portions of the cheese, but the outer layers showed a large increase. A probable explanation is that microorganisms present in large numbers on the surface of the cheese synthesized certain vitamins, including thiamine. Ritter[138] showed that numerous microorganisms present on the surface of soft cheese, especially mold-producing fungi, *Oospora lactis,* yeasts and mycoderms, can produce thiamine in synthetic media not containing this vitamin.

Karlin isolated from cheese three microorganisms (*Empedobacter munsteri, Brevibacterium linens,* and a cornyebacterium) which, when cultured on a simple glucose-mineral solution, synthesized large quantities of all of the B-vitamins.[89]

Values from the literature for the thiamine content of various kinds of cheese are given in Table 7.6.

Riboflavin (Vitamin B-2)

Riboflavin is a dietary essential for man and all animals as well as for many microorganisms. It is a component of two coenzymes, flavin mononucleotide (FMN), and flavin adenine dinucleotide (FAD). As a component of enzymes, riboflavin is involved in the oxidation of glucose, fatty acids, amino acids, and purines. The flavoproteins serve

Table 7.6

CONTENT (MG/KG) OF THIAMINE, RIBOFLAVIN, NIACIN, AND PANTOTHENIC ACID IN VARIOUS CHEESES[a]

Cheese Variety[b]	Thiamine		Riboflavin		Niacin		Pantothenic Acid	
	Average	Range	Average	Range	Average	Range	Average	Range
Very hard								
Parmesan	0.23	0.20–0.26(2)[e]	4.8	2.4–7.1(2)	3.4	1.1 - 7.4 (4)	5.3	— (1)
Hard								
Cheddar	0.30	0.12–0.55(7)	5.0	3.0–8.0(11)	0.92	0.2 - 2.6 (5)	2.7	1.8– 4.0(5)
Edam	0.62	0.40–0.83(2)	4.2	3.9–4.5(3)	0.38	0.38– 0.38(2)	3.2	2.8– 3.5(2)
Gruyère	0.46	0.06–0.86(2)	3.0	2.4–3.5(2)	1.4	1.0 - 1.8 (2)	5.2	4.3– 6.1(2)
Provolone	0.24	— (1)	3.2	— (1)	3.8	1.9 - 5.8 (2)	4.8	— (1)
Swiss	0.43	0.22–0.72(3)	3.2	1.9–6.0(4)	1.7	0.7 - 3.1 (3)	3.0	2.9– 4.4(3)
Semisoft								
Blued	0.27	0.18–0.36(2)	6.0	4.5–7.2(4)	7.8	2.8 -12.5 (4)	12.6	7.8–20.5(3)
Brick	—	—	4.6	4.2–5.1(2)	1.0	0.9 - 1.1 (2)	2.9	2.8– 2.9(2)
Gorgonzola	0.40	0.12–0.68(2)	4.3	4.3–4.4(2)	3.2	— (1)	—	—
Limburger	0.80	0.80–0.80(2)	4.6	3.6–5.6(3)	1.3	0.4 - 2.0 (3)	7.9	3.0–12.8(2)
Münster	—	—	—	—	0.67	— (1)	—	—
Roquefort[e]	0.40	0.30–0.49(2)	5.4	3.5–7.8(4)	5.1	2.9 - 6.6 (4)	9.3	3.2–19.5(3)
Stilton	0.50	0.24–0.75(2)	3.0	— (1)	—	—	—	—
Soft								
Ripened								
Brie	0.60	— (1)	5.9	2.8–9.0(2)	3.8	0.5 - 7.0 (2)	7.4	0.9–14.0(2)
Camembert	0.45	0.40–0.50(2)	6.7	5.0–8.3(4)	8.2	2.8 -11.6 (5)	7.1	0.4–14.0(5)
Unripened								
Cottage	0.26	0.18–0.34(3)	3.3	2.8–4.3(6)	0.92	0.7 -1.15(3)	2.2	1.8– 2.8(3)
Cream	0.24	— (1)	2.6	1.4–5.4(6)	0.81	0.6 -1.0 (3)	2.1	1.4– 2.7(2)
Mozzarella[d]	0.32	— (1)	2.7	— (1)	1.4	0.6 -2.8 (2)	—	—
Neufchâtel	—	—	—	—	0.86	— (1)	—	—
Processed								
Cheddar	0.20	— (1)	5.1	4.3–5.6(4)	0.88	0.8 - 1.0 (3)	6.1	4.3– 7.9(2)
Swiss	0.10	— (1)	3.5	3.0–4.0(2)	0.85	0.7 - 1.0 (2)	2.6	— (1)

a Mean and range of average values obtained from the literature.
b Classified primarily according to Sanders.[140]
c Figures in parentheses indicate number of references consulted.
d May be made from milk of species other than the cow.
e Made from ewe's milk.

essentially as electron carriers, acting through dehydrogenation or hydrogenation of the isoalloxazine ring, as indicated. In some cases, they mediate electron transfer from the pyridine nucleotides to cytochrome c or other one-electron acceptors. In other instances, they catalyze electron transfer directly from the metabolite to molecular oxygen.

Only slight changes can be made in its structure without loss of vitamin activity.

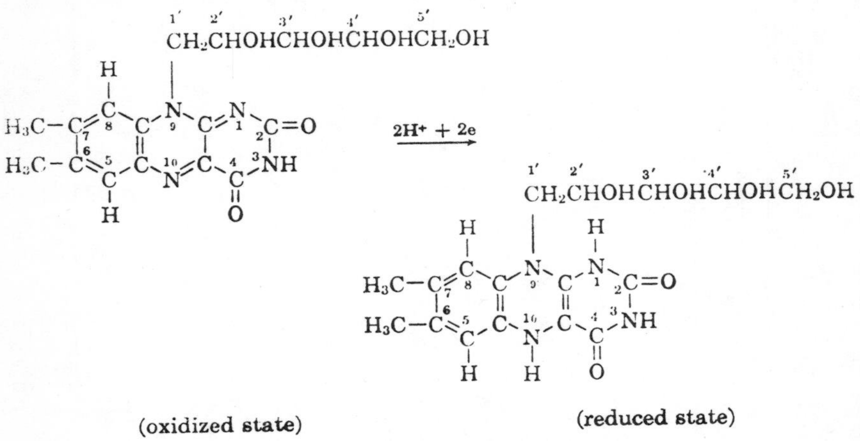

Riboflavin

The most characteristic symptoms of riboflavin deficiency in the human are lesions of the skin and of the eye. A condition known as cheilosis develops, in which the mouth, tongue and lips become abnormally red and lesions appear in the corners of the mouth. Fissures appear around the folds of the nose and ears and scaly desquamations occur. Itching and burning of the eyes, increased sensitivity to light, dimness of vision at a distance, and vascularization and opacity of the cornea are also observed. In experimental animals, severe riboflavin deficiency results in cessation of growth and eventual death.

Riboflavin is the pigment which imparts a greenish-yellow color to whey.

Forms in Milk.—From 54 to 95% of the riboflavin in cow's milk is in the free form. According to Modi and Owen,[116] the rest is present as FAD, while none occurs as FMN. Funai,[45] however, reported 21% in the form of FMN, with 14% as FAD and the remainder in the free form. Anagama *et al.*[2] observed about 25% as FMN and 21% as FAD, the rest being free. In skimmilk, Nagasawa *et al.*[121] found 94 to 95% of the riboflavin free, the remainder being bound about 60% as FMN and 40% as FAD.

In the work of Manson and Modi,[112] all the FAD in raw milk was present in bound form and none in the free form. Indeed, when FAD was added to raw milk, it quickly decomposed into free riboflavin and FMN; almost all of the latter in turn decomposed to give the free vitamin. This decomposition was attributed to phosphatase activity. In pasteurized milk, where there is no phosphatase activity, the decomposition occurs to a much lesser extent, with FMN as the main product.

Apparently the bound riboflavin in milk is for the most part attached to protein as part of an enzyme. The work of Leviton and Pallansch[102] suggests that the sites of binding between the vitamin and casein may be the tyrosine residues of the protein, which attract the ureide-carbonyl and aromatic-amine groups of the riboflavin molecule.

Riboflavin fed to ruminants gives rise to a number of riboflavin metabolites in the milk; such metabolites arise from the degradation of the vitamin by rumen microorganisms.[127]

The riboflavin-containing enzymes diaphorase and xanthine oxidase have been found in milk.

Content in Milk and Milk Products.—The average riboflavin content of fluid whole milk is about 1.74 mg per l (Table 7.4). For most adults, the National Research Council recommends an average daily allowance of 1.5 to 1.7 mg. One quart of milk per day would supply on the average between 95 and 110% of this amount. Milk contributes about 40 to 50% of the total riboflavin in the diet in the United States and many other western countries.

Values for the riboflavin content of various milk products, as taken from the literature, are also given in Table 7.4.

Effect of Season, Diet of Cow, and Breed.—The riboflavin content of milk has generally, but not always, been reported to be higher at certain seasons of the year than at others. The higher content has usually been observed during the spring or summer and has been associated with a change in ration from indoor feeding to fresh pasture feeding, when increases of 20 to 50% occurred. In other instances, however, no significant differences were found between summer and winter rations; in at least one case,[74] a decrease was noted when the change to pasture was made. Such variable results have been attributed to differences in the nature of the winter rations fed to the cows. Since the cow synthesizes riboflavin abundantly in the rumen, any effect of alterations in the diet must be an indirect one, mediated through the rumen microflora.

There are definite differences in the concentration of riboflavin in the milk secreted by cows of different breeds. The milk of Jerseys contains 25 to 50% more riboflavin per unit of weight than that of

Holsteins. Guernseys secrete as much or nearly as much as Jerseys, while Ayrshires secrete about as much as Holsteins. Shorthorn milk contains significantly less riboflavin than Holstein milk.

Change During Lactation.—Colostrum contains about 2 to 4 times the concentration of riboflavin found in later milk. The potency declines very rapidly, however, with successive milkings and by the 3rd to 8th day after calving, it reaches the level present in normal milk. Nagasawa *et al.*[121] found that a larger proportion of the riboflavin is present as FMN, FAD, or esterified riboflavin in colostrum than in normal milk.

Effect of Heat and Other Treatments.—In the absence of light, the riboflavin in milk is quite stable to heat. Pasteurization by either the holding or high temperature-short time method caused either a negligible amount of destruction of the vitamin, or none at all. Other heat treatments, such as those involved in sterilization or in the preparation of evaporated milk, dried milk, or instant nonfat dried milk, were likewise without much effect. The effects on riboflavin when a number of these different processing treatments were applied by British workers to a single large lot of milk are indicated in Table 7.5. Reports of large losses of riboflavin due to heating, which have occasionally been made, are undoubtedly complicated by the effects of light.

Influence of Light and Other Radiation.—When milk is exposed to light, the riboflavin gradually disappears. An increase in temperature or an alkaline pH accelerates the rate of destruction. Loss of the vitamin due to direct sunlight will vary according to such factors as intensity of the light and the size and type of container holding the milk. Thus, for example, when milk was placed in clear glass bottles, 2 hr exposure to sunlight resulted variously, depending upon the above factors, in a destruction of from 20 to 80% of the riboflavin. Loss from milk kept in the shade seldom exceeds 25% of that for comparable samples exposed to direct sunlight; on cloudy or rainy days, even less of the vitamin disappears. Exposure to fluorescent light is less harmful than exposure to sunlight. The rate of disappearance of the vitamin from pasteurized milk is approximately the same as or slightly less than that from raw milk, while it is still less from homogenized milk, which is more opaque.

Such an effect of light could imply a serious loss of riboflavin in the process of delivery of milk to the consumer. Loss on the ordinary delivery truck is not large, except when the milk is on the truck for long periods of time. Wooden cases give better protection than wire cases. In one study 3 to 5% of the vitamin disappeared between the time just before bottling the milk and the time just before delivery to the customer. Destruction during and after delivery when bottles were left on the doorstep averaged 4 to 8%. When milk which has been subjected to light is placed in the dark, no further loss of riboflavin occurs. All the vitamin is preserved in milk kept in highly diffused daylight in normal kitchen containers made of glass, aluminum, or porcelain.

The use of containers other than clear glass bottles increases the retention of riboflavin upon exposure to light. Brown, amber, or ruby glass bottles, or waxed paper cartons, particularly when foil-laminated, lessen considerably or eliminate its destruction.

The wavelengths of light responsible for the destruction of riboflavin in milk are those below 500 to 520 mμ; these rays, however, are incapable of penetrating more than 0.32 in. below the surface of milk. In one test, the band of light in the spectral range 420 to 560 mμ that caused the greatest destruction of the vitamin in liquid skimmilk had a principal wavelength of 445 mμ, which corresponds to a maximum in the absorption spectrum of riboflavin.

Not only does light bring about the destruction of riboflavin itself but, acting through riboflavin, it affects other components of milk. Thus the vitamin is involved in the mechanism of the photochemical oxidation of ascorbic acid (vitamin C). If the riboflavin in milk is first destroyed or removed, the vitamin C is stable to light. The stability of riboflavin toward exposure is affected very little, if at all, by the oxygen tension in the milk.[40]

All riboflavin was retained in milk which had been treated with X-rays at radiation levels which killed 99.5% of the microorganisms present. On the other hand, when milk was subjected to continuous gamma radiation from Co^{60} (80,000 roentgens/hr) 10% of the riboflavin disappeared after 1 hr and as much as 47% after 12 hr.[100] Exposure of liquid milk to one megarad of gamma radiation destroyed about 75% of the vitamin present. With powdered milk, however, application of 2.79 or 5.58 megarads led to no loss of riboflavin.[173] Processing milk with cationic resins to remove radionuclides had no significant effect on the riboflavin content.[49,76]

Influence of Storage.—No riboflavin disappears when milk is kept in a refrigerator in the dark for 24 hr or longer. All the vitamin is retained in dried milk powder stored for up to 16 months. Ice cream maintained for 7 months at about −23°C lost 5.4% of its riboflavin. The riboflavin content of condensed milk kept at 8 to 12°C for 2 yr decreased about 28% of the amount present in the freshly prepared product, while that of condensed milk stored at 10 to 15°C for up to 4 yr similarly declined by about 33%.

Content in Cheese.—In cheese-making, large percentages of riboflavin are lost in the whey. Only between 12 and 34% of the vitamin went into fresh cheese when Cheddar, Cheshire and Trappist cheeses were manufactured.

According to some workers, the riboflavin content of a number of cheeses (Cheddar, Cheshire, Limburger, Blue and Trappist) does not change during curing. Others, however, noted with Cheddar cheese an initial decrease that changed during the second month to a gradual increase, resulting in values at the end of the 6th month that were the same as in the beginning, but leading to slightly higher values by the end of the 12th month.

With other cheeses (Camembert, Liederkranz, Brie, and Limburger), relatively little change takes place in the center of the cheese, but the outside layers show increases in riboflavin content of up to 2 to 4 times that of the central portion. With Gruyere cheese, also, after 5 months, the rind is over twice as rich in the vitamin as the interior. As with thiamine, such changes have been attributed primarily to synthesis of the vitamin by microorganisms on the surface of the cheese.

No destruction of riboflavin takes place in cream cheese during a holding period of up to 14 days or in cottage cheese during a 7 day period. "Regular style" cottage cheese contains more riboflavin than "country style".

Values taken from the literature for the riboflavin content of various kinds of cheese are given in Table 7.6.

Niacin

Niacin activity is exhibited by nicotinic acid and nicotinamide.

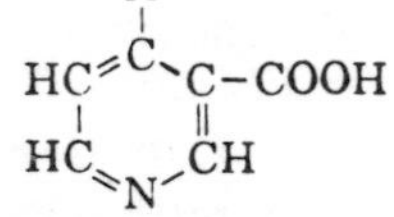

```
      H                          H
      |                          |
HC=C\C-COOH               HC=C\C-CONH2
 |     ||                  |     ||
HC\\N/CH                  HC\\N/CH
 Nicotinic acid             Nicotinamide
```

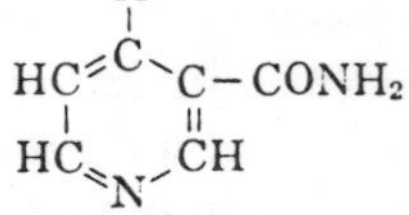

Niacin in coenzyme form plays an important part in tissue respiration. Nicotinamide is a component of two coenzymes—nicotinamide adenine dinucleotide (NAD) and nicotinamide adenine dinucleotide phosphate (NADP). These coenzymes are widely distributed in animal tissues. They participate in a series of biochemical reactions, the effect of which is to transfer electrons from a substrate to another electron acceptor, such as the flavin enzymes, and in the course of which the pyridine nucleotides are alternately reduced and reoxidized. Numerous oxidative biochemical reactions are dependent on these coenzymes, including reactions involved in the Embden-

Meyerhof glycolytic cycle, in the citric acid cycle, and in the breakdown of fatty acids.

All living organisms require niacin, although plants and some animals synthesize it. While the disease pellagra in man usually involves deficiencies of other B vitamins also, niacin overcomes its primary symptoms. These symptoms include epithelial lesions of the digestive tract, with ulcers of the mouth, a fiery red tongue and a soreness and inflammation of the intestinal tract; a dermatitis consisting of a rash which appears simultaneously on those parts of the body normally exposed to the sun; and lesions of the nervous system, which cause mild mental disturbances that may progress eventually to insanity. In dogs, niacin deficiency brings about a disease characterized by a very dark color of the tongue and thus termed "black-tongue".

Forms in Milk.—There is apparently no bound form of the vitamin in milk; 99% of the amount in skimmilk powder consists of nicotinamide,[97] and a similar proportion presumably occurs in whole milk.

Content in Milk and Milk Products.—The average niacin content of fluid whole milk, as determined from the literature, is 0.93 mg per l (see Table 7.4). The recommended dietary allowance of the National Research Council for most adults ranges from 13 to 18 mg per day. It is evident that a quart of milk could supply only a small fraction of these requirements directly. It has been estimated that dairy products contribute about 2% of the total niacin in the American diet.

However, even though the content of niacin in milk is relatively low, milk and milk products are among the most valuable pellagra-preventive foods. This apparent contradiction is explained by the fact that the niacin of milk is fully available, in contrast to the unavailable bound form in which it occurs in certain other foods, such as cereals, and even more by the fact that milk protein is an excellent source of tryptophan (about 500 mg per l of milk). Tryptophan can be used by many animals and man for the synthesis of niacin in the body.

Values derived from the literature for the niacin content of a number of milk products are given in Table 7.4.

Effect of Season and Other Factors.—The niacin content of milk has been observed to be higher during late spring and summer than during winter. This difference was attributed by some workers to pasture feeding. More often, however, no effect of dietary alterations has been observed. There appears to be no difference among breeds in the concentration of niacin secreted in the milk. Colostrum does not contain appreciably more niacin than ordinary milk. During the first 5 months of lactation, a gradual decline in the amount of the vitamin occurs, followed, according to some reports, by an increase thereafter.

Effect of Heat and Other Treatments.—Various heat treatments of milk, such as pasteurization, sterilization, boiling, or those used in the preparation of evaporated milk, condensed milk, dried milk, or instant nonfat dried milk, have little or no effect upon the niacin content (see also Table 7.5).

Exposure of milk to direct sunlight for 2 hr in half-pint bottles is without effect. Subjection of powdered milk to 2.79 or 5.58 megarads of gamma radiation did not result in large losses of niacin.[173] About one-third of the vitamin was destroyed in liquid milk exposed to one megarad; replacement of air in the milk by nitrogen afforded some protection. Processing milk with cationic resins to remove radionuclides decreased the niacin content by 27%.[49,76]

No loss of niacin occurred when frozen milk was held for 21 weeks at −14°C, nor was there any decrease when dried whole milk, packed either in air or inert gas, was stored for 1 yr at room temperature or for 6 months at 37.8°C.

Content in Cheese.—In a study of numerous varieties of cheese,[144] the proteolytic cheeses contained the highest average concentration of niacin; very hard, soft unripened, hard and semihard were next in decreasing order. There were, however, wide variations among cheeses in each of these classes and even among samples of the same variety. As with the other B-vitamins, much of the niacin of milk passes into the whey when cheese is made. Only 10 to 23% of this vitamin was recovered in fresh Cheddar and Trappist cheeses.

The niacin content of Cheddar cheese increased rapidly during the first 3 to 4 weeks of maturation, then tapered off when ripened at 4.4°C, but continued to increase at 10°C and 15.6°C.[124] Addition of lactose to the cheese increased the amount of niacin synthesized until the added lactose was used up.

During ripening of Limburger cheese, the niacin content increases threefold.[152] Burkholder *et al.*[11] found during curing a 2- to 3-fold increase in the vitamin in the core of Camembert, but no change in the interior of Liederkranz and some decrease in the center parts of Brie and Limburger cheeses. The niacin content of the surface layers of all these cheeses increased 10 to 25-fold.

Values from the literature for the niacin content of various kinds of cheese are given in Table 7.6.

Vitamin B-6

Pyridoxine, pyridoxal and pyridoxamine are all active as vitamin B-6. While these three compounds have equal vitamin B-6 activity for the rat, they have different activities for various microorganisms. The biologically active agent is pyridoxal.

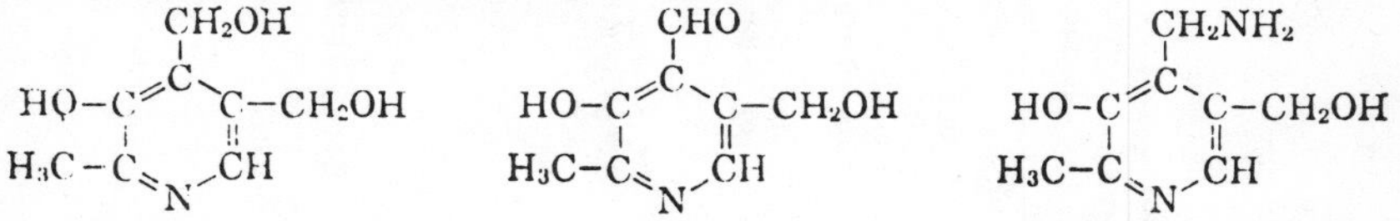

```
      CH2OH                    CHO                      CH2NH2
        |                       |                         |
        C                       C                         C
HO-C=    `C-CH2OH      HO-C=     `C-CH2OH      HO-C=       `C-CH2OH
   |      ||              |       ||              |         ||
H3C-C=   ,CH          H3C-C=    ,CH           H3C-C=      ,CH
      N                       N                        N
  Pyridoxine              Pyridoxal               Pyridoxamine
```

Vitamin B-6 plays an important role in amino acid metabolism in its coenzymatically active form of pyridoxal-5-phosphate. It is required as a coenzyme for transamination reactions in which the amino nitrogen of one amino acid is transferred to the keto analog of another amino acid. Pyridoxal phosphate is also required in the decarboxylation of amino acids and is involved in various other phases of the metabolism of alanine, serine, threonine, glycine, sulfur amino acids, and tryptophan.

Vitamin B-6 is essential for animals, birds, fish, and numerous microorganisms, as well as for man. In the rat, deficiency of this vitamin leads to a decreased growth rate, dermatitis, anemia, and convulsions. Vitamin B-6 deficiency has been produced experimentally in man, but the importance of the vitamin to humans, particularly infants, was not appreciated until recent years, when a convulsive syndrome that occurred in a small percentage of infants fed a commercial sterilized infant formula was found to be promptly alleviated by pyridoxine administration.

Forms in Milk.—Of the vitamin B-6 in whole milk, 14% is in bound form and 86% in free form.[146] The greater part (70 to 95%) is present as pyridoxal, almost all the rest being in the form of pyridoxamine. About 2 to 3% in nonfat dry milk may occur as pyridoxine.[159]

Content in Milk and Milk Products.—There is considerable variation in the vitamin B-6 content of milk as reported in the literature. However, the average of a number of published figures is 0.60 mg per l (Table 7.7). The recommended dietary allowance of the National Research Council for most adults is 2 mg per day. One quart of milk would supply on the average about 30% of this estimated requirement.

Figures taken from the literature for the vitamin B-6 content of various milk products are given in Table 7.7.

Effect of Season, Diet of Cow, and Breed.—The highest concentration of vitamin B-6 in milk occurs in the spring and early summer, while the lowest occurs in winter. The elevation in the spring has been attributed to the change from barn to pasture feeding, acting through an alteration in the microflora of the rumen and a consequent increase in the amount of the vitamin synthesized. No differences

Table 7.7

CONTENT (MG/KG[a]) OF VITAMIN B-6, BIOTIN, FOLACIN, AND VITAMIN B-12 IN MILK AND MILK PRODUCTS[b]

Milk or Milk Product	Vitamin B-6		Biotin		Folacin		Vitamin B-12	
	Average	Range	Average	Range	Average	Range	Average	Range
Whole milk								
Fluid	0.60	0.17–1.90[c](49)[d]	0.030	0.012–0.060(24)	0.059	0.038–0.090(11)[e]	0.0042	0.0024–0.0074(59)
Condensed	0.51	0.42–0.59 (3)	0.040	0.032–0.47	0.161	— (1)	0.0039	0.0031–0.0054(3)
Evaporated	0.69	0.44–1.37 (6)	0.056	0.031–0.090(3)	0.098	0.085–0.111(2)	0.0014	0.0010–0.0019(4)
Dried	3.9	1.7 –7.0 (7)	0.30	0.10 –0.47 (5)	0.50	— (1)	0.026	0.018 –0.038 (6)
Skimmilk								
Fluid	0.41	0.26–0.56 (4)	0.016	0.015–0.016(2)	—	—	0.0037	0.0031–0.0047(5)
Dried	4.4	2.8 –6.8 (9)	0.27	0.14 –0.35 (3)	0.60	0.57 –0.62 (2)	0.034	0.022 –0.045 (9)
Buttermilk								
Fluid	0.35	0.28–0.40 (4)	0.011	— (1)	0.11	— (1)	0.0023	— (1)
Dried	2.4	— (1)	0.29	— (1)	0.40	— (1)	0.019	0.018 –0.02 (1)
Yoghurt	0.36	0.35–0.38 (2)	0.012	— (1)	—	—	0.0012	0.0007–0.0018(5)
Cream								
Half and half	0.38	— (1)	—	—	—	—	—	—
Light table	0.40	— (1)	—	—	—	—	—	—
Medium whipping	0.35	— (1)	—	—	—	—	—	—
Butter	0.04	— (1)	—	—	—	—	—	—
Whey								
Fluid	0.42	0.21–0.77 (3)	0.014	0.013–0.015(2)	—	—	0.0020	0.0015–0.0024(4)
Condensed	1.8	— (1)	0.29	— (1)	—	—	—	—
Dried	4.0	— (1)	0.37	— (1)	0.89	0.88 –0.90 (2)	0.021	0.017 –0.025 (3)

a Mg per l for products designated "fluid."
b Mean and range of average values obtained from the literature.
c Three figures outside this range, one much higher (6.5) and two much lower (each 0.06) have been omitted.
d Figures in parentheses indicate number of references consulted.
e The earlier inaccurate microbiological assays have been excluded.

in the vitamin B-6 content of milk were observed between Jersey and Guernsey cows or between Holstein and Shorthorn cows.

Change During Lactation.—Vitamin B-6 content appears to be elevated somewhat in early colostrum. It reaches its highest concentration, which is about 2 to 3 times normal, at about the 3rd to 14th day of lactation, and then gradually falls to the level in later milk. The low point occurs at about the 4th or 5th month of lactation, the values remaining more or less constant thereafter until at least the 36th week of lactation.

Effect of Heat and Storage.—No significant decreases in vitamin B-6 occur during pasteurization or homogenization of milk or in the manufacture and storage of dried whole milk, dried skimmilk, or instant nonfat dried milk (see also Table 7.5). Large losses of vitamin B-6 have been observed, however, during heat sterilization of milk. In the work of Hassinen *et al.*,[62] sweetened condensed milk, a product not sterilized by heating, retained 78% of the vitamin B-6 of fresh milk. Autoclaving and sterilization brought about greater destruction, evaporated milk retaining only 51 to 64% of the vitamin B-6 present in fresh milk. The decreases in this vitamin took place not only during heating but continued for as much as 10 days during storage thereafter. Beyond that time, further loss was small, even when the milk was held in storage for 1 yr at room temperature.

Hodson[72] explained this apparent decline in vitamin B-6 activity of stored evaporated milk on the basis of a conversion of pyridoxal, first to pyridoxamine and then into an unknown form of vitamin B-6. Bernhart *et al.*[6] suggested that the unknown form of the vitamin was a sulfur-containing compound formed when they autoclaved concentrated milk (25% solids), to which pyridoxal hydrochloride had been added, and allowed it to stand for several weeks at room temperature. They attributed the formation of this compound, which Wendt and Bernhart[163] identified as bis-4-pyridoxal disulfide, to a reaction between pyridoxal and active sulfhydryl groups liberated, particularly from cysteine, during the destruction of milk proteins in the course of heat sterilization. Gregory,[110] however, felt that it had not been shown conclusively that this is the fate of the naturally occurring vitamin in the milk.

Not all workers have noted a loss of vitamin B-6 in the preparation of sterilized and evaporated milk (see Table 7.5). Vitamin B-6 is well-protected by ultrahigh-temperature sterilization of milk; losses of between 0 and 35% were reported, but generally they were small.[51] However, losses of the vitamin due to repeated heat treatments are cumulative, and increase with severity and frequency of heating.[12]

These differences in experimental results may be accounted for, no doubt, by numerous factors, such as variations in sterilizing procedures, previous history of the milk in regard to aeration and exposure to light, and the period and temperature of storage after sterilization.[18] However, removal of dissolved oxygen by flushing milk with nitrogen made no significant difference in the loss of vitamin B-6 due to sterilization.[40]

In tests with infants, a daily intake of approximately 1 l of a mixture containing about 40% evaporated milk and supplying 0.26 mg vitamin B-6 was sufficient in most instances to prevent or cure the convulsive syndrome produced by a deficiency of this vitamin. This amount of vitamin B-6 was about 2 to 3 times the minimum estimated to be required for this protection.

Effect of Light and Other Radiation.—The vitamin B-6 content of milk decreased by 21.4% after 8 hr exposure to daylight in clear glass bottles. In other work, losses of 9 to 31% were found after exposure of milk in similar containers to sunlight for 1 hr and 15 to 46% after 2 hr. Black-lacquered bottles afforded essentially complete protection.

Pedersen[130] found that 50,000 rads of gamma radiation were required to produce losses of vitamin B-6 in liquid milk. Subjection of milk to 1 megarad brought about a decrease of 89% in its vitamin B-6 content. Processing milk with cationic resins to remove radionuclides decreased the vitamin B-6 content by 15%.[49,76]

Content in Cheese.—In a study of a large number of cheese varieties,[144] the proteolytic types had the largest average concentration of vitamin B-6, very hard, hard, semihard, and soft unripened following in decreasing order.

About 45 to 81% of the vitamin B-6 in milk was lost during the manufacture of Cheddar cheese.[41,124] In one investigation,[124] the vitamin B-6 content increased during the first 2 weeks of maturation, decreased for the next 6 weeks, and then increased for 9 months. Addition of lactose to the cheese increased the rate of vitamin B-6 synthesis.

In other studies,[85] involving several varieties of cheese, between 54 and 70% of the vitamin in the original milk passed into the whey. The residual vitamin decreased during the first few days after manufacture, apparently as a result of its consumption by lactic acid bacteria, but then increased on the surface of the cheese and in some cases in the interior, because of synthesis by yeast and mold organisms.

Polansky and Toepfer[131] indicated that over 80% of vitamin B-6 in ripened cheese consists of pyridoxamine, about ¾ of the remainder being present as pyridoxal and the rest as pyridoxine. In cottage cheese, this vitamin occurred about ¼ as pyridozine and the remainder about equally as pyridoxal and pyridoxamine.

Values from the literature for the vitamin B-6 content of various cheeses are given in Table 7.8.

Pantothenic Acid

Pantothenic acid is essential to all living organisms, including man. Only the dextrorotatory form of the compound is effective as a vitamin. It is physiologically active as a part of coenzyme A. This coenzyme plays a vital role in the metabolism of carbohydrates, fatty acids, and nitrogen compounds. It participates in a wide variety of reactions involving the transfer of acetate or 2-carbon units.

$$CH_2OH—C(CH_3)_2—CHOH—CO—NH—CH_2—CH_2—COOH$$

Pantothenic acid

Pantothentic acid-deficient chicks develop keratitis, dermatitis, degenerative changes in the spinal cord, fatty livers, and involution of the thymus. Rats deficient in this vitamin fail to grow normally, and develop porphyrin-stained whiskers, graying of the hair, dermatitis, a "spectacled" eye condition, and necrosis and hemorrhage of the adrenal cortex. Pantothenic acid deficiency has not been observed in the human.

Forms in Milk.—Zook *et al.*[174] found about one-fourth of the pantothenic acid in milk to be bound, and in cheese the proportion was even higher. Thus about 40% of the vitamin in Cheddar cheese was present in the bound state; in cottage cheese about 60% was in this form. In the studies of Brochu *et al.*,[8] bound forms represented 16 to 28% of the total pantothenic acid in milk. Coenzyme A but no phosphopantetheine was detected in milk.[68]

Content in Milk and Milk Products.—The average pantothenic acid content of fluid whole milk, as determined from published figures, is 3.39 mg per l (Table 7.4). The requirement of the human for this vitamin is not known but has been estimated to be approximately 5 to 10 mg per day. On the average, about 35 to 65% of this estimate is supplied by one quart of milk. Values for the pantothenic acid content of various milk products are also given in Table 7.4.

Effect of Diet and Other Factors.—The feed of the cow does not affect the pantothenic acid content of milk.

The effect of breed of cow has also been studied. In none of the experiments, however, were differences between breeds clearly significant.

The amount of pantothenic acid in colostrum at parturition is less

Table 7.8

CONTENT (MG/KG) OF VITAMIN B-6, BIOTIN, FOLACIN, AND VITAMIN B-12 IN VARIOUS CHEESES[a]

Cheese Variety[b]	Vitamin B-6		Biotin		Folacin		Vitamin B-12[c]	
	Average	Range	Average	Range	Average	Range	Average	Range
Very hard								
Parmesan	0.96	— (1)[d]	0.030	0.017 –0.043 (2)	0.073	— (1)	—	—
Hard								
Cheddar	0.74	0.43–1.02(4)	0.022	0.017 –0.033 (3)	0.095	0.05 –0.17 (4)	0.013	0.006 –0.028 (6)
Colby	—	—	0.016	— (1)	—	—	—	—
Edam	0.73	0.60–0.84(3)	0.015	— (1)	0.34	0.16 –0.53 (2)	0.017	0.014 –0.023 (4)
Gruyère	0.78	0.76–0.81(2)	0.013	0.0084–0.017 (2)	0.101	0.098–0.104(2)	0.016	— (1)
Provolone	0.83	— (1)	0.018	— (1)	0.104	— (1)	—	—
Swiss	1.83	0.34–5.6 (4)	0.005	0.0004–0.0094(3)	0.072	0.064–0.08 (2)	0.018	0.009 –0.028 (4)
Semisoft								
Blue[e]	1.60	0.97–2.30(4)	0.046	0.016 –0.076 (2)	0.48	0.36 –0.59 (2)	0.014	— (1)
Brick	0.73	— (1)	0.022	0.016 –0.028 (2)	0.20	— (1)	—	—
Gorgonzola	1.06	— (1)	0.019	— (1)	0.31	— (1)	0.012	— (1)
Limburger	0.54	0.2 –0.89(2)	0.086	0.020 –0.200 (3)	0.58	— (1)	0.010	— (1)
Münster	0.84	0.76–0.93(2)	0.012	0.011 –0.014 (2)	0.33	0.12 –0.53 (2)	0.016	— (1)
Roquefort[f]	1.00	0.97–1.04(2)	0.025	0.015 –0.036 (2)	0.46	0.43 –0.49 (2)	0.014	0.006 –0.027 (4)
Soft								
Ripened								
Brie	3.71	1.52–5.9 (2)	0.062	0.045 –0.080 (2)	0.65	— (1)	0.016	— (1)
Camembert	2.06	1.3 –2.50(4)	0.045	0.023 –0.057 (3)	0.62	0.62 –0.62 (2)	0.013	0.012 –0.014 (3)
Unripened								
Cottage	0.35	0.16–0.54(2)	0.020	— (1)	0.30	0.29 –0.33 (3)	0.0085	0.0059–0.0109(4)
Cream	0.53	— (1)	0.014	0.012 –0.016 (2)	0.14	— (1)	0.0021	0.002 –0.0022(2)
Mozzarella[e]	0.41	0.17–0.64(2)	0.016	— (1)	0.099	— (1)	0.0022	— (1)
Neufchâtel	—	—	0.019	— (1)	0.11	— (1)	—	—
Processed								
Cheddar	0.64	0.45–0.82(2)	0.026	0.017 –0.036 (2)	0.089	0.078–0.10 (2)	0.008	— (1)
Swiss	0.35	— (1)	0.011	— (1)	—	—	0.012	— (1)

[a] Mean and range of average values obtained from the literature.
[b] Classified primarily according to Sanders.[140]
[c] Determined by microbial assay. By a hyperthyroid rat method, Cheddar cheese assayed 0.014 mg/kg (1 reference). For values obtained by employment of an assay using the normal rat, see Table 7.9.
[d] Figures in parentheses indicate number of references consulted.
[e] May be made from milk of species other than the cow.
[f] Made from ewe's milk.

than the level in normal milk. During the first 4 to 14 days after parturition, it increases to a maximum value, which may be 15 to 70% above normal. It then begins a gradual decline, which in one study did not reach its lowest point until about 16 to 20 weeks after parturition. In this case, there was no further change during the remainder of the 36 weeks of lactation observed.

Changes in coenzyme A levels paralleled those of pantothenic acid. Such levels are much higher during second and subsequent lactations than during the first lactation.[68]

Effect of Heat and Other Treatments.—Little or no pantothenic acid is lost during the pasteurization of milk by either the holding or high temperature-short time process. Essentially all the vitamin is retained during the preparation of evaporated milk, condensed milk, sterilized milk, dried whole milk, dried skimmilk, or instant nonfat dried milk (see also Table 7.5). The vitamin is stable in dried whole milk stored at room temperature for 6 months or at 37.8°C for a year.

Exposure of samples of pasteurized milk for 2 hr in half-pint bottles to February noonday sunlight failed to result in any destruction of the vitamin, nor was any loss observed when liquid milk was subjected to 1 megarad of gamma radiation.

Processing milk with cationic resins to remove radionuclides had no effect on the pantothenic acid content.[49,76]

Content in Cheese.—In a study of the pantothenic acid content of many varieties of cheese,[144] the proteolytic types contained the largest amounts of this vitamin. The soft unripened types, on the average, contained less than the hard cheeses, when compared on a fresh-weight basis, but the reverse was true when they were considered on a dry-matter basis.

Large losses of the pantothenic acid of the original milk, particularly of the free form of the vitamin, occur during the manufacture of Cheddar and Cottage cheese.[174] During the curing of Cheddar cheese, the pantothenic acid content decreased for the first few weeks and then gradually increased during the remainder of the maturation period.[124] The pantothenic acid content becomes larger during the ripening of some cheeses (Limburger, Camembert, Brie, etc.) as a result of synthesis by microorganisms. As with other B vitamins, these changes are particularly marked in the outer layers.

Values as given in the literature for the pantothenic acid content of different varieties of cheese are given in Table 7.6.

Biotin

Biotin is an essential nutrient for man and numerous animals, as well as for many microorganisms. Since it can be synthesized by the intestinal flora, however, a deficiency can be produced experimentally in many animals only by feeding raw egg white or sulfa drugs. Biotin participates in many carboxylation and decarboxylation reactions in metabolism. These reactions include the conversion of propionate to methylmalonate, of pyruvate to oxaloacetate, of malate to pyruvate and of oxalosuccinate to alpha-ketoglutarate. Other reactions in which biotin is required include those involved in fatty-acid synthesis and in the metabolism of aspartic and other amino acids.

```
          O
          ‖
          C
      HN3' 2' 1'NH
      |         |
      HC4 ——— 3CH
      |         |   6    7    8    9    10
     H2C5  1  2CHCH2CH2CH2CH2COOH
           S
```

Biotin

Biotin-deficient rats develop a "spectacled" eye condition and, with a more extreme deficiency, a very severe dermatitis. Chicks develop dermatitis and perosis. In man, experimentally induced deficiency of the vitamin leads to dermatitis, depression, lassitude, muscle pains, and anorexia.

Forms and Content in Milk and Milk Products.—Milk is a fairly good source of biotin and one of the substances from which this vitamin was originally isolated. Biotin is present in milk in the free state; there does not appear to be any in bound form.

The average biotin content of fluid whole milk is 30 μg per l (Table 7.7). Considerable variation is found in the biotin content of milk from individual cows. An estimated daily intake of 150 to 300 μg is considered adequate for most people. A quart of milk would supply on the average about 10 to 20% of this. The biotin contents of some milk products are also given in Table 7.7.

Influence of Season and Other Factors.—No influence of season on the biotin content of milk has been observed, except as it may be related to feed. Differences found with variations in diet have been attributed to an effect on the rumen microorganisms. Most studies indicate no significant variations due to breed.

The biotin content of colostrum at parturition is slightly below that of normal milk. It rises to a peak, which in one instance was about twice the normal level, at about the 4th day of lactation, then drops off to the normal level by the 7th day after calving. No real differences occur during the remainder of lactation.

Effect of Heat and Other Treatments.—Loss of biotin during pasteurization of milk by either the holding process or the high temperature-short time method or during sterilization is less than 10%. In the preparation of evaporated, condensed, or dried whole milk, it is not over 10 to 15% (see also Table 7.5). No loss occurs during ultrahigh-temperature sterilization.

Storage of milk in the frozen state at about −14°C for 19 weeks had no influence on the amount of biotin. Similar results were obtained with dried milk maintained either for 1 yr at room temperature as air-packed samples or for 6 months at 37.8°C in air or inert gas.

No destruction was observed when milk samples were exposed to sunlight for 2 hr in half-pint bottles, nor did any occur when liquid milk was subjected to 1 megarad of gamma radiation. Processing milk with cationic resins to remove radionuclides had no significant effect on biotin levels.[49]

Content in Cheese.—In a study[144] of a large number of varieties of cheese, the proteolytic cheeses on the average contained the most biotin, followed in decreasing order by soft, very hard, semihard, and hard types when the comparison was made on a fresh-weight basis.

In the manufacture of Cheddar cheese, about 85 to 90% of the biotin of the original milk is lost; the vitamin content of the curd increased for the first 2 months, then decreased sharply to below the original level at the end of the next 2 months, with little change thereafter.[124] The biotin content increases during ripening of many cheeses (Limburger, Brie, etc.) as a result of synthesis by microorganisms. Often the increase is primarily if not altogether confined to the outer layers, but sometimes it is more general. Ritter[138] showed that numerous microorganisms present on the surface of soft cheese, especially mold-producing fungi, *O. lactis,* yeasts, and mycoderms, can produce considerable quantities of biotin when cultured in media free of this nutrient.

Values for the biotin content of various kinds of cheese are given in Table 7.8.

Folacin

Folic acid and its conjugates with different numbers of glutamic acid residues are active as folacin. The conjugates vary somewhat in their properties and in their potencies for various animals and microorganisms. Most higher animals are capable of utilizing the conjugates as well as folic acid itself.

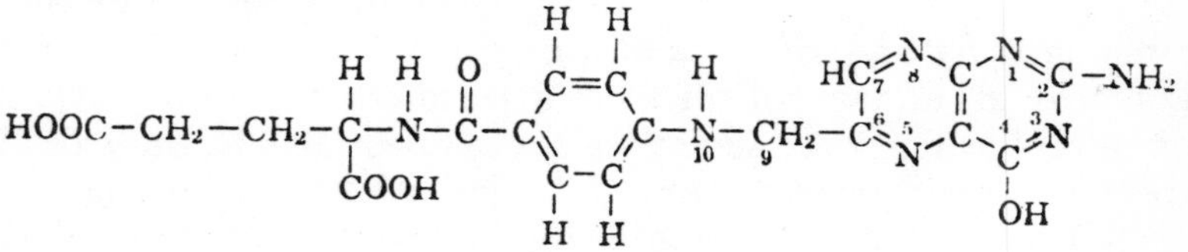

```
                                    H   H
                                    |   |                                N       N
               H   H   O           C=C       H        HC7  8  C  1  C—NH2
               |   |   ||         /    \      |         |6   5  ||   2 |
HOOC—CH2—CH2—C—N—C—C          C—N—CH2—C     C   4  3N
               |                  \    /     10   9          N      C
              COOH                 C—C                              |
                                    |   |                          OH
                                    H   H
```

Folic acid

The physiologically active form of the vitamin is 5,6,7,8,-tetrahydrofolic acid linked in the 5N, 10N or 5N-10N positions with formyl, hydroxymethyl, methyl or formimino groups.

Folacin is a growth factor for various microorganisms and animals. It functions as a coenzyme in reactions involving the incorporation of formate and other single carbon fragments, and thus participates in the metabolism of purines, pyrimidines, and the amino acids methionine, histidine, valine, and serine. In these reactions, folacin serves as an activator and carrier of one-carbon units in their various oxidation levels.

Various animal species differ in their requirement for folacin, depending partly upon the extent to which the vitamin is provided to them by bacterial synthesis in their intestinal tracts. In animals, folacin deficiency causes a low growth rate, anemia, leucopenia, and granulocytopenia. In man, as well as in animals, the vitamin is essential for normal red blood cell development. Diseases such as sprue, nutritional megaloblastic anemia, and megaloblastic anemia of pregnancy respond to administration of folic acid or its conjugates.

Content and Forms in Milk and Milk Products.—Early assays which indicated that milk was a relatively poor source of folic acid proved inaccurate. More recent assays, using an assay organism which responded to bound forms and an assay medium containing ascorbic acid to protect the vitamin from oxidative destruction, indicated a much higher value in milk. According to these recent assays (Table 7.7), milk contains an average of 59 μg per 1. The estimated requirement for the human is 400 μg per day. A quart of milk would supply about 14% of this requirement. As measured by hematological responses of patients, the results of experiments suggested that all milk folacin is available.[46]

Sauberlich and Baumann[141] reported that pasteurized milk contained only a low percentage of 5-formyltetrahydrofolic acid. Collins *et al.*[22] found this form to represent less than 20% of the folacin content of raw milk, while Nambudripad *et al.*[122] obtained a value of about 17% for their sterilized skimmilk samples. The studies of Karlin,[84] however, indicated a much larger proportion of this form in milk.

According to Ford *et al.*[43] folacin in cow's milk occurs predominantly as N^5-methyltetrahydrofolate, as judged by chromatography and differential microbiological assay. The folate in milk is strongly and specifically bound to a minor whey protein, forming a complex of primary molecular weight of about 38,000, but exhibiting concentration-dependent reversible aggregation. The binding protein is present in excess and the milk has the capacity to bind about 50 μg of added folic acid per l. Heating for 10 min at 100°C releases the folate and destroys the binding activity. Folic acid and N^5-methyltetrahydrofolic acid compete for the binding protein and folic acid is preferentially bound. At pH 6 to 8.8, the folate is entirely bound to protein; at pH 5, free folate is present; and at pH 3.6, the complex is wholly disassociated. On restoring the pH to 7.0, the folate and protein recombine.[44] Folic acid is bound much more firmly in milk than in blood serum.[114]

Values for the folacin content of a number of milk products are given in Table 7.7.

Effect of Breed and Stage of Lactation.—As among Guernsey, Holstein, and Jersey cows, no significant effect of breed has been observed on the folacin content of the milk.

The folacin content of colostrum on the day of parturition is from 4 to 36 times the level in normal milk, but the concentration drops very rapidly, reaching normal levels by the 2nd to 3rd day of lactation. The level tends to decrease progressively during the course of lactation.

Effect of Heat and Other Treatment.—Folacin is subject to considerable loss from heat treatment.[12] The effects of successive heat treatments are not simply additive; the first heat treatment magnifies the effect on folacin of subsequent heat treatments. Losses are caused by dissolved oxygen in the milk and can be prevented by excluding oxygen.[40] The loss of folacin is related to the content of reduced ascorbic acid in the milk; as long as any remains, it is oxidized in preference to the folacin. High temperature-short time pasteurization brings about 0 to 12% loss of folacin. Sterilization of milk destroys 40 to 50% of the vitamin; subsequent exposure to sunlight increases the loss considerably. On the other hand, in-bottle sterilization and subsequent storage in the darkness caused no loss of folacin from milk which had been flushed with nitrogen; subsequent exposure to sunlight brought about a 10% loss.

Incubation of fresh cow's milk overnight at pH 6.6 and 37°C produced large increases in folacin.[71] Processing milk with cationic resins to remove radionuclides had no significant effect on folacin.[49,76]

Content in Cheese.—The folacin content of cheese varies considerably from one type to another. It is dependent on the method of manufacture, the degree of maturation, and the microorganisms involved in ripening. The smallest amounts are found in fresh cheese. During ripening, large increases in folacin concentration often occur, particularly at the surface.[83] In work[144] involving many varieties of cheese, the proteolytic cheeses contained on the average the greatest concentration of this vitamin, while the hard cheeses contained the least.

In the manufacture of Cheddar cheese, 60% of the original folacin of milk was retained in the curd;[124] this increased during the first week of ripening, decreased in the following 6 weeks and then increased during the remainder of the 12-month ripening period.

Karlin[88] isolated from cheese a species of empedobacter which, when cultured on a simple glucose-mineral medium, synthesized folacin, about 21% of which was in the form of 5-formyltetrahydrofolic acid. Values for the folacin content of various types of cheese are given in Table 7.8.

Vitamin B-12

Vitamin B-12 activity is exhibited by cyanocobalamin and a number of other corrinoids. The vitamin possesses the most complex structure of any of the vitamins and is unique in being the only one containing a metallic element—cobalt. Cyanocobalamin consists of two major parts—a complex corrin structure closely resembling the porphyrins and a nucleotide-like portion, 5,6-dimethyl - 1 - (α - D - ribofuranosyl) benzimidazole - 3′ - phosphate. The phosphate of the "nucleotide" is esterified with 1-amino-2-propanol, which in turn is joined in amide linkage with a propionic acid side chain on ring D of the large cyclic structure, thus making a bridge between the two major parts. A coordinate link between the cobalt atom and one of the nitrogen atoms on the benzimidazole forms a second bridge. The cyano- group can be replaced by other groups, such as hydroxo-, aquo-, nitroto-, etc., giving rise to a series of cobalamins.

In addition to the cobalamins, a closely related series of analogs also occur. They differ from the cobalamins in most instances only in the nature of the base in the nucleotide portion of the molecule, containing in place of the 5,6-dimethylbenzimidazole a different substituted benzimidazole or a purine (e.g., adenine).

The primary source of vitamin B-12 and its analogs in nature is, almost exclusively, microbial synthesis. Vitamin B-12 is an essential metabolite for a wide variety of organisms. Man and the higher animals apparently are rather specific in their need for cobalamins or for analogs that contain benzimidazole or substituted benzimidazoles as the nucleotide base, although cobalamin with napthimidazole as the base can also be utilized.

Deficiency of vitamin B-12 in animals may produce a variety of pathological conditions. In young animals, marked retardation of growth occurs, accompanied by lowered food consumption. The human disease, pernicious anemia, occurs almost always as the result of faulty absorption of vitamin B-12 from the food.

Study of a wasting disease of ruminants grazing on certain areas in many parts of the world showed that the condition was caused by

Cyanocobalamin

a nutritional deficiency of cobalt, which was lacking in the soil and herbage. It later became evident that the deficiency was primarily one of vitamin B-12. Cobalt given to ruminants by mouth is used by the rumen flora to synthesize vitamin B-12, which then becomes available to the host animal. Under normal conditions, such synthesis is predominantly the sole source of the vitamin for cows and other ruminants with an actively functioning rumen. For example, amounts of animal-active vitamin B-12 in rumen contents of dairy heifers fed chopped alfalfa hay, pelleted hay, hay-grain, or silage were 17 to 663 times those available from the feed.[33] *Selenomonas ruminantium* and *Peptostreptococcus elsdenii* are among the chief functional rumen microorganisms responsible for the production of vitamin B-12 or its analogs in the cow.[34]

The mode of action of vitamin B-12 in biochemical systems has been shown to be coenzymatic, either (1) linked directly or (2) as part of a cobamide coenzyme containing a 5′-deoxyadenosine group covalently bonded to the cobalt atom. The dimethylbenzimidazolecobamide coenzyme, as a constitutent of methylmalonyl-coenzyme A mutase, participates in the metabolism of propionic acid in bacteria and in mammals, the dominant pathway in animals being, in abbreviated form:

$$\text{propionyl CoA} \xrightarrow{\text{propionyl carboxylase}} \text{methylmalonyl CoA} \xrightarrow{\text{MM CoA mutase}} \text{succinyl CoA} \longrightarrow \text{citric acid cycle}$$

MM CoA mutase is particularly important for ruminants since propionic acid produced by rumen microorganisms serves as a major source of energy for these animals.

Studies of the relative occurrence of B-12 coenzymes and vitamins have indicated that, as between the two, the coenzymes are the predominant "natural forms," at least in those organisms that contain high levels of the cobamides. Although in the earlier studies of vitamin B-12, cyanocobalamin was isolated from liver, it now appears that this was mainly an artifact formed by chemical or photolytic decomposition of coenzyme B-12.

Forms in Milk.—The vitamin B-12 activity of milk appears to be due almost entirely to cobalamin. Even though vitamin B-12 analogs predominate over cobalamin in the alimentary tract of ruminants and are present in certain fermented fodders and silage, they occur in milk and in animal tissues only in traces.

Naturally occurring vitamin B-12 is frequently and possibly always bound to a protein or peptide. It occurs bound to protein in the milk of the cow, goat, ewe, rat, sow, and human, and in the colostrum of the cow and goat.

With cow's milk, the distribution of the vitamin appears to be ubiquitous among the various protein fractions, but a higher proportion occurs in whey proteins.[47]

Content in Milk and Milk Products.—The average vitamin B-12 content of milk, as determined from the literature, is 4.2 μg per l (Table 7.7). As would be expected from its water-soluble character, the vitamin occurs in the solids-not-fat fraction of milk. Separated butterfat is devoid of it. Values from the literature for the vitamin B-12 content of some milk products are given in Table 7.7.

Humans are dependent upon dietary vitamin B-12 for their needs. The value of dairy products as a source of this vitamin in human diets was indicated by early studies of British vegetarians. Vegetarians whose diet for many years contained no flesh foods but did include dairy products and a small amount of egg subsisted without developing ill effects. On the other hand, among a group of pure vegetarians ("vegans"), who in the previous 10 years or so had eliminated also the dairy produce and egg from their food and whose diet thus contained no products of animal origin, some individuals developed definite illnesses. Certain symptoms present were similar to some of those which occur in pernicious anemia. The ill effects were apparently due to a deficiency of vitamin B-12, since the symptoms were alleviated by administration of this vitamin; a similar beneficial effect resulted from restoring milk to the diet.

It has been estimated that the daily requirement of most adults is 5.0 μg per day. On the basis of the average value shown in Table 7.7, one quart of milk would furnish about 80% of this amount.

Effect of Diet of Cow, Season, Breed, and Other Factors.—Differences in rations of cows under normal feeding conditions exert, on the whole, only a minor effect on the vitamin B-12 content of milk. Some workers[67,92] observed the vitamin B-12 concentration in cow's milk to be about twice as high early in the pasture feeding period as it was previously during indoor feeding, with a gradual decline to the indoor level.[67] In contrast, other investigators[90] have found, with cows transferred from barn rations to pasture, a noticeable decrease in the vitamin B-12 content of the milk, the level remaining lower during the pasture season than during barn feeding in the fall and winter. Similar findings have been reported of lower values in pasture feeding as compared to indoor feeding in which considerable amounts of A.I.V. silage were fed.[125] In other instances, the vitamin B-12 content of the milk has been observed to be about the same on pasture feeding as on barn feeding or not to undergo seasonal variation. Little if any difference was indicated with different barn rations.[58]

With cattle on farms where the fodder or pasture was deficient in cobalt, supplementing with this element significantly raised the level of vitamin B-12 in the milk[101] or brought about an average, although not significant, increase.[148] In other work with dairy cows, cobalt, added to either pasture or barn rations already containing amounts of this element adequate for normal health and functioning, failed to increase the vitamin B-12 content of the milk.[57]

On the whole, the content of vitamin B-12 in cow's milk seems unrelated to breed of the animal.

The findings of different groups of workers are in accord in indicating a wide range of values for the vitamin B-12 level in milk from cows of the same breed, and much variability among samples collected from the same animal at various times during lactation, even those taken on consecutive days. Differences between samples of milk collected from cows at the morning and at the evening milkings have, on the average, proved to be small and, among various workers, not in the same direction. No relation has been found between the number of previous lactations of individual cows and the vitamin B-12 level in the milk.

Change During Lactation.—The vitamin B-12 content of cow's colostrum is greater than that of the milk later in lactation. The concentration in samples of early colostrum appears to be about 3 to 6 times that in milk. During the first few days postpartum, the level generally falls rapidly, reaching that of normal milk about the 2nd week of lactation or earlier, and remaining at this level without marked deviation through at least 20 to 36 weeks of the milking period.

Effect of Heat and Other Treatments.—Pasteurization causes only slight destruction of the vitamin B-12 in milk. The content is diminished by about 10% by the holding method. The loss appears to be somewhat less with the high temperature-short time process and with the flash method. High-temperature short-time (HTST) pasteurization alone causes a mean loss of 4% of vitamin B-12, while HTST treatment repeated once, or flash pasteurization followed by one or two HTST treatments, causes in general not more than 10% destruction.[12] No significant loss occurs in the manufacturing of dried whole milk,[4] but a 20% reduction occurs in the vitamin B-12 content of reconstituted dried skimmilk compared to unpasteurized fluid milk.[123]

More drastic heat treatment, however, may lead to considerable or even nearly complete destruction of this vitamin. Laboratory studies indicated destruction of about half of the vitamin B-12 in the original milk during the condensing process in making evaporated milk, only 13% remaining after sterilization of the product.[60] In the preparation of evaporated milk by a normal commercial procedure (filled cans held at a maximum temperature of 113°C for 15 min) 83% of the vitamin was destroyed; no further change in vitamin B-12 content occurred upon storing the evaporated milk at 4, 20 or 37°C during a 12-month period.[52] In Table 7.5 are shown the comparative losses of vitamin B-12 resulting when a number of processing treatments were applied by British workers to one bulk of raw milk. It is apparent that vitamin B-12 was nearly completely destroyed during in-bottle sterilization of fluid milk and, as mentioned above, during in-can sterilization of evaporated milk; large loss also occurred in the presterilization treatment of the latter product. It is noteworthy that sterilization by the U.H.T. process alone, when not followed by in-bottle resterilization, caused not much more destruction than did high temperature-short time pasteurization.

Studies of the factors which influence the heat destruction of vitamin B-12 in milk[40] showed that this process is essentially oxidative in nature and that it is linked directly or indirectly with the oxidative degradation of ascorbic acid. Aeration of milk before in-bottle sterilization greatly increased the loss of both vitamin B-12 and ascorbic acid, whereas prior deaeration by flushing with nitrogen or carbon dioxide considerably reduced this heat destruction. This very likely explains the markedly diminished loss of vitamin B-12 observed during in-bottle sterilization of milk which had been subjected to HTST pasteurization previous to a 24-hr storage period, since there would have been considerable loss of ascorbic acid during pasteurization and subsequent storage.[12]

Comparison of the direct process of UHT sterilization, involving injection of steam into milk with consequent complete removal of the oxygen, with the indirect process, in which heat is transferred through stainless steel with an accompanying partial removal of oxygen, showed that both methods prevented large losses of vitamin B-12. However, there was little difference between the two processes (13% loss for the direct and 4% for the indirect process).[13] During storage of sterilized milk for up to 90 days at room temperature, about 30 to 40% of the vitamin was lost. However, in one experiment, milk sterilized by the direct process and containing a large proportion of its original ascorbic acid content and virtually no residual oxygen lost about 60% of its vitamin B-12 content after 60 days storage.[42]

Storage of raw cow's milk in a household-type refrigerator at about 0°C for as long as 3 days caused no detectable change in vitamin B-12 potency.[61] Bottled milk gassed with nitrogen or carbon dioxide and then sterilized by autoclaving suffered a further loss of more than half the residual vitamin B-12 upon exposure to sunlight for 1 hr; a slower rate of loss occurred thereafter.[40]

The few studies which have been reported on the influence of ultraviolet light on the vitamin B-12 content of milk indicate no appreciable destruction. Fresh raw liquid milk suffered no appreciable loss of vitamin B-12 when subjected to doses of 0.25, 0.5, or 1.0 megarad of gamma radiation applied at the rate of 1.0 megarad per hr, a dose of at least 1.0 megarad being required for "commercial" sterility. Processing of milk in plants employing ion-exchange resin to remove cationic radionuclides resulted in no significant change in the vitamin B-12 content of the milk.[49,76]

Cheese.—As pointed out previously, the vitamin B-12 in milk occurs bound to protein. Coagulation of cow's milk with acid or rennet, or a combination of the two, releases a considerable proportion of the vitamin from such binding.[22] When skimmilk at 35°C was coagulated by acidification to pH 4.6 with glacial acetic acid, the ratio of the concentration (w/v) of vitamin B-12 in the whey to that in the original skimmilk was significantly smaller than when skimmilk was coagulated at 40°C with 0.5% of a rennet extract for 45 min (44 versus 61% on the average). Acidification to pH 4 resulted in a higher level of vitamin B-12 in the whey than did acidification to pH 4.6. Little destruction of vitamin B-12 occurs in the coagulation of skimmilk by rennet, at least when used alone. The fraction of the vitamin B-12 level in the original milk that is found in whey varies to some extent with the kind of cheese being made.

With four varieties of cheeses, Karlin[86] determined the vitamin content (dry-matter basis) of the curd immediately after manufacturing, and of the interior and rind at intervals during curing and storage. A rather marked drop in the vitamin B-12 level was observed during the first 24 to 48 hr of curing, a period coinciding with the growth of lactic acid bacteria. The vitamin B-12 content of the inner portion of the cheeses with a moldy rind and/or interior—Coulommiers and Roquefort—continued to decrease somewhat during curing, with a loss, comparing the final content to that of the curd, of 38% for Coulommiers and 44% for Roquefort. However, the vitamin B-12 content of the inner portion of Gruyère cheese increased 80% during ripening compared to the initial level. This enrichment is undoubtedly due to the activity of vitamin B-12-synthesizing microorganisms present. A similar and greater increase (180%) was observed in the rind of Gruyère.

Nilson and co-workers[124] reported that 11% of the vitamin B-12 present in the original milk was retained in Cheddar cheese curd, and that the content of this vitamin in the cheese declined sharply during the first week of ripening and thereafter increased slowly up to 9 months and then very rapidly from the 9th to 12th month. The increase during these last 3 months appeared to be greater at the curing temperatures of 15.6 and 10.0°C than at 4.4°C. On the other hand, Hartman and co-workers[59] found that the vitamin B-12 content of natural Cheddar cheese made from pasteurized milk did not change significantly from that of the original pressed curd at any time during 6 months curing at 12.8°C, or during 18 months curing at 4.4°C. Related studies[60] indicated that about half the vitamin B-12 present in the original milk went into Cheddar cheese, while the remainder was accounted for in the whey. In the manufacture of this variety of cheese, it would appear from the foregoing results and those of other workers that between 10 and 50% of the total vitamin B-12 in the original milk remains in the cheese.

Little is known of the form in which this vitamin occurs in various cheeses. That in Cheddar and Svecia appears to be present mainly in bound form.[14]

Values obtained from the literature for the content of vitamin B-12 in different varieties of cheese are shown in Table 7.8; these values were determined by microbial assay. Comparative determinations by microbial and by rat assay have been made[59] on aliquots of the same samples of various cheeses. As indicated in Table 7.9, higher values were generally obtained by the rat method with natural Cheddar, natural Swiss, and the products made from these cheeses, but not with Cottage cheese. An explanation of the divergent results obtained by the two methods is not readily apparent, although several possibilities have been considered.[22,59,87] The higher vitamin B-12 content of natural Swiss (fresh or dry-matter basis) assayed in this study,[59] as compared to Cheddar cheese, is undoubtedly due to biosynthesis of the vitamin by propionic acid bacteria that are involved in the manufacture of this product. On the dry-matter basis, Cottage cheese was among those in the upper range of vitamin B-12 potency. No relation was noted between mildness or sharpness of Cheddar cheese and vitamin B-12 content. From the values obtained (fresh or dry basis) in this investigation, it appears that some of the vitamin is lost in making process Cheddar and process Swiss from the respective natural products.

Table 7.9
COMPARATIVE MICROBIAL AND RAT ASSAYS OF VITAMIN B-12 IN CHEESE AND CHEESE PRODUCTS[a]

Variety of Cheese or Cheese Product	Average Vitamin B-12 Content, μg/Kg, Fresh Basis	
	Rat Assay	Microbial Assay
Natural Cheddar	20.8	10.3
Process Cheddar	12.0	8.0
Process Cheddar cheese food or spread	7.3	5.7
Natural Swiss	36.2	20.8
Process Swiss	18.4	12.2
Cottage	8.0	8.8

a Data from Hartman *et al.*[59]

VITAMIN C

Fully biologically active forms of vitamin C are L-ascorbic acid and dehydro-L-ascorbic acid. Some analogs of ascorbic acid are partially active as the vitamin.

Vitamin C is required in the formation and maintenance of intercellular material, such as the collagen of fibrous tissue and the matrices of bone, cartilage, and dentin. Some evidence suggests that it may be concerned with the conversion of proline and lysine to hydroxyproline and hydroxylysine, respectively, which are uniquely present in the peptide chain of collagen. The ease with which vitamin C can be reversibly oxidized explains its participation in numerous reactions.

Scurvy, resulting from severe vitamin C deficiency, is characterized by pains in the joints, weak brittle bones, loose fragile teeth, swollen, bleeding, and ulcerated gums, and hemorrhages in various tissues. More frequently, vitamin C deficiency occurs in a milder form as a subclinical deficiency, resulting in a loss of weight, vague pains and other indefinite symptoms.

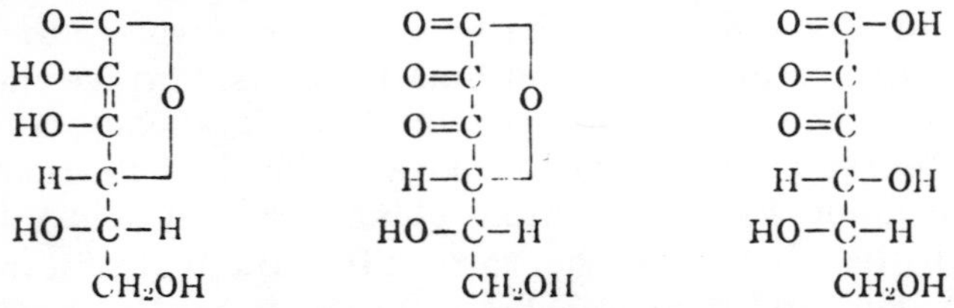

```
   O=C──┐             O=C──┐            O=C−OH
     |  |               |  |              |
  HO−C  |              O=C |              O=C
     ‖  O               |  O              |
  HO−C  |              O=C |              O=C
     |  |               |  |              |
   H−C──┘              H−C─┘            H−C−OH
     |                  |                 |
  HO−C−H             HO−C−H            HO−C−H
     |                  |                 |
    CH2OH              CH2OH             CH2OH
```

L-Ascorbic acid | Dehydro-L-ascorbic acid | Diketo-L-gulonic acid

Forms in Milk

Milk as secreted in the udder and immediately after removal from the udder contains vitamin C only in the form of L-ascorbic acid. However, this reduced form of the vitamin is slowly converted to the reversibly oxidized but still biologically active form, dehydroascorbic acid, which in turn is further slowly oxidized irreversibly to the biologically inactive diketogulonic acid. This acid itself is unstable, particularly when milk is exposed to high temperatures and to a lesser extent when it is exposed to sunlight or to copper; the oxidation proceeds to oxalic acid and threonic acid.[158]

Market milk contains vitamin C in the form of both ascorbic acid and dehydroascorbic acid. The relative percentages of the two forms of the vitamin present will depend upon the previous history of the milk, including such factors as its age, exposure to light, copper content, heat treatment, and temperature of storage.

Content in Milk

On the average, fresh raw milk contains more than 20 mg of vitamin C per l (see Table 7.10). However, a similar average computed for market milk amounts to only about half this value, individual laboratory estimates dropping to as low as 2.4 mg per l. The recommended daily dietary allowances of the National Research Council for this vitamin for adults range from 55 to 60 mg. Potentially, milk could supply a sizable fraction of these requirements, but actually, in many if not most cases, it supplies very little. These losses in vitamin C take place at many stages in processing and distributing milk. Even so, dairy products are estimated to contribute about 5% of the total vitamin C in the American diet.

Table 7.10

CONTENT (MG-KG)[a] OF VITAMIN C, VITAMIN E, AND CHOLINE IN MILK AND MILK PRODUCTS[b]

Milk or Milk Product	Vitamin C		Vitamin E		Choline	
	Average	Range	Average	Range	Average	Range
Whole milk						
Fluid	20.9[c]	15.7– 27.5(49)[d]	1.00	0.20– 1.84(16)	137	43– 285(10)
Condensed	2.6	4 – 58 (7)	3.0	– (1)	344	– (1)
Evaporated	11	4 – 18 (6)	2.2	1.4 – 3 (3)	246	— (1)
Dried	81	26 –120 (7)	7.5	5 –10 (2)	862	394–1070(4)
Skimmilk						
Fluid	19	9 – 25 (4)	—	—	48	— (1)
Dried	98	53 –170 (3)	0.4	0.2 – 0.5 (2)	1182	410–1700(6)
Buttermilk						
Fluid	12	9 – 14 (2)	—	—	—	—
Dried	0	— (1)	—	—	2059	1808–2310(2)
Yoghurt	6.2	0 – 10.9(5)	—	—	6	— (1)
Butter	0	0 – 0 (4)	24	15 –31 (16)	183	20– 400(3)
Ice cream	3	0 – 11 (4)	3	— (1)	24	— (1)
Whey						
Fluid	13	11 – 15 (2)	—	—	—	—
Dried	—	—	—	—	1356	700–2011(2)

[a] Mg per l for products designated "fluid."
[b] Mean and range of average values obtained from the literature.
[c] Fresh milk; average for market milk (18 references): 10.5 (2.4 to 20.5) mg per l.
[d] Figures in parentheses indicate number of references consulted.

Source in Milk

The cow requires no external source of vitamin C since it is synthesized in her body. The vitamin can be formed in the microsomes of the liver and possibly in intestinal and kidney tissues also. Studies have indicated that various hexose sugars, including fructose, mannose, glucose, and galactose, can serve as precursors.

Effect of Season and Other Factors

There is no true correlation between season and the vitamin C content of milk, nor does the ration have an effect on the amount secreted in milk. Differences in the concentration of the vitamin in the milk of cows of various breeds are not great, and probably not significant.

The vitamin C content of colostrum is apparently about 10 to 60% higher than that of normal milk, but it drops within a few days after parturition to normal levels or slightly below. The changes that occur during subsequent states of lactation are somewhat uncertain.

Effect of Light and Other Radiation

Exposure of milk to light brings about oxidation of the vitamin C. If the exposure is not too long in duration, only mild and reversible oxidation to dehydroascorbic acid occurs, and the biological potency is thus retained. Long exposures, however, cause further and irreversible oxidation with loss in biological potency. Visible light in the blue and violet range of the spectrum, as well as ultraviolet radiation, is responsible. Yellow and red light have almost no effect. Protecting milk from light of wavelengths less than 500 mμ prevents this oxidation.

While milk may be and often is exposed to light at many stages in its handling and processing, perhaps the longest period of exposure comes between the time it leaves the dairy and the time it is taken into the house by the consumer. Losses of vitamin C while milk is carried on regular enclosed delivery trucks are not usually in excess of those that normally occur when milk is stored in the dark. However, a survey by Josephson *et al.*[79] of 13 milk routes in 4 large cities indicated that, after delivery to the home, 20% of all milk remained exposed to light for more than 30 min before being taken in by the housewife. In experimental tests, exposure of milk in clear-glass bottles to sunlight for 30 min leads to almost complete disappearance of the reduced ascorbic acid and half or more of the total vitamin C.[20,73,79] Even exposure in the shade or to room light conditions and temperature causes large decreases in vitamin C content. Losses are less in quart and half-gallon bottles than in those of smaller capacity.[79]

When other types of containers are substituted for clear-glass bottles, some degree of protection for the vitamin is obtained. Red, brown, or ruby-glass bottles, and paper cartons are among those that have been found effective. Blue bottles are not very helpful; black-colored containers afford the most protection.

Exposure to fluorescent light is less harmful than exposure to daylight; yet here too, considerable losses can still occur. The amount of destruction is directly related to the radiant power emitted by the light between 400 and 500 mμ. Fluorescent tubes emitting yellow or red light protect the vitamin better than those giving white or green light. Paper cartons afford some protection, the most helpful being those that transmit the least light; brown or plastic containers are also of value.

The presence in milk of oxygen in the form of dissolved air is necessary for light-activated oxidation, and removal of this gas from the milk will prevent destruction of vitamin C by light. Oxygen may be removed by subjecting milk to vacuum treatment or by flushing it out with other gases, such as nitrogen. The primary difficulty seems to lie in keeping the milk from becoming recontaminated with air during the bottling operation. Moreover, a bad flavor may develop when milk flushed with nitrogen is exposed to sunlight.[64]

Riboflavin is also necessary in milk as an energy receptor and as a sensitizer for the destruction of ascorbic acid by light. Its removal from milk by destruction or absorption will prevent oxidation from taking place.

Treatment with gamma rays from Co^{60} (80,000 roentgens/hr), used to sterilize milk, caused 77% destruction in 1 hr and 100% in 6 hr.[100] Pedersen[130] observed considerable losses with only 5000 rads of gamma radiation. Subjection of pasteurized milk (fortified with ascorbic acid) to 0.374 megarad led to the loss of 15% of the vitamin.[160] In other work, however, exposure of liquid milk to 0.125 megarad destroyed 89%, while one megarad destroyed 93%; replacement of air in the milk with nitrogen afforded some protection, particularly at the lower radiation level.

Relation to Oxidation-Reduction Potential of Milk

An inverse relation between the oxidation-reduction potential and the vitamin C content has been observed in both fresh milk and that refrigerated for 5 days. Restoration of the vitamin C content in stored milk to its original value by adding ascorbic acid also restores the oxidation-reduction potential of the milk to its former value. Moreover, the oxidation-reduction potential determines the amount of oxidation that takes place. A low E_h almost completely inhibits the oxidation of vitamin C, while a high potential rapidly destroys the vitamin.

Effect of Heat

Losses of vitamin C in milk from heat treatment are secondary to those from light. In the absence of contaminating metals, the destruction of this vitamin by heat does not result from an increase in the rate of conversion of ascorbic acid to dehydroascorbic acid, but rather from a very great increase in the rate of destruction of the latter form of the vitamin. Dehydroascorbic acid is quite thermolabile and is rapidly destroyed when milk is heated.

Pasteurization of milk by the holding method (61 to 65°C for 30 min) generally results in a decrease in vitamin C of about 20%. Destruction by the flash method (85°C) or the high temperature-short time method (71 to 73°C, 15 sec) is much less. Greater losses from heating occur when milk becomes contaminated with certain metals, especially copper, from containers or processing equipment.

In a comparison of milk heated for 30 min at holding pasteurization temperatures with that heated for a similar period at high temperature-short time pasteurization temperatures, vitamin C was more stable at the higher temperatures, the dividing line being about 73°C.[48,54] Josephson and Doan[80] found that heating milk above about 77°C led to formation of sulfhydryl compounds; as these are reducing substances, this resulted in a lowered oxidation-reduction potential and apparently in the protection of vitamin C against oxidation. Similarly, Burgwald and Josephson[10] observed that high temperature-short time pasteurization produced sulfhydryl compounds which, they felt, protected the vitamin from oxidation. The protective effect of high temperatures has been attributed by Stribley *et al.*[150] to the formation of compounds of copper with protein and with the sulfhydryl compounds produced by heat, thus rendering the copper unavailable for the destruction of ascorbic acid.

In the usual sterilization procedures, losses of vitamin C are quite severe, ranging from 43 to 100%, but these can be reduced sharply by the ultrahigh-temperature sterilization process (heating for a few seconds at 135 to 150°C). Even with this process, however, all the dehydroascorbic acid and 20% of the ascorbic acid is destroyed.[42] Direct heating by steam injection is more protective than indirect heating since it removes the oxygen from the milk.

Effect of Metal Contamination

Contamination of milk with certain metals catalyzes the oxidation of ascorbic acid. Of all metals tested, copper has the most pronounced catalytic action; 5 parts per billion of copper had more effect than 100 parts per billion of iron or 1000 parts per billion of either chromium or nickel.[166] The action of copper apparently is to catalyze the conversion of ascorbic acid to dehydroascorbic acid,[171] and not the irreversible oxidation of dehydroascorbic acid itself.

Effect of Storage

Once activated by light, the destruction of vitamin C continues rapidly even after storage in the dark at low temperatures. When milk was held in 2 refrigerated farm bulk holding tanks at about 3 to 9°C for 36 hr, during which time warm milk from successive milkings was added to the tanks, the content of reduced ascorbic acid decreased by 50 to 75%, as compared to only 25% for milk held for the same period in 10-gal milk cans at about 4°C.[105] In milk kept at refrigerator temperatures, loss of vitamin C averaged 20 to 30% after 1 day, 25 to 50% after 2 days, and 50 to 75% after 3 days. As in the case of destruction by light, loss of vitamin C during storage can be decreased or eliminated by removal of oxygen. Further loss with storage of sterilized milk processed by the ultrahigh-temperature method depends upon whether residual oxygen remains in the milk. The vitamin is stable if no oxygen is present, but quickly disappears if small amounts of oxygen are dissolved in the milk. Unfortunately, a cooked flavor persists in the milk if no oxygen remains.

Content in Milk Products

In the preparation of dried milk powder, 20 to 30% of the vitamin C was lost. Storage for a year in inert gas pack at ordinary temperatures led to a further 20% decrease. The vitamin is better preserved in dried milk packed with inert gas than in that packed with air. The second processing involved in the production of instant nonfat dried milk reduced its vitamin C content below that of regular nonfat dried milk.

In the manufacture of evaporated milk, Doan and Josephson[32] noted a loss of from 50 to 90%, the average being 75%. In the next 2 months, there was a further average decrease of 5%. Between 3 weeks and 3 months after manufacture, the evaporated milk reached a state of stability, little destruction of the vitamin occurring thereafter.

Some of the vitamin is also destroyed in making condensed milk. In one instance, the loss amounted to about 20%; in another, it was 31%, increasing over a 2-yr storage period to 73%.

There is little or no vitamin C in cheese, though small amounts have been reported in a few varieties. Clayton and Folsom[21] reported none in Cheddar cheese. In the work of Dearden *et al.*,[27] the vitamin disappeared rapidly in the process of making Cheddar, Cheshire, or Stilton cheese. About 60 to 95% was lost in draining the first whey, and further loss occurred in subsequent wheys. Less than 2% of that present in the original milk was recovered in the third curd. None survived the maturation period. Randoin[134] found some vitamin C in freshly made cheese but relatively little or none in ripened cheese.

According to most workers, ice cream contains little or no vitamin C. Destruction of the vitamin during manufacture has been attributed to rapid oxidation by the air dispersed throughout the product.

Values taken from the literature for the vitamin C content of several milk products are given in Table 7.10.

Beginning in 1964, Canada permitted the addition of vitamin C to evaporated milk, but required that if this were done, the milk must contain at least 70 mg per l when reconstituted to the original moisture content. On the basis of their experiments, Bullock *et al.*[9] suggested that about 280 mg ascorbic acid per kg should be added to evaporated milk to ensure meeting the legal standard.

VITAMIN D

Vitamin D activity is possessed by a number of structurally similar compounds. All can be formed by irradiation of certain provitamin steroids characterized by two double bonds which are always in the same positions in the ring system. Steroids of this type exhibit certain ultraviolet absorption bands. Conversion of the provitamin into the active vitamin by ultraviolet irradiation involves rupture of ring B in the steroid system, so that the structure loses its true sterol identity and an intermediate product is formed. If irradiation is carried too far, the vitamin activity is lost and toxic substances are formed.

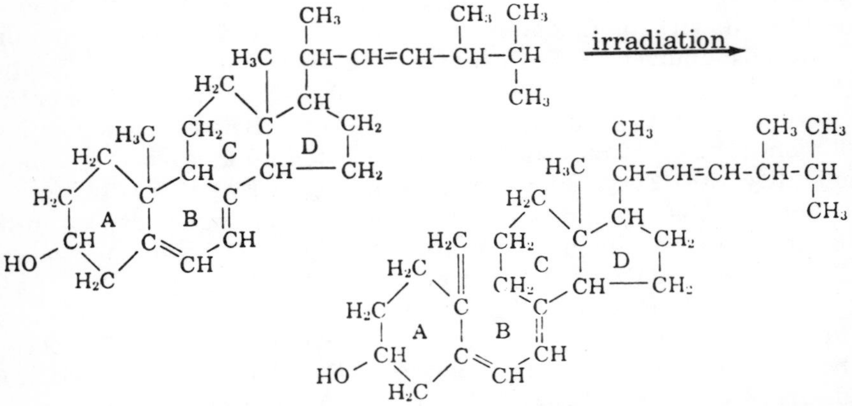

```
                 CH3          CH3  CH3
                  |            |    |
            H3C  CH-CH=CH-CH--CH          irradiation →
         H2C  |   |                |
          /  \|  CH               CH3
   H3C  CH2   C     CH2                     CH3            CH3 CH3
H2C  |   |  C |  D  |                        |              |   |
H2C    C   CH   CH---CH2            H3C  CH-CH=CH-CH-CH
 | A  |  B  C                        H2C  |   |               |
 CH   C    ||                         /  \|  CH              CH3
HO  H2C  CH  CH              H2C  CH2    C    CH2
                          H2C  ||  |  C  |  D  |
                        H2C    C  CH2   CH---CH2
                         | A  |  B   C
                         CH   C      ||
                       HO  H2C  CH   CH
```

Ergosterol   Ergocalciferol

The two principal forms of the vitamin are ergocalciferol (vitamin D_2) produced by irradiation of ergosterol, and cholecalciferol (vitamin D_3) produced by irradiation of 7-dehydrocholesterol. Ergosterol is the provitamin present in plants and yeast; 7-dehydrocholesterol is found in the higher animals and man.

All compounds with vitamin activity have the same nucleus but differ in the constitution of the side chain. Slight modifications in the side chain can greatly affect the physiological response. Thus both ergocalciferol and cholecalciferol are active for the rat, but only the latter is effective in the chick.

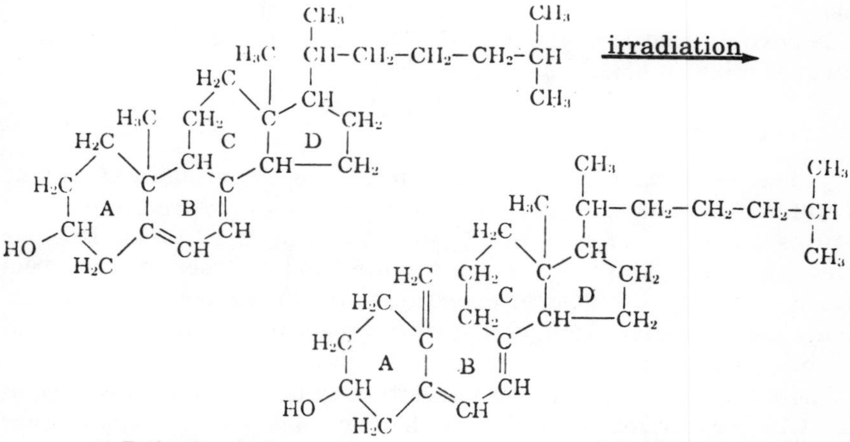

irradiation →

7-Dehydrocholesterol Cholecalciferol

The physiologically active form of vitamin D is 25-hydroxycholecalciferol, formed in the liver from cholecalciferol. A similar compound, 25-hydroxyergocalciferol, is formed from ergocalciferol. 25-Hydroxy-vitamin D causes the cells lining the intestine to form a protein that promotes calcium absorption. Parathyroid hormone is involved in the process. The net effect is to increase the plasma calcium and phosphorus levels, which in turn leads to normal bone formation.

Deficiency of the vitamin involves, as indicated above, a failure to deposit calcium salts in the skeleton, resulting in faulty bone metabolism and consequently leading to rickets in children and, less frequently, to osteomalacia in older people. Both these disease conditions are characterized by soft and misshapen bones. Species vary considerably in their tendency to develop rickets. A vitamin D deficiency is difficult to produce experimentally in the rat when the calcium-phosphorus ratio in the diet is correct, but the vitamin D-deficient chick develops severe rickets regardless of the mineral balance of the diet.

Forms in Milk

Both the principal forms of vitamin D—ergocalciferol and cholecalciferol—can appear in milk. Cholecalciferol is formed by solar radiation acting upon the skin of cows exposed to sunlight and then transmitted to the mammary gland, where it is secreted in the milk. Ergocalciferol occurs in irradiated plant products fed to cows and thus in the milk from these cows. During irradiation of milk itself, cholecalciferol is formed, while either form of the vitamin may occur in the concentrates added to milk depending upon the source of the concentrate used. It is possible that some of the other forms of the vitamin may also at times occur in small quantities in milk.

Content in Milk and Butter

As is the case with most other natural foods, the vitamin D content of unfortified milk is quite low. Averages calculated from published figures are 13.0 (range: 3 to 23) and 31.2 (range: 12 to 43) I.U. per quart (13.7 and 33.0 I.U. per l) for winter and summer milk, respectively. Unfortified butter also contains low amounts of the vitamin. Averages calculated from published figures are 0.15 and 0.48 I.U. of vitamin D per gm of winter and summer butter, respectively.

These figures are quite low in relation to the dietary allowance of 400 I.U. per day recommended by the National Research Council for infants, children, adolescents, and pregnant and lactating women. However, vitamin D-enriched milk has been made generally available in many countries, including the United States, Canada, Great Britain, Sweden, Germany, Czechoslovakia, Switzerland, and the U.S.S.R. Milk has been used as a vehicle for furnishing additional amounts of the vitamin, since this food is so universally consumed by infants and children. Moreover, milk is an excellent source of the antirachitic minerals, calcium and phosphorus, and contains these elements in a proportion that is highly desirable, since it is similar to that existing in normal bone. Vitamin D-enriched milk generally contains 400 I.U. per quart in the United States. Nonfat dry milk, purchased for export in connection with the Food for Peace Program, was enriched with 500 I.U./100 gm.[23]

In recent years, there has been some concern lest an excessive intake of vitamin D might cause illness in young children. Thus it was recommended that the vitamin D content of enriched dried milk, distributed in Great Britain under welfare programs, be decreased from 1000 to 350 I.U. per 100 gm.[93]

Effect of Breed and Stage of Lactation

Guernsey milk is more potent in vitamin D than Shorthorn or Holstein milk; similarly, Jersey milk is more potent than Holstein milk. Such differences, however, are related to the percentage of fat in the milk. On the basis of the vitamin D content of butterfat, no differences among breeds have been observed in most experiments, though Thompson *et al.*[157] obtained lower values with Jerseys than with Ayrshires and even less with Holsteins, especially with summer butter.

The vitamin D content of colostrum is about 3 to 10 times that of normal milk. It declines quickly to normal values by about the 4th or 5th day of lactation.

Effect of Sunlight

Although the vitamin D content of milk depends partly on the diet of the cow, it depends also on the amount of sunlight to which the cow has been exposed.[65] The concentration of this vitamin in milk reaches a maximum during the summer months and a minimum during the winter, the potency during the summer being from 2 to as much as 9 times that in winter.

Methods of Increasing the Vitamin D Content of Milk

The vitamin D content of milk can be increased in the following ways: (1) by feeding the cows substances rich in vitamin D; (2) by irradiating the milk; or (3) by adding vitamin D concentrates directly to the milk.

Natural feeds provided to cows vary considerably in vitamin D content. Fresh pasture is a poorer source of the vitamin than are winter rations, though the latter vary in this respect. Very little vitamin D is supplied by unfortified grain concentrates, but cows obtain fair amounts from hay. Although green plants as such contain little or none of this vitamin before they are cut, vitamin D is formed in dead plant tissue, such as dead stems or leaves or partially injured leaves that are present on the plants.[117,153] The vitamin D content is increased by as much as 5 to 18 times[66] while hay crops lie exposed to the sun during curing. The content in sun-cured prairie hay, alfalfa hay or timothy hay is sufficient to produce a noticeable increase in the vitamin D content of the milk; even barn-dried hay and wilted silage contain appreciable amounts of the vitamin. The use of special feeds has also been found of value in increasing the vitamin D content of milk. Generally speaking, however, the increase brought about by the ingestion of natural feeds is not enough to be of much practical significance. Large dosages of the vitamin must be fed.

A number of materials which have been made vitamin D-active by irradiation have been fed to cows with resultant increases, some very large, in the vitamin D content of the milk. For a number of years, vitamin D-enriched milk, known as "metabolized milk," was produced commercially by feeding cows irradiated yeast containing not less than 1.6 million U.S.P. units per lb. The milk from such cows generally had a potency of about 430 U.S.P. units per quart.

The vitamin D potency of milk can likewise be increased by ultraviolet irradiation. The sterol in milk which is activated by this procedure is 7-dehydrocholesterol, derived from the cholesterol contained in the butterfat component of milk. In commercial practice, when this method was used, vitamin D milk with a potency of 135 and later 400 I.U. per quart was usually produced.

The method used in the United States today for enriching milk with vitamin D is that of adding concentrates directly to the milk or milk products. Such concentrates are available commercially either as ergocalciferol or cholecalciferol. Crystalline cholecalciferol, obtained by irradiation of 7-dehydrocholesterol, is homogenized in a medium of butter oil and condensed dry milk solids, standardized and sterilized. Alternatively, ergocalciferol is prepared by irradiation of ergosterol and either (1) treated as above, or (2) dispersed in an oil carrier plus an emulsifying agent or in propylene glycol solutions. The potencies of such concentrates may range from 4,000 to 200,000 I.U. per ml. These concentrates are added to fluid milk before pasteurization or to evaporated milk before homogenization. Nonfat dry milk can be satisfactorily enriched by either (1) homogenizing a coconut fat carrier of ergocalciferol into the condensed milk before drying, or (2) blending a dry beadlet containing ergocalciferol into the dry milk just before packaging.

Effect of Heat and Other Treatment

Vitamin D is very stable in milk. It is not affected by pasteurization, boiling or sterilization. Moreover, storage of fluid milk for 30 months at −17.8°C, of irradiated evaporated milk for 2 to 3 yr, or of frozen butter for over 2 yr resulted in little or no loss of the vitamin. The vitamin D potency of enriched dried milk declined little, if at all, during storage for 1 yr at 24°C.

VITAMIN E

Vitamin E activity is exhibited by a group of substances known as the tocopherols, of which at least 8 occur in nature. The most potent of them and the one generally used as a standard is α-tocopherol.

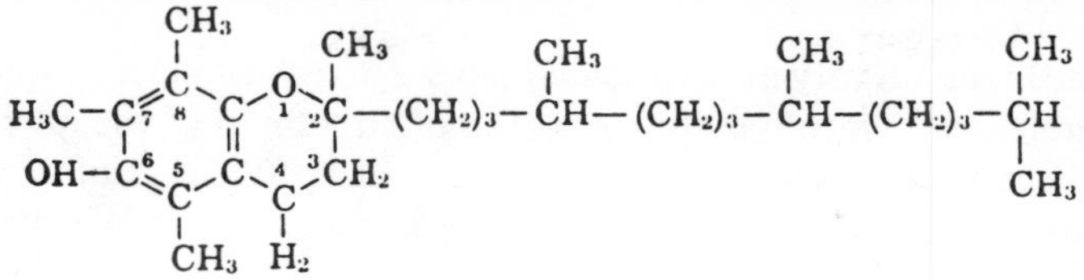

```
           CH3
            |       CH3           CH3            CH3            CH3
            C        |             |              |              |
H3C—C7 8  C   O 1  2C—(CH2)3—CH—(CH2)3—CH—(CH2)3—CH
    |     ||        |                                            |
OH—C6 5  C  4  3CH2                                             CH3
       C        C
       |        |
      CH3      H2
```

α-Tocopherol

The other well-known tocopherols—β, γ, and δ—appear to have, respectively, about 1/3 to 1/2, 1/12 to 1/4 and 1/100 the biological potency of the α form. Other tocopherols reported to occur in nature are ϵ-tocopherol, ζ_1-tocopherol (from wheat bran), ζ_2-tocopherol (from rice), and η-tocopherol. The formula for α-tocopherol is shown above. The compound with hydrogen in place of methyl groups attached at positions 5, 7, and 8 is called tocol; therefore, α-tocopherol is 5,7,8,-trimethyl tocol; β-tocopherol is 5,8-dimethyl tocol; γ-tocopherol, 7,8-dimethyl tocol; δ-tocopherol, 8-methyl tocol; ζ_2-tocopherol, 5,7-dimethyl tocol; and η-tocopherol, 7-methyl tocol. ϵ-Tocopherol and ζ_1-tocopherol are unsaturated analogs of β- and α-tocopherol, respectively.

The specific biochemical function of vitamin E in the body has not been determined. Many of its effects appear to be due to its properties as a biological antioxidant. The requirement for this vitamin increases with the intake of polysaturated fatty acids. Selenium affords a partial protection for some of the disorders associated with a deficiency of vitamin E. Feeding polysaturated fat increases the requirement for vitamin E as a protection against oxidation of the unsaturated fatty acids in the tissues.

Vitamin E has been shown to be essential for many different animals, though the deficiency symptoms are not the same in all species. In the rat and a few other animals, deficiency brings about testicular degeneration in the male and fetal resorption in the female. Muscular dystrophy develops in a larger number of species; at times the cardiac and smooth muscles as well as the skeletal muscles are affected. Vitamin E deficiency has not been observed in the adult human, but there is some indication that an inadequacy of the vitamin may be involved in the development of kernicterus and retrolental fibroplasia in the premature infant.

Forms and Distribution in Milk

α-Tocopherol appears in most instances to be the only tocopherol present in milk. Kanno *et al.*[82] reported that γ-tocopherol made up about 5% of the total vitamin E content of milk. In one instance,[157] a small proportion of ζ-tocopherol was found in milk samples; it was raised to nearly 30% of the tocopherols present by feeding brewers' grain to cows.

Mammals selectively absorb and deposit α-tocopherol in their tissues. Thus, when cows were fed supplements of this form of tocopherol, the vitamin E content of the milk was increased by 172%, but when 90% of the supplement consisted of the γ and δ forms, the increase amounted to only 55%.[133]

Vitamin E is associated with the fat in milk. The concentration of vitamin E in the lipid of the milk fat globule membrane is 3 times that in the lipid inside the fat globule; the tocopherol associated with the membrane is more easily oxidized than that in the interior.[37]

Content in Milk and Milk Products

Milk contains only low amounts of vitamin E. Figures derived from the literature average 1.00 mg per l (see Table 7.10). The content of tocopherol in milkfat, as computed from values obtained by 26 laboratories, averaged 0.023 mg per gm, with a range of 0.005 to 0.035 mg. The National Research Council has recommended a daily dietary allowance of 25 to 30 I.U. of vitamin E for most adults. A quart of milk would supply about 5 to 6% of these amounts.

According to Davidov *et al.*,[26] the transfer of vitamin E from milk to cheese does not exceed 30%. The content of the vitamin in most cheeses does not change during ripening, except for cheeses of the Roquefort type; with the latter, an increase from 4.9 to 7 mg per kg was observed between the first and third month, possibly due to synthesis by mold organisms.

Values obtained for the vitamin E content of cheese include (mg/kg): Cheddar, 10; Edam, 3; Gruyère, 3; and Roquefort, 6.5. Values for the vitamin E content of other milk products are given in Table 7.10.

Effect of Breed

In comparing 4 breeds of cows, Krukovsky *et al.*[99] reported that the milkfat from Guernseys contained the most tocopherol, while that from Holsteins contained the least. Milkfat from Brown Swiss and Jersey cows gave intermediate values. On the other hand, Parrish *et al.*[129] observed no appreciable differences in the tocopherol content of colostrum and early milk of Holstein, Ayrshire, Guernsey, and Jersey cows, nor did Thompson *et al.*[157] observe differences between the Ayrshire, Holstein and Jersey breeds.

Influence of Season and Diet of Cow

The average vitamin E content of milk produced during the summer months is higher than in milk produced during the winter months.

In most cases, summer butterfat has also been found to contain a greater concentration of tocopherol, the level amounting to about 1.3 to 2.5 times that in winter. Seasonal variations reflect changes in the diet of cows from one time of the year to another. Fresh pasture apparently contains more tocopherol than winter rations. The latter also vary in their tocopherol contents. Supplementation of the ration with tocopherol also increases the content of vitamin E in milk and butterfat, but the increase is much greater in winter than in summer.

Substantial amounts of vitamin E are destroyed in the preintestinal tract of ruminants; the amounts destroyed increase as readily fermentable carbohydrate is added to the ration.[1]

Change During Lactation

The vitamin E content of colostrum averages from 2.5 to 7 times that of normal milk, the highest levels being attained in either the first or second milking after calving. The concentration subsequently declines rapidly to a normal level, which is reached by about the 8th to 21st day of lactation. A further slow decrease throughout the lactation period was observed by one group of workers.

Effect of Heat and Other Treatments

Heat and other processing treatments involved in the production of pasteurized milk, butter, condensed milk, or evaporated milk do not appear to bring about any significant loss of vitamin E.

In a study of the effect of gamma radiation on raw whole milk, a dose of 80,000 roentgens per hr destroyed 29% of the vitamin E in 1 hr, 40% in 3 hr, and 61% in 6 hr.[100] In other experiments,[130] 20,000 rads of gamma radiation were needed to produce losses. Liquid milk exposed to one megarad lost 51% of its vitamin E.[94] The replacement of air in the milk by nitrogen reduced this loss by about one-half. Irradiation of dried whole milk with a dose of 0.1 Mrad destroyed 36% of vitamin E.[31]

Storage of condensed milk at 10 to 15°C for up to 4 yr did not alter the vitamin E content, according to one group of workers. In another instance, however, a 17% decline in potency was observed after 2 yr.

VITAMIN K

Vitamin K activity is exhibited by members of a group of compounds having similar biological activity but somewhat different structures. The presence of the quinone ring structure in vitamin K is essential for its biological activity. Menaquinone, [2-methyl-1,4-napthoquinone],

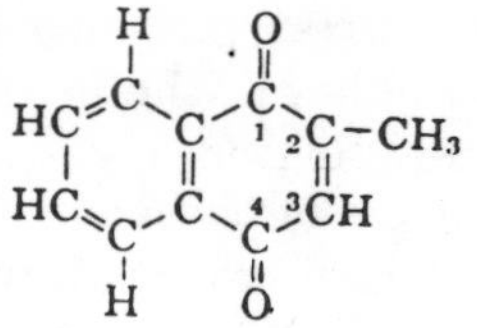

$$
\begin{array}{c}
\text{H} \quad \text{O} \\
\text{HC}=\text{C}-\text{C}-\overset{1}{\text{C}}-\overset{2}{\text{C}}-\text{CH}_3 \\
\text{HC}=\text{C}-\text{C}-\overset{4}{\text{C}}=\overset{3}{\text{CH}} \\
\text{H} \quad \text{O}
\end{array}
$$

Menaquinone

a nutritionally active form of the vitamin, is the usual standard for expressing potency. It does not occur in nature, however, but is made synthetically. Phytylmenaquinone (Vitamin K_1), a naturally occurring form, differs from menaquinone in that it has a phytyl side chain attached at position 3, replacing hydrogen:

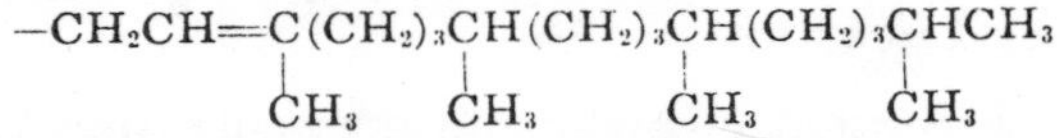

$$-CH_2CH{=}\underset{CH_3}{C}(CH_2)_3\underset{CH_3}{CH}(CH_2)_3\underset{CH_3}{CH}(CH_2)_3\underset{CH_3}{CHCH_3}$$

Prenylmenaquinone (Vitamin K_2), another form of the vitamin that does occur in nature, has similarly attached at position 3 a farnesylgeranylgeranyl side chain:

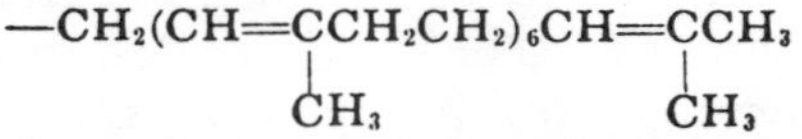

$$-CH_2(CH{=}\underset{CH_3}{C}CH_2CH_2)_6CH{=}\underset{CH_3}{C}CH_3$$

Many other derivatives of menaquinone also have vitamin K activity.

Prenylmenaquinone exists with various numbers of prenyl units on its side chain. By one test, prenylmenaquinone-4 (that is, containing 4 prenyl units) was more active than prenylmenaquinones-3, 5 or 6.

The chief function of vitamin K lies in the regulation and maintenance of the normal level of prothrombin in the blood. Subnormal prothrombin levels, which occur in vitamin K deficiency, lead to a retardation in clotting time of the blood and often to subcutaneous and intramuscular hemorrhages. This deficiency in the human occurs mostly in newborn infants. At birth, the blood prothrombin is low and, until the intestinal flora is established, continues to decrease because of insufficient supplies of vitamin K in the diet. Hemorrhagic disease sometimes results. Daily doses of $1\mu g$ vitamin K are sufficient to prevent the development of this syndrome.

Forms in Milk

The form of vitamin K in milk has been said[29] to depend upon the form available in the food of the cow, which for the most part would be phytylmenaquinone. However, large quantities of vitamin K are synthesized in the rumen. Rumen contents proved to be a good source of this vitamin even when the ration fed to the cow was practically devoid of it.[108] Vitamin K isolated from the bovine rumen was composed of prenylmenaquinones-10,11,12, and 13.[113] Presumably this synthesized vitamin would be transferred to the milk, though not necessarily in exactly the same form.

Content in Milk and Milk Products

The concentration of vitamin K is very low in milk and milk products. Only traces were found in pasteurized milk, buttermilk and butteroil.[120] Cultured buttermilk was more potent, though it was not a good source. Available quantitative figures for milk are small in number and widely different. Dam *et al.*[124] found an average of 80 (range 0 to 160) μg/l in winter milk, Goldman and Deposito[50] 60 μg/l in homogenized milk, while Jeans *et al.*[77] observed only 0.2 μg/l. The daily requirement for the infant has been estimated as approximately 1 μg.

Anderson *et al.*[3] found some indication that part of the vitamin K in milk is destroyed during both pasteurization and evaporation.

Effect of Diet of Cow

It has been stated[29] that if the cow receives an adequate amount of vitamin K in her ration, appreciable amounts are found in her milk. Little quantitative information is available, however, and in view of the large amounts of the vitamin synthesized in the rumen, it seems questionable whether variations in the amount of vitamin K in the diet would have much effect. When menaquinone was fed to cows (100 mg/cow/day) none could be found in the milk.[69,122] However, large relative increases over the amounts of vitamin K normally present in milk might have occurred and yet remained unobserved, since it would appear that the chemical methods used could not detect quantities below 50 to 100 μg per 100 ml.

ESSENTIAL FATTY ACIDS

When rats are placed on a diet devoid of fat but otherwise complete, they fail to grow, and develop characteristic skin lesions. The development of this syndrome can be prevented by the addition to the diet of certain highly unsaturated fatty acids. These compounds are termed essential fatty acids since, unlike most fatty acids, they cannot be synthesized by the animal, at least not at a rate sufficient to meet its needs. Symptoms of this deficiency have been observed in dogs, chickens, mice, hamsters, guinea pigs, and insects.

The function of the essential fatty acids in the body is not too well understood but they may be of importance in the structure of cell membranes, for the normal transport of blood lipids and in connection with enzyme systems.

Essential fatty acid activity is exhibited by a group of naturally occurring polyunsaturated fatty acids, the most important of which are linoleic, linolenic, and arachidonic acids.

$$CH_3(CH_2)_4CH{=}CHCH_2CH{=}CH(CH_2)_7COOH$$

Linoleic acid (*cis,cis*-9,12-octadecadienoic acid)

$$CH_3CH_2CH{=}CHCH_2CH{=}CHCH_2CH{=}CH(CH_2)_7COOH$$

Linolenic acid (*cis,cis,cis*-9,12,15-octadecatrienoic acid)

$$CH_3(CH_2)_4CH{=}CHCH_2CH{=}CHCH_2CH{=}CHCH_2CH{=}CH(CH_2)_3COOH$$

Arachidonic acid (*cis,cis,cis,cis*-5,8,11,14-eicosatetraenoic acid)

Linoleic and arachidonic acids restore growth and cure the skin lesions in essential fatty-acid deficiency, but linolenic, although it restores growth, has no effect on the skin condition and thus does not possess full essential fatty acid activity.

Forms and Content in Milk

Milk and butterfat do not contain large percentages of the essential fatty acids, although all three of the principal members of the group have been reported to occur in milkfat. Most estimates have not identified linoleic, linolenic and arachidonic acids as such, but rather represent determinations of dienoic, trienoic, and tetra- or polyenoic acid content. Thus they include not only the essential fatty acids but also inactive isomers with different double bond positions, *trans* forms, and the like. The following represents averages of figures taken from 60 papers, for dienoic, trienoic, and tetra- or polyenoic acids in milkfat: 2.6 (range: 0.9 to 5.8), 1.3 (range: 0.5 to 3.3), and 0.7 (range: 0.1 to 1.6) percent, respectively, of the total fatty acids measured. Between one-fourth and one-half of the dienoic acids, but only traces of the trienoic and tetraenoic acids are present as conjugated forms; the conjugated forms have no essential fatty acid activity. The nonconjugated dienoic acids of butterfat contain a mixture of *cis-cis* and either *cis-trans* or *trans-trans* isomers, while the conjugated forms have been identified as *cis-trans* and *trans-trans* isomers. The *cis-trans* and *trans-trans* isomers are without biological activity. There seems no doubt that some ordinary *cis-cis* linoleic acid is present in milkfat, but the percentage seems to be in question. Hilditch[70] found only minute quantities but White and Brown[164] reported that this substance made up about ⅔ to ¾ of the dienoic acids. By biological assay, which is probably the best test of activity, the total essential fatty acid content of butterfat calculated as linoleic acid was determined in one case[30] to be between 2.3 and 3.6%, as compared to slightly higher values (3.0 to 5.0%) for a hydrogenated vegetable fat; in another instance[154] it was observed to vary between 0.6 and 1.3% for samples of summer butterfat.

The essential fatty acid requirement of the infant has been estimated to be about 1.4% of calories.

Effect of Diet of Cow and Other Factors

Attempts to increase the levels of the essential fatty acids in milkfat by feeding rations containing large amounts of these acids to the animal have either not been successful at all or at most have increased the levels to only a small extent. This relative lack of success can be explained by the fact that unsaturated fatty acids in the diet are partially or completely hydrogenated in the rumen. The pathway of hydrogenation of linoleic acid by a pure culture of a rumen organism, *Butyrvibrio fibrisolvens*, has been studied.[91] In the first steps, the *cis*-12 bond of linoleic acid is converted to a *trans*-11 bond, and this compound in turn changed to a *trans*-11-octadecanoic acid. The latter is then eventually hydrogenated to stearic acid.

While pasture contains considerable amounts of linolenic acid, the transfer of cows from barn feeding to pasture does not result in an increased level of this substance in milk, but in an increased content of oleic acid and stearic acid, increases in the monounsaturated fatty acid amounting to 10 to 15%.[145] A small percentage of the additional essential fatty acids apparently does get through to the milk, however, for the levels of dienoic and trienoic acids have been reported to be higher during the summer pasture period than during the winter.

Intravenous infusion into cows of vegetable oils containing high levels of essential fatty acids will quickly increase the levels of these acids in the milk, though the level soon declines again once the infusion ceases. In recent work,[142] Scott *et al.* have shown that vegetable oils homogenized with sodium caseinate and treated with formaldehyde will be protected against hydrogenation in the rumen when fed to cows. Feeding of such formaldehyde-treated casein-oil particles will raise the polyunsaturated fat level of the milk to 20 to 30% of the total fatty acids. Similar results have been obtained in studies at Beltsville.

Effects of Heat and Other Treatment

Heating milk at 130°C for 20 sec was reported to cause the loss of 34% linoleic acid, 13% linolenic acid and 7% arachadonic acid. Resterilization in bottles at 118°C for 14 min increased these losses to 37, 21 and 35%, respectively. On the other hand, no decreases were observed in linoleic or archadonic acids when milk was dried. Some loss in the dried product occurred after storage for 7 to 8 months at room temperature.

Colostrum does not appear to contain significantly more of the essential fatty acids than normal milk.[143]

No differences were observed among the Guernsey, Holstein and Jersey breeds in regard to the essential fatty acids secreted in their milk.[151]

OTHER FACTORS

p-Aminobenzoic Acid (PABA)

Evidence was presented in the past which indicated that PABA is essential for growth and lactation in rats, mice, and certain other species. Part or all of the activity of PABA is undoubtedly due to its function as a precursor of folic acid, of which it is a structural constituent. This is particularly true with regard to its role in higher animals. Whole milk contains an average of about 10 μg PABA per 100 ml, with a range of 4 to 15 μg; about 20% of the total is present in bound form.

Bifidus Factor

György *et al.*[55] isolated a strain of *Lactobacillus bifidus* (designated as variant Pennsylvanicus) that requires an unknown growth factor present in human milk. Eluates from human milk increase the growth rate of rats fed rations composed to simulate approximately the composition of human milk and thus containing low protein and high lactose levels. The concentration of the *L. bifidus* factor is very low in the milk of ruminants as compared to human milk. Cow's milk has only about 1/40 to 1/50 the activity of human milk. In both, the activity is associated with the skimmilk fraction.

Carnitine (Vitamin B_T)

Fraenkel reported that the growing larvae of the mealworm, *Tenebrio molitor,* required an unidentified nutrient found in yeast. Without this substance, their growth rate was reduced, and 80% of them kept at a temperature of 25°C died between the 4th and 6th week of age. Fraenkel called the unidentified factor vitamin B_T. Skimmilk and dried whey were very good sources of the factor, being fully active in one experiment in proportions of 0.3 and 0.15%, respectively, in the diet, while in another test, dried whey gave a maximum effect at a level of 0.075 or 0.1%. The active compound was isolated from both condensed whey solids and liver extract[15] and was identified as L-carnitine, a substance previously known as a component of muscle. Both carnitine (0.062 micromole/ml) and acetylcarnitine (0.057 micromole/ml) were present in the milk of normal cows. The level of carnitine was approximately half and that of acetylcarnitine approximately double these levels in ketotic cows.[36] No effect of breed was observed. There is considerable question whether a dietary source of carnitine is required by higher animals.

Choline

Choline is required by all animals and a few microorganisms as well. It is needed, though, in such comparatively large amounts that its classification as a vitamin has at times been questioned. One of its primary functions in animals is to serve as a structural component of fat and nerve tissue.

The average choline content of fluid whole milk, as determined from published figures is 137 mg per l (Table 7.10). The daily choline intake of an adult eating mixed diets has been estimated as between 500 and 900 mg per day. A quart of milk would supply about 15 to 30% of this amount. The choline contents of some milk products are given in Table 7.10. Assays for the choline content of cheese gave 220 mg/kg for Parmesan cheese (1 study) and 335 mg/kg for Cheddar cheese (average of 2 papers).

According to Kahane and Levy,[81] the content of water-soluble choline in milk is from 25 to 40% of the total. About 25 to 55% of the water-soluble form is in the free state; the remainder is in a combined form, possibly as the phosphoric glycerol ester of choline. Frequently when milk is allowed to stand, some of the water-insoluble form decomposes into the combined water-soluble form, but not into free choline. The water-insoluble choline in milk occurs mainly in the form of lecithin and sphingomyelin, lecithin making up by far the greater amount. Phospholipids form a colloidal complex with milk proteins and hence, during the separation of cream from milk, a large part of the choline is retained in the skimmilk. In one experiment,[81] skimmilk contained over 7 times as much choline as was found in cream.

No differences have been observed in the choline content of milk from cows of different breeds. Some workers reported that the choline content of milk from cows on pasture is greater than that of milk from cows on barn feed, but others found no relation to type of feeding or season.

The choline content of colostrum at the first milking following parturition is about 5 to 8 times as high as in normal milk, but decreases quite rapidly with successive milkings until, at the 6th milking, it is down practically to the normal level.

Pasteurization brings about little or no decrease in the choline content of milk. Sterilization was reported to cause a loss of 20%. Little change was observed in the preparation of dried milk.

Myo-Inositol

Myo-Inositol has been reported necessary for the proper growth not only of mice but also of chicks, rats, and other animals. It prevented retarded growth and loss of hair in mice, corrected a "spectacled" eye condition and a type of fatty liver in rats, and stimulated gastrointestinal motility in dogs.

The average myo-inositol content of milk, as determined from the literature, is 110 mg per l, laboratory averages ranging from 60 to 180 mg. About 75% of the total in milk is in the free form. Although no dietary need has been established for humans, Williams[167] estimated that the human requirement, assuming no intestinal synthesis, was about 1 gm per 2,500 Calories of diet. One quart of milk furnishes about 10% of this amount. No difference was observed between the myo-inositol content of Jersey milk and that of Guernsey milk. Dried whole milk contained about 140 mg per 100 gm, while one sample of Cheddar cheese assayed about 25 mg per 100 gm.[19]

Lipoic Acid

Lipoic acid (6,8-dithioctanoic acid; 6-thioctic acid; protogen) is required for the growth of several microorganisms. The evidence for a dietary requirement by animals is somewhat contradictory. In early work, before lipoic acid had been identified, dried buttermilk was reported to have about 2% of the biological activity of the standard liver press-cake digest, which was a potent source of the factor.

In milk, lipoic acid is associated with the fraction containing the fat globule membrane and apparently is tightly bound to protein components of this fraction.[7]

Orotic Acid and Vitamin B-13

An unidentified growth factor was reported as having been isolated in a relatively high degree of purity from distillers dried solubles by Novak and Hauge in 1948. This factor, termed vitamin B-13, stimulated the growth of rats when fed at a level of 2 μg per day and gave a maximum response at 10 μg per day. A similar growth response with rats was obtained later by Manna and Hauge[111] from orotic acid, although it was necessary to use much larger amounts of this substance to obtain the same result as with vitamin B-13. These authors suggested that orotic acid might be a decomposition product of vitamin B-13. Vitamin B-13 concentrates, however, were reported later not to contain any orotic acid.[25]

Workers in Italy reported a high mortality of young rats from mothers fed a ration chiefly of plant products and containing 5% of exhaustively washed casein. This deficiency could be prevented by using crude casein in place of the extracted casein, or, in large measure, by feeding orotic acid. Moruzzi *et al.* [119] noted that orotic acid acted as a growth factor in chickens when they were deficient in vitamin B-12 or methionine. It appeared that, in some instances, orotic acid could partially substitute for vitamin B-12.

The average orotic acid content of fluid whole milk, as determined from published figures (11 papers) is 73 μg/ml. Crude casein has been reported to contain about 26 μg per gm; dried nonfat milk, 950 μg per gm; and dried delactosed whey, 2600 μg per gm. It has been suggested that these levels may be excessively high.[126] No striking differences were observed in the orotic acid content of milk from Guernsey, Holstein and Brown Swiss cows. Diet of the cow is apparently only a minor factor, though there is some indication that the orotic acid content is slightly higher in summer milk than winter milk. Morning milk has been reported to contain less than evening milk. Orotic acid is lowest in colostrum. It was found, in one instance, to reach a peak a few days after parturition and then decline to normal levels.

Coenzyme Q (Ubiquinone)

Coenzyme Q activity is exhibited by a group of quinones that are active in electron transport in respiration and coupled oxidative phosphorylation. There are two enzyme sites, one for the succinoxidase system and the other for the NADH-oxidase system. The various active compounds vary in the number of units in the sidechain of their molecule. Coenzymes Q_1 to Q_{10} have been found in nature, the latter being the predominant mammalian form. Hexahydrocoenzyme Q has been found to prevent exudative diathesis in chicks, necrosis of the gizzard and other deficiency symptoms in turkey poults, reproductive failure in rats associated with vitamin E deficiency, genetic dystrophy in mice, and nutritional dystrophy in rabbits.

The conventional human diet has been estimated to contain about 5 to 10 mg per day of these substances, but there is no evidence to indicate a dietary requirement by animals. Page *et al.*[128] found 3 μg per gm in butter. Neither homogenized milk nor vitamin-free casein contained amounts measurable by assay. In butter, the coenzyme Q present was Coenzyme Q_{10}.

Table 7.11

VITAMIN CONTENT (MG/L) OF MILK OF SPECIES OTHER THAN THE COW

Species	Vitamin A[a]	Thiamine[b]	Riboflavin	Niacin	Vitamin B-6	Pantothenic Acid	Biotin	Folacin	Vitamin B-12[c]	Ascorbic Acid[d]
Ruminants										
Buffalo	2000(12)[f]	0.58(4)	1.43(5)	1.28(4)	0.25(1)	2.4 (3)	0.106(2)	—	0.0036(3)	21(5)
Eland	—	0.45(1)	1.65(2)	—	0.65(1)	3.50(1)	0.035(1)	—	0.0035(1)	—
Giraffe[g]	2533(1)	0.54(1)	1.53(1)	2.10(1)	0.54(1)	2.18(1)	0.009(1)	—	0.011 (1)	—
Goat[h]	1912(5)	0.41(9)	1.84(11)	1.87(7)	0.07(2)	3.44(4)	0.039(2)	0.008 (2)	0.0007(7)	15(9)
Okapi	—	1.20(1)	10.00(1)	0.80(1)	5.30(1)	21.00(1)	0.030(1)	—	0.090 (1)	—
Reindeer[i]	4144(1)	1.83(2)	9.03(2)	1.49(2)	0.90(2)	6.63(2)	0.14 (2)	—	0.012 (2)	—
Sheep[j]	1439(10)	0.69(8)	3.82(10)	4.27(6)	—	3.64(2)	0.093(1)	0.0024(2)	0.0064(9)	43(6)
Nonruminants										
Ass	—	0.62(2)	1.04(2)	1.98(3)	—	—	—	—	0.0011(1)	58(2)
Dog	8667(1)	0.05(1)	6.11(3)	7.8 (2)	0.08(1)	3.09(2)	0.104(2)	0.0024(2)	0.0072(2)	—
Guinea Pig	1834(2)	0.59(2)	2.60(2)	11.10(1)	—	4.78(1)	0.045(1)	0.800 (1)	—	333(3)
Horse[k]	800(4)	0.30(7)	0.33(7)	0.58(5)	0.21(3)	3.02(4)	0.022(1)	0.0012(2)	0.0012(4)	104(11)
Human[l]	1898(16)	0.16(27)	0.36(16)	1.47(11)	0.10(7)	1.84(9)	0.008(7)	0.038 (5)	0.0003(9)	43(36)
Mink	—	1.65(1)	3.1 (1)	16.48(1)	0.18(1)	3.71(1)	0.041(1)	—	0.0041(1)	—
Mouse[m]	—	5.7 (1)	10.4 (1)	41 (1)	0.14(1)	23 (1)	0.32 (1)	—	—	—
Pig[n]	1036(6)	0.70(7)	2.21(8)	8.35(8)	0.40(5)	5.00(8)	0.014(4)	0.0039(1)	0.0016(6)	140(6)
Rabbit	—	1.15(1)	3.6 (1)	6.8 (1)	2.8 (1)	14.2 (1)	0.31 (2)	—	0.08 (1)	—
Rat	4333(1)	1.49(1)	1.12(2)	18.1 (1)	0.79(1)	5.70(1)	0.085(1)	0.179 (2)	0.0345(3)	8(2)
Rhinoceros	—	0.57(3)	0.22(2)	0.54(2)	0.12(2)	3.40(2)	0.005(1)	—	0.003 (2)	17(1)
Whale	7194(1)	1.16(2)	0.96(2)	20.4 (1)	1.10(1)	13.1 (1)	0.050(1)	—	0.0085(1)	70(1)

[a] I.U./1; camel milk: 2692 (2); elephant milk: <1000 (1).
[b] Elephant milk: 0.75(1).
[c] Musk ox milk: 0.0034(1); camel milk: 0.0024(2); caribou milk: 0.0043.
[d] Camel milk: 64(5); elephant milk: 40(2).
[e] Vitamin E: 2.24 mg/l (2).
[f] Numbers in parentheses indicate number of references consulted.
[g] Vitamin E: 0.34 mg/l (1).
[h] Choline: 150 mg/l (1); vitamin D: 23.7 IU/l (1); inositol: 210 mg/l (1).
[i] Vitamin D: 21? IU/l (1); vitamin E: 3.66 mg/l (1).
[j] Choline: 43 mg/l (1); vitamin E: 15.8 mg/l (1).
[k] Choline: 35 mg/l (1); vitamin E: 0.82 mg/l (1); inositol: 180 mg/l (1).
[l] Choline: 90 mg/l (1); vitamin D: 22 IU/l (8); vitamin E: 6.55 mg/l (6); inositol: 330 mg/l (1).
[m] Inositol: 1200 mg/l (1)

MILK OF SPECIES OTHER THAN THE COW

This chapter has been primarily concerned with the milk of the cow. For purposes of comparison, the vitamin content of the milk of other species is given in Table 7.11. The values listed represent averages of those found in the literature. In some instances, these figures are based on a small number of samples and hence are of limited value so far as generalizations concerning the particular species are concerned.

While much less work has been done on the factors influencing the content of these vitamins in the milk of species other than the cow, it seems clear that the levels are influenced by many of the same factors, such as stage of lactation, which bring about variations in the vitamin content in the milk of the cow. Moreover, in the case of nonruminants, since they lack the intercession of the ruminal microflora, the concentrations of the B vitamins in their milk are also greatly affected by the variations in the amounts of these vitamins in their diets.

REFERENCES

1. ALDERSON, N. E., MITCHELL, G. E., JR., LITTLE, C. O., WARNER, R. E., and TUCKER, R. E., J. Nutr., *101,* 655 (1971).
2. ANAGAMA, Y., AND KUZUYA, Y., Res. Bull. Fac. Agr. Gifu Univ., *29,* 247 (1970).
3. ANDERSON, H. D., ELVEHJEM, C. A., and GONCE, J. E., JR., J. Nutr., *20,* 433 (1940).
4. ANTENER, I., Intern. Z. Vitaminforsch., *29*, 357 (1959).
5. BARTH, J., AVANTS, J. K., BRUCKNER, B. H., and EDMONDSON, L. F., J. Agr. Food Chem., *18,* 324 (1970).
6. BERNHART, F. W., D'AMATO, E., and TOMARELLI, R. M., Arch. Biochem. Biophys., *88,* 267 (1960).
7. BINGHAM, R. J., HUBER, J. D., and AURAND, L. W., J. Dairy Sci., *50,* 318 (1967).
8. BROCHU, E., RIEL, R., and VÉZINA, C., "Les Bactéries Lactiques et Les Laits Fermentés", Rech. Agron. No. 3, Le Conseil des Recherches Agricoles, Ministèr de L'Agriculture, Quebec (1959).
9. BULLOCK, D. H., SINGH, S., and PEARSON, A. M., J. Dairy Sci., *51,* 921 (1968).
10. BURGWALD, L. H., and JOSEPHSON, D. V., J. Dairy Sci., *30,* 371 (1947).
11. BURKHOLDER, P. R., COLLIER, J., and MOYER, D., Food Res., *8,* 314 (1943).
12. BURTON, H., FORD, J. E., FRANKLIN, J. G., and PORTER, J. W. G., J. Dairy Res., *34,* 193 (1967).
13. BURTON, H., FORD, J. E., PERKIN, A. G., PORTER, J. W. G., SCOTT, K. J., THOMPSON, S. Y., TOOTHILL, J., and EDWARDS-WEBB, J. D., J. Dairy Res., *37,* 529 (1970).
14. CALLIERI, D. A., Acta Chem. Scand., *13,* 737 (1959).
15. CARTER, H. E., BHATTACHARYYA, P. K., WEIDMAN, K. R., and FRAENKEL, G., Arch. Biochem. Biophys., *38,* 405 (1952).

16. CARY, C. A., GEDDES, W. F., GUILBERT, H. R., HATHAWAY, I. L., PETERSON, W. H., SALMON, W. D., SNEDECOR, G. W., and ZSCHEILE, F. P., JR., U.S. Dept. Agr. Misc. Publ. *636* (1947).
17. CHANDA, R., and OWEN, E. C., J. Agr. Sci., *42,* 403 (1952).
18. CHAPMAN, H. R., FORD, J. E., KON, S. K., THOMPSON, S. Y., ROWLAND, S. J., CROSSLEY, E. L., and ROTHWELL, J., J. Dairy Res., *24,* 191 (1957).
19. CHELDELIN, V. H., and WILLIAMS, R. J., Texas Univ. Publ. *4237,* 105 (1942).
20. CHILSON, W. H., MARTIN, W. H., and PARRISH, D. B., J. Dairy Sci., *32,* 306 (1949).
21. CLAYTON, M. M., and FOLSOM, M. T., J. Home Econ., *32,* 390 (1940).
22. COLLINS, E. B., and YAGUCHI, M., J. Dairy Sci., *42,* 1927 (1959).
23. COULTER, S. T., and THOMAS, E. L., J. Agr. Food Chem., *16,* 158 (1968).
24. DAM, H., GLAVIND, J., LARSEN, H., and PLUM, P., Acta Med. Scand., *112,* 210 (1942).
25. DANSI, A., DAL POZZO, A., ROTTA, L., and ZANINI, C., Boll. Chim. Farm., *101,* 380 (1962).
26. DAVIDOV, R. B., GUL'KO, L. E., and ERMAKOVA, M. A., "Principal Vitamins in Milk and Milk Products," Moscow, 1956.
27. DEARDEN, D. V., HENRY, K. M., HOUSTON, J., KON, S. K., and THOMPSON, S. Y., J. Dairy Res., *14,* 100 (1945).
28. DEJONG, S., Enzymologia, *10,* 253 (1942).
29. DEUEL, H. J., JR., "The Lipids, Their Chemistry and Biochemistry," Vol. 1, Interscience Publishers, Inc., New York, 1951.
30. DEUEL, H. J., JR., GREENBERG, S. M., ANISFELD, L., and MEINICK, D., J. Nutr., 45, 535 (1951).
31. DIEHL, J. F., Z. Lebensmittel-Untersuch. u. Forsch., *142,* 1 (1970).
32. DOAN, F. J., and JOSEPHSON, D. V., J. Dairy Sci., *26,* 1031 (1943).
33. DRYDEN, L. P., and HARTMAN, A. M., J. Dairy Sci., *54,* 235 (1971).
34. DRYDEN, L. P., HARTMAN, A. M., BRYANT, M. P., ROBINSON, I. M., and MOORE, L. A., Nature, *195,* 201 (1962).
35. EATON, H. D., JOHNSON, R. E., MATTERSON, L. D., and SPIELMAN, A. A., J. Dairy Sci., *32,* 587 (1949).
36. ERFLE, J. D., FISHER, L. J., and SAUER, F., J. Dairy Sci., *53,* 486 (1970).
37. ERICKSON, D. R., DUNKLEY, W. L., and SMITH, L. M., J. Food Sci., *29,* 269 (1964).
38. EVANS, E. V., IRVINE, O. R., and BRYANT, L. R., J. Nutr. *32,* 227 (1946).
39. FIGUEIRA, F., MENDONCA, S., ROCHA, J., AZEVEDO, M., BUNCE, G. E., and REYNOLDS, J. W., Am. J. Clin. Nutr., *22,* 588 (1969).
40. FORD, J. E., J. Dairy Res., *34,* 239 (1967).
41. FORD, J. E., and GREGORY, M. E., Ann. Rept., Natl. Inst. Res. Dairying, Reading, England, 1957, 109.
42. FORD, J. E., PORTER, J. W. G., THOMPSON, S. Y., TOOTHILL, H., and EDWARDS-WEBB, J., J. Dairy Res., *36,* 447 (1969).
43. FORD, J. E., SALTER, D. N., and SCOTT, K. J., J. Dairy Res., *36,* 435 (1969).
44. FORD, J. E., SALTER, D. N., and SCOTT, K. J., Proc. Nutr. Soc., *28,* 39A (1969).
45. FUNAI, Y., Tokushima J. Exptl. Med., *2,* 201 (1955).
46. GHITIS, J., and TRIPATHY, K., Am. J. Clin. Nutr., *23,* 141 (1970).
47. GIZIS, E., KIM, Y. P., BRUNNER, J. R., and SCHWEIGERT, B. S., J. Nutr., *87,* 349 (1965).
48. GJESSING, E. C., and TROUT, G. M., J. Dairy Sci., *23,* 373 (1940).
49. GLASCOCK, R. F., and BRYANT, D. T. W., J. Dairy Res., *35,* 269 (1968).
50. GOLDMAN, H. I., and DEPOSITO, F., Am. J. Diseases Children, *111,* 430 (1966).
51. GREGORY, M. E., and BURTON, H., J. Dairy Res., *32,* 13 (1965).
52. GREGORY, M. E., HENRY, K. M., and KON, S. K., J. Dairy Res., *31,* 113 (1964).
53. GREGORY, M. E., and KON, S. K., Acta Biochim. Polon., *11,* 169 (1964).
54. GUTHRIE, E. S., Cornell Univ. Agr. Expt. Sta., Mem. *340,* 3 (1955).
55. GYÖRGY, P., KUHN, R., NORRIS, R. F., ROSE, C. S., and ZILLIKEN, F., Am. J. Diseases Children, *84,* 482 (1952).

56. HALL, H. S., Biennial Rev., National Inst. Res. Dairying, Reading, England, 61, 1970.
57. HARTMAN, A. M., and DRYDEN, L. P., Arch. Biochem. Biophys., *40,* 310 (1952).
58. HARTMAN, A. M., DRYDEN, L. P., and CARY, C. A., J. Am. Dietet. Assoc., *25,* 929 (1949).
59. HARTMAN, A. M., DRYDEN, L. P., and HARGROVE, R. E., Food Res., *21,* 540 (1956).
60. HARTMAN, A. M., DRYDEN, L. P., MOORE, L. A., and HODGSON, R. E., 14th Intern. Dairy Congr., Proc., *1,* No. 2, 103, Rome (1956).
61. HARTMAN, A. M., DRYDEN, L. P., and RIEDEL, G. H., J. Nutr., *59,* 77 (1956).
62. HASSINEN, J. B., DURBIN, G. T., and BERNHART, F. W., J. Nutr., *53,* 249 (1954).
63. HAYES, B. W., MITCHELL, G. E., JR., LITTLE, C. O., and BRADLEY, N. W., J. Animal Sci., *25,* 539 (1966).
64. HENDRICKX, H., and DEMOOR, H., Meded. Landbouwhogesch. Gent., *28,* 1189 (1963).
65. HENRY, K. M., HOSKING, Z. D., THOMPSON, S. Y., TOOTHILL, J., EDWARDS-WEBB, J. D., and SMITH, L. P., J. Dairy Res., *38,* 209 (1971).
66. HENRY, K. M., THOMPSON, S. Y., McCALLUM, J. W., and STEWART, J., 15th Intern. Dairy Congr., Proc., *1,* 172, London (1959).
67. HEUS, J. G. DE, and MAN, T. J. DE, Voeding, *12,* 361 (1951).
68. HIBBITT, K. G., J. Dairy Res., *31,* 105 (1964).
69. HIGGINBOTTOM, C., J. Dairy Res., *22,* 48 (1955).
70. HILDITCH, T. P., "The Chemical Constitution of Natural Fats," 3rd Ed., John Wiley and Sons, Inc., New York, 1956.
71. HODSON, A. Z., J. Nutr., *38,* 25 (1949).
72. HODSON, A. Z., J. Agr. Food Chem., *4,* 876 (1956).
73. HOLMES, A. D., and JONES, C. P., J. Nutr., *29,* 201 (1945).
74. HOLMES, A. D., JONES, C. P., and WERTZ, A. W., Am. J. Diseases Children, *67,* 376 (1944).
75. HOUSTON, J., KON, S. K., and THOMPSON, S. Y., J. Dairy Res., *11,* 155 (1940).
76. ISAACKS, R. E., HAZZARD, D. G., BOOTH, J., FOOKS, J. H., and EDMONDSON, L. F., J. Agr. Food Chem., *15,* 295 (1967).
77. JEANS, P. C., and MARRIOT, W. M., "Infant Nutrition," 4th Ed., C. V. Mosby Co., St. Louis, 1947.
78. JONES, I. R., WESWIG, P. H., BONE, J. F., PETERS, M. A., and ALPAN, S. O., J. Dairy Sci., *49,* 491 (1966).
79. JOSEPHSON, D. V., BURGWALD, L. H., and STOLTZ, R. B., J. Dairy Sci., *29,* 273 (1946).
80. JOSEPHSON, D. V., and DOAN, F. J., Milk Dealer, *29,* No. 2, 35 (1939).
81. KAHANE, E., and LÉVY, J., Lait, *25,* 193 (1945).
82. KANNO, C., YAMAUCHI, K., and TSUGO, T., J. Dairy Sci., *51,* 1713 (1968).
83. KARLIN, R., Ann. Nutr. Aliment., *12,* No. 5, 115 (1958).
84. KARLIN, R., Ann. Nutr. Aliment., *14,* No. 4, 53 (1960).
85. KARLIN, R., Intern. Z. Vitaminforsch., *31,* 176 (1961).
86. KARLIN, R., Intern. Z.Vitaminforsch., *31,* 326 (1961).
87. KARLIN, R., Ann. Nutr. Aliment., *15,* No. 5, 103 (1961).
88. KARLIN, R., Compt. Rend., *256,* 1164 (1963).
89. KARLIN, R., Ann. Nutr. Aliment., *19,* C661 (1965).
90. KARLIN, R., and PORTAFAIZ, D., Compt. Rend. Soc. Biol., *154,* 1810 (1960).
91. KEPLER, C. R., TUCKER, W. P., and TOVE, S. B., J. Biol. Chem., *245,* 3612 (1970).
92. KOETSVELD, E. E. VAN, Nature, *171,* 483 (1953).
93. KON, S. K., Food Agr. Organ., U.N., FAO Nutr. Studies, *17* (1959).
94. KON, S. K., Federation Proc., *20,* No. 1, Pt. III, 209 (1961).
95. KON, S. K., and HENRY, K. M., J. Dairy Res., *18,* 317 (1951).
96. KON, S. K., and PORTER, J. W. G., Proc. Nutr. Soc., Engl. Scot., *12,* xii (1953).

97. KREHL, W. A., HUERGA, J. DE LA, ELVEHJEM, C. A., and HART, E. B., J. Biol. Chem., *166,* 53 (1946).
98. KRUISHEER, C. I., and HERDER, P. C. DEN, 13th Intern. Dairy Congr., Proc., *3,* 1354, The Hague (1953).
99. KRUKOVSKY, V. N., WHITING, F., and LOOSLI, J. K., J. Dairy Sci., *33,* 791 (1950).
100. KING, H.-C., GADEN, E. L., JR., and KING, C. G., J. Agr. Food Chem., *1,* 142 (1953).
101. LAGANOVSKII, S. YA., Dokl. Vses. Konf. Po Molochn. Delu, Moscow, 1958, 421.
102. LEVITON, A., and PALLANSCH, M. J., J. Dairy Sci., *43,* 1713 (1960).
103. LEWIS, J. M., COYLAN, S. Q., and MESSINA, A., Pediatrics, *5,* 425 (1950).
104. LING, E. R., KON, S. K., and PORTER, J. W. G., "Milk: The Mammary Gland and its Secretion." Edited by S. K. Kon and A. T. Cowie, Vol. 2, Chapt. 17, Academic Press, New York, 1961.
105. LISKA, B. J., and CALBERT, H. E., J. Milk Food Technol., *21,* 252 (1958).
106. McDOWALL, F. H., and McGILLIVRAY, W. A., J. Dairy Res., *30,* 47, 59 (1963).
107. McDOWELL, A. K. R., and McDOWALL, F. H., J. Dairy Res., *20,* 76 (1953).
108. McELROY, L. W., and GOSS, H., J. Nutr. *20,* 527 (1940).
109. McGILLIVRAY, W. A., J. Dairy Res., *24,* 352 (1957).
110. McGILLIVRAY, W. A., and GREGORY, M. E., J. Dairy Res., *29,* 211 (1962).
111. MANNA, L., and HAUGE, S. M., J. Biol. Chem., *202,* 91 (1953).
112. MANSON, W., and MODI, V. V., Biochim. Biophys. Acta, *24,* 423 (1957).
113. MATSCHINER, J. T., J. Nutr., *100,* 192 (1970).
114. METZ, J., ZALUSKY, R., and HERBERT, V., Am. J. Clin. Nutr., *21,* 289 (1968).
115. MITCHELL, G. E., JR., LITTLE, C. O., and HAYES, B. W., J. Animal Sci., *26,* 837 (1967).
116. MODI, V. V., and OWEN, E. C., Nature, *178,* 1120 (1956).
117. MOORE, L. A., THOMAS, J. W., JACOBSON, W. C., MELIN, C. G., and SHEPHERD, J. B., J. Dairy Sci., *31,* 489 (1948).
118. MORFEE, T. D., and LISKA, B. J., J. Dairy Sci., *54,* 1082 (1971).
119. MORUZZI, G., VIVIANI, R., and MARCHETTI, M., Biochem. Z., *333,* 318 (1960).
120. MUELLER, W. S., and WERTZ, A. W., J. Dairy Sci., *28,* 167 (1945).
121. NAGASAWA, T., KUZUYA, Y., and SHIGETA, N., Japan. J. Zootech. Sci., *32,* 235, 240 (1961).
122. NAMBUDRIPAD, V. K. N., LAXMINARAYANA, H., and IYA, K. K., 14th Intern. Dairy Congr., Proc., *1,* No. 2, 388, Rome (1956).
123. NEUJAHR, H. Y., JONASSON, M., and LUNDIN, H., Milk Dairy Res., Alnarp, Rept., *59,* (1960).
124. NILSON, K. M., VAKIL, J. R., and SHAHANI, K. M., J. Nutr., *86,* 362 (1965).
125. NURMIKKO, V., and VIRTANEN, A. I., 14th Intern. Dairy Congr., Proc., *1,* No. 2, 873, Rome (1956).
126. OKONKWO, P. O., and KINSELLA, J. E., Am. J. Clin. Nutr., *22,* 532 (1969).
127. OWEN, E. C., and WEST, D. W., Brit. J. Nutr., *24,* 45 (1970).
128. PAGE, A. C., JR., GALE, P. H., KONIUSZY, F., and FOLKERS, K., Arch. Biochem. Biophys., *85,* 474 (1959).
129. PARRISH, D. B., WISE, G. H., and HUGHES, J. S., J. Dairy Sci., *30,* 849 (1947).
130. PEDERSEN, A. H., Nord. Mejeritidsskr., *25,* No. 3, 40 (1959).
131. POLANSKY, M. M., and TOEPFER, E. W., J. Agr. Food Chem., *17,* 1394 (1969).
132. PORTER, J. W. G., "Digestive Physiology and Nutrition of the Ruminant," page 226, Edited by D. Lewis, Butterworths, London, 1961.
133. QUAIFE, M. L., SWANSON, W. J., DJU, M. Y., and HARRIS, P. L., Ann. N.Y. Acad. Sci., *52,* 300 (1949).
134. RANDOIN, L., Bull. Soc. Chim. Biol., *23,* 358 (1941).
135. RASMUSSEN, R. A., AGUSTO, R. G., and MASSEY, C. H., J. Agr. Food Chem., *12,* 413 (1964).
136. RECHCIGL, M., JR., KRUKOVSKY, V. N., and SCHULTZ, L. H., J. Dairy Sci., *38,* 594 (1955).

137. REINART, A., and NESBITT, J. M., 14th Intern. Dairy Congr., Proc., *1,* No. 2, 934, Rome, (1956).
138. RITTER, W., 12th Intern. Dairy.Congr., Proc., *5,* 35, Stockholm (1949).
139. SAMPATH, S. R., ANANTAKRISHNAN, C. P., and SEN, K. C., Indian J. Dairy Sci., *8,* 129 (1955).
140. SANDERS, G. P., U.S. Dept. Agr., Agr. Handbook, *54,* (1953).
141. SAUBERLICH, H. E., and BAUMANN, C. A., J. Biol. Chem., *176,* 165 (1948).
142. SCOTT, T. W., COOK, L. J., and MILLS, S.C., J. Am Oil Chem. Soc., *48,* 358 (1971).
143. SENFT, B., and KLOBASA, F., Milchwiss., *25,* 391 (1970).
144. SHAHANI, K. M., HATHAWAY, I. L., and KELLY, P. L., J. Dairy Sci., *45,* 833 (1962).
145. SHORLAND, F. B., WEENINK, R. O., and JOHN, A. T., Nature, *175,* 1129 (1955).
146. SIEGEL, L., MELNICK, D., and OSER, B. L., J. Biol. Chem., *149,* 361 (1943).
147. SINHA, S. P., Intern. Z. Vitaminforsch., *33,* 262 (1963).
148. SKERMAN, K. D., and O'HALLORAN, M. W., Australian Vet. J., *38,* 98 (1962).
149. SPROULE, W. H., HAMILTON, F. W., LACKNER, C. E., JACKSON, S. H., DRAKE, G. H., and MOFFAT, M., Can. Dairy Ice Cream J., No. *12,* 25 (1946).
150. STRIBLEY, R. C., NELSON, C. W., JR., CLARK, R. E., and BERNHART, F. W., J. Dairy Sci., *33,* 573 (1950)
151. STULL, J. W., and BROWN, W. H., J. Dairy Sci., *47,* 1412 (1964).
152. SULLIVAN, R. A., BLOOM, E., and JARMOL, J., J. Nutr., *25,* 473 (1943).
153. THOMAS, J. W., and MOORE, L. A., J. Dairy Sci., *34,* 916 (1953).
154. THOMASSON, H., Intern. Z. Vitaminforsch., *25,* 62 (1953).
155. THOMPSON, S. Y., J. Dairy Res., *35,* 149 (1968).
156. THOMPSON, S. Y., and ASCARELLI, I., Ann. Rept. Natl. Inst. Res. Dairying, Reading, England, 1957, 100.
157. THOMPSON, S. Y., HENRY, K. M., and KON, S. K., J. Dairy Res., *31,* 1 (1964).
158. TOBIAS, J., and HERREID, E. O., J. Dairy Sci., *42,* 428 (1959).
159. TOEPFER, E. W., POLANSKY, M. M., RICHARDSON, L. R., and WILKES, S., J. Agr. Food Chem., *11,* 523 (1963).
160. VINCENT, O., J. Hyg. Epidemiol. Microbiol. Immunol. (Prague), *5,* 248 (1961).
161. VIRTANEN, A. I., Science, *153,* 1603 (1966).
162. WALD, G., Nature, *219,* 800 (1968).
163. WENDT, G., and BERNHART, F. W., Arch. Biochem. Biophys., *88,* 270 (1960).
164. WHITE, M. F., and BROWN, J. B., J. Am. Oil Chem. Soc., *26,* 385 (1949).
165. WHITE, R. F., EATON, H. D., and PATTON, S., J. Dairy Sci., *37,* 147 (1954).
166. WHITNAH, C. H., RIDDELL, W. H., and CAULFIELD, W. J., J. Dairy Sci., *19,* 373 (1936).
167. WILLIAMS, R. J., J. Am. Med. Assoc., *119,* 1 (1942).
168. WING, J. M., J. Dairy Sci., *52,* 479 (1969).
169. WISEMAN, H. G., SHEPHERD, J. B., and CARY, C. A., 12th Intern. Dairy Congr., Proc., *1,* 61, Stockholm (1949).
170. WODSAK, W., Nahrung, *9,* 167 (1965).
171. WOESSNER, W. W., WECKEL, K. G., and SCHUETTE, H. A., J. Dairy Sci., *23,* 1131 (1940).
172. WORKER, N. A., and McGILLIVRAY, W. A., J. Dairy Res., *24,* 85 (1957).
173. ZIPORIN, Z. Z., KRAYBILL, H. F., and THACH, H. J., J. Nutr., *63,* 201 (1957).
174. ZOOK, E. G., MacARTHUR, J. J., and TOEPFER, E. W., U.S. Dept. Agr., Agr. Handbook, *97* (1956).
175. ZSCHEILE, F. P., HENRY, R. I., WHITE, J. W., JR., NASH, H. A., SHREWSBURY, C. L., and HAUGE, S. M., Ind. Eng. Chem., Anal. Ed., *16,* 190 (1944).

CHAPTER 8

Robert Jenness†
W. F. Shipe, Jr.
J. W. Sherbon‡

Physical Properties of Milk *

INTRODUCTION

Physically, milk is a rather dilute emulsion, colloidal dispersion, and solution. Its physical properties are essentially those of water modified to some extent by the concentration and state of dispersion of the solid components.

In the dairy industry measurements of the physical properties of milk and dairy products are made to secure data necessary for design of dairy equipment (e.g., heat conductivity and viscosity) to determine the concentration of a component or group of components (e.g., specific gravity to estimate solids-not-fat or freezing point to determine added water), or to assess the extent of a chemical or physical change (e.g., titratable acidity to follow bacterial action or viscosity to assess aggregation of protein micelles or fat globules). The great advantage of physical measurements for such purposes is their speed and simplicity as well as their potentialities for automation.

Use of a physical property to measure concentrations or changes in degree of dispersion demands knowledge of the contribution of the several components to that property. Furthermore the natural range of variation of the property in milks or products is of major interest. The precision and suitability of possible methods of measurement are also of prime importance.

In this chapter, several physical properties will be discussed in terms of (a) general physical principles, (b) objectives of study in the dairy field, (c) methods of measurement, (d) contributions of milk components, (e) normal range of values and extent of natural variations, and (f) effects of processing treatments.

Primarily, this chapter deals with the physical properties of milk itself (and is even confined to milk of the bovine species). Most of the principles discussed are, however, applicable in some degree to the physical properties of various milk products.

* Revised from Chapter 8 in the 1st Edition by R. Jenness, W. F. Shipe, Jr., and C. H. Whitnah.

† Robert Jenness, Department of Biochemistry, College of Biological Sciences, University of Minnesota, St. Paul, Minnesota.

‡ W. F. Shipe, Jr. and J. W. Sherbon, Department of Dairy and Food Science, New York State College of Agriculture, Cornell University, Ithaca, New York.

ACID-BASE EQUILIBRIA

The equilibria involving protons and the substances which bind them are among the most important in dairy chemistry. The ionized and ionizable components of milk are in a state of rather delicate physical balance. Certain treatments which alter the state of dispersion of proteins and salts are reflected in the status of the protons. Thus the intensity (pH) and capacity (buffer power) factors of the acid-base equilibria have come to be much used in processing control.

Principles involved in these equilibria are presented in detail in numerous works and will not be repeated here. For a thorough and modern treatment including definitions of pH scales and methods of measurement, the reader is referred to such works as those of Edsall and Wyman[58] and Bates.[14] Applications to milk are discussed by Jenness and Patton.[109] It should suffice here to present some of the basic relationships in equation form.

$$pH = \log \frac{1}{a_{H_3O^+}} = \log \frac{1}{f_H[H_3O^+]}$$

where $a_{H_3O^+}$ = activity of hydronium ion, $[H_3O^+]$ = concentration of hydronium ion in moles per liter, f_H = activity coefficient of hydronium ion.

For many purposes it is sufficiently accurate to use $[H_3O^+]$ instead of $a_{H_3O^+}$. It is often written simply $[H^+]$. For a weak acid HA dissociating into H^+ and A^- the dissociation constant, K_A, is given by the expression:

$$K_A = \frac{[H^+][A^-]}{[HA]}$$

and hence:

$$pH = pK_A + \log \frac{[A^-]}{[HA]}$$

where $pK_A = \log 1/K_A$.

The measure of buffer capacity dB/dpH is the slope of the titration curve (pH plotted vs. increments of base added) at any point.

$$\frac{dB}{dpH} = 2.303 \left[\frac{K_A C[H^+]}{(K_A + [H^+])^2} + [H^+] + [OH^-] \right]$$

where C = total concentration of weak acid, or, for a close approximation between pH's of 3 to 11 and values of C of 0.01 to 0.10*M*, the last two terms may be neglected.

Maximum buffering occurs when:

$$[H^+] = K_A$$

Thus

$$\left(\frac{dB}{d\text{pH}}\right)_{\text{max}} = \frac{2.303C}{4} = 0.576C$$

In applications to milk dB/dpH is evaluated experimentally since calculations from the concentrations of buffer salts present are extremely involved.

The pH of cow's milk is commonly stated as falling between 6.5 and 6.7, with 6.6 the most usual value. It should be emphasized, however, that this value applies only at temperatures of measurement near 25°C. The pH of milk exhibits a greater dependence on temperature than that of such buffers as phosphate which is the principal buffer component of milk at pH 6.6. Miller and Sommer[129] reported a specimen with pH of 6.64 at 20°C decreasing to 6.23 at 60°C. Over the same temperature range a phosphate buffer decreases only from pH 6.88 to 6.84 (Bates[14]). Likewise Dixon[55] observed that the pH of milk decreases by about 0.01 unit/°C between 10 and 30°C, and emphasized the importance of careful temperature control in making pH measurements. The marked temperature dependence of the pH of milk probably is attributable to insolubilization of calcium phosphate as the temperature is raised and its solution as temperature is lowered.

Differences in pH and buffering capacity among individual lots of fresh milk reflect compositional variations arising from the functions of the mammary gland. In general the pH is lower (down to pH 6.0[125]) in colostrum and higher (up to 7.5[156]) in cases of mastitis than in normal milk of mid-lactation. Colostrum and mastitis milk are known to differ radically in proportions of the proteins and certain salts.[12,33,69,145,146,224] However no attempt seems to have been made to relate these compositional features precisely to pH and buffering capacity.

Published titration curves for milk[29,37,125,230,247,248,254] have dealt most intensively with the range between pH 4 and 9. There is little available information on buffering below pH 4. Over the range from pH 4 to 9 milk exhibits a pronounced maximum buffering between pH 5 and 6, the position of this maximum depending on the manner of conducting the titration.[254] The low buffer capacity in the region of the phenolphthalein end point (pH 8.3) contributes to the practicality of the well-known procedures for determination of "titratable acidity."

Table 8.1 presents data for the pH and titratable acidity observed in two studies on milks of different breeds. There appears to be a systematic difference between the two sets of data in regard to titratable acidity, but both lie within the range usually observed.

Table 8.1

TITRATABLE ACIDITY AND PH OF MILK

Breed	No. of Samples	Titratable Acidity Range	Titratable Acidity Mean	Mean pH
		Percent	Lactic Acid	
Ayrshire	229[1]	0.08–0.24	0.160	—
Holstein	297[1]	0.10–0.28	0.161	—
	606[2]	—	0.133	6.71
Guernsey	153[1]	0.12–0.30	0.172	—
	384[2]	—	0.151	6.65
Jersey	132[1]	0.10–0.24	0.179	—
	1062[2]	—	0.149	6.66

[1] Caulfield and Riddell[33]
[2] Wilcox and Krienke[253]

In principle it would be logical to combine plots of the buffer index curves of each of the buffer components of milk and thus to obtain a plot which could be compared with that actually found for milk. It is not difficult, of course, to conclude that the principal buffer components are phosphate, citrate, bicarbonate, and proteins, but quantitative assignment of the buffer capacity to these components proves to be rather difficult. Primarily this problem arises from the presence of calcium and magnesium in the system. These alkaline earths are present as free ions; as soluble undissociated complexes with phosphate, citrate, and casein; and as colloidal phosphates associated with the casein. Thus precise definition of the ionic equilibria in milk becomes rather complicated. Analytical methods are not available for all the several kinds of ions and complexes present. Concentrations of some of them must be calculated from the dissociation constants, whose values in turn are dependent on the ionic strength of the system. Nevertheless, some progress has been made with this approach since the situation was reviewed by Clark.[37] Methods for estimating the concentrations of ionic calcium and magnesium are now available.[36,115,132,215] Some of the principal dissociation constants have been evaluated, particularly those for:

$$\text{Ca Cit}^- \rightleftharpoons \text{Ca}^{++} + \text{Cit}^{\equiv} \quad pK = 4.64 - 3.64\sqrt{\mu},$$

and

$$\text{CaHPO}_4 \rightleftharpoons \text{Ca}^{++} + \text{HPO}_4^{=} \quad K = 1.5 \times 10^{-2}$$

Calculations by Boulet and Marier,[24] in which association of calcium and magnesium with both phosphate and citrate are included, give values for calcium-ion concentration in reasonable agreement with determined concentrations. It thus appears that a close approximation of the complete ionic composition of milk could now be made.

Calcium and magnesium influence the titration curves of milk because as the pH is raised they precipitate as colloidal phosphates, and as the pH is lowered colloidal calcium and magnesium phosphates are solubilized. Since these changes in state are sluggish and the composition of the precipitates depends on the conditions,[25] the slope of the titration curves and the position of the maximum buffering will depend on the speed of titration.

Three approaches have been used in attempting to account for the buffer behavior of milk in terms of the properties of its components. These are calculation, fractionation, and titration of artificial mixtures. Whittier[248,249] derived equations for dB/dpH in calcium phosphate and calcium citrate solutions, taking into account available data on dissociation constants and solubility products. Presumably this approach could be extended to calculate the entire buffer curve. It demands precise knowledge of the dissociation constants of the several buffers, the dissociation of the calcium and magnesium complexes and the solubility products of the calcium and magnesium phosphates under the conditions of a titration of milk.

Fractionation of milk and titration of the fractions has been of considerable value. An early attempt to assign the contributions of the components to the titratable acidity (i.e., pH 6.6 to 8.3) was made by Rice and Markley.[172]

Much interest has centered in a scheme of fractionation with oxalate to precipitate calcium and rennet to remove the calcium caseinate phosphate micelles.[101,123,161] As formulated by Ling[123] the scheme involves titrations of milk, oxalated milk, rennet whey and oxalated rennet whey to the phenolphthalein end point. From such titrations Ling calculated that the caseinate contributed about 0.8 meq, the whey constituents 1.0 meq, and the protons released by precipitation of calcium phosphate during the titration 0.4 meq of the total titer of 2.2 meq per 100 ml in certain milks that he analyzed. These data are consistent with calculations based on the concentrations of phosphate and proteins present.[109]

Titrations of artificially prepared mixtures containing phosphate, calcium, citrate, and sometimes proteins have been employed to study the precipitation of calcium phosphate and the inhibitory effect of citrate thereon.[26,59,254] The technique is valuable for basic studies because the composition of the system can be controlled.

In dairy processing operations the pH and buffering power of milk are influenced (aside from the action of microorganisms) by heat treatments that may be applied. Moderate heating such as pasteurization produces small shifts in pH and buffering by expulsion of CO_2 and by precipitation of calcium phosphate with release of hydrogen ions (see Pyne[160]). The drastic heat treatments used in sterilization produce acids by degradation of lactose.[71,72,109] The rate is slow below 90°C but increases[251] markedly above 100°C.

Concentration of milk lowers the pH. At concentrations of 30 and 60 gm solids per 100 gm water the pH's are about 6.2 and 6.0, respectively.[59,103]

During slow freezing the pH of milk has been observed to fall to values as low as 5.8, whereas little change in pH occurred during fast freezing.[222] During storage of frozen milk at − 7 or − 12°C, the pH decreased to a minimum of about pH 6.0 and increased gradually therafter. These effects of freezing and frozen storage are considered to be caused by insolubilization of salt constituents.

OXIDATION-REDUCTION EQUILIBRIA

Whether a reversible oxidation-reaction involves a transfer of oxygen, of hydrogen, of both, or of neither, there is a transfer of electrons between atoms or molecules. The addition of electrons is synonymous with reduction and the withdrawal of electrons constitutes oxidation. On this basis, and the law of mass action, the following basic equation can be derived (Clark[38]).

$$E_h = E_0 - \frac{RT}{nF} \ln \frac{[\text{Red}]}{[\text{Ox}]}$$

where E_h = oxidation reduction potential, E_0 = standard oxidation reduction potential of the system, R = gas constant, T = absolute temperature, n = number of electrons transferred per molecule, F = the Faraday, [Red] = concentration of reduced form, and [Ox] = concentration of oxidized form.

At 25°C and one electron transfer the equation becomes

$$E_h = E_0 + 0.06 \log \frac{[\text{Ox}]}{[\text{Red}]}$$

The standard potential E_0, is obviously the value of the potential at equal concentrations of the oxidized and reduced forms. Its value is an index of the relative position of the system on the scale of potential.

In considering oxidation-reduction equilibria in milk, the principal interest is in the potentials of the system present relative to one another and those that may be superimposed. Hence, it does not seem necessary to discuss effects of differences in n on the slopes of curves of E_h plotted against percentage reduction or of the relations between pH and E_h. A discussion of these relationships is found in the monograph by Clark.[38] In the following discussion the symbol $E_0{}'$ is used to designate the potential of a system containing equal concentrations of oxidant and reductant at a specified pH value. The curves in Fig. 8.1 are for systems present in milk and for indicators added to milk for the purpose of measuring bacterial activity. These curves indicate the relationship among these systems at different pH values.

In a fluid such as milk, containing several oxidation-reduction systems, the effect of each system on the potential depends on several factors. These include the reversibility of the system, its $E_0{}'$ value or position on the scale of potential, the ratio of oxidant to reductant, and the concentration of the active components of the system. Only a reversible system gives a potential at a noble-metal electrode, and this measured potential is an intensity factor analogous to the potential measured on a hydrogen electrode in determining H-ion concentrations.

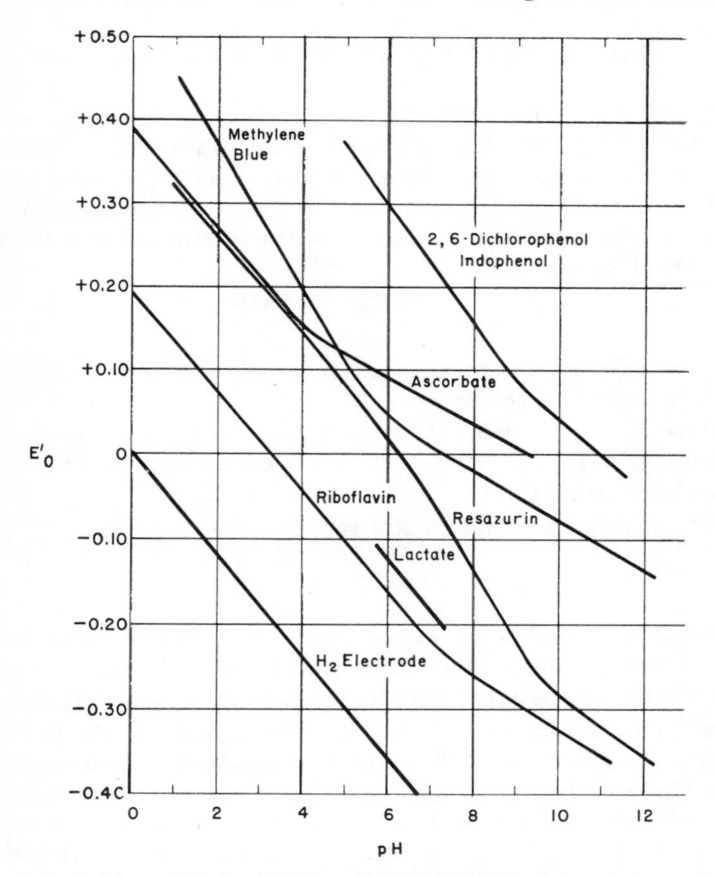

FIG. 8.1 RELATION OF OXIDATION-REDUCTION POTENTIAL OF VARIOUS SYSTEMS TO PH

The quantity factor in oxidation-reduction is the overall concentration of active substance, [Ox] + [Red]. Two solutions of the same system having the same ratio of reductant to oxidant have the same potential, but may have different quantity factors, as [0.1]/[0.2] and [0.8]/[1.6], in which instance the second will be able to oxidize nearly eight times as much reducible substance in a system of low potential as the first. If two reversible systems are combined, the potentials of the two systems change to a common value intermediate between the two initial potentials, part of the oxidant of the system initially more positive being reduced to its reductant form by a reaction whereby part of the reductant of the system initially more negative is oxidized to its oxidant form. The value of the final potential will depend on the relative concentrations of the two systems.

Oxidation-reduction systems exhibit resistance to change of potential when the concentrations of oxidant and reductant are close to equal. This phenomenon, analogous to buffer action in acid-base equilibria, is known as poising.

Fresh milk, as ordinarily produced, exhibits a potential at a gold or platinum electrode of between +0.20 and +0.30 volt. That dissolved oxygen is a major factor in the establishment of this potential has been shown in several ways. Milk drawn from the udder anaerobically reduces methylene blue, indicating that its potential is more negative than of the methylene blue system.[106] When such a milk is exposed to oxygen, its potential rises to a point such that it is positive to that of the methylene blue system. Washing oxygen-containing milk with oxygen-free gas, or allowing growing *Streptococcus lactis* or lactobacilli to remove free oxygen from milk, causes the potential to change in the negative direction.[59,85] Bubbling air or oxygen through these milks will restore the positive potential. This leaves the systems involving ascorbate, lactate and riboflavin as those that may be responsible for the values of oxidation-reduction in oxygen-free milk and may participate in the stabilization of the potential of oxygen-containing milk. The relative positions of these systems are shown in Fig. 8.1. The concentration of ascorbic acid in milk is sufficient for it to exert an appreciable effect, and the system is reversible. The oxidant of the system, dehydroascorbate, readily undergoes further oxidation, but irreversibly; hence the ratio of concentration of ascorbate to that of dehydroascorbate will remain large until the system disappears from the milk, and this system will tend to stabilize the potential at approximately 0.0 volt. The lactate-pyruvate system is an irreversible one, activatable by enzymes and mediators and of a highly negative normal potential,[11] but it is present in fresh milk in such minute quantities that, even if it were activated, its effect could be only very slight.

The riboflavin system is an active reversible one and of highly negative normal potential; but, since its concentration is low and it is present in fresh milk entirely in the oxidized form, its influence would be slight and would not be exerted in the direction of negative potentials. It seems logical to conclude that the ascorbate-dehydroascorbate system is the principal one stabilizing the potential of oxygen-free milk at a value near 0.0 volt and is the system that functions along with the oxygen system to stabilize the potential of oxygen-containing milk in the zone of +0.20 to +0.30 volt.

When milk undergoes fermentation by *Streptococcus lactis,* the oxidation-reduction potential of the milk changes[66] with time, typically as shown in Fig. 8.2. This pattern has been observed and recorded by a number of investigators. Curves characteristic of fermentation by other organisms that may be present in milk differ somewhat,[66] but the tendency in general is for the potential to be changed in a negative direction. The rapid change of potential shown in Fig. 8.2 occurs only after the dissolved oxygen has been consumed by the bacteria and may be identified by the change in color of certain dyes added to the milk. These dyes are oxidants of oxidation-reduction systems. Since the time elapsing before these dyes are reduced to the colorless reductant form is roughly proportional to the number of bacteria present, this "reduction time" is an index of the degree of bacterial contamination.

The effects of heat treatment of milk on the oxidation reduction potential have been studied to a considerable extent.[59,73,85,112] A sharp decrease in the potential coincides with the liberation of sulfhydryl groups by denaturation of the protein, primarily β-lactoglobulin. Minimum potentials are attainable by deaeration and high

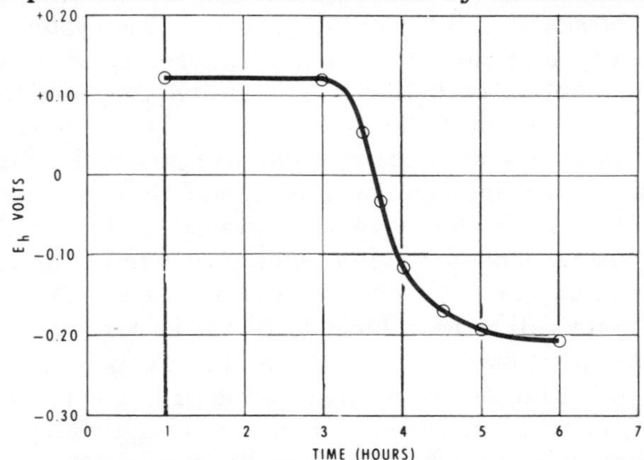

FIG. 8.2 DECREASE IN OXIDATION-REDUCTION POTENTIAL OF MILK DURING INCUBATION WITH A STRAIN OF *S. lactis* AT 25°C

temperature-short time heat treatments.[99] Such treatments also produce dried milks of superior stability against oxidative flavor deterioration.[85]

Using potentiometric titration at 37°C with sodium hydrosulfite ($Na_2S_2O_4$), Nilsson and co-workers[137] demonstrated large variations in the poising effect of milk specimens from individual cows and bulked supplies. The poising index, defined as equivalents of $Na_2S_2O_4$ per l of milk required to attain the fully reduced level, divided by the fall in E_h from the fully oxidized to the fully reduced state varied from 0.00488 to 0.0229 equiv. liter $^{-1}$ volt $^{-1}$ for 58 specimens. Highest poising indexes were found in specimens taken early in lactation, but no correlation of poising with the concentration of any natural component was made.

DENSITY

The density of such a system as milk is a resultant of the densities of its components complicated by variations in the ratio of solid to liquid fat and in the degree of hydration of the proteins. Thus the density of a given specimen of milk is determined by its percentage composition, by its temperature, and by its previous history of temperature fluctuations and processing treatments.

Interest in the density of milk has been concerned mostly with its use, in conjunction with a fat test, to estimate the content of total solids. Equations of the form: $T = AF + BD$ expressing the relation between total solids (T), fat (F), and density (D) have been derived.[109,151,176] Such derivations assume constant values for the densities of the fat and of the mixture of solids-not-fat which enter into the calculation of the coefficients, A and B. Since milkfat has a high coefficient of expansion and also contracts as it solidifies slowly, the temperature of measurement and the previous history of the product must be carefully controlled (see Sharp and Hart.[187]) Variations in composition of fat and in the proportions of lactose, proteins, and salts probably influence the equations much less than do variations due to physical state of the fat. Many studies have compared total solids calculated by such equations with gravimetric results.[95,177,231] In case of disagreement, proposals to modify the equations have been made. This amounts, of course, to including in the equation factors appropriate to the series of specimens analyzed and also to compensate for systematic errors in determination of fat, total solids, and density.

Densities of liquid dairy products as varied as milk, whey, evaporated milk, sweetened condensed milk, and freshly frozen ice cream have been measured (a) by weighing a given volume as in pycnometer; (b) by determining the extent to which an object sinks, as with hydrometers, lactometers,[44,79,62,105,162,204] or a series of beads of graded density;[44,202,203] (c) by hydrostatic weighing of an immersed bulb, as with a Westphal or analytical balance;[126,140] (d) by measuring the volume of a given weight of product as in a dilatometer;[201] and (e) by measuring the distance that a drop of product falls in a density gradient column.[28,207] The choice of method for a given purpose demands that a balance be struck between precision on the one hand and speed and convenience on the other. Description and critical appraisal of the several methods are given by Bauer and Lewin.[16]

Data reported in the literature for density and specific gravity of milk and its fractions and products are somewhat difficult to summarize on a comparative basis because measurements were made at various temperatures. Much of the older data is given as specific gravity at 15.5°C/15.5°C, where the value for fresh whole mixed herd milk seldom lies outside the range of 1.030 to 1.035, and 1.032 is often quoted as an average value. Skimmilk at this temperature has a specific gravity of about 1.036 and evaporated whole milk about 1.0660. Table 8.2 gives results at 20°C/20°C for milk from cows of various breeds.

The density of milk decreases as the temperature is raised. This fact is universally recognized in precautions for temperature control in density measurements and in factors for converting readings taken at one temperature to their equivalents at another. Whole homogenized milk has been found by hydrostatic weighing in a supercooled state to have a maximum density at −5.2°C.[243] Furthermore, the temperature of maximum density increased linearly with extent of dilution with water up to +4°C. An additional complication is that the temperature of maximum density gradually decreases as the sample is held at low temperatures. Since the only study of this point[243] has been made on whole milk, neither the lowering of the temperature of maximum density nor the aging effect can be specifically assigned to the fat or to the nonfat solids.

Much work agrees that the densities of skimmilk and whole milk decrease from the maximum at −5.2°C to about +40°C.[104,200,201,236,239]

For the temperature range 1 to 10°C, Watson and Tittsler[233] derived the relation:

$$\text{Density} = 1.003073 - 0.000179\,t - 0.000368\,F + 0.003744\,N$$

where : t = temperature in °C

F = percent fat

N = percent nonfat solids

Table 8.2

SPECIFIC GRAVITY OF MILK FROM COWS OF VARIOUS BREEDS

Breed	No. of Cows	No. of Samples	Specific Gravity at 20°C./20°C. Range	Mean	Standard Deviation
Ayrshire	14	208	1.0231–1.0357	1.0317	0.0022
Brown Swiss	17	428	1.0270–1.0366	1.0318	0.0016
Guernsey	16	321	1.0274–1.0398	1.0336	0.0018
Holstein	19	268	1.0268–1.0385	1.0324	0.0018
Jersey	15	199	1.0240–1.0369	1.0330	0.0024

Data of Overman, *et al.*[141]

This was based on analysis of 101 specimens of mixed raw milk.

Data obtained by Short[200] in the range 10 to 45°C for whole milk of 3% fat and 8.7% solids-not-fat and skimmilk of 0.02% fat and 8.9% solids-not-fat were fitted to empirical equations of the form: density $= a + bt + ct^2 + dt^3$, where $t =$ temperature in °C. Average effects of varying the percentages of fat and solids-not-fat were also determined. All these data are collected in Table 8.3. It should be emphasized that these coefficients are valid only up to about 45°C. In this range not only the absolute density but also the density relative to water (i.e., specific gravity at t°C/t°C) decreases.[174,201,236,239] This phenomenon occurs with skimmilk and whole milk, but not with a 5% lactose solution; it thus appears to represent changes (perhaps degree of hydration) in the proteins. The decrease in relative density (specific gravity) is of the order of 0.00005 per °C in the range from 10 to 40°C.

Table 8.3

COEFFICIENTS a, b, c, d IN THE EQUATION $D - 1 = a - bt + ct^2 - dt^3$, WHERE D = DENSITY (GM/ML), AND t = TEMPERATURE(°C.)[a]

	$10^4 \times$ a	b	c	d
Whole milk	350	3.58	0.049	0.0010
Skimmilk	366	1.46	0.023	0.0016
Ave. difference per 1% fat[b]	4.8	0.39	0.0061	0.00002
Ave. difference per 1% S.N.F.[b]	3.8	0.08	−0.010	−0.00004

[a] Data of Short.[200]

[b] Interpolations of measurements on mixtures of cream and skimmilk from a single sample were apparently used to determine these averages. Such mixtures would have a constant ratio of each constituent of S.N.F. and would not apply to other samples (e.g., with a different protein/lactose ratio).

Data in the literature are not in agreement as to the relation of the density of milk to temperature above 40°C. Of course the absolute density decreases, but although some pycnometric data indicate that the specific gravity remains virtually constant[236,239] other data obtained with a dilatometer indicate a pronounced increase in specific gravity from 40 to 90°C.[201] The increase found in the latter work is of the order of 0.00005 per °C, and is thus of sufficient magnitude to make verification and interpretation important. Density measurements may be of use in detecting changes in state of milk proteins.

Relationships among fat content, temperature and density of creams have been developed.[17,34,150] Phipps[150] has devised a nomogram covering the range up to 50% of fat and temperatures from 40 to 80°C.

Short[201] reported a small increase (average of six determinations = 0.000065) in density at 20°C as a result of homogenizing whole milk but not skimmilk, and a small decrease (average of 10 determinations = 0.000094) as a result of sterilizing either whole milk or skimmilk for 1 hr at 95°C. These effects varied considerably among samples and thus can be said to be virtually negligible.

As water is removed in the manufacture of concentrated milk products, their density increases. If concentration produces no changes in physical state of the constituents (e.g., change in degree of hydration of proteins or insolubilization of salts) the density of a concentrated product (Dc) could be computed from the density (Di) and total solids (Ti) of the initial milk and the total solids of the concentrate (Tc) by the following equation:[108]

$$Dc = \frac{Di \ D_{H_2O} \ Ti}{Tc(D_{H_2O}\text{-}Di) + DiTi}$$

where D_{H_2O} is density of water. For specific gravity (S_c) at $t°/t°$:

$$S_c = \frac{Dc}{D_{H_2O}} = \frac{DiTi}{Tc(D_{H_2O}\text{-}Di) + DiTi}$$

Monjonnier and Troy[131] present data for concentrated whole and skimmilk which conform approximately to this equation, but data of sufficiently high precision to detect small deviations due to changes in protein hydration or salt precipitation are not available.

VISCOSITY

A review by Swindells and co-workers[211] discusses the theoretical basis of viscosity and methods of measurement. Viscosity may be defined by the following equation:

$$\eta = F/(dV/dx)$$

where η is the coefficient of viscosity, F = force in dynes cm^{-2} necessary to maintain a unit velocity gradient between two parallel planes separated by unit distance, and dV/dx = velocity gradient in sec^{-1} perpendicular to the planes.

The unit of viscosity, the poise, is defined as the force in dynes cm^{-2} required to maintain a relative velocity of 1 cm per sec between two parallel planes 1 cm apart.

Fluidity, ϕ, is the reciprocal of viscosity. The unit commonly used for milk is the centipoise (10^{-2} poise)

$$\phi = 1/\eta$$

Kinematic viscosity, ν, is viscosity divided by density.

$$\nu = \eta / \rho$$

In dealing with solutions and colloidal dispersions use is often made of the following quantities:

Relative viscosity: $\eta_{rel} = \eta$ solution/ηsolvent

Specific viscosity: $\eta_{sp} = \eta_{rel} - 1$

Reduced viscosity: $= \eta_{sp}/c$, where c is the concentration of solute

Intrinsic viscosity: $= [\eta] = \lim_{c \to 0} (\eta_{sp.}/c)$.

Fluids for which the viscosity coefficient η depends only on temperature and pressure and is independent of rate of shear are called "Newtonian." A plot of shear stress versus rate of shear for such fluids is a straight line passing through the origin. Behavior of this type is exhibited by gases, pure liquids, and solutions of materials of low molecular weight. On the other hand, many colloidal dispersions and solutions of high polymers where the molecular species is large show marked deviation from Newtonian behavior, the rate of shear depending on the shear stress (plasticity) or the duration of shear (thixotropy) or both.

Skimmilk and whole milk do not differ appreciably from Newtonian behavior, but cream, concentrated milks, butter, and cheese exhibit varying degrees of non-Newtonian behavior. The literature on these products up to 1953 is summarized and reviewed in the monograph edited by Scott-Blair.[185]

The three types of viscometers that have been used in most studies of the viscosities of dairy products are: coaxial cylinders (e.g., McMichael, Couette, and Brookfield), falling spheres (e.g., Hoeppler) and capillary tubes (e.g., Ostwald). A discussion of viscometer designs for measurements of non-Newtonian fluids[126] considers falling sphere and capillary tube viscometers unsuitable because corrections for non-uniform shear rates are tedious and/or not sufficiently exact. The two conditions proposed for minimizing uncertainty are uniform shear rate and a consistent procedure regarding rate and duration of shear. These points are also emphasized by Potter *et al.*[154] in a study of the applicability of coaxial cylinder viscometers to various concentrated milk products. A general discussion of falling-sphere viscometers[235] presents conditions for and limitations of their use. A "mobilometer,"[127] which has some features of both the coaxial cylinder and falling-sphere types, has been used to measure viscosities of evaporated milk. A sealed microviscometer of the falling-sphere type in which the specimen can be sterilized has been proposed for studies of changes occurring in sterilization and storage of concentrated milk products.[121]

The value of η of 1.0019 ± 0.0003 for water at 20°C[210] and Cragoe's equation[39] $\log(\eta_t/\eta_{20}) = (1.2348(20 - t) - 0.001467(t - 20)^2)/(t + 96)$ for other temperatures seem slow in receiving the recognition which they deserve from their careful determination. Calibration of nearly all viscometers is based on earlier similar determinations.

The viscosity of milk and dairy products depends on the temperature and on the amount and state of dispersion of the solid components. Representative values at 20°C are: whey 1.2 centipoise, skimmilk 1.5 cp, and whole milk 2.0 cp. From these it is evident that the caseinate micelles and the fat globules are the most important contributors to the viscosity. Specific data are given in Table 8.4.

Many workers (see Andrade[3]) have proposed equations of the following general type to relate viscosity of fluids to temperature: $\eta = A e^{B/T}$, where A and B are constants, e is the base of natural logarithms and T is the absolute temperature. Caffyn,[31] in a careful study of the viscosity of homogenized milk, found that plots of $\log \eta$ vs. $1/T$ were not linear over the range 20 –80°C but exhibited a sharp break at about 40°C, which he attributed to melting of the fat. However, plots over the range 5 –80°C of $\log \eta$ vs. $1/T$ for *skimmilk* (data of Eilers *et al.*[59] and of Whitaker *et al.*[239] plotted by R. Jenness[108]) also show distinct breaks at about 30 and 65°C. Therefore it appears that equations of this sort do not apply to the milk system over such a wide range of temperature. Cox[43] examined available data including Caf-

Table 8.4

VISCOSITY OF MILK, SKIMMILK AND WHEY

No. of Samples	Temp.	Whole Milk	Viscosity of Skimmilk	Whey	Ref.
	°C.	c.p.	c.p.	c.p.	
180	27	1.45			159
14	25		1.42		59
62	24		1.47		239
9	24			1.16	239

fyn's[31] and fitted an empirical equation of the form: $\eta_t - \eta_{20} = b_1(t-20) + b_2(t-20)^2 + b_3(t-20)^3$, to relate absolute viscosity of whole milk at any temperature, η_t, to temperature, t. Average values for b_1, b_2, and b_3 were, respectively, -5.895×10^{-4}, -10.089×10^{-6}, and -6.886×10^{-8}. This fits certain data reasonably well but the coefficients fitted for one sample do not necessarily apply to another.

Cox and co-workers[44] surveyed available literature on the relation between composition and viscosity of cow's milk. Although data from any one study could be fitted to an empirical equation of the form: $\eta = A + B_1p + B_2q + B_3p^2$, where p and q are the respective percentages of fat and solids-not-fat, the coefficients B_1, B_2, and B_3 differed markedly from one set of data to another. Viscosity increased with increasing concentrations of both fat and solids-not-fat, of course, but a consistent general relationship was not obtained. Other work[169] has indicated higher coefficients of correlation.

Whitnah[241] found high positive correlations between viscosity at 4°C of milks from individual cows and their fat and protein contents. Surprisingly, the correlation between viscosity and the content of solids other than fat or protein was negative, which may reflect an inverse effect of lactose on viscosity.[241]

The viscosity of colloidal systems depends on the volume occupied by the colloidal particles. The simple equation of Einstein: $\eta_{rel} = 1 + 2.5\theta$, where θ is the fraction of the total volume occupied by the dispersed phase, was derived on the assumption of rigid spherical particles of equal size and appears not to fit the situation for skimmilk or whole milk. Eilers[59] applied an empirical equation

$$\eta_{rel} = \left[1 + \frac{1.25\, VC_v}{1\text{-}1.35\, VC_v}\right]^2$$

to skimmilk. When the viscosity of raw skimmilk relative to that of rennet whey was used, this equation gave a value for the "rheological concentration" or fractional volume occupied, VC_v, of 0.088 for the caseinate particles. C_v is the concentration of caseinate on a dry basis and V is the "voluminosity", which is a composite factor including both hydration and electrical effects. Since C_v was 2.9 gm/100 ml in the skimmilk dealt with, the voluminosity of the caseinate particles was 3.0 ml/gm at 25°C. (i.e., 1 gm of dry caseinate appears to occupy about 3.0 ml). Whitnah[244] applied an equation furnished by Ford to determine voluminosity at various temperatures of caseinate fractions obtained by fractional centrifugation of skimmilk. This yielded values close to 3 ml/gm at 25°C but voluminosity appeared to increase sharply below and decrease moderately above this temperature. Eilers [59] plot of viscosity of skimmilk relative to that of whey at various temperatures also shows a sharp decline from 5 to 30°C, reflecting a decrease in voluminosity of the caseinate micelles. Above 30°C, the decrease is less marked until about 65°C, where the whey proteins begin to be denatured. From this point the viscosity of whey relative to water increases, but that of skimmilk relative to whey remains relatively constant.

Changes in the caseinate micelles produced by either raising or lowering the pH result in an increase in viscosity.[59,157] For example, the viscosity is approximately doubled by addition of 10 ml. 1.4 to 3.8*N* ammonia to 90 ml milk. Addition of alkali (to pH's up to 11.7), urea (up to 4.8 *M*), and calcium complexing agents to concentrated (22.7% solids) skimmilk causes a marked transient increase of several fold in viscosity, followed by a sharp decline.[18,19] This has been interpreted as resulting from a swelling of the micelles, followed by their disintegration.

Various measurements of viscosity as a function of concentration of solids in diluted and concentrated milk[13,54,118] reveal a curvilinear relationship. Eilers[59] calculated voluminosity at several solids concentrations (skimmilk) showing that voluminosity of the caseinate particles does not change, but that the apparent voluminosity of the total solids decreases as concentration is raised. This may merely signify that the equations used, which had been derived for particles of equal size, do not fit the situation of diverse particle sizes found in skimmilk.

Torssell *et al.*[217] attempted to develop a mathematical relationship among viscosity, temperature, and total solids content for skimmilk and whole milk. After showing that Walther's equation:

$$m = \frac{\log\log(\nu_{t1} + 0.8) - \log\log(\nu_{t2} + 0.8)}{\log T_2 - \log T_1}$$

applies in that a straight line was obtained by plotting $\log\log \nu + 0.8$ vs. log T (i.e., that m is a constant), they found that a plot of m vs. solids percent is also a straight line. These relationships held for skimmilk up to 36.5% solids and for whole milk up to 41.7% solids. Undoubtedly the specific plots would differ from milk to milk.

Phipps[150] derived the following equations for creams up to 50% fat at temperatures of 40 to 80°C:

$$\log_{10}\eta = A\ (F + F^{5/3}) + \log_{10}\eta_o$$

where:

$$A = 1.2876 + 11.07 \times 10^{-4}\, t$$

$$\log_{10}\eta_o = 0.7686\ (10^3/T) - 2.4370$$

η = viscosity of cream in cp

η_o = viscosity of skimmilk in cp

F = fat content of cream

t = temperature in °C

T = temperature in °K

A nomogram relates the three variables, viscosity, temperature and fat content.

Heating of either skimmilk or whole milk at temperatures up to about 65°C results in a temporary decrease of a few hundredths of a centipoise in the viscosity measured after quickly cooling to room temperature or lower. Data of Whitnah *et al.*[245] and Eilers[59] agree in indicating that this effect is due to a reversible change in the caseinate micelles. The recovery from the viscosity decrease produced by heating was shown by Whitnah *et al.*[246] to be an approximately linear function of the logarithm of age after heating.

The increase in viscosity when milk and particularly concentrated milk is heated sufficiently to aggregate the proteins is discussed in Chapter 11. The effects of homogenization and clustering of the fat globules on their contribution to the viscosity of the product are dealt with in Chapter 10.

SURFACE AND INTERFACIAL TENSION

The area of contact between two phases is called the "interface," or, especially if one of the phases is gaseous, the "surface." The properties of interfaces and surfaces are determined by the number, kind, and orientation of molecules located in them. A widely used measure is energy per unit area. It is the work required to extend the surface by unit area or force required per unit length, is expressed as ergs cm^{-2} or dynes cm^{-1}, and is often symbolized by the Greek letter gamma (γ).

Surface-active solutes accumulate in the interface between two phases in accordance with their concentration and their ability to reduce the interfacial tension. The relation may be expressed by the Gibbs equation as follows (see Bikerman[22]):

$$\mathrm{T} = \frac{C}{\mathrm{RT}} \cdot \frac{d\gamma_1}{dc}$$

where T = excess concentration of solute in interface over that in bulk of solution, C = concentration of solute in bulk of solution, and $d\gamma_1/dc$ = rate of change of interfacial tension with change in concentration of solute.

This reversible and ideal relationship predicts that the more effective depressants of interfacial tension will tend to accumulate in the interface to the exclusion of others. Actually, in many cases the amount of material concentrated at the interface is greater than would be predicted by the Gibbs equation and the system is not reversible or only sluggishly so.

In milk, the important interfaces are those between the liquid product and air and between the milk plasma and the fat globules contained therein. Studies of the surface tension (liquid/air) have been made to ascertain the relative effectiveness of the milk components as depressants; to follow changes in surface-active components as a result of processing; to follow release of free fatty acids during lipolysis; and to attempt to explain the foaming phenomena characteristic of milk. Interfacial tensions between milkfat and solutions of milk components have been measured in studies of the stabilization of fat globules in natural and processed milks.

The chief methods of measuring surface and interfacial tension are described and critically evaluated by Harkins and Alexander.[84] They may be classified as dynamic and static. In the former, measurements are make on freshly formed surfaces during the span of time required for equilibruim to be established. Such methods enable the rate of orientation of molecules in the interface to be followed. The method of vibrating jets is such a dynamic method.[84] It has been applied to milk in two studies which agree that the surface tension did not change during the observed age of the jet. In the first,[120] it was concluded the surface tension in the jet was equal to that of water and thus that surface orientation had not occurred. The second[242] concluded that the surface tension equalled the static value for milk—i.e., that the change had been completed during the early part of the first wave. The large decrease in surface tension during the interval 0.001–0.01 sec for diluted milk, or for milk from which the proteins had been precipitated, seems to justify the second conclusion. Orientation in newly formed surfaces and interfaces in milk may thus be considered to be extremely rapid in comparison to time intervals involved in dairy-processing operations.

By far the majority of information on surface and interfacial phenomena in milk has been obtained with various static methods.[240] Five of the principal ones involve, respectively, determination of: (1) height of rise of liquid in a capillary; (2) weight or volume of drops formed by liquid flowing from a capillary tip (sometimes considered "semi-dynamic"); (3) force required to pull a ring or plate out of a surface; (4) maximum pressure required to force a bubble of gas through a nozzle immersed in the liquid; and (5) shape of a pendant drop hanging from a capillary. Each of these parameters may be related theoretically and practically to the surface tension.[84] The method involving pulling of a ring or plate from the surface is undoubtedly the most widely employed. It is rapid, simple and capable of accuracy of ±0.25% or better when the force is measured with an analytical balance.[84] Apparatus employing tension balances of various degrees of sensitivity are also available (Du Noüy balances). The critical discussion by Herrington[96] of certain aspects of the measurement of surface tension emphasizes some of the variables and necessary precautions.

The surface tension of milk is of the order of 50 dynes cm^{-1} at 20°C, compared to that of water 72.75 dynes cm^{-1} at the same temperature. The milk proteins, milkfat, phospholipids, and free fatty acids are the principal surface-active components determining the surface properties of milk.

Using the ring method at 20°C, Sharma[186] reported values of 42.3 to 52.1 dynes cm^{-1} (mean 46.8 ± 2.3) for 51 specimens of milk from individual Indian cows, and Parkash[144] found an average of 46.02 ± 1.14 dynes cm^{-1} for 100 specimens. Values of 47.5 to 48.0 dynes cm^{-1} at 20°C were found for individual cow specimens in France by Calandron and Grillet.[32] Mohr and Brockmann[130] in Germany reported 51.0 dynes cm^{-1} for skimmilk, 46.7 for whole milk and 44.8 for cream of 0.04, 2.4 and 34.0% fat, respectively, from a single original lot.

The effectiveness of surface and interfacial tension depressants can be compared by plots of concentration versus tension. Various dilution studies of milk, skimmilk, wheys, and solutions of milk proteins reveal that caseinate and the proteins of the lactalbumin fraction (β-lactoglobulin, α-lactalbumin, and bovine serum albumin) are powerful depressants, while the proteins of the immunoglobulins fraction are somewhat less so.[6,60,107,111,205] In skimmilk and rennet whey the concentrations of proteins are far above the levels where their effects are additive; both products have nearly the same surface tension, but that of whey drops more markedly with dilution.[6] Minor protein fractions that were especially effective depressants but have not been fully characterized were prepared by Ansbacher *et al.*[4] from casein by elution with salt solution, and by Aschaffenburg[6] from the serum from heated skimmilk. The protein and protein-phospholipid complex from the surface of the milkfat globules is one of the most powerful and significant depressants of tension at both the milk/air and plasma/fat interfaces.[107,130,143] The fact that the surface tension of whole milk lies a few dynes cm^{-1} below that of skimmilk may be due to the presence of these substances, as well as to traces of free milkfat in the surface. Surface tension decreases with increasing fat content up to about 4% fat, but does not decline to any extent with further increase.[232] The protein-phospholipid complex is undoubtedly largely responsible for the very low surface tension of sweet cream buttermilk.

As temperature is raised in the range 10 to 60°C, the surface tension of skimmilk and whole milk decreases.[130,232] This decrease is comparable in magnitude to that observed in the surface tension of water, which decreases by about 10 dynes cm^{-1} over this range.

It has frequently been observed that the surface tension of milk that has been held at 5°C and brought to 20°C is lower (2 to 3 dynes cm^{-1} than that of milk cooled to 20°C and measured immediately.[130,188] Sharp and Krukovsky[188] also demonstrated that skimmilk separated at 60°C has a higher (2 to 3 dynes cm^{-1}) surface tension than that separated at 5°C. It thus appears as though a surface-active substance is released from fat globules at low temperature. This behavior is opposite that of the fat globule "agglutinin", also studied by Sharp and Krukovsky,[188] which is adsorbed on solid fat globules. The peculiar phenomenon regarding surface tension has not been explained satisfactorily.

Free fatty acids released by lipolysis of milkfat greatly depress the surface tension of milk. As a matter of fact surface tension has been used to some extent as an objective index for the development of hydrolytic rancidity.[57,96,98,212] Its value for this purpose is somewhat limited by natural variations due to other causes. Of course it must be recognized also that the shorter-chain fatty acids contribute more to rancid flavor and the longer ones are more effective surface-tension depressants.

Homogenization of raw whole milk or cream stimulates lipolysis and thus leads to a decrease in surface tension, but if the product has been previously pasteurized the effect of homogenization is an increase in surface tension.[218,232,234] The reason for such an increase is not known, but suggestions have been made that it results from denaturation or other changes in the lipoprotein complex, or to reduction in the amount of protein available to the milk/air interface because of adsorption on the extended fat surface. The latter explanation seems unlikely in view of the very slight effect of five-fold dilution of the surface tension of skimmilk. Another suggestion that might be advanced is that homogenization reduces the amount of free fat in the product.

Heat treatment of milk has little effect on surface tension except that sterilizing treatments cause an increase of a few dynes cm^{-1} coinciding with grain formation.[136] This effect undoubtedly results from denaturation and coagulation of the proteins so that they are no longer effective surface-active agents.

FREEZING POINT

The freezing point of milk, like that of any aqueous system, depends on the concentration of water-soluble components. The mathematical relationship between depression of the freezing point and concentration of solute was determined by Raoult[169] and is expressed in the equation:

$$T_f = K_f M$$

where T_f = difference between freezing points of solvent and solution, K_f = molal depression constant (1.86°C for water), and M = molal concentration of solute. As Raoult pointed out, this relationship is valid only for dilute solutions of undissociated solutes. The freezing point is, of course, an indirect measure of the osmotic pressure.

In chemical research the freezing point of a solution composed of known weights of solute and solvent affords a means of determining M, the molal concentration, and hence the molecular weight of the solute. In the dairy field, however, the objective of freezing-point measurements is virtually restricted to determination of the water content of the product in order to detect the illegal addition of water. Its value for this purpose rests on the fact that the freezing point of authentic bovine milk varies within very narrow limits. Since the depression of freezing points is directly proportional to the number of particles in solution, it is obvious that it is primarily determined by the major constituents of low molecular weight, the lactose and the salts, and is nearly independent of variations in the concentrations of colloidal micelles and fat globules. Furthermore a complementary relationship exists between lactose and sodium chloride in milk such that the osmotic pressure and hence the freezing point is maintained within a narrow range.

The determination of freezing point demands meticulous attention to detail. The general principal employed is to supercool a sample slightly, to induce crystallization, and then to observe the maximum temperature attained. The temperature of the cooling bath must be controlled; otherwise the rate of heat loss will be greater than the rate of heat transfer to the bath by the heat of fusion of the solution, and the observed freezing point will be too low. Control of the temperature of supercooling and seeding techniques is extremely important; if it varies, the amount of solvent that crystallizes out and consequently the observed freezing point also varies. Attempts have been made to apply a correction factor to the observed freezing point to enable calculation of the "true freezing point." A review of the factors involved is given in Richmond's "Dairy Chemistry."[47] However, from a practical standpoint it is not necessary to determine the true freezing point; rather a high degree of reproducibility between samples and analysts is sought.

In 1921, Horvet[102] published a method and description of apparatus for freezing-point determinations, feeling that earlier methods were not sufficiently standardized. Although the Horvet method has been universally accepted as official, a number of modifications have been developed, such as replacing the ether cooling system with mechanical refrigeration, and addition of mechanical stirring and tapping devices.[199,214] Furthermore, cryoscopes employing thermistors in place of mercury-in-glass thermometers have been developed.[193,194] In 1960, the Association of Official Agricultural Chemists approved the thermistor type cryoscope as an alternate official apparatus. These cryoscopes have become more popular than the Hortvet because of their speed and ease of operation.

Regardless of the method used, determination of the freezing points of solutions is empirical. In recognition of this fact, Hortvet[102] emphasized the need for using standardized equipment and techniques. As an additional means of compensating for variability between analysts, his procedure involves measurement of the difference between the freezing point of a standard solution and the freezing point of milk. Theoretically, any systematic errors should be reflected in both observed values, and thus be eliminated. This feature of the Hortvet technique places special emphasis on the calibration procedure which apparently has not been recognized by all analysts. Of course, the calibration procedure will compensate for systematic errors only when variations in technique have exactly the same effect on the observed results for both the standard solutions and milk. Hortvet's choice of sucrose solutions as calibration standards was unfortunate, since it is difficult to obtain reproducible results with sucrose solutions and they readily undergo microbial decomposition. Salt (NaCl) solutions have been approved as secondary freezing-point standards on the basis of a collaborative study in 1960,[196] which revealed that salt and sucrose standards gave comparable results. Salt standards sterilized in sealed ampoules kept for over a year without a change in freezing point. Henningson[91,92] studied the factors affecting results obtained with thermistor type cryoscopes. His studies led to the development of specific directions pertaining to cooling, seeding, and reading.[92]

The freezing point of bovine milk is usually within the range −0.530 to −0.570°C. The average value is close to −0.540°C.[45,47,93,175,195] It is easy to calculate that lactose and chloride are the principal constituents responsible for this depression. Thus in a milk containing 12.5% solids, 4.75% lactose, and 0.1% chloride, the molal concentration of lactose is $(4.75 \times 1000)/(342 \times 87.5) = 0.159$, and the corresponding depression of freezing point, assuming ideal behavior, is $0.159 \times 1.86 = 0.296$°C. Actually sugars do not behave entirely ideally even at these concentrations. For example, data of Whittier[249] indicate a molal depression by lactose of 2.02°C instead of 1.86°C. The concentration of chloride is 0.032 molal, and assuming that each chloride ion is accompanied by a monovalent ion of opposite charge (i.e., Na^+ or K^+), the depression expected is $1.86 \times 2 \times 0.032 = 0.119$°C. These values are in close agreement with data of Cole *et al.*,[40,41] who measured the effects of adding increments of lactose and potassium chloride to milk. The sum of the contributions of lactose and chloride (0.415°C in the example given) represents 75 to 80% of the entire depression. Assignment of the contributions of the other solutes is not readily made, since information is lacking on distribution of calcium, magnesium, phosphate, and citrate among various ions and complexes. The degree of departure from ideal behavior is also not known.

Variations in freezing point of milk as it is drawn from the udder must reflect variability in the physiological operations of that gland. The exact magnitude of the physicological limits has not been established. It has been assumed that the osmotic pressure of milk is dependent on the osmotic pressure of blood. This assumption has been supported by the observation of Wheelock *et al.*[238] that milk is in osmotic equilibrium with blood flowing through the udder. They noted that freezing-point values for milk agreed more closely with those for mammary-venous blood than for jugular-venous blood. Tucker[221] also found that the freezing points of milk and mammary venous blood were highly correlated, whereas Peterson and Freeman[148] found that there was no significant correlation between the freezing points of milk and jugular blood. Little is known as to whether variability in freezing points reflects primarily deviations from the complementary relationship between lactose and chloride or variations in content of other osmotically active components. Rees[170] has presented some evidence in favor of the latter view, indicating that the concentration of the nonchloride fraction, principally the soluble acid phosphates, was the primary cause of variations.

Environmental factors associated with variations in freezing point have been studied by several investigators. Articles published prior to 1960 have been reviewed,[47,175,195] but since then several additional articles[48,51,52,53,68,90,181] have been published. Some of the variations have been attributed to seasonal effects, feed, water intake, stage of lactation, breed of cow, and time of day (i.e., morning vs. evening milk).

In some cases the effects of these factors have been shown to be interrelated. For example, the differences between the freezing point of morning and evening milk have been shown to be affected by the time of feeding and watering. Geographical differences that have been reported may primarily be due to differences in breed of cattle and in feeding practices. Although variations have been shown to occur, results obtained in England,[206] Australia,[219,220] India,[46] and the United States[9,93,113] all exhibit approximately the same range.

Handling treatments of milk between the time of drawing and the freezing-point determination may be expected to alter the freezing point if they change the net number of osmotically active particles in solution. Effects of some such treatments are reviewed by Shipe.[195] Of course, microbial decomposition with the production of such water-soluble components as lactic acid from lactose will lower the freezing point. Storing samples at low temperatures or freezing them has been reported[50,61,89] to raise freezing points slightly. Likewise, heating has been observed to raise the freezing point by some authors, but not by others. Undoubtedly chilling or heating may produce aggregation of dissolved salts or transfer of dissolved materials to the colloidal caseinate micelles or the fat globules. Such effects would raise the freezing point. Since the changes may be slowly reversible after treatment, the magnitude of change in freezing point would vary with the length of time between treatment and determination of freezing point. However it would be difficult to follow such sluggishly reversible equilibria by means of the freezing point because the changes are so small. Vacuum treatment of milk has been shown[49,197] to raise its freezing point, presumably as a result of removal of carbon dioxide. If water is lost during the vacuum treatment, the freezing point will be lowered, thereby partially compensating for loss of carbon dioxide.

There has been considerable controversy over the interpretation of freezing-point values. In 1970 the Association of Official Analytical Chemists[94] adopted an official interpretation which specified that milk with a freezing point of −0.525°C or below may be *presumed* to be water-free. Detailed procedures have been given for confirming the absence or presence of added water. The choice of the −0.525°C value as an upper limit is based on a statistical evaluation of data from a 1968 cooperative North American survey[93] of freezing points of authentic samples.

ELECTRICAL CONDUCTIVITY

The specific electrical resistance of an electrolyte solution is defined as the resistance of a cube 1 cm in length and 1 sq cm in cross-sectional area.

$$\rho = \frac{\alpha R}{l}$$

where ρ = specific resistance in ohm cm, α = cross-sectional area in cm^2, l = length in cm, and R = measured resistance in ohms ($R = E/I$). Specific conductance, K, is the reciprocal of specific resistance.

$$K = \frac{1}{\rho} = \frac{l}{\alpha R}$$

Measurements of conductance are made with cells of glass in which the solution is contained between platinum electrodes. The resistance of such a cell filled with electrolyte is measured. If the cell were of uniform and measurable length and cross section the conductivity could be computed directly from the equation above. In practice, however, the resistance of a solution of known conductivity (a KCl solution for example) is measured in the cell and a cell constant computed as the product KR. The specific conductance of the electrolyte being measured can then be computed

$$K = \frac{\text{cell constant}}{R}$$

K is expressed in units of reciprocal ohms per centimeter (i.e., $ohm^{-1}\ cm^{-1}$). Methods of measurement including precautions in the selection and use of conductivity cells and the appropriate electrical circuitry are discussed by Shedlovsky.[189]

A chronological summary of literature up to 1954 on the conductivity of milk has been made by Schulz.[183] Conductivity has been considered as a possible index of mastitis infections, of added water, of added neutralizers, and as a means of controlling solids concentration and composition in dairy processing. The specific conductance of cow's milk, reflecting its concentration and activity of ions, is of the order of 0.005 $ohm^{-1}\ cm^{-1}$ at 25°C. Most normal samples fall within the range 0.0040 to 0.0055 $ohm^{-1}\ cm^{-1}$. Higher values usually represent mastitic infections which increase the concentration of sodium and chloride in the milk.[67,152,183]

Temperature control is important in the measurement, since the conductivity of milk increases by about 0.0001 $ohm^{-1}\ cm^{-1}$ per °C rise in temperature.[70,133,152] Increased dissociation of the electrolytes and decreasing viscosity of the medium with increasing temperature are undoubtedly responsible for this effect. An investigation[209] of the viscosity and electrical conductivity of skimmilk from cows and buffaloes failed to reveal a simple relationship. The authors attributed the lack of linear correlations to variations in casein structure and its hydration.

The sodium, potassium, and chloride ions of milk are the greatest contributors to its electrical conductivity, since they are present in the highest concentration. Schulz and Sydow[184] proposed that "chloride-free conductivity" may be a more sensitive index of certain changes and adulterations than conductivity itself. This is the difference between total conductivity and that of a sodium chloride solution having the same chloride content as the milk sample. In tests on 41 specimens of mixed raw milk conductivity averaged 0.00485, chloride conductivity 0.00305, and "chloride-free conductivity" 0.00180 $ohm^{-1} cm^{-1}$. The correlation between chloride content and conductivity has been confirmed by Puri and Parkash.[158] These workers reported that there was no significant difference between the conductivity of cow and buffalo milk, whereas Pal[142] claims that adulteration of buffalo milk with cow milk causes a detectable change in conductivity.

The fat globules of milk reduce the conductivity by occupying volume and by impending the mobility of ions. Thus the conductivity of whole milk is less than that of skimmilk by about 10%, and that of cream varies with fat content.[70,133,155] Homogenization of milk does not measurably influence conductivity.[155] The conductivity of whey and ultrafiltrate is slightly greater than that of skimmilk.[183,184] A possible relationship between the electrical conductivity and physical stability of evaporated milk and concentrated infant milk products has been reported.[82] Samples of poor physical stability tended to have relatively low conductivity values compared to the more stable products.

The production of acidity by bacterial action of course increases the conductivity of milk. An increase of about 0.0001 $ohm^{-1} cm^{-1}$ per Soxhlet-Henkel degree has been noted.[178,216] (The Soxhlet-Henkel degree, °SH, is the ml of $N/4$ NaOH required to titrate 100 ml milk to the phenolphthalein pink). Conductivity can be used to detect added neutralizers.

The influence of dilution and concentration on the conductivity of milk is complicated by their respective effects in promoting and repressing dissociation of salt complexes and solubilization of colloidal salts. Data of various workers[42,184,203,217] indicate that, as milk is concentrated, a maximum is reached in conductivity. With skimmilk the maximum is about 0.0078 $ohm^{-1} cm^{-1}$ and lies at about 28% solids content. Concentration beyond this point results in a decrease in conductivity.

Direct conductivity measurements do not provide a satisfactory index of added water in milk. However, it has been reported[168] that measurement of conductivity in nonaqueous solvents can be useful in detecting adulteration. The conductivities of adulterated milks with two different solvent systems were correlated with the percentage of added water in the milk. One solvent system consisted of 10 ml acetone, 90 ml methanol plus 3 gm sodium chloride, and the other contained 2.65 gm formic acid in 100 ml acetone.

HEAT CAPACITY AND THERMAL CONDUCTIVITY

The heat capacity of a substance is the quantity of heat required to raise the temperature of a unit mass through a unit range. It is usually expressed in terms of cal g^{-1}°C^{-1}. Strictly speaking, of course, heat capacity at any temperature, T, is the limiting value of dQ/dT as dT approaches zero, where dQ is the amount of heat required to raise the temperature from T to $T + dT$. Heat capacity is normally evaluated over a temperature range (dT) of several degrees. The term "specific heat" is used almost interchangeably with "heat capacity". It is the ratio of heat capacity to that of water at 15°C (0.99976 cal g^{-1}°C^{-1}), and thus is dimensionless. The numerical value of the specific heat is nearly the same as that of heat capacity. The heat capacity of air-free water at 1 atm pressure is within 1% of 1 cal g^{-1}°C^{-1} over the range 0 to 100°C.[140]

Determination of heat capacity is best made with a calorimeter incorporating an electric heater. The energy input and resultant temperature rise both are measured. Procedures and precautions for calorimetry are discussed thoroughly by Sturtevant.[208]

The heat capacity of skimmilk has been carefully measured by Phipps,[149] who compared his results with those of earlier workers. Skimmilk exhibits a small but definite linear increase in heat capacity between 0 and 50°C from about 0.933 to 0.954 cal g^{-1}°C^{-1}. There is a marked decrease in heat capacity as the total solids content of the sample is increased.[163] Dried skimmilk products have heat capacities of 0.28 to 0.32 cal g^{-1}°C^{-1} in the 18 to 30°C temperature range.[30]

The heat capacity of milkfat in either the solid or the liquid state is about 0.52 cal g^{-1}°C^{-1} and its latent heat of fusion is about 20 cal/gm.[256] Thus the heat capacity of milk and cream depend strongly upon the fat content. Furthermore, in temperature ranges in which the melting of fat occurs, the apparent heat capacity is the sum of the "true" heat capacity and the energy absorbed by melting of the fat. Thus, the results will vary widely, depending upon the proportion of the fat that was solid at the start of the determination, which in turn depends upon the composition of the fat and the temperature history of the sample. Many workers in the field have observed these effects.[105,138,149,174,192] The apparent heat capacity of fat-containing dairy products has a maximum at 15 to 20°C and often shows a second inflection at about 35°C.[149,190]

Thermal conductivity is the rate of heat transfer by conduction through unit thickness across unit area of substance for a unit difference of temperature.

$$\lambda = \frac{Q\ d}{A\ t\ (T_2 - T_1)}$$

where Q amount of heat is transferred through the sample of cross sectional area A and thickness d in time t, with a temperature differential of $T_2 - T_1$.

Thermal conductivity can be determined using either equilibrium or dynamic methods. Equilibrium methods involve a heated surface, a thin layer of sample, and a cooled surface. The energy required to maintain a steady state and the temperature difference is measured and used in the calculations. Dynamic methods are based on thermal diffusivity, which is obtained from the curvature of heating or cooling plots at various depths within the product. Procedures and applications of thermal conductivity measurements to foods have been reviewed.[147,171,255]

Thermal conductivity, λ, is expressed in cal $cm^{-1}sec^{-1}°C^{-1}$ or in kcal $m^{-1}hr^{-1}°C^{-1}$. The value of λ for water increases from about 0.48 to 0.58 kcal $m^{-1}hr^{-1}°C^{-1}$ between 0 and 100°C. The thermal conductivity of milk decreases slightly between 0 and about 37°C, then increases; but assuming a linear increase over the temperature range 0 to 100°C is usually sufficient.[23,64,171,255] Typical values for λ are 0.46 kcal $m^{-1}hr^{-1}°C^{-1}$ at 37°C and 0.53 kcal $m^{-1}hr^{-1}°C^{-1}$ at 80°C.[147,171,255] There is a marked decrease in λ with increases in either fat or total solids,[117,119,204] but the magnitude of the change is temperature-dependent.[64] Thermal conductivity of dried dairy products depends upon bulk density as well as composition.[63,139,171]

REFRACTIVE INDEX

The refractive index of a substance is defined as the ratio of the speed of light in a vacuum to its speed in that substance. One consequence of refraction is to change the direction of a light ray as it enters or leaves the substance. Measurement of this bending gives a direct measure of refractive index, n. Specifically $n = \sin i / \sin r$, where i is the angle of the ray to the surface as it approaches (incidence) and r is the exit angle (refraction). Principles involved and a detailed critique of methods of measurement are presented by Bauer and coworkers.[15] Since refractive index varies with sample temperature and the wavelength of light, these must be controlled and specified. Thus n_D^{20} refers to the index at 20°C with the D line of the sodium spectrum (589.0 and 589.6 nm).

The refractive index of water is $n_D^{20} = 1.33299$. The value of n_D^{20} for cow's milk generally falls in the range 1.3440 to 1.3485. Buffalo milk is similar to cow's milk,[100] while human, goat, and ewe milks appear to have higher[167] values of refractive index. Since the refractive increments contributed by each solute in a solution are additive, much consideration has been given to the possible use of refractive index as a means of determining total solids or added water in milk. The refractive index of milk itself is somewhat difficult to determine because of the opacity, but by using a refractometer such as the Abbe instrument, which employs a thin layer of sample, it is possible to make satisfactory measurements, particularly with skimmilk products and sweetened condensed milk.[124,173]

The relation between solids content (on a basis of weight per unit volume) and refractive index is linear and the contributions of the several components are additive.[76] However, the individual components of milk differ in specific refractive increment, $\Delta n/(\rho c)$, where ρ is the density of the sample and c is the weight/weight concentration of the component. Thus the relation between per cent solids and refractive index will vary between lots of milk. Goulden[76] reported the following specific refractive increments (ml g^{-1}): casein complex 0.207, soluble proteins 0.187, and lactose 0.140. The total contribution to the refractive index for a milk containing 2.34% casein complex, 0.83% soluble proteins, and 4.83% lactose becomes 0.00500 + 0.00159 + 0.00695, or 0.01354. The residue of 0.95% contributed 0.00166 to the total difference between the refractive indices of water and the milk. Similar data have been reported by Rangappa.[164,166] The refractive index of milk fat is 1.4537 to 1.4552 at 40°C; it has the same refractive index in bulk and in globules.[227] The fat does not contribute to the refractive index of whole milk because refraction occurs at the interface of air and the continuous phase.[76] Sterilization does not alter the refractive index[5] nor does subsequent storage.[35]

Clarification by removal of casein with such agents as calcium chloride, acetic acid, copper sulfate, or rennin has often been employed to obtain a serum more suitable for refractometric measurements. Obviously the composition, and hence the refractive index, of such sera will depend on the method of preparation. Furthermore, some of the serum proteins may be precipitated with the casein by some of the agents used, particularly if the milk has been heated. Refractive index measurements of such sera are not generally considered as satisfactory as freezing point for detection of added water.[47,134,182,213,225] Menefee *et al.*[128] reported a close relation between total solids in evaporated and condensed products and the refractive index of serum prepared therefrom by the copper sulfate method. Of course, a different proportionality constant would hold for each type of product.

The estimation of casein in milk by refractometric techniques appears to hold some promise. The casein may be precipitated, washed and redispersed to yield a solution suitable for refractometry.[27,180] Another method involves computation from the difference between the refractive indices of two samples, one made alkaline to dissolve the casein and the other treated with copper sulfate to precipitate it.[83] Heating the milk would cause the serum proteins to precipitate with the casein. The total solids of co-precipitate preparations can be determined by refractive index using 2.5N NaOH as a dispersant.[56]

The refractive constant or specific refractive index computed by the Lorenz-Lorentz formula:

$$K = \frac{n^2 - 1}{n^2 + 2} \times \frac{1}{\rho}$$

(where n is refractive index and ρ is density) has sometimes been used for milk or milk sera.[162,165,252] It is independent of temperature but not entirely independent of concentration. Milk has a value of K of about 0.2075.

LIGHT ABSORPTION AND SCATTERING

Absorption of electromagnetic radiation by a substance occurs when the radiation has the same frequency as some motion or transition at molecular levels, and when the molecule has either a permanent or induced dipole. The frequency of absorption is controlled by the mass and force involved. At short wavelengths, such as the far ultraviolet, the energy is absorbed by transformation of electrons to higher energy levels. At long wavelengths, such as the infrared, the transitions are in the vibrational and rotational states of molecules. Transitions associated with various regions of the spectrum are shown in Table 8.5. In all regions, the amount of radiation absorbed is proportional to the number of absorption centers. This is commonly expressed in Beer's Law:

$$\log_{10} Io/I = A = abc$$

where Io and I are the intensities of the incident and emergent radiation respectively, A is the absorbance (formerly called optical density), a is the proportionality constant "absorptivity", b is the sample thickness through which the radiation is passed, and c is the concentration of sample in solvent (w/v). If more than one species is present which absorbs radiation of a given wavelength, the absorbances (A) are additive. It should be obvious that absorptivity (a) depends upon the units of b and c and the wavelength of the radiation used.

Fluorescence and phosphoresence are the reemission at longer wavelengths of absorbed radiation with shorter or longer delays, respectively. The intensity of the emitted radiation follows Beer's Law.

The text by Pomeranz and Meloan[153] contains a good introduction

Table 8.5

SUMMARY OF TRANSITIONS INTERACTING WITH RADIATION

Spectral Region	Wavelength	Molecular transitions involved
X-ray	0.001–10nm	Inner electrons
UV-vacuum	10–200nm	Sigma electrons
UV-far	200–290 nm	n and pi electrons
UV-near	290–400 nm	Conjugated systems
Visible	400–800 nm	Highly conjugated systems
IR	0.7–60 μm	Vibrational and rotational
Microwave	cm	Rotational
NMR	0.3–30 m (1–100 megocycles/sec)	Nuclear

to spectroscopy, and the Weissberger[237] series contains full information on all aspects.

The foregoing discussion applies to substances in true solutions or continuous phases. Dispersed particles also scatter light if the particle size and the wavelength of the radiation are of the same order of magnitude. Colloids and emulsions scatter ultraviolet and visible radiation quite effectively. On the other hand, electrons scatter x-rays. The preferred angles of scattering may be observed, as in x-ray diffraction, or the attenuation of the incident radiation may be studied, as in turbidimetry. Scattering of visible light by emulsions has been described adequately by Goulden[74] and by Walstra.[228]

Milk is a colloidal dispersion of proteins and an emulsion of fat in an aqueous solution of lactose, salts, and other compounds. Thus it not only absorbs light at many wavelengths because of the large number of compounds present, but also scatters it as a result of the presence of particles of various sizes. The well-known absorptions by proteins in the 220 to 380 nm region can be distinguished, as can absorption in the 400 to 520 nm region by fat pigments.[202] Scattering is decreased as the wavelength increases;[77,202] thus in the infrared region most of the attenuation is due to absorption. Various specific absorptions can be seen in the near infrared and infrared, most notably those by OH groups near 2.84 μm, CH_2 groups near 3.45 μm, C=O groups at 5.74 μm, and NH_2 groups at 6.56 μm. Since water strongly absorbs infrared radiation, milk is opaque to a major portion of this region.[75]

Recent interest in light absorption, fluorescence and scattering by milk is largely quantitative rather than qualitative in nature. Direct analysis of milk by spectrophotometric techniques offers definite advantages of speed, simplicity, and capabilities for automation. Simple reactions with specific milk components can be used in many cases where direct spectrophotometry is impossible.

Goulden[75] described a method for fat, protein, and lactose in milk based on absorption of infrared energy at specific wavelengths. The "Infrared Milk Analyzer" or IRMA, as it is known commercially, is based on his technique.[79] The difference in absorbances of a homogenized sample of milk and pure water is measured at 5.8, 6.5, and 9.6 μm for fat, protein, and lactose, respectively. Homogenization is used to reduce fat-globule sizes to less than 2 μm to eliminate light-scattering rather than to produce a specific size distribution; therefore the nature of the homogenization is less critical than with light-scattering techniques. The IRMA method correlates well with chemical methods for the various components,[1,21] again with the caution of proper calibration and operation.

The most obvious fluorescent compound in milk is riboflavin, which absorbs strongly at 440–500 nm and emits fluorescent radiation with a maximum at 530 nm. It is readily determined quantitatively in whey by measurement of the fluorescence.[257] Proteins also fluoresce strongly because of their content of aromatic amino acids. Part of the ultraviolet radiation absorbed at 280 nm is emitted as fluorescent radiation at longer wavelengths. A prominent maximum near 340 nm is attributable to tryptophan residues in the protein. Use of fluorescence for quantitation of milk proteins was proposed by Konev and Kozunin,[260] and the technique has been modified and evaluated by several groups.[258,259,261,262] It seems to be subject to somewhat less accuracy than desired because of difficulties in disaggregating the caseinate particles and in standardizing instruments. It also involves a basic uncertainty due to natural variations in the proportions of individual proteins which differ in tryptophan content.

Both the fat and protein of milk scatter light, the amount of scattering depending upon the number and size of particles, the wavelength of the incident radiation, and the difference in refractive index between the different kinds of particles and the solvent.[8,78,87,110,229] Measurement of fat in milk by light-scattering was first described by Haugaard and Pettinati.[88] Homogenization is used to achieve uniform fat-globule size distributions in different samples. Protein particles are solubilized at high pH with disodium ethylenediaminetetraacetate (EDTA), which also serves to dilute the milk sufficiently to eliminate disturbing multiple scattering effects. The commercial version of this method, the "Milko-Tester", utilizes white light in a special photometric system to determine attenuation due to scattering, and thus the fat content of the milk. Dilution of the milk with the EDTA solution before homogenization was found to improve the results.[2,80] With proper attention to instrument calibration and operation, this method compares favorably to more traditional methods of measuring the fat content of milk.[80,198]

Nakai and Le[135] have used a different approach to the measurement of fat by light-scattering. They dissolved both the fat and protein particles with acetic acid, measured the protein content by the absorbance at 280 nm, then reformed a fat emulsion by adding a solution of urea and imidazole. The turbidity was measured at 400 nm and was found to be independent of the initial fat globule size distribution.

Lin *et al.*[122] used inelastic scattering of plane-polarized light of 632.8 nm wavelength from a He-Ne laser to determine the diffusion coefficient and thereby the hydrodynamic radii of monodisperse caseinate micelle fractions from milk. The cumulative distribution curve of the weight fraction of micelles revealed that about 80% of the casein occurs in micelles with radii of 50 to 100 nm and 95% between 40 and 220 nm with the most probable radius at about 80 nm. This method has the advantage that the micelles are examined in their natural medium.

Addition of specific compounds to milk has been used to allow spectrophotometric measurement of lactose as the osazone,[226] and fat by fluorescence.[10,114] The dye-binding method for measuring protein in milk is based on the ability of sulfonic acid dyes to complex with the basic amino acid residues of milk proteins at low pH.[65] Dye-binding correlates well with Kjeldahl,[191] but variations are caused by the different compositions of the different milk proteins.[7,223]

REFERENCES

1. ADDA, J., BLANE-PATIN, E., JEUENET, R., GRAPPIN, R., MOCQUOT, G., POUJARDIEU, B., and RICORDEAU, G., Lait, *48,* 145 (1968).
2. AEGIDUO, P. E., U.S. Pat. 3,402,623 (1969).
3. ANDRADE, E. N., DAC., Proc. Roy. Soc. (London), *215A,* 36(1952).
4. ANSBACHER, S., FLANIGAN, G. E., and SUPPLEE, G. C., J. Dairy Sci., *17,* 723 (1934).
5. ARMANDOLA, P., and BREZZI, G. Latte, *38,* 1013 (1964). Cited in DSA, *27,* [1899] (1965).
6. ASCHAFFENBURG, R., J. Dairy Res., *14,* 316 (1945).
7. ASHWORTH, U. S., J. Dairy Sci., *49,* 133 (1966).
8. ASHWORTH, U. S., J. Dairy Sci., *52,* 263 (1969).
9. BAILEY, E. M., J. Assoc. Offic. Agr. Chem., *5,* 484 (1922).
10. BAKHIREN, N. F., and BUTOV, G. P., Nauchno-tekk Byull, Elecktref, sel'sk. Khoz., *1968,* (2), 34 (1968). Cited in DSA, *32,* [414] (1970).
11. BARRON, E. S. G., and HASTINGS, A. B., J. Biol. Chem., *107,* 567 (1934).
12. BARRY, J. M., and ROWLAND, S. J., Biochem. J., *54,* 575 (1953).
13. BATEMAN, G. F., and SHARP, P. F., J. Agr. Res., *36,* 647 (1928).
14. BATES, R. G., "Determination of pH, Theory and Practice." John Wiley and Sons, Inc., New York, 1964.
15. BAUER, N., FAJANS, K., and LEWIN, S. Z. Chapter 18 in Weissberger, A. "Technique of Organic Chemistry," Vol. 1, 3rd Ed., Interscience Publishers, Inc., New York, 1959.

16. BAUER, N., and LEWIN, S. Z. Chapter 4 in Weissberger, A. "Technique of Organic Chemistry". Vol. 1, 3rd Ed., Interscience Publishers, Inc., New York, 1959.
17. BEARCE, H. W., J. Agr. Res., *3,* 251 (1914).
18. BEEBY, R., and KUMETAT, K., J. Dairy Res., *26,* 248 (1959).
19. BEEBY, R., and LEE, J. W., J. Dairy Res., *26,* 258 (1959).
20. BELL, G. A., Aust, J. Dairy Tech., *17,* 80 (1962).
21. BIGGS, D. A., Conv. Proc. Milk Industry Found. *1964,* 28 (1964). Cited in DSA, *27,* [3288] (1965).
22. BIKERMAN, J. J., "Surface Chemistry", 2nd Ed., Academic Press, New York, 1958.
23. BOGDANOV, S., and GOCHIYAEV, B. Mol. Prom., *22,* (6) 16 (1961). Cited in DSA, *24,* [241] (1962).
24. BOULET, M., and MARIER, J. R., J. Dairy Sci., *43,* 155 (1960).
25. BOULET, M., and MARIER, J. R., Arch. Biochem. Biophys., *93,* 157 (1961).
26. BOULET, M., and ROSE, D., J. Dairy Res., *21,* 227 (1954).
27. BRERETON, J. G., and SHARP, P. F., Ind. Eng. Chem., Anal. Ed., *14,* 872 (1942).
28. BRUNNER, J. R., JACOBS, G., and MADDEN, D. M., Mich. Agr. Exp. Sta. Quart. Bull. *42,* 232 (1959).
29. BUCHANAN, J. H., and PETERSON, E. E., J. Dairy Sci., *10,* 224 (1927).
30. BUMA, T. J., and MEERSTRA, J., Neth. Milk Dairy J., *23,* 124 (1969).
31. CAFFYN, J. E., J. Dairy Res., *18,* 95 (1951).
32. CALANDRON, A. and GRILLET, L., Lait, *44,* 505 (1964).
33. CAULFIELD, W. J., and RIDDELL, W. H., J. Dairy Sci., *19,* 235 (1936).
34. CHEKULAEVA, L. V., Trudy Vologodsk Moloch Inst., *1953,* 12,233. Cited in CA, *54,* [7918] (1960).
35. CHIOFALO, L., and IANNUZZI, L., Zootec. e Vita, *6,* 32 (1963). Cited in *DSA, 27,* [1287] (1965).
36. CHRISTIANSON, G., JENNESS, R., and COULTER, S. T., Anal. Chem., *26,* 1923 (1954).
37. CLARK, W. M., Chapter 6 in Rogers "Fundamentals of Dairy Science", 2nd Ed., Reinhold Publishing Corp., New York, 1934.
38. CLARK, W. M., "Oxidation-Reduction Potentials of Organic Systems". Williams and Wilkins Co., Baltimore, 1960.
39. COE, J. R. and GODFREY, T. B., J. Appl. Phys., *15,* 625 (1944).
40. COLE, E. R., DOUGLAS, J. B., and MEAD, M., J. Dairy Res., *24,* 33 (1957).
41. COLE, E. R., and MEAD, M., J. Dairy Res., *22,* 340 (1955).
42. COSTE, J. H., and SHELBOURN, E. T., Analyst, *44,* 158 (1919).
43. COX, C. P., J. Dairy Res., *19,* 72 (1952).
44. COX, C. P., HASKING, Z. D., and POSENER, L. N., J. Dairy Res. *26,* 182 (1959).
45. DAHLBERG, A. C., ADAMS, H. S., and HELD, M. E., Nat. Research Council Publication *250,* (1953).
46. DASTUR, N. N., DHARMARAJAN, C. S., and RAO, R. V., Indian J. Vet. Sci., *22,* 123 (1952).
47. DAVIS, J. G., and MACDONALD, F. J., "Richmond's Dairy Chemistry," 5th Ed., Charles Griffin and Co., London, 1953.
48. DEMOTT, B. J., J. Milk Food Technol., *29,* 319 (1966).
49. DEMOTT, B. J., J. Milk Food Technol., *30,* 253 (1967).
50. DEMOTT, B. J., and BURCH, T. A., J. Dairy Sci., *49,* 317 (1966).
51. DEMOTT, B. J., HINTON, S. A., and MONTGOMERY, M. J., J. Dairy Sci., *50,* 151 (1967).
52. DEMOTT, B. J., HINTON, S. A., SWANSON, E. W. and MILES, J. T., J. Dairy Sci., *51,* 1363 (1968).
53. DEMOTT, B. J., MONTGOMERY, M. J., and HINTON, S. A., J. Milk Food Technol., *32,* 210 (1969).
54. DEYSHER, E. F., WEBB, B. H., and HOLM, G. E., J. Dairy Sci., *27,* 345 (1944).
55. DIXON, B., Aust. J. Dairy Tech., *18,* 141 (1963).
56. DUNKERLEY, J. Int. Dairy Congr. Proc. 15th, *IE,* 430 (1970).

57. DUNKLEY, W. L., J. Dairy Sci., *34,* 515 (1951).
58. EDSALL, J. T. and WYMAN, J., "Biophysical Chemistry", Vol. 1 Academic Press, New York, 1958.
59. EILERS, H., SAAL, R. N. J., and WAARDEN, M., VAN DEN, "Chemical and Physical Investigations on Dairy Products", Elsevier Publishing Co., New York, 1947.
60. EL-RAFEY, M. S., and RICHARDSON, G. A., J. Dairy Sci., *27,* 1, 19 (1944).
61. ENGLAND, C. W., and NEFF, M., J. Assoc. Offic. Agr. Chem., *46,* 1043 (1963).
62. ERB, R. D., MANUS, L. J., and ASHWORTH, U. S., J. Dairy Sci., *43,* 607 (1960).
63. FARRALL, A. W., HELDMAN, D. R., WANG, P. Y., OJHA, T. P., and CHEN, A. C., Int. Dairy Congr. Proc. 18th, *IE,* 269 (1970).
64. FERNADEZ-MARTIN, F., and MONTES, F., Int. Dairy Congr. Proc. 18th, *IE,* 471 (1970).
65. FRAENKEL-CONRAT, H. and COOPER, M., J. Biol. Chem., *154,* 239 (1944).
66. FRAZIER, W. C. and WHITTIER, E. O., J. Bact., *21,* 239 (1931).
67. FREDHOLM, H., Nord. Jordburgs. Forsk., 195 (1942). Cited in CA, *38,* [3029] (1944).
68. FREEMAN, T. R., BUCY, J. L. and KRATZER, D. D., J. Milk Food Technol., *34,* 212 (1971).
69. GARRETT, O. F. and OVERMAN, O. R., J. Dairy Sci., *23,* 13 (1940).
70. GERBER, V., Z. Untersuch. Lebensm., *54,* 257 (1927).
71. GOULD, I. A., J. Dairy Sci., *28,* 379 (1945).
72. GOULD, I. A. and FRANTZ, R. S., J. Dairy Sci., *28,* 387 (1945).
73. GOULD, I. A., and SOMMER, H. H., Mich. Agr. Exp. Sta. Tech. Bull. 164 (1939).
74. GOULDEN, J. D. S., Brit. J. Appl. Phys., *12,* 456 (1961).
75. GOULDEN, J. D. S., Nature, *191,* 905 (1961).
76. GOULDEN, J. D. S., J. Dairy Res., *30,* 411 (1963).
77. GOULDEN, J. D. S., Dairy Sci. Abst., *27,* 469 (1965).
78. GOULDEN, J. D. S. and SHERMAN, P., J. Dairy Res., *29,* 47 (1962).
79. GOULDEN, J. D. S., SHIELDS, J. and HASERLL, R., J. Soc. Dairy Technol., *17,* 28 (1964).
80. GRAPPIN, R., and JUENET, R., Lait, *50,* 233 (1970).
81. HANKINSON, D. J., and GAUNT, S. N., J. Dairy Sci., *48,* 616 (1965).
82. HANSEN, P. M. T., and LINDAMOOD, J. B., J. Dairy Sci., *54,* 759 (1971).
83. HANSSON, E., Svenska Mejereitid., *49,* 277 (1957).
84. HARKINS, W. D., and ALEXANDER, A. E., Chapter 14 in Weissberger, A., "Technique of Organic Chemistry," Vol. 1, 3rd Ed. Interscience Publishers, Inc., New York, 1959.
85. HARLAND, H. A., COULTER, S. T., and JENNESS, R., J. Dairy Sci., *35,* 643 (1952).
86. HASHIMOTO, Y., ARIMA, S., and SAITO, Z., Anim. Husb., Tokyo, *18,* 921 (1964). Cited in DSA, *27,* [1584] (1965).
87. HAUGAARD, G., J. Dairy Sci., *49,* 1185 (1966).
88. HAUGAARD, G., and PETTINATI, J. D., J. Dairy Sci., *42,* 1255 (1959).
89. HENDERSON, J. L., J. Assoc. Offic. Agr. Chem., *46,* 1030 (1963).
90. HENNINGSON, R. W., J. Assoc. Offic. Agr. Chem., *46,* 1036 (1963).
91. HENNINGSON, R. W., J. Assoc. Offic. Anal. Chem., *49,* 511 (1966).
92. HENNINGSON, R. W., J. Assoc. Offic. Anal. Chem., *50,* 533 (1966).
93. HENNINGSON, R. W., J. Assoc. Offic. Anal. Chem., *52,* 142 (1969).
94. HENNINGSON, R. W., J. Assoc. Offic. Anal. Chem., *53,* 539 (1970).
95. HERMANN, L. F., U.S. Dept. Agr. Marketing Res. Div. (1954).
96. HERRINGTON, B. L., J. Dairy Sci., *37,* 775 (1954).
97. HERRINGTON, B. L., J. Dairy Sci., *43,* 1521 (1960).
98. HETRICK, J. H., and TRACY, P. H., J. Dairy Sci., *31,* 881 (1948).
99. HIGGINBOTTOM, C., and TAYLOR, M. M., J. Dairy Res. *27,* 245 (1960).
100. HOFI, A. A., RIFAAT, I. D., and KHORSHID, M. A., Indian J. Dairy Sci., *19,* 118 (1966).
101. HORST, M. F. ter, Neth. Milk Dairy J., *1,* 137 (1947).

102. HORTVET, J., Ind. Eng. Chem., *13,* 198 (1921).
103. HOWAT, G. R., and WRIGHT, N. C., J. Dairy Res., *5,* 236 (1934).
104. HUTCHINSON, R. C., J. Aust. Inst. Agr. Sci., *6,* 205 (1940).
105. JACK, E. L., and BRUNNER, J. R., J. Dairy Sci., *26,* 169 (1943).
106. JACKSON, C. J., J. Dairy Res., *7,* 31 (1936).
107. JACKSON, R. H., and PALLANSCH, M. J., J. Agr. Food Chem., *9,* 424 (1961).
108. JENNESS, R., Unpublished derivation (1962).
109. JENNESS, R. and PATTON, S., "Principles of Dairy Chemistry", John Wiley and Sons, New York, 1959.
110. JEUNET, R., and GRAPPIN, R., Lait, *50,* 654 (1970).
111. JOHNSTON, J. H. ST., Biochem. J., *21,* 1314 (1927).
112. JOSEPHSON, D. V., and DOAN, F. J., Milk Dealer, *29,* 35 (1939).
113. KLEYN, D. H., and SHIPE, W. F., Am. Milk Rev., *19,* 26 (1957).
114. KONEV, S. V., and KOZLOVA, G. G., Int. Dairy Congr., Proc. 18th, *IE* 84 (1970).
115. KREVELD, A. VAN, and VAN MINNEN, G., Neth. Milk Dairy J., *9,* 1. (1955).
116. LAMB, R. C., McGILLIARD, L. D., and BRUNNER, J. R., J. Dairy Sci., *43,* 491 (1960).
117. LEIDENFROST, W., Fette, Seifen, Anstrichmitt., *61,* 1005 (1959).
118. LEIGHTON, A., and KURTZ, F., J. Dairy Sci., *18,* 105 (1930).
119. LEPILKIN, A., and BORISOV, V., Mol. Prom., *27,* (5) 12 (1966). Cited in DSA, *28,* [3090] (1966).
120. LEVITON, A., and LEIGHTON, A., J. Dairy Sci., *18,* 105 (1935).
121. LEVITON, A., and PALLANSCH, M. J., J. Dairy Sci., *43,* 1389 (1960).
122. LIN, S. H. C., DEWAN, R. K., BLOOMFIELD, V. A., and MORR, C. V., Biochemistry, *10,* 4788 (1971).
123. LING, E. R., J. Dairy Res., *7,* 145 (1936).
124. LUDINGTON, V. D., and BIRD, E. W., Food Res., *6,* 421 (1941).
125. McINTYRE, R. T., PARRISH, D. B., and FOUNTAINE, F. C., J. Dairy Sci., *35,* 356 (1952).
126. McKENNELL, R., Anal. Chem., *32,* 1458 (1960).
127. MAXCY, R. B., and SOMMER, H. H., J. Dairy Sci., *37,* 60 (1954).
128. MENEFEE, S. G., and OVERMAN, O. R., J. Dairy Sci., *22,* 831 (1939).
129. MILLER, P. G. and SOMMER, H. H., J. Dairy Sci., *23,* 405 (1940).
130. MOHR, W., and BROCKMANN, C., Milchwiss, Forsch., *10,* 72 (1930).
131. MOJONNIER, T. S., and TROY, H. C., "Technical Control of Dairy Products" Mojonnier Brothers, Chicago, 1922.
132. MULDOON, P. J., and LISKA, B. J., J. Dairy Sci., *52,* 460 (1969).
133. MÜLLER, W., Milchwiss. Forsch., *11,* 243 (1931).
134. MUNCHBERG, F., and NARBUTAS, J., Milchwiss. Forsch., *19,* 114 (1937).
135. NAKAI, S., and LE, A. C., J. Dairy Sci., *53,* 276 (1970).
136. NELSON, V., J. Dairy Sci., *32,* 775 (1949).
137. NILSSON, G., CARLSON, C., and LAU-ERIKSSON, A., Lantbruks-hogskolans Ann., *36,* 221 (1970).
138. NORRIS, R., GRAY, I. K., McDOWELL, A. K. R., and DOLBY, R. M., J. Dairy Res., *38,* 179 (1971).
139. O'BRIEN, L., and LAWRENCE, A. J., Aust. J. Dairy Tech., *24,* 106 (1969).
140. OSBORNE, N. S., STIMSON, H. F., and GINNINGS, D. C., J. Res. Nat. Bur. Stand., *23,* 197 (1939).
141. OVERMAN, O. R., GARRETT, O. F., WRIGHT, K. E., and SANMANN, F. D., Ill. Agr. Exp. Sta. Bull. *457* (1939).
142. PAL, R. N., Indian J. Dairy Sci., *16,* 92 (1963).
143. PALMER, L. S., J. Dairy Sci., *27,* 471 (1944).
144. PARKASH, S., Indian J. Dairy Sci., *16,* 98 (1963).
145. PARRISH, D. B., WISE, G. H., HUGHES, E. S., and ATKESON, F. W., J. Dairy Sci., 31, 889 (1948).
146. PARRISH, D. B., WISE, G. H., HUGHES, E. S., and ATKESON, F. W., J. Dairy Sci., 33, 457 (1950).

147. PEEPLES, M. L., J. Dairy Sci., *45,* 297 (1962).
148. PETERSON, R. W., and FREEMAN, T. R., J. Dairy Sci., *49,* 806 (1966).
149. PHIPPS, L. W., J. Dairy Res., *24,* 51 (1957).
150. PHIPPS, L. W., J. Dairy Res., *36,* 417 (1969).
151. PIEN, J., Lait, *33,* 129, 241 (1953).
152. PINKERTON, F., and PETERS, I. I., J. Dairy Sci., *41,* 392 (1958).
153. POMERANTZ, V. and MELOAN, C. E., "Food Analysis: Theory and Practice". Avi Pub. Co., Inc., Westport, Conn., 1971.
154. POTTER, F. E., DEYSHER, E. F., and WEBB, B. H., J. Dairy Sci., *32,* 452. (1949).
155. PRENTICE, J. H., J. Dairy Res., *29,* 131 (1962).
156. PROUTY, C. C., J. Dairy Sci., *23,* 899 (1940).
157. PURI, B. R., and GUPTA, H. L., Indian J. Dairy Sci., *8,* 78 (1955).
158. PURI, B. R., and PARKASH, S., Indian J. Dairy Sci., *16,* 47 (1963).
159. PURI, B. R., PARKASH, S., and TOTAJA, K. K., Indian J. Dairy Sci., *17,* 181 (1963).
160. PYNE, G. T., J. Dairy Res., *29,* 101 (1962).
161. PYNE, G. T., and RYAN, J. J., J. Dairy Res., *17,* 200 (1950).
162. RAMAKRISHNAN, C. V., and BANERJEE, B. N., Indian J. Dairy Sci., *5,* 25 (1952).
163. RAMBKE, K., and KONRAD, H., Nahrung, *14,* 475 (1970). Cited in DSA, *33,* [3133] (1971).
164. RANGAPPA, K. S., Nature, *160,* 719 (1947).
165. RANGAPPA, K. S., Biochim. Biophys. Acta, *2,* 207 (1948).
166. RANGAPPA, K. S., Biochim. Biophys. Acta, *2,* 210 (1948).
167. RANGAPPA, K. S., Indian J. Dairy Sci., *17,* 137 (1964).
168. RAO, D. S., SUDHEENDRANATH, C. S., RAO, M. B. and ANANTAKRISHNAN, C. P., Int. Dairy Congr., Proc. 18th *IE,* 88 (1970).
169. RAOULT, F. M., Ann. Chim. Phys., *2,* (6) 66 (1884).
170. REES, H. V., Research Serv. Bull., Tasmanian Dept. Agr., *1,* 1952
171. REIDY, G. A., "I. Methods for determining thermal conductivity and thermal diffusivity of foods. II. Values for thermal properties of foods gathered from the literature." Dept. Food Science, Michigan State Univ., Lansing, 1968.
172. RICE, F. E., and MARKLEY, A. L., J. Dairy Sci., *7,* 468 (1924).
173. RICE, F. E., and MISCALL, J., J. Dairy Sci., 9, 140 (1926).
174. RISHOI, A. H., and SHARP, P. F., J. Dairy Sci., 21, 399 (1938).
175. ROBERTSON, A. H., J. Assoc. Offic. Agr. Chem., *40,* 618 (1957).
176. ROEDER, G., Milchwissenschaft, *8,* 125 (1953).
177. ROWLAND, S. J., and WAGSTAFF, A. W., J. Dairy Res., *26,* 83 (1959).
178. RUGE-LENARTOWICZ, R., Roczn, Zakl. Hig. Warsz., *5,* 91(1954). Cited in DSA, 17, [613] (1955).
179. RUTZ, W. D., WHITNAH, C. H., and BAETZ, G. D., J. Dairy Sci., *38,* 1312 (1955).
180. SCHOBER, R., CHRIST, W., and NICLAUSE, W., Lebensm. Untersuch. u. Forsch., *99,* 299 (1954).
181. SCHOENEMANN, D. R., FINNEGAN, E. J., and SHEURING, J. J., J. Dairy Sci., *47,* 683 (1964).
182. SCHULER, A., Milchwiss. Forsch., *19,* 373 (1938).
183. SCHULZ, M. E., Kieler, Milchwiss. Forschb., *8,* 641 (1956).
184. SCHULZ, M. E., and SYDOW, G., Milchwissenschaft, *12,* 174 (1957).
185. SCOTT-BLAIR, G. W., "Foodstuffs; Their Plasticity, Fluidity and Consistency." North Holland Pub. Co., Amsterdam, 1953.
186. SHARMA, R. R., Indian J. Dairy Sci., *16,* 101 (1963).
187. SHARP, P. F., and HART, R. G., J. Dairy Sci., *19,* 683 (1936).
188. SHARP, P. F., and KRUKOVSKY, V. N., J. Dairy Sci., *22,* 743 (1939).
189. SHEDLOVSKY, T., Chap. 45 in Weissberger, A., "Technique of Organic Chemistry", Vol. 1, 3rd Ed. Interscience Publishers, Inc., New York, 1959.
190. SHERBON, J. W., in Porter, R. S., and Johnson, J. F. "Analytical Calorimetry." Plenum Press, New York, pp 173–180, 1968.

191. SHERBON, J. W., J. Assoc. Offic. Anal. Chem., *53,* 862 (1970).
192. SHERBON, J. W., and COULTER, S. T., J. Dairy Sci., *49,* 1326 (1966).
193. SHIPE, W. F., J. Dairy Sci., *39,* 916 (1956).
194. SHIPE, W. F., J. Assoc. Offic. Agr. Chem., *41,* 262 (1958).
195. SHIPE, W. F., J. Dairy Sci., *42,* 1745 (1959).
196. SHIPE, W. F., J. Assoc. Offic. Agr. Chem., *43,* 411 (1960).
197. SHIPE, W. F., J. Assoc. Offic. Agr. Chem., *47,* 570 (1964).
198. SHIPE, W. F., J. Assoc. Offic. Anal. Chem., *52,* 131 (1969).
199. SHIPE, W. F., DAHLBERG, A. C., and HERRINGTON, B. L., J. Dairy Sci., *36,* 916 (1953).
200. SHORT, A. L., J. Dairy Res., *22,* 69 (1955).
201. SHORT, A. L., J. Soc. Dairy Technol., *9,* 81 (1956).
202. SHUGLIASHVILI, G. V., CHARUEV, N. G., and ABRAMASHVILI, V. I., Trudy nauchnoissled. Inst. Automatiz, proiz. Protessov. Prom. *1967* (4), 91 (1967) Cited in DSA, *32,* [1312] (1970).
203. SOROKIN, YU, Mol. Prom., *16,* (3), 38 (1955). Cited in DSA, *17,* [741] (1955).
204. SPELLS, K. E., Phys. Med. Biol., *5,* 139 (1960).
205. SPREMULLI, G. H., Univ. Mich. Microfilms, Ann Arbor, Mich., Pub. 510 (1942).
206. STUBBS, J. R., and ELSDON, G. D., Analyst, *59,* 146 (1934).
207. STULL, J. W., TAYLOR, R. R., and GHLANDER, A. M., J. Dairy Sci., *48,* 1019 (1965).
208. STURTEVANT, J. M., Chap. 10 in Weissberger, A. "Technique of Organic Chemistry," Vol. 1, 3rd Ed. Interscience Publishers, Inc., New York, 1959.
209. SUDHEENDRANATH, C. S., and RAO, M. B., Dairy Congr. Proc. 18th *IE,* 89 (1970).
210. SWINDELLS, J. F., COE, J. R. and GODFREY, T. B., J. Res. Nat. Bur. Stand., *48,* 1 (1952).
211. SWINDELLS, J. F., ULLMAN, R., and MARK, H., Chap. 12 in Weissberger, A. "Technique of Organic Chemistry". Vol. 1, 3rd Ed. Interscience Publishers, Inc. New York, 1959.
212. TARASSUK, N. P., and SMITH, F. R., J. Dairy Sci., *23,* 1163 (1940).
213. TELLMANN, E., Milchw. Forsch., *15,* 294 (1933).
214. TEMPLE, P. L., Analyst, *62,* 709 (1937).
215. TESSIER, H., and ROSE, D., J. Dairy Sci., *41,* 351 (1958).
216. TILLMANS, J. and OBERMEIER, W., Z. Untersuch. Nahr. u. Genussm., *40,* 23, 31 (1920).
217. TORSSELL, H., SANDBERG, U., and THURESON, L. E., Int. Dairy Congr. Proc. 12th, *2,* 246 (1949).
218. TROUT, G. M., HALLORAN, C. D. and GOULD, I. A., Mich. Agr. Exp. Sta. Bull. *145,* (1935).
219. TUCKER, V. C., Queensl. J. Agr. Sci. *20,* 161 (1963).
220. TUCKER, V. C., Aust. J. Dairy Tech. *25,* 126 (1970).
221. TUCKER, V. C., Aust. J. Dairy Tech. *25,* 137 (1970).
222. VAN DEN BERG, L., J. Dairy Sci., *44,* 26 (1961).
223. VANDERZANT, C., and TENNISON, W. R., Food Technol., *15,* 63 (1961).
224. VANLANDINGHAM, A. H., WEAKLEY, C. E., MOORE, E. N., and HENDERSON, H. O., J. Dairy Sci., *24,* 383 (1941).
225. VLEESCHAUWER, A. DE, and WAEYENBERGE, K., Meded. Landb. Hoogesch. Opzoeksta Gent., *9,* 56 (1941).
226. WAHBA, N., Analyst, *90,* 432 (1965).
227. WALSTRA, P., Neth. Milk Dairy J., *19,* 1 (1965).
228. WALSTRA, P., Neth. Milk Dairy J., *19,* 93 (1965).
229. WALSTRA, P., J. Dairy Sci., *50,* 1839 (1967).
230. WATSON, P. D., J. Dairy Sci., *14,* 50 (1931).
231. WATSON, P. D., J. Dairy Sci., *40,* 334 (1957).
232. WATSON, P. D., J. Dairy Sci., *41,* 1693 (1958).
233. WATSON, P. D., and TITTSLER, R. P., J. Dairy Sci., *44,* 416 (1961).

234. WEBB, B. H., J. Dairy Sci., *16,* 369 (1933).
235. WEBER, W., Kolloid Z., *147,* 14 (1956).
236. WEGENER, H., Milchwissenschaft, *8,* 433 (1953).
237. WEISSBERGER, A., "Technique of Organic Chemistry; Chemical Applications of Spectroscopy". Vol. 9, Interscience Publishers, Inc., New York, 1956.
238. WHEELOCK, J. V., ROOK, J. A. F., and DODD, F. H., J. Dairy Res., 32, 79 (1965).
239. WHITAKER, R., SHERMAN, J. M., and SHARP, P. F., J. Dairy Sci., *10,* 361 (1927).
240. WHITNAH, C. H., J. Dairy Sci., *42,* 1437 (1959).
241. WHITNAH, C. H., J. Agr. Food Chem., *10,* 295 (1962).
242. WHITNAH, C. H., CONRAD, R. M., and COOK, G. L., J. Dairy Sci., *2,* 406 (1949).
243. WHITNAH, C. H., MEDVED, T. M., and RUTZ, W. D., J. Dairy Sci., *40,* 856 (1957).
244. WHITNAH, C. H., and RUTZ, W. D., J. Dairy Sci., *42,* 227 (1959).
245. WHITNAH, C. H., RUTZ, W. D., and ALEXANDER, T. G., J. Dairy Sci., *42,* 221 (1959).
246. WHITNAH, C. H., RUTZ, W. D., and FRYER, H. C., J. Dairy Sci., *39,* 356 (1956).
247. WHITTIER, E. O., J. Biol. Chem., *83,* 79 (1929).
248. WHITTIER, E. O., J. Biol. Chem., *102,* 733 (1933).
249. WHITTIER, E. O., J. Phys. Chem., *37,* 847 (1933).
250. WHITTIER, E. O., J. Biol. Chem., *23,* 283 (1938).
251. WHITTIER, E. O., and BENTON, A. G., J. Dairy Sci., *10,* 343 (1927).
252. WIEGNER, G., Z. Nahr. Genussm., *20,* 70 (1910).
253. WILCOX, C. J., and KRIENKE, W. A., J. Dairy Sci., *47,* 638 (1964).
254. WILEY, W. J., J. Dairy Res., *6,* 71, 86 (1935).
255. WOODAMS, E. E., and NOWREY, J. E., Food Technol. Champaign, *22,* 150 (1968).
256. YONCOSKIE, R. A., J. Am. Oil Chem. Soc., *46,* 49 (1969).

ADDITIONAL REFERENCES

257. American Assoc. Vitamin Chemists, "Methods of Vitamin Assay." *2nd* ed., p. 162 Wiley (Interscience), New York, 1951.
258. BAKALOR, S., Aust. J. Dairy Tech., *20,* 151 (1965).
259. FOX, K. K., HOLSINGER, V. H., and PALLANSCH, M. J., J. Dairy Sci., *46,* 302 (1963).
260. KONEV, S. V., and KOZUNIN, I. I., Dairy Sci. Abst. *23,* [103] (1961).
261. KOOPS, J. and WIJNAND, H. P., Neth. Milk Dairy J., *15,* 333 (1961).
262. PORTER, R. M., J. Dairy Sci., *48,* 99 (1965).

CHAPTER 9

H. M. Farrell, Jr.
M. P. Thompson†

Physical Equilibria: Proteins

INTRODUCTION

Milk is a complex biological fluid, secreted by mammals explicitly for the nourishment of their young. Through the centuries, evolution has produced this stable, fluid, concentrated source of lipid, protein and carbohydrate. Because of its unusual stability (for a biological fluid) milk has become a valuable foodstuff, a commodity, yet many of the problems which arise in the processing of milk stem from the biochemical nature of its components. In dealing with skimmilk, the retention of the unique properties of the casein-protein complex during processing is of the utmost importance.

Chemically, the skimmilk system can be classified as a lyophilic colloid because the protein complexes of skimmilk, which constitute the dispersed phase, are in the correct size range, interact with and are stabilized by the solvent, and do not spontaneously coagulate. The milk protein complex is stable to the earth's gravitational field, yet can be separated from the liquid phase by centrifugation. Milk, then, can be considered a biocolloid; the properties of the dispersed phase (the casein-protein complex) and the dispersion medium (the milk serum) will be discussed.

THE CASEIN MICELLE

For better or for worse, the term "micelle" has been applied to the dispersed phase of milk, the casein-protein complex. The electron micrograph of Figure 9.1 shows a number of typical casein micelles. The nature of the casein micelle has been investigated in many laboratories, and with good reason, for this complex is the essence of a large number of problems encountered in dairy technology, whether it be the preservation of the stability of milk, the curd tension of cheese, or the production of a synthetic engineered food. Hence, understanding those forces

---

*Revised from Chapter 9 in the 1st edition by T. L. McMeekin and M. L. Groves.

† H. M. Farrell Jr. and M. P. Thompson Eastern Regional Research Center, Agricultural Research Service, U.S. Department of Agriculture. Philadelphia, Pennsylvania, 19118

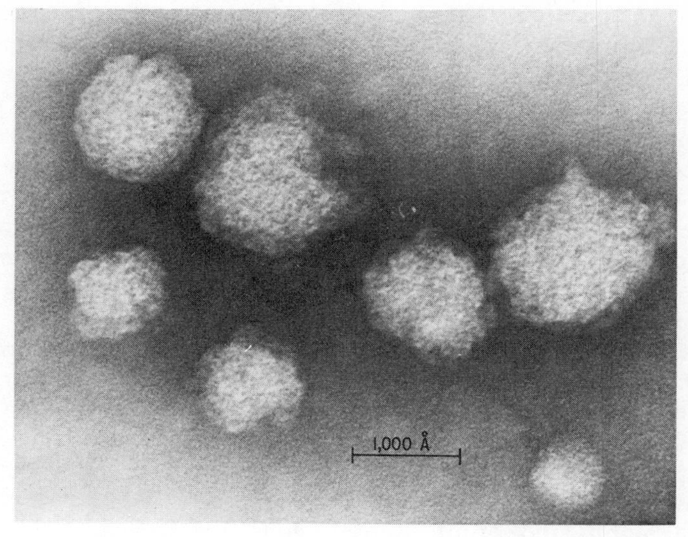

Courtesy of R. J. Carroll

FIG. 9.1. CASEIN MICELLES OF BOVINE SKIMMILK

Fixed in 1% glutaraldehyde and negatively stained with phosphotungstic acid.

which hold the casein micelle together, and *a posteriori* those forces which cause disruption of the casein protein complex, is of paramount importance.

Protein Components of the Casein Micelle

Until the 1930's the milk protein complex was considered to be composed of the rather "homogeneous protein," casein. Then, Linderstrøm-Lang[47] and Mellander[51] demonstrated the heterogeneity of bovine casein. The latter worker termed the electrophoretically distinct fractions α-, β- and γ- casein. From that time until the late 1950's, many methods of fractionation were developed and various casein fractions were isolated and characterized (see Chapter 3). The most significant fractionation was accomplished by Waugh and von Hippel;[102] when they discovered that the α-casein fraction is a mixture of α_{s1}-casein and κ-casein. Indeed, a sample of casein from pooled milk, subjected to gel electrophoresis in urea and mercaptoethanol, yields up to 20 casein components. The demonstration of genetic polymorphism in the β-casein fraction by Aschaffenburg,[1] followed by the work of Thompson *et al.*,[91] on the genetics of the α_{s1}-fraction began to introduce a unifying concept to the field. The contention of Groves and Gordon[33] that the γ-, R-, S-, and TS-fractions are but degradation products of β-casein leads one to speculate that the other minor casein fractions reported in the literature, such as m- and λ-caseins and the proteose peptone fraction,[93] are also degradation products of one or more of the major casein components. From all the work on the characterization of casein, three major components of the casein protein complex have been described, namely α_{s1}-, β- and κ-casein. Without a doubt the major protein of the casein complex is the α_{s1}-fraction. The exact margin by which this fraction exceeds β- and κ-casein seems to be open to debate depending in part on the method of quantitation. However, a good estimate obtained by several methods[70,71] would be α_{s1}, 50%, β, 33%, and κ, 15%.

The names of the various fractions used here are in accord with the A.D.S.A. committee, whose reports[72,87] have done much to order the field of milk protein nomenclature. Thompson[87] has recently reviewed the methods available for the detection of the various known genetic polymorphs of the milk proteins, and Farrell and Thompson[24] have reviewed their occurrence in various breeds and the possible biological significance of milk protein genetic polymorphism.

α_{s1}-Casein is the best characterized component of the casein system. It is a single-chain polypeptide, of known sequence, with 199 amino acid residues and a molecular weight of 23,600 daltons.[31,52,64] The molecule contains 8 phosphate residues, all of which exist as the phosphomonoesters of serine.[52,64] Seven of these phosphoserine residues are clustered in an acidic portion of the molecule bounded by residues 43 and 80 (the second fifth of the molecule from the N → C terminal end). This highly acidic segment contains 12 carboxylic acid residues as well as 7 of the phosphate residues; as postulated by Waugh,[104] it also contains the largest segment of the molecule's net negative charge. Theoretically, from a knowledge of the complete sequence, one can calculate the charge frequency, net charge, and hydrophobicity for various segments of the molecule. These data for α_{s1}-casein are presented in Table 9.1.

The hydrophobicity shown in Table 9.1 was calculated using the method of Bigelow,[6] and can be taken as a quantitative measure of the apolarity of a segment of a molecule or of the molecule itself. The data in Table 9.1 show noncoincidence of high charge frequency and apolarity in the segments shown; however, there exist local areas of charge surrounded by an apolar environment. The proline content of α_{s1}-casein is high;[31,52] these residues appear to be evenly distributed, and proline residues are known to disrupt helical and beta structures. Thus, the sequence data confirm the physical-chemical data, which indicated that the α_{s1}-molecule has little or no recognizable secondary structure such as α-helix or β-structure.[34] The high degree of hy-

Table 9.1

PROFILE OF THE α_{s1}-CASEIN MOLECULE DERIVED FROM ITS PRIMARY STRUCTURE[a]

Residues Considered	Net Charge[b]	Charge Frequency[b,c]	Average Hydrophobicity[c]
1 → 40	+3	0.25	1340
41 → 80	−22½	0.75	641
81 → 120	0	0.35	1310
121 → 160	−1	0.23	1264
161 → 199	−2½	0.14	1164

[a] Adapted from Grosclaude *et al.*,[31] Mercier *et al.*,[52] and Ribadeau-Dumas.[64]
[b] Some error as to assignment of these values may exist, since the exact placement of all amides is not known, serine phosphate = − 2, histidine = + ½.
[c] Calculated as described by Bigelow.[6]

drophobicity exhibited by the segment containing residues 100 → 199 is probably responsible, in part, for the pronounced self-association of the α_{s1}-casein monomer in aqueous solution.[73,75,82,104] This self-association approaches a limiting size under most conditions of ionic strength;[73,75,82,104] the highly charged phosphopeptide region can readily account for this phenomenon through charge repulsions.[104] It is noteworthy that while the self-association of α_{s1}-casein is mostly hydrophobic in nature, and hence temperature-dependent, some ionic bonding, as postulated by Schmidt,[73,75] must occur in the reaction in addition to hydrophobic interactions. At calcium-ion concentrations of 5 to 10mM,[90] α_{s1}-casein forms an insoluble precipitate. The solubility of α_{s1}-casein in aqueous Ca^{2+} solutions has been studied by Waugh *et al.*[103] and by Thompson *et al.*[90] With the exception of the rare genetic variant α_{s1}-A, the calcium solubility of α_{s1}-casein is temperature-independent.[90] Thus, the major protein component of milk is insoluble under normal conditions of pH, ionic strength, and temperature.

β-Casein is the second most abundant milk protein. The molecule is a single chain with 5 phosphoserine residues and a molecular weight of 24,500 daltons.[66,67] In aqueous solution, β-casein has been characterized as a random coil[56] with little or no secondary structure.[34] In aqueous solution, β-casein undergoes an endothermic self-association which reaches a maximum or limiting size depending upon the ionic strength.[58,79,104] An almost complete sequence is available for β-casein.[66,67] The proline content of β-casein is rather evenly distributed, which explains in part why the molecule lacks any secondary structure. The charge frequency, hydrophobicity and net charge for the various segments of β-casein are presented in Table 9.2. Analysis of these data indicates that β-casein is much more "soap-like" than α_{s1}-casein.

Table 9.2

PROFILE OF THE β-CASEIN MOLECULE DERIVED FROM ITS PRIMARY STRUCTURE[a]

Residues Considered	Net Charge[b]	Charge Frequency[b,c]	Average Hydrophobicity[c]
1 → 43	−16	0.65	783
44 → 92	−3½	0.13	1429
93 → 135	+2	0.23	1173
136 → 177	+3	0.07	1467
178 → 209	+2	0.06	1738

[a] Derived from the data of Ribadeau-Dumas *et al.*[64,67]
[b] Some error as to assignment of these values may exist, since the exact placement of all amides is not known serine phosphate = − 2, histidine = + ½.
[c] Calculated as described by Bigelow.[6]

The N-terminal portion of the β-casein molecule (residues 1 → 40) contains the phosphoserine residues and carries essentially all the protein's net charge, while the C-terminal half of the molecule (actually residues 136 → 209) contains many apolar residues (as demonstrated by its high hydrophobicity). The N-terminal concentration of charge and the highly hydrophobic C-terminal may account for the temperature-dependence of the self-association of β-casein[58] since hydrophobic interactions are temperature-sensitive.[43] Like α_{s1}-casein, β-casein is insoluble at room temperature in the presence of Ca^{2+} at concentrations below those encountered in milk. However, the precipitation from solution of β-casein is temperature-dependent, and the calcium-β-caseinate complex is soluble at 1°C up to 400 mM calcium-ion concentration.[90] Again, this temperature-dependence is probably due to the charge distribution of the β-casein monomer.

κ-Casein, the third major component of the milk protein complex, differs from α_{s1}- and β-casein in that it is soluble over a very broad range of calcium-ion concentrations.[102] It is this calcium solubility which led Waugh, upon discovering the κ-fraction, to assign to it the role of casein micelle stabilization.[102] It is also the κ-casein fraction that is most readily cleaved by rennin;[37,40,42] the resulting products are termed para-kappa and the macropeptide. It would appear that κ-casein is the key to micelle structure because it stabilizes the calcium-insoluble α_{s1} and β-caseins, and is the primary site of attack by the enzyme rennin. Ironically, κ-casein is at present the least well characterized fraction with regard to its primary structure and association properties. κ-Casein is unique because it is the only major component of the casein complex which contains cystine (or possibly cysteine). The occurrence of free sulfhydryl groups in the milk protein complex has been reported[3] by some workers, but not by others.[41] Hence, the degree of disulfide bonding which occurs in κ-casein is uncertain, since no free –SH groups have been found in isolated κ-casein fractions.[41,42] Woychik *et al.*[107] reported the reduced molecular weight of κ-casein in 5*M* guanidine hydrochloride to be 17,000 to 19,000. Swaisgood and Brunner[81] reported a value of ≃19,000 for reduced κ-casein. The latter authors[80,81] reported a nonreduced molecular weight in the order of 60,000 for the lightest component, indicating at least a disulfide-linked trimer of the isolated κ-casein; but the weight average molecular weight of their preparation was ≃ 110,000. Cheeseman[17] studied the effect of the binding of the detergent sodium dodecyl sulfate on casein by gel filtration. He concluded that the majority of the κ-casein occurs as a disulfide-linked aggregate, which eluted at the void volume of Sephadex G-200, even in the presence of detergent.

κ-Casein is also the only major component of the casein complex that contains carbohydrate.[54] All the carbohydrate associated with κ-casein is bound to the macropeptide,[37,40,54] which is the highly soluble portion formed by rennin hydrolysis. In addition to being a glycoprotein, κ-casein contains one[37] or two[40] phosphate residues per reduced monomer and, as noted before, is soluble in Ca^{2+}, although it binds this ion.[21] Hill and Wake[37] postulated that κ-casein is an amphiphile or "soap-like" molecule on the basis of what is known of the primary structure of the macropeptide. This C-terminal one-fourth of the molecule, although it is quite hydrophilic and accounts for all the net charge of the κ-casein molecule, contains only a portion of the total number of charged residues. Jollès *et al.*[40] have isolated peptides from the supposedly hydrophobic para-kappa fraction, and have demonstrated that local areas of charge do exist in the para-kappa portion of the molecule. In addition, if κ-casein were a total amphiphile, as suggested, its physical associations should be highly temperature-sensitive. While its properties are temperature-dependent, it falls between β- and α_{s1}-casein in this respect.[7]

Forces Responsible for the Stability of the Casein Micelle

In 1929 Linderstrøm-Lang[47] postulated, as a result of his studies on casein, that the colloidal milk complex should be composed of a mixture of calcium-insoluble proteins stabilized by a calcium-soluble protein. The latter protein would be readily split by rennin, destabilizing the colloid and allowing coagulation to occur. As we have seen, such fractions do exist: α_{s1}- and β-caseins are indeed calcium-insoluble, while κ-casein is not only soluble in the presence of calcium ion, but is readily split by rennin. In addition, Waugh and co-workers[102,103] have demonstrated that α_{s1}- and κ-casein complexes can be reformed from the isolated fractions as measured by sedimentation velocity experiments. Recently, Pepper[59] demonstrated this interaction of α_{s1}- and κ-casein by gel filtration, and studied the concentration-dependence of the interaction. The complexes formed by the interaction of the isolated α_{s1}- and κ-caseins aggregate to form simulated casein micelles upon addition of Ca^{2+} in $0.01M$ imidazole buffer, pH 6.7. As viewed by electron microscopy,[8] these synthetic micelles are virtually identical with fresh milk micelles except for their increased size. The precise mechanism of formation of the natural casein micelles is as yet uncertain, although several theories have been advanced based on the study of synthetic micelles; these theories will be reviewed later. In the course of the discussion of casein micelle structure and formation, a brief summary of the types of bonding forces responsible for the stabilization of protein structure will be given.

Hydrophobic Interactions.—One of the most significant contributions to our understanding of protein stability was made by Kauzman,[43] who elucidated the nature of hydrophobic interactions in proteins. These interactions come about because water exhibits a decreased entropy as a result of the occurrence of apolar amino-acid residues within the solvent. If these apolar residues are forced out of the water and into the interior of a protein molecule, where they can interact with other apolar groups, a small quantity of stabilization energy is gained per residue transferred from the solvent. Several model systems based on the energy of transfer of amino acids from water to ethanol have been studied and yield confirmatory results.[6,83]

These hydrophobic interactions are highly temperature-sensitive, being minimal below 5°C and maximal at higher temperatures. In a recent review article Klotz[44] pointed out that for proteins whose crystallographic structure is known, many apolar side-chains do exist fully or partially exposed to the solvent and therefore exhibit surface patches which are available for interactions with other protein molecules.

From the amino-acid analysis of the α_{s1}-, β- and κ-casein,[29,32,42] it is quite apparent that large numbers of apolar residues occur in these proteins. Furthermore, from the primary structures now available (Tables 9.1 and 9.2), it is clear that these hydrophobic residues are somewhat clustered for α_{s1}- and β-caseins, as well as for κ-casein.[37] According to the calculations of Hill and Wake,[37] the caseins rank among the most hydrophobic proteins of those tabulated by Bigelow.[6] It is not unexpected then, that the casein micelle should be stabilized by hydrophobic bonding. Several investigators[16,23,69,79] have noted that β- and κ-caseins, and α_{s1}-casein to some extent, diffuse out of the micelle at low temperatures. As the temperature decreases, hydrophobic stabilization energy also decreases, and these molecules (β- and κ-casein) are able to diffuse out of the micelle. These observations are consistent with the known primary structure of β-casein (Table 9.2) and the postulated structure of κ-casein.[37] The interactions of β-casein are more temperature-dependent, which indicates that it is probably more "soap-like" than κ-casein. While all the authors cited above agree that β-, and to a lesser extent κ- and α_{s1}-caseins, can be removed from the casein micelle at 1°C, some question arises as to the exact amount released. Rose[69] reported high values for β-casein (up to 30%), while Downey and Murphy's values[23] (up to 15%) are lower. The latter workers, however, pointed out that the stage of lactation and health of the animal play a role in the percentage of cold soluble casein present. All those cited above concur that the α_{s1}-fraction does not diffuse from the micelle to as great an extent as the other two caseins.

The rare α_{s1}-A genetic variant, however, does exhibit highly temperature-dependent interactions. The α_{s1}-A gene is the result of the sequential deletion of up to 13 amino acid residues[29,52,89] bounded by residues 13 and 27, and the majority of these deleted amino acids are apolar.[89] The net result of this deletion is to bring the charged phosphorus-rich area closer to the N-terminal region, making this α_{s1}-genetic variant more like β-casein in its charge distribution, and the physical and solubility properties of α_{s1}-A mirror those of β-casein.[90] Thus, the stability of the casein micelle is due in part to hydrophobic interactions, although some ionic bonding must occur between the α_{s1}- and κ-caseins; the α_{s1}-A deletion probably does not permit the formation of the ionic bonds characteristic of α_{s1}-casein; as a result, micelles containing this protein are less stable to heat, cold, and processing conditions.

Dissociating agents, such as sodium dodecyl sulfate, guanidine-hydrochloride, and urea, all of which are thought to act primarily on hydrophobic interactions, tend to disrupt casein micelle structure in the same fashion, as evidenced by electron microscopy.[12] These solvents reduce the micelle to small subunits approaching 100 Å in diameter. The temperature-dependent properties of the hydrophobic interactions may also explain why milk can withstand moderate to high temperatures, but does not survive extremely low temperatures, such as freezing or ultrahigh temperatures.

Electrostatic Interactions.—It has been pointed out[9,44] that essentially all of the ionic side-chains in the proteins, whose crystallographic structure is known, are fully exposed to the solvent. Thus ionic bonding between negatively charged carboxylic acid residues and positively charged groups contributes little to the stability of a monomeric protein. Notable exceptions to this rule may occur when an ion pair can be formed within a hydrophobic environment;[83] the interactions of subunits of a protein may provide just such an environment. Physical-chemical evidence for the role of ionic bonding in subunit interactions is abundant, while crystallographic evidence is limited to hemoglobin,[26] although several subunit enzymes are currently under study.[35] Conversely, electrostatic interactions between carboxylate residues and divalent metal ions can impart reasonable structural stability to a protein. Calcium stabilizes staphylococcal nuclease[35] and increases the heat stability and reactivity of trypsin.[78] Many metallo-enzymes derive a good deal of their stabilization from specific metal coordination complexes.[22]

The role of inter- and intra-molecular ionic bonds among the α_{s1}-, β- and κ-caseins in stabilization of micelle structure is difficult to assess. Many potential sites for strong ion-pair bonds within an apolar environment exist, as deduced from consideration of the known sequences; and such bonds may play a role in micelle subunit interactions. Pepper *et al.*[61] demonstrated that carbamylation of 5 of 9 lysine residues of κ-casein destroyed the ability of κ-casein to stabilize α_{s1}-casein, thus demonstrating that ionic interactions may play a role in micelle structure. Furthermore, Hill[36] modified the arginine side-chains of the caseins and found differences in coagulation by rennin.

The estimated calcium content of milk is around 30 mM[19,105] far above the concentrations of Ca^{2+} required to precipitate the isolated α_{s1}- and β-caseins at room temperature.[90] The role of the phosphate residues in calcium binding has been investigated by the enzymatic dephosphorylation of α_{s1}-casein. Pepper and Thompson[60] and Bingham *et al.*[8] demonstrated that dephosphorylated α_{s1}-casein was still precipitated by calcium and showed decreased stabilization by κ-casein. The latter authors postulated that two nonphosphate calcium-binding sites occur in α_{s1}-casein, and it is the binding to these sites which induces precipitation of the dephosphorylated casein. Investigation of the κ-casein stabilized, dephosphorylated α_{s1}-casein by electron microscopy[8] showed larger but fewer micelle-like structures. In milks containing α_{s1}-A,[90,92] such large micelles are poorly solvated and less stable. Thus the formation of micelle-like structures is not totally dependent upon the formation of calcium-phosphate bonds between caseins; however, the resulting micelles may be less stable. Indeed, removal of calcium from micelles by chelating agents such as EDTA and fluoride[12] leads to disruption of casein micelle structure, as evidenced by electron microscopy.

The total number of charged groups of the casein monomers (Tables 9.1 and 9.2) reveals that in the formation of a casein micelle, not all of these ionic groups can occupy a surface position. This would indicate either that much energy is used to bury these groups or that the structure is porous and available to the solvent, water. The latter proposition is born out by the experimental evidence. Ribadeau-Dumas and Garnier[65] noted that carboxypeptidase A is able to remove, quantitatively, the carboxyl-terminal residues from the α_{s1}-, β- and κ-caseins of native micelles, demonstrating that this enzyme (M.W. 40,000) is able to penetrate into the center of the casein micelle. Thompson *et al.*[90,92] have shown that the casein micelle is a highly solvated structure with an average of 1.90 gm water per gm protein. They also noted (Fig. 9.2) a strong positive correlation between the degree of solvation and heat stability.[92] The degree of solvation of the micelle, and hence the heat stability of the milk, hinges upon a variety of factors[62,68,92] not the least of which is the calcium:phosphate ratio. Increases in the calcium content of milk causes decreased heat stability,[62,68,99] possibly by altering the degree of solvation of the casein micelle. Thus the micelle emerges as a highly solvated porous structure. Environments which tend to decrease solvent interaction lower the stability of the micelle, which in turn destabilizes the milk. These interactions relate back to the proposition that the ionic residues of the individual casein

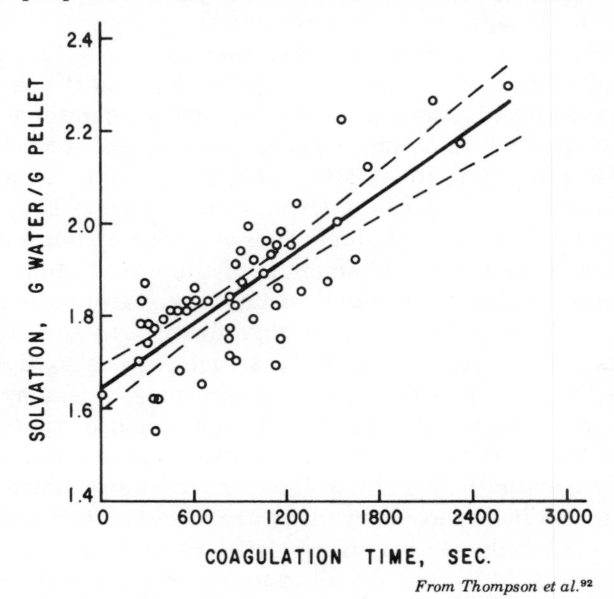

From Thompson et al.[92]

FIG. 9.2. PLOT OF THE SOLVATION OF THE CASEIN PELLETS (GRAMS WATER/GRAM PELLET) VERSUS HEAT COAGULATION TIME IN SECONDS

Dotted lines represent 95% confidence limits.

monomers cannot be totally buried, but must be exposed to solvent.

The better early measurements of the monomer molecular weights of the isolated casein fractions were obtained at pH 11 to 12.[50,102] At these pH values the positively charged lysine residues and a portion of the arginine residues have been neutralized, thus increasing the charge repulsions of the carboxyl and phosphate residues. However, prolonged exposure to high pH may produce degradation, as pointed out by Noelken.[55] These same effects operate in the casein micelle; as the pH of milk is brought to 11 to 12, the micelle structure is disrupted, with accompanying changes in turbidity and viscosity. Presumably, exposure to high pH for long periods of time, for example, in the production of sodium caseinate, may cause degradation, and hence alter the characteristics of the product.

Hydrogen Bonding Secondary and Tertiary Structure.—Many globular proteins, such as myoglobin, are stabilized by a high degree of α-helical structure.[9] In addition to the fibrous proteins, the so-called β- or pleated sheet structure has been detected by X-ray crystallography in globular proteins, notably lactate dehydrogenase[35] and others.[9,26] These secondary structures are stabilized by the formation of hydrogen bonds along the polypeptide backbone. Many proteins have been shown to contain significant amounts of secondary structure, as determined by spectral methods, such as circular dichroism, optical rotatory dispersion, and infrared spectroscopy.[95] However, in at least one case these methods have proved inaccurate in predicting the amount of secondary structure of a protein.[5] Therefore, the spectral methods can provide a good estimate of the amount of secondary structure, but they are subject to error. In many cases, then, some degree of stabilization is achieved by the formation of α-helical or β-structure, but not all stable proteins contain considerable amounts of these conformations. Other bonding forces (noted above) and perhaps even "sterically restricted" random structures may contribute significantly to the stabilization of a protein. The formation of α- and β-structures is also highly dependent upon the amino acid side-chains (proline, for example, breaks helical structures). In fact, the state of ionization of the side-chains and the solvent used play a role in the formation of α-helix.[9,26]

Spectral investigations of the isolated caseins have shown that these proteins possess little secondary structure. Herskovits[34] demonstrated by optical rotatory dispersion, using Moffit-Yang, Drude and Shechter-Blout analyses, that in aqueous solutions neither the individual casein components (α_{s1}-, β- or κ-) nor whole sodium caseinate exhibit an appreciable degree of α-helical content. Noelken and Reibstein[56] concluded that β-casein exhibits a random coil conformation in both aqueous solution and in 6*M* guanidine-HCl. Evidence has been accumulated[58,73,82,104] that α_{s1}- and β-caseins are intermediate between a totally random and a globular protein in conformation. The above observed properties of the caseins are in good agreement with the high incidence of proline scattered throughout the α_{s1}- and β-caseins as derived from analyses of their sequences. Since little or no secondary structure occurs in the individual casein components, one would expect that the degree of stabilization contributed to the casein micelle by α-helix or β-structure would be quite low.

Theoretically, hydrogen bonds between ionizable side-chains accessible to the solvent, water, contribute to a limited degree to the stabilization of monomeric proteins.[9,44] These groups are already hydrogen-bonded to water and the water-residue hydrogen bond would have to be broken before a residue-residue hydrogen bond could be formed. Nevertheless, once two subunits of a protein begin to interact, these surface groups may no longer be totally hydrated, and hydrogen bonds could form between monomers as a result of the altered environment.

Hydrogen bonding between casein monomers in the casein micelle may occur. Subunit interactions, at present, have not been sufficiently detailed by crystallographic evidence to support or rule out these types of bonds, but some intra-chain hydrogen bonds do occur in monomeric proteins.[9] It is also possible that some hydrogen bonding may occur in the self-association[73,75] of α_{s1}-casein. Certainly, in the formation of the highly aggregated casein micelle, such bonds between the various casein components would be possible.

The Role of Disulfide Bonds.—The folding about of helical segments, pleated sheet areas and unordered structures of a polypeptide chain is referred to as tertiary structure. The tertiary structure of proteins can be locked in place by the formation of disulfide bonds between distal cysteine residues. In fact, non-identical polypeptide chains can be held together by disulfide bonding, as in the case of γ-globulins. Evidence has been presented that for several proteins,[9] the disulfide bridges do not cause the formation of secondary and tertiary structure, but tend to stabilize the preformed conformations. Proteins such as lysozyme and RNase with a relatively high degree of disulfide bonding are quite stable, but not all stable proteins necessarily contain disulfide bonds.

As noted above, κ-casein is the only major component of the casein-protein complex that contains cystine (or cysteine). The occurrence of free sulfhydryl groups in the native casein complex has been reported by some workers,[3] but not by others.[41] Hence, the degree of disulfide cross-linkages, which normally occur in the casein micelle, is difficult to estimate. Swaisgood and Brunner[80] reported that a good approximation of the minimum size of κ-casein would be a disulfide-linked trimer, but for the most part their evidence[80,81] would indicate a greater degree of cross-linking. However, Woychik *et al.*[106] demonstrated that reduced and alkylated κ-casein stabilized α_{s1}-casein against calcium precipitation as well as native κ-casein. It appears that while the disulfide bridges of the casein micelle may contribute to the overall stability of the casein micelles, they are neither the driving force for micelle formation nor the central feature of the formed micelle.

Colloidal Calcium Phosphate.—The total calcium content of skimmilk has been estimated to be 30 mM,[19,105] but the calcium-ion content of serum, prepared by ultrafiltration or centrifugation of skimmilk is only ~ 2.9 mM.[10,19] Specific ion electrode studies yield a value of 2.5 mM calcium (II) for skimmilk.[20] Thus, more than 90% of the calcium content of skimmilk is in some way associated with the casein micelles. Subsequent washing of the micelles removes only a small portion of the calcium and other salts. The mineral contents of washed micelles prepared by centrifugation[19] and of "primary micelles" prepared by gel filtration[10] are compared in Table 9.3; both methods appear to yield similar calcium and phosphate contents. The existence of this so-called "colloidal calcium phosphate" was postulated as early as 1915 by Van Slyke and Bosworth,[100] who concluded that the nonprotein-bound colloidal calcium phosphate was present in a 1:1 molar ratio which approximates discalcium phosphate. Later workers[19,105] have calculated that the colloidal calcium phosphate more closely resembles tri-calcium phosphate with a Ca:PO_4 molar ratio of 1.5. Calculation

Table 9.3

TOTAL MINERAL COMPOSITION OF CASEIN MICELLES
mMOLES/100 GM CASEIN[a]

	Washed Micelles[b] by Centrifugation	Micelles by Gel Filtration[10]	Unwashed Micelles by Centrifugation[b]	Unwashed Micelles by Gel Filtration[10]
Calcium	69.6	68.9	71.0	79.0
	64.1			
Magnesium	4.2	3.3	4.5	6.9
Sodium	4.5			
Potassium	6.2	6.2		
Casein (PO_4)	22.2	28.2		
	23.2			
Inorganic (PO_4)	28.9	21.8	47.8	43.2
	27.8			
Citrate	1.6	0.0	6.2	4.7

[a] Casein N × 6.4.
[b] Adapted from Table 70, McMeekin and Groves, Chapter 9, "Fundamentals of Dairy Chemistry," 1st Edition.

of such a ratio after subtracting casein-bound calcium is subject to inherent error. Binding studies[21] on the isolated β- and κ-caseins show a good 1:1 correlation between calcium ions bound and phosphate residues, while α_{s1}-casein appears to have[8,21] 1 to 2 nonphosphate calcium binding sites. The application of these results allows the calculation of a Ca/PO_4 molar ratio from Table 9.3. The ratio obtained for washed micelles and micelles prepared by gel filtration are 1.6 and 1.8, respectively. The latter value differs from that calculated by Boulet *et al.*[10] because they assumed a 2:1 casein phosphate:calcium ion ratio. Thus, attempting an exact assignment of calcium to either the casein fraction or the colloidal calcium phosphate fraction can cause discrepancies in the calculated ratio. It must be realized that these data are average values based on average distributions of the caseins and the minerals. Not only do the mineral content and the casein distribution vary from one individual milk to another, but the various micelle fractions within a single sample are probably not of uniform composition.

It is clear from Table 9.3 that there are two distinct forms of ions associated with the casein micelle—an outer system, perhaps in the form of a charged double layer,[10] and an inner system not easily washed away. As noted above, the casein micelle is a highly porous, well solvated system, and the occlusion of ions within this network is not unexpected; however, some actual complex formation between the colloidal calcium phosphate and the casein cannot be ruled out. If one examines the pK's of phosphoric acid, it would seem most likely that the associating species of phosphate would be $(HPO_4)^{2-}$. Termine and Posner[85] studied the *in vitro* formation of calcium phosphate at pH 7.4, and concluded that an amorphous calcium phosphate phase (with a Ca:PO_4 molar ratio of 1.5) formed prior to the transition to crystalline apatite. In a subsequent study,[86] it was shown that casein and some other macromolecules enhanced the stability of the amorphous calcium phosphate and, in fact, retarded the amorphous → crystalline transition. It would appear then that conditions should favor the formation of an amorphous-calcium phosphate-caseinate complex in milk. The exact nature of this complex (or occlusion) is as yet undetermined, though its role in casein micelle stabilization is well documented.

Pyne and McGann[63] demonstrated that the colloidal calcium phosphate content of milk decreases as the pH is lowered from 6.7 to 5.0 at 5°C. If a small sample of pH 5.0 milk is then dialyzed at 5°C against several large volumes of the original milk, the pH returns to 6.7, but the colloidal calcium phosphate is no longer present. Milk brought to essentially zero colloidal calcium phosphate concentration at pH 5 and dialyzed back to 6.7 in this manner has been termed colloidal calcium phosphate-free milk (CPF milk). In a later study McGann and Pyne[48] investigated the properties of CPF milk as compared to the original nontreated milk. The CPF milk is translucent as compared to ordinary milk, and has a greatly increased viscosity. Addition of Ca^{2+} up to $\sim 1M$ has little effect on normal milk at 25°C, provided the increase in pH is not compensated for. CPF milks, however, are precipitated at added calcium-ion concentrations of only 25 mM. There is no apparent difference between CPF and normal milks with regard to the primary phase of rennin attack as measured by release of soluble nitrogen but, interestingly, the CPF milks are slightly more heat-stable. Finally, McGann and Pyne[48] noted that, at low temperatures, β-casein is more firmly bound to rennin-clotted normal milk than to rennin-clotted CPF milks. Jenness *et al.*[39] noted a marked increase in serum or nonmicellar casein, accompanied by an increased translucence as the colloidal calcium phosphate content of milk was reduced by the addition of EDTA. Rose[69] noted that, while Ca^{2+} addition generally decreases the serum casein content of milk, lowering the pH of milk to 5.3 and the subsequent release of Ca^{2+} actually increase the serum casein content. This result led Rose[69] to speculate that the colloidal calcium phosphate aids in maintaining micelle stability. CPF milks and normal milks were compared by Downey and Murphy[23] with respect to their elution volumes on gel chromatography (Sepharose 2B) in a synthetic milk serum. The normal casein micelles eluted at V_o yielded a molecular weight of $> 10^8$, but CPF micelles eluted at a volume consistent with a molecular weight of $\sim 2 \times 10^6$. However, this result could also be explained by a marked change in shape (frictional ratio).

All the discussion presented above indicates that colloidal calcium phosphate is involved in maintaining the structural integrity of the casein micelle. Occlusion of amorphous apatite or possible complexation of the mineral must occur, but the exact mechanism by which stabilization is achieved is as yet unknown.

Casein Micelle Structure

From the above discussion one might conclude that the individual caseins have been studied in sufficient detail to yield an intimate knowledge of the structure of the casein micelle. This is not the case, and nearly as many models have been proposed as there are investigators. Let us briefly consider why this situation exists. Electron microscopy (Fig. 9.3) of the casein micelles of bovine milk indicates an average diameter of $\sim$1,400 Å for the spherically shaped micelles. Thus, the volume occupied by a micelle would be of the order of $\sim 1.4 \times 10^9$ A^3. For comparison, the β-lactoglobulin monomer[96] occupies a volume

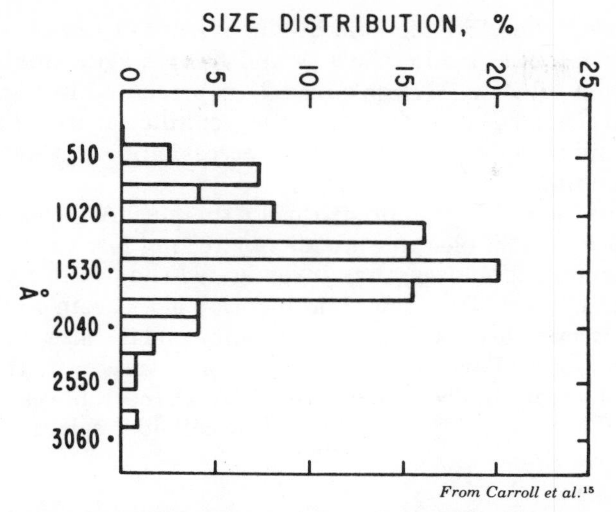

From Carroll et al.[15]

FIG. 9.3. DETERMINATION OF THE SIZE DISTRIBUTION OF GLUTARALDEHYDE-FIXED CASEIN MICELLES FROM SKIMMILK

of ~2.4 × 10^4 $Å^3$. Theoretically, more than 50,000 β-lactoglobulin-like monomers could be arranged into a sphere the size of a casein micelle. Molecular weight measurements for the micelle range from 10^7 to 10^9.[15,23] A speculative calculation, based on 23,000 average M.W. for the casein monomers $[(3\alpha_{s1} + 2\beta + 1\kappa)/6]$ and employing only 25,000 monomers yields a micelle molecular weight of 6 × 10^8. This would indicate low-density packing of the casein monomers, which is consistent with the high hydrations, the random structures and the high negative charge densities of the caseins, as compared to β-lactoglobulin. It is therefore understandable that the mechanism of assembly of this aggregate of around 25,000 monomers has not been fully elucidated. For the purpose of discussion, we shall group the various proposed models into three classes.

Coat-Core Models.—The first class of models to be discussed actually contains two diametrically opposed theories. The model proposed by Waugh and his co-workers[70,104] is primarily based upon their studies of the solubilities of the caseins in Ca^{2+} solutions. The model, in essence, describes the formation of low weight ratio complexes of α_{s1}- and κ-casein in the absence of calcium. Upon addition of calcium ion, the α_{s1} - or β-caseins, depicted as monomers with a charged phosphate loop in Fig. 9.4a, begin to aggregate to a limiting size (the caseinate core). In the presence of the low weight ratio α_{s1}-κ-complexes, precipitation of the casein is prevented by the formation of a monolayer of these low-weight α_{s1}-κ-complexes which envelops the core aggregates. This coat has the κ-casein monomers spread out on the surface, and the micelle size is therefore dictated by the amount of κ-casein available. In the absence of κ-casein, the α_{s1}- and β-cores agglutinate and precipitate from solution. Waugh's model, as presented in Fig. 9.4a, has a good deal of appeal since it explains the lyophilic nature of the colloidal casein complex, as well as the ready accessibility of κ-casein to the enzyme rennin.

Parry and Carroll[57] attempted to locate this surface κ-casein proposed by Waugh by use of electron microscopy. Using ferritin-labeled anti-κ-casein-immunoglobulins, they investigated the possiblity of surface κ-casein and found little or no concentration of κ-casein on the surface of the casein micelles. Based on these results, and the size of the isolated κ-casein complex, Parry concluded that the κ-casein might serve as a point of nucleation, about which the calcium-insoluble caseins might cluster and subsequently be stabilized by colloidal calcium phosphate

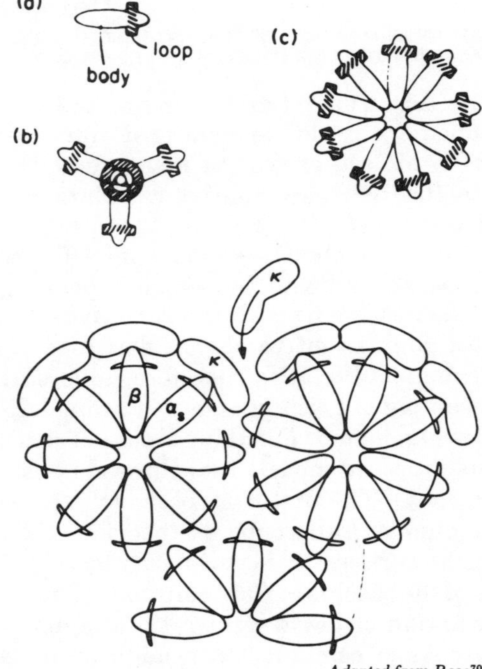

Adapted from Rose[70]

FIG. 9.4a. WAUGH'S PROPOSED MODEL FOR THE CASEIN MICELLE

(a)—Monomer model of α_{s1}- or β-casein with charged loop. (b)—A tetramer of α_{s1}-casein monomers. (c)—planar model of a core polymer of α_{s1}- and β-caseins. The lower portion shows how κ-casein might coat core polymers.

(see Fig. 9.4b). The action of rennin on the micelles was accounted for by demonstrating that serum κ-casein can participate in coagulation and may be involved in the formation of bridges between micelles.

The models of Parry and Waugh both predict a nonuniform distribution of κ-casein and in a sense are based upon nucleation about a core (Parry's core = κ-casein; Waught's core = α_{s1}-, β-calcium caseinate). It is important to note that both models predict no particular stoichiometry for the casein components and demonstrate no subunit structures composed of all three casein components. Secondly, Waugh's model does not incorporate any colloidal calcium phosphate which, as noted above, plays an important role in casein micelle structure and stability.

Finally, Ashoor *et al.*[2] have recently demonstrated that papain which had been cross-linked by glutaraldehyde into a large insoluble polymer caused proteolysis of all three major components of isolated casein micelles. The α_{s1}-, β- and κ-caseins were all cleaved proportionately by the enzyme superpolymer. Therefore, all three components must occupy surface positions on the micelle in relatively the same proportions in which they occur in milk. This result would seem to rule out any preferential localization of κ-casein.

Internal Structure Models.—The second class of models to be discussed are based upon the known properties of the isolated casein

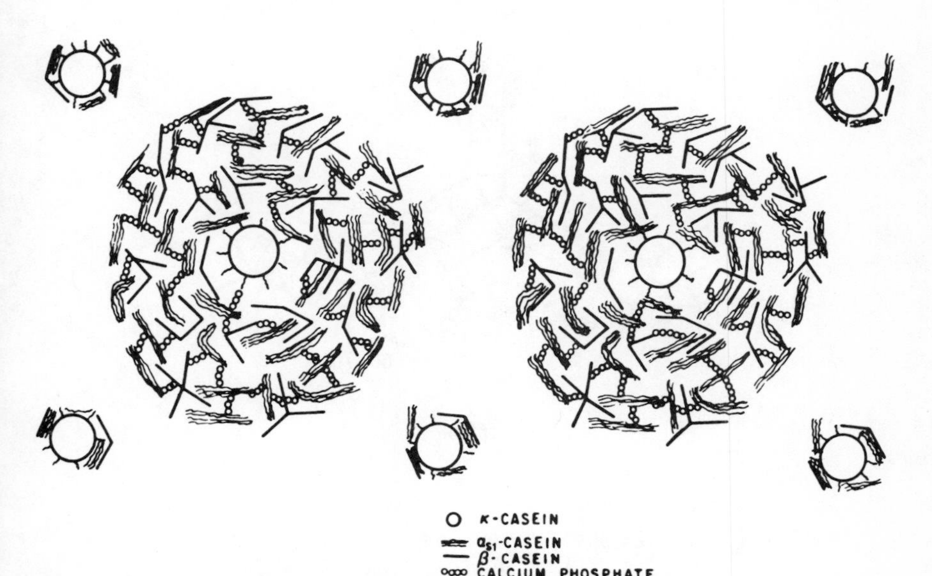

FIG. 9.4b. CASEIN MICELLE MODEL PROPOSED BY PARRY AND CAROLL[57] DEPICTING THE LOCATION OF κ-CASEIN IN THE MICELLE

components, which in turn cause or direct the formation of the internal structure of the casein micelle.

Garnier and Ribadeau-Dumas have proposed a model for the casein micelle[27] which places a good deal of emphasis on κ-casein as the keystone of micelle structure. Trimers of κ-casein are linked to three chains of α_{s1}- and β-casein which radiate from the κ-casein node (a Y-like structure), as shown in Fig. 9.5a. These chains of α_{s1}- and β-casein may connect with other κ-nodes to form a loosely packed network. Garnier and Ribadeau-Dumas favor this type of network because it yields an open, porous structure, and they have demonstrated[65] that carboxypeptidase-A with a molecular weight of ~40,000 is able to remove the C-terminal amino acids of all the casein components. The model satisfies the demonstrated porosity, but places great steric restraints upon κ-casein, which possesses no α-helical or other prominant secondary structures. In addition, studies by Cheeseman[17] and others[80,81] indicate that while disulfide-linked trimers of κ-casein do occur, the majority of the κ-casein may form aggregates of higher as well as lower orders. Finally, the model assigns no definite role to calcium caseinate interactions, and ignores the possibility of colloidal calcium phosphate involvement in stabilization of the micelle.

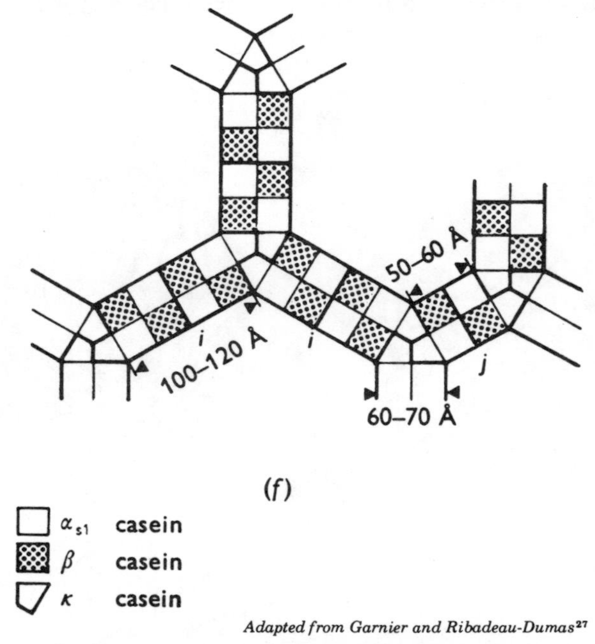

Adapted from Garnier and Ribadeau-Dumas[27]

FIG. 9.5a. STRUCTURE OF THE REPEATING UNIT OF THE CASEIN MICELLE

Rose used the known endothermic polymerization of β-casein as the basis for his micelle model.[70] In this model β-casein monomers begin to self-associate into chain-like polymers to which α_{s1}-monomers become attached (Fig. 9.5b), and κ-casein in turn interacts with the α_{s1}-monomers. The β-casein of the thread is directed inward, the κ-outward, but as these segments coalesce, a small amount of κ-casein is inevitably placed in an internal position. As the micelle is formed, colloidal calcium phosphate is incorporated into the network as a stabilizing agent. The model is appealing in that it accounts for the occurrence of some overall stoichiometry of the various casein components, while demonstrating the role of colloidal calcium phosphate in micelle stabilization. The choice of β-casein as the basis for micelle formation is, however, questionable since Waugh *et al*[104] have shown that α_{s1}- and β-caseins tend to form mixed polymers randomly; secondly, β-casein is quite structureless in solution; and finally, synthetic micelles can be formed from simple α_{s1}- and κ-casein complexes in the absence of β-casein.

Subunit Models.—The final class of models to be discussed is that which proposes subunit structure for the casein micelle. Shimmin and Hill[77] first postulated such a model based upon their study of ultrathin cross-sections of embedded casein micelles by electron microscopy. They predicted a diameter of 100 Å for the subunits of the casein micelle.

Morr[53] studied the disruption of casein micelles and proposed that the α_{s1}-β- and κ-monomers may be aggregated by calcium into small

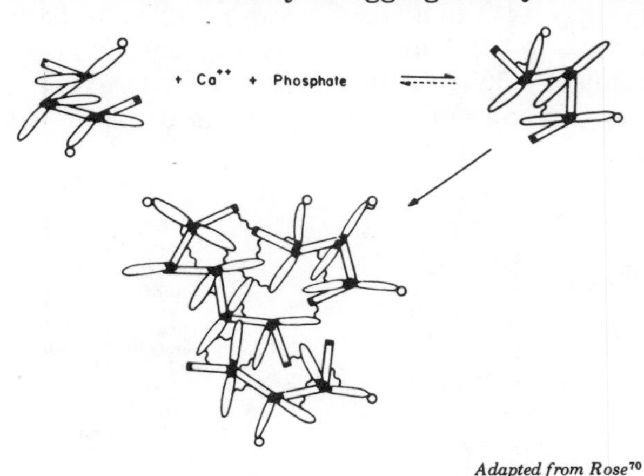

Adapted from Rose[70]

FIG. 9.5b. SCHEMATIC REPRESENTATION OF THE FORMATION OF A SMALL CASEIN MICELLE

The rods represent β-casein, the more eliptical rods represent α_{s1}-casein and the S-shaped lines depict apatite chain formation. The circles represent κ-casein.

subunits in much the same fashion as Waugh[104] had suggested for the entire micelle. Morr's subunits, as estimated by sedimentation velocity, have a diameter of ~ 300 Å. The subunits are stabilized by hydrophobic bonding and calcium caseinate bridges, and these subunits, in turn, are aggregated into micellar structure by colloidal calcium phosphate. Morr's model is summarized in Fig. 9.6. The average subunit size, postulated by Morr, is somewhat larger than that of Shimmin and Hill.

The hypothesis of Shimmin and Hill,[77] that sections of the casein micelles contain particles of ~100 Å in diameter, was invoked by Carroll *et al.*[13] and by Farrell and Thompson,[24] who observed by electron microscopy particles of ~100 Å diameter in the Golgi vacuoles of lactating rat mammary gland. These particles were uniform in size and appeared to form thread-like structures which, in turn, coalesced into the spherically shaped casein micelle (Fig. 9.7a, b). Subsequently, Beery *et al.*[4] reported similar observations in bovine mammary tissue. The biosynthesis of the casein micelle from small subunits was correlated with the disruption of casein micelles by dissociating agents by Carroll *et al.*[12] Using EDTA, urea, sodium lauryl sulfate, and sodium fluoride to disrupt micelles, the latter workers found particles of ~ 100 ± 20 Å diameter; and they noted that micelle assembly from subunits should lead to a rather uniform distribution of α_{s1}-, β- and κ-caseins both on the surface and in the interior of the casein micelle. Schmidt and Bucheim[74] dialyzed milk free of calcium in the cold and also used high pressure to disrupt casein micelles; in both cases they obtained subunits of 100 Å diameter. Subsequently, Pepper[59] reported a Stokes radius of ~50 Å for first-cycle (Ca^{2+} -free) casein as determined by gel filtration. The first-cycle casein, after gel filtration,

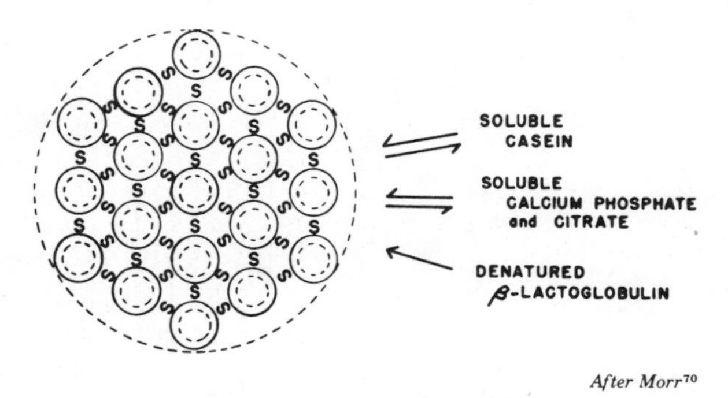

After Morr[70]

FIG. 9.6. STRUCTURE OF THE CASEIN MICELLE

The S-shaped lines represent calcium phosphate linkages between small spherical complexes of the α_{s1}-, β-, and κ-caseins.

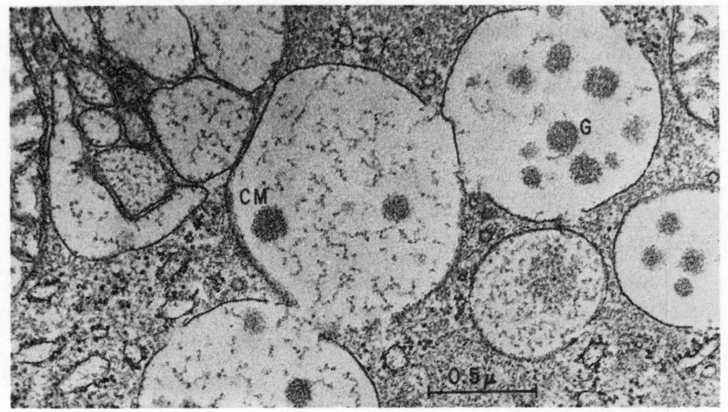

From Carroll et al.[13]

FIG. 9.7a. FORMATION OF CASEIN MICELLES (CM) WITHIN GOLGI VACUOLES (G) OF LACTATING RAT MAMMARY GLAND

Initially, thread-like structures with some degree of periodicity appear, then more compact micelles seem to occur. Sections of the gland were fixed in buffered OsO_4, Epon embedded, and stained with uranyl acetate and lead citrate.

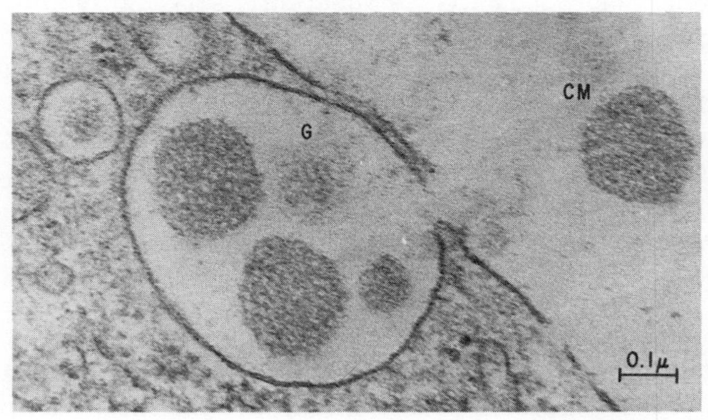

From Carroll et al.[13]

FIG. 9.7b. A GOLGI VACUOLE ABOUT TO DISCHARGE ITS CONTENTS INTO THE ALVEOLAR LUMEN

The golgi vacuole shown appears to impinge upon the plasma membrane. A casein micelle is already present in the lumen.

contained qualitatively all the major casein fractions. Therefore, the question yet to be resolved is whether or not these casein micelle subunits observed by all the above exhibit any stoichiometry in terms of their α_{s1}-, β-, and κ-casein content.

It has long been recognized that at least the α_{s1}-and κ-casein components occur in close association in the "α-casein complex," β-casein being more loosely connected to the micellar complex. Furthermore, the total micellar casein exhibits an overall ratio of 3 α_{s1}-:2β-:1 κ-casein. The apparent uniformity of first-cycle (Ca^{2+}-free) casein and the subunits of the Golgi vacuoles would favor some consistent stoichiometry, but there exists the reported correlation between micelle size and κ-casein content[69,71,104] which would argue against uniform subunit composition. Thus, the existence of some type of subunit structure appears certain, and the question to be decided now is the nature of these reported subunits.

From the biosynthetic point of view, the build-up of the micelle from subunits is quite attractive, as it brings the casein components into the region of assembly with minimal interactions. Addition of calcium ion could cause the polymerization of casein subunits into longer chains, which could be stabilized into micellar spheres by the deposition of colloidal calcium phosphate. The assembly of the casein micelle from preformed subunits need not be as specific as the mechanism of assembly of tobacco mosaic virus. In the latter case, its structured RNA core[11] plays a vital role in directing correctly the assembly of the virus particle from its preformed subunits, whereas in the case of the micelle, amorphous apatite apparently serves this function. In attempting to solve the problem of casein micelle structure, it should be born in mind that the biological function of the micelle is efficient nutrition. Hence, the interactions which yield this product, the casein micelle, need not be as specific as those which result in the formation of a virus or an enzyme.

Proteolytic Action and Micelle Models.—The action of rennin on the casein micelle is primarily hydrolysis of the highly sensitive phenylalanine-methionine peptide bond of κ-casein. Sequence data[40] show that 3 to 8 residues from this bond in either direction is the unusual sequence proline-proline, which perhaps accounts for the high susceptibility of this specific bond. Splitting of the bond[42] causes formation of the rather insoluble para-κ-casein and the highly hydrophilic (and carbohydrate-rich) peptide termed GMP or MP (glycomacropeptide or macropeptide). As a result of the action of rennin, the micelles coagulate or clot. All the casein micelle models must account for this phenomenon.

Waugh's model solves this problem most readily; since all the κ-casein is on the surface, the cleavage of GMP results in the loss of charge repulsions which normally prevent coagulation. Parry pointed out the possible role of serum κ-casein in the rennin reaction, and Garnier and Ribadeau-Dumas noted the possibility that rennin could penetrate a highly porous system. Recently, Ashoor *et al.*[2] have shown that papain, which had been insolubilized by glutaraldehyde cross-linkages, caused proteolysis of all three major components of casein micelles, demonstrating that not only κ- but α_{s1}- and β-casein occupy surface positions. Because of the high incidence of apolar residues in all the caseins and the large number of monomers, it is inconceivable that all the apolar chains would be buried and, therefore, many hydrophobic patches probably exist on the surface of the micelle. Normally, coagulation is prevented by the charged surface groups contributed by all three components. When a sufficient number of the κ-casein-GMP fragments have been removed by rennin action, coagulation or clotting could occur, with the addition of serum para-κ-bridges, as postulated by Parry.[57] In addition, internal hydrolysis of all casein components (and this too occurs) would cause the porous micelle to dehydrate on loss of internal macropeptide, leading to additional destabilization of the system. This latter phenomenon may be as important as those occurring at the surface.

What has been noted above in regard to the specific enzyme rennin would also be true of any protease which would be added to, or naturally occur in, milk. Chen and Ledford[18] have studied the "milk protease" and found it to be trypsin-like. Limited proteolysis by this or by other as yet uncharacterized proteases or their reactivation upon storage may be the cause of many of the problems encountered in the storage of processed whole or concentrated milks and dairy products.

THE MILK SERUM PHASE

The Serum Proteins

The proteins of the dispersed phase of milk, the casein micelles, account for up to 74% of the total protein of skimmilk (Table 9.4). The serum or whey proteins are those which remain in solution after

Table 9.4

AVERAGE COMPOSITION OF WARM SKIMMILK PROTEIN[a]

	gm/100 gm Milk[b]	% Total Protein[b]
Colloidal casein	2.36	74
Serum casein	0.26	8
β-Lactoglobulin	0.29	9
α-Lactalbumin	0.13	4
Bovine serum albumin	0.03	1
Total immunoglobulins	0.06	2
Other proteins	0.06	2

[a] Averaged from data of several sources.[69,71,72,93]
[b] All values normalized to 3.2 gm total protein/100 gm milk.

the micelles are removed; these proteins are not incorporated into colloidal complexes. Depending upon the method of removal of the casein micelles, varying amounts of serum casein will be included in this fraction. The major noncasein serum proteins are: β-lactoglobulin, α-lactalbumin, bovine serum albumin and the milk immunoglobulins.[72]

β-Lactoglobulin.—β-Lactoglobulin is the major serum protein; it accounts for up to 50% of the noncasein protein of skimmilk and exhibits quite unique structural features. β-Lactoglobulin has a monomer molecular weight of 18,000 daltons; Frank and Braunitzer[25] have published a rather complete amino-acid sequence for the molecule. Although X-ray crystallography[30] has been done on β-lactoglobulin, high resolution data has not been obtained. However, from physical data and chemical modification experiments, much has been deduced concerning the topography of the molecule,[99] as well as its rather unique mode of association. The 18,000 dalton monomer occurs only below pH 3.5 and above pH 7.5. Between pH 3.5 and pH 7.5, β-lactoglobulin exists as a dimer with a molecular weight of 36,000; its conformation has been well characterized.[98] At reduced temperatures, between pH 3.7 and 5.1, the dimers of the β-lactoglobulin A genetic variant form a specific octamer with 422 symmetry.[96] β-Lactoglobulin B does not undergo this self-association to an appreciable extent, and the C genetic variant does not octamerize at all.[96,98] The conformational transitions, which bring about these associations, have been well documented, and the analogy of these interactions to allosteric control mechanisms has been noted.[97] The precise biological function of this molecule is unknown, though several hypotheses have been advanced.[24]

Because of its ready availability and unique properties, β-lactoglobulin has been used as a model protein in many studies. However, the major noncasein protein of milk, β-lactoglobulin, also plays an important part during processing as the result of two of its features. In the first place, β-lactoglobulin has well-defined secondary, tertiary and quaternary structures which are susceptible to denaturation. Secondly, one free sulfhydryl group occurs per 18,000 monomer.[46] Thus the integrity of the β-lactoglobulin molecule must be retained in part, in order to prevent coagulation as the result of irreversible denaturation; and conditions which affect free sulfhydryl groups or cause disulfide interchange with other proteins must be avoided.

Protein denaturation has been reviewed by Tanford.[84] Environmental influences that cause reversible denaturation, when carried to an extreme, generally produce some irreversible denaturation (see Coagulation Chapter 11). In discussing the phenomenon of denaturation, Tanford[84] outlined the contributions of conformational, inter-chain, protein-solvent, and electrostatic interactions to the free energy of the transition from the native (N) to the denatured (D) state. He considered both the influence of environmental factors, such as temperature and pressure, and also of binding effects on the N → D transition. Increases of temperature and pressure, as well as the binding of denaturants (e.g., guanidine, urea, and detergents), the binding of inorganic ion, and the binding of hydrogen ion (pH) may facilitate the transition to the denatured state. Tanford further notes that each type of denaturing agent leads to a particular denatured state. In each case, the resulting polypeptide chain has more exposed apolar groups, which increases the probability of aggregation unless these areas are stabilized by the solvent (e.g., urea and guanidine). Denaturation in aqueous solution often leads to aggregation and, finally, precipitation or coagulation. It is this latter event which one wishes to prevent in the processing of dairy products. Carroll *et al.*[14] demonstrated by the use of electron microscopy that denaturation and subsequent precipitation of whey proteins is the basis for the gelation of high-temperature, short-time, sterilized, concentrated milk. Hence, it would appear that environmental influences, which tend to preserve the native structure of β-lactoglobulin and the other whey proteins, would favor stability; conversely, denaturation and subsequent coagulation would lead to instability. Factors which influence the denaturation of β-lactoglobulin have been reviewed by Tilley[94] and McKenzie.[49] Generally speaking, heat, pH above 8.6,[94] and increased calcium ion concentration[108] tend to increase the N → D transition of β-lactoglobulin.

A second source of instability in β-lactoglobulin is its unique sequence[24] -Cys-Cys-. It has been reported[49] that residue 69 is the free sulfhydryl group of β-lactoglobulin, while residue 70 is involved in a disulfide bond. Hence, exposure to strong base or heat may promote disulfide interchange and lead to denaturation. McKenzie[49] has summarized the work of several groups who have demonstrated that sulfhydryl blocking reagents increase the stability of β-lactoglobulin. The free sulfhydryl group of β-lactoglobulin has been implicated in the formation of complexes with κ-casein upon heating.[109] Townend *et al.*[99] pointed out that in the native state this residue is partially buried; hence, some degree of denaturation must occur in order to totally expose this sulfhydryl residue. Since molecules such as carboxypeptidase and myoglobin[65] can penetrate the casein micelle, β-lactoglobulin may likewise do so, thus inducing formation of disulfide-bonded complexes with κ-casein upon heating. Such complexes would alter the stability of the casein micelle.

α-Lactalbumin.—The best-characterized milk protein is α-lactalbumin. Gordon[28] has recently reviewed the properties of α-lactalbumin, which accounts for up to 25% of whey protein and ~4% of total milk protein (Table 9.4). α-Lactalbumin has been assigned a unique biochemical role as the specifier protein of the lactose synthetase system. The complete sequence of α-lactalbumin is known[101] and the molecule exhibits a strong structural relationship to lysozyme. α-Lactalbumin is quite stable, with 4 disulfide cross-links for a monomer molecular weight of 14,000. Kronman[45] has demonstrated that α-lactalbumin is denatured at low pH and then undergoes an association reaction, which requires somewhat elevated protein concentration and may not occur in milk products. However, in whey concentrates, acid denaturation of α-lactalbumin may play an important role in the loss of its functional properties. Hunziker and Tarassuk[38] have reported that the free sulfhydryl of β-lactoglobulin can promote complex formation with α-lactalbumin through disulfide reactions.

Other Serum Proteins.—Serum albumin and immunoglobulins[72] occur in skimmilk to a limited extent; in conjunction with various enzymes,[76] they may account for up to 4% of total milk protein. All these proteins contain a significant amount of native structure and are susceptible to various forms of denaturation. In fact, limited protein denaturation is a desired result in terms of the required inactivation of many of the enzymes which have been noted to occur in milk. The reactivation of proteolytic and other enzymes, upon storage of processed dairy products, is undoubtedly the source of many problems. Hence, it appears as though conditions must be controlled so as to adequately inactivate enzymes, and reduce bacterial contamination without causing severe denaturation and consequent coagulation of the major whey proteins.

The Serum Caseins.—Not all the casein secreted by the lactating mammary cells is incorporated into casein micelles. Rose[69] and Downey and Murphy[23] have studied the occurrence and distribution of the serum caseins. Rose[69] found that in warm milk the serum caseins, on the average, account for about 10% of the total casein, and the serum casein contained all the major components of the micelle in varying proportions, but always in the order β->κ->α_{s1}-casein. In cold milks the percentages of β- and κ-casein in the serum increase as the serum casein increases. Rose has concluded that serum casein does not appear to be in true equilibrium with micellar casein,[69] and this conclusion has been supported by Downey and Murphy[23] and Boulet *et al.*[10]

Serum caseins may play an important role in the destabilization of casein micelles. The serum caseins are not stabilized by colloidal calcium phosphate; they would be readily attacked by proteolytic enzymes and would be more susceptible to other environmental factors. The serum caseins would also be more likely to interact with β-lactoglobulin or other whey proteins. Parry and Carroll[57] have proposed that serum κ-casein is more readily attacked by rennin and plays an important role in the rennin clotting reaction.

Salt Content of Milk Serum

The salt content of milk serum, separated by several methods, has been determined by Davies and White.[19] Their average results for two milks, obtained by separating milk serum by diffusion, centrifugation, and clotting with rennet, are compared in Table 9.5. These values are in essential agreement with each other and with the very extensive previous results on the composition of the nonprotein aqueous phase of milk by the same authors.[105] Recently,[20] calcium ion-specific electrodes have been used to determine the free Ca^{2+} concentration of skimmilk.

Table 9.5

DISTRIBUTION OF SALTS (IN MG/100 GM MILK) BETWEEN DISSOLVED AND COLLOIDAL STATE IN MILK[a]

	Dissolved				Colloidal State
	Total in Milk	Diffusate (20°C)	Rennet Whey	Centrifuged Serum	(Average 3 Methods)
Total calcium	114.2	38.1	39.9	40.9	74.6
Ionized calcium	10.3[c]	11.7	11.6	11.9	—
Magnesium	11.0	7.4	7.8	8.1	3.3
Sodium	50	46	47	47	3.3
Potassium	148	137	143	141	8.0
Total phosphorus	84.8	37.7	37.4	37.9	47.1
Inorganic phosphorus	—	31.8	30.8	31.8	—
Citric acid	166	156	152	154	12.0
Chloride	106	106.5	106.2	105.6	—
Total nitrogen	—	20.7	124.6	110.7	—
Casein nitrogen	364	0	21.6	6.8	—
Lactose[b]	4800	4800	4800	4800	—

[a] Adapted from Davies and White.[19]
[b] Average of two separated milks, corrected for bound water.
[c] By specific ion electrode calculated from the data of Demott.[20]

Equilibria Between the Colloidal and the Serum Phases

Because of the complex nature of the milk system, the question of whether or not true equilibria occur is difficult to assess. The Handbook of Chemistry and Physics defines equilibrium as " . . . . . the state of affairs in which a reaction and its reverse reaction are taking place at equal velocities, so that the concentration of reactants is constant." Such conditions, in terms of physical equilibria, must exist for some milk components, but do not occur for others. For the sake of discussion, let us assume that the casein micelle is a porous, highly hydrated complex, with some degree of subunit structure, and that the micelle contains colloidal calcium phosphate, and is surrounded by a double layer of ions. Clearly, the outer layers of ions can be in equilibrium with those of the solvent, and these ions are removed readily by gel filtration.[10] The bulk of the inorganic ions of milk are definitely not in equilibrium with the environment as they are occluded in the dispersed phase and are not readily dialyzed away[19,105] or removed by gel filtration.[10] The work of Rose[69] indicates that, at room temperature and above, the serum and micellar caseins are also not in equilibrium. However, at lower temperatures,[16,69,79] β-casein readily dissociates from the micelle and enters the serum phase, together with κ-casein and some α_{s1}-casein. These fractions could be in equilibrium because, as the milk is warmed, these caseins return to the colloidal phase; however, whether true microscopic reversibility occurs or not is questionable, as the opportunity for hysteretic effects is enormous in the milk system. Even at low temperatures, a large part of the β-casein, and most of the α_{s1}-and κ-caseins, are not removed from the casein micelle. The serum caseins, while they are apparently not in equilibrium with the micellar casein, are probably in equilibrium with the ions of the serum phase. The addition of calcium decreases the net amount of serum casein,[69] and added calcium has also been reported to decrease the heat stability of milk.[62] Hence, effects on the serum proteins may destabilize the entire system. Also, the serum proteins, such as β-lactoglobulin, must be in equilibrium with the ionic environment; again added salts appear to affect the stability of these serum proteins.[108] If the micelle is sufficiently porous to admit carboxypeptidase,[65] β-lactoglobulin and the other serum proteins should equilibriate within the micellar phase as well. The water of the milk system, as well as the lactose, should be in equilibrium between the dispersed and the serum phase, but some internal water of hydration may not be, as the occluded calcium phosphate undoubtedly affects the hydration, and hence the heat stability,[92] of the milk system.

Generally speaking, many equilibrium situations occur in skimmilk as well as several distinct systems that are not in equilibrium. The innate stability of this biological fluid may depend on the correct balance between these states. Conditions which perturb the equilibrium states temporarily affect the milk system; but overall the nonequilibrium states, which include colloidal calcium phosphate, the serum caseins, and the water of hydration, appear to contribute most to the stability of milk. Hence, environmental influences which disrupt the nonequilibrium states tend to cause greater destabilization of milk.

REFERENCES

1. ASCHAFFENBURG, R., Nature, *192,* 431 (1961).
2. ASHOOR, S. H., SAIR, R. A., OLSON, N. F., and RICHARDSON, T., Biochim. Biophys. Acta, *229,* 423 (1971).
3. BEEBY, R., Biochim. Biophys. Acta, *82,* 418 (1964).
4. BEERY, K. E., HOOD, L. F., and PATTON, S., J. Dairy Sci., *54,* 911 (1971).
5. BEYCHOK, S., Ann. Rev. Biochem., *37,* 448 (1968).
6. BIGELOW, C. C., J. Theoret. Biol., *16,* 187 (1967).
7. BINGHAM, E. W., J. Dairy Sci., *54,* 1077 (1971).
8. BINGHAM, E. W., FARRELL, H. M., JR., and CARROLL, R. J., Biochemistry, *11,* 2450 (1972).
9. BLOW, D. M., and STEITZ, T. A., Ann. Rev. Biochem., *39,* 63 (1970).
10. BOULET, M., YANG, A., and RIEL, R. R., Can. J. Biochem., *48,* 816 (1970).
11. BUTLER, P. J. G., Nature, *233,* 25 (1971).
12. CARROLL, R. J., FARRELL, H. M., JR., and THOMPSON, M. P., J. Dairy Sci., *54,* 752 (1971).
13. CARROLL, R. J., THOMPSON, M. P., and FARRELL, H. M., 28th Annual EMSA Proceedings. (1970), p. 150.
14. CARROLL, R. J., THOMPSON, M. P., and MELNYCHYN, P., J. Dairy Sci., *54* 1245 (1971).
15. CARROLL, R. J., THOMPSON, M. P., and NUTTING, G. C., J. Dairy Sci., *51,* 1903 (1968).
16. CARROLL, R. J., THOMPSON, M. P., BRUNNER, J. R., and KOLAR C., J. Dairy Sci., *50,* 941 (1967).
17. CHEESEMAN, G. C., J. Dairy Res., *35,* 439 (1968).
18. CHEN, J. H., and LEDFORD, R. A., J. Dairy Sci., *54,* 763 (1971).
19. DAVIES, D. T., and WHITE, J. C. D., J. Dairy Res., *27,* 171 (1960).
20. DEMOTT, B. J., J. Dairy Sci., *51,* 1008 (1968).
21. DICKSON, I. R., and PERKINS, D. J., Biochem. J., *124,* 235 (1971).
22. DIXON, M., and WEBB, E. C., "Enzymes," 2nd Ed., p. 672, Academic Press, New York, 1964.
23. DOWNEY, W. K., and MURPHY, R. F., J. Dairy Res., *37,* 361 (1970).
24. FARRELL, H. M., JR., and THOMPSON, M. P., J. Dairy Sci., *54,* 1219 (1971).
25. FRANK, G., and BRAUNITZER, G., Hoppe-Seyler's Z. Physiol. Chem., *348,* 1691 (1967).
26. FRIEDEN, C., Ann. Rev. Biochem., *40,* 653 (1971).
27. GARNIER, J., and RIBADEAU-DUMAS, B., J. Dairy Res. *37,* 493 (1970).
28. GORDON, W. G., in "Milk Proteins," Vol. 2, p. 331, Ed. by H. A. McKenzie, New York, 1971.
29. GORDON, W. G., BASCH, J. J., and THOMPSON, M. P., J. Dairy Sci., *48,* 1010 (1965).
30. GREEN, D. W., and ASCHAFFENBURG, R., J. Molec. Biol., *1,* 54 (1959).
31. GROSCLAUDE, F., MERCIER, J. C., and RIBADEAU-DUMAS, B., Eur. J. Biochem., *14,* 98 (1970).

32. GROVES, M. L., and GORDON, W. G., Biochim. Biophys. Acta, *194,* 421 (1969).
33. GROVES, M. L., and GORDON, W. G., Federation Proc., *30,* 1274 (1971).
34. HERSKOVITS, T. T., Biochemistry, *5,* 1018 (1966).
35. HESS, G. P., and RUPLEY, J. A., Ann Rev. Biochem., *40,* 1013 (1971).
36. HILL, R. D., J. Dairy Res., *37,* 187 (1970).
37. HILL, R. J., and WAKE, R. G., Nature, *221,* 635 (1969).
38. HUNZIKER, H. G., and TARASSUK, N. P., J. Dairy Sci., *48,* 733 (1965).
39. JENNESS, R., MORR, C. V., and JOSEPHSON, R. V., J. Dairy Sci., *49,* 712 (1966).
40. JOLLÈS, J., JOLLÈS, P., and ALAIS, C., Nature, *222,* 668 (1969).
41. JOLLÈS, P., ALAIS, C., and JOLLÈS, J., Arch. Biochem. Bisphys., *98,* 56 (1962).
42. KALAN, E. B., and WOYCHIK, J. H., J. Dairy Sci., *48,* 1423 (1965).
43. KAUZMAN, W., Adv. Protein Chem., *14,* 1 (1959).
44. KLOTZ, I. M., Arch. Biochem. Biophys., *138,* 704 (1970).
45. KRONMAN, M. J., ANDREOTTI, R. E., and VITOLS, R., Biochemistry, *3,* 1152 (1964).
46. LARSON, B. L., and JENNESS, R., J. Dairy Sci., *34,* 483 (1951).
47. LINDERSTRØM-LANG, K., Compt. rend. trav. lab. Carlsberg, *17,*no. 9, 1 (1929).
48. McGANN, T. C. A., and PYNE, G. T., J. Dairy Res., *27,* 403 (1960).
49. McKENZIE, H. A., "Milk Proteins," Vol. 2, p. 257, Academic Press, New York, 1971.
50. McKENZIE, H. A., and WAKE, R. G., Aust. J. Chem., *12,* 734 (1959).
51. MELLANDER, O., Biochem. Z., *300,* 240 (1939).
52. MERCIER, J. C., GROSCLAUDE, F., and RIBADEAU-DUMAS, B., Eur. J. Biochem., *23,* 41 (1971).
53. MORR, C. V., J. Dairy Sci., *50,* 1744 (1967).
54. NITSCHMANN, Hs., and HENZI, R., Helv. Chim. Acta, *42,* 1985 (1959).
55. NOELKEN, M., Biochim. Biophys. Acta, *140,* 537 (1967).
56. NOELKEN, M., and REIBSTEIN, M., Arch. Biochem. Biophys., *123,* 397 (1968).
57. PARRY, R. M., JR., and CARROLL, R. J., Biochim. Biophys. Acta, *194,* 138 (1969).
58. PAYENS, T. A. J., BRINKHUIS, J. A., and VAN MARKWIJK, B. W., Biochim. Biophys. Acta, *175,* 434 (1969).
59. PEPPER, L., Biochim. Biophys. Acta, *278,* 147 (1972).
60. PEPPER, L., and THOMPSON, M. P., J. Dairy Sci., *46,* 764 (1963).
61. PEPPER, L., HIPP, N. J., GORDON, W. G., Biochim. Biophys. Acta, *207,* 340 (1970).
62. PURI, B. R., ARORA, K., and TOTEJA, K. K., Indian J. Dairy Sci., *22,* 85 (1969).
63. PYNE, G. T., and McGANN, T. C. A., J. Dairy Res., *27,* 9 (1960).
64. RIBADEAU-DUMAS, B., personal communication.
65. RIBADEAU-DUMAS, B., and GARNIER, J., J. Dairy Res., *37,* 269 (1970).
66. RIBADEAU-DUMAS, B., GROSCLAUDE, F., and MERCIÉR, J. C., Eur. J. Biochem., *14,* 451 (1970).
67. RIBADEAU-DUMAS, B., BRIGNON, G., GROSCLAUDE, F., and MERCIÉR, J. C., Eur. J. Biochem., *25,* 505(1972).
68. ROSE, D., J. Dairy Sci., *44,* 430 (1961).
69. ROSE, D., J. Dairy Sci., *51,* 1897 (1968).
70. ROSE, D., Dairy Sci. Abstr., *31,* 171 (1969).
71. ROSE, D., DAVIES, D. T., and YAGUCHI, M., J. Dairy Sci., *52,* 8 (1969).
72. ROSE, D., BRUNNER, J. R., KALAN, E. B., LARSON, B. L., MELNYCHYN, P., SWAISGOOD, H. E., and WAUGH, D. F., J. Dairy Sci., *53,* 1 (1970).
73. SCHMIDT, D. G., Biochim. Biophys. Acta, *207,* 130 (1970).
74. SCHMIDT, D. G., and BUCHHEIM, W., Milchwissenshaft, *25,* 596 (1970).
75. SCHMIDT, D. G., and VAN MARKWIJK, B. W., Biochim. Biophys. Acta, *154,* 613 (1968).
76. SHAHANI, K. M., J. Dairy Sci., *49,* 907 (1966).
77. SHIMMIN, P. D., and HILL, R. D., J. Dairy Res., *31,* 121 (1964).
78. SIPOS, T., and MERKEL, J. R., Biochemistry, *9,* 2766 (1970).
79. SULLIVAN, R. A., FITZPATRICK, M. M., STANTON, E. K., ANNINO, R., KISSEL, G., and PALERMITI, F., Arch. Biochem. Biophys., *55,* 455 (1955).

80. SWAISGOOD, H. E., and BRUNNER, J. R., J. Dairy Sci., *45,* 1 (1962).
81. SWAISGOOD, H. E., and BRUNNER, J. R., Biochem. Biophys. Res. Commun., *12,* 148 (1963).
82. SWAISGOOD, H. E., and TIMASHEFF, S. N., Arch. Biochem. Biophys., *125,* 344 (1968).
83. TANFORD, C., "Physical Chemistry of Macromolecules," p. 131, John Wiley and Sons, New York, 1961.
84. TANFORD, C., Adv. Protein Chem., *24,* 1 (1970).
85. TERMINE, J. D., and POSNER, A. S., Arch. Biochem. Biophys., *140,* 307 (1970).
86. TERMINE, J. D., PECKAUSKAS, R. A., and POSNER, A. S., Arch. Biochem. Biophys., *140,* 318 (1970).
87. THOMPSON, M. P., J. Dairy Sci., *53,* 1341 (1970).
88. THOMPSON, M. P., In "Milk Proteins" Ed. by H. A. McKenzie, Vol. 2, p. 117, Academic Press, New York, 1971.
89. THOMPSON, M. P., FARRELL, H. A., JR., and GREENBERG, R., Comp. Biochem. Physiol., *28,* 471 (1969).
90. THOMPSON, M. P., GORDON, W. G., BOSWELL, R. T., and FARRELL, H. M., JR., J. Dairy Sci., *52,* 1166 (1969).
91. THOMPSON, M. P., KIDDY, C. A., PEPPER, L., and ZITTLE, C. A., Nature, *195,* 1001 (1962).
92. THOMPSON, M. P., BOSWELL, R. T., MARTIN, V., JENNESS, R., and KIDDY, C. A., J. Dairy Sci., *52,* 796 (1969).
93. THOMPSON, M. P., TARASSUK, N. P., JENNESS, R., LILLEVIK, H. A., ASHWORTH, U. S., and ROSE, D., J. Dairy Sci., *48,* 159 (1965).
94. TILLEY, J. M. A., Dairy Sci. Abstr., *22,* 111 (1960).
95. TIMASHEFF, S. N., and GORBUNOFF, M. J., Ann Rev. Biochem., *35,* pt. 1, 13 (1967).
96. TIMASHEFF, S. N., and TOWNEND, R. E., Nature, *203,* 517 (1964).
97. TIMASHEFF, S. N., and TOWNEND, R., In "Protides of the Biological Fluids: Proceedings of the 16th Colloquium, Bruges, 1968," Ed. by H. Peeters, p. 33, Pergamon, New York, 1969.
98. TOWNEND, R., KUMOSINSKI, T. F., and TIMASHEFF, S. N., J. Biol. Chem., *242,* 4538 (1967).
99. TOWNEND, R., HERSKOVITS, T. T., TIMASHEFF, S. N., and GORBUNOFF, M. J., Arch. Biochem. Biophys., *129,* 567 (1969).
100. VAN SLYKE, L. L., and BOSWORTH, A. W., J. Biol. Chem., *20,* 135 (1915).
101. VANAMAN, T. C., BREW, K., and HILL, R. L., J. Biol. Chem., *245,* 4583 (1970).
102. WAUGH, D. F., and VON HIPPEL, P. H., J. Am. Chem. Soc., *78,* 4576 (1956).
103. WAUGH, D. F., SLATTERY, C. W., and CREAMER, L. K., Biochemistry, *10,* 817 (1971).
104. WAUGH, D. F., CREAMER, L. K., SLATTERY, C. W., and DRESDNER, G. W., Biochemistry, *9,* 786 (1970).
105. WHITE, J. C. D., and DAVIES, D. T., J. Dairy Res., *25,* 236 (1958).
106. WOYCHIK, J. H., Arch. Biochem. Biophys., *109,* 542 (1965).
107. WOYCHIK, J. H., KALAN, E. B., and NOELKEN, M. E., Biochemistry, *5,* 2276 (1966).
108. ZITTLE, C. A., DELLAMONICA, E. S., RUDD, R. K., and CUSTER, J. H., J. Am. Chem. Soc., *79,* 4661 (1957).
109. ZITTLE, C. A., THOMPSON, M. P., CUSTER, J. H., and CERBULIS, J., J. Dairy Sci., *45,* 807 (1962).

CHAPTER 10

J. R. Brunner*

Physical Equilibria in Milk: the Lipid Phase

INTRODUCTION

A detailed discussion of the chemical characteristics of the components of the lipid phase of cow's milk is presented in Chapters 4 and 5. Factors related to the dispersion and physical state of the lipid phase are the subject matter of this chapter. A definitive differentiation between physical and chemical properties represents a formidable task, since changes in the chemical characteristics of the lipid phase frequently influence its physical attributes. The following subject areas have been selected for particular emphasis: (a) characteristics of the fat emulsion; (b) physical equilibria in milkfat; (c) nature of the fat-globule surface; and (d) the lipid phase in processed dairy products.

Obviously, a detailed treatment of all aspects of these areas has not been attempted. The following references are recommended for supplemental reading: "The Milkfat Globule Membrane,"[187] "Melting and Solidification of Fats,"[11] "Lipid Chemistry,"[120] and "Principles of Dairy Chemistry."[164]

CHARACTERISTICS OF THE FAT EMULSION

At the temperature of milk at the time of its secretion, milkfat exists as a microscopic, immiscible emulsion of liquid fat in aqueous phase of milk plasma. Because of the surface forces inherent in this system, the fat particles take the form of finely divided spheres, stabilized by a third phase oriented at the interfacial surface, usually referred to as the fat-globule membrane. At temperatures low enough to initiate crystallization in the fat phase, the higher-melting glycerides appear to crystallize in the peripheral layers of the fat globule. As crystallization proceeds, the composition of the interfacial surface changes, the once spherical fat globules become distorted, and the fat phase assumes characteristics of a solid suspension.

* J. R. Brunner, Department of Food Science and Human Nutrition, Michigan State University, East Lansing, Michigan 48823

Size and Distribution of the Fat Emulsion

Being of microscipic dimension, it is not surprising that fat globules were first observed by van Leeuwenhoek,[213] who compared them to similar-appearing globules he had observed in blood. Bouchardt and Quevenne[27] were among the first of the early investigators to recognize that the fat globules vary in size, reflecting species, breed, and individual characteristics. They measured fat globules as large as 22.2 μm and suggested that globules as small as 0.1 μm existed. However, most of the fat globules found in normal, unagitated cow's milk are smaller than 4 μm and seldom do they exceed 10 μm in diameter. Globules smaller than 1.0 μm in diameter are difficult to identify and enumerate.

The classical procedure for evaluating the size and distribution of fat globules in milk utilizes microscopic counting techniques. Samples of milk are prepared for observation by dilution (50- to 200-fold) with a 50/50 mixture of glycerol and water or a 2 to 10% solution of gelatin.[48,205] Thus, adequate dispersion of the fat globules is assured and Brownian activity is suppressed. The specimen is mounted and observed at magnifications appropriately selected to identify the full range of globule sizes. Where quantitative counting is desired, known quantities of milk must be examined. Campbell[48] found that a haemocytometer, designed for the microscopic counting of blood cells, adequately served this purpose. Babcock[9] used thin, calibrated capillary tubes to hold his milk samples. Size is estimated by comparison with a calibrated ocular micrometer. To assist in the identification and measurement of small globules, Campbell attached a projector to the eyepiece of the microscope and projected the microscopic field onto a previously calibrated screen. In other studies, counts were made from photographs of the microscopic fields.[3,60,278,288] Although this technique would appear to offer advantages over direct ocular observations, van Kreveld[205] concluded that measurements made from photomicrographs were less adequate. However, recent improvements in electronic photograph scanning techniques as proposed by Andreev[3] offer renewed possibilities for further development of the procedure.

Several hundred fat globules representing numerous microscopic fields are measured and tabulated as frequency data into conveniently selected size groups (e.g., 0.5 to 1.0 μm, 1.0 to 1.5 μm, etc). These size-distribution profiles illustrate the distribution of the fat phase relative to fat-globule size. But since the volume of a sphere is a function of the cube of its diameter ($\pi d^3/6$), the number or percentage of different size globules provides an inadequate image of the distribution of the fat content. The relationship of globular diameter to volume is illustrated by the data of van Kreveld[205] (see Fig. 10.1.) Obviously, as the volume expression predicts, the bulk of the fat in milk is represented by the larger globules. Although more numerous, the small globules (<1 μm) represent but a small portion of the fat.

The fat globules in different milks are frequently compared on the basis of their average diameter and volume. The most nearly representative values are derived from their respective size-distribution profiles and are computed as follows:

$$\bar{d} = \Sigma nd/N$$

and

$$\bar{V} = \Sigma \pi nd^3/6N$$

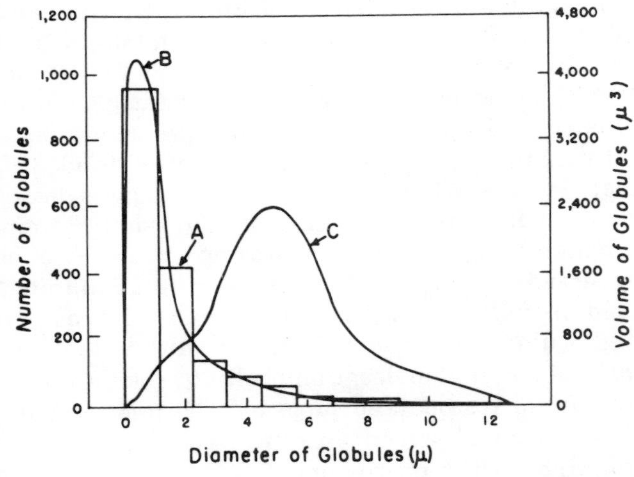

Data of van Kreveld[205]

FIG. 10.1. SIZE AND VOLUME DISTRIBUTION OF FAT GLOBULES IN COW'S MILK

A—Histogram of counting. B—Smoothed, size-distribution curve. C—Volume-distribution curve.

where d = mean average diameter, $\bar{V}$ = mean average volume, n = number or percentage of globules in each size class, N = total number of globules observed, and d = diameter at midpoint of size class.[164]

The accuracy of the values obtained by these computations reflects the reliability of the data represented in the size-distribution profile. Campbell[48] and van Kreveld[205] observed a skewed distribution favoring the small globules in individual size categories. Campbell concluded, as a result of statistical testing, that for 1 μm classes the correction required to compensate for the skewed distribution characteristics was small and consistent enough to ignore. van Kreveld[206,207] made a critical evaluation of the microscopic counting technique and concluded that it provided remarkedly reproducible measurements. He attributed this experimental reliability to the fact that fat globules exhibit a regularity of spatial arrangement due to an inherent repelling charge which seemed to be effective at distances ranging up to 50 μm. Walstra[407] reported a range of from 6 to 25 μm (mean free distance of 12 μm) as the distance separating individual fat globules in whole milk.

The size-distribution characteristics of the fat globules in milk have been evaluated by methods other than microscopic counting and sorting. van Dam and Sirks,[69] Puri *et al.*,[308] and Dolby[80] classified the globules into size groups by an application of the Stokes formula. Their results compared surprisingly well with those obtained by direct microscopic observations. Puri utilized a micro-modification of Robinson's[326] technique for determining the size-distribution of soil particles. The time (V = velocity cm/sec) required for individual fat globules to rise a given distance was predicted from the gravity equation (see p. 774). Fresh milk was agitated gently and poured into a wide-mouth, flat-bottom jar and allowed to stand undisturbed for the duration of the experiment. Portions of milk were withdrawn by means of a microsampling pipet at precisely measured distances from the bottom of the jar. Sampling corresponded to the time estimated for fat globules of a specific size to rise to the position of sampling. The fat content of these samples represented that portion of fat contained in globules smaller than the globules for which the sampling time was computed. Although Puri's data were expressed in terms of size-distribution curves, they represented, more precisely, the volume-distribution characteristics of the lipid phase. He neglected to account for the influence of fat-globule clustering in his computations, or may have assumed a minimum contribution, since sampling occurred near the bottom of the container. About 24 hr. were required to complete the determination. Dolby reduced the sorting time by employing centrifugal force to speed the rising velocity of the fat globules.

In recent years, fat globule size-distribution characteristics have been estimated with an electronic counting device known as the Coulter Counter.[21] Fat globules suspended in a suitable electrolyte, e.g., 2% KI in a 10^{-3} molar phosphate buffer containing 0.1% formaldehyde and small percentages of antibiotic, are passed through a precisely calibrated aperture, e.g., 30 or 50 μm in diameter. Each passing globule displaces an equivalent amount of electrolyte, momentarily changing the resistance between electrodes located on both sides of the aperture.

This variation in resistance is reflected in a voltage change which is electronically amplified, sorted and counted. The pulse intensity is proportional to the size of the globule passing through the aperture. By means of an impulse selector, the number of various fat-globule size-distribution characteristics can be estimated in a matter of minutes.

Although Whittlestone[425] was the first to apply this technique to cow's milk, his fat-globule size-distribution data were of a relative nature because of the absence of a calibrating emulsion. Cornell and Pallansch[61] developed a reliable technique and used it to study fat-globule size-distribution patterns in various milk products (see Fig. 10.2). From these data they calculated the fat content of the original milk that compared favorably with analyses obtained by the Mojonnier method. Kernohan and Lepherd[173] employed a similar technique to study the size-distribution of fat globules in cow's milk during the course of milking. Walstra and Oortwijn[410] reported that, when specific precautions are taken, size distribution data obtained with the Coulter Counter compare favorably with those obtained by microscopic and turbidimetric techniques.

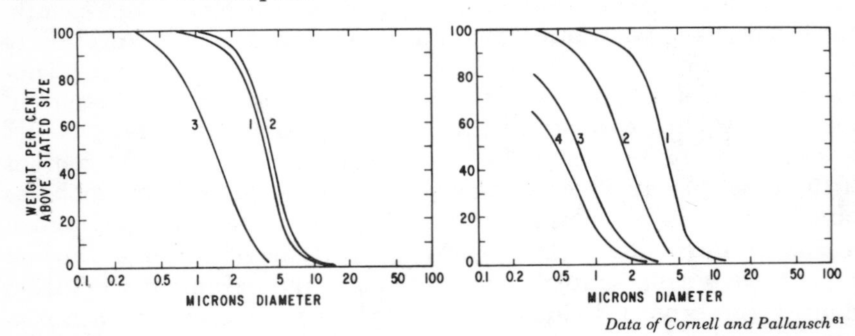

FIG. 10.2. FAT GLOBULE SIZE DISTRIBUTION PATTERNS OBTAINED WITH COULTER COUNTER

LEFT: 1—Raw milk. 2—Raw cream. 3—Skimmilk. RIGHT: 1—Raw milk. 2—Homogenized at 70 kg/cm². 3—Homogenized at 176 kg/cm². 4—Homogenized at 282 kg/cm².

During the past decade, turbidimetric and spectroturbidimetric techniques have been developed for estimating the fat content of milk as well as for assessing the parameters of fat-globule size-distribution patterns.[7,70,106-108,129,130] Walstra and his colleagues made significant contributions to both the theory and practice of these procedures.[401,402,405,406,411] Employing their method, the optical density of diluted milk is measured over a range of wavelengths, i.e., 0.38 to 1.7 μm. From these data and other information, including the concentration and refractive indexes of the fat phase, a specific turbidity spectrum is calculated and plotted. This plot is compared with plots derived from assumed globule size-distribution patterns or patterns obtained with other techniques, e.g., Coulter Counter or microscopic counting. Once these preliminary calculations are made, the spectroturbidimetric method becomes a rapid, simple approach to the assessment of fat-globule size-distribution characteristics. And, from readings made at a single wavelength and the fat content of the milk, volume/surface ratio, average globule diameter, and total surface area of the globular phase can be determined. With special handling, the technique presents a practical means for measuring the fat content of milk.[403] In a recent paper, Walstra *et al.*[411] compared direct microscopic counting, photomicrography, electronic counting (Coulter Counter) and spectroturbidimetry as methods for determining the various dimensional parameters of fat phase in milk (see Fig. 10.3).

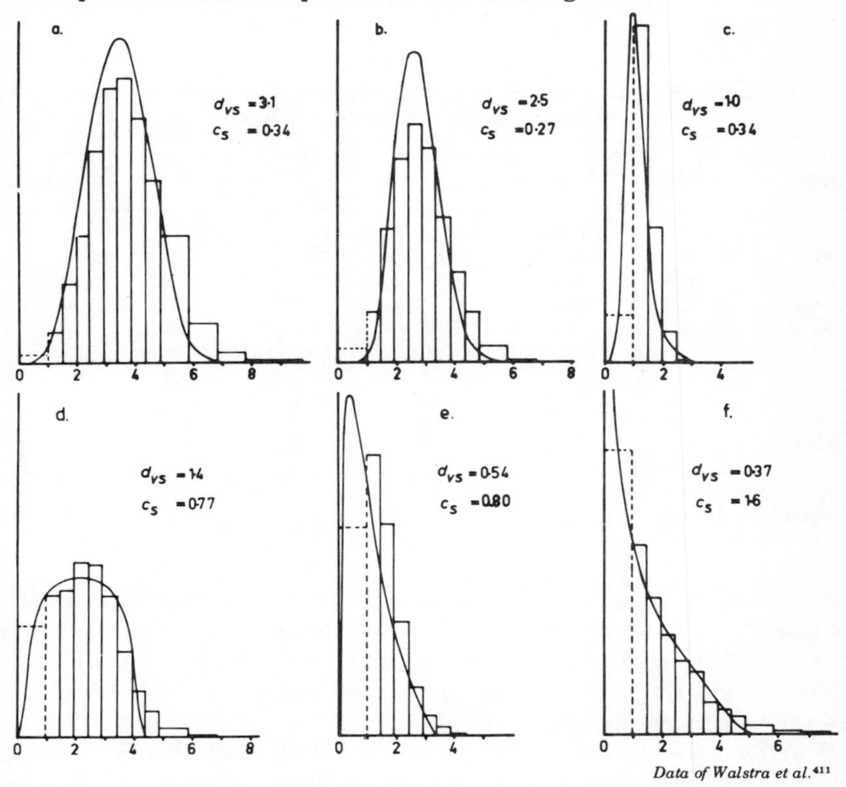

Data of Walstra et al.[411]

FIG. 10.3. COMPARISON OF FAT GLOBULE SIZE DISTRIBUTION PATTERNS OBTAINED WITH COULTER COUNTER (HISTOGRAMS) AND BY SPECTROTURBIDIMETRY (CURVES)

Abcissa—Globule diameter (μm). Ordinate—Fat volume per , m class. a and b—Fresh milk. c—Clarified milk. d and e—Homogenized milk (50 kg/cm²). f—Ultrasonated milk.

Following Babcock's[9] introduction of a quantitative microscopic technique, numerous investigations were initiated for the purpose of evaluating the influence of natural and environmental factors on the size and distribution of fat globules in milk. Campbell[48] reviewed many of these early studies and concluded that most investigators were in agreement concerning the influence of breed, stage of lactation, and fractional milking; but the influence of environmental factors such as feed, climate, and condition were less definitive.

Characteristically, the Channel Isle breeds produce milk containing more fat and a higher proportion of large fat globules than do the other breeds. The range in size from the smallest to the largest globule is approximately equal in all milks. The data in Table 10.1 show charac-

Table 10.1

DISTRIBUTION OF THE FAT PHASE IN THE MILK OF FOUR BREEDS OF COWS ACCORDING TO THE SIZE OF FAT GLOBULES[a]

Breed	Average Size of Fat Globules in Distribution Classes					
	0–2.4μm	2.4–4.8μm	4 8–7.2μm	7.2–9.6μm	9.6–12.0μm	12–14.4μm
	Percentage of Fat Globules in Each of the Groups[b]					
Jersey	8.1	38.3	32.1	18.1	5.3	1.1
Guernsey	6.5	38.9	35.0	14.4	4.4	0.7
Ayrshire	14.6	54.0	23.4	6.2	1.6	0.2
Holstein	14.5	54.6	24.5	5.1	1.1	0.2
	Percentage of Total Fat in Each of the Groups					
Jersey	0.1	11.3	26.1	30.7	23.9	7.9
Guernsey	0.1	11.3	33.2	29.7	25.7	—
Ayrshire	0.3	34.0	41.6	17.8	6.3	—
Holstein	0.3	38.3	50.1	11.3	—	—

[a] Data from van Slyke.[350]
[b] Approximately 1,000 globules counted.

teristics of the fat-globule distribution in milks of four breeds according to size and fat content. As the lactation period advances, the fat globules become relatively smaller and more numerous, except for a short period during the early months, where the reverse relationship applies. These relationships are apparent in Table 10.2.

Van Dam and Sirks[69] measured the size and number of fat globules in milks obtained at 3, 8 and 12 weeks following parturition. Their data are presented as size-frequency curves in Fig. 10.4. The curves constructed from data representing the first two observations show two maxima. As the lactation period advanced, the two-frequency maxima merged into a single peak. The size-distribution patterns constructed from Campbell's[48] data reveal a similar shift in the frequency profiles for Guernsey milks (Fig. 10.5).

Table 10.2

RELATIONSHIP BETWEEN THE MONTH OF LACTATION AND THE AVERAGE RELATIVE SIZE AND NUMBER OF FAT GLOBULES IN MILK OF DIFFERENT BREEDS OF DAIRY COWS[a]

Month of Lactation	Breed of Cows							
	Jersey		Guernsey		Ayrshire		Holstein	
	S[b]	N[c]	S	N	S	N	S	N
First	1,104	49	928	61	687	62	—	—
Second	1,098	53	1,063	46	580	66	640	53
Third	1,228	45	954	52	624	70	576	67
Fourth	1,097	86	659	76	426	85	256	132
Fifth	1,149	56	839	64	384	93	396	95
Sixth	846	76	737	70	399	94	595	66
Seventh	1,017	62	584	103	322	117	340	94
Eighth	733	80	568	89	298	140	310	123
Ninth	715	83	408	120	241	168	384	132
Tenth	571	103	426	104	248	162	284	168

[a] Data of van Slyke.[350]
[b] Average relative size.
[c] Average relative number.

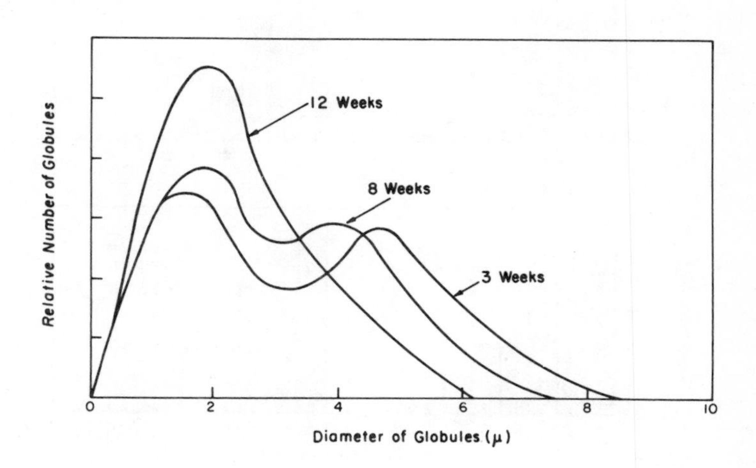

Data of van Dam and Sirks[69]

FIG. 10.4. SIZE-DISTRIBUTION CURVES OF FAT GLOBULES IN MILK AT VARIOUS STAGES OF LACTATION

These graphs do not show a corresponding behavior for Holstein milks. This difference in the shapes of the fat-globule size-distribution profiles may represent a breed characteristic or merely the inability

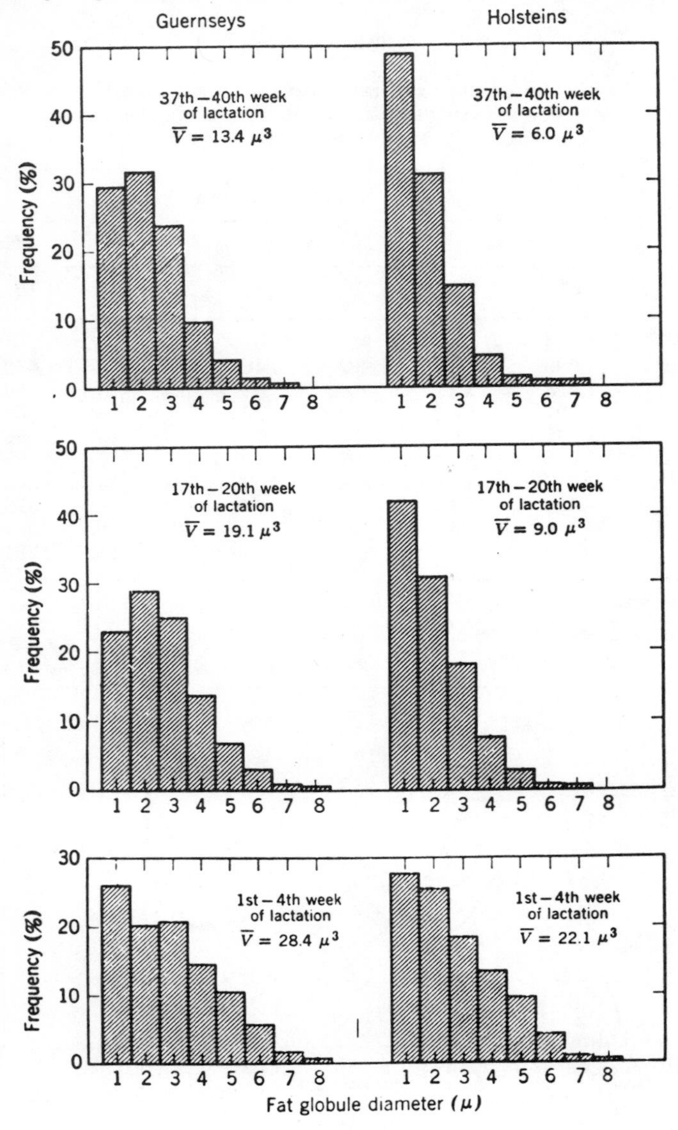

Data of Campbell[48] as presented by Jenness and Patton[164]

FIG. 10.5. PERCENTAGE FREQUENCY DISTRIBUTIONS OF FAT GLOBULE SIZE FOR GUERNSEY AND HOLSTEIN COWS AT VARIOUS STAGES OF LACTATION

of the investigator to measure globules smaller than 0.5 μm in diameter. Collier's[57,58] studies caused him to conclude that milk from successive lactations of the same animal contained increasing proportions of small fat globules. The change was most pronounced from the first to the second lactation period. Contrary to these observations, Whittlestone[424] was unable to detect any definite trend in the size-distribution characteristic with respect to stage of lactation.

Babcock[9] observed that the relative sizes of the fat globules in morning and evening milks are not similar unless the milking intervals are approximately equal. The first milk drawn during a single milking contained fewer and smaller fat globules than that collected at the end of the milking.[57,58,116,257,320,340] The dramatic increase in the relative size of the fat globules toward the termination of milking corresponds to an increase in the fat content of the milk. Whittlestone[424] observed no difference between the size of fat globules in middle milk and strippings. He concluded that the small globules observed in the foremilk result from a filtering action on the part of the small ductules in the mammary gland. Clusters of fat globules, including large fat globules, were retarded in the alveoli and released later in the milking period. Kernohan and Lepherd[173] suggested that this filtering of large fat globules may contribute more to the rise in fat percentage during milking than has previously been recognized.

The manner in which environmental influences, such as feed, temperature, and condition, affect the size and number of fat globules in milk is not entirely clear. Woll[432] and Hunziker[150] concurred in their conclusions that dry feeds favor the production of small fat globules while succulent feeds favor larger globules. Campbell[48] stated that feeding practices might produce an increase, a decrease, or have no effect, but that a drastic change from one type of feed to another usually results in changes in the size and number of fat globules produced. Injury, sickness and changes in condition usually were accompanied by a sharp increase in globular size. King[177] observed cows during confinement and noted that an elevation in body temperature resulted in a decrease in the yield of milk which contained more fat and larger globules. Campbell concluded that the size of the fat globules is determined by multiple factors, contributing to milk production rather than those which control fat production.

One cubic centimeter of normal milk contains approximately 1.5 to 3.0 x 10^9 fat globules of all sizes.[97] Although the milk of different breeds varies in fat content, the numbers of globules are approximately similar. Usually, milk of high fat content contains a higher proportion of large fat globules. Obviously, the surface area of the globule phase must be enormous and is greatest in milk containing a high proportion of large globules. But, since the volume of a sphere is a function of the cube of its diameter ($\pi d^3/6$) and the surface a function of its square (πd^2), it is also obvious that the specific surface (i.e., surface per unit of fat) is greater for milks containing large proportions of small fat globules. The data comprising the size-distribution profiles provide the information required to evaluate the surface-volume parameter (e.g., $S/V = 6\Sigma nd^2/\Sigma nd^3$). When the diameter is expressed in micrometers (μm), the surface in μm^2 and the volume is μm^3, the specific surface (S/V) has the dimensions of $\mu m^2/\mu m^3$, or reciprocal micrometers (μm^{-1}).

Campbell[48] reported specific surface values for individual Guernsey and Holstein milks of 1.26 μm^{-1} and 1.61 μm^{-1}, respectively. With these values and an assumed fat content of 5% for Guernsey milk and 3% for Holstein milk, Jenness and Patton[164] computed the hypothetical globular surface in liter quantities of milk:

For Guernsey;

$$5\% \text{ fat is equivalent to } \frac{5 \times 1.032 \times 1000}{100 \times 0.90} = 57 \text{ cm}^3 \text{ fat per l.}$$

$$\text{Total fat surface} = 57 \text{ cm}^3 \times 1.26 \times \frac{10^{-8}\text{cm}^2}{10^{-12}\text{cm}^3}$$

$$= 720{,}000 \text{ cm}^2 \text{ per 1, or about 700 sq ft per l.}$$

For Holstein;

$$3\% \text{ fat is equivalent to } \frac{3 \times 1.032 \times 1000}{100 \times 0.90} = 34 \text{ cm}^3 \text{ fat per l.}$$

$$\text{Total fat surface} = 34 \text{ cm}^3 \times 1.61 \times \frac{10^{-8}\text{cm}^2}{10^{-12}\text{cm}^3}$$

$$= 550{,}000 \text{ cm}^2 \text{ per l.}$$

Consequent to the realization that the fat globules of cow's milk vary in size and distribution, investigators sought to determine if corresponding compositional differences also existed. Gutzeit[116] carried on the first extensive investigation of the composition of large and small fat globules. He found no differences in the properties of fat obtained from cream and skimmilk. At the same time, he observed that the fat globules in the first milk drawn were smaller than those in the strippings. Consequently, he collected samples of milk at the beginning, middle, and end of separate milkings, but was unable to demonstrate any differences in the properties of the isolated fat. However, studies of a similar nature conducted by van Dam and Holwerda[68] revealed differences in composition between large and small fat globules: cream fat from mixed milk had a lower refractive index than the fat of its skimmilk.

Mulder[257] reviewed the studies of the earlier workers and found sufficient conflict between their results to induce him to reinvestigate the question of compositional differences between fat globules of different size. He observed that the fat obtained from first drawn milk, representing a high proportion of small fat globules, had slightly higher iodine and refractometer numbers than the fat of milk drawn near the conclusion of milking. Furthermore, he was able to demonstrate a difference in composition between cream fat and skimmilk fat obtained from mixed milk, as well as from the milk representing the fractional milkings. A summary of his data is presented in Table 10.3. It was immaterial to the general trend of the experimental results whether the fat analyzed was obtained by solvent extraction or by churning. Despite this evidence of compositional differences, Mulder did not feel justified in drawing the conclusion that small globules differ in composition from large ones. Instead, he postulated that the differences in composition between the large globules themselves might be greater than the differences between average large and average small globules.

Sommer[354] argued that there was no valid reason for assuming that differences in globule sizes imply a difference in composition of the fat phase, or that differences in composition cause the wide distribution in sizes. He conjectured that globular size results from chance, and that the rate and concentration at which glycerides are formed determines their size.

Tverdokhleb[393] reported that butterfat obtained from mechanically separated cream containing predominantly large fat globules was characterized by a high content of oleic and other long-chain fatty acids. Otherwise, the fats obtained from large- and small-globule creams were similar in chemical and physical characteristics. Brunner (unpublished) found that small fat globules ($<1.0\ \mu m$ in diameter) contain slightly higher concentrations of C_{18} unsaturated fatty acids, especially $C_{18:2}$ and $C_{18:3}$. These differences persisted, even though the proportion of C_{18} unsaturated acids in the milkfat varied (see

Table 10.3

CHARACTERISTICS OF LARGE AND SMALL FAT GLOBULES[a]

Sample	Beginning of Lactation		End of Lactation	
	First Milk, Small Globules	Last Milk, Large Globules	First Milk, Small Globules	Last Milk Large Globules
		Refractometer Number[a]		
Cow No. 4	43.8	43.0	43.3	42.4
Cow No. 55	44.0	42.5	43.7	43.2
		Iodine Value[a]		
Cow No. 4	42.5	39.4	—	—
Cow No. 55	41.8	35.7	—	—

	Refractometer number[a]	Iodine Value[a]	Melting Point, °C[a]
Cream fat (large globules)	43.2	38.5	34.5
Skimmilk fat (small globules)	43.6	40.2	33.7

	Fatty Acid Composition,[b] %			
	Winter Fat		Summer Fat	
	Small Globules	Large Globules	Small Globules	Large Globules
$C_{18:1}$	22.9	18.6	34.4	30.9
$C_{18:2}$	4.3	2.3	5.4	4.9
$C_{18:3}$	+ +[c]	+	+ +	+

[a] Compiled from the data of Mulder.[257]
[b] Compiled from data of Brunner *et al.* (unpublished)
[c] Unmeasured; small globules contained more linolenic acid than did large globules.

Table 10.3). The small fat globule preparation was obtained by repeated centrifugation of the supernatant layer of fresh, raw milk. The lipid phase was solvent-extracted and freed of its phospholipid moiety by silicic acid chromatography. The glyceride fraction was methylated and analyzed by gas chromatography.

An exception to the composition pattern reported in the above table was noted by Kernohan *et al.*,[174] who found higher concentrations of the C_{18} unsaturated fatty acids in the fat obtained in the strippings, usually a higher proportion of large fat globules. Although no compositional data were given, Walstra and Borggreve[409] noted differences in the refractive indices of individual fat globules from a single milking. Fat globules with diameters between 3.5 and 6.5 μm were examined.

The Creaming Phenomenon

The rising of fat globules and subsequent formation of a cream layer in normal cow's milk represents one of the fundamental physical properties of the fat emulsion. Despite the numerous investigations which have focused on this process, a clearcut conception of the fundamental principles involved has not been established. The creaming property is no longer of practical significance to the dairy industry. Most of the milk processed for the fluid market is either heat-treated or homogenized—processes which retard or eliminate the formation of a cream layer—and packaged in opaque paper containers. Furthermore, the traditional emphasis on the cream line and fat content of milk is shifting to the protein components. Nevertheless, a satisfactory explanation of the creaming phenomenon remains as an intriguing challenge to dairy scientists.

The fat globules in milk, being lighter than the plasma phase, rise to form a cream layer in from 20 to 30 min under optimum conditions of quiescent storage. The rate of rise (V) of individual globules may be estimated from Stokes' law for the rate of settling of spherical particles:

$$V = \frac{r^2 2 (d_1 - d_2) g}{9 \eta}$$

where r = radius of fat globules, d_1 = density of the plasma phase, d_2 = density of the fat phase, g = the gravitational constant, and η = specific viscosity of the plasma phase. Troy and Sharp[392] measured the rates of rise of individual globules varying in size from 1.8 to 41 μm in diameter. Their data, presented in Fig. 10.6, demonstrate the validity of using Stokes' equation to estimate the rising velocity of fat globules.

Obviously, the appearance of a cream layer, if formed as a result of the rising of individual globules, would require approximately 50 hr instead of the usual matter of minutes. Most investigators concur in the concept that fat globules do not rise singly, but as loose clusters at considerably greater velocities.[69,313] Troy and Sharp[392] measured the velocity of rise of fat globule clusters ranging in size from 10 to ≃800μm in diameter and concluded that clusters rise sufficiently fast to account for the rapid formation of the cream layer (see Fig. 10.7).

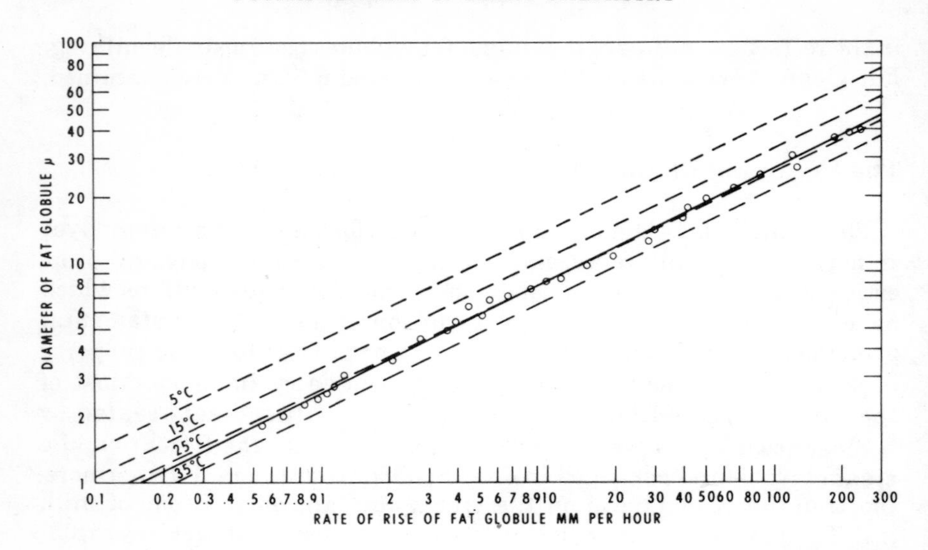

From Troy and Sharp[261]

FIG. 10.6. RATE OF RISE OF INDIVIDUAL FAT GLOBULES OF VARIOUS SIZES

The broken lines express the values calculated from Stokes' equation. The circles represent experimental values obtained at 24°C.

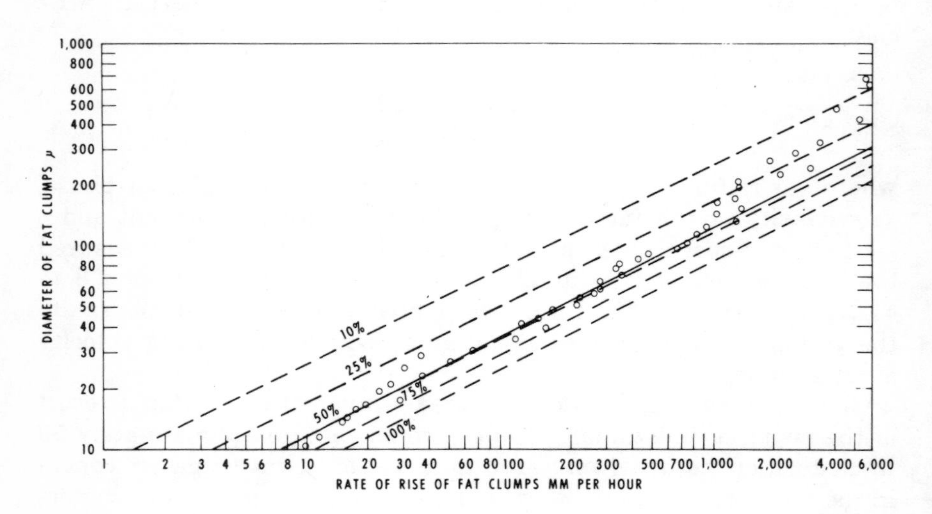

From Troy and Sharp[392]

FIG. 10.7. RATE OF RISE OF FAT GLOBULE CLUSTERS OF VARIOUS SIZES

The broken lines express the values calculated from Stokes' equation. The circles represent experimental values obtained at 24°C.

The rising velocities were compared with values calculated for spherical clusters of similar size, but containing assumed proportions of fat, e.g., 10 to 100% fat. The comparisons indicated that clusters smaller than ≃150 μm in diameter contain approximately 50% fat. As the size of the clusters increased, the velocity of rise was lower than predicted.

The opportunity for cluster formation is provided, apparently, by the disparity in the size of the fat globules present in milk.[118,313] During the first moments after setting, the individual globules commence to rise at rates proportional to the square of their radii (from Stokes' law). The larger globules, rising several times faster than the smaller ones, collide with their smaller counterparts. If conditions are favorable for clustering, they cohere and continue to rise at an increased rate commensurate with their increased radii. Dunkley and Sommer[87] observed that the size of the rising clusters was related to the creaming property of milk. Once the clusters attained an optimum size, they continued to rise without further aggregation. This apparent limitation seems to reflect a state of equilibrium between the forces contributing to cluster formation and the frictional forces encountered in the upward sweep. From all indications, Brownian activity plays only a minor role in the formation of fat-globule clusters.

The depth of the cream layer and the rapidity of its formation relate to (a) the absolute size of the fat globules, (b) the composition of the milk, (c) the extent to which the milk has been agitated, and (d) the temperature history of the milk. Other conditions being equal, large fat globules and high fat content contribute to maximum clustering and high cream volume. Dahlberg and Marquardt[64] pointed out that there is no significant difference in the cream volumes of high- and low-fat milks when compared on the basis of fat content. On the other hand, Bottazzi *et al.*[26] observed that the cream volume of milk from individual cows is not related to fat content and varies with stage of lactation. Mild agitation during the initial stages of creaming enhances the formation of fat-globule clusters. Violent agitation is detrimental, if not inhibiting, to cluster formation. Cooling the milk to low temperatures (0 to 5%C) promotes the rapid formation of a deep cream layer. If disturbed, the cream layer will not reform to the same extent unless the milk is rewarmed and again cooled.

Although the formation of fat-globule clusters has been implicated as a principal factor in the creaming process, the underlying mechanism of their formation is only partially understood. Babcock[10] was the first to postulate that the creaming of cow's milk results from an agglutination-type reaction, and noted that the addition of bovine blood and/or colostrum to milk enhanced the process. Numerous workers have observed that components in the milk plasma, and more specifically in the serum proteins, are involved in the clustering phenomenon.[65,69,87,164,279,335] Hekma and his colleagues concluded that the creaming factor present in blood is inactivated, as are bacterial agglutinins, when heated to 65°C. They suggested that the clustering of fat globules is caused by an "agglutinin" common to both blood and milk.[67,135] Brouwer[29] fractionated the blood globulins into euglobulin and pseudoglobulin fractions and demonstrated the superiority of the euglobulins as a cluster-promoting agent.

Hekma[134] reported that mechanically separated skimmilk behaved as though it contained more of the creaming agent than did skimmilk obtained by gravity creaming. The fact that the surface tension of milk could be reversibly lowered by cooling and raised by warming caused Mohr and Brockman[248] to postulate that there are differences in the nature of the material adsorbed on the surface of warm and cold fat globules. It remained for Sharp and Krukovsky[346] to prove the postulate. They demonstrated that a creaming factor, "agglutinin", is closely associated with the fat phase in cold milk, but is eliminated into the plasma phase upon warming. Seemingly, the association-dissociation equilibrium is correlated with the physical state of the fat, if not dependent on it. Agglutinin was isolated from cold-separated cream and classified as an euglobulin fraction of the milk proteins.[87] Samuelsson *et al.*[335] concluded that the agglutinin factor contains at least two functional components, one heat-labile and the other agitation-labile. Normal creaming will occur if one portion of the available agglutinin has been inactivated by agitation (homogenization) and the other by heating, but not if all the agglutinin has undergone one or the other of these treatments. Further, the activity of agglutinin was favored by factors causing its gradual precipitation, e.g., dilution, low-temperature aging in the range of 5 to 10°C, and mild heat treatment. Similar observations were reported by Kenyon and Jenness.[171]

The clustering of fat globules and the subsequent formation of a cream layer appears to depend principally on the adsorption of the euglobulin proteins at the fat-globule surface. Just how the adsorbed euglobulins contribute to the cluster formation remains unanswered. Possibly they serve to reduce the fat/plasma interfacial tension, as Jenness and Patton[164] have suggested, thus permitting the globules to approach one another to form clusters. The fact that euglobulin-rich skimmilk foams copiously was offered as evidence in support of the interfacial surface activity of euglobulin. Actually, the surface tension of euglobulin-poor skimmilk is lower than that of its euglobulin-rich counterpart. Sharp and Krukovsky[346] conjectured that as euglobulin is adsorbed at the surface of cold fat globules, a surface-active component of the native fat surface is displaced into the plasma. Horwitz *et al.*[147]

reported a rise in the interfacial tension of fat globules in agitated milk which they attributed to the displacement of a surface-active component from the globular surface. Jackson and Pallansch[160] demonstrated that euglobulin is not as interfacially active as, for instance, α-lactalbumin or β-lactoglobulin. Thus, the removal of euglobulin from the plasma might explain the observed reduction in the surface tension of the cold-separated skimmilk. Furthermore, the native interfacial protein (mucoprotein fraction) is far more interfacially active than the euglobulin fraction. Therefore, it seems improbable that the clustering of fat globules can be explained solely on the basis of an increase in the interfacial activity resulting from the adsorption of euglobulin at the interfacial surface.

Several investigations, notably Sharp and Krukovsky,[346] Dunkley and Sommer,[87] and others[123,172,355] have suggested that the euglobulins appear to act in much the same manner as bacterial agglutinins. It is unfortunate that this comparison has been made, since it implies the operation of an antibody-antigen interaction. No evidence exists to suggest that the fat globule interfacial material contains antigenic components to the euglobulin fraction; and it is doubtful if any such intimation was intended. Dunkley and Sommer's conception of a physical interaction between the euglobulin and components of the native fat surface seems to be more acceptable. The theory they proposed to explain the clustering of fat globules embraces the following considerations:

(1) partial dehydration of the adsorbed membrane on the fat globule effected by a specific polar adsorption of the euglobulin;
(2) aggregation of fat globules resulting from the adsorption of a single euglobulin molecule jointly by two fat globules;
(3) maintenance of the surface potential of the fat globules below the critical level by the presence of salts.

It is of interest to note that in thin layers of milk the fat globules exert a repulsive force which serves to maintain their individuality. This behavior has been attributed to the electrokinetic potential of the fat globules.[207] Numerous workers have attempted to relate the electrokinetic potential characteristics with the creaming phenomenon, and have concluded that the surface charge on the fat globules, although variable, shows no relationship to the creaming properties of milk. Dahle and Jack[65] observed creaming in recombined milk prepared from heated fat globules and normal skimmilk. Conversely, creaming was absent when heated skimmilk and nonheated globules were combined. They concluded, therefore, that a component of the milk plasma and not the charge on the fat globule is responsible for the creaming phenomenon. Dunkley and Sommer[87] anticipated that the addition of cations (e.g., Na^+, Fe^{+++}, and Al^{+++}) in equivalent chloride concentrations would enhance creaming as a result of their contributions to a reduction in the zeta-potential at the globule surface, and conversely, that the addition of citrate and phosphate ions would suppress creaming. Contrary to expectations, the cations suppressed creaming while the anions had no apparent effect. Also, it was noted that creaming was suppressed when the pH of milk was lowered. Thus, they concluded that the globule potential plays an insignificant role in the creaming phenomenon.

Payens[297,300] demonstrated that euglobulins added to milk associate with the fat globules, especially at lower temperatures, and concluded that they had no effect on the zeta-potential of the fat globules. Thus, again, it seems unlikely that euglobulins function by reducing the surface potential. Interestingly, β-lactoglobulin was also adsorbed on the fat globule, but had no effect. Thus, Payens concluded that agglutination proceeds by the formation of euglobulin bridges between fat globules, essentially in accordance with the model proposed previously by Dunkley and Sommer.

At this point, a distinction should be made between the electrokinetic potential of fat globules and the forces involved in protein interactions. Essentially, the electrokinetic potential represents the electrostatic resultant, i.e., zeta-potential, of the various charged groups inherent in the composition of the fat-globule surface. The magnitude and sign of the charge may be shown by the migration velocity of a particle when placed in an electric field. Both the fat globules and proteins in milk exhibit a negative electrokinetic potential.[156,309,386] Specifically, the euglobulins are less negative than the fat-globule surface, as demonstrated by their isoelectric points at pH$\simeq$6.0 and 4.0, respectively. However, the similarity of sign does not preclude their potential ability to participate in various types of protein-protein interactions. Here, the nature of the individual charged groups rather than the overall net change is of particular significance.

The predominance of evidence implicates milk euglobulins as a principal participant in the creaming phenomenon. Apparently, the euglobulins exist in milk in a very delicate equilibrium between the dispersed state and a tendency to aggregate. Samuelsson *et al.*[333] concluded that conditions which favor the gradual aggregation of the agglutinins enhance the creaming ability of milk. Dunkley and Sommer[87] observed that euglobulins aggregate and precipitate in cold whey. Interestingly, the aggregate was dispersed again by warming the whey. A similar behavior is noted when a solution of euglobulins is dialyzed against distilled water. Apparently, the forces acting to maintain the proteins in a state of dispersion are suppressed enough by cooling or dilution to permit molecular interactions.

The same forces may be operating in the formation of fat-globule clusters, since both the cooling and dilution of milk enhance its creaming properties. However, it is apparent that modified euglobulins also exhibit a specificity for the fat-globule surface; otherwise, adsorption on the casein micelles would be indicated. The fact that euglobulins are relatively more positive than the fat-globule surface may account for an interaction between the oppositely charged groups of the respective components. This interaction would be suppressed by the addition of cations or a decrease in pH, but not significantly altered by the addition of anions. On the other hand, a structural specificity may exist between euglobulins and membrane proteins. Whichever the case may be, the role played by close-in apolar forces, i.e., van der Waals' forces, may be significant to the entire interacting system. It is a matter for conjecture whether activated euglobulins first adsorb, to the fat surface or interact between themselves prior to adsorption. Once adsorbed at the interfacial surface, euglobulins seem to serve as a "net" that holds fat globules in a loose matrix characteristic of fat-globule clusters. As aging occurs, the clusters become tighter. When the temperature of creamed milk is raised, the euglobulins return to their normal state of dispersion and the clusters disappear. It is unlikely that the physical state of the fat would be more than coincidental to the reversible interactions characterizing the role of euglobulin in the creaming phenomenon.

The failure of heated or agitated (homogenized) milk to cream has generally been attributed to the irreversible denaturation of euglobulin.[87,164,214] However, the results of recent experiments seem, in part, to refute this concept. Payens[297] was unable to detect any difference in the physical properties and clustering ability of euglobulin isolated from colostrum prior to and following homogenization. In a subsequent study, Koops *et al.*[199] observed normal creaming when washed cream was added to a homogenized model system containing milk dialyzate and euglobulin. However, when micellar casein or κ-casein was added to the system, creaming was eliminated, despite the observation that iodine-131-labeled euglobulin was adsorbed on the fat-globule surface. The adsorbed euglobulin was accompanied by small amounts of casein, particularly κ-casein. Koops *et al.* suggested that a euglobulin-casein (κ-casein) complex was formed during homogenization which was capable of adsorbing on the fat surface but unable to effect clustering, or that homogenization induced the adsorption of κ-casein on the fat surface, screening adsorption sites for euglobulin.

As previously indicated, the classic nomenclature employed by dairy scientists differentiated the immunoglobulin fraction of cow's milk protein into a water-soluble portion (pseudoglobulin) and a water-insoluble portion (euglobulin). However, elucidation of the immunoglobulin system of bovine blood in recent years demands the adoption of a contemporary and consistent system of nomenclature.[47,327] This heterogeneous family of large molecular weight proteins is classified primarily as IgG, a 4-chain molecule consisting of 2 light chains and 2 heavy-chain polypeptides which sediments at approximately 7S; IgA, usually a dimer sedimenting as a 13S molecule; and IgM, a pentamer sedimenting as a 19–20S molecule. IgG, the most abundant immunoglobulin found in milk, is further classified into subclasses IgG1 and IgG2, which are antigenically different and can be separated by gel electrophoresis, anion-exchange chromatography and ethanol fractionation.

These components can be correlated with the classic immunoglobulin fractions described by Smith.[351] His pseudoglobulin and plasma T-globulin contain mostly IgG1. The pseudoglobulin preparation also contains "secretory IgA". His serum γ-globulin contains both IgG1 and IgG2, and his euglobulin consists of IgG2-like globulins, electrophoretically slower migrating IgG1 globulins, IgA and IgM. Thus, it is apparent that the euglobulin fraction, classically related to the creaming phenomenon of cow's milk, contains most, if not all, of the uniquely structured immunomacroglobulins, i.e., the 13S-IgA and the 19S-IgM.

The work of Payens,[297–300] Gammack and Gupta,[103,104] Stadhouders and Hup,[355] and Franzen[98] demonstrated that the 19S macroglobulin (IgM) found in the classic euglobulin fraction is a cryoglobulin, i.e., aggregates at low temperatures. It was further shown that this component is the cluster-promoting agent in cow's milk. Gammack and Gupta observed an augmentation of the fat-clustering phenomenon when lipoproteins isolated by centrifugation from skimmilk, presumably of fat globule membrane origin, were added to model systems containing a mixture of IgM and IgA plus agglutinin-poor cream. This observation supports the earlier report of Hansson,[123] who noted that creaming in raw milk was enhanced by the addition of phospholipids. Stadhouders and Hup suggested that the cryoglobulin fraction contains at least two species, one which agglutinates fat globules and another which attaches bacteria to fat globules. Antibodies responsible for the agglutination of bacteria were found in the cryoglobulin supernatant and were presumed to be of the IgG-7S species. They suggested that cryoglobulins function through nonspecific interactions, whereas bacterial agglutination involve a specific interaction.

The recent observations of Jenness and Parkash[163] that goats' milk lacks the fat-clustering factor—also absent in buffaloes' and sows' milk—is of interest. This may indicate that not all species of the 19S macroglobulin are cryogenic, reflecting subtle but specific differences in molecular composition and structure between the active and inactive forms of IgM.

Thus, after more than a century of experimentation, we have arrived at the threshold of understanding this interesting phenomenon of fat-globule clustering. Its resolution depends upon elucidation of the cryo-aggregation activity of the 19S-IgM and its concomitant interaction with fat globule membrane-associated lipoproteins.

PHYSICAL EQUILIBRIA IN MILKFAT

The melting and solidification characteristics of milkfat not only are of practical importance to the processing of dairy products but represent intriguing examples of the physical equilibria that operate in the fat phase. Of principal interest are the factors related to the physical state of milkfat and its crystalline forms.

Milkfat is composed principally of mixed triglycerides differing widely in physical characteristics. The composition and chemical structure of these glycerides are discussed in Chapter 4. Greenbank[111] characterized 11 fractions obtained by crystallization from absolute ethanol (Table 10.4). Melting points ranged from −118°C for the noncrystalline residue to 56.0°C for the first fraction precipitated. Approximately 20% of the fractionated sample melted at temperatures higher than the melting point of the intact milkfat, whereas another 25% melted at temperatures below 0°C. Weckel and Stein[416] separated butterfat into 4 fractions by successive crystallization at 30, 25 and 20°C. They observed that the yield of the various fractions was materially affected by the type of ration fed to the cows and the season in which the fat was produced. The butterfat and the fractions separated from it differed in consistency and melting characteristics. The relative hardness of these fractions was consistent with the degree of unsaturation.

Milkfat exists in the liquid state at the temperature of freshly secreted milk (39 to 40°C). In view of the physical heterogeneity of the glyceride complement, it is apparent that the components are, to a limited extent, mutually soluble. As the temperature of the milk is lowered, the fat commences to solidify, forming what Mulder[258]

Table 10.4

CHEMICAL AND PHYSICAL PROPERTIES OF DIFFERENT FRACTIONS OF BUTTERFAT[a]

Temperature Precipitated °C[b]	Fraction Precipitated %	Melting Point °C.	Saturated Acids %	Unsaturated Acids %
45	3.87	56.0	92.3	4.7
40	0.81	52.0	90.1	5.1
35	7.22	49.0	88.2	7.0
30	3.73	47.0	86.7	8.4
25	1.43	41.4	83.3	8.6
20	3.09	40.1	82.9	13.7
15	17.08	30.8	79.0	15.9
10	26.02	27.7	73.6	22.3
0	10.51	20.2	67.2	27.6
−5	6.71	−4.2	39.2	55.5
−10	11.26	−10.1	37.5	56.0
Soluble	7.83	−11.8	36.6	56.9
Original fat	—	34.3	57.8	37.1

[a] Data of Greenbank.[111]
[b] Precipitated from absolute alcohol (ethanol); initial conc. = 1500 gm. fat/30 liters of solvent.

describes as mixed crystals, i.e., more than one glyceride type in the crystal lattice. The nature and magnitude of the solidified phase reflects (a) the rate and depth of the cooling process, (b) the state of dispersion of the lipid phase, and (c) the compositional characteristics of the glyceride components. Additionally, there is a distinct possibility that a portion of the triglyceride, i.e., the high-melting glyceride fraction, exists in the form of liquid crystals and as such orients at the peripheral area of the fat globules in close proximation with the membrane layer.[179,187]

Melting and Solidification

Melting point, melting interval, and *solidification point* are the commonly determined characteristics of milkfat. The melting point is represented by the temperature at which previously solidified fat transforms completely into the liquid state. If melting occurs over a temperature range it is called the melting interval. Unless precautions are taken to assure that completely liquefied fat is cooled rapidly to low temperatures ($\simeq -10$°C), variable melting characteristics are noted. Similarly, variations in the solidification temperature, i.e., first appearance of the solid phase, of milkfat reflect the cooling procedure. Rahn and Sharp[315] reported solidification points of 19.7 and 23.6°C for samples of the same fat cooled by immersion at 14 and 20°C, respectively. Mulder[258] cited one example in which the solidification and melting points were separated by approximately 19°C. Richardson[318] reported that the melting and solidification temperatures were separated by only a few degrees when liquid fat was cooled slowly. Ideally, these two values should coincide. Mohr and Baur[247] observed melting points of 33.9 and 40.5°C for samples of the same fat which had been solidified at −10°C and approximately 20°C, respectively. The corresponding melting intervals were 1.4 and 14.4°C.

These experimental observations indicated that the solidified fractions, although formed from the same fat, were of different composition. Mulder,[258] utilizing the principles of phase rule chemistry, formulated a "mixed crystal" hypothesis to explain the solidification and melting of milkfat. In view of the array of glyceride components comprising milkfat, no accurate representation of the phase diagram could possibly be determined. Nevertheless, by assuming that all triglycerides follow a basic phase pattern, Mulder constructed a hypothetical partial phase diagram, (Fig. 10.8). The usefulness of this concept as a means of visualizing the behavior of fat when cooled and heated is best illustrated by application.

Assume that milkfat of composition y is cooled to t_1 and allowed to solidify. The solid fat has a composition equivalent to a_1. If the fat had been cooled to a lower temperature, t_2, the solid phase would have a composition corresponding to b_1 instead of a_1. Now, suppose the melting point of the fat which was solidified at $t_2(b_1)$ is to be determined. If heating could be accomplished at a slow enough rate to allow for the establishment of a continuous equilibrium between the liquid and solid phases, then, at t_1 the composition of the solid and liquid phases would correspond to a_1 and a_2, respectively. This would be similar to the equilibrium in the sample cooled to t_1. However, in performing a melting-point determination, heating occurs relatively rapidly. Therefore, the solid-liquid phases are not in continuous equilibrium with each other. Consequently, the melting points for a_1 and b_1 are represented by t_3 and t_4, respectively.

Mulder concluded, that "the lower the temperature at which the fat is caused to solidify (within the solid-liquid region) the lower is the melting point. The dependence of the 'solidification point' on the temperature of cooling can be explained in the light of the great tendency of butterfat to pass into a super-cooled state." Accordingly, a sample of fat cooled rapidly to t_1 will form crystals of composition a_1 and the solidification point lies at t_1.

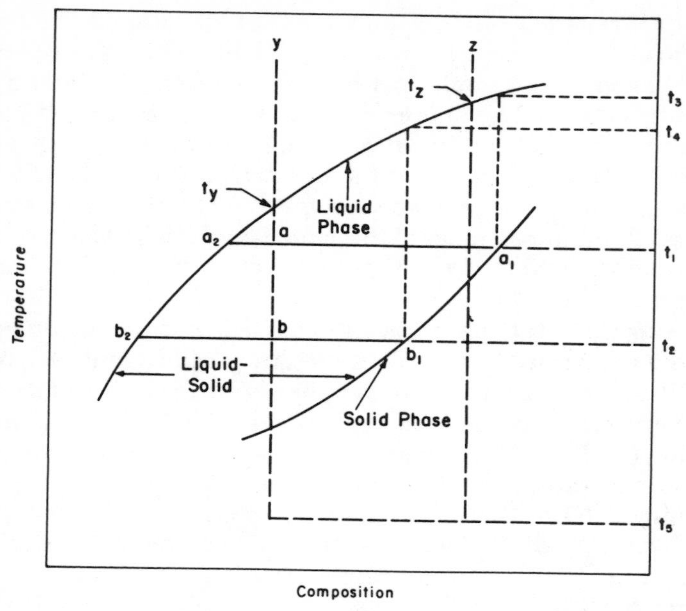

Adapted from Mulder[258]

FIG. 10.8. HYPOTHETICAL REPRESENTATION OF A PARTIAL PHASE DIAGRAM DESCRIBING THE MELTING AND SOLIDIFICATION BEHAVIOR OF MILKFAT

These considerations hold only for rapidly cooled samples of fat. If the cooling pattern is slow or stepwise, part of the fat will have solidified before reaching the desired temperature. Thus, the solid phase would consist of a heterogenous array of mixed crystals, all characterized by different solidification and melting points. Heterogeneous crystallized fat is characterized by indefinite melting and solidification points and wide melting intervals. More consistent values for these physical characteristics can be obtained if the fat is cooled rapidly to temperature t_5 below the solid-liquid zone, where all the fat is crystallized as homogeneous mixed crystals. Referring to Fig. 10.8 and assuming that two samples of milkfat, corresponding to y and z, are cooled to the same temperature t_1, their crystalline phases will be similarly composed. Thus, similar melting points will result. If, on the other hand, the samples are cooled rapidly to a temperature t_5 in the solid phase zone and held until solidification is complete, they will melt at widely diverse temperatures, i.e., t_z and t_y.

Obviously, the behavior of milkfat during melting or solidification cannot be characterized adequately by its melting or solidification points. For in a sense, these values characterize the resultant of all the forces involved in the phase change, but do not provide an evaluation of the component forces. For this purpose, there are three methods in general use, namely; calorimetric, differential-thermal analysis (DTA), and dilatometric. The first two methods utilize the heat of fusion property of fats to follow changes in physical state, whereas the dilatometric method is based on the melting dilation property. Since these methods yield quantitative data, they can be used not only for the purpose of detecting changes in physical state but also as a means of assessing the extent of these changes. All three methods can be applied to fat in mass or, as in the case of milk and cream, in its natural state of dispersion. The application of these methods to the study of the physical state of milkfat follows.

Calorimetric.—One of the simplest calorimetric techniques is the so-called method of mixtures, whereby the heat capacity of one material is determined by mixing in predetermined proportions with a material of known heat capacity at a different temperature. Rishoi and Sharp[322,232] utilized this technique to determine the "apparent" specific heat of creams cooled in different manners. Later Rishoi[321] measured the heat of solidification of cream fat which was reported at 18.7 cal/g of fat at $\simeq 0°C$. The technique is not particularly suited to the detection of changes in physical state.

An adaptation of the classic Nernst[272] calorimetric technique was employed by Jack and Brunner[155] as a means for estimating the proportions of liquid and solid fat in cream. The heat applied to the calorimeter by means of an electric current was determined from Joule's law: $H = EIt/4.186$, where H = calories generated, E = potential difference between terminals of heating element, I = current passed through the heating element, t = time (sec) of current flow, and 4.186 = Joule's constant.

Samples of cream ranging from 15 to 35% fat were cooled in ice water and held at 0°C for 14 to 16 hr to achieve physical equilibrium within the fat. Portions of the stabilized creams were transferred to the calorimeter and heated slowly to 60°C. By subtracting the heat requirements for the skimmilk portion (0.943 cal/gm at 0°C to 0.966 cal/gm at 60°C), the heat required to raise the fat phase over the temperature range was determined. A few samples were treated with NaCl and observed from −10°C. These data were plotted vs. temperature as a heat-content curve (Fig. 10.9).

The heat required to melt the fat was estimated from the point of "half fusion" at $\simeq$20 cal/gm fat.

Two basic assumptions were made to arrive at this value. The first was that all the cream fat is solidified at −10°C. Mulder and Klomp[264] cooled cream to −18°C and observed that the melting started at from

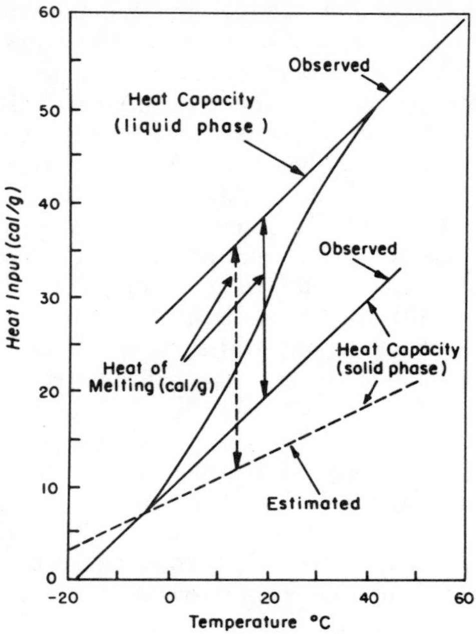

Reconstructed from the data of Jack and Brunner[155]

FIG. 10.9. CURVES SHOWING THE HEAT REQUIRED TO RAISE 1 GM OF MILKFAT FROM ≈ −10° TO 60°C Solid lines represent experimentally determined values (i.e., heat capacities of liquid and solid fat equal 0.5 cal./gm/°C; heat of melting equals 19.5 cal./gm). Broken lines represent estimated corrections to compensate for an assumed lack of complete solidification (i.e., assuming a value of 0.3 cal./gm/°C heat capacity of liquid fat, the heat of melting was estimated at ≈ 25 cal./gm).

−2 to −4°C. On the other hand, Hannewijk and Haighton[122] expressed the opinion that even at −18°C the fat was not entirely solidified. They cooled butterfat to −70°C and observed that melting began at ≃−30°C. Similarly deMan and Wood[239] observed melting as low as −35°C. Hannewijk's values for the proportions of solid fat at a given temperature are presented in Table 10.5.

The second assumption, namely that the heat capacity of solid and liquid fat are similar (≃0.50 cal/gm/°C fat) also appears to have been erroneous. Although 0.5 cal/gm/°C seems to be a good value for liquid fat—Phipps[305] reported a value of 0.52 cal/gm/°C—Bailey[11] suggested that it is too high for the solid phase. Thermodynamic considerations would appear to support this contention. Further, arguments to support

Table 10.5

PERCENTAGES OF SOLID PHASE IN MILKFATS COOLED BY DIFFERENT PROCEDURES[a]

Temperature, °C	Rapidly Cooled to −70°C, Nonstabilized		Slowly Cooled to −70°C, Stabilized	
	Summer[b]	Winter[c]	Summer	Winter
−40	100	100	100	100
−30	100	100	98	99
−20	98	97	91	94
−10	94	93	78	82
−5	91	91	71	74
0	84	84	62	64
5	73	76	52	55
10	59	62	43	45
15	42	46	31	34
20	22	24	17	18
25	14	14	12	13
30	7	11	6	5
35	0	0	2	1
40	0	0	0	0

[a] Data of Hannewijk and Haighton.[122]
[b] Iodine value = 41.3.
[c] Iodine value = 34.5.

a lower value for the heat capacity of solid fat may be derived from the fact that the ratio between the specific heat and coefficients of expansion of a substrate is nearly constant at different temperatures. Since the coefficient of expansion for liquid and solid fat are approximately 0.85 and 0.54mm³/gm/°C, respectively, it may be assumed that a similar ratio exists for the specific heat values (i.e., 0.5 and $\simeq$0.3 cal/gm/°C, respectively). Steiner[358] reported heat capacities for solid and liquid cocoa butter of 0.51 and 0.37 cal/gm/°C, respectively. More recently, Yoncoskie[436,437] reported similar values for milkfat of 0.521 and 0.423 cal/gm/°C.

If the considerations discussed above are applied to the data of Jack and Brunner, as indicated by the dotted lines in Fig. 10.9, a higher value for the heat of melting would be obtained, e.g., $\simeq$25 cal/gm. This value agrees with that reported by Yoncoskie, i.e., 25 cal/gm, for milk fat tempered at 0°C for 25 hr. Sherbon[347] reported heat of melting values ranging from 23.05 to 32.73 cal/gm for samples of milkfat cooled at different rates and held at −40°C for various periods of time. The highest values resulted from slow cooling rates and extended tempering times. It is obvious from the hypothetically corrected melting curve that the heat of melting varies over the melting range, as has been observed for other mixed triglycerides. Consequently, the heat of melting measured from the point of half-fusion represents an average value. The variations arise not only from the presumed differences in the heat capacity of solid and liquid fat, but from different heat of fusion values for the triglyceride types composing the fat. It requires considerably less heat to melt unsaturated fats than saturated fats.

The proportions of solid and liquid fat in a given sample of cream may be *estimated* from the thermal characteristics:

$$\text{Percentage of solid fat} = \frac{H - C_{T_1}, C_{T_2},}{F_{1/2}}$$

where H = calories required to raise the temperature of 1 gm fat from T_1 to T_2 ; C_{T_1, T_2} = heat required to raise 1 gm fat (liquid and/or solid) from T_1 to T_2 [roughly estimated by 0.5 $(T_2 - T_1)$]; and $F_{1/2}$ =

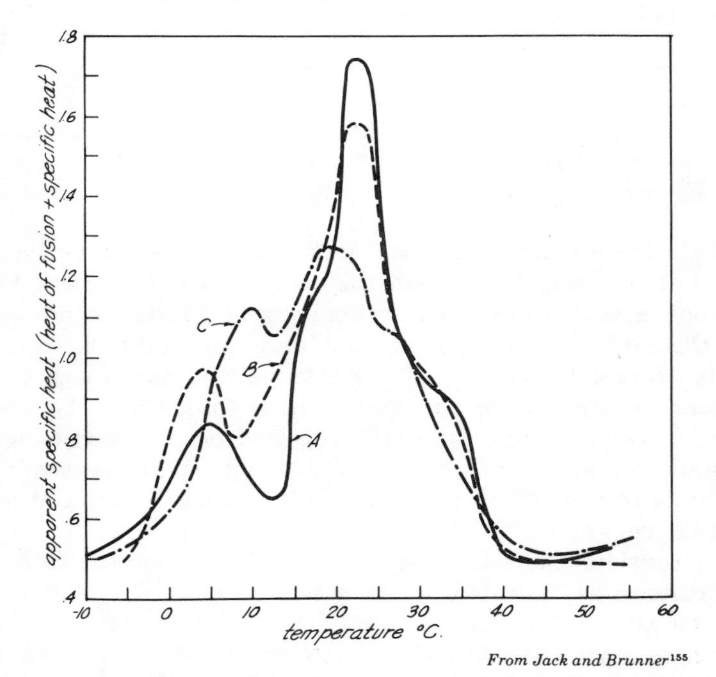

From Jack and Brunner[155]

FIG. 10.10. APPARENT SPECIFIC HEAT VALUES OF MILKFAT OVER THE RANGE OF −10° TO 60°C

Curves A and B–Values for milkfat in cream cooled in brine and held for 16 hr at −10°C. Curve C–Cream cooled in ice water and held for 50 hr at 0°C.

heat of melting measured at point of half-fusion, i.e., $\simeq 25$ cal/gm fat (estimated value).

The melting of milkfat constitutes a solubility phenomenon as well as the phenomenon of fusion. Therefore the heat requirement for melting is probably a combination of heat of fusion and heat of solution. Although the effects of variations in the physical state within the fat are apparent as inflection points on the melting curve, they are best illustrated by plotting the derivation values of the melting curve vs. temperature (Fig. 10.10).

Curves A and B represent samples treated identically and show the same melting characteristics, differing only in intensity; the sample represented by curve C was prepared differently and has different melting characteristics; yet slopes in the region of what is apparently complete solidification, and also where the fat is completely liquid, are essentially the same. In addition, the areas subtended by each curve are equivalent to the heat required for melting 1 gm fat. In these data, the areas under all three curves are almost identical, being equivalent to 19.2 calories for A, 19.8 calories for B, and 19.2 calories for C.

Phipps,[305] by means of adiabatic calorimetry, studied the melting characteristics of fat in creams which were cooled in various ways. The derivative curves shown in Fig. 10.11 illustrate the effects of direct cooling and holding at different temperatures on the relative extent

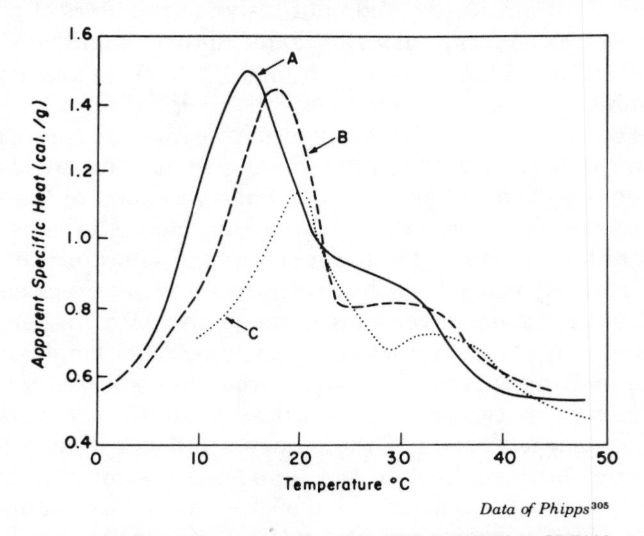

FIG. 10.11. APPARENT HEAT CURVES FOR THE FAT IN CREAM COOLED TO, AND STORED AT, DIFFERENT TEMPERATURES
Curve A—Stored at 0°C. Curve B—5°C. Curve C—10°C.

and nature of the solidified phase. The major peak in each of the three curves was attributed to the principal latent heat effect associated with the melting of the fat. At the higher storage temperatures the magnitude of the effect was reduced, since less fat had solidified. Also, the maximum melting occurred at higher temperatures, no doubt commensurate with the composition of the solidified phase. A second melting peak observed between 25 and 35°C also shifted to higher temperatures as the temperature of storage increased. Phipps' observations were essentially similar to those reported by Brunner and Jack.[36] Phipps suggested that this peak represented an "independent" fraction of fat composed of glycerides of higher melting characteristics than the remainder. Mulder,[258] on the other hand, attributed this characteristic to the presence of a metastable crystalline form.

Differential Thermal Analysis (DTA).—Standard calorimetric procedures, although quite sensitive, are relatively tedious and require precise calibration and temperature control. Most of these requirements are obviated by the technique of DTA. Here, the difference in temperature (Δt) between a sample and an inert reference material is recorded as a function of temperature. The differential scanning calorimeter (DSC) produces thermograms in which the area delineated by the output curve is directly proportional to the total amount of energy transferred in or out of the sample. The ordination is, therefore, proportional to the rate of heat transfer at any given time. Thus, it is possible to obtain both quantitative and qualitative parameters of phase transition, as well as heat capacities for materials over a temperature range of −190°C to 600°C. Sample size is from 0.1 to 100 mg and sensitivities are encountered down to 2 μcal/sec.[117,119,122,211,218,270]

Until recently, this technique has not been employed extensively to follow the behavior of milkfat during phase transitions. Jack[154] was among the first to apply an elementary version of the technique to demonstrate wide variation in the melting characteristics of milkfat cooled at different rates. The slower the rate of solidification the wider was the melting range. As the cooling rate was increased, melting occurred within a narrower range, suggestive of a higher degree of homogeneity in the fat crystals. Hannewijk[121,122] employed DTA to study the melting curves of stabilized and nonstabilized milkfat. The melting curves for two samples of winter butterfat are shown in Fig. 10.12. The characteristics of these curves indicate that considerable quantities of fat remained in the liquid state even after storage at −5°C for ≃15 hr. The solidification of the liquid fraction upon deeper cooling to −70°C is shown by the dip at −13°C. The broken line, representing the melting curve for fat cooled directly to −70°C, shows different

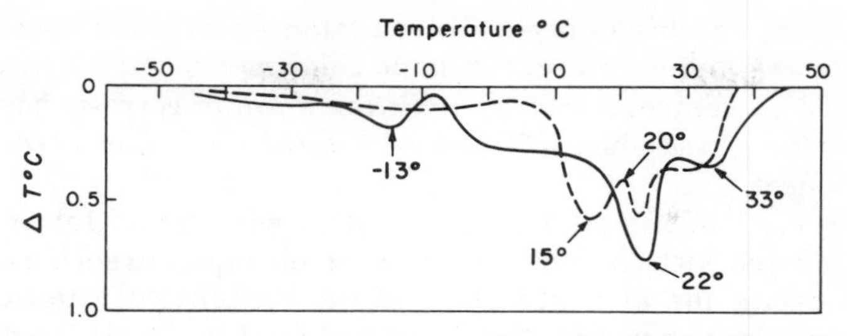

Data of Hannewijk and Haighton[122]

FIG. 10.12. MELTING CURVES FOR MILKFAT DERIVED BY DIFFERENTIAL-THERMAL ANALYSES

Solid line represents winter butterfat cooled slowly to −5°C, tempered for 15 hr, followed by rapid cooling to −70°C. Broken line is for butterfat cooled rapidly to −70°C.

melting characteristics. Employing the same technique, Cantabrana and deMan[49] obtained both heating and cooling curves for milkfat which had been treated and modified in various ways. Varied thermal treatments resulted in differences in the DTA thermograms reflecting the formation of mixed crystals of different glyceride groups. Removal of a portion of the high-melting glyceride and the randomization of the fat by interesterification produced marked changes in the thermograms.

Yoncoskie[436,438] employed DTA to determine the heat capacities of milkfat at −43°C and 70°C, reporting values of 0.423 and 0.52 cal/gm/°C, respectively. The heat of melting was calculated from the thermograms and reported as 21.4 cal/gm. By means of DSC-derived cooling curves, Roos and Tuinstra[327] were able to detect the adulteration of butter with beef tallow, establishing a "limit of detection" at 10% added tallow. Antila *et al.*,[4] employing the DTA technique, observed that milkfat exhibited 4 endothermic melting and 2 exothermic crystallization peaks, and that the characteristics of the melting curve were influenced by the rate of cooling and the duration of crystallization.

Ladbrooke and Chapman[211] prepared an informative and useful review on the application of DTA and DSC techniques to the characterization of lipids, proteins, and biological membranes. Interestingly, the presence of cholesterol in biological membranes seems to initiate crystallization of the polar lipid moiety. They concluded that cholesterol plays a dominant role in stabilizing membrane structures.

Dilatometer.—vanDam[66] was among the first to employ the dilatomic technique to study the physical properties of butterfat. It is exactly analogous to the calorimetric method, except that it senses volume changes rather than thermal properties for characterizing phase transformations. Consequently, dilatometric curves (specific volume plotted vs. temperature) resemble calorimetric curves (heat content plotted vs. temperature). The *melting dilation* corresponds to the *heat of melting,* and the *coefficient of cubical expansion* corresponds to *specific heat.*

The reservoir of the dilatometer is filled with liquid fat or cream which is covered with a nonfreezing immiscible liquid in such a manner that the expansion or contraction of the contents of the bulb are indicated by the rise or fall of the confined fluid in a calibrated small-bore tube. The expansion in volume (i.e., specific volume $\simeq mm^3/gm$) is plotted vs. temperature (Fig. 10.13). In this form the expansion curve resembles the heat-content curve obtained by the calorimetric method (see Fig. 10.9). As the solid fat is heated from *a* to *b* an increase in volume occurs as a result of the expansion of the solid fat. Upon

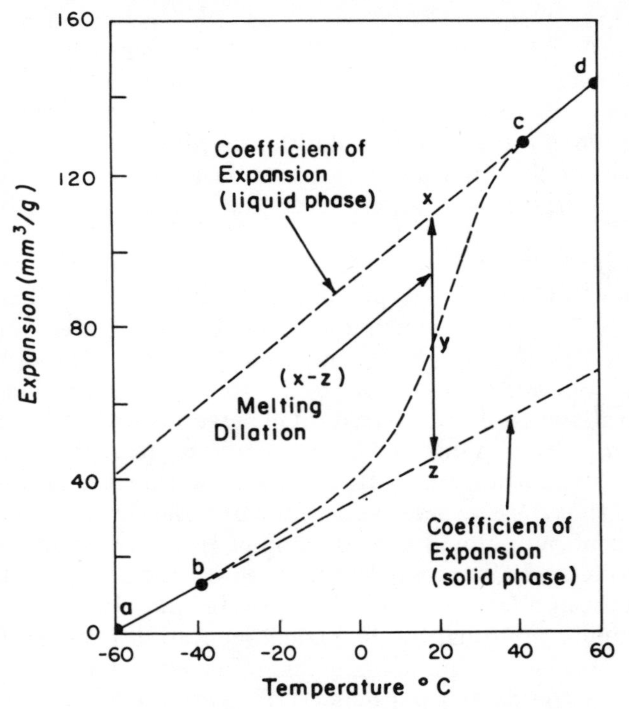

Adapted from data of de Man and Wood[239]

FIG. 10.13. CURVE SHOWING THE THERMAL EXPANSION OF MILKFAT DURING TRANSITION FROM A SOLID (−60°C) TO A LIQUID (60°C) PHASE

Coefficients of expansion for liquid (40–60°C) and solid (−50 to −40°C) fat were 0.85 and 0.58 $mm^3/gm/°C$, respectively. The isothermal melting dilation was 65.1 mm^3/gm at 20°C.

the initiation of melting (b) the system begins to expand more rapidly. At c melting is complete and the subsequent expansion from c to d results from the expansion of liquid fat. By convention the line ($x - z$), drawn through the point of half-fusion perpendicular to the abscissa, represents the average value for the isothermal dilation of the sample of fat under observation. The melting dilation varies throughout the entire melting range, a fact which must be considered when using the values to estimate the proportions of solid and liquid phase at any temperature during the melting. The magnitude of the specific melting dilation also reflects the compositional characteristics of the fat. The glycerides from the fat which melt at lower temperatures, i.e., principally due to the unsaturated glycerides, give slightly lower melting dilations than the high-melting fractions.[11,122]

Values for the coefficients of expansion of solid and liquid phase and the melting dilation of milkfat are listed in Table 10.6. The values obtained by Hannewijk and Haighton[122] appear to be the most plausible, especially for the expansion coefficient of solid fat. Their low value of 0.45 $mm^3/gm/°C$ was determined from butterfat which had been cooled to $-70°C$ in 48 hr and was thus "stabilized", to assure a state of complete solidification. The melting dilation values reported by various investigations seem to reflect various degrees of solidification of the fat. In fact, Hannewijk suggested that Mulder's[257] determinations were obtained with incompletely solidified fat, since $-18°C$ was not considered to be a low enough temperature to assure complete solidification.

The percentage of solid phase present in a sample of fat at any given temperature T during the melting process may be computed as follows:

$$\text{Percentage of solid fat} = E/D_{sT} \times 100,$$

where E = experimentally observed expansion due to melting at temperature T, and D_{sT} = specific melting dilation at temperature T. Values for D_{sT} can be determined experimentally by measuring the magnitude of $x - z$ (Fig. 10.13) over the melting range,[239] or by calculations as follows:

$$D_{sT} = D_{s0} + \Delta D_s \cdot T,$$

where D_{sO} = isothermal melting dilation at 0°C, ΔD = 0.40, and T = the observed temperature.[122] To *estimate* the percentage of solid fat in a given sample of cream at temperature (T_1), the following equation applies:

$$\text{Percentage of solid fat} = \frac{E - M_{T1,\ T2}}{D_{s1/2}}$$

Table 10.6

COEFFICIENTS OF THERMAL EXPANSION FOR SOLID AND LIQUID MILKFAT AND DILATION DUE TO MELTING

Coefficient of Cubical Expansion $Mm^3/Gm/°C$		Melting Dilation Mm^3/Gm	Remarks	Investigator
Solid Fat	Liquid Fat			
0.88	0.88	≈50 ave. value over melting range	Cream cooled to −18°C	Mulder and Klomp[264]
At −16° to −2°C 0.45	At≈50°C 0.84 + 0.00056T	≈66 at 0°C	Fat cooled to −70°C in 48 hr.	Hannewijk and Haighton[122]
0.57		≈57 at °C	Fat cooled to −70°C in 30 min.	
		≈74 at 20°C		
0.58	0.85	62 at 10°C	Fat cooled to −50°C	de Man and Wood[239]
At −50° to −40°C	At 40° to 60°C	65 at 20°C 68 at 30°C		

where E = experimentally observed expansion/gm fat from T_1 to T_2 $M_{T2,T2}$ = thermal expansion of 1 gm fat between temperature T_1 and T_2 [roughly estimated by $0.8(T_2 - T_1)$]; and $D_{s1/2}$ = melting dilation measured at point of half fusion (≃70 mm/gm fat).

Tverdokhleb[396] stated that the dilatometric technique provides the best estimate of the amount of solidified fat in cream and butterfat. The following formula was derived for this purpose:

$$T = \frac{\frac{V_{t_2} - V_t}{P} - K_{Zh}(t_2 - t)}{100 - (t_2 - t)(K_{Zh} - K_T)} \times 100,$$

where V_{t_2} and V_t are dilatometer readings in mm³ at t_2(liquid fat) and at t (observed temperature); P is the weight of the fat in gm; and K_{Zh} and K_T are 0.83 and 0.30 (coefficients of thermal expansion of liquid and solid fat, respectively, in mm³/gm/°C).

The melting characteristics of the solid fat may be amplified, as was done in the case of the calorimetric heat-content curve, by plotting the derivative values (apparent specific dilation = mm³/gm/°C) vs. temperature. Similarly, the area subtended under the curve and above the line connecting the coefficients of dilation for liquid and solid fat represents the average melting-dilation, probably equivalent to $D_{s1/2}$.

Tverdokhleb[394-398] employed dilatometry to determine the extent of solidification in milkfat and fractions thereof when subjected to selected cooling temperatures and holding times. He concluded that there is no uniformity in the composition of the glyceride fractions and that effective solidification—65 to 70%—of milkfat occurs within the first 30 min. Maximum solidification was achieved in 2 to 3 hr at 2 to 8°C and 3 to 5 hr at higher temperatures.

DeMan[235] utilized the dilatometric technique to assess the kinetics of milkfat crystallization. The plots of crystallization data shown in Fig. 10.14 are representative of a first-order reaction. Thus, the rate content k can be determined from the following equation:

$$t = \frac{2.303}{k} \log Co - \frac{2.303}{k} \log (Co - C)$$

where Co represents the initial concentration of solidified fat and C represents the concentration at time t. Then, crystallization half-time can be determined from the following equation:

$$t_{0.5} = \frac{2.303}{k} \cdot \log 2$$

These two parameters for different samples of milkfat and modified milkfat are listed in Table 10.7. The rate constants under conditions of rapid cooling are similar for all samples. However, when a slower

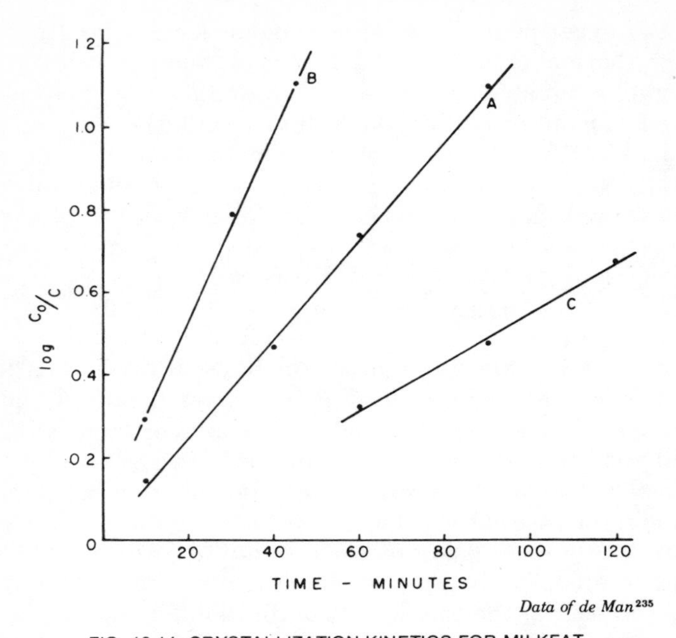

Data of de Man[235]

FIG. 10.14. CRYSTALLIZATION KINETICS FOR MILKFAT

A—Cooled in water at 0°C. B—Interesterifield milkfat cooled at 0°C. C—Cooled in air at 5°C. C_0 and C represent initial concentration of crystallized fat and the concentration at time *t*, respectively.

rate of cooling was employed, the number of crystallization centers became rate-determining, e.g., samples containing higher concentrations of HMG—thus, more crystallization centers—yield higher rate constants. The energy of activation of the crystallization process was calculated from the slope of a plot of log k vs. $1/T$ and found to be 11.0 kcal/mole for September fat.

Factors Affecting Melting and Solidification.—Mulder[257] employed the dilatometric method to follow the melting behavior of fat in creams which had been cooled by various procedures, presumably equivalent to processes encountered when preparing cream for churning. He utilized the mixed-crystal hypothesis to interpret the experimental observations. A summarization of the conclusions drawn from these interesting studies follows:

(1) Cream fat allowed to solidify at low temperatures exhibits maximum melting at lower temperatures than fat allowed to crystallize at higher temperatures, and to a point, seems to be independent of the glyceride composition. However, the proportions of liquid and solid fat reflect the compositional characteristics.

Table 10.7

CRYSTALLIZATION HALF TIMES ($t_{0.5}$) AND RATE CONSTANTS (k) OF MILK FATS AND MODIFIED MILK FATS UNDER DIFFERENT COOLING CONDITIONS.[a]

Description of sample and cooling conditions	$t_{0.5}$ (sec)	k $10^{-4} \cdot sec^{-1}$
June, water 0°C	720	9.6
September, water 0°C	720	9.6
December, water 0°C	720	9.6
September, water 20°C	2850	2.4
September, air 5°C	4235	1.6
September, air 15°C	5940	1.2
Interesterified Sept. water 0°C	690	10.0
Interesterified Sept. water 20°C	600	11.6
Interesterified Sept. air 5°C	2790	2.5
Interesterified Sept. air 15°C	3360	2.1
Sept. minus 2.2% HMG*, water 0°C	720	9.6
Sept. minus 2.2% HMG, air 15°C	6750	1.0
Sept. minus 6.5% HMG, water 0°C	900	7.7
Sept. minus 6.5% HMG, air 15°C	18000	0.4

*) High melting glycerides
a Data of deMan[235]

(2) More fat is caused to pass into the solid state by direct cooling than by stepwise cooling to a specified temperature. The characteristic crystals, formed from each phase in the stepwise cooling process, are not greatly altered by cooling to the next lower temperature. Consequently, the pre-history of semisolid fat cannot be negated by simply holding the cream at a lower temperature.
(3) Changes in the proportion of solid fat occur as a result of recrystallization at temperatures slightly higher than the original crystallization temperature. Although a solid-liquid equilibrium is achieved rather slowly, the proportion of solid fat diminishes concurrently with an increased crystallization, characteristic of the higher temperature.

In general, these observations were supported by dilatomeric studies of deMan and Wood,[239] and by the calorimetric studies of Phipps[305] and others.[11,397,398]

The state of dispersion of the fat phase seems to influence its rate of solidification as well as its rate of melting. Mulder[257] showed that crystallization was retarded in homogenized cream, (Fig. 10.15). As the temperature was lowered to −18°C, however, crystallization in the finely dispersed globules proceeded rapidly until as much of the fat was crystallized as in the free fat samples. Phipps' calorimetric

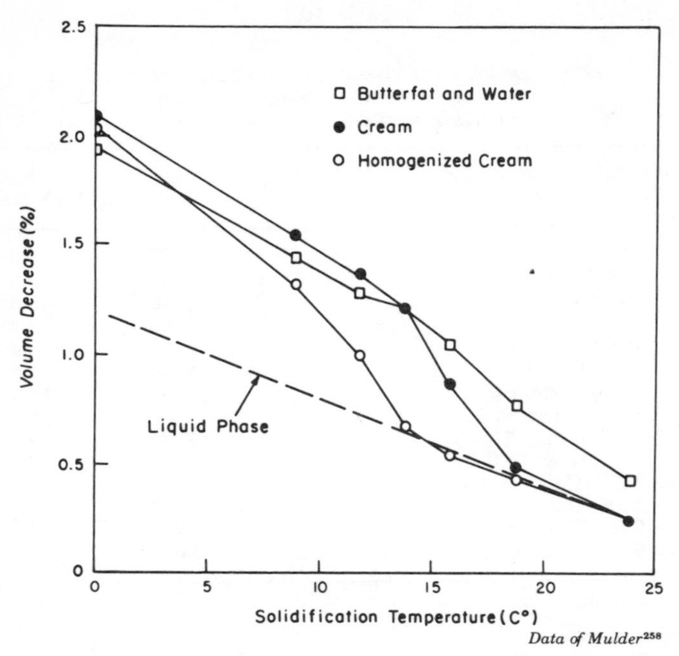

FIG. 10.15. SOLIDIFICATION CURVES FOR MILKFAT IN VARIOUS STATES OF DISPERSION

All samples were cooled from a liquid state (50°C) to the indicated temperatures and equilibrated for 18 hr.

data indicate that a more homogeneous crystallization resulted when the fat phase was finely dispersed.

The composition of a fat influences the proportions of liquid and solid phases observed at any temperature.[235,397,398] Highly unsaturated fats, such as summer fat, crystallize and melt at lower temperatures. Bailey[11] stated that unsaturated fats do not form mixed crystals with the high-melting glyceride fraction as readily as they do with their own kind. This behavior may explain the appearance of the tailing peak characteristically noted in melting and dilation curves. Both deMan[233] and Weihe and Greenbank[417] observed that interesterification modified the physical characteristics of milkfat. A harder fat showing a higher softening point resulted. The effect seemed to be most pronounced in the higher-temperature range and was interpreted to represent an increase in the high-melting glyceride fraction. deMan interpreted this observation to support the proposition that glycerides in milkfat are not formed as a result of random interchange of the fatty acid moieties but that there is some degree of order in their formation. Isomerization caused a lesser increase in the softening point than interesterification but resulted in a greater increase in hardness. The increase in hardness was also apparent in the lower-melting fraction, indicating that the presence of *trans*-unsaturated acyl radicals caused the fat to be harder than if all the unsaturation were present in the *cis*-form. The relationships of chemical composition and cooling treatments on the relative hardness of milkfat are shown in Table 10.8.

Table 10.8

THE EFFECT OF COOLING TREATMENT AND CHEMICAL MODIFICATIONS ON THE HARDNESS OF MILKFAT PRODUCED IN DIFFERENT SEASONS[a]

Fat Sample		Penetration Hardness, 10^{-1} gm.			
		10°C	15°C	20°C	25°C
Untreated, June	Slowly cooled	103	74	0	—
	Rapidly cooled	262	163	46	0
Dec.	Slowly cooled	144	108	0	—
	Rapidly cooled	329	240	36	0
Interesterified, June	Slowly cooled	271	245	158	113
	Rapidly cooled	344	322	118	10
Dec.	Slowly cooled	288	204	152	115
	Rapidly cooled	310	231	111	17
Isomerized, June	Slowly cooled	445	257	124	93
	Rapidly cooled	486	305	107	0
Dec.	Slowly cooled	512	369	167	148
	Rapidly cooled	550	440	157	19

[a] Data of deMan.[233]

POLYMORPHISM

The ability of glycerides to exhibit more than one crystalline structure is described by the term polymorphism and is characterized by a multiple-melting point. For an informative presentation on this subject the reader is referred to Bailey's[11] monograph, "Melting and Solidification of Fats." The exact nature of polymorphism in triglycerides has been a controversial subject, especially between the schools of thought represented by Malkin and his colleagues, on one hand, and by Lutton and his co-workers, on the other.

Clarkson and Malkin[55] presented X-ray diffraction and melting-point data from which they concluded that three polymorphic forms could exist in simple triglycerides. The two higher-melting forms were crystalline in nature. They believed that the lower-melting form, resulting from sharp chilling of the glyceride, is to be noncrystalline (γ-form). The polymorphic forms were called γ, α, and β in order of increasing stability. Later, Malkin and Meara[231] observed a fourth form with a melting point between the α- and β-forms which they designated as β'. Lutton[226] denied the existence of the γ-form in favor of a third crystalline form, claiming that Malkin had erroneously associated the melting point and the X-ray diffraction data. He observed that the polymorphic form which showed an X-ray pattern similar to Malkin's γ-form possessed a melting point similar to the α-form. Furthermore, he could not demonstrate the presence of a noncrystalline form. Subsequently, Clarkson and Malkin[56] modified their original contention that a noncrystalline form did exist, and at the same time acknowledged that a third crystalline form had been observed. In general, the existence of the noncrystalline form has not been verified. In fact, Chapman[50] presented evidence obtained by infrared spectrometry which seems to refute the existence of a vitreous form. The accepted terminology to designate the various crystalline forms of triglycerides is, according to Lutton, α, β', and β, in order of increasing stability.

The schematic representation of polymorphic transitions shown in Fig. 10.16 illustrates the possible transition encountered with saturated triglycendes. The particular transition occurring in a sample depends significantly upon the rate of cooling or heating; the greater the rate the longer the jump between forms. Recently, Buchheim[39] investigated the crystalline parameters of trilaurate by electron diffraction. Dif-

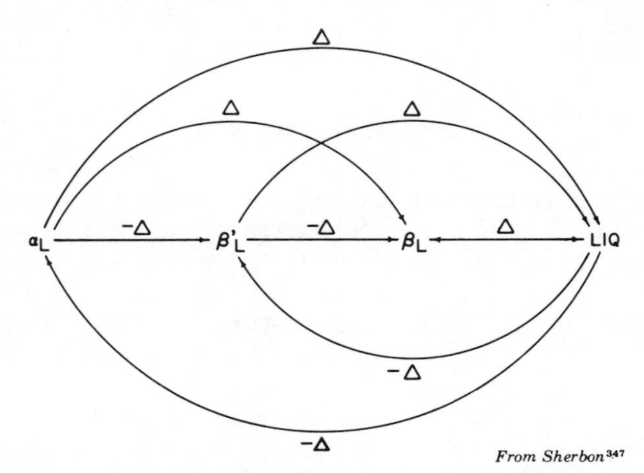

From Sherbon[347]

FIG. 10.16. DIAGRAMMATIC REPRESENTATION OF POLYMORPHIC TRANSITIONS IN GLYCERIDES

fraction patterns of single crystals of α-, β'-, and β-modifications revealed a single α-crystalline form, two β'-modifications, and two (B_{II} and B_{III}) of the five β-modifications previously detected by X-ray diffraction. He proposed a hexagonal elementary cell for the α-form, orthorhombic cells for the two β'-modifications, and triclinc cells for the B_{II} and B_{III} modifications. Vergelesov and Ivanouskaya[400] employed continuous recordings of X-ray spectra to assess crystallization kinetics and polymorphic transitions in milkfat held at 2 to 8°C during the entire process. In model mixtures of triglycerides, polymorphic changes proceed from the α-modification to the more stable β'- and β-modifications. On rapid cooling, α- and possibly γ-modifications were formed which began shifting to the more permanent β'-modification within 13 to 30 min following the initiation of cooling. The transition was complete within the first 2 hrs.

Because milkfat is composed of a wide range of triglycerides, melting characteristics and polymorphic transitions are not as sharply delineated as similar characteristics observed for model systems of more homogenous triglycerides. Krautwurst[204] prepared 4 milkfat fractions by precipitation from acetone which differed in composition. X-ray and DTA studies revealed that these fractions differed in crystal structure and capacity to assume crystalline modifications. These properties were most evident in the high-melting fraction.

Polymorphism in Milkfat

Clear-cut polymorphic behavior is not apparent in fats which consist essentially of mixed triglycerides. In milkfat, with its array of mixed triglycerides, the transformation characteristics may be obscured. In view of Mulder's[257] mixed-crystal hypothesis, it was considered doubtful that milkfat would display detectable polymorphic behavior. The disappearance of one crystalline form upon melting and the subsequent formation of a new form can be rationalized on this basis. Yet, Mulder reported that butterfat exhibits a double-melting phenomenon which he attributed to the transition of a metastable crystalline form at 19°C into a stable, higher-melting form. He was able to observe this transition with a polarizing microscope.

Thomas[379] observed three melting points in high-melting fractions crystallized from milkfat. Two melting points were recorded for the low-temperature fractions and for whole fat. Conceivably the temperatures employed were not low enough to induce the α-form, or the transition from α and β' was so heterogeneous that it was not observed. Thomas observed that the transition of the unstable β'-form to the more stable β-form occurred within a relatively short time in butter, probably not exceeding 24 hr. Employing the technique of DTA, Tverdokhleb[395] observed clear-cut, polymorphic transformations in butterfat. The "vitreous" or γ-form, produced by rapid cooling to low temperatures, was unstable and crystallized at 3 to 13°C. The endothermic effect associated with this transformation was recorded at 12.4 to 14.5°C. The metastable α-form was formed by cooling rapidly to 5°C. Its transformation into the β'-form was noted at 17 to 22°C, the temperature at which maximum melting occurs. Butterfat held at 15 to 18°C crystallized in the β'-form, which showed an endothermic effect on transforming into the stable β-form at 28 to 31°C. The stable β-form was formed by cooling at temperatures in excess of 20°C.

Belousov and Vergelesov[20] and Tverdokhleb[397] observed that the temperatures at which phase transformations occurred depended upon the chemical composition of the milkfat, being higher for the high-melting than for the low-melting fractions. The phase transformation commenced within 30 min after rapid cooling of the cream and continued for 2 to 4 hr. Tverdokhleb[398] observed that, depending on the rate of cooling, the solidification of milkfat may be accompanied by the formation of various crystal forms and fractional crystallization of separate glyceride groups, or by the formation of mixed crystals. Under appropriate conditions of incubation, crystallized milkfat may form polymorphic modifications in which monotropic transformations (i.e., $\gamma \rightarrow \alpha \rightarrow \beta' \rightarrow \beta$) are apparent.

The relatively frequent reference to the γ- or vitreous form of milkfat in the dairy literature is somewhat confusing in view of Lutton's[226] and Chapman's[50] work. Indeed, Belousov and Vergelesov's[19,20] X-ray patterns of γ- and α-forms derived from a high-melting milkfat fraction showed identical short spacing of 4.12 Å, and their respective crystals appeared identical when examined by polarizing microscopy. A possibility exists that the formation of a solid solution is being misinterpreted as a polymorphic transition.

The rapid cooling of liquid milkfat to −20°C was conducive to small-crystal formation (i.e., 1 to 2 μm in diameter). Slow cooling to 15°C resulted in the formation of crystals up to 40 μm in diameter (see Fig. 10.17). deMan[234] observed that the formation of small crystals was accompanied by an increase in the proportion of solid fat. X-ray diffraction technique was employed to study the crystalline habits. These data, presented in Fig. 10.18, indicate that slow cooling resulted in the formation of both the metastable β' (3.8 and 4.2 Å) and stable β-modifications (4.6 Å), whereas rapid cooling resulted in the formation of the β'-form. The presence of the α-form (4.4 Å), if present, could

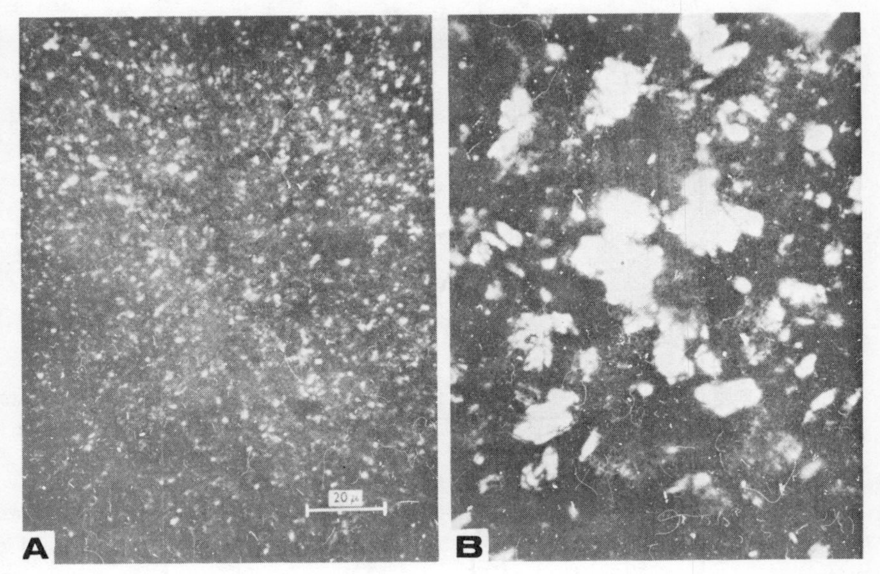

From de Man[234]

FIG. 10.17. POLARIZED LIGHT PHOTOMICROGRAPH OF MILKFAT CRYSTALS
A—Rapidly cooled fat. B—Slowly cooled fat.

have been observed by the strong β'-spacing at 4.2 Å. In the same series of experiments, deMan modified the original milkfat by interesterification, a procedure which resulted in the production of a higher portion of high-melting glycerides. The modified fat formed small crystals when cooled slowly, whereas rapid cooling produced large crystals. This is the exact opposite of the crystalline behavior of the original fat. In both cases, the crystals were the β'-modifications. These observations illustrate the significance of glyceride-types in the solidification and crystalline habit of fats. Knoop[195] utilized the X-ray diffraction technique to show the difference in the crystalline habits of winter and summer milkfats. An analysis of these data indicates that crystallization is more definitive in winter fat than in summer fat. This observation is congruous with the higher proportion of high-melting glycerides in winter butter which seem to be intimately involved in the initiation of crystallization.

Woodrow and deMan[434] employed infrared spectroscopy and X-ray diffraction to monitor the occurrence of polymorphic forms of milkfat. Slow cooling of isolated milkfat resulted in the formation of the β' and β-forms. Rapid cooling produced the α-modification which, upon holding the sample at 5°C, transformed to the β' and β-forms. A high-

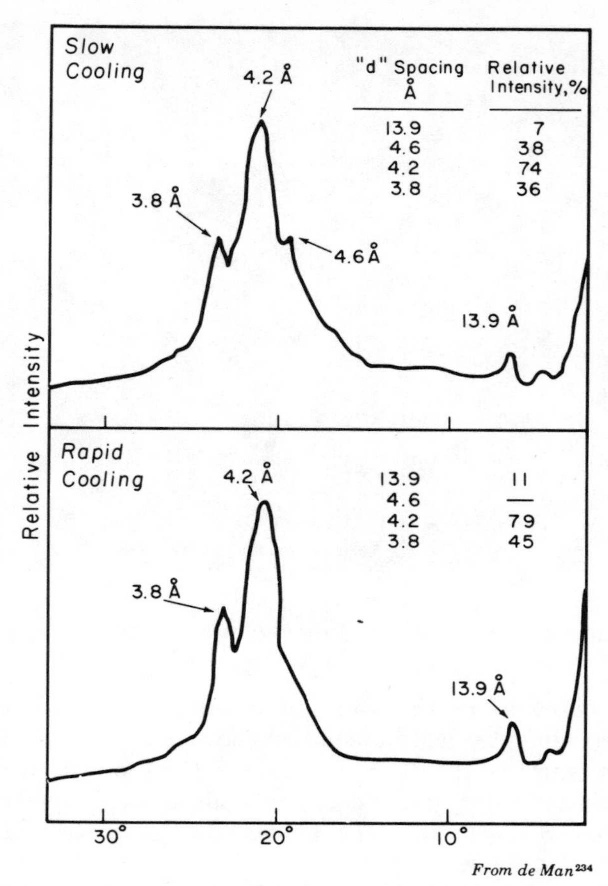

From de Man[234]

FIG. 10.18. X-RAY DIFFRACTION PATTERNS OF SLOWLY COOLED (A) AND RAPIDLY COOLED (B) MILKFAT

Patterns cover a range of 2θ angles from 2° to 34°C.

melting fraction crystallized from acetone existed in the stable β-form. The IR spectra shown in Fig. 10.19 illustrate the identification of the various polymorphic forms of milkfat. Rapid cooling of the milkfat to 0°C resulted in an absorption band at 720 cm^{-1}, indicating the presence of the α-form. Upon tempering to 5°C, an absorption doublet at 718 and 727 cm^{-1} became apparent, indicative of the more stable β'-form. Similar measurements performed with the high-melting glyceride fraction were even more discernible, a ramification of its relatively more homogeneous composition and crystallization behavior. These studies demonstrated that infrared spectroscopy is a useful and convenient method for assessing the crystalline state of milkfat.

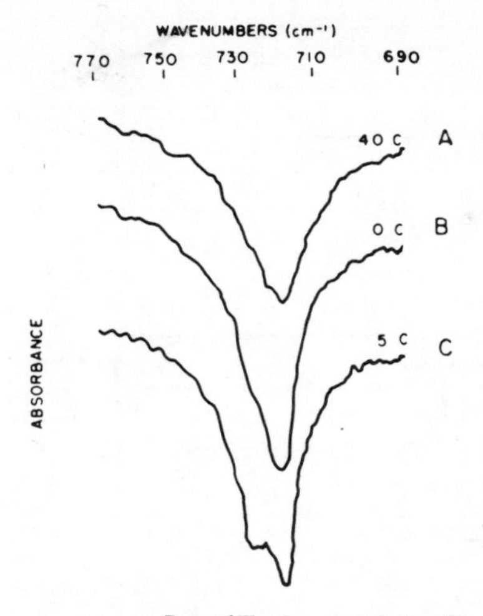

Data of Woodrow and de Man[434]

FIG. 10.19. INFRARED SPECTRA OF MILKFAT

A—Milkfat at 40°C. B—After rapid cooling to 0°C. C—After holding at 5°C for ten hr.

The manner in which the glyceride molecules are arranged in the various crystalline forms of milkfat is relatively obscure. Again, the excellent works of Malkin[55,56,230] and Lutton[227-229] offer some insight into the question (see Fig. 10.20). Malkin proposed a double chain-length structure in which the fatty acids do not all extend from the same side of the glycerol group, similar in orientation to a "tuning fork." Lutton, on the other hand, studied the existence of a triple-chain length structure. He suggested that there was a sorting of long chains from short chains to permit their respective alignment. For unsymmetrical glycerides he proposed a "chair" structure.

These configurations are represented by the long spacings in X-ray diffraction patterns. The long spacings of symmetrical glycerides in the β-modification suggest that they conform to Malkin's double-chain, tuning fork arrangement. But, as the glycerides become more unsymmetrical, the long spacings suggest that there is an ordered alignment into a triple-chain, chair structure. In the lower-melting modifications (γ-, α-, and β'-) the triple-chain arrangement is less common. Information on the long spacings in the milkfat is sparse. deMan[234] estimated values of 41.2 Å and 41.8 Å for milkfats which were cooled slowly and rapidly, respectively.

The nature of crystallization within individual fat globules has not

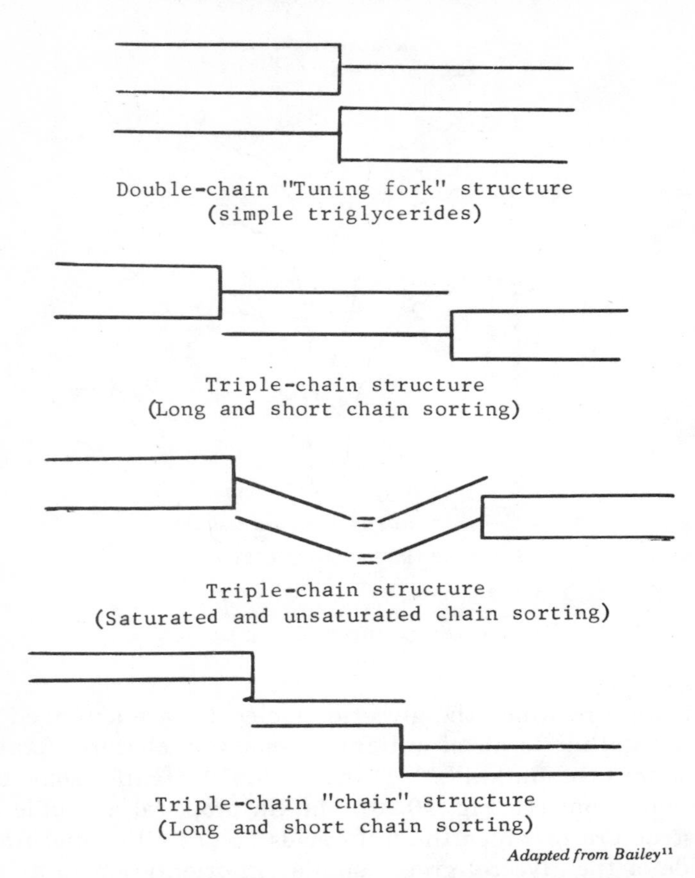

Adapted from Bailey[11]

FIG. 10.20. DIAGRAMMATIC REPRESENTATION OF MOLECULAR ARRANGEMENTS IN TRIGLYCERIDE CRYSTALS

been extensively studied. Most of the techniques previously considered, i.e., calorimetry, dilatometry and X-ray crystallography, are best applied to the free glycerides or, when applied to milk or cream, yield values reflecting the entire lipid phase. King's[179,180] use of the polarizing microscope to observe birefringence in the peripheral layer of individual fat globules provided the best information available. The phenomenon of birefringence, most easily observed in the intact fat globules residing in butter, is attributed to the presence of needle-shaped glyceride crystals (probably the high-melting fraction) oriented tangentially to the globule boundary. Walstra,[404] in a carefully conducted experiment, was able to detect birefringence in the fat globules of milk as well as in butter, cream, and model systems. Buchheim[41,42]

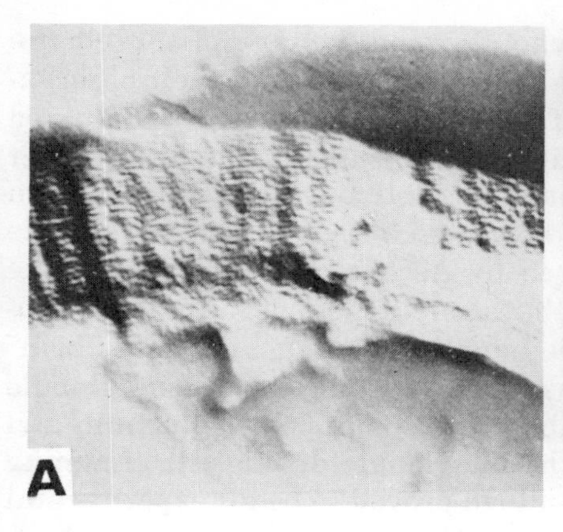

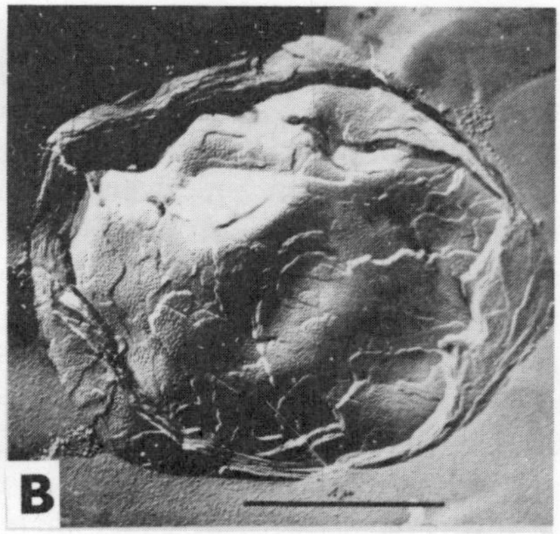

Data of Buchheim[41,42]

FIG. 10.21. ELECTRON MICROGRAPH OF FREEZE-ETCHED FAT GLOBULES SHOWING LAMINAR CRYSTALLIZATION IN THE PERIPHERAL REGION

A—Mixture of tripalmitin (crystallized region) and tricaprin (liquid region). B—Normal fat globule.

extended the dimensions of these observations by applying electron microscopy to freeze-etched specimens. Figure 10.21, Frame A, shows the laminar crystallization of fat at the peripheral layer of simulated fat globules composed of tripalmitin and tricaprins; the tripalmitins constitute the crystalline layer. In similar studies with milk fat globules (Frame B) he observed a crystalline layer about 5.0 μm thick at the periphery of the globule, and suggested that it was composed predominantly of the high-melting glyceride fraction. Crystallization in the inner core of the fat globule was not as ordered.

By measuring the ultrasonic extinction of fat at 284 MHz, Wauschkuhn and Knoop[412] were able to assess the physical state and structure of globular milkfat subjected to various temperature treatments. The results were comparable with information obtained by classic X-ray diffraction studies performed on free milkfat.

NATURE OF THE FAT-GLOBULE SURFACE

One hundred and fifty years after van Leeuwenhoek's identification of microscopic fat globules in cow's milk, Ascherson[6] observed the emulsion-stabilizing membrane surrounding the fat globules. He theorized that the "haptogenic membrane", as he called it, was formed as a result of the capillary condensation of the albumin and its subsequent aggregation at the fat/plasma, interfacial surface. His ideas were generally accepted and supported by his contemporaries.[366]

In 1885, Babcock published his classic studies on the fat globules of milk and challenged those who claimed to have observed the "membrane" surrounding the fat globule. He attributed their observations to a "lack of skill in the use of the microscope." Babcock compared the fat globules of milk with artificial emulsions of the oil-in-water type, stabilized by an adsorbed phase, possibly by a thin film of serum proteins. He was not explicit in how his concept of the fat globule surface differed from that proposed by Ascherson, nor did he make reference to Ascherson's earlier publication in his report. In 1889, Babcock[10] reported that he had identified small percentages of "lacto-fibrin" in milk and that it functioned as a true membrane on the globule surface. He believed that lacto-fibrin was a natural agglutinin, and responsible for the clumping of the fat globule during the creaming of milk. Although Babcock's lacto-fibrin proved to be an experimental artifact,[133] his conception of the agglutination phenomenon, a half-century prior to the discovery of a fat globule agglutinin in milk, remains as a monumental contribution. Furthermore, Babcock's investigations stimulated an extended effort on the part of scientists to determine the nature of the material serving to stabilize the fat emulsion of milk.

Early attempts to identify the material adsorbed on the surface of fat globules utilized the techniques of differential staining and microscopic observations. The fact that milk contains numerous proteins, each capable of stabilizing the fat phase, represented a serious complication to clear-cut experimental results. Storch,[365] aware of this experimental deficiency, "washed" the fat globules free of plasma proteins by repeated dilution and re-separation by means of a mechanical separator. Following the addition of ammoniacal picro-carmine, he was able to detect what appeared to be a stained envelope surrounding the fat globules. A solvent-extracted residue of the washed cream exhibited mucin-like properties, and was considered to differ from the major proteins of milk.

Various modifications of the washing procedure have been employed to prepare "plasma-free" fat globules. Abderhalden and Völtz[1] and Titus *et al.*[385] employed the technique of gravity creaming through a column of water to avoid the loss of components loosely bound to the globule surface. Both groups of investigators concluded that casein is the principal protein associated with fat globules. The technique has been criticized on the basis that significant quantities of plasma proteins are incorporated into the rising clusters of fat globules. By means of a unique washing procedure, Hattori[128] obtained fat globules essentially free of plasma proteins. Milk and chloroform-saturated water were mixed in a 1:9 ratio. The chloroform was taken up by the fat globules, causing them to swell and settle to the bottom of the mixing column. The globule layer was withdrawn, remixed with fresh chloroform-saturated water, and again allowed to separate. The procedure was repeated several times until finally the globule phase was extracted with ether. Hattori called the proteinaceous residue "haptein" and observed that its chemical composition was different from the other known proteins in milk.

Not until Palmer and Samuelsson[282] detected high concentrations of phospholipids in the unextracted fat globule "membrane" was it definitely established that substances other than proteins were involved. The contemporary knowledge concerning the composition of the fat globule surface stems principally from studies of specimens isolated from washed cream.[2, 12, 13, 22, 32-34, 35, 38, 51, 52, 83, 84, 124, 126, 128, 131, 132, 137, 138, 140-142, 149, 157-159, 161, 162, 255, 256, 282, 285, 316, 319, 331, 368-371, 380-383, 428, 429] Some investigators, notably Mulder and his colleagues, preferred to estimate the interfacial concentration of a particular component by determining the partition of the component between cream and skimmilk phases.[259,265,268] Components residing at the fat surface are concentrated in the cream phase when milk is separated. The fat-soluble components, such as carotenoids, which are preferentially oriented at the interfacial surface, may be detected in a somewhat similar manner. These substances exist in relatively high concentrations in the small fat globules of skimmilk where the surface volume ratio exceeds that for larger globules of the cream phase.[422]

Isolation of the Fat Globule Membrane

Palmer and Samuelsson[282] were the first to isolate and partially identify the intact fat globule membrane material. The essential features of Storch's washing procedure were utilized to prepare a plasma-free, washed cream which was cooled and churned to release the fat membrane material into the buttermilk. The membrane sol was concentrated from the combined buttermilk and butter serum by either (a) acidification to the isoelectric point (pH 3.9 to 4.0) with glacial acetic acid, or (b) pervaporation in Visking cellulosic casings. When desired, the concentrates were dried over a desiccant or by lyophilization. Lipids were removed by reflex extraction of the concentrate or dried material with absolute ethyl alcohol for 36 hr followed by absolute ether for 48 hr.

Brunner *et al.*[32] isolated the membrane protein moiety from pasteurized milk by a similar procedure but incorporated a significant modification to avoid exposure of the protein to warm alcohol. Instead, ethyl alcohol was added at 0 to 5°C, followed by several washings with ethyl ether at the same temperature. Finally, the alcohol-free preparation was washed with warm ethyl ether. By this procedure, membrane protein was essentially freed of its lipid moiety. Subsequently, Herald and Brunner[138] concentrated the membrane material by salting-out in 2.2*M* ammonium sulfate. The salted-out sol was collected by centrifugation at 25,000 × *g* and extracted with a cold (0 to 5°C) mixture of ethyl alcohol in ethyl ether 35% alcohol, v/v). The delipidated proteins were dispersed in water or isotonic saline and classified by centrifugation at 25,000 × *g* for 1 hr into 2 protein fractions. The protein collected in the pellet was reddish-brown and quite insoluble in the usual aqueous buffer system. It was designated as the "insoluble" or pellet fraction and represented the first clear-cut evidence that the membrane material contained more than one protein. The author and his colleagues detected losses in enzymatic activity, as the crude membrane was processed through the purification steps, which was essentially apparent following the salting-out step. As these studies progressed, more emphasis was directed to the composition of the lipid moiety and, eventually, to the nature of the lipoprotein constituting the fat globule membrane. A flow diagram showing the procedures employed in the author's laboratory for the preparation of principal fractions of fat globule membrane material is outlined in Fig. 10.22.

Obviously, any attempt to isolate and characterize the material identified with the fat globule interfacial surface presupposes elimination of plasma proteins from the system. Although the globule-washing technique represents the best approach to this objective, it has been justifiably criticized for at least three deficiencies which the author feels compelled to review. The first is that components loosely associated at the interfacial layer are washed away. No doubt this is true for those components, principally plasma proteins, that are oriented peripherally through secondary forces. But it could be argued that such components do not constitute an integral part of the native membrane system. Sharp and Krukovsky's[346] discovery that a fat globule agglutinin factor is reversibly adsorbed at the fat-globule surface in cold milk exemplifies a case in point.

Another important issue is the reduction in the ionic concentration of the aqueous phase resulting from the addition of water. In this environment of decreased ionic activity, the normally suppressed ionic groups of the plasma proteins become more active, enhancing their interacting potential at the fat-globule surface. Although this allegation is valid in principle, it is doubtful if associations formed from this cause could survive the dissociating effects of the mechanical forces inherent in the washing procedure.

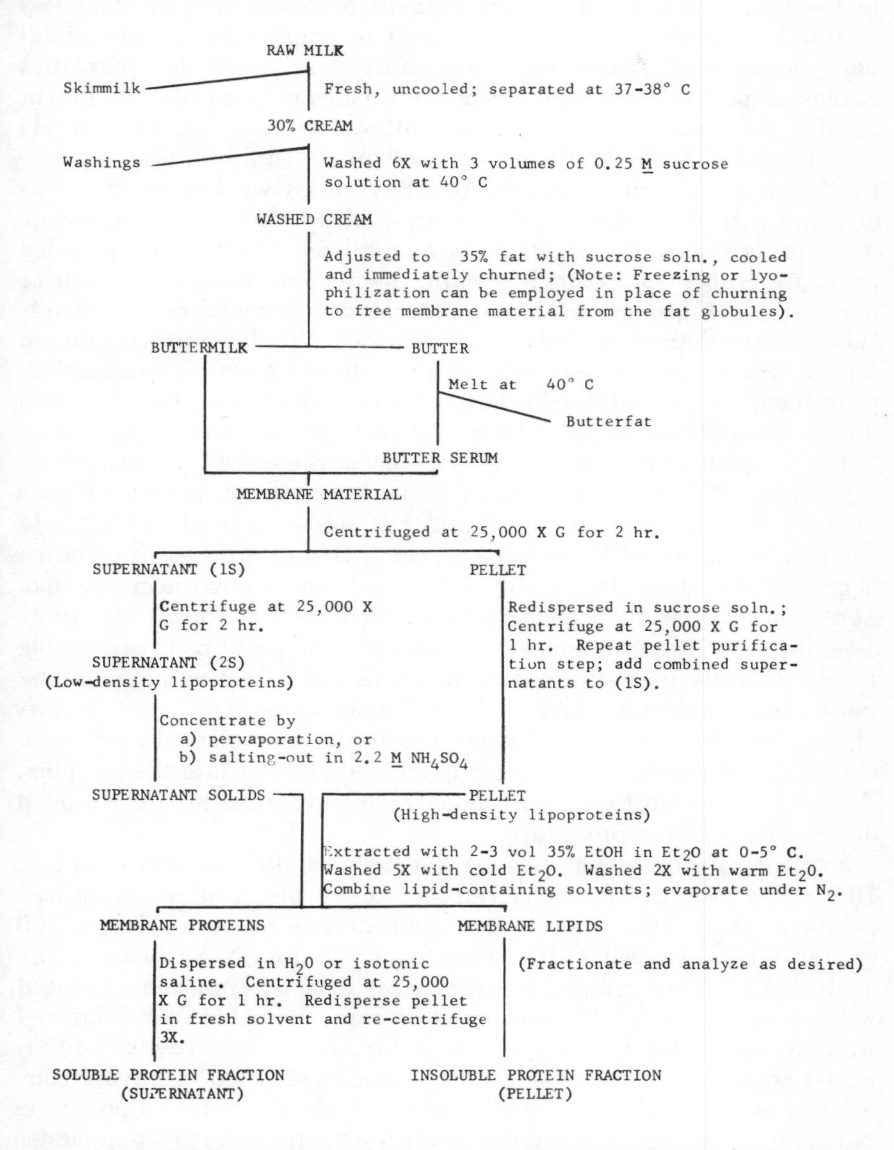
RAW MILK
Skimmilk
Fresh, uncooled; separated at 37-38° C
30% CREAM
Washings
Washed 6X with 3 volumes of 0.25 M sucrose solution at 40° C
WASHED CREAM
Adjusted to 35% fat with sucrose soln., cooled and immediately churned; (Note: Freezing or lyophilization can be employed in place of churning to free membrane material from the fat globules).
BUTTERMILK
BUTTER
Melt at 40° C
Butterfat
BUTTER SERUM
MEMBRANE MATERIAL
Centrifuged at 25,000 X G for 2 hr.
SUPERNATANT (1S)
PELLET
Centrifuge at 25,000 X G for 2 hr.
Redispersed in sucrose soln.; Centrifuge at 25,000 X G for 1 hr. Repeat pellet purification step; add combined supernatants to (1S).
SUPERNATANT (2S)
(Low-density lipoproteins)
Concentrate by
a) pervaporation, or
b) salting-out in 2.2 M NH_4SO_4
SUPERNATANT SOLIDS
PELLET
(High-density lipoproteins)
Extracted with 2-3 vol 35% EtOH in Et_2O at 0-5° C.
Washed 5X with cold Et_2O. Washed 2X with warm Et_2O.
Combine lipid-containing solvents; evaporate under N_2.
MEMBRANE PROTEINS
MEMBRANE LIPIDS
Dispersed in H_2O or isotonic saline. Centrifuged at 25,000 X G for 1 hr. Redisperse pellet in fresh solvent and re-centrifuge 3X.
(Fractionate and analyze as desired)
SOLUBLE PROTEIN FRACTION
(SUPERNATANT)
INSOLUBLE PROTEIN FRACTION
(PELLET)

FIG. 10.22. PROCEDURE FOR THE ISOLATION OF FAT GLOBULE MEMBRANE LIPOPROTEINS AND CONSTITUENT COMPONENTS

Finally, the question of the similarity of the material isolated to that which was lost during preparation represents a more formidable criticism of the method. Losses of membrane material which occur in the first 2 or 3 washings are difficult to assess. The gradual loss of small fat globules makes it difficult to employ the phospholipid-phosphorus: lipid ratio with any assurance. Also, detectable quantities of plasma proteins are retained in the washed cream until the fourth or fifth washings. On occasion the author has encountered unstable milk in which the fat emulsion exhibited gross instability after 3 washings. Conversely, some fat globules have been washed as many as 20 times without an appreciable loss in stability. Obviously, membrane material is lost as the emulsion is destabilized. Only fresh, uncooled milk, preferably from a single milking, should be employed as a source material. Thompson *et al.*[383] demonstrated that membrane preparations obtained from uncooled and previously cooled milk contained 56 and 32% protein, respectively. Also, the lipid moiety of the uncooled-milk membranes contained a larger proportion of phospholipids than did the membrane prepared from previously cooled milk.

The washed-cream isolation procedure was reassessed in the author's laboratory.[369] These studies revealed that the washing operation should be restricted to the minimum number of steps required to eliminate the plasma proteins. Three washes, consisting of 3 times the volume of cream, should result in a 300- to 400-fold decrease in plasma components. Starch-urea-gel electropherograms showed that 3 washings adequately reduced the concentration of plasma proteins to acceptable levels. With the exception of the skimmilk and first-washing residues, most of the material added to the gel slots remained in the vicinity of the slot. Generally, fat-globule membrane specimens do not enter electrophoretic gels unless specially treated to dissociate the complex. Thus it is apparent that a portion of the membrane complex is eroded during the washing procedure.

Evidence of this erosion is apparent from the data presented in Fig. 10.23. The protein content of the membrane preparations stabilized at about 41 to 43%, whereas the yield decreased from 1.02 gm/100 gm fat (for 3X-washed) to 0.89 gm/100 gm fat (for 5X-washed)—a loss of about 13%. Erosion of the membrane material was markedly reduced when a sucrose (0.25*M*)-saline solution was substituted for deionized water as the washing medium. Presumably, the increased density differential between the aqueous and lipid phases resulted in a more complete recovery of the lipid phase, particularly the denser lipoprotein particles. Washing with sucrose-containing solutions is recommended for the preparation of membrane specimens in which enzymatic activities are of principal concern.[83,84] Analyses of this nature should

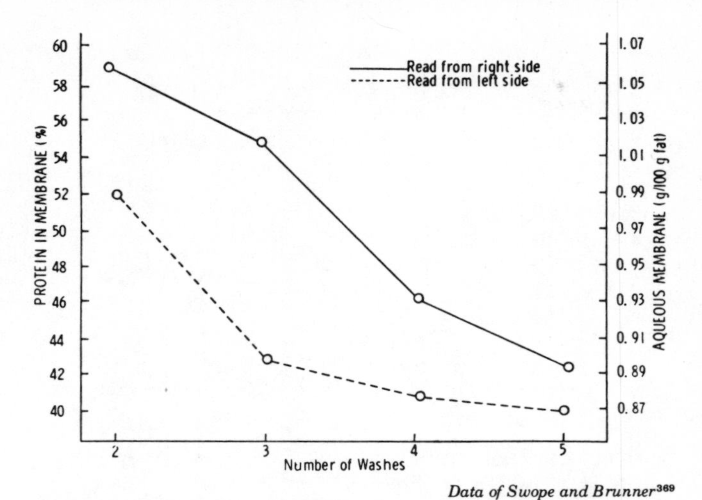

Data of Swope and Brunner[369]

FIG. 10.23. YIELD AND COMPOSITION OF THE AQUEOUS FRACTION RECOVERED FROM ALIQUOTS OF A 2- to 5-TIMES WASHED CREAM

be performed with membrane preparations obtained within 3 to 4 hr after milking. On the other hand, when specimens are prepared for additional fractionation, composition analysis and other involved manipulations, the control of bacterial growth is better achieved when deionized water is used as the washing medium. Ordinarily, for preparations of this nature, a mixture of penicillin-G and streptomycin sulfate is added to the membrane suspension at concentration levels of 46 mg/l and 75 mg/l, respectively.

For 3 preparations in which deionized water was used for the washing medium, the recovery of membrane material was amazingly reproducible, ranging from 1.4 to 1.6 gm/100 gm fat. Similarly, its distribution between the aqueous phase and the butter serum, as well as the composition of these fractions, was quite uniform, as indicated in Fig. 10.24. Notably, the serum membrane fraction contained a higher concentration of lipids than its counterpart obtained from the buttermilk phase. These data are consistent with those reported by Jenness and Palmer,[161,162] who found 2 to 3 times more phospholipid in butter-serum membrane than in buttermilk membrane.

Further evidence of erosion of the fat-globule membrane resulting from preparative manipulations was demonstrated by the studies of Chien and Richardson,[52,53] who assessed the quantitative and compositional nature of lipoprotein fractions desorbed from thrice-washed fat globules subjected to milk agitation. About 55% of the membrane material, which they designated as the "outer layer", was affected. No evidence was presented concerning the amount of membrane mate-

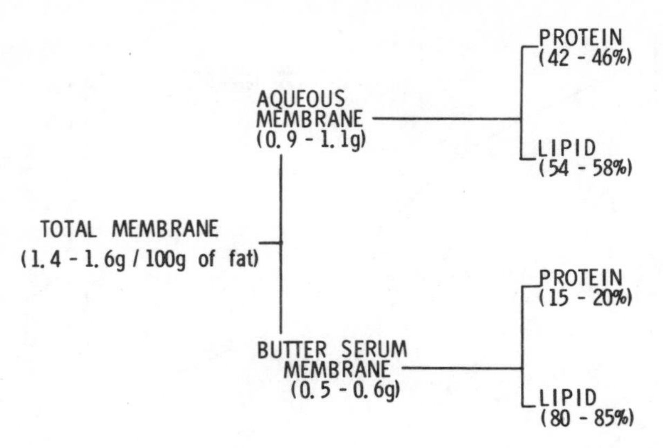

Data of Swope and Brunner[369]

FIG. 10.24. DISTRIBUTION AND COMPOSITION OF FAT GLOBULE MEMBRANE MATERIAL FOLLOWING THE CHURNING OF THRICE-WASHED CREAM

rial lost during the preliminary washings. An indication that membrane material is lost from the fat globule during the cream-washing process was provided by Koyama,[202,203] who reported that cream washed 4 times possessed only 13% of the alkaline phosphatase activity found in unwashed cream.

Although the churning of washed cream represents the most widely used method for obtaining fat-globule membrane specimens, several investigators have resorted to a freeze-thaw technique to liberate the membrane layer, which was subsequently collected as a pellet by differential or density-gradient sedimentation.[2,83,149,168,169] Recognizing the problems inherent in the wash-churning procedure, Saito and Igarashi[331] devised a relatively mild procedure for the preparation of fat-globule membrane specimens (fractions). Cooled whole milk, containing 30% (w/v) sucrose, was overlayed with a small amount of 0.05% (w/v) saline and centrifuged at room temperature at $900 \times g$ for 15 min. The layer of fat globules which collected at the top of the centrifuge tube was removed and twice extracted with cold (1°C) saline (1*M*), followed by centrifugation at $1700 \times g$ for 15 min. The fatty mass was removed from the tube and the aqueous layer was subjected to further centrifugal classifications. All fractions were extracted with a chloroform-methanol mixture to yield their respective protein and lipid moieties.

Chemical and Physical Characteristics

In general, the fat globule/plasma interfacial material consists of a mixture of proteins and lipids which seem to exist in an ordered complex or as discrete lipoprotein particles, or a combination of both. Numerous enzymes, a number of which are plasma membrane-specific, have been identified in freshly prepared specimens.

The membrane complex, when liberated from the fat-globule surface, exists as an opaque, dispersed sol. Palmer and Samuelsson[282] observed that it could be aggregated and concentrated in its iso-electric zone of pH 3.9 to 4.1. Herald and Brunner[138] concentrated the sol by salting-out in 2.2*M* ammonium sulfate. Brunner and Thompson[37] centrifugally fractionated (25,000 $\times$ *g* for 2 hr) the lipoprotein complex into "high- and low-density" fractions. Alexander and Lusena's[2] studies showed that approximately 25% of the sodium desoxycholate-dissociated material remained in the supernatant following centrifugation at 105,000 $\times$ *g* for 5 hr. Bailie and Morton[12,13] collected two-thirds of the membrane material by differential centrifugation at 50,000 $\times$ *g* for 2 hr, and presented evidence to establish the microsomal nature of this material.

Fresh membrane preparations have a pinkish appearance, presumably caused by the presence of the haemo-protein, cytochrome b_5.[240] In its oxidized state, as in aged milk, this component is denatured and becomes dark brown. Therefore, its presence in the undenatured form serves as a convenient indicator of representative membrane preparations. Furthermore, lyophilized membrane specimens, both the lipoprotein complex and corresponding delipidated fractions, have atypical physical properties. Thus only freshly prepared specimens should be studied.

Proteins.—Palmer and Samuelsson[282] isolated the protein moiety and classified it as globulin-like. But, unlike Storch,[365] they were unable to detect the presence of glycoproteins. In subsequent studies Palmer and Wiese[285] and Wiese and Palmer[429] reported values for nitrogen (12.22%), sulfur (0.96%), and phosphorus (0.33%). They attributed the phosphorus content to the presence of residual phospholipids. However, phosphorus has generally been observed as a constituent of membrane protein preparations. Harwalker and Brunner[unpublished] observed that the tenaciously bound residual lipid, following extraction of the membrane complex with an alcohol-ether mixture (35% alcohol, v/v), consisted principally of neutral lipids rather than phospholipids. Herald and Brunner[138] separated the defatted protein moiety into approximately equal fractions by centrifugation at 25,000 $\times$ *g* for 30 min. The pellet-protein was reddish-brown and insoluble in 25% sulfuric acid, 8*M* urea, and conventional buffer solutions. Agents capable of solubilizing a suspension of this protein included strong basic solutions, sodium sulfide, mercaptoethanol, peracetic acid, sodium dodecyl sulfate, and other anionic and nonionic detergents. The elemental compositions of the "soluble", "insoluble", and unfractionated membrane-proteins are compared in Table 10.9. The membrane proteins, especially the soluble fraction, are characterized by relatively low nitrogen contents. The soluble fraction contains a carbohydrate moiety (6 to 14%) consisting of hexose, hexosamines and sialic acid (N-acetyl neuraminic acid); thus it may be classified as a glycoprotein. In buffered solutions at about neutral values of pH and 0.1 ionic strength, this fraction has mucoidal appearance. At higher ionic strengths (i.e., 0.6*M* KCl) the protein is better dispersed and its solutions are clear.

Table 10.9

COMPOSITION OF FAT GLOBULE MEMBRANE PROTEIN[a]

Component	Total Protein Fraction[b]	Soluble Protein Fraction[c]	Insoluble Protein Fraction[d]
Yield, gm/100 gm fat	0.3–0.86	—	—
Nitrogen, %	10.9–12.4	11.1–11.7	13.8
Phosphorus, %	0.32–0.62	0.46–0.78	0.23–0.29
Sulfur, %	0.73–2.58	0.70–0.93	0.94
Trace elements			
Cu, μg/g	20–135	132–135	82
Fe, μg/g	326–600	510	91
Mo, μg/g	72	5	125
Mn, μg/g	11	11	11
Zn, μg/g	20	17	22
Carbohydrate moiety			
Hexose	—	2.7–5.9	trace
Hexosamines	—	1.1–3.9	trace
Sialic acid (NANA)	—	2.0–4.0	—

[a] Ranges of values include all values reviewed by author.
[b] Proteins account for from 30 to 60% of the membrane preparations; remaining material consists principally of phospholipids and associated lipids.
[c] Designated as glycoprotein[38,381] or mucoprotein.[159]
[d] Designated as pseudokeratin.[138]

The biological specificity of the membrane protein was first demonstrated by Palmer and Lewis.[280] Coulson and Jackson[62] confirmed the antigenic properties of the interfacial mucoprotein presumed to be a major component of the soluble protein fraction. These observations, in addition to the compositional characteristics of the membrane proteins, establish the "uniqueness" of the membrane proteins in milk. Recently, R. M. Dowben* detected the presence of an actin-like component, extracted from the fat globule membrane with 0.6*N* KCl, which interacts with bovine myosin.

*Personal communication

Palmer and Powell[281] demonstrated that the membrane phospholipid-protein complex was interfacially more active in a butteroil-water system than were the plasma proteins; further, that the membrane material extracted from the butter phase of churned washed cream was more active than the corresponding material found in the buttermilk phase. This difference in the interfacial activity of the two membrane preparations was attributed to the relatively higher concentration of phospholipids in the preparation obtained from butter serum. Jenness and Palmer[161,162] reported protein/phospholipid ratios of 1.0 to 2.0 and 2.4 to 3.8, respectively, for butter serum and buttermilk membrane preparations. Jackson and Pallansch[160] observed that the soluble membrane protein exerted more interfacial activity at a butteroil-water interface than did other milk proteins.

Electrophoretic characteristics of the membrane proteins have been reported.[33,35,38,159,316,337,338] In general, moving-boundary electrophoretic patterns exhibited a single diffuse gradient peak in alkaline buffers (e.g., veronal pH 8.0 to 8.6; protein conc: 0.5 to 1.0%). In acid buffers (e.g., glycine-HCl, pH 2.0 to 2.5) two or three principal boundaries were apparent. In contrast, the patterns reported by Jackson *et al.*[159] showed but one diffuse boundary in buffers ranging from pH 2.0 to 8.0 and for solutions containing 2.0% protein. Isoelectric points at pH 3.6 to 5.0 were reported for the major boundary. Whether the apparent heterogeneity of this protein fraction represents a mixture of individual proteins or merely reflects the presence of polymeric species of a single protein has not been clarified. Thompson[380] and Brunner and Thompson[38] presented electrophoretic evidence demonstrating that the unextracted, low-density lipoprotein fraction was more heterogeneous than its apoprotein, and that the lipoprotein showed a higher electrophoretic mobility. It was postulated that the glycoprotein exists in various states of association with the lipid moiety. Thus the apparent homogeneity in the moving-boundary electrophoretic patterns was attributed to the incomplete removal of the tenaciously bound lipid moiety.

Sedimentation-velocity patterns of both the soluble apoprotein and the nonextracted, low-density lipoprotein fraction showed evidence of heterogeneity, especially when observed at levels of concentration in excess of 1.0%.[34,38,126,159,316,380] Sedimentation coefficients ($S_{20,w}$) ranging from about 8 to 17 in veronal buffer, pH 8.6, were observed for the low-density lipoprotein fraction. The soluble protein fraction had somewhat lower values ($\simeq$ 8S to 10S), with more apparent homogeneity. Again, the experimental data indicated prevalent intermolecular activity. Jackson's[159] purified mucoprotein fraction ($S_{20,w} = 4.7$), although apparently homogeneous when centrifuged at 25,900 rpm, showed evidence of molecular polydispersity when a rotor speed of 59,700 rpm was employed. Harwalker and Brunner[126] estimated a molecular weight in excess of 1×10^6 daltons for the soluble protein fraction by sedimentation-equilibrium methods. A $S_{20,w}$ of 10 was reported for this preparation in veronal buffer, pH 8.6. Recognizing that the soluble protein probably exists as an intermolecular complex under these experimental conditions, they reexamined the protein in various dissociating systems, e.g., sodium dodecyl sulfate (SDS), SDS plus sodium disulfide, guanidine-HCl plus 2-mercaptoethanol and 67% acetic acid. In the presence of these agents, the protein was dissociated into two species having sedimentation coefficients ($S_{20,w}$) ranging from 2 to 4, with corresponding molecular weights in the order of 2×10^5 to 4×10^5 daltons. Trautman[388] plots of sedimentation-equilibrium data revealed the presence of a small species with an average molecular weight estimated at 2×10^4 daltons, presumably the monomeric species. These studies accentuated further the significance of intermolecular association in the membrane protein and its lipid moieties. Presumably, apolar forces, hydrogen bonding, electrostatic charges and intermolecular disulfide bonds play contributing roles in these interactions.

Physical characterization of the insoluble membrane protein fraction has been more difficult because of its insolubility in the commonly employed aqueous buffer systems. A limited insight into its properties has been achieved by solubilization with one or more of the protein-dissociating agents referred to previously. Tiselius patterns of this protein fraction, treated with disulfide cleaving agents (e.g., sodium disulfide, 2-mercaptoethanol and highly alkaline solutions) showed a single, sharp boundary with a mobility of −4.0 Tiselius units. Sodium dodecyl sulfate adsorbs on the protein, thus making it more hydrophilic. Two fast-moving boundaries were observed for the insoluble protein treated in this manner[138,380] However, when these agent-treated specimens were subjected to ultracentrifugal analysis, they sedimented at low speeds, precluding any meaningful measurements. Obviously, the insoluble protein fraction exists as a highly associated complex, suggesting that it may function as a structural component in the membrane complex.

Membrane lipoprotein fractions as well as solvent-extracted protein fractions showed no electrophoretic migration in starch-urea-gels at pH 9.0.[126,369] The migration of agent-treated specimens was slightly improved, but inadequate for characterization. Disc electrophoresis in 7.5% polyacrylamide gels was more satisfactory, especially when dissociating agents were incorporated into the gels. With the exception of the insoluble protein fraction, which did not move into the gel under any condition, the various membrane preparations (i.e., low- and high-density lipoproteins and their respective soluble apoproteins) showed limited resolution, varying with the type of dissociating agent employed. Amidox-10L, a nonionic detergent, functioned effectively in this capacity. Two distinct zones just behind the ion front and a diffuse zone near the origin were observed. In all cases, appreciable amounts of the specimens remained undissociated in the sample slot. In general, the usual techniques of gel electrophoresis are not satisfactory for studying the membrane system.

More recently, Keenan *et al.*[168] employed the solvent system of Takayama *et al.*,[373] i.e., phenol acetic acid-water in a ratio of 2:1:1 (w/v/v). They succeeded in resolving the fat-globule membrane into approximately 8 zones (one principal zone) by electrophoresis in polyacrylamide gels (7.5% acrylamide, 35% acetic acid and 5*M* urea); acetic acid (10%) was used in both buffer reservoirs. Similar results were obtained with this system in the author's laboratory for both the lipoproteins and their respective solvent-extracted apoproteins. Apparently, the presence of phenol in the solvent system is the key to the success of this technique.

Available information relative to the amino-acid composition of the membrane protein is limited and inconclusive (Table 10.10). The complexity of the protein system and the difficulty encountered in obtaining discrete components have been contributing factors to this lack of definitive data. Hare *et al.*[124] and Brunner *et al.*[32] published partial analyses for the membrane proteins that were estimated by microbiological assay. In subsequent studies, Herald and Brunner[138] reported amino-acid composition for the soluble and insoluble protein fractions that were estimated by paper chromatographic techniques. At best these data can be regarded only in terms of general qualitative composition of heterogeneous fractions. Therefore, it seems unwise at this time to make comparisons between the various sets of data. Chien and Richardson[51,52] reported compositional data for 5 lipoprotein fractions obtained by physical manipulation of washed fat globules. The amino-acid compositions, determined by ion-exchange chromatographic techniques, of the protein moieties of these fractions were essentially similar to those previously reported. Values were included for proline (3.7 to 4.6%) and phenylalanine (3.8 to 5.9%) which had not been reported previously. These investigators concluded that the lipoprotein fractions contain various mixtures of both soluble and insoluble proteins. Also of interest was their observation that all 5 fractions contained from 1.0 to 1.4% glucosamine and from 0.5 to 2.5% galactosamine, further demonstrating the ubiquitous distribution of the soluble glycoprotein components throughout the membrane complex. Saito and Igarashi[331] reported amino-acid compositions for the apoprotein

Table 10.10

AMINO ACID COMPOSITION OF FAT-GLOBULE MEMBRANE PROTEIN (WEIGHT %)

Amino acid residue	Membrane[a] proteins (unfractionated)	Membrane[b] proteins (unfractionated)	Insoluble[b] protein fraction	Soluble[b] protein fraction	Membrane[c] proteins	Membrane[d] proteins
Alanine	—	3.2	3.8	2.4	3.5–4.2	4.3–5.9
Arginine	5.0	7.0	8.2	5.5	3.5–5.1	6.1–7.3
Aspartic acid	—	6.6	7.6	5.4	6.1–7.3	9.1–10.1
½ Cystine	1.5	2.4	2.3	2.7	—	trace–3.5
Glutamic acid	—	9.5	10.8	7.8	7.8–10.6	13.5–15.6
Glycine	3.0	3.2	3.9	2.3	2.8–4.0	4.4–5.1
Histidine	1.7	4.0	4.1	3.9	1.9–2.2	1.1–4.0
Isoleucine	3.5	4.5	4.6	4.4	3.2–4.3	4.1–7.0
Leucine	9.0	7.5	8.2	6.7	5.7–7.7	8.9–9.4
Lysine	6.1	8.0	8.6	7.7	4.5–6.0	6.4–9.0
Methionine	2.1	2.3	3.0	1.4	1.3–2.4	0.0–0.5
Phenylalanine	—	—	—	—	3.8–5.9	5.7–7.2
Proline	—	—	—	—	3.7–4.6	5.3–6.7
Serine	—	4.7	5.3	3.9	3.6–4.3	6.0–8.6
Threonine	6.4	5.0	5.3	4.5	4.2–4.9	5.7–6.1
Tryptophan	0.9	2.1	1.9	2.4	—	—
Tyrosine	—	4.5	4.4	4.6	2.8–3.9	2.8–3.6
Valine	5.4	5.4	6.7	3.6	—	5.2–6.4
Nitrogen	12.34	12.61	11.10	13.80	—	—

[a] Hare *et al.*[124]
[b] Herald and Brunner.[138]
[c] Chien and Richardson,[53] range of values reported for 5 lipoprotein fractions.
[d] Saito and Igarashi,[331] range of values reported for 4 fractions.

moieties of 4 lipoprotein fractions extracted from the membrane complex with 1*M* saline. These were essentially similar in composition and, except for the relatively higher concentrations of glutamic acid, compared favorably with values previously reported.

More recent analyses were given by Brunner and his colleagues[370,371] for protein moieties of membrane fractions prepared by differential centrifugation of the membrane sol—i.e., 7,500S, 230S and 35S pellets—and a glycoprotein isolated from the 35S pellet by extraction with 0.6*N* KCl, (Table 10.11). It is apparent that the protein constituting the 35S fraction, as well as the glycoprotein isolated therefrom, have somewhat different amino-acid profiles from the proteins constituting the faster-sedimenting fractions. These data, together with the reported difference in carbohydrate composition, reflect again the heterogeneous nature of the membrane protein system.

Table 10.11

AMINO ACID COMPOSITION
OF DELIPIDATED MEMBRANE FRACTIONS

Amino Acid Residue	Pellet fraction[a,b,c] 7,500 S	230 S (gm/100 gm protein)	35 S	Glycoprotein[d]
Lysine	6.6	6.6	5.8	6.9
Histidine	2.1	2.3	2.7	3.4
Arginine	6.6	6.5	5.5	4.4
Aspartic Acid	7.7	7.6	9.0	6.7
Threonine	4.7	4.9	5.8	6.2
Serine	5.4	5.6	5.2	5.7
Glutamic Acid	11.3	11.0	10.6	10.6
Proline	5.1	5.4	4.3	2.7
Glycine	3.2	3.3	3.5	2.3
Alanine	3.9	3.9	4.0	2.4
½ Cystine	1.5	1.4	1.9	0.9
Valine	4.5	4.4	4.5	2.6
Methionine	1.9	1.4	1.2	1.3
Isoleucine	4.4	4.3	4.5	3.5
Leucine	8.2	8.1	7.4	7.2
Tyrosine	3.6	3.5	3.9	2.5
Phenylalanine	5.3	5.5	5.3	3.6
Tryprophan	2.2	2.3	2.4	—
Amino Acid residue (%)	88.1	88.1	87.5	73.0
Total carbohydrate (%)	6.6	8.4	10.2	≃ 19.0
P as H_2PO_3	0.4	0.4	0.6	—
Total (%)	95.1	96.9	98.3	92.0

a Pellet protein residues following lipid extraction, Swope and Brunner.[370]
b Weight percentage of each amino acid residue based on a nitrogen content of 14.1% for the 7,500 S, 230 S and 35 S fractions.
c Amino acid content based on duplicate analyses of 20- and 70-hr. digests.
d Glycoprotein extracted from 35 S protein with 0.6N KCl, 20 hr. digest; Swope *et al.*[371]

Enzymes.—Numerous minor proteins, many of which are enzymes, have been identified in milk.[246,343,423] The origin of most of these enzymes is not entirely clear, and its determination is complicated by the possibility that some may be derived from microorganisms, leucocytes and various cell organelles. Assuming that the observed enzymatic activities of fat-globule membrane specimens are endogenous, a remaining question concerns their origin, whether it is by vascular transfer or as products of the mammary gland. Both routes must be considered. Aldolase, which has been identified with the fat-globule membrane, has the same order of activity in blood as in milk.[306] Furthermore, milk ribonuclease is identical with pancreatic ribonuclease, suggesting that its presence in milk is due to "leakage" from the vascular system.[24,25,63,151] Conversely, Folly and White,[92] studying the effects of thyroxine on milk secretion and alkaline phosphatase activity, could find no evidence to suggest that milk alkaline phosphatase originates from blood.

That many enzymes are associated with the fat membrane is well documented.[12,13,54,84,131,138,241,296,439] Kitchen *et al.*[193] detected greater total activities for alkaline and acid phosphatase, catalase, xanthine oxidase, aldolase and ribonuclease in skimmilk than in buttermilk. And, with the exception of ribonuclease, all showed higher specific activities in the buttermilk fraction. Morton's studies made it apparent that many of the enzymes associated with the membrane material are actually microsomal constituents existing as lipoprotein complexes.[12,13,240,254,256] He identified alkaline phosphatase and xanthine oxidase (the two principal enzyme components), DPNH-diaphorase, DPNH-cytochrome *c* reductase, and a haemo-protein (cytochrome b_5). Furthermore, he compared the enzymatic activities of milk microsomes with those of microsomes isolated from mammary tissue (Table 10.12). Differences in enzymatic activities between milk and gland microsomes were attributed to residence time in the milk system. This effect is exemplified by the lower activities of cytochrome *c* reductase and diaphorase in milk microsomes when compared to their corresponding activities in the gland microsomes. Of particular interest was the activation of cellular xanthine oxidase when exposed to milk plasma. Numerous investigators have observed the activation of milk xanthine oxidase by agitation, freezing, homogenization and other manipulations which affect the physiochemical state of the enzyme lipoproteins.[113,114,175,176,324] Morton concluded that milk and mammary gland microsomes are identical.

A recent study by Dowben *et al.*[84] with carefully processed fat-globule membrane specimens revealed activities for the enzymes listed in Table 10.13. Additionally, phosphatidic acid phosphatase and a small amount

Table 10.12

COMPARATIVE ENZYMATIC ACTIVITIES OF MICROSOMAL PARTICLES ISOLATED FROM THE MAMMARY GLAND AND MILK FROM THE COW[a]

Cow No.	Particles	Alkaline Phosphatase[b]	Xanthine Dehydrogenase[c]	DPNH-Diaphorase[c]	DPHN-Cytochrome Reductase[d]
1	Milk microsomes	1.14	1.150	0.13	0
	Gland microsomes	3.42	0.040	0.58	3.28
2	Milk microsomes	9.35	0.670	0.11	0.088
	Gland microsomes	2.85	0.046	1.58	7.20
2	Gland microsomes	18.4	0.3	11.0	52.5
	Gland microsomes incubated in milk serum	29.9	0.9	0.18	0.04

a Data of Bailie and Morton.[12]
b Inorganic phosphate liberated: μ moles/mg./hr. at 38°C.
c 2,6 dichlorophenolindophenol reduced: μ moles/mg./hr. at 19°C.
d Cytochrome *c* reduced: μ moles/mg./hr. at 19°C.

Table 10.13

ENZYMATIC ACTIVITIES ASSOCIATED WITH THE FAT GLOBULE MEMBRANE[a]

Enzyme	Number of determinations	Average	(range)
Alkaline phosphomonoesterase	8	0.63	(0.21–0.92)[b]
Acid phosphomonoesterase	8	0.087	(0.061–0.14)[b]
Phosphodiesterase (pH 8.9)	6	0.134	(0.092–0.161)[b]
Phosphodiesterase (pH 5.0)	6	0.008	(0.004–0.016)[b]
NADH-cytochrome *c* reductase	6	0.0068	(0.0031–0.0106)[c]
Cholinesterase	4	0.171	(0.139–0.227)[d]
Mg^{2+}-activated ATPase	8	0.74	(0.29–0.94)[b]
(Na^{+}-k^{+}-Mg^{2+})-activated ATPase	8	0.91	(0.54–1.18)[b]
Glucose-6-phosphatase	3	0.064	(0.056–0.074)[b]
Xanthine oxidase	4	1.304	(1.008–1.656)[e]
Aldolase	4	0.016	(0.012–0.024)[f]
Phosphatidic acid phosphatase	1	appreciable	
Diglyceride kinase	1	low	

a Data of Dowben *et al.*[84]
bμmoles Pi liberated/mg protein/hr
cμmoles cytochrome *c* reduced/mg protein/hr
dμmoles acetyl-β-methylcholine hydrolyzed/mg protein/hr
eμg xanthine oxidized/mg protein/hr
fμmoles fructose diphosphate split/mg protein/hr

of diglyceride kinase were found. No succinate-cytochrome *c* reductase, cytochrome *c* oxidase, β-glucuronidase, lipase, aspartate oxoglutarate aminotransferase or lactate dehydrogenase activities were detected. Centrifugally classified (100,000 $\times$ *g* for 1 hr) membrane lipoprotein fractions were assayed for selected enzymatic activities which are shown in Table 10.14.

Table 10.14

DISTRIBUTION OF ENZYMES BETWEEN FAT MEMBRANE FRACTIONS[a]

	Membrane fraction		
Enzyme	Pellet	Interfacial fluff	Supernatant
Alkaline phosphomenoesterase[b]	0.362	0.496	0.163
Alkaline phosphodiesterase[b]	0.157	0.129	0.307
Mg^{2+}-activated ATPase[b]	0.023	0.124	0.1308
(Na^{+}-$k^{+}$$Mg^{2+}$)-activated ATPase[b]	0.048	0.148	0.016
Xanthine oxidase[c]	0.670	1.526	0.087
Aldolase[d]	0.0000	0.0043	0.067
Total protein	30.3%	39.7%	22.0%

a Data of Dowben *et al.*[84]
b μmoles Pi/mg protein/hr.
c μg xanthine oxidized/mg protein/min
d μmoles fructose diphosphate split/mg protein/hr

Herald and Brunner[137,138] reported that alkaline phosphatase was concentrated in the soluble protein fraction, while xanthine oxidase was a component of the insoluble fraction of solvent-extracted membrane material. Zittle *et al.*[439] observed that about 85% of the original activities of xanthine oxidase and alkaline phosphatase were lost from the membrane preparations recovered following 4 washing cycles. They hypothesized that the xanthine oxidase-lipoprotein complex occurs in a range of particulate sizes, but that the phosphatase-lipoprotein particulate is more uniform in size. It was suggested that the enzymes occur in specific complexes or in different proportions in the same complex.

Tarassak and Frankel[376] proposed that one of the lipase systems of milk is preferentially adsorbed on the fat-globule surface when milk is cooled. They believed that this lipase is different from the more prevalent plasma lipase, which is normally associated with micellar casein. Recently, Olivecrona and Lindahl[275] observed that a serum-activated lipoprotein lipase was preferentially concentrated in the buttermilk fraction of churned cream and that other lipase systems were essentially absent. The absence of lipase in the buttermilk confirms the previous observation by Dowben *et al.*[83] On the other hand, Saito and Igarashi[331] and Quigley *et al.*[311] detected lipase activity in fat-globule membrane preparations.

Matsushita *et al.*[241] demonstrated phosphodiesterase activity in the microsomal fraction obtained from washed fat globule prepared according to the procedure of Morton. The enzyme hydrolyzed ribonucleic acid to 5′-mononucleotides. Recently, it was demonstrated that the fat-globule membrane contains a 5′-nucleotidase capable of hydrolyzing the nucleotide-5′-monophosphate.[148,296] Also, thioctic acid, a coenzyme in the redox system of the oxidative decarboxylation reaction and active in sulfhydryl-disulfide interchange as a sulfhydryl oxidizing agent, has been isolated from fat-globule membranes.[23]

The enzyme complement of the membrane material explains, in part, the distribution of trace elements given in Table 10.9. Of particular significance are the relatively high concentrations of molybdenum and iron, which are known constituents of xanthine oxidase. Herald and Brunner[138] reported that these elements are preferentially located in the insoluble membrane protein fraction. Swope *et al.*[372] noted that the liberation of a greenish-yellow component (riboflavin) from membrane specimens was enhanced by addition of acid to pH 4.0, or by addition of the anionic detergent dodecyl sulfate, and by freeze-thawing or freeze-drying; also that 92 to 96% of the membrane riboflavin exists as flavin-adenine dinucleotide, a coenzyme of xanthine oxidase. From the Mo, Fe and flavin-adenine dinucleotide (FAD)—all constituents of xanthine oxidase—contents of the membrane protein, Swope *et al.*[369] estimated the enzyme content of the membrane protein at about 8%. Interestingly, the riboflavin content of human, mares', cow's, sheep and goats' milk parallels the activity of xanthine oxidase therein.[223,276,325] Because FAD is loosely bound in the enzyme complex and may be susceptible to degradation by plasma enzymes, it could be that xanthine oxidase is the source of riboflavin in milk.

The relatively high concentration of copper was not unexpected, since it is recognized that membrane protein exhibits an affinity for this element.[8,175,176,265,333,334] However, it is not entirely certain whether all the membrane copper represents an adsorbed constituent. Roussos and Morrow[330] reported an active xanthine oxidase in which molybdenum was replaced by copper.

Lipids.—Palmer and Wiesse[285] identified lecithin, cephalin and sphingomyelin as the principal membrane phospholipids, accounting for 19% of the total membrane material. In subsequent studies, Palmer and his colleagues reported the presence of a high-melting glyceride fraction (HMG), which constituted about 50% of the lipid moiety.[163,164,285] Thompson *et al.*[383] separated membrane lipids into principal classes by silicic acid chromatography and found, in addition to high concentrations of phospholipids and neutral triglycerides, significant quantities of mono- and diglycerides, free fatty acids, cholesterol, cholesterol esters and, in some preparations, carotenoids and a squalene-like component (Table 10.15). The compositional variations between the two preparations reported here reflect differences in preparative procedures. Membrane preparation No. 2 was washed with ethyl ether to remove loosely bound lipid components prior to extraction with an ethanol-ethyl ether mixture. Thompson *et al.* concluded that the carotenoids, squalene and free fatty acids and, possibly a portion of the neutral glycerides are associated loosely in the membrane structure. Earlier studies revealed that a large portion of the neutral triglycerides consisted of a high-melting glyceride fraction (HMG), which crystallizes as a waxy-appearing material from an ethanolic solution of the neutral triglycerides at 21 to 23°C.[164,382,431] More recent studies by Huang and Kuksis,[149] Hladík,[140] and Richardson and Guss[319] substantially support the qualitative distribution of membrane lipids as reported here. Principal discrepancies were the absence of mono-

Table 10.15

COMPOSITION OF THE LIPID FRACTION OF THE FAT GLOBULE MEMBRANE[a] (IN PERCENT)

	Sample No. 1[b]		Sample No. 2[c]	
Component	Percentage of Membrane Lipids	Percentage of Whole Membrane	Percentage of Membrane Lipids	Percentage of Whole Membrane
Carotenoids	0.45	0.30	0	0
Squalene	0.61	0.40	0	0
Cholesterol esters	0.79	0.54	0.63	0.27
Triglycerides	53.41	36.12	49.98	21.88
Free fatty acids plus other triglycerides	6.30	4.26	—	—
Cholesterol	5.17	3.50	3.64	1.59
Diglycerides	8.14	5.49	10.58	4.63
Monoglycerides	4.66	3.14	6.45	2.82
Phospholipids	20.35	13.76	28.72	12.57
Totals	99.88	67.51	100.00	43.76

a Data of Thompson *et al.*[383]
b Isolated from a pooled milk source, previously cooled. Membrane preparation contained 32.5% protein.
c Isolated from freshly drawn, warm milk. Washed-cream "buttermilk" was washed with Et_2O prior to fractionation. Membrane preparation contained 56.2% protein.

glycerides and lower concentrations of free fatty acids and cholesterol. These variations were rationalized on the basis of differences in the preparative and analytical procedures employed (see Table 10.16).

Table 10.16

COMPARISON OF MEMBRANE SPECIMENS OBTAINED BY DIVERSE PROCEDURES [a]

	Phospho-lipids	Neutral-lipids	Protein	Mass ratio protein/phospholipids
	(%)			
Membrane material (1)	23	31	47	2.0
Membrane material (2)	13.8	53.7	32.5	2.4
Total membrane fraction (4)	13.8	47.7	38.5	2.7
DOC-released lipoprotein particles (3)	42	14	44	1.0
Water-insoluble residue material of inner membrane (3)	16	61	23	1.4

a From Peereboom[302]
(1) Prepared from butter serum by J. Koops[198]
(2) Prepared from butter serum with fractionation into protein and lipid fractions by Thompson *et al.*[383]
(3) As calculated from the values given by S. Hayashi and Smith[131]
(4) According to H. C. Chien and T. Richardson[52,53]

Hladík and Forman[141] determined by gas chromatographic analysis that the total membrane lipids contained lower concentrations of C_4 to C_{12} and higher concentrations of C_{14}, C_{16} and C_{18} fatty acids than did core fat from the same source. In a subsequent study, they determined the fatty-acid distribution in the major lipid classes of the membrane lipid moiety (see Table 10.17). These data demonstrate significant differences in the fatty-acid composition of the various lipid classes constituting the membrane complex. The relatively high concentrations of C_{10} to C_{12} residues and low concentrations of C_{16} to C_{18} residues in the cholesterol ester fraction differ markedly from the values reported by Keenan *et al.*[168] Although preparations from different animals on different feeding regimes might account for some variation in composition, it is difficult to reconcile the divergence of recorded data on this basis. Thus, the observed differences appear to be related to the purity of the specimens or the analytical technique employed.

The fatty-acid composition of membrane triglycerides is of interest because of the controversy involving the existence of a high-melting glyceride fraction (HMG) as an integral part of the membrane complex. Thompson *et al.*[382] observed that a HMG fraction crystallized from

Table 10.17

FATTY ACID COMPOSITIONS OF THE PRINCIPAL LIPID CLASSES CONSTITUTING THE FAT GLOBULE MEMBRANE[a]

Fatty Acid Residue[b]	Lipid Classes										
	Cholesterol Esters		Triglycerides	Diglycerides	Free Fatty Acids	Phospholipids					
						Total	PC	PE	PI	SP	PS
Reference	142	168	142[b]	142[b]	142	142[b]	168	168	168	168	168
	Weight %										
8 :0	—	—	—	—	4.2	—	—	—	—	—	—
8 :1	—	—	—	—	3.5	—	—	—	—	—	—
9 :0	—	—	—	—	8.0	—	—	—	—	—	—
10 :0	12.2	5.1	2.5	—	7.7	1.0	—	—	—	—	—
10 :1	11.2	—	—	—	—	—	—	—	—	—	—
11 :0	17.1	1.8	—	—	1.6	—	—	—	—	—	—
12b:0	—	—	—	—	1.3	—	—	—	—	—	—
12 :0	8.3	3.8	3.5	3.7	19.4	—	—	—	—	—	—
12 :1	—	—	—	2.8	—	—	—	—	—	—	—
12 :2	—	—	—	1.1	1.1	—	—	—	—	—	—
13 :0	6.0	—	—	—	—	—	—	—	—	—	—
13 :1	—	1.3	—	—	—	—	—	—	—	—	—

14 :0	6.2	9.7	12.0	9.7	9.7	5.6	2.7	6.0	2.0	8.0	5.0
14 :1	—	1.7	2.1	1.8	—	—	—	—	—	—	—
14 :2	—	—	—	1.8	—	—	—	—	—	—	—
15 :0	4.7	2.4	2.7	2.9	1.3	—	—	—	—	—	1.0
15 :1	2.8	3.2	—	—	—	—	—	—	—	—	—
16[b]:0	—	—	1.2	1.7	—	—	—	—	—	—	—
16 :0	9.4	24.0	33.6	33.3	17.9	25.2	32.0	8.0	23.0	29.0	17.0
16 :1	3.8	9.6	4.5	5.3	2.7	2.4	—	—	—	—	—
16 :2	3.9		2.1	3.7	1.7	1.5	—	—	—	—	—
17 :0	2.8	1.2	1.2	2.1	1.4	—	—	—	—	—	2.0
17 :1		1.7	—	—	—	—	—	—	—	—	—
18 :0	1.5	5.3	9.6	9.0	5.4	5.0	14.0	15.0	18.0	5.0	15.0
18 :1	4.3	25.8	16.6	11.7	9.1	43.2	33.0	40.0	31.0		27.0
18 :2	—	1.2	2.3	2.7	1.9	4.7	6.0	8.0	2.0	—	10.0
18 :3	—	—	—	—	—	—	—	—	—	—	2.0
20 :0	1.2	—	—	—	—	—	2.0	5.0	3.0	—	—
20 :1	1.0	—	—	—	2.3	—	—			—	—
20 :4	—	—	—	—	—	—	7.0	21.0	12.0	—	18.0
22 :0	—	—	—	—	—	2.4	—	—	—	11.0	—
22 :1	—	—	—	—	—	4.6	—	—	—	—	—
23 :0	—	—	—	—	—	—	—	—	—	28.0	—
24 :0	—	—	—	—	—	2.2	—	—	—	14.0	—

[a]Compiled from data of Hladik and Forman[142], representing three determinations.
[b]Number before colon represents the number of carbons; number after colon represents the number of double bonds. Only residues constituting one per cent or more of the total are reported (see original articles for details).

membrane triglycerides and a corresponding fraction from butter oil were similar in composition and physical characteristics. HMG is characterized by its waxy appearance, high melting range (50 to 53°C) and its relatively high concentrations of myristic, palmitic and stearic acid residues, which constitute ≃ 86% of the glyceride fraction. Wolf and Dugan[431] elucidated the structure of this fraction as 71% GS_3, 26% GS_2U and 3% GSU_2. Whether the HMG fraction constitutes an integral part of the membrane lipid or is merely an artifact of the preparative procedure remains unanswered. Vasić and deMan[399] examined the lipid extracted from membrane material prepared at 37°C and found that it possessed a fatty-acid composition similar to that of milk fat; they conclude that the HMG fraction is not a normal constituent of the membrane. They hypothesized that, at temperatures generally employed in the preparation—i.e., the churning step—of membrane material, the HMG fraction crystallizes on the inner surface of the membrane. This interpretation is harmonious with microscopic observations of a birefringent layer (crystallized fat) at the surface of fat globules.[41,187,404] King[187] proposed that HMG combines with the membrane phospholipids either as a solid solution or, upon churning cooled cream, in a crystalline state. The chemical and physical properties of these two lipid components favor an interaction through apolar associations in their long-chain fatty-acid residue. Contrary to Vasić and deMan's observation, Harwalker and Brunner[unpublished] observed that the HMG content of a membrane specimen obtained from washed fat globules by agitation at 40°C contained slightly more HMG and less phospholipids than specimens produced by churning at lower temperatures. Also, that the ratio of phosphatidyl serine to phosphatidyl choline was higher in the latter preparation.

The fatty-acid composition of the membrane triglycerides—prepared by churning—reported by Hladík and Forman[142] (see Table 10.17) show no unusual values to indicate the presence of high concentrations of HMG fractions. On the other hand, Kärkkäinen[165] reported that a triglyceride fraction extracted from a fat-globule membrane preparation with petroleum ether contained 13.5% myristic, 48% palmitic and 28% stearic fatty acid residues—values indicative of the presence of the HMG fraction. Keenan *et al.*[169] observed quantitative differences in the fatty-acid profiles of the neutral glycerides extracted from secretory plasma membrane and fat-globule membrane preparations, the latter showing significantly higher concentrations of palmitic and stearic acid residues. Their electron micrographs of fat-globule membranes revealed a diffuse, electron-dense layer on the inner side of the membrane which was attributed to tenaciously adsorbed HMG derived from the core fat. Also, Huang and Kuksis[149] reported that the triglycerides of membranes derived from winter milk contained proportionally more palmitic and stearic acid and less oleic acid than the triglycerides of the corresponding fat cores. In both the above experiments, membrane specimens were prepared by a freeze-thaw technique.

The phospholipids of milk have been extensively classified and characterized (see Chapter 4). Most of the available data were obtained from phospholipids derived from churned cream buttermilk. Thus, the phospholipids characterized represent the major portion of those associated with the fat-globule membrane. These have been identified as phosphatidyl choline, phosphatidyl ethanolamines, sphingomyelins, phosphatidyl serines, phosphatidyl inositols, and, to a lesser extent, cerebrosides and plasmalogens. No appreciable percentages of fatty acids below C_{10} are found; the principal saturated acids are palmitic and stearic (see Table 10.17). The unsaturated acids $C_{18:1}$ and $C_{18:2}$ account for about 50% of the residues of phosphatidyl cholines, serines, and ethanolamines. The sphingomyelins have relatively low concentrations of unsaturated acids; the saturated acids C_{16}, C_{18}, C_{22} and C_{23} account for about 80% of the total. The cerebroside fractions contain glycolipid components A and B, corresponding to blood serum glycocerebroside (ceremide-monohexoside), respectively.[252,253] Hladík and Michalec[143] confirmed the presence of these two glycolipids in the membrane lipoprotein complex. The monohexosides accounted for 6.5% and the dihexosides 2.5% of the total phospholipids. The major sphingosine bases were C_{18}-sphingosine and probably C_{16}-sphingosine and C_{16}-dihydrosphingosine. The minor component was C_{18}-dihydrosphingosine. No marked differences were observed in the fatty acid composition, the major constituents having chain lengths of $C_{22:0}$, $C_{23:0}$ and $C_{16:0}$. Only small proportions of hydroxy acids were present. Galanos and Kapoulos[101,102] identified at least 16 different glycophospholipids, 2 cerebrosides and 3 gangliomide-like mucolipids in fat-globule membranes. Kayser and Patton[166] isolated glycosyl- and lactosyl-ceremides from skimmilk and the fat globule membrane, which differed in fatty-acid composition; the skimmilk components contained primary C_{18} and shorter-chain fatty acids, whereas the membrane components were characterized by acid containing 20 to 25 carbons. The significance of this compositional difference was related to their respective role in membrane myelination and stability.

The phospholipid moiety in fat-globule membrane preparations *per se* is composed principally (75 to 80%) of sphingomyelin, phosphatidyl choline and phosphatidyl ethanolamine in about equal proportions, with phosphatidyl serine and phosphatidyl inositols in lower concentrations.[149,168,293] The distribution of fatty-acid residues in the membrane phospholipids is listed in Table 10.17.

The distribution of phospholipids in various fractions of milk is illustrated in Table 10.18. These data reveal several interesting relation-

Table 10.18

PHOSPHOLIPID CONTENT OF MILK AND MILK PRODUCTS(%)[a]

Product	Phospholipids in product	Fat in product	Phospholipids in fat
Whole milk	0.0337	3.88	0.869
Skim milk	0.0169	0.09	17.29
Cream	0.1816	41.13	0.442
Buttermilk	0.1819	1.94	9.378
Butter	0.1872	84.8	0.2207

a From Kurtz.[210]

ships: first, that about 50% of the phospholipids are found in skimmilk, probably as components of small fat globules and lipoprotein particles; secondly, that most of the phospholipids associated with cream are recovered in the aqueous phase (buttermilk) after churning. Although we might expect skimmilk and buttermilk lipoproteins to be similar and of common origin, existing experimental evidence suggests that there are specific differences. Patton *et al.*[290,295] reported a lipogenic agent in milk which has the capacity for glyceride biosynthesis. This material has physical and compositional similarities to the membrane complex. However, its precise relationship to the membrane is unclear, since the biosynthetic activity was confined almost entirely to a lipoprotein fraction obtained from milk plasma. In further experiments to clarify the nature of the skimmilk lipoprotein, Patton and Keenan[293] reported that skimmilk lipoprotein (42% of total milk phospholipid) and fat-globule membrane lipoprotein contained the same phospholipids in essentially the same proportions, with similar fatty-acid compositions. However, upon infusion of ^{14}C-palmitate into the mammary gland, more extensive labeling was detected in the skimmilk phospholipids. Thus, it was concluded that fat-globule membrane is an unlikely source of skimmilk lipoprotein. Gammack and Gupta[103] isolated a lipoprotein fraction from skimmilk which exhibited a specific activity for alkaline phosphatase in the range of activities reported for fat-globule microsomes.[255] They concluded that lipoprotein particles in the skimmilk are derived from the same cellular membranes that cover the surface of fat globules. In a subsequent study, they reported that the skimmilk lipoprotein fraction is a necessary constituent for the normal creaming of cow's milk, suggesting that at least a portion of the skimmilk lipoproteins are different from those associated with the fat-globule membrane.[104]

Recently, Buchheim[44] produced electron micrographs of lipoprotein particles obtained by ultracentrifugation of whole milk (human) and a suspension of fat-globule membranes (Fig. 10.25). The whole milk particles illustrated in Frame A, ranging in size up to 500 nm, are smooth-surfaced and spherical. Particles sedimented from the membrane sol (Frame B) are more uniform in size—50 to 125 nm—rough-surfaced and less spherical. The clustered particles obvious in Frame A represent micellar casein. Whether all the small lipoprotein particles apparent in Frame A are plasma lipoproteins or dislodged fat-membrane lipoproteins cannot be ascertained. Nor is it unequivocally certain that the two large bodies are not small fat globules with densities exceeding unity. Because of the experimental difficulties encountered in effecting a clear separation of serum and membrane-adsorbed lipoprotein particles, the question regarding the origin of lipoprotein particles in milk remains essentially unanswered.

Molecular Organization

The composition of the fat-globule interfacial material has been reasonably well elucidated. However, the organization of these components in the membrane complex is not completely understood. Certainly, as their nature might indicate, it is a foregone conclusion that intermolecular associations contribute to the structural characteristics of the complex. These involve the interactions between proteins and lipids, principally, but also include those of the lipid-lipid and protein-protein types. Gurd[115] presented an excellent review of complex biological structures and the forces contributing to their formation and stability. The salient features of these concepts as applied to the fat-globule membrane were discussed by Brunner.[31,126]

The classic model for the fat-globule membrane, as proposed by King,[187] consists of a radially oriented surface layer of polar phospholipids and other less polar lipids—e.g., cholesterol, vitamin A—which, in association with molecules of high-melting glycerides, extends into the peripheral area of the fat core. In association with the hydrophilic polar heads of the phospholipids, he envisioned a closely oriented layer of protein, extending into the plasma phase a distance commensurate with the residual electrostatic charge. Although these

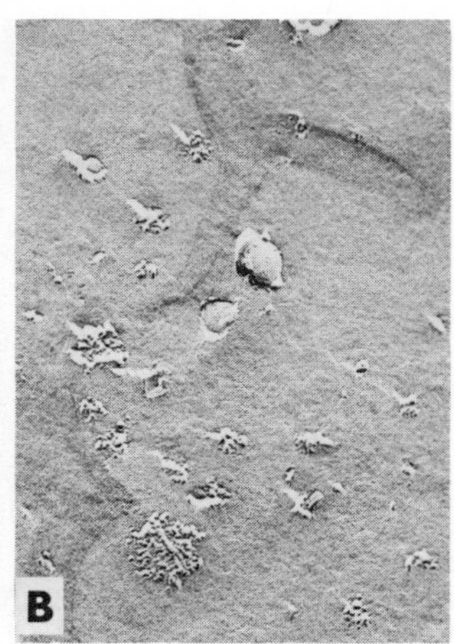

Data of Buchheim[44]

FIG. 10.25. ELECTRON MICROGRAPHS OF FREEZE-DRIED LIPOPROTEIN PARTICLES FROM HUMAN SKIMMILK (A) AND WASHED CREAM BUTTERMILK (B)

Central particle in Frame A is ≃500 nm in diameter.

components, and others, are known to be present in the membrane complex, this model no longer represents a tenable concept for the endogenous membrane structure. Instead, the studies of Bargmann,[15-17] Buchheim,[40,44] Dowben,[83,84] Keenan,[168,169] Knoop,[194] Morton,[12,13,256] Patton,[146,291] Steward,[359-361] and Swope[370] demonstrated that the interfacial material consists of an array of fragmented plasma membranes and attendant microsomal and lipoprotein particles. Characteristics pertinent to the organization of this complex are considered in the following paragraphs.

Alexander and Lusena[2] studied the effect of dissociating agents on the membrane complex. They classified desoxycholate-dissociated membrane into various fractions by differential centrifugation according to the scheme shown in Fig. 10.26. Two high-density fractions (A and D) contained 32% lipid, one-half of which was phospholipid and accounted for 30% of the membrane lipoprotein. The low-density fraction (B) amounted to 36% of the membrane preparation and contained 65% lipid, one-quarter of which was phospholipid. The remaining material was self-sedimenting at 105,000 × *g* for 6 hr and contained 70% lipid, one-half of which was phospholipid. Xanthine oxidase and alkaline phosphatase activities were predominantly associated with the high- and low-density fractions, respectively. No activity was detected in the supernatant fraction. In all fractions except D, 60 to 80% of the phospholipids was lecithin, with small percentages of

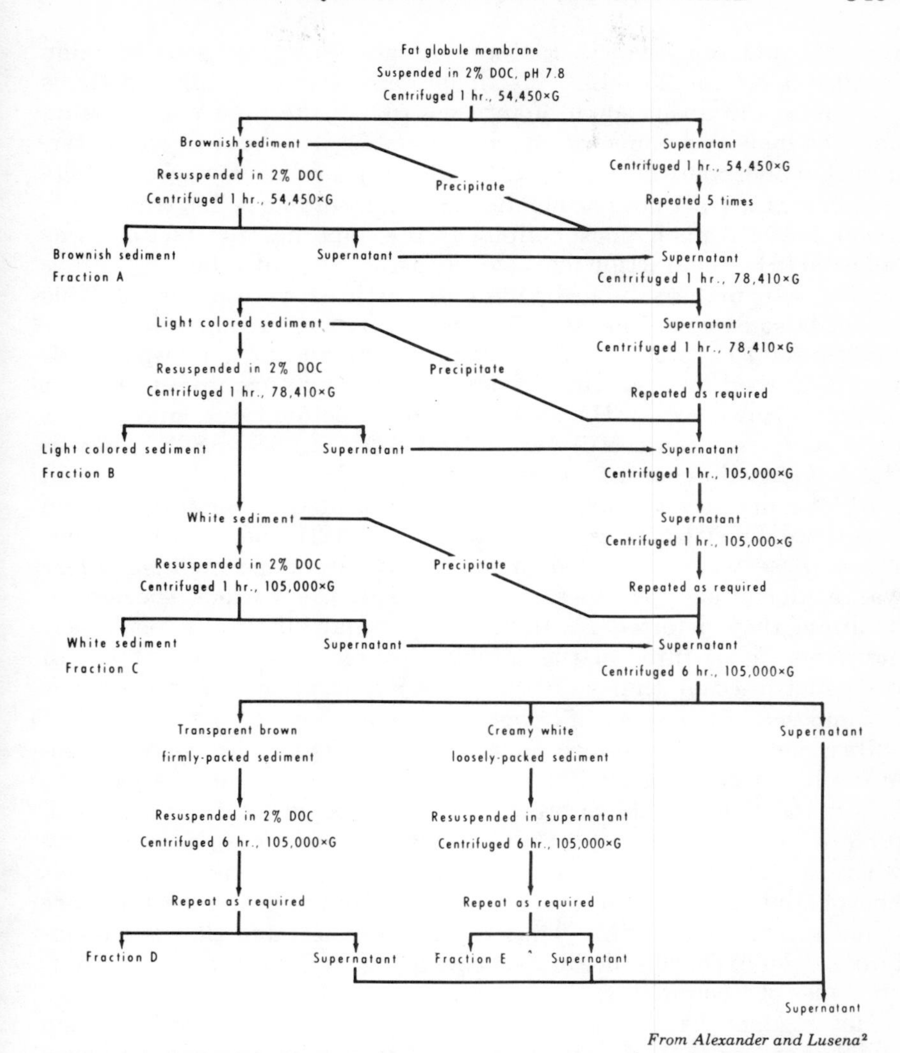

From Alexander and Lusena[2]

FIG. 10.26. FRACTIONATION OF LIPOPROTEIN FRACTIONS FROM DESOXYCHOLATE-DISSOCIATED FAT GLOBULE MEMBRANE MATERIAL

lysolecithin, sphingomyelin, lysocephalin, phosphatidyl ethanolamine, and serine. The phospholipids in Fraction D were composed of lecithin (40%), phosphatidyl ethanolamine (30%), and phosphatidyl serine (20%). The supernatant fraction contained slightly higher levels of lysolecithin, lysocephalin and sphingomyelin.

By somewhat similar procedures, Hayashi and Smith[131] found that about 45% of the membrane lipoproteins were released by desoxycholate and remained in the supernatant following centrifugation at 50,000 × g for 90 min. This lipoprotein fraction contained lipids and protein in about equal quantities, 76% of the lipid moiety being phospholipids. A sedimentation coefficient ($S_{20,w}$) of about 13 was determined in phosphate buffer, pH 7.0, indicating a relatively high particle weight for the fraction. The nondissociated, insoluble pellet, containing about half as much phospholipid as the supernatant fraction, was believed to be the fraction most clearly associated with the fat globules and on which the soluble lipoprotein particles were adsorbed. This model is somewhat like Morton's,[256] who proposed that microsomes containing 70% protein, 15% neutral lipids and 15% phospholipids were adsorbed on a continuous layer of protein surrounding the fat droplets. However, as Hayashi and Smith pointed out, microsomes have sedimentation coefficients in the order of 50S to 80S, whereas the dissociated lipoprotein particles were in the order of 13S. Hayashi *et al.*[132] centrifugally classified as the dissociated lipoprotein fraction into three density classes, i.e., < 1.13, $1.13-1.21$ and > 1.21 gm/cm^3. When these were examined in an analytical ultracentrifuge, there was evidence of gross interaction between the various species. In addition, they detected xanthine oxidase and alkaline phosphatase activities in all three of the soluble lipoprotein species, but not in the pellet fraction, thus supporting their concept that the membrane is composed of two types of lipoprotein complexes.

Brunner and Thompson[38] centrifugally classified untreated membrane lipoproteins at 25,000 × g for 2 hr into a mucoidal-appearing supernatant and a reddish-brown pellet. The supernatant carried lipoproteins containing about 60% lipids, 23% of which was phospholipids, whereas the pellet contained 11% lipids, of which 38% was phospholipids. All the carotenoids, squalene and cholesterol esters were found in the supernatant. Thus it was demonstrated that the membrane, even in the absence of dissociating agents, is composed of different types of lipoproteins.

Heterogeneity in the membrane structure was verified by Chien and Richardson,[52,53] who obtained 5 discrete lipoprotein fractions, based upon relative desorbability from the membrane, by stirring washed cream until there was a phase inversion. Fractionation was accomplished by centrifugation at 50,000 × g for 90 min, applied first to the stirred sample and then to the combined aqueous phase obtained from the churned cream. A schematic representation of the samples collected is shown in Fig. 10.27. The first step is yielded supernatant (*Os*) and pellet (*Op*) fractions which were considered to represent the loosely bound or "outer layer" of the membrane complex. These fractions accounted for 55% of the membrane material. The *Os* fraction was

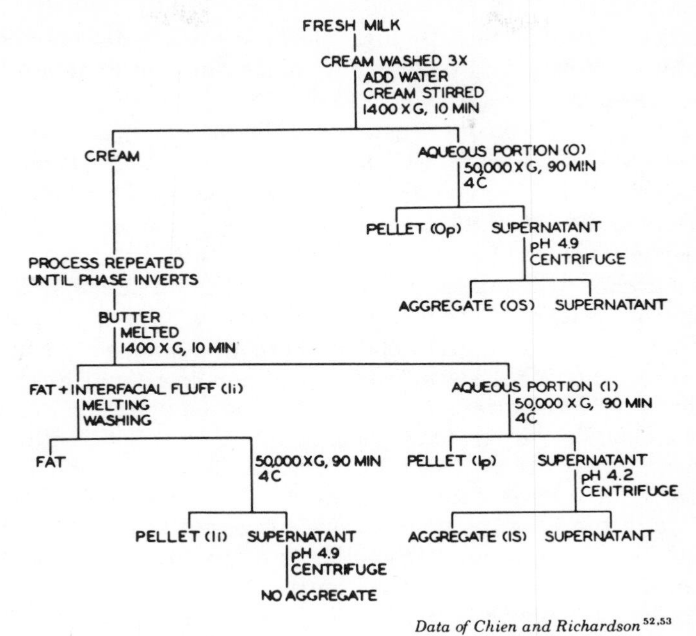

Data of Chien and Richardson[52,53]

FIG. 10.27. FRACTIONATION OF FAT GLOBULE MEMBRANE LIPOPROTEINS

fluffy-white and high in lipids (87%), while the *Op* fraction was reddish-brown and low in lipids (40%). The second step yielded three fractions, i.e., pellet (*Ip*), supernatant (*Is*) and interfacial fluff (*Ii*), which were considered as the inner layer of the membrane complex. The pellet, supernatant and interfacial-fluff lipoprotein fractions contained 43%, 72% and 40% lipids, respectively. All fractions contained phospholipids and neutral lipids in which myristic, palmitic, stearic and oleic acids were the predominant residues. Similarities within the amino acid and hexosamine compositions of the fractional protein moieties were interpreted as an indication that all 5 lipoprotein fractions contained a similar mixture of protein moieties.

Additional evidence that lipoproteins constitute an integral part of the membrane material was provided by the study of Berlin *et al.*[22] A microsomal fraction obtained from a membrane suspension by centrifugation at 14,000 × *g* for 45 min was disrupted in an Aminco-French pressure cell at 8,000 psi. A low-density lipoprotein was isolated in low yields from the homogenate by differential centrifugation (flotation) in NaBr-containing solutions of varying specific gravities. The final purification step was accomplished by centrifugation in solution (sp. gr. 1.064) at 114,000 × *g* for 4 hr. The lipoprotein was removed from the top one milliliter of the tube, just below the surface, dialyzed against NaBr solution of density 1.063 gm/cm^3, and examined in an analytical centrifuge. A flotation coefficient (S°f) of 26.2 and an average molecular weight of 3.9 x 10^6 daltons were estimated. The sedimentation diagrams showed numerous aberrant sedimenting boundaries, a characteristic usually attributed to molecular polydispersity within the solute. This liproprotein analyzed 12.8% protein and 87.13% lipid, of which 52.02% was phospholipids and 35.11% neutral lipids.

Recent studies by Swope and Brunner[370,371] again demonstrated the heterogeneous nature of the membrane complex. Typical preparations of fat-globule membrane obtained by churning washed cream were subjected to centrifugal classification and compositional analysis. A representative distribution of particle sizes based upon sedimentation conditions designed to yield pellets of > 7500S to < 35S is illustrated in Table 10.19. No more than 5 to 10% of the membrane material could qualify as water-soluble components (i.e., < 35S). Obviously, the smaller units survived the globular-washing step in preparation, and apparently were dissociated from the complex during the churning step. Conceivably, the observed distribution in particle size of the membrane complex may have been a ramification of agitation-induced dissociation. This being the situation, all the centrifugally classified fractions should have similar composition if the membrane was homogeneous, which, however, was not the case. For example, the proportion of total lipids in these sedimented fractions increased with decreasing values of S. And, within the lipid moieties, there occurred decreasing proportions of phospholipids with decreasing values of S. Variations in the protein complement of these fractions were exem-

Table 10.19

COMPOSITION AND APPEARANCE OF MEMBRANE FRACTIONS[a]

Pellet fractions[b]	Yield	Constituents				Appearance
		Protein[c,d]	Lipid	Phospholipid	Cholesterol	
	(g)	(%)				
7,500S	1.77	82.5	17.5	11.1	0.7	White
230S	4.44	70.5	29.5	15.7	0.9	White-reddish-brown
35S	3.13	44.7	55.3	23.7	1.5	Reddish-brown

a Data of Swope and Brunner.[370]
b Prepared from three liters of washed cream containing 50% fat.
c Estimated by difference between yields and mass of lipids.
d Delipidated proteins contained 6.6%, 8.4% and 10.2% carbohydrate for the 7,500S, 230S and 35S fractions, respectively.

plified by increasingly higher carbohydrate contents with decreasing values of S.

Electron microscopy has contributed dimensional aspects to the nature of the membrane material. Schwarz and Fischer[342] were the first to observe the fat-globule membrane prepared for electron microscopy. From these observations they proposed that the membrane is composed of three layers: an inner layer of protein, a middle layer of phospholipids resembling small beads, and a poorly defined outer layer. Morton's[256] electron micrographs of fat-globule membrane fractions revealed the presence of numerous spherical particles, 30 to 200 nm in diameter, which he identified as microsomes. He postulated that these lipoprotein particles are adsorbed on a protein layer surrounding the fat globules and account for the phospholipids known to be constituents of the membrane complex. No data were presented to characterize the "continuous protein" layer. In opposition to aspects of this model were the observations of Alexander and Lusena,[2] who reported that 25% of the membrane phospholipids were devoid of enzymatic activity. Therefore, it seems unlikely that all the phospholipids exist as "microsomal" constituents.

Ultrasectioned specimens of defatted washed cream were observed by Dowben *et al.*[83] (Fig. 10.28). Tripartite structures approximately 9 nm thick and typical of the "unit membrane" concept, are distinguishable in selected segments of the peripheral layer. In other segments, the membrane is less ordered and appears as a concentrated layer of spherical particles.

Steward and Levine[359-361] examined fat globules in lactating mammary tissue and in freshly secreted milk, (Fig. 10.29). An electron-dense interfacial layer $\simeq$15–25 nm thick was observed surrounding developing intercellular fat droplets. Upon secretion, the fat globule acquired an additional layer of plasma membrane which shows evidence of disintegration during its residence in milk. Keenan *et al.*[169] showed somewhat similar micrographs of the fat-globule membrane, but attributed the inner layer of electron-dense material to a tightly bound layer of triglycerides, possibly the high-melting fraction (see Fig. 10.29). Both groups of investigators were observing glutaraldehyde-fixed OsO_4-post-fixed specimens mounted in epoxy-type resins and sliced into thin sections for electron microscopy.

Swope and Brunner[370] presented micrographs of glutaraldehyde and OsO_4-fixed fat-globule membrane fragments which were prepared for electron microscopy by shadowing with platinum-carbon (Fig. 10.30). These micrographs show more detail of the macrostructural characteristics of the membrane and its surface than do sectioned specimens. They postulated that the membrane consisted of a highly organized

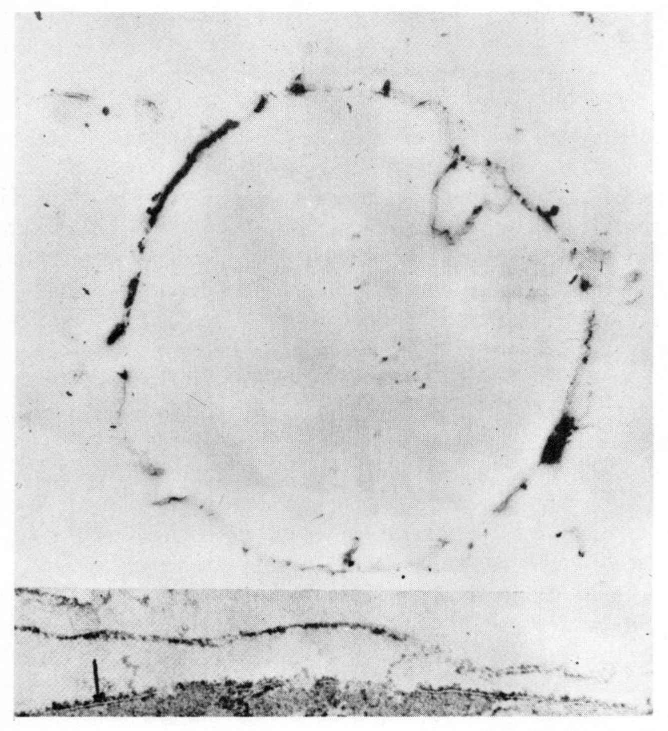

Data of Dowben et al.[84]

FIG. 10.28. ELECTRON MICROGRAPH OF FAT GLOBULE MEMBRANES

A typical tripartite membrane structure of about 9 nm cross-sectional dimension is evident (arrow) in the lower frame.

molecular layer, smooth on one side and rough on the other, on which lipoprotein particles of various sizes are loosely adsorbed. No designation was made as to whether the lipoprotein particles are adsorbed on the outer or inner side of the basic membrane structure. In a recent review, Peereboom[302] suggested that the biomembrane model as proposed by Staehelin,[356] in which a central bimolecular layer is covered with a layer of separate lipoprotein particles, is analogous to the model proposed by both Morton[256] and Hayashi and Smith.[131] In most aspects this concept is compatible with the Swope-Brunner model.

Buchheim,[40,44] employing the specimen preparation technique of freeze-etching, observed both the inner and outer surface of fat-globule membranes (human milk). His electron micrographs provide evidence in support of the model just proposed, namely, that the fat-globule membrane consists of a relatively continuous basic structure

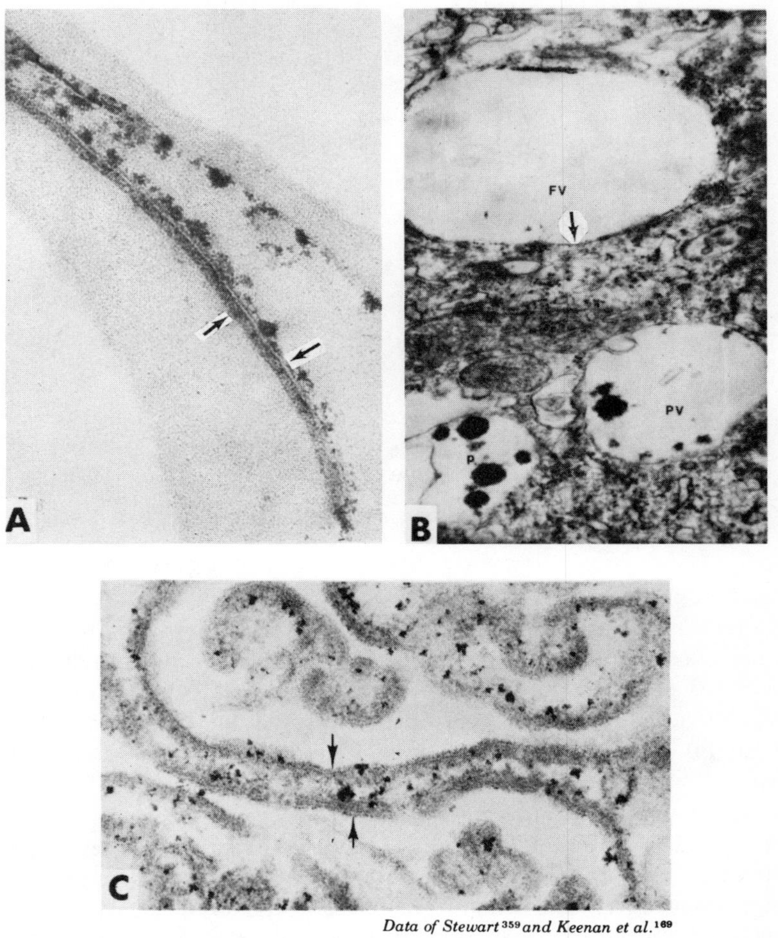

Data of Stewart[359] *and Keenan et al.*[169]

FIG. 10.29. ELECTRON MICROGRAPHS OF LUMINAL (FRAMES A AND C) AND INTRACELLULAR (FRAME B) FAT GLOBULE MEMBRANES

Frames A and C show typical tripartite membrane structures and a closely associated layer which Keenan (Frame C) attributed to high-melting glyceride. Stewart (Frame A) suggested that this associated layer was composed of cytoplasmic components observed at the interfacial layer between cytoplasm and developing intracellular fat droplets (Frame B). He attributed the more diffuse layer in Frame A to associated glycerides.

on which lipoprotein particles appear to be adsorbed. In this instance, the lipoprotein particles are adsorbed predominantly on the underside of the membrane. In Fig. 10.31, Frame A illustrates the outer surface of the membrane. There is evidence of a layer-like disintegration on the membrane surface, possibly resulting from manipulations inherent in the freeze-etching technique, as suggested by Staehelin.[356] The mem-

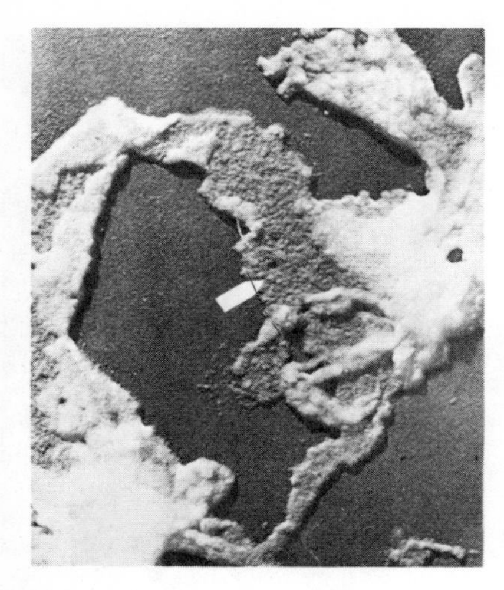

Data of Swope and Brunner[370]

FIG. 10.30. ELECTRON MICROGRAPHS OF PLATINUM-SHADOWED FRAGMENTS OF FAT GLOBULE MEMBRANE

Specimen was collected in the 7500 S pellet of washed-cream buttermilk.

brane appears smooth and continuous, with bulges showing in areas of sub-surface bodies, presumably lipoprotein particles or aggregates of included cytoplasmic material. The crater-like pockets apparent in Frame B presumably represent vacated sites on the underside of the membrane which were originally occupied by lipoprotein particles about 80 nm in diameter. Frame C shows a cross-sectional surface through a fat globule. The lipoprotein particles are apparent as bulges in the peripheral layer. Frame D represents a conventionally fixed—i.e., glutaraldehyde and OsO_4, mounted in epoxy—and sectioned fat globule. Note the electron-dense areas in the peripheral layer corresponding to the lipoprotein particles so vividly apparent in the freeze-etched specimens. Also, the lack of definitive tripartite membrane structures suggests that the basic membrane has undergone structural reorganization. The diffuse layer immediately adjacent to the membrane may represent tenaciously adsorbed glycerides, possibly the high-melting fraction.

The interesting observation here, of course, concerns the location of the lipoprotein particles on the inner surface of the membrane. The ease with which these particles can be recovered from manipulated membrane specimens suggests to the author that they either break through the basic membrane or become exposed when the membrane degrades from its endogenous structure. On the other hand, the apparent absence of lipoprotein particles on the outer surface of the membrane may be a ramification of the freeze-etch technique employed

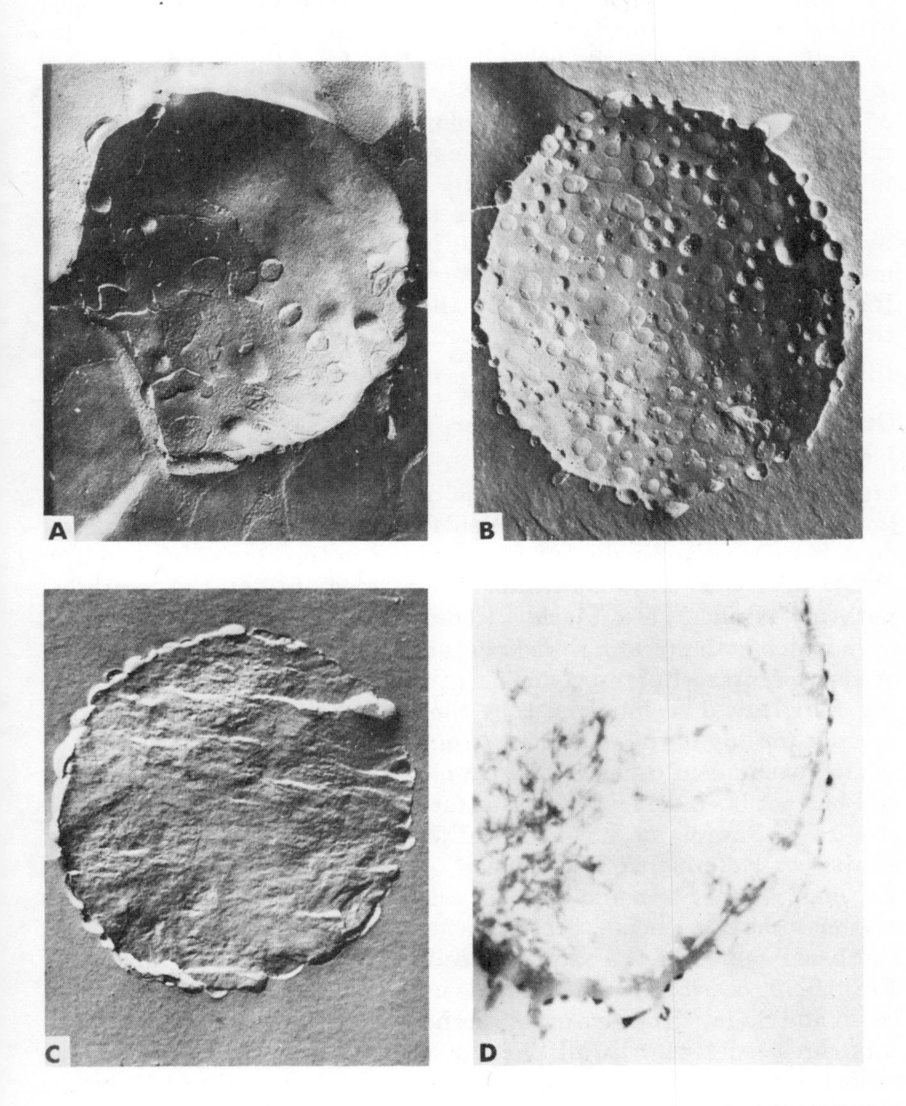

Data of Buchheim[44]

FIG. 10.31. ELECTRON MICROGRAPHS OF FREEZE-ETCHED (FRAMES A, B, AND C) AND SECTIONED (FRAME D) FAT GLOBULE MEMBRANES

A—Outer surface, showing smooth membrane structure. B—Inner surface showing adsorbed lipoprotein particles and vacated sites (particles are about 80 nm in diameter). C—Cross-section of fat globule showing lipoprotein particles in peripheral layer. D—Conventionally fixed (glutaraldehyde—OsO_4) and sectioned fat globule showing lipoprotein particles in vicinity of membrane.

for the preparation of the specimens for electron microscopy. Staehelin[356] pointed out that this procedure could remove such particles as well as portions of the basic membrane from the chipped surface of the frozen specimen. Knoop[194] estimated a thickness of 20 to 25 nm for the membrane of alveolar fat globules, of which 6 to 8 nm was attributed to cytoplasmic components. However, she was unable to detect a unit membrane structure surrounding freshly milked fat globules, observing that the surface layer was 5 to 10 nm thick. She hypothesized that the cell membrane layer and the bordering cytoplasmic layers are rapidly degraded by enzymatic activity in milk plasma. Indeed, Morton's[12,13] observation that mammary microsomal particulates were altered when exposed to milk seems to support this interpretation.

Origin of the Fat Globule Membrane

The origin of the intracellular fat droplet and its encapsulating membrane precludes an understanding of lipid synthesis. Suffice it to mention that the fatty acids of milk lipids arise from two distinct sources: from blood plasma lipids, and by synthesis in the mammary gland.[91,93,105,294,304] One of the significant intermediates in the synthetic pathway is phosphatidic acid (α-, β-diacyl-α-glycerophosphoric acid) from which both glycerides and phospholipids are derived through the activity of intracellular enzymes. Graham *el al.*[109] discovered, employing radiological techniques, that the phospholipids of blood do not account for any of the phosphorus components of milk.

The mechanism by which fat droplets are formed in the secreting cell has not been described satisfactorily. However, from the classical studies of Bargmann[15-17] and Wellings,[419,420] and the more recent studies of Helminen,[136] Keenan,[167-170] Kinsella and McCarthy,[192] Patton,[18,146,291] Stein and Stein,[357] and Steward,[359-361] certain events appear to be sequential. The formation of fat droplets is initiated in the basal region of the secretory cell, as illustrated in Frame A of Fig. 10.32, possibly at the endoplasmic reticulum, as suggested by Stein and Stein.[357] The observation that fat droplets of varying diameters can be detected in all parts of the cell as well as in secreted milk suggests that droplet size is governed by some kind of limiting structure, possibly adsorbed phospholipids[168] or ergatoplasmic vacuolar membrane.[357] Indeed, a limiting interfacial boundary is evident in Frame B, showing the fat droplets in the course of their intercytoplasmic movement. Helminen and Ericsson[136] concluded that the intracellular fat droplets are surrounded with a membrane or membrane-like condensation of ground cytoplasm. Kinsella and McCarthy[192] observed the secretion of fat droplets from mammary gland cells in culture.

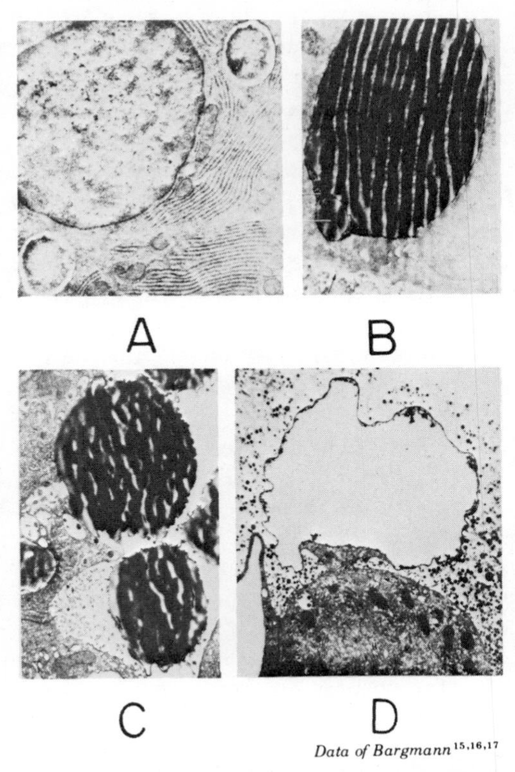

Data of Bargmann[15,16,17]

FIG. 10.32. ELECTRON MICROGRAPHS SHOWING EVENTS IN THE FORMATION AND SECRETION OF FAT GLOBULES AND MILK PROTEINS

A—Formation of fat globules in the basal region of the cell. B—Fat globule (small globule attached to large one in lower part of picture) in intracytoplasmic movement. Note the limiting structure surrounding the globule. C—Fat globules being "pinched off" into alveolar lumen. Protein-packed Golgi vacuoles about to empty their contents are evident. D—Solvent-extracted specimen showing the fat globule membrane on a globule about to be "pinched off."

Interestingly, they were able to detect clear vacuolized areas in cells lacking fat droplets. At the apical area of the cell, Frames C and D, the fat droplet protrudes into the alveolar lumen and is constricted off, carrying with it a portion of the apical plasma membrane and in some instances, small amounts of cytoplasmic material commonly referred to as "signets".[359,433] As pointed out by Patton,[288] the presence of these cytoplasmic crescents around a relatively small number of secreted fat globules does not necessarily support the mechanism of apocrine secretion. Instead, the general consensus supports the proposition that little cytoplasmic material is secreted in this manner. A definitive bit of evidence in support of this concept was contributed by Patton *et al.*,[292] who failed to detect the phospholipid cardiolipin (presumably a mitochondrian-specific component) in the polar lipid fraction of milk.

Closely allied to the mechanism of fat secretion is the secretion of milk proteins. It is a matter of record that only minor constituents of the milk protein system originate from the vascular system (e.g., serum albumin and the immune globulins in particular). Dixon *et al.*[72] observed that the acinar epithelium allows the passage of certain blood proteins (γ-globulins), while restricting the passage of others. These pass through the cytoplasm and discharge into the acinar lumen by an undisclosed process. In contrast, proteins synthesized in the mammary gland (caseins, β-lactoglobulin, α-lactalbumin, etc.) seem to be packed and released to the lumen by discrete action of the Golgi apparatus.[15-17,18,136,168,169,292,419,420] This mechanism involves the evacuation of Golgi vacuoles through the plasma membrane. When the vacuolar membrane contacts the acinar plasma membrane, it "blends" to form a continuous membrane layer in such a manner that the outer surface of the vacuolar membrane becomes the inner surface of the plasma membrane. Therefore, it seems apparent that the Golgi vacuolar membrane becomes a part of the same plasma membrane which envelops the fat droplet during secretion (see Frames C and D, Fig. 10.32). This mechanism relates the secretion of milk fat and the loss of plasma membrane with the secretion of proteins and the replenishment of plasma membrane.

In addition to the evidence provided by electron microscopy, recent biochemical investigations have revealed that fat-globule membrane and secretory cell plasma membrane are similar, and that plasma membrane is derived from the rough endoplasmic reticular membrane *via* the Golgi apparatus. Keenan *et al.*[168] compared the phospholipid composition of the fat-globule membrane, plasma membrane and nascent fat droplets (intracellular fat) obtained from secretory cell homogenate (Table 10.20). These analyses illustrate compositional similarities between fat-globule membranes and plasma membranes. Nascent fat contains a significantly lower concentration of sphingomyelins and higher concentrations of phosphotidyl choline than either of the membrane specimens. Patton and Fowkes[291] reported that the polar lipids associated with nascent fat droplets were comprised essentially of phosphatidyl choline—consistent with its role in triglyceride synthesis—and that phosphatidyl ethanolamine was completely lacking. Additionally, the carotenoids—components found in cell debris and organelle fractions as well as in the fat-globule mem-

Table 10.20

PHOSPHOLIPID DISTRIBUTION IN THE PLASMA MEMBRANE, MILK FAT GLOBULE MEMBRANE, AND NASCENT FAT FROM THE LACTATING COW[a,b]

	Cow A			Cow B		
Lipid component	Plasma membrane	Milk fat globule membrane	Nascent fat	Plasma membrane	Milk fat globule membrane	Nascent fat
Sphingomyelin	22.0 ± 4.1	21.4 ± 2.9	10.1 ± 3.0	27.0 ± 2.4	22.4 ± 2.5	10.3 ± 1.5
Phosphatidyl choline	27.0 ± 4.2	25.7 ± 3.0	38.8 ± 1.7	30.8 ± 3.8	31.7 ± 1.3	42.9 ± 3.9
Phosphatidyl serine	8.4 ± 0.4	14.0 ± 2.3	11.5 ± 1.0	8.5 ± 0.8	9.1 ± 0.6	6.2 ± 2.0
Phosphatidyl inositol	13.2 ± 3.4	11.1 ± 2.2	9.7 ± 2.6	12.3 ± 2.6	10.3 ± 0.8	8.2 ± 1.1
Phosphatidyl ethanolamine	29.4 ± 1.9	27.8 ± 1.3	22.0 ± 0.8	21.3 ± 2.8	27.2 ± 0.4	21.7 ± 3.6
Lysophosphatidyl choline	—	—	7.9 ± 2.9	—	—	10.7 ± 3.2

a Data of Keenan *et al.*[168]
b Values are means plus and minus standard deviations, n = 3. Expressed as % of total lipid phosphorus.

brane—were absent in nascent fat. Thus, it appears that some lipid components (and proteins) not associated with intracellular fat droplets become so at the time of their secretion.

Keenan *et al.*[168,169] demonstrated electrophoretic similarities between the protein moieties of fat-globule membranes and plasma membranes. Negative-contrast electron micrographs of plasma membrane and fat-globule membrane revealed significant morphological differences (Fig. 10.33). Plasma membrane showed extensive vesiculation whereas intact fat-globule membrane appeared plate-like. The latter condition was attributed to the adsorption of high-melting triglycerides at the membrane interface. Further support for the structural function of fat-globule membrane phospholipids was provided by the experiments of McCarthy and Patton,[242] who infused ^{14}C-palmitic acid into a lactating mammary gland. Radioactivity in the milk phospholipids (fat-globule membrane) persisted for a longer period of time than observed for either the glyceride or cholesterol ester moieties.

Evidence in support of the membrane-flow concept was provided by the very significant studies of Keenan and Morré.[167,250] Detailed lipid analyses of rough endoplasmic reticulum (Golgi apparatus) and plasma membrane-rich fractions obtained from rat-liver homogenate revealed that Golgi membranes contained concentrations of phosphatidyl choline and sphingomyelin intermediate between that of endoplasmic reticulum and plasma membranes. Endoplasmic reticulum was highest in phosphatidyl choline and lowest in sphingomyelin. Levels of lysophosphatidyl choline, phosphatidyl serine, phosphatidyl inositol and phosphatidyl ethanolamine were relatively constant between fractions. Lysophosphatidyl ethanolamine was detected in Golgi apparatus and plasma membrane but not in endoplasmic reticulum. Similarily, gel electrophoretic patterns of the three fractions revealed that Golgi apparatus contained a protein profile intermediate between that of endoplasmic reticulum and plasma membrane. These results suggest a partial derivation of the Golgi membrane from endoplasmic reticulum and reutilization of these membranes in the production of secretory vesicles and, thus, the plasma membrane (Fig. 10.34). The increasing proportion of cholesterol and sphingomyelin encountered as the membrane flow proceeds from the endoplasmic reticulum to the plasma membrane may be the principal factor contributing to the progressive changes in the physical properties of these membrane types. Patton[289] proposed that this transition renders membranes more lipophilic, less permeable and less easily degraded—a condition which favors membrane-related phenomena.

The observation that fat-globule membrane contains glycoproteins composed of hexose, hexosamines and sialic acid,[52,53,159,370,371,381] and

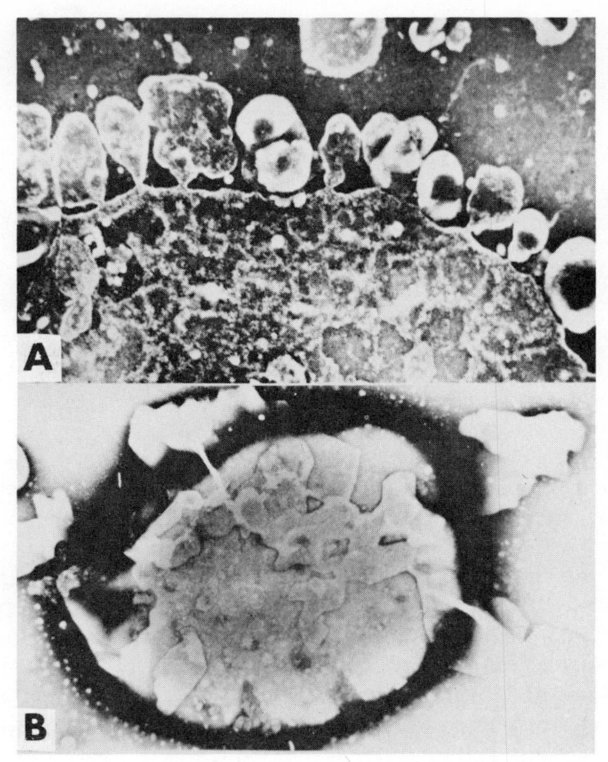

Data of Keenan et al.[168]

FIG. 10.33. NEGATIVE CONTRAST ELECTRON MICROGRAPHS OF (A) BOVINE MAMMARY PLASMA MEMBRANE AND (B) INTACT FAT GLOBULES

Note the numerous vesticles attached to the plasma membrane and the plate-like structure of the fat globule membrane.

glycolipids,[99,143,168,252,253] as well as the Golgi-specific enzyme lactose synthetase,[28] further supports the concept that fat-globule membrane material originates in Golgi cisternal membrane.[167-169,250] On the basis of the concept of membrane flow, it is feasible to rationalize the presence of membrane-specific enzymes,[83,84,296] microsomes,[256,357] and deoxyribonucleic acid,[368] as constituents of fat-globule membranes. Even the presence of milk plasma lipoproteins[293] possessing a phospholipid distribution similar to fat-globule membranes, but of different structure and activity, may be the result of Golgi synthesis. That these lipoproteins are not of vascular origin was shown by Von Murell *et al.*[269] Utilizing antigenic techniques, they found that only traces of blood lipoproteins could be detected in milk.

The enzymatic studies of Bailie and Morton,[12,13] Dowben *et al.*,[83,84]

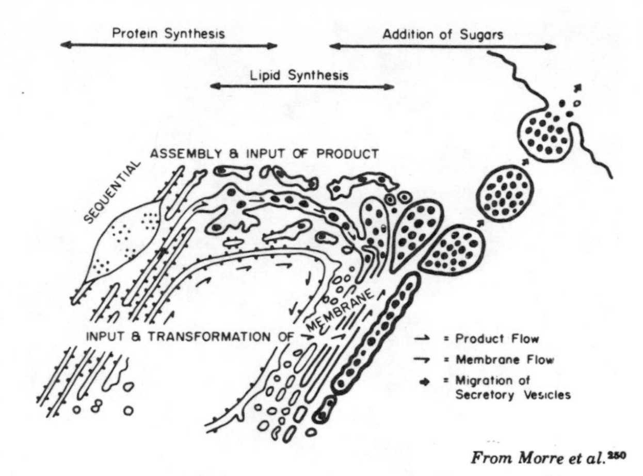

From Morre et al.[350]

FIG. 10.34. SCHEMATIC REPRESENTATION OF EVENTS INVOLVING THE TRANSITION OF ENDOPLASMIC RETICULUM MEMBRANE TO PLASMA MEMBRANE

and Patton and Trams[296] demonstrated the presence of many enzyme systems previously observed in microsomal fractions (endoplasmic reticulum), hepatic plasma membrane and erythrocytic membrane. Especially significant for this interpretation was the identification of xanthine oxidase, alkaline phosphomonoesterase, phosphodiesterase, 5′-nucleotidase, nucleotide pyrophosphatase, NADA-cytochrome c reductase, glucose-6-phosphatase, true cholinesterase and Mg^{2+} -activated ATPase. The presence of (Na^+, K^+-Mg^{2+})-activated ATPase is of particular interest because this enzyme has been identified with a variety of tissues which show selective permeability to cations, and is believed to be involved in "active transport" across cell membranes.[349] Aldolase, an enzyme found in high-speed supernatants of tissue homogenates, was present in the fat membrane preparations, but lactate dehydrogenase and aspartate-oxoglutarate aminotransferase—other high-speed supernatant enzymes—were not detected. In addition, acid phosphomonoesterase, found in erythrocytic membranes, was also detected. Lipase, β-glucuronidase, succinicdehydrogenase and cytochrome c oxidase were not found. Although the enzyme complement of the fat-globule membrane is predominantly of cellular membrane origin, cytoplasmic components also are associated with the preparations. However, there is no evidence to suggest the presence of mitochondrial components. Dowben *et al.*[84] noted that the agglutination and hemolysis of bovine erythrocytes by fat-membrane antisera provide additional evidence in support of the concept that the fat-globule membrane is a cell-membrane derivative. Further, they suggested that the observed selective permeability of fat-globule membranes to potassium ion is not a function of osmosis but is due to active transport by enzymes.

Summarizing the above discussions, the author concludes that all available evidence overwhelmingly supports a plasma membrane origin for the fat-globule membrane. Morphological deviations from typical plasma membrane, often encountered in isolated specimens, are not explainable in terms of composition. The high-melting glycerides which generally accompany isolated specimens may, in part, contribute to this characteristic. In addition to the enveloping membrane layer, there is evidence that low percentages of occluded or adsorbed cytoplasmic components and lipoprotein particles are present at the membrane-fat droplet interface. And, in milk, the fat-globule membrane seems to undergo disintegration and reorganization, forming granular-like structures which are eroded in heterogeneous fractions upon physical manipulation.

THE LIPID PHASE IN PROCESSED DAIRY PRODUCTS

Freshly produced milk represents a perishable raw material which must be stabilized against spoilage and converted into consumable stable products by a multiplicity of processing procedures. The effects of these treatments on the physical and chemical characteristics of the lipid phase become apparent as (1) alterations in the nature of the fat-globule membrane, (2) changes in the state of dispersion, or (3) variations in the physical state of the fat (i.e., the extent and nature of the solidified phase).

Manifestations of shifts in the chemical and physical equilibria inherent in the lipid phase are apparent immediately following secretion of fat globules into the alveolar lumen. The original membrane covering seems to disintegrate in the presence of the plasma phase.[169,194] Bailie and Morton[12,13] observed that mammary gland microsomes show a reduction in lipid and nucleic acid content following a period of incubation in milk plasma. Apparently there are enzyme systems native to the plasma phase which are capable of degrading the nucleic acid-bearing lipoproteins, e.g., phosphodiesterase and ribonuclease.

The complex lipoproteins composing the fat-globule membrane and those associated with it are sensitive to the usual agents which tend to disorganize native proteins and lipids. Consequently, it is conceivable that all processing procedures exert a denaturing influence on the organization of the membrane material. Actually, many of these alterations enhance the physical and chemical stability of the lipid phase.

Some of the effects of processing procedures as related to characteristics of the lipid phase of milk are discussed in the following sections.

Procedures Related to Milk Production

In recent years, milking procedures and methods of handling raw milk have undergone revolutionary changes. The entire procedure from the cow to the processing plant has yielded to mechanization. Pipeline milking systems, refrigerated bulk-tanks, and every-other-day pickup have supplemented the 10-gallon can and daily pickup operations of the past. Many of the old problems inherent in the handling of raw milk were eliminated (e.g., high cost of production and relatively high *S. lactis* counts). However, new problems were engendered and these are considered here.

Rancid flavor, sometimes accompanied by partial destabilization of the fat emulsion, may be one of the prevalent manifestations of the pipeline milking system. Usually the problem is alleviated by redesigning the pipeline to eliminate "risers" and other points in the system where milk is agitated violently. Then, too, the problem may be aggravated or even originated in the cooling tank. Occasionally, localized zones of warming are encountered before final cooling is achieved. This situation is encountered most frequently in tanks where both the nature of agitation and the cooling capacity are inadequate.

Rancid flavor results from the activity of lipases on milkfat. Krukovsky and Herrington[208] observed that lipolysis in cold, raw milk is accelerated by warming cold milk to 30°C and recooling to < 10°C. They postulated that the rate of lipolysis depends on the crystalline state of the fat. Apparently, solididified fat is less available to the enzyme system than is liquid fat.[209] It was assumed that the activation of lipase depended more upon changes in the nature of the substrate than on changes in the enzyme. Alterations in the endogenous membrane material and readsorption of plasma-components (i.e., micellar casein) at the fat/plasma interface must be considered as contributing factors to this phenomenon. Numerous workers have shown that the lipases are associated with components of micellar casein, most likely the κ-casein component.[85,95,96,100]

Tarassuk and Frankel[376] presented evidence to the effect that cow's milk contains at least two lipases, designated by them as "membrane lipase" and "plasma lipase". The membrane lipase participates in the phenomenon of "spontaneous" activation often encountered in the milk of cows in advanced lactation. This enzyme moiety appears to be irreversibly and actively adsorbed at the fat-globule surface when milk is cooled (i.e., no temperature activation treatment is required). The nature of its behavior indicates to this writer that membrane lipase may exist as a lipoprotein complex of microsomal origin. The plasma-casein is associated with the so-called "activated" or "induced" type of lipolysis. This type of lipase is activated by the agitation or foaming of milk, or by manipulating the temperature of milk through the solidification of milkfat.[375]

Whether lipolysis results from the activation of the plasma lipase system or is merely a consequence of its relocation in proximity with its substrate (i.e., the fat globule) remains unanswered. Packard[277] observed that anionic surfactants substantially reduce the surface tension of milk at concentrations known to activate the lipase system and liberate large proportions of the membrane phospholipids. The fact that anionic detergents exert a dissociating effect on molecular complexes makes Packard's observations particularly noteworthy. Conceivably, a structural change in the lipase molecule and/or a modification of the fat-globule membrane are prerequisites for the activation of plasma lipases. Certainly the adsorption of lipase on the fat-globule surface or its orientation into foam lamellae represent circumstances which contribute to the modification of protein structure. Similarly, procedures which cause agitation, foaming, or temperature fluctuation in raw milk result in a modification of the fat-globule membrane structure.

Lundstedt[225] suggested that agitation of milk may cause a release of phospholipids from the membrane into the plasma. He postulated that the phospholipids are adsorbed on the casein micelles, resulting in a reduction in curd tension. However, Palmer and Tarassuk[283] were unable to demonstrate a reduction in the curd tension of milk upon addition of phospholipids *per se*. In a subsequent study, they concluded that free fatty acids generated by the action of the lipase systems are responsible for the observed reduction in curd tension.[284] Greenbank and Pallansch[113] demonstrated that phospholipids were first released from the fat phase by homogenization, and readsorbed as the pressure was increased. Harwalker *et al.*[127] determined that the released phospholipids are components of fragmented fat-globule membranes. It is conceivable that these membrane fragments then interact with the casein micelles through an apparently specific mechanism involving κ-casein.[199] Thurston *et al.*[384] pointed out that the release of phospholipids from the fat surface accounts for the observed reduction in the susceptibility of agitated milk to copper-induced oxidation.

The tendency for fat globules to aggregate and form clumps in agitated raw milk was discussed by King.[181] The incorporation of air into milk or cream causes an extensive increase in the air/plasma interfacial surface. At temperatures where the fat exists in a partially liquid state (i.e., $\simeq$10–20°C), part of the fat-globule membrane is rubbed free and small quantities of lipid fat are extruded. Free fat and fat globules orient at the surface of the incorporated air bubbles where a portion of the free fat "cements" the fat globules into small clumps. Eventually the bubbles collapse, leaving clumps of fat globules and numerous small droplets of fat. Apparently, these clumps are held together through forces inherent in the liquid fat phase and are not reversibly formed and dissociated. In this respect, they differ from the previously described fat-globule clusters associated with the creaming phenomenon.

At normal values of pH (≃6.7), the fat-globule membrane exists in a relatively stable, hydrophilic condition. An increase in the pH enhances this condition, but predisposes the fat globule to subdivision. A decrease in pH contributes to its stability and the tendency for fat globules to clump.[81,274] Thus, the bacteriological quality of the milk becomes important in terms of the physical stability of the lipid phase, not to mention the public health implications. Stone[364] demonstrated that a lecithinase-producing strain of *Bacillus cereus* was capable of destabilizing the fat globule membrane and, thus, the lipid phase. Furthermore, members of the psychrophilic flora, which have become increasingly troublesome in contemporary raw milk supplies, elaborate lipolytic and proteolytic enzymes.

Centrifugal Separation

The factors involved in centrifugal separation of the lipid phase from the plasma phase of milk are represented in terms of Stokes' law and the law of centrifugal force. Mechanical separators operating at from 6,000 to 10,000 rpm develop forces ranging from 5,000 to 10,000 gravities, depending upon the size of the bowl. If these values are substituted into the Stokes' equation (see p. 487), the resulting velocity represents the speed at which fat globules leave the plasma phase into the cream phase. To facilitate partitioning of the two phases, the bowl of the separator is filled with a series of closefitting, A-shaped disks. The heavier, skimmilk phase is channeled to the outside of the bowl, while the lighter cream phase moves toward the center. A separating-disk located at the top of the disk assembly prevents remixing of the phases.

The efficiency with which a mechanically sound separator performs depends on two principal factors: (1) the fat content and relative size of the fat globules; (2) the temperature of the milk. Rahn[314] concluded from his studies of the size of fat globules in skimmilk that globules less than 1 μm in diameter are not removed, and that those in the 1 to 2 μm class are but slightly affected. When the temperature of milk is raised to 40°C, the properties of milk represented by the factors in Stokes' law are affected in such a manner that the efficiency of separation is maximized.[344] The design of commercial separators has been improved sufficiently to permit the practice of separating cold milk (5 to 10°C) with relatively small losses of fat in the skimmilk. However, to achieve this, it has been necessary to reduce the flow of milk to allow for a longer residence time in the separator bowl. The relative inefficiency encountered in cold milk separation arises principally from the increased viscosity of milk, the decreased density-differential between the lipid and plasma phases, and to a limited degree, a decrease in the radii of the partially solidified fat globules.

Agitation effects encountered during the separation of milk cause the release of membrane materials, which seems to be accentuated at the higher temperatures. Additional losses are encountered when quantities of incorporated air are present in the milk to contribute to the destabilization and disintegration of the fat globule. In fact, the result of such a condition is an excessive loss of small fat globules into the skimmilk phase. Cold separation, although less efficient, is less destructive to the native membrane structure. Apparently, the partial solidification of the lipid phase exerts a stabilizing effect. A portion of the plasma immunoglobulins, including the cryoglobulin IgM, remain with the cream when the milk is separated at low temperatures, but are desorbed and recovered in the skimmilk when separation is accomplished at temperatures above the melting zone of milkfat.

The heavier components of milk are collected in the separator bowl; they include such materials as large casein micelles, leucocytes, bacterial cells, extraneous foreign bodies, and quantities of enzyme-rich lipoproteins.

Lipatov *et al.*[219] observed a reduction in globule mean diameters which increased with increasing separator temperature.

Heating and Vacuum Processing

Pasteurization and allied forms of heat treatment are the foci of most milk-processing procedures. The objectives of all applications of heat treatment are (a) destruction of pathogenic organisms, and (b) preservation of the desirable characteristics of milk and its products pending their consumption. Pasteurization (61.8°C for 30 min, or at 71.8°C for 15 sec) represents a minimum heat treatment designed to destroy all the pathogenic bacteria and most of the nonpathogenic organisms. Fortunately, most of the degradative enzyme systems are partially or totally inactivated by this treatment. Products processed in this manner are generally nonsterile and have limited shelf-life, even when stored at refrigerator temperatures. Where longer "shelf-life" is desirable, higher temperatures must be employed (at 79.6 to 87.8°C for 20 to 40 sec). Ultimately, ultra-high pasteurization (93.4°C for 3 sec to ≃149.5°C for 1 sec) in conjunction with aseptic packaging, or in-can sterilization (116.7°C for 12 to 15 min, 129.6°C for 3 to 5 min) are employed to achieve complete destruction of all bacteria and spores. The shelf-life of products processed in this manner is usually dependent upon factors contributing to the chemical and physical stability of the components, principally the lipid and protein systems.

Minimum pasteurization processes do not materially alter the physical behavior of the lipid phase. Cream layer formation in the nonhomogenized product proceeds normally, and only small portions of the serum proteins are denatured. As the amount of heat applied to milk is increased, more serum proteins are denatured and creaming is retarded or completely eliminated. The fat globules fail to cluster and must rise, according to Stokes' law, as individual globules, forming tough layers of tightly packed fat globules after about 4 hr. The extent of creaming in processed milk reflects the relative denaturation of the serum proteins. Again, the significance of the role played by the immunoglobulin fraction in the creaming phenomenon becomes apparent.

Apparently, high-temperature processes contribute to alterations in the nature of the fat-globule membrane. Lowenstein and Gould[221,222] reported that fat globules in milk heated to 82°C for 15 min lost membrane material, and that the loss of proteins exceeded the loss of lipids. They indicated that phospholipids and protein were lost in the same ratio, which suggests that lipoproteins were displaced from the globule surface as a unit and not as free phospholipids. Also, they showed that in re-made milks (i.e., butter oil plus skimmilk) the proteins of heat-treated skimmilk (82°C for 15 min) were adsorbed at the fat/plasma interfacial surface in lower concentrations than the proteins from nonheated (40°C for 0 min) skimmilk. However, the proportion of adsorbed fat was greater in the case of the heated proteins. In opposition to these results, Hladík[139] demonstrated that the loss in membrane material in milk heated at temperatures ranging from 63°C to sterilization was ramified in a loss of membrane lipids, principally the triglyceride fraction. He postulated that the phospholipids remained closely associated with the protein moiety, which remained unaltered. Jackson[157] found that the amount of membrane material recovered from homogenized milk heated to 80°C for 30 min was less than that from homogenized milk heated to 70 or 60°C. Homogenization was performed at 2,500 psi and 58°C (Table 10.21).

Radema[312] determined the displacement of phospholipids from the fat-globule surface resulting from various processing treatments.

Table 10.21

EFFECT OF HEAT TREATMENT ON THE YIELD OF MEMBRANE MATERIALS ISOLATED FROM HOMOGENIZED MILK[a]

Observation	30 Min. Heat Treatment		
	60°C.	70°C.	80°C.
Index of heat treatment (whey protein N)	8.1	6.7	4.1
Volume of washed homogenized cream, ml	420	610	730
Fat in washed cream, %	46.4	49.9	55.1
Solids in washed cream, %	52.2	55.1	59.8
Nitrogen in washed cream, mg. %	165	126	116
Protein, % (%N × 6.38)	1.4	0.6	0.2
Grams of protein/100 gm. fat	2.27	1.60	1.31

a Data of Jackson[157]

To estimate the proportions of fat-bound and plasma phospholipids, samples of treated milk were separated into cream and skimmilk fractions in a laboratory centrifuge. Values for total lipid and phospholipid-phosphorus were determined and the following assumptions made:

> (1) the fat in the cream contains the same proportion of phosphatides as does the fat in the separated milk; (2) the aqueous phase of the cream contains the same proportion of phosphatides as does the aqueous phase of the separated milk; thus the partition of the phosphatides between the fat and the aqueous phase can be derived.

Obviously, this technique lacked the capability of determining whether phospholipids *per se* or the lipoprotein complex was being displaced. Nevertheless, it was demonstrated that mild agitation of light cream at 20°C did not cause release of phospholipids, but that higher temperatures (100°C), alone or accompanied by stirring, did cause displacement of phospholipids into the plasma phase. This displacement was maximized when cream was pasteurized following its separation from pasteurized milk. Fat globules minus their stabilizing surface are unstable and tend to coalesce. In the past, this condition often resulted in the formation of a cream-plug at the top of bottled milk.

In an interesting experiment involving the reactivation of heat-denatured alkaline phosphatase, Peereboom[302,303] proposed that enzyme-containing lipoprotein particles are released into the aqueous phase by pasteurization. When they were recombined with raw cream, renaturation of the denatured enzyme occurred. This phenomenon was attributed to the influence of the conformation of the native enzyme and associated lipoproteins present in the membrane. However, he concluded that indigenous and reactivated enzymes are different, as evidenced by their electrophoretic migration in gels.

A large portion of the indigenous copper in milk is associated with the fat globules. King[191] stated that milks having high contents of natural copper oxidize more readily than those with low concentrations. Peereboom[302] related the copper content of milk with the lipoprotein particulates adsorbed on the fat-globule membrane. He postulated that the release of these copper-containing lipoproteins by high-temperature processing of cream would result in a lower content in the resulting butter. Thus, less oxidative deterioration in the lipid phase would occur.

The violent agitation which occurs in high-velocity heating systems from which air has not been excluded, and in recirculating-type single-effect evaporators employed to concentrate milk contributes to the displacement of phospholipids into the plasma phase.[112] Many high-temperature pasteurization systems utilize a vacuum chamber on the high side of the heating section to (a) facilitate removal of feed flavors from the milk, and (b) provide initial cooling of the hot milk. Dolby[81] observed that the fat globules in cream were split by direct steam injection in the final heating stage of a Vacreator. The flash-boiling which occurred in the vacuum-cooling unit caused a partial clumping. Consequently, processors have found it advantageous to reroute the flow of milk so that homogenization occurs after the vacuum treatment.

Leviton and Pallansch[216] implicated the lipid phase as a factor contributing to the heat and storage stability of high-temperature short-hold sterilized, concentrated milk, but noted that the behavior of the lipid phase varied in different milks. In unstable milk they observed that the fat globules aggregated not only with themselves but with the protein suspension, thus accelerating gel formation. The variable behavior of the lipid phase was attributed to its ability to function as a protein when involved in gel formation or as a truly inert emulsoid when not involved.

Homogenization

Homogenization has become a standard industrial process, universally practiced as a means of stabilizing the fat emulsion against gravity separation. The normal characteristics of milk are profoundly altered by the homogenization process. Although these effects are most obvious in the characteristics of the lipid phase, subtle but significant changes in the characteristics of the plasma proteins are effected which seem to contribute materially to the overall characteristics of homogenized milk. In many aspects, homogenized milk represents a denatured form of its nonhomogenized counterpart. Much of the information pertaining to homogenized milk was compiled prior to 1950 and summarized by Trout.[389]

Essentially all homogenized milk is produced by mechanical means. Milk is forced through small passages under pressure (e.g., 2,000 to 2,500 psi) at velocities of approximately 600 to 800 ft/sec. The disintegration of the original fat globules is achieved by a combination of factors, namely, shearing, impingement, distention, and cavitation. The net result is reduction of the fat globules to approximately 1 μm in diameter, which is accompanied by a 4- to 6-fold increase in the fat/plasma interfacial surface area. Herein lies one of the basic causes for the altered characteristics of homogenized milk. The newly created fat globules are no longer completely covered with the original membrane material, but instead are coated with a mixture of proteins adsorbed from the plasma phase.

Brunner *et al.*[32-34] partially characterized the proteins isolated from the lipid/plasma interface of homogenized milk. They concluded that the proteins thus obtained differ from the proteins of nonhomogenized-milk membrane material. From electrophoretic and ultracentrifugal characterization data, they tentatively identified the "membrane-proteins" of homogenized milk as plasma proteins. In a subsequent study, Jackson and Brunner[158] fractionated this protein complement and reported the presence of high concentrations of casein and lesser quantities of serum proteins. The casein fraction appeared to be complexed with an unidentified lipid fraction. This modification in the electrophoretic pattern of homogenized milk was also noted by Tobias and Serf.[387] Yields of membrane-protein approximating 2.3 gm/100 gm fat were obtained. This value represented approximately 4 times the protein recovered from nonhomogenized milk membrane preparations (i.e., 0.5 to 0.8 gm/100 gm fat), and was commensurate with the increase in the interfacial surface produced by homogenization.

Fox *et al.*[94] studied a fat-protein complex produced by the homogenization of milk. The complex particles were isolated from the sediment of milk samples centrifuged at 150,000 × *g* for 90 min at 40°C. Presumably, these particles did not form a part of the fat-plasma interfacial layer; but under the influence of the centrifugal field employed this assumption may not have been valid. Nevertheless, Fox showed that casein was the protein moiety of the complex and that it was probably associated with the fat fraction through apolar bonding forces. Interestingly, they postulated that the amount of fat bound to casein is a function of the micellar or aggregate size, and that only one size of casein micelle exists at concentrations of 37% T.S. and above. They postulated further that the casein micelle is activated at the moment it passes through the valve of the homogenizer, predisposing it to interaction with the lipid phase.

The association of casein with the fat globule membrane in homogenized milk was substantiated by Buchheim's[43] elegant microscopy (Fig. 10.35). The micrograph of this freeze-etched specimen shows casein subunits oriented at the lipid/plasma interface—probably "pulled out" casein micelles—as well as casein micellar "bridges" between adjacent fat globules.

As a consequence of the increased dispersion of the lipid phase and its resurfaced, interfacial boundary, many interesting and profound changes in the characteristics of milk result from homogenization. Homogenized milk is whiter in appearance, bland in flavor, less heat-stable and more sensitive to light-induced deterioration, less susceptible to copper-induced oxidative changes, and has greater foaming capacity and lower curd tension than nonhomogenized milk. Above all, the lipid phase in properly homogenized milk exists as a stabilized emulsoid. These properties represent ramifications of the state of lipid phase and are discussed here.

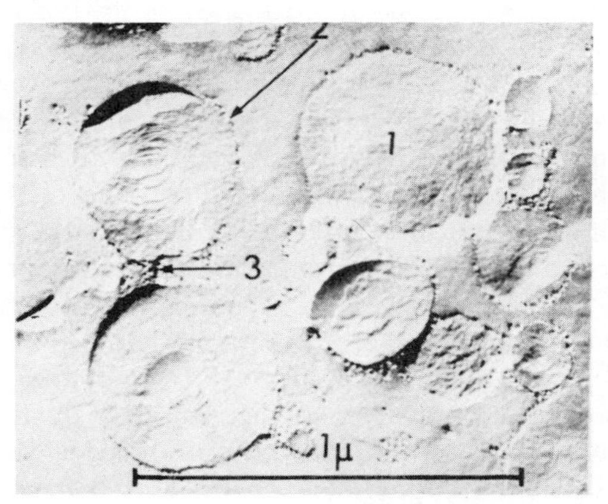

Data of Buchheim[43]

FIG. 10.35. ELECTRON MICROGRAPH OF FREEZE-ETCHED SPECIMEN OF HOMOGENIZED MILK

Micrograph shows (1) fat globules to which casein subunits are adsorbed (2) and joined together by casein micelles (3).

Dispersion of the Lipid Phase.—Rahn and Sharp[315] stated that roughly 85% of the fat globules in homogenized milk are less than 2 μm in diameter and that all are under 3 μm. Wittig[430] claimed that in properly homogenized milk, which he called "micromized", the fat globules should be less than 1 μm in diameter. Wiegner[427] reported that the average-size fat globule was subdivided into approximately 1,200 small globules by homogenization. On the other hand,

Sommer[352] calculated that 216 globules, 1 μm in diameter, would be formed from an original fat globule 6μm in diameter. Jenness and Patton[164] estimated the increase in surface area based upon the change in the surface-volume ratio of the lipid phase resulting from homogenization. They used as an example a sample of Guernsey milk having an S/V ratio of 1.26 μm^{-1} and assumed a uniform yield of 1 μm-diameter globules upon homogenization (S/V = 6.00 μm^{-1} for 1-μm diameter globules). It follows that the increase in surface area was estimated at 4.8-fold (i.e., 6.00/1.26). They pointed out that it is erroneous to calculate the increase in surface from the mean diameter ($\bar{d} = \Sigma\, \pi d/N$) prior to and following homogenization. The calculation should be made from summation data derived from the fat-globule size-distribution patterns.

Of the several types of forces which occur in the homogenizer and considered significant to the physical phenomenon of fat dispersion, shear and cavitation have been most generally investigated. In a recent study, Walstra[408] suggested that energy density is the most important factor. Dispersion of the lipid phase depends on the dissipation of input energy in the shortest possible time; thus it is a function of valve design. With properly designed valves, eddy currents in order of 0.1 μm in diameter—much smaller than fat globules—are produced which disrupt the fat globule.

The significant factor involved in the noncreaming of homogenized milk is not so much the observation that the fat globules are relatively small, but that the state of their dispersion remains fixed. There is no obvious tendency for the fat globules in properly homogenized milk to form loose clusters similar to those observed in nonhomogenized milk. This interesting behavior has been attributed to the homogenization-denaturation of the creaming factor (i.e., the immunoglobulin fraction of the milk plasma protein). In this respect Dunkley and Sommer[87] observed that there was a lack of creaming in milk reconstituted from nonhomogenized cream and homogenized skimmilk, but that creaming occurred in the combination of unclumped homogenized cream and nonhomogenized skimmilk. Samuelsson *et al.*[335] reported that fat-globule agglutinin contains two functional components, one of which was agitation-labile and the other heat-labile. Sharp[345] and Sommer[353] stated that fat-globule clustering and creaming in homogenized milk could be induced by the addition of agglutinins (immunoglobulins). Despite the preponderance of evidence which implicates the immunoglobulin fraction as the clustering factor, it seems reasonable to attribute a portion of the nonclustering characteristics of homogenized milk to the nature of the "new" fat globule surface (i.e., adsorbed plasma proteins). The forces that operate to maintain the dispersion of casein micelles may function to disperse the fat globules.

The physical state and concentration of the fat phase at the time of homogenization contribute to the size and dispersion of the resulting fat globules. Homogenization of cold milk, in which the fat is essentially solidified, is virtually ineffective. Processing at temperatures conducive to partial solidification of milk fat (i.e., 30 to 35°C) results in incomplete dispersion of the fat phase, frequently accompanied by extensive clumping and a "chalk-like" flavor defect.[76,251] Homogenization is most efficient when the fat phase is in a liquid state and in concentrations normal to milk. Products of high fat content are more likely to show evidence of fat clumping, especially when the concentration of serum solids is low with respect to the fat content.[73,76] In fact, the oil-type emulsion in high-fat cream ($\simeq$ 80%) can be "broken" by homogenization at temperatures in the neighborhood of 18 to 20°C.

Two-stage, repeated and high-pressure homogenization procedures cause formation of small fat globules.[74,112,215,413,415] The dispersion of the lipid phase increases temperature of homogenization and is commensurate with the decreasing viscosity of milk at higher temperatures. Greenbank and Pallansch[112] concluded that the agitation effects inherent in the homogenization process cause a displacement of the phospholipids from the fat globule surface into the plasma phase. However, at homogenization pressures around 8,000 psi, the phospholipids were readsorbed. No evidence was presented to indicate whether the phospholipids were displaced *per se* or as a lipoprotein complex. Koops and Tarassak[201] concluded that homogenization is no more effective in reducing the amount of fat-bound phospholipids than other processing treatments (i.e., cooling, separation and heating), but that homogenization reduces by 2 to 3 times the amount of phospholipids per unit of globular surface. This fact may, in part, explain the inhibition of copper-induced oxidized flavor by homogenization. Greenbank and Pallansch[113] observed that xanthine oxidase inactivated by heating milk to 85°C for 15 sec was partially reactivated following concentration to 50% total solids and homogenization at 4,500 psi. Treatments which seem to reactivate the enzyme system also are capable of deactivation if applied at high intensities.[113,307,324]

Viscosity.—With one notable exception, most investigators have observed an increase in the viscosity of homogenized milk. Trout *et al.*[391] felt that variations in homogenizing pressures and temperatures accounted for the discrepancies observed. In this respect, the extent of clumping seemed to be a significant factor in the viscosity determinations. Whitnah *et al.*[421] studied the viscosity changes in milk (4% fat) by homogenization at pressures ranging from 15 to 3,500 psi and temperatures of 4 to 49°C. In all cases they observed an increase which was independent of temperature but increased at higher pressures of homogenization (e.g., 7.1% at 1,000 psi to 15% at 3,500 psi). They concluded that the causes for changes in the viscosity of milk as a result of homogenization are obscure, possibly independent of the changes in fat-particle size. However, in view of more contemporary findings, increases in viscosity can be attributed to newly formed protein-lipid complexes.

Appearance and Flavor Characteristics.—The fine state of dispersion in which fat phase exists in homogenized milk provides a greater opportunity for both the absorption and reflection of incident light. Thus, homogenized milk appears whiter than nonhomogenized milk. The bland flavor characteristic of homogenized milk represents another result of the dispersed state of the fat phase. Small fat globules coated with plasma proteins, principally casein, impart a tactile sensation not unlike that produced by the components of the skimmilk phase. A chalk-like flavor accompanied by extensive fat-clumping and an increase in apparent viscosity are encountered in homogenized milk processed at low temperatures.[251]

Homogenization predisposes milk to enhance lipolytic and photochemical activity. Apparently, plasma lipase is activated by the agitation effects of homogenization and orients at the fat/plasma interfacial surface (e.g., casein-associated lipase). Unless inactivated by heat treatment, lipolysis proceeds rampantly, causing a bitter, soap-like flavor. Even then, precautions should be taken to eliminate possible contamination of homogenized milk with raw milk.

The photocatalyzed flavor-defects which occur in homogenized milk can result from oxidative effects in the lipid phase as well as from degradation of the protein components. "Light-activated" flavor is the terminology usually employed to describe the protein-associated flavor defect whereas "oxidized" flavor is reserved for flavor defects of lipid origin.[86] Patton[287] attributed the light-activated flavor defect to methional produced as a result of the photocatalyzed reaction between the amino acid methionine and riboflavin. The availability of increased concentrations of methionine as a result of homogenization seems feasible. Iwaida and Tsugo[152] reported that homogenized skimmilk showed a reduction in casein nitrogen. An increase in the proteose-peptone content indicated a release of similar materials from the casein complex. Their results indicate that the micellar structure of the casein complex is drastically altered. When adsorbed and extended on the fat-globule surface, this modified casein, or its cleaved fraction, could represent a potentially active substrate. Weinstein *et al.*[418] demonstrated typical light-activated flavor in solutions of a minor protein fraction isolated from whey. In retrospect, this particular fraction contained most of the low molecular weight cleavage products from the action of rennin on casein.[38] Apparently, copper ion and ascorbic acid catalyze the development of this flavor development.[86]

Samuelsson[332] suggested that not all the components of the light-induced flavor defect emanate from free methionine. He showed by means of di- and tripeptide models that radiation in the range of 300 to 400 nm did not hydrolyze peptides but did split the C-S bond of methionine to form RSH, R, SR_2, $RSSR_2$, H_2S_2, formaldehyde, acetaldehyde and propanal, all of which contribute to the flavor defect.

Finley and Shipe[88] isolated a low-density lipoprotein from cream by differential centrifugation at $42{,}000 \times g$ for 2 hr at 30° which they believed to be a principal source of light-induced flavor. Upon exposure to a fluorescent light source, the protein portion of this lipoprotein fraction was partially degraded, liberating tryptophan, tryosine, lysine and methionine. The lipid portion was partially oxidized, as indicated by a reduction in oleic and linoleic acid residues and the formation of a series of 24-dinitrophenylhydrazine-reacting products. These changes were different from those produced by the addition of copper ion.

Thurston *et al.*[384] and others[77,377,378,390] have demonstrated that homogenized milk is resistant to copper-induced oxidized flavor resulting from the oxidation of components of the lipid phase. Tarassuk *et al.*[377,378] suggested that this increased resistance results from a dilution of the susceptible reactive sites (phopholipids) on the increased fat-globule surface. They observed that the susceptibility to copper-catalyzed oxidation decreased with increasing homogenization pressures (i.e., increased surface-volume ratio).

Foaming.—The enhanced foaming capacity of homogenized milk has not been explained adequately. Jenness and Patton[164] proposed that the release of foam-promoting substance from the fat-globule membrane could be responsible. However, milk plasma contains an abundance of surface-active components, and therefore it seems unlikely that a small amount of membrane material could be this specific. Alternatively, the enhanced foaming capacity might be attributed to (a) increased dispersion of the fat phase and (b) adsorption of plasma proteins at the fat/plasma interfacial surface. The smaller globules could move through the foam lamellae at will without imparting localized sites of mechanical stress. Further, the nature of the adsorbed surface is compatible with the foam-stabilizing plasma protein. Essentially, homogenized milk behaves more like skimmilk in its foaming characteristics.

The larger fat globules of nonhomogenized milk, coated with lipoproteins and small quantities of mono- and diglycerides together with free fat, contribute to the instability of milk foams. This is especially so at temperatures which permit the membrane substrate to move into the air/plasma interfacial surface. Brunner[30] demonstrated the antifoaming characteristics of mono- and diglycerides. Ross[329] identified an antifoamant as an agent which is immiscible with the continuous phase and capable of orienting itself into the foam lamellae in such a manner as to prevent its formation or cause its collapse.

Holden *et al.*[145] observed a seasonal variation in the phospholipid content of milk, being greater during the winter months. This variation was correlated with the static foaming capacity of milk which was minimized during the winter. This seasonal variation in whole milk phospholipids reflected primarily variations in the lipid phase.

Surface Tension.—The surface tension of pasteurized milk is increased by homogenization.[75,391] Webb[414] observed that surface tension increased as the homogenizing pressure was increased. Doan considered this change in surface tension to be due to removal of surface-active proteins from the plasma phase by adsorption on the increased fat surface. The studies of Jackson and Pallansch[160] indicated that the plasma proteins are active at the oil/plasma interface. Nevertheless, it is difficult to perceive how the adsorption of a relatively small amount of plasma protein could cause a change in the net surface tension, unless, of course, selective adsorption of the most surface-active components occurs. Aschaffenburg[5] observed that the proteose-peptone fraction, present in low concentrations in milk plasma, is extremely surface-active. It could be that the supply of this highly surface-active glycoprotein fraction in the plasma phase is nearly depleted by adsorption on the fat surface in homogenized milk, thus causing a rise in surface tension corresponding to the composition of the plasma phase. On the other hand, homogenization causes changes in the physical nature of the plasma proteins and the original fat globule membrane, either of which could account for the observed increase in surface tension.

Homogenization of improperly pasteurized milk or the contamination of homogenized milk with small quantities of raw milk activates the plasma lipase, causing partial hydrolysis of the lipid phase. The cleavage products, consisting of free fatty acids and mono- and diglycerides, contribute to the bitter flavor and lower surface tension of rancid milk, respectively.

Curd Tension.—The clot formed after the addition of rennet or pepsin to homogenized milk is less firm than that formed in nonhomogenized milk. The "softness" of the clot reflects the efficiency and pressure of homogenization. Doan[75] showed that the curd tension of pasteurized milk (4.2% fat) was reduced from 55 g to $\simeq$ 20 g by homogenization at 2,000 psi. Further increases in pressure up to 4,000 psi did not bring about additional reduction in curd tension. It seems apparent that the lipid phase is intimately involved, as the homogenization of skimmilk causes no obvious changes in its clot-forming characteristics.

The reduction in curd tension has been attributed to (a) increase in the number of globules serving as points of weakness in the gel, and (b) the adsorption of casein on the increased fat surface, causing a reduction in concentration in the plasma phase. Both factors may be operative and their net effect would be additive. It is inconceivable that a reduction of 4 to 8% in the casein content of the plasma phase could, independently, account for so drastic a reduction in curd tension.[76] However, the immobilization of an additional quality of κ-casein by the immunoglobulin fraction, as indicated by Koops *et al.*,[199] could be a contributing factor to this phenomenon.

Interestingly, Iwaida and Tsugo[153] observed that high homogenization pressures ($\simeq$5,700 psi) caused additional reduction of the curd tension of milk, which they attributed to an alteration in the casein particles themselves. The effect was also demonstrated in homogenized skimmilk.

Temperature treatments drastic enough to initiate denaturation in the serum proteins cause a collateral reduction in curd tension. At intermediate temperatures ($\simeq$70°C) the effects of heat and homogenization are additive. However, at higher temperatures ($>$82°C) of heat treatment, homogenization produces no additional reduction in curd tension.[76] Numerous workers have demonstrated heat-induced intermolecular interactions between the sulfhydryl and disulfide-bearing proteins, e.g., β-lactoglobulin and casein.[125,245,440] In these complexes, casein is unavailable to the primary reaction of rennin; thus a lower curd tension is to be expected. There is evidence that casein and the protein of the fat-globule membrane undergo a similar interaction.

Free fatty acids inhibit the formation of a rennet clot.[284] Conceivably, the intermediate chain-length (C_8-C_{14}) fatty acids interact with either the substrate or the rennin molecules, thus impeding the formation of a clot or the enzymatic activity, much in the same manner as sodium dodecyl sulfate inhibits the activity of rennin on κ-casein.

Heat Stability.—The stability of milk to heat treatment (e.g., time required to show evidence of precipitation at 120°C) is reduced by homogenization. In general, the stability is decreased by increasing pressures of homogenization.[75,76,415] Single-stage homogenization, high fat content, and other factors contributing to fat clumping cause a decrease in the heat stability of milk. Likewise, the addition of divalent cations or a reduction in pH enhances the instability characteristic.

Apparently, the adsorption of plasma proteins on the fat-globule surface must be considered a major contributing force. Denaturation of the heat-labile proteins predisposes the entire plasma system to protein-protein and protein-lipid interactions, and the aggregation of protein-coated fat globules into larger particles is thus enhanced.

Freezing and Dehydration

Numerous studies related to the properties of frozen and dehydrated milk and milk products appear in the literature devoted to dairy products technology. Only the fundamental aspects of the effect of these processes on the chemical and physical stability of the lipid phase are considered here. Native lipoproteins and induced lipid-protein complexes tend to dissociate when subjected to sharp-freezing and dehydration. In many respects, freezing and dehydration represent identical treatments. The water phase is either immobilized or removed, leaving the dissolved and suspended components in a state of dehydration. Alteration in the state of bound water is of paramount significance to the destabilization of the fat emulsion.

Alexander and Lusena[2] and others[149,168,169] found that freezing washed cream to $-10°C$ for 20 hr resulted in the release of the fat-globule membrane, much as is accomplished by churning. Thompson *et al.*[383] employed freeze-drying to dissociate the membrane complex. They extracted the lipid moiety with nonpolar solvents. Dissociation of the nonlyophilized lipoprotein normally requires the use of ethanol or butanol in combination with ethyl ether.[138] Apparently, destabilization of the fat globule membrane contributes to many of the flavor and physical defects encountered in frozen and dehydrated products.

Browning reactions similar to the classic Maillard reaction between carbonyl groups of sugar and free amino acids have been reported for the interaction of phospholipids and sugars.[144,212] These sugar-amino condensations involve addition of the amino groups to the aldehyde group of the sugar. Water is eliminated to produce N-substituted aldosylamines; then, through an Amadori-like rearrangement, the N-substituted ketosylamine is formed, which degrades to colored melanoidin substances.

Lipid-protein copolymerization reactions of the type described by Tappel[374] may conceivably play an important role in the shelf-life of dry whole milk. This reaction involves polymerization of peroxidized polyunsaturated fatty acids and copolymerization of lipid oxidation products with proteins. These reactions could very well be catalyzed by the heme-containing cytochromes or other trace metal containing proteins of the membrane and milk plasma. The products of the reaction are insoluble and dark in color. In the same tone, Nishida and Kummerow[273] reported that interaction of the hydroperoxide of methyl linoleate caused destabilization of the low-density β-lipoprotein of blood. Such a reaction in milk could contribute to destabilization of the membrane lipoprotein.

Despite the general conception that phospholipids contribute significantly to oxidative deterioration, they do serve as synergistic antioxidants. The ionized phosphate group enters into the initial stages of the oxidative process, involving the unsaturated bonds in fatty-acid molecules. At a time when these charged phosphate groups can no longer stabilize the hydroperoxides being formed, the system deteriorates rapidly and rampant oxidative deterioration becomes apparent.

Drum-processed whole milk powder contains considerably more unemulsified fat ($\simeq$65 to 80% of total fat) than its spray-processed counterpart ($\approx$5% fat). The unemulsified fat, usually referred to as "free" fat, represents the lipids extracted with nonpolar solvents (e.g., 50/50 mixture of ethyl and petroleum ethers). The composition of the free fat varies with the amount recovered. Where the fraction constitutes a major portion of the lipid phase, the composition is characteristic of the milkfat. In spray-processed whole milk powder, small quantities of free phospholipids are apparent. The triglyceride fraction of free fat contains from 5 to 10% of the high-melting fraction.[189] The presence of free fat in powdered milk appears to be significant to the dispersibility of dry milk particles. The solvent-rinsed powder is highly dispersible, indicating that the hydrophobic free fat interferes with the ability of water to reach the hydrophilic surface.

Shipstead and Tarassuk[348] postulated that free fat is the first fat to undergo oxidative deterioration in whole milk powder. Swanson and Sommer[367] and Koops and Pette[200] presented evidence to the effect that oxidation is initiated in the phospholipid fraction and proceeds to the triglycerides. Litman and Ashworth[220] isolated a scum-like material from reconstituted, aged whole milk. They observed that scum development was related to the initial free fat content of the powder. It was postulated that development of the defect is the result of a fat-protein complex involving free fat and unstable protein. King[184] reported that free fat showed low fluorescence when stained with fluorochromes, which he later ascribed to high content of crystalline fat, probably the high-melting fraction.[189]

King[184] studied the distribution of fat globules in dehydrated milks by fluorescence microscopy. The fat particles in spray-dried whole milk were small and uniformly distributed. However, in aged samples, where the adsorption of moisture had induced lactose crystallization, appreciable percentages of coalesced fat were observed. It was spread around the air cells and on the periphery of the powder particles. Small particles were joined together through "fat bridges." Crystallization was noticed in the free fat, but not in the globules. Fewer air cells were observed in whole milk powder than in nonfat dry milk, probably because of the antifoam characteristics of the lipid phase at drying temperatures (≈82°C). In contrast to spray powders, roller-dried powder contained large globules, fragments, and patches of fat in and on the dried particle. The fat was easily extruded from these particles.

The physical state of the lipid phase has been related to the wettability of milk powder. Milk powders in which the fat phase exists in a liquid state are more readily wetted than if partial or complete solidification of the fat phase has occurred.[45,271,310] Baker and Samuels[14] employed surfactants in the processing of milk powder to determine if increased interfacial activity would enhance its wettability. They concluded that increased interfacial activity is ineffective unless the fat is in a liquid state.

Chien *et al.*[51] studied the effects of dilution, freezing and thawing, and drying by various procedures on the dispersibility of fat-globule membrane dispersions. Dilution and freeze-thawing favored aggregation of the membrane fragments; drying by the roller process markedly decreased dispersibility; drying by a low-temperature vacuum process only slightly increased dispersibility, whereas freeze-drying and spray-drying processes increased dispersibility. Knoop and Wortman,[197] by means of electron microscopy, observed degradation of the fat globule membrane in cream stored for one year at −15°C, leading to destabilization of the lipid emulsion.

Churning and the Physical Characteristics of Butter

The concentration of the lipid phase of milk into butter or oil has been practiced by man since the beginning of recorded history. The *art* was passed from generation to generation and was practiced by those fortunate enough to possess milk-producing animals. Not until the latter part of the past century, and largely due to the invention of the centrifugal cream separator, was buttermaking removed from the home. Since then, more research has been devoted to the problems and manufacture of butter than any other milk product. Much of the voluminous literature related to these studies was ably reviewed and presented in McDowall's two volumes.[244] In this section only the essential rudimentary concepts of churning and the structure of butter are considered.

Prior to the last decade, most of the world's butter was manufactured by the batch-process in which cream (30 to 35% fat) was churned and worked in wooden or metal churns. Butters produced in this manner were relatively uniform in appearance and physical characteristics. Since World War II, "continuous" procedures have been introduced to achieve increased manufacturing efficiencies. The characteristics of butters produced by these procedures vary somewhat between processes and from the conventional batch-processed butter. Detailed information on the technology of continuous buttermaking has been presented by Wiechers and de Goede[426] and in reviews by McDowall[243] and King.[182]

Regardless of the manufacturing method employed, the essential feature of churning involves destabilization of the lipid-phase emulsion by means of mechanical agitation. The various theories propounded to explain this phenomenon for both types of manufacturing processes have been summarized by King.[182,183 190]

Conventional Buttermaking.—The "auto-flotation" theory advanced by van Dam and Holwerda[68] and elaborated by King[183] and others[217,336,341] represents a contemporary viewpoint of the sequence of events which occur during the churning process. Briefly summarized, this theory states that churning is initiated by agitation and the incorporation of numerous small air bubbles. Partially denuded fat globules gather at the fat/plasma interface where they form small clumps "centered" by liquid fat exuded from the fat globules. A portion of this hydrophobic liquid fat spreads over the surface of the air bubbles, causing them to collapse. The monomolecular layer of film fat is dispersed in the plasma phase as colloidally dispersed fat particles, accounting for about 25% of the lipids found in the buttermilk. The aggregation of fat globules continues, promoted by the forces of churning and reflotation until butter granules are formed. About one-half of the fat-globule membrane material is liberated into the buttermilk phase.[164]

Although the flotation theory places a great deal of emphasis on foam formation in the early stages of churning, foaming is not prerequisite to the agglomeration of fat globules. However, the physical state of the fat at the temperature of churning is of primary importance. Brunner and Jack[36] demonstrated that the churnability of cream is not entirely dependent on the degree of solidification in the fat phase, but to a greater extent on the distribution of the liquid and solid phases. They postulated that the fat on the surface of the fat globules is neither totally liquid nor totally solid, but exists in a semisolid, sticky condition, conducive to fat-globule agglomeration.

King,[178] employing polarized and incident light microscopy, observed that fat clumps are surrounded with a liquid fat phase. He later suggested that the structural integrity of clumped fat globules is maintained by radial orientation of the rod-like molecules of the higher-melting butterfat fractions.[180] This "shell-crystal" possesses plastic properties, possibly related to the mesomorphic substance (liquid crystals), and is microscopically distinguishable as an optically active birefringent layer. Sometimes the formation of the crystal-layer is induced by the forces of churning and coincides with the exudation of free fat from the globules. Occasionally, the layer-crystal segregates into needle crystals oriented tangentially to the globule surface. Prior to the working of butter granules in the churn, approximately 80% of the fat phase exists as globular fat. As the butter granules are worked, more and more of the fat globules are crushed, causing the release of liquid fat. This free fat, which varies from approximately 54 to 98% in butter, constitutes the continuous phase. Fat globules, and finely dispersed droplets of the aqueous phase, represent the dispersed phase. The distribution of continuous and dispersed phases in butter was represented diagrammatically by King[185] (see Fig. 10.36). This concept was verified by the studies of von Gravel,[110] who was able to obtain photographs of the microstructure of butter showing essentially the same features. Wortmann's[435] electron micrographs of sour cream butter show essentially a continuous lipid phase, bearing remnants of the fat-globule membrane. Occasional fat globules with partial membrane can be detected (Fig. 10.37).

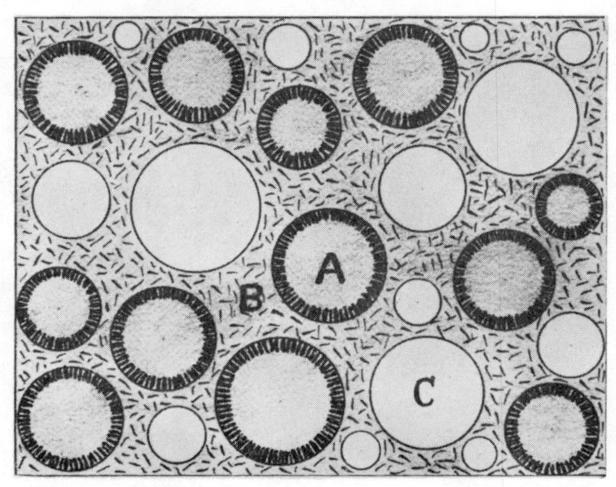

FIG. 10.36. DIAGRAMMATIC REPRESENTATION OF THE STRUCTURE OF BUTTER

A—Fat globule with crystalline peripheral layer consisting of radially oriented molecules of higher melting glycerides. B—Free fat and small fat crystals in the continuous fat phase. C—Water droplets.

Mulder and his colleagues[260-268] compared the dispersion of the aqueous phase during the working of butter with the stirring of an emulsion of high viscosity. In stirred emulsions, some of the dispersed droplets are further dispersed, while others collide and coalesce. Whichever process predominates depends on velocity gradient, viscosity, number and size of droplets, surface tension, etc. Obviously, if the aqueous phase is to be uniformly and finely dispersed, the shearing forces ($\eta dv/dx$) must exceed the stabilizing forces ($2\gamma/r$), i.e., $\eta(dv/dx) > 2\gamma/r$, where η is viscosity of butter, dv/dx = velocity gradient, γ = surface tension of droplet, and r = radius of droplet.

Accordingly, the vigorous working of firm butter would serve to increase the state of dispersion in the aqueous phase. The appearance of butter is related to the extent of this dispersion. Finely dispersed droplets cause the butter to be opaque and whitish in color. Fewer and larger droplets cause the butter to have a clear, deep yellow appearance.

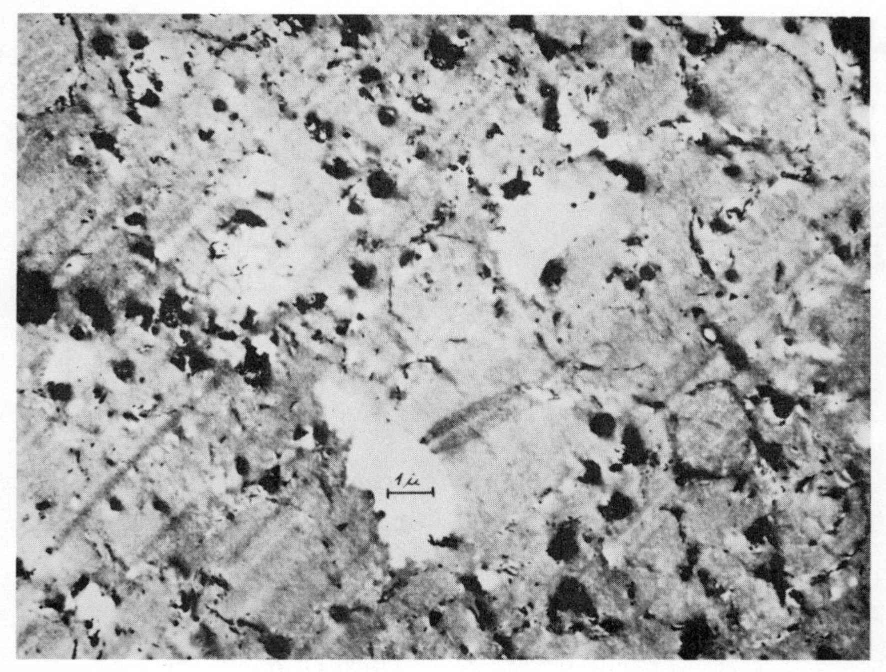

Data of Wortmann[435]

FIG. 10.37. ELECTRON MICROGRAPHS OF ULTRATHIN SLICES OF SOUR CREAM BUTTER

Fat globules present in the butter exhibit only remnants of the original membrane.

Continuous Buttermaking.—The continuous processes may be divided into two principal groups: (a) methods involving the accelerated churning of cream of normal composition, e.g., the Fritz (German) and Senn (Swiss) processes; and (b) methods that utilize reseparated high-fat cream, e.g., the Alfa (German and Swedish), New Way (Australian), Cherry-Burrell's "Gold 'n Flow" (U.S.A.), and Creamery Package (U.S.A.) processes.

The theories employed to explain the churning and working of butter in the conventional, batch-type procedure are applicable to the processes listed in the first group. According to King,[183] fat globules in the Fritz machine are oriented at the air/cream interface of a thin layer of cream as it moves through the churning cylinder. The violent agitation which occurs in this thin layer of cream is akin to the agitation which takes place in a churn. Whereas conventional churning and working requires about 1½ hr, a similar sequence is accomplished in a matter of minutes in the Fritz process. Excessive losses of fat in the buttermilk and poor control of moisture have been troublesome. Reputedly, the newer modifications of this process have overcome most of these difficulties and the butter produced is indistinguishable from the conventionally processed product.

The Alfa and New Way processes utilize a two-unit operation. Normal cream is reseparated in an air-tight separator to $\simeq$80% fat prior to its introduction into a "transmutator", where it is chilled and agitated. Although high-fat cream represents an oil-in-water type emulsion, its stability has been impaired. Upon chilling, the cream becomes highly viscous and the emulsion commences to break. Agitation of the chilled cream causes additional destabilization of the fat globules, and ensuing liberation of the liquid fat which, again, becomes the continuous phase in the butter. The agglomeration step seems to occur within the cream phase rather than at the air/plasma interface, as was the case in conventional churning. Because of the low extruding temperatures (i.e., 9 to 14°C for the Alfa process and 1 to 3°C for the New Way Process) and the inherent capacity of liquid fat to exist in a supercooled state, crystallization continues for some time following manufacture. A finely dispersed aqueous phase and a relatively high curd content are characteristic of these butters. Alfa butter contains more fat globules in the continuous phase (11 to 31%) than New Way butter ($\simeq$6%), and more "intact" fat globules appear in the moisture droplets. These characteristics reflect temperature differences in the worker-chiller.

The Cherry-Burrell and Creamery Package processes are unique in that the cream is "broken" prior to chilling and working. In the Cherry-Burrell process, the first cream is reheated and agitated to achieve partial destabilization of the lipid phase. Complete destabilization is accomplished in a re-separation step. The Creamery Package process utilizes homogenization to destabilize the reseparated cream. Since in both processes the lipid and aqueous phases are separated, precautions must be exercised in the remixing step to assure homogeneity in the finished product. The standardized mixture is fed to a series of chill-workers and finally to working units where the congealed mixture is worked to the desired consistency for printing. In this respect, the processes resemble those used in the margarine industry. These butters are characterized by a complete absence of globular fat and by large crystals or crystal aggregates in the continuous phase.

Information presented in Table 10.22 serves to summarize the physical characteristics of butter manufactured by the various processes.

Factors Affecting the Consistency of Butter.—The term "consistency" is used in its broadest sense and is indicative of the summation of the rheological properties of butter, e.g., hardness, firmness, structure, texture, etc. Numerous investigators have assessed the factors which contribute to the consistency characteristics. deMan[232] listed the seemingly pertinent factors as (a) composition of the milkfat, (b) temperature treatment of the cream, (c) mechanical treatment of the butter, (d) thermal treatment of the butter, and (e) chemical modification of the fat.

The consistency of butter depends to a large extent on the fatty-acid composition of the fat phase, and therefore is affected by seasonal changes in the nature of the fat. Dolby[78] estimated that about 80% of the variations in the hardness of butter are due to compositional changes in the fat. Cows on winter rations usually produce a harder fat than those on pasture. However, as deMan[232] pointed out, the control of fat composition by feeding practices represents a highly impractical means of regulating the consistency of butter. Consequently only those factors under the control of the buttermaker are considered feasible measures.

Numerous investigators have demonstrated that the consistency of butter can be influenced by temperature manipulation in the cream prior to churning. Most of these procedures have evolved from empirical experimentation; consequently the fundamental changes occurring in the physical nature of the fat were not elucidated. Of current interest in the butter industry is the so-called "8–19–16" or "Swedish" method.[249,317,336,441] Immediately following pasteurization cream is cooled to 8°C and held for 2 hr to initiate crystallization. A tempering period follows in which the cream is warmed to 19°C and held for approximately 2 to 6 hr. Then the cream is cooled to 16°C and held for about 15 hr before final cooling to the churning temperature.

Table 10.22

COMPARISON OF THE STRUCTURAL PROPERTIES OF CONVENTIONAL BUTTER AND BUTTER MADE BY CONTINUOUS METHODS[a]

Characteristic	Conventional Butter	Continuous Methods			
		Fritz Process	Alfa Process	New Way Process	Cherry-Burrell Process
Microscopic structure	Fat globules and water droplets in free fat[b]				Water droplets in free fat
Dielectric constant[c]	5–8	5–7	6–8		
Globular fat	2–46%, 3.5–4 μ diam.	12–17%, 3.5–4.2 μ diam.	11–31%, 3.3–4.1 μ diam.	≈ 6% 4.9 μ. diam.	Few globules
Free fat	54–98%	83–88%	69–89%	94%	Almost 100%
Aqueous phase, diam. of droplets	<1–> 30 μ	<1–> 15 μ	<1–<7[d]	<1–<7 μ	<1–30 μ
Gaseous phase (% volume)	3%	≈ 7%	<1%	—	N_2 added to 4–5%
Consistency	S1. tendency to openness	Tendency to crumbliness	Close texture, "layering"	Close texture	—

a From King[186]
b Some moisture droplets may contain fat globules.
c Dielectric constant of 80°C. cream is about 130, butterfat 3.
d Less susceptible to microbiological deterioration, more susceptible to oxidation.

Proponents of this procedure feel that the initial deep chilling, followed by the tempering period, is basic to its successful application. On the other hand, Dolby[82] felt that the initial cooling does not influence the hardness of butter but does result in reduction of fat losses. The tempering at 19°C was considered to be responsible for the benefits derived, which were influenced by the tempering time. The manner in which cream is cooled seems to be as significant as the temperatures employed (see Table 10.5). As previously discussed in this chapter, rapid cooling causes more crystallization than slow or stepwise cooling. Rapid cooling causes increased hardness in butter.[79,237,238]

Tverdokhleb[393] observed that conventionally churned butter exhibits two crystalline phases—α and β'—in which the β' phase predominates. In continuously processed butter the α-modification prevails, but may contain some γ-modifications. Phase transformations continue in this type of butter following printing. Tverdokhleb concluded that the structure and consistency of butter are determined by the nature of the phase distribution, degree of fat stabilization, and correlation between the amount of solid and liquid fat. The higher the concentration of the high-melting fraction the greater the degree of fat solidification and stabilization, and the harder the butter.

In addition to its effect on the dispersion of the aqueous phase, the working of butter influences its hardness. deMan and Wood[238] studied the effect of commercial printing machines on the consistency of butter. They observed that the extent of hardness reduction was dependent on the initial hardness of the butter; hard butter was softened to a lesser extent than softer butter. The working of butter in commercial plasticizing machines in preparation for printing causes a reduction in hardness and an improvement in moisture distribution.

Fundamentally, the hardness of butter depends on the physical state of the fat. Also, it seems reasonable to assume that the nature of the crystalline structure within the solid phase exerts some effect. Size and distribution of the crystalline phase and the manner in which the liquid phase is enmeshed in the crystalline matrix seems to be relevant to the hardness characteristic. It is a common observation that butter continues to increase in hardness after its removal from the churn. This phenomenon is known as "setting", and may be caused by either continued crystallization of the liquid fat or thixotropic changes. The former effect results in an increase in the proportion of solid fat, while the latter involves changes in the spatial arrangement of the particles. Setting is temporarily arrested by storage at low temperatures but resumes when the temperature is raised. Apparently, the liquid phase plays an important role in the thixotropic process. Thixotropy in butter was reviewed by deMan and Wood.[236 238]

Modifying milkfat chemically by random or directed interesterifications, isomerization, or hydrogenation causes an increase in hardness.[232-234,417] Interesterification and hydrogenation increase the proportion of higher-melting glycerides. Isomerization from the *cis* to *trans* form of oleic results in an increased melting point of elaidic acid. Aside from the fact that these modifications do not improve the consistency of butter, they represent expensive procedures and are limited in application to processes capable of utilizing de-emulsified fat, e.g., the Cherry-Burrell and Creamery Package continuous buttermaking processes.

Numerous excellent papers describing the interrelationships of butterfat composition, polymorphism, crystallization and processing treatments on the physical properties and consistency of butter have emanated from the Russian workers Tverdokhleb, Belousov and Vergelesov (see Ref. No. 59 for abstracts of these papers). King's[190] review is recommended for supplemental reading.

De-Emulsification of Milkfat.—Obviously, the methods employed to manufacture butter represent practical approaches to the destabilization of the milkfat emulsion. Numerous continuous methods, somewhat akin to the procedures used in the American continuous methods, are particularly suited to large-scale production of anhydrous butter oil.[89] Final refinement of the milkfat is achieved by repeated washing and reseparation steps, followed by vacuum dehydration. Oils prepared in this manner are practically phospholipid-free and exhibit excellent storage characteristics.

Patton[286] and Stine and Patton[362,363] studied the de-emulsification of milkfat by heating cream to 82°C for 15 min in the presence of 10% or less of selected surfactants. Oils obtained in this manner were identical with butter oil except that they were lower in phospholipid content. Of the different classes of surfactants studied, acids, alcohols, and amines were more effective as de-emulsifiers than aldehydes, ketones, and esters. In general, optimum results were obtained with the C_3 to C_5 homologues. The mechanism of de-emulsification was ascribed to the interfacial activity of the surfactants.

King[188] observed that the addition of surface-active substances to whole milk either induced fat globules, clumps, and lenses to rise to the surface, or cleared the surface of all forms of fat. The latter effect was observed for sorbitan or polyoxyethylene sorbitan esters of higher fatty acids. In the homologous series of monohydric alcohols studied, methyl and ethyl alcohols induced only a moderate rising action, whereas the *n*-propyl to *n*-decyl members were highly active. He proposed that the hydrophobic properties of the fat globules are caused by the adsorption of molecules of alcohol on the outer layer of the membrane protein. Further, it was suggested that the lower alcohols (C_3 to C_5) participated in the disorganization of the membrane complex through their effect on the polar bonds between phospholipids and proteins. However, apolar bonding also makes a significant contribution to the stability of the membrane structure. Therefore in view of the known dissociating effects of surface-active, anionic detergents on the apolar structure of plasma membranes, it seems reasonable to assume that the fat-globule membrane complex is dissociated in a similar manner.

Butterfat has been separated into high- and low-melting point fractions by partial crystallization. Separation of the oil phase from the crystallized fat phases was achieved by either centrifugal separation or pressure filtration.[90,339] Suggested uses for the high-melting fraction include; hard butter for tropical regions; hard coatings for chocolate or ice cream and incorporation into spray-dried products; the low-melting fraction can be incorporated during conventional churning to improve the spreadability of butter.[71]

Loo[224] produced a wettable dried whole milk by working low- and high-melting fractions with skimmilk prior to drying so that in the final product the fractions were separated by the SNF phase.

REFERENCES

1. ABDERHALDEN, E., and VÖLTZ, W., Hoppe-Seyler's Z. physiol. Chem., *59,* 13 (1909).
2. ALEXANDER, K. M., and LUSENA, C. V., J. Dairy Sci., *44,* 1414 (1961).
3. ANDREEV, A., Milchwissenschaft, *23,* 476 (1968). Cited in Dairy Sci. Abstr., *30,* 616 (1968).
4. ANTILA, V., LUOTO, K., and ANTILA, M., Meijeritieteellimen Aikakauskiaja, *25,* 31 (1965).
5. ASCHAFFENBURG, R., and OGSTON, A. S., J. Dairy Res., *14,* 3161 (1946).
6. ASCHERSON, F. M., A translation by Emil Hatschek in "Foundations of Colloid Chemistry" (1840). E. Benn Ltd., London (1925).
7. ASHWORTH, V. S., J. Dairy Sci., *34,* 317 (1951).
8. AULAKH, J. A., and STINE, C. M., J. Dairy Sci., *49,* 707 (1968).
9. BABCOCK, S. M., 4th Ann. Rept. N.Y. (Geneva) Agr. Expt. Sta. (1885).
10. BABCOCK, S. M., Wis. Agr. Expt. Sta. Bull., *18* (1889).
11. BAILEY, A. E., "Melting and Solidification of Fats," Interscience Pub. Inc., New York, 1950.
12. BAILIE, MARGOT J., and MORTON, R. K., Biochem. J., *69,* 35 (1958).
13. BAILIE, MARGOT J., and MORTON, R. K., Biochem. J., *69,* 44 (1958).
14. BAKER, B. E., and SAMUELS, E. R., J. Dairy Sci., *44,* 407 (1961).
15. BARGMANN, W., Z. Tierzüchung. Zuchtüngbiol., *76,* 416 (1962).
16. BARGMANN, W., FLEISCHAUURER, K., and KNOOP, A., Z. Zellforsch. Mikroskop. Anat., Abt. Histochem., *53,* 545 (1961).
17. BARGMANN, W., and KNOOP, A., Z. Zellforsch. Mikroskop. Anat., Abt. Histochem., *49,* 344 (1959).
18. BERRY, K. E., HOOD, L. F., and PATTON, S., J. Dairy Sci., *54,* 911 (1971).
19. BELOUSOV, A. P., and VERGELESOV, V. M., Molochn. Prom., *21,* 27 (1960).

20. BELOUSOV, A. P., and VERGELESOV, V. M., 16th Intern. Dairy Congr., Proc., Copenhagen, *B,* 122 (1962).
21. BERG, R. H., Am. Soc. Testing Mater., Spec. Tech. Pub., *234,* 245 (1958).
22. BERLIN, E., LAKSCHMANON, S., KLIMAN, P G., and PALLANSCH, M. J., Biochemistry, *6,* 1388 (1967).
23. BINGHAM, R. J., HUBER, J. D., and AURAND, L. W., J. Dairy Sci., *50,* 318 (1967).
24. BINGHAM, ELIZABETH W., and ZITTLE, C. A., Biochem. Biophys. Res. Comm., *7,* 408 (1962).
25. BINGHAM, ELIZABETH W., and ZITTLE, C. A., Arch Biochem. Biophys., *106,* 235 (1964).
26. BOTTAZZI, V., VELLAGLIO, F., and MONTESCANI, G., Sci. Tech. Latt. -Casear, *20,* 219 (1968). Cited in Dairy Sci. Abstr., *31,* 341 (1969).
27. BOUCHARDAT, M., and QUEVENNE, T. A., DuLait (1857). Cited by Campbell, Ref. no. 48.
28 BRODBECK, U., EBNER, K. E., J. Biol. Chem., *241,* 762 (1966).
29. BROUWER, E., Proefzuivelboerderij Hoorn, *1924,* 18 (1924). Cited by Dunkley and Sommer, Ref. no. 87.
30. BRUNNER, J. R., J. Dairy Sci., *33,* 741 (1950).
31. BRUNNER, J. R., J. Dairy Sci., *45,* 943 (1962).
32. BRUNNER, J. R., DUNCAN, C. W., and TROUT, G. M., Food Res., *18,* 454 (1953).
33. BRUNNER, J. R., LILLEVIK, H. A., TROUT, G. M., and DUNCAN, C. W., Food Res., *18,* 463 (1953).
34. BRUNNER, J. R., DUNCAN, C. W., TROUT, G. M., and MACKENZIE, MAXIME, Food Res., *18,* 469 (1953).
35. BRUNNER, J. R., and HERALD, C. T., J. Dairy Sci., *41,* 1489 (1958).
36. BRUNNER, J. R., and JACK, E. L., J. Dairy Sci., *33,* 267 (1950).
37. BRUNNER, J. R., and THOMPSON, M. P., J. Dairy Sci., *44,* 1170 (1961).
38. BRUNNER, J. R., and THOMPSON, M. P., J. Dairy Sci., *44,* 1224 (1961).
39. BUCHHEIM, W., Kiel. Milchwirtsch. Forschungsber., *22,* 3 (1970.
40. BUCHHEIM, W., Naturwissenschaften, *12,* 672 (1970).
41. BUCHHEIM, W., Milchwissenschaft, *25,* 65 (1970).
42. BUCHHEIM, W., Milchwissenschaft, *25,* 223 (1970).
43. BUCHHEIM, W., Molkerei- Z., *25,* 461 (1971).
44. BUCHHEIM, W. Personal communication (1971).
45. BULLOCK, D. H., and WINDER, W. C., J. Dairy Sci., *41,* 708 (1958).
46. BAKER, B. E., and SAMUELS, E. R., J. Dairy Sci., *44,* 407 (1961).
47. BUTLER, J. E., J. Dairy Sci., *52,* 1895 (1969).
48. CAMPBELL, M. H., Vt. Agr. Expt. Sta. Bull., *341* (1932).
49. CANTABRANA, F., and DEMAN, J. M., J. Dairy Sci., *47,* 32 (1964).
50. CHAPMAN, D., Chem. Rev., *62,* 433 (1962).
51. CHIEN, H. C., RICHARDSON, T., and AMUNDSON, C. H., J. Dairy Res., *33,* 217 (1966).
52. CHIEN, H. C., and RICHARDSON, T., J. Dairy Sci., *50,* 451 (1967).
53. CHIEN, H. C., and RICHARDSON, T., J. Dairy Sci., *50,* 1868 (1967).
54. CHRZASZCZ, T., and GORALOWNA, C., Biochem. Z., *166,* 172 (1925).
55. CLARKSON, C. E., and MALKIN, T., J. Chem. Soc., *1934,* 666 (1934).
56. CLARKSON, C. E., and MALKIN, T., J. Chem. Soc., *1948,* 985 (1948).
57. COLLIER, P., 10th Ann. Rept. N.Y. (Geneva) Agr. Expt. Sta. (1891).
58. COLLIER, P., 11th Ann. Rept. N.Y. (Geneva) Agr. Expt. Sta. (1892).
59. Comm. Bur. Dairy Sci. Tech., Shinfield, Reading, England. "Crystalline Milk Fat," Annotated Bibliography, no. 16, (1960–71).
60. COOPER, W. F., NUTTALL, W. N., and FREAK, G. A., J. Agr. Sci., *4,* 331 (1911).
61. CORNELL, D.G., and PALLANSCH, M. J., J. Dairy Sci., *49,* 1371 (1966).
62. COULSON, E. J., and JACKSON, R. H., Arch Biochem. Biophys., *97,* 378 (1962).
63. COULSON, E. J., and STEVENS, H., Arch. Biochem. Biophys., *107,* 336 (1964).

64. DAHLBERG, A. C., and MARQUARDT, J. C., N.Y. (Geneva) Agr. Expt. Sta. Tech. Bull., *157* (1929).
65. DAHLE, C. D., and JACK, E. L., J. Dairy Sci., *20,* 605 (1937).
66. VAN DAM, W., Verslag. Landbouwk.Ondersoek., *16,* 1 (1915). Cited by Dunkley and Sommer, Ref. no. 87.
67. VAN DAM, W., HEKMA, E., and SIRKS, H. A., Proefquivelboerderij Hoorn, *1922,* 81 (1922). Cited by Dunkley and Sommer, Ref. no. 87.
68. VAN DAM, W., and HOLWERDA, B. S., Verslag. Landbouwk. Onderzoek., 40, 175 (1934). Cited by Dunkley and Sommer, Ref. no. 87.
69. VAN DAM, W., and SIRKS, H. A., Verslag. Landbouwk. Onderzoek., *26,* 106 (1922). Cited by Dunkley and Sommer, Ref. no. 87.
70. DECKOFF, L. P., and REES, L. H., Milk Dealer, *46,* 61 (1957).
71. DIXON, B. D., MAITLAND, V. J., 18th Intern. Dairy Congr., Proc., Melbourne, *1E,* 245 (1970).
72. DIXON, F. S., WEIGLE, W. O., and VAGQUEZ, J. U., Lab. Invest., *10,* 216 (1961).
73. DOAN, F. J., 41st Ann. Rept. Pa. Agr. Expt. Sta. Bull., *230* (1928).
74. DOAN, F. J., J. Dairy Sci., *14,* 527 (1931).
75. DOAN, F. J., J. Milk Tech., *1,* 2 (1938).
76. DOAN, F. J., and MINISTER, C. H., Pa. Agr. Expt. Sta. Tech. Bull., *287* (1933).
77. DOAN, F. J., and MYERS, C. H., J. Dairy Sci., *26,* 893 (1936).
78. DOLBY, R. M., J. Dairy Res., *16,* 336 (1949).
79. DOLBY, R. M., J. Dairy Res., *21,* 67 (1954).
80. DOLBY, R. M., J. Dairy Res., *24,* 68 (1957).
81. DOLBY, R. M., J. Dairy Res., *24,* 372 (1957).
82. DOLBY, R. M., Austral. J. Dairy Tech., *14,* 103 (1959).
83. DOWBEN, R. M., and BRUNNER, J. R., J. Clin. Invest., *45,* 1006 (1966).
84. DOWBEN, R. M., BRUNNER, J. R., and PHILPOTT, D. E., Biochim. Biophys. Acta, *135,* 1 (1967).
85. DOWNEY, W. K., and MURPHY, R. F., J. Dairy Res., *37,* 47 (1970).
86. DUNKLEY, W. L., FRANKLIN, J. P., and PANGBORN, R. M., J. Dairy Sci., *45,* 1040 (1962).
87. DUNKLEY, W. L., and SOMMER, H. H., Wis. Agr. Expt. Sta. Res. Bull., *15* (1944).
88. FINLEY, J. W., and SHIPE, W. F., J. Dairy Sci., *54,* 15 (1971).
89. FJAERVOLL, A., Dairy Ind., *35,* 424 (1970).
90. FJAERVOLL, A., Dairy Ind., *35,* 502 (1970).
91. FOLCH, J., J. Biol. Chem., *146,* 35 (1942).
92. FOLLEY, S. J., and WHITE, P., Proc. R. Soc., Ser B, *120,* 330 (1936).
93. FOLLEY, S. J., Biochem. Soc. Symp., *9,* 52 (1952).
94. FOX, K. K., HOLSINGER, VIRGINIA, CAHA, JEANNE, and PALLANSCH, M. J., J. Dairy Sci., *43,* 1396 (1960).
95. FOX, P. F., and TARASSUK, N. P., J. Dairy Sci., *51,* 826 (1968).
96. FOX, P. F., YAGUCHI, M., and TARASSUK, N. P., J. Dairy Sci., *50,* 307 (1967).
97. FLEISCHMANN, W., "Lehrbuchder Milchwirtschaft," 1st Ed., M. Heinsius, Bremen (1893). Cited by Campbell, Ref. no. 48.
98. FRANZEN, R. W., "Isolation and Characterization of Bovine Lacteal Immunoglobulins," Thesis, Michigan State University (1971).
99. FUJINO, Y., SAEKI, T., Japan J. Zootech. Sci., *40,* 349 (1969).
100. GAFFNEY, P. J., HARPER, W. J., and GOULD, I. A., J. Dairy Sci., *45,* 646 (1962).
101. GALANOS, D. S., and KAPOULOS, V. M., Biochim. Biophys. Acta, *98,* 278 (1965).
102. GALANOS, D. S., and KAPOULOS, V. M., Biochim. Biophys. Acta, *98,* 298 (1965).
103. GAMMACK, D. B., and GUPTA, B. B., Biochem. J., *103,* (3) 72 (1967).
104. GAMMACK, D. B., and GUPTA, B. B., 18th Intern. Dairy Congr. Proc., Melbourne, *1E,* 20 (1970).
105. GARTON, G. A., J. Lipid Res., *4,* 237 (1963).
106. GOULDEN, J. D. S., Dairy Ind., *23,* 558 (1958).
107. GOULDEN, J. D. S., J. Dairy Res., *25,* 228 (1958).
108. GOULDEN, J. D. S., and SHERMAN, P., J. Dairy Res., *29,* 47 (1962).

109. GRAHAM, W. R., JONES, T. S. G., and KAY, H. D., Proc. R. Soc. Ser. B., *120,* 330 (1936).
110. VON GRAVEL, L., Z Lebensm.-Untersuch., *104,* 1 (1956).
111. GREENBANK, G. R., 13th Intern. Dairy Congr., Proc., The Hague, *3,* 1269 (1953).
112. GREENBANK, G. R., and PALLANSCH, M. J., J. Dairy Sci., *44,* 1597 (1961).
113. GREENBANK, G. R., and PALLANSCH, M. J., J. Dairy Sci., *45,* 958 (1962).
114. GUDNASON, G. V., "Studies on the Distribution of Xanthine Oxidase between the Fat and Skimmilk Phases of Milk and the Effect of Aging on the Heat Stability of Enzymes", Thesis, Cornell University (1961).
115. GURD, R. R. N., in "Lipid Chemistry", D. J. Hanahan, Ed., John Wiley and Sons, Inc., New York (1960).
116. GUTZEIT, E., Landwirtsch. Jahrb. Schweiz., *25,* 539 (1895).
117. HAIGHTON, A. J., and HANNEWIJK, J., J. Am. Oil Chem. Soc., *35,* 344 (1958).
118. HAMMER, B. W., Iowa Agr. Expt. Sta. Res. Bull., *31* (1916).
119. HAMPSON, J. W., and ROTHBART, H. L., J. Am. Oil Chem. Soc., *46,* 143 (1969).
120. HANAHAN, D. J., "Lipid Chemistry", John Wiley and Sons, Inc., New York (1960).
121. HANNEWIJK, J., Chem. Weekblad, *51,* 419 (1955).
122. HANNEWIJK, J., and HAIGHTON, A. J., Neth. Milk Dairy J., *11,* 304 (1957).
123. HANSSON, E., 12th Intern. Dairy Congr., Proc., Stockholm, *2,* 30 (1949).
124. HARE, J. H., SCHWARTZ, D. P., and WEESE, S. J., J. Dairy Sci., *35,* 615 (1952).
125. HARTMAN, G. H., JR., and SWANSON, A. M., J. Dairy Sci., *48,* 1161 (1965).
126. HARWALKER, V. R., and BRUNNER, J. R., J. Dairy Sci., *48,* 1139 (1965).
127. HARWALKER, V. R., LUNDEEN, G. A., LINDQUIST, KARIN, and BRUNNER, J. R., J. Dairy Sci., *47,* 665 (1964).
128. HATTORI, K., J. Pharm. Soc. (Japan) No. *516,* 123 (1925). Cited in Chem. Abstr., *19,* 2380 (1925).
129. HAUGAARD, G., J. Dairy Sci., *49,* 1185 (1966).
130. HAUGAARD, G., and PETTINATI, J. D., J. Dairy Sci., *42,* 1255 (1959).
131. HAYASHI, S., and SMITH, L. M., Biochemistry, *4,* 2550 (1965).
132. HAYASHI, S., ERICKSON, D. R., and SMITH, L. M., Biochemistry, *4,* 2557 (1965).
133. HEKMA, E., Proefzuivelboerderij Hoorn, *1922,* 1 (1922). Cited by Dunkley and Sommer, Ref. no. 87.
134. HEKMA, E., Proefzuivelboerderij Hoorn, *1924,* 36 (1924). Cited by Dunkley and Sommer, Ref. no. 87.
135. HEKMA, E., and SIRKS, H. A., Proefzuivelboerderij Hoorn, *1923,* 88 (1923). Cited by Dunkley and Sommer, Ref. no. 86.
136. HELMINEN, H. J., and ERICSSON, J. L. E., J. Ultrastruct. Res., *25,* 193 (1968).
137. HERALD, C. T., and BRUNNER, J. R., J. Dairy Sci., *40,* 446 (1957).
138. HERALD, C. T., and BRUNNER, J. R., J. Dairy Sci., *40,* 948 (1957).
139. HLADÍK, J., 17th Intern. Dairy Congr., Proc., München, *B* (2), (1966).
140. HLADÍK, J., Sb. Vysoke Skoly Chem.-Technol. v Praze, *E20,* 57 (1968).
141. HLADÍK, J., and FORMAN, L., Sb. Vysoke Skoly Chem.-Technol. v Praze, *E15,* 69 (1967).
142. HLADÍK, J., and FORMAN, L., Sb. Vysoke Skoly Chem.-Technol. v Praze, *E21,* 33 (1968).
143. HLADÍK, J., and MICHALEC, C., Acta Biol. Med. Ger., *16,* 696 (1966).
144. HODGE, J. E., J. Agr. Food Chem., *1,* 928 (1953).
145. HOLDEN, T. F., ACETO, N. C., DELLAMONICA, E. S., and CALHOUN, M. J., J. Dairy Sci., *49,* 346 (1966).
146. HOOD, L. F., and PATTON, S., J. Dairy Sci., *51,* 928 (1968).
147. HORWITZ, C., FREMLIN, J. H., and FARR, R. F., Phys. Med. Biol, *10,* 385 (1965).
148. HUANG, C. M., and KEENAN, T. W., J. Dairy Sci., *54,* 770 (1971).
149. HUANG, T. C., and KUKSIS, A., Lipids, *2,* 453 (1967).
150. HUNZIKER, O. F., Ind. Agr. Expt. Sta. Bull., *150* (1912).
151. IBUKI, F., MORI, T. MATSUSHITA, S., and HATA, T., Agr. Biol. Chem. (Japan), *29,* 635 (1965).
152. IWAIDA, M., and TSUGO, T., Agr. Biol. Chem. (Japan), *25,* 130 (1961).

153. IWAIDA, M., and TSUGO, T., 16th Intern. Dairy Congr., Proc., Copenhagen, *A*, 707 (1962).
154. JACK, E. L., Am. Dairy Sci. Assoc., Western Div., 27th Ann. Meet., Proc, 68 (1941).
155. JACK, E. L., and BRUNNER, J. R., J. Dairy Sci., *34,* 169 (1943).
156. JACK, E. L., and DAHLE, C. D., J. Dairy Sci., *20,* 551 (1937).
157. JACKSON, R. H., "Proteins of the Homogenized Milk Fat Globule Membrane," Thesis, Michigan State University (1959).
158. JACKSON, R. H., and BRUNNER, J. R., J. Dairy Sci., *43,* 912 (1960).
159. JACKSON, R. H., COULSON, E. J., and CLARK, W. R., Arch, Biochem. Biophys., *97,* 273 (1962).
160. JACKSON, R. H., and PALLANSCH, M. J., J. Agr. Food Chem., *9,* 424 (1961).
161. JENNESS, R., and PALMER, L. S., J. Dairy Sci., *28,* 611 (1945).
162. JENNESS, R., and PALMER, L. S., J. Dairy Sci., *28,* 653 (1945).
163. JENNESS, R., and PARKASH, Sat, J. Dairy Sci., *54,* 123 (1971).
164. JENNESS, R., and PATTON, S., "Principles of Dairy Chemistry," John Wiley and Sons, Inc., New York (1959).
165. KÅRKKÅINEN, V. J., Fat and Oil Chem., 4th Scand. Symp. Fats Oils, (1965) *1966,* 293 (1966).
166. KAYSER, S. G., and PATTON, S., Biochem. Biophys. Res. Comm., *41,* 1572 (1970).
167. KEENAN, T. W., and MORRÉ, D. J., Biochemistry, *9,* 19 (1969).
168. KEENAN, T. W., MORRÉ, D. J., OLSON, DIANE E., YUNGHANS, W. N., and PATTON, S., J. Cell Biol., *44,* 80 (1970).
169. KEENAN, T. W., OLSON, DIANE E., and MOLLENHAUER, H. H., J. Dairy Sci., *54,* 295 (1971).
170. KEENAN, T. W., SAACKE, R. G., and PATTON, S., J. Dairy Sci., *53,* 1349 (1970).
171. KENYON, A. J., and JENNESS, R., J. Dairy Sci., *41,* 716 (1958).
172. KENYON, A. J., JENNESS, R., and ANDERSON, R. K., J. Dairy Sci., *49,* 1144 (1966).
173. KERNOHAN, ELIZABETH A., and LEPHERD, E. E., J. Dairy Res., *36,* 177 (1969).
174. KERNOHAN, ELIZABETH A., WADSWORTH, J. C., and LASCELLAS, A. K., J. Dairy Res., *38,* 65 (1971).
175. KIERMEIER, F., and SOLMS-BARUTH, H., Z. Lebensm. Untersuch.-Forsch., *130,* 291 (1966).
176. KIERMEIER, F., and SOLMS-BARUTH, H., Z. Lebensm. Untersuch.-Forsch., *132,* 12 (1966).
177. KING, J. O. L., J. Dairy Res., *23,* 105 (1956).
178. KING, N., Milchw. Forsch., *14,* 114 (1932).
179. KING, N., Milchwissenschaft, *4,* 250 (1949).
180. KING, N., Neth. Milk Dairy J., *4,* 30 (1950).
181. KING, N., Dairy Ind., *16,* 727 (1951).
182. KING, N., J. Dairy Res., *14,* 225 (1952).
183. KING, N., J. Dairy Res., *15,* 589 (1953).
184. KING, N., J. Dairy Res., *22,* 205 (1955).
185. KING, N., Dairy Ind., *20,* 311 (1955).
186. KING, N., Dairy Ind., *20,* 409 (1955).
187. KING, N., "The Milk Fat Globule Membrane," Comm. Agr. Bur., Franham Roy. Bucks, England (1955).
188. KING, N., J. Dairy Res., *24,* 43 (1957).
189. KING, N., Austral. J. Dairy Tech., *15,* 77 (1960).
190. KING, N., Dairy Sci. Abstr., *26,* 151 (1964).
191. KING, R. L., J. Dairy Sci., *46,* 267 (1963).
192. KINSELLA, J. E., and McCARTHY, R. D., Biochim. Biophys. Acta, *164,* 518 (1968).
193. KITCHEN, B. J., TAYLOR, G. C., and WHITE, I. C., J. Dairy Res., *37,* 279 (1970).
194. KNOOP, A., Personal Communication (1962).
195. KNOOP, E., Kiel. Milchs. Forschber., *10,* 207 (1958).
196. KNOOP, E., and WORTMANN, A., Milchwissenschaft, *14,* 106 (1959).

197. KNOOP, E., and WORTMANN, A., Deutsch. Molkerei-Ztg., *80,* 491 (1959).
198. KOOPS, J., "Koelhuisgebreken van boter," Thesis, University of Wageningen (1963).
199. KOOPS, J., PAYENS, T. A. J., and MAGOT, M. F. K., Neth. Milk Dairy J., *20,* 296 (1966).
200. KOOPS, J., and PETTE, J. W., 14th Intern. Dairy Congr., Proc., Rome, 2 (1), 168 (1956).
201. KOOPS, J., and TARASSUK, N. P., Neth. Milk Dairy J., *13,* 180 (1959).
202. KOYAMA, S., Japan J. Zootech Sci., *31,* 88 (1960).
203. KOYAMA, S., Japan J. Dairy Sci., *16,* A14 (1967).
204. KRAUTWURST, J., Kiel. Milchwirtsch. Forschungsber., *22,* 225 (1970).
205. VAN KREVELD, A., Rec. Trav. Chim., *61,* 29 (1942).
206. VAN KREVELD, A., Rec. Trav. Chim., *61,* 41 (1942).
207. VAN KREVELD, A., Rec. Trav. Chim., *65,* 321 (1946).
208. KRUKOVSKY, V. N., and HERRINGTON, B. J., J. Dairy Sci., *22,* 137 (1939).
209. KRUKOVSKY, V. N., and SHARP, P. F., J. Dairy Sci., *23,* 1109 (1940).
210. KURTZ, F. E., in "Fundamentals of Dairy Chemistry," Ed. Webb, B. H., and Johnson, A. H., the Avi Pub. Co., Westport (1965).
211. LADBROOKE, D. B., and CHAPMAN, D., Chem. Phys. Lipids, *3,* 304 (1969).
212. LEA, C. H., J. Sci. Food Agr., *8,* 1 (1957).
213. VAN LEEUWENHOEK, A., Phil. Trans., *9,* 102 (1674).
214. LEMBKE, A., and KAUFMANN, W., 13th Intern. Dairy Congr. Proc., The Hague, *3,* 1394 (1953).
215. LEVITON, A., and PALLANSCH, M. J., J. Dairy Sci., *42,* 20 (1959).
216. LEVITON, A., and PALLANSCH, M. J., J. Dairy Sci., *44,* 633 (1961).
217. LEVOWITZ, D., and VANDERMEULEN, P. A., J. Dairy Sci., *20,* 657 (1937).
218. LEVY, P. F., Am. Lab., *2,* 46 (1970).
219. LIPATOV, N N., MIKHAILOUSKY, E. A., ZOLOTIN, YU. P., and USKOV, V. I., Moloch. Prom., *31* (9), 30 (1970). Cited in Dairy Sci. Abstr., *33,* 77 (1971).
220. LITMAN, I. I., and ASHWORTH, U.S., J. Dairy Sci., *40,* 403 (1957).
221. LOEWENSTEIN, M., and GOULD, I. A., J. Dairy Sci., *37,* 644 (1954).
222. LOEWENSTEIN, M., and GOULD, I. A., 17TH Intern. Dairy Congr., Proc., Müchen, *B,* 399 (1966).
223. LONG, C., In "Biochemists' Handbook", C. Long, Ed., D. Van Norstrand Co., Inc., Princeton, N. J. (1961).
224. LOO, C. C., Fat Crystallization, U.S. Pat. 3, 301, 682 (1967).
225. LUNDSTEDT, E., Milk Plant Monthly, *23* (7), 32 (1934).
226. LUTTON, E. S., J. Am. Chem. Soc., *67,* 524 (1945).
227. LUTTON, E. S., J. Am. Chem. Soc., *68,* 676 (1946).
228. LUTTON, E. S., J. Am. Chem. Soc., *70,* 248 (1948).
229. LUTTON, E. S., J. Am. Oil Chem. Soc., *27,* 276 (1950).
230. MALKIN, T., "Progress in the Chemistry of Fats and other Lipids," *2,* 1 (1954). Pergamon Press, Ltd., London.
231. MALKIN, T., and MEARA, M. L., J. Chem. Soc., *1939,* 103 (1939).
232. DEMAN, J. M., Dairy Ind., *26,* 37 (1961).
233. DEMAN, J. M., J. Dairy Res., *28,* 81 (1961).
234. DEMAN, J. M., J. Dairy Res., *28,* 87 (1961).
235. DEMAN, J. M., Milchwissenschaft, *18,* 67 (1963).
236. DEMAN, J. M., and WOOD, F. W., Dairy Ind., *23,* 265 (1958).
237. DEMAN, J. M., and WOOD, F. W., J. Dairy Sci., *41,* 360 (1958).
238. DEMAN, J. M., and WOOD, F. W., J. Dairy Sci., *42,* 56 (1959).
239. DEMAN, J. M., and WOOD, F. W., J. Dairy Res., *26,* 17 (1959).
240. MARTIN, E. M., and MORTON, R. K., Nature, *176,* 111 (1955).
241. MATSUSHITA, S., IBUKI, F., MORI, T., and HATA, T., Agr. Biol. Chem., (Japan) *29,* 436 (1965).
242. McCARTHY, R. D., and PATTON, S., Biochim. Biophys. Acta, *70,* 102 (1963).

243. McDOWALL, F. H., Food Ind., *10,* 909 (1947).
244. McDOWELL, F. H., "The Buttermaking Manuel," Two volumes, New Zealand University Press, Wellington (1953).
245. McGUZAN, W. A., ZEHREN, V., ZEHREN, V. C., and SWANSON, A. M., Science, *120,* 435 (1954).
246. McKENZIE, H. A., Advan. Protein Chem., *22,* 55 (1967).
247. MOHR, W., and BAUR, J., Vorratspfl. Lebensm. Forsch., *2,* 383, 509 (1939). Cited by Mulder, Ref. no. 258.
248. MOHR, W., and BROCKMANN, C., Milchw. Forsch., *10,* 90 (1930).
249. MOHR, W., MOHR, E., and PETERS, K. H., Molkerei. Koserei-Ztg., *6,* 326 (1955).
250. MORRÉ, D. J., KEENAN, T. W., and MOLLENHAUER, H. H., 1st Intern. Symp. Cell Biol. Cytopharmacol., Venice. Raven Press, New York (1969).
251. MOORE, A. V., and TROUT, G. M., Mich. Agr. Expt. Sta. Quart. Bull., *29,* 177 (1947).
252. MORRISON, W. R., and HAY, J. D., Biochim. Biophys. Acta, *202,* 460 (1970).
253. MORRISON, W. R., and SMITH, L. M., Biochim. Biophys. Acta, *84,* 759 (1964).
254. MORTON, R. K., Biochem. J., *55,* 786 (1953).
255. MORTON, R. K., Nature, *171,* 734 (1953).
256. MORTON, R. K., Biochem. J., *57,* 231 (1954).
257. MULDER, H., Verslag. Landbouwk. Onderzoek., *51* (2), 39 (1945).
258. MULDER, H., Neth. Milk Dairy J., *7,* 149 (1953).
259. MULDER, H., Neth. Milk Dairy J., *11,* 197 (1957).
260. MULDER, H., and DEN BRAUER, F. C. A., Neth. Milk Dairy J., *10,* 199 (1956).
261. MULDER, H., and DEN BRAUER, F. C. A., Neth. Milk Dairy J., *10,* 230 (1956).
262. MULDER, H., DEN BRAUER, F. C. A., and WELLE, TH. G., Neth. Milk Dairy J., *10,* 206 (1956).
263. MULDER, H., DEN BRAUER, F. C. A., and WELLE, TH. G., Neth. Milk Dairy J., *10,* 214 (1956).
264. MULDER, H., and KLOMP, R., Neth. Milk Dairy J., *10,* 123 (1956).
265. MULDER, H., and KOPPEJAN, C. A., 13th Intern. Dairy Congr., Proc., The Hague, *3,* 1402 (1953).
266. MULDER, H., MENGER, J. W., and KOOPS, J., Neth. Milk Dairy J., *11,* 263 (1957).
267. MULDER, H.,and MENGER, J. W., Neth. Milk Dairy J., *12,* 1 (1958).
268. MULDER, H., and ZUIDHOF, T. A., Neth. Milk Dairy J., *12,* 173 (1958).
269. VON MURALT, G., GUGLER, E., and ROULET, D. L. A., In "Immunoelectrophoretic Analysis", P. Graber and P. Burtin, Eds., p. 261, Elsevier, New York (1964).
270. MURPHY, C. B., Anal. Chem., *30,* 867 (1958).
271. NELSON, W. E., and WINDER, W. C., J. Dairy Sci., *44,* 1156 (1968).
272. NERNST, W., Ann. Physik., *36,* 395 (1911).
273. NISHIDA, T., and KUMMEROW, F. A., J. Lipid Res., *1,* 450 (1960).
274. NUGENT, R. L., J. Phys. Chem., *36,* 449 (1932).
275. OLIVECRONA, T., and LINDAHL, A., Acta Chem. Scand., *23,* 3587 (1969).
276. OWEN, E. C., and HART, L. I., Proc. Nutr. Soc., *21,* 15 (1962).
277. PACKARD, V. S., "A Study of Some Factors Involved in Milk Lipase Action." Thesis, University of Minnesota (1960).
278. PAECH, W., Z. Lebensm. Untersuch.-Forsch., *142,* 47 (1970).
279. PALMER, L. S., and ANDERSON, E. O., J. Dairy Sci., *9,* 1 (1926).
280. PALMER, L. S., and LEWIS, J. J., Arch. Pathol., *16,* 303 (1933).
281. PALMER, L. S., and POWELL, M. E., J. Dairy Sci., *27,* 471 (1944).
282. PALMER, L. S., and SAMUELSSON, E.-G., Proc. Soc. Expt. Biol. Med., *21,* 537 (1924).
283. PALMER, L. S., and TARASSUK, N. P., J. Dairy Sci., *19,* 323 (1936).
284. PALMER, L. S., and TARASSUK, N. P., J. Dairy Sci., *23,* 861 (1940).
285. PALMER, L. S., and WIESE, HILDA, F., J. Dairy Sci., *16,* 41 (1933).
286. PATTON, S., J. Dairy Sci., *35,* 324 (1952).
287. PATTON, S., J. Dairy Sci., *37,* 446 (1954).

288. PATTON, S., Nature, *228,* 97 (1970).
289. PATTON, S., J. Theor. Biol., *29,* 489 (1970).
290. PATTON, S., DURDAN, A., and McCARTHY, R. D., J. Dairy Sci., *47,* 489 (1964).
291. PATTON, S., and FOWKES, F. M., J. Theor. Biol. *15,* 274 (1967).
292. PATTON, S., HOOD, L. F., and PATTON, J. S., J. Lipid Res., *10,* 260 (1969).
293. PATTON, S., and KEENAN, T. W., Lipids, *6,* 58 (1971).
294. PATTON, S., and McCARTHY, R. D., J. Dairy Sci., *46,* 916 (1963).
295. PATTON, S., McCARTHY, R. D., and DIMICK, P. S., J. Dairy Sci., *48,* 1389 (1965).
296. PATTON, S., and TRAMS, G., Federation Proc., *14,* 230 (1971).
297. PAYENS, T. A. J., Kiel. Milchwirtsch. Forschungsber., *16,* 457 (1964).
298. PAYENS, T. A. J., Milchwissenschaft, *23,* 325 (1968).
299. PAYENS, T. A. J., and BOTH, P., J. Immunol., *7,* 869 (1970).
300. PAYENS, T. A. J., KOOPS, J., and MOGOT, M. F. K., Biochim. Biophys. Acta, *94,* 576 (1965).
301. PEEREBOOM, J. W. S., Neth. Milk Dairy J., *20,* 113 (1966).
302. PEEREBOOM, J. W. C., Fette-Seifen Anstrichmittel, *71,* 314 (1969).
303. PEEREBOOM, J. W. C., Fette-Seifen Anstrichmittel, *72,* 299 (1970).
304. PENNINGTON, R. J., Biochem. J., *51,* 251 (1952).
305. PHIPPS, L. W., J. Dairy Res., *24, 51* (1957).
306. POLIS, B. D., and SHMUKLER, H. W., J. Dairy Sci., *33,* 619 (1950).
307. POLONOVSKI, M., BANDU, L., and MEUZIL, E., LeLait, *29,* 1 (1949).
308. PURI, B. R., LAKHANPAL, M. L., and GUPTA, S. C., Indian J. Dairy Sci., *5,* 189 (1952).
309. PURI, B. R., and TOTEJA, K. K., Indian J. Dairy Sci., *21,* 119 (1968).
310. PYNE, C. H., and COULTER, S. T., J. Dairy Sci., *44,* 1156 (1961).
311. QUIGLEY, T. W., ROE, C. E., and PALLANSCH, M. J., Federation Proc., *17* (1), 292 (1958).
312. RADEMA, L., 14th Intern. Dairy Congr., Proc., Rome, *1*(2), 403 (1956).
313. RAHN, O., MOLKERIE-Z. Hildesh., *38,* 1321 (1924).
314. RAHN, O., Milchw. Forsch., *2,* 382 (1925).
315. RAHN, O., and SHARP, P. F., "Physik der Milchwirtschaft," Paul Parey, Berlin (1928).
316. RAMACHANDRAN, K. S., and WHITNEY, McL. R., J. Dairy Sci., *42,* 854 (1960).
317. REINHART, A., and NESBITT, J. M., Can. Dairy Ice Cream J., *38,* 54 (1959).
318. RICHARDSON, G. A., J. Dairy Sci., *19,* 749 (1936).
319. RICHARDSON, T., and GUSS, P. L., J. Dairy Sci., *48,* 523 (1965).
320. RIESSIG, G., Milchw. Forsch., *19,* 273 (1937).
321. RISHOI, A. H., J. Dairy Sci., *35,* 1125 (1952).
322. RISHOI, A. H., and SHARP, P. F., J. Dairy Sci., *21,* 399 (1938).
323. RISHOI, A. H., and SHARP. P. F., J. Dairy Sci., *21,* 683 (1938).
324. ROBERT, L., and POLONOVSKI, J., Disc. Faraday Soc., *20,* 54 (1956).
325. RODKEY, F. L., and BALL, E. G., J. Lab. Clin. Med., *31,* 354 (1946).
326. ROBINSON, G. W., J. Agr. Sci., *12,* 302 (1922).
327. ROOS, J. B., and TUINSTRA, L. G. M. Th., Neth. Milk Dairy J., *23,* 37 (1969).
328. ROSE, DYSON, BRUNNER, J.R., KALAN, E. B., LARSON, B. L., MELNYCHYN, P., SWAISGOOD, H. E., and WAUGH, D. F., J. Dairy Sci., *53,* 1 (1970).
329. ROSS, S., Chem. Ind., 64, 757 (1949).
330. ROUSSOS, G. G., and MORROW, B. H., Appl. Spectr., *22,* 769 (1968).
331. SAITO, Z., and IGARASHI, Y., Bull. Fac. Agr. Hirosaki Univ., *1969* (15), 50 (1969).
332. SAMUELSSON, E. -G., 16th Intern. Dairy Congr., Proc., Copenhagen, A, 552, (1962).
333. SAMUELSSON, E. -G., Milchwissenschaft, *21,* 235 (1966).
334. SAMUELSSON, E. -G., Milk and Dairy Res., Alnap, Sweden, Rept. no. 76 (1967).
335. SAMUELSSON, E. -G., BENGTSSON, G., NILSSON, S., and MATTSSON, N., Svenska. Mejeritidn., *46,* 163 (1954).
336. SAMUELSSON, E. -G., and PETTERSON, K. I., Svenska Mejeritidn., *29,* 65 (1937).

337. SASAKI, R., and KOYAMA, S., 14th Intern. Dairy Congr., Proc., Rome, *2,* 381 (1956).
338. SASAKI, R., and KOYAMA, S., 14th Intern. Dairy Congr., Proc., Rome, *2,* 390 (1956).
339. SCHAAP, J. E., BERESTEYN, E. C. H. VAN, Alg. Zuivelbl., *63,* 257 (1970). Cited in Dairy Sci. Abstr., *33,* 73 (1971).
340. SCHOLZ, J., Milchw. Forsch., *19,* 203 (1937).
341. SCHULMAN, J. H., and SMITH, T. D., Kolloidzchr., *126,* 20 (1952).
342. SCHWARZ, G., and FISCHER, O., Milchw. Forsch., *18,* 53 (1936).
343. SHAHANI, K. M., J. Dairy Sci., *49,* 407 (1966).
344. SHARP, P. F., J. Dairy Sci., *11,* 259 (1928).
345. SHARP, P. F., J. Dairy Sci., *23,* 771 (1940).
346. SHARP, P. F., and KRUKOVSKY, V. N., J. Dairy Sci., *22,* 743 (1939).
347. SHERBON, J. W., "Analytical Calorimetry," Plenum Press, New York (1968).
348. SHIPSTEAD, H., and TARASSUK, N. P., Abst. Am. Chem. Soc., Mar., 5A (1953).
349. SKOU, J. C., Physiol. Rev., *45,* 569 (1965).
350. VAN SLYKE, L. L., 10 Ann. Rept., N.Y. (Geneva) Agr. Expt. Sta. (1891).
351. SMITH, E. L., J. Biol. Chem., *165,* 655 (1946).
352. SOMMER, H. H., "The Theory and Practice of Ice Cream Making," 5th Ed. The Author, Madison, Wis. (1946)
353. SOMMER, H. H., "Market Milk and Related Products," 2nd Ed., The Author, Madison, Wis. (1946).
354. SOMMER, H. H., Milk Dealer, *41* (1), 58 (1951).
355. STADHOUDERS, J., and HUP, G., Neth. Milk Dairy J., *24,* 79 (1970).
356. STAEHELIN, L. A., J. Ultrastruct., *22,* 326 (1968).
357. STEIN, O., and STEIN, Y., J. Cell Biol., *34,* 251 (1967).
358. STEINER, E. H., J. Sci. Food Agr., *6,* 777 (1955).
359. STEWART, P. S., "A Study of the Formation and Structure of the Milk Fat Globule Membrane," Thesis, University of Guelph, (1970).
360. STEWART, P. S., and IRVINE, D. M., J. Dairy Sci., *53,* 279 (1970).
361. STEWART, P. S., and IRVINE, D. M., 18th Intern. Dairy Congr., Proc., Melborne, *1E,* 71 (1971).
362. STINE, C. M., and PATTON, S., J. Dairy Sci., *35,* 655 (1952).
363. STINE, C.M., and PATTON, S., J. Dairy Sci., *36,* 516 (1953).
364. STONE, M. JEAN, J. Dairy Res., *19,* 311 (1952).
365. STORCH, V., Translated and communicated by H. Farber, Analyst, *22,* 197 (1897).
366. STURTEVANT, E. L., New York State Agr. Soc., Trans., *32* (1872–76).
367. SWANSON, A. M., and SOMMER, H. H., J. Dairy Sci., *23,* 201 (1940).
368. SWOPE, F. C., and BRUNNER, J. R., J. Dairy Sci., *48,* 1705 (1965).
369. SWOPE, F. C., and BRUNNER, J. R., Milchwissenschaft, *23,* 470 (1968).
370. SWOPE, F. C., and BRUNNER, J. R., J. Dairy Sci., *53,* 691 (1970).
371. SWOPE, F. C., RHEE, K. C., and BRUNNER, J. R., Milchwissenschaft, *23,* 744 (1968).
372. SWOPE, F. C., BRUNNER, J. R., and VADEHRA, D. V., J. Dairy Sci., *48,* 1707 (1965).
373. TAKAYAMA, K., MACLERMAN, D. H., TZAGOLOFF, A., and STONER, C. D., Arch. Biochem. Biophys., *114,* 223 (1966).
374. TAPPEL, A. L., Arch. Biochem. Biophys., *54,* 266 (1959).
375. TARASSUK, N. P., and FRANKEL, E. N., J. Dairy Sci., *38,* 438 (1955).
376. TARASSUK, N. P., and FRANKEL, E. N., J. Dairy Sci.,*40,* 418 (1957).
377. TARASSUK, N. P., KOOPS, J., and PETTE, J. W., Neth. Milk Dairy J., *13,* 258 (1959).
378. TARASSUK, N. P., and KOOPS, J., J. Dairy Sci., *43,* 93 (1960).
379. THOMAS, A., "Polymorphism in Butterfat," Thesis, University of Minnesota (1950).
380. THOMPSON, M. P., "Structure of the Milk Fat Globule Membrane," Thesis, Michigan State University (1960).
381. THOMPSON, M. P., and BRUNNER, J. R., J. Dairy Sci., *42,* 369 (1959).

382. THOMPSON, M. P., BRUNNER, J. R., and STINE, C. M., J. Dairy Sci., *42,* 165 (1959).
383. THOMPSON, M. P., BRUNNER, J. R., LINDQUIST, KARIN, and STINE, C. M., J. Dairy Sci., *44,* 1589 (1961).
384. THURSTON, L. M., BROWN, W. C., and DUSTMAN, R. B., J. Dairy Sci., *19,* 671 (1936).
385. TITUS, R. W., SOMMER, H. H., and HART, E. B., J. Biol. Chem., *76,* 237 (1928).
386. TJEPKEMA, R., and RICHARDSON, T., J. Dairy Sci., *49,* 1120 (1966).
387. TOBIAS, J., and SERF, R. M., J. Dairy Sci., *42,* 550 (1959).
388. TRAUTMAN, R., J. Phys. Chem., *60,* 1211 (1956).
389. TROUT, G. M., "Homogenized Milk, A Review and Guide," Michigan State University, East Lansing (1950).
390. TROUT, G. M., and GOULD, I. A., Mich. Agr. Expt. Sta., Quart. Bull., *21,* 21 (1938).
391. TROUT, G. M., HALLORAN, C. P., and GOULD, I. A., Mich. Agr. Expt. Sta. Tech. Bull., *145* (1935).
392. TROY, H. C., and SHARP, P. F., J. Dairy Sci., *11,* 189 (1928).
393. TVERDOKHLEB, G. V., LLA Raksti, *6,* 485 (1957). Cited in Dairy Sci. Abstr., *22,* 531 (1960).
394. TVERDOKHLEB, G. V., Moloch. Prom., *18* (12), 26 (1957). Cited in Dairy Sci. Abstr., *20,* 257 (1958).
395. TVERDOKHLEB, G. V., Izv. Vysshikh. Uchebn. Zavedenii, Pischevaya Tekhnol., No. *6,* 25 (1959). Cited in Dairy Sci. Abstr., *22,* 531 (1960).
396. TVERDOKHLEB, G. V., Izv. Vysshikh. Uchebn. Zavedenii, Pischevaya Tekhnol., No. *1,* 122 (1961). Cited in Dairy Sci. Abstr., *24,* 217 (1962).
397. TVERDOKHLEB, G. V., 16th Intern. Dairy Congr., Proc., Copenhagen, *B,* 145 (1962).
398. TVERDOKHLEB, G. V., 16th Intern. Dairy Congr., Proc., Copenhagen, *B,* 155 (1962).
399. VASIĆ, J., and DEMAN, J. M., 17th Intern. Dairy Congr., Proc., München, *C,* 167 (1966).
400. VERGELESOV, V. M., and IVANOUSKAYA, L. S., Izv. Vysshikh. Uchebn. Zavedenii, Pischevaya Tekhnol., No. *4,* 160 (1970). Cited in Dairy Sci. Abstr., *33,* 331 (1971).
401. WALSTRA, P., Neth. Milk Dairy J., *19,* 93 (1965).
402. WALSTRA, P., Neth. Milk Dairy J., *19,* 266 (1965).
403. WALSTRA, P., J. Dairy Sci., *50,* 1839 (1967).
404. WALSTRA, P. W., Neth. Milk Dairy J., *21,* 166 (1967).
405. WALSTRA, P., J. Colloid Interface Sci., *27,* 493 (1968).
406. WALSTRA, P., Neth. Milk Dairy J., *23,* 238 (1969).
407. WALSTRA, P., Neth. Milk Dairy J., *23,* 245 (1969).
408. WALSTRA, P., Neth. Milk Dairy J., *23,* 290 (1969).
409. WALSTRA, P., and BORGGREVE, G. J., Neth. Milk Dairy J., *20,* 140 (1966).
410. WALSTRA, P., and OORTWIJN, H., J. Colloid Interface Sci., *20,* 424 (1969).
411. WALSTRA, P., OORTWIJN, H., and DEGRAAF, J. J., Neth. Milk Dairy J., *23,* 12 (1969).
412. WAUSCHKUHN, P., and KNOOP, E., Milchwissenschaft, *25,* 70 (1970).
413. WEBB, B. H., J. Dairy Sci., *14,* 508 (1931).
414. WEBB, B. H., J. Dairy Sci., *16,* 369 (1933).
415. WEBB, B. H., and HOLM, G. E., J. Dairy Sci., *22,* 363 (1939).
416. WECKEL, K. G., and STEIN, J. A., Milk Dealer, *45,* 154 (1956).
417. WEIHE, H. D., and GREENBANK, G. R., J. Dairy Sci., *41,* 703 (1958).
418. WEINSTEIN, B. R., and TROUT, G. M., J. Dairy Sci., *34,* 554 (1951).
419. WELLINGS, S. R., GREENBAUM, B. W., and DEOME, K. B., J. Nat. Cancer Inst., *25,* 423 (1960).
420. WELLINGS, S. R., DEOME, K. B., and PITELKA, D. R., J. Nat. Cancer Inst., *25,* 393 (1960).

421. WHITNAH, C. H., RUTZ, W. D., and FRYER, H. C., J. Dairy Sci., *39,* 1500 (1956).
422. WHITE, R. F., EATON, H. D., and PATTON, S., J. Dairy Sci., *39,* 1500 (1956).
423. WHITNEY, R. McL., J. Dairy Sci., *43,* 1303 (1958).
424. WHITTLESTONE, W. G., J. Dairy Res., *21,* 50 (1954).
425. WHITTLESTONE, W. G., Austral. J. Dairy Tech., *17,* 108 (1962).
426. WIECHERS, S. E., and DEGUEDE, B., "Continuous Buttermaking", North-Holland Pub. Co., Amsterdam (1950).
427. WIEGNER, G., Kolloid-Z., *15,* 105 (1914).
428. WIESE, HILDA F., and PALMER, L. S., J. Dairy Sci., *15,* 37 (1932).
429. WIESE, Hilda F., and PALMER, L. S., J. Dairy Sci., *17,* 29 (1934).
430. WITTIG, A. B., 12th Intern. Dairy Congr., Proc., Stockholm, *3,* 118 (1949).
431. WOLF, D. P., and DUGAN, L. R., J. Am. Oil Chem. Soc., *41,* 139 (1964).
432. WOLL, F. W., J. Agr. Sci., *6,* 446, 515 (1892).
433. WOODING, F. B. P., PEAKER, M., and LINZELL, J. L., Nature, *226,* 762 (1970).
434. WOODROW, I. L., and DEMAN, J. M., J. Dairy Sci., *51,* 996 (1968).
435. WORTMANN, A., Kiel Milchwirtsch. Forschungsber. *15,* 341 (1963).
436. YONCOSKIE, R. A., J. Am. Oil Chem. Soc., *44,* 446 (1967).
437. YONCOSKIE, R. A., "Analytical Calorimetry", Plenum Press, New York (1968).
438. YONCOSKIE, R. A., J. Am. Oil Chem Soc., *46,* 49 (1969).
439. ZITTLE, C. A., DELLAMONICA, E. S., CUSTER, J. H., and RUDD, J. H., J. Dairy Sci., *39,* 528 (1956).
440. ZITTLE, C. A., THOMPSON, M. P., CUSTER, J. H., and CERBULIS, J., J. Dairy Sci., *45,* 807 (1962).
441. ZOTTOLA, E. A., WILSTER, G. H., and STEIN, R. W., J. Dairy Sci., *44,* 41 (1961).

CHAPTER 11

R. M. Parry, Jr.†

Milk Coagulation and Protein Denaturation *

INTRODUCTION

This chapter is concerned with the complex interactions which occur among the many milk components and proteins, and the resulting effect on the properties of milk in the fluid, concentrated and dried state. In several cases these effects are cumulative, but no quantitative method of estimating the contribution of each component to observed changes has been forthcoming. Empirical observations have given some indication of the path of these changes leading to destabilization. These changes will be discussed in three sections: first, the equilibrium (or nonequilibrium) nature of fluid milk; second, the changes induced by heating this system; and third, the changes manifested by concentrating the solids of milk.

COLLOIDAL STABILITY OF MILK

Milk is a polyphasic secretion containing emulsified fat, colloidal casein micelles, and dissolved protein, lactose and salts, and it undergoes continual change in the mammary gland lumen prior to milking. Most of these changes are attributable to reactions on the enzymatic level. Subsequent handling procedures, such as cooling, heating, homogenization, and concentration can disturb the colloidal-dissolved state equilibrium. Therefore, the age and treatment history of milk is very important to understanding and predicting its properties.

Coagulation of the milk system is often attributed to destabilization of the casein micelles. However, study of this phenomenon reveals that the end result may be a summation of many minute changes in the colloid system. The interactions of caseins and salts that are responsible for these changes will be discussed in this section.

---

* Revised from chapter 11 in the 1st edition by Leon Tumerman and Byron H. Webb.

† R. M. Parry, Jr., Eastern Regional Research Laboratory, Agricultural Research Service, U.S. Department of Agriculture, Philadelphia, Pennsylvania 19118.

Casein Dispersion

Chapter 3 deals with the various casein components and their genetic variability. Our concern here is with the behavior of the three caseins which constitute over 90% of those present in normal milk, namely, α_{s1}—, β- and κ-casein. Chapter 9 discusses the various proposed models of how these proteins interact in milk to form micelles. This discussion assembles observations on the behavior of these proteins in milk, and those made on the purified components, to elucidate the factors influencing changes occurring in the milk system.

The casein micelle population is estimated to be in the order of 10^{12} particles per cubic centimeter of milk with an average free path of approximately 3,600 Å between particles.[254] The micelles are in constant kinetic motion, and because they are closely packed in milk, the entire colloidal dispersion may be immobilized by cohesion of a relatively small number of the particles. Unrestrained growth of the casein micelles, therefore, leads to gelation of milk. However, in normal milk the micelles, as a separate entity, are markedly stable to extremes of temperature and concentration. The structural stability and relative inertness of the micelles can be demonstrated by centrifuging milk to sediment the micelles into a pellet. This dense mass can be redispersed and the micelles will be identical with the natural system in size distribution when examined by electron microscopy. This property of reversibility after close approach (concentration) and dilution attests to the internal rigidity of the colloidal particle, which has been attributed to its protein-protein and protein-salt interactions.[142,193]

Figure 11.1 is an electron microscope photograph of skimmilk. The casein micelles are recognized as the rough-surfaced spherically shaped particles. These colloidal casein particles range in size from 500 to 2,500 Å and show a high degree of light scattering, which gives rise to the characteristic "milky" appearance to the fluid.

Numerous studies have been made of micelle population in milk by the electron microscope.[20,23,117,164,196,202,214] Figure 11.2 is a micelle size distribution nomograph of glutaraldehyde-fixed skimmilk. Some disagreement as to the size distribution of casein micelles is reported in the literature; the disagreement arises from the method(s) of specimen preparation. The glutaraldehyde procedure of Carroll *et al.*[20] gives good fixation with minimal shrinkage of the particles. It is in reasonable agreement with workers who have separated micelles into size classes by differential centrifugation and who have made size measurements by light-scattering techniques.[35,52,134] Figure 11.2 shows that the most frequently occurring casein micelle size is around 1,300 Å.

The appearance of the micelle in the electron microscope has led

From Parry and Carroll[172]

FIG. 11.1. ELECTRON MICROSCOPE PICTURE OF SKIMMILK MICELLES, GLUTARALDEHYDE FIXED AND SHADOWED WITH PLATINUM

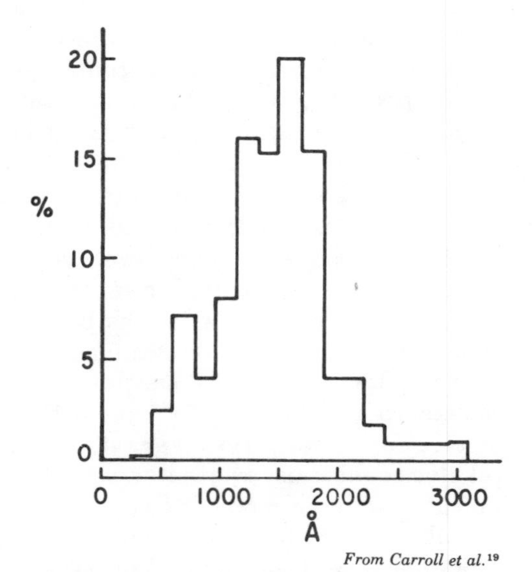

From Carroll et al.[19]

FIG 11.2. RELATIVE FREQUENCY OF OCCURRENCE OF CASEIN MICELLES VERSUS SIZE IN RAW MILK

several workers to suggest that the aggregate is composed of subunits.[150,158,209,214,244] Although this could explain the size distribution and rough-surfaced appearance of the micelle, studies of casein composition as a function of size indicate that the composition of the micelles is far from constant.[134,193,223]

Considerable variation can be found in the literature on the quantitation of α_{s1}-, β- and κ-casein present in the micelle. The problem seems to center on the difficulty of accurately determining κ-casein. It has been found that the dye-binding capacity of this protein is considerably less than that of α_{s1}- or β-casein, which are approximately equal in this respect.[144] This unequal binding of dye to the caseins yields very low values for κ-casein when quantitated by polyacrylamide gel electrophoresis. Quantitation of sialic acid, a carbohydrate component of κ-casein, has been shown to be unreliable, since the carbohydrate content varies from 0 to 10%.[139,194,255] The use of the known sulfur content of κ as an analytical determinant has been recently questioned by the report that the minor α_s components (α_{s3}, α_{s4} and α_{s5}) also contain sulfur;[82] the reader is referred to Chapter 3 for further discussion of these proteins. Casein quantitation has been reported also on the basis of the carboxyl end-terminal amino acid released by carboxypeptidase A.[187,197]

A detailed study of three methods available for casein quantitation has been made by Rose *et al.*[197] This is the best study available to date and compares the sialic acid and sulfhydryl tests to a new method using analytical anion-exchange cellulose chromatography (DEAE). This technique would appear to give the best quantitative results and should be the least affected by compositional variations that are prominent in other techniques.

Table 11.1 shows the results of the above three casein analyses on a sample of whole isoelectric casein and various centrifugal cuts used for size separation of the micelles. It is apparent from this table that specific identification of the procedure used to separate casein is necessary, because of large compositional variations between pH 4.6 precipitated casein and the micellar protein obtained by centrifugation. This is due to the partial solubility of κ-casein at pH 4.6. Reported values for the isoelectric point of κ are 3.7[262] and 4.1[194] as determined by free-boundary electrophoresis. However, its behavior in urea-polyacrylamide gel electrophoresis, where it moves slower than either α_{s1} or β, would indicate a charge more in line with its isoionic point of pH 5.2.[262] The solubilization of κ can be seen when "washing" acid casein at pH 4.0, a preparative treatment often used to remove a proteolytic enzyme associated with casein. If the reported lower isoelectric points were those of the native protein, solubilization and hence loss of the protein would probably not occur to such an extent. The results of column gradient isoelectric focusing on the caseins[104] in 7*M* urea and mercaptoethanol have shown the isoelectric point of κ to be 5.95 to 6.53, which would explain the high solubility at pH 4 in dilute acetic acid.

Table 11.1 also shows the differences in casein composition among different micelle sizes. This unusual characteristic of an aggregated system was first observed by Sullivan *et al.*[223] Such an inverse relationship between micelle size and κ content holds true regardless of the method chosen for determination of the κ; that is, when κ-casein is examined by methods previously mentioned, they all indicate less κ in the biggest micelles than in the smallest, even though the methods vary considerably in the absolute amount determined.

An often neglected casein fraction, which may play an important role in milk stability, is serum casein. This nonmicellar casein does not sediment with the micelles (100,000 $\times$ *g* for 60 min) and has been reported to be 3 to 10% of the total casein in uncooled milk.[188,193] The noncentrifugable or serum casein increases in cooled milk (stored overnight at 4°C) to approximately 20% of the total casein.[193] The composition of this fraction, shown in Table 11.1, indicates that the content of κ rises to a third of the total casein in fresh noncooled milk. It has also been reported that micelles, which were centrifuged and redispersed 7 times in a milk salt-simulating buffer, lose their ability to be clotted by rennin.[172] This would suggest that fluid milk instability problems may be attributable to the serum κ-casein destabilization acting as an initiator of clot formation.

Table 11.1

CASEIN COMPOSITION[a]

Centrifugation Time (min.)[b]	Proportion				Sulfhydryl[d] per 25,000 gm	Sialic Acid (%)[d]
	α_{s1}	β	$\kappa+\gamma$	Minor		
0 (Isoelectric Casein)	50.00	31.8	14.8	2.5	—	—
0–7½	47	34	16	3	0.4	0.28
7½–15	46	32	18	4	0.45	0.35
15–30	45	31	20	4	0.52	0.46
30–60	42	29	26	3	0.68	0.65
Serum[c]	39	23	33	5	0.98	0.88

[a] Data obtained Rose *et al.*[197]
[b] All centrifugations were done in a Spinco No. 30 rotor, 30,000 rpm, 100,000 $\times$ *g* max.
[c] Casein not sedimented after 60 min centrifugation.
[d] SH-groups and sialic acid are reported for freeze-dried material, uncorrected for moisture.

Effects of temperature on the casein distribution between the micelles and serum phase have been examined by Rose,[193] Murphy *et al.*,[158] and Downey and Murphy.[31] These authors have reported that dissociation (and hence solubilization of β-casein into the serum phase) occurs readily upon cooling of milk to 5°C. This is in agreement with Sullivan *et al.*[224] and Payens *et al.*,[173,174] who have studied the temperature-dependent association of purified β-casein. They found that β-casein has a monomer molecular weight of 26,000 daltons at 5°C; increasing the temperature to 25°C causes rapid aggregation and eventual precipitation.

β-Casein accounts for approximately 50% of the casein released upon cooling, the remainder being evenly divided between α_s and κ. The total percentage of casein solubilized shows variation between cows; i.e., ranges are reported from 10 to 21%[31] and from 14 to 44%[193] by different investigators. Downey and Murphy found very high serum or nonmicellar casein contents in late lactation cows and levels up to 55% of total casein from a cow having clinical mastitis. The fact that migration of casein from the "micellar state" to the serum phase does not occur in the same casein ratios as found in the micelle would appear to be significant in understanding the manner by which these proteins associate inside the micelle. Interestingly, the micelles, when repeatedly "washed" by centrifugation and redispersion in a milk salt-simulating buffer,[101] do not show any size change when examined in an electron microscope.[172] No work has described the properties of micelles which have been rewarmed to 30°C although serum casein does return to its previous low level of about 6%. One could imagine that micelles, after such a treatment (similar to that found in fluid milk processing), might have some alteration in their surface properties due to the large amount of previously solubilized β-casein that may well precipitate on or near the micelle surface.

Salt Dispersion

The nonprotein components play a critical role in the physical stability of milk. The inorganic components are a complex mixture, dispersed in both the micelles and the serum phases. The work of Davies and White[28] first measured the salt content of both phases and studied the changes which occur upon cooling. They found serum phase increase of all salts except chloride ion when milk cooled from 20 to 3°C. The serum or filtrate obtained at 3°C becomes "cloudy" on rewarming to 37°C. This indicates equilibrium changes in the salt balance by formation of complex colloidal salts of sufficient particle size to scatter visible light. The ions most frequently thought to undergo complexation are calcium, $HPO_4^=$, $H_2PO_4^-$ and citrate; the resultant insoluble salts of calcium phosphate probably have structures resembling apatite.[12] This rigid salt structure, also referred to as colloidal calcium phosphate, maintains micelle structure in spite of the loss of a significant percentage of casein when milk is cooled. This was demonstrated by Jenness *et al.*,[102] who found complete micelle disruption upon removal of the colloidal calcium phosphate.

The soluble (diffusible) calcium has been estimated by means of phase separation, including renneting, ultrafiltration and dialysis. The total calcium concentration is approximately 30 m*M* per liter, of which 10m*M* is diffusible and 2 to 3 m*M* is ionized.[22,212,243] Calculating from the dissociation constants of calcium citrate and calcium phosphate, Smeets[215] verified the average ionic calcium concentration of 2.75 m*M* per liter by using the muroxide method.

It is well known that variations in free Ca^{2+} can alter the heat stability of milk. The increase in serum calcium found on cooling could adversely affect heat stability. The formation of colloidal calcium salts, in treatments such as forewarming, might reverse this destabilizing influence.

About two-thirds of the calcium in milk is located in the micelle. It occurs covalently bound to the phosphate esters of serine (or possibly threonine) complexed with $HPO_4^=$, $H_2PO_4^-$ and citrate,[195] and also is associated with acidic groups on the proteins.[30] From the work of McGann and Pyne[142] and Jenness *et al.*,[102] we know that when chelating agents are added to milk (e.g., EDTA or citrate), disruption of micelles occurs, apparently into small units of casein measuring 100 to 125 Å in diameter.[17,39,158,209,214] The occurrence of these subunits and their relationship to micelle structure are discussed in more detail in Chapter 9. Micelle disruption into subunits also has been found when isoelectric casein (pH 4.6 precipitated) is redispersed at neutral pH.[166] These treatments, which totally remove colloidal calcium phosphate, indicate that the most important role of this salt is in maintaining micelle integrity. Hence, when examining the effects of milk manufacturing on product instability, it is important to consider not only protein dispersion, but also the role and the ratio of serum to colloidal salt.

Boulet *et al.*[13] have reported an unusual approach to describe micelle salt dispersion. Using Sephadex G-150, they separated whole milk into three fractions: (1) micelles, (2) whey protein, and (3) serum salts, lactose and low molecular weight components. Analysis of the calcium, phosphate and citrate showed that "fraction one" micelles contained no citrate and had lost 22% of the calcium, 50% of the magnesium, and 47% of the phosphate contained in "whole" micelles of milk. These authors concluded that the loss of these salts on gel filtration is due to the removal of labile ions adsorbed in the micelle, probably in the form of a diffuse double layer. The removal of this layer of ions does not destroy the integrity of the micelle, since rechromatography of micelles gives the same salt content. The "fraction one" micelles have a calcium-to-inorganic-phosphate ratio between 2.52 and 4.12, which is much higher than the ratio of any known calcium phosphate salt. This indicates that more than one type of salt form must be responsible for the internal structure of the micelle. Boulet *et al.*[13] suggest that approximately half the Ca^{++} is in apatite form, with the remainder in primary salt interactions between casein carboxyls and serine phosphate.

Yamauchi *et al.*[257] studied the behavior of $^{45}Ca^{++}$ added as the chloride salt to milk, to determine the distribution and exchange of this cation between soluble and colloidal phases of milk. They found that of the calcium in the colloidal phase about 40% was not exchanged after 48 hr, terming it "hard-to-exchange" calcium. This calcium was absent, or nearly so, in colloidal phosphate-free milk and calcium caseinate phosphate dispersion. The speed of exchange in the latter system was faster than in milk. They suggest that hard-to-exchange calcium is present in a part of the colloidal phosphate portion of casein micelles. This nonexchangeable property is typical of hydroxyapatite and is in reasonable agreement with Boulet's suggestion of 50% apatite.

EFFECT OF HEAT ON FLUID MILK

The preceding section summarizes some of the critical observations necessary to understanding the colloidal milk dispersion, particularly those factors involving casein and salt. The introduction of irreversible changes in the milk system, such as processing treatments, may decisively affect the stability of the milk colloids. Heat sterilization of concentrated milk, for example, may simultaneously cause partial dephosphorylation of casein, denaturation of the serum proteins, interactions among the lactose, casein and serum constituents, an increase in acidity derived from multiple sources, and some irreversible changes in salt equilibria. All have a bearing on the coagulation process in varying degrees. Such induced changes, superimposed on the natural variation in milk composition, contribute to some of the instability of the colloidal dispersion. These factors frequently obscure the underlying cause of coagulation, and anomalous results are not uncommon. Hence, no satisfactory correlation has yet been made between the heat stability of milk and its analytical composition; nor is it possible on the basis of available data to predict with accuracy the heat stability of a concentrate from the original fluid milk. Although preheat treatment of fluid milk imparts resistance to clotting by enzymes, as well as greater heat stability on subsequent concentration, it lowers the stability of both fluid and concentrated milks to freezing. Similarly, various forms of phosphate that act as calcium sequestrants effectively retard heat coagulation and are, therefore, indispensable to the manufacture of evaporated milk.

Protein Denaturation

The term denaturation as applied to proteins has been subject to various connotations in milk literature. Native protein molecules are known to be folded into well-defined, more or less rigid, three-dimensional structures. For most proteins this structure is compact and globular, as exemplified by lysozyme, α-lactalbumin and β-lactoglobulin. In some proteins the native structure is rod-like; or it is a rod with globular appendages as in the case of myosin. The caseins, however, are known to be essentially random coils in comparison with other protein secondary structures.[80,165,175,207,210,225]

The native structure of a protein remains stable over a fairly wide range of external conditions, but its internal organization into α-helical or β-structures and/or disulfide bonds can be permanently disrupted by changes in physical or chemical environment. This process is irreversible denaturation. In many instances however, denaturation is a reversible process, showing that certain structural changes reflect changes in the stabilities of various possible conformations of the protein molecule. Examples of denaturing agents important to the dairy industry are heat and acidity.

Denaturation is complicated by the fact that not all proteins behave similarly in the presence of the same disruptive agent (conversely, widely different effects can be found with different agents[228]) and many possible molecular configurations may occur between the native and irreversibly denatured forms.[99] Furthermore, as the protein approaches this denatured state, its association behavior towards other proteins is markedly altered. This has been observed in β-lactoglobulin-κ-casein systems when they are heated.

Casein, while considered to be "denatured" in the chemical sense of having very little, if any, α-helix or β-structures, does have a "native" structure, which is primarily due to self-association with other caseins. This structure, therefore, is mostly of the quaternary type arising from hydrophobic, electrostatic and, to a limited extent, hydrogen bonding between protein molecules.[207,210] Thus, the effect of temperature and pH can drastically affect casein association and result in micelle alteration, to an even greater extent than observed with compact globular proteins.

The serious voids in our understanding of the equilibria and compositional changes in milk at high temperatures hardly allow for speculation on the mechanism of heat coagulation. This mechanism is further obscured by variations in composition and heat stability of different milks, and even of milks drawn from different quarters of the udder of a single cow. Consequently, no satisfactory correlations have yet been established between heat stability and the analytical composition of normal fluid milks or their concentrates.[5]

In the strictest sense, casein is not a heat-coagulable protein. Its dispersion in normal fluid milk is very stable to heat and may resist coagulation for as long as 14 hr at boiling temperatures, and 1 hr at 130°C. The heat coagulation of casein in milks of normal stability occurs largely as a result of compositional changes in the milk, caused by sustained exposure to high temperatures. Of major significance among these effects are the increase of acidity, the conversion of soluble calcium and phosphates to colloidal forms, and interactions (denaturation, hydrolytic cleavages) between the protein components. The impact of such heat effects on colloidal stability is amplified in concentrated milks where coagulation tends to increase logarithmically with milk solids concentration.

Salt and pH

Salt balance and acidity are regarded as two of the most important factors in the heat stability of milk. The low heat stability of colostral milks is ascribed to a higher level of ionic calcium.[199,251] Noncolostral milks of inordinately low heat stability (Utrecht abnormality) are of normal composition with respect to total calcium and acidity, but evidence a notably higher ionic calcium activity.[211] Milks of normal heat stability have a calcium-ion concentration ranging from 2.0 to 4.0 millimolar in the ultrafiltrate, whereas the ionic calcium values in milks exhibiting the Utrecht abnormality range from 4 to 7 millimolar.[11] This defect can be simulated in milks of normal heat stability by the addition of calcium salt ($CaCl_2$) to elevate the ionic calcium concentration above 4 millimolar.[185] Adjustment to a more alkaline pH, or addition of calcium sequestrants, constitute effective corrective measures for this particular heat-stability defect.[211,212]

Heat treatment is known to cause reduction of both the total soluble and ionic calcium. Under pasteurizing conditions, the reduction is slight, but significant losses of soluble calcium and phosphorus occur above 76°C.[7,81,231] Values reported for changes in salt distribution in heated milk show large discrepancies, which may be caused by inadequate precautions against subsequent shifts in equilibrium and

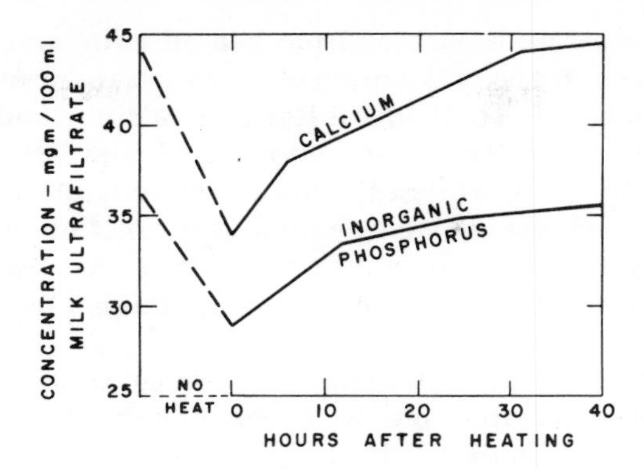

From data by Hilgeman and Jenness[81]

FIG. 11.3. ULTRAFILTERABLE CALCIUM AND PHOSPHORUS IN MILK HEATED AT 78° C FOR 30 MIN, MEASURED INITIALLY AND AFTER COOL-AGING AT 5°C

by variations in the time lapse between heat treatment and analysis. Figure 11.3 shows the initial loss of about 25% of the soluble calcium in milk heated to 78°C for 30 minutes.[81] When the milk is aged at 5°C, gradual reversion toward the original soluble calcium level occurs over a period of 24 to 48 hr. Soluble phosphorus undergoes a similar change. This decrease in soluble calcium and phosphorus upon heat treatment and its reversion on cool aging has also been confirmed by analysis of calcium and phosphorus in centrifuged milk[38] as well as in ultrafiltration.[198]

The concentration of both soluble calcium and magnesium is increased in evaporated milk, but less than that of potassium and sodium, which implies precipitation of some of the Ca^{++} and Mg^{++} cations, possibly as the phosphate salts. This departure from the anticipated activity of calcium and magnesium ion becomes even more pronounced with increasing milk solids concentration. The combined effect of added disodium phosphate stabilizer and heat sterilization may lower the ionic calcium concentration in evaporated milk by 20 to 40%. The addition of 0.15% disodium phosphate to raw skimmilk lowers the ratio serum Ca^{++}/total Ca^{++} from 0.317 to 0.216, which is further reduced to 0.116 following heat treatment at 88°C for 15 min.[38]

Since changes in the electrolyte composition of heated milk are undoubtedly involved in the mechanism of heat coagulation, their measurement at elevated temperatures is critical. This is particularly true in view of the evidence that heat-induced changes in the salt balance revert on cooling, and may escape detection. Analyses by

Rose[198] of ultrafiltrate separated from heated milk samples establish that the changes in the salt balance are much more extensive than had been anticipated. Total ultrafilterable calcium and phosphorus, separated from milk at 94°C, are reduced by 50 and 18%, respectively, over raw milk controls, while calcium-ion concentration shows a 60% reduction. These changes, whose extent is a function of temperature level, generally attain equilibrium within 5 min. Reversion of the salt balance to 75 to 90% of the value of the original unheated milk requires cool aging at 5°C for 22 hr.

Reduction of the soluble and ionic calcium concentration in heated milk is partially attributable to a conversion of soluble calcium phosphate to the colloidal state, as discussed above. Evenhuis and Vries[45,46] suggest that crystallization of calcium phosphate to hydroxyapatite occurs in heated milk, a process that is greatly accelerated if a large surface area for nucleation is provided, for example, in the form of inactive yeast cells.[15] The precipitation and crystallization process begins at 60°C increasing rapidly at higher temperatures. If the phosphate in milk does crystallize to hydroxyapatite on heating, the Ca/P ratio of the precipitate would be 1.67, a value in good agreement with the analysis of van der Burg.[15] Approximately 48% of the casein-bound calcium and magnesium is precipitated with phosphate in milk treated with yeast cells for 20 min at 20°C. This precipitation of half the cations from casein is in accord with the conclusion of Boulet *et al.*[13] and Yamauchi *et al.*,[257] who similarly noted that approximately half the calcium is "exchangeable." The calcium in raw rennet whey is, however, readily precipitated with the denatured serum proteins, and does not require the use of a yeast cell surface. This precipitated calcium phosphate is highly resistant to dissolution on cooling. The precipitation of soluble calcium as the apatite crystal would leave the milk unsaturated with respect to calcium and thus enhance its heat stability. The beneficial effects of forewarming have similarly been related to heat modification of the salt equilibrium. It has also been suggested[15,115,183] that the advantage of strong preheat may originate with detachment of some of the colloidal phosphate from its complex with casein, a change noted in the polarograms of barium caseinate-barium phosphate complexes.[181]

Jenness and Parkash[103] have noted that part of the differences in heat stability and pH stability curves between individual milks could be eliminated by dialysis against bulk milk. This dialysis did not equalize the concentration of ultrafilterable calcium, magnesium or phosphorus of the test samples and the bulk milk. However, removal of colloidal calcium phosphate by acidification and then neutralization usually resulted in increased heat stability of both unconcentrated and concentrated (2:1) skimmilk, but the pH stability curve was of the same type as that of the original skimmilk.

The influence of titratable acidity or pH upon heat coagulation has long been a recognized correlation. The general relationship between adjusted pH and spontaneous coagulation temperatures of raw skimmilk, as defined by Miller and Sommer,[149] is shown in Figure 11.4. A high order of sensitivity to pH is evident for values between 6.2 and 6.4. At pH values below 6.4, added phosphate tends to displace the curve toward higher stability, while calcium has the reverse effect. Such heat stability curves have been found to exhibit a maximum at a pH that appears to be a specific characteristic for each milk.

Thus the colloidal and serum salt levels are critical in maintaining casein integrity, and it has been pointed out that this delicate equilibrium can be disturbed by lowering or raising the temperature. The stability of milk after exposure to a processing treatment is similarly sensitive to pH.

The acidity of milk increases with temperature, partially as a result of changes in the buffer capacity of the milk salts and the expulsion of CO_2 on heating. Miller and Sommer[149] observed that the pH of skimmilk decreases approximately 0.1 pH for each 10°C temperature rise. Rose *et al.*[198] reported that the hydrogen-ion concentration of a 94°C milk ultrafiltrate is at least twice that of a comparable 25°C ultrafiltrate. Concentration of milk is accompanied by a significant decrease in pH, which may contribute to the greater heat-susceptibility of condensed milks. Under prolonged heat treatment at elevated temperatures additional acidity is developed as a result of further changes in the milk. This acidity may be derived from thermal decomposition of the lactose to organic acids, interaction of lactose with the milk proteins, hydrolytic dephosphorylation of casein and displacement of the calcium-phosphate equilibrium.

Of the total acidity developed in milk heated at 120°C for 90 minutes to the point of coagulation, Pyne and McHenry[183] attribute one-half to lactose decomposition, one-third to casein dephosphorylation and the residual acidity to phosphate equilibria. Such heat-developed acidity contributes materially to the heat coagulation of milk.[182,183] In a study of 26 samples of fluid skimmilk it was noted that the pH at the time of coagulation ranged from 5.5 to 6.0, the more rapidly coagulating samples generally developing the most acidity.[183] The level of acidity developing before the heated milk coagulates appears to be inversely related to the calcium-ion activity, suggesting that the developed acidity may enhance the coagulating effect of ionic calcium. On the average, fluid skimmilks with an "effective calcium-ion concentration" in excess of 4.8 m*M* per l coagulated within 20 min at 130°C.

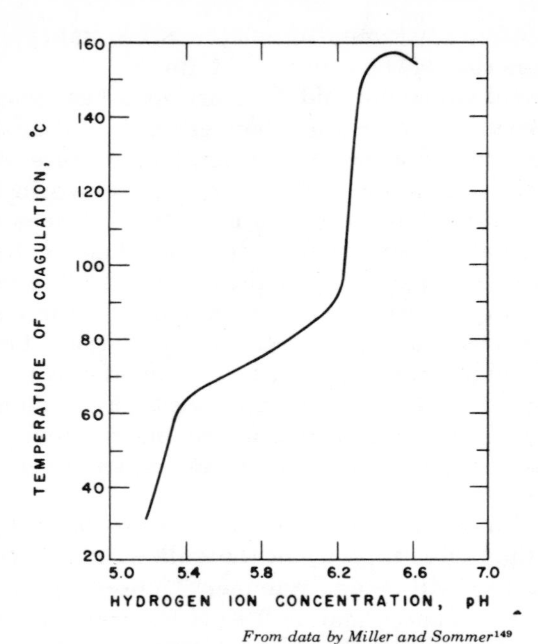

From data by Miller and Sommer[149]

FIG. 11.4 THE EFFECT OF pH ON THE COAGULATION TEMPERATURE OF SKIMMILK

At coagulation, the pH had been depressed only 0.4 unit. Milks containing 4 m*M* serum calcium survived 30 min at 130°C, at which time the pH fell 0.7 unit. At a calcium-ion concentration of 3.7 m*M*, the average milk withstood coagulation at 130°C for periods up to 60 min, with a concomitant pH decrease of 1.0 unit.

Rose's observations that a minor change in the pH of milk can dramatically affect heat stability has increased our understanding of this phenomenon.[190,191] He showed that milk from individual cows will vary over a narrow pH range of 6.5 to 6.9, and measurements of heat stability at 140°C indicate notably increased stability at about pH 6.6 to 6.7. Therefore, heat stability is related to the difference between the pH of the milk and the pH of maximum stability. The heat stability-pH curve is significantly altered by addition of pure β-lactoglobulin, but not by α-lactalbumin or globulin mixtures.

Casein Alteration

The influence of heat on the natural state of milk proteins is substantial. This is eventually observed as heat coagulation; however, many other changes precede this phenomenon. We shall consider here the heat effects on casein and then explore the whey protein-casein interaction phenomenon.

The hydrolytic cleavage of peptide and phosphate bonds in casein can result from either enzymatic action or elevated heat treatment of milk. The liberation of inorganic phosphate from casein has been previously alluded to in connection with acidity development in heated milk. Structural changes of this type exert an influence on the stability of the caseinate complex, and it is essential to consider their contribution to the coagulation of milk.

Minor changes occur in the nonprotein nitrogen (NPN) during pasteurization or HTST sterilization of milk.[105,213] Belec and Jenness[6] found that α-casein solutions heated at 135°C hydrolyze slightly more rapidly than those of whole or β-casein, attaining a nitrogen release level of 5% in 20 min, and approximately 15% in 60 min. This would indicate that the α_s and κ fractions are the most susceptible protein fractions, while the intact micelle does impart some protection against nitrogen release. Fox *et al.*[53] found that after heating fresh raw skimmilk 30 min at 55 to 95°C, gelation occurred within 24 hr at room temperature, and was accompanied by an increase in nitrogen soluble in 5% trichloroacetic acid. They were able to rule out bacterial action as the cause, and identified the peptides released as similar to those set free by the action of rennin on milk. This indicates that κ-casein's macropeptide may be the labile group in this hydrolysis.

Under rigorous heat treatment, casein yields substantial hydrolysis of nitrogen and phosphorus. Howat and Wright[92,93] found 15% of the nitrogen released as small fragments in neutral sodium or calcium caseinate solutions heated for 5 hr at 120°C, as well as complete dephosphorylation of sodium caseinate, and 85% dephosphorylation of calcium caseinate. Coagulation of 3% calcium caseinate solutions at pH 6.9, heated in the range 90 to 115°C, was accompanied by a proportionate dephosphorylation of the casein.[95] This was taken as presumptive evidence that coagulation and phosphate release may be related reactions. The reaction velocity for both dephosphorylation and coagulation increases threefold for each 10°C rise in temperature. However, phosphate does not appear to be liberated in evaporated milk aged 6 to 9 months at room temperature.[92] Studies of casein dephosphorylation by Belec and Jenness[6] simulated normal heat-processing conditions. Dephosphorylation of both sodium caseinate and skimmilk casein, heated 110 to 140°C, conforms to first-order kinetics with an energy of activation of 25 to 29 kcal/mole. The rate of phosphate release is almost identical for α- and β-casein and is independent of pH in the range 6.0 to 7.0. Twofold concentration of skimmilk increases the rate of dephosphorylation, while preheating at 90°C for 10 min decreases the rate by 20 to 25%.

Work on NPN release has been confounded, however, by the identification of milk protease which may survive the heat treatment and/or be reactivated on storage.[21] A moderate increase of NPN in HTST-sterilized milk during storage has been attributed to gradual breakdown of β-casein; the greatest increase occurred at a storage temperature of 37°C.[105] Reactivation of a β-casein-specific protease in milk is suggested as a possible cause for the hydrolysis noted.

Reactivation of alkaline phosphatase after whole milk was sterilized (141°C for 7 sec) has been observed.[37,118] The reactivation was apparently dependent upon the presence of casein[37] in the heated mixture and upon subsequent storage at room temperature,[118] whereupon 50% recovery of the original milk level was noted. However, no evidence of lipase reactivation in milk has been found following sterilization.[37,241] It is conceivable that reactivation of a naturally occurring milk protease occurs after sterilization. This may be true, since specific combination of the enzyme with its substrate (β-casein) or an inhibitor would enhance its resistance to heat denaturation. Since low percentages of this enzyme could have a detrimental effect on long-term storage stability of sterilized milk, a brief review of the information available on this enzyme follows.

Although the first description of the presence of protease in milk was made in 1897,[2] the enzyme still remains one of the least characterized milk enzymes. Considering the observations made by several investigators,[71,113,222] it appears certain that the enzyme does not originate from a bacterial protease but is a natural component of milk. Warner and Polis[242] found that most proteolytic activity in milk is precipitated with casein when milk is acidified. Zittle[261] extended this method by washing the acid curd with water acidified to pH 4.0 and by precipitating the solubilized protease with ammonium sulfate. Yamauchi *et al.*[258] were able to achieve 70-fold purification over unfractionated protease. When they examined some of the properties of this enriched fraction, they found[109] that the protease had a sharp pH optimum at 8.0 with an accompanying shoulder at 6.5; it showed tryptic and chymotryptic substrate specificity, and the most effective inhibitor was diisopropyl fluorophosphate. They also found that in the purified state the protease could be destroyed by heating at 80°C for 10 min, and that when it was present in aseptically canned milk, considerable proteolysis of the casein occurred after storage for 30 days at 30°C. Murthy *et al.*[159] have observed by free-boundary electrophoresis that casein in milk sterilized by UHT treatment undergoes change during extended storage, and noted that this alteration would be due to the enzymic action of remaining or reactivated protease. Several investigators[21,109,148] have found that β-casein is most susceptible to proteolysis, followed by α_{s1}- and κ-casein. Chen and Ledford[21] achieved a 180-fold purification of milk protease, and this preparation gave mostly trypsin-like activity. Their data suggest, indirectly, that milk protease activity may be involved in the age thickening and gelation of HTST-sterilized milk. As more information is obtained on this milk protease its direct link to storage instability problems may be proved, and hopefully the required processing treatment can then be designed to destroy it. This defect could cause greater problems as manufacturers look for longer shelf life for their manufactured milk products.

Serum Protein Denaturation

The denaturation of the milk serum or whey proteins modifies the course of milk coagulation and the rheological properties of the curd formed by acid or enzymes. When milk is heated to temperatures that denature serum proteins, cooked flavor, increased heat stability following concentration, reduced colloidal stability when frozen, and resistance to clotting by rennin ensue. Such changes are of practical significance in the production of cheese, milk powder, and concentrated milk products.

The serum proteins constitute approximately 0.6% of milk or 20% of total milk protein. The composition of the serum protein fraction, based on electrophoretic analysis,[123] is approximately 55% β-lactoglobulin, 12% α-lactalbumin, 10% "proteose-peptone," and 10 to 15% casein, the remainder being composed of the globulins and enzymes. Serum proteins are subject to more marked variation than casein and are particularly affected by the stage of lactation. When denatured, whey proteins lose their solubility in the isoelectric region, or in salt solutions, and coprecipitate with casein.

The heat stability of concentrated milk is materially increased if the fluid milk is subjected to heat treatment before condensing. This forewarming process is essential to the stability of evaporated milk during sterilization; much of the benefit has been ascribed to heat modification of the serum proteins,[184,189,217] as well as to favorable changes in the salt balance. It has been generally maintained that serum proteins are not extensively denatured in pasteurized fluid milk, whereas their denaturation in concentrates is extensive. Attention has also been drawn to the role of the serum proteins in heat coagulation of colostral and mastitic milks.[34,47,48,217,251] In considering the heat effects upon these proteins, we will first consider studies on the purified proteins, and then look into complex mixed systems of casein-whey protein interactions.

The effect of heat on proteins should be considered as a two-stage process: (1) the secondary and tertiary structure is altered, causing denaturation, and (2), the proteins aggregate, which may lead to coagulation. β-Lactoglobulin (β-lg) is a well-characterized protein which has been shown to undergo such a process. Several reviews (Sawyer[205] and McKenzie[145]) have been published on heat denaturation of β-lg. Briefly, the mechanism for β-lg appears first as a dissociation from dimer (36,000) to monomer (18,000), the monomer increasing as the temperature is increased from 20 to 45°C.[58] Dupont[32,33] studied the optical rotation of β-lg A and B variants and found that they went through a transition above 40°C. The change in rotation was reversible, but at higher temperatures it was followed by a slow, irreversible one. This irreversible change (or aggregation) was found to yield at least two different products,[171,204] one being small in size (3.7S) with sulfhydryl groups involved, and the other yielding large aggregates (29S) that did not have sulfhydryl bridges. Apparently, the genetic type of β-lg will change the relative proportion of the type of aggregation product formed (Table 11.2).

Table 11.2*

AGGREGATION OF GENETIC VARIANTS OF β-LACTOGLOBULIN DURING HEAT TREATMENT

Genetic variant	% aggregation not involving SH/SS reactions	% aggregation involving SH/SS reactions
β-lg A	37	63
β-lg B	55	45
β-lg C	72	28

* From Sawyer.[204]

The genetic variants of β-lg vary in their susceptibility to heat denaturation. The most heat stabile is β-lg A, followed by the B and C variants.[204] Gough and Jenness[61] have found that the behavior of the isolated genetic type of β-lg paralleled the heat stability of the milk, i.e., β-lg B milks were less stable than β-lg A milks. No work on the heat stability of the less common β-lg C milk has been reported. Predictably, it should be the least heat-stable. Rose[192] was unable to detect heat-stability differences between type AB and B milk, but did find a significant correlation between (a) the amount of β-lg in 20 milk samples and (b) the maximum and minimum heat stability.

The loss of solubility of heat-denatured serum proteins, when acidified or saturated with sodium chloride, has served as a useful criterion for kinetic studies of the overall denaturation process in milk. Early methods of analysis, based on acid precipitation, adequately demonstrated that the threshold temperature for initiation of serum-protein denaturation is slightly above the general region of pasteurization conditions. For complete denaturation, heat treatments at 77.5°C (1 hr), 80°C (30 min), and 90°C (5 min), have been cited.[112,201] Subsequent refinements introduced in the measurement of serum-protein denaturation rates enabled Rowland[199] to obtain smooth curves for serum-protein denaturation as a function of temperature, with values ranging from 10.4 to 28.0% in milk heated at 63 to 70°C for 30 min. Although the energy of activation and the order of reaction could not be precisely calculated, Rowland demonstrated that the thermal coef-

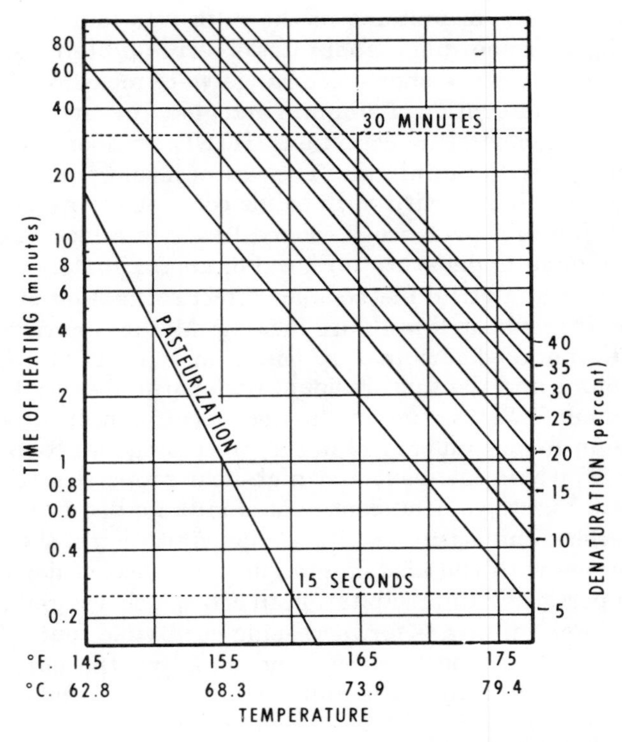

Data from Harland et al.[68]

FIG. 11.5. THE HEAT DENATURATION OF THE SERUM PROTEINS IN SKIMMILK

ficient of 1.5 per 1°C, in this temperature range, was of the high order of magnitude generally observed for protein denaturation. Similar time-temperature equivalents for denaturation of the combined serum proteins in skimmilk, determined by Harland *et al.*[68] using the salt-precipitation method, are shown in Fig. 11.5. In the temperature range 62 to 80°C the relationship of temperature to time for a constant level of serum-protein denaturation is semilogarithmic. A temperature increase of 7.5°C reduces the time required for a fixed level of denaturation ten-fold ("Z" value of 13.5). Again, the heat treatments required for pasteurization of milk are clearly below serum protein denaturing conditions.

In the high-temperature short-time range (Fig. 11.6), the relationship for equivalent serum protein denaturation becomes curvilinear.[70] At 80 to 90°C, the Z value is approximately 19 and increases further at higher temperatures. The percentage of total serum proteins denatured is about 10% lower than indicated in Fig. 11.6, since a maximum of 90.3% of the serum proteins in the milk was found denaturable. Comparable, but somewhat higher rates of serum-protein denaturation in the high-temperature short-time range were measured by Hetrick and Tracy.[79] Denaturation of albumin and globulin cannot be detected when the time-temperature conditions (HTST) for phosphatase inactivation is restricted to a minimum. The rate of serum-protein denaturation varies but slightly with milk solids concentrations in the range 9 to 40%.[68] The first evidence of cooked flavor in milk coincides with an albumin-globulin denaturation level of approximately 58%.

Quantitative analytical methods for direct estimation of denatured serum proteins in heated milk are lacking. Methods currently in use rely on differential determination based on the total initial serum-protein content and the final residual undenatured protein following heat treatment.[67] Estimation of the heat treatment of milk products constitutes an important control in the processing of cottage cheese[146] and bread,[122] where low-heat and high-heat properties respectively, are required. Excessive heat treatment of milk for use in conventional cottage cheese manufacture results in coprecipitation of the denatured serum proteins with the acid casein curd, and loss of desirable curd tension properties.[154] Coprecipitation in cottage cheese milk offers an important potential means for increasing curd yield; but special care must be used in setting, cooking, and handling the curd. Detailed information on coprecipitable manufacturing can be found in the excellent review by Muller.[157] Milk processed for bread baking must receive a sufficiently high heat treatment to inactivate unidentified factors responsible for gluten breakdown, extreme dough extensibility, and

poor loaf volume.[122] For satisfactory bread loaf volume, the milk must be subjected to a minimum heat treatment of 69 to 74°C for 30 min.[122] Milk processed for commercial bread baking is generally heated in the range of 85 to 100°C.

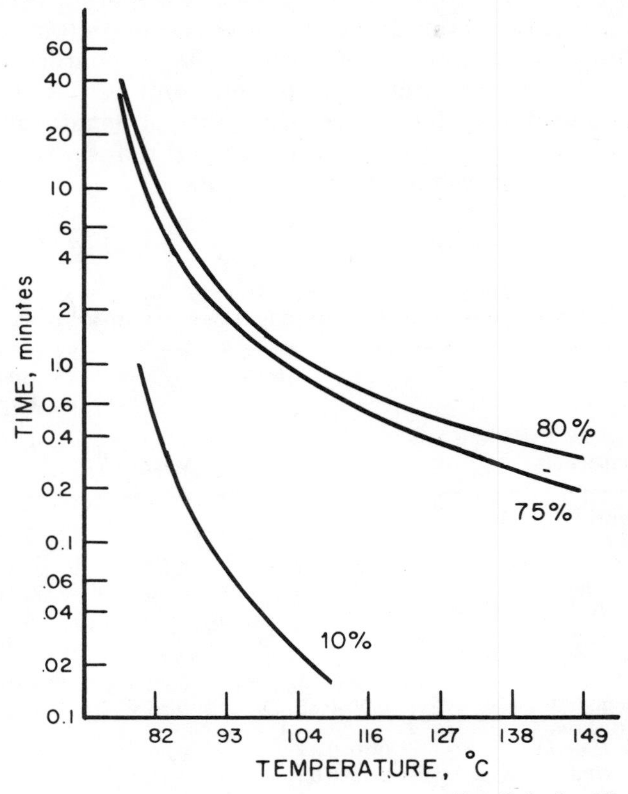

From Harland et al.[70]

FIG. 11.6. SERUM PROTEIN DENATURATION, MEASURED BY THE HARLAND-ASHWORTH METHOD, IN SKIMMILK HEATED UNDER HIGH-TEMPERATURE–SHORT-TIME CONDITIONS

Fahrenheit temperatures on the original graph have been converted to the nearest centigrade value.

The use of denatured serum protein values to assess the heat treatment of milk products is most satisfactory when the original total serum-protein content is known, because the serum-protein content of milk is subject to considerable variation. An extensive survey of the serum proteins in milk by Harland *et al.*[69] disclosed sizeable variations in both concentration and denaturability. The range of such variations, which appear to be related primarily to regional and breed differences rather than to seasonal effects, is indicated in Table 11.3, reproduced from values published by Harland *et al.*[69] Wide variations were observed in the precipitability of the serum proteins in 81 milk samples heated at 74°C for 30 min. A close correlation between the amount of protein denatured and the total serum-protein content is evident. The absolute amount of serum protein denatured shows a closer correlation to the total serum-protein content than do the percentage denaturation values. Such variability of denaturation rates probably reflects distribution differences in the serum proteins, as well as in the ionic composition of the milk.

Table 11.3

VARIATIONS IN THE DISTRIBUTION AND HEAT LABILITY OF THE SERUM PROTEIN IN COMMERCIAL BULKED MILK[a]

Component[b]	Range	Mean	Standard Deviation (Single Determinations)
Serum protein N (Rowland method)			
Total, mg/ml	0.82–1.48	1.04	0.12
Heat-labile, mg/ml	0.60–1.10	0.76	0.08
Serum protein N (Harland Ashworth method)			
Total, mg/ml	0.62–0.91	0.76	0.06
Heat-labile, mg/ml	0.55–0.85	0.66	0.06
Sulfhydryl (iodosobenzoate), meq/ml	0.19–0.44	0.30	0.16
Casein N, mg/ml	3.49–6.02	4.31	0.47
Nonprotein N, mg/ml	0.23–0.42	0.31	0.04
Total protein N, mg/ml	4.52–7.28	5.35	0.54
Total nitrogen, mg/ml	4.82–7.70	5.66	0.56
Solids nonfat, %	8.11–10.55	9.25	0.47

a Survey of 81 samples. From Harland, Coulter, and Jenness.[69]
b Data calculated on basis of skimmilk.

The denaturation curves for the individual serum-protein components, obtained by quantitative electrophoretic analysis of the serum from heated milk, place the immune globulins, serum albumin, β-lg and α-lactalbumin (α-la), in order of increasing resistance to heat denaturation. The denaturability of the serum-protein components, measured electrophoretically by Larson and Rolleri,[123] is reproduced in Fig. 11.7. The complete serum-protein denaturation curves and the total extent of denaturation for the heat treatments indicated conform to the values reported elsewhere.[68,122,200] Heat treatment of skimmilk at 70°C for 30 min denatures only 6% of the α-la, but 32% of the β-lg, 52% of the serum albumin, and 89% of the immune globulins;

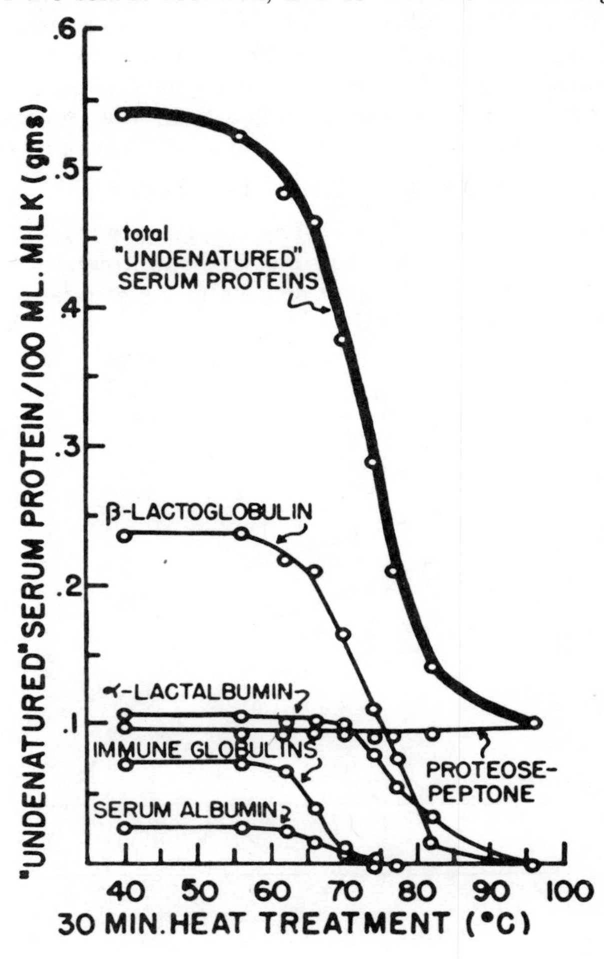

From Larson and Rolleri[123]

FIG. 11.7. THE DENATURATION OF THE TOTAL AND INDIVIDUAL SERUM PROTEIN COMPONENTS IN MILK HEATED AT VARIOUS TEMPERATURES FOR 30 MIN

this represents, cumulatively, 29% of the total serum proteins. Upon elevated-temperature treatment, the composition of the undenatured proteins in milk serum will vary; the most significant increase observed is in the most heat-resistant component, α-la. Over one-half of the total α-la is able to withstand 77°C for 30 min, a heat treatment sufficient to completely denature the serum albumin and immune globulin fractions.

An important extension of the Larson and Rolleri study[123] has appeared which adds to kinetic data on the heat denaturation of β-lg and α-la. Lyster[136] studied the heat effects on whey protein using sensitive immunological techniques developed by Larson *et al.*[119,124] He found that the denaturation of α-la is a first-order reaction; between 90 and 155°C the kinetic constant k_1 in sec^{-1} is given by the equation

$$\log k_1 = 7.15 - 3.60\ (10^3/T)$$

where T is the temperature in °K. The denaturation of β-lg in skimmilk is second-order with respect to time, and the kinetic constant k_2 in $lg^{-1}\ sec^{-1}$ is given by two equations, valid for different temperature ranges. Between 68 and 90°C

$$\log k_2 = 37.95 - 14.51\ (10^3/T);$$

between 90 and 135°C

$$\log k_2 = 5.98 - 2.86\ (10^3/T).$$

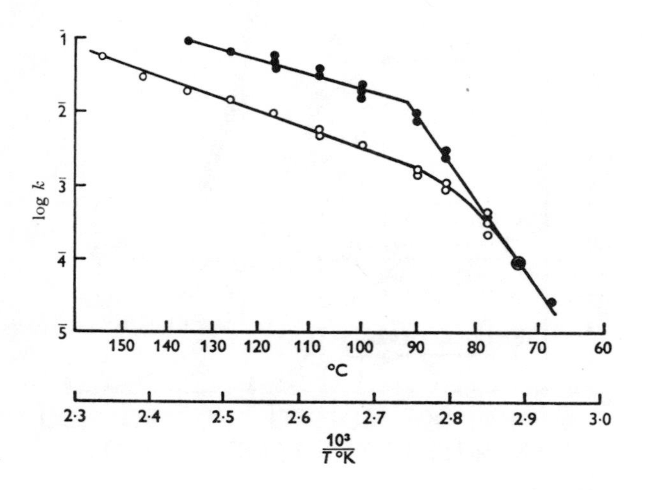

From Lyster[136]

FIG. 11.8. ARRHENIUS PLOT OF THE KINETIC CONSTANTS o———o, α-LACTALBUMIN; •———•, β-LACTOGLOBULIN

Fig. 11.8 is the Arrhenius plot of those equations for β-lg and α-la. The denaturation rates decreased when a reagent specific for sulfhydryl groups was added, suggesting that such groups are involved in the denaturation of both proteins. The validity of the equations derived from the immuno diffusion method can be seen from Table 11.4, where the data of Larson and Rolleri[123] are compared with the calculated values of Lyster.[136]

Table 11.4

DENATURATION OF β-LACTOGLOBULIN AND α-LACTALBUMIN IN MILK AFTER HEATING FOR 30 MIN AT VARIOUS TEMPERATURES

Temperature °C	β-Lactoglobulin Found[1]	β-Lactoglobulin Calculated[2]	α-Lactalbumin Found[1]	α-Lactalbumin Calculated[2]
62	92	98	—	—
66	89	93	—	—
70	70	81	—	—
74	47	59	75	80
77	32	38	51	64
82	7	14	32	32
96	0	1	0	1

[1] Determined electrophoretically.[123]
[2] Calculated from kinetic equations.[136]

Lyster[136] was able to detect differences in the β-lg denaturation among milks with genetic types of A, B and AB. The constant k_2 for type AB is a value close to that for mixed herd milk, and for types A and B the constants were 50% lower and higher, respectively. He concluded that in practice the genetic variation may be safely ignored, since most bulk milk contains equal proportions of the two variants.[1]

An important manifestation of protein denaturation is the increased reactivity of specific groups resulting from structural modification of the protein molecule. The activation of sulfhydryl groups is one of the most readily detectable of the chemical changes accompanying denaturation.[180] An excellent quantitative method has been reported for the determination of sulfhydryl and disulfide groups in native and heat-denatured milk proteins.[137] β-Lg is known to be the most influential sulfhydryl-containing component of the serum proteins.[98,120] From the amino-acid analysis of β-lg,[145] it is known that for each 36,000 molecular weight dimer, it contains two –SH groups and four –S–S–linkages. The sulfhydryl group activation by heat has been associated with the origin of cooked flavor, the antioxygenic properties of heated milk, and aggregation with the caseins.[25,73,98,260,265]

Several reports[66,121,129] have been made on the rate and extent of reaction of the –SH groups of β-lg with various reagents and denaturing conditions. It was found that these groups were for the most part unreactive when the protein was in the native state; however, a marked increase in reactivity was found after the primary denaturation phase as induced by heat, urea, guanidine or alkali at pH 12. Apparently, this primary denaturation phase is identical to the reversible dimer-to-monomer dissociation referred to previously. Prolonged exposure to denaturation conditions will cause even more extensive protein unfolding, leading to cleavage of disulfide groups and/or other exchange reactions which further complicate the picture.[66,145]

The aggregation and precipitation of heat-denatured β-lg is also highly dependent on pH and the presence of calcium ion. Zittle *et al.*[263,264] observed a progressive decrease in the calcium-ion concentration required to precipitate heated β-lg as the pH of its solution was lowered. It was surmised that the calcium interacts at negatively charged carboxyl sites on the protein, thus reducing the net charge to zero, causing isoelectric precipitation. α-La is also known to aggregate in the presence of calcium, magnesium and barium after heating to 100°C.[116] Although calcium-ion concentration in milk is adequate to precipitate β-lg, denatured serum proteins in heated milk do not coagulate to any appreciable extent. This is generally construed as evidence of a heat-induced interaction between β-lg and casein.

Casein-whey Protein Interaction

Considerable data on heat-induced denaturation and interaction of whey protein and casein components of skimmilk have been accumulated from numerous investigators using different milk protein systems under different experimental conditions.[4,55,72,73,97,153,155-160,179,234,259,260] It is known that purified β-lg and κ-casein strongly interact when heated together or when heated in skimmilk. This interaction is mediated by a thiodisulfide interchange, and the principal chemical linkage responsible for stability of the complex is the disulfide bond.

Studies of the isolated proteins have shown that complexation is inhibited when a sulfhydryl blocking reagent or a reducing agent is present. Limited heat aggregation occurs when previously heated and recooled β-lg is mixed with κ-casein.[179] McKenzie *et al.*[143] found that when heat-denatured β-lg and κ-casein were mixed at room temperature, no interaction took place. When the mixture was heated at 75°C for 7.5 hr, about 50% of the κ-casein had interacted. The presence of N-ethylmaleimide (a thiol blocking agent) during the heat treatment of β-lg prior to heat treatment of the mixture did not affect the result. The complex was found to be relatively stable and did not dissociate at low concentrations. High concentrations of a dissociating agent (6*M* guanidine hydrochloride) did not disrupt the complex, but addition of mercaptoethanol caused complete dissociation. They concluded that the formation of intermolecular disulfide bonds is a major factor in stabilizing the complex, but that some form of less specific association may also take place involving physical entanglement of polypeptide chains. This observation is in line with Lyster's[136] report of three forms of denatured β-lg; one form involves disulfide interchange and the other two probably involve hydrophobic, hydrogen and ionic bonding which would alter β-lg's sensitivity to ionic strength and salt composition.

Morr and Josephson[153] identified two whey protein-casein aggregates formed after heating skimmilk. One was identified as protein particles aggregated through thio-disulfide bonds; the second type was formed by gross aggregation of the former, probably by a calcium-binding mechanism. It was surmised that the large aggregate was prevented from precipitation in skimmilk by interaction with colloidal casein and involved nonspecific calcium salt linkages.

Morrissey,[156] however, examined sodium caseinate and colloidal phosphate-free systems and found that neither the micellar structure of casein nor the initial colloidal phosphate of milk plays an important role in the heat stability of milk. He attributed the observed minimum heat stability of milk at pH 6.9 to heat-induced precipitation of calcium phosphate on some type of caseinate/β-lg complex, which is then "sensitized" to the coagulating action of calcium salts. Because of the complexity of skimmilk, no quantitative study has been made to predict the relative importance of the thiol-disulfide reaction of β-lg with κ-casein, as compared to its nonspecific salt-induced aggregation with the entire casein complex. It would appear that both reactions are critical to an understanding of this system. When this question is finally resolved, a clearer picture of milk heat stability will emerge and better methods for efficient coprecipitate manufacture should be possible.

The relationship between milk heat stability and pH was investigated by Tessier and Rose.[233] They found that the milk from 32 individual cows could be classified into two types, as seen in Fig. 11.9. Type A milk had a maximum and minimum in the pH range of 6.4 to 7.1, whereas type B had only a gradually increasing heat stability throughout this pH range. Approximately two-thirds of the milks studied were of type A. Milk from three cows varied between A and B during one

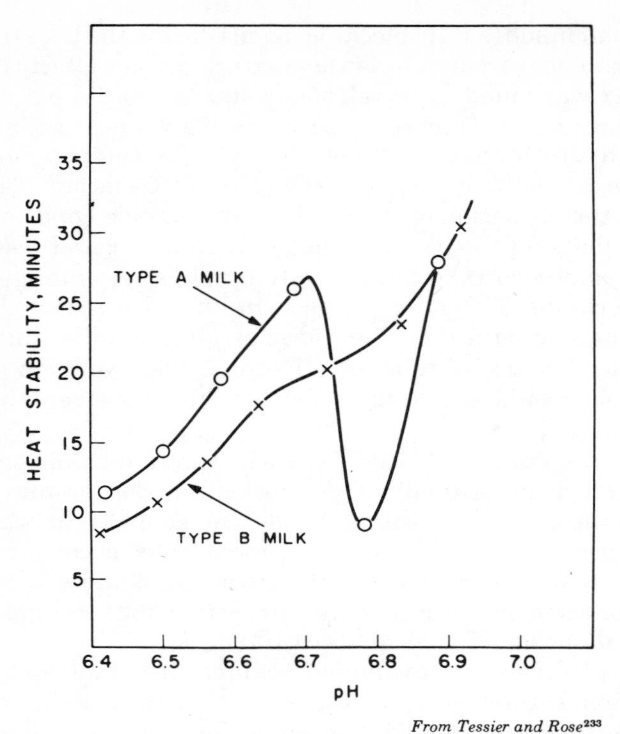

From Tessier and Rose[233]

FIG. 11.9. pH HEAT STABILITY CURVES OF INDIVIDUAL COW MILKS

o————o = Type A milk. x........x = Type B milk.

lactation period, but most cows consistently produced one type of milk. All bulk milks examined were type A. Addition of κ-casein to a type A milk reduced or eliminated the minimum in the curve, but addition of similar amounts of β- or α_s-casein had little effect. However, the addition of β-lg to type B milks induced a minimum in the heat stability-pH plot, converting it to type A.

Thus we can see that skimmilk heat stability is dependent on the relative amount of κ-casein and β-lg present in the milk. Since added κ-casein did remove the minimum found with type A milks, the role of the serum κ-casein, mentioned earlier, must be considered as a possible factor in heat stability. Unfortunately, no information is available on variation of the serum κ-casein content in milk. It is known that the whey proteins will vary considerably in certain abnormal milks (such as milks from cows with mastitis infections).

Reports have indicated that subclinical mastitis is widespread.[57,64]

The work of Feagen *et al.*[48] has shown that milk drawn from infected quarters has a much lower heat stability than normal milk. The heat stability of mixed subclinical mastitic milk and uninfected milk could not be predicted from the heat stabilities of either milk. However, mixing infected milk with normal milk invariably lowered the heat stability of skimmilk powder; the extent of alteration was dependent upon the proportion of subclinical mastitic quarters present.[47] The authors did not test for changes in salt composition or protein distribution which should occur as a result of the infection.

Lactose Reactions

Thermal decomposition of lactose, as a partial source of acidity in heated milk, has been well documented. The rate of acid formation has been found proportional to the lactose concentration, and the formation of formic and lactic acids in heated milk has been correlated with the degree of lactose decomposition. Formic acid constitutes approximately 57% of the acid generated by milk heated at 116°C for 2 hr[62,63] and 75% at 100°C for 6 hr.[152] The lactic acid produced in milk heated sufficiently to decompose 25% of the lactose accounts for less than 5% of the total acidity.[62] In addition to formic and lactic acids, butyric, propionic, acetic and pyruvic acids have been identified chromatographically in heated milk.[111,151,152] The presence of stabilizing salts appreciably increases the rate of acid formation during heating.[151,152] The addition of 0.5% disodium phosphate may triple total acid formation because of an increase in the formic, lactic, and acetic acid components.

As a consequence of sugar decomposition reactions,[26,76,176] acidity gradually develops in evaporated milk and dried milk products during prolonged storage at moderate temperatures. At 38°C storage, 50 to 75% of the titratable acidity increase in evaporated milk is due to formic acid. Lactic acid formation during storage is negligible.

The development of acidity in heated or aged sterile milk is accompanied by a brown discoloration, attributed to the formation of a poorly defined class of high molecular weight pigments termed melanoidins. The bulk of experimental evidence supports the view that this browning reaction results from Maillard type condensation of lactose with the milk proteins rather than from caramelization of the lactose. Ramsey *et al.*[186] postulated that a Maillard interaction between the aldose group of the lactose with the free amino groups of casein is the primary source of color in heated milk. The browning of fluid milk and development of acid by thermal decomposition of lactose are not significant below 80°C, under the usual pH conditions of milk.[16,65]

The occurrence of an interaction between lactose and the proteins in milk has been established on the basis of (a) radioisotope tracer studies and (b) changes in the concentration of free amino nitrogen and reducing sugar. Patton and Flipse[177] concluded, from the distribution of C^{14}, that the formation of formic acid and maltol from lactose proceeds from carbon atom number 1. Thus, the reducing end of the glucose moiety appears to be the primary source of sugar decomposition in heated milk. The reactive aldehyde group of lactose is believed to interact with the free amino groups found on the protein; about 90% of the total free amino groups are the ϵ-amino groups of lysine. Ferretti *et al.*[50A] were able to isolate and identify 80 compounds formed from heating a casein/lactose model system. Many of the complex mixture of products formed were readily accounted for solely on the basis of a Maillard-type interaction.

COAGULATION OF CONCENTRATED MILKS

Changes in Milk Induced by Concentration

The stability of the caseinate-phosphate complex to various coagulating agents, including heat, rennin, and alcohol, declines rapidly with increasing milk solids concentration. Heat coagulation at a constant temperature tends to increase logarithmically with solids concentration.[86,247] Similarly, storage thickening and gelation of heat-sterilized concentrates, particularly high temperature-short time processed, are accelerated at higher milk solids concentrations.[3,18,77,230] The factors controlling the stability of concentrated milk are even less well understood than those involved in fluid milk stability. The heat stability of evaporated milk bears no apparent relationship to that of its related raw milk.[247]

The pH of fluid milk declines with concentration from an average initial value of 6.6 to approximately 6.2 at 26% total solids after batch sterilization. The decrease in pH with increasing milk solids concentration is substantially linear in the 9 to 40% skim solids range.[93] The pH of nonfat dry milk reconstituted to a 9% solids level is approximately 6.6, while the pH of a 23% solids solution is 6.2 and that of a 40% solution is 6.0.[93] Redilution of concentrated milk with water approximately restores the initial pH. The reversibility of this pH change is presumed to arise primarily from the interconversion of soluble phosphate and tricalcium phosphate.

The activity of calcium and magnesium increases during the condensing of milk, but at a lower rate than that of sodium and potassium, implying formation of insoluble or undissociated salts of calcium and magnesium with increasing milkconcentration. Van Kreveld and Van Minnen,[239] using an equilibrium exchange resin method, found 6% of the calcium and 16% of the magnesium of milk to be in the ionic state. The proportions declined to approximately 2 and 5% respectively in sterilized, evaporated milk. The relative activity of the calcium ion decreases more readily than that of the magnesium ion during concentration. However, magnesium-ion activity appears to be depressed preferentially during sterilization. Ion activities in sweetened condensed milk are significantly lower than in sterilized evaporated milk, an effect attributed to the lower ionizing power of concentrated sugar solutions.

Changes in the viscosity of skimmilk that result from concentration are reversible on redilution of the milk. However, heat sterilization of concentrated milk leads to large viscosity increases that are not fully reversed by redilution, affording evidence of casein micelle aggregation. Hostettler and Imhof[88,89] have demonstrated by electron microscopy the formation of large, stable, irregular agglomerates in unsweetened condensed milk and drum-dried powder. The formation of such aggregates constitutes the basis for the thickening and gelation of condensed milks during sterilization and storage. Micellar aggregation is less evident in sweetened condensed milk and almost nonexistent in spray-dried milk.

This early work was expanded by Carroll *et al.*[19] and Schmidt,[206] who studied the electron microscopic characteristics of sterilized concentrated milks. Casein micelles from concentrated skimmilk are much larger than micelles from fresh skimmilk. Fig. 11.10 is a nomograph of the size distribution of 3:1 HTST skimmilk. Comparison of this distribution with Fig. 11.2, which depicts the micelle dispersion in fresh milk, shows that the average size of casein micelles in fresh sterilized concentrated skimmilk is about twice as large. Micelle diameters in the concentrate range from 1,500 to 4,000 Å, with 80% between 1,800 and 3,200 Å. These values are in good agreement with Schmidt,[206] who reported micelle diameters in the range of 2,000 to 2,500 Å.

When 3:1 HTST concentrated skimmilk with added polyphosphate was studied at weekly intervals for 17 weeks, no sign of gelation was observed. In the concentrate without added polyphosphate, however, Carroll *et al.*[19] observed the gelling or aggregation of micelles after 9 weeks, as seen in Fig. 11.11a. After 13 weeks of storage at room temperature (Fig. 11.11b) increased aggregation was evident. A prominent feature of this aggregation is the presence of linkages or bridges between the protein particles. Bridging of micelles from HTST concentrates, but not from conventionally sterilized milks, has also been reported by Schmidt and Buchheim.[208] At 17 weeks storage, gelation

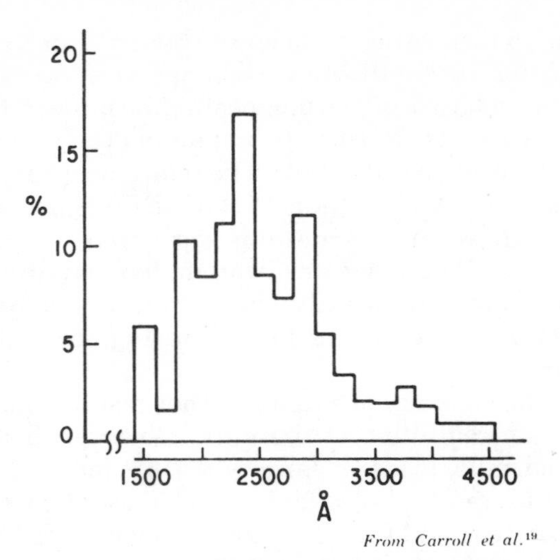

From Carroll *et al.*[19]

FIG. 11.10. RELATIVE FREQUENCY OF OCCURRENCE OF CASEIN MICELLES VERSUS SIZE IN HTST 3:1 CONCENTRATED SKIMMILK

was complete. Fig. 11.11c shows the tight packing of the casein micelles in the gel and the areas of bridging between the micelles. At higher magnification (Fig. 11.11d), the bridging of micelles is depicted more vividly. It appears that the particulate material, also present in ungelled samples, may comprise the bridging material in the gelled samples. It is important to note that the micelles have not lost their individuality, but their surfaces are much more textured than those observed at the beginning of the storage period.

The importance of the particulate material in the gelation phenomenon can be seen in the electron micrographs of Fig. 11.12. Skimmilk was concentrated 2:1 at low temperature (Fig. 11.12a) and then heated at 100°C for 15 min (Fig. 11.12b). An increase in the amount of particulate material surrounding the micelles is seen after heat treatment, and close inspection shows that chain-like aggregates have formed; also, an increase in both average size and electron density of micelles resulted from heat treatment. In Figs. 11.12c and 11.12d the samples received the same treatment, except that the casein micelles were removed by sedimentation. Before heating, the supernatant consists of discrete particles 150 to 200 Å in diameter. These particles have been observed by several investigators, and it has been suggested that they represent small micellar units,[19] although, as was discussed earlier (page 608), they are known to have an abnormally high κ-casein content compared to the casein composition of the micelle. When this superna-

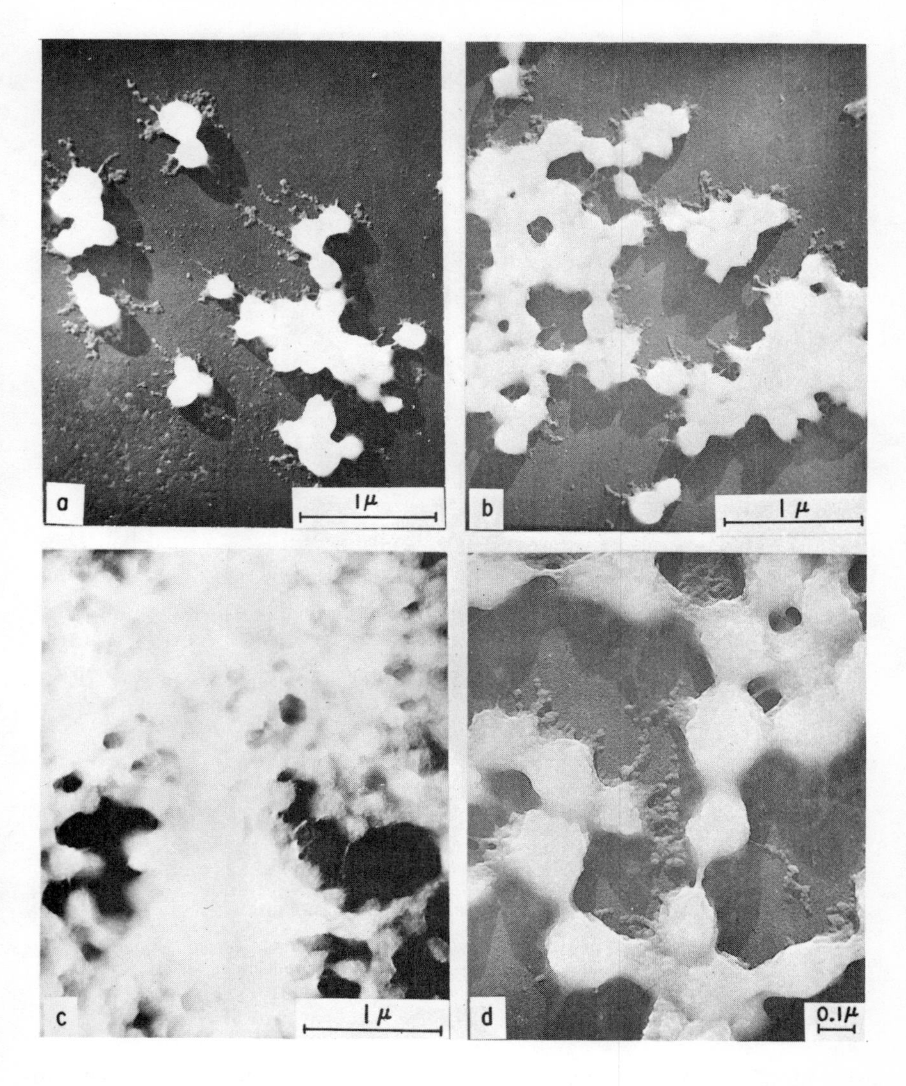

From Carroll et al.[19]

FIG. 11.11. EFFECT OF STORAGE ON HTST 3:1 STERILE CONCENTRATED SKIMMILK

a—9 weeks. b—13 weeks. c—17 weeks. d—15 weeks at higher magnification.

tant was heated at 100°C, a flocculant precipitate was formed, which appeared (Fig. 11.12d) as large masses similar to the bridging material connecting the micelles in gelled concentrated milk (Fig. 11.11d).

Electron microscopy of micelles of concentrated skimmilk shortly

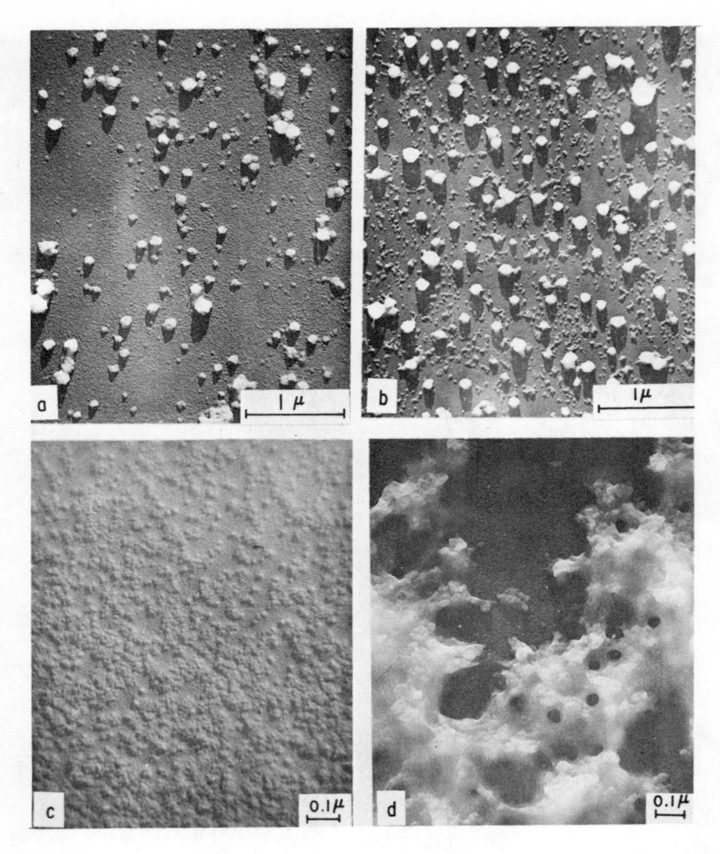

From Carroll et al.[19]

FIG. 11.12. ELECTRON MICROGRAPHS OF SKIMMILK

a—2:1 concentrated skimmilk prepared in laboratory. b—Skimmilk (a) heated at 100°C for 15 minutes. c—Supernatant from skimmilk (a) after casein micelles removed by centrifugation. d—Skimmilk (c) heated at 100°C for 15 minutes.

after sterilization has shown them to be smooth-surfaced and electron-dense. The micelles lose this appearance during prolonged storage, and the micelles in the gel are more textured than those in the fresh concentrate. This could be explained by the interaction of the "κ-casein rich" supernatant with β-lg by the thiol-disulfide mechanism, forming a protein particle which could associate with the micelle itself or form the chain-like aggregates seen in the electron microscope. These highly coalescent particles associating through hydrophobic and charge interactions could be stabilized by the serum calcium, thus forming electron-dense micelles. Fox *et al.*[54] have reported that concentrated milks prepared by conventional evaporation procedures and by HTST developed a sediment of calcium phosphate which increased with time. This would indicate that the complex of calcium-κ-casein β-lg casein micelle loses its calcium during storage causing: (1) formation of an insoluble salt sediment, (2) less opacity or electron density of the micelles, and (3) promotion of micelle aggregation or gelation mediated through the highly coalescent surface proteins which are now unsatisfied ionically because of the loss of cations.

Effect of Concentration on Heat Coagulation

The coagulation of milk by heat is a function of the milk solids concentration, as well as of the temperature and time of heating. The way in which increases in concentration of nonfat milk solids cause the product to coagulate at successively lower temperatures is shown in Fig. 11.13. The relationship is fenerally linear with respect to temperature and logarithmic with respect to heating time.[23,162,247,251] Cole and Tarassuk[23] reported finding some deviation from this straight-line relationship when they studied milks heated in the temperature range 110 to 160°C. About 12 hr of heating at 100°C is necessary to coagulate fresh milk. At 130°C coagulation occurs in approximately 1 hr, while at 150°C the reaction occurs in about 3 min. Though there are wide variations among milks, the time of coagulation of an average evaporated milk (total solids 26%), prepared from milk of good quality, may increase from 10 min at 131°C to 60 min at 114.5°C and to 7,500 minutes at 80°C.

The time of coagulation at a definite temperature varies with the forewarming treatment to which the milk is subjected before concentration. Preliminary heating or forewarming applied to a milk not subsequently concentrated lowers its heat stability. Forewarming continues to cause lowering in heat stability of milks that are concentrated up to a 13% nonfat-solids level. If a milk is further concentrated to 14% or more solids-not-fat, its heat stability is markedly increased by a proper forewarming treatment.[247] Thus there is a critical solids level, about 13.5%, below which forewarming of raw milk decreases stability and above which it raises stability. Hence, forewarming is an important manufacturing step in the evaporated milk industry, where a milk of at least 18% solids-not-fat is to be produced.

No heat stability test is known which, when used on raw milk, will accurately predict the stability of its concentrated product. It is, therefore, not possible to predetermine the forewarming to apply to a fresh milk, so that after concentration it will respond to a sterilization process

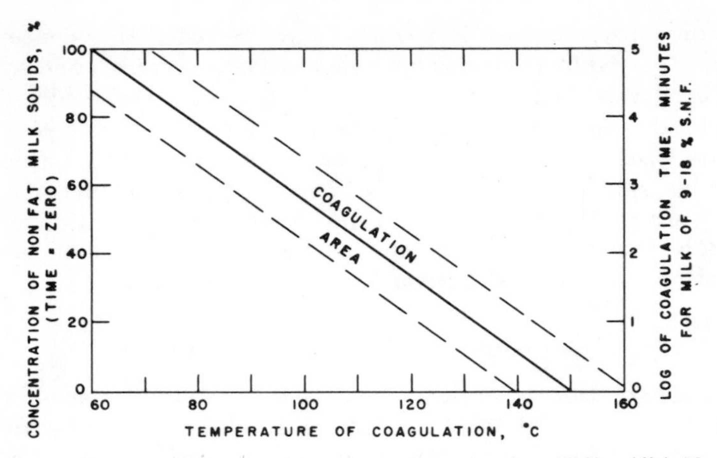

Drawn from data of Webb and Holm[247]

FIG. 11.13. THE GENERAL RELATIONSHIP BETWEEN THE HEAT COAGULATION AREA OF MILKS, THEIR CONCENTRATIONS, AND THE TIME AND TEMPERATURE OF HEATING

by producing a required body. The evaporated milk industry determines heat stability and regulates the body of the finished product by making pilot-batch tests and by addition of stabilizing salts. Tests to determine approximate heat stability, such as alcohol,[27] phosphate,[207] or protein stability test,[221,252] are sometimes used on raw milk. A decision as to the optimum forewarming conditions is usually made on the basis of the behavior of the milk received the previous day.

Marked variations in stability occur with variations in forewarming temperature and time. Rapid improvement in resistance of the concentrate to heat coagulation is brought about by increasing forewarming temperature. Milk to be concentrated and sterilized as evaporated milk is usually forewarmed in a hot well by direct steam injection to raise it to 95°C, where it is held for 10 to 20 min. The stability thus imparted to the milk will usually enable it to withstand heat sterilization as a 26% solids evaporated milk. Greater heat stability can be obtained by the use of high-temperature forewarming carried out by injecting steam directly into the milk, or by forcing the milk through a pressure-heating system, usually in the form of a tubular heater. As the temperature of forewarming is raised, shorter holding times are necessary to attain maximum heat stability. For the manufacture of a 3:1 sterilized milk of 35% solids content, a high-temperature forewarming treatment within the range of 115°C for 2 min to 138°C for 15 sec will usually be required. The forewarming of sweetened condensed milk, discussed later, must be conducted differently from that of evaporated milk.

The rapid decrease in the heat stability of milk with increasing concentrations makes it impractical to attempt to sterilize a product which has been concentrated to a ratio substantially higher than 3:1 (27% nonfat solids). In commercial practice evaporated milk concentrated 2:1 can be sterilized under several temperature-time combinations, such as 115°C for 16 min, 130°C for 2 min, and 140°C for 3 sec. Milks concentrated to 3:1 can sometimes be adjusted by forewarming and salt-balance procedures so that they will withstand the heat necessary for sterilization.

Statistical analysis has shown that the addition of carrageenan to sterilized milk concentrate, and the interaction between carrageenan and forewarming, were highly significant in maintaining storage stability and flavor in the product.[226] The daily processing of such a concentrate by the usual sequence of processing steps would be difficult, and with some milk supplies, almost impossible. The relatively low heat stability of concentrated milks may be avoided in high-temperature sterilization processes by sterilizing the milk prior to aseptic homogenization and canning.[40,220]

Viscosity of Sterilized Milk Concentrates

Viscosity control during processing and storage is an important consideration in dairy-product manufacture. Pre-coagulation thickening occurs in the manufacture and storage of evaporated, plain and sweetened condensed milks, ice-cream mix, and various specialty products. Thickening of milk precedes its coagulation by heat, and increases rapidly with increasing concentration. The actual process of thickening occurs in a relatively short time, just prior to separation of the curd. Coagulation is easily determined by visual appearance of curd aggregates, but thickening is a less perceptible change. Since thickening is a transient state, it is often difficult to stabilize a milk concentrate at any desired viscosity. Evaporated milk must be treated so that the sterilization heat brings it almost to, but never beyond, the point of coagulation. Here the thickening reaction produces a desirable creamy body, although some thinning occurs during storage. The rate of thickening during sterilization in the can at 116°C although variable, becomes greatest in the 10 min preceding coagulation.[29] Milks of high heat stability, which reach the end of the sterilization period before entering the thickening phase, do not develop the high viscosity shown by milks of lower heat stability.

Evaporated milk is subjected to conditions of agitation that vary considerably in different sterilization processes. A heavy-bodied milk will not form under the severe agitation sometimes used to attain rapid heat exchange. However, Gammack and Weckel,[56] using an intermittent hold, Ball,[3] by addition of lactic acid to raw milk, and Tarassuk *et al.*,[227,229] by a proteolytic enzyme treatment before concentration, produced evaporated milks which (after sterilization) had higher viscosities than untreated control samples.

In the case of sweetened condensed milk, excessive thickening developed during manufacture will lead to objectionable gelation in storage. Thickening may be caused by improper forewarming, high acidity in the milk, high concentration of solids, high storage temperature, and other factors.

Sweetened condensed milk should be forewarmed at a lower temperature than evaporated milk to avoid storage thickening. The effect of time and temperature of forewarming of milk on the viscosity of its sweetened condensed product has been studied by several investigators.[178,218,248,250] Representative values are shown in Table 11.5. Although forewarming at 71°C gives a low-viscosity product, this temperature does not effectively destroy many bacteria, yeasts, molds, and enzymes, the presence of which often cause body and flavor defects during storage. It is commercial practice to forewarm the milk at about 82°C. This yields a product that has sufficient viscosity to retard separation of fat and crystallized lactose during storage, but will not gel when held at room temperature or lower for several weeks. When high-temperature short-time heating equipment is available, forewarming at 115°C for less than 1 min may be used to produce low-viscosity sweetened condensed milk for shipment to warm climates. Thickening is usually measured and expressed as relative viscosity; generally no attempt is made to determine plasticity.

Table 11.5

EFFECT OF TIME AND TEMPERATURE OF FOREWARMING ON THE VISCOSITY OF SWEETENED CONDENSED SKIMMILK[a]

Forewarming Conditions		Viscosity after Storage at 15.5°C in Poises		
°C	Min	1 Day	24 Days	58 Days
71	10	25	40	95
82	10	360	534	853
95	10	570	846	—
115.5	0.5	20	30	69

[a] The data are those of Webb and Hufnagel[248] except that the values at 71°C for 10 min were estimated from references cited in the text.

Completion of the sterilizing process does not mark the end of the effects of time and temperature on the stability of the concentrated milk system. Changes continue, but at a rate dependent not only on previous heat treatments, but also on time and temperature of storage. Most products thicken with age, but evaporated milk made by the usual longhold sterilization method thins to a basic viscosity during the first few days of storage.[29] In contrast, high-temperature short-time (HTST)-sterilized milk (135°C for 30 sec) thickens until it gels in storage. The rate of thickening decreases as the processing heat increases.[8] The viscosity of sweetened condensed milk increases logarithmically with the temperature of storage and arithmetically with storage time.[249] For conditions of constant viscosity, time varies logarithmically with storage temperature. Whether the gelation of HTST-sterilized milks in storage represents an extension, at a slow rate, of the heat-coagulation reaction is uncertain.[230] Thickening during the storage of HTST-concentrated milks proceeds rapidly after an initial increase in viscosity. The more severe the heat treatment to which samples are subjected, the more stable are the milks against thickening. Samples of 2:1 milks processed in a tubular heater at 135°C for 30 sec and subsequently heated in cans at 115°C for 2, 6, 12 and 18 min, started to thicken at 32, 36, 41, and 51 weeks, respectively.[8]

Many workers[3,18,77,230,250] have shown in various ways that the onset of gelation is hastened by lessening exposure of the milk to heat during the sterilization process, as in HTST proceeding. Milks sterilized by irradiation without heat gel quickly in storage.[83,84,85] This gel-forming property of irradiated milk is probably related to the same mechanism that causes gel formation in milks processed with a minimum of heat during sterilization. There is no correlation between long-continuing fluidity and high heat stability in evaporated milk.[8] Heat appears to stabilize milk against gelation in storage more effectively when applied to milks of high solids content (26 to 34%) than to milk of normal concentration.[230,250] Heat applied to milk concentrates at a solids level of 32 to 45%, followed by dilution to 26%, imparts to the diluted milk a greater resistance toward age-thickening.[220] Thickening is retarded by low-temperature storage.[56]

There has been intensive effort to devise a sterilization process for preparation of a beverage-quality concentrated milk at the 2:1 or 3:1 level by manipulation of time and temperature of heating. The objective has been the production of a concentrate with a minimum of cooked flavor, discoloration, fat separation, thickening and staling in flavor during storage.[9,14,18,40,161,220,226,250] Of these defects, thickening and gelation have been the most obvious and objectionable. An acceptable sterile concentrate can be prepared by the use of optimum processing conditions, followed by storage at 10 to 15°C. A typical process sequence would include forewarming at 115°C for 2 min, concentration not to exceed 3:1, sterilization at 135°C for 30 sec, cooling to at least 98°C, aseptic homogenization at 4,000 psi, cooling to 20°C, and aseptic canning. Passing the uncooled canned milk through a hot-water bath at 72°C for 10 min further delays gelation but increases cooked flavor. Both flavor and physical stability are benefited by refrigerated storage. The proposal has also been made that raw milk be sterilized, then aseptically homogenized, concentrated, and packaged.[14] This procedure avoids coagulation problems during processing but does not delay storage thickening.

One process for retarding gel formation adjusts the calcium-sodium ion ratio of the milk by use of an ion-exchange resin.[219] Calbert and co-workers[18,60] have sought to stabilize HTST-processed milk by holding it after sterilization at 94°C to permit critical viscosity development. The incipient coagulum is next destroyed by homogenization and it is then not expected to reform during storage. One problem in the use of this procedure is to so adjust the degree of thickening that, when the product is finally smoothed out by homogenization, neither sediment formation nor gelation will occur in storage.

Polyphosphates effectively retard age thickening and gelation in HTST-sterilized milk concentrates and sweetened condensed milk.[132] Early work indicated these salts had a rather uncertain stabilizing effect, but optimum percentages of polyphosphate glass, having an average of 4.8 phosphorus atoms per chain, now appear to afford effective stabilization.[131]

The polyphosphates appear to slowly hydrolyze to orthophosphate, which has a thickening effect, thus retarding lactose crystallization. Much of the apparent stabilization may be due to the protein-peptizing effect of the polyphosphates. Tumerman and Guth[238] found that by the addition of 0.25 to 1.0% NaCl to milk before or after concentration, the freezing point was lowered and lactose crystallization was suppressed, giving increased stability to frozen concentrates. Samel and Muers[203] found that polyvalent anions, including phosphate, destabilize sweetened condensed milk because of the formation of insoluble calcium salts, which increase viscosity on aging. Leviton *et al.*[132] places an optimum concentration for a sterilized concentrate at 0.6 lb polyphosphate per 100 lb milk solids, thereby extending storage life of a 3:1 concentrate at 21°C from 50 to 347 days. The kind of phosphate salt used as a stabilizer is important, since the common Na_2HPO_4 and some other phosphates increase gel formation during storage.[36,132]

Kaleb *et al.*[106,107,108] have investigated the properties of milk gels obtained by heating 50% solutions of nonfat dry milk as a possible new food product. They reported that maximum gel firmness is obtained by heating the solutions to 100°C for 10 min; other factors, such as protein concentration and temperature of testing, are important in determining firmness. Compared to fibrin, the concentration of milk proteins must be much higher to form gels, and unlike gelatin, milk gels are irreversible. The nutritional value of thermally induced milk gels apparently changes little when heating at 100°C for 10 min, but higher temperatures and longer times will dramatically decrease quality.[106] Similarly, the nutritional quality of milk decreases considerably on storage of concentrated or dry milk or by heating.[42,43,167]

Salt Balance and Heat Stability of Concentrated Milks

Sommer and Hart[216,217] first showed that a critical balance between the natural acidic and basic salt components of milk appeared necessary to provide maximum stability to heat coagulation. In most cases inadequate resistance to coagulation is related to the presence of an excess of calcium and magnesium. Addition of phosphates, citrates, or carbonates to such milks was shown to improve their heat stability appreciably. It is generally assumed that the salt balance directly affects the heat stability of the casein, but this balance may operate indirectly on the casein through its effect on denaturation and interaction with the serum proteins.[163] The relation between the mineral composition of milk and its heat stability has been studied intensively, but the distribution and role of milk-serum electrolytes at high temperatures remains obscure.[44,183,198,251]

When milk is preheated and concentrated, the effect of acid and stabilizing salts may be the same or opposite to that observed in the original raw milk. Evaporated milk may respond to small changes in pH in the same manner as fluid milk, but its pH sensitivity in the presence of added β-lacto-globulin is much less than that of fluid milk.

Three types of concentrated milks are shown in the lower curves of Fig. 11.14. One kind of milk is stabilized by addition of orthophosphate or citrate; another is stabilized by addition of calcium or magnesium; and a third is destabilized by salt additions.[87,247] The three types of milk are usually derived from a raw milk, of which the curve shown in upper Fig. 11.14 is typical.

Concentrated milk III, stabilized by small percentages of $CaCl_2$ or other chloride including hydrochloric acid, is comparatively rare, apparently being secreted by from 10 to 20% of the cows of a normal herd. If a Type III raw milk is placed in storage, there will be a gradual shift in the heat stability curve of its evaporated product through Type

II to Type I, although little change may be noted in the stability curve of the unevaporated sample. The shift from Type III to Type I will be accelerated by the development of lactic acid during storage,[245] but it may also proceed without the accompaniment of a measurable change in pH.

There is no clear division among the three types of milk, since they merge into one another, and numerous curves can be obtained showing a difference in degree of variation. There may be a shift from type to type as a result of a change in forewarming temperature, as shown in Fig. 11.15.[246] Thus a milk forewarmed at 95°C was heat-stabilized by calcium after concentration. When higher forewarming temperatures were used, both calcium and phosphate destabilized the milk, but some stability was obtained at certain critical levels of phosphate.

Many different salts may be used with substantially the same results as are obtained with $CaCl_2$ and Na_2HPO_4.[247] Differences in the valency of the ions concerned generally account for the differences in ionic concentration found necessary to produce a given result. The chlorides of H, Na, K, Ca, Mg, Ba, Al and Th furnish a source of strong cations, while the sodium and potassium citrates and orthophosphates may be used when strong anions are required. In general, calcium has a greater impact than phosphate on heat stability. The addition of small percentages of phosphate to a milk to increase heat stability is useful if a phosphate-stabilized milk is encountered (curve I, Fig. 11.14). However, if the milk is stabilized by calcium (curve III), the addition of phosphate does not increase stability, but at the same time it effects no marked lowering of stability. This fact accounts for the success of phosphate salts in evaporated milk manufacture.

The addition of salts to milk before forewarming and condensing modifies the heat stability curve of evaporated milk. If a small proportion of calcium chloride is added to a fresh milk which normally has a stability curve of a type II milk, Fig. 11.14, the resulting curve is shifted toward curve I. If a phosphate salt is added to the fresh milk, the heat stability curve is shifted toward curve III. This behavior is to be expected, since it is the normal adjustment arising from an excess of either a strongly positive or negative ion for this particular milk.

From the foregoing discussion it can be seen that the effect of salts on the heat stability of milk is largely empirical but of paramount importance in the manufacture of evaporated milk. The stabilizing salts that have been approved by the Food and Drug Administration[51] are calcium chloride, sodium citrate, and disodium phosphate. The hearing record shows that these salts are needed in evaporated milk manufacture to control heat coagulation in certain types of milk and

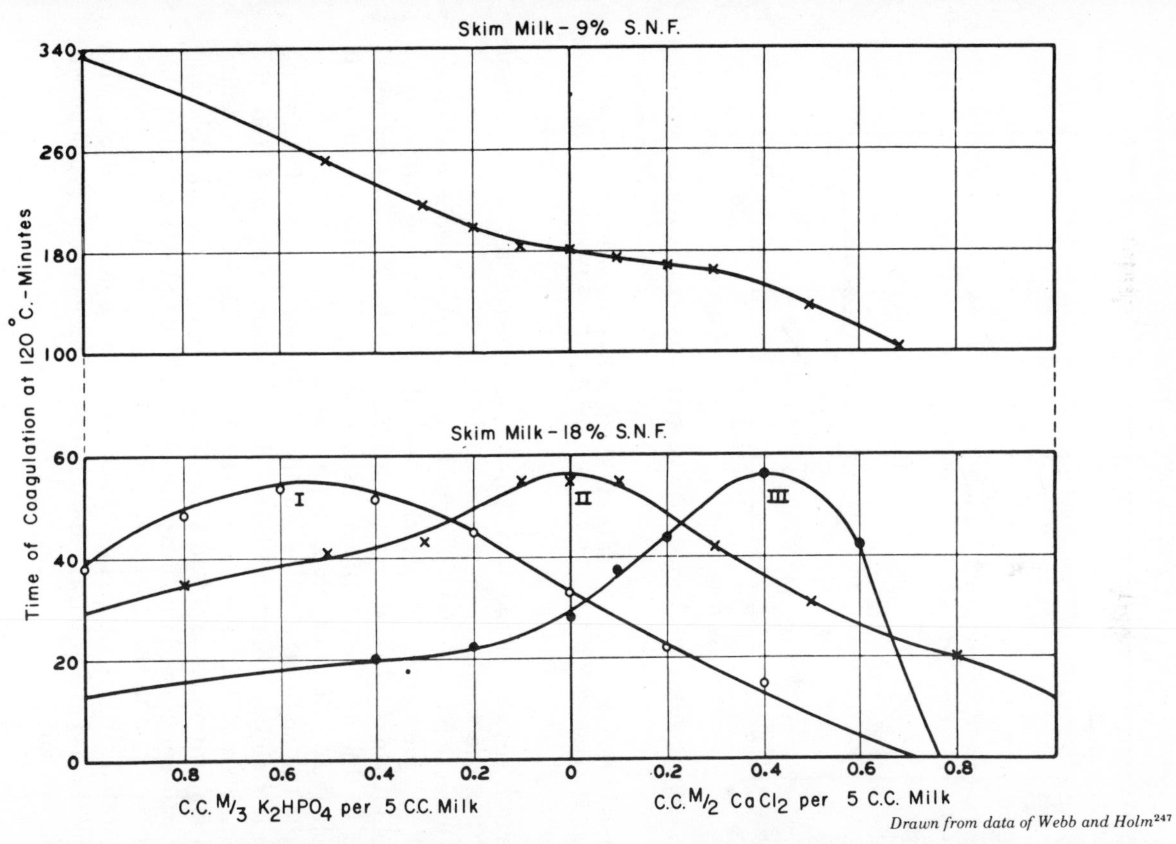

Drawn from data of Webb and Holm[247]

FIG. 11.14. THE EFFECT OF THE ADDITION OF VARYING AMOUNTS OF DIFFERENT ELECTROLYTES UPON THE HEAT STABILITY OF NORMAL AND EVAPORATED SAMPLES OF THE THREE DIFFERENT TYPES OF MILK

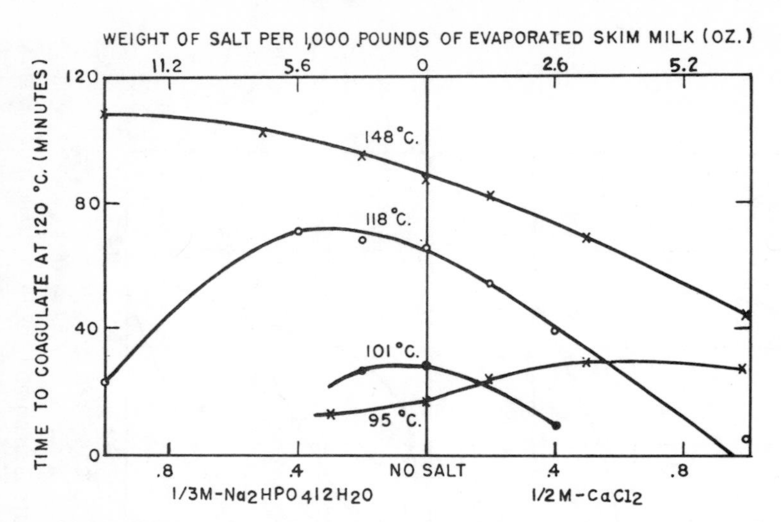

From Webb and Bell[246]

FIG. 11.15. THE EFFECT OF STABILIZING SALTS AND FOREWARMING TREATMENTS UPON THE HEAT STABILITY OF EVAPORATED SKIMMILKS

at certain seasons of the year. Their use is permitted to the extent of 0.1% of the weight of the evaporated product.

An exception to the stabilizing influence of orthophosphate has been noted. Stewart *et al.*[220] advocate use of forewarming rather than phosphates to obtain stabilization in HTST sterilization. The principal defect in this product, gel formation in storage, can be minimized if phosphate addition is avoided in processing.

While the addition of salts to milk is a simple method of changing its heat stability, this can also be done by removing ions from the system. Skimmilk treated to remove 60% of its calcium may be added in an amount equal to 0.5 to 2.5% of the batch of original milk, and the mixture processed into evaporated milk.[168,169] Stabilization by this means is equal to the use of from 2 to 7.3 oz of disodium phosphate. Evaporated milk of 40% solids can be stabilized against heat coagulation by the use of a mineral ion-exchanged milk as a 10% replacement for normal solids.[110]

Effect of Nonionic Substances

Nonionic and inert materials, such as fats, starches, sugars, and vegetable pulps, usually promote protein coagulation during heating of milk systems in which they are dispersed. Suspended particles appear to adsorb protein, finer dispersions tending to concentrate the protein with increasing effectiveness. This accumulation of protein on the interphase tends to lower protein stability by encouraging local coagulation, which quickly destabilizes the entire system. Ground casein and vegetable pulps, decolorizing carbon, and ground filter paper are among the substances that have been shown to lower heat stability.[253] Sugars, including lactose, sucrose, and dextrose, usually promote heat coagulation, although dextrose under certain conditions exerts a stabilizing influence.[253] Lactose removal from concentrated milk by dialysis, crystallization and centrifugation, or hydrolysis has been shown to effectively prevent age-thickening.[41]

The adverse effect of sugars and starches on the heat stability of milk protein systems is significant in the manufacture of certain foods processed at high temperatures. Sterilized cream-style soups, sauces, and creamed vegetables must have a smooth body, free from lumps and visible curd. Wheat flour, corn or potato starch are common thickeners which promote coagulation of the protein. Since coagulation in most products of this type cannot be avoided, the material must be handled so that a soft, smooth, gel type of coagulum is produced. Milkfat in its normally dispersed state affects the heat stability of the system but slightly.[86] Homogenization, however, increases the dispersion of the fat and creates new surfaces for adsorption, simultaneously concentrating the protein on these surfaces, forming a fat-protein complex. Clumping of the newly formed globules often occurs, and this further concentrates and destabilizes the protein. It follows, therefore, that those factors which influence the degree of fat clumping will also modify the heat stability of the product. The effect of homogenization on heat stability is slight at low fat levels, but becomes appreciable with increasing fat content.[96]

The specific conditions of homogenization have an important bearing on the heat stability of fluid milk products. Preheating temperature, homogenization pressure, fat and solid-not-fat concentrations, and salt equilibria were shown years ago to affect the fat clumping and coagulability of milks and creams subjected to homogenization.[246A] The general nature of these relationships is shown in Fig. 11.16. Rehomogenization, or the use of a second-stage valve, wherein the second pressure is lower than the first, breaks up the larger clumps and consequently increases heat stability over that observed after a single homogenization.[141] Increasing homogenization temperatures favor heat stability of concentrated milks when homogenization is practiced before forewarming and condensing.[141]

The feathering of homogenized creams when added to coffee is a form of heat coagulation.[246A] Coffee cream is usually homogenized to retard fat separation and to impart a smooth, creamy body. The effect of homogenization on the heat stability of 20 and 30% cream is shown in Fig. 11.16. The salt content of the cream or of the coffee, chiefly the presence of relatively high levels of calcium in either, is an important factor.[236] Acid development before or after processing quickly renders cream susceptible to feathering. King[114] lowered the feathering value of cream by adding to it 5% of skimmilk treated with an ion-exchange resin to lower its calcium content.

In the processing of milk, coffee cream, and evaporated milk, the destabilizing effect of homogenization is incidental to the primary objective of retarding fat separation in storage. The interrelationships of

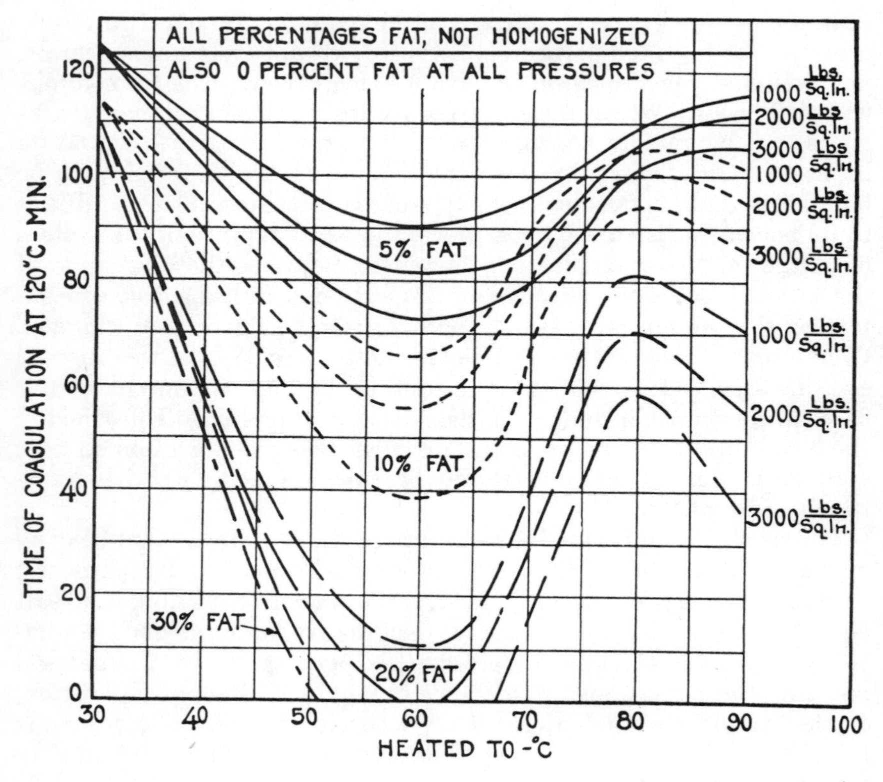

From Webb and Holm[246A]

FIG. 11.16. VARIATIONS OF COAGULATION TIME OF MILK OF NORMAL SOLIDS-NOT-FAT CONTENT WITH CHANGES IN FOREWARMING TEMPERATURE, FAT CONTENT, AND HOMOGENIZATION PRESSURES

No homogenization below 50°C.

viscosity, heat stability, conditions of homogenization, and storage on fat separation in evaporated milk have been studied. Trout has reviewed work on the homogenization of milk from all aspects.[237]

Solubility of Dried Milks

Effect of Heat.—The native properties of milk components are substantially unmodified by moderate milk-drying conditions. In freeze-dried milk, the equilibrium ratio of α- and β-lactose and the salt distribution remain essentially intact. The normal size dispersion of the caseinate phase and its clottability by rennin are substantially recovered on reconstitution of the dried product. Depending on the preheat conditions, dryer design and temperature of operation, the properties of spray-dried powder may vary significantly. The initial temperature of an evaporating milk droplet, in a spray dryer with parallel air flow, does not appreciably exceed the unit bulk temperature, and can be effectively held to temperatures below 60°C.[24,140] As the falling rate period is approached in the course of further evaporation, the temperature rises to a final value determined by the final temperature of the drying gas and the residence time in the dryer. Under properly controlled spray-drying conditions, the changes in milk protein structure and solubility are minor. Spray drying does not denature the serum protein significantly, and the level of serum-protein denaturation in dry milk is substantially equal to that of the condensed milk from which it is processed.[68]

Loss of solubility of dry milk during processing or storage is largely a manifestation of changes in the stability of the caseinate-phosphate complex. The solubility of spray-dried powder is unaffected by preheat treatments of the fluid milk that cause extensive serum-protein denaturation.[138] Exposure of dry milk to a dry-heat treatment sufficient to cause total insolubilization of the caseinate phase may have negligible effect on the serum protein. In a maximally insolubilized nonfat dry milk, the total casein, comprising 35% of the solids, may be rendered insoluble, while the solubility of the serum components remains intact. Jenness and Coulter[100] suggest that preheat may exert a stabilizing influence against coagulation during drying, similar to the benefit of preheat in evaporated-milk sterilization.

While preheat conditions do not significantly affect the initial solubility of dried whole milk, powders processed from high-preheat milk tend to develop insolubility, during subsequent storage at 37°C, more rapidly than low-preheat powders. The free fat in dry whole milk decreases somewhat on storage and appears to be related to solubility decrease and film formation in the reconstituted milk.[135]

Heat damage in nonfat dry milk can be reduced in the spray-drying step. Atomizing equipment which gives the narrowest possible particle size distribution is superior, since in sprays having a broad size distribution the solids from smaller droplets tend to burn long before sufficient water is removed from the larger drops. Also, condensing equipment that can only operate efficiently with one or more stages held at 85°C or higher is undesirable, unless it can be demonstrated that the holding time of the milk is short.[170]

Under normal spray-drying conditions the casein solubility is spared. At excessively high exit-air temperatures, the casein is spray-dried milk powders may insolubilize at a rate that is approximately logarithmic with the temperature.[235] The conditions that prevail during roller drying of milk alter the solubility of the casein extensively. The film of milk on the heated drum has a residence time of several seconds, during which it attains progressively increasing solids concentration upon continuous evaporation. At solids level above 60%, the casein is particularly vulnerable to the high temperature of the drum surface. Coagulation is essentially instantaneous at solids concentrations exceeding 80%. Wright[256] determined that each increase of 1% in milk solids lowered by 1°C the temperature required to effect a constant degree of insolubilization for a fixed heating period. The maximum quantity of protein rendered insoluble by any moist heat treatment approximates 75% of the total. At a fixed moisture level, the time required to produce a constant degree of insolubility is a logarithmic function of the temperature. This relationship is shown in Fig. 11.17. The rate of protein insolubilization increases nearly fivefold for each 10°C rise in temperature. The slope of this curve is similar to that derived from measurements on the heat stability of evaporated milk. Similarly, the coagulation time for milk powder at 86.7% solids is extrapolable from the heat-stability curves of concentrated milks of 28 to 38% solids levels, suggesting that the heat coagulation of evaporated milk and the insolubility of dried milk may have a common origin. At moisture levels below 13%, the rate of insolubilization of heated milk is sharply reduced. Nevertheless, considerable solubility loss is sustained by milk powder with a residual moisture below 3%, when subjected to dry-heat treatment (see Fig. 11.17).

The "solubility value" of the protein in milk powder is a function of the temperature and energy applied to effect its redispersion in water to colloidal dimensions.[91] The solubility of dry-heated powder varies with the temperature of reconstitution, whereas the protein in high-moisture powders is irreversibly insolubilized.[92,256] With increasing temperature of reconstitution, full solubility of the protein

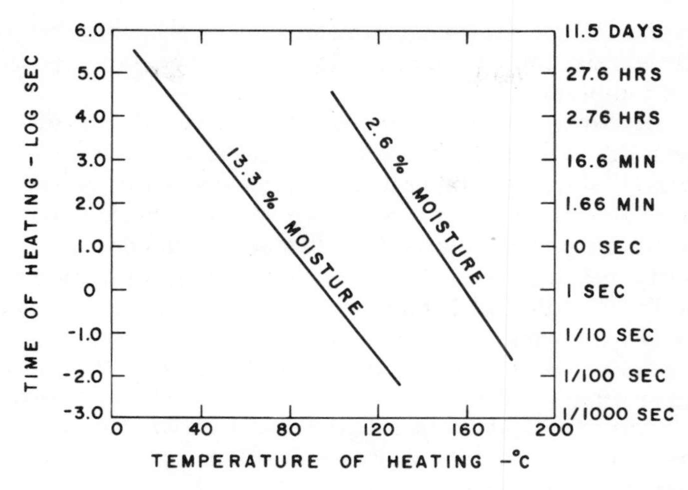

From data by N. C. Wright[256]

FIG 11.17. THE TIME-TEMPERATURE RELATIONSHIP FOR THE DEVELOPMENT OF 50% PROTEIN INSOLUBILITY IN NONFAT MILK POWDER HEATED AT 2.6 and 13.3% MOISTURE

Solubility measured at 20°C.

in dry-heated milk can be recovered. The solubility of milk powder in 20 and 50°C water is therefore a useful criterion for differentiating dry heat from moist heat insolubility. Howatt and Wright[91,92] demonstrated that protein insolubility in roller-dried milk is partially reversed at elevated temperatures of reconstitution. The momentary final contact of the dry-milk film with the hot surface of the drum during roller drying, therefore, accounts for a significant proportion of the total insolubility of roller-dried milk powder. Wright[256] considers dry-heat insolubility a physical, rather than a chemical, change in the casein, since the rate of insolubilization is directly proportional to the time of heating, and is independent of the quantity of unchanged casein. The casein in milk powder treated with absolute alcohol undergoes similar insolubilization, that is also reversible at higher temperatures of reconstitution. Dehydration of critical water, such as loss of water of imbibition, is therefore suggested as a possible cause of dry-heat insolubility.

The solubility changes in dry-heated milk powder are wholly related to destabilization of the caseinate complex, with no detectable alteration in the serum protein and nonprotein nitrogen components.[256] The insolubles of dry milk manufactured by different processes are generally of similar composition and comprise casein together with calcium and phosphorus in a ratio suggesting tricalcium phosphate.[246] In roller-dried whole milk, some fat appears to be associated with the insoluble caseinate complex.

Effect of Storage Conditions on Stability of Dried Milks.—The stability of dry milk during storage is critically affected by moisture content and storage temperature. High moisture levels, due to inadequate dehydration or reabsorption of atmospheric moisture, promote insolubilization at relatively mild storage temperatures. The rate of solubility loss is a function of both moisture concentration and temperature.[25,26,75,76] Below 5% moisture, solubility changes are relatively insignificant at normal storage temperatures. In a study of the solubility changes in hydrated nonfat dry milk, Henry *et al.*[75,76] found that the temperature coefficient of the reaction leading to insolubility exceeds a value of 5 at temperatures above 20°C. Consequently, the solubility of moist powders may remain unchanged for long storage periods at 20°C, but falls rapidly at 37°C. The solubility of nonfat dry milk falls as the humidity, time or temperature is increased. Similarly, the nutritional value of nonfat dry milk has been shown to decrease on storage.[42,43,167] The insolubility is reversible on reconstitution in 50°C water, resembling the differential solubility of dry heated milk powder or roller-dried powder. The initial solubility of nonfat dry milk can be retained for 700 days at 37°C storage if the moisture level does not exceed 4.7%.

The occurrence of solubility loss in powder of 7.6% moisture is coincident with crystallization of the lactose, as a result of which the equilibrium relative humidity increases from 42 to 55%. The change in activity of the water due to lactose crystallization may contribute to the overall reactions leading to solubility loss. The change in stability is nonbacterial, and the rate appears to be unrelated to the gas atmosphere in the container. The insoluble component is predominantly casein, which can be totally insolubilized during prolonged storage at 37°C, in powder hydrated to 7.6% moisture. Solubility of the lactalbumin and lactoglobulin components is similarly impaired. Changes in the distribution of soluble nitrogen in hydrated milk powder are shown in Fig. 11.18.[76]

A number of significant changes occur in high-moisture milk powder, concurrent with the loss of solubility. The lactose is gradually bound by the protein, accompanied by a parallel reduction of free amino nitrogen. The pH decreases steadily, and the characteristic changes associated with the Maillard reaction between sugars and amino nitrogen become evident, including development of brown discoloration and production of carbon dioxide, reducing substances, and fluorescing compounds. Stale and caramelized flavors also develop rapidly in milk powder under conditions of high humidity and elevated storage temperatures, and a significant loss in biological value of the protein is incurred. Changes in nonfat dry milk hydrated to 7.6% moisture and aged 100 days at 37°C include a nearly total loss of protein solubility, a pH decrease of 0.4 unit in the reconstituted milk, crystallization of 80% of the lactose, and destruction of 70% of the original amino nitrogen and approximately 6% of the lactose.[76]

Most of the deteriorative changes in moist milk powder are attributed to a 1 : 1 interaction between the free amino groups of the milk proteins, largely the ε-amino groups of the lysine residues and the potential aldehyde group of lactose.[76] The initial complex is soluble and colorless,

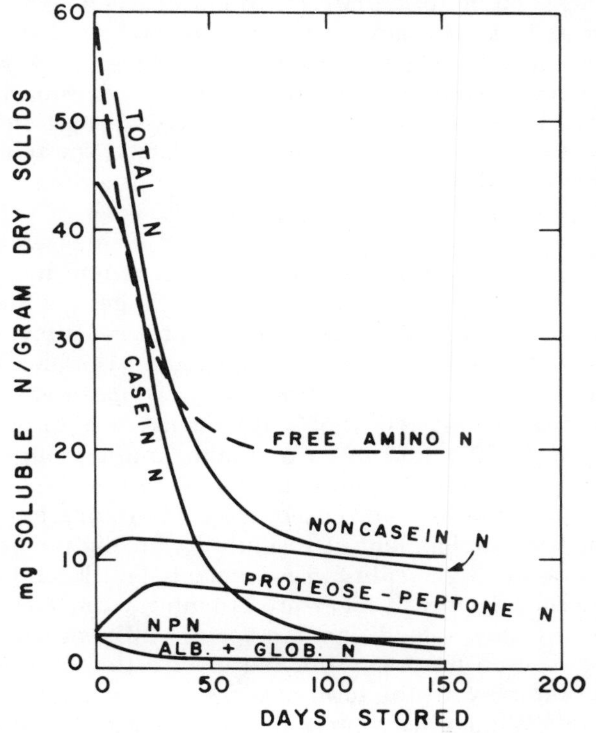

Data from Henry et al.[76]

FIG. 11.18. CHANGES IN THE DISTRIBUTION OF THE SOLUBLE NITROGEN IN NONFAT MILK POWDER CONTAINING 7.3% MOISTURE, STORED AT 37°C

Simultaneous changes in free amino nitrogen, determined by the Van Slyke method and expressed as mgm per gm of protein nitrogen, are shown in the dashed curve.

and as much as 65 to 75% of the reactive amino nitrogen may be bound before appreciable solubility or color change becomes evident.[76,126] The reaction rate is largely determined by the activity of the moisture. The temperature coefficient exceeds 6 for milk powders of 7.6% moisture, decreasing to a value of 2 at moisture levels below 5%. The progressive loss of amino nitrogen during storage is paralleled by a decrease in lactose and an increase in the weight of the nondialyzable fraction. The complexed sugar becomes irreversibly bound and, through a series of undefined degradative changes, yields the brown pigments characteristic of the Maillard reaction. The reaction in moist powder consumes a maximum of 70% of the total free amino nitrogen as determined by Van Slyke analysis.

While the loss of protein solubility in dry milk is generally attributed to the sugar-protein interaction, the mechanism is obscure. Henry *et al.*[76] speculate that insolubility may finally arise from induced denaturation of the protein molecule as a consequence of its complex formation with lactose, or subsequent degradation reactions.

On the basis of amino nitrogen binding capacity, the relative order of reactivity of the various sugars toward casein is xylose > arabinose > glucose > maltose > lactose > fructose.[133] Lea *et al.*[125,127,128] established that the ϵ-amino group of lysine is the primary reaction site in the casein-glucose system. The decrease in amino nitrogen is at a maximum in the 65 to 70% relative humidity range, and the temperature coefficient at 15 to 25°C is 5.4. The greater reactivity of glucose induces more rapid insolubilization of the caseinate complex in addition to intense discoloration. In freeze-dried preparations of various sugar-milk protein mixtures stored at 55% R.H., sucrose is unreactive, but glucose inactivates 70 to 80% of the available amino groups, although at a concentration of only 1/6 equivalents.[126]

The interaction of lactose and casein is a contributing but not essential factor in the development of insolubility in high-moisture milk powder. The already high acidity in concentrated milk solids is further increased as a consequence of sugar-protein interaction, and this secondary change may have an additional important influence on solubility. Insufficient attention has been directed toward the observation that the caseinate complex, in the total absence of sugar, will insolubilize at 55% R.H.[59,74] While glucose appears to accelerate the development of insolubility, casein in the absence of sugar may insolubilize more rapidly than casein in the presence of lactose. Furthermore, sucrose protects the solubility of the casein, and the insolubilizing effect of glucose in the presence of sucrose or lactose is largely suppressed.

From studies on water absorption by milk powders and their components, Berlin *et al.*[10] have shown that the first water absorbed at low relative humidity is by the milk proteins. As the vapor pressure increases, lactose becomes the principal water-absorption site until sufficient is acquired to convert it into the nonhygroscopic α-hydrate form. At relative humidities above 50%, the lactose binds little water, but rapid hydration of the salts occurs. It may be during this stage that the concentration of the phosphates, particularly the potassium form, in the salts becomes high enough to destabilize the casein micelles and render them insoluble.

Lactose is actually essential to the stability of the caseinate complex under many conditions, and its crystallization in milk products during frozen storage is highly detrimental to solubility. Furthermore, Gerlsma[59] observed that the caseinate complex, substantially freed from lactose by centrifugal separation, loses approximately half of its capacity to redisperse on spray drying. Restoration of lactose, or addition of glucose, sucrose, or sorbitol, effectively protects the solubility of the caseinate during spray drying. Therefore, crystallization of lactose in high-moisture milk powder must exert a more substantial influence on the deterioration of protein solubility than that attributed merely to its effect on the activity of the water in the system. As a protective factor, lactose conceivably moderates the destabilizing influence of calcium ion, either by direct complex formation[78] or by lowering the ionizing power of the calcium salts in its solutions.[239] The concentrated, amorphous lactose matrix in moist milk powder, prior to its crystallization, may further act as a physical barrier against micellar aggregation, as in frozen concentrated milk, thereby preserving the ability of colloid to redisperse on reconstitution. Interaction between lactose and casein, particularly under conditions involving extensive degradation of the sugar-protein complex, would inevitably supplement the deterioration of casein solubility.

REFERENCES

1. ASCHAFFENBURG, R., J. Dairy Res., *35,* 447 (1968).
2. BABCOCK, S. M., and RUSSEL, H. L., Wisconsin Agri. Exp. Sta., 14th Ann. Rep. 161 (1897).
3. BALL, C. O., Food Res., *20,* 351 (1955).
4. BEEBY, R., XVII Int. Dairy Cong. B 95 (1966).
5. BEEBY, R., HILL, R. D., and SNOW, N. S., "Milk Proteins." Vol. II, p. 422, H. A. McKENZIE, ed., Academic Press, Inc., New York, 1971.
6. BELEC, J., and JENNESS, R., J. Dairy Sci., *44,* 1163 (1961).
7. BELL, R. W., J. Biol. Chem., *64,* 391 (1925).
8. BELL, R. W., CURRAN, H. R., and EVANS, F. R., J. Dairy Sci., *27,* 913 (1944).
9. BELL, R. W., and WEBB, B. H., J. Dairy Sci., *26,* 579 (1943).
10. BERLIN, E., ANDERSON, B. A., and PALLANSCH, M. J., J. Dairy Sci., *51,* 1912 (1968).

11. BOOGAERDT, J., Nature, *174,* 884 (1954).
12. BOULET, M., XVII Int. Dairy Cong. B 111 (1966).
13. BOULET, M., YANG, A., and RIEL, R. R., Can. J Biochem., *48,* 816 (1970).
14. BOYD, J. M., U.S. Patent 2,827,381, March 18, 1958.
15. BURG, P., VAN DER, Neth. Milk Dairy J., *1,* 69 (1947).
16. BURTON, H., J. Dairy Res., *21,* 194 (1954).
17. CALAPAJ, G. G., J. Dairy Res., *35,* 1 (1968).
18. CALBERT, H. E., and SWANSON, A. M., XV Int. Dairy Cong. Proc., *1,* 442 (1959).
19. CARROLL, R. J., THOMPSON, M. P., and MELNYCHYN, P., J. Dairy Sci., *54,* 1245 (1971).
20. CARROLL, R. J., THOMPSON, M. P., and NUTTING, G. C., J. Dairy Sci., *51,* 1903 (1968).
21. CHEN, J. H., and LEDFORD, R. A., J. Dairy Sci., *54,* 763 (1971).
22. CHRISTIANSON, G., JENNESS, R., and COULTER, S. T., Anal. Chem., *26,* 1923 (1954).
23. COLE, W. C., and TARASSUK, N. P., J. Dairy Sci., *29,* 421 (1946).
24. COULTER, S. T., J. Dairy Sci., *38,* 1180 (1955).
25. COULTER, S. T., JENNESS, R., and CROWE, L. K., J. Dairy Sci., *31,* 986 (1948).
26. COULTER, S. T., JENNESS, R., and GEDDES, W. F., Adv. Food Res., *3,* 45; E. M. Mrak and G. F. Stewart, ed., Academic Press, Inc., New York, 1951.
27. DAHLBERG, A. O., and GARNER, H. S., U.S. Dept Agri. Bull. 944 (1921).
28. DAVIES, D. T., and WHITE, J. C. D., J. Dairy Res., *27,* 171 (1960).
29. DEYSHER, E. F., WEBB, B. H., and HOLM, G. E., J. Dairy Sci., *27,* 345 (1944).
30. DICKSON, I. R., and PERKINS, D. J., Biochem. J., 113, 7P (1969).
31. DOWNEY, W. K., MURPHY, R. F., and AHERNE, S. A., Biochem J. *115,* 21P (1969).
32. DUPONT, M., Biochim. Biophys. Acta, *94,* 573 (1965).
33. DUPONT, M., Biochim. Biophys. Acta, *102,* 500 (1965).
34. DUMONT, P., Éc. Nationale vét. Lyon: Thèse No. 19 (1965). Cited in DSA *28,* 2001 (1966).
35. D'YACHENKO, P. F., and ALEKSSVA, N. Yu, Trudy vses. Nauchno-issled. Inst. moloch Prom., *27,* 3 (1970). Cited in DSA *33,* 472 (1971).
36. EDMONDSON, L. F., J. Dairy Sci., *42,* 910 (1959).
37. EDMONDSON, L. F., AVANTS, J. K., DOUGLAS, F. W., JR., and EASTERLY, D. G., J. Dairy Sci., *49,* 708 (1966).
38. EDMONDSON, L. F., and TARASSUK, N. P., J. Dairy Sci., *39,* 123 (1956).
39. EGGMAN, H., Milchwissenschaft, *24,* 479 (1969).
40. ELLERTSON, M. E., and PEARCE, S. J., U.S. Patent 2,822,277, Feb. 4, (1958).
41. EL-NEGOUMY, A. M., and BOYD, J. C., J. Dairy Sci., *48,* 23 (1965).
42. ERBERSDOBLER, H., Milchwissenschaft, *25,* 280 (1970).
43. ERBERSDOBLER, H., Fraftfutter, *53,* 172, 228 (1970).
44. EVENHUIS, N., Neth. Milk Dairy J., *11,* 225 (1957).
45. EVENHUIS, N., and VRIES, T. R. DE, Neth. Milk Dairy J., *9,* 146 (1955).
46. EVENHUIS, N., and VRIES, T. R. DE, Neth. Milk Dairy J., *10,* 101 (1956).
47. FEAGAN, J. T., GRIFFIN, A. T., and LLOYD, G. T., J. Dairy Sci., *49,* 933 (1966).
48. FEAGAN, J. T., GRIFFIN, A. T., and LLOYD, G. T., J. Dairy Sci., *49,* 940 (1966).
49. FEAGAN, J. T., GRIFFIN, A. T., and LLOYD, G. T., J. Dairy Sci., *49,* 1010 (1966).
50. FERRETTI, A., FLANAGAN, V. P., and RUTH, J. M., J. Agr. Food Chem., *18,* 13 (1970).
50A. FERRETTI, A., FLANAGAN, V. P., J. Agr. Food Chem., *19,* 245 (1971).
51. Food and Drug Admin., Fed. Register, *5,* 2442 (July 2, 1940).
52. FOX, K. K., Proc. Sixth Milk Concen. Conf., Davis, Calif. 45 (1963).
53. FOX, K. K., HOLSINGER, V. H., POSATI, L. P., and PALLANSCH, M. J., J. Dairy Sci., *50,* 1032 (1967).
54. FOX, K. K., HOLSINGER, V. H., POSATI, L. P., and PALLANSCH, M. J., J. Dairy Sci., *50,* 1032 (1967).
55. FREIMUTH, U., and KRAUSE, W., Nahrung, *12,* 597 (1968).

56. GAMMACK, D. B., and WECKEL, K. G., J. Dairy Sci., *40,* 1570 (1957).
57. GARDINER, M. R., and MUNCH-PETERSON, E., Australian J. Dairy Tech., *20,* 171 (1965).
58. GEORGES, C., GUINAND, S., and TONNELAT, J., Biochim. Biophys. Acta, *59,* 737 (1952).
59. GERLSMA, S. Y., Neth. Milk and Dairy J., *11,* 83 (1957).
60. GIROUX, R. N., CALBERT, H. E., and SWANSON, A. M., J. Dairy Sci., *41,* 710 (1958).
61. GOUGH, P., and JENNESS, R., J. Dairy Sci., *45,* 1033 (1962).
62. GOULD, I. A., J. Dairy Sci., *28,* 379 (1945).
63. GOULD, I. A., and FRANTZ, R. S., J. Dairy Sci., *29,* 27 (1946).
64. GRAY, D. M., and SCHALM, O. W., J. Am. Vet. Med. Assoc., *136,* 195 (1960).
65. GRIMBLEBY, F. H., J. Dairy Res., *21,* 207 (1954).
66. HABEEB, A. F. S. A., Can. J. Biochem. Physiol., *38,* 269 (1960).
67. HARLAND, H. A., and ASHWORTH, U. S., J. Dairy Sci., *28,* 879 (1945).
68. HARLAND, H. A., COULTER, S. T., and JENNESS, R., J. Dairy Sci., *35,* 363 (1952).
69. HARLAND, H. A., COULTER, S. T., and JENNESS, R., J. Dairy Sci., *38,* 858 (1955).
70. HARLAND, H. A., COULTER, S. T., TOWNLEY, V. H., and JENNESS, R., J. Dairy Sci., *36,* 568 (1953).
71. HARPER, W. J., ROBERTSON, J. A., JR., and GOULD, I. A., J. Dairy Sci., *43,* 1850 (1960).
72. HARTMAN, G. H., JR., and SWANSON, A. M., J. Dairy Sci., *48,* 1161 (1965).
73. HARTMAN, G. H., JR., SWANSON, A. M., and STAHMANN, M. A., J. Dairy Sci., *48,* 780 (1965).
74. HAUGAARD, G., TUMERMAN, L., and SILVESTRI, H., J. Am. Chem. Soc., *73,* 4594 (1951).
75. HENRY, K. M., KON, S. K., LEA, C. H., and SMITH, J. D. H., 12th Int. Dairy Cong. Proc. *2,* 166 (1949).
76. HENRY, K. M., KON, S. K., LEA, C. H., and WHITE, J. C. D., J. Dairy Res., *15,* 292 (1948).
77. HERREID, E. O., Rept. Proc. Conference on Milk Concentrates, Eastern Utilization and Research Div., U.S. Dept. Agr., Philadelphia, p. 38 (1955).
78. HERRINGTON, B. L., J. Dairy Sci., *17,* 805 (1934).
79. HETRICK, J. H., and TRACY, P. H., J. Dairy Sci., *33,* 410 (1950).
80. HERSKOVITS, T. T., Biochemistry, *5,* 1018 (1966).
81. HILGEMANN, M., and JENNESS, R., J. Dairy Sci., *34,* 483 (1951).
82. HOAGLAND, P. D., THOMPSON, M. P., and KALAN, E. B., J. Dairy Sci., *54,* 1103 (1971).
83. HOFF, J. E., SUNYACH, J., PROCTOR, B. E., and GOLDBLITH, S. A., Food Tech., *14,* 24 (1960).
84. HOFF, J. E., SUNYACH, J., PROCTOR, B. E., and GOLDBLITH, S. A., Food Tech., *14,* 27 (1960).
85. HOFF, J. E., WERTHEIM, J. H., ROYCHOUDHURY, R. N., DEOLALKER, S. T., PROCTOR, B. E., and GOLDBLITH, S. A., Food Tech., *12,* 648 (1958).
86. HOLM, G. E., DEYSHER, E. G., and EVANS, F. R., J. Dairy Sci., *6,* 556 (1923).
87. HOLM, G. E., WEBB, B. H., and DEYSHER, E. F., J. Dairy Sci., *15,* 331 (1932).
88. HOSTETTLER, H., and IMHOF, K., Milchwissenschaft, *6,* 351, 400 (1951).
89. HOSTETTLER, H., and IMHOF, K., XIII Int. Dairy Cong. Proc., *2,* 423 (1953).
90. HOSTETTLER, H., and IMHOF, K., Landwirtsch. Jahrb. Schweiz., *66,* 307 (1952).
91. HOWAT, G. R., SMITH, J. A. B., WAITE, R., and WRIGHT, N. C., J. Dairy Res., *10,* 498 (1939).
92. HOWAT, G. R., and WRIGHT, N. C., J. Dairy Res., *4,* 265 (1933).
93. HOWAT, G. R., and WRIGHT, N. C., J. Dairy Res., *5,* 236 (1934).
94. HOWAT, G. R., and WRIGHT, N. C., Biochem. J., *28,* 1336 (1934).
95. HOWAT, G. R., and WRIGHT, N. C., Biochem. J., *30,* 1413 (1936).

96. HUMBERT, E. S., BRUNNER, J. R., and TROUT, G. M., Food Tech., *10,* 134 (1956).
97. HUNZIKER, H. G., and TARASSUK, N. P., J. Dairy Sci., *48,* 733 (1965).
98. HUTTON, J. T., and PATTON, S., J. Dairy Sci., *35,* 699 (1952).
99. IKAI, A., and TANFORD, C., Nature, *230,* 100 (1971).
100. JENNESS, R., and COULTER, S. T., Quartermaster Food and Container Institute. Surveys of Prog. on Military Subsistence. Problems Ser. I. Food Stability No. 6, Dry Milk Products, p. 17 (1955).
101. JENNESS, R., and KOOPS, J., Neth. Milk Dairy J., *16,* 153 (1962).
102. JENNESS, R., MORR, C. V., and JOSEPHSON, R. V., J. Dairy Sci., *49,* 712 (1966).
103. JENNESS, R., and PARKASH, S., J. Dairy Sci., *50,* 952 (1967).
104. JOSEPHSON, R. V., Diss. Abstr., *32,* 359-B (1971).
105. KADABA, L. R., Ph.D. Thesis, Univ. of Illinois, Urbana (1957). Cited in DSA *17,* [2971] (1957).
106. KALEB, M., ANDERSON, G. H., and SIMS, R. P. A., J. Dairy Sci., *55,* 1073 (1972).
107. KALEB, M., EMMONS, D. B., and VOISEY, P. W., J. Dairy Sci., *54,* 638 (1971).
108. KALEB, M., VOISEY, P. W., and EMMONS, D. G., J. Dairy Sci., *54,* 178 (1971).
109. KAMINOGAWA, S., YAMUCHI, K., and TSUGO, T., Jap. J. Zootech. Sci., *40,* 559 (1969).
110. KEENEY, M., and JOSEPHSON, D. V., J. Dairy Sci., *30,* 539 (1947).
111. KERN, J., WEISER, H. H., HARPER, W. J., and GOULD, I. A., J. Dairy Sci., *37,* 904 (1954).
112. KIEFERLE, F., and GLOETZL, J., Milch. Forsch., *11,* 62 (1930).
113. KIERMEIER, F., and SEMPER, G., Z. Lebensmittel. U. Forsch., *111,* 282 (1960).
114. KING, N., XII Int. Dairy Cong. Proc., *2,* 36 (1949).
115. KOMETIANI, P.A., Milch. Forsch., *12,* 433 (1931).
116. KOYAMA, S., SUZUKI, K., INOURA, N., Bull. Coll. Agr. vet. Med. Nihon. Univ., *19,* 9 (1964). Cited in DSA *27,* 1894 (1965).
117. KNOOP, E., and WORTMANN, A., Milchwissenschaft, *15,* 273 (1960).
118. KRESHECK, G. C., and HARPER, W. J., XVII Int. Dairy Cong. B277 (1966).
119. LARSON, B. L., and HAGEMAN, E. C., J. Dairy Sci., *46,* 14 (1963).
120. LARSON, B. L., and JENNESS, R., J. Dairy Sci., *33,* 896 (1950).
121. LARSON, B. L., and JENNESS, R., J. Am. Chem. Soc., *74,* 3090 (1952).
122. LARSON, B. L., JENNESS, R., GEDDES, W. F., and COULTER, S. T., Cereal Chem., *28,* 351 (1951).
123. LARSON, B. L., and ROLLERI, G. D., J. Dairy Sci., *38,* 351 (1955).
124. LARSON, B. L., and TWAROG, J. M., J. Dairy Sci., *44,* 1843 (1961).
125. LEA, C. H., and HANNAN, R. S., Biochim. Biophys. Acta, *3,* 313 (1949).
126. LEA, C. H., J. Dairy Res., *15,* 369 (1948).
127. LEA, C. H., and HANNAN, R. S., Biochim. Biophys. Acta, *4,* 518 (1950).
128. LEA, C. H., and HANNAN, R. S., Nature, *165,* 438 (1950).
129. LESLIE, J., BUTLER, L. G., and GORIN, G., Arch. Biochem. Biophys., *99,* 86 (1962).
130. LEVITON, A., and PALLANSCH, M. J., J. Dairy Sci., *44,* 633 (1961).
131. LEVITON, A., PALLANSCH, M. J., and WEBB, B. H., XVI Int. Dairy Cong. Proc. B: Sec. V.2, 1009 (1962).
132. LEVITON, A., VESTAL, J. H., VETTEL, H. E., and WEBB, B. H., XVII Int. Dairy Cong. Proc. E, 133 (1966).
133. LEWIS, V. M., and LEA, C. H., Biochim. Biophys. Acta, *4,* 532 (1950).
134. LIN, S. H. C., DEWAN, R. K., BLOOMFIELD, V. A., and MORR, C. V., Biochemistry, *10,* 4788 (1971).
135. LITMAN, I. I., and ASHWORTH, U. S., J. Dairy Sci., *40,* 403 (1957).
136. LYSTER, R. L. J., J. Dairy Res., *37,* 233 (1970).
137. MANNING, P. B., HEINSELMAN, A. L., JENNESS, R., and COULTER, S. T., J. Dairy Sci., *52,* 886 (1969).
138. MANUS, L. J., and ASHWORTH, U. S., J. Dairy Sci., *31,* 935 (1948).

139. MARIER, J. R., TESSIER, H., and ROSE, D., J. Dairy Sci., *46,* 373 (1963).
140. MARSHALL, W. R., JR., Quartermaster Food and Container Institute Surveys of Progr. on Military Subsistence, Problems Ser. I. Food Stability, No. 6, Dry Milk Products Symposium, p. 90 (1955).
141. MAXCY, R. B., and SOMMER, H. H., J. Dairy Sci., *37,* 1061 (1954).
142. MCGANN, T. C. A., and PYNE, G. T., J. Dairy Res., *27,* 403 (1960).
143. MCKENZIE, G. H., NORTON, R. S., and SAWYER, W. H., J. Dairy Res., *38,* 343 (1971).
144. MCKENZIE, H. A., "Milk Protein," Adv. Protein Chem., *22,* 55 (1967).
145. MCKENZIE, H. A., ed. "Milk Proteins." Vol. 2. Academic Press, Inc. New York, 1971.
146. MCMEEKIN, T. L., J. Milk Food Tech., *15,* 57 (1952).
147. MACKINLAY, A. G., and WAKE, R. G., Biochim. Biophys. Acta, *104,* 167 (1965).
148. MIKOLAJCIK, E. M., J. Dairy Sci., *51,* 457 (1968).
149. MILLER, P. G., and SOMMER, H. H., J. Dairy Sci., *23,* 405 (1940).
150. MORR, C. V., J. Dairy Sci., *50,* 1744 (1967).
151. MORR, C. V., GOULD, I. A., and HARPER, W. J., J. Dairy Sci., *38,* 594 (1955).
152. MORR, C. V., HARPER, W. J., and GOULD, I. A., J. Dairy Sci., *40,* 964 (1957).
153. MORR, C. V., and JOSEPHSON, R. V., J. Dairy Sci., *51,* 1349 (1968).
154. MORRIS, H. A., COULTER, S. T., COMBS, W. B., and HEINZEL, L. R., J. Dairy Sci., *34,* 487 (1951).
155. MORRISSEY, P. A., Ir. J. Agri. Res., *8,* 201 (1969).
156. MORRISSEY, P. A., J. Dairy Res., *36,* 343 (1969).
157. MULLER, L. L., Dairy Sci. Abstr., *33,* 659 (1971).
158. MURPHY, R. F., DOWNEY, W. K., KEARNEY, R. D., Biochem. J., *115,* 22P (1969).
159. MURTHY, L., HERREID, E. O., and WHITNEY, R. MCL., J. Dairy Sci., *41,* 1324 (1958).
160. NAKANISHI, T., TAKAHASHI, K., and IMAGAWA, T., Jap. J. Dairy Sci., *17A,* 28 (1968).
161. NELSON, V., J. Dairy Sci., *31,* 415 (1948).
162. NELSON, V., J. Dairy Sci., *32,* 775 (1949).
163. NELSON, V., J. Dairy Sci., *35,* 388 (1952).
164. NITSCHMANN, H., Helv. Chim. Acta, *32,* 1258 (1949).
165. NOELKEN, M., Biochim. Biophys. Acta, *140,* 537 (1967).
166. ODAGIRI, S., and NICKERSON, T. A., J. Dairy Sci., *48,* 1157 (1965).
167. OSNER, R. C., and JOHNSON, R. M., J. Food Tech., *3,* 81 (1968).
168. OTTING, H. E., U.S. Patent 2,490,599, December 6, 1949.
169. OTTING, H. E., CHRYSLER, L. H., and ALMY, E. F., U.S. Patent 2,473,493, June 14, 1949.
170. PALLANSCH, M. J., in "Byproducts from Milk," by Webb, B. H. and Whittier, E. O. eds., 2nd Ed., p. 124, Avi, Westport, Conn. 1970.
171. PANTALONI, D., C. R. Acad. Sci., Paris, *259,* 1775 (1964).
172. PARRY, R. M., JR., and CARROLL, R. J., Biochim. Biophys. Acta, *194,* 138 (1969).
173. PAYENS, T. A. J., and HEREMANS, K., Biopolymers, *8,* 335 (1969).
174. PAYENS, T. A. J., and VAN MARKWIJK, B. W., Biochim. Biophys. Acta, *71,* 517 (1963).
175. PAYENS, T. A. J., and SCHMIDT, D. G., Biochim. Biophys. Acta, *109,* 214 (1965).
176. PATTON, S., J. Dairy Sci., *35,* 1053 (1952).
177. PATTON, S., and FLIPSE, R. J., Science, *125,* 1087 (1957).
178. PEEBLES, D. D., U.S. Patent 2,565,085, August 21, 1951.
179. PURKAYASTHA, R., TESSIER, H., and ROSE, D., J. Dairy Sci., *50,* 764 (1967).
180. PUTMAN, F. W., "The Proteins," H. Neurath and K. Bailey, eds. Vol. I, Part B, p. 807, Academic Press, Inc., New York, 1953.
181. PYNE, G. T., XII Int. Dairy Cong. Proc., *2,* 229 (1949).
182. PYNE, G. T., XIII Int. Dairy Cong. Proc., *3,* 1032 (1953).
183. PYNE, G. T., and McHENRY, K. A., J. Dairy Res., *22,* 60 (1955).

184. RAMSDELL, G. A., and HUFNAGEL, C. A., XIII Int. Dairy Cong. Proc., *3,* 1025 (1953).
185. RAMSDELL, G. A., JOHNSON, W. T., JR., and EVANS, F. R., J. Dairy Sci., *14,* 93 (1931).
186. RAMSEY, R. J., TRACY, P. H., and RUEHE, H. A., J. Dairy Sci., *16,* 17 (1933).
187. RIBADEAU-DUMAS, B., and GARNIER, J., C. R. Acad. Sci., Paris, *268,* 2504 (1969).
188. RIBADEAU-DUMAS, B., and GARNIER, J., J. Dairy Res., *37,* 269 (1970).
189. ROGERS, L. A., DEYSHER, E. F., and EVANS, F. R., J. Dairy Sci., *4,* 294 (1921).
190. ROSE, D., J. Dairy Sci., *44,* 430 (1961).
191. ROSE, D., J. Dairy Sci., *44,* 1405 (1961).
192. ROSE, D., J. Dairy Sci., *45,* 1305 (1962).
193. ROSE, D., J. Dairy Sci., *51,* 1897 (1968).
194. ROSE, D., BRUNNER, J. R., KALAN, E. B., LARSON, B. L., MELNYCHLYN, P., SWAISGOOD, H. E., and WAUGH, D. F., J. Dairy Sci., *53,* 1 (1970).
195. ROSE, D., and COLVIN, J. R., J. Dairy Sci., *49,* 351 (1966).
196. ROSE, D., and COLVIN, J. R., J. Dairy Sci., *49,* 1091 (1966).
197. ROSE, D., DAVIES, D. T., YAGUCHI, M., J. Dairy Sci., *52,* 8 (1969).
198. ROSE, D., and TESSIER, H., J. Dairy Sci., *42,* 969 (1959).
199. ROWLAND, S. J., J. Dairy Res., *5,* 46 (1933).
200. ROWLAND, S. J., J. Dairy Res., *8,* 1 (1937).
201. RUPP, P., U.S. Dept. Agri. Bur. An. Ind., Bull. 166 (1913).
202. SAITO, Z., and HASHIMOTO, Y., J. Fac. Agri. Hokkaido Univ., *54,* 17 (1964).
203. SAMEL, R., and MUERS, M. M., J. Dairy Res., *29,* 269 (1962).
204. SAWYER, W. H., J. Dairy Sci., *51,* 323 (1968).
205. SAWYER, W. H., J. Dairy Sci., *52,* 1347 (1969).
206. SCHMIDT, D. G., Neth. Milk Dairy J., *22,* 40 (1968).
207. SCHMIDT, D. G., Ph.D. Thesis, Utrecht, Netherlands (1969).
208. SCHMIDT, D. G., and BUCHHEIM, W., Milchwissenschaft, *23,* 505 (1968).
209. SCHMIDT, D. G., BUCHHEIM, W., Milchwissenschaft, *25,* 596 (1970).
210. SCHMIDT, D. G., PAYENS, T. A. J., MARKWIJK, B. W. VAN, and BRINKUIS, J. A., Biochem. Biophys. Res. Commun., *27,* 448 (1967).
211. SEEKLES, L., and SMEETS, W. T. G. M., Neth. Milk Dairy J., *1,* 7 (1947).
212. SEEKLES, L., and SMEETS, W. T. G. M., Lait, *34,* 610 (1954).
213. SHAHANI, K. M., and SOMMER, H. H., J. Dairy Sci., *34,* 1035 (1951).
214. SHIMMIN, P. D., and HILL, R. D., Aust. J. Dairy Technol., *20,* 119 (1965).
215. SMEETS, W. T. G. M., Neth. Milk Dairy J., *9,* 249 (1955).
216. SOMMER, H. H., and BINNEY, T. H., J. Dairy Sci., *6,* 176 (1923).
217. SOMMER, H. H., and HART, E. B., Wis. Agri. Expt. Sta. Res. Bull. 67 (1926).
218. STEBNITZ, V. C., and SOMMER, H. H., J. Dairy Sci., *18,* 805 (1935).
219. STEWART, A. P., JR., JOHNSON, R. A., and ANDERSON, T. T., U.S. Patent 3,008,840, November 14, 1961.
220. STEWART, A. P., JR., JOHNSON, R. A., and WILCOX, D. F., U.S. Patent 2,886,450, May 12, 1959.
221. STORRS, A. B., J. Dairy Sci., *25,* 19 (1942).
222. STORRS, A. B., and HULL, M. E., J. Dairy Sci., *39,* 1097 (1956).
223. SULLIVAN, R. A., FITZPATRICK, M. M., and STANTON, E. K., Nature, *183,* 616 (1959).
224. SULLIVAN, R. A., FITZPATRICK, M. M., STANTON, E. K., ANNINO, R., KISSEL, G., and PALERMITI, F., Arch. Biochem. Biophys., *55,* 455 (1955).
225. SWAISGOOD, H. E., and TIMASHEFF, S. N., Abstr. Pap. Am. Chem. Soc. C165, 152nd Meeting, 1966.
226. SWANSON, A. M., MAXWELL, G. E., and ROEHRIG, P. C., J. Dairy Sci., *51,* 932 (1968).
227. TAMSMA, A. F., and TARASSUK, N. P., J. Dairy Sci., *39,* 26 (1956).
228. TANFORD, C., Adv. Protein Chem., *23,* 121 (1968).
229. TARASSUK, N. P., and NURY, M. S., J. Dairy Sci., *35,* 857 (1952).

230. TARASSUK, N. P., and TAMSMA, A. F., J. Agri. Food Chem., *4,* 1033 (1956).
231. TER HORST, M. G., Neth. Milk Dairy J., *4,* 246 (1950).
232. TESSIER, H., and ROSE, D., J. Dairy Sci., *41,* 351 (1958).
233. TESSIER, H., and ROSE, D., J. Diary Sci., *47,* 1047 (1964).
234. TESSIER, H., YAGUCHI, M., and ROSE, D., J. Dairy Sci., *50,* 941 (1967).
235. TOWNLEY, V. H., M.S. Thesis, Univ. of Minn. (1950); Ph.D. Thesis, Univ. of Minn., Minneapolis (1953).
236. TRACY, P. H., and RUEHE, H. A., Ill. Agr. Expt. Sta. Bull. 352, 569 (1930).
237. TROUT, G. M., "Homogenized Milk," p. 30. Michigan State College Press, East Lansing, 1950.
238. TUMMERMAN, L., and GUTH, J. H., U.S. Patent 3,210,201 (1965).
239. VAN KREVELD, A., and VAN MINNEN, G., Neth. Milk Dairy J., *9,* 1 (1955).
240. WAITE, R., and WHITE, J. C. D., J. Dairy Res., *16,* 379 (1949).
241. WALLANDER, J. F., and SWANSON, A. M., J. Dairy Sci., *50,* 949 (1967).
242. WARNER, R. C., and POLIS, E., J. Am. Chem. Soc., *67,* 529 (1945).
243. WAUGH, D. F., "Milk Proteins," Vol. II. H. A. McKenzie, ed. Academic Press, Inc., New York, 1971.
244. WAUGH, D. F., and NOBEL, R. W., Proc. Fed. Amer. Soc. Expt. Biol., *24,* 418 (1965).
245. WEBB, B. H., J. Dairy Sci., *11,* 471 (1928).
246. WEBB, B. H., and BELL, R. W., J. Dairy Sci., *25,* 301 (1942).
246A. WEBB, B. H., and HOLM, G. E., J. Dairy Sci., *11,* 243 (1928).
247. WEBB, B. H., and HOLM, G. E., J. Dairy Sci., *15,* 345 (1932).
248. WEBB, B. H., and HUFNAGEL, C. F., U.S. Dept. Agri. Bur. Dairy Ind., BDIM-*Inf*-47 (1947).
249. WEBB, B. H., and HUFNAGEL, C. F., J. Dairy Sci., *31,* 21 (1948).
250. WEBB, B. H., and WHITTIER, E. O., Eds. "Byproducts from Milk," 2nd ed. p. 83, Avi, Westport Conn., 1970.
251. WHITE, J. C. D., and DAVIES, D. T., J. Dairy Res., *25,* 267 (1958).
252. WHITNEY, R. McL., PAULSON, K., and MURTHY, G. K., J. Dairy Sci., *35,* 937 (1952).
253. WHITTIER, E. O., and WEBB, B. H. "Byproducts from Milk," p. 191, Reinhold Publishing Corp., New York, 1950.
254. WIEGNER, G., Z. Nahr. Genussm., *27,* 425 (1914).
255. WOYCHIK, J. H., KALAN, E. B., and NOELKEN, M. E., Biochemistry, *5,* 2276 (1966).
256. WRIGHT, N. C., J. Dairy Res., *4,* 122 (1932).
257. YAMAUCHI, K., YONEDA, Y., KOGA, Y., TSUGO, T., Agri. Biol. Chem. Tokyo, *33,* 907 (1969).
258. YAMAUCHI, K., KAMINOGAWA, S., TSUGO, T., Jap. J. Zootech. Sci., *40,* 551 (1969).
259. YOSHINO, U., TAKEMOTO, S., YAMAUCHI, K., TSUGO, T., Jap. J. Zootech. Sci., *40,* 299 (1969).
260. YOSHINO, U., TANAKA, K., YAMAUCHI, K., TSUGO, T., Jap. J. Zootech. Sci., *40,* 476 (1969).
261. ZITTLE, C. A., J. Dairy Sci., *48,* 771 (1965).
262. ZITTLE, C. A., and CUSTER, J. H., J. Dairy Sci., *46,* 1183 (1963).
263. ZITTLE, C. A., and DELLAMONICA, E. S., J. Dairy Sci., *39,* 514 (1956).
264. ZITTLE, C. A., DELLAMONICA, E. S., RUDD, R. K., and CUSTER, J. H., J. Am. Chem. Soc., *79,* 4661 (1957).
265. ZWEIG, C., and BLOCK, R. J., J. Dairy Sci., *36,* 427 (1953).

CHAPTER 12

C. A. Ernstrom*
Noble P. Wong†

Milk-clotting Enzymes and Cheese Chemistry

C. A. Ernstrom

Part I. Milk-clotting Enzymes and their Action

Milk-clotting enzymes have been used since antiquity for the manufacture of cheese and other foods,[437] and have been obtained from a great many microbial, plant and animal sources.[588] Until very recently the calf gastric enzyme, rennin (EC 3.4.4.3) in the form of a crude extract, paste or powder called rennet, was used almost exclusively in commercial cheese making. However, during the decade 1962–1972 there developed a substantial shortage of calf stomachs (vells), and cheese makers resorted to rennet substitutes to meet the needs of an expanding cheese industry.[202] It is questionable whether calf vells available in the United States in 1972 could have filled more than one-third of the country's need for milk-clotting enzymes. However, rennin still remains the enzyme of choice, and the standard against which all others are evaluated.[202]

Most published reviews of enzyme-catalyzed clotting of milk have been concerned with rennin and its action.[51,54,125,172,194,195,197,199,229,338,358,430,618] This is understandable since a significant shift to the use of nonrennin milk-clotting enzymes has occurred only in recent years. However, several reviews have treated the subject of rennet substitutes and their use in cheese making.[29,71,232,318,329,434,579,581]

Nearly all proteolytic enzymes will clot milk under proper conditions.[54] Therefore, it is not surprising that milk-clotting enzymes have been obtained from virtually every class of living organism. The so-called "sweet curdling" of milk is attributed to milk-clotting enzymes secreted by bacteria.[204,212,240] Milk-clotting enzymes are also produced

*C. A. Ernstrom, Head, Department of Nutrition & Food Sciences, Utah State University, Logan, Utah 84322
†Noble P. Wong, Dairy Foods Nutrition Laboratory, Nutrition Institute, Agricultural Research Service, U.S. Department of Agriculture, Beltsville, Md.

by fungi,[238,239,274,745] higher plants,[48,276,354,411,467,693,747,756,772] and various animal organs.[695]

United States Federal Standards for cheese have been modified to allow the use of ". . . sufficient rennet, or other safe and suitable milk-clotting enzyme that produces equivalent curd formation, or both . . ." in the manufacture of standardized varieties of cheese.[178,179]

An FAO Committee of the United Nations has proposed the term "rennet", preceded by the name or source of a particular enzyme as a nomenclature scheme for all milk-clotting enzyme preparations, e.g., plant rennet, microbial rennet, pepsin rennet.[202] Historically the term rennet has been applied only to the crude extract from bovine calf stomachs. Also, U.S. Cheese Standards distinguish between rennet and other safe and suitable milk-clotting enzymes. In this discussion the term "rennet" will be used to designate only the calf-stomach extract. Other enzyme preparations will be identified by their common or usual name, or by the term "protease" preceded by an identifying source, e.g., *Mucor miehei* protease.

CLOTTING ENZYMES FROM HIGHER PLANTS

It has been suggested that enzymes from higher plants might be useful in cheese making, and some have been tried.[97,167,375,377] However, attempts to use them usually have met with disappointment. Most plant proteases are strongly proteolytic and cause extensive digestion of the curd.[7,276] This has resulted in reduced yields, bitter flavors and pasty-bodied cheese.[97,167,377,713]

Papain (EC 3.4.4.10)

The latex of the plant *Carica papaya* yields papain and several other proteases.[23] The properties of papain were reviewed by Arnon[23] in 1970. It has powerful milk-clotting activity,[116] but is also highly proteolytic. It requires a free sulfhydryl group for its catalytic activity, is reversibly inactivated in air at low cysteine concentrations, but can be reactivated in the presence of high levels of cysteine.[23] Crude papaya latex contains both papain and chymopapain (EC 3.4.4.11) that have been crystallized separately.[312,388] Milk-clotting time has been used for assessing the activities of both enzymes,[299,312,327,388] although assay methods based on protein digestion or hydrolysis of specific peptides are preferred.[23] Hinkel and Alford[299] and Balls and Hoover[40] supported the idea that the milk-clotting power of papain represents its proteolytic activity. However, Skelton[646] found no direct relationship between milk-clotting and protein-digesting activities. He suggested that the milk-clotting and general proteolytic components of papain are different. This view was also supported by Pozsarne-Hajnal *et al.*[540]

Ficin (EC 3.4.4.12)

Milk-clotting enzymes (ficin) are present in the latex of several species of the genus *Ficus* (fig), but the best source is *Ficus carica*.[376,735,771] The properties of ficin have been reviewed by Liener and Friedenson.[414] Like papain, ficin is a sulfhydryl enzyme, and may be inactivated by such thiol reagents as mercuric iodide, N-ethyl maleimide, and iodoacetic acid. Zuckerman-Stark[773] suggested that fig latex contains two factors: a milk-clotting factor and a protein-digesting factor. He found that 2×10^{-4} to $1 \times 10^{-3}M$ iodoacetic acid inhibited proteolysis by ficin, but only slightly affected its milk-clotting activity.[773] Pozsarne-Hajnal *et al.*[540] reported that $2 \times 10^{-4}M$ iodoacetic acid inhibited the curd-digesting action of both ficin and papain without affecting their clotting activities. They also found that it required a $5 \times 10^{-3}M$ concentration to affect bromelain similarly. Cheese made with ficin developed bitter flavors which decreased in intensity during curing.[377]

Bromelain

Bromelain (EC 3.4.4.24) from pineapple has been considered as a possible substitute for rennet,[276] and milk-clotting tests have been used for measuring its activity.[154] However, it also has proved too proteolytic for use by the cheese industry.[540] A recent discussion of bromelain enzymes was published by Murachi.[480]

Other Plant Proteases

An enzyme extract from *Withania coagulans* was suggested by Dastur *et al.*[117] as a coagulant for cheese making. Kothavalla and Khubchandani[375] used it in the manufacture of Surati and Cheedar cheese. No unusual flavors were reported, but fat losses were high.

Christen and Virasoro[96] in 1935 studied the milk-clotting properties of extracts from the flower petals of *Cynara cardunculus* (cardoon). These extracts are used by Portuguese farmers for the manufacture of Serra cheese from sheep's milk.[711] Vieira de Sa' and Barbosa[711,712,713] found that the milk-clotting activity of cardoon extracts varied more with pH and substrate (cow's or sheep's milk) than rennet. When used for making Edam cheese, the extract was much more proteolytic than rennet, and the cheese was acid and bitter. The quality of Roquefort cheese was good, but yields were low when cardoon extract was used. No differences were observed between rennet and cardoon extract when used in the manufacture of Serra cheese from sheep's milk. Cheddar cheese made with cardoon extract was extremely bitter and pasty at 30 days of age.[167]

CLOTTING ENZYMES FROM BACTERIA

The search for suitable rennet substitutes has led to the investigation of a number of extracellular proteases produced by bacteria. Milk-clotting enzymes from *Streptococcus liquifaciens,*[651] *Micrococcus caseolyticus,*[140] *Bacillus cereus,*[92,459,650] *Bacillus polymyxa,*[235] *Bacillus mesentericus,*[367] *Bacillus coagulans,*[485] and *Bacillus subtilis*[314,387,482,547,548,685] have received considerable attention. Recent patents have been issued for the production and use of milk-clotting enzymes from bacteria, particularly from *Bacillus subtilis.*[236,237,332,333,334,481,483] Melachouris and Tuckey[459,460] found that enzymes from *B. cereus* rapidly degraded whole casein, α-casein and particularly β-casein. Puhan[547] reported that high proteolytic activity and nonspecific hydrolysis characterized the action of a protease from *B. subtilis.* Irvine *et al.*[314] reported that Cheddar cheese made with a *B. subtilis* protease had acceptable flavor, but that yields were low due to excessive proteolysis. Puhan and Steffan[548] found that a protease from the same organism was unsuitable for the production of hard cheese.

There have been a few reports of acceptable cheese made with bacterial coagulants.[235,482,650,651] However, results have not been consistently favorable, and no bacterial protease has found its way into commercial use. Bacterial protease preparations are complex,[547] and the occasional report of success may reflect the fact that suitable coagulants exist as part of crude mixtures containing other highly proteolytic enzymes.

CLOTTING ENZYMES FROM FUNGI

Fungal proteases have been investigated extensively in the search for rennet substitutes. Sannabadthi and Srinivasan[612] studied proteases from *Aspergillus nidulans, Aspergillus glaucus, Syncephalastrum racemosum,* and *Cladosporium herbarum.* All had higher proteolytic activity than rennet and functioned at higher temperatures. Knight[362] tested proteases from 39 molds, and obtained encouraging results with a protease from *Byssochlamys fulva.* Reps *et al.*[566] also reported encouraging results with a protease from the same organism. The milk-clotting characteristics of proteases from *Aspergillus candidus,*[83] *Aspergillus niger,*[510] *Rhizophus oligosporous,*[724] three strains of *Basidiomycetes,*[347] and other fungi[546] have been investigated.

In recent years patents have been issued for the production of milk-clotting enzymes from *Endothia parasitica,*[614,615] *Mucor pusillus* var. *Lindt,*[18,19] and *Mucor miehei.*[27,81,82] Proteases from these organisms have been subjected to considerable investigation, and in the United States have been accepted as "safe and suitable" substitutes for rennet in the manufacture of all standard varieties of cheese.[179,180,181]

Protease from Endothia Parasitica (EP Protease)

A protease preparation from *Endothia parasitica* (EP protease) has been used with varying success for cheese making,[16] and was reported to accelerate the ripening of Cheddar cheese.[637] It was more proteolytic than rennet during the curing of Edam, Tilsit and Butter cheese,[680] and produced bitterness when the amount used exceeded 70% of the activity normally used in the form of rennet. When it was used for making Camembert cheese, proteolysis was greater and yields were lower than when rennet was used.[446] Morris and McKenzie[479] found that at three months of age Cheddar cheese made with 75% EP protease and 25% rennet was slightly inferior in flavor and body to control cheese made with rennet. Bitterness was characteristic of Cheddar cheese made with the EP protease.[28,152,491] Taleggio cheese made with EP protease was harder and more bitter than the rennet control.[569]

Even though bitterness has been characteristic of several cheese varieties made with EP protease, when used in Emmental cheese where high cooking temperatures (51.7 to 54.4°C) are employed, the cheese was of excellent quality.[66,556,580] Cooking temperatures are usually reached when the curd is still relatively sweet, and the enzyme least stable. Therefore, the protease activity is probably destroyed during cooking, and is unavailable to affect the cheese during curing. Sardinas[616] found that EP protease was completely destroyed in 5 min at 60°C at pH 4.5, where it has maximum stability. Above pH 6.0 it was much less stable. Larson and Whitaker[397] and Sequi and Rotini[634] found that the heat stability of the enzyme decreased with increasing pH above 4.5. However, Reps *et al.*[567,568] showed that EP protease in sweet milk was more heat-stable than rennin.

Whitaker[736] reported that EP protease has its maximum stability between pH 3.8 and 4.8. Below pH 2.5, activity losses were associated with an increase in ninhydrin reaction groups, which suggested autolysis of the molecule. Above pH 6.5, activity was rapidly lost;

this was accompanied by decreased solubility and no increase in ninhydrin reaction groups.

Alais and Novak[6] separated an EP protease preparation into three fractions with different degrees of proteolytic activity, but were unable to separate highly active milk-clotting activity from proteolytic activity.

Changes of pH in milk did not affect the milk-clotting activity of EP protease as much as they did the activity of rennin.[5,396,567] Also, the initial coagulation of milk by EP protease was reportedly slower and the subsequent increase in curd firmness faster than when comparable levels of rennet were used.[557,574]

Hagemeyer *et al.*[254] crystallized EP protease in the form of needles, Sardinas[616] crystallized it as plates, and Moews and Bunn[469] prepared the enzyme in three different crystalline forms. Hagemeyer *et al.*[254] reported that the isolectric point of EP protease was below pH 4.6, since it was negatively charged and strongly adsorbed on DEAE-cellulose in 0.2*M* acetate buffer at that pH. Sardinas[616] reported that his preparation of the enzyme was cationic at pH 4.5 on cellulose paper. Hagemeyer *et al.*[254] concluded that EP protease is not a sulfhydryl enzyme because none of the sulfhydryl reagents they used inhibited its activity. Neither did it resemble chymotrypsin or trypsin so far as its reactivity with α-N-tosyl-L-lysine phenylalanine chloromethyl ketone (TPCK) and α-N-tosyl-L-lysine chloromethyl ketone (TLCK) was concerned. Incubation of the enzyme at pH 4.6 and OC with sodium tetrathionate ($1 \times 10^{-3}M$), $HgCl_2$ ($1 \times 10^{-3}M$), N-ethylmaleimide ($2 \times 10^{-3}M$), iodoacetamide ($2 \times 10^{-3}M$), *p*-chloromercuribenzoate ($1 \times 10^{-3}M$), and cysteine ($1 \times 10^{-2}M$) had no effect on the activity of the enzyme. Addition of 1×10^{-2} M EDTA to the substrate actually increased its activity by 25%. EP protease produced 18 peptides from the oxidized B chain of insulin compared to 12 for rennin.[736] Amino acid composition and molecular weights are given in Table 12.1.

Protease from Mucor Pusillus var. Lindt (MP Protease)

A milk-clotting protease from *Mucor pusillus* var. *Lindt* (MP protease) was isolated from soil by Arima *et al.*[20] It had a ratio of milk-clotting to proteolytic activity that was closer to that of rennin than previously tested fungal proteases.[319] The crude enzyme was purified by column chromatography, and its homogeneity established by ultracentrifugal and electrophoretic analyses at several pH values.[320] The enzyme was crystallized in three different forms, all with the same specific activity.[21] It differed serologically from rennin, and was less readily inhibited by normal blood serum.[321]

MP protease has given satisfactory results as a rennet substitute in the manufacture of a number of cheese varieties, and is currently enjoying considerable commercial acceptance.[489] Zwaginga *et al.*[775] compared MP protease with calf rennet in 7 different factories making Edam and Gouda cheese from both raw and heat-treated milk. Taste panels showed a slight preference for the rennet cheese. Van den Berg *et al.*[703] compared MP protease with rennet in 190 commercial batches of cheese involving 7 different varieties or styles. Graders showed a preference for Gouda cheese made with rennet, but preferences for Broodkaas and Edam made with MP protease were equal to preferences for the rennet controls. Bitterness was encountered more frequently in the experimental than in the control lots. Martens[436] reported that Gouda cheese made with MP protease was equal in quality to that made with 50-50 pepsin-rennet blends. Richardson *et al.*[575] made normal Brick, Cheddar, Mozzarella and Parmesan cheese with MP protease. They found slight bitterness in the Cheddar cheese only after 14 months of age. Kikuchi and Toyoda[356] frequently found bitter flavors in cheese made with several microbial proteases, which included MP protease.

The clotting activity of MP protease was more sensitive to pH changes between 6.4 and 6.8 than was rennin activity, but much less sensitive than the activity of porcine pepsin.[575] The same authors reported that $CaCl_2$ added to milk affected the clotting activity of MP protease more than it did that of rennin. However, the addition of $CaCl_2$ to milk reduced the pH, and since no pH adjustments were made to correct for this, it cannot be concluded that the observed effects were due entirely to calcium chloride. Calcium ions had no effect on the proteolytic activity of MP protease.[22,762] Richardson *et al.*[575] found that MP protease was more stable than rennin between pH 4.75 and 6.25. Iwasaki *et al.*[319] also reported that it was more stable to pH changes and to heat than rennin. They found that MP protease clotted milk optimally at 56°C as opposed to rennin at 44°C. Richardson *et al.*[575] concluded that MP protease and rennin are not compatible in liquid solution because the fungal enzyme appeared to destroy rennin activity. Mickelsen and Fish[466] found MP protease much less proteolytic than EP protease on whole casein, α_s-casein and β-casein at pH 6.65. However, MP protease was more proteolytic than rennin.

MP protease released 80.5% of the sialic acid from κ-casein as part of a macropeptide, but no free sialic acid was liberated.[760] Photooxidation of MP protease modified the histidine side-chains of the enzyme. This modification correlated reasonably well with a loss in the enzyme's ability to release sialic acid peptides from κ-casein. It was suggested that histidine may be associated with the active site.[760] Unlike rennin, MP protease can split several dipeptides, including glutamylphenylalanine, phenylalanyl-tyrosine, phenylalanyl-leucine, tyrosyl-leucine, and glutamyl-tyrosine.[761] Reported amino acid composition and molecular weights are given in Table 12.1 All varieties of *Mucor pusillus* may not produce proteases suitable for cheese making. Babel and Somkuti[32] found that a protease from a *Mucor pusillus* (not variety Lindt) gave cheese with a rancid flavor and a coarse, mealy texture. The rancid flavor, however, must have been due to lipase contamination of the sample. Some early preparations of *Mucor pusillus* var. *Lindt* also contained lipase activity.[575]

Protease from Mucor Miehei (MM Protease)

A protease derived from *Mucor miehei* (MM protease) was tested in Europe as a possible rennet substitute before it was approved for use in the U.S.[181] When compared with rennet in the manufacture of 15 lots of Emmental cheese, yield and quality were the same as the rennet controls.[358,448] No bitter flavors were reported, and it was concluded that MM protease is completely satisfactory for Emmental cheese manufacture. Thomasow *et al.*[681] used MM protease in the manufacture of Edam, Tilsit and Butter cheese. Yields and qualities of the experimental and control cheeses were similar. Cheese made with MM protease ripened faster, but bitterness was not evident. Domiati cheese made with MM protease had satisfactory organoleptic properties, gave a slightly higher yield, and developed more soluble nitrogen during curing than rennet cheese.[255] Prins and Nielsen[545] reported that large-scale tests of MM protease in Cheddar cheese manufacture resulted in cheese of excellent quality that was judged to be as good as rennet cheese even after an extended ripening period.

Edelsten and Jensen[151] separated an MM protease preparation into three fractions by gel filtration. All three had clotting activity, but gave evidence of different temperature optima. A purified preparation which gave no evidence of heterogeneity in electrophoresis, sedimentation, diffusion or gel filtration on Sephadex G-200 was prepared by Ottesen and Rickert.[511] Amino acids accounted for 90% of the solid matter; 6% was attributed to carbohydrates and 4% to salts. Molecular weights and amino acid composition of the purified protease are given in Table 12.1. It is of interest that within the concentration range 0.5 to 1.5% at pH 5.25 there was no concentration dependence of the sedimentation coefficient, which was $S^{\circ}_{20,\omega} = 3.35$ Svedberg units. Thus, the enzyme did not appear to form dimers or polymers within this concentration range. This is in marked contrast to rennin, whose aggregating tendencies are definitely concentration-dependent.[39,143] *Mucor miehei* protease was crystallized by Sternberg,[655] who found that the milk-clotting and proteolytic activities of the enzyme were inhibited by $HgCl_2$, $4 \times 10^{-2}M$ mercaptoethanol, and alkyl dimethylbenzyl ammonium chloride. Their studies also showed that activity was not metal-dependent, and that the enzyme did not have serine or SH groups associated with the active site. In the presence of aluminum sulfate, MM protease lost some of its ability to split the essential bond for milk clotting, but its general proteolytic activity was not impaired.[655]

The enzyme has a broad stability maximum between pH 4.0 and 6.0[655] Ottesen and Rickert[511] found that MM protease is remarkably stable. After 8 days incubation at 38°C between pH 3.0 and 6.0, over 90% of the activity was retained. The enzyme lost no activity during 11 hr incubation at pH 6.0 in 8*M* urea. Sternberg[655] reported its pH optima for proteolytic activity as 4.0 on urea-denatured hemoglobin, 4.5 on bovine serum albumin, and 3.5 on acid-denatured hemoglobin. Its proteolytic activity on the B chain of oxidized insulin was different from that of rennin and pepsin.[577]

CLOTTING ENZYMES FROM ANIMAL ORGANS

Several animal proteases have been investigated for their milk-clotting potential, but only pepsin (porcine and bovine) and rennin have been of interest to the cheese industry.

Porcine Pepsin (EC 3.4.4.1)

Limited amounts of porcine pepsin were used as rennet substitutes during World War I.[426,463,656,657] However, it was not until about 1960 that interest in this enzyme for cheese making was renewed. Since then, numerous workers have recommended it as a satisfactory substitute for at least part of the rennet in making many varieties of cheese.[79,120,162,503,659,679] Early reports of bitterness in Cheddar cheese made with pepsin[120,717] have not been substantiated.[79,426,457,679] Even though the quality of pepsin cheese has been good, the Research Committee of the National Cheese Institute[486] recommended that it not be used as a complete substitute for rennet. Difficulties with slow coagulation can be expected when porcine pepsin is used in milk that is not sufficiently acid at the time of setting.[42,163,656] If setting is slow and the curd cut when too soft, fat losses may be high.[79,162]

Between pH 6.3 and 6.8 the milk-clotting activity of porcine pepsin decreases much more rapidly than that of rennin.[698] In fact, at pH 6.8, usual levels of pepsin may not clot milk.[170] Emmons[159] suggested that part of this activity loss may be due to pH-inactivation of the pepsin in the milk. Pyatnitskii and Pyatnitskaya[550] reported that pepsin coagulation time depends on the percentage of pepsin added to the milk and the rate of inactivation of the enzyme in the milk. Pepsin becomes increasingly unstable at pH values above 6.0, even though the minor pepsins B and C are reportedly stable at pH 6.9 and 25°C.[594] It appears that pepsin activity during cheese making is reduced or eliminated during and subsequent to coagulation of the milk. Wang[725] was unable to detect the presence of any milk-clotting activity in fresh Cheddar cheese curd made with porcine pepsin. Green[245] concluded that most, if not all, porcine pepsin used as a coagulant is inactivated during Cheddar cheese making. Ketting and Paulay[352] found that the clotting activity of pepsin, unlike that of rennin, decreased disproportionately with decreasing enzyme concentration. Loss of enzyme activity in the milk could have accounted for this. Emmons[159] experienced significant inactivation when commercial pepsin and pepsin-rennet mixtures were diluted with high-pH hard water 10 min before adding them to the cheese vat.

Mickelsen and Ernstrom[465] reported that mixtures of porcine pepsin and rennet were stable between pH 5.0 and 6.0, but that pepsin activity was lost from the mixture above pH 6.0. This loss was shown to be entirely due to pepsin instability. Below pH 5.0 rennin activity was destroyed by pepsin.

A procedure for the preparation of crude commercial pepsin from autolyzed hog stomachs was described by Rajagopalan *et al.*[555] The gastric mucosa were accumulated and stored in the frozen state. They were then brought to pH 2.7 to 2.9 with HCl and incubated at about 50°C for several hours. The digestion mixture was allowed to settle, and the clear supernatant liquid chilled to 4°C and concentrated under reduced pressure. The mucin was precipitated with alcohol and removed, after which pepsin was precipitated by the addition of more alcohol. The pepsin was then filtered off and vacuum-dried. Cerna *et al.*[78] also described a method for extracting porcine and bovine pepsin from their respective tissues. A method for the preparation of homogeneous pepsin from crystalline pepsinogen was described by Rajagopalan *et al.*[555]

Crude porcine pepsin contains a major pepsin, minor pepsins B, C, and D,[593] and gastricsin (EC 3.4.4.22).[91] The minor pepsins appear to arise from separate zymogens, although the possibility that they represent degradation products of pepsin has not been completely discounted.[594] Pepsins B and C clot milk, but much less readily than the major pepsin, and are relatively stable up to pH 6.9.[595] Pepsin D is highly unstable above pH 6.0, but is twice as active as the major pepsin in clotting milk.[710] Ryle[592] reported that some animals make pepsinogens for all four of these enzymes, but others lacked one or more of the zymogens at the time of slaughter.

Gastricsin was originally isolated from human gastric juice, and shown to have proteolytic properties different from those of human pepsin.[576] On Berridge substrate[53] human gastricsin and human pepsin exhibited milk clotting activities about 45% as great as crystalline rennin.[667] Later, Chiang *et al.*[91] demonstrated the presence of gastricsin in extracts of hog stomach mucosa as well as in crude commercial pepsin. Porcine gastricsin had 15% greater milk-clotting activity than human gastricsin and 8% less than porcine pepsin.[664] The pH optima for hemoglobin digestion were 3.0 and 2.0 for porcine gastricsin and pepsin, respectively.[91] Porcine gastricsin exhibited 6% more proteolytic activity than human gastricsin on hemoglobin substrate.[664] Also, on hemoglobin substrate pepsin was only 74% as active as gastricsin, while on egg albumin pepsin hydrolyzed the substrate more than twice as fast as gastricsin.[666] Gastricsin, like porcine pepsin, is stable in acidic solutions, but both lose activity rapidly in neutral and alkaline solutions.[91,664] The percentage of gastricsin present in crude porcine pepsin extracts has not been reported. However, it must contribute to the milk-clotting activity of pepsin solutions used in cheese manufacture. Since its pH stability is similar to that of pepsin, it is probably subject to the same stability problems as porcine pepsin during cheese making.

Porcine pepsin is secreted by hog stomach mucosa as catalytically inactive pepsinogen[279,284] with a molecular weight of 40,400.[738] Herriott[282] reported that pepsinogen was stable in neutral and slightly alkaline solutions, but underwent reversible denaturation above 55°C at pH 7.0 and at room temperature at pH 11.0. This change was accompanied by the reversible removal of two protons.

The conversion of pepsinogen to pepsin is catalyzed by pepsin below pH 5.0;[280] 7 to 9[709] peptide bonds are hydrolyzed during the formation of pepsin, which results in splitting off about 20% of the molecule. However, it is likely that cleavage of only one of these bonds is necessary to release the active enzyme.[709] One of the polypeptides released during activation is a pepsin inhibitor (MW 5000). The inhibitor is bound to pepsin between pH 5.0 and 6.0 and inhibits both the milk-clotting and protein-digesting power of the enzyme.[281] At lower pH values the inhibitor dissociates from the enzyme and activity is restored. The inhibitor is destroyed by pepsin between pH 2.0 and 5.0, with a rate maximum near pH 4.0.[281]

Pepsin is a single-chain protein,[709] and optical rotary dispersion measurements suggest that the molecule exhibits very little helical coiling[330,527] even though it is folded to produce a globular particle.[150] The forces tending to hold the folds of the protein together may include hydrogen bonds, hydrophobic bonds, electrostatic bonds and disulfide bonds.[282,346] There are only three disulfide bonds in pepsin, and at least one of these is not essential to enzymatic activity.[67,282] Hydrogen bonds probably do not play an important role in stabilizing the molecule because 4*M* urea has little effect on inactivating the enzyme.[526,654] Pepsin contains a high percentage of amino acids with hydrophobic side-chains,[527] and the contribution of hydrophobic bonding to protein stability may be substantial.[282] The number of ionizable groups in pepsin which contribute electrostatic charges is high. Herriott[282] pointed out that the isoelectric point of pepsin, if one exists, must be below pH 1.0 because even at that pH the protein is not cationic.[283,686]

Bovine Pepsin

The presence of pepsin in rennet extract has been frequently implied.[50,125,131,774] However, in 1961, Linklater[420] reported that pepsin accounted for only 0 to 6% of the milk-clotting activity of commercial rennet extracts. He used porcine pepsin as a reference standard, and considered that bovine pepsin, when present in rennet, was the same as porcine pepsin. Pepsins from the two species are now known to be different.[345] Bovine pepsin has increased in use as a coagulant because of the practice of extracting stomachs from older calves and adult cattle.[202,503] Leitch[410] reported that the enzymatic secretion of the young calf was predominately rennin, but that at the age of 5 months, the calf stomach yielded pepsin almost to the exclusion of rennin.

In approving pepsin as a "safe and suitable" substitute for rennet, only porcine pepsin was considered by the U.S. Food and Drug Administration.[487] This raises questions concerning the definition of rennet, and the legal acceptability of bovine pepsin as a milk coagulant in the United States. In spite of this, substantial amounts of bovine pepsin are present in a high percentage of rennet extracts.[638]

Green[245] reported that Cheddar cheese made entirely with bovine pepsin was only slightly inferior to that made with calf rennet, and Fox and Walley[211] found no significant difference between Cheddar cheese made with bovine pepsin and rennet.

Crystalline bovine pepsin was prepared from gastric juice by Northrup[497] in 1933. Chow and Kassell[93] isolated bovine pepsinogen and showed that it differs considerably in amino acid composition from porcine pepsinogen. Like porcine gastric extracts, those from the bovine species contain a major pepsinogen and several minor pepsinogens, which upon activation at pH 2.0 produce a corresponding series of pepsins.[456] All the pepsins reportedly had N terminal valines and C terminal alanines, and differed only in their organic phosphate content.[456] Lang and Kassell[392] reported that the bovine pepsins did not differ from each other in enzymatic activity. However, Antonini and Ribadeau-Dumas[17] reported the presence of a pepsin I and a pepsin II in activated beef stomach extracts. Inactivation at alkaline pH values revealed that pepsin I behaves much like rennin, whereas pepsin II resembles porcine pepsin. Douillard[147] found that bovine pepsin I was rapidly inactivated in 7.2*M* urea at pH 6.3 and 10.3°C, while bovine pepsin II was stable. Even in 8.6*M* urea at pH 3.25 and 27.5°C bovine pepsin II lost activity slowly. Bovine pepsin has about 25 and 40% of the activity of porcine pepsin toward N-acetyl-L-phenylalanyl-L-diiodotyrosine and N-acetyl-L-phenylalanyl-L-tyrosine, respectively, and shows little activity toward benzyloxycardonyl-L-histidyl-L-phenylalanyl-L-tryptophan ethyl ester, which is a good substrate for porcine pepsin.[392] On hemoglobin substrate, bovine pepsin had only 60 to 70% of the activity of porcine pepsin.[392] The same authors reported that pepsin from the two species had similar milk-clotting activities. Fox[205] found that the milk-clotting activity of bovine pepsin is less pH-dependent than that of porcine pepsin, and could coagulate milk up to pH 6.9. Fox[205] suggested that bovine pepsin has proteolytic properties more like those of rennin, and is less subject to pH denaturation than porcine pepsin. This was verified by Emmons and Elliott,[160] who reported that at pH 6.35 and 30°C porcine pepsin was destroyed, while bovine pepsin was denatured not at all or very little. Shovers *et al.*[638] reported the existence of a calf pepsin that was electrophoretically distinguishable from adult bovine pepsin. The amino acid composition and molecular weights of some milk-clotting enzymes are shown in Table 12.1.

RENNET AND RENNIN

Rennet Manufacture

Cheese makers usually secure enzyme coagulants from dairy supply laboratories, but methods for making whey rennet are described by Sammis[597] and Peter,[528] and a procedure for preparing homemade rennet is given by Todd and Cornish.[689]

Early attempts[59,201] to obtain rennet from living fistulated calves yielded solutions too low in activity to be commercially useful. An operation for producing abomasal fistulas in calves for collecting rennet was described by Nair *et al.*[484] Using this technique, Ganguli[223] and others[441,442] collected rennet twice daily until the calves were 3 to 4 months old. During the collection period 30 to 40 times as much rennet was collected per calf as could be obtained by the slaughter method.

Table 12.1

AMINO ACID COMPOSITION AND MOLECULAR WEIGHTS OF SOME IMPORTANT MILK CLOTTING ENZYMES. (MOLES PER 10^5G PROTEIN)

	Porcine Pepsin (555)	Bovine Pepsin (392)	Rennin (192)	EP Protease (736)	MM Protease (511)	MP Protease (763)
Lys	2.9	0	24.0	30.9	22.5	37.3
His	2.9	3.0	14.1	8.1	4.3	4.6
Arg	5.9	9.0	16.7	4.5	15.2	13.4
½ Cystine	17.5	18.0	24.6	4.3	9.8	7.5
Asp	118.1	107.9	103.3	68.5	110.0	142.5
Thr	72.6	74.9	61.8	133.3	72.4	69.3
Ser	126.1	131.9	80.1	117.1	89.4	72.2
Glu	75.4	83.9	93.3	38.9	61.1	64.4
Pro	46.5	45.0	43.7	35.2	43.2	46.1
Gly	99.6	98.9	75.4	100.3	83.0	109.5
Ala	46.0	42.0	41.1	76.3	66.5	53.6
Val	58.0	68.9	66.3	59.2	63.3	77.8
Met	11.7	9.0	21.4	1.2	15.1	10.4
Ile	68.2	83.9	50.6	46.4	46.8	38.6
Leu	83.4	56.9	64.0	49.9	49.5	48.4
Tyr	46.8	48.0	51.8	51.7	48.7	41.8
Phe	40.7	42.0	45.0	34.9	52.6	61.8
Trp	17.6	15.0	12.4	8.5	7.6	8.0
Molecular Weights	34,163 (555)	33,367 (392)	30,700 (192)	37,500 (254)	38,200 (511)	29,000 (763)
	32,700 (738)		30,400 (132)	34,000 (616)		30,600 (763)
	35,000 (165)		34,000 (132)			32,500 (22)
	36,212 (65)		31,000 (13)			29,690 (22)
	32,930 (506)		33,000 (143)			
			40,000 (628)			

Pang[517] reported that milk-clotting activity is present in the abomasum of bovine fetuses as early as the 6th month of development, and increases in potency as the fetus approaches full term. The total recoverable milk-clotting activity from fetal abomasa during the 6th, 7th, 8th and 9th month amounted to 2, 7, 12 and 31% respectively of that found in high-quality vells from milk-fed calves slaughtered at 3 to 4 days of age. It was not determined whether the large increase in activity between a 9 month fetus and 3 to 4 day old milk-fed calf was strictly related to the birth process or was stimulated by milk consumption.

Herian and Krcal[278] investigated the production of an enzyme coagulant from lamb vells. It took 3 to 4 lamb vells to equal the weight of one calf vell, and the yield of activity per kg of tissue was half that of calf vells. They reported that extracts from lamb vells were more effective in coagulating sheep's milk than cow's milk.

Calves' stomachs (vells) are generally prepared in their abattoir in one of two ways: dry-blown or flat-salted. In the dry-blown process, used mostly in Europe and New Zealand, the contents of the stomach are expressed and the intact vell tied, inflated with air, and dried. The flat-salted process is used in the United States. Here the vells are open at the slaughter house, washed, and packed in dry salt for shipment to the rennet processor,[94] who cleans and dries them prior to extraction.[538] The dried vells in both processes are shredded, mixed with an inert filler and extracted with sodium chloride solutions. Recent developments in the flat-salted process have made possible the extraction of undried vells even though the extraction rate is slower than with dried vells.[489]

Crude rennet extract contains active rennin as well as an inactive precursor (prorennin). Addition of acid to the extract facilitates conversion of the prorennin to rennin and allows the extract to reach maximum activity. Rand and Ernstrom[560] recommended activating rennet extracts at pH 5.0. Even though activation at lower pH values was more rapid, poor stability of rennin below pH 5.0 in the presence of sodium chloride resulted in reduced yields.[464]

Activated extract may or may not be clarified with alum before filtering.[538,704] Finally, it is standardized with respect to activity, salt concentration, pH, and color. Liquid rennet is usually preserved by a high (14 to 20%) salt content and by addition of preservatives such as sodium benzoate and propylene glycol. The microflora in rennet has been investigated,[9,363,682] and generally found to contribute little to the curing or quality of cheese. However, poor sanitation and inadequate quality control during rennet manufacture can result in undesirable contamination of cheese milk.[490] Uncontrolled microbiological growth can also cause rapid deterioration of the clotting strength in rennet extracts. Rennet should be kept under refrigeration and at a pH between 5.5 and 5.9 for maximum stability.[464]

Rennet powders are prepared from precipitates obtained by acidifying activated extracts or by saturating with sodium chloride, or both.[538,683] Gelatin is frequently added to assist in obtaining more complete precipitation and recovery of the rennin from solution.

Specially prepared rennet pastes are used in the manufacture of some varieties of Italian cheese.[265,270,535,536] Historically these pastes have been made by macerating fresh calf stomachs and their contents. However, they can be made by the addition of clean ground vells to partially dried coagulated milk.[631,682] Lipase activity is a desideratum in making some Italian varieties of cheese, and if not present in the rennet, must be added separately. Lipase enzymes for this purpose are extracted from goat, lamb, or calf gullets.[266] When lipase extracts are used, satisfactory results may be obtained by using liquid rennet instead of rennet paste.[445]

Isolation of Rennin

Rennin has been purified and crystallized.[50,187,259,500] Both needle-shaped and block-shaped crystals have been reported, but Foltmann[187] found that needle-shaped crystals were transformed into rectangular plates upon storage at −15°C. A method for the preparation of crystalline rennin from rennet extract was reported by Ernstrom,[169] and has been used in a number of laboratories with consistently good results. The procedure is outlined below and in Table 12.2.

(1) Rennet extract at room temperature was adjusted to pH 5.0 with 3*N* HCl, and saturated with sodium chloride. A precipitate formed (Precipitate I) which was separated from the supernatant by centrifuging for 45 min at 5,000 *g* (max.) in an International centrifuge with a No. 845 angle head.

(2) Precipitate I was dissolved in distilled water, and diluted to one-half the original volume. The solution was adjusted to pH 6.3 with 0.1*N* NaOH* and filtered.

(3) Steps 1 and 2 were repeated until 4 salt precipitations had been made (Precipitates I, II, III, and IV). The dissolved Precipitate IV was diluted to the same volume as the solution of Precipitate III.

(4) Potassium alum (1.0 gm per 1000 ml rennin solution), dissolved in a small amount of water, was added to the solution of Precipitate IV, and immediately neutralized to pH 6.3 with 0.1*N* NaOH*. An aluminum hydroxide gel formed which was removed by centrifuging for 15 min at 5,000 *g* (max.). The alum supernatant was retained.

(5) Rennin in the alum supernatant was precipitated by adjusting to pH 4.6 with 0.1*N* HCl and saturating with sodium chloride, and separated by centrifuging in a refrigerated centrifuge for 45 min at 5,000 *g* (max.) (Precipitate V). The supernatant was discarded.

(6) Precipitate V was dissolved in a minimum of 0.05*M* sodium phosphate buffer at pH 6.8 and dialyzed at 2°C against Berridge's[50] salting-in buffer (50 gm $MgCl_2 \cdot 6H_2O$ plus 15 gm $CH_3COONa \cdot 3H_2O$ per 1, adjusted to pH 5.4 with 10*N* H_2SO_4).

(7) Crystals formed after a few days and continued to deposit for about 4 weeks. Crystal formation was materially hastened by adding a few seed crystals from a previous batch.

(8) The crystals were washed twice in the crystallizing buffer and twice at pH 5.6 in 0.05*M* acetate buffer containing 0.3% sodium chloride.

* An improved yield can be obtained if 0.1*M* Na_2HPO_3 is used in place of 0.1*N* NaOH to neutralize the solutions.

Table 12.2

ANALYSES OF FRACTIONS DURING PURIFICATION AND CRYSTALLIZATION OF RENNIN

Fraction	Volume	Activity	Total Activity	Activity Recovered	Nitrogen	Specific Activity
	(ml)	RU/ml	(RU)	(%)	(mg/ml)	(RU/mg N)
Original extract	3,870	100	387,000	100	5.01	20
Solution of						
Precipitate I	1,960	150	294,000	76	0.96	156
Precipitate II	955	263	251,000	65	1.07	246
Precipitate III	550	394	216,700	56	1.48	266
Precipitate IV	477	459	218,900	57	1.56	294
Alum supernatant	499	265	132,200	34	0.78	340
Solution of						
Precipitate V	72	1,718	123,700	32	4.81	357
Crystals[a]			62,270	16		401

[a] A portion of the crystals was dissolved in 0.2 N acetic acid, dialyzed against 0.05 M phosphate buffer at pH 6.8, and recrystallized by dialyzing at 2°C against Berridge's[50] buffer. The specific activitv of the recrystallized rennin was 399 RU/mg N (From Ernstrom[169]).

Heterogeneity of Rennin and Prorennin

Heterogeneity of crystalline rennin has been observed by free-boundary electrophoresis in low ionic strength buffers,[169] by paper electrophoresis,[524] by immunoelectrophoresis,[622] and by polyacrylamide gel electrophoresis.[24]

Foltmann[189] successfully fractionated crystalline rennin by chromatography on DEAE cellulose, and obtained three active fractions with relative specific activities of 125, 100, and 55–60. They were designated rennins A, B and C in order of decreasing specific milk-clotting activity. He found no indication that rennins A and B were heterogeneous, but considered rennin C to be a mixture containing some degradation products.[193]

Rennin originates in the stomach mucosa of the calf as an inactive zymogen called prorennin,[360] and is converted to active rennin in the acid environment of the stomach. Foltmann[186,190,191] and Rand[558] purified prorennin by a combination of salt fractionation and chromatography on DEAE cellulose. Activation of chromatographically purified prorennin resulted in an increase in inert material and a mixture of rennins A, B, and C that corresponded to the same fractions found in crystalline rennin.[190]

Using DEAE cellulose chromatography and elution with a linear phosphate concentration gradient, Foltmann[191] partially separated prorennin into two fractions. Rechromatography established the existence of two components that were designated prorennins A and B. Activation of the individual prorennins resulted in rennins which appeared in the same positions in the chromatograms as rennins A and B isolated from crystalline rennin.[194] Bundy *et al.*[73] and Shukri[641] were unable to show chromatographic heterogeneity of prorennin by following Foltmann's procedure. However, it was suggested that prorennin heterogeneity might be due to genetic variants not present in some starting materials.[73]

Asato and Rand[24] verified the heterogeneity of prorennin and rennin by polyacrylamide gel electrophoresis. One minor and two major prorennins, and two minor and two major rennins were identified. Proteolytic or potential proteolytic activity was demonstrated for each electrophoretic component. This led them to suggest that each zymogen gave rise to a separate rennin.

Activation of Prorennin

Prorennin is converted to active rennin by limited proteolysis accompanied by a molecular weight reduction of approximately 14%.[74,199] The rate of conversion increases markedly with decreasing pH below

5.0.[560] Foltmann[194] reported a maximum yield of rennin when prorennin was rapidly activated at pH 2.0 in the absence of salt. Rand and Ernstrom[560] obtained essentially the same activities in the absence of sodium chloride by rapid activation at pH 2.0 and by slow activation at pH 5.0. However, at pH 2.0 the newly formed rennin was not stable, and activity deteriorated when activation mixtures were held at that pH. At pH 5.0 activated rennin was quite stable and good yields were recovered. Also at pH 5.0, NaCl concentrations up to *2M* increased the rate of activation. Foltmann[191] reported that very low salt concentrations enhanced activation at pH 2.0 However, at pH 2.0 chloride salts have a devastating effect on the stability of rennin,[464] and when prorennin is activated at that pH in the presence of sodium chloride, it must be neutralized immediately to prevent activity losses that are approximately proportional to the salt concentration.[560] Since commercial rennet extracts contain substantial proportions of sodium chloride, maximum yields can usually be realized by activating slowly at pH 5.0.[560]

In addition to rate of activation and yield of rennin, the course of the activation reaction is affected by pH and salt (NaCl) concentration. If milk-clotting activity is plotted against activation time at pH 5.0, the course of activation resembles an S-shaped curve. If activation is carried out in the presence of preformed rennin, the S-shape disappears and the initial rate of the activation process increases with increasing concentration of preformed rennin.[194]

Rand and Ernstrom[560] reported that activation of prorennin at pH 5.0 was predominantly autocatalytic, particularly in the presence of 1.7*M* sodium chloride (See Fig. 12.1). As the pH decreased below 5.0, the course of activation appeared to deviate increasingly from autocatalysis. Foltmann[194] demonstrated that at pH 4.7 the course of activation was not a purely autocatalytic reaction. At low pH, activation became very rapid, and Rand and Ernstrom[560] found it difficult to measure accurately the rate of increase in activity at pH 2.0, even at low temperatures. However, Foltmann[194] reported activation rates at pH 2.0 in the absence of salts. He noted that, at the beginning of the reaction, activation proceeded at a faster rate than predicted for a first-order reaction, and suggested that it was reminiscent of second-order kinetics. By adding glycerol to the reaction mixture, Shukri[640,641] was able to reduce markedly the rate of activation at both pH 5.0 and pH 2.0 without affecting the apparent course of the reaction or the final level of activity. Activation mixtures contains 45 to 55% glycerol at pH 2.0 produced activation rates closely resembling first-order kinetics. A plot of log substrate concentration vs. time produced straight lines except for the initial points, as shown in Fig. 12.2. The

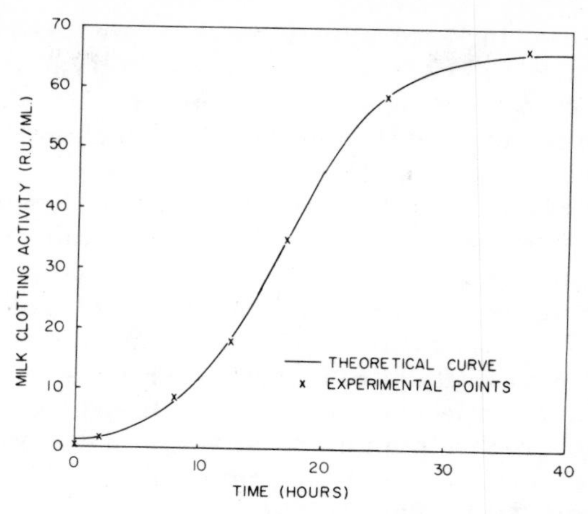

From Rand and Ernstrom[560]

FIG. 12.1. ACTIVATION OF PRORENNIN AT pH 5.0 IN 1.7 M SODIUM CHLORIDE COMPARED TO A THEORETICAL CURVE CALCULATED FROM THE AUTOCATALYTIC EQUATION GIVEN BY HERRIOTT[279]

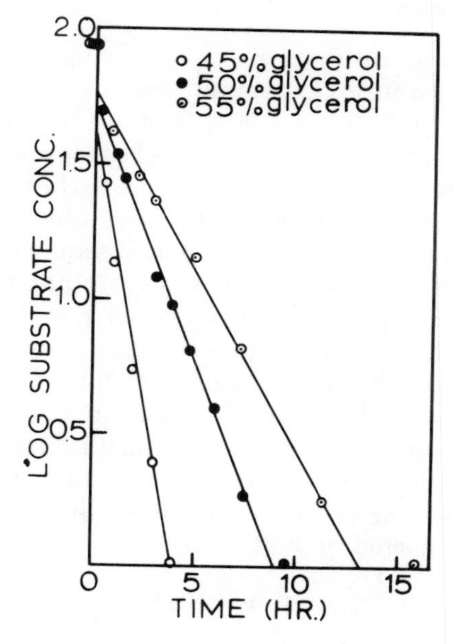

From Shukri[640]

FIG. 12.2. A PLOT OF LOG SUBSTRATE (PRORENNIN) CONCENTRATION VERSUS TIME FOR THE ACTIVATION OF PRORENNIN AT pH 2.0 AND 25°C IN THE PRESENCE OF 45, 50, AND 55% GLYCEROL

more rapid initial increase in activity than predicted by first-order kinetics was in agreement with Foltmann's observation.[194]

There appear to be at least two mechanisms by which prorennin is converted to rennin, and the importance each plays in an activation mixture depends largely on pH. Foltmann[199] suggested the possibility of an irreversibly activated rennin produced by limited proteolysis of prorennin, and also a reversibly active prorennin that manifests itself only at low pH values when electrostatic charges on acidic carboxyl groups have been neutralized. He explained that the prorennin molecule may be stabilized in its inactive form by a peptide fragment split off during autocatalytic activation, or by strong electrostatic charges within the molecule that are neutralized at low pH. However, no evidence of an active prorennin has yet been demonstrated.

The conversion of prorennin to rennin can be catalyzed by pepsin, especially at pH 2.0.[194] That this conversion can be due entirely to peptic activity was established by Rand and Ernstrom,[561] who activated prorennin with pepsin at pH 5.5 and 6.0 were autocatalytic activation was negligible.

Structure of Prorennin and Rennin

Activation of prorennin involves the splitting of peptides from the N-terminal end of prorennin with simultaneous reduction in molecular weight from about 36,000 to 31,000. The N-terminal amino acid in prorennin is alanine,[74,190] while glycine is N-terminal on rennin.[194,331] Jirgensons *et al.*[331] found leucine or isoleucine in the C-terminal position on rennin, which is consistent with the more recent findings of Foltmann[196] that similar C-terminal amino-acid sequences exist in rennin and prorennin, and that isoleucine is C-terminal in both. Bundy *et al.*[73] reported asparagine in the C-terminal position on rennin, but this has not been substantiated. However, they isolated from activation mixtures a large peptide (MW 4000) containing N-terminal alanine, which was presumably N-terminal in prorennin.

Structural data suggest that prorennin must be a single-chain protein containing 3 disulfide bridges which remain in the rennin portion of the molecule following activation. The N-terminal amino acid sequence of prorennin was reported by Foltmann[198] as

N-ala-glu-ile-thr-arg-ile-pro-leu-tyr-lys-gly-lys-ser-leu-arg-

Hydrolysis during activation exposes the N-terminal end of rennin whose sequence also has been determined.[196]

N-gly-glu-val-ala-ser-val-pro-leu-thr-asn—
-tyr-ile-lys-gly-phe-tyr-gln-ser-asp-leu-tyr—

The core sequence of amino acids in rennin has not been reported, but the sequence in the region of each disulfide bridge was given by Foltmann and Hartley.[200] They found that in a chain of 270 amino-acid residues, 2 of the three disulfide bridges were in the form of small loops of 5 and 6 residues respectively.

Bridge A

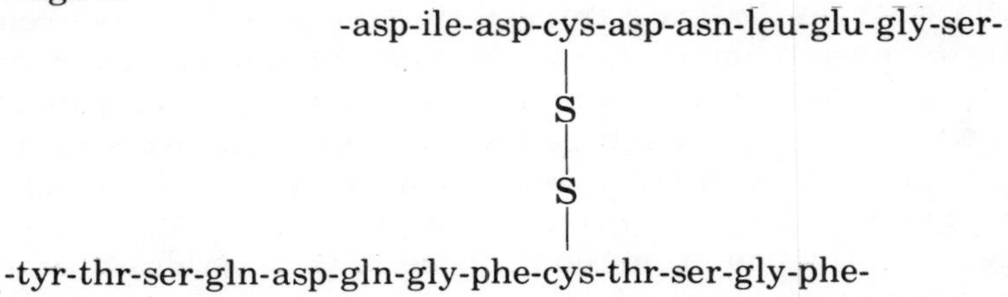

```
                    -asp-ile-asp-cys-asp-asn-leu-glu-gly-ser-
                                 |
                                 S
                                 |
                                 S
                                 |
-tyr-thr-ser-gln-asp-gln-gly-phe-cys-thr-ser-gly-phe-
```

Bridge B

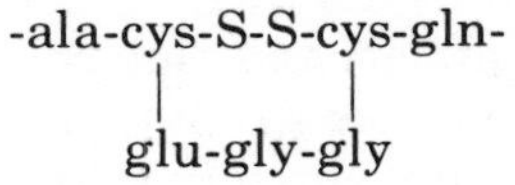

```
-ala-cys-S-S-cys-gln-
     |        |
     glu-gly-gly
```

Bridge C

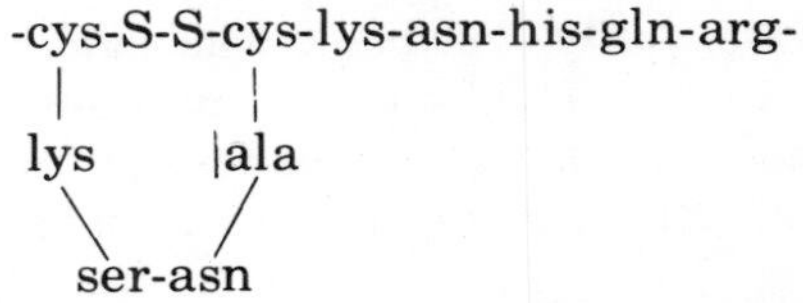

```
-cys-S-S-cys-lys-asn-his-gln-arg-
 |        |
lys      ala
  \      /
  ser-asn
```

The C-terminal amino-acid sequence of rennin was reported by Foltmann.[196,199]

-ile-leu-gly-asp-val-phe

ala-arg-asp-phe-val-ser-tyr-tyr-glu-arg-ile

asn-asn-leu-val-gly-leu-ala-lys-ala-ile-COOH

The first 19 C-terminal residues shown here also have been identified in a C-terminal peptide split from prorennin.[198]

These structural studies were made on whole crystalline rennin, but Foltmann[199] pointed out that since the 3 cystine peptides are substantially different, and since the amino-acid composition of the individual rennins are quite similar, the 3 S-S bridges must be present in each major rennin component.

A number of similar structural features and amino-acid sequences exist among rennin, porcine pepsin, human pepsin and human gastricsin.[145,199,310,665] However, Cheeseman,[86,87] Shukri,[641] and Ottesen and Rickert[511] found that, unlike porcine pepsin, rennin is readily inactivated by urea. Shukri[641] also noted that prorennin is rapidly denatured by urea, and that once denatured, it could not be activated. This suggests that, unlike pepsin,[526,654] hydrogen bonds may play an important role in maintaining the secondary and tertiary structure of rennin in an active configuration. Cheeseman[87] concluded that activity loss in urea solutions was not related to the extent of unfolding of the molecule, but to the exposure of specific groups that were susceptible to the urea environment. Rennin was not inhibited by the pepsin inhibitors *p*-bromophenacyl bromide and α-diazo-*p*-bromoacetophenone, which suggested that the active site does not possess a reactive carboxyl group, as in pepsin. However, some evidence was found for a tyrosine group that was important to enzyme activity.[87]

During the activation of prorennin none of the peptides liberated inhibit the activity of rennin.[73]

Sedimentation studies by Baldwin and Wake[39] demonstrated that rennin in solution forms dimers and high polymers, and that the extent of association decreases as ionic strength increases. When high concentrations of rennin were sedimented in solutions at 0.02 ionic strength, the Schlieren patterns showed two partly resolved peaks. At low rennin concentration only one peak was evident. When the ionic strength was increased to 0.32, one peak was observed at all rennin concentrations. Similar properties of association were noted by Djurtoft *et al.*[143]

The isoelectric point of rennin is in the region of 4.5.[259,628]

Stability of Prorennin and Rennin

Kleiner and Tauber[360] demonstrated that, in contrast to rennin, prorennin is stable above pH 7.0. At room temperature it is usually stable from pH 5.3 to 9.0.[199] However, Rand[558] found that while crude prorennin preparations were quite stable at pH 8.0, highly purified solutions lost potential activity at that pH. Exposure of prorennin to 6*M* urea completely and rapidly destroyed its ability to become activated.[641]

Foltmann[187] found that the optimum pH for rennin stability is between 5.3 and 6.3, and that the enzyme is moderately stable at pH 2.0. As the pH was raised from 6.3 to neutrality the activity of the enzyme was destroyed at an increasing rate. A region of instability near pH 3.5 was also noted. This was confirmed by Mickelsen and Ernstrom,[464] who also found that at that pH rennin was even less stable in the presence of sodium chloride (See Fig. 12.3). Foltmann[187] suggested that the region of instability near pH 3.5 may be due to self-digestion, since this was near the pH optimum for the proteolytic activity of rennin. This proved correct when Mickelsen and Ernstrom[464] noted that activity losses at pH 3.8 paralleled an increase in ninhydrin

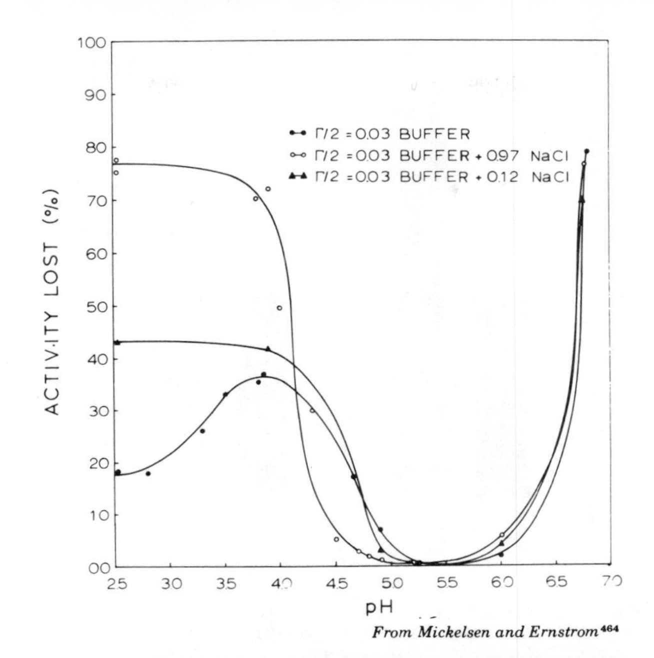

From Mickelsen and Ernstrom[464]

FIG. 12.3. EFFECT OF pH ON THE STABILITY OF RENNIN DURING 96 HR INCUBATION AT 30°C IN BUFFERS OF $\Gamma/2=1.0$ (0.03 DUE TO BUFFER, 0.97 DUE TO NaCl), $\Gamma/2=0.15$ (0.03 DUE TO BUFFER, 0.12 DUE TO NaCl), AND $\Gamma/2=0.03$ (NO NaCl)

color development. Limited proteolysis of rennin A at pH 3.5 resulted in its conversion to rennin C with a decrease in specific activity.[199]

Schober *et al.*[623] suggested on the basis of ninhydrin color development that activity losses at pH 7.0 are also due to autolysis. However, the proteolytic activity of rennin is very low at that pH. Foltmann[199] suggested the possibility that inactivation at pH 7.0 may be a combination of alkali denaturation and autolysis such that some of the rennin molecules undergo an unfolding, which makes them easily accessible to the small amount of proteolytic activity still present at pH 7.0. Activity losses above pH 6.0 were accompanied by precipitation of protein from solution and appeared to be highly temperature-dependent.[464]

The poor stability of rennin below pH 5.0 at high ionic strengths was shown to be at least partly ion-specific.[464] This is illustrated in Figure 12.4. Chloride ion was the most destructive, while lactate appeared to improve stability.

Hill and Laing[291] inactivated rennin by photooxidation, and attributed this to the destruction of histidine. They also reported destruction of rennin activity by coupling dimethylaminonaphthalene

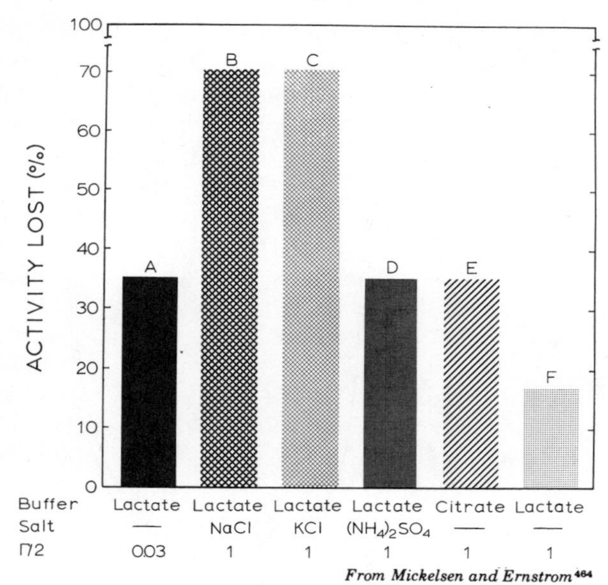

From Mickelsen and Ernstrom[464]

FIG. 12.4. EFFECT OF SPECIFIC IONS ($\Gamma/2=1.0$) ON LOSS OF ACTIVITY IN RENNIN SOLUTIONS INCUBATED AT 30°C FOR 96 HR (pH 3.8)

sulfonyl chloride (dansyl chloride) to rennin at pH 8.0 and 3°C.[292] They later attributed this to dansylation of a lysine residue.[293] Rennin was not inactivated by dansyl chloride at pH 6.5. Even though rennin is unstable at pH 8.0, the authors reported little inactivation of their controls, presumably because of the low temperature used. Rickert[578] reported that dansylation of rennin resulted in a 70 to 75% loss of proteolytic activity, and that subsequent treatment with 4*M* NH_4 OH at pH 6.25 increased the activity 2.5-fold, with the release of 1 dansyl moiety per mole of rennin. He felt it unlikely that the dansyl group was attached to a lysine residue in rennin because NH_4 OH would not release the dansyl moiety from dansylated polylysine. He suggested that inactivation of rennin by dansylation resulted from coupling with histidine, which Hill and Laing[291] had previously shown to be necessary for rennin activity.

Rennin can be digested by other proteolytic enzymes.[624,671] This fact has assumed considerable importance since blends of rennin with other milk-clotting enzymes, particularly porcine pepsin, have become commercially available to the cheese industry. Mickelsen and Ernstrom[465] found that blends of rennin and porcine pepsin were most stable at pH 5.5. Below pH 5.0 rennin activity was destroyed by pepsin, and above pH 5.5 the pepsin became progressively less stable. There was no evidence that rennin affected the pepsin activity.

Tsugo and Yamauchi[696,697] demonstrated the inhibitory action of blood serum on the milk-clotting activity of rennin. Horse and pig sera were most effective of those studied.[697] Human serum from patients with cancer and some other diseases inhibited rennin activity more than normal human serum.[733] Tauber[670] found that human serum inhibited pepsin much more than it did rennin, whereas Tsugo and Yamauchi[697] found that horse serum was more inhibitory toward rennin than pepsin. The specific inhibitor in blood serum was not identified, but its effect was destroyed by heat, alcohol, formalin, acetone, and mercuric chloride. It was precipitated from solution by 0.4 to 0.5 saturation with ammonium sulfate. Its effect resulted from direct inactivation of the enzyme because it had no effect on the nonenzymatic phase of milk clotting.[697] Furthermore, when blood serum was added directly to rennin solutions (pH 5.3), their activities were reduced within 10 min to a level proportional to the amount of serum added.[697]

MEASURING MILK-CLOTTING ACTIVITY

An ideal test for measuring milk-clotting activity has never been devised. In practice, activity is determined by the rapidity with which the enzyme clots milk under a set of specified conditions. This differs from the usual procedure in enzyme chemistry where one measures the rate at which the products of an enzyme-catalyzed reaction appear, or conversely, the rate at which the substrate disappears. Milk clotting is a complex process, involving both a primary enzymatic phase in which κ-casein is altered and loses its ability to stabilize the remainder of the caseinate complex, and a secondary nonenzymatic phase in which aggregation of the altered caseinate takes place. The secondary phase is particularly susceptible to variation in milk composition and to the presence of added salts. Natural variations in the composition of milk result in differences in clotting times, which make it difficult to establish a standard milk substrate for the clotting test. The milk-clotting activity of a protease apparently is based on its ability to catalyze the hydrolysis of a particularly susceptible phenylalanyl-methionine bond in κ-casein.[137,335] However, at pH values where it will coagulate milk, rennin does not readily hydrolyze simple peptides containing such a bond.[286] Determination of the conditions necessary for the specific enzyme attack on κ-casein is the object of considerable research activity, and it is unlikely that a standard unit of milk-clotting activity will be established until a simple substrate is found on which the primary action of milk clotting enzymes can be followed with precision. Some progress has been made toward finding such a substrate;[285,286] and it is hoped that this work will continue.

Aside from the above-mentioned difficulties, there are also problems of accurately observing the end point of coagulation, and of standardizing the conditions of the clotting test. It is customary to observe visually the formation of a clot, or rather the sudden fracture of a film of milk on the wall of a bottle or test tube. Apparatus for measuring clotting time in this way have been described by Sommer and Matsen[649] (Fig. 12.5), Berridge,[52] and Bakker *et al.*[37], and have been used for many years in rennet control laboratories.

A blood clot timer adapted for measuring milk-clotting time by deMan and Batra[138] has been used in industry control laboratories for standardizing rennet solutions.[636] It is faster than visual methods, requires less substrate, and has an automatic end-point detector. The ratio of enzyme solution to substrate is higher than is used with most other clotting tests. Therefore, care must be exercised to prevent differences in pH or salt concentration in the enzyme solution from affecting the clotting time.

Everson and Winder[174] reported that the clotting point of milk was marked by a sharp increase in the velocity of ultrasonic energy passing through the sample. They used this principle to develop a device and testing procedure that gave a reproducible recorded end point.

Thomasow[678] used a thrombelastograph for measuring the clotting time of milk. This device also measured the elasticity of the coagulum as it formed, and gave information about the time of solidification of the curd. Clotting times measured with this device were longer than those determined visually.

Nitschmann and Bohren[493] suggested that the liberation of nonprotein nitrogen (NPN) from casein might serve as a measure of primary rennin action. They pointed out, however, that this method is based

From Sommer and Matsen[649]

FIG. 12.5. APPARATUS FOR MEASURING THE CLOTTING TIME OF MILK

on differences in the solubility of casein degradation products in 12% trichloroacetic acid (TCA) and is less accurate than other methods of enzyme assay where newly liberated end groups are titrated.

Douillard and Ribadeau-Dumas[148] used the clotting time of κ-casein solutions for measuring the activity of rennin, porcine pepsin, and bovine pepsin. They indicated that careful control of ionic strength, pH, and κ-casein concentration was necessary to obtain an accuracy of 1% within a single laboratory.

Scott-Blair and Oosthuizen[632] suggested the measurement of viscosity change in milk as an index of clotting time. They showed that viscosity during the course of rennin action first dropped to a minimum, then rose as incipient clotting started. The initial or decreasing viscosity phase followed zero kinetics for a useful period, during which time rennin assays could be made. They found that plots of changes in specific viscosity against time during the action of rennin on caseinates from a variety of sources produced the same slope. However, the slope varied with different rennet extracts, which they attributed to the presence of varying percentages of proteolytic enzymes other than rennin.[633] Falder[175] reported a procedure for detecting the clotting point of milk by an indirect measure of viscosity. However, viscosity methods seem to have no advantage in accuracy or precision over visual methods.[199] Optical methods for measuring the clotting of milk were reported by Claesson and Nitschmann[99] and Claesson and Claesson.[98]

Niki *et al.*[492] reported a steady increase in sensitivity to calcium salts during renneting of calcium caseinate solutions, and found that the change could be measured spectrophotometrically at 610 nm. They reported a linear relationship between calcium sensitivity and NPN liberation, and suggested this as a basis for measuring milk-clotting activity.

The course of any enzyme-catalyzed reaction is influenced by a number of factors. However, regardless of the shape of an enzyme-time reaction curve, the reaction velocity is always proportional to enzyme concentration.[488] Therefore, for any given amount of reaction, the product of enzyme concentration and time is constant. This relationship for the clotting of milk by rennin was first proposed by Storch and Segelke;[661] it may be expressed as $T \times C = K$, where T = clotting time, C = enzyme concentration, and K = constant. This expression assumes that the clotting point marks the end of a given amount of enzyme action, and that the rate of the nonenzymatic stage of coagulation is always constant. These assumptions, of course, are not valid. Grimmer and Kruger[249] proposed a logarithmic expression for relating clotting time to enzyme concentration. They, as well as others,[14,95] found that K in the Storch-Segelke expression decreased as rennin concentration increased when milk was used as the substrate, but that the converse was true when the substrate was calcium caseinate. However, Holter[301] found that K decreased with decreasing rennin concentration in both milk and calcium caseinate substrates at 25 and 32°C. The same was true of nonfat dry milk reconstituted in water and in 0.01M calcium chloride.[168] At 40°C, Holter[301] observed that K increased with decreasing rennin concentration, but this was attributed to the instability of the enzyme at that temperature.[349] According to Holter,[301] the behavior of the clotting system may be described by the Storch-Segelke relationship provided an additional term is inserted. He proposed the expression $C(T - X) = K$, where X denotes the time required for the second stage of the reaction, i.e., the clotting phase. Foltmann[188] has rewritten Holter's equation in the form $T = K\frac{1}{C} + t$, where T is the observed clotting time, C the rennin concentration, and t the time lag between the end of the primary enzymatic reaction and the onset of clotting.

The expressions of Holter and Foltmann assume that the enzymatic and nonenzymatic phases of milk clotting are sequential and not overlapping. However, it should be pointed out that they do overlap.[553] Foltmann[188] followed the liberation of nonprotein nitrogen by rennin as a measure of the primary enzyme reaction, and found that at 25°C the end of the primary reaction and the clotting point approached one another as the pH was adjusted from 6.7 to 6.3. Below pH 6.3, clotting occurred before the enzymatic reaction was completed. Garnier and Chevalier[227] demonstrated that aggregation of casein starts before "proteolysis" is terminated, although the amount of aggregation corresponds to the percentage of substrate hydrolyzed, and that "proteolysis" continues after aggregation ceases.

The problem of standardizing the rennin assay was discussed at length by Berridge.[53] Inasmuch as milks differ greatly in their susceptibility to rennin action, analytical consistency may be approached, at least within one laboratory, by using a standard supply of nonfat dry milk. The material should be prepared and stored in sealed containers at low temperature so the original properties of the powder are retained as nearly as possible. Berridge[53] suggested reconstituting 60 gm low-heat, nonfat dry milk in 500 ml 0.01M $CaCl_2$. Clotting time is considerably shorter in this substrate than in nonfat dry milk reconstituted in water. Berridge's substrate also gives more reproducible results since it is less subject to the many factors that cause variation in the clotting time of normal milk. Nonlinearity between clotting time and the reciprocal of rennet concentration is reduced, but not entirely removed. Following reconstitution the clotting time of the substrate continues to increase during storage at 2°C. Therefore, in order to achieve a constant clotting time, it is helpful to allow the reconstituted substrate to age for about 20 hr before use.[168] Different samples of reconstituted nonfat dry milk vary in their clotting times. Therefore, when a new batch of nonfat dry milk is introduced into the laboratory it must be checked against the old with an enzyme solution of known activity. To avoid errors due to variability in the clotting properties of milk substrates, commercial suppliers are careful to maintain standard rennet solutions that are always tested with unknown samples on the same substrate. An arbitrary activity value is assigned the standard, and activities of unknown solutions are expressed in relation to that of the standard.[169] In making such comparisons, the enzymes are diluted so they give approximately the same clotting time. This eliminates significant errors due to deviation from the enzyme-time relationship, and permits calculation of the activity of unknown samples with reference to the standard as follows:

R.U./ml = +*100 $(t_s/t_u) \times (D_u/D_s$; where R.U./ml = rennin units per ml of unknown, t_s = clotting time of standard, t_u = clotting time of unknown, D_u = dilution of unknown, and D_s = dilution of standard*.

Standard solutions are renewed frequently to maintain uniform activity.

This system has worked relatively well in the rennet industry, but changes in strength of the standard over a period of years is a problem that is difficult to assess. Furthermore, the lack of a uniform standard for rennet activity makes it difficult for research workers in different laboratories to compare their results.

The British Standards Institution has attempted to assist with this problem by providing freeze-dried samples of rennet powder of standard activity.

The problem of standardizing the milk-clotting activity of enzyme solutions has been compounded by the introduction of new nonrennin coagulants. From the standpoint of the cheese industry, prime concern is with activity of the enzymes in the cheese vat. When rennet was the sole coagulant, unknown rennet extracts could be standardized against standard rennet solutions on Berridge's or any other reproducible substrate. Even though clotting times varied with different milk samples, changes in the properties of milk always affected the activity of the enzyme in the same way. However, when a nonrennin enzyme coagulant is standardized against a standard rennet solution, it will not have the same activity as rennet when used in cheese milk having a different pH or composition from the substrate in which it was stan-

*The above expression assumes an arbitrary activity of 100 rennin units per ml for the standard.

dardized.[170,256,575] When dealing with porcine pepsin, an additional problem is created by inactivation of the enzyme during dilution with hard water[159] and during coagulation of the cheese milk.[550] The situation has become even more complicated by the marketing of enzyme mixtures such as rennet-porcine pepsin, rennet-bovine pepsin, MP protease-porcine pepsin and rennet-porcine pepsin-bovine pepsin.

Micro Tests for Milk-clotting Enzymes

Gorini and Lanzavecchia[241] described a very sensitive substrate for measuring the milk-clotting activity of low concentrations of proteolytic enzymes. Their substrate, buffered at pH 5.8, consisted of 1 gm nonfat dry milk dissolved in a mixture of 70 ml 6.6 $\times$ $10^{-2}M$ cacodylic acid, 30 ml 6.6 $\times$ $10^{-2}M$ triethanolamine, and 1 ml 3*M* $CaCl_2$. By adjusting the pH of the substrate to 5.7[725] the sensitivity was increased, and Reyes[570] was able to use it to measure the residual rennin in curd and whey.

Elliott and Emmons[155] described a passive indirect hemagglutination test, and a corresponding inhibition test for measuring residual rennin in cheese. They also produced high titer antisera for EP protease and MP protease, and suggested that these enzymes could also be quantitatively detected in cheese.

Lawrence and Sanderson[406] proposed another micro-method for measuring rennin and other proteolytic enzymes. Measurement of concentration was based on the rate of radial diffusion of the enzyme through a thin layer of caseinate-agar gel. The limit of diffusion was marked by a zone of precipitated casein. Cheeseman[85] found that for high enzyme concentrations the milk-clotting assay was superior to the casein-agar diffusion test. Diffusion tests appear to hold real potential for measuring low concentrations of milk-clotting enzymes, but it will still be necessary to correlate the results with milk-clotting activity.

Detecting Enzymes in Mixed Coagulants

Casein-agar diffusion tests have proved useful for qualitative identification of several milk-clotting enzymes.[134,224,572] Richardson[572] was able to detect 0.25% EP protease in rennet, porcine pepsin, and rennet-porcine pepsin blends. MP protease was detectable at levels of 0.5% in rennet, 0.25% in porcine pepsin, and 1% in 50-50 rennet-porcine pepsin blends.

Shovers *et al.*[638] successfully separated several animal and microbial milk-clotting enzymes by electrophoresis on polyacrylamide gels at pH 6.2 to 6.3. Active components were readily identified by layering milk on the slab and noting the spots where clotting occurred. The system was sensitive enough to separate and identify as little as 5% of a clotting enzyme in the presence of another enzyme having a different migration rate.

Shovers and Kornowski[639] reported a quantitative method for measuring porcine pepsin in rennin-pepsin mixtures. The procedure was based on the complete destruction of porcine pepsin at pH 7.3 in 30 min at 30°C. As a small amount of rennin may also be destroyed at that pH, a purified rennin control is run simultaneously with the unknown mixture. However, Shovers[636] reported that rennin as well as bovine pepsin and the commercially available microbial coagulants are completely stable under these test conditions. Clotting activities are determined before and after exposure of the sample to pH 7.3. After correcting for loss of rennin activity, if any occurs, the remaining loss is assumed to be porcine pepsin.

Emmons and Elliott[160] used the same principle of differential pH stability to estimate bovine pepsin, porcine pepsin and rennin in mixtures of these enzymes. Porcine pepsin was estimated by comparing the rate of inactivation of the unknown mixture at pH 6.36 with that of crystalline porcine pepsin. Rennin and bovine pepsin are quite stable at that pH. Thus, if the crystalline pepsin lost 50% of its activity at that pH and the unknown lost 25%, the unknown sample contained 50% porcine pepsin. Rennin was estimated by comparing the activity remaining in a mixture after 30 min at pH 7.65 and 30°C with that of a purified rennin solution. Under these conditions both porcine and bovine pepsins are rapidly inactivated. If the unknown had 25% of its activity remaining, and the purified rennin 50% of its activity, the unknown would have contained 50% of rennin-like activity. Bovine pepsin, possessing intermediate stability, was estimated by difference.[160] Douillard[147] established two sets of conditions involving temperature, pH, and urea concentration for distinguishing between rennin, bovine pepsin I and bovine pepsin II in mixtures of these enzymes. In 7.2*M* urea at pH 6.3 and 10.3°C only bovine pepsin I was destroyed. In 8.6*M* urea at pH 3.25 and 27.5°C, both rennin and bovine pepsin I were quickly destroyed, while the activity of bovine pepsin II decreased very slowly.

ENZYMATIC COAGULATION OF MILK

The only obvious change that occurs when milk is treated with a coagulating enzyme is the formation of a visible clot. Chemical analyses reveal no gross change in composition.[753] The clot is formed by the caseinate fraction of milk, and the presence in the curd of lactose, whey proteins, and fat is due to mechanical occlusion. The coagulum can be dispersed by the addition of alkali or x dialysis against a concentrated solution of sodium chloride.

The main physiological function of rennin is to digest milk in the acid environment of the calf's stomach. The same function can be performed by pepsin in other animals and in humans. Rennin is apparently absent from the gastric secretion of normal human infants.[433] In spite of the pH at which they function physiologically these enzymes rapidly clot fresh milk at pH 6.6 to 6.8 where their general proteolytic activities are extremely low. Therefore, the enzymatic action necessary to cause milk clotting is apparently very specific and restricted.

The early theories of Hammarsten[257] and Van Slyke and Bosworth [706] that attempted to explain the clotting of milk by rennin were based on the belief that casein was homomolecular. They proposed that rennin converted native casein into a new form called paracasein, which clotted in the presence of calcium ion. Hammarsten's idea may be represented by the following scheme:

$$\text{Casein} \xrightarrow{\text{(rennin)}} \text{Paracasein} + \text{Soluble whey albumin}$$
$$\downarrow Ca^{++}$$
$$\text{Insoluble Clot}$$

Early English workers used the terms caseinogen and casein in place of casein and paracasein.[262] Since recognition of the heterogeneity of casein, it has become common practice to use the term paracasein to identify not only rennin-treated whole casein, but also individual casein fractions that have been acted on by rennin, e.g., para-α-casein[88,417] or para-κ-casein.[430]

As stated in the previous section, milk clotting occurs in two separate phases; a primary phase in which the enzyme attacks κ-casein to destroy its stabilizing capacity, and a secondary nonenzymatic phase in which the destabilized system clots in the presence of calcium ion. Some authors[4,338] have also considered a tertiary phase, which is a slow gradual hydrolysis of casein components following clotting. This, however, is not essential to the clotting process.

Primary Enzymatic Phase

Calcium-sensitive α_s-caseins account for 45 to 55% of skimmilk protein, and along with the β-caseins (25 to 35% of skimmilk protein) and κ-casein (8 to 15% of skimmilk protein) make up nearly all the protein in the caseinate micelles of milk.[586]

Waugh and Von Hippel[730] demonstrated that κ-casein functions in a significant way to stabilize caseinate micelles against precipitation, and that the stabilizing action of κ-casein is specifically destroyed by rennin. Zittle[766] and Zittle and Walter[770] verified the stabilizing effect of κ-casein on calcium-sensitive α_s- and β-caseins. Earlier, Linderstrøm-Lang[415] had suggested that the caseinate system of milk was stabilized by a rennin-sensitive protective colloid.

Wake[718] and Garnier[226] showed that the nonprotein nitrogen ("soluble whey albumin" described by Hammarsten) that was liberated from casein by the action of rennin came specifically from κ-casein (see Fig. 12.6). Thus, the scheme proposed by Hammarsten was more accurately described by Mackinlay and Wake[430] as

$$\kappa\text{-casein} \xrightarrow{\text{(rennin)}} \underset{\text{(insoluble)}}{\text{Para-}\kappa\text{-casein}} + \underset{\text{(soluble)}}{\text{macropeptide}}$$

The effect of rennin on the other casein components during milk clotting is minor.[718]

When isolated κ-casein is treated with rennin at pH 6.7 in the absence of calcium, para-κ-casein precipitates,[84] and 23 to 30% of the total nitrogen remains in solution.[47,718] When rennin acts on whole casein, calcium is needed to form a clot or precipitate.[768] In the absence of calcium, para-κ-casein, which by itself is insoluble, must interact with the calcium-sensitive caseins to keep from precipitating.[430] When κ-casein, as part of the caseinate micelle, is attacked by a milk-clotting enzyme, its ability to stabilize the micelle is destroyed. In the presence of calcium ion the calcium-sensitive caseinate fractions along with insoluble para-κ-casein form a clot. However, the mechanism by which clot formation occurs with natural milk micelles is far from clear, and will be discussed later.

Since the splitting of soluble nonprotein nitrogen from κ-casein destroys its stabilizing power, the source of this stabilizing activity has been sought in the soluble products. Wake[718] found that when a solution of isolated κ-casein was treated with rennin, aggregates of para-κ-casein were easily removed by centrifugation, leaving a clear solution containing 23% of the κ-casein nitrogen. No additional precipitate formed when the pH was adjusted to 4.7. However, in 12% trichloroacetic acid (TCA) only 8% of the original nitrogen remained in solution.

The heterogeneous nature of the soluble nitrogenous material split from casein by rennin was suggested by Alais *et al.*[4] when they reported that the amount of NPN recovered from rennin-treated casein varied with the concentration of TCA used as a precipitating agent. The 12% TCA-soluble fraction was almost electrophoretically homogeneous,[1] and had a molecular weight of about 6000 to 8000.[494] Nitschmann

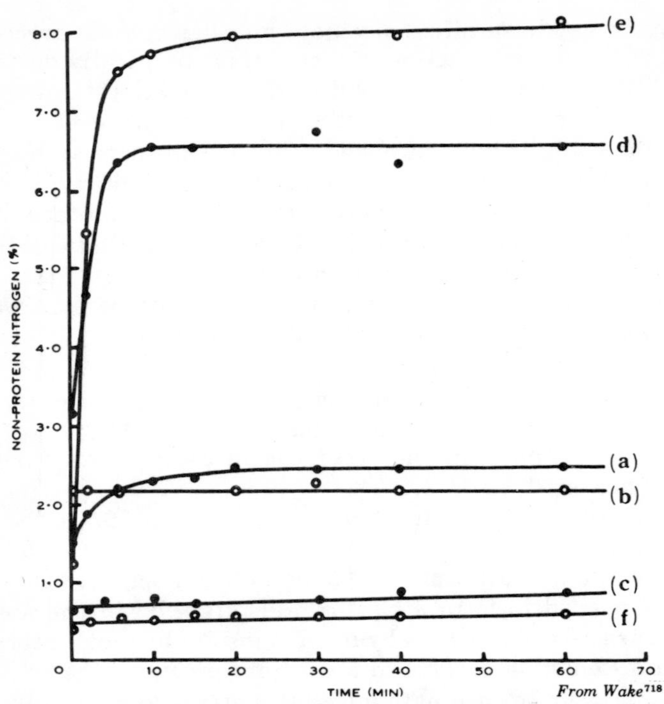

From Wake[718]

FIG. 12.6. LIBERATION OF NONPROTEIN NITROGEN (SOLUBLE IN 12% TCA) FROM MILK PROTEIN FRACTIONS BY RENNIN

All experiments were conducted at pH 6.7 and 30°C using 1μg rennin N/ml. a—2% first cycle casein (largely whole casein). b—Total whey protein. c—2% second cycle casein-fraction P (mostly α_s- and β-caseins). d—1% second cycle casein-fraction S (mostly κ- and β-caseins). e—1% κ-casein. f—2% α-casein.

et al.[495] further characterized this fraction as a glycomacropeptide that was extremely hydrophilic and contained 30% carbohydrate. Alais and Jolles[2] found that the glycomacropeptide contained 28.7% carbohydrate consisting of galactosamine 7.0%, galactose 7.4% and N-acetyl neuraminic (sialic) acid 14.3%. Wheelock and Sinkinson[734] and Sinkinson and Wheelock[644] have shown that the galactosamine is N-acetylated.

It was tempting to think of the strongly hydrophilic properties of the carbohydrate in the glycomacropeptide as accounting for the stabilizing power of κ-casein, since all the carbohydrate appears with the soluble products following rennin action.[340,430,432] However, Gibbons and Cheeseman[233] were able to remove, by neuraminidase action, 97% of the neuraminic acid from κ-casein without affecting its stabilizing power or its response to rennin.

Further progress on this problem awaited development of a method for reducing the disulfide bonds in κ-casein and forming S-carboxymethyl (SCM) or S-carboxyamidomethyl (SCAM) κ-casein.[139,429,752] This procedure led to a demonstration of the heterogeneity of κ-casein through the separation of individual κ-casein molecules normally held together in large aggregates by S-S bonds. Rennin appeared to attack SCM κ-casein in the same way it did κ-casein, viz., para-SCM-κ-casein was insolubilized, and over 25% of the nitrogen remained in solution.[429]

Electrophoretic studies of κ-casein isolated from pooled milk and treated to break disulfide bonds revealed the presence of 7 to 8 discrete bands[429,620,752] and possibly more.[428] All the components corresponding to these bands were attacked by rennin.

Using SCM κ-casein, Mackinlay and Wake[429] demonstrated that the heterogeneity of the soluble products resulting from rennin action arose mainly from the heterogeneity of the original κ-casein with respect to its carbohydrate content, and to two amino-acid differences in the macropeptide moiety.[294] Electrophoretic separation of SCM κ-casein also showed the existence of carbohydrate-free as well as carbohydrate-rich κ-caseins. Both had essentially the same micelle stabilizing power and both were equally sensitive to attack by rennin.[429] Thus the term "macropeptide" is more appropriate for the total NPN split from κ-casein than "glycomacropeptide." The carbohydrate material had little influence on the stabilizing properties of κ-casein, but its presence in the macropeptide reduced its solubility in 12% TCA.[429] This is consistent with the fact that one-third of the NPN liberated by rennin was soluble in 12% TCA.

Since the carbohydrate moiety has been eliminated as an important factor in the stabilizing function of κ-casein, more attention has been given to other properties of the macropeptide. Hill and Wake[295] pointed out that amino acids in the N-terminal two-thirds of κ-casein are strongly hydrophobic, and those in the C-terminal third are hydrophilic. They suggested that the macropeptide portion of κ-casein may provide a solvated coat which functions as the stabilizer at the surface of caseinate micelles.

The presence of low molecular weight dialyzable material that sometimes appears in the soluble portion of reaction mixtures following rennin attack[341,494] is probably not the result of primary enzyme attack, but of minor nonspecific proteolysis.[342]

Thus, the heterogeneity of reaction products resulting from the action of rennin on κ-casein appears to arise from the heterogeneity of κ-casein rather than from the primary action of rennin. It is likely, therefore, that the essential enzymatic action in milk clotting involves the rupture of a single specific bond in κ-casein, although Mackinlay and Wake[429] reported the possibility of major and minor para derivatives of κ-casein that can arise from the attack of rennin at two different sites on the κ-casein molecule.

If the enzymatic liberation of a macropeptide from κ-casein results from the hydrolysis of a peptide bond, a new α-amino end group would appear either on the para-κ-casein or the macropeptide. Early studies by Nitschmann *et al.*[495] and Jolles *et al.*[340] were unable to detect the presence of free α-amino groups in the macropeptide fraction. Wake[718] found measurable levels of dinitrophenyl (DNP) derivatives of aspartic and glutamic acids in both κ-casein and para-κ-casein, but no differences were detected between N-terminal amino acids in κ- and para-κ-caseins. Garnier[225] and Garnier *et al.*[228] followed the course of rennin action on κ-casein with a pH stat and found the same liberation of protons over the entire pH range 5.4 to 7.4. This argued against hydrolysis of a single peptide bond. Failure to find evidence of a new N-terminal amino acid led several workers to suggest that the rennin-sensitive linkage is not a peptide bond.[47,149,341]

Jolles *et al.*[339] reported that the macropeptide released by rennin contained phosphoserine, but no aromatic amino acids, no sulfur amino acids or arginine. Based on analysis with carboxypeptidase, they reported that valine, alanine, threonine, and serine were set free from the C-terminal end of the macropeptide in relative amounts of 1, 1, 0.5, and 0.4, respectively, and that the same amino acids were C-terminal on κ-casein. This provided additional evidence that the macropeptide occupies a C-terminal position on κ-casein.[340] However, they too were unable to find a new N-terminal amino acid following rennin action. Pujolle *et al.*[549] agreed that valine was C-terminal, but were in only partial agreement on the remaining sequence.

The reduction of κ-casein with lithium borohydride by Jolles *et al.*[341] split κ-casein into a soluble fraction which resembled the macropeptide and a precipitate which resembled para-κ-casein. Since phenylalanol was detected in the precipitate, it was suggested that phenylalanine occupied the C-terminal position on para-κ-casein. Phenylalanine or leucine had been previously shown by carboxypeptidase analysis to be C-terminal in para-κ-casein.[340]

After limited acid hydrolysis of the macropeptide, Delfour *et al.*[137] identified a particularly labile N-terminal methionine that had previously escaped detection because of destruction during hydrolysis. de Koning *et al.*[133] also established methionine as N-terminal in the macropeptide. Later, Delfour *et al.*[136] elucidated the amino-acid sequence of the N-terminal octadecapeptide in the macropeptide. This again showed methionine in the N-terminal position. The question of whether the rennin-sensitive bond is a phenylalanyl-methionine peptide bond was clarified when Jolles *et al.*[335] isolated from a tryptic digest of κ-casein, a trideca peptide containing a phe-met bond and having an amino-acid sequence corresponding to the C-terminal end of κ-casein and the N-terminal end of the macropeptide. Additional evidence based on amino-acid sequences[336,337] further supported a phenylalanyl-methionine peptide bond as the primary site of rennin attack on κ-casein. An amino-acid sequence adjacent to the suggested site of rennin attack was proposed by Jolles *et al.*[336] and is illustrated in Table 12.2.

The great susceptibility of the phe-met bond in κ-casein to attack by proteolytic enzymes, $LiBH_4$[341] and heat[3] cannot be attributed to the nature of the bond itself. The same bond in natural[335] and synthetic[285] peptides was not as subject to attack by rennin as was κ-casein. Hill[285,286] reported that rennin split the phe-met bond in the methyl ester pentapeptide, H-ser-leu-phe-met-ala-o-methyl at pH values below 6.3, but would not attack other peptides of this series when serine was missing. Even so, the rate of rennin attack on κ-casein was about 1000 times greater than on the pentapeptide ester. The amino-acid sequence of the pentapeptide ester duplicated that found in κ-casein near the phe-met bond by Jolles *et al.*,[340] Jolles,[338] and Hill.[285] However, in more recent publications by Jolles *et al.*[336,337] the positions of leucine and serine have been reversed (see Table 12.3).

Rennin attack on κ-casein appears to be somewhat dependent on the presence of intact histidine near the susceptible bond.[290] Hill[288] considers that both histidine and serine side-chains in κ-casein act to make the phe-met bond susceptible to enzyme attack. These residues appear to have catalytic effects rather than serving as enzyme binding sites.[286]

The rennin-susceptible bond in κ-casein also appears to be the principal target for nonrennin milk-clotting enzymes. Dennis and Wake[139] reported that rennin, pepsin and chymotrypsin all destroyed the stabilizing power of κ-casein in the same way, and gave rise to the same insoluble products. Habermann *et al.*[251] had previously reported that pepsin liberated the same "glycomacropeptide" from κ-casein as rennin. Evidence for a similar attack on the rennin-sensitive bond in κ-casein has been reported for pepsin, chymotrypsin, a microbial protease,[132] EP protease[396] and MP protease.[760] The general proteolytic activities of these enzymes vary, but their milk-clotting activities seem to result from the same specific action on κ-casein.

Table 12.3

PRESENT KNOWLEDGE OF AMINO ACID SEQUENCE OF COW κ-CASEIN AND THE RENNIN SUSCEPTIBLE BOND

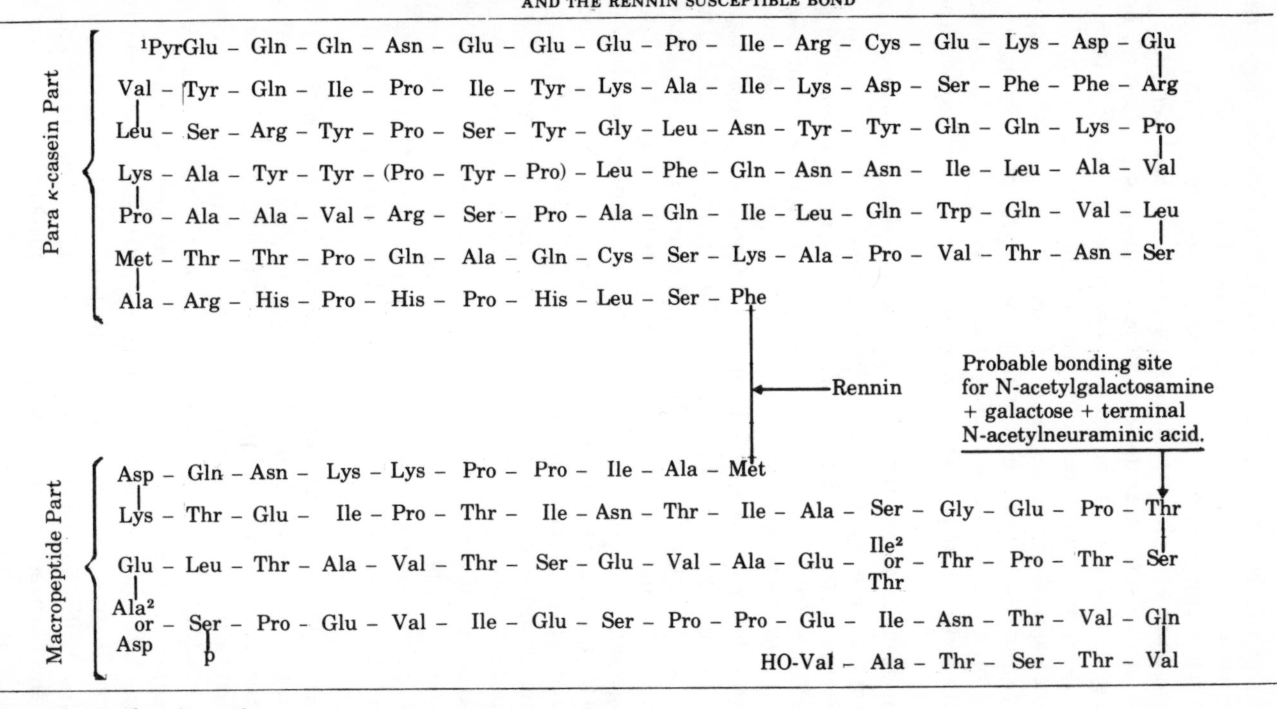

Para κ-casein Part

[1]PyrGlu – Gln – Gln – Asn – Glu – Glu – Glu – Pro – Ile – Arg – Cys – Glu – Lys – Asp – Glu
Val – Tyr – Gln – Ile – Pro – Ile – Tyr – Lys – Ala – Ile – Lys – Asp – Ser – Phe – Phe – Arg
Leu – Ser – Arg – Tyr – Pro – Ser – Tyr – Gly – Leu – Asn – Tyr – Tyr – Gln – Gln – Lys – Pro
Lys – Ala – Tyr – Tyr – (Pro – Tyr – Pro) – Leu – Phe – Gln – Asn – Asn – Ile – Leu – Ala – Val
Pro – Ala – Ala – Val – Arg – Ser – Pro – Ala – Gln – Ile – Leu – Gln – Trp – Gln – Val – Leu
Met – Thr – Thr – Pro – Gln – Ala – Gln – Cys – Ser – Lys – Ala – Pro – Val – Thr – Asn – Ser
Ala – Arg – His – Pro – His – Pro – His – Leu – Ser – Phe

Rennin (arrow pointing to the Phe–Met bond)

Probable bonding site for N-acetylgalactosamine + galactose + terminal N-acetylneuraminic acid. (arrow pointing to Thr)

Macropeptide Part

Met – Ala – Ile – Pro – Pro – Lys – Lys – Asn – Gln – Asp
Lys – Thr – Glu – Ile – Pro – Thr – Ile – Asn – Thr – Ile – Ala – Ser – Gly – Glu – Pro – Thr
Glu – Leu – Thr – Ala – Val – Thr – Ser – Glu – Val – Ala – Glu – Ile[2] or Thr – Thr – Pro – Thr – Ser
Ala[2] or Asp – Ser(P) – Pro – Glu – Val – Ile – Glu – Ser – Pro – Pro – Glu – Ile – Asn – Thr – Val – Gln
HO-Val – Ala – Thr – Ser – Thr – Val

[1]Pyrrolidonecarboxylic acid.
[2]Genetic variant κA-casein contains Thr and Asp in residues 136 and 148, κB-casein contains Ile and Ala.
Sources: Jolles *et. al*,[336a] Jolles *et. al*,[337a] Jolles *et. al*,[337b] and Mercier *et. al.*[362a]

THE NONENZYMATIC PHASE OF MILK CLOTTING

Milk clotting that follows rennin action is not well understood. Much of the problem arises from confusion over the structure of caseinate micelles in milk and the relationship of that structure to known facts about clotting. The properties and structure of micelles prepared from isolated casein components have been thoroughly reviewed by Waugh.[728] Much valuable information has been gained from such artificial systems, but as pointed out by the same author, the natural micelle system has not been studied in detail. Artificially prepared caseinate micelles do not enjoy the same degree of stability as natural milk micelles.[451] In fact, Waugh[728] commented that one of the striking properties of the milk micelle system is its extraordinary stability. Several models have been proposed for the structure of natural micelles,[230,357,474,518,525,585,729] but only limited attempts have been made to explain in detail their enzyme-catalyzed destabilization, aggregation to form a clot, and subsequent syneresis.

As the stabilizing power of κ-casein is destroyed by enzyme action, the caseinate micelles in milk become progressively more susceptible to clotting in the presence of calcium ion. A clot will not form in the absence of calcium. Whole paracasein is very sensitive to slight variations in calcium near the concentration needed for clotting. Pyne[553] found that a variation of 0.2 millimole per l was critical in a total of 3 to 4 millimoles per l at any temperature where coagulation of whole paracasein would occur. He also noted that the calcium requirements for coagulation were affected by the presence of other ions, and that strontium and magnesium are needed in roughly twice the concentration of calcium and barium for equally rapid second-stage clotting. As increasing concentrations of calcium chloride are added to milk, rennin coagulation time decreases, reaches a minimum, then increases again.[573]

Pyne and McGann[554] presented evidence that caseinate micelles as they exist in milk are associated with a colloidal calcium-phosphate-citrate complex. These micelles have remarkable stability to calcium ion, much more so than caseinate in colloidal-phosphate-free milk or artificial caseinate sols prepared from isolated caseins.[451] This observation is difficult to explain when one considers that colloidal calcium phosphate in rennin-treated systems renders the system less stable to ionic calcium at levels such as occur in milk.[553] The great stability of unrenneted caseinate micelles may be explained if one assumes that the colloidal calcium phosphate plays a structural role in the micelle that enhances the stabilizing effectiveness of κ-casein. A structural role for colloidal calcium phosphate in micelles has been suggested by McGann and Pyne.[452] Waugh[728] questions whether inorganic colloidal calcium phosphate plays such a role. Some of the proposed micelle models do not consider it an essential part of the structure,[230,729] while others do.[474,518,525,585]

Green and Crutchfield[247] found that both native and rennin-treated caseinate micelles moved toward the anode in electrophoresis. However, rennin treatment reduced their electrophoretic mobility and zeta potential, indicating a reduction in the magnitude of the negative charge.

Normally, aggregation of caseinate micelles begins before enzyme action is complete; thus the primary and secondary phases of coagulation overlap.[553] Berridge[49] was able to separate the two phases by taking advantage of their different temperature coefficients. He found that the enzyme reaction proceeded at a reduced but reasonable rate at 0°C without producing clotting. By quickly raising the temperature of rennin-treated milk to various levels, he was able to measure the clotting time as a function of temperature, and determine the temperature coefficient for the nonenzymatic phase of milk clotting. He reported it to be 1.3 to 1.6 per °C. Pyne[553] found the coefficient to be even greater (2 per °C) in calcium caseinate solutions. Morganroth[473] noted earlier that rennin would not clot milk at temperatures below 8°C regardless of the length of the incubation period. The difference between temperature coefficients of the enzymatic and nonenzymatic stages of coagulation has been useful as a basis for estimating the time of the nonenzymatic stage at various temperatures.[49,553] It also served as the basis for a continuous curd-forming process in which cold milk was treated with rennet for an extended period, then quickly brought to a temperature where the curd formed instantly.[57,58]

Several investigators have used electron microscopy to study the sequence of physical changes that occur during the coagulation of milk.[307,308,313,529] Initially, short thread-like structures appeared to join the micelles. Later these formed into large fibrils and the caseinate particles tended to agglomerate and eventually change into a cross-linked network of fibrous structures. The final coagulum was described as an irregularly arranged structure of paracaseinate particles in a three-dimensional thread-like network. The milk-fat globules and whey were trapped within the paraceasein structure.

It was suggested[525] that the role of calcium may be to cross-link destabilized caseinate micelles during clot formation. However, experiments have not revealed a difference between the calcium-binding capacity of casein and paracasein.[755,768] Green[244] found that SCM κ-casein, SCM κ-casein containing 2.5 dimethylaminonapthalene sulfonyl residues per mole, and rennin-treated dimethylaminonaphthalene sulfonated SCM κ-casein all bound calcium to the same extent at 30°C and pH 6.5. This added support to evidence that clotting does not involve formation of calcium bridges between micelles.

Hill and Laing[290] inhibited coagulation of paracasein by photooxidation of histidine side chains. Coagulation of rennin-altered whole casein was also inhibited by blocking a few ϵ-amino groups on lysine and 1 or 2 arginine side-chains.[287,289]

Beeby *et al.*[46] suggested that both coagulation and syneresis were the result of interactions in which a number of different groups, including imidazo, guanidine, amine, ester phosphate, colloidal phosphate and ionic calcium, play a part.

Any explanation of the clotting mechanism must be based on an assumed structure of the caseinate micelle in milk. While a number of micelle models have been suggested, three have attracted considerable attention. They are basically quite different and require different explanations of the clotting phenomenon. It is apparent that much work is needed to resolve the experimental problems that have led to such diverse micelle models and their corresponding postulates for the secondary phase of milk clotting.

Coagulation Based on the Micelle Model of Waugh and Noble[729]

Known as the coat-core model, the micelle is portrayed as containing a core of pure calcium α_S-caseinate covered by a uniform coat composed of units of an interaction product consisting of calcium α_S- and κ-caseinate. (See Fig. 12.7.) Coat units can also be constructed of β- and κ-caseinates. The coat units are oriented at the surface of the micelle so that the macropeptide portion of κ-casein is exposed. As the macropeptide is strongly hydrophilic and has great affinity for the milk serum,[295] there is little tendency for surfaces of different micelles to interact with one another.[729] On the other hand, there is strong interaction between the hydrophobic para-κ-portion of κ-casein and calcium α_s- or β-caseins to form coat units. These coat units are in turn strongly interacted with core calcium α_s-caseinate. It is assumed that the micelles are solid spheres and that the average micelle size is determined by the amount of κ-casein available to form coat units.[663] Waugh[728] points out that the calcium-dependency of coat formation must be greater than that required for precipitation of calcium α_s-caseinate.

Milk-clotting enzymes would have easy access to the exposed macropeptides, and following their removal by enzyme action, clotting would occur as a result of interaction of the para-κ or para-coat surfaces. Waugh[728] considered that before two micelles could cohere, a certain

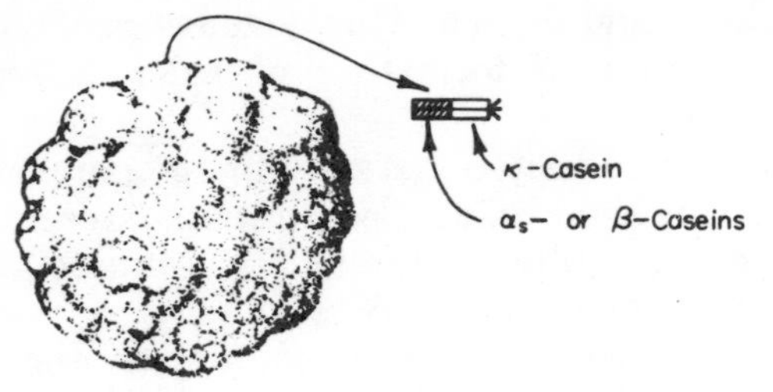

From Waugh[728]

FIG. 12.7. SOLID SPHERE MILK MICELLE MODEL PROPOSED BY WAUGH AND NOBLE[729]

fraction of the coat units within small surface areas on two micelles must be modified. Since some of the κ-casein is in solution and not part of the micelles,[518] it too may enter into clot formation when transformed into para-κ-casein. Polymerization of micelles was attributed to interaction between "para-coat" surfaces without further rearrangement of micellar structures. Strands that have been observed between micelles could be fingers of "para-coat" and possibly free para-κ-casein. In the absence of calcium, para-κ-caseinate can be prevented from precipitation by a small excess of α_s- or β-casein. Since α_s-, β- and para-κ-casein interact very strongly, it is believed that the destabilizing effect of calcium ion is due to its effect on the electrostatic free energy of the micelles.[727]

Coagulation Based on the Micelle Model of Garnier and Ribadeau-Dumas[230]

This micellar model is visualized as having an open sponge-like structure with a uniform distribution of the three different casein subunits (α_s-, β-, and κ-caseins) throughout the micelle. It is composed of α_s- and β-casein polymers held together at the nodes by trimers of κ-casein. The number of α_s- and β-casein subunits in the branches may vary.

Evidence for this model was based on the complete accessibility of all casein subunits to large molecular weight reagents such as carboxypeptidase A[571] and presumably rennin. When rennin was linked to Agarose to form large particles, it rapidly hydrolyzed κ-casein in solution, but could not clot milk.[246] Also Ashoor *et al.*[25] used papain polymerized to glutaraldehyde to form particles larger than casein micelles that were capable of hydrolyzing the micelles progressively from the outside toward the center. The ratios of α_s-, β- and κ-caseins hydrolyzed were similar in both soluble and micellar caseins. These studies suggested a uniformity of composition for micelles with respect to casein subunits.

In this sponge-like model, milk-clotting enzymes could attack κ-casein nodes in the interior as well as on the exterior of the micelle. With κ-casein transformed into para-κ-casein, it could still bind to α_s- and β-caseinate branches and maintain the micellar structure. Through collision of micelles, para-κ-casein trimers on exterior surfaces interact hydrophobically to initiate aggregation of micelles. The micelles would then coalesce to form aggregates. Garnier and Ribadeau-Dumas[230] suggested that mechanical treatment such as cutting, stirring, cheddaring, milling, or even Brownian movement may tend to distort and disrupt the micelles, causing interaction between trimers of para-κ-casein within the micelles. The micelles would collapse and become permanently distorted, as shown in Fig. 12.8. The product

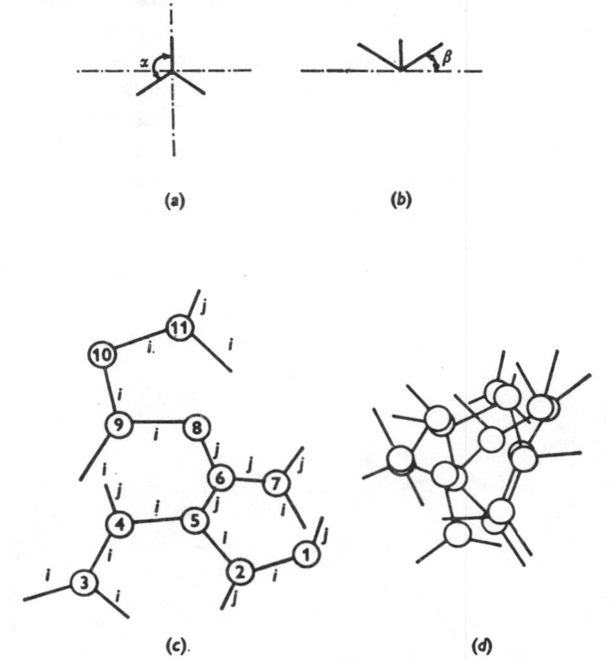

FIG. 12.8. SPONGE-LIKE MICELLE MODEL PROPOSED BY GARNIER AND RIBADEAU-DUMAS[230]

a and b—Assembly of one trimer of κ-casein and 3 copolymers of α_{s_1}-, β-caseins (dash). a—Top projection. b—Side projection. Angles $\alpha = 120°$ and $\beta = 35°$. c—Scheme of micelle model of 161 subunits. "i" corresponds to a mixed octamer, 4α_s- and 4 β-casein subunits, and "j" to a mixed tetramer, 2α_s-and 2 β-casein subunits. The numbered circles correspond to trimers of κ-casein forming the nodes of the network of the micelle. d—Clot of collapsed micelle after transformation of κ-casein into para-κ-casein by rennin. Double circles mean association of 2 trimers of para-κ-casein. Note compactness of the new network at the origin of syneresis.

would become a closely packed network of copolymers of α_S-, β- and para-κ-caseinate, which would readily explain the rapid syneresis of curd that takes place following clot formation. No precise role for calcium or colloidal calcium phosphate was proposed for the model.[230] Green[245] suggested that calcium ion may function by increasing the amount of casein in the micellar form and/or by binding electrostatically to charged groups on the caseins to reduce their net negative charge.

Coagulation Based on the Model of Parry and Carroll[518]

In this model the caseinate micelle consists of polymers of α_S- and β-casein formed around a κ-casein nucleus and held together by colloidal calcium phosphate. The κ-casein in the micelles is not responsible for micelle stability. Parry and Carroll[518] estimated that 30% of the κ-casein in milk is not present in the micelle, but exists as soluble κ-casein associated with small amounts of α_S- and β-casein (see Fig. 12.9). It is proposed that rennin attacks only the small soluble κ-casein aggregates, converting them into highly insoluble para-κ-casein. These in turn, interact strongly with micellar surfaces to build up a 3-dimensional network of micelles connected by interlocking strands of insoluble para-κ-casein. Presumably a given concentration of calcium ion would be needed to maintain sufficient colloidal calcium phosphate in the micelle. Green[244] questioned this model because when milk micelles were separated from the serum, pretreatment of the serum

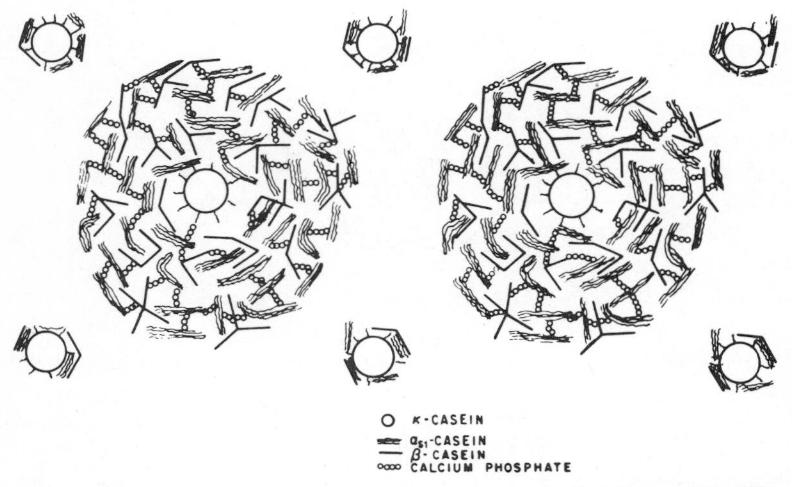

From Parry and Carroll[518]

FIG. 12.9. MILK MICELLE MODEL PROPOSED BY PARRY AND CARROLL SHOWING NON-MICELLAR κ-CASEIN

with rennet did not reduce the clotting time when the micelles were resuspended in the treated serum. When micelles were washed 3 times with milk dialyzate, the clotting time did not increase, but in fact decreased. Since these results were opposite to those reported by Parry and Carroll,[518] additional information is needed before proper conclusions can be drawn about the feasibility of this micelle system.

PROTEOLYTIC ACTION OF MILK-CLOTTING ENZYMES

The influence of milk-clotting enzymes on cheese curing has always been difficult to measure in the presence of the many changes brought about by microorganisms. Rennet certainly contributes to proteolysis during curing, but reduction of rennet to half the normal amount caused a relatively small decrease in soluble nitrogen when compared to the total percentage present in cured cheese.[168] Yamamoto *et al.*[754] made Cheddar cheese with twice the normal proportion of rennet, and found that it increased soluble nitrogen, improved the texture, but did not promote flavor development. The effect of the amount of rennet used in Cheddar cheese making was most noticeable on body and texture only during the very early stages of curing.[168,457] Young Cheddar cheese made with porcine pepsin contained less soluble nitrogen and exhibited a more curdy body than rennet cheese of similar age.[457] However, after curing it was satisfactory for market, and contained reasonable levels of soluble nitrogen.[457] The likelihood that little, if any, porcine pepsin survives cheese making [245] suggests that the proteolytic activity of milk-clotting enzymes assists, but is not essential to cheese curing. On the other hand, too much or the wrong kind of proteolytic activity can adversely affect cheese quality.

Determination of nonprotein nitrogen liberation in milk or cheese is probably not a very critical measure of the proteolytic activity of milk-clotting enzymes. Lindqvist and Storgards[417,419] and Cerbulis *et al.*[77] found that the proteolytic products of rennin action on casein were largely insoluble in 12% trichloroacetic acid (TCA). Fox[206] found that changes in the starch-urea gel electrophoresis patterns of rennin-treated casein provided a more critical evaluation of rennin proteolysis than did increases in nonprotein nitrogen. His electrophoretic studies indicated that the pH optimum for rennin hydrolysis of casein was 5.8. Changes in nonprotein nitrogen suggested a lower pH optimum. Below pH 5.8 the rate of proteolysis and nature of the proteolytic products changed. Rennin proteolysis of α_S-casein was more pH-dependent than that of β-casein. Fox[208] also found that the rate of casein proteolysis by rennin increased as the state of aggregation of caseinate micelles decreased. He considered that colloidal calcium phosphate plays an important aggregating role in micelle structure, and that the proteolytic action of rennin on both α_s- and β-caseins increases with decreasing amounts of micellar colloidal calcium phosphate. Similarly, Ledford *et al.*[407] reported that soluble casein is more rapidly degraded by rennin than micellar casein. Cheddar cheese normally reaches pH 5.0 to 5.2 following manufacture, and according to Pyne and McGann[554] should be nearly devoid of colloidal calcium phosphate. Also the changes brought about during and after coagulation would further tend to disrupt the micellar structure and make the casein more susceptible to enzyme attack.[206]

Lindqvist and Storgards[416] discussed the differences of opinion among research workers concerning the relative resistance to proteolysis of α- and β-caseins. The same authors[417,418] studied changes in electrophoretic patterns of α- and β-caseins during hydrolysis with crystalline rennin. Their patterns indicated that hydrolysis of α-casein over a period of 25 days followed a stepwise degradation, while hydrolysis of β-casein resulted in a profusion of degradation products. Nonprotein nitrogen (soluble in 2% TCA) was liberated 3 times faster from α- than from β-casein. Yoshino[757] found no free amino acids in a fraction soluble in 12% TCA following hydrolysis of casein by rennin. The same author also found no evidence that rennin could hydrolyze β-casein.[758] Prolonged action of rennin on α-casein resulted in a split of the α-casein peak found in electrophoresis, accompanied by the appearance of a slow-moving component (slower than γ casein).[759] The α-casein used in the above experiments included both α_s- and κ-caseins. El Negonmy[156] pointed out that the action of rennin on κ-casein produced several degradation products that on starch gel electrophoresis overlapped those from α_s-casein, and made it impossible to determine their origin. He found that both α_{s1}- and β-caseins yielded specific products whose patterns on starch gel electrophoresis did not alter when rennin action was extended from 3 to 14 hr. The $\alpha_{s2,3,4}$ components were unaffected by rennin during a 14-hr treatment, while 45 to 48% of β-casein was solubilized during 3 hr of rennin action.

Ledford *et al.*[408] examined the proteins in several varieties of cheese by polyacrylamide gel electrophoresis, and found that α_{s1}-casein is much more readily hydrolyzed than β-casein. In fact, α_{s1}-casein was degraded in all varieties examined, whereas β-casein remained largely intact in Camembert, Swiss, Cheddar and Brick cheese. Complete degradation of both α_{s1}- and β-caseins was evident in Blue and Muenster. The proteolytic effects of microorganisms were included in these experiments, although one starterless cheese was examined. Ledford *et al.*[407] found that rennin hydrolyzed α_s-casein much more rapidly than it did

β-casein at all pH values from 7.0 to 5.5. Little activity was discernible above pH 7.0, and activity on both fractions increased with decreasing pH. At pH 6 proteolytic activity on α_s-casein was several times greater than on β-casein. Fox and Walley[210] demonstrated that at pH 6.5 α_{s1}-casein (3%) was optimally degraded by rennin in the presence of 5% sodium chloride, and by pepsin at somewhat higher levels (5 to 10%). Salt concentrations of 15 to 20% somewhat inhibited the proteolytic activity of rennin on α_{s1}-casein, but 50% of the substrate was still hydrolyzed in the presence of 20% NaCl. The same authors found that proteolysis of β-casein by rennin and pepsin was completely inhibited in the presence of 10% NaCl, and significantly reduced by 5% salt. Rennin hydrolyzates of β-casein were very bitter, but hydrolyzates of α_{s1}-casein were not bitter.[210] It was suggested that the effectiveness of NaCl in controlling the development of bitter flavors in Cheddar cheese may be due to its inhibitory effect on the proteolysis of β-casein.[210] Creamer[103] explained the differences between rates of degradation of α_{s1}-*and* β-caseins in Cheddar cheese on the basis of water activity (a_w). In relatively dilute aqueous solutions, α_{s1}- *and* β-caseins were degraded by rennet at comparable rates. In solutions containing 30% sucrose (a_w = 0.983), the rate of β-casein hydrolysis at pH 6.0 was only one-fifth that of α_{s1}-casein. In sugar concentrations of 60% (α_w = 0.963) β-casein was only slightly hydrolyzed, and α_{s1}-casein underwent initial hydrolysis about 20 times slower than when sugar was absent. Whether the a_w affected the enzymes or the substrate was not determined.

It was previously mentioned that the action of most milk-clotting enzymes probably involves the hydrolysis of a specific bond in κ-casein. However, beyond that action, the enzymes vary in their activity on κ-casein and other proteins. Lawrence and Creamer[404] found that the rate of turbidity development during the action of rennin, chymotrypsin, porcine pepsin, trypsin, EP protease, subtilisin and MP protease on 1% κ-casein solutions varied substantially. Also, the turbidity obtained with some of the enzymes depended on calcium-ion concentration. The para-κ-casein coagulum formed was not stable, but was solubilized to various degrees by different enzymes. They concluded that the coagulated products resulting from the action of these enzymes on κ-casein were not all the same. Creamer *et al.*[104] reported that β-casein as well as κ-casein contains a single bond that was quite easily hydrolyzed by rennin, but that pepsin showed no such specificity.

Brandl[69] reported very slight proteolytic activity of rennin in calcium caseinate suspensions at pH 5.5 and 6.5, with greater activity exhibited by EP protease and MP protease. Hansen[261] found that the proteolytic activities of rennin, porcine pepsin, rennet, and bovine pepsin were about the same in 2.8% whole casein solutions at pH 5.5, but that activities of EP protease, MP protease and MM protease were several times greater. The ratios of milk clotting to proteolytic activity in nonfat dry milk reconstituted in 0.01*M* $CaCl_2$ were about the same for rennet and MP protease, but much lower for EP protease due to its very high proteolytic activity.[15] Differences in proteolysis in Cheddar cheese made with EP protease, rennet, MP protease and MM protease were examined electrophoretically by Edwards and Kosikowski,[153] who found that α-casein was attacked least readily by EP protease, followed by the Mucor proteases and EP protease in that order. Mickelsen and Fish[466] did not concur with these results; they reported that EP protease was most proteolytic on both α_s- and β-caseins, and that both EP protease and MP protease produced more nonprotein nitrogen than rennet or porcine pepsin when incubated in casein solutions at pH 6.5, and also when mixed with fresh cheese pastes.

Green[245] pointed out that the activity of protease enzymes varies with pH, and that this, as well as their ability to survive the cheese-making operation, must be considered when evaluating their proteolytic effects on cheese curing.

Proteolytic pH optima reported for several common milk-clotting enzymes were: 1.8[496] and 1.35 to 2.0[392] for porcine pepsin; 3.8 for liberation of nonprotein nitrogen from casein by rennin;[187] 5.8 for degradation of electrophoretic components of casein by rennin;[206] 4.0 and 4.5 for proteolysis of hemoglobin and -casein, respectively, by MP protease;[22] 2.0 and 2.5 for proteolysis of hemoglobin and casein, respectively, by EP protease;[396] approximately 3.3, 4.1, and 4.5 for proteolysis of acid-denatured hemoglobin, urea-denatured hemoglobin and casein, respectively, by MM protease;[655] and 1.35 and 1.35 to 2.0 for proteolysis of N-acetyl-L-phenylalanyl-L-diiodotyrosine and N-acetyl-L-phenylalanyl-L-tyrosine, respectively, by bovine pepsins.[392]

The proteolytic specificities of rennin and porcine pepsin were originally compared on the B chain of oxidized insulin by Fish.[183] His work was re-evaluated by Bang-Jensen *et al.*[41] with the discovery of some major sites of rennin attack that were not previously reported. They concluded that there were no differences between the proteolytic specificities of the individual rennin fractions, but that characteristic differences existed between the specificities of rennin and porcine pepsin. They also showed that pepsin rapidly inactivated ribonuclease, whereas rennin did not. An extremely broad specificity for the proteolytic action of EP protease on the B chain or oxidized insulin was reported by Larsen and Whitaker.[396] The proteolytic action of MM protease on the B chain of insulin was reported by Rickert and is shown in Table 12.4.

Table 12.4

BONDS HYDROLYZED IN THE B CHAIN OF OXIDIZED INSULIN BY THREE MILK-CLOTTING ENZYMES

1 2 3 4 5 6 7 8 9 10 11 12 13 14 15 16 17 18 19
phe–val–asp–glu–his–leu–cy–gly–ser–his–leu–val–glu–ala–leu–tyr–leu–val–cy–

20 21 22 23 24 25 26 27 28 29 30
–gly–glu–arg–gly–phe–phe–tyr–thr–pro–lys–ala

Enzyme	1–2	4–5	11–12	13–14	14–15	15–16	16–17	18–19	23–24	24–25	25–26	26–27
MM protease[577]	Λ+					Λ+	Λ+			Λ+	Λ+	Λ
Rennin[41]	Λ		Λ−	Λ+	Λ?	Λ	Λ?	Λ+		Λ	Λ	
Porcine pepsin[611]	Λ−	Λ	Λ+	Λ	Λ	Λ	Λ+		Λ	Λ+	Λ+	

Λ+ Bonds hydrolyzed rapidly.
Λ Bonds hydrolyzed moderately fast.
Λ− Bonds hydrolyzed slowly.
Λ? Hydrolysis uncertain.

Other Enzymatic Activity.

Phosphoamidase activity of rennin was suggested by Holter and Li[301a] and supported by Dyachenko.[149] However, their results appear to have been due to contaminants in their preparations and not to rennin.[183] Mattenheimer *et al.*[443] found no phosphatase activity attributable to purified rennin, but did find it in some lots of commercial rennet extract.

Bakri and Wolfe[38] reported that lysozyme was able to catalyze the coagulation of casein micelles in a manner similar to the action of rennin. The action of both rennin and lysozyme showed similar pH optima and calcium dependencies for clotting casein.

Czulak[107] suggested that bitter flavors associated with acid cheese may be due to the inability of certain strains of starter bacteria to hydrolyze bitter peptides produced by rennin. Lawrence and Gilles[405] advocated a reduction in the level of rennet used by New Zealand cheese factories because the use of less rennet reduced the incidence of bitterness in Cheddar cheese when certain "fast" starters (HP, ML_8, BA_1, E_8, and ML_1) were used. However, 2- to 3-fold increases in usage of rennet did not increase bitterness when "slow" starters (AM_1, AM_2 and US_3) were used.

INSOLUBLE CLOTTING ENZYMES

Insoluble enzyme derivatives in which the active protein is chemically bonded to a high molecular-weight polymer can be easily removed from reaction mixtures after enzymatic action is complete. They are also suited to continuous or repeated usage. Crook *et al.*[105] summarized the principal methods for the preparation of cellulose-insoluble enzymes and have described the properties of bromelain and its insoluble cellulose derivatives.

Green and Crutchfield[246] proposed the use of insolubilized enzymes to clot milk. The enzymatic reaction would be carried out below 10°C, after which the enzyme could be separated from the milk. The milk would then be warmed to permit second-stage clotting to proceed. They suggested that this would allow reuse of the enzyme, help overcome the world shortage of rennet, and permit the use of different enzymes for clotting milk and curing cheese. They prepared active insoluble derivatives of rennin and chymotrypsin by forming rennin-Agarose (Sepharose 2B Pharmacia), rennin-aminoethylcellulose and chymotrypsin-Agarose. These preparations never exhibited more than 20% of the specific activity shown by the corresponding soluble enzymes. This was attributed to steric hindrance. They also found that specific activity decreased as the particle size of the substrate increased. In milk they could not show that clotting activity was due to the insoluble derivatives because of contamination with free soluble enzymes.

Ferrier *et al.*[182] used porcine pepsin covalently coupled to 40 to 60-mesh porous glass in a temperature-controlled column. Skimmilk acidified to pH 5.6 or 5.9 was passed through the enzyme column at about 15°C, which allowed sufficient enzyme activity to alter the κ-casein and yet prevent coagulation of the skimmilk. After passage through the column, the skimmilk was clotted at 30°C and underwent syneresis at 45°C to form a typical skimmilk curd. In contrast to Green and Crutchfield,[246] they were able to show that enzyme activity was entirely due to insoluble enzymes. However, they were unable to explain a progressive reduction in activity of the pepsin-glass during continuous flow of skimmilk through the column. Porous glass also accumulated colloidal material from skimmilk which was deleterious to its use in enzyme columns.

OTHER FACTORS ASSOCIATED WITH THE CLOTTING PROCESS

A number of additives and conditions may accelerate or retard the visible clotting of milk. They may also alter the properties of the coagulum. Their effect may be on the enzymatic or nonenzymatic stage, or both. Sodium chloride added to milk reduces clotting tendencies and weakens the curd.[645] Di- and trivalent cations usually hasten the clotting of milk by rennin. Rudiger and Wurster[591] reported that $CaCl_2$ has its maximum effectiveness on clotting time at a concentration near 0.142%.

Natural Variations in Clotting Properties of Milk

Milk from animals with subclinical mastitis produces a weaker curd and clots more slowly than normal milk.[649] It has been recognized that milk from healthy animals also varies in its ability to clot with rennin. Disease-free milk with poor clotting properties occasionally causes cheese-making problems. While it appears to be an individual characteristic of some cows,[90] the defect has occurred spasmodically in herds that usually produce milk with normal clotting properties.[450] Mocquot *et al.*[468] observed that slow clotting occurred during periods of maximum milk secretion, but they could establish no definite relationship between milk yield and clotting ability. The addition of acid or calcium chloride to slow-clotting milk may cause an improvement, but some milk does not respond well to that treatment.[450]

Attempts have been made to determine whether differences in the clotting properties of milk are associated with the caseins or with other components of milk. In some instances the results have been contradictory. Chevalier *et al.*[89] removed the caseinate micelles from fast- and slow-clotting milks by centrifugation, and resuspended them in serum prepared from mixed milk. They found that the resuspended caseinate from slow-clotting milk retained its clotting characteristics, as did the caseinate from milk that clotted well. The same authors reported similar results when acid-precipitated caseins were isolated from fast- and slow-clotting milks and reconstituted into an artificial milk. However, no data were presented on their work with acid-precipitated caseins. They concluded that differences in the nature of the caseinate accounted for variations in the clotting properties of their samples. On the other hand, Hostettler and Ruegger[309] found that casein precipitated by acid from fast- and slow-clotting milks had similar clotting characteristics when reconstituted in solutions of the same hydrogen- and calcium-ion concentrations. When reconstituted nonfat dry milk was dialyzed against either fast- or slow-clotting milks, its clotting ability approached that of those milks.[552] Mocquot *et al.*[468] mixed the sediment from centrifuged "slow-renneting" milk with the serum from normal milk and found that the clotting time was intermediate between that of the two original milks. Kosikowski and Mocquot[374] reported that "slow rennet" milks are characterized by low mineral salt content (both calcium and phosphorus), smaller caseinate particle size, and higher moisture content in the caseinate deposits centrifuged from the milk. They also pointed out that curd made from "slow-rennet" milk has, in most instances, a higher moisture content than curd from "fast-rennet" or normal milk, and that this in turn is reflected in the ability of the curd to expel whey during cheese making.

Disruption of the natural fat globule membrane in raw milk results in decreased pH, decreased surface tension, and decreased susceptibility to rennin clotting.[668] In the absence of intact natural membrane material, milkfat appears more susceptible to hydrolysis by milk lipases. Tarassuk and Richardson[669] and Palmer and Tarassuk[516] found that certain free fatty acids (lauric, myristic, and palmitic), when present in milk aged at reduced temperatures, inhibited or completely stopped clotting of the milk by rennin at 35°C. The inhibiting effect was overcome by warming the milk to a temperature above the melting point of the inhibiting fatty acid, then cooling to 35°C and adding rennin. Capric and oleic acids also inhibited clotting by rennin[515], but lower molecular-weight fatty acids did not.[669] It was suggested that solid fatty acids may be adsorbed on caseinate particles in milk and thus block their accessibility to rennin.[669] This, however, does not explain inhibition by oleic acid. Palmer and Hankinson[515] could not completely overcome the inhibiting effect of oleic acid by heating milk. Jenness and Patton[328] considered that the inhibition is more probably caused by free fatty acids binding some of the calcium ion as insoluble salts.

Homogenization hastens rennin coagulation of milk and cream,[731] but reduces firmness of the curd.[692]

Temperature History of Milk

When raw milk is cooled and held at low temperatures its clotting time, measured at 20 to 35°C, increases progressively for several hours.[562] Fox[207] found that in milk from individual cows, the increase in rennin clotting time due to cool-aging ranged between 10 and 200%, occasional samples showing increases up to 700%. Bulk herd milks showed increases ranging from 9 to 60%. The effect of cool-aging was reversed by moderate heating (40°C for 10 min)[207] or by pasteurizing.[562]

The effect of temperatures from 0 to 62°C on the clottability of milk by rennin appears to be largely reversible. Pyne[551] showed that heating milk produced a temporary transfer of some of the soluble calcium and phosphate to the colloidal calcium-phosphate caseinate complex. He considered that this caused an increased sensitivity of the paracaseinate to calcium ion which was responsible for the reduced rennin clotting time. Subsequent reduction in temperature caused a reverse effect as calcium and phosphate moved from the colloidal to the soluble state.[119] According to Pyne,[551] dissolution of the colloidal calcium phosphate decreased the sensitivity of the paracaseinate to precipitation by calcium ion.

Fox[207] showed that the increase in rennin clotting time was inversely related to the ratio of colloidal to soluble phosphate and to the total calcium content of the milk. Addition of low levels of $CaCl_2$ was effective in reducing the susceptibility of milk to cool-aging effects.

Changes in the state of the calcium and phosphate as a result of heating and cooling milk would be expected to alter the pH. This could be an additional factor responsible for the cool-aging effect. Hadlund[253] reported that inoculation of milk with 0.25 to 1.0% lactic culture prior to cold storage at 4 to 5°C reduced the effects of cool-aging on the rennet clotting time by 40 to 55%.

After milk has been heated to 65 to 100°C for 30 min, there is an immediate increase in rennin clotting time, followed by a further increase as the milk is "cool-aged."[539] No doubt heat treatments of this order cause precipitation of colloidal calcium phosphate with an accompanying decrease in pH. On subsequent cooling there would be some dissolution of these salts and an increase in pH. However, the immediate increase in rennin clotting time following exposure of milk to such temperatures suggests that the milk proteins are altered to a degree that more than offsets the effect of precipitation of calcium phosphate. Pyne[551] observed the cool-aging effect in calcium phosphocaseinate sols heated to 85°C for 36 min, but not in simple calcium caseinate sols. Kannan and Jenness[343] found that β-lactoglobulin was involved in both the immediate and delayed prolongation of clotting time caused by heating milk to 85°C for 30 min. Part of their studies were carried out on milk systems practically free of whey proteins. There were prepared by centrifuging the caseinate micelles from skimmilk and resuspending them in protein-free milk serum prepared by dialyzing distilled water against a large quantity of milk. In contrast to the work of Pyne, they found little or no change in rennin clotting time of whey-protein-free caseinate sols upon heating or upon subsequent holding at reduced temperatures. The reason for this difference was not apparent unless it can be attributed to differences in the way their caseinate sols were prepared. When β-lactoglobulin was added to caseinate sols prior to heating, Kannan and Jenness[343] found that the immediate prolongation of coagulation time and subsequent hysteresis followed a course typical for heated skimmilk. They suggested that heating to 85°C for 30 min caused changes involving β-lactoglobulin that interfered with the primary attack of rennin on casein. Zittle *et al.*[769] demonstrated that the rennin coagulation of κ-casein was considerably retarded after a heat-induced interaction between κ-casein and β-lactoglobulin. Hindle and Wheelock[297] agreed that heat sterilization of milk affected the primary phase of the clotting reaction when judged by clotting ability of the milk, but found that it did not interfere with the ability of rennin to release carbohydrate-containing macropeptide material from κ-casein. The same authors[298] later showed that heat sterilization inhibited the release of carbohydrate-free macropeptide material from κ-casein by rennin. They suggested that when whole milk is heated, formation of a complex between β-lactoglobulin and κ-casein takes place more readily with species of κ-casein which lack carbohydrate.

Factors Affecting Curd Firmness

Factors that affect clotting time also influence curd firmness, but the effects are not always parallel. It is generally considered that high acidity favors the formation of a firm curd, but Kelley[349] demonstrated that curd firmness increased with decreasing pH to a maximum at pH 5.8, below which it decreased rapidly (see Fig. 12.10). The clotting time, however, continued to decrease. Like coagulation time, curd firmness varies considerably with breed and individuality of animals.[349] Dilution of milk with whey reduces curd firmness, but to a lesser degree than with water.[349] The effect of dilution on curd firmness is much more pronounced than on clotting time. As the clotting temperature of milk is increased, curd firmness also increases up to 41 to 42°C. Kelley[349] reported that rennin curd formed above 40°C was more rubbery than that formed at lower temperatures, and the curd knife could not make a clean cut. He felt that such curd firmness measurements do not have the same meaning as those taken at lower temperatures.

Certain free fatty acids have a depressing effect on the formation of curd by rennin.[516,668] The fat content of milk has no appreciable effect on clotting time or curd firmness provided the casein content is held constant. Reduced curd firmness observed when fat is increased by standardizing milk with high-testing cream is attributed to a reduction in the percentage of casein.[349] When milk is homogenized before

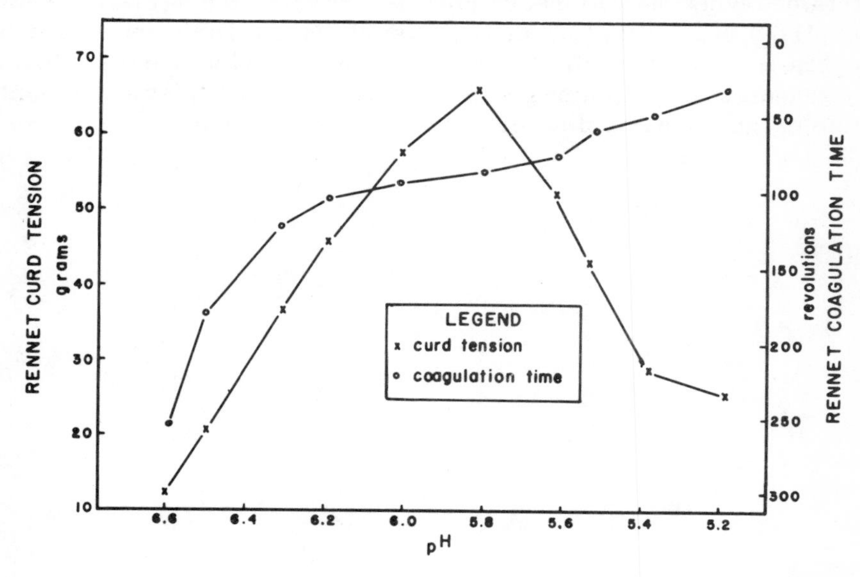

From Kelley[349]

FIG. 12.10. INFLUENCE OF pH ON RENNIN CLOTTING TIME AND CURD FIRMNESS OF MILK

setting with rennet, there is an increase in the retention of fat in the curd.[692] Curd firmness is also reduced, but it can be restored by increasing the solids-not-fat content of the milk.[447] Addition of $CaCl_2$ to milk increases curd firmness, but Kelley[349] obtained no further increase in curd firmness with the addition of more than 0.07% $CaCl_2$.

Factors Affecting Syneresis of Curd

A remarkable property of curd formed by milk-clotting enzymes is the rapid syneresis and expulsion of whey that follows clotting. A micellar explanation of this phenomenon was proposed by Garnier and Ribadeau-Dumas.[230] Lawrence[398] found a linear relationship between temperature and rate of syneresis for about 30 min, after which the effect of temperature was less pronounced. He also reported that the rate of syneresis was considerably increased by stirring, and to a lesser extent by removal of whey. Syneresis was repressed by addition of whey to the curd. Patel *et al.*[519] reported that rate of heating did not have a significant effect on syneresis of curd. Increasing agitation and addition of $CaCl_2$ had slight effects, while temperature and pH had significant effects. Wallace and Aiyar[722] subjected artificial milk systems to dinitrophenylation, and found that it significantly retarded syneresis of rennet-casein gels. The greatest effect was evident at pH 6.8, but it was also noted at lower pH values. Dinitrophenylation appeared to affect only the β-casein in the system, which led the authors to suggest that free amino groups in β-casein probably play a dominant role in the syneresis of rennet-clotted casein systems.

Noble P. Wong

Part II. Cheese Chemistry

INTRODUCTION

Cheese is a highly nutritious and palatable food. It consists of the casein and usually nearly all the fat, insoluble salts, and colloidal material, together with part of the moisture or serum of the original milk in which are contained lactose, whey proteins, soluble salts, vitamins, and other minor milk components. Data on the composition of common varieties of cheese are given in Chapter 2.

In cheese manufacture, the essential solids of milk are reduced to a concentrated form. The milk is coagulated, usually by the addition of rennet or acid, or both. A bacterial culture is customarily added as a starter for acid development. Curdling results from the coagulation of the casein, entrapping as much as 90% of the fat. The curd is cut into cubes and excess moisture is expelled from the cubes by continued action of rennet, by the development of acid, by stirring, by a moderate or high degree of heating and, in the case of the harder varieties, by pressing the curd. The curd is shaped in forms, salt is applied on the surface if it has not been added to the curd in the vat, and the cheese is stored in curing rooms to ripen at a definitely controlled temperature and humidity. Thus, by the action of rennet and specific fermentative agents and the use of mechanical means, milk is transformed into a less perishable and more concentrated food product, which varies in type according to the kind and composition of milk used, the specific agents added, the method of manufacture, and the curing conditions.

Although more than 400 varieties of cheese have been described under different names,[605] many of them are similar, and probably there are only about 14 distinct types of cheese. No two of these are made by the same method; the details of setting the milk, cutting, stirring, heating, draining, pressing, salting, and curing are modified to produce the characteristic flavor, body, and texture peculiar to each kind. The various types are described and data on the composition of common varieties are given in Chapter 2. The details of the technical procedures for the manufacture of different cheeses are beyond the scope of this book. However, suggested references for Cheddar and other varieties are Van Slyke and Price,[708] Wilster,[744] Kosikowski,[369] Wilson and Reinbold,[742] and Davis.[123] Cottage cheese manufacture is described in detail by Emmons[157,158] and Emmons and Tuckey.[173]

CHEMISTRY OF CHEESE-MAKING

Rennet and Acid Curds

The rennet used in the commercial manufacture of cheese is a liquid rennet extract. Unsuccessful attempts to obtain good Cheddar cheese with crystalline rennin led to the conclusion that rennet extract contains needed substances other than rennin.[55,56] Recently, cheese of good quality has been made with crystalline rennin.[421] The characteristics, properties, and activities of rennin and other enzymes in relation to the coagulation of milk in cheese-making have been discussed fully in Part I of this chapter. The coagulum produced by rennet has considerable elasticity and undergoes shrinkage, thereby causing elimination of whey. Shrinkage increases as acidity develops. Curd formed by the action of acid alone, as the pH approaches that of the isoelectric point of casein (pH 4.6), is not elastic but gelatinous and fragile, and tends to shatter more and contract less than that formed by rennet. The rennet coagulum encloses with it most of the fat and insoluble salts of the milk, while in curd formed by acid the insoluble salts are rendered soluble by the acid and are largely lost in the whey. Since about one-fourth of the phosphorus is held in organic combinations and since the calcium becomes soluble more rapidly than the phosphorus and other ash components, the Ca/P ratio and also the percentage of calcium are comparatively low in cheese made under conditions of high acidity during drainage, particularly if the acid curd is washed.

Normally, from 60 to 65% of the calcium and from 50 to 60% of the phosphorus in milk is retained in Cheddar cheese,[144,316,444] but values of 80% and 38%, respectively, have been reported.[449] About 58% of the calcium and 59% of the phosphorus is retained in Brick cheese and 46% of the calcium and 43% of the phosphorus in Blue cheese.[317] Retention in Cottage, Neufchâtel, and Cream cheese is much less, being about 20 and 37% for calcium and phosphorus, respectively.[449] Cottage cheese has been found to contain 0.08% calcium and 0.23% phosphorus,[231] 0.085 and 0.146%, respectively,[317] and from 0.09

to 0.128 and 0.134 to 0.186%, respectively.[449] Rennet and acid types of Cottage cheese did not show significant differences in calcium content.[449] Because the reaction of the rennet curd is close to neutrality, conditions are more favorable for the development of numerous groups of organisms than in acid curd, where the lower pH restricts such growth. Consequently, with rennet-curd manipulations and the use of special ripening agents, curing temperatures and humidities, it is possible to manufacture cheese with a great range of variations in composition and distinctive differences in flavor, body, and texture.

Hard and Soft Cheese

Moisture content is one of the principal factors influencing the firmness of cheese. In hard cheeses it varies from about 30 to 40% and, under suitable storage conditions, these types may be kept for a year or even longer. Semisoft cheeses contain from 39 to 50% moisture. Soft cheeses contain from 50 to about 75% moisture, with a maximum limit of 80% for certain varieties. Soft, high-moisture types are more perishable than hard cheeses. In the manufacture of hard cheeses, the curd is cut finer, usually cooked at a higher temperature, and formed at a greater pressure than in the manufacture of the soft varieties. Lactic acid fermentation causes the lactose to disappear from the cheese, usually within a few days after manufacture. However, the soft ripened varieties contain more whey and therefore more lactose, and the lactose does not disappear entirely from soft cheeses until after 1 or 2 weeks. In the ripening of hard cheeses, the casein is progressively changed into more soluble compounds, such as polypeptides and amino acids, together with a low percentage of ammonia, until at the end of ripening, approximately one-third of the casein has been converted into a water-soluble form.

Cheddar and similar types of hard cheese retain about 80% of the milk calcium, and soft cheeses usually retain about 20%; the phosphorus content varies with the proportion of protein in the cheese, approximately 38% of the milk phosphorus being retained in soft and Cheddar-type cheeses.[449] A study of Cheddar cheese hardness and some associated physical characteristics revealed an inverse relationship between total calcium content and cheese hardness.[732]

A considerable proportion of the paracasein of soft cheeses is converted into soluble forms on curing, and this contributes to the softness; amino acids are liberated from the proteins more rapidly than in hard cheeses, even though the curing temperature is lower; in most soft varieties a greater percentage of ammonia is liberated than in hard varieties. The rate and extent of proteolysis increase with increased moisture content. Proteolysis is also increased by surface ripening that occurs in some soft and semisoft varieties. The products of proteolysis contribute to the intensity and type of flavor. The interior of hard cheeses is entirely anaerobic.[55] The oxidation-reduction potential of hard cheeses decreases rapidly during the first day and remains low.[121] Ripening occurs relatively uniformly throughout hard cheeses, but in soft cheeses having a surface microflora, ripening is more rapid at the surface and proceeds from the surface inward.

Temperature of Coagulation

The temperature employed for coagulation of the milk is between 20 and 40°C, but for most kinds of cheese is usually between 22 and 35°C. For example, milk for Cheddar cheese is usually "set" (rennet added) at 30 to 31°C; Swiss at 31 to 34°C; and Cottage at 21 to 22°C for an overnight method, or at 31 to 32°C for a short-time method. The quantity of rennet and the setting temperature have been correlated with the development of acidity to yield the desired curd characteristics for the kind of cheese being made. The relative efficiency of the coagulation process at various temperatures between 20 and 50°C with fixed proportions of rennet in milk and a fixed time period of coagulation has been determined.[185] The results indicate that the optimum relative quantity of coagulum formed is at 41°C.

Rennet extract contains both rennin and pepsin. On the basis of time required for coagulation, the milk-coagulating power of rennet increases with temperature up to 45°C, but that of pepsin reaches a maximum at 35°C.[694] Equal-quantity mixtures of rennet and pepsin have been used in making Cheddar cheese.[79,162] One investigator found no differences in flavor and body and texture of the two types of cheese, whereas the other observed that cheese made with 100% rennet has slightly less bitter flavor and stronger Cheddar flavor.

Milk is generally coagulated at a lower temperature for soft cheese than for hard cheese. Marked differences in the texture of the curd can be produced by variations in coagulating temperatures. Neufchâtel, Cream, and Cottage cheeses set at from 21 to 27°C have a soft, pliable, jelly-like curd; Cheddar curd formed at 30°C is decidedly firm; and Limburger curd set at 33 to 35°C is distinctly tough and rubbery. Experiments have shown that the elasticity of the coagulum increases in direct proportion to the temperature up to 41°C.[8]

The combined actions of the rennet, temperature, and acid affect the rate of formation, the firmness, the elasticity, and other properties of the resulting curd and the rate and extent of whey expulsion. Failure of any one to function normally may delay drainage and result in defective cheese. The curdling period for hard cheeses, until the curd is ready to be cut, is usually from 25 to 40 min; some of the softer cheeses such as Camembert and Roquefort require from 1 to 2 hr. When the curdling extends over a long period, droplets of moisture collect on the surface, increase in size, and finally unite to form a layer of expelled moisture. The solidifying action of the rennet practically ceases soon thereafter.

Firm-curd milk is said to produce better quality Cheddar cheese than milk in which the coagulated curd is soft.[296] Milk from mastitis-infected cows coagulates relatively slowly with rennet, forms soft curd, and usually has a relatively high pH; the pH is a major factor in influencing the rennet coagulation time, while the percentage of casein is more important in influencing the rennet curd tension.[607] In cheese made from soft-curd milk, the moisture content is comparatively high, the body is weak and mealy, the fat loss in the whey is high, and the yield of cheese is low.

Cutting the Curd

The purpose of cutting is to permit a large proportion of the whey to escape from the curd. The curd is ready to be cut when it will break cleanly, without shattering, ahead of a thermometer that is inserted at an angle and raised slowly. Cutting the curd too soon or. using a curd breaker instead of knives, or cutting it too fine, increases the loss of fat in the whey.

The moisture content of the curd is reduced by cutting the curd into comparatively small cubes.[601,355] When different vats of Cheddar curd were cut into cubes ¼, ⅜, ½, and ¾ in. thick and drained for 2½ hr, the curd contained 50, 53, 58, and 70% of moisture, respectively, showing that cutting into small cubes promotes drainage. Reduction in curd particle size below ¼ in. encouraged moisture expulsion but also increased fat losses.[177]

For Swiss cheese, the curd is cut finer than for Cheddar, resulting in increased whey expulsion during cooking and decreased drainage during pressing.[366] Cutting too fine causes retention of moisture by diminishing the outflow of whey between the particles, resulting in slow and often insufficient drainage during the pressing period. Inadequate drainage results in the retention of excess lactose, causing excessive acidity and subsequent ripening defects. Harping Swiss cheese curd finely causes the cheese to contain about 1% more moisture than harping coarsely, and results in a loss of solids and a slight decrease in yield.[606,613] The presence of so-called cheese dust, caused by harping too finely, tends particularly to cause the retention of whey and may lead to defective fermentation, over-setting, and gasiness, often localized in the form of pin-hole areas or blow-holes.[366]

The curd particles vary in size from that of a hempseed to that of a cherry, and in some cases even larger, depending on the variety of cheese being made. In the manufacture of certain kinds of soft cheese, such as Brie and Camembert, coagulation occurs very slowly, and therefore the curd is not dipped until from 1 to 3 hr after setting. In the manufacture of Cream and Neufchâtel cheeses, the coagulum is poured directly upon drain cloths without preliminary cutting. In the manufacture of many varieties the curd is reduced gradually, by means of a so-called harp instead of curd knives, into small pieces that will drain uniformly without excessive loss of fat. The specific procedure employed in reducing the sizes of the particles influences development of desirable properties of the type of cheese being made. Uniform-sized curd pieces are important in all cutting procedures; lack of uniformity results in variations in firmness, elasticity, and moisture content, and leads to defects in texture and quality.

A thin, continuous coating forms quickly on the surface of each curd particle soon after cutting, and the inner portion contains more whey and is softer. This elastic film retains fat as well as moisture within the curd. When the curd particles are broken, as by excessive or rapid stirring during the early stages, more of the fat is lost in the whey. Some fat loss is unavoidable. Cheddar cheese whey should not contain more than 0.3% of fat. If the curd particles are divided more than usual, the whey may contain from 0.6 to 0.9% fat. The whey may be separated and the resultant cream used for manufacture of butter.

After the curd has been cut and permitted to settle to promote expulsion of whey, it is slowly stirred occasionally, but frequently enough to prevent the particles from cohering and from forming lumps, which may retain excessive amounts of whey and cause defects in the cheese. As the lactic fermentation progresses and the rate of acid formation increases, the curd becomes particularly sticky and tenacious. Subsequent steps in the procedure serve to harden the curd particles and cause them to contract as whey is expelled.

Cooking the Curd and Whey

Heating or cooking is employed in making many kinds of cheese. In the case of Swiss, for example, a culture of lactobacilli and heat-resistant streptococci is added, and the kettle contents are heated to a temperature between 51 and 54°C with continued stirring. For Cottage cheese, the curd is heated to 43 to 52°C. The curd for certain plastic-curd Italian-type cheeses, such as Mozzarella and Provolone, is usually heated to between 55 and 60°C after it has matted, and is worked while hot until it becomes very elastic, shiny, stringy, and free from mechanical openings. The curd for Cheddar cheese is ordinarily heated to only about 38°C. The activity of the usual lactic acid-producing starter, which contains predominantly *Streptococcus lactis* and *Streptococcus cremoris,* is retarded at higher temperatures. However, in a simplified short-time procedure for making Cheddar-type,[723] a starter containing *Streptococcus durans* is also added. This culture is more resistant to heat. With it, the curd is heated to a temperature of 43°C, or even as high as 46°C. The resulting cheese is compact and especially free from mechanical openings.

Heating hastens expulsion of whey from the curd, increases elasticity, changes the texture so that it is more compact and has fewer openings, and alters the bacterial flora. Stirring at a high temperature causes a desirable physical condition of the curd which permits the whey to filter off through and among the cheese particles.[710] The adhesive properties of the granules are so diminished during heating that whey movement between the particles is facilitated in the cheese on the press. The specific gravity of the granules increases from about 1.056 to 1.073 during cooking.[710] Cooking at a high temperature decreases the moisture content[508,710,355,177] and causes the cheese to ripen less rapidly.[508] The cooking temperature of Swiss cheese curd stops the growth of *S. lactis,*[215] and rennet action is greatly retarded, if not wholly inhibited. During heating, the combined effects of heat and acid cause changes in the properties of the rennet curd and it tends to lose its soft, sticky properties. Cheddar curd becomes firmer and more rubbery during cooking, whereas curds that are heated at higher temperatures, such as those of Parmesan, Swiss, and especially Provolone-type (plastic-curd) cheeses, acquire increased plasticity.

Acidity and Hydrogen-Ion Concentration

The coagulating activity of rennet varies with the acidity of the milk. An increase in acidity increases the solubility of calcium salts, and the resulting decrease in pCa accelerates the coagulation of the casein. The rate and extent of acid development influence the texture and subsequent ripening of the cheese. Cheese made with excessive acid development, especially that made from overripe milk, is likely to have so-called acid defects, i.e., bitter flavors, a "short", brittle body that frequently becomes pasty, and a tendency to form splits or cracks. The correlation of acidity with the properties of the milk and curd has been customarily made by determination of titratable acidity of the milk and whey, and this method is still employed. The changes in H-ion concentration can be satisfactorily determined by use of a pH meter. However, if the sample is mixed with water, the ratio of the sample to water should not be greater than 1:2, since dilution of the curd causes a definite shift of the pH reading.[70] It is well known that different samples of milk have different buffering capacities, depending on the concentration of casein, salts, and other components including acids, and that the acids present are only partly dissociated, i.e., only partly effective. For these reasons titration values do not give as true an indication of effective acidity as is obtained by determination of the hydrogen-ion concentration or pH value.

The initial acidity of the milk as well as the subsequent rate and extent of formation of acid in the curd and whey is of major importance in cheese-making. Certain kinds of cheese require comparatively fresh milk, as in the cases of Limberger, Brick, Bel Paese, and Brie, while certain other kinds, such as Cheddar, Camembert, and Blue, can be made from milk that has developed a slight acidity. The development of acid by lactic acid-forming bacteria is essential during manufacture and in the subsequent ripening of cheese. However, excessive bacterial growth and activity in the milk, which sometimes occur under unsanitary conditions and without adequate cooling before delivery, is decidedly harmful. In studies on more than 500 commercial Swiss cheeses, Rogers *et al.*[583] found that milk lacking in ripeness (acid development) and with an average pH greater than 6.51 at dipping, only about 12% of the cheeses were of good quality; with milk ripened slightly and the pH between 6.36 and 6.51 at dipping, 75% of the cheeses were of good quality; and with riper milk, average pH below 6.36 at dipping, 50% of the cheeses were of good quality. In the case of Cheddar cheese made from good-quality pasteurized milk, Wilson[738a] emphasized that control of the development of acidity has more influence on the quality of the cheese than any other factor. Pimblett[537] observed that increased rate of acid production, especially in the early stages of manufacture due to excess starter inocula (up to 5%), influenced appreciably the quality of the cheese. The development of acidity can be controlled within the most favorable limits by using pasteurized milk, by varying the proportion of starter according to the characteristics of the milk, and by permitting the milk to ripen for 1 hr after addition of starter.[422,738a,740] Chapman *et al.*,[80] however, concluded that the use of unripened milk (rennet added 15 min after starter) gave greater control over acid development and moisture expulsion throughout cheese-making than did ripened milk in which the starter was allowed to develop 0.02% lactic acid over and above original acidity. A pH value between 5.40 and 5.50 at the time of milling the curd is desirable for making the best cheese.[740]

The hydrogen-ion concentration is higher (pH lower) in the curd of cheese than in the whey during the period of rapid acid formation; but later, as the rate of acid development decreases, the pH value of the curd and whey tend to become the same.[70] About 95% of the bacteria present while the whey is draining are contained in the curd cubes and not in the whey,[602] indicating that early lactic acid fermentation occurs primarily within the cubes. The acidity within the cubes of Cheddar cheese curd increases faster than that of the surrounding whey, and the whey gains most of its acidity from the curd.[602] Removal of part of the whey from the vat soon after cutting does not affect the rate of separation of whey from curd, but the acidity of the remaining whey increases more rapidly as a result of such withdrawal.

The curd of Cheddar cheese does not show much tendency to cohere immediately after it is cut. With increased acidity and temperature, however, it becomes cohesive and, when the whey is removed, the pieces of curd unite into a single mass so that the original cubes cannot be detected. This fusion of the pieces of curd is known as "matting". It is a part of the "cheddaring" process and is an important step contributing to the desired body and texture. The acidity increases rapidly in the vat and the loss of calcium and other acid-soluble salts is therefore greater than in Swiss and certain other varieties in which the curd is heated more and dipped sooner after setting.

Cheddar cheese of inferior quality results from excessive H-ion concentration or titratable acidity at any given stage during manufacture from either raw milk[70] or pasteurized milk.[740] The acid and bitter flavors that occur frequently in raw-milk cheese are largely a result of overdevelopment of acidity during manufacture.[303]

Typical data[422,740] showing acidity changes in the manufacture of Cheddar cheese of good quality from pasteurized milk are shown in Table 12.5.

The pH value in Cheddar cheese decreases to between 4.95 and 5.3, preferably between 5.05 and 5.20, during the first few days, and then increases only slightly for a few months, finally increasing more rapidly to approximately 5.3 to 5.5 in 1 year;[70] meanwhile, during ripening, the titratable acidity increases from about 0.7% in 1 day to between 1.0 and 1.25% in 1 year.

For Swiss cheese, the making process takes less time and the heating is higher than for Cheddar; the acidity of the whey is less, being usually between 0.10 and 0.12%. The acidity increases only slightly during the making process, e.g., from an average of pH 6.57 to an average of 6.48,[215] and at least part of the decrease in pH is due to physicochemical rather than to bacteriological causes. Because of the high cooking temperature, bacteria do not grow appreciably in the kettle. However,

Table 12.5

TYPICAL CHANGES IN ACIDITY DURING CHEDDAR CHEESE MANUFACTURE

Stage in Making Process	Time After Adding Starter, Hr	Acidity	
		Percent	pH
Before adding starter	0	0.165	6.55
Adding rennet	1	0.175	6.50
Cutting	1½	0.12[a]	6.40
Draining whey	3¼	0.14[a]	6.15
Milling	5½	0.48[a]	5.40
Dressing	7	—	5.30
End of pressing	24	—	5.15

[a] Titration values obtained on whey.

acidity develops rapidly as the cheese cools and drains on the press, largely as a result of the activity of *Streptococcus* ***thermophilus***.[213] After several hours the increase of acidity is carried on by *Lactobacillus bulgaricus* (*Lactobacillus helviticus*) or *Lactobacillus lactis*, closely related types, which thrive as the temperature decreases and which are stimulated by the earlier production of acids by streptococci.[688] The acidity in cheese of good quality reaches a range usually between pH 5.0 and 5.3, or preferably between 5.10 and 5.15, in 1 day.[75,214] When the pH value is below 5.0, the excessive acidity has a very inhibitory effect on *Propionibacterium shermanii*;[687] thus propionic acid fermentation, which is largely responsible for the formation of eyes, fails to proceed normally. Excessive acidity also causes the cheese to be brittle and causes cracks to develop instead of eyes. When the pH is higher than approximately 5.3 at 21 hr, the cheese usually develops too many eyes. This is a common defect known as "oversetting". Rindless, block, Iowa Swiss-type cheese has good eye formation when the pH is 5.05, but it develops too many eyes when the pH is above 5.2.[130]

With Swiss cheese, the rapid development of acidity on the press promotes the drainage of whey and also increases the firmness of the cheese.[726] Meanwhile, insoluble calcium and other salts become more soluble. The acidity of the cheese on the press develops more rapidly near the surface than in the interior, because of the effect of surface cooling on the bacterial flora. Excessive acid development at the surface may be detrimental to proper drainage of whey, leading to defects in ripening.[75] During the ripening of Swiss cheese the pH increases gradually, reaching approximately 5.75 in 9 months; after the 3rd week the pH increases more rapidly near the rind than in the interior of wheel type cheese,[365] but not in the wrapped type.[658]

The rate of lactose fermentation in Emmentaler cheese is greatly reduced as the cooking temperature is increased in the range of 48 to 60°C.[508] The growth of lactic acid-forming bacteria is greatly reduced or completely inhibited; thus the rate of decrease in pH is greatly reduced. However, the rate of acid development is also dependent on the type, quantity, and activity of the starter used.

In surface-ripened cheeses, such as Limburger, Brick, Brie, Camembert, Tilsiter, and numerous others, acidity at the rind decreases during ripening until the reaction becomes almost neutral, while in the inside of the ripened cheese the pH value is usually between 5.2 and 5.6. In Tilsiter cheese, pH values exceeding 5.7 to 5.82 have been found to be associated with defective flavors.[626] Yeasts growing on the surface of Limburger cheese and acting on the proteins and lactates increase the pH to above 5.85 at the surface, after which *Bacterium linens* becomes established and outgrows the yeasts, causing an additional decrease in acidity and additional ripening.[351] In the case of Brick cheese, successive growth of yeasts, micrococci, and *B. linens* reduces the acidity at the surface, e.g., from pH 5.05 to 4.5, in 2 months, producing a pungent aroma which diffuses into the cheese.[394] Fatty acids higher than butyric and acetic are formed by the micrococci associated with the development of the characteristic Brick cheese aroma.[425]

The gradual increase in pH during cheese ripening is caused by destruction of the lactic acid, formation of nonacidic decomposition products and weaker or less highly dissociated acids, including acetic and carbonic, and liberation of alkaline products of protein decomposition.

Lactose is the source of lactic acid in cheese, and in and its fermentation products are soluble in the whey. Relatively rapid development of acidity during manufacture and pressing causes rapid drainage and low moisture content, and therefore tends to cause the pH in young cheese to be higher than normal. This effect has been noted in Swiss cheese[75,213] in which too rapid drainage with high pH in young cheese, causes eyes to form too early and in too great numbers. Too slow drainage, leaving more lactose and resulting in low pH in young cheese, causes deficient eye formation and curd-splitting defects. Among different kinds of ripened cheeses, pH values in young cheeses are higher in hard, low-moisture types than in soft, high-moisture types.

In the manufacture of some cheeses, such as Colby, Monterey, and washed- and soaked-cured varieties of American-type cheese, cool water is added to the curd in the vat. Water is sometimes added in making Swiss cheese. Washing with cool water increases moisture in the curd and dilutes the acid formed, thus increasing the pH of freshly made cheese.Washing also makes the cheese softer, increases openness of texture, and increases perishability. Washing has also proved effective in preventing seaminess in Cheddar.[108]

The curd strength of Cottage cheese made from nonfat dry milk is closely related to pH, but not to titratable acidity.[164] Starter cultures differ greatly in their relations between pH and titratable acidity, presumably because they differ in their production of weekly ionized acids such as carbonic and acetic.

Homogenized Milk in Cheese Making

The use of homogenized milk in cheese making has been reviewed by Peters.[531] Generally the benefits of homogenized milk in the manufacture of cheeses are (1) lower fat losses in the whey, (2) higher yields of cheese, (3) lower shrinkage losses during ripening, and (4) reduced fat leakage at elevated temperatures.

Homogenization of the milk, cream, or mix is a common practice in the manufacture of Neufchâtel and Cream cheese and some types of cheese spreads. Cream cheese made from homogenized milk has a finer grain and a closer texture than that made from unhomogenized milk, and the curd tends to retain more moisture. The process reduces the loss of fat in the whey and is beneficial for some cheese in producing a soft, smooth, stable body that does not lose fat by leakage at room temperature.[435] The immediate effect of homogenization is chiefly on the fat. The size of the fat globules is reduced and their surface area is increased from 10 to 30 times, thus greatly increasing the possibility of hydrolysis of the fat due to lipase activity.[146] The process reduces the rennet curd tension (softens the curd) of whole milk but not that of skimmilk.[607] The pressure is increased up to about 2,000 psi as the homogenization temperature is raised. The curd-softening effect may be caused by increased adsorption of casein on the greater area of the newly formed fat surfaces.[447] Besides causing softer curd, homogenization causes the cheese to contain relatively more moisture and, in the case of hard cheese, to have a relatively brittle body which tends to form cracks. These effects upon the composition and body, when not desirable, can be minimized by separating the milk and homogenizing only the cream,[741] or by adding low-heat nonfat dry milk or concentrated skimmilk.[447]

Homogenization of raw milk produces rancidity and bitterness in Neufchâtel and Cream cheeses.[439] This effect occurs rapidly and is caused principally by the great increase of surface area of the fat globules as a result of homogenization. Homogenization may also cause some activation of lipase, which acts upon the fat and causes hydrolytic rancidity by liberating fatty acids. This defect is overcome by pasteurization before or at the time of homogenization, with heating conditions adequate to destroy or inactivate the lipase. For this purpose, a pasteurizing temperature of 63°C or higher and a homogenizing pressure ranging from 1,500 to 4,000 psi (usually 2,000 to 2,500) has been recommended for processing the cream to be used in making Cream cheese.[114,563,590,764]

Homogenization of the milk for Blue cheese causes considerable improvement in its ripening and flavor development.[391] Growth of the essential mold, *P. roqueforti,* is accelerated, the cheese is lighter in color, the body is softer, the amount of volatile acidity and the acidity of the fat are increased, and the desirable, sharp flavor develops more rapidly. The increased lipolysis results in a marked increase in liberation of fatty acids and subsequent formation of methyl ketones, with a corresponding increase in the typical piquant flavor. With homogenized milk, a distinctly rancid odor, with a bitter flavor suggesting butyric acid, may develop within a few hours during manufacture, but in a few weeks rancidity generally disappears and the desired sharp, peppery flavor becomes predominant.[391] In a procedure described by Babel,[30] the milk is separated and only the cream is homogenized at 29°C and 2,500 psi.

For the manufacture of Cheddar cheese, a procedure was patented[741] in which the milk is separated and the resultant cream heated to 63°C and homogenized at 1,000 to 2,500 psi, to prevent leakage of fat from the cheese in case it is held at a high temperature. Pasteurization essentially inactivates the lipolytic enzymes, and homogenization of the cream instead of the milk results in cheese of better body. Increasing the homogenizing pressure results in progressive decrease in the elasticity of curd, slower whey expulsion, less loss of fat in the whey, less color intensity, and less oiling-off.[530] Cheese of better body and texture was obtained with 500 and 1,000 psi. In the manufacture of Italian cheeses for grating, such as Parmesan and Romano, some lipolysis is desirable to increase the flavor. Flavor development in a grating-type cheese can be accelerated by separating the milk and homogenizing the cream at 2,500 psi at 43°C.[115] The lipase in raw skimmilk liberates free fatty acids from the fat present in homogenized cream, thereby increasing the piquant flavor.

Early research[596] showed that homogenization of raw milk caused serious defects in Brick, Limburger, and Swiss cheeses. The curd was fragile and easily shattered. The cheese developed cracks and flavor defects. However, less fat loss, increased yields, and less fat leakage, but lower quality, have been reported[533] for Swiss cheese made from homogenized, pasteurized milk. A homogenizing pressure of 500 psi gave better overall results than pressures of 0, 1,000, or 2,000 psi.

Clarified Milk in Cheese Making

Clarification of milk is very beneficial in improving the eye formation and increasing the elasticity of Swiss cheese. Experiments conducted by the U.S. Department of Agriculture showed that clarifying milk for Swiss cheese resulted in less "over-setting," larger and fewer eyes, and improved body and texture. Of 241 wheels of Swiss cheese made in factories from clarified milk, 77.6% graded Fancy, whereas only 30.3% graded Fancy of 109 wheels made from nonclarified milk.

The effects of clarification on milk include decrease in the tendency of the fat to form aggregates upon standing; removal of a large proportion of the leucocytes from mastitis milk; increase in the rate of growth of starter organisms; increase in concentration of oxygen and decrease of carbon dioxide; increase in the rate at which the oxidation-reduction potential decreases; and slight decrease in the stability of milk to alcohol.[440] The effects on cheese include marked decrease in number and increase in size and uniformity of eyes; increase in the firmness of the cheese and in the incidence of the glaesler (curd-splitting) defect; increase in acid formation; decrease in moisture content and in yield of cheese; and increase in fat loss in the whey. The beneficial effects of clarification are most pronounced in the case of mastitis milk. No conclusive evidence has been offered pointing to any one predominating factor that would explain the improvement in eye formation due to clarification. Rather, it appears to result from a combination of factors, among which removal of extraneous and cellular material, more uniform distribution of bacteria, and also the effects of agitation in dispersing the fat and decreasing its tendency to form clusters, appear to be important.[250,440,507]

Clarification may affect the composition of cheese. For example, data on 150 pairs of experimental Swiss cheese, analyzed before curing, showed that clarification had the following results: decreased the average percentage of moisture from 39.19 to 38.56; decreased the average yield from 9.82 to 9.65%; and increased the average percentage of fat in the whey from 0.62 to 0.67.[660]

Clarification of milk produces some improvement in the body, texture and flavor of Cheddar cheese.[101,184,305,743] The beneficial effects are greater with milk of good quality than with milk of poor quality.[101]

Pasteurized Milk in Cheese Making

Although the primary purpose of pasteurization is to destroy pathogenic microorganisms, pasteurization in cheese making improves the quality of the product by killing undesirable gas- and flavor-forming microorganisms and inactivating some of their enzymes.

The development of pasteurization of milk in manufacturing Cheddar cheese has been reviewed.[422] Many states have adopted regulations requiring that Cheddar and various other cheeses either be made from pasteurized milk or be cured for a definite length of time to permit any pathogenic organisms that may possibly be present to die or become inactive. Federal definitions and standards of identity for the various kinds and groups of cheeses, issued by the Federal Food and Drug Administration,[701] require that if the milk used is not pasteurized the cheese must be cured for not less than 60 days at a temperature not lower than 1.67°C. It is of vital concern, in the interest of the health of consumers, that all unripened cheese and those which are consumed within a short period, e.g., 2 months, should be made from pasteurized milk.

The Federal definitions and standards for cheeses and cheese products, and also regulations of state and local health agencies, require the use of the phosphatase test as a criterion of pasteurization. The activity of the alkaline phosphatase enzyme of milk is almost entirely destroyed by the specified heating conditions of pasteurization, and determination of the residual phosphatase activity in milk or cheese provides a reliable index of the adequacy of heating. Following the development of earlier phosphatase tests for milk, a test applicable to cheese and other dairy products was developed by Sanders and Sager.[604,608] It has been adopted as an official method and is described in textbooks on methods of analysis.[10,26] Modified procedures have also been developed.[371,431,455]

It has become obvious that the pasteurizing temperature for cheese milk should not exceed by more than a few degrees the minimal temperatures specified above for each respective holding period. For example, with a holding period of 15 sec, the temperature should never exceed 74°C and preferably not 73°C.[422] If the milk is overheated, there is noticeable softening of the rennet curd in milk[564] and of the resulting cheese.[621] The making process is delayed,[353] and the curd is weak and retains excessive moisture. A longer rennet coagulation time is accompanied by a linear decrease in the percentage of nitrogen in the whey.[470,471] The whey from such pasteurized-milk cheese filters slowly and leaves a gelatinous precipitate,[470] indicating a change in the physical properties of the serum proteins, which apparently is a result of partial denaturation by heat. The less complete rennet coagulation of milk after overheating may be partly due to flocculation of serum proteins on the casein micelles.[471]

With high-grade milk, pasteurization has relatively little effect on the rates of protein decomposition and flavor development in Cheddar cheese.[470,609] Ultra-high temperatures used for pasteurization, however, increase protein degradation, decrease rates of amino-acid liberation, and produce sulfide and cooked flavors and weak and crumbly bodies.[458] With low-grade milk, however, protein decomposition and flavor development are much faster in raw-milk cheese than in pasteurized-milk cheese, but quality, including flavor, is distinctly lower in raw-milk cheese. Wilson *et al.*[740] emphasized the importance of quality of milk and pointed out that the detrimental effects of inferior milk can be partly but not entirely overcome by pasteurization, which destroys the lactic acid- and gas-producing bacteria and thus permits the cheese maker to control the fermentation. Although pasteurization destroys the harmful bacteria, it does not entirely remove the effects which these bacteria produced prior to pasteurization. Certain bacterial enzyme systems are much more heat-resistant than the bacteria themselves.[502] Consequently, pasteurization may kill nearly all the bacteria, but some of their enzymes may survive such heating and greatly influence cheese ripening. Many commercial cheese makers prefer to use "heated" milk (60 to 68.5°C for 15 sec or less) instead of pasteurized milk for making Cheddar cheese.[374] They believe that such a heat treatment is sufficient to control undesirable gas- and flavor-forming bacteria and does not adversely affect the development of the desired characteristic sharp flavor. Price and Call[543] made a comparison of the effects of raw and heated milk on the quality and ripening of Cheddar cheese. They concluded that the initial choice of degree of heat treatment of milk for Cheddar cheese manufacture depends on the quality of raw milk-plant sanitation and the market for strong or milk-flavored cheese.

The yield of Cheddar cheese made from pasteurized milk is from 2.5 to 4.0% more than that from raw milk,[260,315,353,422,534,541,544,598] chiefly because of retention of moisture. Several workers have reported less loss of fat and nonfat solids into the whey of cheese made from pasteurized milk than from raw milk.[315,534,541,598] However, Price and Prickett[544] concluded that the differences in retention of fat and nonfat are not great enough to be significant. In making Neufchâtel and Cream cheeses, pasteurization increases fat loss unless the milk is homogenized, and it improves the flavor and keeping quality.[435] Heating at 80°C for 10 min has been recommended. Slow coagulation and mealy, pasty, soft, and fragile defects occur in Cottage cheese made conventionally from overheated skimmilk.[102,165] However, cheese of good quality has been made from milk heated to 79°C for 30 min, to denature whey proteins and thus increase the yield, when the percentage of rennet was increased and the curd was cut under optimum conditions.[165] Increases in cheese yield of 10 to 20% can be obtained in the manufacture of Cottage cheese, Cheddar, and other varieties as a result of substantial retention of the whey proteins by subjecting the cheese milk to heat treatment in a range of 85°C for 15 min to 151.6°C for 0.7 sec.[498]

Pasteurized milk is generally considered unsatisfactory for the manufacture of Emmentaler or Swiss cheese.[217,242] The heating causes changes in the casein which adversely affect the formation of eyes, and there is a tendency for anaerobic organisms to increase and cause defective eye formation.[242] By modifying the manufacturing procedure, however, Emmentaler cheese of good quality has been made from pasteurized milk.[218,746,749] In the case of milk of poor quality, pasteurization, together with the use of proper starters and a modified making process, is said to bring about a decided improvement in the quality of the cheese.[219]

Use of Salts and Other Compounds in Cheese Making

In order to promote coagulation, calcium chloride may be added to milk which otherwise would not coagulate normally. The definitions and standards for cheeses, issued by the Food and Drug Administration,[701] provide for the optional addition of not more than 0.02% calcium chloride to milk for numerous kinds of cheese. The addition of small percentages of calcium chloride causes improved rennet coagulation, as reported in early research on making Limburger cheese[359] and Cheddar cheese[599] from pasteurized milk. Many cheesemakers add it to stimulate rennet coagulation in pasteurized milk and in milk that is deficient in calcium. Often, it is added to normal milk to increase curd strength.

The addition of calcium chloride to raw milk results in faster curdling, improved draining of whey, clearer whey, increased retention of milkfat and other solids, greater yield of cheese, and some saving of rennet.[361,542,702] When too little rennet is used with calcium chloride, protein breakdown is slow and curdiness persists in Cheddar cheese, even though coagulation may seem to be normal.[171] It is believed[361] that the increased precipitation of the curd by calcium chloride is due to at least two factors: first, an effect favoring the formation of insoluble phosphates and citrates of calcium at the expense of phosphates and citrates of sodium and potassium, in which the casein is relatively soluble; and secondly, a depressing effect on the dissociation of calcium caseinate, with which it has a common ion.

Recently, buffer salts, trisodium citrate, and sodium chloride have been added to cheese in conjunction with increased proportions of lactic acid cultures to accelerate ripening.[462] Addition of calcium chloride to skimmilk used in making Cottage cheese has been recommended[691,708] in order to compensate for possible calcium deficiency in the milk, to remedy coagulating defects resulting from excessive heat during pasteurization, and to aid in firming the curd. However, there is considerable evidence that this is neither necessary nor beneficial.[165,277,684]

The use of hydrogen peroxide in cheese milk as an aid in controlling the bacterial flora and thereby improving fermentation and cheese quality has been described[589,209,476] and has found some utility, particularly for Swiss cheese manufacture. Because of the liberation of free oxygen, this compound is germicidal to anaerobes and to at least some members of the coliform group, thus helping to control gassy defects. Clarified, raw milk is treated with 0.02 to 0.05% hydrogen peroxide, heated to 52°C for 25 sec in a high-temperature short-time plate-type pasteurizer, cooled to setting temperature (30 to 34.5°C), and a small percentage of catalase added to destroy the remaining peroxide.[589] The milk is then made into cheese in the usual way.

Attempts have been made, mostly in Europe, to reduce or control the early gassy defect in cheese by means of potassium nitrate (saltpeter) and other oxidizing salts that have preservative properties. The addition of 0.04% potassium nitrate to milk inhibited the growth of the coliaerogenes group, inhibited or prevented formation of gas (carbon dioxide and hydrogen) in cheese, and did not inhibit the desirable lactic-acid bacteria.[750] At a concentration of 10 to 20 gm/100 kg milk, potassium nitrate gave adequate protection against butyric acid fermentation in most cases.[221] Other research[222,714] showed that potassium chlorate is much more effective than either potassium nitrate or potassium nitrite in controlling gassiness in cheese, and is less inhibitory to the normal lactic fermentation. A concentration of 0.002% added to milk was sufficient. However, none of these salts was effective in controlling early gas defects in cheese caused by lactose-fermenting yeasts.

Sorbic acid or its sodium or potassium salt may be added to sliced or cut Cheddar cheese at levels not to exceed 0.3% by weight calculated as sorbic acid.[701] The efficacy of sorbic acid as a fungistat has been adequately demonstrated[627,647] and methods for its analysis described.[390,461,737]

Pasteurized Process Cheese

Cheese may be classified as either natural or processed. Natural cheese is made directly from milk. Process (or processed) cheese is made from natural cheese by grinding finely and mixing together, with heating and stirring, several lots of cheese of the same variety or two or more varieties, together with an added emulsifying agent and specified optional ingredients, into a homogeneous mass, and packaging the resultant product while hot. The kinds of cheese commonly processed include principally Cheddar and related American-type cheeses, and also Swiss, Brick, and Limburger. Other cheese products of this general type include the following, all of which are pasteurized: blended cheese, in which emulsifying salts and acidifying chemicals permitted in pasteurized process cheese are not used; process cheese food, which may contain skimmilk or whey or dehydrated products derived therefrom, as well as certain other optional ingredients, though at least 51% of the weight of the cheese food must consist of the cheese ingredient; and process cheese spread, which is softer and contains a greater percentage of moisture and a lower percentage of fat. The complete specifications of ingredients, composition, and other details are given in the definitions and standards of identity of cheeses, issued by the Food and Drug Administration.[701]

The age of natural cheese used in making the blend has significant effects upon the body, texture, and flavor of process cheese. Cheese less than about a week old cannot be stabilized in processing to prevent loss of fat; but with aged cheese, the use of an emulsifier prevents this instability.[672] Predominance of young cheese in the blend causes the body to be firm and rubbery, but too much aged cheese yields a weak body and grainy texture.[118,513,672,673] The higher pH value and the increased proportion of water-soluble protein in aged cheese may contribute to these effects. A suitable blend consists of 2 or 3 parts of young cheese, about 2 months old, and 1 part of aged cheese, 4 to 7 months old.[672]

The Federal definitions and standards[701] specify that pasteurized process cheese be heated for not less than 30 sec at not less than 65.6°C. Temperatures as high as 70 to 72°C are sometimes used. The duration of heating with agitation at such temperatures in a horizontal cooker may be about 5 and 3 min, respectively, or longer in a kettle-type cooker. The product flows into the packages while hot, and thus the total heating period is greatly prolonged. Process cheese in sealed containers is therefore less perishable than natural cheese. Further ripening does not take place. Process cheese is not sterile, however, and may contain several hundred or even a few thousand viable microorganisms per gram. From 0.5 to 1% salt (sodium chloride) is usually added, which helps prevent microbial action. The normal, alkaline phosphatase present in natural cheese made from raw milk is inactivated by the heat of processing.

In the initial heating there is a slight separation of fat. The melted fat is reincorporated as the temperature reaches about 60°C,[673] and is finely emulsified during subsequent heating and agitation, so that leakage of fat does not occur. The product has been described as a fat-in-water suspensoid, in which small fat particles carry negative charges and an emulsoid sol of hydrated casein and emulsifier is adsorbed at the fat-water interface, preventing loss of charge and fat separation.[513] The smaller the fat particles, the higher is the proportion of stabilizing material adsorbed and the greater are the stability and viscosity. Physically, the stabilizing effect is comparable to the effects of homogenization in stabilizing the fat in milk.

Process cheese heated to 71°C is much firmer than that heated to 65°C.[672] As the temperature is increased, the paracasein becomes plastic and stringy, so that while still hot it can be drawn out in threads and forms a homogeneous, stable mass. In the presence of potassium hydroxide (monovalent cation), the solubility of casein increases as the temperature is raised; but with calcium hydroxide (divalent cation) present, casein does not become more soluble upon heating.[582] A shifting of equilibria in the direction of higher complexes is assumed. The heat of processing also apparently causes an increase in the proportion of bound water in process cheese.

The function of emulsifying salts is to stabilize the product so that fat does not separate and to improve the smoothness of body and texture.[675] The Federal definitions and standards[701] for cheeses specify numerous emulsifiers, principally sodium and potassium salts of phosphates, citrate, and tartrate, in such quantity that the weight of the solids of the emulsifying agents does not exceed 3% of the weight of the pasteurized process cheese. In practice, sodium citrate and disodium phosphate, or a mixture of these, have been generally found to be very effective. The percentage added is often about 2 rather than a maximum of 3. Sodium citrate is said to be most satisfactory.[512,514] potassium citrate may cause an objectionable flavor. The best emulsifier should contain a monovalent cation combined with a polyvalent anion.[252] The relative emulsifying properties of sodium acetate, disodium tartrate, and trisodium citrate are 1:20:100. Citrates and phosphates act as peptizing agents and protein solvents and promote the formation of an adsorbed layer of casein at the surface of each fat globule, preventing the escape of fat. The emulsifier acts as a solvent for cheese protein, and stabilizes the emulsion by adsorption at the fat-water interface.[513]

Control of acidity is an important factor influencing the quality and stability of process cheese.[118] Citric, phosphoric, and other acids may be added to increase acidity. However, according to Federal definitions and standards,[701] the quantity added shall not increase the acidity of process cheese to below pH 5.3, and that of process cheese spread to below pH 4.0. Templeton and Sommer found a pH range of 5.6 to 6.1 preferable for the best body of process cheese,[673] and recommended addition of citric acid to obtain this range. Later,[676] they reported that such values, obtained with suspensions of 1 part of cheese and 10 parts of water, were approximately 0.45 pH unit higher than those with undiluted cheese. The maximum peptizing effect of sodium chloride on paracasein is said to occur between pH 5.5 and 6.0.[635] However, pH values in this range indicate that the acidity is not sufficient to prevent spoilage. Van Slyke and Price[708] recommended a minimum acidity represented by pH approximately 5.4 in these products, to improve the keeping quality. Increased acidity is a factor in inhibiting the activity of microorganisms that may cause spoilage.

Cheddar and Dariworld cheeses having a high pH are too firm, regardless of age, for use in making cheese spreads.[505] The suitability of cheese of normal pH improves as it ages to 60 days.

Among the defects in process cheese, the development of gas and swelling may occur occasionally if the product is not adequately refrigerated. This fermentation is stopped by a processing temperature of 71 to 74°C in the case of unusual contamination with gas-forming microorganisms.[512] In the case of cheese spreads, which are more perishable, pasteurization at temperatures of 71 to 82°C has been recommended to destroy putrefactive organisms.[674] However, large numbers of cocci and spore-forming bacilli were found in process cheese spread made from Dariworld cheese, pasteurized at 85°C for 1 min, and stored at 32°C, but not in spread stored at 21°C.[504] Obviously certain kinds of bacteria can survive the higher temperatures and subsequently grow at a high storage temperature.

Process cheese in which Rochelle salt is used as an emulsifier may be "sandy" as a result of the presence of gritty particles that have been identified as calcium tartrate.[648] "Sandiness" caused by lactose crystals has been observed in cheese spreads containing more than 10% whey solids.[674] Bloom, a surface discoloration observed in processed American cheese slices, has been characterized as a fine network of tricalcium citrate tetrahydrate crystals.[619]

Cheese Curing

The terms "curing" and "ripening" have not been definitely defined in descriptions pertaining to cheese. In this discussion, "curing" is considered arbitrarily as referring to the methods and conditions—temperature, humidity, sanitation, etc.—of care and treatment from manufacture to marketing. "Ripening" refers to the chemical and physical changes taking place in the cheese under the conditions of curing. The curing temperatures for most varieties are generally between 4.4 and 12.8°C, with exceptions for certain varieties. Soft, high-moisture varieties are cured at lower temperatures than hard cheeses, since a combination of high moisture content and high temperature promotes development of inherent defects.

Effects of curing temperatures on the quality of Cheddar cheese have been reviewed by Wilson *et al.*,[739] who found that cheese cured from 3 to 4 months at 10°C prior to low-temperature storage had better quality than cheese stored continuously for 6 months at 1°C. They reported that when cheese was made properly from milk of good quality it could be safely cured at temperatures as high as 10°C, with reasonable assurance of development of good flavor. However, cheese made from poor quality milk required a lower curing temperature. Cheddar cheese cured at 17°C[216] had as good flavor as that cured at 7°C, and the rate of proteolysis was from 40 to 100% more rapid at the higher temperature. In the case of Cheddar cheese made from pasteurized milk, Wilson *et al.*[738a] reported that a curing temperature as high as 10°C, or even somewhat higher, ensures suitable ripening; a temperature as low as 1°C is too low for ripening cheese but is suitable for storing ripened cheese and retarding the development of undesirable flavors. Sanders *et al.*[609] found that Cheddar cheese made from pasteurized milk of good quality was as fully ripened in from 3 to 4 months at 15.5°C as that cured for 6 months at 10°C, and the flavor and quality of such cheese was not injured by the higher temperature during that time. Thus, by using good-quality pasteurized milk, some increase in curing temperature can be employed to compensate for the fact that pasteurized-milk cheese generally does not ripen as rapidly as raw-milk cheese. When low-grade milk was used, however, the quality of the cheese cured at 15.5°C was definitely inferior to that of the cheese cured at 10°C. The defects present in cheese cured at the higher temperature, especially in raw-milk cheese, included acid, sour, bitter, and unclean flavors, gassiness, and crumbly, brittle body. In Cheddar cheeses made with twice the normal amount of rennet, increasing the curing temperature from 10°C to 15°C, markedly promoted ripening, but fermented odors and strong tastes developed.[754] Effects of curing treatments for Cheddar cheese, including temperature and humidity, on the flavor, quality, ripening, and yield of cheese have been described by Van Slyke and Price.[708]

Swiss cheese is usually kept for 10 to 14 days in a cold room at a temperature of 12 to 15°C, meanwhile being salted by immersion in a salt-brine tank. During this time it is firm and rather brittle. It is then transferred to a room at 20 to 23°C. Salt is applied to the surface with the relative humidity in the curing room, generally 80-85%, being sufficiently high so that the salt becomes moist and is gradually adsorbed. Meanwhile the cheese becomes sufficiently elastic so that a plug of cheese may be wound around the finger. During this principal ripening period propionic acid fermentation takes place, in which *Propionibacterium shermanii* and related microorganisms form propionic and acetic acids and carbon dioxide from lactic acid and lactates. The propionic acid contributes to the characteristic sweet flavor, and the carbon dioxide collects to form eyes. The cheese tends to become more firm and less pliable as it loses moisture.[365] If the firmness becomes extreme during eye formation, defective slits and cracks and the so-called "glaesler" defect may occur. pH is a critical factor affecting the growth and activity of propionic acid-forming microorganisms. If the pH is below 5.0 during early curing and eye formation, these organisms are strongly inhibited[687] and the eyes fail to develop normally. Low pH also causes the body to be brittle and to lack elasticity, resulting in cracks rather than round eyes. Relatively high pH, e.g., higher than 5.35, causes the over-setting defect, with rapid swelling and gas formation, and the development of an excessive number of small eyes. In abnormal or exceptionally early eye formation the gas contains a relatively large proportion of hydrogen[306] in addition to carbon dioxide. The gas collects in areas whose localities have no necessary relation to the areas where the gas is produced,[100] and "rapid gas formation must tend to the formation of numerous small holes, while slow gas production must admit of the formation of large holes." When the eyes have developed sufficiently, usually after 4 to 6 weeks in the warm room, the cheese is returned to the cold room at a temperature of about 13°C for further but slower ripening and more flavor development. It may be stored later at a lower temperature.

The interior of Swiss cheese is practically anaerobic,[364] and fat decomposition in cheese of this type is said to be negligible.[748] Proteolysis is said to occur somewhat more rapidly in the interior of wheel-type cheese than at the rind.[243,364]

Cheese Ripening

Biochemical aspects of cheese ripening have been discussed and illustrated diagrammatically by Harper and Kristoffersen,[267] and more recently by Schormüller,[625] with particular emphasis on metabolic pathways, intermediate compounds, and degradation products. Microbiological and chemical aspects of Cheddar cheese ripening have been reviewed by Marth,[438] with major emphasis on chemical reactions and components related to flavor. The literature pertaining to the chemical and microbiological aspects of Cheddar cheese flavor has been reviewed by Mabbitt[427] and Fryer.[220] Certain phases of the chemistry of cheese flavors, including amino acids, fatty acids, secondary degradation products, and possible reaction routes, were reviewed by Harper.[264] Microbial lipase activity in cheeses has recently been discussed.[403]

The principal components of milk which undergo chemical and physical changes in the ripening of cheese are carbohydrate, proteins, and fat. High moisture content promotes rapid ripening. Added salt improves the flavor and retards biological activity. Calcium and other salt and mineral components influence enzymatic activity. Vitamins are necessary as growth factors for microorganisms or as coenzymes.[344] Citric acid contributes to the formation of biacetyl. The principal chemical changes are: (1) fermentation of lactose to lactic acid and small quantities of acetic and propionic acids and carbon dioxide; (2) proteolysis; and (3) breakdown of the fat with liberation of fatty acids. Although pasteurization is commonly employed to eliminate undesirable microorganisms and some of their enzymes, and starters are added to produce the desired flora, it is generally understood that the chemical changes are actually induced chiefly by enzymes. These enzymes are derived from (1) microorganisms added as starters, including lactic acid-forming streptococci and lactobacilli, propionic acid-forming bacteria, and certain molds; (2) various microorganisms present in the milk; (3) the rennet; and (4) the milk.

Carbohydrate Changes.—The changes in lactose occur primarily during cheese manufacture and during the first stages of cheese ripening. Regardless of the stage at which lactic acid is formed, most of the lactose originally present has disappeared after 24 hr with only trace amounts of glucose and galactose detectable for the next 7 to 14 days. The glycolysis of lactose to lactic acid requires about 14 enzymatic steps.[267] Formation of lactic acid is essential for proper manufacture, flavor development, normal ripening, and good keeping quality.

The development of lactic acid is the cause of the principal ripening of cheese in an acid medium. The lactic acid developed represses undesirable microorganisms, such as the coli-aerogenes group, butyric, and bacteria which may lead to the formation of acetic acid, carbon dioxide, hydrogen, and butyric acid.[625]

Protein Changes.—Cheddar cheese, which may be considered as an example, undergoes a series of chemical and physical changes during ripening, which cause the body of the cheese to lose its firm, tough, curdy properties and to become soft and mellow. During this progressive proteolysis the paracasein and the minor proteins are gradually converted to simpler nitrogenous compounds, namely proteoses, peptones, amino acids, and ammonia. Meanwhile, the insoluble constituents are to some extent changed to soluble forms.

In early research, the rate and extent of proteolysis was investigated extensively as a measure of ripening. A study of the distributive pattern of lower molecular-weight peptides in cheese indicated that ripening appears in two stages: hydrolysis of protein to proteoses and peptones, then hydrolysis to the lower molecular-weight peptides and amino acids.[690] Van Slyke and Hart[707] found that, in a ripening period of 1 year, water-soluble nitrogen increased to 44.7% of total nitrogen, amino nitrogen increased to 28.4% of the total, and ammonia nitrogen to 5.4% of the total. Meanwhile, the proportion of proteoses and peptones increased at first and then decreased, the decrease being caused by conversion to amino acids and ammonia. Conditions that increased the rate and extent of proteolysis were: (1) increase in ripening temperature; (2) more rennet; (3) higher moisture content; (4) less salt; (5) larger size of cheese; and (6) moderate content of acid. Incubation of cheese milk for 5½ hr at 30°C and pH 6.6 with 1% *Streptococcus cremoris* did not affect the mechanism of protein breakdown, although ripening was accelerated in Edam cheese.[368] Cheese quality was improved by relatively slow ripening. In cheese cured at 15.5°C[705] water-soluble nitrogen, expressed as percentage of total nitrogen, increased from about 4% in fresh cheese to 25% in cheese 4 months old. Nonprotein nitrogen in ripe Cheddar cheese represented as much as 24% of total nitrogen.[350] A study of Cheddar cheese made from milk heated at different temperatures indicated that protein degradation increases as heat treatment increases, and that the rate of amino-acid liberation decreases as heat treatment increases.[458]

Sanders *et al.*[610] found that proteolysis is much more rapid in cheese made from poor quality milk than in that made from good milk, but the desirable flavor and quality of the cheese is directly related to the quality of the milk and not to the rate and extent of proteolysis. Pasteurization of milk of average market quality causes a decrease in the rate of ripening, as indicated by the rate and extent of proteolysis, but the pasteurized-milk cheeses are superior in flavor and quality. In mature Camembert cheese, as much as 70% of total nitrogen may be protein nitrogen and less than 50% soluble nitrogen, of which nearly 25% may be ammonia.[324]

The rate, nature, and extent of protein decomposition during cheese ripening is related to and influenced by the nature and concentration of proteolytic microbial enzymes, increased moisture content,[600] the presence of lactic acid,[765] temperature,[43,216,610,707] and other factors such as pH, oxidation-reduction potential, and salts that affect enzyme activity.

Protein decomposition, peptide formation, and free amino acids in cheese have been reviewed recently by Schormüller.[625] The pattern of amino acids in cheese protein is similar to that in casein.[20] During ripening, amino acids are liberated gradually by decomposition of paracasein and its intermediate products, certain free amino acids appearing within a few hours after manufacture and numerous others within a few weeks.[111,302] With Cheddar cheese, 18 different free amino acids have been identified by chromatographic methods, and others may be present. Amino acids are liberated more rapidly in cheese made with a supplemental culture of *Lactobacillus casei* than in that made with only the usual lactic starter.[72] A definite relationship exists between the percentage of certain free, water-extractable amino acids and intensity of flavor.[271] A direct relationship between cheese flavor and free tryptophan content has been reported.[173] When a mixture of amino acids in the proportions found in ripened Cheddar cheese was added to fresh curd, the preparation did not have the typical flavor of Cheddar cheese, but the amino acids appeared to contribute an important "background" flavor.[427]

Certain amino acids are liberated in Cheddar cheese within a few days after manufacture, and more rapidly in raw milk cheese than in that made from pasteurized milk.[370] Marked increases of many of the amino acids coincide with optimum flavor and body, but their role in flavor development has not been fully established. Free tyrosine in Cheddar cheese results from hydrolysis of the lower peptides, and an appreciable amount is liberated after the content of soluble protein has reached its maximum.[642] Thus, total tyrosine may be more significant than soluble protein values as a measure of ripening. An early loss of serine, but not of threonine, occurs during normal cheese ripening, and a loss of both occurs during abnormal ripening.[625a] A possible relation of DL-serine deamination and hydrogen sulfide production by certain strains of *L. casei* to Cheddar cheese flavor has been suggested.[385] Arginine is of special interest, because several authors have indicated that it has a repulsive, unpleasant, or bittersweet taste and is responsible for abnormal flavor development.[625]

The amino acid patterns of 30 commercial foreign-type cheeses have been investigated;[373] the reported values ranged from 0 to more than 6 mg per gram of cheese. Although the amino-acid contents differed somewhat in different varieties, no foreign-type cheese could be identified solely by its chromatographic amino-acid pattern. Transaminase and also decarboxylase activity in ripening sour-milk cheese has been demonstrated,[625a] these enzymes being elaborated by microorganisms. Free amino acids are formed to some extent by transamination processes in addition to formation directly by proteolysis. The presence of at least 12 to 19 different free amino acids in various kinds of cheese has been demonstrated by chromatography.[662] However, the presence or absence of certain amino acids *per se* did not determine the specific flavor which resulted from other compounds, probably including the peptide fractions. Several varieties of cheese have been investigated by electrophoretic examination of the peptide fraction from which the casein and liberated amino acids had been removed.[662] This fraction yielded a different and characteristic curve for each type of cheese; the curve changed gradually during ripening, indicating that the peptide fraction and changes therein may be very important in cheese ripening, influencing the flavor and other characteristics of each variety. During cheese ripening κ-, β-, and α-caseins are degraded to compounds of lower electrophoretic mobility.[416]

Various amines are formed in cheese ripening by decarboxylation of amino acids. A considerable quantity of tyramine is formed in Cheddar cheese made with a culture of *Streptococcus faecalis,* in addition to the usual lactic starter.[112,113,372] *Streptococcus durans* and numerous other organisms present in milk also form substantial quantities of tyramine.[311] Although the use of supplemental starters containing *S. faecalis* or *S. durans* greatly increases tyramine production, the amount of tyramine is not always related directly to cheese quality. The use of poor-quality raw milk resulted in increased tyramine content and more flavor in the cheese, but the flavor was less characteristic and less desirable. It has been concluded that tyramine itself is not a determining factor in desirable flavor production, but is merely incidental in the ripening process.[110,311] Cheddar cheese may contain agmatine, cadaverine, histamine, putrescine, tryptamine, and tyramine, and α-amino butyric acid.[643] Cheese made from raw milk contained higher concentrations of these compounds than that made from pasteurized milk, but there was little or no relationship between the concentration of amines and unclean flavor. The concentration of γ-amino butyric acid, however, was generally related to the intensity of unclean flavor, indicating abnormal fermentation.

Lipid Changes.—The lipid fraction of Cheddar-type cheeses contributes more to the development of flavor than any other component. Whereas milk proteins and lactose are sources of many flavor percursors for Cheddar cheese, milk fat is perhaps more important because cheese made from skimmilk does not develop a typical flavor.[501] The role of milk fat in the development of Cheddar cheese flavor has been demonstrated.[401,501]

The action of lipolytic enzymes liberating fatty acids by hydrolysis of the fat is important in the ripening of many varieties, particularly with respect to flavor (see Flavor and Flavor Defect, this chapter). Lipases occur principally from three sources. Normal milk contains lipases; microorganisms involved in the ripening process produce lipases, and some rennet preparations exhibit lipolytic action. Milk lipases are strongly activated by homogenization and to some extent by agitation, foaming, and even changes in temperature. Agitation of cheese milk increases lipolysis in the ripening of Cheddar cheese, causing unclean flavors, bitterness, and rancidity.[302] Some workers assume that milk lipase is inactivated during cheese making, but studies have shown that milk with strong lipase activity produces a rancid cheese.[625]

Rancid flavors are also produced in Cheddar cheese by inoculating the milk with lipolytic bacteria, isolated from silage and low-grade milk.[304] In each instance the flavor defects were proportional to the increase in the acid degree of the fat, and were therefore ascribed to liberation of fatty acids.[710] The microorganisms in raw milk are principally representatives of the *Alcaligenes, Achromobacter, Pseudomonas,* and *Serratia* groups. Cultures of these strains added to pasteurized milk increase fat splitting considerably in cheese made from this milk.[653]

Milk lipases are normally inactivated at pasteurization temperatures. Also easily inactivated are the normal lipase-producing organisms of raw milk (*Pseudomonas Achromobactereaceae*); however, some microbial lipases can withstand the usual pasteurization temperatures very well.[652] These enzymes can be very important in altering the course of flavor development during the ripening period.

The principal ripening of Blue (blue-mold, blue-veined) cheese, like that of Gorgonzola, Roquefort, Stilton, and other blue-veined cheeses, is caused by the lipolytic and proteolytic activity of the blue-green mold, *Penicillium roqueforti,* with which the curd is inoculated. There is also some surface-ripening associated with the formation of slime in which yeasts, mold, and micrococci, *Bacterium linens,* and under some conditions, *Bacterium erythrogenes,* have been identified.[272] Surface-ripening progresses from the surface inward, and is accompanied by increased proteolysis, higher volatile acidity, and increased alkalinity, as well as improved flavor, and quality of such ripening is not excessive.[475] The microorganisms and enzymes in the milk and starter, as well as the rennet, also contribute to ripening.

P.roqueforti produces a water-soluble lipase system which hydrolyzes fat to yield fatty acids, among which caproic, caprylic, and capric acids and their salts yield strong flavors and are largely responsible for the sharp, peppery flavor,[106] whereas the free fatty acids in Blue cheese have been measured quantitatively;[11] other work has indicated that they are not the main flavor components.[477] *P. roqueforti* forms heptanone-2 from caprylic acid, and this ketone is an important flavor component of Blue cheese.[258] Quantitative determination of the C_3 to C_{11} methyl ketones in Blue cheese revealed considerable variation, but heptanone-2 was always predominant.[12] Others analyzed Blue, Camembert, and Roquefort for C_3 to C_{15} methyl ketones and found heptanone-2 in highest concentration in all but one sample of Roquefort cheese.[629,630] Steam distillation and ether extraction of Blue cheese yields a concentrate that contains acetone, pentanone-2, heptanone-2, and nonanone-2.[520] These methyl ketones have an aroma similar to that of Blue cheese. Their precursors are considered to be butyric, caproic, caprylic, and capric acids, respectively. The mechanism generally accepted for their formation is oxidation of the fatty acids to the beta-keto acid and decarboxylation of this acid to a ketone having one less carbon than the precursors.[234] Lowered oxygen tension reduces the amount of ketone formed. When present in sufficient concentration these ketones are toxic to mold, and hence they may inhibit overgrowth of mold.[234] Vacuum distillation and chromatographic analysis of numerous samples of blue-mold cheese showed the presence of the above carbonyl compounds in all samples that had a typical aroma.[472] However, the methyl ketone content was not considered to be the only characteristic of Blue cheese flavor. The secondary alcohols, pentanol-2, heptanol-2, and nonanol-2 are formed in Blue cheese, probably by reduction of the methyl ketones, pentanone-2, heptanone-2, and nonanone-2.[323] The chemical oxidation and decarboxylation of fatty acids by molds, yielding methyl ketones, and the role of fungi in Blue, Camembert, and other varieties of cheese have been described by others[30,660] respectively. The research related to the manufacture of blue-veined cheese has been reviewed.[35,36]

Rennet is another source of lipase, rennet paste causing more lipolysis and a better flavor than rennet extract.[31] In some instances rennet paste causes a rancid flavor in early ripening, but this defect usually disappears later. In the manufacture of certain Italian-type cheeses, rennet paste has been found to produce more flavor than rennet extract, since the paste contains lipolytic as well as proteolytic enzymes in addition to rennin. The lipases liberate fatty acids during ripening and thus increase the volatile acidity and enhance the sharp, characteristic flavor.[270] Glandular enzyme preparations obtained from calves and goats are also being used, together with rennet extract, producing lipolysis and proteolysis in the cheese similar to the effects of rennet paste.[176] The glandular preparations contain the same multiple lipase systems as rennet pastes provided they are from the same animal source, but the lipases from calves, kids, and lambs are different.[266] These products were used in a series of investigations to determine their lipolytic activities and their effects on cheese flavor. Purified glandular products produced Provolone cheese similar to that made with crude rennet pastes.[269] The free fatty acid content varied with the animal source of lipase, while the free amino acid content was related to the type of bacterial starter. The concentrations of free butyric and free glutamic acids were related to characteriatic flavor intensity. In Romano cheese the content of free glutamic and of free butyric acids was generally related to flavor intensity.[423] However, other compounds, including higher fatty acids, were recognized as possible contributors to flavor. Later work[424] indicated that the glutamic acid content of Romano cheese appeared not to be related to flavor intensity. The rate of formation and the concentration of free butyric, however, was definitely and directly related to the animal source of lipase, that from kids being more active than that from calves. Pancreatic lipase released predominantly fatty acids C_{12} or higher from synthetic triglycerides. *Aspergillus* lipase released primarily the lower ones, and milk lipase released significant concentrations of both higher and lower fatty acids.[263]

Effect of Salt.—Salt is added to cheese to improve flavor and to control ripening. It also has a desiccating effect[587,707] and thus it increases firmness,[732] especially with respect to rind formation, so that the cheese is subject to less damage in handling. The quantity of salt may differ, within certain limits, for each variety. The salt that affects bacterial growth and ripening is only that in the moisture content of the cheese. Too little salt in Cheddar cheese causes a weak and pasty body, abnormal ripening, and increased shrinkage in curing; and too much salt causes a dry, brittle body and cracking of the rind, with possible mold damage and discoloration.[700] Cheddar cheese with a high salt content (1.7% or more) may have less tendency to become bitter than cheese with less salt, and cheese containing too little salt develops a low pH.[699] Over the range of 1.25 to 5% the salting rate had little effect on the grade score of Cheddar cheese.[478] Variations arising from hand salting, however, can result in inconsistent grading.[499] The high concentration of salt applied to the surface of Limburger cheese suppresses undesirable organisms, but permits the growth of yeasts and *B. linens,* which contribute to surface-ripening.[351] In mold-ripened cheeses, salt restrains the growth of *Oidium lactis* (*Geotrichum candidum*) in Camembert and inhibits it in Roquefort, but does not prevent the growth of the molds active in the ripening of Camembert, Roquefort, and ripened Neufchâtel cheeses.[677] Salt inhibits the growth of certain microorganisms more than others; thus, the salting of Camembert cheese is said[613] to aid in regulating the proper proportions of the organisms which effect the ripening. By its ability to inhibit bacteria, salt decreases the rate of formation of protein decomposition products.[603,707] The presence of salt influences the solubility of nitrogenous compounds in the cheese[603,765] and therefore modifies the body.

Over-salting of Swiss cheese causes the eyes to form slowly and insufficiently, and too little salt is one of the causes of the "over-setting" defect. The outer portion of surface-salted cheese contains more salt than the interior, thus accounting in part for the formation of smaller and often less numerous eyes near the surface. The activity of propionic bacteria, which ferment lactates and produce carbon dioxide for eye formation, is retarded by 0.5% salt and practically stopped by 2.5%.[509] In the case of Brick cheese, which is salted on the surface, excessive salting (more than 2%) causes delayed lactose fermentation, slow ripening, firmness and curdiness of body, unnatural white color, and loss of yield; too little salt causes abnormal fermentation, weak body, and open texture.[76] Salt extracts moisture from Brick cheese, the rate of moisture loss being much greater during salting than either before or after salting.[393] The time of salting is important; salting on the first day after manufacture, rather than later, favors the type of fermentation that improves the flavor and body.

Flavor and Flavor Defects

According to Harper,[264] cheese flavor can be divided into two categories by distillation: the nonvolatile portion responsible for taste, and the volatile fraction responsible for aroma. The nonvolatile part contains the amino acids, the nonvolatile acids, amines, minerals, and salts, while the volatile part consists of fatty acids, aldehydes, ketones, alcohols, amines, esters, hydrogen sulfide, and sulfides.

General reports on Cheddar cheese flavor were recently publisted by Mabbit,[427] Marth,[438] and Fryer.[220] Volatile compounds found so far in Cheddar cheese are fatty acids, aldehydes, methyl ketones, sulfur compounds, primary and secondary alcohols, esters from primary and secondary alcohols, and delta lactones. Newer techniques such as gas chromatography and mass spectrometry have been responsible for recent advances in the isolation and identification of volatile compounds in cheese.[68,129,413,619,751] Although the list of identified compounds is increasing, often too little attention is paid to the significance of these compounds.[203]

Neutral carbonyl compounds identified in Cheddar cheese are formaldehyde, acetaldehyde, acetone, 2-butanone, 2-pentanone, 3-methyl butanal, 2-heptanone, 2-nonanone, 2-undecanone, 2-tridecanone, 2-pentadecanone, diacetyl, *n*-decanal, *n*-dodecanal, and acetoin.[44,45,127,128,380,523] Volatile carbonyls are considered to be significant by some[273] but not so important by others.[274] Methyl ketones, however, are responsible for the characteristic aroma of Blue and Roquefort cheeses.[106,258,520,629,630] From 4 to 11 keto acids and from 2 to 8 neutral carbonyl compounds were found in each of 4 varieties of cheese, namely: Cheddar, Swiss, Blue, and Camembert.[45] For each neutral carbonyl, there was a corresponding possible immediate precursor in the group of acidic carbonyls. Some workers[312] concluded that acidic and neutral carbonyl compounds are not related to Cheddar cheese flavor, but impairment of flavor may be associated with increase in pyruvic acid, decrease in α-ketoglutaric or disappearance of oxalacetic acid.

Ten volatile carbonyls were identified definitely and 4 tentatively in Cheddar cheese.[127] The complete mixture gave a detectable, but not typical, cheese-like aroma. A more cheese-like aroma was obtained from a synthetic mixture of carbonyls, and acetic, butyric, caproic, caprylic, capric, and 3 mercaptopropionic acids. Although the aroma of this mixture was the most typical of any evaluated, it was incomplete.

The importance of fatty acids to the aroma of Cheddar has recently been emphasized.[203] Several workers have studied the free fatty acids of Cheddar,[521] measured their concentrations from butyric to linoleic,[61,141] and noted their changes during ripening.[142,399] There seemed to be no correlation of fatty-acid content with flavor except in rancid samples, where the concentration of fatty acids was much higher.[61,142] In addition to acetic and propionic acid, 2-methyl butyric acid was identified in Swiss cheese and considered to be flavor-significant.[395] The only differences observed between the fatty acids of Cheddar and Swiss, as determined by gas chromatography, were higher proportions of propionic and smaller proportions of isovaleric in the Swiss.[522]

Generally sulfur compounds are thought to contribute to Cheddar flavor. Of the many compounds identified in Cheddar cheese aroma[523] dimethyl sulfide was the only one considered to be obviously and directly important to cheese aroma, except possibly diacetyl and 3-hydroxybutanone. Volatile carbonyl compounds and hydrogen sulfide in Cheddar cheese have been studied in relation to cheese flavor.[719,721] Evidence[720] suggests that Cheddar cheese flavor may be due to a blend of methyl ketones, fatty acids, a single substance such as hydrogen sulfide. Other research[381] showed that the concentrations of ammonia, free amino acids, and free fatty acids in Cheddar cheeses increase continuously during ripening, whereas the concentration of hydrogen sulfide fluctuates. Raw-milk cheese contained more hydrogen sulfide than did pasteurized-milk cheese, but their contents of ammonia, amino acids, and fatty acids were about the same. The flavor of the cheese seemed to be related more to the ratios of free fatty acids and hydrogen sulfide concentrations than to any other compounds or combination of compounds, suggesting that cheese flavor is dependent on the degradation of both fat and protein.

Definite and probably highly significant differences have been found between the concentrations of hydrogen sulfide and free fatty acids in commercial Cheddar cheese made from raw milk and those in cheese made from pasteurized milk.[382] Also the ratio of free fatty acids to hydrogen sulfide was smaller in raw-milk cheese. The intensity of flavor appeared to be related to this ratio.

Results have indicated a relationship between Cheddar cheese flavor and active-SH sulfhydryl groups[60] as determined by the thiamine disulfide method. Using a different method for determining hydrogen sulfide in Cheddar cheese, it was concluded that there is no direct relationship between H_2S concentration and typical Cheddar cheese flavor.[400]

Methanethiol was isolated and identified as a component of Cheddar cheese aroma; however, there was no estimation of its significance.[412] Hydrogen sulfide and methanethiol were involved in the flavor of Trappist, Liederkranz, and Brick cheese, however.[248]

The degradation of certain amino acids, particularly methionine, to aldehydes having one less carbon atom (Strecker reaction) has been considered important in the development of Cheddar cheese flavor.[348] The addition of less than 1 ppm of methional to Cottage cheese curd imparted a Cheddar cheese flavor.

An early transient bitterness in Cheddar cheese has been ascribed to high proteolytic activity of rennet at low pH, with an accumulation of polypeptides including bitter peptones, and inability of bacterial proteinases, at low pH, to convert the polypeptides to amino acids.[107,326] Some starter cultures are less active than others in their ability to degrade polypeptides to amino acids.[109] Many strains of *S. cremoris* produce bitter cheese.[161] Presumably all strains form bitter peptides, and most strains hydrolyze them to nonbitter substances. In general, the degree of bitterness has been related to low pH of cheese, but some strains produce bitter cheese regardless of pH.

The sweetish flavor of Emmentaler cheese was ascribed largely to the amino acid proline,[716] but other amino acids and also propionic and acetic acids were believed to contribute to the characteristic flavor. Others[300] have found that high values for proline and propionic acid are associated with desirable flavor in domestic Swiss cheese, and that butyric acid is detrimental to good flavor. The amino-acid content of Emmentaler cheese increases rapidly during ripening.[325] Various off-flavors have been investigated. The fruity flavor defect in Cheddar cheese was due to ethanol, ethyl caproate, and ethyl butyrate.[60] Phenolic flavor[34] in Gouda was due to *p*-cresol produced by a special group of lactobacilli isolated from inadequately filtered rennet. A catty flavor called Ribes[33] or odor of feline urine[454] was investigated simultaneously by two workers. It was identified by one as a 6-carbon monounsaturated ketone and more conclusively as 2-mercapto-2-methyl-pentane-4-one. Although the compound was not found in cheese, it was shown to arise easily by the reaction of mesityl oxide and hydrogen sulfide, components which are found in cheese with Ribes flavor defect.

The available knowledge of the chemistry of cheese ripening including specific compounds and their relation to flavor seems to justify the conclusion that each variety of cheese, with flavor, body, and texture characteristics different from other varieties, is characterized by a specific type of ripening, and that type of ripening depends on the bacterial flora of the milk and the starter, the composition and enzyme content of the milk, the rennet or other added enzymes, the amount of added salt, and the conditions of manufacturing and curing.

Many flavor-producing substances such as peptides, amino acids, fatty acids, aldehydes, carbonyl compounds, ketones, alcohols, esters, diacetyl, ammonia, and hydrogen sulfide have been isolated from cheese. However, no single one or combination of them possesses a typical cheese flavor. Most authorities believe that typical cheese flavor results from the blending of a variety of specific individual substances in proper proportions. This viewpoint has become known as the "Component Balance" theory of flavor origin.[374] According to this theory, there is a certain background flavor more or less common to all varieties of ripened cheese. Specific varietal flavor results from a blending of specific flavor substances in balanced proportions with the background flavor. Shifts in the "Component Balance" could cause shifts in the quality or the intensity of cheese flavor. A small percentage of diacetyl, 0.05 mg or less per 100 gm, contributes to the desired characteristic flavor of Cheddar cheese, whereas higher content may produce a flavor defect. The great number and variety of flavor-contributing substances in cheese would seem to preclude a possibility that any single substance *per se* is responsible for typical cheese flavor. However, there does remain a possibility that a single substance or a group of substances may be responsible for the typical distinguishing characteristic flavor of each variety.

REFERENCES

1. ALAIS, C., 14th Intern. Dairy Congr. II, 823 (1956).
2. ALAIS, C., and JOLLES, P., Biochim. Biophys. Acta, *51,* 315 (1961).
3. ALAIS, C., KIGER, N., and JOLLES, P., J. Dairy Sci., *50,* 1738 (1967).
4. ALAIS, C., MOCQUOT, G., NITSCHMANN, H., and ZAHLER, P., Helv. Chim. Acta, *36,* 1955 (1953).
5. ALAIS, C., NOVAK, G., Lait, *48,* 393 (1968).
6. ALAIS, C., and NOVAK, G., 18th Intern. Dairy Congr., IE, 279 (1970).
7. ALBONICO, F., GIANANI, L., RESMINI, P., and ZANINI, A., Ind. Agr., *4,* 289 (1966). Cited in DSA *29,* 2854 (1967).
8. ALLEMANN, O., and SCHMID, H., Landw. Jahrb. Schweiz, *30,* 357 (1916).
9. ALLEN, L. A., and KNOWLES, N. R., J. Dairy Res., *5,* 185 (1934).
10. Am. Public Health Assoc., "Standard Methods for the Examination of Dairy Products," 11th Ed., New York, 1960.
11. ANDERSON, D. F., and DAY, E. A., J. Dairy Sci., *48,* 248 (1965).
12. ANDERSON, D. F., and DAY, E. A., J. Agr. Food Chem., *14,* 241 (1966).
13. ANDREWS, P., Biochem. J., *91,* 222 (1964).
14. ANGELI, F., Chim. ind., (Sao Paulo), *22,* 464 (1940).
15. ANNIBALDI, S., BENEVENTI, G. P., and NIZZOLA, L., Scienza Tec. latt-casear, *21,* 27 (1970). Cited in DSA *32,* 3566 (1970).
16. Anonymous, Food Eng., *39,* (5) 88 (1967).
17. ANTONINI, J., and RIBADEAU-DUMAS, B., Biochimie, *53,* 321 (1971).
18. ARIMA, K., and IWASAKI, S., U.S. Pat. 3,212,905 (1965).
19. ARIMA, K., and IWASAKI, S., W. German Pat. Appl. 1,442,118 (1969).
20. ARIMA, K., IWASAKI, S., and TAMURA, G., Agr. Biol. Chem., *31,* 540 (1967).
21. ARIMA, K., YU, J., IWASAKI, S., and TAMURA, G., Appl. Microbiol., *16,* 1727 (1968).
22. ARIMA, K., YU, J., and IWASAKI, S., *in* "Methods in Enzymology", 19, 446, G. E. Perlman and L. Lorand, Ed., Academic Press, New York, 1970.
23. ARNON, R., *in* "Methods in Enzymology", 19, 226, G. E. Perlman and L. Lorand, Ed., Academic Press, New York, 1970.
24. ASATO, N., and RAND, A. G., JR., Anal. Biochem., *44,* 32 (1971).
25. ASHOOR, S. H., SAIR, R. A., OLSEN, N. F., and RICHARDSON, T., Biochim. Biophys. Acta, *229,* 423 (1971).
26. Assoc. Offic. Anal. Chemists, "Official Methods of Analysis", 11th Ed., (Association of Official Analytical Chemists), Washington, D.C., 1970.
27. AUNSTRUP, K., W. German Pat. Appl. 1,517,775 (1970).
28. Australian Div. Dairy Res., CSIRO Annual Report, 7 (1967).
29. BABBAR, I. J., SRINIVASAN, R. A., CHAKRAVORTY, S. C., and DUDANI, A. T., Indian J. Dairy Sci., *18,* 89 (1967). Cited in DSA *28,* 2955 (1966).
30. BABEL, F. J., Econ. Botany, *7,* 27 (1953).
31. BABEL, F. J., and HAMMER, B. W., J. Dairy Sci., *28,* 201 (1945).
32. BABEL, F. J., and SOMKUTI, G. A., J. Dairy Sci., *51,* 937 (1968).
33. BADINGS, H. T., J. Dairy Sci., *50,* 1347 (1967).
34. BADINGS, H. T., STADHOUDERS, J., and VAN DUIN, H., J. Dairy Sci., *51,* 31 (1968).

35. BAKALOR, S., Dairy Sci. Abstr., *24,* 529 (1962).
36. BAKALOR, S., Dairy Sci. Abstr., *24,* 583 (1962).
37. BAKKER, G., SCHEFFERS, W. A., and WIKEN, T. O., Neth. Milk Dairy J., *22,* 16 (1968).
38. BAKRI, M., and WOLFE, F. H., Can. J. Biochem., *49,* 882 (1971).
39. BALDWIN, R. L., and WAKE, R. G., Abstr. of papers, Am. Chem. Soc., *136,* 35c (1959).
40. BALLS, A. K., and HOOVER, S. R., J. Biol. Chem., *737* (1937).
41. BANG-JENSEN, V., FOLTMANN, B., and ROMBOUTS, W., Compt. rend. trav. lab. Carlsberg, *34,* 326 (1964).
42. BARR, G. H., Agr. Gaz. Can., *4,* 660 (1917).
43. BARTHEL, C., SANDBERG, E., and HAGLUND, E., Lait, *8,* 285, 762 (1928).
44. BASSETT, E. W., and HARPER, W. J., J. Dairy Sci., *39,* 918 (1956).
45. BASSETT, E. W., and HARPER, W. J., J. Dairy Sci., *41,* 1206 (1958).
46. BEEBY, R., HILL, R. D., and SNOW, N. S., *in* "Milk Proteins, Chemistry and Molecular Biology", 2, 421, H. A. McKenzie, Ed., Academic Press, New York, 1971.
47. BEEBY, R., and NITSCHMANN, H., J. Dairy Res., *30,* 7 (1963).
48. BERKOWITZ-HUNDERT, R., and LEIBOWITZ, J., Enzymologia, *25,* 257 (1963).
49. BERRIDGE, N. J., Nature, *149,* 194 (1942).
50. BERRIDGE, N. J., Biochem. J., *39,* 179 (1945).
51. BERRIDGE, N. J., *in* "The Enzymes", 1, part 2, 1079, J. B. Sumner and K. Myrback, Ed., Academic Press, New York, 1951.
52. BERRIDGE, N. J., J. Dairy Res., *19,* 328 (1952).
53. BERRIDGE, N. J., Analyst, *77,* 57 (1952).
54. BERRIDGE, N. J., *in* "Advances in Enzymology", 15, 423, E. F. Nord, Ed., Interscience Publishers, New York, 1954.
55. BERRIDGE, N. J., J. Dairy Res., *22,* 384 (1955).
56. BERRIDGE, N. J., 14th Intern. Dairy Congr., *2,* pt. 2, 59 (1956).
57. BERRIDGE, N. J., U.S. Patent 2,997,395 (1961).
58. BERRIDGE, N. J., Dairy Eng., *80,* 130 (1963).
59. BERRIDGE, N. J., DAVIS, J. G., KON, P. M., KON, S. K., and SPRATLING, F. R., J. Dairy Res., *13,* 145 (1943).
60. BILLS, D. D., Diss. Abstr., Sect. B, *27,* 2582 (1967).
61. BILLS, D. D., and DAY, E. A., J. Dairy Sci., *47,* 733 (1964).
62. BILLS, D. D., MORGAN, M. E., LIBBEY, L. M., and DAY, E. A., J. Dairy Sci., *48,* 1168 (1965).
63. BILLS, D. D., WILLITS, R. E., and DAY, E. A., J. Dairy Sci., *49,* 681 (1966).
64. BLOCK, R. J., J. Dairy Sci., *34,* 1 (1951).
65. BLUMENFELD, O. O., and PERLMANN, G. E., J. Gen. Physiol., *42,* 553 (1959).
66. BOLLIGER, O., and SCHILT, P., Schweiz. Milchztg. (Lait Romand) *95,* (59) Wiss. Beil. Nr. *121,* 1029 (1969).
67. BOVEY, F. A., and YANARI, S. S., Abstr. of papers, Am. Chem. Soc., *134,* 33c (1954).
68. BRADLEY, R. L., JR., and STINE, C. M., J. Gas. Chromatog., *6,* 344 (1968).
69. BRANDL, E., Ost. Milchw., *25,* 211 (1970) Cited in DSA *32,* 3905 (1970).
70. BROWN, L. W., and PRICE, W. V., J. Dairy Sci., *17,* 33 (1934).
71. BRUGHERA, F., and SALVADORI, P., Minerva Nipiol, *19,* 182 (1969).
72. BULLOCK, D. H., and IRVINE, O. R., J. Dairy Sci., *39,* 1229 (1956).
73. BUNDY, H. F., ALBIZATI, L. D., and HOGANCAMP, D. M., Arch. Biochem. Biophys., *118,* 536 (1967).
74. BUNDY, H. F., WESTBERG, N. J., DUMMEL, B. M., and BECKER, C. A., Biochemistry, *3,* 923 (1964).
75. BURKEY, L. A., SANDERS, G. P., and MATHESON, K. J., J. Dairy Sci., *18,* 719 (1935).
76. BYERS, E. L., and PRICE, W. V., J. Dairy Sci., *20,* 307 (1937).
77. CERBULIS, J., CUSTER, J. H., and ZITTLE, C. A., J. Dairy Sci., *43,* 1725 (1960).

78. CERNA, E., KNEZ, V., POZIVIL, J., and STRMISKO, J., Prum. Potravin, *17,* 385 (1966). Cited in DSA *29,* 1327 (1967).
79. CHAPMAN, H. R., and BURNETT, J., Dairy Ind., *33,* 308 (1968).
80. CHAPMAN, H. R., and HARRISON, A. J. W., J. Soc. Dairy Technol. *16,* 139 (1963).
81. CHARLES, R. L., GERTZMANN, D. P., and MELACHOURIS, N., U.S. Pat. 3,549,390 (1970).
82. CHARLES, R. L., GERTZMANN, D. P., and MELACHOURIS, N., W. German Pat. Appl. 1,945,447 (1970).
83. CHEBOTAREV, A. I., DUROVA, ZH. I., and PETINA, T. A., Moloch. Prom. *30,* (11) 10 (1969). Cited in DSA *32,* 1014 (1970).
84. CHEESEMAN, G. C., J. Dairy Res., *29,* 163 (1962).
85. CHEESEMAN, G. C., J. Dairy Res., *30,* 17 (1963).
86. CHEESEMAN, G. C., Nature, *205,* 1011 (1965).
87. CHEESEMAN, G. C., J. Dairy Res., *36,* 299 (1969).
88. CHERBULIEZ, E., and BAUDET, P., Helv. Chim. Acta, *33,* 1673 (1950).
89. CHEVALIER, R., MOCQUOT, G., ALAIS, C., and BONNAT, M., Compt. rend. acad. sci. Paris, *230,* 581 (1950).
90. CHEVALIER, R., MOCQUOT, G., ALAIS, C., and BONNAT, M., Compt. rend. acad. sci. Paris, *231,* 249 (1950).
91. CHIANG, L., SANCHEZ-CHIAN, L., MILLS, J., and TANG, J., J. Biol. Chem., *242,* 3098 (1967).
92. CHOUDHERY, A. K., and MIKOLAJCIK, E. M., J. Dairy Sci., *52,* 896 (1969).
93. CHOW, R. B., and KASSELL, B., J. Biol. Chem., *243,* 1718 (1968).
94. Chr. Hansen's Laboratory Inc., "Calve's Stomachs—Their Preparation for Market", Milwaukee, 1958.
95. CHRISTEN, C., and VIRASORO, E., Lait, *12,* 923 (1932).
96. CHRISTEN, C., and VIRASORO, E., Lait, *15,* 354, 496 (1935).
97. CHRISTEN, C., and VIRASORO, E., Anales soc. cient. Argentina, Seccion Santa Fe, 7, 18 (1935).
98. CLAESSON, O., and CLAESSON, E., 18th Intern. Dairy Congr., IE, 42 (1970).
99. CLAESSON, O., and NITSCHMANN, H., Acta Agr. Scand., *7,* 341 (1957).
100. CLARK, W. M., J. Dairy Sci., *1,* 91 (1917).
101. COMBS, W. B., MARTIN, W. H., and HUGGLAR, N. A., J. Dairy Sci., *7,* 524 (1924).
102. CORDES, W. A., J. Dairy Sci., *42,* 2012 (1959).
103. CREAMER, L. K., New Zealand J. Dairy Sci. Tech., *6,* 91 (1971).
104. CREAMER, L. K., MILLS, O. E., and RICHARDS, E. L., 18th Intern. Dairy Congr., IE, 45 (1970).
105. CROOK, E. M., BROCKLEHURST, K., and WHARTON, C. W., in "Methods in Enzymology", 19, 963, G. E. Perlmann and L. Lorand, Ed., Academic Press, New York, 1970.
106. CURRIE, J. N., J. Agr. Res., *2,* 1, 429 (1914).
107. CZULAK, J., Australian J. Dairy Technol., *14,* 177 (1959).
108. CZULAK, J., CONOCHIE, J., and HAMMOND, L. A., Australian J. Dairy Technol., *19,* 147 (1964).
109. CZULAK, J., and SHIMMIN, P. D., Australian J. Dairy Technol., *16,* 96 (1961).
110. DACRE, J. C., J. Dairy Res., *20,* 217 (1953).
111. DACRE, J. C., J. Sci. Food Agr., *4,* 604 (1953).
112. DACRE, J. C., J. Dairy Res., *22,* 219 (1955).
113. DAHLBERG, A. C., and KOSIKOWSKY, F. V., J. Dairy Sci., *31,* 305 (1948); *32,* 316, 630 (1949).
114. DAHLE, C. D., and NAGEOTTE, G. J., Nat. Butter Cheese J., *39* (10) 44; (11) 40 (1948).
115. DAHLE, C. D., and WATROUS, G. H., JR., Nat. Butter Cheese J., *36,* (11) 44 (1945).
116. DASTUR, N. N., Indian Farming, *9,* 451 (1948).

117. DASTUR, N. N., SHAMA SASTRY, K. N., and VENKATAPPIAH, D., Indian J. Vet. Sci., *18,* 233 (1948).
118. DAVEL, H. B., and RETIEF, D. J., Dept. Agr., Union So. Africa, Sci. Bull., 58 (1927); New York Produce Review and Am. Creamery, *65,* 384 (1928).
119. DAVIES, D. T., and WHITE, J. C. D., J. Dairy Res., *27,* 171 (1960).
120. DAVIES, W. L., DAVIS, J. G., DEARDEN, D. V., and MATTICK, A. T. R., J. Dairy Res., *5,* 144 (1934).
121. DAVIS, J. G., J. Dairy Res., *3,* 241 (1932).
122. DAVIS, J. G., Chem. and Ind., *60,* 259 (1941).
123. DAVIS, J. G., "Cheese, Vol. I. Basic Technology," American Elsevier Publ. Co., Inc., 1965.
124. DAVIS, J. G., "Cheese, Vol. II. Annotated Bibliography-Subject Index," American Elsevier Publ. Co., Inc., 1965.
125. DAVIS, J. G., Dairy Ind., *36,* 135 (1971).
126. DAY, E. A., and ANDERSON, D. F., J. Agr. Fd. Chem., *13,* 2 (1965).
127. DAY, E. A., BASSETTE, R., and KEENEY, M., J. Dairy Sci., *43,* 463 (1960).
128. DAY, E. A., and KEENEY, M., J. Dairy Sci., *41,* 718 (1958).
129. DAY, E. A., and LIBBEY, L. M., J. Food Sci., *29,* 583 (1964).
130. DEANE, D. D., and COHENOUR, F. D., J. Dairy Sci., *44,* 451, (1961).
131. de BAUN, R. M., CONNORS, W. M., and SULLIVAN, R. A., Arch. Biochem. Biophys., *43,* 324 (1953).
132. DE KONING, P. J., Ph.D. Thesis, University of Amsterdam (1967). Cited in DSA *29,* 3265 (1967).
133. DE KONING, P. J., VAN ROOIJEN, P. J., and KOK, A., Biochem. Biophys. Res. Comm., *24,* 616 (1966).
134. DE KONING, P. J., VAN ROOIJEN, P. J., and KOK, A., Neth. Milk Dairy J., *23,* 55 (1969).
135. DELFORNO, G., GRUEV, P., Ind. Latte, 6, 216 (1970). Cited in DSA *33,* 3886 (1971).
136. DELFOUR, A., ALAIS, C., and JOLLES, P., Chimia, *20,* 148 (1966).
137. DELFOUR, A., JOLLES, J., ALAIS, C., and JOLLES, P., Biochem. Biophys. Res. Comm., *19,* 452 (1965).
138. DEMAN, J. M., and BATRA, S. C., Dairy Ind., *29,* (1) 32 (1964).
139. DENNIS, E. S., and WAKE, R. G., Biochim. Biophys. Acta, *97,* 159 (1965).
140. DESMAZEAUD, M., and HERMIER, J., Ann. Biol. anim. Biochim. Biophys., *8,* 565 (1969). Cited in DSA *31,* 1798 (1969).
141. DIXON, R. P., and DEMAN, J. M., Can. Inst. Fd. Technol. J., *1,* 51 (1968).
142. DIXON, R. P., and DEMAN, J. M., Can. Inst. Fd. Technol. J., *2,* 127 (1969).
143. DJURTOFT, R., FOLTMANN, B., and JOHANSEN, A., Compt. rend. trav. Lab. Carlsberg, *34,* 287 (1964).
144. DOLBY, R. M., and MCDOWALL, F. H., J. Dairy Res., *6,* 218 (1935).
145. DOPHEIDE, T. A. A., MOORE, S., and STEIN, W. H., J. Biol. Chem., *242,* 1833 (1967).
146. DORNER, W., and WIDMER, A.,Milk Plant Monthly, *21,* (7), 50 (1932).
147. DOUILLARD, R., Biochimie, *53,* 447 (1971).
148. DOUILLARD, R., and RIBADEAU-DUMAS, B., Bull. Soc. Chim. Biol., *52,* 1429 (1970).
149. DYACHENKO, P. F., 15th Intern. Dairy Congr., *2,* 629 (1959).
150. EDELHOCH, H., J. Am. Chem. Soc., *79,* 6100 (1957).
151. EDELSTEN, D., and JENSEN, J. S., 18th Intern. Dairy Congr., IE, 280 (1970).
152. EDWARDS, J. L., JR., Diss. Abstr. Int., B30, 4194 (1970).
153. EDWARDS, J. L., and KOSIKOWSKI, F. V., J. Dairy Sci., *52,* 1675, (1969).
154. EL-GHARBAWI, M., and WHITAKER, J. R., Biochemistry, *2,* 476 (1963).
155. ELLIOTT, J. A., and EMMONS, D. B., Can. Inst. Fd. Tech. J., *4,* 16 (1971).
156. EL NEGOUMY, A. M., J. Dairy Res., *37,* 437 (1970).
157. EMMONS, D. B., Dairy Sci. Abstr., *25,* 129 (1963).
158. EMMONS, D. B., Dairy Sci. Abstr., *25,* 175 (1963).
159. EMMONS, D. B., J. Dairy Sci., *53,* 1177 (1970).

160. EMMONS, D. B., and ELLIOTT, J. A., 15th Ann. Conf. Can. Inst. Food Sci. and Technol., Toronto (1972).
161. EMMONS, D. B., MCGUGAN, W. A., ELLIOTT, J. A., and MORSE, P. M., J. Dairy Sci., *45,* 332 (1962).
162. EMMONS, D. B., PETRASOVITS, A., GILLAN, R. H., and BAIN, J. M., 18th Intern. Dairy Congr., IE, 294 (1970).
163. EMMONS, D. B., PETRASOVITS, A., GILLAN, R. H., and BAIN, J. M., Can. Inst. Food Technol. J., *4,* 31 (1971).
164. EMMONS, D. B., PRICE, W. V., and TORRIE, J. H., J. Dairy Sci., *43,* 480 (1960).
165. EMMONS, D. B., SWANSON, A. M., and PRICE, W. V., J. Dairy Sci., *42,* 1020 (1959).
166. EMMONS, D. B., and TUCKEY, S. L., Cottage Cheese and Other Cultured Milk Products," Pfizer Cheese Monographs, Vol. III, Charles Pfizer & Co., Inc., New York, 1967.
167. ERNSTROM, C. A., Unpublished results. University of Wisconsin (1953).
168. ERNSTROM, C. A., Ph.D. Thesis, University of Wisconsin (1956).
169. ERNSTROM, C. A., J. Dairy Sci., *41,* 1663 (1958).
170. ERNSTROM, C. A., Milk Prod. J., *52,* (5) 8 (1961).
171. ERNSTROM, C. A., PRICE, W. V., and SWANSON, A. M., J. Dairy Sci., *41,* 61 (1958).
172. ERNSTROM, C. A., and TITTSLER, R. P., *in* "Fundamentals of Dairy Chemistry" p. 590, B. H. Webb and A. H. Johnson, Ed., Avi Publishing Co., Westport, 1965.
173. EREKSON, A. B., J. Dairy Sci., *32,* 704 (1949).
174. EVERSON, T. C., and WINDER, W. C., J. Dairy Sci., *51,* 940 (1968).
175. FALDER, A., 18th Intern. Dairy Congr., IE, 328 (1970).
176. FARNHAM, M. G., U.S. Patent 2,531,329 (Nov. 21, 1950).
177. FEAGAN, J. T., ERWIN, L. J., and DIXON, B. D., Australian J. Dairy Technol., *20,* 214 (1965).
178. Fed. Regist., *32,* 410 and 4350 (1967).
179. Fed. Regist., *34,* 8908 and 15841 (1969).
180. Fed. Regist., *34,* 4887 (1969).
181. Fed. Regist., *37,* 6734 (1972).
182. FERRIER, L. K., RICHARDSON, T., OLSON, N. F., and HICKS, C. L., J. Dairy Sci., *55,* 726 (1972).
183. FISH, J. C., Nature, *180,* 345 (1957).
184. FISK, W. W., and PRICE, W. V., N.Y. (Cornell) Agr. Expt. Sta. Bull. 418 (1923).
185. FLEISCHMANN, W., "Lehrbuch der Milchwirtschaft", rev. by Weigmann, H., 7th Ed., p. 666, P. Parey, Berlin, 1932.
186. FOLTMANN, B., Acta Chem. Scand., *12,* 343 (1958).
187. FOLTMANN, B., Acta Chem. Scand., *13,* 1927 (1959).
188. FOLTMANN, B., 15th Intern. Dairy Congr., *2,* 655 (1959).
189. FOLTMANN, B., Acta Chem. Scand., *14,* 2059 (1960).
190. FOLTMANN, B., Acta Chem. Scand., *14,* 2247 (1960).
191. FOLTMANN, B., Compt. rend. trav. lab. Carlsberg, *32,* 425 (1962).
192. FOLTMANN, B., Compt. rend. trav. lab. Carlsberg, *34,* 275 (1964).
193. FOLTMANN, B., Compt. rend. trav. lab. Carlsberg, *34,* 319 (1964).
194. FOLTMANN, B., Compt. rend. trav. lab. Carlsberg, *35,* 143 (1966).
195. FOLTMANN, B., Biochem. J., *115,* 3P (1969).
196. FOLTMANN, B., Phil. Trans. R. Soc., *B257,* 147 (1970).
197. FOLTMANN, B., *in* "Methods of Enzymology", 19, 421, G. E. Perlmann and L. Lorand, Ed., Academic Press, New York, 1970.
198. FOLTMANN, B., FEBS Letters, *17,* 87 (1971).
199. FOLTMANN, B., *in* "Milk Proteins, Chemistry and Molecular Biology", 2, 217, H. A. McKenzie, Ed., Academic Press, New York, 1971.
200. FOLTMANN, B., and HARTLEY, B. S., Biochem. J., *104,* 1064 (1967).
201. FOMIN, D., Molochno-Maslodel'naya Prom., *6,* (9) 16 (1939). Cited in DSA *3,* 16 (1941).

202. Food and Agriculture Organization of the United Nations. Report of the FAO Ad Hoc Consultation on World Shortage of Rennet in Cheese Making. Rome, Italy (1968).
203. FORSS, D. A., and PATTON, S., J. Dairy Sci., *49,* 89 (1966).
204. FOSTER, E. M., NELSON, F. E., SPECK, M. L., DOETSCH, R. N., and OLSON, J. C., "Dairy Microbiology", Prentice-Hall, Englewood Cliffs, 1957.
205. FOX, P. F., J. Dairy Res., *36,* 427 (1969).
206. FOX, P. F., J. Dairy Sci., *52,* 1214 (1969).
207. FOX, P. F., Irish J. Agr. Res., *8,* 175 (1969).
208. FOX, P. F., J. Dairy Res., *37,* 173 (1970).
209. FOX, P. F., and KOSIKOWSKI, F. V., J. Dairy Sci., *45,* 648 (1962).
210. FOX, P. F., and WALLEY, B. F., J. Dairy Res., *38,* 165 (1971).
211. FOX, P. F., and WALLEY, B. F., Irish J. Agr. Res., *10,* 358 (1971).
212. FRAZIER, W. C., J. Dairy Sci., *8,* 370 (1925).
213. FRAZIER, W. C., BURKEY, L. A., BOYER, A. J., SANDERS, G. P., and MATHESON, K. J., J. Dairy Sci., *18,* 373 (1935).
214. FRAZIER, W. C., JOHNSON, W. T., JR., EVANS, F. R., and RAMSDELL, G. A., J. Dairy Sci., *18,* 503 (1935).
215. FRAZIER, W. C., SANDERS, G. P., BOYER, A. J., and LONG, H. F., J. Bacteriol., *27,* 539 (1934).
216. FREEMAN, T. R., and DAHLE, C. D., Penna. Agr. Expt. Sta. Bull. 362 (1938).
217. FREUDENREICH, E. V., and JENSEN, O. S., Centr.-Bl. f. Bakter. u. Parasitenk. II., *6,* 112 (1900).
218. FRISCHLING, K., Suddeut. Molkerei Ztg., *51,* 1244 (1930); Österr. Milchwirtsch. Ztg., *37,* 275 (1930).
219. FRÜHWALD, H., "Die Neuzeitliche Emmentalerkäserei," Kurz and Co., Kempten im Allgäu, 1932.
220. FRYER, T. F., Dairy Sci., Abstr., *31,* 471 (1969).
221. GÁBOR, P., JÁNOS, C., and ISTVÁN, E., Tegipar *17,* 1 (1968). Cited in DSA, *30,* 1556 (1968).
222. GALESLOTT, T. E., Neth. Milk Dairy J., *1,* 33 and 238 (1947).
223. GANGULI, N. C., Indian Farming, *20,* (6) 36 (1970).
224. GANGULI, N. C., and BHALERO, V. R., J. Dairy Sci., *48,* 438 (1965).
225. GARNIER, J., Compt. rend. acad. sci., Paris, *247,* 1515 (1958).
226. GARNIER, J., Proc. Intern. Symp. Enzyme Chem., Tokyo, Kyoto, *2,* 524 (1958).
227. GARNIER, J., and CHEVALIER, R., Compt. rend. acad. sci. Paris, *252,* 350 (1961).
228. GARNIER, J., MOCQUOT, G., and BRIGNON, G., Compt. rend. acad. sci., Paris, *254,* 372 (1962).
229. GARNIER, J., MOCQUOT, G., RIBADEAU-DUMAS, B., and MAUBOIS, J. L., Ann. Nutr. Aliment, *22,* B495 (1968).
230. GARNIER, J., and RIBADEAU-DUMAS, B., J. Dairy Res., *37,* 493 (1970).
231. GARRETT, O. F., J. Dairy Sci., *26,* 305 (1943).
232. GENIN, G., Lait, *48,* 53 (1968).
233. GIBBONS, R. A., and CHEESEMAN, G. C., Biochim. Biophys. Acta, *56,* 354 (1962).
234. GIROLAMI, R. L., and KNIGHT, S. G., Appl. Microbiol., *3,* 264 (1955).
235. GODO SHUSEI, KK, Br. Pat. 1,156,387 and 1,156, 388 (1969).
236. Godo Shusei Co., Japanese Patent 27,714/69 (1969).
237. Godo Shusei Co., Japanese Patent 12,255/70 (1970).
238. GONASHVILI, S. G., Doklady Vsesoyuz. Akad. Sel'skokhoz. Naukim V. I. Lenina, *14,* (12) 32 (1949).
239. GONASHVILI, S. G., Molochnaya Prom., *16,* 27 (1955).
240. GORINI, C., J. Bacteriol., *20,* 297 (1930).
241. GORINI, L., and LANZAVECCHIA, G., Biochim. Biophys. Acta, *14,* 407 (1954).
242. GRATZ, O., Süddeut. Molkerei Ztg., *52,* 1269 (1931).
243. GRATZ, O., and SZANYI, S., Biochem. Z., *63,* 436 (1914).
244. GREEN, M. L., J. Dairy Res., *39,* 55 (1972).
245. GREEN, M. L., J. Dairy Res., *39,* 261 (1972).
246. GREEN, M. L., and CRUTCHFIELD, G., Biochem. J., *115,* 183 (1969).

247. GREEN, M. L., and CRUTCHFIELD, G., J. Dairy Res., *38,* 151 (1971).
248. GRILL, H., JR., PATTON, S., and CONE, J. F., J. Dairy Sci., *49,* 409 (1966).
249. GRIMMER, W., and KRUGER, M., Milchwirtsch. Forsch., *2,* 457 (1925).
250. GUITTONNEAU, G., SAJOUS, P., and DEPEET, R., Lait, *11,* 809 (1931).
251. HABERMANN, W., MATTENHEIMER, H., SKY-PECK, H., and SINOHARA, H., Chimia, *15,* 339 (1961).
252. HABICHT, C., Z. Untersuch Lebensm., *66,* 81 (1933); Milchwirtsch. Forsch., *15,* 347 (1934).
253. HADLUND, G., 18th Intern. Dairy Congr., IE, 520 (1970).
254. HAGEMEYER, K., FAWWAL, J., and WHITAKER, J. R., J. Dairy Sci., *51,* 1916 (1968).
255. HAMDY, A., 18th Intern. Dairy Congr., IE, 350 (1970).
256. HAMDY, A., and EDELSTEN, D., Milchwissenschaft, *25,* 450 (1970).
257. HAMMARSTEN, O., Nova Acta Reg. Soc. Scient., Upsala, 1877. Cited by O. Hammarsten and S. G. Hedin *in* "A Textbook of Physiological Chemistry". Auth. trans. by J. A. Mandell, 7th Ed., John Wiley and Sons, Inc., New York, 1914.
258. HAMMER, B. W., and BRYANT, E. W., Iowa State Coll. J. Sci., *11,* 281 (1937).
259. HANKINSON, C. L., J. Dairy Sci., *26,* 53 (1943).
260. HANSEN, H. C., BENDIXEN, H. A., and THEOPHILIUS, D. R., J. Dairy Sci., *16,* 121 (1933).
261. HANSEN, K., 18th Intern. Dairy Congr., IE, 51 (1970).
262. HARDEN, A., and MACALLUM, A. B., Biochem. J., *8,* 90 (1914).
263. HARPER, W. J., J. Dairy Sci., *40,* 556 (1957).
264. HARPER, W. J., J. Dairy Sci., *42,* 207 (1959).
265. HARPER, W. J., and GOULD, I. A., Butter, Cheese and Milk Prod. J., *43,* (5) 24, (8) 22 (1952).
266. HARPER, W. J., and GOULD, I. A., J. Dairy Sci., *38,* 87 (1955).
267. HARPER, W. J., and KRISTOFFERSON, T., J. Dairy Sci., *39,* 1773 (1956).
268. HARPER, W. J., and KRISTOFFERSON, T., J. Agr. Fd. Chem., *18,* 563 (1970).
269. HARPER, W. J., and LONG, J. E., J. Dairy Sci., *39,* 129 (1956).
270. HARPER, W. J., MASKELL, K. T., and HARGROVE, R. E., Butter, Cheese, and Milk Prod. J., *43,* (12) 20 (1952).
271. HARPER, W. J., and SWANSON, A. M., 12th Intern. Dairy Congr., *2,* 147 (1949).
272. HARTLEY, C. B., and JEZESKI, J. J., J. Dairy Sci., *37,* 436 (1954).
273. HARVEY, R. J., and WALKER, J. R. L., J. Dairy Res., *27,* 335 (1960).
274. HATANO, J., Biochem. Z., *149,* 228 (1924).
275. HAYASHI, H., NEGISHI, T., ITO, S., and FUJINO, Y., Res. Bull. Obihero Zootechn. Univ. Ser. I4, 465 (1967). Cited in DSA, *30,* 338 (1968).
276. HEINICKE, R. M., Science, *118,* 753 (1953).
277. HENSON, J. H., and MILLER, T. B., J. Dairy Res., *22,* 211 (1955).
278. HERIAN, L., and KRCAL, A., Prum. Potrovin *22,* (5) 137 (1971). Cited in DSA *33,* 5455 (1971).
279. HERRIOTT, R. M., J. Gen. Physiol., *21,* 501 (1938).
280. HERRIOTT, R. M., J. Gen. Physiol., *22,* 65 (1938).
281. HERRIOTT, R. M., J. Gen. Physiol., *24,* 325 (1941).
282. HERRIOTT, R. M., J. Gen. Physiol., *45,* Suppl. 57 (1962).
283. HERRIOTT, R. M., DESREUX, V., and NORTHRUP, J. H., J. Gen. Physiol., *23,* 439 (1940).
284. HERRIOTT, R. M., and NORTHRUP, J. H., Science, *83,* 469 (1936).
285. HILL, R. D., Biochem. Biophys. Res. Comm., *33,* 659 (1968).
286. HILL, R. D., J. Dairy Res., *36,* 409 (1969).
287. HILL, R. D., J. Dairy Res., *37,* 187 (1970).
288. HILL, R. D., Personal Communication (1971).
289. HILL, R. D., and CRAKER, B. A., J. Dairy Res., *35,* 13 (1968).
290. HILL, R. D., and LAING, R. R., J. Dairy Res., *32,* 193 (1965).
291. HILL, R. D., and LAING, R. R., Biochim. Biophys. Acta, *99,* 352 (1965).
292. HILL, R. D., and LAING, R. R., Nature, *210,* 1160 (1966).
293. HILL, R. D., and LAING, R. R., Biochim. Biophys. Acta, *132,* 188 (1967).

294. HILL, R. J., NAUGHTON, A. M., and WAKE, R. G., Biochim. Biophys. Acta, *200,* 267 (1970).
295. HILL, R. J., and WAKE, R. G., Nature, *221,* 635 (1969).
296. HILL, R. L., and MERRILL, A. C., Utah Agr. Expt. Bull. 236 (1932).
297. HINDLE, F. J., and WHEELOCK, J. V., Biochem. J., *115,* 19P (1969).
298. HINDLE, E. J., and WHEELOCK, J. N., J. Dairy Res., *37,* 389 (1970).
299. HINKLE, E. T., JR., and ALFORD, C. E., Ann. N.Y. Acad. Sci., *43,* 208 (1951).
300. HINTZ, P. C., SLATTER, W. L., and HARPER, W. J., J. Dairy Sci., *39,* 235 (1956).
301. HOLTER, H., Biochem. Z., *255,* 160 (1932).
301a. HOLTER, H., and LI, S. O., Acta Chem. Scand., *4,* 1321 (1950).
302. HONER, C. J., and TUCKEY, S. L., J. Dairy Sci., *34,* 475 (1951).
303. HOOD, E. G., and GIBSON, C. A., Can. Dairy and Ice Cream J., *27,* (11) 45 (1948).
304. HOOD, E. G., GIBSON, C. A., and BOWEN, J. F., Can. Dairy and Ice Cream J., *28,* (2) 27; *28,* (6) 27 (1949).
305. HOOD, E. G., and HLYNKA, I., Nat. Butter and Cheese J., *37* (1) 38; (2) 39 (1946).
306. HOSTETTLER, H., Landwirtsch Jahrbh., *46,* 609 (1932).
307. HOSTETTLER, H., and IMHOFF, K., Milchwissenchaft, *6,* 351 (1951).
308. HOSTETTLER, H., and IMHOFF, K., Schweiz. Milchztg, *81,* Wiss. Beilage No. *31,* 244 (1955).
309. HOSTETTLER, H., and RUEGGER, H. R., Landwirtsch. Jahrb. Schweitz., *64,* 669 (1950).
310. HUANG, W. Y., and TANG, J., J. Biol. Chem., *245,* 2189 (1970).
311. HUPFER, J. A., JR., SANDERS, G. P., and TITTSLER, R. P., J. Dairy Sci., *33,* 401 (1950).
312. HWANG, K., and IVY, A. C., Ann. N. Y. Acad. Sci., *54,* 161 (1951).
313. IMHOFF, K., and HOSTETTLER, H., Schweiz. Milchztg., *82,* 289 (1956).
314. IRVINE, D. M., PUHAN, Z., and GRUETZNER, V., J. Dairy Sci., *52,* 889 (1969).
315. IRVINE, O. R., BRYANT, L. R., HILL, D. C., and SPROULE, W. H., Can. Dairy and Ice Cream J., *27,* (11) 49 (1947).
316. IRVINE, O. R., BRYANT, L. R., SPROULE, W. H., JACKSON, S. H., CROOK, A., and JOHNSTONE, W. M., Sci. Agr., *25,* 817 (1945).
317. IRVINE, O. R., BRYANT, L. R., SPROULE, W. H., JACKSON, S. H., CROOK, A., and JOHNSTONE, W. M., Sci. Agr., *25,* 833 (1945).
318. IWASAKI, S., Kobunshi, *16,* 1213 (1967). Cited in DSA *30,* 3977 (1968).
319. IWASAKI, S., TAMURA, G., and ARIMA, K., Agr. Biol. Chem. *31,* 546 (1967). Cited in DSA *29,* 3984 (1967).
320. IWASAKI, S., YASUI, T., TAMURA, G., and ARIMA, K., Agr. Biol. Chem., *31,* 1421 (1967). Cited in DSA *30,* 2390 (1968).
321. IWASAKI, S., YASUI, T., TAMURA, G., and ARIMA, K., Agr. Biol. Chem., *31,* 1427 (1967). Cited in DSA *30,* 2391 (1968).
322. JACKSON, W. H., Flavor Research in Cheese. *In* "Flavor Research and Food Acceptance," p. 324, Reinhold Publ. Corp., New York, 1958.
323. JACKSON, H. W., and HUSSONG, R. B., J. Dairy Sci., *41,* 920 (1958).
324. JACQUET, J., and LENOIR, J., Compt. rend., *238,* 2201 (1954).
325. JAGER, H., Milchwiss. Ber., *4,* 72 (1954).
326. JAGO, G. R., Dairy Ind., *27,* 772 (1962).
327. JANSEN, E. F., and BALLS, A. K., J. Biol. Chem., *137,* 459 (1941).
328. JENNESS, R., and PATTON, S., "Principles of Dairy Chemistry", John Wiley and Sons, Inc., New York, 1959.
329. JENSEN, J. S., Maelkeritidende, *82,* 57 (1969).
330. JIRGENSONS, B., Arch. Biochem. Biophys., *74,* 70 (1958).
331. JIRGENSONS, B., IKENAKA, T., and GORGURAKI, V., Makromol. Chem., *28,* 96 (1958).
332. JOHN LABBATT LTD., Br. Pat. 1,202,378 (1970).
333. JOHN LABBATT LTD., Br. Pat. 1,203,371 (1970).
334. JOHN LABBATT LTD., Br. Pat. 1,223,860 (1971).

335. JOLLES, J., ALAIS, C., and JOLLES, P., Biochim. Biophys. Acta, *168,* 591 (1968).
336. JOLLES, J., ALAIS, C., and JOLLES, P., Helv. Chim. Acta, *53,* 1918 (1970).
336a. JOLLES, J., FIAT, A., ALAIS, C., and JOLLES, P., FEBS Letters *30,* 173 (1973).
337. JOLLES, J., JOLLES, P., and ALAIS, C., Nature, *222,* 668 (1969).
337a. JOLLES, J., SCHOETGEN, F., ALAIS, C., FIAT, A., and JOLLES, P., Helv. Chim. Acta *55,* 2872 (1972).
337b. JOLLES, J., SCHOETGEN, F., ALAIS, C., and JOLLES, P., Chimia *26,* 645 (1972).
338. JOLLES, P., Angew, Chemie (Int. Ed.) *5,* 558 (1966).
339. JOLLES, P., ALAIS, C., and JOLLES, J., Biochim. Biophys. Acta, *51,* 309 (1961).
340. JOLLES, P., ALAIS, C., and JOLLES, J., Arch. Biochem. Biophys., *98,* 56 (1962).
341. JOLLES, P., ALAIS, C., and JOLLES, J., Biochim. Biophys. Acta, *69,* 511 (1963).
341A. JOLLES, J., SCHOENTGEN, F., ALAIS, C., FIAT, A., and JOLLES, P., Helv. Chim. Acta, *55,* 2872 (1972).
342. KALAN, E. B., and WOYCHIK, J. H., J. Dairy Sci., *48,* 1423 (1965).
343. KANNAN, A., and JENNESS, R., J. Dairy Sci., *44,* 808 (1961).
344. KARLIN, R., Ann. Nutrition et Aliment *15,* 103 (1961).
345. KASSELL, B., and MEITNER, P. A., *in* "Methods in Enzymology", 19, 337, G. E. Perlmann, and L. Lorand, Ed., Academic Press, New York, 1970.
346. KAUZMANN, W., *in* "Mechanism of Enzyme Action", W. D. McElroy and H. B. Glass, Ed., Johns Hopkins Press, Baltimore, 1970.
347. KAWAI, M., Agr. Biol. Chem., *34,* 162 (1970).
348. KEENEY, M., and DAY, E. A., J. Dairy Sci., *40,* 874 (1957).
349. KELLEY, L. A., Ph.D. Thesis, University of Wisconsin (1951).
350. KELLY, C. D., N.Y. (Geneva) Agr. Expt. Sta., Tech. Bull. 200 and 201 (1932).
351. KELLY, C. D., and MARQUARDT, J. C., J. Dairy Sci., *22,* 309 (1939).
352. KETTING, F., and PAULAY, G., 18th Intern. Dairy Congr., IE, 291 (1970).
353. KIEFERLE, F., and EISENREICH, L., Milchw. Forsch., *16,* 1 (1933).
354. KIERMEIER, F., and SCHMID, M., Z. Lebensmitteluntersuchung und Forsch, *133,* 217 (1967). Cited in DSA *30,* 686 (1968).
355. KIERMEIER, F., and WÜLLERSTORFF, B. V., Milchwissenschaft, *18,* 75 (1963).
356. KIKUCHI, T., and TOYODA, S., 18th Intern. Dairy Congr. IE, 285 (1970).
357. KIRCHMEIER, O., 18th Intern. Dairy Congr., IE, 25 (1970).
358. KIURU, K., Suom. Kem., *42A,* 237 (1969). Cited in DSA *32,* 2186 (1970).
359. KLEIN, J., and KIRSTEN, A., Milchw. Ztg., *27,* 785, 803 (1898).
360. KLEINER, I. S., and TAUBER, H., J. Biol. Chem., *96,* 755 (1932).
361. KNAYSI, G., and NELSON, J. D., J. Dairy Sci., *10,* 396 (1927).
362. KNIGHT, S. G., Can. J. Bot., *12,* 420 (1966). Cited in DSA *29,* 2498 (1967).
363. KNOWLES, N. R., J. Dairy Research, *7,* 63 (1936).
364. KOESTLER, G., Landw. Jahrb. Schweiz., *43,* 1065 (1929).
365. KOESTLER, G., Landw. Jahrb. Schweiz., *46,* 51 (1932).
366. KOESTLER, G., Landw. Jahrb. Schweiz., *47,* 156, 1121 (1933).
367. KONDRATENKO, M., BODURSKA, I., and MANAFOVA, N., in "V pamet na Akademik Ignat Emanuilov trudove vurkhu izbrani problemi na mikrobiologiyata, enzimologiyata i biokhimiyata", pp. 375, Sofia, Bulgaria. Cited in DSA *32,* 1232 (1970).
368. KONING, P. J., DE, Int. Dairy Congr., 1962 B, 362 (1962).
369. KOSIKOWSKI, F., "Cheese and Fermented Milk Foods", Edwards Brothers, Inc., Ann Arbor, Mich., 1966.
370. KOSIKOWSKI, F. V., J. Dairy Sci., *34,* 235 (1951).
371. KOSIKOWSKI, F. V., J. Dairy Sci., *34,* 1151 (1951).
372. KOSIKOWSKI, F. V., and DAHLBERG, A. C., J. Dairy Sci., *31,* 293 (1948).
373. KOSIKOWSKI, F. B., and DAHLBERG, A. C., J. Dairy Sci., *37,* 167 (1954).
374. KOSIKOWSKI, F. V., and MOCQUOT, G., "Advances in Cheese Technology," FAO Agr. Studies No. 38, Food and Agr. Org. of the U.N., Rome, 1958.
375. KOTHAVALLA, Z. R., and KHUBCHANDANI, P. G., Indian J. Vet. Sci., *10,* 284 (1940).

376. KRISHNAMURTI, C. R., and SUBRAHMANYAN, V., Indian J. Dairy Sci., *1,* 27, 106 (1948); *2,* 19, 43 (1949).
377. KRISHNASWAMY, M. A., JOHAR, D. S., SUBRAHAMANYAN, V., and THOMAS, S. P., Food Technol., *15,* 482 (1961).
378. KRISTOFFERSEN, T., Proc. Flavor Chem. Symp. (Campbell Soup Co.) 201 (1961).
379. KRISTOFFERSEN, T., J. Dairy Sci., *50,* 279 (1967).
380. KRISTOFFERSEN, T., and GOULD, I. A., 15th Intern. Dairy Congr., Proc., *2,* 720 (1959).
381. KRISTOFFERSEN, T., and GOULD, I. A., J. Dairy Sci., *43,* 1202 (1960).
382. KRISTOFFERSEN, T., GOULD, I. A., and HARPER, W. J., Milk Prods. J., *50* (5) 14 (1959).
383. KRISTOFFERSEN, T., GOULD, I. A., and PURVIS, G. A., J. Dairy Sci., *47,* 599 (1964).
384. KRISTOFFERSEN, T., MIKOLAJCIK, E. M., and GOULD, I. A., J. Dairy Sci., *50,* 292 (1967).
385. KRISTOFFERSEN, T., and NELSON, F. E., J. Dairy Sci., *38,* 1319 (1955).
386. KROGER, M., and PATTON, S., J. Dairy Sci. *47,* 296 (1964).
387. KRUGER, W., and KRENKEL, K., Nahrung, *12,* 149 (1968). Cited in DSA *30,* 2053 (1968).
388. KUNIMITSU, D. K., and YASUNOBU, K. T., in "Methods in Enzymology", 19, 244, G. E. Perlmann and G. Lorand Ed., New York, 1970.
389. KUZDZAL-SAVOIE, S., and KUZDZAL, W., Lait, *47,* 461 (1966). Cited in DSA *29,* 2478 (1967).
390. LaCROIX, D. E., and WONG, N. P., J. Assoc. Offic. Anal. Chemists', *54,* 361 (1971).
391. LANE, C. B., and HAMMER, B. W., Iowa Agr. Expt. Sta., Res. Bull. 237 (1938).
392. LANG, H. M., and KASSELL, B., Biochemistry, *10,* 2296 (1971).
393. LANGHUS, W. L., and PRICE, W. V., J. Dairy Sci., *24,* 873 (1941).
394. LANGHUS, W. L., PRICE, W. V., SOMMER, H. H., and FRAZIER, W. C., J. Dairy Sci., *28.* 827 (1945).
395. LANGLER, J. E., and DAY, E. A., J. Dairy Sci., *49,* 91 (1966).
396. LARSON, M. K., and WHITAKER, J. R., J. Dairy Sci., *53,* 253 (1970).
397. LARSON, M. K., and WHITAKER, J. R., J. Dairy Sci., *53,* 262 (1970).
398. LAWRENCE, A. J., Aust. J. of Dairy Tech., *166,* 169 (1959).
399. LAWRENCE, R. C., J. Dairy Res., *30,* 161 (1963).
400. LAWRENCE, R. C., J. Dairy Res., *30,* 235 (1963).
401. LAWRENCE, R. C., N. Z. J. Dairy Technol., *2,* 55 (1967).
402. LAWRENCE, R. C., Dairy Sci. Abstr., *29,* 1 (1967).
403. LAWRENCE, R. C., Dairy Sci. Abstr., *29,* 59 (1967).
404. LAWRENCE, R. C., and CREAMER, L. K., J. Dairy Res., *36,* 11 (1969).
405. LAWRENCE, R. C., and GILLES, J., N. Z. J. Dairy Sci. Tech., *6,* 30 (1971).
406. LAWRENCE, R. C., and SANDERSON, W. B., J. Dairy Res., *36,* 21 (1969).
407. LEDFORD, R. A., CHEN, J. H., and NATH, K. R., J. Dairy Sci., *51,* 792 (1968).
408. LEDFORD, R. A., O'SULLIVAN, A. C., and NATH, K. R., J. Dairy Sci., *49,* 1098 (1966).
409. LEE, D., and RYLE, A. P., Biochem. J., *104,* 742 (1967).
410. LEITCH, R. H., Congr. intern. tech. et chim. ind. Agr. Compt. rend. 5th Congr., *2,* 307 (1937).
411. LENNOX, F. G., and ELLIS, W. J., Biochem. J., *39,* 465 (1945).
412. LIBBEY, L. M., and DAY, E. A., J. Dairy Sci., *46,* 859 (1963).
413. LIEBICH, H. M., DOUGLAS, D. R., BAYER, E., and ZLATKIS, A., J. Chrom. Sci., *8,* 355 (1970).
414. LIENER, I. E., and FRIEDENSON, B., in "Methods in Enzymology", 19, 261, G. E. Perlmann and L. Lorand Ed., Academic Press, New York, 1970.
415. LINDERSTRØM-LANG, K., Compt. rend. trav. lab. Carlsberg, *17,* 9 (1929).
416. LINDQVIST, B., and STORGARDS, T., 15th Intern. Dairy Congr. Proc., *2,* 679 (1959).

417. LINDQVIST, B., and STORGARDS, T., Acta Chem. Scand., *13,* 1839 (1959).
418. LINDQVIST, B., and STORGARDS, T., Acta Chem. Scand., *14,* 757 (1960).
419. LINDQVIST, B., and STORGARDS, T., 16th Intern. Dairy Congr., B, 665 (1962).
420. LINKLATER, P. M., Ph.D Thesis, University of Wisconsin (1961).
421. LINKLATER, P. M., and ERNSTROM, C. A., J. Dairy Sci., *44,* 1621 (1961).
422. LOCHRY, H. R., SANDERS, G. P., MALKAMES, J. P., JR., and WALTER, H. E., U.S. Dept. Agr., Circ. 880 (1951).
423. LONG, J. E., and HARPER, W. J., J. Dairy Sci., *39,* 138 (1956).
424. LONG, J. E., and HARPER, W. J., J. Dairy Sci. *39,* 245 (1956).
425. LUBERT, D. J., and FRAZIER, W. C., J. Dairy Sci., *38,* 981 (1955).
426. LUCAS, P. S., Oregon Agr. Expt. Sta. Bull. 155, (1918).
427. MABBITT, L. A., J. Dairy Res., *22,* 224 (1955).
428. MacKINLAY, A. G., HILL, R. J., and WAKE, R. G., Biochim. Biophys. Acta, *115,* 103 (1966).
429. MacKINLAY, A. G., and WAKE, R. G., Biochim. Biophys. Acta, *104,* 167 (1965).
430. MacKINLAY, A. G., and WAKE, R. G., in "Milk Proteins, Chemistry and Molecular Biology", 2, 175, H. A. McKenzie Ed. Academic Press, New York, 1971.
431. MAHON, J. H., ANGLIN, C., and CHAPMAN, R. A., J. Assoc. Offic. Agr. Chemists, *38,* 482 (1955).
432. MALPRESS, F. H., Biochem. J., *80,* 19P (1961).
433. MALPRESS, F. H., Nature, *215,* 855 (1967).
434. MANN, E. J., Dairy Ind., *32,* 761 (1967).
435. MARQUARDT, J. C., J. Dairy Sci., *10,* 309 (1927).
436. MARTENS, R., Meded. Rijksfac. Landb. Wet. Gent., *34,* 317 (1969). Cited in DSA *33,* 2258 (1971).
437. MARTH, E. H., Milk Prod. J., *44,* (10) 30 (1953).
438. MARTH, E. H., J. Dairy Sci., *46,* 869 (1963).
439. MATHESON, K. J., and CAMMACK, F. R., U.S. Dept. Agri. Bull. 669 (1918).
440. MATHESON, K. J., SANDERS, G. P., BURKEY, L. A., and CONE, J. F., J. Dairy Sci., *27,* 483 (1944).
441. MATHUR, M. P., NAIN, C. K., WALLI, T. K., NATH, I., and GANGULI, N. C., 18th Intern. Dairy Congr., IE, 290 (1970).
442. MATHUR, M. P., PRABHAKARAN, R. J. V., NAIR, P. G., and BHALERAO, V. R., Indian J. Dairy Sci., *23,* 252 (1970).
443. MATTENHEIMER, H., NITSCHMANN, H., and ZAHLER, P., Helv. Chim. Acta, *35,* 1970 (1952).
444. MATTICK, E. C. V., J. Dairy Res., *9,* 233 (1938).
445. MATTIOLI, I., CORBERI, E., and BERGAMASCHI, P., Anali. Microbiol., *20,* 31 (1970). Cited in DSA *33,* 5646 (1971).
446. MAUBOIS, J. L., and MOCQUOT, G., Lait, *49,* 497 (1969). Cited in DSA *32,* 872 (1970).
447. MAXCY, R. B., PRICE, W. V., and IRVINE, D. M., J. Dairy Sci., *38,* 80 (1955).
448. MAYR, A., Dte, Molk-Ztg. *92,* 544 (1971). Cited in DSA *33,* 3305 (1971).
449. McCAMMON, R. B., CAULFIELD, W. J., and KRAMER, M. M., J. Dairy Sci., *16,* 253 (1933).
450. McDOWELL, F. H., DOLBY, R. M., and McDOWELL, A. K. R., J. Dairy Res., *8,* 31 (1937).
451. McGANN, T. C. A., Ph.D Thesis, University College, Cork, Ireland (1960).
452. McGANN, T. C. A., and PYNE, G. T., J. Dairy Res., *27,* 403 (1960).
453. McGUGAN, W. A., Can. Fd. Ind., *34* (9) 44 (1963).
454. McGUGAN, W. A., and EMMONS, D. B., J. Dairy Sci., *50,* 1495 (1967).
455. McGUGAN, W. A., and HOWSAM, S. G., J. Dairy Sci., *47,* 139 (1964).
456. MEITNER, P. A., and KASSELL, B., Biochem. J., *121,* 249 (1971).
457. MELACHOURIS, N. P., and TUCKEY, S. L., J. Dairy Sci., *47,* 1 (1964).
458. MELACHOURIS, N. P., and TUCKEY, S. L., J. Dairy Sci., *49,* 800 (1966).
459. MELACHOURIS, N. P., and TUCKEY, S. L., J. Dairy Sci., *50,* 943 (1967).
460. MELACHOURIS, N., and TUCKEY, S. L., J. Dairy Sci., *51,* 650 (1968).

461. MELNICK, D., and LUCKMANN, F. H., Food Res. *19,* 20 (1954).
462. MENSHIKOV, N., 17th Intern. Dairy Congr. D 565 (1966).
462a. MERCIER, J. C., URO, J., RIBADEAU-DUMAS, B., and GROSCLAUDE, F., Eur. J. Biochem. *27,* 535 (1972).
463. MERKER, H. M., J. Dairy Sci., *2,* 482 (1919).
464. MICKELSEN, R., and ERNSTROM, C. A., J. Dairy Sci., *50,* 645 (1967).
465. MICKELSEN, R., and ERNSTROM, C. A., J. Dairy Sci., *55,* 294 (1972).
466. MICKELSEN, R., and FISH, N. L., J. Dairy Sci., *53,* 704 (1970).
467. MISHRA, M., Indian Vet. J., 42, 134 (1965). Cited in DSA *30,* 1374 (1968).
468. MOCQUOT, G., ALAIS, C., and CHEVALIER, R., Ann. Technol. Agr. (Paris) *3,* 1 (1954).
469. MOEWS, P. C., and BUNN, C. W., J. Mol. Biol., *4,* 395 (1970).
470. MOIR, G. M., J. Dairy Res., *1,* 149 (1930); *2,* 176 (1931).
471. MOIR, G. M., J. Dairy Res., *2,* 68 (1930); *3,* 80 (1931).
472. MORGAN, M. E., and ANDERSON, E. O., J. Dairy Sci., *39,* 253 (1956).
473. MORGANROTH, J., Centr. Bakteriol Parasitenk. Abt. I., *26,* 349 (1899).
474. MORR, C. V., J. Dairy Sci., *50,* 1744 (1967).
475. MORRIS, H. A., COMBS, W. B., and COULTER, S. T., J. Dairy Sci., *34,* 209 (1951).
476. MORRIS, H. A., and JEZESKI, J. J., J. Dairy Sci., *47,* 681 (1964).
477. MORRIS, H. A., JEZESKI, J. J., COMBS, W. B., and KUROMOTO, S., J. Dairy Sci., *46,* 1 (1963).
478. MORRIS, T. A., Qd. J. Agr. Sci., *19,* 93 (1962). Cited in DSA, *25,* 54 (1963).
479. MORRIS, T. A., and McKENZIE, I. J., 18th Intern. Dairy Congr., IE, 293 (1970).
480. MURACHI, T., *in* "Methods in Enzymology", 19, 273, G. E. Perlmann and L. Lorand Ed., Academic Press, New York, 1970.
481. MURRAY, E. D., and GRUETZNER, V. E., U.S. Pat., 3,543,563 (1970).
482. MURRAY, E. D., and KENDALL, M. S., U.S. Pat., 3,482,997 (1969).
483. MURRAY, E. D., and PRINCE, M. P., U.S. Pat., 3,507,750 (1970).
484. NAIR, P. G., SINGH, M., and BHALERAO, V. R., Indian Vet. J., *42,* 250 (1965).
485. NAKAYAMA, T., Jap. Pat., 6387 (1965). Cited in DSA *30,* 1174 (1968).
486. National Cheese Institute, Research Committee, Milk Prod. J., *51,* (7) 43 (1960).
487. NATIONAL CHEESE INSTITUTE, Supplement to 1968 petition for amendment of a regulation under Sec. 401, Fed. Food, Drug and Cosmetic Act, 1969.
488. NEILANDS, J. B., and STUMPF, P. K., "Outlines of Enzyme Chemistry", 2nd Ed., John Wiley and Sons Inc., New York, 1958.
489. NELSON, J. H., Personal communication (1972).
490. NENKOV, M., Vet. Med. Nauki Sof., *8,* (3) 49 (1971). Cited in DSA *33,* 5247 (1971).
491. New Zealand Dairy Research Inst. Annual Report, 24, (1967).
492. NIKI, R., CHANG, K. K., and ARIMA, S., J. Fac. Agric. Hokkaido University, *55,* 421 (1968). Cited in DSA *32,* 2670 (1970).
493. NITSCHMANN, H., and BOHREN, H. W., Helv. Chim. Acta, *38,* 1953 (1955).
494. NITSCHMANN, H., and HENZI, R., Helv. Chim. Acta, *42,* 1985 (1959).
495. NITSCHMANN, H., WISSMANN, H., and HENZI, R., Chimia, *11,* 76 (1957).
496. NORTHRUP, J. H., J. Gen. Physiol., *5,* 263 (1922).
497. NORTHRUP, J. H., J. Gen. Physiol., *16,* 615 (1933).
498. NOZNICK, P. P., and BUNDUS, R. H., U.S. Patent 3,316,098 (1967).
499. O'CONNOR, C. B., Dairy Ind. *33,* 625 (1968).
500. OEDA, M., and KASAI, B., Bull. Nippon Vet. and Zootech. College, No. 3 (1954).
501. OHREN, J. A., and TUCKEY, S. L., J. Dairy Sci., *52,* 598 (1969).
502. OLIVER, W. H., J. Gen. Microbiol., *7,* 329 (1952); Ibid., *8,* 38 (1953).
503. OLSON, N. F., Dairy Record, *71,* (8) 7 (1971).
504. OLSON, N. F., and PRICE, W. V., J. Dairy Sci., *44,* 1394 (1961).
505. OLSON, N. F., VAKALERIS, D. G., PRICE, W. V., and KNIGHT, S. G., J. Dairy Sci., *41,* 1005 (1958).

506. OREKHOVITCH, V. N., SHPIKITER, V. O., and PETROVA, V. I., Dokl. Akad. Nauk. S.S.S.R., *11,* 401 (1956).
507. ORLA-JENSEN, S., Ann. Agr. Suisse, *7,* 14 (1906).
508. ORLA-JENSEN, S., Rev. gen. Lait, *5,* 464 (1906); Ann. Agr. Suisse, *7,* 20, 253 (1906).
509. ORLA-JENSEN, S., Landw. Jahrb. Schweiz., *20,* 437 (1906).
510. OSMAN, H. G. ABDEL-FATTAH, A. F., and MABROUK, S. S., J. Gen. Microbiol., *59,* 131 (1969).
511. OTTESEN, M., and RICKERT, W., Compt. rend. trav. lab. Carlsberg, *37,* 301 (1970).
512. PALMER, H. J., and SLY, W. H., Dairy Ind., *6,* 241 (1941).
513. PALMER, H. J., and SLY, W. H., Dairy Ind., *8,* 427 (1943).
514. PALMER, H. J., and SLY, W. H., J. Soc. Chem. Ind., *63,* 363 (1944).
515. PALMER, L. S., and HANKINSON, C. L., J. Dairy Sci., *24,* 429 (1941).
516. PALMER, L. S., and TARASSUK, N. P., J. Dairy Sci., *23,* 861 (1940).
517. PANG, S. H., M. S. Thesis, Utah State University (1969).
518. PARRY, R. M., JR., and CARROLL, R. J., Biochim.Biophys. Acta, *194,* 138 (1969).
519. PATEL, M. C., LUND, D. B., and OLSON, N. F., J. Dairy Sci., *55,* 913 (1972).
520. PATTON, S., J. Dairy Sci., *33,* 680 (1950).
521. PATTON, S., J. Dairy Sci., *46,* 856 (1963).
522. PATTON, S., J. Dairy Sci., *47,* 817 (1964).
523. PATTON, S., WONG, N. P., and FORSS, D. A., J. Dairy Sci., *41,* 857 (1958).
524. PAYENS, T. A. J., 16th Intern. Dairy Congr., B, 410 (1962).
525. PAYENS, T. A. J., J. Dairy Sci., *49,* 1317 (1966).
526. PERLMANN, G. E., Arch. Biochem. Biophys., *65,* 210 (1956).
527. PERLMANN, G. E., Proc. Nat. Acad. Sci., *45,* 915 (1959).
528. PETER, A., "Praktsche anleitung zur fabrikation und behandlung des Emmenthaler Kase", 6th Ed., K. J. Wyss Erbern, Bern, 1930.
529. PETER, J. E., and DIETRICH, J. W., Tex. J. Sci., *6,* 442 (1954).
530. PETERS, I. I., J. Dairy Sci., *39,* 1083 (1956).
531. PETERS, I. I., Dairy Sci. Abstr., *26,* 457 (1964).
532. PETERS, I. I., and MOORE, A. V., J. Dairy Sci., *41,* 70 (1958).
533. PETERS, I. I., and NELSON, F. E., Milk Prod. J., *51* (11), 14 (1960).
534. PHILLIPS, C. A., J. Dairy Sci., *11,* 292 (1928).
535. PHILLIPS, C. A., Nat. Butter Cheese J., *34,* (6) 13 (1943).
536. PHILLIPS, C. A., RICHARDSON, G. A., and TARASSUK, N. P., J. Dairy Sci., *25,* 728 (1942).
537. PIMBLETT, J., Int. Dairy Congr. B 721 (1962).
538. PLACEK, C., BAVISOTTO, V. S. and JADD, E. C., Ind. Eng. Chem., *52,* 2 (1960).
539. POWELL, M. E., J. Dairy Sci., *19,* 305 (1936).
540. POZSARNE-HAJNAL, K., VAMOSNE-VIGYAZO, L., NONN-NE-SAS, H., and HEGEDUSNE-VOLGYESI, E., Elelmiszertudomany, *3,* (2) 55 (1969).
541. PRICE, W. V., J. Dairy Sci., *10,* 155 (1927).
542. PRICE, W. V., J. Dairy Sci., *10,* 373 (1927).
543. PRICE, W. V., and CALL, A. O., J. Milk Fd. Technol., *32,* 304 (1969).
544. PRICE, W. V., and PRICKETT, P. S., J. Dairy Sci., *11,* 69 (1928).
545. PRINS, J., and NIELSEN, T. K., Process Biochem. *5,* (5) 34 (1970).
546. PUHAN, Z., Milchwissenschaft, *23,* 331 (1968).
547. PUHAN, Z., J. Dairy Sci., *52,* 889 (1969).
548. PUHAN, Z., and STEFFEN, C., Schweiz. Milchztg, *93,* (56) Wiss Beil. Nr. 114, 937 (1967). Cited in DSA *29,* 3818 (1967).
549. PUJOLLE, J., RIBADEAU-DUMAS, B., GARNIER, J., and PION, R., Biochem. Biophys. Res. Comm., *25,* 285 (1966).
550. PYATNITSKII, N. P., and PYATNITSKAYA, I. N., Izv. vyssh. ucheb. Zaved., Pishch. Tekhnol, *1969,* 156 (1969). Cited in DSA *31,* 4313 (1969).
551. PYNE, G. T., Biochem. J., *39,* 385 (1945).

552. PYNES, G. T., Chem. and Ind. (London), *13,* 302 (1953).
553. PYNE, G. T., Dairy Sci. Abstracts, *17,* 532 (1955).
554. PYNE, G. T., and McGANN, T. C. A., J. Dairy Res., *27,* 9 (1960).
555. RAJAGOPALAN, T. G., MOORE, S., and STEIN, W. H., J. Biol. Chem., *241,* 4940 (1966).
556. RAMET, J. P., ALAIS, C., and WEBER, F., Lait, *49,* 40 (1969).
557. RAMET, J. P., and SCHLUTER, A. S., Revue lait, fr Industrie lait (*280*) 730 (1970). Cited in DSA *33,* 1612 (1971).
558. RAND, A. G., M. S. Thesis, University of Wisconsin (1961).
559. RAND, A. G., Ph.D. Thesis, University of Wisconsin (1964).
560. RAND, A. G., and ERNSTROM, C. A., J. Dairy Sci., *47,* 1181 (1964).
561. RAND, A. G., and ERNSTROM, C. A., J. Dairy Sci., *51,* 1756 (1968).
562. RAPP, H., and CALBERT, H. E., J. Dairy Sci., *37,* 637 (1954).
563. REICHART, E. L., Neb. Agr. Expt. Sta. Bull. 303 (1936).
564. REID, W. H. E., and FLESHMAN, C. L., Dairy World 11 (12) 20 (1933).
565. REITER, B., FRYER, T. F., SHARPE, M. E., and LAWRENCE, R. C., J. Appl. Bacteriol., *29,* 231 (1966).
566. REPS, A., POZNANSKI, S., KOWALSKA, W., Bull. Acad. Pol. Sci., C1 II Ser. Sci. Biol., *17,* 535 (1969). Cited in DSA *32,* 1676 (1970).
567. REPS, A., POZNANSKI, S., and KOWALSKA, W., Milchwissenschaft, *25,* 146 (1970).
568. REPS, A., POZNANSKI, S., SURAZYNSKI, A., and CHOMINIEC, E., Przegl. mlecz, *20,* (6) (1971). Cited in DSA *33,* 5457 (1971).
569. RESMINI, P., VOLONTERIO, G., SARACCHI, S., and BOZZOLATI, M., Industrie Agr., *9,* 6 (1971). Cited in DSA *33,* 3904 (1971).
570. REYES, J., M. S. Thesis, Utah State University (1971).
571. RIBADEAU-DUMAS, B., and GARNIER, J., J. Dairy Res., *37,* 269 (1970).
572. RICHARDSON, G. H., J. Dairy Sci., *53,* 1373 (1970).
573. RICHARDSON, G. H., and ERNSTROM, C. A., Unpublished results (1971).
574. RICHARDSON, G. H., GANDHI, N. R., DIVATIA, M. A., and ERNSTROM, C. A., J. Dairy Sci., *54,* 182 (1971).
575. RICHARDSON, G. H., NELSON, J. H., LUBNOW, R. E., SCHWARBERG, R. L., J. Dairy Sci., *50,* 1066 (1967).
576. RICHMOND, V., TANG, J., WOLF, S., TRUCCO, R. E., and CAPUTTO, R., Biochim. Biophys. Acta, *29,* 453 (1958).
577. RICKERT, W., Compt. rend. trav. lab. Carlsberg, *38,* 1 (1970).
578. RICKERT, W., Biochim. Biophys. Acta, *220,* 628 (1970).
579. RITTER, W., Dte. Molk-Ztg., *91,* 2222, (1970). Cited in DSA *33,* 3880 (1971).
580. RITTER, W., Dte. Molk-Ztg. (Kempten-Allgau F48, 2222) (1970).
581. RITTER, W., SCHILT, P., and BLANC, B. H., Schweiz. Milchztg. (Lait Romand), *95,* (51) 442 (1969). Cited in DSA *31,* 3251 (1969).
582. ROBERTSON, T. B., J. Biol. Chem., *5,* 147 (1908).
583. ROGERS, L. A., HARDELL, R. E., and FEUTZ, F., J. Dairy Sci., *22,* 43 (1939).
584. ROSE, D., Dairy Sci. Abstracts, *25,* 45 (1963).
585. ROSE, D., Dairy Sci. Abstracts, *31,* 171 (1969).
586. ROSE, D., BRUNNER, J. R., KALAN, E. B., LARSON, B. L., MELNYCHYN, P., SWAISGOOD, H. E., and WAUGH, D. F., J. Dairy Sci., *53,* 1 (1970).
587. ROSENGREN, L. F., and HAGLUND, E., Centr. Bakt. Parasitenk., II, Abt. 45, 156 (1915).
588. ROTINI, O. T., 13th Intern. Dairy Congr., *3,* 1406 (1953).
589. ROUNDY, Z. D., Milk Prod. J. *52,* No. 7, 12 (1961).
590. ROUNDY, Z. D., and PRICE, W. V., J. Dairy Sci., *24,* 235 (1941).
591. RUDIGER, M., and WURSTER, K., Biochem. Z., *216,* 367 (1929).
592. RYLE, A. P., *in* "The Council for International Organization of Medical Sciences Symposium", (P25), H. Munro Ed., Blackwell, Oxford, 1964.
593. RYLE, A. P., Biochem. J., *98,* 485 (1966).

594. RYLE, A. P., in "Methods in Enzymology", 19, 316, G. E. Perlmann and L. Lorand Ed., Academic Press, New York, 1970.
595. RYLE, A. P., and PORTER, R. R., Biochem. J., *73,* 75 (1959).
596. SAMMIS, J. L., Wis. Agr. Expt. Sta. Bull. 241, 16 (1914).
597. SAMMIS, J. L., "Cheese Making" 12th Ed., The Cheese Makers Book Co., Madison, 1948.
598. SAMMIS, J. L., and BRUHN, A. T., Wis. Expt. Sta. Res. Bull. No. 27 (1912).
599. SAMMIS, J. L., and BRUHN, A. T., U.S. Dept. Agr. Bur. An. Ind. Bull. 165 (1913).
600. SAMMIS, J. L. and GERMAINE, L., Butter Cheese J., *20,* No. 39, 13 (1929).
601. SAMMIS, J. L., LAABS, F. W., and SUZUKI, S. K., Wis. Agr. Expt. Sta. Circ. Information *20,* 4 (1911).
602. SAMMIS, J. L., SUZUKI, S. K., and LAABS, F. W., U.S. Dept. Agr. Bur. An. Ind. Bull. 122 (1910).
603. SANDBERG, E., HAGLUND, E., and BARTHEL, C., Lait, *10,* 1 (1930).
604. SANDERS, G. P., J. Assoc. Offic. Agr. Chem. *31,* 306 (1948).
605. SANDERS, G. P., "Cheese Varieties and Descriptions," U.S. Dept. Agr. Handbook 54 (1953).
606. SANDERS, G. P., FARRAR, R. R., HARDELL, R. E., FEUTZ, F., and BURKEY, L. A., J. Dairy Sci., *23,* 905 (1940).
607. SANDERS, G. P., MATHESON, K. J., and BURKEY, L. A., J. Dairy Sci., *19,* 395 (1936).
608. SANDERS, G. P., and SAGER, O. S., J. Dairy Sci., *30,* 909 (1947).
609. SANDERS, G. P., TITTSLER, R. P., and WALTER, H. E., U.S. Dept. Agr. Bur. Dairy Ind., BDIM-Inf-29 (1946).
610. SANDERS, G. P., WALTER, H. E., and TITTSLER, R. P., J. Dairy Sci., *29,* 497 (1946).
611. SANGER, F., and TUPPY, H., Biochem. J., *49,* 481 (1951).
612. SANNABADTHI, S. S., SRINIVASAN, R. R., 18th Intern. Dairy Congr., IE, 278 (1970).
613. SANSONETTI, F., Lait, *10,* 627, 1109 (1930).
614. SARDINAS, J. L., U.S. Patent 3,275,453 (1966).
615. SARDINAS, J. L., W. German Patent Appl. 1,442,140 (1968).
616. SARDINAS, J. L., Appl. Microbiol., *16,* 248 (1968).
617. SCARPELLINO, R., and KOSIKOWSKI, F. V., J. Dairy Sci., *45,* 343 (1962).
618. SCEDROV, O., and LESIC, L., Kemija Ind., *13,* 763 (1964). Cited in DSA *28,* 1972 (1966).
619. SCHARPF, L. G., JR., and KICHLINE, T. P., Food Technol., *23,* 835 (1969).
620. SCHMIDT, D. G., Biochim. Biophys. Acta, *90,* 411 (1964).
621. SCHNECK, A., and ELGER, A., Milchwirtschaft Zentr., *60,* 205 (1931).
622. SCHOBER, R., and HEIMBURGER, N., Milchwissenschaft, *15,* 561 (1960).
623. SCHOBER, R., HEIMBURGER, N., and PRINTZ, I., Milchwissenschaft, *15,* 506 (1960).
624. SCHOBERL, A., and RAMBACHER, P., Biochem. Z., *305,* 223 (1940).
625. SCHORMÜLLER, J., Adv. Food Res., *16,* 231, (1968).
625a. SCHORMÜLLER, J., GLATHE, M., and HUTH, H., Z. Lebensm. Untersuch.
626. SCHULTZ, M., Molkerei Ztg. (Hildesheim), *44,* 527 (1930).
627. SCHULTZ, M. E. and THOMASOW, J., Milchwissenschaft *25,* 330, (1970).
628. SCHWANDER, H., ZAHLER, P., and NITSCHMANN, H., Helv. Chim. Acta, *35,* 553 (1952).
629. SCHWARTZ, D. P., and PARKS, O. W., J. Dairy Sci., *46,* 989, (1963).
630. SCHWARTZ, D. P., PARKS, O. W., and BOYD, E. N., J. Dairy Sci., *46,* 1422 (1963).
631. SCOTT, E. C., and McDONALD, G. W., U.S. Patent 2,482,520 (1949).
632. SCOTT BLAIR, G. W., and OOSTHUIZEN, J. C., J. Dairy Res., *28,* 165 (1961).
633. SCOTT BLAIR, G. W., and OOSTHUIZEN, J. C., Nature, *191,* 697 (1961).

634. SEQUI, P., and ROTINI, O. T., Agrochimica, *14,* 379 (1970).
635. SHARP, P. F., and McINERNEY, T. J., J. Dairy Sci., *19,* 573 (1936).
636. SHOVERS, J., Personal Communication (1968).
637. SHOVERS, J., and BAVISOTTO, V. S., J. Dairy Sci., *50,* 942 (1967).
638. SHOVERS, J., FOSSUM, G., and NEAL, A., J. Dairy Sci., *55,* 675 (1972).
639. SHOVERS, J., and KORNOWSKI, R., J. Dairy Sci., *51,* 937 (1968).
640. SHUKRI, N. A., M. S. Thesis, University of Wisconsin (1965).
641. SHUKRI, N. A., Ph.D. Thesis, Utah State University (1969).
642. SILVERMAN, G. J., and KOSIKOWSKI, F. V., J. Dairy Sci., *38,* 950 (1955).
643. SILVERMAN, G. J., and KOSIKOWSKI, F. V., J. Dairy Sci., *39,* 1134 (1956).
644. SINKINSON, G., and WHEELOCK, J. V., Biochem. J., *115,* 19P (1969).
645. SIRRY, I., and SHIPE, W. F., J. Dairy Sci., *41,* 204 (1958).
646. SKELTON, G. S., Enzymologia, *40,* 170 (1971).
647. SMITH, D. P., and ROLLIN, N. J., Food Res., *19,* 59 (1954).
648. SOMMER, H. H., J. Dairy Sci., *13,* 288 (1930).
649. SOMMER, H. H., and MATSEN, H., J. Dairy Sci., *18,* 741 (1935).
650. SRINIVASAN, R. A., Indian Sci. Abstr. *3,* 7184 (1966).
651. SRINIVASAN, R. A., ANANTHARAMIAH, S. N., and ANANTAKRISHNAN, C. P., Indian J. Dairy Sci., *21,* 149 (1968). Cited in DSA, *31,* 3431 (1969).
652. STADHOUDERS, J., DEVRIES, E., and MULDER, H., 15th Intern. Dairy Congr., London 2, 709 (1959).
653. STADHOUDERS, J., and MULDER, H., Neth. Milk Dairy J., *14,* 141 (1960).
654. STEINHARDT, J., J. Biol. Chem., *123,* 543 (1938).
655. STERNBERG, M. Z., J. Dairy Sci., *54,* 159 (1971).
656. STEUART, D. W., J. Board Agr., *24,* 313 (1917).
657. STEVENSON, C., J. Agr. (N.Z.), *14,* 32 (1917).
658. STINE, J. B., U.S. Patent 2,494,636 (Jan. 17, 1950).
659. STINE, J. B., Personal Communication (1962).
660. STOKOE, W. N., Biochem. J., *22,* 80 (1928).
661. STORCH, V., and SEGELKE, T., Milchztg., *3,* 997 (1874). Cited by C. Porcher in Le Lait, *10,* 47 (1930).
662. STORGÅRDS, T., and LINDQUIST, B., 13th Intern. Dairy Congr., Proc., *2,* 625 (1953).
663. SULLIVAN, R. A., FITZPATRICK, M. M., STANTON, E. R., ANNINO, R., KISSEL, G., and PALERMITI, F., Arch. Biochem. Biophys., *55,* 455 (1955).
664. TANG, J., *in* "Methods in Enzymology", 19,406, G. E. PERLMANN and L. LORAND ED., Academic Press, New York, 1970.
665. TANG, J., and HARTLEY, B. S., Biochim. J., *118,* 611 (1970).
666. TANG, J., MILLS, J., CHIANG, L., and DE CHIANG, L., Ann. N.Y., Acad. Sci., *140,* 688 (1967).
667. TANG, J., WOLF, S., CAPUTTO, R., and TRUCCO, R. E., J. Biol. Chem., *234,* 1174 (1959).
668. TARASSUK, N. P., and PALMER, L. S., J. Dairy Sci., *22,* 543 (1939).
669. TARASSUK, N. P., and RICHARDSON, G. A., J. Dairy Sci., *24,* 667 (1941).
670. TAUBER, H., J. Biol. Chem., *107,* 161 (1934).
671. TAUBER, H., and KLEINER, I. S., J. Biol. Chem., *104,* 259 (1934).
672. TEMPLETON, H. L., and SOMMER, H. H., J. Dairy Sci., *13,* 203 (1930).
673. TEMPLETON, H. L., and SOMMER, H. H., J. Dairy Sci., *15,* 29 (1932).
674. TEMPLETON, H. L., and SOMMER, H. H., J. Dairy Sci., *15,* 155 (1932).
675. TEMPLETON, H. L., and SOMMER, H. H., J. Dairy Sci., *19,* 561 (1936).
676. TEMPLETON, H. L., and SOMMER, H. H., J. Dairy Sci., *20,* 231 (1937).
677. THOM, C., CURRIE, J. N., and MATHESON, K. J., Conn. (Storrs) Agr. Expt. Sta. Bull. 79, 387 (1914).
678. THOMASOW, J., Milchwissenschaft, *23,* 725 (1968).
679. THOMASOW, J., Milchwissenschaft, *26,* 276 (1971).
680. THOMASOW, J., MROWETZ, G., and SCHMANKE, E., Milchwissenschaft, *25,* 211 (1970).

681. THOMASOW, J., MROWETZ, G., and SCHMANKE, E., Kieler Milchw. Forsch. Ber., *23,* 57 (1971). Cited in DSA *33,* 3304 (1971).
682. THONI, J., Ann. Agric. Suisse, *46,* 708 (1945).
683. THORNLEY, B. D., and HILTON, S., U.S. Patent 2,337,947 (1943); 2,352,037 (1944).
684. THURSTON, L. M., and GOULD, I., JR., J. Dairy Sci., *16,* 467 (1933).
685. TIPOGRAF, D. Y., VESELOV, A. I., LeVAN, N., and MOSICHEV, M. S., Prikl, Biokhim. Mikrobiol., *2,* 45 (1966). Cited in DSA *28,* 2238 (1966).
686. TISELIUS, A., HENSCHEN, G. E., and SVENSSON, H., Biochem. J., *32,* 1814 (1938).
687. TITTSLER, R. P., and SANDERS, G. P., J. Dairy Sci., *36,* 574 (1953).
688. TITTSLER, R. P., WOLK, J., and HARGROVE, R. E., J. Dairy Sci., *37,* 638 (1954).
689. TODD, A., and CORNISH, E. C. V., J. Board of Agr., *24,* 307 (1917).
690. TOKITA, F., and NAKANISHI, T., Milchwissenschaft, *17,* 198 (1962).
691. TRACY, P. H., and RUEHE, H. A., Ill. Agr. Expt. Sta., Cir. 445 (1936).
692. TROUT, G. M., "Homogenized Milk," Michigan State College Press, East Lansing, 1950.
693. TSOULI, J., Compt. rend, hebd. Seanc. Acad. Sci., Ser. D. Paris, *270,* 396 (1970). Cited in DSA *32,* 1799 (1970).
694. TSUGO, T., Japan J. Zootech. Sci., *23,* 178 (1953); Dairy Sci. Abstr., *15,* 993 (1953).
695. TSUGO, T., and KOBAYASHI, M., Jap: J. Zootech. Sci., *26,* 173 (1955).
696. TSUGO, T., and YAMAUCHI, K., 13th Intern. Dairy Congr., *4,* 641 (1953).
697. TSUGO, T., and YAMAUCHI, K., J. Agr. Chem. Soc. (Japan) *29,* 457 (1955).
698. TSUGO, T., and YAMAUCHI, K., 15th Intern. Dairy Congr., *2,* 636 (1959).
699. TUCKEY, S. L., and RUEHE, H. A., J. Dairy Sci., *23,* 517 (1940).
700. TUSTIN, E. B., JR., Nat. Butter Cheese J., *37* (9), 44 (1946).
701. U.S. Food and Drug Administration, "Cheeses and Cheese Products, Definitions and Standards"; Federal Food, Drug, and Cosmetic Act, Part 19, Title 21, as amended (1971).
702. VAILLANT, E., Lait, *4,* 7 (1924).
703. VANDEN BERG, G., ALDERLIESTE, P. J., ROBBERTSEN, T., and ZWAGINGA, P., Officeel Org. K. ned. Zuivelbond, *62,* 1032 (1970). Cited in DSA *33,* 2758 (1971).
704. VAN DER BURG, B., and VAND DER SCHEER, A. F., Congr., intern. Tech. et chim. ind. agr. Compt. rend. 5th Congr., *2,* 321 (1937).
705. VAN SLYKE, L. L., and BOSWORTH, A. W., N.Y. (Geneva) Agr. Expt. Sta. Tech. Bull. 4 (1907).
706. VAN SLYKE, L. L., and BOSWORTH, A. W., N.Y. (Geneva) Agr. Expt. Sta. Tech. Bull. 37 (1914).
707. VAN SLYKE, L. L., and HART, E. B., N.Y. (Geneva) Agr. Expt. Sta., Bull., 236 (1903).
708. VAN SLYKE, L. L., and PRICE, W. V., "Cheese," Orange Judd Publ. Co., Inc., New York, 1949.
709. VAN VUNAKIS, H., and HERRIOTT, R. M., Biochim. Biophys. Acta, *23,* 600 (1957).
710. VAS, K., Kiserlet. Közlemenyek, *33,* 377 (1930); Milchwirtsch. Forsch., *11,* 519 (1931).
711. VIEIRA DE SA, F., and BARBOSA, M., 18th Intern. Dairy Congr., IE, 286 (1970).
712. VIEIRA DE SA, F., and BARBOSA, M., 18th Intern. Dairy Congr., IE, 287 (1970).
713. VIEIRA DE SA, F., and BARBOSA, M., 18th Intern. Dairy Congr., IE, 288 (1970).
714. VIEIRA DE SA, F., and MATOS, AGUAS, J. P., 17th Intern. Dairy Congr., D, 659 (1966).
715. VIRASORO, E., Anales inst. invest. cienty. y tecnol. (Univ. nac. litoral, Santa Fe, Arg.), *7,* 81 (1938).
716. VIRTANEN, A. I., KREULA, M. S., and NURMIKKO, V. T., 12th Intern. Dairy Congr., Proc., *2,* 268 (1949).
717. VIVIAN, A., Hoards Dairyman, *34,* (19) 473 (1903).

718. WAKE, R. G., Aust. J. Biol. Sci., *12,* 479 (1959).
719. WALKER, J. R. L., J. Dairy Res., *26,* 273 (1959).
720. WALKER, J. R. L., J. Dairy Res., *28,* 1 (1961).
721. WALKER, J. R. L., and HARVEY, R. J., J. Dairy Res., *26,* 265 (1959).
722. WALLACE, G. M., and AIYAR, K. R., 18th Intern. Dairy Congr., IE, 48 (1970).
723. WALTER, H. E., SADLER, A. M., MALKAMES, J. P., and MITCHELL, C. D., U.S. Dept. Agr., Agr. Research Service, ARS-73-11 (1956).
724. WANG, H. L., RUTTLE, D. I., HESSELTINE, C. W., Can J. Microbiol., *15,* 99 (1969).
725. WANG, J. C. T., M. S. Thesis, Utah State University (1969).
726. WATSON, P. D., Ind. Eng. Chem., *19,* 1272 (1927); J. Dairy Sci., *12,* 289 (1929).
727. WAUGH, D. F., Personal Communication (1971).
728. WAUGH, D. F., in "Milk Proteins, Chemistry and Molecular Biology", 2, 3, H. A. McKenzie Ed., Academic Press, New York, 1971.
729. WAUGH, D. F., and NOBLE, R. W. JR., J. Am. Chem. Soc., *87,* 2246 (1965).
730. WAUGH, D. F., and VON HIPPEL, P. H., J. Am. Chem. Soc., *78,* 4576 (1956).
731. WEBB, B. H., J. Dairy Sci., *14,* 508 (1931).
732. WEIK, R. W., Diss. Abstr. *25,* 4638 (1965).
733. WEST, P. M., and HILLIARD, J., Proc. Soc. Expt'l Biol. Med., *71,* 169 (1949).
734. WHEELOCK, J. V., and SINKINSON, G., Biochim. Biophys. Acta, *194,* 597 (1969).
735. WHITAKER, J. R., Food Tech., *13,* (2) 87 (1959).
736. WHITAKER, J. R., *in* "Methods in Enzymology", 19, 436, G. E. Perlmann and L. Lorand Ed., Academic Press, N. Y., 1970.
737. WILAMOWSKI, G., J. Assoc. Offic. Anal. Chem., *54,* 663 (1971).
738. WILLIAMS, R. C., JR., and RAJAGOPALAN, T. G., J. Biol. Chem., *241,* 4951 (1966).
738a. WILSON, H. L., U.S. Dept. of Agr., Bur. Dairy Ind., BDIM-947 (1942).
739. WILSON, H. L., HALL, S. A., and JOHNSON, W. T., JR., J. Dairy Sci., *24,* 169 (1941).
740. WILSON, H. L., HALL, S. A., and ROGERS, L. A., J. Dairy Sci., *28,* 187 (1945).
741. WILSON, H. L., and JOHNSON, W. T., JR., U.S. patent, 2,127,453 (Aug. 16, 1938).
742. WILSON, H. L., and REINBOLD, G. W., "American Cheese Varieties", Pfizer Cheese Monograph, Vol. II, Chas. Pfizer & Co., Inc., New York.
743. WILSTER, G., Iowa State Coll. J. Sci., *4,* 181 (1930).
744. WILSTER, G. H., "Practical Cheesemaking", 11th Ed. O.S.C. Cooperative Association, Corvallis, Oregon, 1959.
745. WINDLAN, H., and KOSIKOWSKI, F. V., J. Dairy Sci., *39,* 917 (1956).
746. WINKLER, W., Osterr. Milchwirtsch. Ztg., *36,* 285 (1929).
747. WINNICK, T., DAVIS, A. R., and GREENBERG, D. M., J. Gen. Physiol., *23,* 275, 289 301 (1940).
748. WINTERSTEIN, E., and THÖNY, J., Z. physiol. Chem., *36,* 28 (1902).
749. WOERLE, H., Österr. Milchwirtsch. Ztg., *36,* 309 (1929).
750. WOJTKIEWIEZ, A. F., Zentr. Bakt. Parasitenk. II Abt., *87,* 349 (1933).
751. WOJTOWICZ, M. B., and McGUGAN, W., Qualitas Pl. Mater. Veg., *11,* 281 (1964). Cited in DSA *27,* 2289 (1965).
752. WOYCHIK, J. H., Arch. Biochem. Biophys., *109,* 542 (1965).
753. WRIGHT, N. C., Biochem. J., *18,* 245 (1924).
754. YAMAMOTO, T., TAKAHASHI, K., YOSHINO, M., and YOSHITAKE, M., Bull. Nat. Inst. Anim. Ind. Chiba., *18,* 53 (1968). Cited in DSA *32,* 1915 (1970).
755. YAMAUCHI, K., J. Agr. Chem. Soc. Japan, *33,* 1134 (1960).
756. YASHODA, K. M., Current Sci. (India), *10,* 23 (1941).
757. YOSHINO, U., Nippon Nogei-Kagaku Kaishi, *32,* 736 (1958).
758. YOSHINO, U., Agr. Chem. Soc. Japan, *33,* 487 (1959).
759. YOUNG, J. O., and ASHWORTH, U. S., J. Dairy Sci., *43,* 856 (1960).
760. YU, J., LIU, W., TAMURA, G., and ARIMA, K., Agr. Biol. Chem., *32,* 1482 (1968).

761. YU, J., OSAWA, H., TAMURA, G., HONG, Y. M., and ARIMA, K., Korean J. Food Sci. Tech., *2,* 43 (1970). Cited in DSA *33,* 5641 (1971).
762. YU, J., TAMURA, G., and ARIMA, K., Agr. Biol. Chem., *32,* 1048 (1968).
763. YU, J., TAMURA, G., and ARIMA, K., Biochim. Biophys. Acta, *171,* 138 (1969).
764. ZAKARIASEN, B. M., and COMBS, W. B., J. Dairy Sci., *24,* 543 (1941).
765. ZAYKOWSKY, J., and SLOBODSKA-ZAYKOWSKA, N., Biochem. Z., *159,* 199 (1925).
766. ZITTLE, C. A., J. Dairy Sci., *44,* 2101 (1961).
767. ZITTLE, C. A., CERBULIS, J., PEPPER, L., and DELLA MONICA, E. S., J. Dairy Sci., *42,* 1897 (1959).
768. ZITTLE, C. A., DELLA MONICA, E. S., and PEPPER, L., Arch. Biochem. Biophys., *81,* 187 (1959).
769. ZITTLE, C. A., THOMPSON, M. P., CUSTER, J. H., and CERBULIS, J., J. Dairy Sci., *45,* 807 (1962).
770. ZITTLE, C. A., and WALTER, M., J. Dairy Sci., *46,* 1189 (1963).
771. ZUCKERMAN-STARK, S., and LEIBOWITZ, J., Enzymologia, *23,* 71 (1961).
772. ZUCKERMAN-STARK, S., and LEIBOWITZ, J., Enzymologia, *25,* 252 (1963).
773. ZUCKERMANN-STARK, S., Enzymologia, *32,* 380 (1967).
774. ZVYAGINTSEV, V. I., GUDKOV, A. V., TOLKACHEV, A. N., and BUZOV, I. P., Moloch. Prom., *32,* 19 (1971). Cited in DSA *33,* 3644 (1971).
775. ZWAGINGA, P., ALDERLIESTE, P. J., and ROBBERTSEN, T., Rapp. ned. Inst. Zuive londerz, *74,* 19 (1969). Cited in DSA *31,* 3259 (1969).

CHAPTER 13

Elmer H. Marth[†]

Fermentations*

INTRODUCTION

This chapter presents the primary chemical principles, laws, and generalizations which govern fermentations in milk and milk products. In selecting topics, richness and profundity in chemical content have been primary considerations. Although stress is placed on intermediary metabolism, the utilitarian aspects of dairy fermentations have not been ignored. A balance is sought between the theoretical and the practical.

Fermentation is a term with many shades of meaning, each of which defines limited biochemical changes brought about by microorganisms or their enzyme systems. Prescott and Dunn[429] reviewed the changes of meaning which the term has undergone since its derivation as a descriptive word for a gentle and "boiling" condition observed in wine making. They define fermentation in a broad sense as "a process in which chemical changes are brought about in an organic substrate, whether carbohydrate or protein or fat or some other type of organic material, through the action of biochemical catalysts known as 'enzymes' elaborated by specific types of living microorganisms." Fermentations encountered in dairy technology fall within the scope of this definition; some are desirable and others undesirable. Desirable fermentations leading to useful intermediate or end products are usually brought about by inoculating milk or its by-products with pure or mixed cultures; the undesirable ones form substances deleterious to milk or its products and are caused by organisms naturally in milk, or by contaminants.

---

* Revised from Chapter 13 in the 1st edition by Abraham Leviton and Elmer H. Marth.

† Elmer H. Marth, Department of Food Science, University of Wisconsin, Madison, Wisconsin 53706.

A more restrictive definition also has been proposed. According to Elsden,[114] "Fermentation may be defined as a biological process in which chemical energy is made available for growth by oxidative reactions, the ultimate hydrogen acceptors for which are substances other than O_2." This definition, rather than the broader one, is more suitable to discuss intermediate products of carbohydrate metabolism in yeast and muscle. Reactions which produce the intermediates are well established and furnish the theoretical basis to study component reactions in bacterial fermentations. Elsden has enunciated the thesis that the primary mechanism found in muscle and yeast, the operation of which causes formation of pyruvic acid, is also found in bacteria which ferment carbohydrates, and that differences observed in end products result from the ability of bacteria to synthesize different hydrogen acceptors from pyruvic acid.

THEORIES AND METABOLIC PROCESSES OF FERMENTATION

When one organism with simple nutritional requirements predominates and overgrows others, the fermentation in milk parallels that of a pure culture. The usual effect, however, is one in which variable results are obtained because of differences in bacterial strains, in the milk supply and in its treatment, and in external conditions such as temperature, oxygen supply, acidity, etc.

Details of fermentations are important and empirical formulations are indispensable in the art and practice. These will be discussed in the second and third sections of this chapter. This section is devoted to generalizations resulting from studies of isolated systems containing components of cell-free extracts and their specific substrates. Reactions, reaction products, and reaction mechanisms observed in such systems occur in dairy fermentations, but their occurrence is not confined to any particular fermentation medium or substrate.

Topics will be discussed in the following order: the Embden-Meyerhof pathway to pyruvic acid; the pentose phosphate pathway to pyruvic acid; lactic acid fermentation; alcohol fermentation; glycerol fermentation; propionic acid fermentation; oxidative decarboxylation of pyruvic acid; *Escherichia coli* fermentation; *Enterobacter* type fermentation; butyric and related fermentations; fat metabolism; citric acid cycle; "energy-rich" bonds and their formation; and polysaccharide synthesis and assimilation. The order listed is a logical one, for each topic flows from from the preceding one.

Embden-Meyerhof Pathway

The Embden-Meyerhoff pathway constitutes a generalization of great importance, as significant in bacterial processes as it is in yeast fermentation and in glycolysis. The pathway may conveniently be divided into six parts, the first leading to the formation of fructose-1-6-diphosphate; the second to the phosphotrioses, phosphodihydroxyacetone, and 3-phosphoglyceraldehyde; the third to 3-phosphoglycerate; the fourth to 2-phosphoglycerate; the fifth to phosphoenolpyruvate; and the sixth to pyruvic acid. Reactions are given below.

Formation of Frutose-1,6-Diphosphate.—The first stage comprises the following series of three reactions:[48,80,182,183,389,457,567,573]

$$\underbrace{CHOH \cdot (CHOH)_3 \cdot CH}_{O} \cdot CH_2OH \underset{\text{hexokinase}}{\overset{\text{ATP}}{\rightleftharpoons}} \quad \text{(R1)}$$

$$\underbrace{CHOH \cdot (CHOH)_3 \cdot CH}_{O} \cdot CH_2OPO_3H_2$$

6-phosphoglucose (glucose-6-phosphate)

$$\underbrace{CHOH \cdot (CHOH)_3 \cdot CH}_{O} \cdot CH_2OPO_3H_2 \underset{\text{isomerase}}{\overset{\text{phosphohexose}}{\rightleftharpoons}}$$

$$CH_2OH \cdot \overbrace{COH \cdot (CHOH)_2 \cdot CH}^{O} \cdot CH_2OPO_3H_2 \quad \text{(R2)}$$

6-phosphofructose (fructose-6-phosphate)

$$CH_2OH \cdot \underbrace{COH \cdot (CHOH)_2 \cdot CH}_{O} \cdot CH_2OPO_3H_2 \underset{\text{ATP}}{\overset{\text{phosphohexokinase}}{\rightleftharpoons}}$$

$$CH_2OPO_3H_2 \cdot \underbrace{COH \cdot (CHOH)_2 \cdot CH}_{O} \cdot CH_2OPO_3H_2 \qquad (R3)$$

1,6-phosphofructose (fructose-1-6-diphosphate)

A more efficient mechanism to convert fructose-6-phosphate to fructose-1-6-diphosphate has been proposed.[328]

$$2 \text{ fructose-6-phosphate} \rightleftharpoons \text{fructose diphosphate} + \text{glucose} \qquad (R4)$$

The Second Stage: Phosphotriose Formation.—[21,368-370,537]

$$CH_2OPO_3H_2 \cdot \underbrace{COH \cdot (CHOH)_2 \cdot CH}_{O} \cdot CH_2OPO_3H_2 \xrightleftharpoons{\text{aldolase}}$$

$$CH_2OPO_3H_2 \cdot CO \cdot CH_2OH + CHO \cdot CHOH \cdot CH_2OPO_3H_2$$

phosphodihydroxyacetone 3-phosphoglyceraldehyde (R5)

$$CHO \cdot CHOH \cdot CH_2OPO_3H_2 \xrightleftharpoons{\text{isomerase}} CH_2OH \cdot CO \cdot CH_2OPO_3H_2 \qquad (R6)$$

phosphodihydroxyacetone

The Third Stage: 3-Phosphoglycerate Formation.—The third stage consists of two steps, the first leading to 1,3-diphosphoglycerate and the second to 3-phosphoglycerate, thus:[59,371,386,541]

$$CH_2OPO_3H_2 \cdot CHOH \cdot CHO + DPN^+ + H_3PO_4 \xrightleftharpoons[\text{dehydrogenase}]{\text{phosphoglyceraldehyde}}$$

$$CH_2OPO_3H_2 \cdot CHOH \cdot CO_2{\sim}PO_3H_2 + DPNH + H^+$$

1,3-diphosphoglyceric acid (R7)

$$COO \sim PO_3^{2-} \cdot CHOH \cdot CH_2OPO_3^{2-} + ADP^{2-} \underset{\text{kinase}}{\overset{\text{3-phosphoglycerate}}{\rightleftharpoons}} \qquad (R8)$$

$$COO^- \cdot CHOH \cdot CH_2OPO_3^{2-} + ATP^{3-}$$

3-phosphoglycerate

The Fourth Stage: 2-Phosphoglycerate Formation.—[507]

(R9)

$$COOH \cdot CHOH \cdot CH_2OPO_3H_2 \underset{\text{mutase}}{\overset{\text{phosphoglycero-}}{\rightleftharpoons}} COOH \cdot CHOPO_3H_2 \cdot CH_2OH$$

2-phosphoglyceric acid

(R10)

$$COOH \cdot CHOH \cdot CH_2OPO_3H_2 + COOH \cdot CHOPO_3H_2 \cdot CH_2OPO_3H_2 \underset{\text{apomutase}}{\overset{\text{phosphoglycero-}}{\rightleftharpoons}}$$

2, 3-phosphoglyceric acid

$$COOH \cdot CHOPO_3H_2 \cdot CH_2OH + COOH \cdot CHOPO_3H_2 \cdot CH_2OPO_3H_2$$

The Fifth Stage: Phosphoenolpyruvate Formation.—[333,542]

$$COOH \cdot CHOPO_3H_2 \cdot CH_2OH \underset{+H_2O}{\overset{\text{enolase, } -H_2O}{\rightleftharpoons}} COOH \cdot CO \sim PO_3H_2 : CH_2 \qquad (R11)$$

phosphoenolpyruvic acid

The Sixth Stage: Pyruvate Formation.—[38,47,290,297,372]

(R12)

$$COOH \cdot CO \sim PO_3H_2 : CH_2 + ADP \underset{\text{kinase}}{\overset{\text{pyruvate}}{\rightleftharpoons}} COOH \cdot CO \cdot CH_3 + ATP$$

pyruvic acid

Pyruvate is a key intermediate whose importance will become clear in the ensuing discussion. Several enzymes and coenzymes have been involved in the steps leading to its formation. Many of the enzymes have been isolated in crystalline form. Two important coenzymes make their appearance for the first time: one is adenosine triphosphate (ATP), the other diphosphopyridine nucleotide (DPN), currently known as nicotinamide-adenine-dinucleotide (NAD). The structural formulas are given below:

```
N=C—NH2                                adenosine triphosphate
|  |
HC  C—N
||  ||    \CH
N—C—N/———CH·(CHOH)2·CH·CH2OHPO3~HPO3~HPO3
          |_____O_____|
```

```
 ______________
| N=C—NH2      |
|  |  |        |  adenine       _________________________________
|HC  C—N       |  grouping     |     ______O______                |
| ||  ||   \CH |               |    |             |               |
| N—C—N/———————|———————————————|—CH·(CHOH)2·CH·CH2·CH2OPO2⁻       |
|______________|               |                        |         |
 ______________                |                        O         |
|HC=CH         |               |                        |         |
| |   |        |               |                        |         |
|HC  N+————————|———————————————|—CH·(CHOH)2·CH·CH2·CH2OPO2⁻       |
| ||  ||       |               |    |_____O_______|               |
|HC—CH         |               |__________________________________|
|______________|
 nicotinamide                          D-ribose grouping
  grouping
                 diphosphopyridine nucleotide
```

Pentose Phosphate Pathway

It is generally accepted that the Embden-Meyerhof pathway is not unique for bacteria, and that among others there are two branches (the oxidative and the nonoxidative) of a metabolic route which conjoins with the Embden-Meyerhof scheme at a number of points, and is known as the pentose phosphate pathway. The oxidative pathway is utilized by many heterolactic bacteria, including several species that produce biacetyl, the aromatic substance in starter cultures. The pathway comprises the following steps: oxidation of 6-phosphoglucose to the α-lactone of 6-phosphogluconic acid in the presence of Warburg's (Zwischenferment) 6-phosphoglucose dehydrogenase,[81,403,539] and the coenzyme triphosphopyridine nucleotide (TPN),* also known as coenzyme II; hydrolysis of the lactone to 6-phosphogluconic acid in the presence of a

* TPN is related structurally to DPN above containing an additional phosphate group bound presumably to the adenosine part of the molecule.

lactonase; oxidative decarboxylation of 6-phosphogluconic acid to ribulose-5-phosphate in the presence of 6-phosphogluconic dehydrogenase (decarboxylating) and TPN; isomerization of ribulose-5-phosphate to ribose-5-phosphate in the presence of phosphoriboisomerase (this step is necessary only when ribose-5-phosphate is the substrate);[72-74,224,326,481] epimerization of ribulose-5-phosphate to xylulose-5-phosphate in the presence of phosphoketo-pentoepimerase; and phosphorolytic cleavage of xylulose-5-phosphate to acetyl phosphate and 3-phosphoglyceraldehyde in the presence of phosphoketolase.[197,228] With formation of phosphoglyceraldehyde, the pentose and the Embden-Meyerhof pathways merge enroute to production of pyruvic acid. The active two-carbon compound, acetyl phosphate, is an extremely important intermediate which will be discussed further elsewhere in this chapter.

Nonoxidative Pentose Phosphate Pathway

This pathway is distinguished by its requirement for "active glycolaldehyde" donors and acceptors. In pentose synthesis, glyceraldehyde-3-phosphate is the acceptor and sedoheptulose-7-phosphate and fructose-6-phosphate are important donors, thus:[222]

$$\underset{\text{ribulose phosphate}}{CH_2OH \cdot CO \cdot (CHOH)_2 \cdot CH_2OPO_3H_2} \xrightleftharpoons{\text{transketolase}} \underset{\text{enzyme}}{\overset{H_2COH}{\overset{|}{HC{=}O}}} + \underset{\text{phosphoglyceraldehyde}}{CHO \cdot CHOH \cdot CH_2OPO_3H_2} \quad (R13)$$

$$\text{ribulose-5-phosphate} \xrightleftharpoons{\text{isomerase}} \text{ribose-5-phosphate} \quad (R14)$$

$$\underset{\text{ribose-5-phosphate}}{CHO \cdot (CHOH)_3 \cdot CH_2OPO_3H_2} + \underset{\text{enzyme}}{\overset{H_2COH}{HC{=}O}} \rightleftharpoons \underset{\text{sedoheptulose-6-phosphate}}{CH_2OH \cdot CO \cdot (CHOH)_4 \cdot CH_2OPO_3}$$

and in a similar manner, fructose-6-phosphate serves as a donor to the acceptor glyceraldehyde-3-phosphate yielding ribulose-5-phosphate and erythrose-4-phosphate. Not only is sedoheptulose formed in a transketolase-catalyzed reaction, but it is also formed in a transaldolase mediated reaction between fructose-6-phosphate and erythrose-4-phosphate.

The preparation and properties of transketolase have been described in a number of publications.[95,225,436,437,497] Transaldolase has been obtained from yeast in a highly purified state.[222]

Lactic Acid Fermentation

Knowledge of the several metabolic pathways is a prerequisite to a clear understanding of bacterial fermentations, including lactic acid fermentation. Lactic acid, the most common of all substances produced microbially in milk, can develop naturally (although uncontrollably) in raw milk since lactic acid bacteria are usually present. Lactic acid is formed in the manufacture of virtually all cultured products, such as buttermilk, yogurt, and kefir among cultured milks, and cottage, Cheddar, Swiss, and brick, among the cheeses.

Description of the Lactic Acid Bacteria.—Lactic acid bacteria are gram-positive, nonspore-forming, nonmotile, and almost always catalase-negative. The homofermentative types produce lactic acid from sugar in yields ranging from 80 to 98%, and small quantities of other products. Included among these types are species of the genus *Streptococcus,* and the subgenera, *Streptobacterium* and *Thermobacterium*. *Streptococcus* species produce dextro-lactic acid in concentrations up to approximately 1%; *Streptobacterium* species produce, at an optimum temperature of 30°C, either dextro- or inactive lactic acid in concentrations up to approximately 1.5%; *Thermobacterium* species produce, at optimum temperatures of 40°C or higher levo- or inactive lactic acid in concentrations up to approximately 3%. *Lactobacillus bulgaricus** in yogurt and *Streptococcus lactis* in cultured buttermilk and cheese are important homofermentative types encountered in dairy technology.

The heterofermentative lactic acid bacteria ferment glucose to form CO_2, alcohol, and acetic acid, in addition to lactic acid. Comprising this group are species of *Lactobacillus,* which produce inactive lactic acid, and species of *Leuconostoc,* which usually develop alcohol, CO_2, and limited amounts of lactic and acetic acid. Levo-lactic acid is always produced, and simultaneously dextro-lactic acid is sometimes formed. About 25% of glucose may be converted to CO_2. The flavor-producing organisms of butter and cultured buttermilk, *Leuconostoc citrovorum* and *Leuconostoc dextranicum,* belongs to this group. Based on the observation that alcohol and carbonic and acetic acid are formed in significant quantities by *Lactobacillus bifidus* and members of the genus *Pediococcus,* these bacteria also could be considered as heterolactics.

---

* In bacteriological nomenclature, the names of the genera are spelled out wherever they occur for the first time. Thereafter they are identified by their first letter.

However, studies with radioactive glucose indicate that the aforementioned bacteria ferment glucose according to the Embden-Meyerhof scheme, as do the homofermentative types.[56,246] Hence *L. bifidus* and members of the genus *Pediococcus* have been classified with the homolactics.

The distinction between homofermentative and heterofermentative lactic acid bacteria is not a hard-and-fast one. In the experiments of Friedemann,[130] a strain of the homofermentative *Streptococcus faecalis* produced 74, 14, 6, and 7 m*M*, respectively, of lactate, formate, acetate, and ethanol per 50 m*M* glucose, whereas the heterofermentative *Leuconostoc mesenteroides* produced 81 m*M* lactate, 4 m*M* acetate, and 10 m*M* ethanol. Clearly the distinction here is not so much in the ratio between lactate and other products as in the ratio between formate and ethanol. Alkaline media favor formation of products other than lactic acid. *S. faecalis* var. *liquefaciens* produces 87% acid at pH 5, but only 61% at pH 9.[165]

The homolactics are quite responsive to altered substrates and conditions of fermentation.[163] Use of an oxidized substrate diverts the lactic fermentation to dismutation and to acetylmethylcarbinol (acetoin) formation; use of reduced substrate diverts it to a coupled lactate-succinate fermentation. Alkaline glucose dissimilation favors formation of formate, acetate, and ethanol.

Platt and Foster[420] observed that *Streptococcus cremoris,* grown anaerobically in a glucose medium subject to no pH control, produced acetic and formic acid, CO_2, and ethanol in addition to lactic acid, and in a medium held at pH 7.0 more products appeared. *S. lactis* produced some acetoin but no formic acid in the absence of pH control, and products altered in character at pH 7.0. Thus lactic, acetic, and formic acid, CO_2, ethanol, biacetyl, acetoin, and 2,3-butanediol were found in a medium sparged with nitrogen. *Streptococcus thermophilus* formed the combined products developed by *S. cremoris* and *S. lactis* in the absence of pH control, and the products were altered in kind at pH 7.0. Harvey[189] noted that *S. lactis* and *S. cremoris* also could develop acetaldehyde and acetone.

A surprising result was obtained by Steele *et al.,*[501] who, while studying capsular polysaccharide synthesis by Group A streptococci, observed that a typically heterolactic instead of a homolactic fermentation was obtained when galactose served as a carbon source.

Species of the genus *Lactobacillus* may be homofermentative, belonging then to the subgenus *Thermobacterium* or the subgenus *Streptobacterium;* or they may be heterofermentative, belonging then to the subgenus *Betabacterium.* An interesting organism is *Lactobacillus pentosus.* It converts pentoses into lactic and acetic acid in equimolar proportions. Other pentose-decomposing organisms break down the sugar and produce relatively greater quantities of lactic acid, thus:

$$6C_5H_{10}O_5 \rightarrow 8C_3H_6O_3 + 3C_2H_4O_2$$

lactic acid acetic acid

(R15)

The *L. pentosus* fermentation has been studied with 1-^{14}C-labeled L-arabinose and xylose. Cleavage occurred between 2-C and 3-C. Labeled carbon was found in the methyl group of acetic acid, indicating that a 2-keto-pentose was the precursor. Studying this fermentation has contributed much to our understanding of intermediary metabolism.

Intermediates in the Homofermentative Lactic Fermentation.—The metabolic route followed to produce lactic acid by homolactic bacteria is believed to parallel that followed in glycolysis, i.e., in the production of lactic acid from glucose in muscle. The conclusiveness with which parallel pathways to pyruvic acid in muscle and yeast metabolism have been demonstrated supports the hypothesis that production of lactic acid in a homolactic fermentation proceeds according to the Embden-Meyerhof mechanism. This means that lactic acid fermentation follows the pathway.of alcohol fermentation up to the point at which pyruvic acid is produced, and diverges at this point. Lacking pyruvic apocarboxylase but possessing lactic apodehydrogenase,[367] the lactic acid bacteria utilize reduced coenzyme I to reduce pyruvic to lactic acid.

Gibbs *et al.*[142] carried out investigations with *Lactobacillus casei* and 1-^{14}C-labeled glucose. The lactate resulting from the fermentation contained all its radioactivity in the methyl group, a result predicted by the Embden-Meyerhof scheme. Kuhn and Tiedemann investigated the metabolism of *L. bifidus* and the conversion of 1-^{14}C-labeled glucose into radioactive products. It is known that in the *L. bifidus* fermentation of glucose, acetic and lactic acids are formed in approximately equal quantities, and CO_2 is not found. Kuhn and Tiedemann[291] found radioactivity equally distributed between acetic and lactic acid. However, in conformity with the Embden-Meyerhof scheme, activity was associated with the methyl, but not with the carboxyl group. Incorporation of $CH_3 \cdot {}^{14}COOH$ into the medium resulted in formation of lactic acid with labeled α-carbon. The acetic and lactic acid appeared to possess a common precursor. Cell preparations showed aldolase and lactic dehydrogenase activity. If, in glucose fermentation, homofermentative lactic acid bacteria follow the Embden-Meyerhof route, the required enzymes should be present in these bacteria. Aldolase, a key enzyme, has been found in extracts prepared from *S. faecalis*, *S. lactis*, a so-called strain of *L. citrovorum* (Actually a *Pediococcus* species), *Lactobacillus parabifidus, L. bulgaricus, Lactobacillus delbrückii, Lactobacillus plantarum, Lactobacillus arabinosus, Lactobacillus leichmannii, L. bifidus,* and *Microbacterium lacticum.*[57,291,525] Hexokinase appeared in extracts of *L. bulgaricus* and *M. lacticum;*[468,525] and phosphohexoseisomerase, phosphohexokinase, phosphotrioseisomerase, phosphoglyceraldehyde dehydrogenase, phosphoglyceromutase, pyruvatekinase, and enolase appeared in extracts of *M. lacticum.*[525] Lactate dehydrogenase and phosphoglyceraldehyde dehydrogenase have been found in extracts of the heterofermentative *L. mesenteroides,* yet the key enzymes belonging to the Embden-Meyerhof scheme, aldolase and phosphotrioseisomerase, have not been found.[57,103,104]

Intermediates in the Heterolactic Fermentation.—Von Baeyer observed that oxygen in organic compounds tends to move toward the carbon richest in oxygen. In glucose, the carbon of the aldehyde group, 1-C, is richest in oxygen. The rule of von Baeyer is not followed in the Embden-Meyerhof scheme. The CO_2 produced in the alcohol fermentation, and the carboxyl group produced in the lactic fermentation, derive not from the oxygen-rich 1-C, but rather from 3-C and 4-C. This situation prevails because isomerization reactions occur, mediated by phosphoglucoisomerase and phosphotriose isomerase. It has already been pointed out that the homolactic fermentation follows the Embden-Meyerhof pathway. However, no unique pathway exists among the heterolactics. Metabolism of glucose by *L. bifidus* appears to follow the Embden-Meyerhof scheme; however, metabolism of glucose by *L. mesenteroides* clearly follows the oxidative pentose phosphate route, and moreover, one conforming to the von Baeyer rule. With a strain of *L. mesenteroides,* 1-^{14}C-labeled glucose gave rise to labeled CO_2, and 3,4-^{14}C-labeled glucose gave rise to carbinol-C-labeled ethanol and carboxyl-C-labeled lactate in equal proportions.[103,164] The methyl carbon of ethanol and the α and β-carbons of lactate were unlabeled, and were presumed to arise from the 2,5,6,-carbons of glucose. The overall reaction is expressed thus:

$$C_6H_{12}O_6 \rightarrow CO_2 + C_2H_5OH + CH_3 \cdot CHOH \cdot CO_2H \qquad \text{(R16)}$$

One mole of glucose always yielded one mole each of CO_2, ethanol, and lactate. Aldolase and isomerases in cell extracts were not demonstrable. However, enzymes linking the reactions leading from 3-phosphoglyceraldehyde to D(–) lactic acid, as in the Embden-Meyerhof scheme, were found. Furthermore, 6-phosphoglucose dehydrogenase appeared in *L. mesenteroides.*[104]

These findings may be explained by the reaction pattern of the pentose phosphate scheme. Actually, an impetus to explore this alternate route to pyruvic acid was provided by the work on the *Leuconostoc* fermentation. On examination of the reactions comprising this pathway, it may readily be seen that the carbon in CO_2 would derive from the 1-C of glucose; the carboxyl carbon of lactic acid and the carbinol carbon of ethanol would derive from the 3,4-C of glucose; and finally the methyl carbon of lactic acid and of ethanol would derive from the 5,6-C of glucose.

Optical Configuration of Lactic Acid.—Certain lactobacilli, which normally produce DL (±) lactic acid, will, when grown in a niacin-deficient medium, synthesize the D(−) acid.[261] Since DPN, a niacin derivative, is a cofactor of racemase, it appears that racemase activity would be seriously reduced in media deficient in niacin; and hence the DL (±) acid would not accumulate in normal proportions. Occurrence of specific lactic dehydrogenases for synthesis of D(−) and DL(±) lactic acid accounts for the observation that the configuration of lactic acid often serves to distinguish between species.

Associative Growth.—Lactic acid bacteria are extremely fastidious in their growth requirements. Often species grown in association will thrive under conditions in which one or more of the bacteria, if cultured individually, will fail to grow. Exceedingly interesting experiments to illustrate the relationship between symbiotic growth and the requirement for specific metabolites have been described by Nurmikko,[405] and independently by Koft and Morrison.[270] Using a medium which was complete only for a strain of *S. lactis* but deficient in phenylalanine for *L. citrovorum, L. arabinosus, Lactobacillus fermenti,* and *L. mesenteroides,* and deficient in folic acid for *L. citrovorum* and *S. faecalis,* Nurmikko compartmentalized individual species in one of six cells, while at the same time he provided for exchange of dialyzable components according to the scheme given in this chapter. Under these conditions all species thrived; the missing growth factors for any one organism were supplied by one or more of the remaining species.

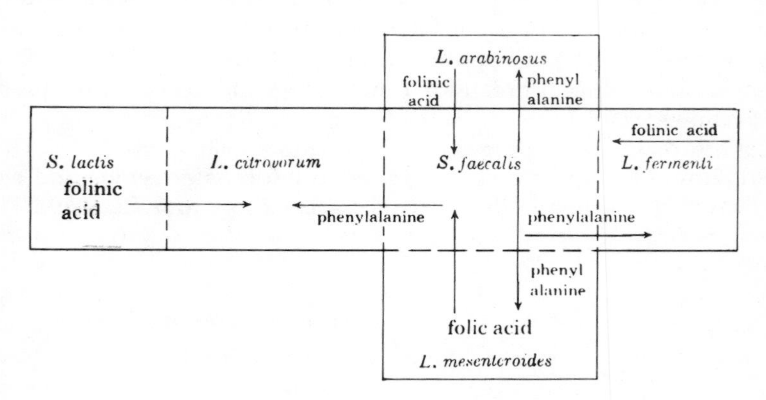

Propionic Acid Fermentation

Propionibacteria are gram-positive, catalase-positive, nonspore-forming, nonmotile, facultative anaerobic, and rod-shaped. Eleven species of propionibacteria are presently recognized. Some of these organisms are utilized in industrially important processes where they: (a) produce the characteristic flavor and eyes in Swiss (Emmental)-type cheeses, (b) synthesize vitamin B_{12} (discussed later in this chapter), or (c) produce propionic acid which can be used commercially.[209]

Growth of Propionibacteria.—Nutritional requirements of these bacteria are complex. Hettinga and Reinbold,[209] in their review of this subject, indicate that amino acids added to a medium are beneficial but not essential for growth, whereas certain vitamins (pantothenic acid and biotin), minerals, and unknown constituents of yeast extract are needed for growth and metabolism. Growth of propionibacteria can be inhibited by certain concentrations of calcium or sodium propionate, calcium or sodium lactate, acetate, or sodium chloride; and by glucose when improperly heated.

Propionibacteria can grow at 2.8 to 7.0°C, though slowly, and at 40°C.[209] Optimal activity seems to occur at 24 to 27°C, although many investigators grow the bacteria at 30°C. These bacteria grow best at pH values ranging from 6.5 to 7.0. Growth and production of propionic acid is essentially inhibited at pH 5.0. Excess light can inhibit the propionibacteria and can reduce the vitamin B_{12} and flavin content in their cells. Certain lactobacilli, if grown in association with propionibacteria, are believed to stimulate them, whereas micrococci, under similar conditions, are thought to be inhibitory. Lactate agar and reduced oxygen tension are commonly used when propionibacteria are cultivated in the laboratory.

Production of Propionic Acid.—Propionibacteria can produce propionic acid from several different substrates. Thus, for example, these bacteria can utilize the lactic acid in Swiss cheese to form propionate. The general equation for this conversion is:[210]

$$3 \text{ lactate} \rightarrow 2 \text{ propionate} + \text{acetate} + CO_2 + H_2O \qquad \text{(R17)}$$

They also can produce propionate from glucose and the general equation then becomes[210]

$$1.5 \text{ glucose} \rightarrow 2 \text{ propionate} + \text{acetate} + CO_2 + H_2O \qquad \text{(R18)}$$

Production of propionic acid involves numerous enzymes and reactions, which have been summarized by Hettings and Reinbold[210] as follows:

(a) Glycolysis

$$1.5 \text{ glucose} + 3NAD^+ + 3P_i \rightleftharpoons 3\text{P-enolpyruvate} + 3NADH + 3H \qquad \text{(R19)}$$

(b) Pyruvokinase

$$\text{pyruvate} + NAD^+ + \text{CoA} \rightleftharpoons 3 \text{ pyruvate} + 3 \text{ ATP} \qquad \text{R20)}$$

(c) Pyruvic dehydrogenase

$$\text{pyruvate} + NAD^+ + \text{CoA} \rightleftharpoons \text{acetyl-CoA} + NADH + H^+ + CO_2 \qquad \text{(R20a)}$$

(d) Phosphotransacetylase
acetyl-CoA + P_i ⇄ acetyl-P + CoASH (R21)
(e) Acetyl kinase
acetyl-P + ADP ⇄ acetate ATP (R22)
(f) Transcarboxylase
2 pyruvate + 2 methylmalonyl-CoA ⇄ 2 oxaloacetate + 2 propionyl-CoA (R23)
(g) Phosphoenolpyruvic carboxytransphosphorylase
3 P-enolpyruvate + 3 CO_2 + 3P_i ⇄ 3 oxaloacetate + 3PP_i (R24)
(h) Malic dehydrogenase
2 oxaloacetate + 2NADH + 2H ⇄ 2 malate + $2NAD^+$ (R25)
(i) Fumarase
2 malate ⇄ 2 furmarate + $2H_2O$ (R26)
2NADH + $2H^+$ + $2P_i$ + 2ADP + 2FP ⇄ 2NAD + 2ATP + FPH_2 (R26a)
(j) Fumarate reductase
2 fumarate + FPH_2 ⇄ 2 succinate + 2FP (R26b)
(k) CoA transferase
2 succinate + 2 propionyl – CoA ⇄ succinyl – CoA + 2 propionate (R26c)
(l) Methylmalonyl isomerase
2 succinyl-CoA ⇄ 2 methylmalonyl-CoA(b) (R27)
(m) Methylmalonyl racemase
2 methylmalonyl = 2 methylmalonyl-CoA(b) ⇄ 2 methylmalonyl-CoA(a) (R27a)

In addition to the above, presence of lactic dehydrogenase allows the propionibacteria to utilize lactate rather than glucose.

The reactions of propionic acid fermentation which lead to production of propionate, acetate, and CO_2 can be summarized by the following scheme.[210] Methylmalonyl-CoA (a) and (b) here and above are the isomers. FP in both locations is flavoprotein and FPH_2 is reduced flavoprotein.

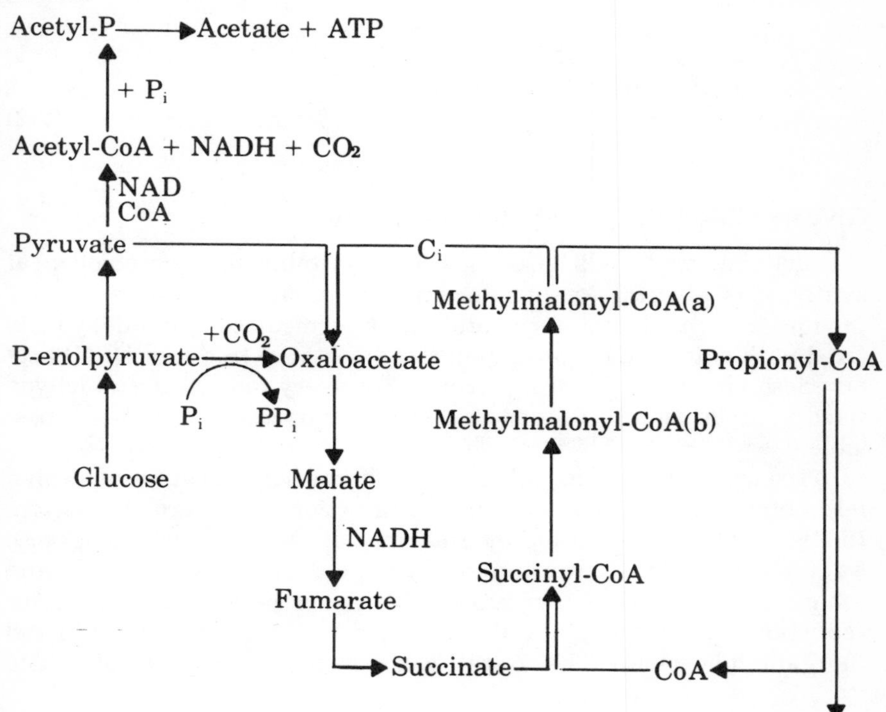

Alcohol (Yeast) Fermentation

Results of work on alcohol fermentation contributed greatly to elucidating the Embden-Meyerhof metabolic pathway. This pathway, up to the point of pyruvate formation, is now known to be shared by many kinds of bacteria.

Alcohol fermentation is of more than theoretical interest in dairy technology. Whey is an excellent substrate for alcohol production, and alcohol is an important component of some cultured milks. Alcohol fermentation may be modified to produce potentially important lactose-fermenting yeasts.

Alcohol and homolactic fermentations, in part, follow a common course, departing from it with the production of the important intermediate pyruvic acid. Possessing the enzyme carboxylase and the coenzyme diphosphothiamine,[334] yeast acts to decarboxylate pyruvic acid, with the production of acetaldehyde and CO_2. The acetaldehyde is reduced forthwith in the presence of alcohol dehydrogenase and reduced coenzyme I (DPNH) to ethyl alcohol, thus:[155,279,289,435]

$$CH_3 \cdot CO \cdot COOH \xrightarrow{\text{pyruvic carboxylase}} \underset{\text{acetaldehyde}}{CH_3 \cdot CHO} + CO_2 \qquad \text{(R31)}$$

$$CH_3 \cdot CHO \underset{-2H}{\overset{+2H}{\rightleftharpoons}} \underset{\text{ethyl alcohol}}{CH_3 \cdot CH_2OH} \qquad \text{(R32)}$$

Glycerol (Modified Alcohol) Fermentation

Dairy raw materials also should be suitable for microbiological synthesis of glycerol—a substance needed in situations such as wars produce. If, in alcohol fermentation, intermediate acetaldehyde is fixed with sulfite, the spared reducing substance (reduced DPN) acts on triose phosphate to form glycerol. For every mole of acetaldehyde fixed, production of one mole of glycerol becomes possible.[79,120,167,230,298,391,393-395,536,575]

In the presence of added alkali,[390,392,396] the normal course of alcohol fermentation is also modified in the direction of glycerol formation. In a weak alkaline solution, the intermediate acetaldehyde undergoes an intermolecular Cannizaro oxidation-reduction, acetic acid and ethanol are formed, and reduced DPN again becomes available for reduction of 3-phosphoglyceraldehyde, to ultimately yield glycerol and inorganic phosphate. The overall reaction in the presence of sulfite is:

$$C_6H_{12}O_6 + Na_2SO_3 + H_2O \rightarrow \underset{\text{glycerol}}{C_3H_8O_3} + \underset{\text{fixed aldehyde}}{C_2H_4O \cdot NaHSO_3} + NaHCO_3 \quad (R33)$$

glucose

The overall reaction in the presence of alkali is:

$$2C_6H_{12}O_6 + H_2O \longrightarrow CO_2 + \underset{\text{acetic acid}}{C_2H_3O_2} + \underset{\text{ethanol}}{C_2H_5OH} + \underset{\text{glycerol}}{2C_3H_8O_3} \quad (R34)$$

Another approach for glycerol synthesis has been suggested by Wallerstein and Stern[535] and Beloff and Stern.[28] They proposed the use of selective inhibitors of alcohol dehydrogenase or carboxylase to prevent either reduction of acetaldehyde to alcohol or formation of acetaldehyde from pyruvate; thus the fermentation is directed toward formation of glycerol in the absence of large quantities of a fixing or trapping reagent.

Oxidative Decarboxylation of Pyruvic Acid

Pyruvic acid is clearly an intermediate whose existence makes possible many parallel reactions. With formation of this acid, the main pathway in glucose dissimilation begins to branch. Even greater branching occurs after oxidative decarboxylation of pyruvic acid. An active two-carbon fragment is formed, and this compound enters into a series of parallel reactions which account for many of the products of fermentation. Before considering other fermentations, it is profitable to consider formation of this active intermediate.

In fermentations with heterolactic bacteria, acetic acid and CO_2 are commonly produced in appreciable quantities. With *L. delbrückii,* homofermentative under anaerobic conditions, pyruvic acid under aerobic conditions is converted into acetic acid and CO_2, thus:

$$CH_3 \cdot CO \cdot COOH + \tfrac{1}{2}O_2 \longrightarrow CH_3 \cdot COOH + CO_2 \quad (R35)$$

Anaerobically, pyruvic acid may undergo dismutation to form lactic, acetic, and carbonic acid. Lipmann,[327] working with an enzyme solution prepared from *L. delbrückii,* showed that phosphate is essential for oxidative decarboxylation. This led eventually to isolation of acetyl phosphate as a product of the oxidation. Later the same conditions were found to hold when acetyl phosphate was formed by various anaerobic and facultative anaerobic organisms, such as *E. coli, Clostridium butylicum, Clostridium saccharobutyricum, Clostridium sporogenes, Enterobacter aerogenes,* and *S. faecalis.*

Acetyl phosphate can serve as an acetyl or acetate donor provided that an activator known as coenzyme A is present.[329,330] The structure of the compound appears to be:[100]

```
                                                          ┌ ─ ─ ─ ─ ─ ─ ─ ─ ┐
CH2·C(CH3)2·CHOH·C—N—CH2·CH2·C—|—N—CH2·CH2·SH   |
 |                ||  |           ||  |  |              |
 O                O   H           O   |  H              |
 |                                    └ ─ ─ ─ ─ ─ ─ ─ ─ ┘
—O—P→O                                 thioethanolamine
   |               |                         group
   O               O
   |               |
—O—P→O        —O—P→O
   |               |
   O               O
   |               |
CH2·CH·CH·CHOH·CH-adenine
     |_____O_____|
```

It contains adenylic acid and a pyrophosphate bridge cross-linking the adenylic 5′-position with the 4′-position of pantothenic acid. The pantothenic acid is peptide-linked via its terminal carboxyl group with thioethanolamine, and this derived moiety is known as pantetheine. Pantetheine and its oxidized disulfide form, pantethine, are growth factors required by *L. bulgaricus* and *Lactobacillus acidophilus*. The bacterial (*Proteus morganii*) requirement for pantetheine is closely related to its involvement in coenzyme A.[109,211,404]

Oxidative decarboxylation of pyruvic acid consists of four stages.[274] In the first stage, pyruvate reacts (see R36) with lipothiamide-pyrophosphate (LTPP), which is a complex of lipoic acid and thiamine pyrophosphate (TPP):[407,439,440]

```
      CH2
     /   \
  CH2    CH · (CH2)4 · CONH—(thiamine pyrophosphate) + CH3 · CO · COO⁻ ⇌
   |      |
   S——————S
```

(R36)

$$CH_3 \cdot COS—CH_2 \cdot CH_2 \cdot CHS^- \cdot (CH_2)_4 \cdot CONH—(TPP) + CO_2$$

In the second stage, the acetyl group is transferred from the acetyl lipothiamidepyrophosphate complex to coenzyme A. This reaction is mediated by a transacetylase, thus:

$$CoA—SH + CH_3 \cdot CO—S—(LTPP)—S^- \rightleftharpoons CoA—SOC \cdot CH_3 + HS—(LTPP)—S^- \quad (R37)$$

In this stage also, LTPP is regenerated. Hydrogen from reduced lipothiamidepyrophosphate is transferred to DPN_{ox} (oxidized DPN). Hydrogen from DPN_{red} (reduced DPN) is then transferred to O_2 via a flavoprotein enzyme to form H_2O_2, as in the oxidative decarboxylation of pyruvate by *L. delbrückii*. Alternatively, the hydrogen may be transferred to pyruvate via lactic dehydrogenase, as in the anaerobic oxidative decarboxylation of pyruvic acid, and lactic acid is formed.

In the third stage acetyl phosphate is formed and in the fourth and final stage, acetate and ATP are generated, thus: (R38)

$$CH_3 \cdot COS \sim (CoA) + H_3PO_4 \rightleftharpoons CH_3 \cdot CO_2 \sim H_2PO_3 + CoA{-}SH$$

$$CH_3 \cdot CO_2 \sim H_2PO_3 + ADP^{=} \rightleftharpoons ATP^{=} + CH_3COO^{-} + H^{+} \quad (R39)$$

Reaction (R38) is mediated by phosphotransacetylase.

Fermentations of the Escherichia coli Type

Escherichia is one genus of the tribe *Eschericheae,* which also includes *Enterobacter* and *Klebsiella.* Bacteria in the tribe *Eschericheae,* if acetylmethylcarbinol is not produced and if the methyl red test is positive, belong to the genus *Escherichia.* If acetylmethylcarbinol is produced, if the methyl red test is negative, and if citrates may be used as the sole source of carbon, bacteria in the same tribe belong to the genus *Enterobacter.* Bacteria in the genus *Escherichia* ferment many substrates including lactose with acid and gas production.[161] Carbon dioxide and hydrogen are produced from glucose in approximately equal volumes. The coliforms when grown in milk and milk products may bring about flavor defects and gasiness, for example, early gas formation in cheese.

Quantitatively, the end products of fermentations with *E. coli* are likely to be variable and are influenced by pH (see Table 13.1). The percentage of alcohol produced does not appear to be affected by changes in pH.[516] Formic acid represents free acid and that which is decomposed according to the equation:

$$HCOOH \rightleftharpoons H_2 + CO_2 \quad (R40)$$

Production of lactic acid diminishes, and that of acetic and formic acid increases, as pH increases. This points to the likelihood that pyruvic acid is a common source of these acids but not of ethanol.

In the presence of $CaCO_3$, it is generally believed that *E. coli* ferments glucose to yield 1 part carbon dioxide, 1 part hydrogen, 1 part acetic acid, 1 part alcohol, 2 parts lactic acid, and ½ part succinic acid. Some of the *Enterobacteriaceae,* such as *Salmonella typhi,* yield little H_2 gas and correspondingly large quantities of formic acid.

The Formic Acid Fermentation.—Formic acid production is a characteristic of *E. coli.* Harden deduced in 1901 that hydrogen is derived from formic acid (see Tikka[516]). Some bacteria produce formic acid but not hydrogen. Mixed cultures produce formic acid from pyruvic acid, thus:[388]

$$CH_3 \cdot CO \cdot COOH + H_2O \rightarrow \underset{\text{acetic acid}}{CH_3 \cdot COOH} + \underset{\text{formic acid}}{HCOOH} \quad (R41)$$

Table 13.1

FERMENTATION OF GLUCOSE BY *E. coli* AT DIFFERENT pH LEVELS[a]

Glucose Gm	Initial H-ion Conc., pH	Lactic Acid Mg	Lactic Acid %	Acetic Acid Mg	Acetic Acid %	Formic Acid Mg	Formic Acid %	Ethanol Mg	Ethanol %	H_2 Ml
4.0	7.1	815	20.4	724	18.1	994	16.2	645	21.0	45
4.0	7.1	860	21.5	812	20.3	736	12.0	676	22.0	46
2.0	6.4	926	46.3	90	4.5	85	2.8	325	21.1	181
2.0	6.4	816	40.8	120	6.0	113	3.7	296	19.3	146
2.0	7.4	82	4.1	588	29.4	622	20.2	340	22.1	29
2.0	7.6	54	2.7	683	34.1	822	26.8	325	21.1	37

[a] Data by Tikka.[516]

Jones *et al.*[244] studied the reaction with labeled $CH_3 \cdot {}^{14}COOH$ and labeled pyrophosphate. Their findings may be summarized in the following series of reactions:

$CH_3 \cdot CO \cdot COOH + HS(CoA) \rightleftharpoons CH_3 \cdot COS(CoA) + HCOOH$ (R42)
ad—P∼PP + enz · OH ⇌ enz∼P—ad + pyrosphosphate (R43)
enz∼P—ad + HS(CoA) ⇌ enz∼S(CoA) + adenylic acid (R44)
$CH_3 \cdot COOH$ + enz∼S(CoA) ⇌ $CH_3 \cdot CO{\sim}S(CoA)$ + enz · OH (R45)
$CH_3 \cdot CO{\sim}S(CoA)$ + phosphate ⇌ $CH_3 \cdot CO(O{\sim}PO_3H_2)$ + HS(CoA) (R46)

AD-P∼PP (see R43) is bound by the apoenzyme, and modifies it, with the release of pyrophosphate and formation of a protein-adenylic acid complex. The overall reaction leading from pyruvic acid to formic and acetic acid is:

$CH_3 \cdot COOH + HCOOH + ATP \rightleftharpoons$
$CH_3 \cdot CO \cdot COOH$ + pyrophosphate + adenylic acid (R47)
pyruvic acid

Formation of H_2.—Reversibility of the conversion of formic acid to H_2 and CO_2 (see R40) has been demonstrated with *E. coli* extracts and radioactive $NaH^{13}CO_3$. There is a clear implication that molecular hydrogen can be utilized by *E. coli* to effect reduction of carbonic acid. The question arises: How is this brought about?

Stephenson and Stickland[503] believe that three enzymes are required in hydrogen formation. Two of the three are necessary to account for the following reactions, thus:

$$HCOO^- + H_2O \xrightleftharpoons{\text{dehydrogenase}} HCO_3^- + 2H^+ + 2e^- \quad (R48)$$

$$2H^+ + 2e^- \xrightleftharpoons{\text{hydrogenase}} H_2 \quad (R49)$$

Hydrogen is not always produced in the presence of both dehydrogenase and hydrogenase. This is attributed to the absence of an intermediate electron carrier named "hydrogenlyase."[503] "Hydrogenlyase," although it does not develop in suspensions of *E. coli* in glucose, does develop in broth suspensions containing 1% formate. Synthesis of "hydrogenlyase" is observed after 1 hr. Growth is not required.

Formation of Ethanol.—It is generally agreed that the ethanol produced by *E. coli* does not arise in the same manner as that produced by yeast. Dawes and Foster[99] have proposed the following scheme:

pyruvic acid → acetyl CoA → acetaldehyde → ethanol (R50)

When *E. coli* fermented glucose, the concentration of ethanol increased during the first 60 min and decreased thereafter. Acetaldehyde was identified as a true intermediate. With the use of cell suspensions of a pantothenate-requiring mutant, it was shown that CoA was a cofactor in ethanol production. Ethanol dehydrogenase and acetaldehyde dehydrogenase were prepared in partially purified forms.

Formation of Lactic Acid.—The evidence is strongly in favor of parallel metabolic pathways for homolactics and *E. coli* in the production of lactic acid from glucose. The presence of lactic dehydrogenase in *E. coli* has been studied by Quastel and Whetham.[433]

Enterobacter Type Fermentation

Biochemically, the genus *Enterobacter* differs from *Escherichia* in that the latter produces CO_2 and H_2 in equimolar ratios, and the former produces from 5 to 8 times more CO_2. Furthermore, acetoin is produced in the *Enterobacter* but not in the *Escherichia* type fermentation.[444]

Species of the genus *Enterobacter* are facultative anaerobes. *E. aerogenes* is normally found on grains and plants and to various degrees in the intestinal canal of man and animals. *E. cloacae* is found in human and animal feces, sewage, soil, and water. Differentiation of *Enterobacter* from *Escherichia* permits one to question whether contamination of a milk supply is of animal or soil origin.

Formation of Acetoin.—Acetoin may be obtained by reduction of biacetyl. The flavor of butter has been attributed to biacetyl. Its production in butter making has been ascribed to the action on citrate of *L. citrovorum* and *L. dextranicum,* two bacteria unrelated to *Enterobacter* species. Nevertheless, production of biacetyl in the intermediary metabolism of *Enterobacter* species bears on the general question of its origin. It should be remembered that *Enterobacter,* like the *Leuconostoc* species, can utilize exogenous pyruvate as the sole source of carbon.[56]

Slade and Werkman[487] explained the action of *Enterobacter cloacae* (*Enterobacter indologenes*) cells on carboxy-carbon-labeled acetate and pyruvate, thus:

$$\underset{\text{active aldehyde}}{CH_3 \cdot {}^{13}\overset{\overset{\displaystyle H}{|}}{C}{-}O{-}} + \underset{\text{Pyruvic acid}}{CH_3 \cdot CO \cdot COOH} \rightleftharpoons \underset{\alpha\text{ - acetolactic acid}}{CH_3{-}\overset{\overset{\displaystyle OH}{|}}{\underset{\underset{\displaystyle CH_3{-}{}^{13}C{=}O}{|}}{C}}{-}COOH} \quad \text{(R51)}$$

$$\begin{array}{l} \quad\quad\ \ OH \\ \quad\quad\ \ | \\ CH_3 - C - COOH \\ \quad\quad\ \ | \\ CH_3 - {}^{13}C \\ \quad\quad\ \ \| \\ \quad\quad\ \ O \end{array} \rightleftharpoons CH_3 \cdot CHOH \cdot {}^{13}\underset{\underset{O}{\|}}{C} \cdot CH_3 + CO_2 \quad (R52)$$

acetoin

The active aldehyde is formed from acetate through a reversal of the reaction leading from pyruvate to acetate. Two enzymes are required—one to catalyze condensation to α-acetolactic acid and the other to foster decarboxylation. The labeled molecule in the first of the two reactions (R51) is either acetaldehyde or a closely related derivative formed by the reduction of labeled acetate.

With *Enterobacter* species not utilizing acetaldehyde, the following intermediate reaction occurs, thus:

$$2CH_3 \cdot CO \cdot COOH \rightarrow CO_2 + \begin{array}{l} \quad\quad\ \ OH \\ \quad\quad\ \ | \\ CH_3 - C - COOH \\ \quad\quad\ \ | \\ CH_3 - C{=}O \end{array} \quad (R53)$$

pyruvic acid α-acetolactic acid

Juni[245] found enzymes both for reactions (R51) and (R52) in cell-free preparations from *E. aerogenes, Bacillus subtilis, Staphylococcus aureus,* and *Serratia marcescens.* The enzyme apparatus of *E. coli* containing only α-acetolactic decarboxylase is an exception to the rule that the enzymes mentioned above always occur together.

Cell suspensions of *S. faecalis* promoted oxidation of pyruvate to acetate and CO_2, or anaerobically to lactate, acetate and CO_2, whereas cell-free extracts catalyzed dismutation to acetoin and CO_2 at a pH optimum of 6.1.[108]

The detailed pathway to acetoin is probably not universal. Animal tissues can metabolize acetaldehyde to acetoin in the absence of pyruvate; yeast can metabolize acetaldehyde only in the presence of pyruvate; and finally *Enterobacter* species cannot utilize acetaldehyde.

Martius and Lynen[355] proposed the following mecahism to account for the requirement of phosphate in *Enterobacter* metabolism and for utilization of acetaldehyde by *Bacillus polymyxa:*

(R54)

$$CH_3 \cdot COO \sim H_2PO_3 + CH_3 \cdot \underset{\displaystyle H}{\underset{|}{C}} = O \rightarrow CH_3 \cdot CO \cdot CO \cdot CH_3 + H_3PO_4$$

acetylphosphate acetaldehyde biacetyl

Biacetyl rather than acetoin is the primary product. In succeeding steps, biacetyl is reduced to acetoin, and acetoin to 2,3-butylene glycol.

In the so-called acetoin fermentation very little acetoin and biacetyl are actually found; 2,3-butylene glycol, the reduction product, accumulates.

Oxidation of 2,3-butylene glycol to biacetyl, nonenzymatically, proceeds in alkaline solution. Biacetyl condenses with peptone constituents or α-naphtol and creatine, to yield the bright pink product of the Voges-Proskauer reaction, which is positive for *Enterobacter* and characteristically differentiates *Enterobacter* from *Escherichia.*

The products of typical *Enterobacter* and *Escherichia* fermentations are compared in Table 13.2.[445]

Industrially, studies of 2,3-butylene glycol (2,3-butanediol) fermentation have centered around *E. aerogenes,* and *B. polymyxa.*[33,34,162,387] Results may bear on possible uses for whey.

Butyric Acid and Related Fermentations

Of microorganisms producing butyric acid, the *Clostridium* species are technologically the most important, giving rise in the butyl alcohol-acetone fermentation to industrial solvents and riboflavin. In dairy technology the *Clostridium* species cause late gassing and off-flavors in cheese.

Table 13.2

PRODUCTS OF TYPICAL *Enterobacter* AND *Escherichia*[a] FERMENTATIONS

Product	Yield, Moles per 100 Moles Glucose *E. indologenes*	*E. coli*
CO_2	172	44
H_2	36	43
Formic acid	18	2
Acetic acid	0.5	44
Ethanol	70	42
Lactic acid	3	84
Succinic acid	0	29
2,3-Butylene glycol	66.5	

[a] Data from Reynolds and Werkman.[445]

The *Clostridium* species are gram-positive, spore-forming rods, either strictly anaerobic or microaerophilic. Bacteria in the genus differ in their fermentations. *Clostridium butyricum,* synonymous with *C. saccharobutyricum,* typically produces the following from 100 moles fermented glucose: 233 moles hydrogen, 196 moles carbon dioxide, 43 moles acetic, and 75 moles butyric acid. Other species of the genus producing large quantities of butyric acid are: *Clostridium amylobacter, Clostridium pasteurianum,* and *Clostridium lactoacetophilum,* which, as the name implies, is active toward lactate in the presence of acetate.

Closely related to butyric is the butylic fermentation. Butylic clostridia—*C. butylicum, Clostridium acetobutylicum,* and others—can continue butyric fermentation to a more advanced stage, leading to production of butanol and other volatile solvents.

Butyric Acid Fermentation.—In the formation of butyric acid, C_3 compounds may be utilized (glycerol by *C. acetobutylicum,* pyruvate by *C. butylicum,* and lactate by *C. lactoacetophilum*). Consequently it appears that butyric acid formation is not a direct result of the splitting of a hexose into C_4 and C_2 constituents.

In the light of the demonstrated importance of coenzyme A in the oxidation of butyric acid, Barker[19] formulated the following series of reactions:

$$2\ CH_3\cdot CO\cdot COO^- + 2CoA \underset{+2CO_2}{\overset{-2CO_2}{\rightleftharpoons}} \underset{\text{acetyl-CoA}}{2\ CH_3\cdot CO\cdot CoA} \underset{+H_2O}{\overset{-H_2O}{\rightleftharpoons}} \underset{\text{acetoacetyl-CoA}}{CH_3\cdot CO\cdot CH_2\cdot CO\text{—}CoA} \quad (R55)$$

$$CH_3\cdot CO\cdot CH_2\cdot CO\text{—}CoA \underset{-2H}{\overset{+2H}{\rightleftharpoons}} \underset{\text{vinylacetyl-CoA}}{CH_2{=}CH\cdot CH_2\cdot CO\text{—}CoA} \text{ or } \underset{\text{crotonyl-CoA}}{CH_3\cdot CH{=}CH\cdot CO\text{—}CoA} + H_2O \quad (R56)$$

$$\text{vinylacetyl—CoA} \underset{-2H}{\overset{+2H}{\rightleftharpoons}} \underset{\text{butyryl—CoA}}{CH_3\cdot CH_2\cdot CH_2\cdot CO\text{—}CoA} \quad (R57)$$

$$\text{butyryl—CoA} \rightleftharpoons \text{butyrate} + \text{CoA} \quad (R57a)$$

Attempts to demonstrate the conversion of pyruvate to butyrate met with difficulties. *C. butylicum* converts pyruvate into acetate, CO_2 and H_2 at an optimum pH of 5.0. Cell-free extracts, in the presence of phosphate, form acetyl phosphate, CO_2, and H_2.

The mechanism postulated for conversion of acetate to butyrate also explains conversion of lactate to butyrate by *C. lactoacetophilum.* Acetate is required and the condensation again involves CoA, and is mediated by an independent condensing apoenzyme. One mole of added acetate is consumed per mole of butyric acid formed.

In the formation of butyric acid from lactic acid by *Butyribacterium rettgeri,* CO_2 serves as a precursor of additional acetate, and is required, thus:

$$CH_3 \cdot CHOH \cdot COOH + CO_2 + 4H \rightarrow 2CH_3 \cdot COOH + H_2O \quad (R58)$$

Energetically, the reaction involving utilization of CO_2 in the formation of a -C-C- bond proceeds because lactate is simultaneously oxidized to yield CO_2.[20]

Butyric fermentation and conversion of pyruvate to acetate by *C. acetobutylicum* are inhibited by CO.[256,288] Inactivation of an iron-containing protein complex appears to be involved.[538]

Butanol Fermentation.—When acetone and butanol are produced commercially by the action of *C. acetobutylicum* on corn meal, the following products are formed in the early stages of fermentation: lactic and acetic acids, CO_2, and H_2. A so-called "break" is reached after 13 to 17 hr in an active fermentation, coinciding with development of maximum titratable acidity. Then the acid titer decreases rapidly, and at the same time production of volatile solvents begins and increases rapidly. Acids are precursors of the solvents.

Studies by Koepsell *et al.*[268] with cell-free extracts of *C. butylicum,* and by Stadtman and Barker,[500] with *Clostridium kluyveri* may be summarized, thus:

$$CH_3 \cdot COOPO_3^- + CH_3 \cdot (CH_2)_2 \cdot COO^- \leftrightharpoons CH_3COO^- + CH_3 \cdot (CH_2)_2 \cdot COOPO_3^- \quad (R59)$$

$$C_3H_7 \cdot COOPO_3^- + 2H_2 \rightarrow C_3H_7 \cdot CH_2OH + HPO_4^= \quad (R60)$$

The yields were as high as 77% of theoretical. In view of later findings it would seem that butyryl-CoA rather than butyryl phosphate is the direct precursor.

Eliasburg[113] observed that an increase in pressure of hydrogen from atmospheric to 20 atmospheres reduced the yield of butyric acid from 30 to 15%, and correspondingly increased the yield of butanol from 0.3 to 8%.

Acetone and Isopropanol Fermentations.—Addition of acetate to an active maize fermentation liquor results in increased yields of acetone.[442] With resting-cell suspensions of *C. acetobutylicum,* 2 moles acetate yield 1 mole acetone and 1 mole CO_2. The reaction proceeds by way of acetoacetic acid, and the energy requirement for condensation of acetic acid is met in the simultaneous decomposition of glucose or pyruvate.[97] Acetoacetic decarboxylase, prepared from suspensions of *C. acetobutylicum,* has properties resembling somewhat those of riboflavin phosphate.[98] Formation of acetone is adequately represented by the following reactions:

$$2CH_3 \cdot COOH \rightleftharpoons CH_3 \cdot CO \cdot CH_2 \cdot COOH + H_2O \quad (R61)$$

acetic acid   acetoacetic acid

$$CH_3 \cdot CO \cdot CH_2 \cdot COOH \rightleftharpoons CH_3 \cdot CO \cdot CH_3 + CO_2 \quad (R62)$$

acetone

In isopropanol fermentation by *C. butylicum,* addition of labeled acetone results in formation of labeled isopropanol. Adding 2 moles acetic acid yields 1 mole isopropanol and 1 mole CO_2. Clearly, isopropanol must result from the reduction of acetone.

Production of ethanol, to the extent of 2 to 3% of glucose fermented, always accompanies formation of butanol. Ethanol derived from fermentations in which media contain labeled acetate always contains labeled carbon, and hence ethanol derives from acetate via coenzyme A and an appropriate dehydrogenase system.

Respiratory Processes

The terminal products of respiration are CO_2 and H_2O. In fermentation the ultimate acceptor of electrons and protons is some organic molecule, for example DPN, and in respiration it is oxygen. Calculations based on available redox potentials (see Table 13.3) show that, for each two electrons transferred, the free-energy change accompanying respiration exceeds that accompanying fermentation by 53.4 kcal/mole of product. Respiration therefore sets free larger stores of energy than does fermentative oxidation.

What mechanism underlies the transfer of hydrogen to oxygen? There is no single unique mechanism, yet the number of mechanisms is quite limited. Transfer occurs in a sequence of steps in which electrons and protons move through a series of acceptors to molecular oxygen. Organic substrates under conditions of culture yield electrons and protons to appropriate enzyme systems known as dehydrogenases. These then effect transfer of their acquired complement of electrons and protons either to the so-called cytochrome system or to coenzymes, the latter of which are bound with some measure of firmness to the protein moiety of dehydrogenases. Reduced coenzymes can then transfer electrons and protons to flavoprotein enzymes; and thus in regaining their oxidized state they may again accept hydrogen. The detailed mechanism underlying the transfer of hydrogen from one coenzyme to another is obscure. Several schemes have been postulated showing how energy-rich bonds

Table 13.3

OXIDATION-REDUCTION POTENTIALS[a]

Oxidation-Reduction System Oxidized Compound	Reduced Compound	Temp., °C	pH	E_h Volt
Coenzyme I	Reduced coenzyme	30	7.0	−0.282
Riboflavin	Leuco-riboflavin	20	7.0	−0.186
Heme, ferri	Heme, ferro	30	8.18	−0.188
Pyruvate	Lactate	35	7.01	−0.180
Malate	Succinate	37	7.0	−0.094
Ferric ion	Iron	25	0.0	−0.36
Cytochrome b, Fe^{3+}	Cytochrome b, Fe^{2+}	20	7.4	−0.04
α-Ketoglutarate	Glutamate		7.0	−0.030 calc.
Methylene blue	Leuco-methylene blue		7.0	+0.011
Cytochrome a, Fe^{3+}	Cytochrome, Fe^{2+}	20	7.4	+0.262
Oxygen	Water	25	7.0	+0.815
Iodine	Iodide	25	0.0	+0.5345
Oxygen	Hydrogen peroxide	25	0.0	+0.682
Oleate	Palmitate		7.0	+0.025
Fumarate	Succinate		7.0	0.00
Resazurin	Hydroresorufin		7.0	+0.051
Acetaldehyde	Ethanol	20	7.0	−0.200
Cystine	Cysteine	25	7.0	−0.14
Gluconate	Glucose		7.0	−0.45
Carbon dioxide	Formate	30	7.0	−0.42
Hydrogen ion	Hydrogen	All temps	0.00	0.000

[a] From Anderson and Plaut.[7]

can be formed with reduced DPN, flavin-adenine-dinucleotide and ferricytochrome a_3.[148,160]

Flavoproteins

Flavoproteins constitute a system of important oxidases and dehydrogenases containing as part of their molecular structure either riboflavin mononucleotide (FMN) or more commonly flavin-adenine-dinucleotide (FAD).

The view long held that flavoprotein enzymes are limited to the role of terminal oxidases has now been abandoned. Some flavoproteins contain metal which participates as an essential link in an electron transport system. Thus, xanthine oxidase, familiar to dairy chemists because it is associated with the fat globule membrane in substantial amounts, is a metalloflavoprotein. Evidence that it is an iron-containing molybdo-flavoprotein has come from many sources.[70,105,449,518] Whitely and Ordal investigated the role of xanthine oxidase and hydrogenase, both from *Micrococcus lactilyticus,* in the transport sequence associated with the following reaction:[556]

$$\text{hypoxanthine} \rightleftharpoons \text{xanthine} + H_2 \quad \text{(R63)}$$

Electron transport proceeds according to the sequence shown below:

hypoxanthine → X
xanthine oxidase (an iron-containing molybdo-flavoprotein)→
hydrogenase (an iron-containing molybdo-flavoprotein)→
hydrogen (R64)

X is a heat-labile coupling compound, probably one containing iron.

Other metalloflavoproteins that have been studied include: reduced DPN-cytochrome c-reductase, which contains 4 Fe atoms for each flavin molecule;[344] butyryl-coenzyme A dehydrogenase, which contains Cu;[156] aldehyde oxidase and reduced DPN-nitrate-reductase, which contain molybdenum;[398] and fumaric hydrogenase and acyl-CoA dehydrogenase, which contain Fe.[87]

Xanthine oxidase causes slow reduction of methylene blue in "sterile" raw milk. The reduction time may be shortened appreciably by introduction of xanthine, hypoxanthine, or other suitable substrates. Thermal inactivation of the enzyme requires temperatures as high as 85°C. Milks heated at this or higher temperatures suffer significant losses in xanthine oxidase activity.[156,223,422]

Among other oxidases, one finds glucose oxidase, 6-phosphoglucose oxidase, cytochrome c oxidase, nonspecific D- and L-amino oxidases,[35] and the specific D-aspartic oxidase. The following two reactions are mediated with impure and pure preparations, respectively, of L- and D-amino oxidases, thus:

$$R\cdot CH(NH_2)\cdot COOH + \tfrac{1}{2}O_2 \rightarrow R\cdot CO\cdot COOH + NH_3 \quad \text{(R65)}$$

Pure preparations behave differently, thus:

$$R\cdot CH(NH_2)\cdot COOH + O_2 + H_2O \rightarrow R\cdot CO\cdot COOH + NH_3 + H_2O_2 \quad \text{(R66)}$$

Oxidative decarboxylation of amino acids, which in microbial metabolism leads to formation of flavor-producing aldehydes and ketones, may be important in the manufacture of mold-ripened cheeses such as Roquefort, and of cultured milks such as yogurt.

Less widespread among microorganisms than the amino-acid oxidases are the diamino-oxidases, which catalyze the oxidation of diamines to ammonia, aldehydes, and hydrogen peroxide. These oxidases are apparently flavoproteins possessing a prosthetic group of the FAD type.

Respiration of Lactic Acid Bacteria.—Bertho and Glück[32] showed that catalase-free, facultative anaerobic, homolactic bacteria underwent a forced respiration in the presence of oxygen to yield hydrogen peroxide. Later, Warburg and Christian[540] proved that, in Bertho's experiments, the presence of Warburg's yellow ferment was responsible for the entire fermentation. In the respiration of *L. delbrückii* and *L. acidophilus* with the substrate glucose, one-half of the oxygen consumed is utilized in CO_2 production, the other one-half appearing in hydrogen peroxide. In later stages, the fermentation product, pyruvic acid, reacts with hydrogen peroxide, and the ratio between CO_2 formed and O_2 consumed increases. Respiration of homolactics proceeds through the intervention of flavoprotein oxidases, foremost among which is 6-phosphoglucose oxidase. This phenomenon manifests itself in such a way that the color of a suspension of certain bacteria can be transformed from white to yellow by the admission of air, and alternatively it may be transformed from yellow to white by the exclusion of air.

The respiration is unphysiologic, and unless the medium contains some means of dissipating peroxide and preventing its accumulation, inhibition occurs. The "viridans" group of streptococci, and *L. bulgaricus* and *L. acidophilus,* produce a green pigment, a peroxidation product of hemoglobin, when grown on blood agar.[311,361]

Added peroxide inhibits growth of the aerobes *Pasteurella (Yersinia) pestis* and *Pasteurella (Yersinia) septicum,* and hematin reverses the effect. The function of hematin, blood or catalase is to reverse the inhibition brought about by the product of respiration, hydrogen peroxide.

Accumulation of H_2O_2 accounts for the observation that strong aeration initially inhibits growth of starter organisms. Although lactic acid organisms do not as a rule contain catalase, they do contain peroxidases which tend to dissipate the initial accumulation of H_2O_2. Thus Dolin has shown that *S. faecalis,* in oxidizing reduced DPN, produces H_2O_2 which in turn is reduced in the presence of a flavin-containing peroxidase.[107] Hydrogen peroxide produced by lactic acid bacteria strongly inhibits growth of anaerobes.[555]

The uniqueness of the flavoproteins in the transfer of electrons has been postulated to reside in the structure of the flavin moiety, which is conducive to formation of semiquinones, i.e., intermediate free radicals.[374]

Cytochrome System

Dehydrogenases may transfer electrons and protons to the cytochrome system or they may transfer hydrogen to coenzymes. Reduced coenzymes are reoxidized in respiration by way of flavoenzymes, which in turn are reoxidized either directly by oxygen, or indirectly by way of the cytochrome system. The cytochrome system, comprising the cytochromes and cytochrome oxidase, is extremely important.

In the dehydrogenation of succinate to fumarate and **water**, the following sequence is observed in the transfer of single electrons:

$$\text{succinate} \xrightarrow{e \mp H^+} Fe^{+++}\text{-cytochrome b} \xrightarrow{e \mp H^+}$$

$$\text{Slater's factor} \xrightarrow{e \mp H^+} Fe^{+++}\text{-cytochrome c} \xrightarrow{e \mp H^+}$$

$$Fe^{+++}\text{-cytochrome a} \xrightarrow{e \mp H^+} Fe^{+++}\text{-cytochrome oxidase-}O_2 \quad (R67)$$

This cytochrome system is absent in anaerobes such as the clostridia, and in homolactic bacteria. It is present in yeasts, *Bacillus subtilis,* and *Pseudomonas* species, among others. *E. coli* appears to contain modified cytochromes and a modified cytochrome oxidase, distinguished by their absorption bands.[488]

Dehydrogenases

The ordinary dehydrogenases, with some exceptions, transfer 2 hydrogens and 2 electrons from a substrate to a codehydrogenase, usually coenzyme I or coenzyme II. One exception is succinic dehydrogenase, which, according to some evidence, is a flavoprotein. A second exception is the formic acid dehydrogenase of bacteria. This bacterial dehydrogenase seems to require cytochrome b; and of the same character are yeast and bacterial lactic acid dehydrogenases. A third exception is the pyruvic acid oxidizing system, which contains lipothiamide-pyrophosphate among its ensemblage of coenzymes as the primary coenzyme involved in the oxidation of pyruvic acid.

Besides biological acceptors, there are numerous nonbiological ones, such as methylene blue and resazurin. They furnish a basis for various tests designed to characterize the quality of milk and milk products.

Citric Acid Cycle

In the oxidative decarboxylation of pyruvic acid, acetyl-CoA is one of the reaction products. In terminal respiration this compound, together with oxalacetate, initiates a series of enzymatically mediated reactions which proceed cyclically with regeneration of oxalacetic acid, consumption of 2 molecules of water, and formation of 2 molecules of carbon dioxide. Thus the oxidation is executed in a stepwise fashion such that the energy released is made available to the cell, thereby satisfying its energy requirements for growth, reproduction, movement, and other activities concerned with survival. The series of reactions known variously as the citric acid cycle, Krebs cycle, or tricarboxylic

oxalacetic acid — $COOH \cdot CH_2 \cdot CO \cdot COOH$ $\underset{-2H}{\overset{+2H}{\rightleftarrows}}$ malic acid — $COOH \cdot CH_2 \cdot CHOH \cdot COOH$ $\underset{+H_2O}{\overset{-H_2O}{\rightleftarrows}}$ fumaric acid — $COOH \cdot CH{=}CH \cdot COOH$ $\underset{-2H}{\overset{+2H}{\rightleftarrows}}$ succinic acid — $COOH \cdot CH_2 \cdot CH_2 \cdot COOH$

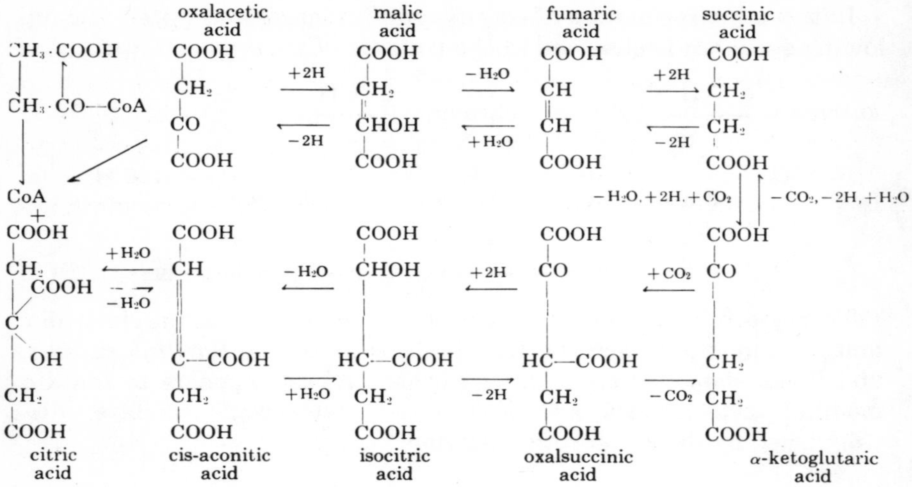

$CH_3 \cdot COOH \rightleftarrows CH_3 \cdot CO{-}CoA$; $CH_3 \cdot CO{-}CoA$ + oxalacetic acid → CoA + citric acid

succinic acid $\underset{-CO_2, -2H, +H_2O}{\overset{-H_2O, +2H, +CO_2}{\rightleftarrows}}$ α-ketoglutaric acid

citric acid — $COOH \cdot CH_2 \cdot C(OH)(COOH) \cdot CH_2 \cdot COOH$ $\underset{-H_2O}{\overset{+H_2O}{\rightleftarrows}}$ cis-aconitic acid — $COOH \cdot CH{=}C(COOH) \cdot CH_2 \cdot COOH$ $\underset{+H_2O}{\overset{-H_2O}{\rightleftarrows}}$ isocitric acid — $COOH \cdot CHOH \cdot HC(COOH) \cdot CH_2 \cdot COOH$ $\underset{-2H}{\overset{+2H}{\rightleftarrows}}$ oxalsuccinic acid — $COOH \cdot CO \cdot HC(COOH) \cdot CH_2 \cdot COOH$ $\underset{-CO_2}{\overset{+CO_2}{\rightleftarrows}}$ α-ketoglutaric acid — $COOH \cdot CO \cdot CH_2 \cdot CH_2 \cdot COOH$

FIG. 13.1 THE KREBS OR CITRIC ACID CYCLE

acid cycle takes place in animal tissues.[283,284] In the metabolism of microorganisms, operation of such a cycle, although the subject of much investigation and controversy, has gained acceptance.[508,553] The citric acid cycle of intermediates is given in this chapter.

An interesting development is the discovery of the importance of the citric acid cycle in the metabolism of amino acids. In oxidative deaminations mediated by α-amino acid oxidases and dehydrogenases, α-oxoacids are formed, together with NH_3. Transamination reactions can effect exchange between amino and ketonic groups. The citric acid cycle contains three α-oxoacids (oxalacetic, oxalsuccinic, and α-ketoglutaric), and therefore it can accommodate the deamination products of certain α-amino acids; or conversely it can serve as a source of α-amino acids and subsequently proteins. Two patterns of ^{14}C distribution in amino acids of *E. coli* have been found.[1,362,456] These acids were derived from one of the following: $^{14}CO_2$, $^{14}CH_3$-COOH, $CH_3{}^{14}COOH$, aspartic acid and glutamic acid both labeled uniformly, and 1-^{14}C-labeled glutamic acid. Grown in the presence of ^{12}C-glucose and $^{14}CO_2$, *E. coli* yielded cellular proteins, which when degraded showed: (a) that 20 to 40% of aspartic acid activity was located in C-1 and the balance in C-4; and (b) that more than 97% of glutamic acid activity resided in C-1. When $CH_3{}^{14}COOH$ was used, analysis of the degraded proteins showed that 80+% of the activity resided in C-5, and 5+% in C-1, of glutamic acid. The values were those which would obtain if the citric acid cycle was operative. Specific activity measurements and competition experiments showed that biosynthetic relationships exist among most of the amino acids. Two separate acetate fixation mechanisms were believed to be in operation in amino-acid synthesis—one in which leucine was a product, and a second in which glutamic acid or a substance in equilibrium with it, α-ketoglutaric acid, was an early product. With respect to the two families of amino acids, the members of each of which followed a definite and specific radioactivity pattern based on the specific activity of either glutamic or aspartic acid, the following genesis was postulated:

glutamic acid → glutamic-γ-semialdehyde → proline
glutamic acid → N-acetyl-glutamic acid → N-α-acetylornithine
ornithine → citrulline → arginine

I

diaminopimelic acid ↓ ; methionine ↑

lysine ← aspartic acid → homoserine → threonine
α-ketobutyric acid → α-keto-β-methylvaleric acid
↓↑ ; ↓↑
α-aminobutyric acid ; isoleucine

II

Organisms which can meet all their carbon requirements from 2-carbon compounds such as alcohol and acetate possess a mechanism known as the glyoxylate cycle for achieving a net synthesis of C_4-dicarboxylic acids from acetic acid.[275-277] Thus acetate enters the cycle, reacts with oxalacetate to form citrate; citrate is transformed to *cis*-aconitate, which in turn is changed to isocitrate; succinate leaves the cycle in a reaction in which glyoxylate is also formed; glyoxylate combines with activated acetate at this point to yield malate, which in turn is oxidized to oxalacetate, thus completing the cycle. One mole succinate is produced for each 2 moles acetate entering the cycle. Intermediates removed from the citric acid cycle can be regenerated in the glyoxylic acid scheme. The modified citric acid cycle occurs in species of *Pseudomonas*, in molds, and in many strains of *E. coli*.

Pasteur Effect

The aerobic dissimilation of carbohydrates represents a more complete oxidation of substrate than the anaerobic one. The much greater store of energy released becomes available through the instrumentality of "energy-rich" phosphate bonds for those assimilatory processes of the cell which require energy. This means that for a given amount of substrate, more cells will be produced in aerobic dissimilation.

Pasteur was the first to generalize on this behavior, and consequently the action of air in effecting increased cellular development, or conversely, diminished glycolysis, became known as the Pasteur effect. The effect is peculiar to facultative organisms; and if we wish we may utilize the Pasteur effect to define and limit this group, i.e., the facultative microorganisms are those which exhibit a Pasteur effect.

Current views would have the phosphorylation reactions occurring during respiration compete with the coupled dehydrogenation and phosphorylation of phosphoglyceraldehyde.[240,338,341] The supply of inorganic phosphorus required for phosphorylation of phosphoglyceraldehyde becomes exhausted, and glucose fermentation is inhibited. The lower uptake of glucose during respiration, as compared with that during fermentation, is attributed to the lack of adenosine triphosphate at the site of glucose phosphorylation. Such adenosine triphosphate as is formed accumulates in the mitochondria, and is therefore not readily available for glucose phosphorylation.

Fat Metabolism

Although fats as a source of carbon and energy are not as important in microbial as in tissue metabolism, the fact remains that microorganisms may contain within their cytoplasm fat bodies which, in mycobacteria and in fungi such as *Geotrichum candidum* (*Oospora lactis*), constitute a substantial proportion of the dry microbial mass. We may presume that fats constitute a reservoir of carbon and energy upon which the highly aerobic organism may draw under starvation conditions. Many of the flavors observed in highly flavored cheeses, such as Roquefort, Gorgonzola, Camembert, and Limburger, have their origin in the breakdown of lipids.

Fat normally occurs dispersed as globules of microscopic dimensions, and, unless lipolyzed, will remain unavailable to the cell. Many microorganisms have been found to contain lipases.[6,78] Collins and Hammer[78] isolated 159 lipolytic bacterial cultures from various sources—all of which could conceivably contaminate a milk supply. Lipolytic organisms often present in milk include *Achromobacter lipolyticum, Pseudomonas fragi, Pseudomonas fluorescens,* and *Achromobacter lipidis*.[6,112,175,229] Also, yeast of the genus *Torula* and most molds possess lipolytic activity.[458]

Lipolytic activity and the important part it plays in flavor and texture development in cheese has been the subject of extensive investigations.[63,176,218,219,418,522] Girolami and Knight[147] found that resting cells of *Penicillium roqueforti* oxidized fatty acids in the presence of phosphate and magnesium to respective methyl ketones having one less carbon atom. Jackson and Hussong, while studying Blue cheese, isolated several secondary alcohols—the oxidation products of pentanone-2, heptanone-2, and nonanone-2.[232]

Lipolysis makes available to the organism glycerol and a variety of fatty acids, saturated and unsaturated, of different chain lengths. Glycerol, it may be presumed, is decomposed according to the Embden-Meyerhof scheme, entering the pathway to pyruvic acid at the point corresponding to triose formation in the decomposition of glucose. It is dehydrogenated, a reversal of the step leading to its formation from phosphoglyceraldehyde, and ultimately its 3-carbon "backbone" appears in pyruvic acid. The oxidation of glycerol makes available for subsequent reduction reactions 4 hydrogen atoms compared with the 2 made available in the oxidation of trioses. The difference reflects the lower degree of oxidation of glycerol. Generally substrates possessing a relatively low degree of oxidation give rise to end products which *in toto* represent a relatively low degree of oxidation. The degree of oxidation of n moles of an organic substance with the generalized formula $C_xH_yO_z$ may be represented by the number $n(y - 2z)$.[314] In anaerobic decomposition, the number representing the degree of oxidation of the substrate must be equal to the sum of the numbers representing the degree of oxidation of the end products.

Metabolism of Fatty Acids.—A long time elapsed before the observations of Knoop[266] on the basic principles underlying β-oxidation of fatty acids were extended to show that the CO_2 involved in β-oxidation is formed in the citric acid cycle.[42,299,382,559] This meant, by analogy, that an active 2-C compound might play an important role in fatty-acid oxidation. This intermediate is acetyl-CoA, and an early indication of its role in fat metabolism was suggested by the studies of Barker,[19] Burton,[54] and Lynen[342] in connection with the reversible series of reactions leading from acetyl-CoA to butyric acid in butyric acid fermentation.

We shall now consider the general mechanism for oxidation of fatty acids. In the first stage, an acyl-coenzyme A compound is formed. The reaction may require "sparking" by acetyl-coenzyme A, or it may take place in the presence of ATP and coenzyme A. In the presence of acyl dehydrogenase and flavin-adenine-dinucleotide, dehydrogenation occurs, and saturated acyl-CoA is converted to the corresponding unsaturated acyl-CoA compound, with the double bond located between the α and β carbon atoms. Hydration in the presence of enoyl hydratase then yields L-β-oxyacyl-CoA, which in turn is oxidized in the presence of β-oxyacyl dehydrogenase and DPN+ to yield β-ketoacyl-CoA. This compound then undergoes thiolysis in the presence of CoA and β-ketoacylthiolase. Acetyl-CoA and an acyl-CoA compound are formed.

The chain length of the new acyl compound is 2 carbon atoms shorter than the original one. Acetyl-CoA is available for renewal of the cycle, and in each such renewal acetyl-CoA and an acyl-CoA compound bereft of 2 carbon atoms are formed. Acetyl-CoA becomes available, as a result of such β-oxidation, for condensation with oxalacetate to yield citrate; thus "active acetate" enters the citric acid cycle, and what emerges are the oxidation products, carbon dioxide and water. Unless oxalacetate is renewed to replace the quantity transformed into precursors of amino acids, "active acetate" will not again enter the citric acid cycle. Normally, oxalacetate is renewed as one result of the dissimilation of carbohydrate. Pyruvate or phosphopyruvate, resulting from the fermentation of carbohydrates, reacts with carbon dioxide, and oxalacetate is formed. If for some reason the link between carbohydrate and fatty acid metabolism is disturbed, 2 molecules of acetyl-CoA combine in the presence of β-ketothiolase to yield acetoacetyl-CoA, and this compound in turn yields acetoacetic acid upon enzymatic hydrolysis.

Biosynthesis of Fats.—Carbohydrates comprise the primary source of material and energy for biosynthesis of fats. Decomposition of carbohydrates results in formation of pyruvic acid and 2 molecules of reduced coenzyme I. Oxidative decarboxylation of pyruvic acid then yields acetyl-CoA, carbon dioxide, and a second molecule of reduced coenzyme I. With formation of acetyl-CoA, the stage is set for synthesis of higher fatty acids by way of the fatty-acid cycle. The sequence of reactions leading to β-oxidation of fatty acids is reversed. Two molecules of acetyl-CoA condense to yield acetoacetyl-CoA, which in turn accepts hydrogen to form L-β-oxybutyryl-CoA. This oxy-compound then undergoes dehydration followed by reduction to yield butyryl-CoA. Acetyl-CoA enters the cycle for the second time, and reacts with butyryl-CoA. The hydrogenation, dehydration, and reduction steps, in the order indicated, are repeated, and an acyl-CoA compound is formed in which the acyl component now contains 6 carbon atoms. This process, repeated often enough, leads to formation of a pool of acyl-CoA compounds with acyl components having various chain lengths, each chain of which is unbranched and contains an even number of carbon atoms. Biosynthesis of fatty acids having an odd number of carbon atoms requires a primary condensing unit that contains an odd number of carbon atoms. This primary unit is propionyl-CoA. It is formed in the rumen of ruminants by bacteria acting on cellulose. Propionyl-CoA is resorbed from the blood and made available for synthesis. As one might expect, the fat of ruminants contains fatty acids having an odd number of carbon atoms.

In the synthesis of fats, the "energy-rich" acyl-CoA compounds are seldom hydrolyzed. Rather, the energy associated with these compounds is utilized in the synthesis of ester bonds. Esterification of 3-phosphoglycerol with two molecules of acyl-CoA leads to formation of CoA and phosphatidic acid, a precursor of fats, thus:

$$2R \cdot COSCoA + CH_2OH \cdot CHOH \cdot CH_2OPO_3H_2 \rightarrow CH_2OR \cdot CHOR \cdot CH_2OPO_3H_2 + 2HSCoA \quad (R68)$$

Kornberg and Pricer[278] studied enzymes in liver involved in this transacylation, and observed that the relative activity of various acyl-CoA substrates plotted against chain length is maximum for fatty-acid derivatives containing 16 to 18 carbon atoms. This accounts for the predominance of palmitic, stearic, and oleic acid in the fat of animal tissues.

Coenzyme A in fat metabolism appears to act in a multifunctional manner. Compound formation with fatty acids renders more reactive the hydrogen atoms in the neighborhood of the carboxyl group. Spontaneous condensation of 2 molecules of acetic acid to acetoacetic acid under physiological conditions is thermodynamically impossible. Condensation of 2 molecules of acetyl-CoA to acetoacetyl-CoA is possible. Therefore, both condensation and the reverse reaction can occur on the same enzyme surface only with activated acyl groups, such as acyl-CoA provides. Compounds of the insoluble higher fatty acids and coenzyme A are soluble in water, and thus the acyl groups are more accessible to chemical attack.

Characteristic differences between thioesters and the corresponding oxygen esters have been discussed by Lynen.[339] The much larger nuclear charge on the sulfur atom increases the positively charged character of the carbon associated with it in thioesters, and as a consequence, it would be readily susceptible to attack by nucleophilic substituents (Lewis bases) and carbanion-containing molecules.

Malonyl-coenzyme A.—A remarkable development stems from the observation of Klein *et al.*[262,263] that homogenates of anaerobically grown *Saccharomyces cerevisiae* cells incorporated 5 times as much acetate into fatty acids whenever CO_2 was not removed from the aerobic environment of the experiment. A related observation by Gibson *et al.* specifying a bicarbonate requirement for synthesis of long-chain fatty acid was reported almost simultaneously.[144] Commenting on these observations, Lynen[339] proposed that reaction between bicarbonate and acetyl-CoA would lead to formation of a malonyl-CoA intermediate, and that it was malonyl-CoA rather than the acetyl compound which was active in the synthesis of β-keto acids. A decarboxylation would attend the condensation of malonyl- and acetyl-CoA, and the accompanying energy release, Lynen reasoned, would shift the thiolase equilibrium in favor of the β-keto acid. Moreover, the observation that biotin is necessary for synthesis of palmitic acid, and the likelihood that biotin is involved in all biochemical carboxylation mechanisms supported Lynen in his thesis. This interesting hypothesis was confirmed by Lynen and his collaborators; and malonyl-CoA was isolated in experiments with crude enzymes prepared from cell-free yeast homogenates and systems containing unlabeled acetate and radioactive bicarbonate.

Phosphate Bond Energy

The energy available to the microorganism for performance of its vital functions (assimilation, reproduction, movement, etc.) is derived from free-energy changes attending oxidation of the substrate. Energy is to a large extent not wasted as kinetic energy, but is stored as molecular energy as a result of coupled reactions, bimolecular or unimolecular, leading to formation of "energy-rich" phosphate bonds. Lipmann,[328] in 1941, elucidated the function of compounds containing these bonds thus: "During various metabolic processes, phosphate is introduced into compounds not merely or at least not solely to facilitate their breakdown, but as a prospective carrier of energy."

Three types of compounds have been observed to contain "energy-rich" phosphate bonds. Pyrophosphates contain such a bond, thus: $-\mathrm{P}-\mathrm{O}\sim\mathrm{P}$. The slurred line indicates an "energy-rich" bond. The mixed anhydride type, embracing a carboxy-acid and phosphoric acid, is denoted by

$$-\overset{\Vert}{\mathrm{C}}-\mathrm{O}\sim\mathrm{P},$$

and includes types derived from the fundamental type by substitution of bivalent groups for carboxyl oxygen. Finally, a type is found in which phosphoric acid is linked with guanidine, or derivatives of guanidine, thus:

$$-\overset{|}{\mathrm{N}}-\underset{\underset{|}{\mathrm{N}}}{\overset{}{\underset{\Vert}{\mathrm{C}}}}-\overset{|}{\mathrm{N}}\sim\mathrm{P}.$$

The designation of these bonds as "energy-rich" to distinguish them from bonds of the simple ester type signifies that they are relatively easily hydrolyzed. Inorganic pyrophosphates, for example, dissociate almost completely, although slowly, into orthophosphates even at room temperature. Thermodynamic considerations lead one to regard the "energy-rich" bond as relatively labile; considerations of reaction kinetics, on the contrary, lead one to associate a relatively high activation energy requirement with the severance of this bond. Lipmann[330] explained that the phosphate bond is so well suited for its biochemical duties because the relatively high activation energy requirement serves to retard rapid hydrolysis and to impart a measure of stability to thermodynamically labile compounds.

Coupling of Oxidation with Phosphorolysis

Formation of "energy-rich" phosphate bonds results from the coupling of an oxidation-reduction reaction with phosphorylation. This reaction may be intramolecular, as for example in the formation of phosphoenolpyruvic acid from 2-phosphoglyceric acid (see R11); or it may be intermolecular, usually involving Co I or Co II, as in the phosphorylative oxidation of 3-phosphoglyceraldehyde to 1,3-diphosphoglyceric acid (see R7).

If the entire output of energy in aerobic glucose oxidation is converted into phosphate-bond energy, each bond possessing 15 kilocalories, some 45 phosphorylations would result. The ratio between the quantity of inorganic phosphate converted to organic phosphate and the amount of oxygen consumed in oxidative phosphorylations—the P/O ratio—is an important index of the efficiency by which the energy of oxidative processes is converted into available energy. The true P/O ratio was estimated to be 3.0 in the complete oxidation of pyruvate by way of the citric acid cycle.[27,406] In addition to phosphorylation of substrate at any level, phosphorylation of intermediates in the stepwise approach to the final oxidant occurs. Direct experiments have established that phosphorylation is coupled to the aerobic oxidation of reduced diphosphorpyridine nucleotide.[302]

Inorganic ions and permeability and structure of mitochondria are important factors in oxidative phosphorylation.[260,296,300,301]

The mechanism for storage in and transfer of energy through "energy-rich" phosphate bonds is not unique. In oxidation of pyruvic acid, an anhydride bond rich in energy is formed between an "active" acetyl group and a sulfhydryl group of coenzyme A. Such compounds readily enter into reactions with inorganic phosphate to yield "energy-rich" phosphate bonds; acetylphosphate, for example, readily forms in the reaction between acetyl-coenzyme A and inorganic phosphate.

It has been frequently demonstrated that "energy-rich" compounds are utilized in cellular growth. Such assimilatory reactions as peptide, urea, and glutamine synthesis, and indirectly polysaccharide formation by way of hexoses, have been studied.[34,71,190,492]

Coenzyme Q

A group of quinones (related to napthoquinones and vitamin K) occur widely among microorganisms, especially aerobes. These act as coenzymes by undergoing cyclic oxidation and reduction during substrate oxidation in mitochondria, and are involved in oxidative phosphorylations.[88,312,379] The coenzyme has been characterized as a group of 2,3-dimethoxy-5-methylbenzoquinones, thus:[566]

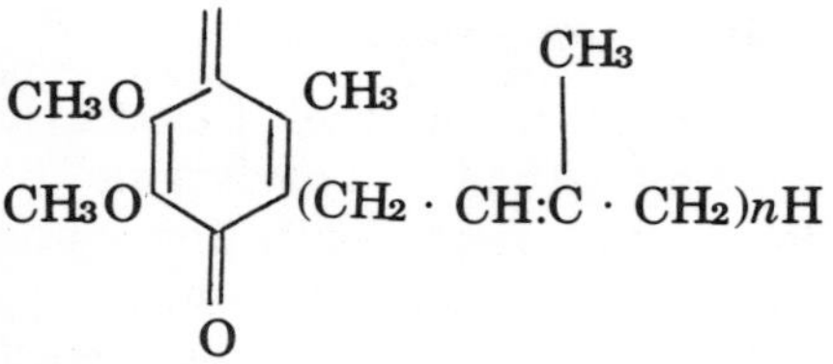

$$CH_3O,\ CH_3O,\ CH_3\text{ (ring, with two =O)}\ (CH_2 \cdot CH{:}\underset{}{\overset{CH_3}{C}} \cdot CH_2)_nH \qquad n = 6 - 10$$

Members of the coenzyme Q group are designated as Q_6 . . . . . . Q_{10} depending on the value of n.

Coenzyme Q is very likely an additional member of the respiratory chain concerned with oxidation of succinate.[89,193,194,303] During phosphorylative oxidation coenzyme Q is concomitantly involved in the energization of inorganic phosphate, with formation of an energy-rich phosphate bond.[193,194] Model compounds such as the quinol phosphates acquire a phosphorylating function on oxidation.[68] Currently, the structural type of phosphate based on a 1,2-pyran ring skeleton is regarded as more significant than the type related to dihydrocoenzyme Q_{10} monophosphate. Folkers *et al.* consider that the 5-phosphomethyl-6-chromanol type I shown below can be oxidized to system II, which then participates in oxidative phosphorylation by generating metaphosphate (HPO_3).[121,531]

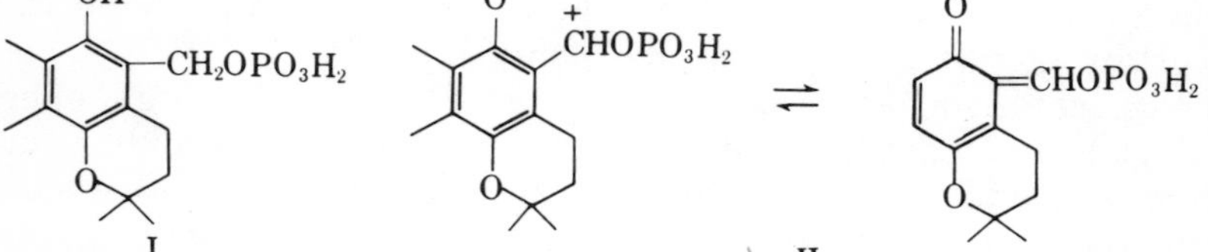

I: OH, $CH_2OPO_3H_2$, O

II: O^-, $\overset{+}{C}HOPO_3H_2$, O $\rightleftharpoons$ O, $=CHOPO_3H_2$, O

Polyscharide Formation

The condition known as ropy or slimy milk, the result of abnormal fermentation, is sometimes encountered in milk in epidemic proportions. Its immediate cause is the formation of gums of a polysaccharide character, and mucins. The mechanism of polysaccharide formation is discussed herewith; practical aspects are considered later.

Free-Energy Changes.—The free-energy change accompanying hydrolysis of sucrose is approximately −6,500 calories per mole; the corresponding free-energy change in hydrolysis of lactose is approximately 2.5 times as great. Hydrolysis is favored, and to reverse the reaction under physiological conditions, i.e., in neutral dilute aqueous solution, synthesis of lactose or sucrose must be coupled with an energy-yielding reaction. Much the same condition prevails in hydrolysis of such polysaccharides as glycogen and starch.

Synthesis of Polysaccharides.—There are two known pathways for synthesis of polysaccharides; one requires mediation of phosphorylase and phosphate, the other does not. The phosphorolytic pathway will be considered first.

Polysaccharide phosphorylases have been found in animal tissues, yeast and bacteria.[62,191,199] These enzymes may catalyze either scission or formation of α-1,4-glucosidic bonds at the nonreducing end of a glycogen or starch chain. In synthesis, these enzymes foster a metathetical reaction between glucose-1-phosphate and the polysaccharide chain. Scission occurs between the carbon and oxygen of the phosphorylated hexose group, the hexose moiety is transferred to the nonreducing end of the polysaccharide chain, and a proton from the chain end is transferred to the phosphate moiety. No net change in free energy is involved, since the free-energy changes attending hydrolysis of both glucose-1-phosphate and starch or glycogen are approximately equal. As the H-ion concentration of an equilibrated solution decreases, the concentration of the stronger acid, glucose-1-phosphate, increases. The reaction requires the presence of a "primer" (a small quantity of starch, glycogen, or dextrin on which chain expansion can take place). The chain prolongation eventually ceases, and the addendum separates from its "primer."[82]

Transglycosidases.—Transglycosidases are enzymes which promote transfer of hexose units from di- to monosaccharides, from di- to oligosaccharides, and from di- to polysaccharides.[198] In these reactions, the free energy from hydrolysis of disaccharides is utilized to effect synthesis of oligo- and polysaccharides. Lactase possesses a transgalactosidylase function, which will be described in the section on lactose metabolism.

Glucose Donor-Uridine Diphosphate Glucose.—In a series of papers, Leloir and his collaborators have described several reactions in which uridine diphosphate-glucose (UDP-glucose) acts as a glucose donor to produce the following: trehalose, sucrose, sucrose phosphate, bacterial cellulose, and glycogen. There is evidence, according to these investigators, that in animal tissues, glycogen is synthesized from UDP-glucose and not from glucose-1-phosphate.[307-310] The occurrence and postulated occurrence of UDP-glucose in all lactose-fermenting microorganisms suggests that this compound may be involved in microbial polysaccharide synthesis.

Assimilation—Material and Energy Balance

Assimilatory processes, of which polysaccharide synthesis is an example, are endoergonic processes whereby the cell reproduces the macromolecular species and essential metabolites required for its renewal and survival. These processes are commercially significant in production of cells, as in yeast manufacture and in production of cellular components such as fats, polysaccharides, and vitamins. They are important in sewage disposal where a balance must be maintained between cellular synthesis and cellular losses (endogenous respiration).

General Considerations.—The efficiency of an assimilatory process is usually described in terms of several coefficients: conversion, fat, protein, and carbohydrate coefficients. These denote in turn the weight of cellular dry substance, weight of fat, weight of protein, and weight of carbohydrates (essentially polysaccharides) which are obtained per 100 gm of an active source of carbon present in the medium. For comparison purposes, if glucose is the source of carbon, the determination is best made when glucose utilization is complete.

The consensus at present considers the active 2-carbon compound acetyl-CoA as the building block for the preponderance of fat and protein found in nature, although there are some, notably Clifton, who regarded a 1-carbon fragment, an "active" formaldehyde as the precursor.[69] Both schemes would make available for assimilation two-thirds of the total glucose present; the remaining third would be used in fermentation.[125]

Maximal Value for the Conversion Coefficient.—Dry yeast on an average contains between 45 and 47% carbon;[511] 100 gm glucose contains 67 gm assimilable sugar, the carbon content of which is approximately 27 gm. The maximum quantity of dried yeast containing 45 to 47% carbon that can be realized from 100 gm glucose (the maximum conversion coefficient) lies therefore, between 57 and 60. Roughly, about 50% of the dry cellular material produced under normal conditions of culture is protein; hence the maximum value of the protein coefficient lies in the neighborhood of 30.

Maximum Value for the Fat Coefficient.—If the energy available for synthesis is assumed constant, it is possible to draw conclusions concerning the maximum value of the fat coefficient.[450,480] Assuming that the microorganism is yeast and that it is composed of fat, protein, ash, and carbohydrates with respective heats of combustion relative to glucose of 2.5, 1.7, 0, and 1, and with respective percentages of fat, protein, ash, and carbohydrate in the dry yeast of F, P, A, and 100—F—P—A, and assuming that the conversion coefficient is E, one may conclude, because of the assumed constancy of the available energy, that the following relationship would hold, thus: $2.5F + 1.75P + 100—F—P—A \leqslant (77 \times 100/E)$. The heat of combustion of normal yeast containing about 50% protein and 5% fat is approximately 77% that of the assimilable sugar; hence 77 occurs in the right-hand member of the equation. The equation shows that if the fat content of yeast is increased at the expense of either protein or carbohydrate, the maximum value of the conversion coefficient, i.e., the yield of dry cells per 100 gm glucose, must decrease. Moreover, as the fat content of cells increases, so must the maximum value of the fat coefficient.

Theoretically, if all the assimilable sugar is converted into glycerol, water, and fat with the tripalmitin composition, the resulting cellular mass would contain 29.8 gm fat for each 100 gm processed sugar.[123] Since, however, an appreciable part of each cell consists of protein and other nonfat components, fat coefficients greater than 15 are not observed in practice.[168]

FERMENTATIONS IN MILK

Fermentations in milk modify its properties favorably or adversely, giving rise to such products as starter culture, cultured beverages and cheese, on the one hand, and to spoiled and degraded products on the other. Some fermentations lead to development of antibiotics which may be desirable or undesirable, depending on circumstances. Fermented milk products are unique in the sense that the acquired organoleptic properties (reflecting a distinctive blending of flavor-imparting components and body qualities) depend on unique characteristics of certain components of milk, for example, curdling properties of the calcium caseinate-phosphate complex, blandness of lactose, and flavor characteristics of lipolysed milkfat. This section is devoted to the following topics: milk (raw and heated) as a fermentation medium, lactose and its dissimilation, citrate and its dissimilation, nitrogenous compounds in milk, proteolysis, cultured milks, and undesirable fermentations in milk.

Milk as a Fermentation Medium

Milk is a versatile fermentation medium. It contains the materials required even by such nutritionally exacting organisms as the lactobacilli. It is not a universal medium, however, and contains some bacteriostatic substances, among which are the mild, easily destroyed lactenins of historic interest.[242,243] The disaccharide lactose is the chief source of carbon, and thus the application of milk is limited to fermentations for which lactose-fermenting organisms are available; or at best, additional and uneconomic processing becomes necessary to convert lactose to fermentable monosaccharides. In propionic acid fermentation, supplementing milk or whey with degraded protein is necessary for maximal fermentation rates to be achieved. Milk normally is deficient in manganese with respect to requirements of lactobacilli for maximal growth rates. It normally is also deficient in iron with respect to requirements of certain species of *Clostridium* for production of volatile solvents and riboflavin. Milk contains only a trace of cobalt, which does not suffice for maximal microbial synthesis of vitamin B_{12}.

On the other hand, milk is a rich source of vitamins and organic growth factors. It is especially rich in riboflavin and orotic acid. Supplementation of milk media with vitamins or extracts is only rarely necessary. Proteins of milk contain in superabundance all the amino acids essential for bacterial growth. Proteoses and amino acids are present in sufficient quantities to promote the onset of most fermentations.

The heat treatment milk undergoes before most fermentations changes its characteristics as a culture medium. Milk autoclaved 15 min, but not longer, is superior to milk heated at 80°C for 10 min for culturing lactic streptococci.[127] Heating milk at 62 to 72°C for 30 to 40 min serves to stimulate growth of starter cultures.[157-159] Heating at 72°C for 45 min, at 82°C for 10 to 45 min, or at 90°C for 1 to 45 min inhibited growth. Drastic heating beyond the limits mentioned again stimulated growth of lactic streptococci. Believed to be responsible for initial stimulation were the following: oxygen expulsion, destruction of inhibitors, partial protein hydrolysis, and serum protein denaturation. Inhibition was associated with formation of toxic volatile sulfides, and the succeeding stimulation with heat-induced disappearance of these sulfides.[157-159]

Lactose, fat, and citric acid comprise the important fermentable (C, H, O) compounds of milk. Fat metabolism has been considered in a general way and will not be discussed further. Citric acid metabolism has been considered only indirectly in connection with the citric acid cycle. Lactose metabolism follows essentially the same course as glucose metabolism. There are, however, distinct features involved in its transformation to the required intermediate compound, glucose-1-phosphate.

Lactose and Lactase.—Lactose, milk's chief source of energy for microbial metabolism, is a disaccharide constituting about 40% of the solids of whole milk. Strictly speaking, lactose is α-d-glucopyranose-1,4-β-d-galactopyranose.

In bacterial metabolism lactose is first hydrolyzed to glucose and galactose. The enzyme which brings this about, lactase or properly, β-D-galactosidase, has been found in lactose-fermenting yeasts, molds, *E. coli, E. cloacae,* and the intestinal wall of the calf.[238] The older literature refers to lactase as an endoenzyme in "true lactics"[409] and to the presence of lactase in *L. bulgaricus,*[41] *S. lactis* var. *taette,*[166] and *E. coli,*[532] among others.

Lactase functions both as a β-D-galactosidase and as a transgalactosidylase; that is, lactase not only brings about hydrolytic cleavage of lactose, but also transfers galactosidyl residues to various acceptors, including water, glucose, galactose, lactose, and oligosaccharides. In solutions containing high concentrations of lactose, the enzymatic reactions may even favor formation of oligosaccharides at the expense of the monosaccharides, glucose and galactose.[11,415,416,454,455,533,534]

One enzyme, and not two, is involved in formation of monoses and oligosaccharides.[415,416] Using a commercial preparation from lactose-fermenting yeasts, Pazur studied the products and postulated the following sequence of reactions:

$$\text{glucose} \underset{1,4}{\rule{1em}{0.4pt}} \text{galactose} + \text{lactase} \rightarrow \text{galactose-lactase} + \text{glucose} \qquad \text{(R69)}$$
$$\text{galactose-lactase} + \text{glucose} \rightarrow \text{glucose} \overset{1,6}{\rule{1em}{0.4pt}} \text{galactose} + \text{lactase} \qquad \text{(R70a)}$$
$$\text{galactose-lactase} + \text{galactose} \rightarrow \text{galactose} \overset{1,6}{\rule{1em}{0.4pt}} \text{galactose} + \text{lactase} \qquad \text{(R70b)}$$
$$\text{galactose-lactase} + \text{glucose} \overset{1,4}{\rule{1em}{0.4pt}} \text{galactose} \rightarrow \text{glucose} \overset{1,4}{\rule{1em}{0.4pt}} \text{galactose} \overset{1,6}{\rule{1em}{0.4pt}} \text{galactose} + \text{lactase} \qquad \text{(R70c)}$$
$$\text{galactose-lactase} + \text{glucose} \overset{1,6}{\rule{1em}{0.4pt}} \text{galactose} \rightarrow \text{glucose} \overset{1,6}{\rule{1em}{0.4pt}} \text{galactose} \overset{1,6}{\rule{1em}{0.4pt}} \text{galactose} + \text{lactase} \qquad \text{(R70d)}$$
$$\text{galactose-lactase} \xrightarrow{+H_2O} \text{galactose} + \text{lactase} \qquad \text{(R70e)}$$

Existence of this mechanism was supported by results of experiments with the following: labeled lactose, lactose plus labeled glucose, and lactose plus labeled galactose. In agreement with Pazur's conclusions were those of Wallenfels *et al.*,[533,534] who worked with preparations from molds, *E. coli,* and calves' intestines.

Roberts *et al.*,[454,455] found at least 11 oligosaccharides among the products of the reaction; with concentrated lactose solutions, high yields of oligosaccharides were obtained.

Conversion of 1-Phosphogalactose to 6-Phosphoglucose.—Following a suggestion of Kosterlitz,[280,281] Caputto *et al.*[60] noted that this conversion was accelerated by two factors, one thermolabile, the other stable. The thermostable factor present in purified yeast extract was uridinediphosphate-glucose; the thermolabile factor was identical with purified phosphoglucomutase, the enzyme catalyzing conversion of 1-, to 6-phosphoglucose. Uridine-diphosphate-glucose (UDPG) has been isolated and has been assigned the following formula:[60]

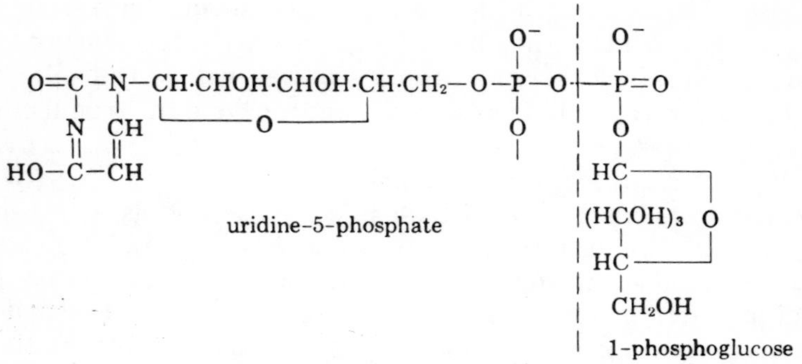

```
                                                   ¦
                                             O⁻    ¦  O⁻
                                             |     ¦  |
O=C—N—CH·CHOH·CHOH·CH·CH₂—O—P—O—¦—P=O
  |  |    |_______________|                   |     ¦  |
  N  CH          O                            O     ¦  O
  ‖  ‖                                        |     ¦  |
HO—C—CH                                             ¦  HC———┐
                                                    ¦  |     |
            uridine-5-phosphate                     ¦ (HCOH)₃ O
                                                    ¦  |     |
                                                    ¦  HC———┘
                                                    ¦  |
                                                    ¦  CH₂OH
                                                    ¦ 1-phosphoglucose
```

The structure was confirmed by splitting off glucose and one inorganic phosphate group during acid hydrolysis of milk.

Twenty-five per cent of UDPG in a crude enzyme preparation is converted on incubation to uridine-diphosphate-galactose. Based on this observation, the following scheme was proposed as a working hypothesis:[306]

$$\text{galactose} + \text{ATP} \rightleftharpoons \text{galactose-1-phosphate} + \text{ADP} \qquad \text{(R71a)}$$

$$\text{galactose-1-phosphate} + \text{UDP-glucose} \qquad \text{(R71b)}$$

$$\rightleftharpoons$$

$$\text{glucose-1-phosphate} + \text{UDP-galactose}$$

$$\text{UDP-galactose} \rightleftharpoons \text{UDP-glucose} \qquad \text{(R71c)}$$

Reaction 71a is catalyzed by a galactokinase present in galactose-adapted yeasts and in *L. bulgaricus*. The occurrence of step b is largely a matter of conjecture. Step c is catalyzed by an enzyme named galactowaldenase. The reaction is essentially a Walden inversion, at C-4, of glucose. Concerning its mechanism, Leloir[306] concluded "It is impossible at present to correlate the knowledge provided by organic chemistry with the enzymatic reactions . . ."

Hansen and Craine[181] and Rutter and Hansen[468] separated the enzyme system containing galactowaldenase from that containing phosphoglucomutase. The waldenase was active only with respect to α-galactose, and α-glucose and β-sugars were not attacked. The equilibrium between UDP-glucose and UDP-galactose was such that 73 to 79% of the equilibrium mixture consisted of UDP-glucose.

Adaptation of Organisms to Lactose.—Many microorganisms contain enzymes necessary to transform lactose to 6-phosphoglucose; these are said to possess the enzymes constitutively. Other microorganisms can be adapted to grow on lactose on the basis of models furnished by the presence of lactose, galactose, or lactose-like and galactose-like compounds; and organisms in which the ability to ferment lactose is acquired are said to possess the enzymes adaptively.[75] Model substances, presence of which is required for induced enzyme synthesis, are called inductors. Inductors, if present in a medium, will induce synthesis of all the enzymes involved in their decomposition, whether some participating enzymes are present constitutively or not.

Elucidation of the mechanism of induced synthesis is receiving much consideration.[75] Sources of energy and amino acids are required. Sometimes the energy can be supplied only under oxidative conditions, as in the synthesis of the galactozymase system in yeast. Adaptive enzymes are not usually perpetuated in serial transfers unless the essential substrate is always present. Organisms trained to ferment lactose over a long series of transfers appear to lose this ability after a few transfers in media devoid of the substrate.

Direct Versus Indirect Lactose Utilization.—Studies of relative fermentation rates of disaccharides and component monosaccharides have always lent themselves to interpretation on the basis of selective permeation effects. An alternative view considers differences in rates as evidence that disaccharides can be dissimilated directly.

Lactose is fermented by various yeasts at a uniform and more rapid rate than are equimolar mixtures of glucose and galactose.[304,385,460,549] Moreover, Myrbäck *et al.* observed that, if toluene suppressed fermentation, hydrolysis of lactose in the presence of toluene proceeded more rapidly than fermentation in its absence.[385] To account for differences in fermentation rates, it seemed reasonable to postulate that lactose could be dissimilated directly. Moreover, in acid solution hydrolysis was suppressed but fermentation was not. Myrbäck argued that lactase was localized in the cell wall and hence was subject to influence by the pH of the medium, whereas the organism's apparatus for direct dissimilation was located in the interior, where the pH might conceivably differ locally from that of the medium.

The theory of direct dissimilation disregards the occurrence of specific permeability effects and the effect of toluene on permeability.[141] The consensus now considers lactase function to be indispensable.

Concerted attacks on problems pertaining to selective permeation of the cell favor the idea of stereospecific permeation systems which determine whether neutral molecules will have access into the cell and at what rate.[70,448 491] Suspensions of *E. coli* containing the inducible enzyme lactase will hydrolyze lactose but not thiogalactosides (lactose analogues in which the oxygen atom of the glycosidic linkage is replaced by sulfur); thus thiogalactoside can accumulate within the cell to an intracellular concentration of galactoside which may exceed by 100-fold or more its concentration in the external medium. To account stoichiometrically and kinetically for the rates and amounts of accumulation, a stoichiometric and a catalytic model were compared. Only the catalytic or permease model, which assigned to the specific permease sites the role of catalyzing accumulation of galactosides within the cell rather than serving as final acceptors, proved able to account successfully for the observed reaction kinetics. To account quantitatively for accumulation of galactoside in the cell, it was necessary to take into consideration the evidence that cells genetically or otherwise devoid of permease were unable to metabolize lactose, although they posessed lactase activity. Thus the correct model appeared to be one in which the permeation barrier in the absence of permease possesses a high degree of impermeability toward carbohydrates, and in which the catalytic permeases if present would be associated with the barrier (not necessarily the cell wall). Hence the rate of permeation into the cell would depend on external concentration, on leakage rate, and above all on the activity of the permease. The rate of exit from the cell would depend on the concentration of free substrate within the barrier and on the leakage rate.

Citric Acid

The importance of citric acid is out of all proportion to its low concentration (0.07 to 0.4%) in milk. Citric acid is an essential substrate for the desirable aroma-producing organisms of milk products. Its decomposition gives rise to the flavors characteristic of butter. Bosworth and Prucha[36] observed that citric acid disappears from milk during the normal souring process. Kickinger[259] showed that the citrates in milk disappear as a result of the action of bacteria which survived heating to the boiling point.

Hastings *et al.*[192] observed that the breakdown of citrates in milk and in synthetic media does not follow parallel courses. The difference brought about by use of synthetic media was marked if citrates were the sole source of carbon. According to these workers, the nature of the available carbon determines the course of citrate metabolism. Citric acid is fermented by many organisms which may find their way into milk, for example, by *E. coli, E. cloacae, E. aerogenes, L. citrovorum, L. dextranicum, B. subtilis, Proteus vulgaris, L. casei, L. acidophilus, L. bulgaricus,* alkali-forming bacteria, and lactose-fermenting yeasts. Citrate metabolism is discussed in the following paragraphs.

Fermentation by Starters

Lactic acid bacteria are used widely as starters in the cheese industry to initiate the desired fermentation that leads to acid formation. These bacteria are also useful in butter manufacture, creating acid conditions which inhibit the growth of undesirable organisms and contributing to production of desirable flavor and aroma. They also serve the same purpose in the manufacture of cultured buttermilk, sour cream, and margarine. The predominant organisms present in the usual lactic starter are *S. lactis* and/or *S. cremoris*. Certain aroma-producing bacteria, *L. citrovorum* and *L. dextranicum,* may be present in smaller numbers. *S. cremoris, Streptococcus diacetilactis,* and occasionally *Pediococcus* species are relied upon in middle European countries for aroma development.[246]

Formation of Lactic Acid by Starters.—*S. lactis* and *S. cremoris* are the acid-producing species in mesophilic starters. These organisms are facultative anaerobes; they grow at 0°C but not at 45°C. They are homofermentative, producing large amounts of dextrolactic acid and much smaller amounts of acetic acid. Their growth requirements are comparatively complex. Niven[402] studied 21 strains of *S. lactis.* No less than 14 amino acids were required for prompt growth. Although *S. lactis* can attack casein, it seems to thrive better on simpler compounds, and in milk this preference is met by a small but sufficient supply of proteoses and peptones and to a lesser degree, by a wide selection of amino acids. Growth and acid production in milk by many lactic-acid bacteria may be inhibited by developed rancidity and by the presence of antibiotics, sanitizers, other chemicals, and bacteriophage. The subject has been discussed in many reviews.[3,13,292,348,350,506]

If spontaneously soured milk is allowed to stand, especially at a comparatively high temperature, it undergoes a second lactic fermentation induced by nonspore-forming bacilli of the *L. bulgaricus* type. Many of these organisms are quite tolerant to lactic acid, and are known to produce 3.25% acid when grown for one month at 29°C. The pH as a consequence may drop to 3.

In the ripening of a lactic culture, lactic acid is produced to the extent of 0.75 to 1.0%, causing the pH of milk to fall to 4.3 to 4.7. This acid is chiefly D-lactic acid but, because of the activity of *L. citrovorum* and particularly of *L. dextranicum,* some l-lactic acid is also produced from lactose. A small percentage of acetic acid is produced by *S. lactis* from lactose and a much larger proportion by the *Leuconostoc* species from citric acid.

Production of Biacetyl.—Production of biacetyl in starter cultures is an associative one. Neither *S. lactis* nor *L. citrovorum* (unless pH is artifically reduced), in the absence of the other, will produce efficiently either biacetyl or acetoin.[264] Biacetyl is usually accompanied by larger proportions of acetoin. Neither is a stable compound in culture, and they tend to be reduced to the more stable 2,3-butylene glycol (2,3-butanediol) in a reaction favored by optimum conditions for the activity of the culture. At 21.1°C and in neutralized cultures, both biacetyl and acetoin tend to disappear quite rapidly.[174] Maximum production, but with relatively poor yields of biacetyl and acetoin, occurs in the pH range 4.1 to 4.4 in pure cultures acidified with sulfuric acid. Maximim yields in cultures acidified with citric acid are much higher, and are obtained in the pH range 3.7 to 3.9.[373]

Cox[86] found that the aroma-producing organisms of milk first produced biacetyl and then destroyed it. He also found that, although a relatively low pH value in the range 4.4 to 5.5 retarded growth, it promoted biacetyl formation. Michaelian *et al.*[373] noted that the yield of biacetyl could be increased if oxygen was bubbled through acidified milk containing a pure culture. Addition of pure acetoin to the medium effected no further increase in yield; hence the observed increase was not a result of nonenzymatic oxidation.

Prill and Hammer[432] found the following conditions to be conducive to production of biacetyl in good yields by starter cultures: addition of citric acid, shaking and aeration, holding at low temperatures, and lowering the pH of a ripened culture with sulfuric acid. Brewer *et al.*[43] found that agitation of pure cultures in air at pressures of 30 psi resulted in increases in yield as high as several hundred per cent.

In butter making, the practice of cooling ripened cultures and holding them for 24 hr in the presence of small proportions of citric acid is recommended if such cultures are intended for inoculation of cooled cream and if further growth is not desired.

Production of acetoin and biacetyl occur by two separate mechanisms, and formation of biacetyl is not simply caused by microbial oxidation of acetoin.[77] Consequently, measurement of biacetyl plus acetoin is not a true indication of the content of biacetyl present, nor is it indicative of the biacetyl-producing capacity of the culture.[77]

Biacetyl formation involves: (a) production of pyruvic acid from citrate, (b) decarboxylation of pyruvic acid to hydroxyethylthiamine pyrophosphate, and (c) reaction of hydroxyethylthiamine pyrophosphate and acetyl-CoA to produce biacetyl. Details of the reactions follow:[77]

$$\underset{\text{citric acid}}{\begin{array}{c}CH_2COOH\\|\\HO\text{-}C\text{-}COOH\\|\\CH_2COOH\end{array}} \xrightarrow[Mn^{++}\text{ or }Mg^{++}]{\text{Citratase}} \underset{\text{acetic acid}}{\begin{array}{c}CH_3\\|\\COOH\end{array}} + \underset{\text{oxalacetic acid}}{\begin{array}{c}COOH\\|\\C=O\\|\\CH_2\\|\\COOH\end{array}} \quad \text{(R72a)}$$

$$\underset{\text{oxalacetic acid}}{\begin{array}{c}COOH\\|\\C\\|\\CH_2\\|\\COOH\end{array}} \xrightarrow[\text{acid decarboxylase}]{\text{oxalacetic}} CO_2^{-} + {}^{-}\underset{\text{pyruvic acid}}{\begin{array}{c}COOH\\|\\C=O\\|\\CH_3\end{array}} \quad \text{(R72b)}$$

Supplying citrate (R72a) to the culture enables it to form pyruvic acid (R72b) without simultaneously forming reduced NAD, as it would when converting pyruvic to lactic acid. Since pyruvic acid accumulates, it is available for synthesis of biacetyl or other compounds.[77] Use of citrate by the organism requires (a) citrate permease for transporting citrate into the cell, (b) citratase for splitting citrate into acetic acid and oxalacetic acid, and (c) oxalacetic acid decarboxylase to form pyruvic acid from oxalacetic acid.

Use of the pyruvic acid thus formed is dependent on its decarboxylation to hydroxyethylthiamine pyrophosphate, a compound which can be designated as acetaldehyde-TPP complex. The reaction requires a divalent metal and thiamine pyrophosphate (TPP), and proceeds as follows:[77]

$$\underset{\text{pyruvic acid}}{\begin{matrix} CH_3 \\ | \\ C = O \\ | \\ COOH \end{matrix}} \xrightarrow[Mg^{+}]{TPP} \underset{\text{acetaldehyde-TPP complex}}{\begin{matrix} CH_3 \\ | \\ C = O \\ | \\ H/TPP \end{matrix}} + CO_2 \qquad (R73)$$

The acetaldehyde-TPP complex thus formed can react with acetyl CoA to form biacetyl according to the following equation:[77]

$$\underset{\text{acetaldehyde-TPP complex}}{\begin{matrix} CH_3 \\ | \\ C=O \\ | \\ H/TPP \end{matrix}} + \underset{\text{acetyl-CoA}}{\begin{matrix} CH_3 \\ \\ C=O \\ | \\ S\text{-}CoA \end{matrix}} \longrightarrow \underset{\text{biacetyl}}{\begin{matrix} CH_3 \\ | \\ C=O \\ | \\ C=O \\ | \\ CH_3 \end{matrix}} + CoA\text{-}SH + TPP \qquad (R74)$$

Microorganisms able to produce biacetyl form only small amounts but normally produce larger amounts of acetoin, presumably because only a limited amount of acetyl-CoA is produced and it is not needed to form acetoin (see below).

Acetyl-CoA used in the reaction to produce biacetyl can be made by microorganisms from the acetaldehyde-TPP complex and oxidized lipoic acid or from acetic acid. Production from acetic acid appears to be of less importance and hence only the equations for the first method are given below.[77]

$$\underset{\text{acetaldehyde-TPP complex}}{CH_3-C(=O)-H/TPP} + \underset{\text{oxidized lipoic acid}}{\text{S—S (LIP)}} \longrightarrow TPP + \underset{\text{S-acetyl dihydrolipoic acid}}{CH_3-C(=O)-C-\text{LIP-SH}} \quad (R75a)$$

$$\underset{\text{S-acetyl dihydrolipoic acid}}{CH_3-C(=O)-S-\text{LIP-SH}} + CoA \cdot SH \longrightarrow \underset{\text{acetyl-CoA}}{CH_3-C(=O)-S-CoA} + \underset{\text{reduced lipoic acid}}{\text{LIP(SH)(SH)}} \quad (R75b)$$

Biacetyl, once formed, can be reduced to acetoin and the acetoin can be reduced to 2,3-butylene glycol. This occurs on extended incubation of cultures and on prolonged storage of cultured products with a resultant loss of flavor and aroma. The reactions proceed as follows:[77]

$$\underset{\text{biacetyl}}{CH_3-C(=O)-C(=O)-CH_3} \xrightarrow[NADH_2]{\text{biacetyl reductase}} \underset{\text{acetoin}}{CH_3-C(=O)-HCOH-CH_3} \underset{NAD}{\overset{NADH_2}{\rightleftharpoons}} \underset{\text{2,3-butylene glycol}}{CH_3-HCOH-HCOH-CH_3} \quad (R76)$$

Although some acetoin can be produced as just outlined, the large quantities associated with some cultures are formed from the acetaldehyde-TPP complex and pyruvic acid according to the following equations:[77]

$$\underset{\text{acetaldehyde-TPP complex}}{\begin{array}{c} CH_3 \\ | \\ C=O \\ | \\ H/TPP \end{array}} + \underset{\text{pyruvic acid}}{\begin{array}{c} COOH \\ | \\ C=O \\ | \\ CH_3 \end{array}} \longrightarrow \underset{\alpha\text{-acetolactic acid}}{\begin{array}{c} CH_3 \\ | \\ C=O \\ | \\ HOC\text{-}COOH \\ | \\ CH_3 \end{array}} + TPP \qquad \text{(R77a)}$$

$$\underset{\alpha\text{-acetolactic acid}}{\begin{array}{c} CH_3 \\ | \\ C=O \\ | \\ HOC\text{-}COOH \\ | \\ CH_3 \end{array}} \xrightarrow[\text{acid decarboxylase}]{\alpha\text{-acetolactic}} \underset{\text{acetoin}}{\begin{array}{c} CH_3 \\ | \\ C=O \\ | \\ HCOH \\ | \\ CH_3 \end{array}} + CO_2 \qquad \text{(R77b)}$$

Such bacteria as *S. diacetilactis, S. faecalis, L. citrovorum, E. aerogenes, P. fluorescens, Lactobacillus brevis,* and *Serratia marcescens* can produce acetoin by the mechanism outlined in equations R77a and R77b. Yeasts and *E. coli* form acetoin from the acetaldehyde-TPP complex and free acetaldehyde. Although these microorganisms produce some α-acetolactic acid, they lack the decarboxylase enzyme needed to form acetoin. Instead they produce valine and pantothenic acid from α-acetolactic acid. Acetoin produced from either pyruvic acid or acetaldehyde reacting with the acetaldehyde-TPP complex can be reduced to 2,3-butylene glycol, as illustrated in equation R76.

Nitrogenous Compounds in Milk

The chief nitrogenous compounds in milk are proteins, among which there are two major groups—the caseins and the whey proteins. Caseins are a complex of proteins with a high degree of affinity for one another. In milk they are associated into a complex called the calcium caseinate-phosphate complex which exists as discrete spherical particles ranging in diameter from 10 to 300 nm. The casein system of proteins coagulates at a pH of approximately 4.6 produced by an acid(s), and at the pH of milk in the presence of suitable proteinases, notably rennin, pepsin, or certain microbial proteinases. The whey system of proteins (about 20% of the total) comprise chiefly β-lactoglobulin, α-lactalbumin, and proteins of the immunoglobulin fraction. β-Lactoglobulin, the most abundant of the whey proteins, is denatured when milk is heated above 70°C. The denatured protein tends to associate itself with other whey proteins and with the calcium caseinate-phosphate complex. Following coagulation of caseins with acid and whey proteins with heat, there remain in the resulting serum a heat-stable proteose-peptone fraction, a variety of amino acids (in some measure reflecting in their relative amounts the amino-acid composition of the proteins in milk), and minute percentages of nonprotein nitrogenous compounds such as urea, ammonia, creatine, creatinine, uric acids, etc.

All the free amino acids needed for effective growth of *S. lactis* are present in milk except phenylalanine and possibly cystine. Glutamine, which is stimulatory, has not been observed. Variation in milk of the nonprotein nitrogen component (peptide fraction) brings about differences in the growth rates of lactic streptococci.[493]

Proteolysis

Metabolism of amino acids has been briefly discussed in the section on the citric acid cycle. Some organisms can use ammonia as the sole source of nitrogen, reversing the process by means of which amino acids are degraded to oxoacids. Many organisms active in milk have the apparatus for hydrolyzing proteins to simpler proteins, polypeptides, and amino acids. Gorini studied the elaboration of rennin-like substances (protein-coagulating enzymes), especially by *S. marcescens*.[151,152] Bacteria-elaborating rennin-like enzymes have the power to "sweet" curdle the casein complex in milk. Thus Hammer and Hussong[180] identified *Bacillus cereus* as the organism responsible for an outbreak of "sweet" curdling in sterile evaporated milk. The coagulum appeared normal to the taste; there was no acid development and no "wheying-off."

Proteinases and peptidases constitute the primary enzyme forms in bacteria responsible for hydrolysis (proteolysis of milk proteins). Proteolysis is important in cheese manufacture and less so in fermented milks. In cheese, proteolysis manifests itself during the long ripening period and is brought about by rennet or other milk-clotting enzymes, proteinases of milk, and probably those of bacteria and molds. Although proteolysis in cheese will be briefly discussed, it is covered more adequately in Chapter 12. Proteolysis is of some significance in the preparation of cultured milks. It obviously plays an important part in the treatment of dairy wastes by aerobic microorganisms. In connection with outbreaks of a bitter flavor defect in evaporated milk, Spitzer and Epple[495] found *Bacillus panis,* a spore-bearing rod with intense proteolytic activity, to be the causative organism.

The role of proteolysis in cheese making may properly be discussed briefly at this point. Proteolysis, as carried out by streptococci and by certain lactobacilli and micrococci, is important in the ripening of Cheddar and some other cheese types; that carried out by *Brevibacterium linens* is important in the ripening of Brick, Limburger, and similar cheese; and that brought about by molds is significant in the ripening of Blue, Roquefort, and Camembert cheese.

Proteolysis by S. lactis.—In cultures of *S. lactis* acid development and proteolysis occur concurrently at 30°C, but not at 32 or 37°C.[188,560] Both endo-, and extracellular proteinases have been isolated from lactic streptococci.[526,527,561] Proteolysis in milk at 32°C is detectable within 4 hr.

The importance of proteolysis in starters is emphasized by the observation that various cultures developed the same amount of acid (0.55%), but produced curd having different curd tension values.[201] The variation may be attributed directly to variations in proteolysis, or indirectly to differences in buffering action brought about by different degrees of proteolysis.[560]

Free amino nitrogen is not associated with bitterness. Several investigators believe that development of bitterness indicates a deficiency in enzymes required to hydrolyze the bitter primary degradation products of cheese proteins.[116]

Proteolysis by L. casei.—Proteolysis develops more rapidly in Cheddar cheese inoculated with *L. casei* than in control cheese.[295] In some instances the difference is evident throughout the ripening period, in others after 8.5 months.[52] Proteolytic strains of *L. casei* added to Cheddar cheese encouraged flavor development and improved flavor in mature cheese.[572]

Proteinases isolated from disrupted cells of *L. casei* exhibited maximal activity at 15 and 38°C, and at a pH near neutrality.[18] Brandsaeter and Nelson[39,40] attribute maximum activity at pH 7.0 to a peptidase, and maximum activity in the pH range 5.5 to 6.5 to a proteinase.

Stereospecific deaminases in *L. casei* brought about deamination of serine at pH 5.4 and 8.1, and asparagine and theronine at pH 8.1. Two deaminases appear to be involved. Thus deamination of DL-serine proceeded equally well at 52°C and pH 7.0 and 46°C and pH 4.6.[285-287]

Proteolytic activity is exhibited by some strains of *L. brevis,*[93] *L. bulgaricus,*[568,569] *Lactobacillus helveticus,*[9] and *Lactobacillus lactis.*[37] The subject has been reviewed by Marth.[346]

Proteolysis by Micrococci.—Micrococci comprise 78% of the nonlactic bacteria in Cheddar cheese.[4] The proteolytic apparatus of *Micrococcus freudenreichii* functions optimally at 30°C and at a pH near neutrality.[18] Analysis of the several sets of conditions for the optimal activity of one-year old Cheddar cheese proteinases supports the view that micrococci may contribute to total proteolytic activity.

Proteolysis by B. linens.—An extracellular proteinase isolated from a two-day old culture of *B. linens* in a liquid medium was active over a pH range between 5.6 and 8.8 and exhibited maximum activity at pH 7.2 to 7.3.[131] Proteolysis was detectable at 0 and 60°C and was optimal at 38°C. The enzyme hydrolyzed α- and β-casein, but not β-lactoglobulin and α-lactalbumin. No evidence of polypeptidase and dipeptidase activity was found.[515] Free amino acids are liberated in Limburger cheese inoculated with *B. linens,*[521] suggesting that under certain conditions enzymes other than proteinases are secreted by the organism, which occurs in the flora of surface-ripened cheeses and which it is believed contributes to their characteristic body and flavor.[132]

Proteolytic Action by Molds.—Blue cheese with inadequate growth of *P. roqueforti,* reflecting inadequate proteolysis, is tough and crumbly instead of soft and smooth.[176] The proteinase of *P. roqueforti* is of the trypsin type with an optimum activity between pH 5.8, and 6.3 and is not inhibited by NaCl at levels found in Blue cheese.[176] Proteinases and related enzymes in *Penicillium camemberti* are chiefly responsible for proteolysis observed in Camembert cheese.[176]

Cultured Milks

The ability of bacteria to affect the flavor and consistency of milk was utilized in making special milk drinks many years before the microorganisms were seen by van Leeuwenhoek. Most of these fermented drinks originated in southern Russia and in countries at the eastern end of the Mediterranean Sea. They were made by producing conditions in milk favorable for the desired fermentation, or by inoculating milk with small amounts of prepared milk. Usually the unwashed containers served to provide the starter flora, especially when skin bottles were used. Under these conditions a mixture of bacteria was present, and flavors resulted which were difficult to reproduce with pure cultures, although the essential bacteria of these fermented drinks were eventually known. In all instances the basic fermentation is a lactic one. Sometimes it occurs in combination with the production of gas, mild alcohol fermentation, and some proteolysis.[83] Kefir may contain about 1.2% alcohol. Some fermented drinks can be made from nonfat dry milk solids and water, and the body and texture of such a product can be improved by addition of 0.35% gelatin.

One of the more popular fermented milks is known in Bulgaria and Turkey as yogurt or yaourt; it is also known in Armenia as matzoon and in Egypt as leben. In this product, the fermentation is brought about entirely by lactic acid bacteria, including *Thermobacterium yoghurt* of Orla-Jensen (probably *L. bulgaricus*), *L. bulgaricus,* and *S. thermophilus. S. lactis* may be present but is not essential; *L. acidophilus* is sometimes added, and then the product is called acidophilus yogurt. Fresh milk is pasteurized, cooled to 40 to 45°C, inoculated, and held at that temperature until ready for use; it is then refrigerated. Rosell[462] recommends an incubation at 45 to 48°C. At this temperature, 3.5% acid develops in 24 to 36 hr. Incubation at 29 to 32°C for 12 to 14 hr is said to produce a yogurt with more distinctive flavor. The body of yogurt may be improved by heating fresh milk at 90°C for 5 min or more, homogenizing the milk after the culture is added, and adding nonfat dry milk (3 parts per 100). Yogurt is made from cow's milk, either pasteurized or partially evaporated, and in some places from sheep or buffalo milk. The Egyptian leben is said to contain a lactose-fermenting yeast which produces mild alcohol fermentation. Winckel[562] described two fermented milks called kajobst and kajovit, which undergo modified yogurt fermentation.

Modern yogurt is almost invariably made from a mixed culture of *S. thermophilus* and *L. bulgaricus*. The two cultures are often maintained separately.

Schulz *et al.*[478,479] identified acetaldehyde as the compound responsible for the characteristic flavor and aroma of yogurt. Addition of 0.001 to 0.005% acetaldehyde to milk soured with *S. thermophilus* imparted a yogurt-like aroma and taste to the product. Four ranges of color intensity were defined, and milks were grouped accordingly. Sour milks possessing well-developed yogurt flavor and aroma contained a relatively high concentration of acetaldehyde, and those lacking these qualities contained less.

A popular American cultured milk is buttermilk. If cultured buttermilk is made with *L. bulgaricus,* it is desirable to temper the sharp acid flavor from the bulgaricus culture by growing a streptococcus in association with it. However, maintenance of the two in mixed culture is difficult; the lactic streptococci cease growing at a pH of 4.0 to 4.2, whereas *L. bulgaricus* decreases the pH considerably below this level. The desired end may be accomplished by growth in separate containers, followed by proportioning and mixing to give a product with the desired flavor and texture. The product made with *L. bulgaricus* is known as Bulgarian buttermilk. A more popular beverage employs either *S. lactis* or *S. cremoris* grown together with either

L. dextranicum or *L. citrovorum*. One function of the streptococcus is to provide a sufficiently acid environment for the "aroma" compounds to be produced maximally.[264,373] If the starter culture is permitted to develop 0.80 to 0.85% acidity, and if incubation is at 21°C, associative growth will produce a buttermilk with excellent flavor and body characteristics.

Kefir is usually made from cow's milk, and is peculiar in that the fermentation is brought about by kefir "grains," which resemble miniature cauliflowers. These "grains" consist of casein, yeasts, and bacteria. The microorganisms include lactose-fermenting *Torula* yeasts, *S. lactis, Betabacterium caucasicum* (probably an *L. brevis* variant), and glycogen-containing rod-shaped kefir bacilli (Henneberg). The grains increase in size in the fermenting milk, and may be strained out, dried, kept for long periods, and used as inocula. The fermentation is usually carried on in closed bottles so that gas is retained and the milk becomes effervescent.

In Russia a milk drink made from unpasteurized mare's milk, and known as kumiss, is used extensively. The fermentation is caused principally by *L. bulgaricus,* lactose-fermenting *Torula* yeasts, and *L. leichmannii*.

Kuban fermented milk, a product of southern Russia, is made from pasteurized milk by combined lactic and alcohol fermentation. The microflora include a lactic streptococcus resembling *S. lactis* var. *hollandicus,* a lactic rod of the *L. bulgaricus* type, and three yeast types.

Taette milk is a sour milk used in the Scandinavian Peninsula. A slime-producing fermentation is induced by a variant of *S. lactis* designated as *S. lactis* var. *taette*. This is possibly identical with *S. lactis* var. *hollandicus,* which has been used to make Edam cheese, and with other streptococci associated with "ropy" milk.

A milk drink known as saya is prepared from fresh unheated milk, ripened at first by *S. lactis* and later by a lactobacillus. In saya milk, considerable carbon dioxide and vigorous proteolysis, are produced. Characteristic of the fermentation is a 6-day ripening period at 11°C. Corminboeuf[83] has described numerous milk beverages, including mazun, groddu, skorup, and tättemjolk.

Undesirable Fermentations in Milk

The flavor and body of cultured milks are distinguished by a delicate balance between the components of the cultured product. Unless conditions of culture are carefully controlled, this balance may not be achieved even when pure cultures are employed. Empirical formulations relating to proper cultural and environmental conditions constitute the art of fermentations. Apart from defects arising from use of improper conditions of culture, other defects may occur in the milk products because of contamination of a milk supply by unwanted organisms, and because many organisms required in fermentations can produce antibiotics.

Flavor defects have been characterized as bitter[331,519] (see section on proteolysis), rancid, ester,[90] cresol-like,[358] barny,[335] doughy,[471] fruity,[216,446] fishy,[172] malty,[178] potato-like,[41] turnip-like,[305] caramel,[255,470] cabbagy,[122] metallic,[14] putrid,[410] and feed.[466,467] Body defects such as sweet curdling in evaporated milk, gasiness in cheese and evaporated and sweetened condensed milk may be brought about by bacterial contaminants. A body defect of rather wide occurrence, which can reach epidemic proportions, is ropy or slimy milk.

Ropy and Slimy Milk.—Ropy or slimy milk of bacterial origin becomes apparent only after milk has been stored, and is therefore distinguishable from the stringy milk associated with mastitis.[117,184] Ropiness may be evident only as a slightly abnormal viscosity, or it may be so pronounced that the affected milk can be drawn out in fine threads a yard long, and in some instances may assume a gel-like consistency. Thickening may be confined to the top layer of the milk.

Among the bacteria causing ropiness are active gelatin liquefiers, including some of the hay bacillus type. More frequently, however, epidemics of ropy milk are caused by some members of the coliform group or the lactic streptococci. The common occurrence of the defect in the presence of certain streptococci has led to the assignment of distinguishing names to these. *S. lactis* var. *taette, S. lactis* var. *hollandicus,* and varieties of the common *S. lactis* are the essential organisms in Swedish ropy milk and in certain Edam cheese starter cultures. Among the organisms associated with development of ropiness are these: *Alcaligenes viscolactis,*[179,207,208,357,472] *S. lactis* var. *hollandicus,*[173,177] certain corynebacteria,[208] and some organisms of the *Escherichia-Enterobacter* group.[353,472,474] Ordinary milk streptococci, such as *S. lactis, S. cremoris,* or *S. thermophilius,* may at times cause ropiness.[208] Certain strains of streptococci, according to Hammer, easily acquire and lose their ability to produce ropiness. Rope-producing strains are more oxygen-exacting than nonrope-producing ones and develop less volatile acid. Induction of rope-producing properties in bacteria by means of bacteriophage has been observed.

The flavor of ropy milk, unless the defect is associated with lactic fermentation, is indistinguishable from that of normal milk; nor is the milk unwholesome in any way.

The immediate cause of the ropy or slimy condition is the bacterial formation of gums or mucins. Gums are the more common cause. These are probably galactans produced by fermentation of lactose. Some of the active peptonizing bacteria produce sliminess by formation of mucins, which are combinations of proteins with a carbohydrate radical. Development of sliminess is closely associated with capsule formation.[46,173] The ability to produce slime in milk is rather general among bacteria, and is readily acquired and lost. Epidemics are caused by some members of the *E. aerogenes* group or the lactic streptococci.

Emmerling,[115] in 1900, and Schardinger,[476] in 1902, observed that *E. aerogenes* produced slime in milk. The empirical formula $(C_6H_{10}O_5)_n$ was assigned to the slimy substance. It dissolved readily in water, yielding a gelatinous solution; it was optically inactive and did not reduce Fehling's solution. Hydrolysis with dilute acids yielded a reducing sugar; oxidation with nitric acid yielded both mucic and oxalic acids. The gummy substance was called arabogalactan. The galactan-producing properties of *Bacterium sacchari* and *Bacterium atherstoni* were studied by Smith,[489] who obtained galactose and arabinose upon hydrolyzing the product and upon oxidation obtained mucic and oxalic acids.

Antibiotic Production in Milk

Production of antibiotic-like substances in cultured dairy products has been associated with homo- and heterolactic streptococci, some of the lactobacilli, and *B. linens*. Production may be unwanted and fortuitous, as in commercial starter cultures, or it may be desired and encouraged.

Nisin.—Elaboration by *S. lactis* of a substance inhibitory to *L. bulgaricus* was reported by Rogers[459] in 1928. The substance later named "nisin" produced by some strains of *S. lactis* is a large polypeptide with a molecular weight of approximately 10,000.[360] Lanthionine and a structural isomer of cystathionine, two sulfur-containing amino acids, were recovered from hydrolyzates by Newton *et al.,* who concluded that nisin resembles the antibiotic subtilin.[397] A partially purified preparation (mol. wt. 7,000) by Cheeseman and Berridge lacked amino or carboxyl end groups, but contained side chains with the epsilon amino group of lysine and the imino group of histidine.[66] Baribo and Foster[17] observed that an endocellular inhibitory substance (probably nisin) was liberated when cultures of *S. lactis* were acidified. Cultures boiled or autoclaved for 10 min at pH 4.8 retained their activity, whereas those heated at pH 7.4 rapidly lost about 50%.

Nisin dissolves in aqueous solution at pH 7, 5.6, and 4.2 to the extent of 75, 1,000, and 12,000 μg per ml, respectively.[195] Solubility is substrate-dependent. Nisin assays are based on its inhibitory action toward *Streptococcus agalactiae* in a tube dilution test.[212,359] One Reading unit is defined as that amount of antibiotic preparation dissolved in *N*/20 HCl which gives the same inhibition as a standard preparation. A 0.1% solution of the standard contained 10,000 Reading units, i.e., it was inhibitory at 1:10,000 dilution. Modified tests using litmus milk and *S. cremoris*,[136] and the one-hour resazurin test and *S. cremoris* have been described.[131] Modified microbiological assays have been reported by several workers.[26,343,514] Nisin is distinguished from chemical preservatives by means of its antibacterial spectrum, particularly with respect to its activity toward some yeasts and lactobacilli.[91] Paper chromatography is useful for identification purposes.[490] The subject of nisin is discussed in greater detail in reviews by Marth[349] and Hurst.[227]

Antibacterial Spectrum.—Hawley reported that various species and strains of the genera *Staphylococcus, Streptococcus, Neisseria, Bacillus, Clostridium,* and *Corynebacterium* are inhibited by nisin.[195] Mattick and Hirsch added actinomycetes, pneumococci, mycobacteria, and *Erysipelothrix* to this list.[359] The nisin concentration required for complete inhibition is organism-specific and ranges from 0.25 to 500 units per ml. Inhibition of *L. casei* by antibiotics from *S. lactis* and *S. cremoris* was observed by Baribo and Foster.[17] Inhibition by nisin of *Propionibacterium* but not of coliform bacteria was reported by Galesloot.[135]

Nisin Inactivation.—*L. plantarum* isolates (thermal death time, 5 min at 65°C) from milk and cheese reduced nisin activity in these substrates.[272] *S. faecalis* and *S. lactis* isolates from raw milk destroyed nisin.[134] Galesloot observed that use of nisin-producing starters in cheese manufacture is uninviting if large numbers of group N streptococci are present.[134] Galesloot also observed that some cultures of *Leuconostoc* were antagonistic. Nisin (but not subtilin) destroying nisinase, an enzyme, has been recovered from some strains of *S. thermophilus*.[5]

Relationship between Starter Cultures and Streptococcal Antibiotics.—When antibiotic-producing and nonproducing strains of *S. lactis* and *S. cremoris* from commercial starter cultures were mixed in equal proportions, the antibiotic-producing strains soon predominated.[226] Domination occurred in only a day or two.[76] Emergence of a predominant strain may be accompanied by a loss of starter activity, and certainly renders the starter more susceptible to complete inactivation by bacteriophage.

Heterolactic Streptococci.—Ritter in 1945 noted that 2 strains of *S. lactis* were inhibited by 5 strains of betacocci (*Leuconostoc* sp.) when grown at 20°C.[453] Later, Mather and Babel[356] found that a creaming mix made up in part of skimmilk cultured with *L. citrovorum*, when added to cottage cheese, inhibited such spoilage organisms as *P. fragi*, *Pseudomonas putrefaciens* or coliforms but not the yeasts *G. candidum* and *Candida pseudotropicalis*. Marth and Hussong[352] showed that filtrates from cultured skimmilk, in which 4 strains of *L. citrovorum* were allowed to ferment citrate, inhibited to different degrees *Staphylococcus aureus*, *E. aerogenes*, *A. viscolactis*, *E. coli*, *P. fragi*, and *P. fluorescens*, but in no instance did they inhibit the yeasts *Torula glutinis*, *S. cerevisiae*, *Saccharomyces fragilis*, or *Mycotorula lipolytica*. Dilution of the filtrates to the level at which they might be present in cottage cheese eliminated the inhibition of all bacteria except 1 of 2 strains of *P. fragi* and 1 of 5 strains of *P. fluorescens*.[351] Collins noted that 3 of 6 strains of *S. diacetilactis* formed an antibacterial substance similar to that produced by *S. cremoris*.[76] Some strains of *S. lactis*, *S. cremoris*, and *S. diacetilactis* were inhibited.

These observations suggest that care must be exercised to combine only suitable strains in the compounding of a mixed-strain starter culture.

Lactobacilli.—Kodama[267] isolated an antimicrobial substance designated "lactolin" from *L. plantarum*, and Wheater,[554] a substance designated as "lactobacillin" from organisms resembling *L. helveticus*. "Lactobacillin" inhibited *C. butyricum* in Gruyere cheese.[215] Cultures and a derived filtrate of *L. helveticus* reduced gas formation by *E. coli*[366] and inhibited *Propionibacterium shermanii* and other propionibacteria.[563] Pasteurized cultures and culture filtrates of *L. acidophilus* inhibited growth of *E. coli*.[411] Concentrated sterile culture filtrates prevented, and growing *L. acidophilus* cultures halted development of *E. coli*, *P. fluorescens*, *Shigella* sp., *Salmonella* sp., and aerobic spore-forming bacilli.[421] "Lactocidin," the antibiotic produced by *L. acidophilus*, was isolated by Vincent *et al.*[529] It has a wide spectrum of antibiotic activity. Winkler reported on the inhibition of propionibacteria by milk cultures of *L. acidophilus*.[563] Some lactobacilli, according to DeKlerk and Coetzer,[101] produce substances inhibitory to other bacteria of the same genus.

Antibiotics in Cultured Milks.—Some cultured milks exhibit antibiotic activity, the causative organisms for which are obscure. Thus acidophilus milk is antagonistic to *E. coli* and bactericidal to *Mycobacterium tuberculosis*;[49] yogurt inhibits *Erysipelothrix rhusiopathiae*,[249] *E. coli*,[248] and human, bovine, and BCG strains of *M. tuberculosis*;[510] kumiss is bacteriostatic or bactericidal to *E. coli*, *S. aureus*, *B. subtilis*,

B. cereus, E. aerogenes, and other organisms;[49,486] kefir is inhibitory to *E. coli, S. aureus,* and *B. subtilis;*[49] and "kuränga" inhibits mycobacteria and organisms in the genus *Bacillus.*[67]

Antibiotics from Surface-Ripened Cheese.—An antimicrobial agent attributed to *B. linens* reportedly appears in surface-ripened cheese stored at 2 to 4°C for 8 wk, and is inhibitory to *S. aureus, B. cereus,* and *Clostridium botulinum.* Strains of *B. linens* yielded culture fluids inhibitory to the germination and outgrowth of *C. botulinum* type A spores.[154,155] Organisms other than *B. linens* on the surface of cheese contribute minor antimicrobial activity.[154] The inhibitory substance from *B. linens* withstood heating at 121°C for 25 min.[154] It properties differed from those of nisin.

Frozen Starter Cultures

Although lactic starter cultures undoubtedly were frozen earlier, research interest in the use of this technique to produce cultures for commercial purposes began during the mid-1950's. Initial attempts involved two approaches: (a) inoculation of milk with the desired culture, incubation in the normal manner so that approximately 0.85% lactic acid was produced, cooling the culture, packaging, and then freezing at approximately −29°C; and (b) inoculation of milk with the desired culture followed immediately by packaging and freezing. In the first procedure, neutralization of acid with sodium hydroxide was claimed to improve both survival of bacteria and activity of survivors.

Several investigators claimed success when frozen ripened (incubated to produce approximately 0.85% titratable acid before freezing) or unripened (incubated briefly or not at all before freezing) cultures were used directly to produce bulk cultures or cheese.[8,61,354,400,465,482-484] Some workers acknowledged that lactic cultures, when frozen as described, lost activity during frozen storage.[447,484] Loss of activity and consequent variability in performance of cultures prepared by these techniques prompted development of alternative procedures to provide frozen cultures which are more uniformly dependable. Even though the procedures just described were not entirely successful for preparing conventional lactic starter cultures, it has been claimed that kefir grains can be preserved by freezing and storage at −18°C. According to Toma and Meleghi,[517] grains held at that temperature for 9 months were easily reactivated and provided normal alcohol and lactic acid fermentation.

Present Method.—Commercial producers of frozen starter cultures now: (a) grow the desired culture in a suitable medium; (b) harvest and concentrate the cells by centrifugation; (c) resuspend cells at a desired concentration in a suitable medium; (d) package the culture, and (e) freeze it at −196°C (liquid nitrogen). The culture is then shipped to the user in containers which hold liquid nitrogen or "Dry Ice" so that a low temperature can be maintained during storage. Quantities of frozen material are packaged so the user can prepare his own bulk culture or can directly inoculate milk that is to be made into a fermented food.

Use of the procedure just described: (a) eliminates the need for carrying starters in a dairy plant and thus reduces the hazards of culture failure; (b) enables the manufacturer of cultures to exert greater control of quality so the user is assured of a pure and active culture; (c) avoids repeated transfer of the culture, thereby minimizing problems of strain dominance, and allows the user to mix pure cultures in the exact proportions desired; and (d) enables the user to store cultures which are nearly instantly active.

Among the first to describe a procedure for preparing frozen concentrated cultures were Lamprech,[293] Foster,[128] and Lamprech and Foster.[294] In their process they grew *S. lactis* or *S. diacetilactis* separately at 25°C in a tryptone-yeast extract-glucose-magnesium phosphate medium. Cells early in the maximum stationary phase (10 to 15 hr incubation) were recovered by centrifugation, resuspended in sterile skim milk to a concentration of 25 to 55 × 10^9 cells per milliliter, and the pH was adjusted to 7.0. The suspensions were then frozen and stored at −20°C. Under these conditions the cultures retained sufficient viability and activity during 10 months of storage to serve satisfactorily as direct inoculum for making buttermilk. Shortly after the reports by Lamprech and Foster,[293,294] Cowman and Speck[84,85] noted less loss of viability and greater retention of proteolytic activity and of the ability to produce acid when concentrates of *S. lactis* suspended in skimmilk were frozen and stored at −196°C (liquid nitrogen) rather than at −20°C. Other investigators also have reported excellent retention of viability and activity by different lactic cultures when they were frozen and stored at −196°C. Included are single- and mixed-strain cultures of *S. lactis,*[24,25,441] mixtures of *L. citrovorum* and *S. diacetilactis,*[530] and single strains of *S. lactis, S. cremoris,* and *S. diacetilactis*.[143,257]

Recently, Vallea and Mocquot[524] reported a process to prepare frozen concentrated cultures of *L. helveticus* and *S. thermophilus*. In their procedure the cultures are grown in cheese whey fortified with papain, yeast extract, manganese sulfate, dried milk, and corn steep liquor; the pH is controlled during growth at 5.0 to 6.5, depending on the organism; cells are recovered by centrifugation and resuspended in skimmilk; and suspensions are frozen and stored at −30°C. Continuous pH control during growth of *S. cremoris* to be used for frozen concentrated cultures was advocated by Peebles *et al.*[417] to obtain maximum cell yields. Higher prefreezing populations (approximately 10^{10}/ml) were obtained when ammonium hydroxide rather than sodium hydroxide was added during growth to neutralize the acid. Concentrates of some *S. cremoris* cultures retained activity for 231 days when frozen and stored at −196°C. Frozen (−196°C) concentrated cultures of *L. citrovorum* grown in a tryptone-yeast extract-glucose-citrate broth medium also retained viability and the ability to produce biacetyl during 30 days of storage in liquid nitrogen.[145] Facilities needed and procedures used to commercially produce frozen (−196°C) concentrated starter cultures have been described by Ziemba.[576]

Other Methods.—The methods of commercial production and distribution of starter cultures just discussed require: (a) liquid nitrogen for freezing and temporary storage; (b) special containers to hold cultures in liquid nitrogen or "dry ice" during distribution, and (c) a distribution system to deliver cultures promptly to the user. Although facilities to handle cultures this way are now available in the U.S., they are not in other parts of the world. Consequently, work is continuing on procedures to preserve lactic cultures by storing them at temperatures in the range of −20 to −40°C. Kawashima *et al.*[253,254] and Kawashima and Maeno[251,252] completed extensive studies in which they stored lactic-acid bacteria at −15 to −20°C. Their results showed that: (a) *S. lactis* retained greater activity than did *L. bulgaricus* after 6 months of storage, although differences disappeared on subculture; (b) satisfactory preservation of yogurt cultures (*L. bulgaricus* and *S. thermophilus*) was obtained when they were grown in sterile skimmilk fortified with 0.5% calcium carbonate and then were diluted with sterile skimmilk before freezing; (c) cultures of *S. lactis, S. cremoris,* or *S. thermophilus* in skimmilk containing 10% total solids exhibited increased resistance to frozen storage and more rapid growth after thawing when L-glutamic acid was added, though corn-steep liquor, yeast extract, and glucose were not beneficial; (d) addition of L-glutamic acid provided no protection to *L. bulgaricus* or *L. acidophilus* during frozen storage, but stimulated their growth when the cultures were thawed.

Stadhouders and van der Waals[498] prepared *S. diacetilactis, L. citrovorum,* and mixtures of the two for freezing by subculturing them 5 times in reconstituted skimmilk and then freezing and storing ripened cultures at −20 or −40°C. After storage for 20 weeks, approximately 50% of the initial activity was lost. In contrast, when the ripened culture was transferred to skimmilk before freezing, from 17 to 31% of the activity was lost, depending on the heat treatment received by the milk before freezing (loss of activity was less in steamed than in pasteurized milk).

When mixed cultures are frozen, there may be changes in the proportion of each organism that will remain viable under different conditions. Jabarit[231] froze a mixed culture composed of *S. lactis* and *L. acidophilus* used to manufacture bioghurt (a German yogurt-like food). When the culture was stored at 0°C, 53% of the viable cells were *S. lactis* and the remainder were *L. acidophilus.* At −30°C 57% were *S. lactis,* but at −35 and −40°C the proportions of the two organisms were approximately equal. The total viable population at 0, −20, and −35°C were 10^{10}, 10^{10}, and 10^9 per ml, respectively.

Production of concentrated cultures and storage at −20 to −40°C also has been advocated by several investigators. Usually, the procedures differ somewhat from those originally proposed by Lamprech and Foster.[294] Two examples of differences are as follows. (a) Growth of lactic streptococci on tryptone-lactose agar, followed by preparation of cell suspensions, recovery of cells by centrifugation, resuspension of cells in glycerol:water (1:1), and storage at −30°C; such cultures when held for up to 8 months did not lose viability or activity and were suitable for use as direct inoculum in cheese making.[2] (b) Growth of cultures in papain-digested milk enriched with yeast extract and lactose, recovery of cells by centrifugation, suspending the concentrate in a glycerol-skimmilk mixture, freezing, and storage at −30°C until the concentrate is used for commercial purposes.[2] Similar processes with slight additional modifications also have been described by Jansen *et al.*[235] and Stadhouders *et al.*[499]

Interaction of Starter Cultures and Food-borne Pathogens

Cultured dairy foods seldom cause food-borne illness in the consumer. If an active starter culture is used, common food-borne pathogens, even if present in the milk, do not grow well and often are inactivated during the fermentation or early during the storage life of the product. Even if some cultured products are recontaminated after manufacture, pathogens do not survive well. Several examples will illustrate this point.

Goel *et al.*[149] added *E. coli* and *E. aerogenes* to commercially prepared yogurt, buttermilk, sour cream, and cottage cheese and then stored the products at 7.2°C. Viable coliforms were nearly always absent from yogurt after 1 or 2 days of storage. They persisted somewhat longer in buttermilk and sour cream, whereas cottage cheese seldom had a deleterious effect on the bacteria. Minor and Marth[378] did similar tests with *S. aureus* and found that when less than 1,000 cells were added per gm of yogurt, sour cream, or buttermilk, viable *S. aureus* seldom could be recovered from the products after 1 or 2 days of storage at 7°C. Use of a higher initial inoculum generally resulted in survival of *S. aureus* for 4 to 5 days, regardless of the product. Park and Marth[413] prepared a series of cultured milks which also contained *Salmonella typhimurium*. Survival of salmonellae in the products stored at 11°C ranged from less than 3 days to more than 9 days, depending on species of starter culture, strain of a given species, level of inoculum used to prepare the cultured product, temperature at which the product was cultured, and amount and speed of acid production. In other studies Park *et al.*[414] noted that *S. typhimurium* survived for up to 7 or 10 months in Cheddar cheese made with a slow acid-producing starter culture and stored at 13 or 7°C, respectively. In contrast, Goepfert *et al.*[150] and Hargrove *et al.*[187] found that *S. typhimurium* survived from 3 to 7 months in Cheddar cheese made with "normal" starter cultures. Additional information on the behavior of salmonellae in fermented dairy foods can be found in a review by Marth.[347]

Results of tests on survival of food-borne pathogens in cultured products suggest that the starter culture is an important factor in determining inhibition of the pathogen in the food. The starter culture also is important in governing growth of the pathogens, if present in milk, during fermentation. Some of the information on *S. aureus* has been reviewed by Minor and Marth.[377]

Reiter *et al.*[443] showed that growth of *S. aureus* in raw, steamed, and pasturized milk was inhibited by a lactic starter culture. When they neutralized the lactic acid as it was produced, inhibition of staphylococcus was still evident. Jezeski *et al.*[239] also observed that growth of *S. aureus* in steamed or sterile reconstituted nonfat dry milk was inhibited by an actively growing *S. lactic* culture. Enterotoxin was detected in *S. aureus-S. lactis* mixed cultures when *S. lactis* was inactivated by bacteriophage, but not when the lactic streptococcus grew normally.

Park and Marth[412] inoculated skimmilk with *S. typhimurium* and with different lactic-acid bacteria. They noted that: (a) *S. cremoris* and *S. lactis* and mixtures of the two repressed growth but did not inactivate *S. typhimurium* during 18 hr incubation at 21 or 30°C when the lactic inoculum was 0.25%. An increase in inoculum to 1% resulted in inactivation of *S. typhimurium* by some of the mixed cultures during incubation at 30°C. Both *S. diacetilactis* and *L. citrovorum* were less inhibitory to *S. typhimurium* than were *S. cremoris* or *S. lactis*. When added at the 1% level, *S. thermophilus* was more detrimental to *S.*

typhimurium at 42°C than was *L. bulgaricus*. Mixtures of these two lactic-acid bacteria, when added at levels of 1.0 and 5.0%, caused virtually complete inactivation of *S. typhimurium* between the 8th and 18th hour of incubation at 42°C. Daly *et al.*[94] observed that *S. diacetilactis* inhibited (in most instances more than 99%) growth of the following spoilage or pathogenic bacteria in milk or broth: *P. fluorescens, P. Fragi; Pseudomonas viscosa, Pseudomonas aeruginosa, Alcaligenes metalcaligenes, A. viscolactis, E. coli, S. marcescens, Salmonella senftenberg, Salmonella tennessee, S. aureus, Clostridium perfringens, Vibrio parahaemolyticus,* and *Streptococcus liquefaciens*. Growth of *S. aureus* in a variety of foods and of *P. putrefaciens* in cottage cheese also was inhibited successfully by *S. diacetilactis*.

The mechanism(s) by which lactic-acid bacteria inhibit or inactivate other bacteria is still open to conjecture. Daly *et al.*,[94] Speck,[494] and Gilliland and Speck[146] have cited evidence which suggests that the following may be involved: (a) production of antibiotics such as nisin, diplococcin, acidophilin, lactocidin, lactolin, and perhaps others; (b) production of hydrogen peroxide by some lactic-acid bacteria; (c) depletion of nutrients by the lactic-acid bacteria which makes growth of pathogens difficult or impossible; (d) production of volatile acids; (e) production of acid and reduction in pH; (f) production of D-leucine; and (g) lowering the oxidation-reduction potential of the substrate.

INDUSTRIAL FERMENTATIONS WITH MILK MEDIA

Among purely industrial fermentations, milk and its derived products have not, for historic and economic reasons, received their full share of attention. Decentralization of casein and cheese manufacture in the early days weakened the competitive position of the low-solids by-product, whey, relative to that of grains and molasses. With changing economic and market trends, it is to be expected that the by-products of milk which are intrinsically suited for many industrial fermentations will become more competitive. In times of unusual demand, such as wars produce, these by-products possess a strong industrial potential.

Whey has been and is being used to manufacture lactic acid. Lactose has been and still is the carbohydrate of choice in the manufacture of penicillin. Whey has been used on a large scale for the microbiological synthesis of riboflavin, butanol, and acetone. Utilization of whey to manufacture alcohol, yeast, and fat has been studied. Whey is a suitable substrate for the microbiological synthesis of vitamin B_{12} in several fermentations. Enzymatic digests of casein are used extensively to manufacture antibiotics, and lend themselves to any fermentation which requires a degraded source of amino acids. Casein and the nitrogenous components of whey give rise to large yields of riboflavin in flavinogenesis by means of the fungus *Eremothecium ashbyii*. Skimmilk is recommended as a medium for the microbiological synthesis of the antibiotic, nisin.

Bacterial oxidations may yield useful products. Vinegar and vinegar substitutes may be obtained from whey in acetic acid fermentations. Lactobionic acid may be obtained in high yields by the action of *Pseudomonas graveolens* on the lactose in whey. Uses of fermented whey as a food or beverage are known or have been suggested. The reader who is interested in a more detailed discussion of these fermentations should consult the discussion by Marth.[345]

Production of Lactic Acid

Production of lactic acid utilizing whey sometimes is industrially important. In this fermentation the culture of choice is *L. bulgaricus* because: (a) it is homofermentative, producing almost theoretical yields of lactic acid; (b) it is thermophilic and, having an optimum growth temperature between 45 and 50°C, it can be grown in pasteurized rather than sterile media with little danger that the medium will become contaminated; (c) it is acid-tolerant and, in a batch process, periodic neutralization of the medium is required with relative infrequency; and finally (d) it grows under either aerobic or anaerobic conditions.

The fermentation with *L. bulgaricus* is likely to be sluggish in whey, and hence the bacterium is sometimes grown in association with a yeast (*Mycoderma*). The function of the yeast is not clearly defined; perhaps it produces stimulatory growth factors for *L. bulgaricus*—an organism which is highly fastidious in its nutritional requirements. In this connection it should be remembered that certain strains of *L. bulgaricus* are unusual in that they cannot use the pyrimidine derivative uracil, but instead require orotic acid; nor can the species in general use pantothenic acid, but instead requires pantetheine; and finally, the species cannot utilize biotin, but instead requires such unsaturated fatty acids as oleic or linoleic. In connection with mineral requirements, *L. bulgaricus,* like other species in this genus, probably requires for maximum growth rates manganese in excess of the quantity usually present in milk. The fermentation proper has been described in some detail.[55,399,408]

Butanol-Acetone-Riboflavin Fermentation

Beside volatile solvents, fermentations with *C. acetobutylicum* yield riboflavin in significant quantities. Optimal content of iron, between 1.5 and 2.0 ppm,[363] and either certain salts of organic acids or calcium carbonate[570,571] is required for maximal synthesis of riboflavin and maximum fermentation rates.

The butanol-acetone fermentation, utilizing whey supplemented with yeast extract, is substantially a butanol fermentation. Approximately 80% of the volatile solvents produced thereby is butanol, 13% is acetone, and 5% is ethanol. Approximately 30% of the lactose fermented is converted to butanol, 5% to acetone, 2% to ethanol, and the balance largely to CO_2 and small quantities of butyric and acetic acids, acetylmethylcarbinol, and hydrogen.[314]

C. acetobutylicum is not too exacting in its growth requirements. Asparagine is needed to effect normal production of solvents in what otherwise would be an acid fermentation. Both biotin and *p*-amino benzoic acid are required in trace quantities, 0.001 and 0.05 μg per ml, respectively. Iron is essential in minimal but variable quantities for attainment of maximal fermentation rates.[313-315,363,365] The requirement for iron varies with the composition of the medium; potassium is also required.[97,98] Trace quantities of manganese sulfate, lithium chloride, strontium chloride, tin chloride, and zinc chloride aid fermentation in whey.[363]

Leviton and Burkey[318] added yeast extract, liver extract, or cornmeal to whey, thus achieving normal fermentation and avoiding addition of iron. The presence of these added solids (1%) in whey assures normal fermentation and high yields of riboflavin.

The physiological state of the organism influences the yield of riboflavin, as well as that of volatile solvent, and consequently requires control.[313] Transfer of inocula when approximately 25% of the gaseous products of fermentation have envolved is conducive to high yields.

Production of Alcohol

Although lactose-fermenting yeasts have been known for some time, their possible utilization to produce ethyl alcohol and yeast from whey received serious study much later. Browne[45] observed that certain *Torula* species yielded more alcohol than might have been expected from statements in the literature. Working with four kefir yeasts, two *Torula* species, one of *Torulopsis,* and one additional yeast species, he obtained alcohol in yields 68 to 80% of theoretical. A maximal yield, 80.3%, based on a theoretical yield of 4 moles alcohol per mole lactose fermented, was obtained with a strain of *Torula cremoris* in a 21.7-hr fermentation at 30 to 32°C.

Continuing Browne's work, Rogosa *et al.*[461] extended the scale of operation, and employed in addition to *Torula* species, *S. fragilis, Saccharomyces lactis, Saccharomyces anamensis, Zygosaccharomyces lactis, Mycotorula lactis,* and *C. pseudotropicalis*. Again *T. cremoris* gave the highest yields. The yield of alcohol averaged 90.73% under laboratory conditions and 84% under pilot-plant conditions.

Alcoholic Beverages

Whey supplemented with malt wort has been used as a raw material to prepare beer. Dietrich[106] added 5.4% malt wort to dilute whey (2.5% solids), precipitated the albumin at 90°C, and filtered the mixture. He then inoculated the filtrate with a strain of the yeast *S. lactis,* and after 5 to 7 days obtained a product with a true beer taste and character.

Whey may be fortified with sucrose, and fermented with yeast to yield an alcoholic whey. Upon freeze concentration a whey liquor with 10 to 69% alcohol was obtained.[119] The alcohol fermentation carried out in whey supplemented with brown sugar yields a whey cordial.[15]

Microbiological Fat Synthesis

Limiting Factors.—Pioneering research showed that an ample and sometimes a critical supply of oxygen is necessary for fat synthesis.[324] It was incorrectly assumed that only in surface culture—more suited for growth of fungi than of yeast—could the necessary supply of oxygen be maintained.[322–325]

Fat was synthesized only within a definite pH range which varied with the medium and the type of organism.[30,451] A high concentration of assimilable carbohydrate and a relatively low concentration of water was conducive to the proliferation of fat-rich cells.[30] A small inoculum, and usually incubation temperatures at the optimum for growth or slightly under, were beneficial. In any given culture, the ratio between cellular fat and protein increased with age. Of utmost importance was the observation that nitrogen is required for fat synthesis.[198,434,477] Since this requirement and that of a high sugar concentration were less exacting for molds than for yeasts, much early research on fat synthesis was devoted to the study of molds.

Inverse relationships were found to apply between the fat coefficient and conversion (protein and carbohydrate) coefficient.[44,200,434] Responding to a nitrogen deficiency, both yeasts and molds synthesized more fat, and less protein and cellular materials. Environmental factors favorable for fat synthesis were those that proved antagonistic for synthesis of protein or carbohydrate.

Fat Synthesis by Yeasts.—Studies with yeasts, particularly those with *Endomyces vernalis,* showed that commercially acceptable yields of fat could be obtained microbiologically.[325] Processes based on surface culture were too costly, and therefore submerged techniques were sought. Nilsson, Enebo, and co-workers,[118,401] working with *Rhodotorula glutinis, Rhodotorula gracilis,* and synthetic media containing invert sugar, pH 4.0 to 4.7, and employing techniques similar to those used in compressed yeast manufacture, obtained yields of dried *R. glutinis* yeast after 3 to 4 days amounting to 31 to 36% of the sugar consumed, and with 25 to 30% fat. Results with *R. gracilis* were even more encouraging. Schulze[480] successfully pursued the problem of producing "fat" yeast on a commercial scale using continuous methods. He first showed that deficiencies of nitrogen and phosphorus in the medium resulted in fat-rich yeasts. Moreover, he found that proliferation of yeast, slow in a nitrogen-deficient medium, proceeded more vigorously when phosphorus was deficient.

Starting with a normal culture isolated from sulfite waste liquor, Schulze, after 3 to 4 weeks of continuous cultivation, obtained a culture containing chiefly fat-rich cells. With this as an inoculum and processing 12,000 l of phosphorus-deficient liquor daily, he obtained 150 kg dried yeast per day, of which 20 to 23% was fat and 25 to 30% protein. The yield based on added reducing substances was approximately 40%, and the generation time was 10 to 12 hr.

Fat Production in Whey Media.—Whey is suitable for growth of fat-rich fungi, and an excellent sulfite waste liquor supplment to grow fat-rich yeasts. There is some question whether whey is sufficiently deficient in phosphorus to support the continuous culture of fat-rich yeasts. Milk serum, and hence whey, contains about 1.4 gm P_2O_5 per l.[12] Yeasts proliferate normally in a medium containing a steady state concentration of 0.7 to 0.8 gm P_2O_5 per l.[480] Hence the P_2O_5 in whey would support a normal, rather than the abnormal, sugar assimilation necessary for fat biosynthesis. However, the possibility exists that calcium or magnesium supplements could be employed to lower the concentration of diffusible phosphate. Some experiments have shown that added calcium salts lead to high fat yields,[271] and this may be related to suppression of phosphate ion.

Whey is superior to other natural media for growth of fat-rich species of the genus *Geotrichum.* Fink and co-workers found only two strains among many which could synthesize fat to a commercially significant degree. These strains of *G. candidum* were distinguished by certain morphological characteristics—a yellow-white convoluted structure of the mat, as compared to the snow-white silk-like texture found in low-fat strains.[124] Almost simultaneously Geffers carried out similar screening tests and noted that only the good producers of fat assimilated lactose.[140]

Although unsupplemented whey is suitable for growing *G. candidum*, fortification with 0.05% KH_2PO_4, 0.02% $MgSO_4$, and 0.1% $(NH_4)_2SO_4$ has been recommended. Fink and co-workers[123,124] obtained 5.7 gm dry material after incubating fortified whey 6 days in Jena culture flasks. The dry substance contained 17.6% crude protein, and 22.5% crude fat.

Fat from *G. candidum* had a pale yellow to pale brown color. Of the two strains studied, one yielded a product with a Vaseline-like consistency, and the other a fat which was liquid at room temperature. The fat characteristics were not only dependent on strain, but also on conditions of culture.

Advantages accruing to the use of mixed cultures were reported by Fredholm,[129] who observed that symbiosis of *G. candidum* and various lactic-acid bacteria such as *S. lactis, S. cremoris,* and *L. citrovorum* yielded 44.7 gm dry cells (42% fat) per 100 gm lactose plus lactic acid, compared with 47.2 gm (25.5% fat) obtained in control experiments. Inoculation with lactic-acid bacteria was followed by 40 hr inoculation with *G. candidum*.

Xylose wort as a basal medium for fat synthesis by *Candida reukauffii* requires supplementation with extracts and mineral salts.[450] Whey is an exceptionally good supplement up to concentrations of 10% in a basal medium (pH 6.8) containing about 2% xylose, 0.2% K_2HPO_4, 0.2% K_2SO_4, 0.05% $MgSO_4$, 0.005% $FeSO_4$, 0.012% $(NH_4)_2HPO_4$, and 1% wheat bran extract. Addition of 10% whey effected an increase in dry yeast production from 3.60 to 10.42 gm per l, an increase of fat from 0.62 to 2.57 gm per l, and an increase of protein from 0.62 to 1.05 gm per l. Testifying to the efficient utilization of sugar in the presence of whey were the exceptionally high conversion and fat coefficients of 55.7 and 13.76, respectively. The fat coefficient closely approached the theoretical maximum of 15, and such high yields were obtained in 48 hr with an inoculum of only 0.1 gm dry cells per l. In experiments with media containing progressively increasing amounts of sugar, all three coefficients—fat, protein, and conversion—decreased progressively, contrary to the expected inverse relationships among the coefficients. However, if time was allowed for nearly all the sugar to be consumed, the anticipated relationships were observed.

The yeast oil had the following characteristics: consistency, fluid; sp. gr., 0.921; unsaponifiable matter, 0.21; iodine no., 70; saponification no., 197. In these characteristics the oil approximated those associated with olive oil and the oils obtained with *E. vernalis* and *Oospora* species.

Fat Synthesis in Whey with Molds and Mold-like Fungi.—Geffers, working with unsupplemented whey and with a selected strain of *Oospora wallroth,* obtained 3 to 5 kg fat and 10 to 12 kg dry substance from 1,000 l of whey after lengthy incubation.[140]

Schulze[480] used whey and a species of *Trichosporon* (synonymous with *Oospora moniliaformis*) and obtained after 5 days incubation at 24°C 47.6 gm dry substance per 100 gm sugar, 12.4 gm crude fat, and 7.6 gm crude protein.

Wix and Woodbine grew *Aspergillus ustus* in whey fortified with 1.14gm NH_4NO_3/l. Incubated on a shaker, the mold used up 96% of the lactose, yielding 17 gm mycelial felt per l with the composition 13% protein and 28% fat.[564] Results with *Penicillium frequentans* were not encouraging.

Lipid Composition.—The higher fatty acids predominate in the fat from both yeasts and molds, and the unsaturated predominate over the saturated acids. Thus *Penicillium javanicum* reportedly contains 60.8% and 30.8% unsaturated and saturated acids, respectively;[543] *G. candidum* 53.1 and 42.8%, respectively;[250] and yeast-fat 50.5 and 15.4%, respectively.[513] Palmitic and stearic acids are the chief saturated acids, and oleic acid is the chief unsaturated acid. The latter was found in fat from yeast, *P. javanicum,* and *G. candidum* to the respective extents of 47.6, 31.7, and 41.2%, and linoleic acid was found to the extents of 2.9, 29.1, and 11.4%. Linolenic acid was not found in yeast-fat, but occurred in samples from *G. candidum* to the extent of 0.12%.

Astonishingly large percentages of phosphatides in yeast grown under controlled conditions have been reported. Of the 2.87% total fat in beer yeast, Salisbury and Anderson found that 58% consisted of phosphatides.[473]

Yeast fat is an excellent natural source for ergosterol and related sterols.[557,558] Ergosterol is also found in the fat from *P. javanicum.* Characteristic of the unsaponifiable fraction of yeast fat is its rather high concentration of the unsaturated hydrocarbon, squalene.[513] Certain yeasts, especially the red pigmented ones, abound in carotenoids.[64,126,336]

Production of Yeast

Conversion of lactose into edible protein both for animal and human consumption is an inviting prospect especially because of modern trends in nutrition which emphasize the relative importance of protein in diets. Principles underlying the microbiological conversion of sugars to protein have been available for many years. It only remained for dairy technology to work out the details for a commercially feasible process.

Demmler,[102] using whey and a mixed culture containing predominantly *Candida utilis,* described a continuous process to produce yeast in high yields. The fermentation was conducted in a Waldhof-type fermentation tank equipped with a rotating sparger. Under normal operating conditions, an average yield ranging from 13 to 15 gm yeast per l of whey was obtained. In addition, 1.24 gm heat-coagulable whey proteins were obtained in association with the yeast. The drum-dried yeast product contained 59.4% protein, 4.7% fat, 26.6% invert sugar, 9.2% ash, 3.17-3.4% P_2O_5, 8.6% moisture, and 0.2% sulfur. The purine content was lower than the average for other yeasts. The drum-dried product was more digestible than the spray-dried, presumably because cell walls are destroyed in the drum-drying operation. However, preliminary heat treatment before spray-processing eliminated this difference.

The lactose-fermenting yeasts contain the following vitamins in milligrams percent on a dry basis: vitamin A, traces; B_1, 12.8; B_2, 4.4; nicotinic acid, 8.3; ascorbic acid, 7.8; and provitamin A, 40.5.[496]

The mixed culture used by Demmler was derived from a process utilizing sulfite pulp waste liquor. The predominant organism, *C. utilis,* could not utilize lactose. Other organisms present in the culture presumably were responsible for the fermentation in whey. These required aeration for both growth and alcohol production. Furthermore, iron, boron, manganese, potassium, magnesium, sulfur, and phosphorus were needed—all of which are present in whey.

Porges *et al.*,[428] studying the problem of waste disposal, obtained interesting data in connection with yeast production. *S. fragilis* was selected as the lactose-fermenting yeast most suitable for further study. In a laboratory batch process, with a 0.1% dispersion of skimmilk solids, *S. fragilis* assimilated 78% of the lactose and 78% of nitrogenous compounds. Addition of 113 ppm nitrogen in the form of ammonium sulfate per 10 parts nonfat milk solids per l increased lactose utilization in a 24-hr period from 30 to 98%, and yeast production in the range of 17- to 42-fold. Proportionate quantities of ammonium sulfate added to solutions containing 25 and 50 gm nonfat milk solids per l brought about respective increases in yeast production ranging between 13- to 22-fold and 12- to 20-fold. With a 12-l Humfeld fermenter, a 1-liter inoculum, and 2.5% nonfat milk solids or 2.5% whey solids, each supplemented with 281 ppm nitrogen, Porges *et al.* obtained yields based on solids present at the start in a 13- to 16-hr period of 40% yeast (including casein solids) from nonfat milk solids, and 20% yeast from whey solids. In a continuous fermentation, employing a 2-l fermenter, 50 gm whey solids, 1,060 ppm nitrogen in the form of ammonium sulfate, and 750 ml starter culture grown *in situ,* they recovered yeast in 24% yield based on original solids present in 2 l whey. Based on available sugar, yields of yeast of 35 and 29% were obtained with *S. fragilis* and *T. cremoris,* respectively, in an experiment with clarified whey.

The yeast fermentation has been brought to a high degree of perfection and placed on a reasonable economic basis.[547-550] Peak oxygen requirements of 100 to 120 ml O_2 per l whey per min, corresponding to a solution rate of 1 lb per min, were realized in both laboratory and plant investigations in which specially designed sprayer-agitation combinations were employed.[545-550] Supplementation of whey in laboratory experiments with 0.5 to 1% ammonium sulfate, 0.5% dipotassium phosphate, and 0.1% yeast extract, together with use of a heavy inoculum constituting 25 to 30% of the weight of sugar present, resulted in both maximal assimilation of available carbon and nitrogen, and maximal assimilation rates. Thus the time was reduced from the usual 12 to 24 hr to 3 to 4 hr without impairment of yeast yield or quality (high protein content). Calculation based on the quantity of lactose and lactic acid carbon convertible to yeast carbon showed that a theoretical yield of 27 gm yeast (containing 45% carbon) per l whey was possible. Actual yields 85% of theoretical were obtained. Stated otherwise, about 0.55 lb dry yeast could be obtained per lb lactose.

Based on laboratory findings, subsequent escalation of operations to a pilot plant scale employing an 800-gal propagator was carried out successfully.[547,548] The following table shows a representative balance sheet for growth of *S. fragilis* in a whey medium.

Exceedingly important in the yeast fermentation are the propagators with their aerator-agitator combinations. These and their operation govern the oxygen absorption rate of the medium which must correspond to the peak oxygen demand of the growing culture. Wasserman and Hampson observed a dependency of the oxygen absorption rate on agitator design and speed and aeration rate.[546] With the Waldhof fermenter, good growth was obtained even when the desired oxygen absorption rate (5 millimoles O_2 per l per min) was not realized.

Of the nitrogenous components of whey, yeast utilizes the ammonia nitrogen and about two-thirds of the heat-noncoagulable organic nitrogenous compounds, to the exclusion of the heat-coagulable nitrogenous substances.[544]

In reproducing its own substance, the yeast cell produces an abundance of nucleic acids. Thus, not all the nitrogen in yeast is protein

Table 13.4

REPRESENTATIVE BALANCE SHEET FOR THE GROWTH OF *S. fragilis* IN WHEY MEDIUM IN PILOT PLANT EQUIPMENT[a]

	Batch 1	Batch 2
Volume whey, gal	450	600
Lactose (4.86% of whey), lb	182	242
Lactose disappearing, lb	182	242
Volume seed yeast, gal	81	150
Final volume in tank, gal	530	800[b]
Gross weight of yeast yield (dry), lb	140	215
Net weight of yeast yield, lb	74	105
Theoretical yeast yield (55% of sugar weight), lb	100	133
% theoretical	74	79

[a] From data of Wasserman, *et al.*[548]
[b] 50 gal water added accidentally.

nitrogen, although calculation of protein concentration is based on this assumption. An estimated 20 to 40% of bacterial nitrogen is considered to belong to nucleoproteins.[29]

Amino Acid and Vitamin Composition of S. fragilis.—*S. fragilis* grown in whey can contain 50% protein (assuming all the nitrogen is protein nitrogen) with an amino-acid composition differing very little from yeasts in general. Of the total amino acids in dried cells an appreciable fraction (28% nitrogen) is extractable when cells are treated successively with trichloroacetic acid, ethyl alcohol, ether-alcohol, and trichloroacetic acid. Histidine is largely extractable, whereas serine and valine remain with the protein fraction. The quantity of amino acids in the extractable, but not the protein, fraction increase if the yeast is grown in media supplemented with 1% $(NH_4)_2SO_4$; in general the quantity and composition of extractables are influenced by yeast strain, composition of the medium, and age of the yeast. The protein fraction is rich in lysine, aspartic acid, and glutamic acid and deficient in sulfur-containing amino acids. Arginine, thereonine, serine, glycine, alanine, valine, isoleucine, and tyrosine are found in respectable concentrations.

Wasserman found the following vitamins and their concentrations in μg per gm in *S. fragilis*: thiamine 24.1, pyridoxine 13.6, riboflavin 36.0, niacin 280.0, folic acid 5.8, pantothenic acid 67.2, *p*-aminobenzoic acid 24.2, biotin 2.0, choline 6,710, and inositol 3,000.

Lactose in the Production of Penicillin

The reason for the startling increase in demand for lactose during

World War II was the discovery that this carbohydrate was uniquely suitable to produce penicillin in high yields. The demand for lactose continued to increase largely in connection with an expanding penicillin industry. However, the competitive position of lactose has been weakened by the discovery that dextrin can be used as a substitute, and that with adequate control of pH other substrates may be used to advantage.

High yields of penicillin are obtained with the molds, *Penicillium notatum* and *Penicillium chrysogenum,* when rapid mycelial growth during the initial phase of the fermentation is followed by a period of slow fermentation. The initial phase can be accelerated by addition of 0.5% glucose to the usual medium containing 2% corn-steep liquor and 2% lactose.[269] Jarvis and Johnson[236] obtained optimum yields with a medium containing 3 parts lactose per 1 part glucose. The advantage accruing from the use of lactose is believed to result from its slow utilization by the fungus. Since the lactic acid in corn-steep liquor is oxidized before lactose, the increase in pH is more rapid; and autolysis of the mycelium is retarded because a longer time is required for exhaustion of nutrients. The inferiority of glucose is related to unfavorable factors resulting from its too-rapid utilization. These are: (a) slow rise of pH during penicillin production; and (b) premature exhaustion of carbohydrate and of ammonia nitrogen, leading to untimely autolysis and a cessation of penicillin synthesis.

Moyer and Coghill[381] were the first to state that yields of penicillin in corn-steep liquor media were relatively small when media containing glucose were compared with those containing lactose. This observation was later confirmed by others.[502,505]

With lactose-glucose media, Jarvis and Johnson[236] found that utilization of acetate proceeded at the same rate as utilization of ammonia, and hence the pH remained constant during their utilization. With the disappearance of glucose, utilization of acetate proceeded at a relatively greater rate, and the pH rose. Lactate was metabolized much less rapidly than ammonia when glucose was decomposed, and only slightly more rapidly during the decomposition of lactose. Applying these observations, Jarvis and Johnson achieved adequate control of pH during fermentation by the simultaneous use of acetate and lactate in lactose-glucose media.

Whey has been utilized with good results to produce penicillin by surface-culture methods.[16] Its commercial utilization as a source of lactose in submerged cultures is objectionable on several grounds: foaming becomes difficult to control, and the enrichment and purification procedures become unduly involved.

Microbiological Synthesis of Nisin

Nisin is distinguished from the vast majority of antibiotics in that it is an assimilable polypeptide that can be tolerated in large dosage by humans, and appears to be without influence on the intestinal flora. It is found in cultured milks made with *S. lactis* cultures as well as in raw milk and in milk products such as cheese. Interest has centered around its ability, when present in cheese, of minimizing (although not in all instances) blowing and inhibiting butyric organisms.[23,58,111,137,213,214,273,438,463,520]

Nisin has been applied successfully in preparing sterile beverage quality chocolate milk. The antibiotic serves as a sterilization aid because it inhibits outgrowth of heat-damaged spores and so permits use of less drastic heat treatments for sterilization.[202]

Skimmilk is a suitable medium to produce nisin.[196,565] It is inoculated with an active strain of *S. lactis,* and after 40 to 48 hr, during which pH values between 4.5 and 5.5 are established, the coagulated proteins containing nisin are separated by centrifugation. This preparation is useful commercially. It may be dried and the nisin extracted with acidified acetone. Methods for further purification are given by Cheeseman and Berridge.[65] In a patent, Hawley and Hall[196] describe a process in which sterilized skimmilk is cultured with *S. lactis* until the titer of nisin at pH 6.0 to 6.3 reaches 1,000 Reading units per ml. Paracasein is precipitated with $CaCl_2$ and rennin, and the resulting whey is adjusted to pH 4.0 to 4.5 with HCl and drained. The combined whey and curd washings adjusted to pH 5.0 are transferred to a circulating system of vertical foam tubes and 0.1% Tween is added. The collected foam contains 40,000 Reading units per ml. Solid nisin is prepared by saturating 500 ml foam with 27 ml acetone. The resulting precipitate is extracted with 500 ml methanol, and the nisin in the extract is precipitated with 1,000 ml acetone. The dried precipitate has an activity of 1.4×10^6 Reading units per gm.

Production of Vitamins

Microbiological Synthesis of Riboflavin.—Three types of microorganisms can synthesize riboflavin in commercially significant quantities. Bacilli of the species *C. acetobutylicum* produce quantities up to 50 mg per l. The yeast *Candida guilleirmondi,* and related species, synthesize it under suitable conditions in quantities exceeding 100 mg per l. The yeast-like fungi *Ashbya gossypii* and *E. ashbyii* are the most productive, and under proper conditions will synthesize riboflavin in quantities up to 2.4 gm per l. The *Eremothecium* and *Ashbya* fermentations have another decided advantage, since flavinogenesis is not inhibited by trace quantities of iron—a problem encountered to an intolerable degree with *C. guilliermondi* and to a lesser degree with *C. acetobutylicum*. Riboflavin synthesis in the acetone-butanol fermentation by means of *C. acetobutylicum* was discussed earlier in connection with production of butanol.

C. guilliermondi and related species will not produce significant quantities of riboflavin in the presence of more than 0.1 ppm iron.[53] This limitation is severe and practically limits the *C. guilliermondi* fermentation to synthetic media.

Of the related species, *E. ashbyii* and *A. gossypii*, only the former utilizes the nitrogenous substances of whey to effect synthesis of riboflavin.[314] With *A. gossypii* and a suitable carbohydrate, luxuriant growth but practically no riboflavin synthesis occurs in whey media.[314]

Whey is not a complete medium for the growth of *E. ashbyii*. Neither lactose nor galactose is broken down by it, nor can it be adapted to lactose.[314] However, whey acidified to 0.1*N* with HCl, heated for 1 hr at 120°C, and neutralized yields a product suitable as a supplement for malt-extract media, whey, mixtures of these, and presumably media containing one or more of several nitrogenous substances. Fullest utilization of the nitrogenous substances of whey requires abundant and efficient aeration, and this entails a serious foam problem, which persists even when foam depressants are added.

With mild agitation and aeration the nonheat-coagulable nitrogenous constituents of whey constitute the most readily utilized nitrogen fraction. Efficient utilization of both carbohydrates and nitrogenous substances is optimal in media that contain relatively small quantities of these substances, that is, under starvation conditions. Of course poor yields on a volume basis are obtained, yet results indicate an advantage would accrue for continuous fermentation employing small, stationary concentrations of carbohydrate and nitrogenous substances.[314] Maximum yields of 200 mg riboflavin per l of whey were obtained by Leviton and Whittier[321] with hydrolyzed, clarified whey diluted with an equal volume of water. Greater yields resulted if hydrolyzed, clarified whey was used to supplement malt extract.[314]

Examination of the experimental and cultural conditions under which these high yields were obtained discloses the use of media rich in dissimilable carbohydrates and nitrogenous substances, together with lengthy and intensive aeration. Phelps[419] reported yields of 650 mg riboflavin per l in media containing 8 ml skimmilk, 1.75% malt extract, and 0.5% cerelose in 25 ml. Sjöström and Håkansson,[485] using the same culture employed by Leviton but modifying it by serial transfers, obtained yields up to 400 mg per l in clarified whey media supplemented with 1% sucrose.

Hendrickx and deVleeschauwer[206] employed higher concentrations of ingredients and vigorous agitation. Average yields with whey containing 1, 2.5, and 5% sucrose were, respectively, 435, 665, and 755 mg per l. However, by using a mixture of equal parts of whey and skimmilk, and 1, 2.5, and 5.0% sucrose, they obtained average yields of 630, 1,150, and 1,575 mg per l, respectively. The maximum yield obtained with 5% sucrose in a whey plus skimmilk medium was 2,375 mg riboflavin per l. Numerous patents have been issued defining various media which produce high yields.[234,380,464,504,509]

Cultural conditions which must be met to obtain reproducible results and high yields are: control of initial pH within the range of 5.4 to 7.0; preparation and transfer of inoculum according to a consistent and uniform plan; use of temperatures between 25 and 35°C; and avoidance of overheating of heat-sensitive components during sterilization. Hendrickx claims that no loss in activity results when inocula are stored 4 to 5 months at low temperatures, and that cultures 12 to 24 hr old possess maximal activity.[205] Potassium benzyl penicillin (25 units or more per ml) added to pasteurized media prevented their infection during fermentation without impairment of yields.[204]

Riboflavin proportions itself between the mycelium and surrounding medium during synthesis.[314] The partition coefficient remains constant up to the point at which the mycelium disappears, and many spores appear extracellularly. At this point, practically all the riboflavin and large quantities of soluble nitrogenous compounds are excreted into the medium.

Cultures, when plated on a yeast-extract-peptone-agar medium, yield pigmented and white colonies.[205,314,452] Serial plate transfers of cells from a pigment-free colony serve to perpetuate the substrain. Only occasionally is a pale yellow colony observed. Serial transfers of cells from pigmented colonies perpetuate a mixture with increasing proportions of pigmented colonies, until a steady state is reached. However, temporary reversions occasionally have been observed in favor of the pigment-free strain. It is advisable, therefore, to inaugurate a fresh series of transfers periodically either from pigmented colonies or from a lyophilized, productive culture. Temporary reversions have also been observed in favor of the pigmented strain, and on occasions extraordinarily high yields are obtained which are difficult to reproduce.

Much work has been done with *A. gossypii*,[430,431] and detailed laboratory, pilot-plant, and plant-scale procedures for the microbiological production of riboflavin have been published.[430]

Microbiological Synthesis of Vitamin B_{12}.—Microbiological synthesis affords the only known means for bulk production of pure vitamin B_{12} and vitamin concentrates containing it. Several reports concerned chiefly with vitamin B_{12} yields in actinomycetes cultures appeared between 1949 and 1951.[138,139,170,171,475]

A strain of *Bacillus megaterium* was found active with suitable substrates, among which whey was one. Garibaldi *et al.*[139] obtained yields of 0.8 mg per l, corresponding to a glucose consumption of 10 g.

A low cobalt-ion concentration was shown by Hendlin and Ruger[203] to limit synthesis of vitamin B_{12}. Cobalt comprises about 4% of the molecule. Working with 13 cultures, including a strain of *Streptomyces griseus,* unidentified rumen and soil isolates, a strain of *Mycobacterium smegmatis,* and *Pseudomonas* species, Hendlin and Ruger[203] found that addition of 1 to 2 ppm of cobalt ion increased yield by 3-fold.

Leviton and Hargrove[185,186,316,317,320] found that bacteria in the genus *Propionibacterium* elaborated vitamin B_{12}-active substances in concentrations equal to or greater than those reportedly obtained with other organisms. The active compound produced was identified as hydroxocobalamine.

They[319,320] compared lactose and glucose as sources of energy in several vitamin B_{12} fermentations. Employing different strains of *B. megaterium* and several unidentified rumen isolates, they found that lactose brought about higher yields and faster fermentation. With *Streptomyces olivaceus* as the organism, and clarified whey as lactose source, Leviton[314] compared lactose and glucose in enzymatically hydrolyzed casein-yeast extract media, in distillers' soluble media, and in ammonium caseinate media. All media were fortified with Co^{++}. Higher yields were obtained with the lactose-containing media.

In laboratory-scale experiments in which *L. casei* was used symbiotically with *Propionibacterium freudenreichii* in the fermentation of whey, the average yield was 2.2 mg per l and the maximum was 4.3 mg per l.[319]

Production of vitamin B_{12} compounds is not species-specific. All species of the *Propionibacterium* genus, when cultivated under the same conditions, produce active substances, yet in different quantities. *P. freudenreichii* and *Propionibacterium zeae* synthesized sufficient quantities to warrant their consideration for commercial exploitation. As propionic acid bacteria are active during Swiss-cheese ripening, it was anticipated, and actually has been demonstrated, that production of vitamin B_{12} in Swiss cheese is influenced by the same factors that influence its production in pure culture, particularly by the cobalt content of milk.[185,186]

Propionic acid bacteria require, for maximal growth rates, a highly degraded source of amino acids. In caseinate media, and even in peptone media, rates are likely to be relatively slow. For maximal yields of vitamin B_{12}, a high degree of anaerobiosis is not required. Because assimilation is largely anaerobic, a high ratio between vitamin concentration and total cell mass is obtained. Thus this fermentation is particularly suitable for preparation of the pure vitamin, since the cell mass contains all the vitamin and furnishes a highly concentrated initial source for further treatment. As a first step in further treatment, harvested cells may be coagulated and then lysed in a 50% by volume acetone solution, or in mixtures of butyl and ethyl alcohols.[317]

Sewage wastes have been shown to contain as much as 4 ppm of vitamin B_{12}.[110,221,375,376] Although frowned on for aesthetic reasons as a source of vitamin B_{12} for human nutrition, wastes from activated sludge processes may well provide the cheapest source for preparation of vitamin B_{12} concentrates used in cattle feed. Symbiotic growth of lactic and acetic acid bacteria has been recommended for producing sour milk products, biologically enriched with vitamin B_{12}.[469] Acetic acid bacteria cultured in whey fortified with cobalt salts led to an 80-fold increase in vitamin B_{12}. Propionic acid bacteria in skimmilk supplemented with dimethylbenzimidazole increased the vitamin content by 300-fold.[96]

In view of work by Barker *et al.* and Weissbach *et al.*, it appears that the natural cobamide produced in bacterial cultures is not vitamin B_{12}, but rather coenzyme B_{12}.[21,22,551] The main difference in composition between the coenzymes derived from *Clostridium tetanomorphum* and *Propionibacterium* cultures and the corresponding vitamins and pseudovitamins is the absence of the cyano, or hydroxo, groups and the presence of an adenine nucleoside in association with cobalt. The nucleoside contains adenine and a sugar-like compound characterized as D-erythro-2,3-hydroxy-Δ-4-pentenal linked to the N-9 position of adenine.[217] In assays with *E. coli* and *Ochromonas malhamensis*, response to the vitamins and corresponding coenzymes is identical. The hydroxocobalamine vitamin B_{12} can be formed from the 5,6-dimethylbenzimidazolyl coenzyme B_{12}. Consideration of the yields reported by Barker *et al.*[22] for the enzyme and by Leviton *et al.*[319] for the vitamin suggests that the vitamin is derived from the coenzyme during preparation of bacterial extracts for analysis.

Berry and Bullerman[31] and Bullerman and Berry[50,51] described a two-stage process for production of vitamin B_{12} by *P. shermanii*. Maintenance of anaerobiosis (first step) during the first one-half of the fermentation is accompanied by formation of the macroring portion of the B_{12} molecule. During the aerobic phase (second step) in the second one-half of the fermentation, the organism attains its maximal population and also attaches the nucleotide portion and thus completes synthesis of the B_{12} molecule. Use of aerobiosis during the second phase of fermentation precludes addition of the B_{12} precursor (5,6-dimethylbenzimidazole) to the medium.

The process, as outlined by Bullerman and Berry involves: (a) preparation of a medium containing 6 to 8% whey solids, 0.5 to 1% yeast extract, and 15 ppm cobalt; (b) adding a 10% inoculum of *P. shermanii* and holding the temperature at 29°C; (c) adjusting the pH daily so it is returned to 6.5 to 7.0; (d) sparging with CO_2 for 84 hr and then with air for 84 hr (1000 ml air/l/min gave maximum yield of B_{12}); and (e) drying the fermented material. The dried product thus obtained contained 365 μg B_{12} per gm, whereas the maximum yield in the unconcentrated liquid approximated 15 μg per ml.

"Oxidative" Fermentations

Whey does not lend itself to direct production of acetic acid by species of the genus *Acetobacter*. Furthermore, use of combined inocula of yeasts and *Acetobacter* species has not proved fruitful. However, Haeseler[169] has described an operable procedure, in which an alcoholic followed by an acetic acid fermentation yielded a vinegar with satisfactory qualities.

Production from whey of a vinegar with as much as 10% acid seems unlikely because of adverse effects from a high salt concentration; production of a 5 to 7% acid vinegar may prove feasible. However, in the process described by Haeseler, a whey vinegar containing only 4% acid was produced. This product, yellow-brown in color, had a malt-vinegar character with only a weak whey taste and a slight saltiness, which were not detrimental. The possibility of slime formation and over-oxidation with whey as a substrate were considered detrimental to the use of quick vinegar processes. A process for making a vinegar substitute from whey has been claimed in a French patent.[523]

Production of lactobionic acid from lactose through bacterial oxidation of the aldehyde group is of some interest because of the properties of this substance. Lockwood and Stodola,[332] using *P. graveolens*, recovered lactobionic acid in 77% yield from a fermentation mixture containing the following components per l: 96 gm anhydrous lactose, 0.62 gm KH_2PO_4, 0.25 gm $MgSO_4 \cdot 7H_2O$, 2.1 gm urea, 28 gm $CaCO_3$, 5 ml corn-steep liquor, and 0.3 ml soybean oil.

Kluyver *et al.*[265] described a *Pseudomonas* species which produces large yields of lactobionic acid. Villecourt and Blachere[528] have reported that *Bacterium anitratum* does not utilize lactose but oxidizes it.

The sequestrant and emulsifying properties of lactobionic acid suggest a commercial potential for this product. In addition, it is a solubilizing agent for calcium salts. Solutions of calcium lactobionate containing up to 70% salt have been prepared. This product may prove valuable in the pharmaceutical trade as a source of calcium.[247]

Other Fermentations Using Whey

Making whey-cheese is, perhaps, one of the earliest fermentations which used whey (or its components) as a substrate. Examples of such cheese include Schottengsied, Primost (Mysost), Ricotta, and Gjetost (made from goat's milk whey).

Whey has been suggested as a culture medium for growth of lactic bacteria. Czulak[92] reported whey could be used to grow *P. roqueforti* and Lundstedt and Fogg[337] found it suitable for growth of *S. diacetilactis*. They noted further that when citrated whey was cultured with *S. diacetilactis* and added to creamed Cottage cheese, a pleasing biacetyl flavor and aroma developed in 2 to 6 days while the cheese was held under refrigeration.

Use of fermented whey as a food has been suggested. Jagielski[233] combined whey and lactose with an appropriate culture and produced a whey kumiss. Later Krul'kevich[258] mixed equal volumes of whey and buttermilk with kumiss yeasts, *L. bulgaricus,* and *L. acidophilus*. The finished product is claimed to resemble kumiss. A condensed whey food comprised, in part, of whey fermented by *L. bulgaricus* and *P. shermanii* has been described in a patent issued to Meade *et al.*[363]

Other uses for whey based on fermentation include production of: (a) lactase enzyme from *S. fragilis* (or other organisms able to utilize whey) as described by Myers and Stimpson[383] and Wendorff *et al.;*[552] (b) a high-vitamin, high-protein product containing little or no lactose and prepared by fermenting whey with an organism able to utilize lactose (e.g., *S. fragilis*), followed by drying the fermented material;[384] and (c) an animal feed suitable for ruminants by fermenting whey with *L. bulgaricus* at a pH of 5.8 to 6.0, concentrating the fermented whey to 30 to 80% solids, and neutralizing the concentrate to pH 7 to 8.[65]

Attempts to improve the quality of whey include those of Johnstone and Pfeffer,[241] who increased its nitrogen content with a nitrogen-fixing strain of *E. aerogenes,* and Davidov and Rykshina,[96] who used whey fortified with $CoCl_2$ and, after fermenting it with acetic acid bacteria, observed an 80-fold increase in vitamin B_{12}.

Addition of a whey paste plus a nisin-producing strain of *S. lactis* to silage has been suggested by Zeilinger and Binder[574] as a means of preventing development of butyric acid bacteria in the fodder.

DAIRY WASTE DISPOSAL

Wet oxidation of dairy waste is one of the most serious and strenuous tasks microorganisms are called upon to perform. The microbiological system must oxidize the carbon and hydrogen of organic compounds to carbon dioxide and water, respectively, and must at the same time conserve its own mass. In other words, the cellular mass must neither increase nor decrease over long periods of time. That this ultimate objective is closely approached in practice testifies to the remarkable power of the metabolic apparatus of microorganisms.

Dairy wastes fall into two categories, one of which may be described as an intrinsic, and the other as a conditional waste. All dairy plants experience losses that are intrinsically a part of plant operation. For example, a small dairy plant that receives 10,000 lb milk daily may produce each working day about 1,250 gal waste with a milk solids concentration of 0.1%. Cheese plants, on the other hand, produce whey as a by-product of cheese making; although whey contains half the nutrients of the milk from which it was derived, it must be treated as a conditional waste—conditional upon the absence of a suitable market for its use. It has been estimated that some 80% of cheese plants have whey-disposal problems. The principles elaborated for the aeration of wastes are quite general and apply to all kinds of dairy wastes. A more detailed discussion on the disposal of dairy wastes can be found in a review by Arbuckle.[10]

Dairy Waste Treatment by Aeration

The magnitude of the chemical or biological oxygen demand of solutions of organic matter determines whether or not these solutions may be safely added to sources of marine life. Chemical oxygen demand (COD) is the amount of oxygen, determined chemically, necessary for the complete oxidation of an organic substance, and is usually reported in parts per million (ppm).[427] It is practically equal for milk wastes to the ultimate biochemical oxidation demand (BOD).

As oxidants either permanganate or dichromate may be employed under standard conditions of concentration, temperature, and time. These reagents have been studied critically; only the results with dichromate were found to reflect accurately the BOD of dairy wastes.[133]

Aeration techniques will be successful only if oxygen can be supplied at a sufficiently high rate to lower the COD to acceptable values.[282]

Extensive investigations on the bio- and chemical oxidation of dairy wastes have shown that each pound of dry organic matter in dairy waste requires about 1.2 lb oxygen for complete oxidation.[219,220,424] During the period of rapid assimilation bacteria need about 37.5% of their complete oxygen requirement, or 0.45 lb; and in the process, 0.52 lb of new cell material is formed per lb waste solids. To oxidize this newly formed sludge, 0.75 lb oxygen is required, the difference between the oxygen required for complete oxidation of 1 lb waste solids and that required for assimilation. During endogenous respiration at 32.2°C, sludge is consumed at an hourly rate of approximately 1%.[224] Thus, if an amount of sludge equal to 0.52 lb of newly formed cells is to be oxidized in the time t_1 no less an amount of sludge than that given below would be required to maintain this condition:

Equilibrium weight of sludge per lb organic matter = $52/t_1$. If the parts of oxygen required to oxidize the organic matter in one million parts of waste volume—the ppm COD—is known, the total oxygen requirement in pounds for any given waste volume, V, in gallons, is easily calculated. The weight of organic solids is equal to 83.3% of the total oxygen requirement (COD) and hence the equilibrium sludge weight is given by the following equation, thus: sludge $= (52 \times V \times \text{ppm COD} \times 8.34 \times 0.833 \times 10^{-6})/t_1$.

If, for example, a waste volume, V, of 10,000 gal with a ppm COD of 1,500 is to be processed in $t = 20$ hr, the equilibrium sludge weight would be 270 lb. The calculation is oversimplified and is about 10% too low assuming as it does, that endogenous respiration and assimilation occur simultaneously during the entire operation. Actually, there is always a retention time during which cellular substance is consumed without replenishment.

The hourly oxygen requirement for sludge respiration is equal to the sludge dissipation rate multiplied by the lb oxygen (1.44) required for the oxidation of each lb ash-free sludge. The hourly oxygen requirement for assimilation is given by the quotient of total oxygen required for assimilation and the time required to introduce the waste. The hourly oxygen requirement during assimilation is equal to the sum of the two aforementioned requirements, and may be expressed in terms of the volume V, of influent, the ppm COD, the feed time, t_2, and the endogenous respiration time t_1, thus: O_2 (lb per hr) $= (5.2V \times \text{ppm COD} \times 10^{-6})t_1 + (3.13V \times \text{ppm COD} \times 10^{-6})/t_2$.

The equation summarizes some of the arguments and data contained in the literature.[426] The aeration device must be designed to furnish the solution with oxygen at the required rate. The tank must be designed to accommodate the milk waste and the sludge. Allowances must be made for a certain proportion of free space (free-board), and settling space. The design, construction and operation of dairy waste disposal units has been described.[423,426]

Processing of Whey Wastes

Whey solids compared with milk solids contain a greater proportion of lactose, and a much smaller proportion of nitrogen. Consequently in the processing of whey wastes even under conditions of adequate aeration, the rate of assimilation may be limited by the COD-nitrogen imbalance. Jasewicz and Porges[237] observed that when sludge (2,000 ppm COD) was used to treat dilute whey waste (1,000 ppm COD) under highly aerobic conditions no additional nitrogen was necessary for complete whey removal, since the essential nitrogen was supplied during endogenous respiration. Addition of ammonium sulfate to aerators was recommended to compensate for the additional load imposed on them when whey is wasted along with the normal load. In pure whey studies it was found that under the laboratory schedule of daily feedings both supplemented and unsupplemented sludges gradually deteriorated, and presented serious bulking problems after three months. This was taken to indicate that supplementation with nitrogen alone was not enough. Pursuing the problem further, Porges and Jasewicz,[425] in a 61-day study of the COD balance in a system to which whey was added 48 times to aerated sludge, observed that whey wastes may be readily treated under certain conditions without nitrogen addition. An average of 75% of the influent whey COD was relieved, when no provisions were made for removal of sludge from the effluent. The sludge accounted for all but 2 to 3% of the effluent COD. Calculation based on a sludge oxidation rate of 6.3% per day showed that dynamic equilibrium would be possible if 100 units of sludge were used to treat 10 units of whey.

REFERENCES

1. ABELSON, P. H., BOLTON, E. T., and ALDOUS, E., J. Biol. Chem., *198,* 165, 173 (1952).
2. ACCOLAS, J. P., AUCLAIR, J., BONILLANE, C., MOCQUOT, G., ROSSEAUX, P., VALLES, E., and VASSAL, L., Proc. 18th Intern. Dairy Congr., 1E, 275 (1970).
3. ALBRIGHT, J. L., TUCKEY, S. L., and WOODS, G. T., J. Dairy Sci., *44,* 779 (1961).
4. ALFORD, J. A., and FRAZIER, W. C., J. Dairy Sci. *33,* 15 (1960).
5. ALIFAX, R., and CHEVALIER, R., J. Dairy Res., *29,* 233 (1962).
6. ANDERSON, J. A., and HARDENBERGH, J. G., J. Bacteriol., *23,* 59 (1932).
7. ANDERSON, L., and PLAUT, G. W. E., in "Respiratory Enzymes," Ed. by Lardy, H. A., p. 76, Burgess Pub. Co., Minneapolis, 1949.
8. ANDERSON, V. B., Milk Dealer, *52* (1), 47 (1962).
9. ANNIBALDI, S., Proc. Intern. Dairy Congr., B, 545 (1962).
10. ARBUCKLE, W. S., "Byproducts from Milk," Ed. by Webb, B. H., and WHITTIER, E. O., p. 405, Avi Publishing Co., Westport, Conn., 1970.

11. ARONSON, M., Arch. Biochem. Biophys., *39,* 370 (1952).
12. ASSOCIATES of LORE A. ROGERS, "Fundamentals of Dairy Science," p. 28, 2nd Ed., Reinhold Publishing Corp., New York, 1935.
13. BABEL, F. J., J. Dairy Sci., *38,* 705 (1955).
14. BAKER, M. P., and HAMMER, B. W., Proc. Iowa Acad. Sci., *32,* 55 (1925).
15. BALDWIN, ANNA E., U.S. Patent 78,640 (1868).
16. BÄR, F., Pharmazie, *1,* 52 (1946).
17. BARIBO, L. E., and FOSTER, E. M., J. Dairy Sci., *34,* 1136 (1951).
18. BARIBO, L. E., and FOSTER, E. M., J. Dairy Sci., *35,* 149 (1952).
19. BARKER, H. A., "Phosphorous Metabolism," Ed. by McElroy, W. D., and Glass, B., Vol. i, p. 204, 241, Johns Hopkins Press, Baltimore, 1951.
20. BARKER, H. A., KAMEN, M. D., and HAAS, V., Proc. Natl. Acad. Sci. U.S., *31,* 355 (1945).
21. BARKER, H. A., SMYTH, R. D., WEISSBACH, H., MUNCH-PETERSEN, A., TOOHEY, J. I., LADD, J. N., VOLCANI, B. E., and WILSON, R. M., J. Biol. Chem., *235,* 181 (1960).
22. BARKER, H. A., SMYTH, R. D., WEISSBACH, H., TOOHEY, J. I., LADD, J. N., and VOLCANI, B. E., J. Biol. Chem., *235,* 480 (1960).
23. BARTUSKOVA-CERNIKOVA, MARTA, Vysoki Skoly Chem.-tech. v. Praze, Potravin. Tech., *6,* 330 (1962).
24. BAUMANN, D. P., and REINBOLD, G. W., J. Dairy Sci., *47,* 674 (1964).
25. BAUMANN, D. P., and REINBOLD, G. W., J. Dairy Sci., *49,* 259 (1966).
26. BEACH, A. S., J. Gen. Microbiol., *6,* 60 (1952).
27. BELITSER, V. A., and TSYBAKOVA, E. T., Biokhimiya, *4,* 516 (1939).
28. BELOFF, R. L., and STERN, K. G., J. Biol. Chem., *158,* 19 (1945).
29. BELOZERSKII, A. N., Cold Spring Harbor Symposia Quant. Biol., *12,* 1 (1947).
30. BERNHAUER, K., Ergeb. Enzymforsch., *9,* 297 (1943).
31. BERRY, E. C., and BULLERMAN, L. B., Appl. Microbiol., *14,* 356 (1966).
32. BERTHO, A., and GLÜCK, H., Ann., *494,* 159 (1932).
33. BLACKWOOD, A. C., NEISH, A. C., BROWN, W. E., and LEDINGHAM, G. A., Can. J. Res., B., *25,* 56 (1947).
34. BLACKWOOD, A. C., and SIMPSON, F. J., Can. J. Res., C., *28,* 613 (1950).
35. BLANCHARD, M., GREEN, D. E., NOCITO, V., and RATNER, S., J. Biol. Chem., *155,* 421 (1944); *161,* 583 (1945).
36. BOSWORTH, A. W., and PRUCHA, M. J., N.Y. (Geneva) Agr. Expt. Sta., Tech. Bull. 14, p. 43 (1910).
37. BOTTAZZI, V., 16th Intern. Dairy Congr., Proc., B, 522 (1962).
38. BOYER, P. D., LARDY, H. A., and PHILLIPS, P. H., J. Biol. Chem., *146,* 673 (1942); *149,* 529 (1943).
39. BRANDSAETER, E., and NELSON, F. E., J. Bacteriol., *72,* 68 (1956).
40. BRANDSAETER, E., and NELSON, F. E., J. Bacteriol., *72,* 73 (1956).
41. BRANNON, J. M., Milk Plant Monthly, *23,* No. 1, 41 (1934).
42. BREUSCH, F. L., Science, *97,* 490 (1943).
43. BREWER, C. R., WERKMAN, C. H., MICHAELIAN, M. B., and HAMMER, B. W., Iowa Agr. Expt. Sta., Research Bull. 233 (1938).
44. BROCK, T. D., Mycologia, *48,* 337 (1956).
45. BROWNE, H. H., Ind. Eng. Chem., News Ed., *19,* 1272 (1941).
46. BUCHANAN, R. E., and HAMMER, B. W., Iowa Agr. Expt. Sta., Research Bull. 22 (1915).
47. BÜCHER, T., Biochim. et Biophys. Acta, *1,* 292 (1947).
48. BUCHNER, E., BUCHNER, H., and HAHN, M., "Die Zymasegärung, Untersuchungen über den Inhalt der Hefenzellen und die biologische Seite des Gärungsproblems," pp. 18, 20, 31, 34, 42, 58, 100, 125, 141, 149, R.Oldenbourg, München, 1903.
49. BUKANOVA, V. I., Gigieni i. Sanit., No. 8, 32 (1952).
50. BULLERMAN, L. B., and BERRY, E. C., Appl. Microbiol., *14,* 353 (1966).

51. BULLERMAN, L. B., and BERRY, E. C., Appl. Microbiol., *14,* 358 (1966).
52. BULLOCK, D. H., and IRVINE, O. R., J. Dairy Sci., *39,* 1229 (1956).
53. BURKHOLDER, P. R. (to Research Corp.), U.S. Patent 2,363,227 (Nov. 21, 1944).
54. BURTON, K., Biochem. J. (London), *59,* 44 (1955).
55. BURTON, L. V., Food Ind., *9,* 571, 634 (1937).
56. BUSSE, M., and KANDLER, O., Nature, *189,* 774 (1961).
57. BUYZE, G., HAMER, J. A. VAN DER, and HAAR, P. G. DE, Antonie van Leeuwenhoek, J. Microbiol. Serol., *23,* 345 (1957).
58. BYLUND, G., NORRGREN, O., and SJOSTROM, G., Svenska Mejeritidn., *46,* 433 (1954).
59. CAPUTTO, R., and DIXON, M., Nature, *156,* 630 (1945).
60. CARDINI, C. E., PALADINI, A., CAPUTTO, R., and LELOIR, L. F., Nature, *165,* 191 (1950).
61. CARDWELL, J. T., and MARTIN, J. H., Milk Dealer, *48* (12), 58 (1959).
62. CARLSON, A. S., and HEHRE, E. J., J. Biol. Chem., *177,* 281 (1949).
63. CHANDAN, R. C., CARRANCEDO, M. G., and SHAHANI, K. M., J. Dairy Sci., 44, 1161 (1961).
64. CHAPMAN, A. C., Biochem. J., *10,* 548 (1916).
65. CHEESEMAN, G. C., and BERRIDGE, N. J., Biochem. J., *65,* 603 (1957).
66. CHEESEMAN, G. C., and BERRIDGE, N. J., Biochem. J., *71,* 185 (1959).
67. CHUZKOVA, Z., Molochnaya Prom., *19,* 34 (1958).
68. CLARK, V. M., HUTCHINSON, D. W., KIRBY, G. W., and TODD, A., J. Chem. Soc., 715 (1961).
69. CLIFTON, C. E., Antonie van Leeuwenhoek, J. Microbiol. Serol. Jubilee Volume Albert J. Kluyver, *12,* 186 (1947).
70. COHEN, G. N., and MONOD, J., Bacteriol. Rev., *21,* 169 (1957).
71. COHEN, P. P., and HAYANO, M., J. Biol. Chem., *172,* 405 (1948).
72. COHEN, S. S., J. Biol. Chem., *177,* 607 (1949).
73. COHEN, S. S., and RAFF, R., J. Biol. Chem., *188,* 501 (1951).
74. COHEN, S. S., and SCOTT, D. B. M., Science, *111,* 543 (1950).
75. COHN, M., Bacteriol. Rev., *21,* 140 (1957).
76. COLLINS, E. B., Appl. Microbiol., *9,* 200 (1961).
77. COLLINS, E. B., J. Dairy Sci., *55,* 1022 (1972).
78. COLLINS, M. A., and HAMMER, B. W., J. Bacteriol., *27,* 487 (1934).
79. CONNSTEIN, W., and LÜDECKE, K., Ber., *52,* 1385 (1919).
80. CORI, C. F., and CORI, G. T., Proc. Soc. Exptl. Biol. Med., *34,* 702 (1936).
81. CORI, O., and LIPMANN, F., J. Biol. Chem., *194,* 417 (1952).
82. CORI, G. T., SWANSON, M. A., and CORI, C. F., Federation Proc., *4,* 234 (1945).
83. CORMINBOEUF, F. G., Sci. Agr., *13,* 466, 596 (1933).
84. COWMAN, R. A., and SPECK, M. L., J. Dairy Sci., *46,* 609 (1963).
85. COWMAN, R. A., and SPECK, M. L., J. Dairy Sci., *48,* 1531 (1965).
86. COX, G. A., J. Dairy Res., *14,* 28 (1945).
87. CRANE, F. L., HAUGE, J. G., and BEINERT, H., Biochim. et Biophys. Acta, *17,* 292 (1955).
88. CRANE, F. L., HATEFI, Y., LESTER, R. L., and WIDMER, C.,Biochim. et Biophys. Acta, *25,* 220 (1957).
89. CRANE, F. L., and LESTER, R. L., Plant Physiol., Suppl. vii (1961).
90. CUNNINGHAM, A., J. Dairy Res., *4,* 197 (1933).
91. CZESZAR, J., and PULAY, G., 14th Intern. Dairy Congr., Proc., *3,* 423 (1956).
92. CZULAK, J., Australian J. Dairy Technol., *15,* 118 (1960).
93. DACRE, J. C., J. Dairy Res., *20,* 217 (1953).
94. DALY, C., SANDINE, W. E., and ELLIKER, P. R., J. Milk Food Technol., *35,* 349 (1972).
95. DATTA, A. G., and RACKER, E., J. Biol. Chem., *236,* 617, 624 (1961).
96. DAVIDOV, R. B., and RYKSHINA, Z. P., Zhivotnovodstvo, *22,* No. 6, 22 (1960); Abstract in Milchwissenschaft, *16,* 434 (1961).

97. DAVIES, R., Biochem. J. (London), *36,* 582 (1942).
98. DAVIES, R., Biochem. J. (London), *37,* 230 (1943).
99. DAWES, E. A., and FOSTER, S. M., Biochim. et Biophys. Acta, *22,* 253 (1956).
100. DECKER, KARL, "Die activierte Essigsaüre," p. 1, F. Enke Verlag, Stuttgart, Germany, 1959.
101. DeKLERK, H. C., and COETZER, J. N., Nature, *192,* 340 (1961).
102. DEMMLER, G., Milchwissenschaft, *5,* 11 (1950).
103. De MOSS, R. D., BARD, R. C., and GUNSALUS, I. C., J. Bacteriol., *62,* 499 (1951).
104. De MOSS, R. D., GUNSALUS, I. C., and BARD, R. C., Bacteriol. Proc., p. 125 (1951).
105. De RENZO, E. C., HEYTLER, P. G., and KALEITA, E., Arch. Biochem. Biophys., *49,* 242 (1954).
106. DIETRICH, K. R., Brauwissenschaft, No. *2,* 26, (1949).
107. DOLIN, M. I., Arch. Biochem. Biophys., *55,* 415 (1955).
108. DOLIN, M. I., and GUNSALUS, I. C., J. Bacteriol., *62,* 199 (1951).
109. DORFMAN, A., BERKMAN, S., and KOSER, S. A., J. Biol. Chem., *144,* 393 (1942).
110. Drug Trade News, Mar. 21, 1952.
111. EASTOE, J. E., and LONG, JOAN E., J. Appl. Bacteriol., *22,* 1 (1959).
112. EIJKMAN, C., Centr. Bakteriol. Parasitenk., I, *29,* 847 (1901).
113. ELIASBURG, P., Biochem. Z., *220,* 259 (1930).
114. ELSDEN, S. R., "The Enzymes," Ed. by Sumner, J. B., and Myrbäck, K., Vol. 2, p. 791, Academic Press, Inc., New York, 1952.
115. EMMERLING, O., Ber., *33,* 2477 (1900).
116. EMMONS, D. B., McGUGAN, W. A., and ELLIOT, J. A., J. Dairy Sci., *43,* 862 (1960).
117. EMRICH, E., Diss. Techn. Hochschule Muenchen (1932); Abstr. in Milchwirtsch. Forsch., *14,* 94 (1932).
118. ENEBO, L., ANDERSON, L. G., and LUNDIN, H., Arch. Biochem., *11,* 383 (1946).
119. ENGEL, E. R., U.S. Patent 2,449,064 (1948).
120. EOFF, J. R., LINDER, W. V., and BEYER, G. F., Ind. Eng. Chem., *11,* 842 (1919).
121. ERICKSON, R. E., WAGNER, A. F., and FOLKERS, K., J. Am. Chem. Soc. *85,* 1534 (1963).
122. FARMER, R. S., and HAMMER, B. W., Iowa Agr. Expt. Sta. Research Bull. *146* (1931).
123. FINK, H., HAEHN, H., and HOERBURGER, W., Chem. Ztg., *61,* 689 (1937).
124. FINK, H., HAEHN, H., and HOERBURGER, W., Chem. Ztg., *61,* 744 (1937).
125. FINK, H., KREBS, J., and LECHNER, R., Biochem. Z., *301,* 143 (1939).
126. FINK, H., and ZENGER, E., Wochschr. Brau., *51,* 129 (1934).
127. FOSTER, E. M., J. Dairy Sci., *35,* 988 (1952).
128. FOSTER, E. M., J. Dairy Sci., *45,* 1290 (1962).
129. FREDHOLM, H., Kgl. Lartbryksakad. Tid., *80,* 341 (1941).
130. FRIEDEMANN, T. E., J. Biol. Chem., *130,* 757 (1939).
131. FRIEDMANN, R., and EPSTEIN, C., J. Gen. Microbiol., *5,* 830 (1951).
132. FRIEDMAN, M. E., NELSON, W. O., and WOOD, W. A., J. Dairy Sci., *36,* 1124 (1953).
133. FRITZ, A., Milchwissenschaft, *15,* 237, 609 (1960).
134. GALESLOOT, T. E., Neth. Milk Dairy J., *10,* 154 (1956).
135. GALESLOOT, T. E., Neth. Milk Dairy J., *11,* 71 (1957).
136. GALESLOOT, T. E., and PETTE, J. W., Neth. Milk Dairy J., *10,* 141 (1956).
137. GARCIA, F. R., Anales Fac. Vet. Letn., *5,* 171 (1959).
138. GAREY, J. C., and DOWNING, J. F., Abstr. papers, 119th Meeting Am. Chem. Soc., p. 22A, Cleveland (Apr. 8, 1951).
139. GARIBALDI, J. A., IJICHI, K., LEWIS, J. C., and McGINNIS, J. (to U.S.A., represented by Secy. of Agr.), U.S. Patent 2,576,932 (Dec. 4, 1951).
140. GEFFERS, H., Arch. Mikrobiol., *8,* 66 (1937).

141. GERHARDT, P., MACGREGOR, D. R., MARR, A. G., OLSEN, B. C., and WILSON, J. B., J. Bacteriol., *65,* 581 (1953).
142. GIBBS, M., DUMROSE, R., BENNETT, F. A., and BUBECK, M. R., J. Biol. Chem., *184,* 545 (1950).
143. GIBSON, C. A., LANDERKIN, G. B., and MORSE, P. M., Appl. Microbiol., *14,* 665 (1966).
144. GIBSON, D. M., TITCHENER, E. B., and WAKIL, S. J., Biochim. et Biophys. Acta, *30,* 376 (1958).
145. GILLILAND, S. E., ANNA, E. D., and SPECK, M. L., Appl. Microbiol., *19,* 890 (1970).
146. GILLILAND, S. E., and SPECK, M. L., J. Milk Food Technol., *35,* 307 (1972).
147. GIROLAMI, R. L., and KNIGHT, S. G., Appl. Microbiol., *3,* 264 (1955).
148. GLAHN, P. E., and NIELSEN, S. O., Nature, *183,* 1578 (1959).
149. GOEL, M. C., KULSHRESTHA, D. C., MARTH, E. H., FRANCIS, D. W., BRADSHAW, J. G., and READ, R. B., JR., J. Milk Food Technol., *34,* 54 (1971).
150. GOEPFERT, J. M., OLSON, N. F., and MARTH, E. H., Appl. Microbiol., *16,* 862 (1968).
151. GORINI, C., J. Bacteriol., *20,* 297 (1930).
152. GORINI, C., Arch. Mikrobiol., *5,* 123 (1933).
153. GRECZ, N., DACK, G. M., and HEDRICK, L. R., J. Food Sci., *26,* 72 (1961); 27, 335 (1962).
154. GRECZ, N., WAGENAAR, R. O., and DACK, G. M., J. Bacteriol., *78,* 506 (1959).
155. GREEN, D. E., HERBERT, D., and SUBRAHMANYAN, V., J. Biol. Chem., *138,* 327 (1941).
156. GREEN, D. E., MII, S., MAHLER, H. R., and BOCK, R. M., J. Biol. Chem., *206,* 1 (1954).
157. GREENE, V. W., and JEZESKI, J. J., J. Dairy Sci., *40,* 1046 (1957).
158. GREENE, V. W., and JEZESKI, J. J., J. Dairy Sci., *40,* 1053 (1957).
159. GREENE, V. W., and JEZESKI, J. J., J. Dairy Sci., *40,* 1062 (1957).
160. GROBE, B., Biochim. et Biophys. Acta, *30,* 560 (1958).
161. GROSS, N. H., and WERKMAN, C. H., Arch. Biochem., *15,* 125 (1947).
162. GUNSALUS, I. C., J. Bacteriol., *48,* 261 (1944).
163. GUNSALUS, I. C., Abstr. papers, 120th Meeting Am. Chem. Soc., p. 4A, New York (Sept. 3, 1951).
164. GUNSALUS, I. C., and GIBBS, M., J. Biol. Chem., *194,* 871 (1952).
165. GUNSALUS, I. C., and NIVEN, C. F., J. Biol. Chem., *145,* 131 (1942).
166. HAACKE, P., Arch. Hyg., *42,* 16 (1902).
167. HAEHN, H., Brit. Patent 488,464 (July 7, 1938).
168. HAEHN, H., and KINTTOF, W., Ber., *56,* 439 (1923).
169. HAESELER, G., Branntweinwirtschaft, Nos. 1 and 2, 6pp. (1947).
170. HALL, H. H., BENJAMIN, J. C., WIESEN, C. F., and TSUCHIYA, H. M., Abstr. papers, 119th Meeting Am. Chem. Soc., p. 22A, Cleveland (Apr. 8, 1951).
171. HALL, H. H., and TSUCHIYA, H. M. (to U.S.A. as represented by Secy. of Agr.), U.S. Patent 2,561,364 (July 24, 1951).
172. HAMMER, B. W., Iowa Agr. Expt. Sta., Research Bull. 38, (1917).
173. HAMMER, B. W., J. Dairy Sci., *13,* 69 (1930).
174. HAMMER, B. W., "Dairy Bacteriology," p. 372, John Wiley & Sons, Inc., New York, 1948.
175. HAMMER, B. W., and BABEL, F. J., J. Dairy Sci., *26,* 83 (1943).
176. HAMMER, B. W., and BABEL, F. J., "Dairy Bacteriology" 4th edition, John Wiley & Sons, New York, 1957.
177. HAMMER, B. W., and BAKER, M. P., Iowa Agr. Expt. Sta., Research Bull. 81 (1923).
178. HAMMER, B. W., and BAKER, M. P., Iowa Agr. Expt. Sta., Research Bull. 99 (1926).
179. HAMMER, B. W., and HUSSONG, R. V., J. Dairy Sci., *14,* 27 (1931).

180. HAMMER, B. W., and HUSSONG, R. V., J. Dairy Sci., *15,* 220 (1932).
181. HANSEN, R. G., and CRAINE, E. M., J. Biol. Chem., *293* (1954).
182. HARDEN, A., "Alcoholic Fermentation," Longmans, Green & Co., New York, 1932.
183. HARDEN, A., and YOUNG, W. J., Proc. Chem. Soc. (London), *21,* 189 (1905); Proc. Roy. Soc. (London), *B77,* 405 (1906).
184. HARDING, H. A., and PRUCHA, M. J., Ill. Agr. Expt. Sta., Bull. 228 (1920).
185. HARGROVE, R. E., and LEVITON, A., Bacteriol. Proc. p. 21 (1952).
186. HARGROVE, R. E., and LEVITON, A. (to U.S.A., as represented by Secy. of Agr.), U.S. Patent 2,715,602 (Aug. 16, 1955).
187. HARGROVE, R. E., McDONOUGH, F. E., and MATTINGLY, W. A., J. Milk Food Technol., *32,* 480, (1969).
188. HARRIMAN, L. A., and HAMMER, B. W., J. Dairy Sci., *14,* 40 (1931).
189. HARVEY, R. J., J. Dairy Res., *27,* 41 (1960).
190. HASSID, W. Z., "Phosphorus Metabolism," Ed. by McElroy, W. D., and Glass, B., Vol. 1, p. 11, Johns Hopkins Press, Baltimore, 1951.
191. HASSID, W. Z., DOUDOROFF, M., and BARKER, H. A., "The Enzymes," Ed. by Sumner, J. B., and Myrbäck, K., Vol. 1, p. 1014, Academic Press, Inc., New York, 1951.
192. HASTINGS, E. G., MANSFIELD, H., and HELZ, G., Proc. Soc. Am. Bact. (1925); Abstr. in J. Bacteriol., *11,* 77 (1926).
193. HATEFI, Y., LESTER, R. L., CRANE, F. L., and WIDMER, C., Biochim. et Biophys. Acta, *31,* 490 (1958).
194. HATEFI, Y., and QUIROS-PEREZ, F., Biochim. et Biophys. Acta, *31,* 502 (1958).
195. HAWLEY, H. B., Food Manuf., *32,* 370, 430 (1957).
196. HAWLEY, H. B., and HALL, R. H. (to Aplin and Barrett Ltd.) Brit. Patent 844,782 (Aug. 17, 1960), U.S. Patent 2,935,503 (May 3, 1960).
197. HEATH, E. C., HURWITZ, J., and HORECKER, B. L., J. Am. Chem. Soc., *78,* 5449 (1956).
198. HEHRE, E. J., J. Biol. Chem., *163,* 221 (1946).
199. HEHRE, E. J., CARLSON, A. S., and NEILL, J. M., Science, *106,* 523 (1947).
200. HEIDE, S., Arch. Mikrobiol., *10,* 135 (1939).
201. HEINEMANN, B., J. Dairy Sci., *40,* 437 (1957).
202. HEINEMANN, B., STUMBO, C. R., and SCURLOCK, A., J. Dairy Sci., *47,* 8 (1964).
203. HENDLIN, D., and RUGER, M. L., Science, *111,* 541 (1950).
204. HENDRICKX, H., Mededel. Landboushogeschool en Opzoekingsstas. Staat Gent., *26,* 134 (1961).
205. HENDRICKX, H., Mededel. Landboushogeschool en Opzoekingsstas. Staat Gent., *26,* 831 (1961).
206. HENDRICKX, H., and VLEESCHAUWER, A. DE, Mededel. Landboushogeschool en Opzoekingsstas. Staat Gent., *20,* 229 (1955).
207. HENNEBERG, W., Molkerei-Ztg. (Hildesheim), *47,* 2369 (1933).
208. HENNEBERG, W., and KNIEFALL, H., Molkerei-Ztg. (Hildesheim), *47,* 1446, 1474, 1492 (1933).
209. HETTINGA, D. H., and REINBOLD, G. W., J. Milk Food Technol., *35,* 295 (1972).
210. HETTINGA, D. H., and REINBOLD, G. W., J. Milk Food Technol., *35,* 358 (1972).
211. HILLS, G. M., Biochem. J. (London), *37,* 418 (1943).
212. HIRSCH, A., J. Gen. Microbiol., *4,* 70 (1950).
213. HIRSCH, A., and GRINSTED, E., J. Dairy Res., *21,* 101 (1954).
214. HIRSCH, A., GRINSTED, E., CHAPMAN, H. R., and MATTICK, A. T. R., J. Dairy Res., *18,* 205 (1951).
215. HIRSCH, A., McCLINTOCK, M., and MOCQUOT, G., J. Dairy Res., *19,* 179 (1952).
216. HISCOX, E. R., and LOMAX, K., Ann. Appl. Biol., *11,* 503 (1924).
217. HOGENKAMP, H. P. C., and BARKER, H. A., J. Biol. Chem., *236,* 3097 (1961).

218. HOOD, E. G., GIBSON, C. A., and BOWEN, J. F., Can. Dairy Ice Cream J., *28,* No. 2, 27 (1949); No. 6, 27 (1949).
219. HOOVER, S. R., JASEWICZ, L., and PORGES, N., Sewage Ind. Wastes, *24,* 1144 (1952).
220. HOOVER, S. R., and PORGES, N., Sewage Ind. Wastes, *24,* 306 (1952).
221. HOOVER, S. R., JASEWICZ, L., PEPINSKY, J. B., and PORGES, N., Sewage Ind. Wastes, *24,* 38, (1952).
222. HORECKER, B. L., and MEHLER, A. H., "Annual Review of Biochemistry," Vol. 24, p. 207, Annual Reviews, Inc., Stanford, 1955.
223. HORECKER, B. L., and HEPPEL, L. A., J. Biol. Chem., *178,* 683 (1949).
224. HORECKER, B. L., and SMYRNIOTIS, P. Z., J. Biol Chem., *193,* 371 (1951).
225. HORECKER, B. L., and SMYRNIOTIS, P. Z., J. Am. Chem. Soc., *74,* 2123 (1952); *75,* 1009 (1953).
226. HOYLE, MARGERY, and NICHOLS, A. A., J. Dairy Res., *15,* 398 (1948).
227. HURST, A., J. Milk Food Technol., *35,* 418 (1972).
228. HURWITZ, J., Biochim. et Biophys. Acta, *28,* 599 (1958).
229. HUSS, H., Centr. Bakteriol. Parasitenk., II, *20,* 474 (1908).
230. I. G. Farbenindustrie, German Patent 727,555 (Oct. 1, 1942).
231. JABRAIT, A., Lait, *49,* 520 (1969).
232. JACKSON, H. W., and HUSSONG, R. V., J. Dairy Sci., *41,* 928 (1958).
233. JAGIELSKI, V., U.S. Patent 117,889 (1871).
234. JAMES, R. M. (to Commercial Solvent Corp.), U.S. Patent 2,498,549 (Feb. 21, 1950).
235. JANSEN, L. A., STADHOUDERS, J., and HUP, G., Voeding en Techniek, *3,* 407 (1969).
236. JARVIS, F. G., and JOHNSON, M. J., J. Am. Chem. Soc. *69,* 3010 (1947).
237. JASEWICZ, L., and PORGES, N., Sewage Ind. Wastes, *30,* 555 (1958).
238. JASEWICZ, L., and WASSERMAN, A. E., J, Dairy Sci., *44,* 393 (1961).
239. JEZESKI, J. J., TATINI, S. R., DE GARCIA, P. C., and OLSON, J. C., JR., Bacteriol. Proc. *12* (1967).
240. JOHNSON, M. J., Science, *94,* 200 (1941).
241. JOHNSTONE, D. B., and PFEFFER, M., Nature, *183,* 992 (1959).
242. JONES, F. S., and LITTLE, R. B., J. Exptl. Med., *45,* 319 (1927).
243. JONES, F. S., and SIMMS, H. S., J. Exptl. Med., *50,* 279 (1929); *51,* 327 (1930).
244. JONES, M. E., LIPMAN, F., HILZ, H., and LYNEN, F., J. Am. Chem. Soc., *75,* 3285 (1953).
245. JUNI, E., Federation Proc., *9,* 396 (1950).
246. KANDLER, O., Milchwissenschaft, *16,* 523 (1961).
247. KASTENS, M. L., and BALDAUSKI, F. A., Ind. Eng. Chem., *44,* 1257 (1952).
248. KATRANDZHIEV, K., Nauch. Trudove, *1,* 46 (1959).
249. KATRANDZHIEV, K., Isvest. Zentral nauch norszled. vetkhig Inst. Zhivoten. Prod. Sofia, *2,* 135 (1962).
250. KAUFMANN, H. P., and SCHMIDT, O., Vorratspflege und Lebensmittelforsch., *1,* 166 (1938).
251. KAWASHIMA, T., and MAENO, M., Jap. J. Zootechnol. Sci., *35,* 48 (1964).
252. KAWASHIMA, T., and MAENO, M. Jap. J. Zootechnol. Sci., *35,* 55, (1964).
253. KAWASHIMA, T., KODAMA, T., and MAENO, M., Jap. J. Zootechnol. Sci., *34,* 218 (1963).
254. KAWASHIMA, T., KODAMA, T., and MAENO, M., Jap. J. Zootechnol. Sci., *34,* 288 (1963).
255. KELLY, C. D., Trans. Roy. Soc. Can., *22,* V, 227 (1928).
256. KEMPNER, W., and KUBOWITZ, F., Biochem. Z., *265,* 245 (1933).
257. KEOGH, B. P., Appl. Microbiol., *19,* 928 (1970).
258. KHRUL'KEVICH, A., Molochanya Prom., *20,* 32 (1959).
259. KICKINGER, H., Biochem. Z., *132,* 210 (1922).
260. KIELLEY, W. W., and KIELLEY, R. K., Federation Proc., *10,* 207 (1951).
261. KITAHARO, K., OBAYASHI, A., and FUKUE, S., Proc. Internat. Symposium

Enzyme Chem., p. 460, Tokyo and Kyoto (1957); Reference in Milchwissenschaft, *16,* 523 (1961).
262. KLEIN, H. P., J. Bacteriol., *73,* 530 (1957).
263. KLEIN, H. P., EATON, N. R., and MURPHY, J. C., Biochim. et Biophys. Acta, *13,* 591 (1954).
264. KLUYVER, A. J., J. Soc. Chem. Ind., *52,* 367T (1933).
265. KLUYVER, A. J., LEY, J. de, and RIJVEN, A., Antonie van Leeuwenhoek J. Microbiol. Serol., *17,* 1 (1951).
266. KNOOP, F., Beitr. Chem. Physiol. Pathol., *6,* 150 (1905).
267. KODAMA. R., J. Antibiotics, *5,* 72 (1952).
268. KOEPSELL, H. J., JOHNSON, M. J., and MEEK, J. S., J. Biol. Chem., *154,* 535 (1944).
269. KOFFLER, H., EMERSON, R. L., PERLMAN, D., and BURRIS, R. H., J. Bacteriol., *50,* 517 (1945).
270. KOFT, B. W., and MORRISON, J. H., J. Bacteriol., *72,* 705 (1956).
271. KONDRAT'EVA, T. M., Microbiology (U.S.S.R.), *9,* 114 (1940).
272. KOOY, J. S., Neth. Milk Dairy J., *6,* 330 (1952).
273. KOOY, J. S., and PETTE, J. W., Neth. Milk Dairy J., *6,* 317 (1952).
274. KORKES, S., CAMPILLO, A. DEL, and GUNSALUS, I. C., Federation Proc., *10,* 210 (1951).
275. KORNBERG, H. L., and KREBS, H. A., Nature, *179,* 988 (1957).
276. KORNBERG, H. L., and MADSEN, N. B., Biochim. et Biophys. Acta, *24,* 651 (1951).
277. KORNBERG, H. L., and MADSEN, N. B., Biochem. (London), *68,* 549 (1958).
278. KORNBERG, A., and PRICER, W. E., JR., J. Biol. Chem., *204,* 345 (1953).
279. KOSHLAND, D. E., JR., and WESTHEIMER, F. H., J. Am. Chem. Soc., *72,* 3383 (1950).
280. KOSTERLITZ, H. W., Biochem. J. (London), *31,* 2217 (1937).
281. KOSTERLITZ, H. W., Biochem. J. (London), *37,* 318 (1943).
282. KOUNTZ, R. R., Purdue Univ., Indus. Waste Conference Proc., *11,* 157 (1956).
283. KREBS, H. A., Biochem. J. (London), *34,* 460 (1940, Harvey Lectures, Ser., 44, 165 (1948–49).
284. KREBS, H. A., and JOHNSON, W. A., Enzymologia, *4,* 148 (1937).
285. KRISTOFFERSEN, T., and NELSON, F. E., Appl. Microbiol., *3,* 268 (1955).
286. KRISTOFFERSEN, T., and NELSON, F. E., J. Dairy Sci., *37,* 635 (1954).
287. KRISTOFFERSEN, T., and NELSON, F. E., J. Dairy Sci., *38,* 1319 (1955).
288. KUBOWITZ, F., Biochem. Z., *274,* 285 (1934).
289. KUBOWITZ F., and LÜTTGENS, W., Biochem. Z., *307,* 170 (1941).
290. KUBOWITZ, F., and OTT, P., Biochem. Z., *314,* 94 (1943).
291. KUHN, R., and TIEDEMANN, H., Z. Naturforsch., *8b,* 428 (1953).
292. KULSHRESTHA, D. C., and MARTH, E. H., J. Milk Food Technol., *33,* 305 (1970).
293. LAMPRECH, E. D., Diss. Abstr., *23,* 1162 (1962).
294. LAMPRECH, E. D., and FOSTER, E. M., J. Appl. Bacteriol., *26,* 359 (1963).
295. LANE, C. B., and HAMMER, B. W., Iowa State Agr. Expt. Sta. Bull. 190 (1935).
296. LARDY, H. A., "A Symposium on Phosphorus Metabolism," Ed. by McElroy, W. D., and Glass, B., Vol. 1, p. 477, Johns Hopkins Press, Baltimore, 1951.
297. LARDY, H. A., and ZIEGLER, J. A., J. Biol. Chem., *159,* 343 (1945).
298. LAWRIE, J. W., "Glycerol and the Glycols," p. 1, Chemical Catalog Co., New York (1928).
299. LEHNINGER, A. L., J. Biol. Chem., *154,* 309 (1944); *164,* 291 (1946).
300. LEHNINGER, A. L., J. Biol. Chem., *178,* 625 (1949).
301. LEHNINGER, A. L., Sci. American, *202,* No. 5, 102 (1960).
302. LEHNINGER, A. L., "Phosphorus Metabolism," Ed. by McElroy, W. D., and Glass, B., Vol. 1, p. 344, Johns Hopkins Press, Baltimore, 1951.
303. LEHNINGER, A. L., and WADKINS, C. L., Ann. Rev. Biochem., *31,* 47 (1962).
304. LEIBOWITZ, J., and HESTRIN, S., Advances in Enzymol., *5,* 87 (1945).
305. LEITSCH, R. H., Scottish J. Agr., *15,* 167 (1932).

306. LELOIR, L. F., "Phosphorus Metabolism," Ed. by McElroy, W. D., and Glass, B., Vol. 1, pp. 67, 80, Johns Hopkins Press, Baltimore, 1951.
307. LELOIR, L. F., and CARDINI, C. E., J. Am. Chem. Soc., *79,* 6340 (1957).
308. LELOIR, L. F., RONGINE DE FEKETE, M. A., and CARDINI, C. E., J. Biol Chem., *236,* 636 (1961).
309. LELOIR, L. F., and GOLDEMBERG, S. H., J. Biol. Chem., *235,* 919 (1960).
310. LELOIR, L. F., OLAVARRIA, J. M., GOLDEMBERG, S. H., and CARMINATTI, H., Arch. Biochem. Biophys., *81,* 508 (1959).
311. LEMBERG, R., LEGGE, J. W., and LOCKWOOD, W. H., Biochem. J. (London), *35,* 328 (1941).
312. LESTER, R. L., CRANE, F. I., and HATEFI, Y., J. Am. Chem. Soc., *80,* 4071 (1958).
313. LEVITON, A., J. Am. Chem. Soc., *68,* 835 (1946).
314. LEVITON, A., unpublished data.
315. LEVITON, A. (to U.S.A., as represented by Secy. of Agr.), U.S. Patent 2,477,812 (Aug. 2, 1949).
316. LEVITON, A. (to U.S.A., as represented by Secy. of Agr.), U.S. Patent 2,753,289 (July 3, 1956).
317. LEVITON, A. (to U.S.A., as represented by Secy. of Agr.), U.S. Patent 2,764,521 (Sept. 25, 1956).
318. LEVITON, A., and BURKEY, L. A., unpublished data (1944).
319. LEVITON, A., and HARGROVE, R. E., unpublished data (1951).
320. LEVITON, A., and HARGROVE, R. E., Ind. Eng. Chem., *44,* 2651 (1952).
321. LEVITON, A., and WHITTIER, E. O., Abstr. in J. Dairy Sci., *33,* 402 (1950).
322. LINDNER, P. (War Commission for Fats and Oils), German Patent 305,091 (Apr. 17, 1920).
323. LINDNER, P. (War Commission for Fats and Oils), German Patent 307,789 (Sept. 2, 1919).
324. LINDNER, P. (War Commission for Fats and Oils), German Patent 320,560 (Apr. 24, 1920).
325. LINDNER, P., Z. Angew. Chem., *35,* 110, 591 (1922).
326. LIPMANN, F., Nature, *138,* 588 (1936).
327. LIPMANN, F., Cold Spring Harbor Symposia Quant. Biol., *7,* 248 (1939).
328. LIPMANN, F., Advances in Enzymol., 1, *99,* (1941).
329. LIPMANN, F., J. Biol. Chem., *160,* 173 (1945).
330. LIPMANN, F., Bacteriol. Rev., *17,* 1 (1953).
331. LOCKHEAD, A. G., Cent. Expt. Farm Reports for 1924 (Canada) Div. Bact., p. 9 (1925).
332. LOCKWOOD, L. B., and STODOLA, F. H., (to U.S.A., as represented by Secy. of Agr.), U.S. Patent 2,496,297 (Feb. 7, 1950).
333. LOHMANN, K., Biochem. Z., *222,* 324 (1930).
334. LOHMANN, K., and SCHUSTER, P., Biochem. Z., *294,* 188 (1937).
335. LUCAS, P. S., Mich. Agr. Expt. Sta., Quart. Bull., *12,* No. 1, 18 (1928).
336. LUCE, E. M., and MACLEAN, I. S., Biochem. J. (London), *19,* 47 (1925).
337. LUNDSTEDT, E., and FOGG, W. B., J. Dairy Sci., *45,* 1327 (1962).
338. LYNEN, F., Ann., *546,* 120 (1941).
339. LYNEN, F., J. Cellular Comp. Physiol., *54* (Suppl. 1) 33 (1959).
340. LYNEN, F., AGRANOFF, B. W., EGGERER, H., HENNING, U., and MÖSLEIN, E. M., Z. Angew. Chem., *71,* 657 (1959).
341. LYNEN, F., HARTMANN, G., NETTER, N. F., and SCHUEGRAF, A., Ciba Foundation Symposium Regulation of Cell Metabolism, p. 256 (1959); Chem. Abs., *54,* 5858a (1960).
342. LYNEN, F., REICHERT, E., and RUEFF, L., Ann., *574,* 1 (1951).
343. MACQUOT, G., and LEFEBVRE, E., J. Appl. Bacteriol., *19,* 322 (1959).
344. MAHLER, H. R., and ELOWE, D. G., J. Biol. Chem., *210,* 165 (1954).
345. MARTH, E. H., "Byproducts from Milk," Ed. by WEBB, B. H., and WHITTIER, E. O., p. 43, Avi Publishing Co., Westport, Conn., 1970.

346. MARTH, E. H., J. Dairy Sci., *46,* 869 (1963).
347. MARTH, E. H., J. Dairy Sci., *52,* 283 (1969).
348. MARTH, E. H., J. Milk Food Technol., *24,* 36, 70 (1961).
349. MARTH, E. H., Residue Rev., *12,* 65 (1966).
350. MARTH, E. H., and ELLICKSON, B. E., J. Milk Food Technol., *22,* 241, 266 (1959).
351. MARTH, E. H., and HUSSONG, R. V., J. Dairy Sci., *45,* 652 (1962).
352. MARTH, E. H., and HUSSONG, R. V., J. Dairy Sci., *46,* 1033 (1963).
353. MARTH, E. H., INGOLD, D. L., and HUSSONG, R. V., J. Dairy Sci., *47,* 1265 (1964).
354. MARTIN, J. H., and CARDWELL, J. T., J. Dairy Sci., *43,* 438 (1960).
355. MARTIUS, C., and LYNEN, F., Advances in Enzymol., *10,* 167 (1950).
356. MATHER, D. W., and BABEL, F. J., J. Dairy Sci., *42,* 1917 (1959).
357. MATTICK, A. T. R., J. Agr. Sci., *16,* 459 (1926).
358. MATTICK, A. T. R., Analyst, *55,* 37 (1930).
359. MATTICK, A. T. R., and HIRSCH, A., Lancet, ii, 5 (1947).
360. MATTICK, A. T. R., and HIRSCH, A., Nature, *154,* 551 (1944).
361. McLEOD, J. W., and GORDON, J., J. Pathol. Bacteriol., *26,* 326, 332 (1923).
362. McQUILLEN, K., and ROBERTS, R. B., J. Biol. Chem., *207,* 81 (1954).
363. MEADE, R. E., POLLARD, H. L., and RODGERS, N. E. (to Western Condensing Co.), U.S. Patent 2,369,680 (Feb. 20, 1945).
364. MEADE, R. E., POLLARD, H. L., and RODGERS, N. E., U.S. Patent 2,369,680 (1945).
365. MEADE, R. E., RODGERS, N. E., and POLLARD, H. L., U.S. Patent 2,433,232 (1947).
366. MEEWES, K. H., and MILOSEVIC, S., Milchwissenschaft, *17,* 678 (1962).
367. MEISTER, A., "Biochemical Preparations," Vol. 2, p. 18, Ed. by Ball, E. G., John Wiley & Sons, Inc., New York, 1952.
368. MEYERHOF, O., and BECK, L. V., J. Biol. Chem., *156,* 109 (1944).
369. MEYERHOF, O., and KIESSLING, W., Biochem. Z., *264,* 40 (1933); 267, 313 (1934).
370. MEYERHOF, O., and LOHMANN, K., Biochem. Z., *271,* 89 (1934).
371. MEYERHOF, O., and OESPER, P., J. Biol. Chem., *170,* 1 (1947).
372. MEYERHOF, O., and OESPER, P., J. Biol. Chem., *179,* 1371 (1949).
373. MICHAELIAN, M. B., HOECKER, W. H., and HAMMER, B. W., J. Dairy Sci., *21,* 213 (1938).
374. MICHAELIS, L., "The Enzymes," Ed. by Sumner, J. B., and Myrbäck, K., Vol. 2, p. 1, Academic Press, Inc., New York, 1951.
375. MINER, C. S., JR., and WOLNAK, B. (to Sewage Commission, Milwaukee, Wis.), U.S. Patent 2,646,386 (July 21, 1953).
376. MINER, C. S., JR., and WOLNAK, B., Abstr. papers, 121st Meeting Am. Chem. Soc., p. 10C, Milwaukee (Mar. 31, 1952).
377. MINOR, T. E., and MARTH, E. H., J. Milk Food Technol., *35,* 77 (1972).
378. MINOR, T. E., and MARTH, E. H., J. Milk Food Technol., *35,* 302 (1972).
379. MORTON, R. M., WILSON, G. M., LOW, J. S., and LEAT, W. M. F., Chem. and Ind., 1649 (1957).
380. MOSS, A. R., and KLEIN, R., Brit. Patent 615,847 (Jan. 12, 1949).
381. MOYER, A. J., and COGHILL, R. D., J. Bacteriol., *51,* 57 (1946).
382. MUNOZ, J. M., and LELOIR, L. F., J. Biol. Chem., *147,* 355 (1943).
383. MYERS, R. P., and STIMPSON, E. G., U.S. Patent 2,762,749 (1956).
384. MYERS, R. P., and WEISBERG, S. M., U.S. Patent 2,128,845 (1938).
385. MYRBÄCK, K., and VASSEUR, E., Z. Physiol. Chem., *277,* 171 (1943).
386. NEGELEIN, E., and BRÖMEL, H., Biochem. Z., *301,* 135; 303, 132 (1939).
387. NEISH, A. C., BLACKWOOD, A. C., ROBERTSON, F. M., and LEDINGHAM, G. A., Can. J. Res., *B25,* 65 (1947).
388. NEUBERG, C., Biochem. Z., *67,* 90 (1914).
389. NEUBERG, C., Biochem. Z., *88,* 432 (1918).
390. NEUBERG, C., and HIRSCH, J., Biochem. Z., *96,* 175; 100, 304 (1919).

391. NEUBERG, C., and HIRSCH, J., Biochem. Z., *98,* 141 (1919).
392. NEUBERG, C., HIRSCH, J., and REINFURTH, E., Biochem. Z., *105,* 307 (1920).
393. NEUBERG, C., and REINFURTH, E., Biochem. Z., *89,* 365; *92,* 234 (1918).
394. NEUBERG, C., and REINFURTH, E., Ber., *52B,* 1677 (1919).
395. NEUBERG, C., and REINFURTH, E., Ber., *53B,* 1039 (1920).
396. NEUBERG, C., and URSUM, W., Biochem. Z., *110,* 193 (1920).
397. NEWTON, G. G. F., ABRAHAM, E. P., and BERRIDGE, N. J., Nature, *171,* 606 (1953).
398. NICHOLAS, D. J. D., and NASON, A., J. Biol. Chem., *207,* 353 (1954).
399. NILSSON, G., Svenska Mejeritidn., *40* (20), 207 (1948).
400. NILSSON, G., and WASS, L., Svenska Mejeritidn., *52,* 643 (1960).
401. NILSSON, R., ENEBO, L., LUNDIN, H., and MYRBÄCK, K., Svensk Kemi Tid., *55,* 41 (1943).
402. NIVEN, C. F., JR., J. Bacteriol., *46,* 573 (1943); 47, 343 (1944).
403. NOLTMANN, E. A., GUBLER, C. J., and KUBY, S. A., J. Biol. Chem., *236,* 1225 (1961).
404. NOVELLI, G. D., and LIPMANN, F., Arch. Biochem., *14,* 23 (1947).
405. NURMIKKO, V., Experientia, *12,* 245 (1956).
406. OCHOA, S., J. Biol. Chem., *151,* 493 (1943).
407. O'KANE, D. J., J. Bacteriol., *60,* 449 (1950).
408. OLIVE, T. R., Chem. Met. Eng., *43,* 480 (1936).
409. ORLA-JENSEN, S., "The Lactic Acid Bacteria," Mem. Acad. Roy. Sci. Let. Danemark (Sec. Sci) VIII, 5, 81–196, 1919.
410. ORLA-JENSEN, S., "Dairy Bacteriology," p. 276, P. Blakiston's Son and Co., Philadelphia, 1931.
411. ORLA-JENSEN, S., ORLA-JENSEN, A. D., and SPUR, B., J. Bacteriol., *12,* 333 (1926).
412. PARK, H. S., and MARTH, E. H., J. Milk Food Technol., *35,* 482 (1972).
413. PARK, H. S., and MARTH, E. H., J. Milk Food Technol., *35,* 489 (1972).
414. PARK, H. S., MARTH, E. H., GOEPFERT, J. M., and OLSON, N. F., J. Milk Food Technol., *33,* 280 (1970).
415. PAZUR, J. H., J. Biol. Chem., *208,* 439 (1953); Science, *117,* 355 (1953).
416. PAZUR, J. H., and GORDON, A. L., J. Am. Chem. Soc., *75,* 3458 (1953).
417. PEEBLES, M. M., GILLILAND, S. E., and SPECK, M. L., Appl. Microbiol., *17,* 805 (1969).
418. PETERSEN, W. H., and JOHNSON, M. J., J. Bacteriol., *58,* 701 (1949).
419. PHELPS, A. S. (to Am. Cyanamid Co.), U.S. Patent 2,473,818 (June 21, 1949).
420. PLATT, T. B., and FOSTER, E. M., J. Bacteriol., *75,* 453 (1958).
421. POLANSKAYA, M. S., LEONOVICH, V. V., and BIBERDIEVA, M. P., Doklady Vesesoyaz Acad. Nauk, 21 (1953).
422. POLONOVSKI, M., NEUZIL, E., BAUDU, L., and POLONOVKSI, J., Compt. rend. soc. biol., *141,* 460 (1947).
423. PORGES, N., Food Technol., *12,* 78 (1958).
424. PORGES, N., J. Milk Food Technol., *19,* 34 (1956).
425. PORGES, N., and JASEWICZ, L., Sewage Ind. Wastes, *31,* 443 (1959).
426. PORGES, N., MICHENER, T. S., JR., JASEWICZ, L., and HOOVER, S. R., "Agriculture Handbook No. 176." Agr. Research Service, Washington, D. C. (1960).
427. PORGES, N., PEPINSKY, J. B., HENDLER, N. C., and HOOVER, S. R., Sewage Ind. Wastes, *22,* 318 (1950).
428. PORGES, N., PEPINSKY, J. B., and JASEWICZ, L., J. Dairy Sci., *34,* 615 (1951).
429. PRESCOTT, S. C., and DUNN, C. G., "Industrial Microbiology," p. 5, 314, McGraw-Hill Book Co., New York, 1949.
430. PRIDHAM, T. B., HALL, H. H., and PFEIFER, V. F., Report. Agr. Research Adm., U.S. Dept. of Agr. (1950).
431. PRIDHAM, T. B., and RAPER, K. B., Mycologia, *44,* 452 (1952).
432. PRILL, E. A., and HAMMER, B. W., J. Dairy Sci., *22,* 67 (1939).
433. QUASTEL, J. H., and WHETHAM, M. D., Biochem. J. (London), *18,* 519 (1924); *19,* 520, 645 (1925).

434. RAAF, H., Arch. Mikrobiol., *12,* 131 (1941).
435. RACKER, E., J. Biol. Chem., *184,* 313 (1950).
436. RACKER, E., "Phosphorus Metabolism," Ed. by McElroy, W. D., and Glass, B., Vol. 1, p. 145, Johns Hopkins Press, Baltimore, 1951.
437. RACKER, E., HABA, G. DE LA, and LEDER, I. G., J. Am. Chem. Soc., *75,* 1010 (1953).
438. RAMSEIER, H. R., Arch. Mikrobiol., *37,* 57 (1960).
439. REED, L. J., Physiol. Rev., *33,* 544 (1953).
440. REED, L. J., and DEBUSK, B. G., J. Biol. Chem., *199,* 873 (1952).
441. REIF, G. D., REINBOLD, G. W., and VEDAMUTHU, E. R., J. Dairy Sci., *50,* 945 (1967).
442. REILLY, J., HICKINBOTTOM, W. J., HENLEY, F. R., and THAYSEN, A. C., Biochem. J. (London), *14,* 229 (1920).
443. REITER, B., FEWINS, B. G., FRYER, T. F., and SHARPE, M. E., J. Dairy Res., *31,* 261 (1964).
444. REYNOLDS, H., and WERKMAN, C. H., J. Bacteriol., *33,* 603 (1937).
445. REYNOLDS, H., and WERKMAN, C. H., Arch. Mikrobiol., *8,* 149 (1937).
446. RICE, F. E., and DOWNS, P. A., J. Dairy Sci., *6,* 532 (1923).
447. RICHARDSON, G. H., and CALBERT, H. E., J. Dairy Sci., *42,* 907 (1959).
448. RICHENBERG, H. V., COHEN, G. N., BUTTIN, G., and MONOD, J., Ann. Inst. Pasteur, *91,* 829 (1956).
449. RICHERT, D. A., and WESTERFELD, W. W., J. Biol. Chem., *209,* 179 (1954).
450. RIPPEL-GALDES, A., PIETSCHMANN-MAYER, K., and KOHLER, W., Arch. Mikrobiol. *14,* 113 (1948).
451. RIPPEL, K., Arch. Mikrobiol., *2,* 72 (1931).
452. RITTER, W., Schweiz. Z. Pathol. Bakteriol., *7,* 370 (1944).
453. RITTER, W., Schweiz. Milchztg., *71,* 255 (1945).
454. ROBERTS, H. R., and McFARREN, E. F., Arch. Biochem. Biophys., *43,* 233 (1953).
455. ROBERTS, H. R., and PETTINATI, J. D., J. Agr. Food Chem., *5,* 130 (1957).
456. ROBERTS, R. B., COWIE, D. B., BRITTEN, R., BOLTON, E. T., and ABELSON, P. H., Proc. Natl. Acad. Sci. U.S., *39,* 1013, 1020 (1953).
457. ROBISON, R., Biochem. J. (London), *16,* 809 (1922).
458. ROGERS, L. A., Centr. Bakteriol. Parasitenk., II, *12,* 395 (1904).
459. ROGERS, L. A., J. Bacteriol., *16,* 32 (1928).
460. ROGOSA, M., J. Biol. Chem., *175,* 413 (1948).
461. ROGOSA, M., BROWNE, H. H., and WHITTIER, E. O., J. Dairy Sci., *30,* 263 (1947).
462. ROSELL, J. M., Can. Pub. Health J., *24,* 344 (1933).
463. ROSELL, J., and MATALLANA, S., Sci. de Alimene., Milano, *2,* 126 (1956).
464. RUDERT, F. J. (to Commercial Solvents Corp.), U.S. Patent 2,374,503 (Apr. 24, 1945).
465. RUDNICK, A. W., JR., and GLENN, W. E., J. Dairy Sci., *43,* 845 (1960).
466. RUEHLE, G. L. A., Mich. Agr. Expt. Sta. Ann. Rept. 174 (1924).
467. RUEHLE, G. L. A., Mich. Agr. Expt. Sta. Tech. Bull. 102 (1930).
468. RUTTER, W. J., and HANSEN, R. G., J. Biol. Chem., *202,* 323 (1953).
469. RYKSHINA, Z. P., Isv. Timiryazev. Selskochog. Akad., *1,* 217 (1960). Abstract in Milchwissenschaft, *16,* 434 (1961).
470. SADLER, W., Trans. Roy. Soc. Can., *20V,* 395 (1926).
471. SADLER, W., Scientific Agr., *10,* 11 (1929).
472. SADLER, W., and MOUNCE, M. J., British Columbia Dept. Agr., Research Bull. 1 (1926).
473. SALISBURY, L. F., and ANDERSON, R. J., J. Biol. Chem., *112,* 541 (1936).
474. SARLES, W. B., and HAMMER, B. W., J. Bacteriol., *25,* 461 (1933).
475. SAUNDERS, A. P., OTTO, R. H., and SYLVESTER, J. C., Abstr. papers, 119th Meeting Am. Chem. Soc., p. 21A, Cleveland (1951).
476. SCHARDINGER, F., Centr. Bakteriol. Parasitenk., II, *8,* 144 (1902).
477. SCHMALFUSS, K., Bodenkunde u. Pflanzenernähr., *5,* 37 (1937).
478. SCHULZ, M. E., and HINGST, G., Milchwissenschaft, *9,* 330 (1954).

479. SCHULZ, M. E., VOSZ, E., and KLEY, W., Milchwissenschaft, *9,* 361 (1954).
480. SCHULZE, K. L., Arch. Mikrobiol., *15,* 315 (1950).
481. SCOTT, D. B. M., and COHEN, S. S., J. Biol. Chem., *188,* 509 (1951).
482. SIMMONS, J. C., and GRAHAM, D. M., J. Dairy Sci., *41,* 705 (1958).
483. SIMMONS, J. C., and GRAHAM, D. M., J. Dairy Sci., *42,* 363 (1959).
484. SIMMONS, J. C., and GRAHAM, D. M., Southern Dairy Prod. J., *64*(4), 80 (1958).
485. SJÖSTRÖM, G., and HÅKANSSON, E. B., 13th Intern. Dairy Congr., Proc., *2,* 697 (1953).
486. SKORODUMOVA, A. M., Voprosy Pitaniya, *15,* 32 (1956).
487. SLADE, H. D., and WERKMAN, C. H., Arch. Biochem., *2,* 97 (1943).
488. SMITH, L., Bacteriol. Rev., *18,* 106 (1954).
489. SMITH, R. G., Centr. Bakteriol. Parasitenk., Abt. II, *10,* 61 (1903).
490. SNELL, N., LJICHI, K., and LEWIS, J. C., Appl. Microbiol., *4,* 13 (1956).
491. SOLS, A., Annual Rev. Biochem., *30,* 217 (1961).
492. SPECK, J. F., J. Biol. Chem., *168,* 403 (1947).
493. SPECK, M. L., J. Dairy Sci., *45,* 1281 (1962).
494. SPECK, M. L., J. Dairy Sci., *55,* 1019 (1972).
495. SPITZER, G., and EPPLE, F. W., J. Dairy Sci., *3,* 1 (1920).
496. SPRINGER, R., Pharmazie, *5,* 113 (1950).
497. SRERE, P., COOPER, J. R., TABACHNICK, M., and RACKER, E., Archiv. Biochem. Biophys., *74,* 295 (1958).
498. STADHOUDERS, J., and van der WALLS, E. B., Misset's Zuivel, *72,* 347 (1966).
499. STADHOUDERS, J., JANSEN, L. A., and HUP, G., Koeltechniek, *62*(2), 29 (1969).
500. STADTMAN, E. R., and BARKER, H. A., J. Biol. Chem., *184,* 769 (1950).
501. STEELE, R. H., WHITE, A. G. C., and PIERCE, W. A., JR., J. Bacteriol., *67,* 86 (1954).
502. STEFANIAK, J. J., GAILEY, F. B., JARVIS, F. G., and JOHNSON, M. J., J. Bacteriol., *52,* 119 (1946).
503. STEPHENSON, M., and STRICKLAND, L. H., Biochem. J. (London), *26,* 712 (1932); *27,* 1517, 1528 (1933).
504. STILES, H. R., (to Commercial Solvents Corp.), U.S. Patent 2,483,855 (Oct. 4, 1949).
505. STONE, R. W., and FARRELL, M. A., Science, *104,* 445 (1946).
506. STORGARDS, J., Milchwissenschaft, *17,* 369 (1962).
507. SUTHERLAND, E. W., POSTERNAK, T. Z., and CORI, C. F., J. Biol. Chem., *179,* 501 (1949).
508. SWIM, H. E., and KRAMPITZ, L. O., J. Bacteriol., *67,* 419 (1954).
509. TABENKIN, B. (to Hoffmann-LaRoche, Inc.), U.S. Patent 2,493,274 (Jan. 3, 1950).
510. TACQUET, A., TISON, F., and DEVULDER, B., Ann. Inst. Pasteur, *100,* 581 (1961).
511. TANIYA, H., Acta Phytochim., *6,* 265 (1932).
512. TATUM, E. L., PETERSON, W. H., and FRED, E. B., J. Bacteriol., *27,* 207 (1934); 29, 563 (1935).
513. TAUFEL, K., THALER, H., and SCHREYEGG, H., Z. Untersuch. Lebensm., *72,* 394 (1936).
514. TEPLEY, M., Prumsyl Potravin, *13,* 270 (1962).
515. THOMASOW, J., Kieler Milchwirtschaft. Forsch. Berichte, *2,* 35 (1950).
516. TIKKA, J., Biochem. Z., *279,* 264 (1935).
517. TOMA, C., and MELEGHI, E., Lucar. Inst. Cercet. Aliment., *2,* 227 (1958).
518. TOTTER, J. R., BURNETT, W. T., JR., MONROE, R. A., WHITNEY, I. B., and COMAR, C. L., Science, *118,* 555 (1953).
519. TRILLAT, A., and SAUTON, B., Compt. rend., *144,* 926 (1907).
520. TSUGO, T., IWAIDA, M., NAGAO, A., KOBAYASHI, Y., TSURUMOTO, K., and HIRAI, K., Shokuhin Eisigaku Zasshi, *3,* 356 (1962).
521. TUCKEY, S. L., and SAHASRABUDKHE, M. R., J. Dairy Sci., *40,* 1329 (1957).

522. TUCKEY, S. L., NELSON, W. O., and HUSSONG, R. V., J. Dairy Sci., *31,* 713 (1948).
523. Usines de Melle (Firmin Boirot, inventor), French patent 942,101 (Jan. 31, 1949).
524. VALLEA, E., and MOCQUOT, G., Lait, *48,* 631 (1968).
525. VANDEMARK, P. J., and WOOD, W. A., J. Bacteriol., *71,* 385 (1956).
526. Van der ZANT, W. C., and NELSON, F. E., J. Dairy Sci., *36,* 1212 (1953).
527. Van der ZANT, W. C., and NELSON, F. E., J. Dairy Sci., *37,* 795 (1954).
528. VILLECOURT, P., and BLACHERE, H., Ann. Inst. Pasteur, *88,* 523 (1955).
529. VINCENT, J. G., VEOMETT, R. C., and RILEY, R. F., J. Bacteriol., *78,* 477 (1959).
530. WAES, G., Revue de l'Agr., *23,* 1097 (1970).
531. WAGNER, A. F., ANO, L., SHUNK, C. H., LINN, B., WOLF, D. E., HOFFMAN, C. F., ERICKSON, R. E., ARISON, B., TRENNER, N. R., and FOLKERS, K., J. Am. Chem. Soc., *85,* 1534 (1963).
532. WAKSMAN, S. A., Abstr. Bacteriol., *6,* 265 (1922).
533. WALLENFELS, K., Naturwiss., *38,* 306 (1951).
534. WALLENFELS, K., BERNT, E., and LIMBERG, G., Ann., *579,* 113; 584, 63 (1953).
535. WALLERSTEIN, J. S., and STERN, K. G., J. Biol. Chem., *158,* 1 (1945).
536. WALMESLEY, R. A. (to Imperial Chem. Industries, Ltd.), U.S. Patent 2,235,056 (Mar. 18, 1941).
537. WARBURG, O., "Wasserstoffübertragende Fermente," p. 51, Berlin, W. Saenger, 1948.
538. WARBURG, O., "Schwermetalle als Wirkungsgruppen von Fermenten," Berlin, W. Saenger, 1948.
539. WARBURG, O., and CHRISTIAN, W., Biochem. Z., *238,* 131 (1931).
540. WARBURG, O., and CHRISTIAN, W., Biochem. Z., *260,* 499 (1933).
541. WARBURG, O., and CHRISTIAN, W., Biochem. Z., *303,* 40 (1939).
542. WARBURG, O., and CHRISTIAN, W., Biochem. Z., *310,* 384 (1942).
543. WARD, G. E., and JAMIESON, G. S., J. Am. Chem. Soc., *56,* 973 (1934).
544. WASSERMAN, A. E., J. Dairy Sci., *43,* 1231 (1960).
545. WASSERMAN, A. E., Appl. Microbiol., *8,* 291 (1960).
546. WASSERMAN, A. E., and HAMPSON, J. W., Appl. Microbiol., *8,* 293 (1960).
547. WASSERMAN, A. E., HAMPSON, J. W., and ALVARE, N. F., J. Water Pollution Control Fed., *33,* 1090 (1961).
548. WASSERMAN, A. E., HAMPSON, J. W., ALVARE, N. F., and ALVARE, N. J., J. Dairy Sci., *44,* 387 (1961).
549. WASSERMAN, A. E., HOPKINS, W. J., and PORGES, N., Sewage Ind. Wastes, *30,* 913 (1958).
550. WASSERMAN, A. E., HOPKINS, W. J., and PORGES, N., Proc. XVth Intern. Dairy Congress, *2,* 1241 (1959).
551. WEISSBACH, H., REDFIELD, B., and PETERKOFSKY, A., J. Biol. Chem., *236,* PC 240 (1961).
552. WENDORFF, W. L., AMUNDSON, C. H., and OLSON, N. F., J. Milk Food Technol. *33,* 451 (1970).
553. WHEAT, R. W., and AJL, S. J., Arch. Biochem. Biophys., *49,* 7 (1954).
554. WHEATER, D. M., HIRSCH, A., and MATTICK, A. T. R., Nature, *168,* 659 (1951).
555. WHEATER, D. M., HIRSCH, A., and MATTICK, A. T. R., Nature, *170,* 623 (1952).
556. WHITELEY, H. R., and ORDAL, E. J., "A Symposium on Inorganic Nitrogen Metabolism," Ed. by McElroy, W. D., and Glass, B., p. 521, Johns Hopkins Press, Baltimore, 1956.
557. WIELAND, H., RATH, F., and BEREND, W., Ann., *548,* 19 (1941).
558. WIELAND, H., RATH, F., and HESSE, H., Ann., *548,* 34 (1941).
559. WIELAND, H., and ROSENTHAL, C., Ann., *554,* 241 (1943).
560. WILLIAMSON, W. T., and SPECK, M. L., J. Dairy Sci., *45,* 164 (1962).
561. WILLIAMSON, W. T., TOVE, S. B., and SPECK, M. L., Bacteriol. Proc., 26 (1962).
562. WINCKEL, M., Z. Volksernähr. Diätkost, *7,* 265 (1932).

563. WINKLER, S., Proc. 13th Intern. Dairy Congr., *3,* 1164 (1953).
564. WIX, P., and WOODBINE, M., J. Appl. Bacteriol., *22,* 175 (1959).
565. WODZAK, W., Deutsche Lebensmittel Rundschau, *5,* 135 (1962).
566. WOLF, D. E., HOFFMAN, C. H., TRENNER, N. R., ARISON, B. H., SHUNK, C. H., LINN, B. O., McPHERSON, J. F., and FOLKERS, K., J. Am. Chem. Soc., *80,* 4753 (1958).
567. WROBLEWSKI, A., J. Prakt. Chem., *64,* 1 (1901).
568. YAMAMOTO, T., ASAO, T., and CHIKUNA, G., Bull. Natl. Inst. Agr. Sci., Japan (Series G6 Animal Husbandry), *2,* 5 (1951).
569. YAMAMOTO, T., ASAO, T., and CHIKUNA, G., Bull. Natl. Inst. Agr. Sci., Japan (Series G6) *28,* 91 (1953).
570. YAMASAKI, I., Biochem. Z., *300,* 160 (1939); *307,* 431 (1941).
571. YAMASAKI, I., and YOSITOME, W., Biochem. Z., *297,* 398 (1938).
572. YATES, A. R., IRVINE, O. R., and CUNNINGHAM, J. D., Can. J. Agr. Sci., *35,* 337 (1955).
573. YOUNG, W. J., Proc. Roy. Soc. (London), *B81,* 528 (1909).
574. ZEILINGER, A., and BINDER, W., Milchwiss. Ber. Wolfpassing, *6,* 197 (1956).
575. ZERNER, E., Ber., *53B,* 325 (1920).
576. ZIEMBA, J. V., Food Engr., *42*(1), 68 (1970).

CHAPTER 14

P. G. Keeney†
Manfred Kroger‡

Frozen Dairy Products *

FREEZING

Various dairy products are intentionally frozen as a means of preservation. Cream, for example, is frozen during the flush season and held for later use in manufactured dairy products, such as ice cream. Concentrated skimmilk, cottage cheese curd, plastic cream, and butter, also, are commonly stored in the frozen state. Furthermore, it is technically possible and sometimes economically feasible to freeze, store and distribute to the ultimate user, fluid milk and concentrated milk as a frozen food. With products held or stored in the frozen state before use, the principal concern is with methods and procedures that will permit frozen storage without altering the characteristics in a way which would be detrimental to the intended use of the products after defrosting.

Intentional freezing is practiced in the manufacture of ice cream, sherbet, and other dairy confections for the purpose of creating a food product to be consumed in the frozen state. The low temperature at which these products are stored does permit them to be held for considerable periods of time before consumption, but this, of course, is not the primary reason for freezing. In contrast to the products mentioned in the previous paragraph, the characteristics of ice cream and frozen desserts, while in the frozen condition, are of first consideration and the properties after melting are of lesser importance.

---

* Authorized for publication May 16, 1972, as Paper No. 4213 in the Journal Series of the Agricultural Experiment Station.
Revised from Chapter 14 in the 1st edition by F. J. Doan and P. G. Keeney.
† P. G. Keeney, Department of Dairy Science, Division of Food Science and Industry, The Pennsylvania State University, University Park, Pa. 16802
‡ Manfred Kroger, Department of Dairy Science, Division of Food Science and Industry, The Pennsylvania State University, University Park, Pa. 16802

Freezing Temperatures and Freezing Effects

The freezing point of milk has been the subject of extensive investigation because it represents an accurate criterion of the presence of added water (see Chapter 8). Only limited attention, however, has been paid to the freezing points of other dairy products, even though such information is often very useful in their handling during and subsequent to manufacture. Determination of the freezing point is relatively simple with dairy products, which are essentially fluid, but is considerably more difficult and less precise with concentrated products low in water content or containing substantial percentages of added components such as sweeteners.

The freezing points of ideal water solutions of known composition may be accurately calculated from the classic freezing point—molecular weight equation, $D = KG/M$, where D = the depression of the freezing point of water, K = 18.6 (the freezing point depression constant of water), G = the grams of dissolved substance in 100 gm water, and M = the molecular weight of the dissolved substance. Such calculations are precise only when applied to dilute solutions containing solutes which do not decompose or ionize in water and do not combine or hydrate with water.

Calculated freezing points for many dairy products are, at best, only approximate because of such factors as the concentration of various solubles in the serum phase; the observed discrepancy between the true and calculated freezing points of sugar solutions, attributable to hydration;[140] and the unknown ionic combinations of the milk salts, their degrees of ionization, and the extent of their combination with proteins. Successful calculation of the freezing points of ice-cream mixes have been made, nevertheless, by utilizing observed freezing points for sucrose solutions (Table 14.1), assuming the same freezing point depression for lactose and employing a value of 78.6 for the "apparent average molecular weight" of the salts of milk. Using these values, good agreement was demonstrated between the calculated and observed freezing points in a series of experimental mixes.[97] Fats, proteins, and colloids in general have a negligible effect on the freezing point of water or water solutions in which they are dispersed.

The method of calculating the freezing point of an ice-cream mix makes use of the sum of the freezing-point depression value for the sugars and the calculated value for the salts. For lactose and sucrose the following equation is utilized to determine the parts of disaccharide sugar per 100 parts water in the mix:

$$\frac{[(\mathrm{MSNF} \times 0.545) + \mathrm{S}]\,100}{W} = \text{parts sucrose equivalent/100 parts water.}$$

MSNF is the percent of milk solids-not-fat (serum solids) in the mix, 0.545 is the proportion of MSNF which is lactose, S is the percent of sucrose in the mix, and W is the percent of water in the mix.

The freezing-point lowering due to the disaccharide sugars (sucrose equivalent) is then obtained by interpolation from the data shown in Table 14.1. The freezing-point depression due to the milk salts is calculated from the equation:

$\frac{MSNF \times 2.37}{W}$ = freezing point depression due to milk salts.[94] MSNF

Table 14.1

FREEZING POINT LOWERING OF SUCROSE SOLUTIONS[a]

Sucrose to 100 Parts Water, Parts	Sucrose, %	Lowering, °C	Lowering Due to 1 Part Sucrose, °C
3.59	3.47	0.21	0.05
6.85	6.41	0.40	0.05
10.84	9.78	0.65	0.06
15.83	13.67	0.95	0.06
19.80	16.53	1.23	0.06
22.58	18.42	1.37	0.06
25.64	20.41	1.58	0.06
28.51	22.19	1.77	0.06
32.22	24.37	1.99	0.06
35.14	26.00	2.15	0.06
37.86	27.46	2.33	0.06
43.72	30.42	2.71	0.06
45.62	31.33	2.82	0.07
50.02	33.35	3.13	0.07
54.74	35.37	3.47	0.07
59.46	37.29	3.81	0.07
64.55	39.23	4.22	0.07
69.74	41.09	4.60	0.07
75.91	43.15	5.07	0.07
82.35	45.16	5.65	0.07
88.67	47.00	6.11	0.07
95.94	48.97	6.76	0.07
102.70	50.65	7.38	0.07
111.30	52.67	8.06	0.07
121.00	54.75	9.02	0.07
131.60	56.82	9.93	0.07
143.10	58.86	10.90	0.07
153.80	60.60	11.69	0.08
165.60	62.35	12.72	0.08
181.70	64.49	13.80	0.08

a From Pickering.[122]

is the percent of milk-solids-not-fat (serum solids) in the mix; 2.37 is a constant, resulting from the depression constant of water, namely 18.6, divided by the "apparent molecular weight" of the salts (78.6), multiplied by the proportion of MSNF which is salts (0.1), multiplied by 100; and W is the percent of water in the mix.

The freezing-point depression due to the sucrose equivalent added to the lowering caused by the milk salts gives the freezing point of the mix.

The above described procedure was developed before the use of corn sweeteners in ice cream became commonplace. Since these are now added to almost all mixes in amounts up to 10% by weight, their contribution must be taken into account when calculating the freezing points of mixes currently being manufactured. Calculation is complicated by the fact that several types of corn sweetener, both in syrup form and dry, are available for use, and that, unfortunately, there is a scarcity of data concerning their freezing behavior. About the best that can be done is to compare dilute solutions of the corn sweeteners with sucrose in respect to depression in freezing point so as to rate them in terms of sucrose equivalent. The appropriate factor can then be incorporated into the equation. Unpublished data by the authors suggest that corn sweeteners can be expressed in terms of sucrose equivalent by lb corn sweetener per 100 lb mix × factor (as listed in Table 14.2).

Table 14.2

FACTORS TO BE USED IN CONVERTING CORN SWEETENERS TO SUCROSE EQUIVALENTS

Type of Corn Sweetener	Factor
Dextrose (8% moisture)	1.90
62 DE Solids	1.16
52 DE Solids	0.93
42 DE Solids	0.80
36 DE Solids	0.63
42 DE High maltose type solids	0.72

The method described herein for calculating freezing point may be applied to evaporated milk, concentrated whole and skimmilk, sweetened condensed milk and similar products, to yield approximate values. It is to be expected that accuracy will decrease with increased concentration of the products.

Milk and Cream.—The freezing point of normal milk is quite constant, varying only about 0.036°C from −0.530 to −0.566°C for all herd milk and most individual cow's milk.[13,68,104,124,161] The freezing points of cream and skimmilk are identical with those of the milk from which they are separated,[39] since the respective sera in such cases are the same. Whey will have the same freezing point as milk when it is obtained without chemical or biological modification.

When the temperature of milk or cream is lowered below the freezing point, ice separates in the form of minute crystals of pure water. On continued holding without agitation, these build up into structures which mechanically hold some of the fat globules and proteins in the interstices of the formation. The soluble ingredients, although held to some degree in the frozen structures, largely diffuse into the unfrozen portions of the product. There, as more ice forms, they concentrate and progressively depress the freezing point.[15]

Under static conditions, products in containers freeze from the outside inward, the top portion congealing early. As ice is formed, the unfrozen portion containing the solubles diffuses toward the center, where it becomes more difficult to freeze because of the concentration. When creaming has preceded freezing, or if it occurs during freezing, the cream layer becomes part of the frozen portion when from 20 to 30% of the mass has solidified.[15]

Movement of the nonfat milk solids into the unfrozen portion of the milk during the freezing process is retarded by the fat. As the fat content is increased, the retarding action becomes more and more pronounced, especially if the product is homogenized.[15] At fat levels of 25% and over, cream freezes homogeneously.[15,172] Thus, frozen cream can be sampled for fat by collecting and melting chips taken from any portion of the solidly frozen mass.[172]

One of the very important effects of freezing milk and cream is the destruction of the fat emulsion.[2,14,29,43] The inadvertent partial freezing of milk or cream may lead to an oily layer of fat on the surface of hot coffee, and it is a contributing factor in such defects as cream plug and serum separation in bottled cream.[151] Demulsification and coalescence of the fat are caused by the internal pressures set up in the freezing milk or cream by expanding ice crystals,[86] and the degree is closely related to the extent of freezing.[43] Even mild freezing, as it sometimes occurs on the tubes of a surface cooler during the cooling process, has an effect on the normal fat emulsion of milk, inasmuch as the subsequent creaming ability of such milk is reduced.[95]

Homogenization prior to freezing almost entirely overcomes the demulsification of the fat of milk[11] as well as of concentrated milk,[44] it partially inhibits it with cream containing less than 30% fat, but has little effect when the fat content is higher.[186] Apparently the smaller fat globules produced by homogenization are less easily deformed and the newly adsorbed surface layers are less prone to rupture by expanding ice crystals. The effect is lost, however, when the fat globule concentration becomes so great as to overcome the advantage of size.[15] In homogenized high-fat cream the available protein is insufficient to stabilize the greatly expanded fat-globule surface; fat-globule clusters are formed,[40] and these are less resistant to the pressures of freezing. The addition of sugar to cream considerably retards fat coalescence due to freezing.[44,105,125,186] This has been attributed to the smaller ice crystals formed and to the lowering of the freezing point, which results in less water being frozen at a given temperature.[43] Increasing the solids-not-fat content also retards fat demulsification.[44,186] Very rapid or "quick" freezing minimizes the destructive effects on the fat emulsion,[93] especially when the cream is agitated,[125] or frozen in thin films[33] or in relatively small containers at very low temperature.[131]

Homogenization of milk increases the amount of immobilized casein adsorbed on the fat-globule surface and, when the fat content is high enough (cream), may actually remove substantially all the casein from colloidal suspension in the serum. During the freezing of homogenized milk there is a definite tendency on the part of all the solids of the milk, including the solubles, to drain away from the ice and to settle.[12] Once milk is solidly frozen, no further movement of solids occurs. When the product is defrosted the settling effect again becomes evident and may proceed to such an extent that a watery, whey-like liquid forms at the top of the container.[67,173] This objectionable condition in home-delivered milk has frequently been the subject of complaint by consumers during winter cold spells. If frozen homogenized milk exhibiting the watery layer after thawing is remixed, it assumes its normal dispersion, and no subsequent separation occurs. Such milk is indistinguishable from unfrozen milk, unless some churning or destruction of the fat emulsion has occurred.[41]

Freezing *per se* has no significant effect on the proteins of milk.[46] Defects cannot be detected after thawing, even though the product may have been solidly frozen. On frozen storage, however, the calcium caseinate-phosphate complex is affected. With time it loses its stability and on thawing appears as a precipitate.[44,186] A water dispersion of normal calcium caseinate and/or a relatively clear milk serum may be prepared by solidly freezing homogenized cream containing 25 to 30% fat followed by thawing on a coarse filter.[186]

It is generally conceded that freezing of milk, cream, and dairy products has little, if any, effect on the nutritional properties.[10,103,110,114] During frozen storage, however, the thiamine and ascorbic acid content of milk may decrease.[110] Various freeze-thaw treatments may activate lipase[191] and xanthine oxidase,[84] presumably through mechanisms involving destabilization of the emulsion.

Evaporated Milk.—The calculated freezing point of evaporated milk, of usual composition, is in the vicinity of −1.38°C. When freezing occurs some changes may be evident after thawing, such as destabilization of the fat emulsion, settling of the solids in the container, and, under drastic conditions, precipitation of casein. In some cases freezing may distort the cans and cause the can seams to weaken, giving rise to leakers on defrosting. Lactose may crystallize in the frozen product and form a sediment which redissolves with difficulty.[70] Freezing does not, however, damage the nutritive properties of evaporated milk.[103]

Sweetened Condensed Milk.—In calculating the freezing point of sweetened condensed milk it must be appreciated that the lactose is present in the crystallized from as well as in solution and that, as the temperature is reduced, more lactose slowly crystallizes. At 18.3°C about 15 gm lactose per 100 gm water are in solution, while all of the added sucrose (solubility 179 gm/100 ml water at 0°C) normally is in solution. Since crystallization of lactose is a relatively slow process, because of the slow reversion of the β to α form, it is probable that, on rapid lowering of the temperature, little additional lactose crystallizes before freezing commences. Thus, calculations of the freezing point, for practical purposes, can be based on the solubility concentration at 18.3°C. If calculations are made on this basis, sweetened condensed milk of average composition has a freezing point in the vicinity of −15.0°C.

Because of its low freezing point, sweetened condensed milk is much less likely to freeze than most other dairy products. Storage at temperatures below normal will, with time, cause more lactose to crystallize, producing a coarse texture and possibly sedimentation, even in the absence of freezing.[70] On occasion, even sucrose may crystallize. With freezing, these effects may be exaggerated and, depending on the degree and time, destabilization of the fat emulsion and casein flocculation may appear, as in other products.

Fresh Concentrated Milk.—Products in this category are merely concentrated by removal of water and held under refrigeration until used. The degree of concentration may vary widely, and the freezing point will be depressed increasingly as the concentration becomes greater. When the composition is accurately known, the approximate freezing point may be calculated as previously described. A concentrated whole milk product containing 10% fat and 23% solids-not-fat would have a freezing point of about −2.0°C, whereas a 4-to-1 concentrated skimmilk with 36% total solids would freeze at −3.13°C.

Cheese.—The freezing points of cheeses vary widely with variety, within the variety, and with age, moisture, and salt content. Freezing points range from about −5.0 to −15°C, depending on the above-mentioned variables.[150,184] Aged cheeses have lower freezing points because of loss of moisture, but more particularly because of increases in soluble substances through protein degradation and possibly fat decomposition.[150] Table 14.3 gives freezing-point data on a variety of cheeses.

Table 14.3

FREEZING POINTS AND MOISTURE CONTENT OF STANDARD VARIETIES OF CHEESE[a]

Name of Cheese	Freezing Point °C	Freezing Point °F	Moisture Content %
Cottage	−1.2	29.8	78.7
Cheddar (processed)	−6.9	19.6	38.8
Limburger	−7.4	18.7	44.4
Picnic Swiss (processed)	−8.1	17.4	—
Brick (processed)	−8.7	16.3	—
Swiss (imported)	−9.6	14.7	—
Swiss	−10.0	12.0	34.4
Cheddar	−12.9	8.8	33.8
Roquefort	−16.3	2.7	39.2

a From Watson and Leighton.[184]

When cheese undergoes extensive freezing, the body and texture become more crumbly and mealy after thawing,[150,181] although some recovery occurs with time.[108] Low-moisture cheeses are less affected by freezing than high-moisture products. Such cheeses as Cottage, Neufchâtel, and Cream are usually seriously damaged and exhibit whey leakage after thawing. Freezing does not significantly affect cheese flavor.[108,148]

Butter.—The freezing point of the water phase in butter ranges from 0°C for unsalted butter to −19.8°C for butter containing 3.5% salt.[109] Freezing has no observable effect on the characteristics or quality of butter. Low-temperature storage, in fact, is advocated; temperatures of −23.3°C are commonly employed because flavor deterioration is definitely retarded under frozen conditions compared with holding at higher temperatures.[133] Inasmuch as the cryohydric point of sodium chloride, −21.2°C, is above common storage temperatures, some salt crystallization occurs under frozen storage conditions, but the crystals readily redissolve on thawing.[133]

Dried Products.—Dried milk and milk products contain so little moisture that it is doubtful whether they exhibit a freezing point. This is understandable, since rapid evaporation of moisture in the drying operation leaves behind a very concentrated, amorphous lactose-water combination, the water of which is very tenaciously held and insufficient to permit lactose crystallization.[174,189]

FROZEN STORAGE

Predictably, dairy products can be stored in the frozen state almost indefinitely without perceptible change of microbial origin. However, chemical and physical interactions among milk components are not completely inhibited and flavor, color, and textural changes will eventually exceed threshold levels. The direction and magnitude of these changes may be greatly influenced by product composition, the manner of processing, and length of storage, as well as by external environmental factors such as exposure to light, storage temperature, temperature constancy, and the humidity, circulatory pattern and composition of the air phase around the frozen product.

While flavor deterioration cannot be rectified by further processing, the physical changes which may have occurred in storage are of little consequence when the thawed product is added as a component of a food to be pasteurized and homogenized, such as ice cream. However, emulsion destabilization, viscosity changes, and sedimentation of protein complexes are important variables which must be controlled when frozen milk and frozen milk concentrates are distributed through retail outlets as substitutes for fresh milk.

Products and Problems

Frozen Cream.—The practice of freezing good-quality cream of 40 to 50% fat for later use in ice-cream manufacture, has been successfully employed for several decades.[31] Provided the cream is properly handled, holding periods of 6 to 10 months are not uncommon at storage temperatures of −17.8 to −26.1°C. The development of oxidized flavor is probably the most troublesome problem, but this can be controlled by avoiding copper contamination, by using high-heat treatments,[62,63,75] and by homogenizing[17,129] the cream. Anti-oxidants also have been found effective for the purpose.[31,162,169]

The churned condition of the fat, noted when frozen cream is thawed, is not a critical factor when the product is to be used in ice cream, because homogenization of the mix re-emulsifies it satisfactorily.

Churning of the fat can be reduced, however, by adding 5 to 15% of sugar to the cream,[43] by use of lower storage temperatures,[20] and of stabilizers.[18] Although homogenization before freezing has little effect in stabilizing the fat emulsion of high-fat cream, mild processing may be employed to assist in preventing development of oxidized flavor.

It has been demonstrated that when cream is frozen very rapidly in a matter of minutes or less, the fat emulsion is not damaged significantly, even after long periods of frozen storage.[20,33,125,131] In such cases, the defrosted product is normal in physical appearance and is practically indistinguishable from fresh table or whipping cream. To produce such cream, special and very rapid freezing techniques must be employed. The freezing of cream in a thin layer on a revolving refrigerated drum is a particularly satisfactory procedure for preserving the natural characteristics of cream after defrosting.[33]

Frozen Fluid Milk.—Pasteurized, homogenized fluid milk has been statically frozen in ordinary fiber containers and held satisfactorily for periods of 3 to 4 months at temperatures of about −23.3°C. When defrosted, the product has normal physical characteristics and an acceptable, though slightly flat and chalky flavor.[137,177,185] Oxidized flavor can be controlled by the procedures mentioned under frozen cream, so that it is not flavor deterioration but instability of the casein which more often limits storage life. Inasmuch as increasing concentrations of casein tend toward decreasing stability, the problem is more critical with concentrated than with fluid milk.

Frozen fluid milk was used very successfully by the military in World War II for furnishing casualties on hospital ships with a product which was generally accepted as fresh milk.[11] It has also been used locally in some places to decrease the number of milk deliveries where customers have ample freezer space.[185] It is being furnished to ships on long voyages as a preserved supply of "fresh milk."[147]

Frozen Concentrated Milk.—From a purely technical viewpoint, milk can be concentrated, homogenized, frozen, and distributed as a frozen food. Keeping periods of 6 to 10 weeks are attained without great difficulty with a 3-to-1 concentrate, if properly processed and handled.[46,170,175] Economically, however, the commercial aspects have not been developed as favorably as has been the case with frozen concentrated orange juice and some other frozen foods.

Oxidized flavor in this product is less of a problem than with frozen fluid milk, but the tendency of the calcium caseinate-phosphate complex to flocculate on defrosting is more serious, and limits the storage life to a shorter period. This loss of stability on the part of casein[187,190] is progressive in frozen storage.[46] At first the flocculation is reversible with heat and agitation, but with continued storage it becomes irreversible.[44,46] Whey proteins contribute to the volume of the precipitate only if they have been denatured by heat treatment applied to the product before freezing.[136,190]

Factors contributing to instability of the casein in frozen storage include: high degree of concentration of the milk; preheating or pasteurizing temperatures in excess of 77°C; storage temperatures above 23.3°C and especially higher than −17.8°C; cooling and/or holding the product under refrigeration between concentrating and freezing; and lengthening storage periods.[46,136,175,187,190]

Casein precipitation during frozen storage is delayed by: lower concentration of the milk; the use of minimum temperatures in pasteurizing or preheating prior to concentration; post-concentration heat treatment of not over 68.7°C for 30 min;[136] slow initial freezing;[135] freezing promptly after concentration; [136,190] addition of sugars or glycerol to the milk;[132,190] and low storage temperatures.[19,45,46] Research on this problem has been reviewed.[192]

The initial amounts of water frozen in skimmilk at various temperatures are shown in Table 14.4, with data for whey and lactose solutions also included. It is evident that at temperatures comparable to those used in the frozen storage of milk very little liquid water remains as solvent for the lactose and soluble salts. It has been estimated that a 10-fold concentration of the salts occurs at −7.5°C prior to lactose crystallization, and that up to a 30-fold concentration develops as lac-

Table 14.4

PERCENTAGE OF WATER FROZEN IN MILK, WHEY, AND LACTOSE SOLUTIONS AT VARIOUS TEMPERATURES[a]

Temperature °C	Skimmilk		Whey		Lactose Solution[c]	
	T.S.[b] 9.3%	T.S. 26.0%	T.S. 6.5%	T.S. 20.3%	5%	15%
−2	75.0	20.0	73.0	30.0	84.1	54.2
−4	87.5	53.0	91.0	58.0	91.2	73.7
−8	92.5	74.0	95.0	77.0	95.8	87.4
−12	94.5	81.0	95.5	83.0	97.0	91.0
−16	95.0	84.5	96.0	86.0	97.6	92.7
−20	95.5	86.0	96.5	88.0	97.9	93.7
−24	96.0	88.0	97.0	90.0	98.1	94.5

a From Tessier, Rose, and Lusena.[166]
b Total solids by oven method.
c Values calculated from molal freezing point depressions of supersaturated solutions.

tose crystallizes.[137] Data in Table 14.5 show analyses of ultrafiltrates from the unfrozen liquid in skimmilk frozen at the above temperature, compared with those from unfrozen skimmilk.[165]

Table 14.5

ANALYSIS OF ULTRAFILTRATES OF SKIMMILK AND THE LIQUID PORTION OF FROZEN ULTRAFILTRATES[a]

Ave. of Three Samples

	Ultrafiltrate of Skimmilk	Ultrafiltrate of Liquid Portion from Frozen Concentrated Milk	Recovery,[b] %
pH	6.7	5.8	—
Chloride	34.9 mM	459 mM	100
Citrate	8.0 "	89 "	84
Phosphate	10.5 "	84 "	61
Sodium	19.7 "	218 "	84
Potassium	38.5 "	393 "	78
Calcium	9.1 "	59 "	50

a From Tessier and Rose.[165]
b Recoveries calculated relative to chloride.

One of the most interesting and significant findings concerning casein precipitation in frozen milk is its temporal relationship to the crystallization of lactose. A number of studies have shown that the casein sol retains its normal colloidal dispersion until some of the lactose crystallizes.[38,166,175,187] Most investigators agree that any factor which accelerates the formation of lactose crystals in the frozen product shortens its storage life, and vice versa. Enzymatic hydrolysis of lactose by lactase in milk before freezing retards both lactose crystallization and casein precipitation in proportion to the extent of hydrolysis.[175] It is also evident that at very low temperatures (below −23.3°C) neither lactose crystallization nor casein flocculation occurs, even after very long storage.[42,175]

Before any lactose crystallizes in frozen milk, it appears that the highly concentrated liquid phase, even with the increase in acidity, is insufficient to cause flocculation of the colloidal casein complex, but it undoubtedly has caused some loss of stability.[165] By remaining in solution, lactose prevents further ice formation and further concentration of the solutes. It may also contribute to such a high viscosity in the unfrozen liquid that diffusion and orientation of ions, molecules, and micelles are virtually suspended.[175] Other added sugars and glycerol seem to act in similar fashion. Since they are more soluble and do not crystallize, their effect is less transitory.[134] When lactose does crystallize, more water freezes, the salts become more concentrated, the acidity increases, more calcium phosphate is rendered colloidal, and the casein becomes flocculated. The precipitation may be a salting-out effect; it may be the result of casein dehydration coupled with the downward pH shift;[154] it may be caused by freshly precipitated calcium phosphate interlinking the casein micelles to form flocculant structures. The latter action might explain the initial reversibility of coagulation with heat and agitation after defrosting.[137,187] Conceivably, all of these phenomena may be involved in the aggregation of the caseinate micelles leading to precipitation.

Lactose, in common with other sugars, exhibits reluctance to crystallize from solution even when considerably supersaturated. As a consequence, milk can be cooled through the freezing range, even very slowly, to low temperatures without promoting crystallization, provided that no nucleation has occurred prior to freezing. At temperatures below − 23°C, when over 80% of the water of concentrated milk and over 95% of that in fluid milk is present as ice (Table 14.6), the lactose is in an amorphous "glass state" and unable to crystallize because of the high viscosity, the paucity of free water, and lack of molecular mobility. In this condition the serum and nonserum phases of frozen milk are extremely stable, and long storage periods are possible. If storage temperatures rise, or higher temperatures are employed, lactose nucleation and crystallization will slowly take place, and the protective influence of the highly supersaturated lactose will be lost. The induction period for crystallization decreases with increasing temperature,[166] but even here it is slowed by the low kinetic energy at these temperatures, the high viscosity and the slow reversion of the β to α forms of lactose.

The calcium caseinate-phosphate complex flocculated during frozen storage of milk has been found to be similar to but not identical with that existing in normal milk. It is associated with more calcium phosphate,[187] and electrophoretic patterns indicate that both the precipitated protein and serum protein fractions are somewhat altered by frozen storage.[37,115] Flocculated casein contains less β-casein than normal acid casein,[116] and has been shown to have characteristics similar to casein salted out of milk by saturation with NaCl.[46] Several workers have suggested that the unfrozen solution, highly concentrated in salts, exerts such an effect in destabilization of the casein sol.[46,137,175] Among the salt ions, calcium seems to be especially active in this connection, possibly due in part to the low cryohydric point of $CaCl_2$[154] and to its specificity in all precipitations of casein.

Calcium-precipitating salts, calcium-sequestering salts, and the removal of calcium by ion-exchange all tend to delay casein flocculation when applied before freezing milk.[46,66,102,190] The beneficial effects of polyphosphates lie in their ability to enhance the dispersibility of flocculated casein rather than any direct effect on protein coagulation or lactose crystallization.[102]

In the light of these considerations, some of the less obvious factors influencing the destabilization and precipitation of casein mentioned previously become more understandable. A prolonged interval between concentration of milk and its freezing permits lactose nucleation,[136] which later may eliminate or greatly reduce the induction period for crystallization. Such nucleation is not likely in fluid milk, and this may be one reason why the frozen fluid product generally has a longer storage life.[190] A moderate heat treatment applied to concentrated milk prior to freezing undoubtedly dissipates any lactose nuclei which may have formed, while temperatures over 77°C applied at any time in the processing denature serum proteins and cause them to be occluded in the flocculation of the casein. Except for reduction of the percentage of soluble calcium in milk, all the factors that promote casein stability in frozen storage are related to maintaining the protective influence of soluble lactose in the liquid phase of the frozen mass or to adding substances, such as sugar, which will function in a similar manner.

In recent years, the chemistry of frozen dairy products, other than ice cream, have received scant attention. Several reports of research findings not previously referred to are worthy of mention.[24,25,47,74,91,139,163,179,185 A,194]

ICE CREAM

The products previously considered in this chapter are generally used in the unfrozen condition. Consequently, the effects of freezing are of interest primarily with respect to the thawed product and how its characteristics differ from the product which has not been frozen. As ice cream is consumed in the frozen state, it is the properties of the frozen product that are of ultimate concern.

Definition

This discussion is oriented toward the frozen dessert legally defined as ice cream. Ice cream is but one type of frozen dessert and ice milk, sherbet, water ice, mellorine, and quiescently frozen confections contribute significantly to the total volume of products generally identified with the ice-cream industry. It should be kept in mind that the chemistry and technology of these products do not differ greatly from that of ice cream.

Ice cream, as defined under U.S. Federal Standards, must contain at least 10% milk fat and 20% total milk solids, except when there is a diluting effect because of the use of such flavoring materials as chocolate, fruits, or nuts. Ice cream may not contain more than 0.5% stabilizer or 0.3% emulsifier, and 1 gal must weigh at least 4.5 lb and contain not less than 1.6 lb food solids. The minimum weight requirement permits the ice-cream manufacturer to whip air into the mix to approximately double the volume (100% overrun). French ice cream or frozen custard must contain at least the equivalent of 1.4% egg yolk solids.

Ice milk is low-fat ice cream containing 2 to 7% milk fat. Sherbet contains twice as much sugar as ice cream, but milk solids are restricted to a maximum of 5%. Water ice is identical with sherbet, except that milk solids are not present. Sherbets and ices may or may not contain an added acidulant.

Mellorine is the name commonly applied to frozen desserts having compositions and properties similar to ice cream and ice milk, except that vegetable fat is used as a replacement for milk fat. Quiescently frozen desserts include a wide variety of products of vaguely defined composition that have been frozen without agitation. Freezing is accomplished by filling a mix into a mold of the desired shape and immersing it in a tank of very cold brine or antifreeze solution. Most stick confections are quiescently frozen. The mix, before being placed in the molds, may be slushed in a freezer to prechill it and to incorporate air as a means of making products of various characteristics.

Composition

Three grades of ice cream usually can be found in most market areas. The chemical composition of these ice creams differs mainly with regard to the fat content. One grade just meets the minimum fat content, often has an overrun that approaches the maximum allowed by law, and usually contains relatively inexpensive flavor ingredients. At the other extreme are the so-called premium ice creams that are high in fat, low in overrun, and usually contain natural flavors. A third grade of ice cream, designed as a compromise between the minimum-cost and premium products, is the type that has dominated the market for many years. Representative formulas for each grade are presented in Table 14.6.

Since ice cream is manufactured with widely differing analytical composition and is flavored with a variety of substances, the average

Table 14.6

COMMON GRADES OF ICE CREAM

	Minimum-Fat %	Regular %	Premium %
Milk fat	10.1	12.0	18.0
Milk solids-not-fat (MSNF)	11.5	11.0	9.5
Sucrose	11.0	12.5	15.5
Corn sweetener solids	7.0	5.0	—
Stabilizer	0.35	0.32	0.25
Emulsifier	-.1	0.1	—
Total solids	40.05	40.92	43.25

chemical composition can only be approximated. Practically all the fat, protein, and minerals are contributed by the milk solids. Lactose, sucrose, corn sweeteners, and to a small extent stabilizers account for the carbohydrate content of ice cream.

Ice Cream Ingredients: Functions and Sources

Milk Fat.—Milk fat is the most important component of ice cream that affects quality. When choosing a formula, an ice-cream manufacturer first selects a fat level and then proportions the other components to blend with it. Milk fat imparts richness, body or substance, and helps to ensure a smooth-textured ice cream. Perhaps the most easily recognized difference between low-fat and high-fat ice cream is the sensation of coldness; low-fat ice cream feels colder in the mouth. Milk fat contributes a subtle flavor quality. It is a good carrier and synergist for added flavor compounds, and it promotes desirable tactile qualities. Most ice-cream manufacturers favor the use of fresh cream as the principal source of fat. Flavor problems are minimized and fresh cream has handling advantages not possessed by stored fat products, which must be thawed or comminuted before being added to the blending vat. Frozen cream, plastic cream, and butter are frequently used as a partial replacement for fluid cream when the latter is in short supply.

Milk Solids-Not-Fat.—The nonfat milk solids (serum solids), besides contributing to the flavor, body, and texture, are essential for the formation and maintenance of small, stable air cells. Concentrated skim and whole milk, and nonfat dry milk are the most important sources of serum solids, supplementing those derived from cream. Whey solids can be used as a partial substitute for skimmilk solids.

Sugars.—The sweetness of ice cream is contributed by sucrose, corn sweeteners, and to a slight degree by lactose. Since components in true solution determine the freezing point, the concentration of sugars prevents ice cream from being frozen solid, even at very low temperatures. Were it not for the sugars, ice cream would freeze hard at a comparatively high temperature,[121] and the product would be entirely lacking in desirable body and texture characteristics.

Most ice cream is now made with some type of corn sweetener as a partial replacement for sucrose. Those products result from the hydrolysis of corn starch by either an acid or an enzyme, or both. The degree of hydrolysis is expressed as the dextrose equivalent, or DE.[28] Complete hydrolysis yields D-glucose (dextrose) which has a DE of 92. When the reaction is stopped at intermediate stages, corn syrups are obtained which consist of dextrose, maltose, and a complex mixture of polysaccharides, some having molecular weights that are quite high.[28] The lower the DE of the corn sweetener, the greater will be the relative proportion of high molecular-weight polysaccharides. These intermediate compounds are noncrystallizable in nature, more or less hydrate or bind water, increase the viscosity of the water phase and permit a higher total solids content in the mix without adding much to the sweetness. They contribute desirable "chewiness" and body to ice cream and also impart protection against the development of coarse-textured ice cream during storage in retail cabinets and in home freezer chests.[51,61,81,195]

When a heavy, chewy body and resistance to heat shock are desired in ice cream, a low-conversion corn syrup is employed having a DE in the range of 28 to 42. When the objective is an ice cream with chewiness, and body and texture characteristics more like that of ice cream containing only sucrose, then intermediate-range (50-57) DE syrups and high-conversion (58-62) syrups are used. High-DE syrups are more economical sources of sweetness than lower-DE products because they are sweeter and less is required per lb of sucrose replacement.[78] Enzymatically hydrolyzed corn syrups in which maltose accumulates in preferences to dextrose are claimed to be superior to regular corn syrups of comparable DE.[48]

Stabilizers.—The primary reason for using a stabilizer in ice cream is to aid in the maintenance of a smooth texture by preventing large ice crystal formation during handling by the manufacturer, dealer, and consumer between manufacture and consumption. Compared to ice cream that does not contain a stabilizer, properly stabilized ice cream will have a heavier body, will not taste as cold, and will melt more slowly and to a creamier consistency.[112]

In many food applications stabilizers are used as thickening, suspending, and emulsifying agents. Their function in ice cream is primarily to control ice-cystal size. All stabilizers increase the viscosity of the unfrozen portion, which restricts molecular migration to crystal nuclei, thereby limiting crystal size. The ability of stabilizers to bind or hold relatively large amounts of water is also advanced to show why a smooth texture is maintained. Experimental support of both explanations is available.[4,147] Because of the lack of reliable physical and chemical tests for evaluating stabilizer effectiveness in ice cream, the compounding of stabilizers is a trial-and-error process involving more art than science.

The most commonly used hydrocolloids are sodium carboxymethylcellulose (CMC), locust bean gum, sodium alginate, propylene glycol alginate, carrageenan and guar gum. Gums occasionally used for special purposes include gelatin and the gum exudates, karaya, tragacanth, and arabic.

Commercial ice-cream stabilizers are usually blends of two or more basic ingredients. CMC, locust bean gum, and guar gum stabilizers, alone or in combination, almost always are blended with a small amount of carrageenan, a sulfated galactose polymer. Carrageenan complexes with milk protein, and when present in mix at a concentration of approximately 0.018% effectively inhibits the serum separation or wheying-off encountered when the above-mentioned gums are used alone.[160] Stabilizers differ in respect to their effect on mix viscosity, ease of dispersion in mix, body and texture, and melting characteristics of ice cream.[32] The particular stabilizer chosen for use will depend on a combination of factors involving price, the type of processing used in making the mix, and on body and texture qualities desired in the finished product.

Emulsifiers.—Emulsifiers are especially desirable when extrusion techniques are used for forming specialty items such as sandwiches, slices, and factory-fill cones. Emulsifiers stiffen ice cream by a mechanism involving the agglomeration of fat globules.[80,90] They favor a smooth texture, since both ice-crystal and air-cell sizes are reduced.[9] Emulsifiers may improve the whipping properties, particularly if stored fats are used and the mix is frozen in a batch or soft ice cream freezer.[176,82] Since overrun and texture problems are rarely encountered in the air-metered continuous freezers which are used in the manufacture of commercial ice cream, the stiffening ability is the most important factor to be considered when selecting an emulsifier.[79]

Both monoglycerides and polysorbate esters are approved for use in ice cream. Monoglycerides are most often obtained as a mono- and diglyceride mixture also containing small proportions of triglycerides and traces of saponification products. The monoglyceride content of commercially available products may vary from 40 to 98%, and since these emulsifiers are usually derived from hydrogenated vegetable and animal fats, palmitic and stearic acid esters predominate.[72] A few commercial products contain high levels of glyceryl monocleate. The effectiveness of such an emulsifier depends on the percentage of monoglycerides and on the particular fatty acids present. There is an inverse relationship between carbon chain length of the fatty acid and the stiffness produced in ice cream; capric and lauric acid monoglycerides are more effective in this respect than glyceryl monopalmitate or monostearate.[87] Monoglycerides containing unsaturated fatty acids impart more stiffness than saturated monoglycerides.[87,90]

Polysorbate emulsifiers provided for in Federal standards are the polyoxyethylene derivatives of sorbitan tristearate and sorbitan monooleate.[50] These emulsifiers are more hydrophilic than the monoglycerides and are used at much lower levels, since they are especially effective in imparting stiffness to ice cream as it leaves the freezer.[87,88]

Mix Processing

Ice-cream mix preparation is a multi-step process involving the blending and dispersing of a variety of liquid and solid ingredients into a homogeneous fluid product that is pasteurized, homogenized, and cooled. Dispersion and activation of the stabilizing ingredient and the control of mix viscosity are probably the most troublesome problems encountered in mix processing.

Stabilizer dispersion seldom presents a problem when mix is vat-pasteurized. However, in HTST or ultra-high heat (UHT) systems, the beneficial effects of heat and time on dispersion and hydration are not fully available prior to passage of the mix through the pasteurizing unit.[112] Improperly dispersed stabilizer may increase the tendency for the mix to coat the hot surfaces of the heat exchanger and cause operating difficulties. Furthermore, the short holding period at high temperatures may be insufficient to completely develop the effectiveness of the stabilizer.[158] To assist in overcoming these difficulties stabilizer manufacturers offer specially processed products, which disperse readily in cold mix and which do not require long holding times at high temperatures for hydration.

Pasteurization.—More heat is required to effectively pasteurize ice-cream mix than milk because of its relatively high total solids content and greater viscosity, which exert a protective effect on some microorganisms.[64] In pasteurizing ice-cream mix by the holding method the official recommendation of the U.S. Public Health Service is a minimum of 68.3°C for 30 min. In commercial plants using vats the practice is to pasteurize in the range of 71.1 to 73.9°C for 30 min.

For the continuous-flow, high-temperature short-time (HTST) pasteurizing equipment used by most large ice cream plants, the U.S. Public Health Service recommendation is 79.4°C for 25 sec.[156A] Continuous HTST pasteurization followed by vacuum deodorization and cooling[132,171] has been used in accordance with the USPHS recommendation of 87.7°C heating.[71]

For UHT pasteurization, the mix is flashed to a temperature of 104 to 150°C with no purposeful or measurable hold. The beneficial effect of high heat on ice-cream mix components (particularly milk proteins) has been recognized,[69] and UHT processing of mix is said to improve the body and texture of ice cream and may allow a reduction in stabilizer usage.[6] Chapter 11 presents the changes that occur in milk with heat treatment.

Homogenization.—Ice cream mix is homogenized to reduce the size of the fat globules and to create a high degree of fat dispersion. The effect of homogenization on mix viscosity is discussed later in this chapter and a discussion of the physical chemistry of homogenization appears in other chapters.

Homogenization was shown many years ago to increase mix viscosity.[111,128] A viscous mix was preferred when freezing was done in batch freezers, because it apparently incorporated air and whipped better than a low-viscosity mix. With modern mix-making equipment, where warm mix leaves the regenerative pasteurizer at 35 to 40°C to be cooled to 4°C over a surface cooler or other continuous cooling device, high viscosity is undesirable. Mix is usually homogenized immediately after pasteurization at a temperature range of 60 to 74°C and at a pressure of 1,500 to 3,000 psi. Optimum homogenization temperature and pressure for a specific operation depend upon several factors, most of which can be controlled by the processor.

Freezing and Hardening of Ice Cream

Freezing.—The function of the ice cream freezer is twofold. The ice cream mix is partially frozen and at the same time air is incorporated or whipped into the mass. Mix is pumped into the freezer together with a metered amount of air. As it passes through the freezer, the temperature is lowered as scraper blades sweep the mix over the refrigerated walls. Refrigeration is supplied by expanding liquid ammonia or freon, usually at −22 to −32°C. Ice in the form of small crystals separates from the water of the mix, thus causing an increase in viscosity of the unfrozen portion. The mass progressively becomes more viscous and finally is discharged from the freezer at −5 to −6°C as a stiff plastic-like stream, suitable for packaging. Continuous agitation, provided by beater blades, disperses air in uniformly distributed, very small bubbles or cells.

The expansion or increase in volume of mix as it is frozen is technically referred to as "overrun." It is expressed in percentage as follows:

$$\% \text{ overrun} = \frac{\text{weight of mix} - \text{weight of ice cream}}{\text{weight of ice cream}} \times 100,$$

where the weight of each product is the weight of the same unit volume.

The result of a proper amount of well distributed air, coupled with small ice crystals, is a fine-grained palatable product. Ice cream of low overrun tends to be dense and soggy; if the overrun is excessively high, the ice cream becomes fluffy and lacks body. The amount of overrun in ice cream ranges from 40 to 100%.

Hardening.—During the time the mix is in the freezer it changes to a flowable semi-frozen mass with approximately half the water frozen into ice.[93] In this condition it is filled into packages and is then ready for further reduction in temperature to freeze the product to a solid mass. This final chilling to below −18°C is known as hardening.

In most small manufacturing plants hardening is accomplished in the cold-storage room in which air at −30°C circulates and sweeps heat away from the surfaces of the ice-cream package. Optimum conditions should bring the core temperature in a half-gallon package of ice cream to −18°C in 4 to 6 hr.[149] In large-volume plants fast-hardening units are employed to at least partially harden packages of ice cream before they are placed in cold storage. Packages are automatically conveyed through the unit and exit after about 1 hr in a hardened condition. Rapid removal of heat is achieved with high-velocity air blasts at −50°C[3] or by conductive transfer of heat through refrigerated contact plates.[157]

The rate at which ice cream hardens depends on a combination of factors, the most important of which are size of container, air temperature, air velocity, and the degree of exposure of container surface.[167,168] Ice cream entering the hardening area contains a vast number of minute ice crystals which act as centers of crystallization for the ice that is to be formed during hardening. Assuming that all the ice that forms during hardening is deposited uniformly upon crystals already present, the increase in size of the existing ice crystals would be insufficient to affect the texture of ice cream significantly. If temperature fluctuations are allowed to occur in the storage room, during transportation, or in dealers' cabinets, small ice crystals will melt when the temperature rises, and this water will be deposited as ice on the larger crystals on refreezing. In this manner of repeated temperature fluctuations, known as heat shocking, a coarse texture results from the development of large individual or agglomerated ice crystals.

The beneficial effects on body and texture of below-normal extrusion temperature at the freezer (−9°C) and immersion-hardening in liquid nitrogen have been demonstrated.[54,56,58,59,138,193]

Ice-cream Mix as an Emulsion and Colloidal Dispersion.—The transmutation of ice-cream mix into ice cream occurs in the freezer under violent conditions of agitation followed by hardening in the quiescent state. The finished product represents a delicately balanced system whose stability and physical structure are dependent on the physicochemical properties of its mix.

Ice-cream mix is essentially an oil-in-water emulsion. The dispersed phase is the milk fat, the continuous phase an aqueous serum consisting of calcium caseinate-calcium phosphate micelles, serum proteins, carbohydrates, and mineral salts. On the basis of particle size, the serum phase is a mixture of a colloidal dispersion and a true solution.

Mix Viscosity.—The viscosity of a quiescently aged mix reaches a maximum which can be reduced by agitation to a lower constant value. This phenomenon of two viscosities, termed apparent and basic,[99] is due to a loose network of casein micelles, fat globules, and stabilizer hydrocolloids individually or collectively aggregated to form a structure. That apparent viscosity is a structural viscosity is evidenced by consistency and yield value determinations,[16,92,98,99,100] the yield value being the force required to break the structure, thus converting the system from plastic to Newtonian. Basic viscosity is not to be regarded as the minimum possible value of viscosity following complete mechanical breakdown of structure, but rather as an arbitrarily useful concept when a standardized method of agitation has been used.[66]

During freezing apparent viscosity is rapidly destroyed, largely disappearing shortly after the mix is introduced into the freezer. Basic viscosity increases as a result of progressive concentration of the mix due to separation of ice crystals, and is a function of the mix solutes and the fundamental emulsion structure of the mix.[35] Apparent viscosity is a physical property of the mix which indicates a potential for change with time and temperature, a factor which may exert an influence on ice-cream structure after freezing.

In homogenized mix basic viscosity is at a minimum if the globules are small and individually dispersed. However, the tendency of subdivided fat globules to clump or aggregate greatly increases the viscosity,[111,128] depending on the number, size and shape of the clumps.[153] The extent to which this phenomenon occurs is influenced by such factors as the proportion of ingredients used in the mix, particularly the MSNF/F ratio, the acidity, presence of emulsifiers, salt balance, and homogenization pressure and temperature.

The increase in mix viscosity due to the formation of clumped fat globules has been attributed to the entrapping of a portion of the continuous phase in the interstices of the clumps. This enclosed portion then functions as if it were fat in its effect on viscosity. The degree of fat clumping may be quantitatively described by a "clumping index" using the viscometric method of Leviton and Leighton,[101] who derived an empirical formula expressing the relationships between the viscosity of a mix free from fat clumps and that of its continuous phase.

Research of relatively recent origin on the rheology of ice cream and ice-cream mix by Sherman and co-workers[73,141,143,144,145,146,188] has resulted in new interpretations of viscosity data for mix. Although their studies were carried out on mixes which contained vegetable fat, interpretations should be equally applicable to milk-fat products.

Rheological and microscopic methods were employed to demonstrate that viscosity increases precipitously as the minimum mean distance between adjacent globules in ice cream mix falls below 1μm.[143] Based on theoretical considerations, the critical distance cited, 1μm, was greater than expected. This was attributed to the effect of an extensive hydrated layer of surface-denatured and coagulated protein adsorbed around the globules during and after homogenization. From viscosity and mean globule size data an average thickness of 0.3μm was calculated theoretically. A rigid interfacial layer of this type would impart flow properties to mix which would resemble those of suspensions of solid spheres in liquid media.

Mix Stability.—According to Sherman,[146] the colloidal stability of dilute oil-in-water emulsions is controlled by two phenomena, flocculation of globules and subsequent coalescence. Flocculated globules are held together by weak forces of attraction and can be redispersed by agitation. Coalescence, on the other hand, is an irreversible process.

Coalescence occurs in two stages provided the mean diameter of the globules is below a critical value.[144] The initial phase is quite rapid, but there is a transition to a slow rate of coalescence when globules with diameters less than 1 μm have disappeared. The energy barrier to coalescence depends on the electrical forces preventing flocculation, on the rate of thinning of the film between globules after they have flocculated, and on the physical properties of the monoglyceride-protein layer.[144] The degree to which the globules in mix flocculate and the coalescence prior to freezing affect the structure of ice cream.

Physical Structure of Ice Cream

Ice cream can be classified as a foam, or a physicochemical system containing a gas (air) dispersed in a liquid, a solid, or a mixture of liquid and solid. According to King[85] "ice cream is a rather complicated foam, the continuous phase representing a partly frozen emulsion, the ice crystals and the solidified fat globules being embedded in the unfrozen water phase." The size of the coarsely dispersed structural components can only be approximated. Typical values reported by Arbuckle[5] are: ice crystals, 50 μm diameter; air cells, 175 μm diameter; distance between ice crystals, 7 μm; distance between air cells, 125 μm. The solidified fat globules, ranging from 0.5 to 2 μm in diameter, may be individually dispersed or may occur in clusters and chains in the unfrozen serum and around the air cells.[9] Microscopic techniques employed in studies of ice-cream texture have been reviewed.[7]

Effect of Mix Composition on Structure of Ice Cream.—Using the petrographic microscope for identification of crystalline and noncrystalline materials as they actually exist in ice cream, it was established that composition of the mix markedly affects the physical structure of ice cream, the fine texture being associated with the presence of a uniformly dispersed system of small, angular-shaped ice crystals.[127] The factors reducing the size of ice crystals (thus favoring a smooth texture in ice cream) can be arranged as follows in decreasing order of effectiveness: fat plus serum solids, fat, serum solids, sugar.[5]

Increasing the fat content reduces the ice crystal size and increases the distance between crystals, but air-cell size and distribution are not affected. Increased fat content produces smaller ice crystals by mechanical obstruction, causing water to freeze out of the mix in smaller crystals instead of allowing it to collect on crystal nuclei and form large ice crystals during freezing.[30]

Raising the serum solids level favors smaller ice crystals and air cells. The distance between ice crystals becomes greater, while air cells are brought closer together. Increased serum solids produce smaller ice crystals, both by mechanical obstruction and the water-binding properties of the extra protein. The increase in distance between ice crystals may be attributed to the lowered freezing point of the mix, resulting in a greater amount of unfrozen material between the crystals.[30]

Partial replacement of sucrose with dextrose increases both ice-crystal size and the distance between crystals.[126] Although published microscopic data are not available, subjective evaluations of texture imply that the substitution of partially hydrolyzed corn starch for sucrose should yield smaller ice crystals.

Adjusting the pH of the mix in the range of 6.1 to 7.5 has no appreciable effect on the microscopic crystalline structure of ice cream,[129] although it is reported that smaller ice crystals are formed at neutral or slightly alkaline pH values.[5,130]

Freezing Equilibria as Functions of Mix Composition and Temperature.—The freezing point of ice cream mix depends on the same factors as those which determine the freezing point of milk, namely, the concentration of the soluble components, such as sucrose, lactose, milk salts, and any others in solution. Fat and protein, although not in solution, exert an indirect effect by replacing water, thus increasing the concentration of the soluble substances in the aqueous phase.

The initial freezing point of the average ice-cream mix is −2.8 to −2.2°C. This temperature essentially reflects the freezing-point depression due to the total sugar content of the mix. When latent heat is removed from water and ice crystals are formed, a new freezing point is established for the remaining solution, since it has become more concentrated with respect to the soluble components. The transfer of sensible heat from the unfrozen solution lowers the temperature to a new freezing point and more water is converted into ice. Thus, the freezing point of the liquid portion of ice cream is continually changing as water is frozen. It is to be expected that as the individual dissolved components become more concentrated in the unfrozen solution, solubility limits may be exceeded and crystallization of dissolved substances may occur. Conclusive proof has not been obtained that sugar, lactose, or any other soluble component crystallizes in normal ice-cream making, and it therefore is believed that they exist in a supersaturated state in the unfrozen portion.[155]

If the sugars and soluble salts remain in solution, it should be possible to calculate the amount of water that will be frozen at any particular temperature during freezing, hardening, and storage of ice cream. For example, suppose it was desired to know the temperature at which 50% of the water would be frozen in a mix containing 12% fat, 11% milk solids-not-fat, 16% sucrose, 0.3% stabilizer, and 60.7% water. The procedure used is the same as that described for calculating the freezing points of milk, ice-cream mix, and other dairy products (see introductory pages of this chapter), except that only 50% of 60.7, or 30.4 parts, of water is used in the formulas.

$$\frac{(\text{MSNF} \times 0.545) + \text{sucrose}}{\text{water}} \times 100 =$$

$$\frac{(11 \times 0.545) + 16}{30.4} \times 100 = 72.3 \text{ parts sugar}$$

Referring in Table 14.1, the equivalent 72.3 parts sugar would depress the freezing point 4.78°C.

Freezing-point lowering caused by milk salts would be:

$$\frac{11 \times 0.1 \times 100 \times 18.6}{30.4 \times 78.6} = 0.86°C.$$

The temperature at which 50% of water is frozen is (−4.78) + (−0.86) = 5.64°C.

Calculating the freezing points of mix at several concentrations and then plotting freezing point versus percentage of water frozen gives a freezing curve for ice cream suitable for predicting the amount of water that will be frozen at any given temperature. The freezing curve for ice cream with a composition conforming to that used in the sample calculation is shown in Fig. 14.1. Calculated values agreed quite well with experimentally measured freezing points for mixes which differed only in water content, 32 to 65%,[97] and with direct measurement of the water/ice equilibrium by dilatometry.[26]

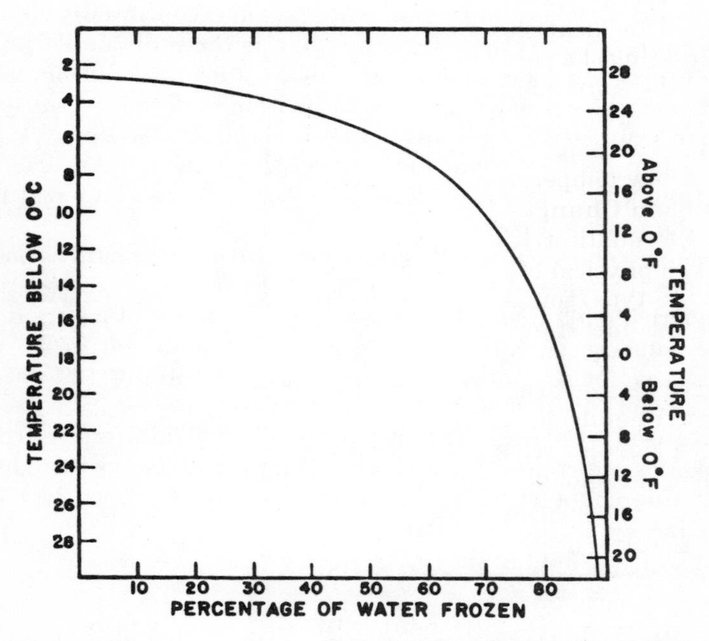

From Berger and White[21]

FIG. 14.1 A TYPICAL FREEZING CURVE FOR ICE CREAM SHOWING THE PERCENTAGE OF WATER FROZEN AT VARIOUS TEMPERATURES

The characteristic freezing curve for ice cream can be used to explain why relatively low drawing temperatures at the freezer and low storage temperatures help to ensure a smooth-textured ice cream.[78] Over 50% of the water must be converted into ice, if the temperature is between −6.1 and −5.6°C, common drawing temperatures for properly operated continuous freezers. This portion of the water is frozen very rapidly, requiring in many instances less than 1 min of freezing time. Fast freezing induces the formation of small ice crystals, a critical prerequisite for a smooth-textured ice cream. Fig. 14.1 reveals that at the higher temperatures a slight temperature change has a relatively large influence on the amount of water existing as ice. For instance, ice cream drawn at −4.4°C, a common drawing temperature for batch freezers, has approximately 38% of the water frozen, whereas 54% is frozen if the ice cream is made in a continuous freezer and drawn at −6.1°C. The fact that continuously frozen ice cream is characteristically smoother in texture than the batch-frozen product can be explained in part on this basis. The freezing-point curve also reveals why a smooth texture can be maintained more satisfactorily at low storage temperatures. A coarse texture which may develop in ice cream is usually attributed to "heat shocking", which involves the alternate thawing and refreezing of a portion of the water because of temperature fluctuations in the hardening room or storage cabinet. Since this is a slow thawing and freezing process, large ice crystal formation is favored. Per degree of temperature change less water and ice will change their states at low temperatures than at higher temperatures.

Emulsion Changes During Freezing.—The desirable structure of ice cream, qualitatively expressed in terms of body and texture, is due to the physical effects of homogenization, whipping, and freezing, and to the type, source, and pretreatment of the mix ingredients. Of paramount importance are the size and state of dispersion of the ice crystals and fat particles. A long-held assumption was that optimum structural characteristics in ice cream depend upon the maintenance of the mix emulsion, consisting of individually dispersed, uniformly small fat globules resulting from homogenization. Since it was found that some fat-globule coalescence is inevitable during freezing and whipping,[80] considerable research attention has been given to an understanding of this process. Terms used to describe the changes in the emulsion include flocculation, clustering, clumping, agglomeration, coalescence, coagulation, and churning. In essence, they reflect the progressive transformation of the fat phase of mix and its eventual concentration at air-serum interfaces in ice cream, which profoundly affect the subjective unctuous qualities of body and texture as well as the objective rheological properties. The importance of these alterations in the mix emulsion during freezing are now widely appreciated; however, much still remains to be learned about the determinative structural features of ice cream.

Positioning of the fat in ice cream has been variously described as an arrangement of small, individually dispersed globules in the lamellae about the air cells;[156] chains of fat globules near the air-serum interface,[1,9] an envelope of free fat around the air bubbles with globules imbedded in it;[21] a denatured protein film at air-serum and globule interfaces which are interconnected;[142,182] and as partially naked small globules which coalesce and migrate to air-bubble interfaces when ice cream is masticated or thawed on the palate.[141] While each of the proposed characterizations is argumentative, all except the first depend upon stripping or rupturing in varying degrees of the emulsifier-protein layer about the globules to enhance inter-globule contact. Thus, the composition and thickness of the material adsorbed at fat-serum interfaces and the nature of the globules enveloped therein are of critical importance to structure in ice cream.

Obviously, both the mix formula and the variable stresses of freezing, mechanical agitation, concentration, and aeration in the freezing cylinder will affect the size distribution of the globules and the degree to which they agglomerate or coalesce. Excess stress can be deleterious, since a churned condition may develop to alter the rheological properties of ice cream. This would be reflected by the presence of butter granules detectable on the palate, by coarse texture, and by a weak, crumbly body. Churning is an especially serious problem under the freezer operating conditions needed to extrude ice cream of a particular geometric design, such as a ribbon for sandwiches and slices, or as a stiff, dry-appearing mass having the qualities necessary for direct-draw cones of soft ice cream. For these products the problem is to induce the correct amount of agglomerated fat for stiffness without causing complete breakdown and churning.

As practiced commercially, freezers are operated at below maximum capacity and the temperature is adjusted downward when there is a special requirement for stiffness and dryness. Reducing the rate of through-put means that mix moves through the freezing cylinder more slowly and is subjected to agitation for a longer period of time, which would favor churning.[53,80] Stiffness is enhanced by lowering the draw temperature to cause a greater percentage of the water to be in the frozen state. However, increased stiffness attending a reduction in temperature can also be attributed to agglomerated fat, which increases as the temperature is lowered.[21,73]

Coalescence and churning tendencies are affected by the solid-to-liquid fat ratio at the time of freezing.[1,21,52] Increased levels of unsaturation resulting from the use of summer butter[52] and fat from cows on special rations[60] make a mix more susceptible to churning, whereas stability improves when milk fat is hydrogenated prior to addition to a mix.[52] The fact that coalescence accelerates as mix storage temperature increases is a corollary of the above.[144]

Resistance to churning is increased by added increments of nonfat milk solids,[87] and/or citrate, phosphate, tetrapyrophosphate, and hexametaphosphate sodium salts.[76,89] Conversely, calcium sulfate accelerates churning, especially when the stabilizing hydrocolloid is locust bean gum.[87]

Emulsifiers, whether monoglycerides or polyoxyethylene sorbitan esters, favor agglomeration and churning during freezing.[87] Similarly, phosphatide-rich egg yolk and buttermilk solids yield a more highly destabilized mix emulsion.[87,90] Hydrolyzed cereal solids, especially the amylose fractions, complex with casein, which is easily stripped from globule surfaces during freezing with a resultant increase in susceptibility to churning.[106] As suggested by electron microscopy,[1] the destabilizing effect of emulsifiers may be the result of a thinner protein layer around the fat globules.

The ingredient and processing variables itemized above have been integrated into the various proposals advanced to characterize the structure of ice cream more fundamentally.

Rheological measurements for subsidence, rate of softening, shape retention, and creep, coupled with data for churned fat and microscopic examinations, have helped to clarify the interrelationships among the components of mix, the freezing process, and changes in the emulsion system.[73,141,142,144,145,146,188] Subsidence and rate of softening relate to the sensations of thickness and richness perceived when ice cream is eaten. The "thick eating" quality, as measured rheologically, intensifies when either emulsifier is incorporated into a mix or the temperature of the extruding ice cream is lowered.[73] These changes also favor destabilization of the emulsion as indicated by data for churned fat.[21,73,87]

Values for churned fat reported in the literature do not provide information about the size of the individual globules in ice cream as such, but they do reveal the extent to which the protective protein layer around the globules is ruptured while mix is being agitated and aerated as it traverses the freezing cylinder. These values relate more to the state of the emulsion in melted ice cream, which would be determined in large measure by the degree to which unprotected globules aggregate and coalesce as ice cream melts on the palate or at ambient temperature.[73]

Similarly, clumps and agglomerates of globules revealed microscopically in melted ice cream would not necessarily be present in the frozen product; nevertheless, their features can still be related to the changes in the emulsion caused by freezing. In a survey of 56 samples of commercial ice cream,[178] globules 3 to 5 μm in diameter were associated in clumps and agglomerates in all but 21 of the melted samples. When clumping was not observed in melted ice cream, the globules were individually dispersed and generally were less than 2 μm in diameter. However, globular size was even smaller in mix before freezing, which led to the conclusion that freezing had altered the fat dispersion even in ice cream having small, individually dispersed globules.

These observations on globule sizes in commercial ice cream fit Sherman's[144] description of the transformations taking place during freezing. Fat coagulation and solidification proceed very rapidly during the first few seconds of freezing, followed by a much slower rate of growth of particles. Since 8 globules of 1 μm diameter must fuse to form a single 2-μm globule, considerable coalescence can occur without the formation of noticeably large particles.[73]

Agitation during freezing disperses the coalesced but still small globules, and further enlargement is hindered and eventually stopped by ice-crystal barriers and the increasing viscosity of the embedding medium.[141] Upon melting, as occurs when ice cream is eaten, these events are reversed. Ice crystals disappear, the aqueous phase becomes more dilute and less viscous, and the foam structure collapses. The globules become more mobile and they coalesce and aggregate into forms of varying shapes and dimensions. These changes during thawing have an important effect on the textural qualities perceived on the palate by the consumer.[141]

Descriptions of the actual structure of ice cream have been developed from studies conducted on unfrozen mix and thawed ice cream, from the use of information collected with model systems, from certain mechanical measurements taken on the frozen product, and with the aid of some imaginative thinking. Much of this can be put in proper perspective by the recently published interpretations of prints showing the submicroscopic structure of ice cream obtained by electron microscopy of freeze-etched samples.[21,23]

Based on electron micrographs, Berger and White[21] suggested that Sherman's[144] characterization of the enveloping membrane as a thick and rigid monoglyceride-protein layer may need modification. Fat

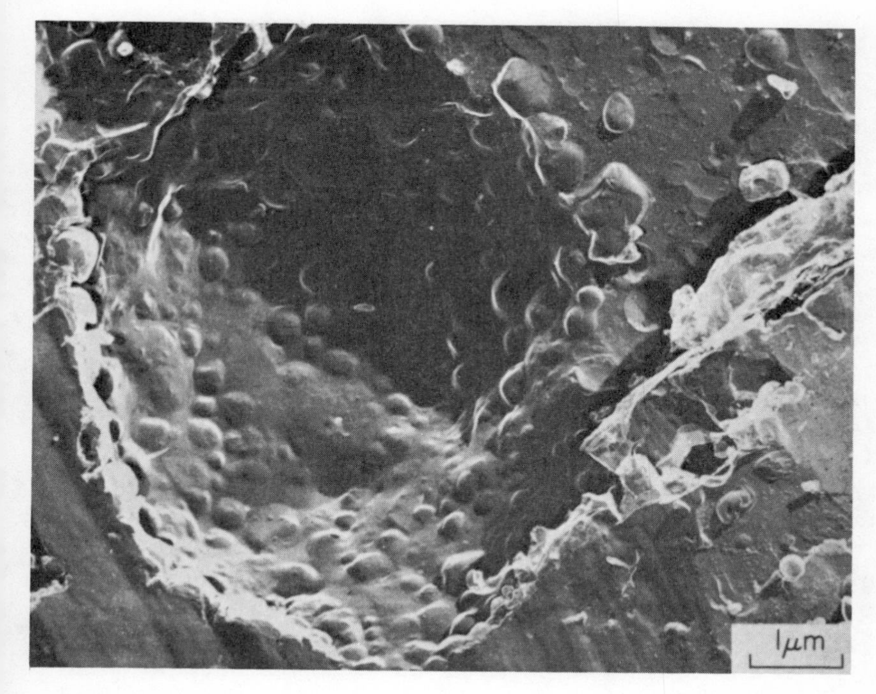

From Berger and White[21]

FIG. 14.2. REPLICA SHOWING INTERIOR OF TYPICAL AIR CELL IN ICE CREAM

globules were more often encased by a thin shell of casein sub-units, one layer thick, which was easily lifted off in the fracture stage of specimen preparation. This indicates that the protein layer is only weakly bound to the fat surface.

Berger and White[21] noted a marked reduction in the number of globules per unit area during freezing, but clumped fat did not appear. They proposed that liquid fat released by rupture of the adsorbed layer coalesces and spreads as a film at the air/serum interface with individual globules cemented in it. The electron micrograph (Fig. 14.2) showing the interior of an air cell in ice cream, supports this interpretation.[21] The smooth surface shown is attributed to fat and is in contrast to the more granular appearance of the continuous phase shown in Fig. 14.3, which reveals casein sub-units in the membrane around globules and in the spaces between globules.

The existence of liquid fat at freezing temperatures is a requisite to most structural interpretations. Interestingly, the lipid fraction

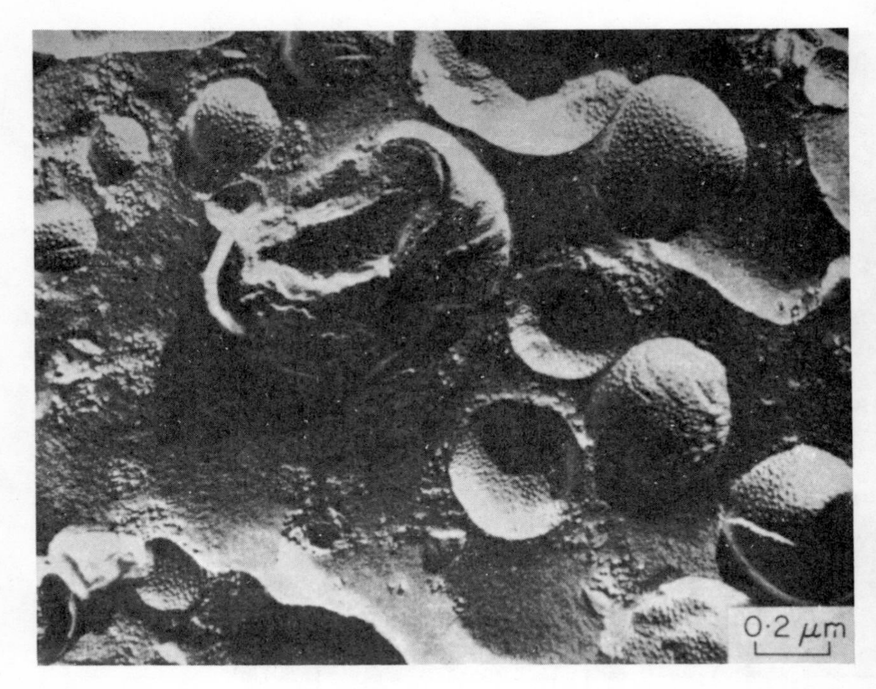

From Berger and White[21]

FIG. 14.3. REPLICA SHOWING FAT GLOBULES WITH CASEIN SUBUNITS IN THE MEMBRANE

recovered from the foam portion of melted ice cream was found to be more unsaturated than the nonfoam-association lipids and contained less monoglyceride.[77,83] Compared to the total extractable lipids of ice cream, the most easily recovered fraction is more unsaturated.[159]

DEFECTS IN ICE CREAM

The scoring system for measuring quality in ice cream adopted by the American Dairy Science Association distributes the 100 total points as follows: 45 for flavor, 30 for body and texture, 15 for bacterial quality, 5 for melting qualities, and 5 for package and color. The emphasis on flavor, and body and texture is justified, since these are the qualities most critically considered by the consumer. A survey involving 4,500 people[123] revealed that "creamy," "smooth," and "velvety" are the terms most often used in describing the texture and body of good-quality ice cream. Poor quality is associated with a "watery," "foamy" product that is not "smooth" and lacks "body." High-quality ice cream has "good flavor," is "rich and sweet" and does not have an "artificial taste."

Flavor Defects

The characterizing flavor of ice cream is identifiable to the consumer by taste, by color, by pieces of flavor ingredient dispersed in the product, and by carton labeling. Economic considerations, problems of availability, the compatibility with the ice-cream manufacturing process, and consumer preferences are factors considered in the selection of flavor ingredients. To meet these requirements, natural and synthetic materials, alone or in combination, are available in a variety of forms to characterize, intensify, and modify the flavor of ice cream.

While the added characterizing flavor ingredient is of primary consideration, the mix itself is an important factor affecting the degree to which consumers like or dislike the flavor of a particular ice cream. Flavor release and flavor masking are terms commonly referred to in qualitative assessments of acceptability. Sequential perception of flavor is dependent in large measure upon the physical or structural features of ice cream, which in turn are controlled by the composition of the mix, processing and freezing variables, and the events between manufacture and ultimate consumption.

Since ice cream is usually consumed within a few weeks after manufacture, the development of undesirable flavors during storage and distribution is not considered a serious problem. However, organoleptic evaluations and chemical measurements have revealed that oxidized and metallic flavors can develop in ice cream exposed to light intensities similar to those found in merchandising cabinets,[107] or when Cu and Fe concentrations are relatively high.[180]

Most flavor defects in ice cream which are not caused by the characterizing flavor can be traced to the use of ingredients of inherently poor quality or to variables related to mix processing.

"Old ingredient," "oxidized," and "cooked" or "scorched" probably are the most common flavor defects involving the milk solids in ice cream. Old ingredient is an ill-defined off-flavor which quite often is microbial in origin. Oxidized flavor, on the other hand, usually can be traced to the use of frozen cream, butter, or plastic cream in mix. Industrial experience indicates that oxidized flavor is least likely to develop in the latter, and research has shown that in regard to flavor, concentrated sweetened cream is preferred to butter and frozen cream as a source of storage fat for ice cream.[8]

Cooked and scorched flavor notes, although not necessarily objectionable in themselves, may tend to mask or modify the characterizing flavor. Generally, these off-flavors are associated with mix pasteurized in HTST and especially UHT systems.[113] Occasionally, cooked flavor in ice cream can be traced to a mix ingredient such as frozen cream used too soon after preparation for storage. A slight flavor preference for low-heat over high-heat skimmilk powder in ice cream has been demonstrated.[65]

Buttermilk and whey solids are legally accepted sources of milk solids-not-fat with certain restrictions for the latter. Buttermilk used soon after spray-drying yielded ice cream having flavor qualities which could not be differentiated from ice cream containing dried skimmilk.[196] However, most dried buttermilk samples stored six weeks before use in ice cream imparted off-flavors. Flavor was not significantly changed when dried whey was used as a partial replacement for regular milk solids-not-fat.[55] A slight preference for electrodialyzed whey over conventional whey solids has been demonstrated.[57]

Considerable work has been carried out, especially in industrial laboratories, to develop sterile mixes for the soft-serve trade. While acceptable physical properties have been realized, flavor problems persist. In addition to a cooked, evaporated milk character, sweetened, sterilized mixes may become hydrolytically rancid when stored at room temperature.[183]

To help maintain smooth texture during the rigors of the distribution system, 6% or more by weight of a corn sweetener may be found in a typical ice cream. At these high levels corn sweeteners tend to mask the characterizing flavor and may impart a syrup, cereal, or caramelized flavor note to the ice cream. A comparative study showed that low-DE corn sweeteners exert a greater flavor-masking effect in ice cream, but hydroxymethylfurfural determinations revealed a greater tendency toward nonenzymatic browning when the mix contained high-DE solids.[49] These findings correlate with known compositional differences among the various corn sweeteners available for use in frozen desserts.

Body and Texture Defects

Body refers to the bite resistance, chewiness, and substance of ice cream. Body is fairly easily standardized by control of mix composition, processing, and freezing. It does not change significantly during storage. The most common body defects are classified as excessive sogginess, fluffy or weak, gummy or sticky, and crumbly. Sogginess usually is associated with low overrun and high total solids. Some consumers consider sogginess in ice cream as a desirable characteristic, and its classification as a defect of body depends on individual preferences.

High levels of corn sweetener solids, particularly low-DE products, impart a heavy, chewy quality to ice cream.

Fluffy body, the opposite of sogginess, is generally caused by insufficient solids and excessive overrun. Gummy or sticky body is usually attributed to high stabilizer levels and in some cases to high levels of milk solids and corn sweetener solids. A crumbly body often results when ice cream contains less than the normal amount of sugar, and it is improperly stabilized.

The texture of ice cream refers to the feeling of smoothness in the mouth as it is consumed. Since small ice crystals are essential for good texture, the ice-cream manufacturer makes every effort to produce small ice crystals by proper stabilization, rapid freezing and hardening, and careful handling during transportation and storage. Texture defects generally develop during storage, particularly when ice cream is in retail cabinets and in home freezer chests, and are due chiefly to fluctuating temperatures.[27]

Sandiness

If lactose crystallizes from solution in ice cream, a very objectionable gritty, or sandy, condition may develop.[22,197] The defect is distinguishable from the coarseness caused by ice crystals in that the sand-like condition persists even after the ice cream has melted. Two factors promote the formation of sandy ice cream; high lactose (MSNF) concentration in the mix and high storage temperatures. Lactose can be held in a supersaturated state in ice cream because of the high viscosity or plasticity of this product.[152] This reasoning has been used to explain why sandiness develops much more frequently at relatively high storage temperatures (above −18°C). The important factors responsible for lactose crystallization are formation of nuclei, the rate of change of beta to alpha lactose, and the deposition of alpha on existing crystal centers.[117] When nuclei are present, the growth of the lactose crystals to a size easily detected in the mouth is rapid, especially at high storage temperatures. In the absence of nuclei, crystallization is slow. At hardening-room temperatures, lactose crystal development is retarded, even in the presence of nuclei, because of the depressed rate of change from beta to alpha and the inability of alpha lactose to migrate and orient due to the high viscosity of the ice cream.[152] Nucleation and crystal growth are favored by the presence of vanilla bean specks,[96] and nut or berry seed fragments.

Several kinds of partially and completely "delactosed" products have been developed so as to make possible an increase in MSNF content without raising the lactose to an excessive level.[42] The danger of sandiness can also be minimized by seeding the mix with lactose or whey solids by a process that is analogous to the seeding of sweetened condensed milk.[36] Induced crystallization of lactose in mix yields ice cream with a firmer body.[119]

Even though ice cream may be supersaturated to lactose, sandiness is not a commonly occurring defect, particularly in recent years. The reduced incidence of sandy ice cream has been attributed primarily to various marine and vegetable gum stabilizers which retard flow of fresh material and interrupt crystal growth by adsorption on crystal surfaces.[96,118] Evidence also suggests that lactose crystallization is inhibited by hydrolyzed starch solids.[198]

Shrinkage

This defect is characterized by the shrinking of the product away from the sides and top of the package. In severe cases the apparent loss in volume may be as high as one-third of the original mass. The shrinkage problem is typified by sporadic, acute, and short-lived seasonal outbreaks in local areas. The defect may suddenly appear and disappear, often with no indication as to the cause or remedy.

Shrinkage results from the loss of air from the ice cream. For air to escape, the film or the lamellae surrounding the air bubbles must be weakened and finally ruptured. Many factors have been implicated, but most should be classified as contributory, rather than actual causes of shrinkage.[120] The state of the whey protein fraction is especially significant since denaturation of this fraction, either by heat or proteolysis, makes ice cream susceptible to shrinkage.[94,164] Undenatured protein, on the other hand, increased the resistance of ice cream to shrinkage. There appear to be differences in milks even among individual cows in regard to shrinkage susceptibility.[164] Surface-tension factors have been associated with the problem, as it has been shown that rancid milk accelerates shrinkage.[34] Other factors that appear to favor shrinkage when it occurs include high fat and MSNF levels, continuous freezers, low drawing temperatures, and fast hardening. The literature on shrinkage has been reviewed by Nickerson and Tarassuk.[120]

REFERENCES

1. ALSAFAR, T., and WOOD, F. W., 17th Int. Dairy Congr. *E/F,* 401 (1966).
2. ANDERSON, E. O., and PIERCE, R. L., Milk Dealer, *18,* (12), 60 (1929).
3. ANDERSON, G., Ice Cream Trade J., *53,* (5), 26 (1958).

4. ANDERSON, R. J., and THOMAS, E. L., 54th Ann. Conv. Int. Assn. of Ice Cream Mfrs., Proc., Chicago, Ill. (1958).
5. ARBUCKLE, W. S., Mo. Agr. Expt. Sta., Res. Bull 320 (1940).
6. ARBUCKLE, W. S., 51st Ann. Conv. Assn. Ice Cream Mfrs., Proc., St Louis, Mo., *2,* 31 (1955).
7. ARBUCKLE, W. S., Ice Cream Trade J., *56,* 62 (1960).
8. ARBUCKLE, W. S., and BELL, R. W., Ice Cream Field, *82,* 40 (1963).
9. ARBUCKLE, W. S., and CREMERS, L. F., Ice Cream Field, *64,* 98 (1954).
10. ASCHAN, L., Ice and Refrigeration, *87,* 321 (1934).
11. BABCOCK, C. J., Roerig, R. N., Stabile, J. N., DUNLAP, W. A., and RANDALL, R., J. Dairy Sci., *29,* 699 (1946).
12. BABCOCK, C. J., STABILE, J. N., RANDALL, R., and WINDHAM, E. S., J. Dairy Sci., *30,* 733 (1947).
13. BAILEY, E. M., J. Assoc. Offic. Agr. Chem., *13,* 198 (1923).
14. BALDWIN, F. B., JR., and COMBS, W. B., Milk Plant Mo., *22,* (1), 18 (1933).
15. BALDWIN, F. B., JR., and DOAN, F. J., J. Dairy Sci., *18,* 629 (1935).
16. BATEMAN, G. M., and SHARP, P. F., J. Dairy Sci., *11,* 380 (1928).
17. BELL, R. W., J. Dairy Sci., *22,* 59 (1939).
18. BELL, R. W., J. Milk Food Tech., *10,* 149 (1947).
19. BELL, R. W., and MUCHA, T. J., J. Dairy Sci., *35,* 1 (1952).
20. BELL, R. W., and SANDERS, C. F., J. Dairy Sci., *29,* 213 (1946).
21. BERGER, K. G., and WHITE, G. W., J. Food Tech., *6,* 285 (1971).
22. BOTHELL, F. H., 5th Ann. Conv., Pacific Ice Cream Mfrs. Assn., Proc., Portland, Oreg., p. 121–125 (1920).
23. BUCHHEIM, W., 18th Int. Dairy Congr. IE, 398 (1970).
24. CHEN, C., and YAMAUCHI, K., Agr. Biol. Chem., *33,* 1333 (1969).
25. CHEN, C., and YAMAUCHI, K., Agr. Biol. Chem., *33,* 1751 (1969).
26. COLE, W. C., Ice Cream Trade J., *34,* (6), 15 (1938).
27. COMBS, W. B., and THOMAS, E. L., Ice Cream Field, *64,* (6), 34 (1954).
28. Corn Industries Research Foundation, "Corn Syrups and Sugars," New York.
29. CVITL, J., Milchwirtsch. Forschgn., *12,* 409 (1931).
30. DAHLBERG, A. C., N. Y. (Geneva), Agr. Expt. Sta. Tech. Bull. *111* (1925).
31. DAHLE, C. D., J. Dairy Sci., *24,* 245 (1941).
32. DAHLE, C. D., and COLLINS, W. F., 43rd Ann. Conv., Int. Assn. of Ice Cream Mfrs., Proc. *50* (1947).
33. DAHLE, C. D., and JOSEPHSON, D. V., Ice Cream Field, *44,* (5), 36 (1945).
34. DAHLE, C. D., HANKINSON, D. J., and MEISER, J. A., 42nd Ann. Conv. Int. Assn. Ice Cream Mfrs., Proc. *2,* 33 (1947).
35. DAVIS, J. G., "A Dictionary of Dairying," p. 515–536, 2nd Ed., Interscience Publishers, Inc., New York, 1955.
36. DECKER, C. W., U.S. Patent 2,641,546, June 6 (1953).
37. DESAI, I. D., and NICKERSON, T. A., Nature, 202, 183 (1964).
38. DESAI, I. D., NICKERSON, T. A., and JENNINGS, W. G., J. Dairy Sci., *44,* 215 (1961).
39. DOAN, F. J., J. Dairy Sci., *10,* 353 (1927).
40. DOAN, F. J., J. Dairy Sci., *12,* 211 (1929).
41. DOAN, F. J., Quart. Rev. Pediatrics, *8,* 194 (1953).
42. DOAN, F. J., J. Dairy Sci., *41,* 325 (1958).
43. DOAN, F. J., and BALDWIN, F. B., JR., J. Dairy Sci., *19,* 225 (1936).
44. DOAN, F. J., and FEATHERMAN, C. E., Milk Dealer, *27,* (3), 33 (1937).
45. DOAN, F. J., and LEEDER, J. G., Food Ind., *16,* 532 (1944).
46. DOAN, F. J., and WARREN, F. G., J. Dairy Sci., *30,* 837 (1947).
47. EL-NEGOUMY, A. M., and BOYD, J. C., J. Dairy Sci., *48,* 23 (1965).
48. EOPECHINO, A. A., and LEEDER, J. G., Ice Cream Review, *50,* 11 (1967); *51,* 14 (1967).
49. EOPECHINO, A. A., and LEEDER, J. G., Food Tech., *23,* 1215 (1969).
50. Federal Register, *25,* 7125 (1960).
51. FINNEGAN, E. J., and SHEURING, J. J., J. Dairy Sci., *47,* 681 (1964).

52. FRAZEUR, D. R., Ice Cream Field, *73,* 32 (1959).
53. FRAZEUR, D. R., Ice Cream Field, *74,* 48 (1959).
54. FRAZEUR, D. R., Ice Cream Trade J., *60,* 72 (1964).
55. FRAZEUR, D. R., Ice Cream Field and Ice Cream Trade J., 149, 22 (1967).
56. FRAZEUR, D. R., and HARRINGTON, R. B., J. Dairy Sci., *47,* 682 (1964).
57. FRAZEUR, D. R., and HARRINGTON, R. B., Ice Cream Field and Ice Cream Trade J., *149,* 40 (1967).
58. FRAZEUR, D. R., and HARRINGTON, R. B., Food Tech., *22,* 910 (1968).
59. FRAZEUR, D. R., and HARRINGTON, R. B., Food Tech., *22,* 912 (1968).
60. FRAZEUR, D. R., and NOLLER, C. H., J. Dairy Sci., *43,* 847 (1960).
61. FRAZEUR, D. R., LISKA, B. J., and PARMELEE, C. E., Ice Cream Review, *45,* 38 (1962).
62. GOULD, I. A., and KEENEY, P. G., J. Dairy Sci., *40,* 297 (1957).
63. GOULD, I. A., and SOMMER, H. H., Mich. Agr. Expt. Sta. Tech. Bull. *164* (1939).
64. HAMMER, B. W., "Dairy Bacteriology," 3rd Ed., p. 356, John Wiley and Sons, New York, 1948.
65. HEDRICK, T. I., ARMITAGE, A. V., and STINE, C. M., Mich Agr. Expt. Sta. Quart. Bull. *47,* 153 (1964).
66. HENING, J. C., J. Dairy Sci., *14,* 84 (1931).
67. HOOD, E. G., and WHITE, A. H., 28th Ann. Conv. Int. Assn. Milk Dealers, Plant Sect. Proc., St. Louis, Mo., p. 68 (1935).
68. HORTVET, J., J. Ind. Eng. Chem., *13,* 198 (1921).
69. HOSTETTLER, H., and IMHOF, K., Chem. Abstr., *48,* 6613 (1954).
70. HUNZIKER, O. F., "Condensed Milk and Milk Powder," 5th Ed., p. 315, La Grange, Ill., 1935.
71. Int. Assoc. Ice Cream Manufacturers, Tech. Bull. *25,* Washington, D. C. (1954).
72. JENSEN, R. G., SAMPUGNA, J., and GANDER, G. W., J. Dairy Sci., *44,* 1057 (1961).
73. JOHN, M. G., and SHERMAN, P., 16th Int. Dairy Congr. *C,* 61 (1962).
74. JOHNSON, A. H., and TUMERMAN, L., 16th Int. Dairy Congr. *B,* 1057 (1962).
75. JOSEPHSON, D. V., and DOAN, F. J., Milk Dealer, *29,* (2), 35 (1939).
76. KEENEY, P. G., J. Dairy Sci., *45,* 430 (1962).
77. KEENEY, P. G., J. Dairy Sci., *45,* 658 (1962).
78. KEENEY, P. G., Ice Cream Field, *82,* 38 (1963).
79. KEENEY, P. G., "Commercial Ice Cream and Frozen Desserts," Circular 525, Pennsylvania State University, 1965.
80. KEENEY, P. G., and JOSEPHSON, D. V., Ice Cream Trade J., *54,* (5), 24 (1958).
81. KEENEY, P. G., and JOSEPHSON, D. V., Ice Cream Trade J., *57,* 28 (1961).
82. KEENEY, P. G., and JOSEPHSON, D. V., 16th Int. Dairy Congr. C, 53 (1962).
83. KEENEY, P. G., and MAGA, J. A., J. Dairy Sci., *48,* 1591 (1965).
84. KIERMEIER, F., and SOLMS-BARUTH, H. GRAF ZU, Z. Lebensmittelunters. u. -Forschung, *130,* 291 (1966).
85. KING, N., Dairy Ind., *15,* 1051 (1950).
86. KING, N., "The Milk Fat Globule Membranc," p. 60, Commonwealth Agricultural Bureau Technical Communication #2 (1955).
87. KLOSER, J. J., and KEENEY, P. G., Ice Cream Trade J., *55,* (5), 26 (1959).
88. KLOTZEK, L. M., and LEEDER, J. G., Ice Cream Review, *50,* 20 (1966).
89. KLOTZEK, L. M., and LEEDER, J. G., Ice Cream Review, *50,* 26 (1966).
90. KNIGHTLY, W. H., Ice Cream Trade J., *55,* (7), 20 (1959).
91. KOZIN, N. I., and PONOMAREVA, G. T., Izv. vyssh. ucheb. Zaved., Pishch. Tekhnol., *1969,* 64 (1969).
92. KURTZ, F. E., J. Phys. Chem., *33,* 1489 (1929).
93. LAGONI, H., and PETERS, K. H., Milchwissenschaft, *16,* (4), 197 (1961).
94. LANDO, J. C., and DAHLE, C. D., Ice Cream Review, *33,* (3), 150 (1949).
95. LEACH, H. J., and MARTIN, W. H., Am. Creamery Poultry Prod. Rev., *77,* (4), 112 (1933).
96. LEEDER, J. G., and OSTROFF, B., 17th Int. Dairy Congr. *E/F,* 409 (1966).

97. LEIGHTON, A., J. Dairy Sci., *10,* 300 (1927).
98. LEIGHTON, A., and KURTZ, F. E., J. Phys. Chem., *33,* 1485 (1929).
99. LEIGHTON, A., and WILLIAMS, O. E., J. Phys. Chem., *31,* 596 (1927).
100. LEIGHTON, A., and WILLIAMS, O. E., J. Phys. Chem., *33,* 1485 (1929).
101. LEVITON, A., and LEIGHTON, A., J. Phys. Chem., *40,* 71 (1936).
102. LEVITON, A., VESTAL, J. H., VETTEL, H. E., and WEBB, B. H., 17th Int. Dairy Congr. *E/F,* 133 (1966).
103. LOUDER, E. A., and SMITH, L. S., J. Dairy Sci., *15,* 113 (1932).
104. MacDONALD, F. J., Dairy Ind., *15,* (4), 413 (1950).
105. MACK, M. J., Mass. Agr. Expt. Sta. Bull. 268 (1930).
106. MAHDI, S. R., and BRADLEY, R. L. S., JR., J. Dairy Sci., *52,* 1738 (1969).
107. MAN, J. M. DE, VUJICIC, VERA, and VUJICIC, I., Canad. Inst. Food Tech. J., *1,* 6 (1968).
108. MARQUARDT, J. C., J. Dairy Sci., *20,* 467 (1937).
109. McDOWALL, F. H., "The Buttermaker's Manual," p. 603, New Zealand Univ. Press, Wellington, 1953.
110. MEYKNECHT, E. A. M., Neth. Milk Dairy J., *15,* 296 (1961).
111. MORTENSEN, M., World's Dairy Congress, Washington, D. C., Proc. *1,* 472 (1923).
112. MOSS, J. R., Ice Cream Trade J., *51,* (1), 40 (1955).
113. MUCK, G., Ice Cream Trade J., *61,* 40 (1965).
114. MUNKWITZ, R. C., BERRY, M. H., and BOYER, W. C., Md. Agr. Expt. Sta. Bull. *344* (1933).
115. NAKANISHI, T., and ITOH, T., Agr. Biol. Chem., *29,* 1099 (1965).
116. NAKANISHI, T., and ITOH, T., J. Agr. Chem. Soc. Japan, *43,* 725 (1969).
117. NICKERSON, T. A., J. Dairy Sci., *39,* 1342 (1956).
118. NICKERSON, T. A., J. Dairy Sci., *45,* 354 (1962).
119. NICKERSON, T. A., and PANGBORN, ROSE M., Food Tech., *15,* 105 (1961).
120. NICKERSON, T. A., and TARASSUK, N. P., J. Dairy Sci., *38,* 1305 (1955).
121. PETER, P. N., J. Phys. Chem., *32,* 1856 (1928).
122. PICKERING, S. V., Ber. der Deut. Chem. Ges., *24,* 333 (1891).
123. POLITZ, A., "Public Attitude and Uses of Dairy Products. Highlights Study No. 10," Published by American Dairy Association, Chicago, 1959.
124. PORCHER, C., and CHEVALIER, G., Le Lait, *3,* 369 (1923).
125. PRICE, W. V., Proc. Int. Assoc. Ice Cream Mfgs., *2,* 17 (1931).
126. REID, W. H. E., and GARRISON, E. R., Mo. Agr. Expt. Sta., Res. Bull. *128* (1929).
127. REID, W. H. E., and HALES, M. W., Mo. Agr. Expt. Sta., Res. Bull. *215* (1934).
128. REID, W. H. E., and MOSELEY, W. K., Mo. Agr. Expt. Sta., Res. Bull. *91* (1926).
129. REID, W. H. E., and SMITH, L. E., Mo. Agr. Expt. Sta., Res. Bull. *340* (1942).
130. REID, W. H. E., DECKER, C. W., and ARBUCKLE, W. S., Mo. Agr. Expt. Sta., Res. Bull. *322* (1940).
131. ROADHOUSE, C. L., and HENDERSON, J. L., Food Ind., *12,* (6), 54 (1940).
132. ROBERTS, W. M., 51st Ann. Conv. Int. Assn. Ice Cream Mfrs. Proc., *2,* 31 (1955).
133. ROGERS, L. A., THOMPSON, S. C., and KIETHLEY, J. R., U.S. Dept. Agr., Bur. Animal Industry, Bull. *48* (1912).
134. ROSE, D., Can. J. Tech., *34,* 145 (1956).
135. ROSE, D., and TESSIER, H., Can. J. Tech., *32,* 85 (1954).
136. ROSE, D., and TESSIER, H., Can. J. Tech., *34,* 139 (1956).
137. ROSE, D., and TESSIER, H., J. Dairy Sci., *42,* 989 (1959).
138. RUDNICK, A. W., ROSS, I. J., and FOX, J. D., J. Dairy Sci., *52,* 883 (1969).
139. SAITO, Z., and IGARASHI, Y., Jap. J. Zootech. Sci., *37,* 478 (1966).
140. SCATCHARD, G., J. Am. Chem. Soc., *43,* 2406 (1921).
141. SHAMA, F., and SHERMAN, P., J. Food Sci., *31,* 699 (1966).
142. SHAW, D., "Rheology of Emulsions," p. 125, Pergamon Press, London, 1963.
143. SHERMAN, P., Food Tech., *15,* 394 (1961).
144. SHERMAN, P., J. Food Sci., *30,* 201 (1965).
145. SHERMAN, P., J. Food Sci., *31,* 707 (1966).
146. SHERMAN, P., J. Texture Studies, *1,* 43 (1969).

147. SHIPE, W. F., ROBERTS, W. M., and BLANTON, L. F., J. Dairy Sci., *46,* 169 (1963).
148. SHUTE, G. C. M., J. Soc. Dairy Tech., *14,* 208 (1961).
149. SMITH, A. C., and DOWD, L. R., Ice Cream Trade J., *57,* 10 (1961).
150. SOMMER, H. H., J. Dairy Sci., *11,* 9 (1928).
151. SOMMER, H. H., "Market Milk and Related Products," 2nd. Ed., p. 456. Madison, Wis., 1946.
152. SOMMER, H. H., 12th Int. Dairy Congress, *3,* 249 (1949).
153. SOMMER, H. H., "Theory and Practice of Ice Cream Making," 6th Ed., p. 242, Madison, Wis., 1951.
154. Ibid., p. 263, 267.
155. Ibid., p. 265.
156. Ibid., p. 319.
156A. SPECK, M. L., GROSCHE, C. A., LUCAS, H. L., and HANKIN, L., J. Dairy Sci., *37,* 37 (1954).
157. STINE, C. M., BARNES, J., and HEDRICK, T. I., Food Tech., *17,* 1059 (1963).
158. STISTRUP, K., and ANDREASEN, J., 16th Int. Dairy Congr. *C,* 19 (1962).
159. STISTRUP, K., and ANDREASEN, J., 16th Int. Dairy Congr. *C,* 29 (1962).
160. STOLOFF, L., "Natural Plant Hydrocolloids," Advances in Chemistry Series No. 11, p. 92, Am. Chem. Soc., Wash., D.C., 1954.
161. STUBBS, J. R., and ELSDON, G. D., Analyst, *59,* 146 (1934).
162. STULL, J. W., HERREID, E. O., and TRACY, P. H., J. Dairy Sci., *30,* 541 (1947).
163. TAMATE, R., and OHTAKA, F., Jap. J. Dairy Sci., *18,* A37 (1969).
164. TARASSUK, N. P., and HUTTON, J. T., 45th Am. Conv. Int. Assn. Ice Cream Mfgs. Proc. *2,* 37 (1949).
165. TESSIER, H., and ROSE, D., Can. J. Tech., *34,* 211 (1956).
166. TESSIER, H., ROSE, D., and LUSENA, C. V., Can. J. Tech., *34,* 131 (1956).
167. TRACY, P. H., and EARL, F. A., Ice Cream Trade J., *56,* 24 (1960).
168. TRACY, P. H., and MCCOWN, C. Y., J. Dairy Sci., *17,* 47 (1934).
169. TRACY, P. H., HASKISSON, W. A., and TRIMBLE, J. M., J. Dairy Sci., *27,* 311 (1944).
170. TRACY, P. H., HETRICK, J., and KRIENKE, W. S., J. Dairy Sci., *33,* 832 (1950).
171. TRACY, P. H., TOBIAS, J., and HERREID, E. O., 46th Ann. Conv. Int. Assn. Ice Cream Mfgs. Proc. *2,* 21 (1950).
172. TRELOGAN, H., and COMBS, W. B., J. Dairy Sci., *11,* 717 (1934).
173. TROUT, G. M., J. Dairy Sci., *24,* 277 (1941).
174. TROY, H. C., and SHARP, P. F., J. Dairy Sci., *13,* 140 (1930).
175. TUMERMAN, L., FRAM, H., and CORNELY, K. W., J. Dairy Sci., *37,* 830 (1954).
176. TURNBOW, G. D., TRACY, P. H., and RAFFETTO, L. A., "The Ice Cream Industry," 2nd Ed., p. 99, John Wiley and Sons, Inc., New York, 1947.
177. U.S. Army Med. Dept. Bull. *85,* 19 (1945).
178. VALAER, E. P., and ARBUCKLE, W. S., Ice Cream Field, *77,* 10 (1961).
179. VAN DEN BERG, L., J. Dairy Sci., *44,* 26 (1961).
180. VANDERZANT, C., and MIAH, A. H., Food Tech., *15,* 515 (1961).
181. VAN SLYKE, L. L., and PUBLOW, C. A., "The Science of Cheese Making," Orange Judd Pub. Co., New York, 1916.
182. WALKER, D. A., J. Dairy Sci., *46,* 591 (1963).
183. WALLANDER, J. F., and SWANSON, A. M., J. Dairy Sci., *48,* 778 (1965).
184. WATSON, P. D., and LEIGHTON, A. J., J. Dairy Sci., *10,* 331 (1927).
185. WEARMOUTH, W. H., Food Manuf., *35,* (3), 97 (1960).
185A. WEBB, B. H., *2,* 224 and *3,* 295 in "The Freezing Preservation of Foods," Tressler, D. K., Van Arsdel, W. B., Copley, M. J. Avi Publishing Co., 1968.
186. WEBB, B. H., and HALL, S. A., J. Dairy Sci., *18,* 275 (1935).
187. WELLS, P. R., and LEEDER, J. G., J. Dairy Sci., *46,* 789 (1963).
188. WHITEHEAD, J., and SHERMAN, P., Food Tech., *21,* 107 (1967).
189. WHITTIER, E. O., and WEBB, B. H., "By-Products from Milk," Reinhold Publ. Corp., New York, 1950.

190. WILDASIN, H. L., and DOAN, F. J., J. Dairy Sci., *34,* 438 (1951).
191. WILLART, S., Svenska Mejeritidn., *56,* 241 (1964).
192. WINDER, W. C., J. Dairy Sci., *45,* 1024 (1962).
193. WISHNETSKY, T., and FINLEY, R. D., Chem. Eng. Prog. Symp. Ser. No. 69 *62,* 80 (1966).
194. YAMAUCHI, K., and CHEN, C., Agr. Biol. Chem., *33,* 1761 (1969).
195. YOUNG, H. B., and MULL, L. E., South. Dairy Prod. J., *67,* 22 (1960).
196. ZIEMER, K. A., AMUNDSON, C. H., WINDER, W. C., and SWANSON, A. M., J. Dairy Sci., *45,* 659 (1962).
197. ZOLLER, H. F., and WILLIAMS, O. E., J. Agr. Res., *21,* 791 (1921).
198. ZUCZKOWA, J., 18th Int. Dairy Congr. *IE* (1970).

Index